UNITEXT

La Matematica per il 3+2

Volume 150

The **UNITEXT - La Matematica per il 3+2** series is designed for undergraduate and graduate academic courses, and also includes advanced textbooks at a research level.

Originally released in Italian, the series now publishes textbooks in English addressed to students in mathematics worldwide.

Some of the most successful books in the series have evolved through several editions, adapting to the evolution of teaching curricula.

Submissions must include at least 3 sample chapters, a table of contents, and a preface outlining the aims and scope of the book, how the book fits in with the current literature, and which courses the book is suitable for.

For any further information, please contact the Editor at Springer: francesca.bonadei@springer.com

THE SERIES IS INDEXED IN SCOPUS

***

UNITEXT is glad to announce a new series of free webinars and interviews handled by the Board members, who rotate in order to interview top experts in their field.

Access this link to subscribe to the events:

https://cassyni.com/events/TPQ2UgkCbJvvz5QbkcWXo3

Valter Moretti

Analytical Mechanics

Classical, Lagrangian and Hamiltonian
Mechanics, Stability Theory, Special
Relativity

 Springer

Valter Moretti
Dipartimento di Matematica
Università di Trento
Trento, Italy

Translated by
Simon G. Chiossi (iD)
Applied Mathematics
Fluminense Federal University
Niterói, Brazil

ISSN 2038-5714   ISSN 2532-3318   (electronic)
UNITEXT
ISSN 2038-5722   ISSN 2038-5757   (electronic)
La Matematica per il 3+2
ISBN 978-3-031-27611-8   ISBN 978-3-031-27612-5   (eBook)
https://doi.org/10.1007/978-3-031-27612-5

This Springer imprint is published by the registered company Springer Nature Switzerland AG
The registered company address is: Gewerbestrasse 11, 6330 Cham, Switzerland

Preface

This textbook aims at introducing readers, primarily students enrolled in undergraduate Mathematics or Physics courses, to the topics and methods of classical Mathematical Physics, together with a mathematical formulation of the *theory of special relativity*. Special attention is devoted to *classical mechanics*, its *Lagrangian formulation*—including an introduction to *Lyapunov stability*, and to the *Hamiltonian formulation*, together with a few important results such as the *Liouville theorem* and *Poincaré's recurrence theorem*. The general purpose is to present the logical-mathematical structure of physical theories, by introducing them in an axiomatic way, and start from a limited number of physical assumptions. To do so, we use the tools of analysis and elementary differential geometry. Certain traditional topics—such as the detailed treatment of rigid bodies—are presented only in their foundational aspects, given that there are several excellent and modern textbooks on analytical mechanics that address this type of subject and more, both in depth and rigorously; for example, [BRSG16] and [Bis16]. We have, on the other hand, insisted on topics that have had a major impact on theoretical and mathematical physics beyond analytical mechanics, such as the Galilean symmetry (equivariance) of classical dynamics and the Poincaré symmetry of relativistic dynamics, or the broad relationship between symmetries and constants of motion, the independence of mathematical objects from choices (think reference frames) or the possibility of describing dynamics in a global way while still working in local coordinates.

This book is the result of 20+ years of teaching undergraduate mathematics and physics at Trento University, starting from the one-year course formerly known as *rational mechanics*, that nowadays is almost inevitably compressed into one semester to fit the current 3-year degree. Based on this experience, the text has been conceived to be flexible and thus adapt to different curricula and to the needs of a variety of students and instructors. Undergraduates may find the abstract formalism daunting at first, but they will definitely reap the rewards of our approach throughout. Instructors may select only certain materials, should they want to simplify matters, or follow the path we set out and take advantage of the more advanced sections.

The author's research work does not dwell on analytical mechanics nor classical mathematical physics, but rather mathematical and foundational aspects of quantum theories, both relativistic and non-relativistic. In a sense, this fact has been greatly beneficial to the preparation of the text (which was born out of lecture notes and has now become a structured, proper textbook). On the one hand, this expertise has prevented the excessive use of abstract formalism, which has become (alas) inadequate to the structure of current undergraduate courses, even though the book has kept to the highest didactical standards, appropriate for undergraduate classes in mathematics and physics. On the other hand, the author's interest in other areas of mathematical physics allowed for a critical understanding of the foundational aspects of classical mechanics, a perspective that permeates the entire text. Here mechanics is not seen as an isolated, albeit important, mathematical chapter of theoretical physics, but rather it is viewed as a source of some of the ideas that lie at the heart of modern theoretical and mathematical physics. For this reason, the mathematical formulation of the elementary notions of space, time, spacetime, frame systems, etc. has been addressed critically. We have provided the most suitable logical-mathematical description and clarified the limits beyond which mathematical formalism ceases to describe physical phenomenology. The author believes that, in mathematical physics, the *physical content* of a mathematical statement should always be made transparent. The mathematical tools employed to describe a physical phenomenon should be efficient, and should neither suggest nor say more than what Physics requires. Based on this, for example, it is imperative to explain why it is crucial that the topology used to describe classical Physics is *Hausdorff*; or to what end one assumes that spacetime has a smooth structure; what is the true physical meaning of the *law of inertia* once we forego the traditional tautological rhetoric; what it means, concretely even though in an idealised context, that a ruler is *ideal*; finally, what we should understand when we say, today, that the time and space of classical Physics are *absolute*, without giving up on the *Galilean invariance* that sanctions the equivalence of the rest spaces and temporal axes of all inertial frame systems. In the same spirit, as we consider the subject of great conceptual importance, the book discusses at length and with different levels of formality the formidable and mysterious relationship between dynamical symmetries and the associated constants of motion. Such a relationship has, on the one hand, proven to be so profound that it has *withstood all the revolutions of twentieth-century physics*, as we shall briefly see when we introduce the Lagrangian formalism in special relativity. On the other hand, this same relationship has become a key tool to *define* fundamental notions; for instance, linear and angular momenta, but also novel, more abstract quantities, contextually inaccessible to classical mechanics, such as *quantum mechanics* and the theory of *elementary particles*.

The first complement wrapping the book up offers a crash course on the theory of ODEs and systems of differential equations, on smooth manifolds as well. This compendium includes the most significant proofs, for instance the local and global existence and uniqueness theorem for ODE systems.

The second complement works out the mathematical structure of special relativity, introduced axiomatically in Chap. 10, starting from the physical principles that underpin it.

The Appendices summarise basic notions of analysis, point-set topology and differential geometry. The final section contains solutions and hints for solving the exercises proposed.

We should point out that the sequence in which various physical and mathematical concepts are presented does not usually correspond to their historical development, nor do the terms we use have the same meaning they had when they were first introduced. For example, Newtonian Mechanics, as discussed in Chap. 3, does not reflect the presentation in Newton's Principia but is the result of a modern approach, itself based on other people's earlier reworkings after Newton, like E. Mach. A critical historical reconstruction of the fundamental concepts of classical mechanics together with much information on the wealth of contributors, from Aristoteles to Hamilton, passing through Galilei, Newton, Euler and Lagrange, can be found in [Bis16]. The concise presentation of special relativity in Chap. 10 is completely geometrical in nature and does not delve into the difficult and genuinely physical problems such as the synchronisation of events at different places,

Prerequisites and Reference Textbooks

The backgrounds required for reading most of the material are differential and integral calculus in one and several variables, elementary notions of Geometry and Linear Algebra, basic facts from point-set Topology and the fundamentals of Physical Mechanics. The core theory of ODE systems should be known (Existence and uniqueness theorems and little else). In any case, the Complement (Chap. 14) contains all the technical results on systems of differential equations that are used in the book including proofs, with minor exceptions. The parts (chapters, section, propositions, remarks etc.) labelled $\boxed{\text{AC}}$ ("Advanced Content") need more sophisticated mathematical tools, especially regarding Differential Geometry. At any arte, all technical notions are briefly recalled before they are invoked, and the Appendices contain a detailed summary with several proofs of the more advanced concepts. The only notions that are not recapped, and appear in a small number of applications in Chap. 12 and Appendix B, regard measure theory, for which we refer to [Rud78].

Among the books with a constructive didactical purpose are the already mentioned [Bis16], and [BRSG16]. As standard references we recommend [Gol50], [Arn92] and [FaMa02] for more advanced material, complements and exercises; their content clearly goes beyond any Italian undergraduate lecture course on Analytical Mechanics. Passing to the advanced texts, beside the classical [AbMa78], an excellent modern book dedicated to the geometric formulation of classical theories, in particular Hamiltonian ones, is the recent [RuSc13]. We finally recommend the abridged [Cardin15] as advanced text for readers interested in modern results on Symplectic Geometry and its applications to classical Mechanics. A superlative introductory textbook on Special (and General) Relativity is Rindler's [Rin06].

Notations and Conventions

(0) Sections, theorems, proofs and exercises marked with $\boxed{\text{AC}}$ ("Advanced Content") are not fundamental to an introductory course, because they refer to more advanced (in particular, mathematical) topics based on the complement and the two appendices on differential geometry at the end of the text. These parts become instead important to those wishing to strengthen the formalism.

(1) – The symbol $:=$ means *equal, by definition, to*;
   – $A \subset B$ includes the possibility $A = B$ (in other books our $\subset$ would be denoted by $\subseteq$);
   – formulas of the type

$$\sum_k a_k + b \,,$$

   where b does *not* depend on the summation index, should be interpreted as

$$b + \sum_k a_k \,,$$

   unless otherwise stated.
   – the Kronecker delta will be written as δ_{ij}, δ^{ij} or δ^i_j indifferently, all having the standard meaning: 0 if $i \neq j$ and 1 if $i = j$.

(2) Vectors in affine and Euclidean spaces or $\mathbb{R}^n$ are written in boldface, e.g. $\mathbf{u}$; the exceptions are vectors in Σ_i^{3N} and vector fields on manifolds (in Appendices A and B in particular) where a normal font is often used. In Chap. 10, dedicated to special relativity, we will not use boldface for vectors in Minkowski spacetime, except for pseudo-orthonormal triples.

(3) Products of vectors.

   – The positive-definite inner product of vectors is always denoted by $\mathbf{v} \cdot \mathbf{u}$ and occasionally by $(\mathbf{v}|\mathbf{u})$.
   – The indefinite inner product of special relativity is indicated by $g(V, F)$.
   – The cross product is written $\mathbf{u} \wedge \mathbf{v}$. The same symbol is used to express the exterior product of forms, and the context should clarify which meaning $\wedge$ has.

(4) Almost everywhere the coordinates and components of (contravariant) vectors are written in the standard tensor notation with upper indices:

$$x^1, x^2, x^3, \ldots, x^n, \qquad \mathbf{v} = \sum_{j=1}^{n} v^j \mathbf{e}_j .$$

Matrices associated with linear operators follow the same tensorial convention; for example, the generic entry of an orthogonal matrix R is written $R^i{}_j$.

(5) In accordance with the standard tensorial notation, summation over certain indices will be expressed by pairs at different height, for instance

$$x'^j = \sum_{k=1}^{3} R^j{}_k x^k, \qquad \frac{dx^j}{dt} = \sum_{k=1}^{2n} S^{jk} \frac{\partial \mathcal{H}(t, \mathbf{x}(t))}{\partial x^k}, \qquad \omega = -\frac{1}{2} \sum_{i,j=1}^{2n} S_{ij} dx^i \wedge dx^j .$$

The only exceptions occur in a few sections in Chap. 5 and Sect. 7.3, where we will exclusively use lower indices to prevent the notation from becoming too cumbersome.

(6) The conventions used for the *symplectic matrix* are as follows:

$$S := \begin{bmatrix} 0 & I \\ -I & 0 \end{bmatrix} ,$$

and

$$S = [S_{ij}]_{i,j=1,\ldots,2n} , \qquad S^{ij} := S_{ij} .$$

(7) We will often denote Lagrangian and Hamiltonian coordinates in the following compact form

$$(t, q) := (t, q^1, \ldots, q^n) , \quad (t, q, \dot{q}) := (t, q^1, \ldots, q^n, \dot{q}^1, \ldots, \dot{q}^n) ,$$

$$(t, q, p) := (t, q^1, \ldots, q^n, p_1, \ldots, p_n) .$$

(8) The wedge product $\wedge$ between p-forms is defined so that:

$$dx^1 \wedge \cdots \wedge dx^p \left(\frac{\partial}{\partial x^1}, \ldots, \frac{\partial}{\partial x^p} \right) = 1 .$$

(9) If M is a differentiable manifold and we define on M a C^k vector field X, a C^k field of differential forms Ω or a function $f : M \to N$ of class $C^k(M; N)$, we implicitly assume that M (and N in the last case) are of class C^r with $r \geq k$.

(10) The *Hamiltonian flow* is denoted by various symbols throughout the book depending on the level of formality and on the context:

   (a) Φ for Hamiltonian systems on $\mathbb{R} \times \mathbb{R}^{2n}$,
   (b) $\Phi^{(Z)}$ for Hamiltonian systems on phase spacetime $F(\mathbb{V}^{n+1})$, where Z is the *Hamiltonian dynamic vector field*,
   (c) φ in the autonomous case on phase space $\mathbb{F}$, regarding the *Poincaré Theorem*,
   (d) $\varphi^{(W)}$ for autonomous Hamiltonian systems on symplectic manifolds, where W is a symplectic dynamic vector field.

   To make things worse $\Phi^{(Z)}$ is also employed to indicate the *dynamic vector field* on the *spacetime of kinetic states* $A(\mathbb{V}^{n+1})$ in the *Lagrangian formulation*. At any rate, there will be no confusion since the two contexts will always be kept distinct.

(11) In Chap. 10, the components of *four-vectors* in Minkowski frames are labelled by Greek indices. For example, V^{μ} with $\mu = 0, 1, 2, 3$. When we consider only the *spatial components*, we will use Roman letters instead, for instance V^{a} with $a = 1, 2, 3$. We shall not make use of the popular *summation convention for repeated indices*.

(12) We shall adopt the following conventions regarding real bilinear forms a : $V \times V \to \mathbb{R}$ on real vector spaces V. a is called **positive semi-definite** when $a(\mathbf{v}, \mathbf{v}) \geq 0$ for every $\mathbf{v} \in V$, and **positive definite** if it is positive semi-definite and $a(\mathbf{v}, \mathbf{v}) = 0$ implies $\mathbf{v} = \mathbf{0}$.

   A recurring situation is that in which a is the bilinear form given by a linear map $A : V \to V$, meaning $a(\mathbf{u}, \mathbf{v}) := \mathbf{u} \cdot A\mathbf{v}$ where $\cdot$ is a positive-definite inner product on V and A is self-adjoint: $A\mathbf{v} \cdot \mathbf{u} = \mathbf{v} \cdot A\mathbf{u}$. Then a is positive semi-definite if and only if the eigenvalues of A are all non-negative; it is positive definite if and only if the eigenvalues of A are strictly positive.

Acknowledgements

A heartfelt and much due thanks go to my colleague Enrico Pagani, who is much bigger expert than me on the subject. Without him this text would never have seen the light of day. Many of the issues dealt with were born out of conversations I had with Enrico over 25 years. The general geometric setup, in particular regarding the Lagrangian formulation of mechanics I have adopted, is a personal reworking, including a few of my own additions, of the critical formulation I have learnt from Enrico.

I thank Gabriele Anzellotti, Giacomo Boris, Stefano Caldini, Alessandro Casalino, Franco Caviglia, Franco Cardin, Claudio Dappiaggi, Matteo De Paris, Giorgio Ercoli, Leonardo Errati, Riccardo Ghiloni, Diego Catalano Ferraioli, Nicola Lombardi, Antonio Lorenzin, Sonia Mazzucchi, Alberto Melati, Giulia Morelli, Simone Murro, Gianluca Nardon, Marco Oppio, Alessandro Perotti, Nicola

Pinamonti, Gianmarco Puleo, Filippo Saatkamp, Lorenzo Sebastiani, Alessandro Sinibaldi and Cesare Straffelini for several suggestions on the content and technical corrections.

I am sincerely indebted to my colleague Nicolò Drago for carefully reading the first Italian edition, pointing out a number of adjustments and discussing with me many parts, eventually contributing to its improvement in a substantial way.

I am very grateful to Antonio Lorenzin and my daughter Bianca Moretti who produced the figures and thus helped to make the text more understandable.

As usual, a big thanks goes to Simon Chiossi for translating the book into English.

Trento, Italy   Valter Moretti
December 2022

General Framework of Analytical Mechanics

Analytical mechanics is first and foremost the mathematical formulation, or better, a collection of mathematical formulations, of classical mechanics produced from the eighteenth century to the present day in order to extend and make rigorous Newton's formulation of mechanics. Analytical mechanics is the attempt to isolate a possibly small number of definitions and axioms, expressible in the thorough language of geometry and analysis, from which one should deduce several properties of physical systems, especially their dynamics. The accent is therefore placed on the rigour and the mathematical conciseness, as is typical in *mathematical physics*, rather than on the physical relevance of results obtained with little care for mathematical rigour, which is the typical approach of *theoretical physics*. (The latter should by no means be regarded with contempt, considering it has produced not only fundamental physical theories, but novel mathematical ideas as well!) At the same time, the rigorous approach has eventually given rise to physically important theoretical results: it suffices to mention *Jacobi's theorem, Noether's theorem, Liouville's theorem* and *Poincaré's recurrence theorem*. Part of the methods and results ensuing from the classical setup have been subsequently subsumed under relativistic theories. The *theory of special relativity* may still be considered, from a certain perspective, a part of (evidently non-classical) analytical mechanics, and as such we treat it in this book.

From the point of view of physics, it is important to stress that the description of reality in terms of *classical* analytical mechanics has clear-cut limits of applicability, and generally speaking, it should be considered an approximation of some other deeper theory. Indeed, it is inadequate in at least two respects.

(1) The classical description no longer holds in the regime of relative velocities comparable to the speed of light/strong gravitational fields/cosmological distances and times. In these contexts, the most satisfying description known at the moment is given by the *Theory of Special Relativity*, which we shall introduce in Chap. 10 mathematically, and in Complement 15 under a more physical lens,

and by the *Theory of General Relativity*,[1] of which classical mechanics is an approximation. As we shall discuss in the sequel, axiomatically, the revolution initiated by relativity has shown how the classical metric structures (lengths and time intervals) are actually *specific to the chosen frame system*, but at the same time they form part of the metric structure of an *absolute* spacetime carrying special symmetry features (at least until we can neglect gravity, or describe it semi-classically) represented by the so-called *Lorentz-Poincaré group*. The ensuing spacetime geometry has proved to be the mathematical language necessary to deal with phenomena of general physical relevance, like the notion of *causality*. The implications of this novel viewpoint have been astonishingly fruitful and have influenced crucially the development of twentieth century physics. Relativity theory has built, jointly with quantum mechanics, the *language itself* and the *paradigm* of one hundred years of research work in theoretical physics. These theories are the foundations of the physical theories of the twenty-first century.

(2) The classical description stops being adequate, roughly speaking, also for microscopic systems (at the molecular scale or smaller). In such contexts the best description is provided by *quantum mechanics* (and at high energies, by *quantum field theory*), of which, once again, classical mechanics is an approximation. Whilst the mathematical language of relativistic theories is still the one of differential geometry, the mathematical language of quantum theories is that of *functional analysis* (*Hilbert spaces* and *operator algebras* in particular). Geometry still exists here, but is almost invariably infinite dimensional. The exception are the contexts such as *quantum information*, where finite-dimensional Differential Geometry still plays an important part.

Classical mechanics, on the other hand, works perfectly well for the most common applications, but not only those. It is remarkable to remind that the *Apollo* missions' enterprise that brought mankind to the moon was entirely conceived within the framework of classical mechanics: all models were built, and all calculations were performed, in a classical regime.

The general scheme of physical theories is far from completed, given that quantum theory and relativity theory do not amalgamate coherently. There are in particular several conceptual problems in trying to reconcile the quantum description with that of general relativity: at present we lack an exhaustive and coherent mathematical description of the physical structure of what is in existence.

---

[1] There remain unresolved issues in cosmology, also within relativistic theories, in particular regarding the so-called problem of *dark energy and dark matter*.

Contents

Chapter 1
The Space and Time of Classical Physics

In this first chapter we will discuss how *Euclidean geometry*, despite its 2300 years of age and its re-elaborations (mainly due to Hilbert) in the twentieth century, provides a perfect mathematical description of the space and time of classical Physics. By this we are referring to the physical space and the physical time as they appear to any possible *observer*, thought of as a collection of instruments (without necessarily being sentient). In the last section, where we extend the notions introduced previously, we will arrive at the concept of *differentiable manifold*, which will be useful in the rest of the book.

Appendix A completes the discussion of the present introductory chapter and summarises a few elementary notions of *point-set Topology*, it studies in more depth the concept of differentiable manifold (which will be further examined in Appendix B) and offers several examples and solved exercises.

1.1 The Mathematical Description of Space and Time in Classical Physics

Every *observer*, viewed as a collection of instruments (as we said, possibly not sentient) places physical events in a three-dimensional space and along a temporal line.

It is not uncommon to read in books that the physical space is $\mathbb{R}^3$. This is incorrect both mathematically and physically. The reason, for starters, is that the structure of $\mathbb{R}^3$ is *not* invariant under displacements (and in general it is not invariant under other physically important transformations called *isometries*, which we will see later), in contrast to the nature of Euclidean geometry's results. This non-translation-invariant structure, physically speaking, does not match the reality of our daily experience. For example, the origin $(0, 0, 0)$ of $\mathbb{R}^3$ or the axes determined by the vectors $(1, 0, 0)$, $(0, 1, 0)$, $(0, 0, 1)$ are privileged structures with no physical

© The Author(s), under exclusive license to Springer Nature Switzerland AG 2023
V. Moretti, *Analytical Mechanics*, La Matematica per il 3+2 150,
https://doi.org/10.1007/978-3-031-27612-5_1

correspondent: there is no physical law that fixes the origin of space, nor any preferred axes. Conversely, the physical laws (at least in inertial frames, which we will discuss in the next chapters) are invariant under displacements, and also under other transformations identified with the group of spatial isometries, as we will see in the sequel. In intuitive terms, to ignore the part of the structure of $\mathbb{R}^3$ that is not translation-invariant means exactly to replace $\mathbb{R}^3$ with a three-dimensional *affine space*.[1] Because of this nature these spaces admit special transformations called *displacements*, which, when the affine space is endowed with a physical meaning and in particular with the usual metric properties, correspond to the physical operations of *rigid displacements* of physical bodies. In order to speak of rigid bodies and invariant transformations other than displacements, like rotations, it becomes necessary to include additional mathematical structures to the notion of affine space, as we will see when discussing the structure of *Euclidean space*.

A similar reasoning applies to the time axis, along which each observer arranges events. In this case, too, there is no privileged point on this ordered set, and physical events obey an invariance property—although this is more difficult to grasp that the analogous feature of physical space—under *temporal displacements*, in the sense that (at least for inertial frame systems, discussed in subsequent chapters) any experiment that can be prepared today can in principle also be prepared tomorrow, giving the same result.

The three-dimensional space and the temporal axis of Classical Physics (but also four-dimensional spacetime in Special Relativity) are first of all *affine spaces*.

Let us recall the definition and main characteristics of affine spaces, and later we will pass to Euclidean space.

1.1.1   Affine Spaces

Definition 1.1 A **(real) affine space of (finite) dimension** n is a set $\mathbb{A}^n$, whose elements are called **points**, equipped with the structures described below.

(1) An n-dimensional real vector space V, called **space of displacements** or **space of free vectors**.
(2) A map $\mathbb{A}^n \times \mathbb{A}^n \ni (P, Q) \mapsto P - Q \in V$ satisfying the following properties:

  (i) for any pair $Q \in \mathbb{A}^n$, $\mathbf{v} \in V$ there is a *unique* $P \in \mathbb{A}^n$ such that $P - Q = \mathbf{v}$;
  (ii) the identity $P - Q + Q - R = P - R$ holds for any triple $P, Q, R \in \mathbb{A}^n$.

If $Q \in \mathbb{A}^n$ and $\mathbf{v} \in V$, then $Q + \mathbf{v} \in \mathbb{A}^n$ denotes the unique point P in $\mathbb{A}^n$ such that $P - Q = \mathbf{v}$. A **line** in $\mathbb{A}^n$ with **origin** P and **tangent vector u** is the map $\mathbb{R} \ni t \mapsto P + t\mathbf{u} \in \mathbb{A}^n$. A **segment** is obtained from a line by restricting t to some interval different from a point.   ◇

---

[1] About this, read the "definition" of affine space in Arnold's fundamental treatise Mathematical Methods of Classical Mechanics [Arn92, p.13].

The affine spaces considered in this book, i.e. the vector spaces of displacements, are exclusively *real* and *finite* dimensional.

Exercise 1.2

(1) Prove that $P - P = \mathbf{0}$ is the zero vector of V, for any $P \in \mathbb{A}^n$.
(2) Prove that, if $Q \in \mathbb{A}^n$ and $\mathbf{u}, \mathbf{v} \in V$ then:

$$(Q + \mathbf{u}) + \mathbf{v} = Q + (\mathbf{u} + \mathbf{v}) . \tag{1.1}$$

(3) Prove that, if $Q, P \in \mathbb{A}^n$ then:

$$P - Q = -(Q - P) . \tag{1.2}$$

(4) Prove that, if $P, Q \in \mathbb{A}^n$ and $\mathbf{u} \in V$ then:

$$P - Q = (P + \mathbf{u}) - (Q + \mathbf{u}) . \tag{1.3}$$

A **local coordinate system** on $\mathbb{A}^n$ is an *injective* mapping $\psi : U \to \mathbb{R}^n$, where $U \subset \mathbb{A}^n$ and $\psi(U) \subset \mathbb{R}^n$ is an open set. Clearly ψ identifies the points of U with the n-tuples in $\psi(U)$ bijectively. The coordinate system is called **global** if $U = \mathbb{A}^n$.

Every affine space $\mathbb{A}^n$ admits a family of global coordinate systems called *Cartesian coordinate systems*, that play a very important role in the theory. These systems are built as follows. Fix a point $O \in \mathbb{A}^n$, called *origin*, and a basis $\mathbf{e}_1, \ldots, \mathbf{e}_n$ of the space of displacements V, called *axes*. By varying $P \in \mathbb{A}^n$ the components $((P - O)^1, \ldots, (P - O)^n)$ of any vector $P - O$, in the chosen basis, define a bijective map $f : \mathbb{A}^n \to \mathbb{R}^n$ allowing to identify the points of $\mathbb{A}^n$ with the points of $\mathbb{R}^n$. This map f, that sets up a correspondence between points P and n-tuples $((P - O)^1, \ldots, (P - O)^n)$, is injective because of (1) and (3) in Exercise 1.2, since the components of a vector in a basis are uniquely determined by the vector. It is also surjective, since if $(x^1, \ldots, x^n) \in \mathbb{R}^n$ then, setting $P := O + \sum_{k=1}^{n} x^k \mathbf{e}_k$, we obviously have $f(P) = (x^1, \ldots, x^n)$.

Definition 1.3 In the affine space $\mathbb{A}^n$ with space of displacements V fix a point $O \in \mathbb{A}^n$ and a basis $\mathbf{e}_1, \ldots, \mathbf{e}_n$ of V. The global coordinate system $(\mathbb{A}^n, f)$, where f associates with $P \in \mathbb{A}^n$ the n-tuple of components of $P - O$ in the basis $\mathbf{e}_1, \ldots, \mathbf{e}_n$, is called **Cartesian coordinate system with origin O and axes $\mathbf{e}_1, \ldots, \mathbf{e}_n$**. Non-Cartesian (local) coordinate systems are called **curvilinear** coordinate systems. ◇

Cartesian coordinate systems are important also because they allow to represent so-called *affine transformations* in an easy way. An affine transformation is a transformation that preserves the affine structure.

Definition 1.4 Let $\mathbb{A}_1^n$ and $\mathbb{A}_2^m$ be affine spaces with spaces of displacements V_1 and V_2 respectively. A map $\psi : \mathbb{A}_1^n \to \mathbb{A}_2^m$ is called an **affine transformation** if the following conditions hold:

(1) ψ is *invariant under displacements*, i.e.

$$\psi(P + \mathbf{u}) - \psi(Q + \mathbf{u}) = \psi(P) - \psi(Q), \quad \text{for any } P, Q \in \mathbb{A}_1^n \text{ and } \mathbf{u} \in V_1;$$

(2) the map $P - Q \mapsto \psi(P) - \psi(Q)$ defines a linear map $V_1 \to V_2$, denoted $d\psi : V_1 \to V_2$.

An affine map $\psi : \mathbb{A}_1^n \to \mathbb{A}_2^n$ is called an **isomorphism of affine spaces** if it is bijective. In that case $\mathbb{A}_1^n$ and $\mathbb{A}_2^n$ are called **isomorphic under** ψ. $\diamond$

Remarks 1.5

(1) The map in (2) is well defined because of (1): there are infinitely many possible choices for the pair of points P, Q giving the same vector $\mathbf{v} := P - Q \in V_1$, but each one defines the same vector $d\psi(\mathbf{v}) := \psi(P) - \psi(Q) \in V_2$ by virtue of (1) (prove this fact).

(2) Note that the inverse of an isomorphism of affine spaces is an affine map (prove it) and hence an isomorphism of affine spaces.

(3) If $\psi : \mathbb{A}_1^n \to \mathbb{A}_2^n$ is an isomorphism of affine spaces, from the Proof of Proposition 1.6 below, $d\psi : V_1 \to V_2$ is a vector space isomorphism.   ∎

An important property of affine transformations is the following.

Proposition 1.6 *An affine transformation between two affine spaces of the same (finite) dimension* $\psi : \mathbb{A}_1^n \to \mathbb{A}_2^n$ *is injective if and only if it is surjective, and therefore in either case bijective.*

Proof If we fix $O \in \mathbb{A}_1^n$, then:

$$\psi(P) = \psi(O) + d\psi(P - O) \quad \forall P \in \mathbb{A}_1^n.$$

From this identity it follows easily that ψ is injective if and only if $d\psi : V_1 \to V_2$ is injective, and ψ is surjective if and only if $d\psi : V_1 \to V_2$ is surjective. The claim then follows immediately from the fact that $d\psi : V_1 \to V_2$ is injective if and only if it is surjective, since the vector spaces V_1 and V_2 have the same (finite) dimension and $d\psi$ is linear. $\square$

Exercise 1.7

(1) Let $(\mathbb{A}^n, f)$ be a Cartesian coordinate system, as in Definition 1.3, with coordinates $x^1, \cdots, x^n$ and $(\mathbb{A}^n, g)$ another Cartesian coordinate system with coordinates $x'^1, \cdots, x'^n$, origin O' and axes $\mathbf{e}'_1, \ldots, \mathbf{e}'_n$, so that

$$\mathbf{e}_i = \sum_j B^j{}_i \mathbf{e}'_j.$$

Prove that the map $g \circ f^{-1}$ is expressed, in coordinates, by the relations:

$$x'^j = \sum_{i=1}^{n} B^j{}_i (x^i + b^i), \tag{1.4}$$

where $(O - O') = \sum_i b^i \mathbf{e}_i$.

(2) Referring to the previous exercise, show that the map $f \circ g^{-1} : \mathbb{R}^n \to \mathbb{R}^n$, in coordinates, is given by:

$$x^i = -b^i + \sum_{j=1}^{n} (B^{-1})^i{}_j x'^j . \tag{1.5}$$

(3) Prove that if $\psi : \mathbb{A}_1^n \to \mathbb{A}_2^m$ is affine, then, for any choice of Cartesian coordinate systems in the two spaces $(\mathbb{A}_1^n, f_1)$ and $(\mathbb{A}_2^m, f_2)$, the *representation of ψ in coordinates*, i.e. the map

$$f_2 \circ \psi \circ f_1^{-1} : \mathbb{R}^n \ni (x_1^1, \ldots, x_1^n) \mapsto (x_2^1, \ldots, x_2^m) \in \mathbb{R}^m$$

has the form

$$x_2^i = c^i + \sum_{j=1}^{n} L^i{}_j x_1^j \quad (i = 1, 2, \ldots, m), \tag{1.6}$$

for suitable coefficients $L^i{}_j$ and c^i depending on ψ and on the coordinate systems. Above, $(x_2^1, \ldots, x_2^m)$ are the Cartesian coordinates of $\psi(P) \in \mathbb{A}_2^m$ and $(x_1^1, \ldots, x_1^n)$ those of $P \in \mathbb{A}_1^n$.

Prove that, conversely, $\psi : \mathbb{A}_1^n \to \mathbb{A}_2^m$ is affine if there exist Cartesian coordinate systems on the two spaces in which ψ has the form (1.6) in coordinates.

(4) Show that affine transformations map straight lines to straight lines. In other words, if $\psi : \mathbb{A}_1^n \to \mathbb{A}_2^n$ is affine and $P(t) := P + t\mathbf{u}$, with $t \in \mathbb{R}$, is the line in $\mathbb{A}_1^n$ with origin P and tangent vector $\mathbf{v} \in V_1$, then $\psi(P(t))$, as $t \in \mathbb{R}$ varies, defines a line in $\mathbb{A}^m$.

For any affine space $\mathbb{A}^n$ the vector space V of displacements acts as a set of transformations $\{T_\mathbf{v}\}_{\mathbf{v} \in V}$ on $\mathbb{A}^n$. The mapping $T_\mathbf{v} : \mathbb{A}^n \to \mathbb{A}^n$ of $\mathbf{v} \in V$ on $\mathbb{A}^n$ is defined in the obvious way as $T_\mathbf{v} : P \mapsto P + \mathbf{v}$. The set $\{T_\mathbf{v}\}_{\mathbf{v} \in V}$ is easily a *group* under composition, called the **group of displacements** of $\mathbb{A}^n$,, since $T_\mathbf{u} T_\mathbf{v} = T_{\mathbf{u}+\mathbf{v}}$. As $T_\mathbf{u} T_\mathbf{v} = T_{\mathbf{u}+\mathbf{v}}$ and $\mathbf{u} + \mathbf{v} = \mathbf{v} + \mathbf{u}$, the group is *Abelian*, i.e. commutative: $T_\mathbf{u} T_\mathbf{v} = T_\mathbf{v} T_\mathbf{u}$ for any pair $\mathbf{u}, \mathbf{v} \in V$.

Let us lay out some of the features of the action of V on $\mathbb{A}^n$ given by the group of displacements.

(1) The map $V \ni \mathbf{v} \mapsto T_{\mathbf{v}}$ is *faithful*, i.e. injective (since $\mathbf{v} = \mathbf{u}$ if $T_{\mathbf{u}} = T_{\mathbf{v}}$). It is a group isomorphism when we view V as Abelian group under addition.

(2) Only $\mathbf{v} = \mathbf{0}$ satisfies $T_{\mathbf{v}}(P) = P$ for some $P \in \mathbb{A}^n$, in other words the action of the group of displacements is *free*; clearly we also have the stronger condition $T_{\mathbf{0}}(P) = P$ for any $P \in \mathbb{A}^n$.

(3) For any pair $P, Q \in \mathbb{A}^n$ there is a displacement $T_{\mathbf{v}}$ such that $T_{\mathbf{v}}(P) = Q$; put in other terms, the group of displacements acts *transitively*.

Remarks 1.8 $\boxed{\text{AC}}$ More generally, given a set S, and a group G with neutral element e and product $\circ$, an **action of G on S** is a map $A : G \times S \ni (g, s) \mapsto A_g(s) \in S$, where $A_g \in \mathcal{G}_S$ and the latter is the group of bijections on S under composition, such that:

(1) $A_e = id$,

(2) $A_g A_{g'} = A_{g \circ g'}$ where $g, g' \in G$.

The action is called **free** if $A_g(s) = s$ for some $s \in S$ implies $g = e$, and **transitive** if for any $s, s' \in S$ there exists $g \in G$ such that $A_g(s) = s'$. At last, an action is called **faithful** if $G \ni g \mapsto A_g \in \mathcal{G}_S$ is injective. Note that an action always determines a group homomorphism $G \to \mathcal{G}_S$, and therefore (a) $G_S := \{A_g\}_{g \in G}$ is a subgroup of $\mathcal{G}_S$ and (b) A is faithful if and only if defines a group *isomorphism* $G \to G_S$. $\blacksquare$

The group of displacements becomes even more interesting when we enhance the affine structure by an inner product, as we will see in the next section.

1.1.2   Euclidean Spaces and Isometries

The structure of affine space is not enough to describe mathematically the nature of the physical space and of the temporal axis. We need additional mathematical structures to, on one hand, account for the experimental possibility of assigning the *physical dimensions*: *lengths, areas, volumes, angles...*, to the bodies that occupy the space, and on the other to keep track of the *duration* of the phenomena that occur. From a purely mathematical point of view affine spaces, if further equipped with the *metric structure* we are about to introduce, are called *Euclidean* because, in the one-, two- and three-dimensional cases they are the spaces of Euclid's geometry. *Three-dimensional Euclidean space* adequately models the physical space of Classical Physics. One-dimensional Euclidean space is a suitable description of the temporal axis of Classical Physics. In this case the metric structure serves to measure time distances.

From a completely abstract point of view—in particular, besides Physics—we may say that when an affine space admits an extra structure that is compatible with the affine structure and that allows to define metric properties, we have a

Euclidean space. Such a metric structure is in practice an inner product on the space of displacements.

Let us recall the relative definitions, reminding that from now on the symbol δ_{ij}, called **Kronecker delta**, is defined by $\delta_{ii} = 1$ and $\delta_{ij} = 0$ if $i \neq j$. The symbols δ_i^k and δ^{ij} mean the same. let us start with a general definition.

Definition 1.9 If V is a real vector space, an **inner product** is a map $s : V \times V \to \mathbb{R}$ that is:

(1) **bilinear**: $s(a\mathbf{v} + b\mathbf{u}, \mathbf{w}) = as(\mathbf{v}, \mathbf{w}) + bs(\mathbf{u}, \mathbf{w})$, if $a, b \in \mathbb{R}$ and $\mathbf{u}, \mathbf{v}, \mathbf{w} \in V$,
     and the analogous requirement when swapping the two arguments;
(2) **symmetric**: $s(\mathbf{u}, \mathbf{v}) = s(\mathbf{v}, \mathbf{u})$ if $\mathbf{u}, \mathbf{v} \in V$;
(3) **positive definite**: $s(\mathbf{v}, \mathbf{v}) \geq 0$ if $\mathbf{v} \in V$, with $s(\mathbf{v}, \mathbf{v}) = 0$ only if $\mathbf{v} = \mathbf{0}$.

A set of vectors $\{\mathbf{e}_i\}_{i=1,\dots,n} \subset V$ is called **orthonormal** if $s(\mathbf{e}_i, \mathbf{e}_j) = \delta_{ij}$ for $i, j = 1, \dots, n$.

The **standard norm**[2] associated with the inner product is defined by $||\mathbf{u}|| := \sqrt{s(\mathbf{u}, \mathbf{u})}$ for any $\mathbf{u} \in V$.   ◇

In the sequel we will almost always write $\mathbf{v} \cdot \mathbf{u}$ to denote the inner product[3] $s(\mathbf{v}, \mathbf{u})$.
Now we are ready to define Euclidean spaces.

Definition 1.10 An affine space $\mathbb{E}^n$ of (finite) dimension n equipped with an inner product:

$$V \times V \ni (\mathbf{u}, \mathbf{v}) \mapsto \mathbf{u} \cdot \mathbf{v} \in \mathbb{R}$$

on the space of displacements V is called **(real) Euclidean space of dimension** n.
Cartesian coordinate systems associated with orthonormal bases for the inner product $\cdot$ are called **orthonormal** coordinate systems.   ◇

Let us now show that the presence of the inner product gives meaning to the metric notions we expect of Physics: distances and angles. We recall for this the definition of *metric space*.

Definition 1.11 A **metric space** is a set M equipped with a function $d : M \times M \to \mathbb{R}$, called **distance**, satisfying:

(1) **symmetry**: $d(P, Q) = d(Q, P)$;
(2) **positivity**: $d(P, Q) \geq 0$, with equality if and only if $P = Q$;
(3) **triangle inequality**: $d(P, Q) \leq d(P, R) + d(R, Q)$ for $P, Q, R \in M$.

A function $f : M \to M'$ between metric spaces M, M' with respective distances d and d' is called an **isometry** if it *preserves the distances*: $d(P, Q) = d'(f(P), f(Q))$ for any pair $P, Q \in M$.   ◇

---

[2] See Sect. 14.2.1 for the general properties of norms.
[3] When dealing with *indefinite* inner products (Chap. 10) we will speak of a *positive definite* inner product each time we refer to Definition 1.9.

Obviously, by (2), isometries are always injective maps. In the case of Euclidean spaces, the presence of the inner product further enriches the affine structure by adding a metric structure, whenever the distance of two points of $\mathbb{E}^n$ is defined as the standard norm of $P - Q$ associated with inner product $\cdot$:

$$d(P, Q) := ||P - Q|| := \sqrt{(P - Q) \cdot (P - Q)}\,. \tag{1.7}$$

Therefore Euclidean spaces are metric spaces in a natural way. Note that the inner product on the space of displacements allows to define the measure of the *angle between two vectors*: the angle α between the vectors $\mathbf{u}$, $\mathbf{v}$, whenever it makes sense, is the unique number α (in $[0, \pi]$) such that

$$||\mathbf{u}||\, ||\mathbf{v}|| \cos \alpha = \mathbf{u} \cdot \mathbf{v}\,.$$

Let us pass to isometries between Euclidean spaces. An affine transformation between two Euclidean spaces that preserves the respective distances is called an **affine isometry**. From Proposition 1.6 it follows immediately that an affine isometry between affine spaces *of the same dimension* is bijective, and hence an isomorphism of affine spaces. Since the notion of isometry is independent of that of affine map, we could ask whether there exist isometries between Euclidean spaces that are not affine. In this respect we have the following remarkable theorem, that for Euclidean spaces of given dimension, identifies isometries with affine isometries. The proof is in the exercises.

Theorem 1.12 *Let $\mathbb{E}_1^n$ and $\mathbb{E}_2^n$ be Euclidean spaces of the same (finite) dimension n. $f : \mathbb{E}_1^n \to \mathbb{E}_2^n$ is an isometry if and only if it is an affine isometry (and in particular an isomorphism of affine spaces).*

An alternative, but equivalent, way to state the above result in more operational terms is given in the solved Exercise 1.16.**3**, which we report below as a theorem.

To reformulate the result let us recall $M(n, \mathbb{R})$ denotes the algebra of real $n \times n$ matrices and $R \in M(n, \mathbb{R})$ is called **orthogonal of order** n when $RR^t = I$ (i.e., in components, $\sum_k R^i{}_k R^j{}_k = \delta^{ij}$).

$$O(n) := \{R \in M(n, \mathbb{R}) \mid RR^t = I\}$$

is a group under matrix multiplication, called **orthogonal group of dimension** n or **rotation group in dimension** n. Exercise 1.14.**3** shows that $O(n)$ may be equivalently defined as

$$O(n) := \{R \in M(n, \mathbb{R}) \mid R^t R = I\}\,.$$

Theorem 1.12 (Reformulation) *Let $\mathbb{E}_1^n$ and $\mathbb{E}_2^n$ be Euclidean spaces of the same dimension with distances d_1 and d_2 respectively. A map $f : \mathbb{E}_1^n \to \mathbb{E}_2^n$ is an isometry if and only if, for a choice (hence any choice) of orthonormal Cartesian coordinates* *in $\mathbb{E}_1^n$ and $\mathbb{E}_2^n$, f is represented in coordinates as $(i = 1, \ldots, n)$*

$$x_2^i = b^i + \sum_{j=1}^{n} R^i{}_j \, x_1^j \, ,$$

where the coefficient matrix $R^i{}_j$ is $n \times n$, real orthogonal and the $b^j \in \mathbb{R}$ are constants. The n-tuple $(x_1^1, \ldots, x_1^n)$ are the coordinates of the generic point $P \in \mathbb{E}_1^n$ and $(x_2^1, \ldots, x_2^n)$ the coordinates of the transformed point $f(P) \in \mathbb{E}_2^n$.

Remarks 1.13

(1) The distance d on a Euclidean space (1.7) enjoys interesting properties. First, it is *invariant under displacements*:

$$d(P + \mathbf{u}, Q + \mathbf{u}) = d(P, Q) \, , \quad \text{for any } P, Q \in \mathbb{E}^n, \ \mathbf{u} \in V \, , \tag{1.8}$$

and therefore *displacements are special types of isometries*. The proof of the invariance is straightforward by the definition of d and recalling property $(P + \mathbf{u}) - (Q + \mathbf{u}) = P - Q$.

A second feature of the Euclidean distance is completely mathematical in nature: by direct computation the map $P, Q \mapsto d(P, Q)^2$ is of class C^∞ (that is, differentiable infinitely many times, including all mixed derivatives of any order) when written in orthonormal coordinates. In fact, if $(x_P^1, \ldots, x_P^n) \in \mathbb{R}^n$ and $(x_Q^1, \ldots, x_Q^n) \in \mathbb{R}^n$ are the orthonormal coordinates of P and Q respectively, it is easy to see that

$$d(P, Q)^2 = \sum_{k=1}^{n} (x_P^k - x_Q^k)^2$$

(the formula holds for any orthonormal coordinate system used to write both arguments of d). The right-hand side is a polynomial in the $2n$ variables x_P^k and x_Q^h and therefore smooth in each, including all possible mixed derivatives of any order. The map $P, Q \mapsto d(P, Q)$ in Cartesian coordinates is instead everywhere continuous, but not C^∞ everywhere, since it admits a singularity exactly for $P = Q$.

(2) As a consequence of Exercise 1.7.**1**, keeping into account that orthogonal matrices are associated with changes of orthonormal bases, the most general transformation rule between the coordinates of different orthonormal coordinate systems takes the form

$$x'^j = \sum_{i=1}^{n} R^j{}_i (x^i + b^i), \tag{1.9}$$

where the real numbers b^i are arbitrarily fixed and the coefficients $R^j{}_i$ determine a given *orthogonal matrix* of dimension n. The formula may be recast, renaming the numbers $\sum_{i=1}^{n} R^j{}_i b^i$ by c^j, as

$$x'^{j} = c^{j} + \sum_{i=1}^{n} R^{j}{}_{i} x^{i} .$$

(3) From the definition of $O(3)$, using Binet's determinant formula and the fact that the determinant of the transpose equals the original determinant, we immediately obtain that if $R \in O(3)$ then $\det R = \pm 1$. Both signs can occur in $O(3)$. In fact I and $-I$ belong to $O(3)$ and have determinant 1 and -1 respectively. The subset of matrices with determinant 1 is denoted by $SO(3)$, and it is easy to see that it is a subgroup of $O(3)$. $SO(3)$ is called **special orthogonal group** (of dimension 3) and its elements are the **proper rotations**. Matrices with determinant -1 cannot form a subgroup since their set does not contain the neutral element (the matrix $I \in SO(3)$). However, if $S \in O(3)$ and $\det S = -1$, then $S = (-I)R$, where $R := (-I)S \in SO(3)$. Hence all rotations with determinant -1, called **improper rotations** or, with a slightly imprecise term, **reflections**, arise from a proper rotation followed by the action of $-I$; the latter is called **inversion** or **parity inversion**. This justifies the notation $-ISO(3)$ for the set of improper rotations (of dimension 3). An important example of improper rotation is the reflection of an object through a mirror. With respect to an orthonormal coordinate system, if we imagine the mirror is the plane $x = 0$, the improper rotation corresponds to the diagonal matrix with -1 as first element and two further 1s. The determinant clearly equals $(-1) \cdot 1 \cdot 1 = -1$.

(4) ⟨AC⟩ There is a fully topological characterisation of $SO(3)$ and $-ISO(3)$ of interest in Physics. Let us view proper and improper rotations as *active transformations*, meaning they act on geometric objects by transforming them (see the next section). As is evident from the physical experience, improper rotations, such as reflections through a mirror, are "discontinuous" physical operations: they cannot be obtained by a sequence of "small modifications" of the initial figure. This intuitive idea has a precise mathematical correspondent that we set out to explain without all the details. The group $O(3)$ is a subset of $\mathbb{R}^9$, since real 3×3 matrices can be thought of as vectors in $\mathbb{R}^9$. If we put on $O(3)$ the induced topology of $\mathbb{R}^9$, the group multiplication and inversion are continuous operations. In this sense $O(3)$ is a *topological group*. This fact is actually true for the whole isometry group of a Euclidean space, but at present we shall only deal with the subgroup of rotations. Staying with subsets in a topological space, note that $O(3)$ is not a connected subset of $\mathbb{R}^9$ because the map $\det : O(3) \to \mathbb{R}$ is continuous and has the disconnected set $\{1, -1\}$ as range, whereas the continuous image of a connected set must be connected. That means the two subsets $SO(3)$ and $-ISO(3)$, apart from being disjoint, disconnect $O(3)$. It can be proved that both are connected, so

$SO(3)$ and $-ISO(3)$ are the (two) connected components of $O(3)$. The former is the one containing the identity. Observe at last that if an improper rotation, such as a reflection, $S \in -ISO(3)$ could be joined to I by a continuous path $\gamma : [a, b] \to SO(3)$ with $\gamma(a) = I$ and $\gamma(b) = S$, i.e. a path of rotations, then I and S would belong to the same connected component of $O(3)$ (the path's image $\gamma([a, b])$ is connected as continuous image of a connected set). But as we have just seen, this is false. More intuitively, we have just explained that improper rotations, like any reflection through a mirror, are "discontinuous" physical operations: they cannot be obtained by a sequence of "small modifications" of the initial figure, that is by a sequence of small rotations starting from the identity. ∎

Exercise 1.14

(1) Prove that any inner product s as per Definition 1.9 satisfies the **Cauchy-Schwarz inequality** :

$$|s(\mathbf{u}, \mathbf{v})| \leq ||\mathbf{u}||\, ||\mathbf{v}||, \quad \mathbf{u}, \mathbf{v} \in V . \tag{1.10}$$

(2) Prove that if $\{\mathbf{e}_1, \ldots, \mathbf{e}_n\}$ is an orthonormal basis of V then $\mathbf{v} = \sum_{j=1}^{n} v^j \mathbf{e}_j$ implies $v^j = \mathbf{v} \cdot \mathbf{e}_j$ if $j = 1, \ldots, n$. If furthermore $\mathbf{u} = \sum_{j=1}^{n} u^j \mathbf{e}_j$ then $\mathbf{v} \cdot \mathbf{u} = \sum_{j=1}^{n} v^j u^j$.

(3) Prove that $O(n) := \{R \in M(n, \mathbb{R}) \mid RR^t = I\}$ may be defined equivalently by $O(n) := \{R \in M(n, \mathbb{R}) \mid R^t R = I\}$. In other words, show that, according to the first definition, $R \in O(n)$ if and only if $R^t \in O(n)$.

(4) Prove that the norm $||\mathbf{u}|| := \sqrt{\mathbf{u} \cdot \mathbf{u}}$ associated with an inner product satisfies the **triangle inequality** :

$$||\mathbf{u} + \mathbf{v}|| \leq ||\mathbf{u}|| + ||\mathbf{v}||, \quad \mathbf{u}, \mathbf{v} \in V ,$$

so the distance on a Euclidean space defined as in (1.7) satisfies the triangle inequality for distances, $d(P, Q) \leq d(P, R) + d(R, Q)$, seen in Definition 1.11.

1.1.3   The Isometry Group of $\mathbb{E}^n$ and the Active and Passive Interpretation

Using directly the characterisation of isometries in orthonormal coordinates proved in Theorem 1.12, for example, one easily proves that for a given Euclidean space $\mathbb{E}^n$, *the collection of isometries $\psi : \mathbb{E}^n \to \mathbb{E}^n$ forms a group under composition*. This group is called the **isometry group of** $\mathbb{E}^n$. Thus the composite of two isometries is an isometry and the inverse of an isometry exists and is an isometry. We can say more about the structure of this group by studying its action[4] on $\mathbb{E}^n$. If we fix an orthonormal coordinate system on $\mathbb{E}^n$, in these coordinates it is easy to show that the most general form of an isometry $\psi : \mathbb{E}^n \to \mathbb{E}^n$ mapping the generic $P \in \mathbb{E}^n$ of coordinates $(x^1, \ldots, x^n)$ to $\psi(P) \in \mathbb{E}^n$ of coordinates $(x'^1, \ldots, x'^n)$, is still given by (1.9), where the real numbers b^i and the matrix R are fixed and depend on ψ. This kind of transformations are called **active**, because they act by "displacing" points in space. As a consequence of that explicit form for the group elements we immediately see that the isometry group includes as subgroups the group of displacements and that of rotations about a given point (the origin of the coordinates chosen by representation (1.9)). The isometries of three-dimensional physical space are therefore exactly the *roto-translations*. Given an orthonormal coordinate system, these are precisely the maps of the form (1.9) interpreted as *active transformations*. That the isometry group contains the displacements, given the affine structure of $\mathbb{E}^n$, implies that the group acts *transitively* on $\mathbb{E}^n$: for any pair of points $P, Q \in \mathbb{E}^n$ there is an isometry (in particular, a displacement) mapping P to Q. The fact that the group contains the rotations about a point implies that the isometry group's action, in contrast to that of displacements alone, is no longer *free*: there exist transformations that fix some $P \in \mathbb{E}^n$ and are not the identity, since any rotation around P fixes P.

As we will see shortly, the isometries of $\mathbb{E}^3$, physically, are the operations we can perform actively on bodies without altering the metric properties.

We stress again that by working in a unique and fixed $\mathbb{E}^n$, we have now interpreted the isometric transformations (1.9) **actively**, i.e. as transformations between points of $\mathbb{E}^n$. However (1.9), exactly as we did in (2) of Remark 1.13, can be interpreted **passively** as well, i.e. with reference to two *distinct* orthonormal coordinate systems in $\mathbb{E}^n$, where we describe the *same* point, which is not "displaced" but just described in different coordinates. Relations (1.9) then describe how the coordinates of one point, in two orthonormal coordinate systems, are related. In this case the map in coordinates is simply the identity map $\mathbb{E}^n \ni P \mapsto P \in \mathbb{E}^n$. The fact that it does not look like the identity when we pass to the coordinates in $\mathbb{R}^n$ is because we are using *two* different systems of coordinates, *both orthonormal*, on the domain and codomain. The two interpretations, active and passive, are both employed in the physical applications.

1.1.4 Invariant Arclengths, Areas and Volumes Under the Isometry Group

In the construction of physical theories not only distances and angles are important, but also other mathematical objects like areas and volumes. We shall not address

---

[4] AC   Abstractly, we may think of the action of the isometry group on $\mathbb{E}^n$ as a *right group action* (faithful by construction) in the mathematical sense of Remark 1.8, as we have done for the group of displacements.

how these notions are defined starting from the structure of Euclidean space, but we will simply illustrate a few general considerations.

Areas and *volumes* of measurable sets, in the sense of the *Peano-Jordan-Riemann measure*, can be constructed starting from the concepts of distance and angle used to define rectangles, parallelepipeds and their surfaces and volumes in $\mathbb{E}^2$ and $\mathbb{E}^3$. The notion generalises to any $\mathbb{E}^n$. Exactly as for distances, the volume measures in any $\mathbb{E}^n$ obtained via the Peano-Jordan-Riemann process are *invariant under the action of the isometries of* $\mathbb{E}^n$. In other terms, if for example we let an isometric transformation $\psi : \mathbb{E}^n \to \mathbb{E}^n$ acts on a set $G \subset \mathbb{E}^n$ with volume $vol(G)$, then $\psi(G)$ has volume equal to $vol(\psi(G)) = vol(G)$. The more sophisticated and powerful approach, based on the theory of the *Lebesgue measure*, requires some additional technical tools; nonetheless, the notion of Euclidean space $\mathbb{E}^n$ naturally detects a unique concept of (Lebesgue) measure that is invariant by the isometry group of $\mathbb{E}^n$ and assigns the usual value to the volumes of n-rectangles $[a_1, b_1] \times \cdots \times [a_n, b_n]$ of $\mathbb{E}^n$.

The assignment of an inner product in V that automatically comes with a Euclidean space allows to define, in a completely similar way to that of $\mathbb{R}^n$, the concept of *length of a rectifiable curve*, for sufficiently regular curves $\gamma : [a, b] \to \mathbb{E}^n$. We will examine this notion in the next chapter for $n = 3$. In the same way, when $n = 3$, the inner product of $\mathbb{E}^3$ allows to define, in analogy to what one does in $\mathbb{R}^3$, the notion of *area of a rectifiable surface*, for sufficiently regular surfaces $P = P(u, v) \in \mathbb{R}^3$, with (u, v) in some open set in $\mathbb{R}^2$. Both notions are *invariant under the isometry group* of the corresponding Euclidean space.

1.1.5 Orientation of Euclidean Spaces and Cross Product

The space of displacements V of a Euclidean space $\mathbb{E}^n$, or more generally of an affine space $\mathbb{A}^n$, is naturally *orientable*. Let us review this important notion. If $\mathcal{B}$ is the collection of all bases of V, and $A, B \in \mathcal{B}$, with $A = \{e_r^{(A)}\}_{r=1,2,\dots,n}$ and $B = \{e_r^{(B)}\}_{r=1,2,\dots,n}$, we write $M(A, B)$ for the $n \times n$ matrix of change of basis, that is, the matrix with coefficients $M(A, B)^j{}_i$ given by

$$e_i^{(A)} = \sum_{j=1}^{n} M(A, B)^j{}_i \, e_j^{(B)} .$$

As $M(A, B)$ is non-singular, the determinant is non-zero and so it can be either positive or negative. The relation

$$\forall A, B \in \mathcal{B} , \quad A \sim B \quad \text{if and only if } \det M(A, B) > 0$$

is an equivalence relation with two equivalence classes. The collection $\mathcal{B}$ naturally decomposes as union of such disjoint equivalence classes. The choice of one equivalence class, containing so-called **positively oriented** bases, is an **orientation** of V and of the associated Euclidean (or affine) space.

Remarks 1.15 In the case of a Euclidean space $\mathbb{E}^3$, such as our physical space, the two equivalence classes of bases contain so-called **right-handed** and **left-handed** bases. The former type includes the triple formed by our right hand's thumb, pointing finger and middle finger, in this order. The other one is similar but on the left hand. *It is customary to choose as positively oriented triples the right-handed ones, and we shall follow this convention.* ∎

In case of the oriented $\mathbb{E}^3$, the Euclidean space's orientation defines an orientation for the rotations about a given axis $\mathbf{u}$. A rotation $R \in SO(3)$ by angle $\theta \in (0, 2\pi)$ about $\mathbf{u}$ is called **positive** if, given a vector $\mathbf{v} \neq \mathbf{u}, \mathbf{0}$, the triple $\mathbf{v}, R\mathbf{v}, \mathbf{u}$ is right-handed. For the oriented $\mathbb{E}^3$ the **cross product** of two vectors $\mathbf{u}, \mathbf{v} \in V$ forming an angle $\alpha \in [0, \pi]$ is, as is well known, the vector $\mathbf{u} \wedge \mathbf{v} \in V$ of modulus $||\mathbf{u}|| ||\mathbf{v}|| \sin \alpha$, normal to $\mathbf{u}$ and $\mathbf{v}$ and oriented so that $\mathbf{u}, \mathbf{v}, \mathbf{u} \wedge \mathbf{v}$ is a right-handed triple (provided none of the vectors has zero modulus).

We remind that the map $\wedge : V \times V \to V$ is linear in the first argument:

$$(a\mathbf{u} + b\mathbf{v}) \wedge \mathbf{w} = a(\mathbf{u} \wedge \mathbf{w}) + b(\mathbf{v} \wedge \mathbf{w}) \quad \text{for any } a, b \in \mathbb{R} \text{ and any } \mathbf{u}, \mathbf{v}, \mathbf{w} \in V,$$

and *skew-symmetric*:

$$\mathbf{u} \wedge \mathbf{v} = -\mathbf{v} \wedge \mathbf{u} \quad \text{for any } \mathbf{u}, \mathbf{v}, \in V,$$

whence also linear in the second argument. The $\wedge$ operation is *not associative* though. Put otherwise, and excluding special choices: $(\mathbf{u} \wedge \mathbf{v}) \wedge \mathbf{w} \neq \mathbf{u} \wedge (\mathbf{v} \wedge \mathbf{w})$. The definition we have given above for the cross product is *equivalent* to the determinant rule:

$$\mathbf{u} \wedge \mathbf{u} = (u^2 v^3 - u^3 v^2)\mathbf{e}_1 - (u^1 v^3 - u^3 v^1)\mathbf{e}_2 + (u^1 v^2 - u^2 v^1)\mathbf{e}_3,$$

as long as the basis $\mathbf{e}_1, \mathbf{e}_2, \mathbf{e}_3$ *of* V *is right-handed and orthonormal, and* $\mathbf{u} = \sum_{j=1}^{3} u^j \mathbf{e}_j$ *and* $\mathbf{v} = \sum_{j=1}^{3} v^j \mathbf{e}_j$.

We leave the proofs of these facts to the reader. There is an alternative definition of cross product based on the notion of *pseudo-vector*, which we will not use.

Exercise 1.16

(1) Let $\mathbb{E}^n$ be a Euclidean space with distance d. Given a point $O \in \mathbb{E}^n$, identify (bijectively) the vectors of the space of displacements V of $\mathbb{E}^n$ with the points of $\mathbb{E}^n$ using $\mathbf{u} \mapsto O + u$. Prove that the inner product $\cdot$ on V may be written in terms of d as:

$$\mathbf{u} \cdot \mathbf{v} = \frac{1}{2} d (O, O + (\mathbf{u} + \mathbf{v}))^2 + \frac{1}{2} d (O, O + (\mathbf{u} - \mathbf{v}))^2. \tag{1.11}$$

(2) Let $\mathbb{E}_1^n$ and $\mathbb{E}_2^n$ be Euclidean spaces of the same dimension n, with distances d_1, d_2 and inner products $(\cdot|\cdot)_1$ and $(\cdot|\cdot)_2$ respectively. Let $\phi : \mathbb{E}_1^n \to \mathbb{E}_2^m$ be an affine transformation. Prove that $d\phi$ preserves the inner product (and therefore the angle between vectors), i.e. for any given $Q \in \mathbb{E}_1^3$:

$$(\phi(P) - \phi(Q)|\phi(P') - \phi(Q))_2 = (P - Q|P' - Q)_1, \ \forall P, P' \in \mathbb{E}_1^3. \tag{1.12}$$

if and only if ϕ preserves distances, i.e.

$$d_2(\phi(P), \phi(Q)) = d_1(P, Q), \ \ \forall P, Q \in \mathbb{E}_1^3. \tag{1.13}$$

(3) Let $\mathbb{E}_1^n$ and $\mathbb{E}_2^n$ be Euclidean spaces of the same dimension with respective distances d_1 and d_2. Show that a transformation $\phi : \mathbb{E}_1^n \to \mathbb{E}_2^n$ is an isometry if and only if, for one (hence very) choice of orthonormal coordinates in $\mathbb{E}_1^n$ and $\mathbb{E}_2^n$, ϕ has the form ($i = 1, \ldots, n$)

$$x_2^i = b^i + \sum_{j=1}^{n} R^i{}_j \, x_1^j, \tag{1.14}$$

where the coefficient matrix $R^i{}_j$ is $n \times n$, real, orthogonal.

(4) Given affine spaces $\mathbb{A}^n$ and $\mathbb{A}^m$, a map $\psi : \mathbb{A}^n \to \mathbb{A}^m$ is called a *diffeomorphism* when it is (1) bijective and (2), if we represent ψ and ψ^{-1} in Cartesian coordinates in $\mathbb{A}^n$ and $\mathbb{A}^m$ as maps $\mathbb{R}^n \to \mathbb{R}^m$ and $\mathbb{R}^m \to \mathbb{R}^n$ respectively, these functions are everywhere differentiable infinitely many times (i.e. of class C^∞). Considering the isometries of two Euclidean spaces, show that:

  (i) the isometries of two Euclidean spaces of the same dimension are diffeomorphisms;

  (ii) inverses of isometries are isometries;

  (iii) the composite of two isometries is an isometry;

  (iv) if $\phi : \mathbb{E}_1^n \to \mathbb{E}_2^n$ is an isometry then $d\phi : V_1 \to V_2$ is a vector space isomorphism that preserves the inner product.

(5) Show that isometries $\phi : \mathbb{E}^n \to \mathbb{E}^n$ form a group, actually a subgroup of diffeomorphisms of the Euclidean space $\mathbb{E}^n$.

(6) $\boxed{\text{AC}}$ Prove Theorem 1.12.

1.2 Space and Time for an Observer: Physical Correspondences

Let us go back to the three-dimensional physical space, viewed as $\mathbb{E}^3$, and the time axis, viewed as $\mathbb{E}^1$, and discuss a few aspects of physical nature.

Physically speaking, distances, inner products and angles formed by segments and vectors in $\mathbb{E}^3$ are measured by fixing a class of *ideal rigid rulers* that we assume are available at every point in space at any time. Translations, given by vectors **v** in $\mathbb{E}^3$, should be thought of as physical displacements of material bodies. Hence there must exist bodies (at least the rigid rulers!) that are metrically invariant under physical displacements, just like the distance d associated with the inner product.

The time axis' structure is motivated similarly: using *ideal clocks* we can translate the time instants of a given interval, but also measure the time distance between two instants using an invariant distance under temporal displacements. In these operations the choice of a time origin does not enter the game, nor is there a physical law prescribing it. In other words, we are formalising the time axis as a 1-dimensional affine space with an inner product on the displacement space,[5] i.e. a space $\mathbb{E}^1$. Further, note that the inner product of $\mathbb{E}^1$ determines a pair of unit vectors in the displacement space. Choosing one of the two as privileged defines the *future time orientation*. More precisely, the time axis should be thought of as an $\mathbb{E}^1$ together with the choice of one of the two normalised vectors.

The above description for the space and time of a given observer is valid both in Classical Physics and Special Relativity (at least for so-called inertial observers). The differences between the two theories emerge when one considers different observers, as we will see in Chap. 10.

Remarks 1.17 The moment we fix an origin in $\mathbb{E}^1$, that space is identified with $\mathbb{R}^1$ under the only (ortho)normal coordinate system associated with the origin and the basis formed by the future-pointing unit vector. The positive real semi-axis of $\mathbb{R}$ defines the future direction in coordinates. In components, the inner product of the space of displacements of $\mathbb{E}^1$ is nothing but the usual product of real numbers. From this point of view the time axis corresponds to an "$\mathbb{R}^1$ with no chosen origin". ∎

1.2.1 Rigid Rulers and Ideal Clocks

We should ask ourselves what it means, physically, to speak of the class of *ideal rigid rulers* and the class of *ideal* clocks. For a ruler to be *ideal* refers to the following property which we take as true. When choosing any two rulers, they coincide if they are at rest in the same position in an arbitrary frame, and this continues to be true after they move (including accelerated motions) once they are returned to relative rest in some frame (possibly different from the initial one). The same criterion applies for ideal clocks used to measure time distances on the temporal axis: two

---

[5] To produce a translation-invariant distance on an affine space it would suffice to have a norm on the space of displacements. But assigning a norm on a *one-dimensional real* vector space is completely equivalent to fixing an inner product that generates the given norm. Such an inner product is unique. If **e** is a vector with $||\mathbf{e}|| = 1$ then the unique inner product generating the norm is $\mathbf{v} \cdot \mathbf{u} = uv$, and clearly $||\mathbf{u}|| = |u|$ where $\mathbf{v} = v\mathbf{e}$, $\mathbf{u} = u\mathbf{e}$.

clocks keep time in the same way when they are at rest in the same place in some frame, and this stays true if the clocks undergo changes but return to relative rest in some frame (possibly other than the initial one, and even if they are no longer synchronised as they were initially).

Remarks 1.18 [AC] The notions of *ideal* rulers and clocks described above are valid in Relativistic Physics as well, though more care is required to construct the corresponding mathematical entities. In general, when relinquishing the classical formulation the notion of space at rest with an observer or a frame is more complicated to describe mathematically, and is no longer conceived in terms of a Euclidean space as just seen. Rather, it is described by a smooth manifold of dimension 3 endowed with a positive-definite metric: a *Riemannian* 3-manifold. Ideal rulers should then be thought of, in this context, as represented by ("infinitesimal") vectors in the *tangent space* to the manifold, moving from point to point by the procedure called *parallel transport* with respect to the *Levi-Civita connection*. Any attempt to interpret rulers as true subsets of the manifold (corresponding to the finite segments in Euclidean spaces) would clash with the absence of a group of isometries acting transitively on a generic Riemannian manifold with non-constant curvature: there would be no way to describe the operations of physical displacement of a ruler from one place to another without altering the dimensions. Finally, note that the geometry of the rest space, in this context, might depend on time. ∎

1.2.2 Existence of Physical Geometry

It is important to emphasise that we cannot use the same measuring instrument at all scales: for example we cannot measure astronomical distances, nor intermolecular ones, with a 1-metre ruler. We need different types of instruments for different scales (and we can also use different instruments at the same scale, too, for example optical instruments instead of rulers). The fact that different measuring instruments of the same geometric quantity working at the same or at distinct scales behave coherently, for instance giving the same result at intermediate scales where we can use instruments of different sorts simultaneously, corresponds to the idea that *there exists a geometry that is independent of the measuring instruments*. This fact, which we should consider an experimental evidence, is far from obvious even if we consider it completely natural because we experience it directly and indirectly every day. An analogous discussion goes for the time axis and ideal clocks.

Because these facts are *experimental*, the existence of a geometry in the aforementioned sense might be disproved when working in unusual conditions. In Physics, in fact, there has been much speculation about the possible discontinuous nature of space and even time at very small scales: *Planck's length and time*

$$\ell_P := \sqrt{\frac{\hbar G}{c^3}} \sim 10^{-33} \text{ cm}, \quad T_P = \frac{\ell_P}{c} \sim 10^{-43} \text{ s},$$

at which some sort of *Quantum Gravity* should hold. These scales are in fact defined using the fundamental constants of quantum theory, gravity and relativity. These are: the *(reduced) Planck constant* $\hbar$, the *gravitational constant G* and the *speed of light c*, respectively. At such scales the classical structure of space and time discussed in this chapter, and also a weaker notion of differentiable manifold (as in general relativistic theories) or even of topological space locally homeomorphic to $\mathbb{R}^4$, would cease to be physically appropriate, perhaps even allowing for a geometric object of other nature. Maybe some version of *non-commutative geometry*. There is no direct or indirect experimental evidence of all of that for the time being. No direct evidence is accessible through our current technology.

1.3   Introduction to the Notion of Differentiable Manifold

The most general and powerful mathematical tool apt to describe the overall properties of three-dimensional physical space, spacetime and the abstract space in which we study the physical systems of classical theories, is the notion of *differentiable manifold*. It was essentially created by B. Riemann in the nineteenth century, and we shall introduce it here as our final mathematical tool. Appendices A and B review other notions and results from Differential Geometry.

1.3.1   Classes of Differentiable Maps

We start with the following technical definition, stated here once and for all and uses in the remainder of the book.

Definition 1.19 Fix integers $n, m = 1, 2, \ldots$ and $k = 0, 1, \ldots$ and a non-empty open set $\Omega \subset \mathbb{R}^n$.

(1) A map $f : \Omega \to \mathbb{R}^m$ is said to be **of class** C^k if all the partial derivatives (mixed ones too) of the components of f exist and are continuous up to order k included. The set of such functions is denoted $C^k(\Omega; \mathbb{R}^m)$ and we define $C^k(\Omega) := C^k(\Omega; \mathbb{R})$.

(2) $f : \Omega \to \mathbb{R}^m$ is said of **class** C^∞ if it is of class C^k for any $k = 0, 1, \ldots$ and we put:

$$C^\infty(\Omega; \mathbb{R}^n) := \bigcap_{k=0,1,\ldots} C^k(\Omega; \mathbb{R}^n) \quad \text{and} \quad C^\infty(\Omega) := C^\infty(\Omega; \mathbb{R}) . \qquad \diamond$$

1.3.2 Local Charts and Differentiable Manifolds

Informally speaking, a differentiable manifold is a set M of arbitrary objects[6] generically called *points*, that can be covered by *coordinate systems*, i.e. injective maps $\psi : U \rightarrow \mathbb{R}^n$, where the number n should not depend on the $U \subset M$ considered, and the union of all domains U is M. Thus the various pieces U of M covered by a corresponding coordinate system ψ can be put, separately, in one-to-one correspondence with portions $\psi(U)$ of $\mathbb{R}^n$. With each point $p \in U$ we bijectively associate its coordinates $(x^1(p), \ldots, x^n(p)) := \psi(p) \in \psi(U) \subset \mathbb{R}^n$. In this way the set M can be treated *locally* as if it were $\mathbb{R}^n$, even if it is not $\mathbb{R}^n$, by choosing to work in the above coordinates. When two coordinate systems (U, ψ) and (V, ϕ) are such that $U \cap V \neq \varnothing$, on the intersection we can use ψ or ϕ indifferently. We then require a simple compatibility condition: the maps $\psi \circ \phi^{-1}$ and $\phi \circ \psi^{-1}$, respectively defined on $\phi(U \cap V) \subset \mathbb{R}^n$ and $\psi(U \cap V) \subset \mathbb{R}^n$ and called **transition functions**, must be *differentiable with continuity* up to some order k independent of U and V.

In the end we will be able to transfer to the set M the mathematical notions, fundamental to Physics, that are originally only defined on $\mathbb{R}^n$. In particular we can make the concept of *differentiable* function precise, or of curve (of any class) defined on M, simply by reading them as functions of the local coordinates and "imagining" of being in $\mathbb{R}^n$. In realty, since each local coordinate system only covers a part of M, to develop the theory we must account for how the various coordinates are patched together in the domains' intersections. In this way one can for example develop the theory of differential equations on M to describe the evolution equations of physical systems constrained to "live" on M: think of a surface M on which a point particle moves, or the spacetime M where the physical system evolving in it, described in terms differential equations, is an electromagnetic field.

To render viable the definition of differentiable manifold that we have thus far introduced in words only, we shall make some topological assumptions. The first requirements are intended to specify the structure of coordinate systems better. We assume from the start that the sets $\psi(U)$ are *open* in $\mathbb{R}^n$ (where the common notion of differentiability we want to use makes sense). Open subsets $V \subset \psi(U) \subset \mathbb{R}^n$ will correspond to subsets $\psi^{-1}(V) \subset U$. It is natural to view each $\psi^{-1}(V) \subset M$ as open in some pre-existing topology on M, which is identified with the topology of $\mathbb{R}^n$ in coordinates. In other words we suppose from the start M is a topological space, that the domains U are open in M in that topology and that the bijections $\psi : U \rightarrow \psi(U)$ are continuous with continuous inverses. Now we may define *coordinate systems* formally.

---

[6] for example the points on a surface in $\mathbb{R}^3$, although this example is very special.

Definition 1.20 Let M be a topological space. A **local coordinate system**—or **local chart**—of dimension n on M is a **local homeomorphism** from M to $\mathbb{R}^n$, i.e. a pair (U, ϕ) with $U \subset M$ and $\phi : U \to \mathbb{R}^n$ such that:

(1) U is open in M,
(2) $\phi(U)$ is open in $\mathbb{R}^n$,
(3) $\phi : U \to \phi(U)$ is a homeomorphism (where $\phi(U)$ has the induced topology of $\mathbb{R}^n$).

For any $p \in U$ the numbers $(x^1(p), \cdots, x^n(p)) = \phi(p)$ are the **coordinates** of p in the chart (U, ϕ).

Two n-dimensional local charts (U, ϕ) and (V, ψ) on M are C^k-**compatible** (or k-**compatible** for short) if the transition functions

$$\phi \circ \psi^{-1} : \psi(U \cap V) \to \phi(U \cap V)$$

and

$$\psi \circ \phi^{-1} : \phi(U \cap V) \to \psi(U \cap V)$$

are both of class C^k, or if $U \cap V = \varnothing$. $\diamond$

Not all topological spaces admit local charts, but some do: every affine space for example, with respect to its natural topology. A topological space admitting C^0-compatible local charts around each point is called **locally Euclidean**.

Back to the general setup, in order for M to be a differentiable manifold we impose two further conditions, which require a technical reminder.

Definition 1.21 Let M be a topological space.

(1) M is called **Hausdorff** if, for any distinct $p, q \in M$, there exist two open sets U_p, U_q such that $U_p \ni p$, $U_q \ni q$ and $U_p \cap U_q = \varnothing$.
(2) M is called **second countable** if there is a **countable basis** of open sets for the topology, i.e. every open set is a countable union of elements of the basis. $\diamond$

$\mathbb{R}^n$ and all affine spaces $\mathbb{A}^n$ with the topology of $\mathbb{R}^n$ induced by any Cartesian coordinate system (the topology thus obtained does not depend on the chosen Cartesian coordinate system) are certainly second countable Hausdorff spaces. In that case the two sets U_p and U_q can be taken to be open balls centred (in coordinates) at p and q respectively, of sufficiently small radii. The countable basis can always be chosen, given Cartesian coordinates, as the set of open balls with rational radius and centred at points with rational coordinates.

Definition 1.22 A **differentiable manifold of dimension** n and **class** C^k, for given $n \in \{1, 2, 3, \cdots\}$ and $k \in \{1, 2, \ldots, \} \cup \{\infty\}$, is a locally Euclidean, Hausdorff and second countable topological space M, whose elements are called **points**, equipped with a **differentiable structure of class** C^k and **dimension** n. The differentiable

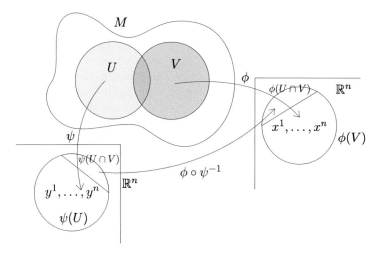

Fig. 1.1 Two local charts on the manifold M

structure is a collection of n-dimensional local charts $\mathcal{A} = \{(U_i, \phi_i)\}_{i \in I}$ satisfying the following:

(1) $\cup_{i \in I} U_i = M$;
(2) the local charts in $\mathcal{A}$ must be pairwise C^k-*compatible*;
(3) $\mathcal{A}$ is **maximal** with respect to (ii): if (U, ϕ) is an n-dimensional local chart on M compatible with every chart of $\mathcal{A}$, then $(U, \phi) \in \mathcal{A}$. $\diamond$

Except for very special manifolds, in general it is not possible to find one chart that covers a manifold entirely. A local chart with domain the entire M is called a **global chart** or a **global coordinate system** (think a Cartesian coordinate system on an affine space). We may say, in general terms, that *being covered by several local charts* is the property that best characterises the idea of a manifold, or at least the basic reason for which this mathematical notion was invented: to handle objects that are locally, but not globally, identifiable with pieces of $\mathbb{R}^n$, and such identification can be made in many, all legitimate, ways (Fig. 1.1).

A collection of local charts $\mathcal{A}$ on the differentiable manifold M that satisfies (1) and (2) but maybe not (3) is called an **atlas** on M **of dimension** n and **class** C^k. It is easy to prove that for any atlas $\mathcal{A}$ on M there is a unique maximal atlas containing it, i.e. a unique differentiable structure containing it. This structure is said to be **induced by the atlas**. Note that two atlases on M such that every chart of one is compatible every chart of the other induce the same differentiable structure on M. Hence to assign a differentiable structure it suffices to prescribe an atlas, one of the many that determine the differentiable structure.

Examples 1.23

(1) The simplest and in some sense most obvious and "useless" example of differentiable manifold, of class C^∞ and dimension n, is any non-empty open subset of $\mathbb{R}^n$ (including $\mathbb{R}^n$ itself) with the natural topology (as well known, this is Hausdorff and second countable) and standard differentiable structure given by the identity map, which by itself defines an atlas. All of this can be extended to affine and Euclidean spaces. Observe first that every affine space $\mathbb{A}^n$ is a second countable, Hausdorff topological space. The topology with these properties comes from fixing any Cartesian coordinate system $\psi : \mathbb{A}^n \to \mathbb{R}^n$ and defining open sets in $\mathbb{A}^n$ to be the empty set and the sets $\psi^{-1}(U)$ where $U \subset \mathbb{R}^n$ is open. It can be proved easily that this collection of sets does not depend on the chosen Cartesian coordinate system. In case $\mathbb{A}^n$ is also a Euclidean space, in principle there is another topology at play: the one induced by the distance d coming from the inner product. But this topology is actually the same one described above, since it is obtained by the procedure we described if ψ is an orthonormal coordinate system. Every affine space $\mathbb{A}^n$ then admits a natural differentiable manifold structure (of class C^∞) that contains in particular the compatible natural global coordinates given by all possible *Cartesian coordinate systems* on $\mathbb{A}^n$. Such a natural differentiable structure is explicitly described in the appendix to Sect. A.6.7. In particular, Euclidean space $\mathbb{E}^3$ is a differentiable manifold where each Cartesian coordinate system (orthonormal or not) is global and belongs to the same C^∞ structure. The various local coordinate systems, such as spherical coordinates, belong to the same differentiable structure.

(2) Consider the unit sphere $\mathbb{S}^2$ of $\mathbb{R}^3$ with centre the origin and topology induced by $\mathbb{R}^3$ (open sets in $\mathbb{S}^2$ are by definition the intersections of $\mathbb{S}^2$ with open sets in $\mathbb{R}^3$). In canonical coordinates x^1, x^2, x^3 on $\mathbb{R}^3$:

$$\mathbb{S}^2 := \left\{ (x^1, x^2, x^3) \in \mathbb{R}^3 \;\middle|\; (x^1)^2 + (x^2)^2 + (x^3)^2 = 1 \right\} .$$

$\mathbb{S}^2$ inherits the structure of a 2-dimensional C^∞ manifold from $\mathbb{R}^3$, by defining an atlas on $\mathbb{S}^2$ made of 6 local charts $(\mathbb{S}^2_{(i)\pm}, \phi_\pm^{(i)})$ ($i = 1, 2, 3$) as follows. Consider the x^i-axis ($i = 1, 2, 3$) and the pair of open hemispheres $\mathbb{S}^2_{(i)\pm}$ with south-north axis given by the x^i-axis. The local charts $\phi_\pm^{(i)} : \mathbb{S}^2_{(i)\pm} \to \mathbb{R}^2$ associate with every $p \in \mathbb{S}^2_{(i)\pm}$ its coordinates on the plane $x^i = 0$. One can prove that $\mathbb{S}^2$ cannot admit a global chart, as opposed to $\mathbb{R}^3$ (or any open subset): if so we would have a continuous map on the compact space $\mathbb{S}^2$ whose image is non-compact, an absurd. *This shows that the class of differentiable manifolds does not reduce to non-empty open subsets of $\mathbb{R}^n$.*

(3) A similar example is the unit circle $\mathbb{S}^1$ in $\mathbb{R}^2$ centred at the origin, equipped with the induced topology of $\mathbb{R}^2$ (open sets in $\mathbb{S}^1$ are by definition the intersections of $\mathbb{S}^1$ with open sets of $\mathbb{R}^2$). This space can be covered with an atlas of 4 local charts (N, ψ_N), (O, ψ_O), (S, ψ_S), (E, ψ_E) where the sets $N, O, S, E \subset \mathbb{S}^1$

are the open semi-circles centred at $(0, 1)$, $(-1, 0)$, $(0, -1)$, $(1, 0)$ respectively, the maps ψ_N and ψ_S are the projections of points on the circle onto the segment $(-1, 1)$ of the x-axis while ψ_O and ψ_E the projections of points on the circle onto the segment $(-1, 1)$ of the y-axis. We have for example $\psi_N \circ \psi_E^{-1} : y \mapsto \sqrt{1 - y^2}$ where y is in $(0, 1) = \psi_E(N \cap E)$. This map, on the given domain, is clearly C^∞ (it would not be at $y = 1$, but the latter is excluded from the domain). Consequently $\mathbb{S}^1$ acquires from $\mathbb{R}^2$ the structure of 1-dimensional C^∞ manifold. In this case too there is no single chart with domain equal to the whole $\mathbb{S}^1$.

(4) Consider $\mathbb{R}^3$ and the equivalence relation $(x, y, z) \sim (x', y', z')$ if and only if $(x, y, z) = (x' + k, y' + h, z' + l)$ with $k, h, l \in \mathbb{Z}$. The ensuing set $\mathbb{T}^3$ roughly corresponds to a unit cube with identified opposing faces. It can be equipped, in a natural manner, with a (C^∞) structure of dimension 3 and there is no global chart for this differentiable manifold. The topology is defined by requiring that the open sets of $\mathbb{T}^3$ are precisely the subsets in $\mathbb{T}^3$ whose pre-image under the canonical projection $\mathbb{R}^3 \ni (x, y, z) \mapsto [(x, y, z)] \in \mathbb{T}^3$ is open in $\mathbb{R}^3$ (hence π is in particular continuous). This topology turns $\mathbb{T}^3$ into a second countable Hausdorff space as is easy to prove. For any point $p \in (0, 1)^3$, a local chart on $\mathbb{T}^3$ containing the point in its domain is the function mapping the point to the usual coordinates (x, y, z) of $\mathbb{R}^3$. For a point q on some face of the cube $[0, 1]^3$, to fix ideas suppose q is on the face $x = 0$ (or equivalently, $x = 1$), a local chart on $\mathbb{T}^3$ containing q is the usual map associating with the point its coordinates on $\mathbb{R}^3$ restricted to the domain $-1/2 < x < 1/2$, $y, z \in (0, 1)$. We must in fact keep into account that the values in $(-1/2, 0)$ for the coordinate x determine points in $\mathbb{T}^3$ because of the identification $x + k \sim x$, where $k \in \mathbb{Z}$.

Remarks 1.24

(1) One might ask what is the purpose of the topological requirements of a differentiable manifold M, i.e. the Hausdorff and second countable properties. Both are technical, and as one can prove with counterexamples, they are independent of the other conditions in the definition. The Hausdorff property warrants for example the uniqueness of limits, and therefore of the solutions to differential equations, such as those describing the evolution of physical systems on a differentiable manifold (this will be the general technical subject of the rest of this book). Asking second countability instead is less useful at an elementary level, but is necessary to extend, for example, the integral calculus of volumes and areas to differentiable manifolds (guaranteeing, together with other conditions, the property of so-called *paracompactness*).

(2) If (U, ϕ) and (V, ψ) are local charts on the C^k manifold M, and supposing $U \cap V \neq \varnothing$, the k-compatibility of local charts (U, ϕ) and (V, ψ) implies that the Jacobian matrix of $\phi \circ \psi^{-1}$, being invertible, has non-zero determinant everywhere. Vice versa, if $\phi \circ \psi^{-1} : \psi(U \cap V) \to \phi(U \cap V)$ is bijective, of class C^k, with non-null Jacobian determinant on $\psi(U \cap V)$, then $\psi \circ \phi^{-1} : \phi(U \cap V) \to \psi(U \cap V)$ is also C^k and therefore the two local charts are

k-compatible. The proof of this fact (Exercise A.21.**5**) relies on the known [KrPa03]:

Theorem 1.25 (Inverse Function Theorem) *Let* $f : D \to \mathbb{R}^n$, *with* $D \subset \mathbb{R}^n$ *open and non-empty, be a* C^k *map, with* $k = 1, 2, \ldots, \infty$ *given. If the Jacobian matrix of* f, *evaluated at* $p \in D$, *has non-zero determinant then there exist open neighbourhoods* $U \subset D$ *of* p *and* V *of* $f(p)$ *such that:*

(1) $f|_U : U \to V$ *is bijective,*
(2) *the inverse* $f|_U^{-1} : V \to U$ *is of class* C^k.

(3) One can prove that if $1 \leq k < \infty$, we can get rid of some charts from the differentiable structure (indeed, infinitely many charts!), so that the remaining set is still an atlas with $k = \infty$. One may also consider *real analytic* manifolds (in symbols, C^ω), where all maps $\phi \circ \psi^{-1}$ and $\psi \circ \phi^{-1}$ are taken real analytic. Affine spaces have a natural differentiable structure of class C^ω. ∎

1.3.3 Differentiable Functions and Curves on a Manifold and Diffeomorphisms

Since a differentiable manifold is locally indistinguishable from $\mathbb{R}^n$, the differentiable structure allows to make sense of the notions of *differentiable map* and *differentiable curve* on a differentiable manifold by reducing to the corresponding definitions in $\mathbb{R}^n$ via the local charts that cover any differentiable manifold. Let us begin with the simplest case.

Definition 1.26 If M is a C^k manifold of dimension m and $f : M \to \mathbb{R}$ a map, we will say that:

(1) f is of **class** C^r, with $0 \leq r \leq k$, whenever the maps $f \circ \phi^{-1}$ are of class C^r as maps from $\mathbb{R}^m$ to $\mathbb{R}$ for any local chart (U, ϕ) on M;
(2) f is **differentiable** (with respect to M) if it is of class C^k. ◇

Now we can pass to curves.

Definition 1.27 If M is C^k manifold of dimension m and $\gamma : I \to M$, with $I \subset \mathbb{R}$ an open interval, we will say that:

(1) γ is a **curve of class** C^r, with $0 \leq r \leq k$, if the maps $\phi \circ \gamma$ are of class C^r as functions from I to $\mathbb{R}^m$ for any local chart (U, ϕ) on M;
(2) γ is **a differentiable curve** (with respect to M) if it is of class C^k.

If the interval $I \subset \mathbb{R}$, different from a single point, contains one or both the endpoints, a map $\gamma : I \to M$ is called **curve of class** C^k (**differentiable**) if it is the restriction of a curve of class C^k (resp. differentiable) defined as in (1) (resp. (2)) on an open interval $J \supset I$. ◇

Fig. 1.2 Representation in coordinates of the function $f : M \to N$

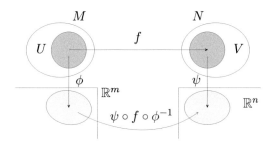

The definition scheme for curves of class C^k suggests the following definition, valid in general for maps between manifolds M and N.

Definition 1.28 Given $f : M \to N$ and two local charts $(U, \phi), (V, \psi)$ on the differentiable manifolds N and M respectively, the function $\psi \circ f \circ \phi^{-1} : \phi(U) \to \psi(V)$ is called **representation in coordinates** of f. One defines similarly the representation in coordinates of a curve.   ◇

The above definition allows to extend the previous definitions to the general case $f : M \to N$.

Definition 1.29 If M is a C^k manifold of dimension m and N a C^h manifold, we will say $f : M \to N$ is of **class** C^r, with $0 \leq r \leq \min\{k, h\}$, written $f \in C^r(M; N)$, if all of its representations in coordinates are of class C^r as maps from $\mathbb{R}^m$ to $\mathbb{R}^n$.   ◇

Remarks 1.30 It is easy to show that for $f : M \to N$ to be C^r, it is enough that the $\psi \circ f \circ \phi^{-1}$ are C^r as the local charts $(U, \phi), (V, \psi)$ vary in the two atlases on M and N respectively, without having to check the condition for *all* possible local charts of the two manifolds.   ∎

A crucial notion is that of *diffeomorphism* between manifolds. This is a map that identifies two differentiable manifolds in the most complete way (up to a given differentiability order), keeping in account the topological and differentiable structures (Fig. 1.2).

Definition 1.31 A **diffeomorphism of order** r or r**-diffeomorphism** between differentiable manifolds M and N of equal dimension, is an injective and surjective C^r map $f : M \to N$ with C^r inverse.

If there is a diffeomorphism of order r between two manifolds, the latter are called **diffeomorphic** to order r under the diffeomorphism.[7]

If $f : M \to N$ is such that $f|_U : U \to f(U) = V$ is a diffeomorphism (of order r) for two open sets $U \subset M$ and $V \subset N$ equipped with the obvious

---

[7] The specification of the order is usually dropped if r coincides with the common differentiability class of M and N.

differentiable structures inherited from M and N respectively, then we say f is a
local diffeomorphism (of order r from U to V).   ◇

Examples 1.32

(1) Every linear map between real vector spaces of finite dimension is automatically
a diffeomorphism between the natural C^∞ structures of the two spaces.

(2) The angular coordinate $\theta \in [0, 2\pi)$ on the unit circle $\mathbb{S}^1$ in $\mathbb{R}^2$ is *not* a
differentiable function for the C^∞ manifold structure the circle inherits from
$\mathbb{R}^2$. Why? The map $z = \sin \theta$ defined for $\theta \in [0, 2\pi)$, instead, is differentiable
on $\mathbb{S}^1$. Why?

(3) $\mathbb{R}^n$ and the open unit ball centred at the origin $B^n \subset \mathbb{R}^n$ are diffeomorphic
differentiable manifolds. In this case the differentiable structures of $\mathbb{R}^n$ and
B^n are the natural ones coming from the standard Cartesian coordinates of
$\mathbb{R}^n$ (restricted to B^n for the ball), which on their own form C^∞ atlases. A
diffeomorphism identifying B^n with $\mathbb{R}^n$ is, for example, the C^∞ bijective map
with C^∞ inverse:

$$f : B^n \ni \mathbf{x} \mapsto \frac{\mathbf{x}}{1 - ||\mathbf{x}||^2} \in \mathbb{R}^n .$$

(4) Every local chart (U, ψ) on a differentiable manifold M defines a local
diffeomorphism of the same order as M, defined on U with values in $f(U)$,
equipped with the natural differentiable structure induced by $\mathbb{R}^m$.

Further differential-geometric notions that will at times be used in the sequel can be
looked up in the Appendix.

Chapter 2
The Spacetime of Classical Physics and Classical Kinematics

In this chapter we introduce the structure of *spacetime* of Classical Physics, the notion of *reference frame* and the fundamental ideas of elementary, absolute and relative *Kinematics*.

2.1 The Spacetime of Classical Physics and Its Geometric Structures

We saw in the previous chapter that Classical Physics' space and time can be modelled by Euclidean spaces, namely endowed with metric structures that represent, in mathematical terms, the measuring instruments. This description of space and time, however, is not accurate enough to account for the physical phenomenology: a more complete description must contemplate the possibility, as clarified by Galileo in the first part of the seventeenth century, of having several equivalent notions of rest space (and time axis) together with the absolute character of the metric properties of physical bodies (and of time intervals). This elementary fact, completely absent in Aristotele's Physics, historically coincided with the turning point that led to the birth of Physics as a modern science. The conceptual tool we will use to account for all of that is *spacetime*. This notion was in truth born in a relativistic context from the work of Einstein and Minkowski in the twentieth century, as we will see in Chap. 10, but can be successfully adapted to Physics' pre-relativistic formulations.

© The Author(s), under exclusive license to Springer Nature Switzerland AG 2023
V. Moretti, *Analytical Mechanics*, La Matematica per il 3+2 150,
https://doi.org/10.1007/978-3-031-27612-5_2

2.1.1   Multiple Rest Spaces and Absolute Metric Structure of Space and Time

The description of the previous chapter lacks an essential phenomenological ingredient that was highlighted by, among other things, Galileo in his 1632 *Dialogue Concerning the Two Chief World Systems*: physical experience teaches us that bodies that are at rest for one observer may not be so for another observer. In general we can say that there exists different notions of rest depending on the observer—where by "observer" we mean here and henceforth a *system of measuring instruments without a conscious activity*, as we have remarked several times already. Every observer has their own personal rest space. On the other hand, there is a second experimental datum of the utmost importance: despite a body might be at rest for one observer and not for another, *the physical dimensions (lengths, areas, volumes, angles) of a given physical body are the same for all observers irrespective of the state of motion each observer assigns to it*. Similarly, although two given events may seem to occur at different (places and) times depending on the observer, *the duration of the time interval between the two events is the same for all observers*. In this sense the metric properties of space and time are *absolute*: they do not depend on the observer. So we should extend any physical modelling to include, in one mathematical description, the co-existence of different rest spaces and the fact that the metric properties of bodies and time intervals are absolute. One of the most interesting solutions, that has revealed itself to be most fertile in the development of physical theories, is to use the concept of *spacetime* already in Classical Physics.

It is well known that all this classical representation of the physical world is only the best approximation, and patently stops being valid when the relative speeds at play are comparable to the speed of light, roughly 300,000 km/s; in this case one needs a relativistic description, which we will introduce in Chap. 10.

2.1.2   The Spacetime of Classical Physics and the World Lines

One fundamental physical postulate (which has become clear at the beginning of the twentieth century by work of Einstein and mathematicians like Minkowski and Poincaré), common to classical and relativistic physical theories, states that everything that occurs is decomposable into *events*. Physically, an event corresponds to the smallest possible spacetime determination, detectable by assigning three spatial coordinates and one time coordinate. These assignments are relative to different observers, here understood as pure *reference frames*. The set of events forms *spacetime*. It should be made clear starting now that the nature of every event and of spacetime itself is independent from—and pre-exists—their representations, given in the single reference frames in terms of spacetime coordinates. Within such a *framework*, everything that occurs must admit a description in terms of *relations* or *coincidences* between events.

We shall also assume that spacetime has a *topological* nature—technically, a *second countable Hausdorff topological space*—and on top of that a *differentiable* nature—technically corresponding to the presence of a *four-dimensional differentiable manifold* structure, in practice a collection of coordinates in the neighbourhood of each event that identify that neighbourhood with a corresponding neighbourhood in $\mathbb{R}^4$. The reason for dimension 4 should be evident: to determine an event in one observer's representation we need four pieces of information, typically three spatial ones and one temporal. Having multiple observers corresponds, at this point still in a vague way, to possibly distinct systems of coordinates on spacetime, exactly in the spirit of a differentiable structure.

The topological structure allows in particular (but not only) to make sense of the ideas of "spatial and temporal proximity". Requiring that the topology is Hausdorff has a profound physical motivation.[1] Because of the experimental errors made by measuring instruments, any spacetime determination is only approximatively accurate. We may locate an event only with a certain approximation. In mathematical terms, we may only detect *neighbourhoods* of events and not the events themselves. These neighbourhoods are determined by the imprecisions, as small as we like but non-zero, of the instruments that measure space and time. *That the spacetime topology is Hausdorff means that we can always distinguish two events so long as the measurements are sufficiently precise, even if not infinitely precise.*

The differentiable structure allows to make sense of the notions of curve and differentiable function (with respect to the coordinates) defined on spacetime. As we shall see shortly, in this way we can introduce the notions of velocity and acceleration, to describe the evolution of point particles. The differentiable structure of spacetime warrants, much more generally, the possibility of describing in terms of *differential equations* the equations governing the spacetime evolution of classical (and relativistic) physical systems, whether discrete (collections of points) or continuous (fluids or fields). The existence and uniqueness of the solutions to these equations (in presence of initial and/or boundary conditions) corresponds to the classical physical postulate called *determinism*. The Existence and uniqueness theorems are related to the topological and differentiable structure of spacetime.

If we consider explicitly the case of classical spacetime as defined below, spacetime includes two further structures to account for the observations of Sect. 2.1.1: *absolute time* and *absolute space*. The word *absolute* refers to *the independence from possible frame systems of any measurement of angles, distances and time intervals* performed on physical bodies. These structures are therefore automatically absolute, i.e. independent of the frame of reference, because we still have not introduced the concept reference frame.

Definition 2.1 (The Spacetime of Classical Physics)   The **spacetime of Classical Physics** is a four-dimensional differentiable manifold denoted by $\mathbb{V}^4$. The points

---

[1] The same argument holds for the Hausdorff property of the physical space $\mathbb{E}^3$ and the physical time axis $\mathbb{E}^1$, though in that case the property showed up *a posteriori*, whereas now we are assuming it a priori.

$\mathbb{V}^4$

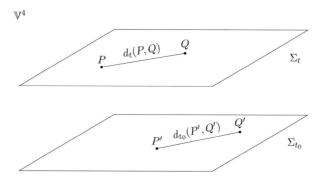

Fig. 2.1 $\mathbb{V}^4$ and two absolute spaces at given times

of $\mathbb{V}^4$ are called **events**. $\mathbb{V}^4$ is equipped with a privileged function defined up to an additive constant, $T : \mathbb{V}^4 \to \mathbb{R}$, called **absolute time**, which is required to be differentiable, surjective and non-singular (Definition A.22). We further assume that:

(1) each of the pairwise-disjoint sets: $\Sigma_t := \{p \in \mathbb{V}^4 \mid T(p) = t\}$, called **absolute space at time** $t \in \mathbb{R}$, has the structure of a three-dimensional Euclidean space (with space of displacements V_t and inner product $(\cdot|\cdot)_t$);

(2) the geometric structures on every Σ_t is compatible with the differentiable structure of $\mathbb{V}^4$, in the sense that, for any $t \in \mathbb{R}$, on a neighbourhood O_e of any event $e \in \Sigma_t$ there is a four-dimensional system of local coordinates x^0, x^1, x^2, x^3 on $\mathbb{V}^4$ defining (the restriction of) *orthonormal Cartesian* coordinates x^1, x^2, x^3 on $O \cap \Sigma_t$ when $x^0 = 0$ (Fig. 2.1).   ◇

Remarks 2.2

(1) $\mathbb{V}^4$ will be of class C^k with $k \in \{1, 2, \ldots, \} \cup \{\infty\}$ to be determined depending on the kind of theory we need; usually $k \geq 2$ is necessary, and often sufficient, for the various formulations of Dynamics. We remind, as explained in Sect. 1.3.3, that for any curve and function on spacetime (such as T), *differentiable* means of the same class C^k as spacetime is.

(2) As mentioned earlier, it is physically important to recall that there is ongoing speculation on the possible discontinuous nature of spacetime itself at very small scales (Planck scales $\sim 10^{-33}$ cm and 10^{-43} s), where some sort of *Quantum Gravity* should hold. At these scales the classical topological-differentiable structure of Definition 2.1 would no longer be physically appropriate.   ∎

Spacetime is essentially a container: parts of it are occupied by the time evolutions of material objects existing in the universe. From the classical viewpoint these objects are made of elementary entities called *point particles* that determine geometric points in every absolute space at a given time, and possess additional physical properties such as a *mass*, which will be discussed later. Physically,

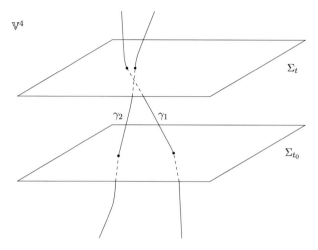

Fig. 2.2 Two world lines in $\mathbb{V}^4$

we should not forget that whether a physical system can or not be treated as a point particle, thus disregarding the possible "inner" structure, may depend on the precision of our measuring instruments and on the chosen degree of accuracy.

The time evolution of a point particle in spacetime constitutes an elementary object called *history* or *world line (of a point particle)* (Fig. 2.2).

Definition 2.3 (World Line or History)   The **history** or **world line** (**of a point particle**) is a differentiable curve

$$I \ni t \mapsto \gamma(t) \in \mathbb{V}^4 \,,$$

where $I \subset \mathbb{R}$ is a non-trivial interval that can be identified with an interval of the absolute time, meaning that for some *constant* $c \in \mathbb{R}$, usually depending on γ, we have

$$T(\gamma(t)) = t + c \,, \ \forall t \in I. \tag{2.1}$$

$\diamond$

Remarks 2.4

(1) A few facts should be pointed out about the absolute time function T. For starters T is defined up to an additive constant, which corresponds to the evident physical fact that we can fix the conventional origin of time as we please. The constant c in Definition 2.3 expresses the possibility to change this origin. That T does not admit dilations means we have made a universal choice of the unit of time. The direction in which T grows determines the future direction. All this agrees with Remark 1.17 and with the formalisation of the temporal axis as $\mathbb{E}^1$ with a choice of privileged unit vector denoting the future direction: the range of T can be thought of as a system of (ortho)normal coordinates oriented towards the future of the time axis $\mathbb{E}^1$. Physically, the existence of an absolute time is equivalent to saying that we posit the existence of *ideal clocks*, available at every point in spacetime, for measuring that time.

(2) The following facts about absolute spaces hold by construction.

(i) The fact T is onto guarantees that $\Sigma_t \neq \emptyset$ for any $t \in \mathbb{R}$.

(ii) By definition, if $t \neq t'$, the associated absolute spaces are disjoint: $\Sigma_t \cap \Sigma_{t'} = \emptyset$.

(iii) T is defined everywhere on $\mathbb{V}^4$: if $p \in \mathbb{V}^4$ then $\Sigma_t \ni p$ for some $t \in \mathbb{R}$. Hence $\mathbb{V}^4 = \cup_{t \in \mathbb{R}} \Sigma_t$.

All in all, spacetime is foliated by pairwise-disjoint absolute spaces and coincides with their union.

(3) Each absolute space Σ_t is a real three-dimensional Euclidean space. Physically, the distance d_t, hence the inner products and angles between segments in Σ_t, is measured by assigning a collection of *ideal rigid rulers* that should be assumed available at each point in spacetime. Translations induced by vectors $\mathbf{v}$ in Σ_t should be thought of as physical displacements of material bodies.

(4) In the definition of world line the requisite $T(\gamma(t)) = t + c$ for any $t \in I$ ensures that absolute time can be used as parameter to describe the world line, but also that a world line cannot cross the same absolute space Σ_t more than once, for any given[2] t (and cannot in particular self-intersect). As a consequence of this constraint, the point particle, of which the world line is the history, cannot "go back in time" (it cannot return to Σ_t if it has gone through it previously). This is a physical requisite that avoids the causal paradoxes of science fiction.

(5) We can imagine to attach ideal clocks to two point particles. These clocks, up to the choice of time origin, will indicate the labels t of the various hypersurfaces Σ_t hit by the world lines at the points. If the points meet at the event $p \in \Sigma_t$, and at such event we synchronise the clocks *manually* so that they both show t at that event, then at every successive encounter the clocks will show the same time $t' > t$ (irrespective of their motion status relative to the encounter). Moreover, if we synchronise two clocks that are placed apart, at some distance, using a third clock that shuttles back and forth between the two, the synchronisation will be preserved in the future and all synchronisation procedures are equivalent. All of this is because ideal clocks simply *read* the absolute time that was declared to exist. We know from Physics that this state of affairs is actually only approximate, and ceases to hold when the relative speeds at play (for example the relative speed between instantaneously close clocks) are close to the speed of light. This means that the axiom postulating absolute time has limited physical validity and should be replaced by something else if we want

---

[2] If $\gamma(t_1), \gamma(t_2) \in \Sigma_t$ with $t_1 < t_2$, then $t_1 + c = T(\gamma(t_1)) = t = T(\gamma(t_2)) = t_2 + c$, so $t_1 = t_2$ which is impossible.

to have a more realistic image of the physical world. Similarly, if we associate rigid rulers with the point particles, every time the rulers are compared at the same event p (even with different relative speeds) they must coincide, since they just *read* the absolute geometry of the absolute space $\Sigma_t \ni p$. In this case, too, the physical reality is rather different when the speeds considered (the relative speed of the two rulers we are comparing) are close to the speed of light, and that means that assuming space, and geometry, absolute does not depict the physical reality faithfully. We shall provide some details on the special relativistic formulation of spacetime in Chap. 10.

(6) $\boxed{\text{AC}}$ The Σ_t are *embedded submanifolds* of dimension $4 - 1 = 3$ by the regular-value theorem (Theorem A.26 in the Appendix) since T is everywhere non-singular. Hence if $p \in \Sigma_t$ there is a local chart (U, ψ) in $\mathbb{V}^4$ with $\psi : U \ni q \mapsto (x^1(q), x^2(q), x^3(q), x^4(q)) \in \psi(U) \subset \mathbb{R}^4$ such that $p \in U$ and the set $U \cap \Sigma_t$ is described in coordinates by the quadruples $(x^1, x^2, x^3, x^4) \in \psi(U)$ with $x^4 = 0$. The map $\psi' : U \cap \Sigma_t \ni q \mapsto (x^1(q), x^2(q), x^3(q)) \in \mathbb{R}^3$ is a local chart on Σ_t around p. All similarly constructed charts around points of Σ_t starting from charts in $\mathbb{V}^4$ are C^k compatible (as restrictions of compatible local charts on $\mathbb{V}^4$) and hence form an atlas that turns Σ_t into a differentiable manifold of dimension 3. This differentiable structure on Σ_t is by construction *induced* by that of $\mathbb{V}^4$.

As for the Euclidean structure of Σ_t, asking that it be "compatible with the global structure of $\mathbb{V}^4$" means that the differentiable structure associated with the affine structure, and the one due to the Σ_t being embedded submanifolds, coincide. In other words the three-dimensional Cartesian coordinates on Σ_t are C^k compatible with the local coordinates induced on Σ_t by $\mathbb{V}^4$. ∎

2.2 Reference Frames

Now we wish to introduce the tools for assigning to each world line a position, a velocity and an acceleration at each time instant. To do so we need to introduce the notion of *reference frame*, which will replace that of *observer* we used until now in an intuitive way. Traditionally this part of Mechanics falls under the name **Kinematics**: the part of Mechanics that deals with the pure description on motions.

A *reference frame* is a means to represent, in some fixed "rest space", what happens in spacetime, as absolute time goes by. We can think of a reference frame as an identification of $\mathbb{V}^4$ with the Cartesian product $\mathbb{R} \times \mathbb{E}^3$, where $\mathbb{R}$ is the *axis of absolute time* (common to all frames, and on which we have chosen an origin possibly depending on the frame) and $\mathbb{E}^3$ is the *frame's rest space* equipped with a Euclidean structure, that should be identified at each instant t with Σ_t. This identification must preserve all the structures of $\mathbb{E}^3$, the affine one and the Euclidean one, so it must occur in terms of isometries; as we know, these are isomorphisms between both the metric structures and the affine structures of the Euclidean spaces.

Definition 2.5 (Reference Frames) A **reference frame**, o **frame** for short (of Classical Physics), $\mathscr{R}$ is a pair $(\Pi_{\mathscr{R}}, E_{\mathscr{R}})$, where

(1) $E_{\mathscr{R}}$ is a three-dimensional Euclidean space (with space of displacements, inner product and distance respectively denoted by $V_{\mathscr{R}}$, $(\cdot|\cdot)_{\mathscr{R}}$ and $d_{\mathscr{R}}$);
(2) $\Pi_{\mathscr{R}} : \mathbb{V}^4 \to E_{\mathscr{R}}$ is a surjective differentiable map such that, for any instant of absolute time $t \in \mathbb{R}$, $\Pi_{\mathscr{R}}|_{\Sigma_t} : \Sigma_t \to E_{\mathscr{R}}$ is a Euclidean isometry (hence, in particular, also an isomorphism of affine spaces that preserves the inner products).

$E_{\mathscr{R}}$ is called **rest space** of $\mathscr{R}$ and its elements **points at rest** for $\mathscr{R}$.   ◇

The definition has as a consequence the following proposition, which will play an important part in the sequel.

Proposition 2.6 *If $(\Pi_{\mathscr{R}}, E_{\mathscr{R}})$ is a frame, the map*

$$\mathbb{V}^4 \ni e \mapsto (T(e), \Pi_{\mathscr{R}}(e)) \in \mathbb{R} \times E_{\mathscr{R}}$$

is injective and surjective and therefore identifies spacetime with the decomposition in space and time given by the Cartesian product $\mathbb{R} \times E_{\mathscr{R}}$ (equivalently, due to the arbitrary additive constant in the definition of T, it identifies spacetime with $\mathbb{E}^1 \times E_{\mathscr{R}}$, where $\mathbb{E}^1$ is the temporal axis with no preferred origin).

Any curve $\mathbb{R} \ni t \mapsto (t, P) \in \mathbb{R} \times E_{\mathscr{R}}$, for any given $P \in E_{\mathscr{R}}$, is a world line in the sense of Definition 2.3 when viewed in $\mathbb{V}^4$ under the above identification of $\mathbb{R} \times E_{\mathscr{R}}$ and $\mathbb{V}^4$. This world line, γ_P, is called world line **of the point P at rest for** $\mathscr{R}$.

Proof Consider the map $\mathbb{V}^4 \ni e \mapsto (T(e), \Pi_{\mathscr{R}}(e)) \in \mathbb{R} \times E_{\mathscr{R}}$. It is injective. There are in fact two possibilities if $e \neq e'$: (1) the two events belong to distinct absolute spaces, and then $T(e) \neq T(e')$, or (2) e and e' belong to the same absolute space Σ_t, in which case they determine distinct points of Σ_t, whence $\Pi_{\mathscr{R}}(e) \neq \Pi_{\mathscr{R}}(e')$ because $\Pi_{\mathscr{R}}|_{\Sigma_t} : \Sigma_t \to E_{\mathscr{R}}$ is bijective (as isomorphism of affine spaces). In either case $e \neq e'$ implies $(T(e), \Pi_{\mathscr{R}}(e)) \neq (T(e'), \Pi_{\mathscr{R}}(e'))$. As regards surjectivity let us observe the following. Given an arbitrary $(t_0, P) \in \mathbb{R} \times E_{\mathscr{R}}$, that the map $\Pi_{\mathscr{R}}|_{\Sigma_{t_0}} : \Sigma_{t_0} \to E_{\mathscr{R}}$ is by assumption surjective means that $\Pi_{\mathscr{R}}^{-1}(P)$ intersects Σ_{t_0} at some e. By construction $(T(e), \Pi_{\mathscr{R}}(e)) = (t_0, P)$. Clearly every curve γ_P described by $\mathbb{R} \ni t \mapsto (t, P) \in \mathbb{R} \times E_{\mathscr{R}} \equiv \mathbb{V}^4$ satisfies $T(\gamma_P(t)) = t$ for any $t \in \mathbb{R}$. The fact that every curve γ_P is differentiable follows directly from the existence of a system of global coordinates on $\mathbb{V}^4$, called *rest orthonormal coordinate system of $\mathscr{R}$* (see Definition 2.8), that is compatible with spacetime's differentiable structure, as proved in Proposition 2.10, and where every curve γ_P has trivial structure $t \mapsto (t, x_P^1, x_P^2, x_P^3)$, where the coordinates x_P^i are kept constant.   □

Remarks 2.7

(1) We have just seen that the points P of the rest space $E_\mathscr{R}$ of a frame $\mathscr{R}$ trace, as they evolve in time, world lines in $\mathbb{V}^4$. Each of these world lines γ_P determines the same point P in rest space. The rest space is instant by instant isometrically identified with Σ_t. Hence *the spatial distance $d_t(\gamma_P(t), \gamma_Q(t))$ between the world lines γ_P and γ_Q of the points P, Q at rest in the frame $\mathscr{R}$ is constant, as absolute time varies.* This is not the only property of the family $\{\gamma_P\}_{P \in E_\mathscr{R}}$ of world lines; there are more, that we list below and are of immediate verification. The curves γ_t are *synchronised*, i.e. the additive constants of the parameter t describing them $(\gamma_P = \gamma_P(t))$ are chosen so that *every* curve intersects each Σ_τ at the value $t = \tau$ of its parameter. *Every* spacetime event is reached by some γ_P. The lines γ_P *never intersect* one another.
(2) Regarding the notion of frame, it makes no physical sense to distinguish two pairs (Π, E) and (Π', E') that satisfy Definition 2.5 but are isomorphic, meaning there exists an isomorphism of Euclidean spaces (i.e. a Euclidean isometry) $\psi : E \to E'$ such that $\Pi' = \psi \circ \Pi$. Therefore we shall assume, in such a case, that *both pairs (Π, E) and (Π', E') define the same frame $\mathscr{R}$.*■

2.2.1 Rest Orthonormal Coordinate Systems of Moving Frames

We would like to introduce special global charts on $\mathbb{V}^4$, associated with a given frame $\mathscr{R}$, that we will refer to as being *at rest* with $\mathscr{R}$. For this, let us fix orthonormal coordinates x^1, x^2, x^3 on the rest space $E_\mathscr{R}$. Consider then the function $\sigma_\mathscr{R} : \mathbb{V}^4 \to \mathbb{R}^4$ that after projecting every spacetime event $p \in \Sigma_t$ on the rest space of $\mathscr{R}$, $E_\mathscr{R}$, associates with p the ordered quadruple of real numbers formed by the value of absolute time $T(p) = t$, possibly redefined by some additive constant c, and by the three Cartesian coordinates of $\Pi_\mathscr{R}(p)$ in the rest space of $\mathscr{R}$:

$$\sigma_\mathscr{R} : p \mapsto \left(T(p) + c, x^1\left(\Pi_\mathscr{R}(p)\right), x^2\left(\Pi_\mathscr{R}(p)\right), x^3\left(\Pi_\mathscr{R}(p)\right) \right) .$$

We claim $\sigma_\mathscr{R}$ is bijective and hence, in particular, it defines a global chart $(\mathbb{V}^4, \sigma_\mathscr{R})$ and so an atlas on $\mathbb{V}^4$. Without loss of generality we will suppose $c = 0$ in the sequel.

Surjectivity Given $(t, x^1, x^2, x^3) \in \mathbb{R}^4$ consider $\Pi_\mathscr{R}|_{\Sigma_t} : \Sigma_t \to E_\mathscr{R}$. This map is bijective as explained in Theorem 2.14. Hence there exists $p \in \Sigma_t$ such that $\Pi_\mathscr{R}(p) \in E_\mathscr{R}$ has coordinates (x^1, x^2, x^3). By construction $\sigma_\mathscr{R}(p) = (t, x^1, x^2, x^3)$.

Injectivity If $p \neq p'$ there are two possibilities. (1) $T(p) \neq T(p')$ so $\sigma_\mathscr{R}(p) \neq \sigma_\mathscr{R}(p')$. (2) $T(p) = T(p') =: t$, and then $p, p' \in \Sigma_t$, but $p \neq p'$ because the points are distinct. Therefore, as $\Pi_\mathscr{R}|_{\Sigma_t} : \Sigma_t \to E_\mathscr{R}$ is bijective hence injective, we will have $\Pi_{\mathscr{R}}(p) = \Pi_{\mathscr{R}}|_{\Sigma_t}(p) \neq \Pi_{\mathscr{R}}|_{\Sigma_t}(p') = \Pi_{\mathscr{R}}(p')$ and so

$$
(x^1 (\Pi_{\mathscr{R}}(p)), x^2 (\Pi_{\mathscr{R}}(p)), x^3 (\Pi_{\mathscr{R}}(p)))
$$
$$
\neq (x^1 (\Pi_{\mathscr{R}}(p')), x^2 (\Pi_{\mathscr{R}}(p')), x^3 (\Pi_{\mathscr{R}}(p'))),
$$

from which $\sigma_{\mathscr{R}}(p) \neq \sigma_{\mathscr{R}}(p')$.

Definition 2.8 (Rest Orthonormal Coordinates of a Frame)   Let $\mathscr{R}$ be a frame in spacetime $\mathbb{V}^4$, x^1, x^2, x^3 orthonormal coordinates in rest space $E_{\mathscr{R}}$ and $\sigma_{\mathscr{R}}$: $\mathbb{V}^4 \to \mathbb{R}^4$ the bijection mapping each spacetime event p to the numbers $(T(p) + c, x^1 (\Pi_{\mathscr{R}}(p)), x^2 (\Pi_{\mathscr{R}}(p)), x^3 (\Pi_{\mathscr{R}}(p)))$, where c is a constant. The global chart $(\mathbb{V}^4, \sigma_{\mathscr{R}})$ is called a **system of rest orthonormal coordinates on $\mathscr{R}$**.   ◇

A fundamental consequence of technical character is the following. Given frames $\mathscr{R}$ and $\mathscr{R}'$ with rest orthonormal coordinates (t, x^1, x^2, x^3) and (t', x'^1, x'^2, x'^3) respectively, the function $\sigma_{\mathscr{R}'} \circ \sigma_{\mathscr{R}}^{-1} : \mathbb{R}^4 \to \mathbb{R}^4$, i.e. the "transformation rule from 'unprimed' coordinates to 'primed' coordinates", always has the form

$$
\begin{cases}
t' = t + c, \\
x'^i = c^i(t) + \sum_{j=1}^{3} R^i{}_j(t) \, x^j, \quad i = 1, 2, 3,
\end{cases}
\tag{2.2}
$$

where $c \in \mathbb{R}$ is a constant, the maps $\mathbb{R} \ni t \mapsto c^i(t)$ and $\mathbb{R} \ni t \mapsto R^i{}_j(t)$ have the same differentiability order as $\mathbb{V}^4$, and for any $t \in \mathbb{R}$ the coefficient matrix $R^i{}_j(t)$ is real 3×3 and orthogonal. The first identity is obvious from the fact that both systems use absolute time as time coordinates up to an additive constant. The second identity is straightforward (see the exercises) using the fact that for any given $t \in \mathbb{R}$, the transformations $\Pi_{\mathscr{R}}|_{\Sigma_t} \circ (\Pi_{\mathscr{R}'}|_{\Sigma_t})^{-1} : E_{\mathscr{R}'} \to E_{\mathscr{R}}$ are Euclidean (hence affine) isometries since they are composites of isometries. Now we may use the result of 3 in Exercise 1.16 for any given time t, which directly gives (2.2).

Remarks 2.9

(1) Observe that there exist infinitely many orthonormal coordinates systems at rest with a given frame $\mathscr{R}$. This multitude is parametrised by the choice of origin $O \in E_{\mathscr{R}}$, the choice of orthonormal basis in $V_{\mathscr{R}}$, but also by the choice of constant in the definition of absolute time.

(2) $\boxed{\text{AC}}$ Physically, frames' rest coordinates are the natural coordinates of spacetime. Spacetime is a manifold precisely because of the existence of various coordinates systems, at rest with different frames, which identify it with $\mathbb{R}^4$. At this point one could then ask whether the differentiable structure induced on $\mathbb{V}^4$ by the rest coordinates of frames truly coincides with the differentiable structure we have assumed to exist a priori on $\mathbb{V}^4$ in Definition 2.1. In other words we would like to know if the rest coordinates of frames just defined are C^k-compatible (as for Definition 1.22) with the local coordinates we already

have by declaring that spacetime is a differentiable manifold. The answer, as it should, is yes, due to the following technical proposition based on absolute time being a non-singular map. The proof is the solution to one of the ensuing exercises.   ∎

Proposition 2.10 $\boxed{\text{AC}}$ *Every system of rest orthonormal coordinates on an arbitrary frame $\mathscr{R}$ of $\mathbb{V}^4$ of class C^k is C^k-compatible with the local charts on $\mathbb{V}^4$ (i.e. it belongs to the differentiable structure of $\mathbb{V}^4$).*

Exercises 2.11

(1) $\boxed{\text{AC}}$ Prove Proposition 2.10.
(2) Show that given frames $\mathscr{R}$ and $\mathscr{R}'$ with rest orthonormal coordinates (t, x^1, x^2, x^3) and (t', x'^1, x'^2, x'^3), the function $\sigma_{\mathscr{R}'} \circ \sigma_{\mathscr{R}}^{-1} : \mathbb{R}^4 \to \mathbb{R}^4$, i.e. the "transformation rule from 'unprimed' to 'primed' coordinates", has the form (2.2) where $c \in \mathbb{R}$ is a constant, the maps $\mathbb{R} \ni t \mapsto c^i(t)$ and $\mathbb{R} \ni t \mapsto R^i{}_j(t)$ are C^k (as $\mathbb{V}^4$) and, for any $t \in \mathbb{R}$, the coefficient matrix $R^i{}_j(t)$ is real 3×3 and orthogonal.
(3) Let $\mathscr{R}$ be a frame with rest orthonormal coordinates (t, x^1, x^2, x^3). Consider another set of (global) coordinates on $\mathbb{V}^4$, (t', x'^1, x'^2, x'^3), related to the first set by (2.2) (where $\mathbb{R} \ni t \mapsto c^i(t)$ and $\mathbb{R} \ni t \mapsto R^i{}_j(t)$ have the aforementioned properties). Prove that:

   (i) there is a frame $\mathscr{R}'$ for which (t', x'^1, x'^2, x'^3) are rest orthonormal coordinates;
   (ii) If $\mathscr{R} = \mathscr{R}'$ the maps $\mathbb{R} \ni t \mapsto c^i(t)$ and $\mathbb{R} \ni t \mapsto R^i{}_j(t)$ are constant. Conversely, if these maps are constant, the frames $\mathscr{R}$ and $\mathscr{R}'$ are, physically, the same frame, meaning they determine the same collection of world lines of rest points, possibly labelled differently for $\mathscr{R}$ and $\mathscr{R}'$ and perhaps with a different origin for the time coordinate.

Remarks 2.12

(1) Since we did not define on $\mathbb{V}^4$ an affine structure, systems of rest orthonormal coordinates of a frame are *not, despite the name*, Cartesian coordinates on $\mathbb{V}^4$!

   We can actually say more. Even if every given frame $\mathscr{R}$ identifies spacetime with the affine space $\mathbb{E}^1 \times E_{\mathscr{R}}$, it is *not* possible to equip spacetime with a preferred *affine structure*, thus turning it into an affine space $\mathbb{A}^4$, for which the rest Cartesian coordinates of each frame (including the time coordinates!), i.e. the Cartesian coordinates of each affine space $\mathbb{E}^1 \times E_{\mathscr{R}}$, are Cartesian coordinates on $\mathbb{A}^4$. The reason lies in the general form of equations (2.2). If we fix a frame $\mathscr{R}$ and consider two events p and q, we may associate with the pair (p, q) a vector $p - q$ with components

$$(t_p - t_q, x_p - x_q, y_p - y_q, z_p - z_q),$$

where we have introduced rest orthonormal coordinates on $\mathscr{R}$. If we change to other rest coordinates on the same $\mathscr{R}$, the transformation rule for the components of $p - q$ is affine, of the form (2.2), with coefficients c^i all zero and $R^i{}_j(t)$ independent of time. However, if we pass to another frame $\mathscr{R}'$ and attempt to associate with the same pair (p, q) a vector $p - q$ of components

$$(t'_p - t'_q, x'_p - x'_q, y'_p - y'_q, z'_p - z'_q)$$

in some rest orthonormal coordinates on $\mathscr{R}'$, we discover that, in general, the components of the vector do not transform affinely when passing to the previous frame. This is due to the time dependence of the coefficients $R^i{}_j(t)$ and $c^i(t)$ in (2.2), which appears because we have changed frames and because we are at liberty to pick non-simultaneous events p and q. To sum up: it is not possible, using rest coordinates on the frames of spacetime, to define a notion of *vector* $p - q$ that is independent of the chosen frame.

(2) $\boxed{\text{AC}}$ A transformation of type (2.2) is completely determined by a triple $(c, \vec{c}, R)$, where $c \in \mathbb{R}$, while $\vec{c} : \mathbb{R} \ni t \mapsto \vec{c}(t) = (c^1(t), c^2(t), c^3(t))^t \in \mathbb{R}^3$ and $R : \mathbb{R} \ni t \mapsto R(t) \in O(3)$ are differentiable maps (both of the same class as $\mathbb{V}^4$). If we compose two such transformations $(c, \vec{c}, R)$ and $(c', \vec{c}', R')$, in this order, we still obtain a transformation of the same type

$$(c'', \vec{c}'', R'') = (c + c', \vec{c}'_c + R'_c \vec{c}, R'_c R) ,$$

where

$$\vec{c}_k(s) := \vec{c}(s + k) , \quad R_k(s) := R(s + k) , \quad \forall s, k \in \mathbb{R} .$$

It is not hard to show that the collection of such triples, equipped with the operation

$$(c', \vec{c}', R') \circ (c, \vec{c}, R) := (c + c', \vec{c}'_c + R'_c \vec{c}, R'_c R)$$

defines a (non-commutative) group. The latter, which we might call the **fundamental kinematic group**, describes every kinematic relation among all possible reference frames of Classical Physics. It is an "infinite-dimensional" group, since its elements are determined by assigning arbitrary functions and not a finite number of coefficients, like for the elements of $O(3)$. When discussing Dynamics, we will see that a central role is played by a finite-dimensional subgroup of the above, called the *Galilean group*, (see Exercise 3.3).

(3) $\boxed{\text{AC}}$ Because every rest coordinate system on a frame $\mathscr{R}$ is a coordinate system of the differentiable structure of $\mathbb{V}^4$, it, too, defines a diffeomorphism between $\mathbb{V}^4$ and $\mathbb{R} \times E_\mathscr{R}$. This result can be stated in differential-geometric terms (see Appendix B) saying that $\mathbb{V}^4$ is a *(fibre) bundle with base* $\mathbb{R}$, *standard fibre* $\mathbb{E}^3$ and *canonical projection* $T : \mathbb{V}^4 \to \mathbb{R}$ given by absolute time. Consequently

spacetime is a *trivialisable fibre bundle* (meaning diffeomorphic to the product of the base manifold times the standard fibre), although the identification is not unique and there is no way to fix a privileged one, precisely because that would entail choosing one frame as more important than the others.

Further, observe that the histories of point particles identify with the *local sections* of the bundle, and the base $\mathbb{R}$ can be taken to be $\mathbb{E}^1$ if we fix a conventional time origin, i.e. once we choose a specific function T. ∎

2.2.2 $\boxed{\text{AC}}$ *An Alternative But Equivalent Definition of Frame*

To give an alternative definition of frame we can put the emphasis on the aforementioned proprieties of world lines of rest points in the frame (viewing these curves are primary entities) rather than on the splitting $\mathbb{E}^1 \times E_{\mathscr{R}}$. This perspective is useful for the theory's subsequent developments (in particular for General Relativity theories). In practice we can imagine a frame $\mathscr{R}$ as a collection $\{\gamma_P\}_{P \in E_{\mathscr{R}}}$ of world lines. *Notice that the index set $E_{\mathscr{R}}$ does not have a Euclidean structure in this context, for the time being.* The world lines in $\{\gamma_P\}_{P \in E_{\mathscr{R}}}$ are viewed as parametrised on the *whole* absolute time and *synchronised*, i.e. the additive constants of the parameter t describing them ($\gamma_P = \gamma_P(t)$) are fixed so that *every* curve intersects each Σ_τ for the value $t = \tau$ of its parameter. The lines of $\mathscr{R}$ should permeate the *entire* spacetime. One also assumes the lines γ_P *do not cross*. In this way every event is reached by a unique world line; the subscript P is a label that distinguishes one world line from the others, and defines the event's "spatial location". The value of the parameter, the absolute time at which γ_P meets the event, defines the event's "temporal location". If the world lines evolve *rigidly*, i.e. preserving the mutual distances of the Σ_t as t varies, we can put on the rest space $E_{\mathscr{R}}$, as is expected in Physics, a metric structure that identifies it with every Σ_t, thus bringing us back to the first model of frame discussed in the previous definition.

Definition 2.13 (Reference Frames, Alternative Definition) A **reference frame** (of Classical Physics), or **frame** for short, $\mathscr{R}$ is a collection of world lines $\{\gamma_P\}_{P \in E_{\mathscr{R}}}$ satisfying the following conditions.

(1) **Temporal maximality**: for any $P \in E_{\mathscr{R}}$, the domain of γ_P is the entire $\mathbb{R}$.
(2) **Synchronisation**: $\gamma_P(t) \in \Sigma_t$ for any $t \in \mathbb{R}$ and any $P \in E_{\mathscr{R}}$ (for a suitable choice of additive constant for absolute time).
(3) **Globality**: $\bigcup_{P \in E_{\mathscr{R}}} \gamma_P(\mathbb{R}) = \mathbb{V}^4$.
(4) **Impenetrability**: $\gamma_P(\mathbb{R}) \cap \gamma_Q(\mathbb{R}) = \varnothing$ if $P \neq Q$ for any $P, Q \in E_{\mathscr{R}}$.
(5) **Rigidity**: $d_t(\gamma_P(t), \gamma_Q(t)) = d_{t'}(\gamma_P(t'), \gamma_Q(t'))$ for any $P, Q \in E_{\mathscr{R}}$ and any $t, t' \in \mathbb{R}$.

The indexing set $E_{\mathscr{R}}$ is called **rest space** of $\mathscr{R}$. ◇

Observe that for any $p \in \mathbb{V}^4$, there is a unique $P \in E_{\mathscr{R}}$ whose world line γ_P passes through p. Thus we have a surjective function $\Pi_{\mathscr{R}} : \mathbb{V}^4 \to E_{\mathscr{R}}$ mapping $p \in \mathbb{V}^4$ to the index $P \in E_{\mathscr{R}}$ as explained. The surjectivity of $\Pi_{\mathscr{R}}$ is straightforward from the definition of $\Pi_{\mathscr{R}}$.

As announced, we shall now prove that any frame in the sense of the definition just given defines a frame as per the first definition.

Theorem 2.14 *Let $\mathscr{R}$ be a frame on $\mathbb{V}^4$ in the sense of Definition 2.13, and consider the map $\Pi_{\mathscr{R}} : \mathbb{V}^4 \to E_{\mathscr{R}}$ associating with $p \in \mathbb{V}^4$ the index P of the unique world line of $\mathscr{R}$ meeting p. The rest space $E_{\mathscr{R}}$ admits the structure of a three-dimensional Euclidean space (with space of displacements, inner product and distance $V_{\mathscr{R}}$, $(\cdot|\cdot)_{\mathscr{R}}$ and $d_{\mathscr{R}}$ respectively), so that:*

(1) $\Pi_{\mathscr{R}}$ is differentiable,
(2) for any $t \in \mathbb{R}$ the restriction $\Pi_{\mathscr{R}}|_{\Sigma_t} : \Sigma_t \to E_{\mathscr{R}}$ is an isometry, so in particular an affine isomorphism that preserves the inner products.

Finally, the pair $(\Pi_{\mathscr{R}}, E_{\mathscr{R}})$ is a frame in the sense of Definition 2.13 and the world lines of the rest points for $(\Pi_{\mathscr{R}}, E_{\mathscr{R}})$ (in the sense of Proposition 2.6) coincide with the γ_P.

Proof Let us fix a value $t_0 \in \mathbb{R}$ and refer to Σ_{t_0}. Any curve γ_P intersects Σ_{t_0} exactly at one point P'. Every point P' of Σ_{t_0} determines a unique world line of $\mathscr{R}$, say γ_P. Hence there is a 1-1 correspondence between the indices P of world lines of $\mathscr{R}$ and the points of Σ_{t_0}, so we may view the indices $P \in E_{\mathscr{R}}$ as points of Σ_{t_0}. Since Σ_{t_0} is a Euclidean space, $E_{\mathscr{R}}$ inherits a Euclidean structure under that identification. With this definition $\Pi_{\mathscr{R}}|_{\Sigma_t} : \Sigma_t \to E_{\mathscr{R}}$ reads $\Pi_{\mathscr{R}}|_{\Sigma_t} : \Sigma_t \to \Sigma_{t_0}$. By construction, if $Q_i \in \Sigma_t$ then, for any given $i = 1, 2$, Q_i and $P_i := \Pi_{\mathscr{R}}|_{\Sigma_t}(Q_i) \in \Sigma_{t_0}$ belong to the same world line of $\mathscr{R}$. Due to the rigidity request of Definition 2.13, $d_{t_0}(P_1, P_2) = d_t(Q_1, Q_2)$. We conclude that $\Pi_{\mathscr{R}}|_{\Sigma_t}$ is an isometry between Σ_t and $\Sigma_{t_0} \equiv E_{\mathscr{R}}$. But then, by (iv) in Exercise 1.16.4, $\Pi_{\mathscr{R}}|_{\Sigma_t} : \Sigma_t \to \Sigma_{t_0}$ is an affine isometry as well, and therefore in particular an affine isomorphism that preserves inner products. Note that in this way $E_{\mathscr{R}}$ inherits a differentiable structure making it diffeomorphic to Σ_{t_0}. The rest of the claim is at this point all obvious apart from the fact that with the choices made $\Pi_{\mathscr{R}}$ is differentiable. To prove that, notice that there are global coordinates (t, x^1, x^2, x^3) on $\mathbb{V}^4$, i.e. rest orthonormal coordinates of $\mathscr{R}$ (see Definition 2.8, or using Definition 2.5), compatible with the differentiable structure of spacetime, as shown in Proposition 2.10 (whose proof holds under Definition 2.5 as well) and in which $\Pi_{\mathscr{R}}$ is trivial: $(t, x^1, x^2, x^3) \mapsto (x^1, x^2, x^3)$. $\square$

Remarks 2.15 In the sequel we shall always refer to Definition 2.5 for frames. ∎

2.3 Absolute Point-Particle Kinematics

We shall now describe the motion of a point particle with respect to a given frame $\mathscr{R}$, introduce the concepts of velocity and acceleration and show how the latter decomposes along a trihedron associated with the curve traced by the point particle in the space $E_{\mathscr{R}}$.

2.3.1 Differentiating Curves in Affine Spaces

If $\mathbb{A}^n$ is an affine space with space of displacements V and $P = P(t)$, $t \in (a, b)$, describes a C^k curve, the vectors ($p \leq k$) for $\tau \in (a, b)$:

$$\frac{d^p P}{dt^p}\Big|_{t=\tau} \in V$$

are defined using the differentiable structure of $\mathbb{A}^n$ induced by the affine one. Given Cartesian coordinates with origin O and axes $\{\mathbf{e}_i\}_{i=1,\dots,n}$, the curve $P = P(t)$ will be defined by a curve $x^i = P^i(t)$ in $\mathbb{R}^n$ of class C^k. Then we put:

$$\frac{d^p P}{dt^p}\Big|_{t=\tau} := \sum_{i=1}^{n} \frac{d^p P^i}{dt^p}\Big|_{t=\tau} \mathbf{e}_i . \tag{2.3}$$

Similarly, if $\mathbf{v} = \mathbf{v}(t) \in V$ with $t \in (a, b)$ describes a curve with values in the space of displacements of an affine space $\mathbb{A}^n$, for $\tau \in (a, b)$ we let

$$\frac{d^p \mathbf{v}}{dt^p}\Big|_{t=\tau} := \sum_{i=1}^{n} \frac{d^p v^i}{dt^p}\Big|_{t=\tau} \mathbf{e}_i , \tag{2.4}$$

provided the right hand exists, where the $v^i(t)$ are the components of $\mathbf{v}(t)$ in the basis $\{\mathbf{e}_i\}_{i=1,\dots,n}$ of V. We wish to prove that the two definitions are well posed, i.e. independent of the choice of Cartesian coordinates in the former case and of basis in the latter. From Exercise 1.7.**1**, the coordinates of $P(t)$, $P'^j(t)$ in a new Cartesian system on $\mathbb{A}^n$ are related to the $P^i(t)$ by the non-homogeneous linear transformation:

$$P'^j(t) = \sum_{i=1}^{n} B^j{}_i (P^i(t) + b^i)$$

where $\mathbf{e}'_j = \sum_k (B^{-1})^k{}_j \mathbf{e}_k$. Consequently

$$\frac{d^p P'^j}{dt^p} = \sum_{i=1}^n B^j{}_i \frac{d^p P^i}{dt^p} \,,$$

and then

$$\sum_j \frac{d^p P'^j}{dt^p} \mathbf{e}'_j = \sum_{i,j,k} (B^{-1})^k{}_j B^j{}_i \frac{d^p P^i}{dt^p} \mathbf{e}_k = \sum_{i,k} \delta^k_i \frac{d^p P^i}{dt^p} \mathbf{e}_k = \sum_i \frac{d^p P^i}{dt^p} \mathbf{e}_i \,.$$

The proof in the other case is analogous.

Remarks 2.16 We leave to the reader to show that the above derivatives can be computed, totally equivalently, using the function that maps vectors to pairs of points in affine spaces. For example, with the above definition of derivative, for the first derivative we have:

$$\frac{dP}{dt}\bigg|_{t=\tau} = \lim_{h \to 0} \frac{1}{h}(P(\tau + h) - P(\tau))$$

for the curve $P = P(t)$, $t \in (a, b)$, provided the two sides exist at the same time. ∎

2.3.2   Elementary Kinematic Quantities

We can now define the fundamental notions of Kinematics: the velocity and acceleration of a point particle in a frame $\mathscr{R}$ where the particle's equation of motion has been prescribed.

Definition 2.17 Consider a frame $\mathscr{R}$ and a point particle described by a world line $I \ni t \mapsto \gamma(t) \in \mathbb{V}^4$. We may represent the latter in rest space $E_\mathscr{R}$ as a curve

$$t \mapsto P_\gamma(t) := \Pi_\mathscr{R}(\gamma(t)) \,,$$

which by construction is differentiable. This curve is called **equation of motion** of the world line γ in the frame $\mathscr{R}$. The **velocity of γ with respect to $\mathscr{R}$, at the instant** τ, is the vector of $V_\mathscr{R}$:

$$\mathbf{v}|_\mathscr{R}(\tau) := \frac{dP_\gamma}{dt}\bigg|_{t=\tau} \,.$$

The **acceleration of** γ **with respect to** $\mathscr{R}$, **at the instant** τ, is the vector of $V_\mathscr{R}$:

$$\mathbf{a}|_\mathscr{R}(\tau) := \frac{d^2 P_\gamma}{dt^2}\Big|_{t=\tau} .$$

$\diamond$

Exercises 2.18

(1) Consider, in a plane in the rest space of the frame $\mathscr{R}$, $E_\mathscr{R}$, polar coordinates r, ϕ ($r > 0$, $\phi \in (-\pi, \pi)$) referred to orthogonal Cartesian coordinates x, y. Write the velocity $\mathbf{v}$ and the acceleration $\mathbf{a}$ with respect to $\mathscr{R}$ of a point constrained to move on the plane using polar coordinates and the unit vectors $\mathbf{e}_r$ (tangent to the coordinate line r and pointing outwards) and $\mathbf{e}_\varphi$ (tangent to the coordinate line ϕ oriented as θ increases). Show in particular that, denoting the first derivative in time by a dot,

$$\mathbf{v} = \dot{r}\,\mathbf{e}_r + r\dot{\phi}\,\mathbf{e}_\varphi , \tag{2.5}$$

$$\mathbf{a} = (\ddot{r} - r\dot{\phi}^2)\,\mathbf{e}_r + (r\ddot{\phi} + 2\dot{r}\dot{\phi})\,\mathbf{e}_\varphi \tag{2.6}$$

(2) Consider, in the rest space of the frame $\mathscr{R}$, $E_\mathscr{R}$, spherical coordinates r, ϕ, θ, ($r > 0$, $\phi \in (-\pi, \pi)$, $\theta \in (0, \pi)$) referred to orthogonal Cartesian coordinates x, y, z. Write the velocity and the acceleration with respect to $\mathscr{R}$ of a point using spherical coordinates and the unit vectors $\mathbf{e}_r$ (tangent to the coordinate line r and pointing outwards), $\mathbf{e}_\varphi$ (tangent to the coordinate line ϕ oriented with the increasing ϕ) and $\mathbf{e}_\theta$ (tangent to the coordinate line θ oriented with the increasing θ). In particular show that:

$$\mathbf{v} = \dot{r}\,\mathbf{e}_r + r\dot{\theta}\,\mathbf{e}_\theta + r\dot{\phi}\sin\theta\,\mathbf{e}_\varphi , \tag{2.7}$$

$$\mathbf{a} = (\ddot{r} - r\dot{\theta}^2 - r\dot{\phi}^2\sin^2\theta)\,\mathbf{e}_r + (r\ddot{\theta} + 2\dot{r}\dot{\theta} - r\dot{\phi}^2\sin\theta\cos\theta)\,\mathbf{e}_\theta$$
$$+(r\ddot{\phi}\sin\theta + 2\dot{r}\dot{\phi}\sin\theta + 2r\dot{\phi}\dot{\theta}\cos\theta)\,\mathbf{e}_\varphi . \tag{2.8}$$

(3) Consider, in the rest space of the frame $\mathscr{R}$, $E_\mathscr{R}$, cylindrical coordinates r, ϕ, z ($r > 0$, $\phi \in (-\pi, \pi)$, $z \in \mathbb{R}$) referred to orthogonal Cartesian coordinates x, y, z. Write the velocity and the acceleration with respect to $\mathscr{R}$ of a point using cylindrical coordinates and the unit vectors $\mathbf{e}_r$ (tangent to the coordinate line r and pointing outwards), $\mathbf{e}_\varphi$ (tangent to the coordinate line ϕ oriented with the increasing ϕ) and $\mathbf{e}_z$ (tangent to the coordinate line z with positive orientation). Show in particular that, denoting the first derivative in time by a dot,

$$\mathbf{v} = \dot{r}\,\mathbf{e}_r + r\dot{\phi}\,\mathbf{e}_\varphi + \dot{z}\,\mathbf{e}_z , \tag{2.9}$$

$$\mathbf{a} = (\ddot{r} - r\dot{\phi}^2)\,\mathbf{e}_r + (r\ddot{\phi} + 2\dot{r}\dot{\phi})\,\mathbf{e}_\varphi + \ddot{z}\,\mathbf{e}_z \tag{2.10}$$

(4) $\boxed{\text{AC}}$ Exercises 2.18.**2** and 2.18.**3** can be solved, as regards the acceleration, in a quicker way using the covariant derivative induced by the Levi-Civita connection of $\mathbb{E}^3 = E_{\mathcal{R}}$ associated with the inner product. In general local curvilinear coordinates x^1, x^2, x^3, if $(x^1(t), x^2(t), x^3(t))$ are the coordinates of $P(t)$ at time $t \in \mathbb{R}$, then

$$\mathbf{a} = \sum_{i=1}^{3} \left(\frac{d^2 x^i}{dt^2} + \sum_{j,k=1}^{3} \Gamma^i{}_{jk} \frac{dx^j}{dt} \frac{dx^k}{dt} \right) \frac{\partial}{\partial x^i} . \tag{2.11}$$

where the *Christoffel symbols* $\Gamma^i{}_{jk}$ are, as usual, defined via the metric tensor $\sum_{ik} g_{ik} dx^i \otimes dx^k$ (and its inverse, of components g^{rs}), as:

$$\Gamma^i{}_{jk} = \sum_{r=1}^{3} \frac{1}{2} g^{ir} \left(\frac{\partial g_{rj}}{\partial x^k} + \frac{\partial g_{kr}}{\partial x^j} - \frac{\partial g_{jk}}{\partial x^r} \right) . \tag{2.12}$$

The vectors $\frac{\partial}{\partial x^i}$ form the basis related to the unit vectors by $\mathbf{e}_i = (g_{ii})^{-1/2} \frac{\partial}{\partial x^i}$. For example, for the spherical coordinates $x^1 = r, x^2 = \theta, x^3 = \phi$, we have: $\mathbf{e}_r = \frac{\partial}{\partial r}$, $\mathbf{e}_\theta = r^{-1} \frac{\partial}{\partial \theta}$, $\mathbf{e}_\varphi = (r \sin \theta)^{-1} \frac{\partial}{\partial \phi}$. Explain why this result holds.

2.3.3   Kinematics for Point Particles Constrained to Stationary Curves and Surfaces

Often the motion of point particles is *constrained*, i.e. forced to occur in fixed geometric sets, which may be curves or surfaces. Here we will deal with the situation where the motion happens along curves or on surfaces that do not move with respect to a frame. A technically relevant aspect is to develop mathematical techniques to decompose the particle's velocity and acceleration along bases of vectors associated with those sets. When discussing the dynamics of constrained point particles, we will see that it will be physically important to be able to decompose the particle's acceleration into a *tangent component* and a *normal component* to the surface or curve on which the motion occurs.

Consider a curve Γ, of equation $P = P(u)$ with $u \in (a, b)$, *at rest* with respect to the frame $\mathcal{R}$. In other words, the curve's equation, in rest orthonormal coordinates on $\mathcal{R}$, does not contain time explicitly (the parameter u is not time). We will demand that the curve is C^1 (so that, in particular, it is *rectifiable* [Apo91I]) and **regular**, i.e. the tangent vector is everywhere non-zero along Γ:

$$\frac{dP}{du} \neq 0 , \quad \text{for } u \in (a, b).$$

Suppose a point particle is constrained to move on the curve. Note that Γ has a preferred parametrisation that we set out to introduce now. Fix a point $O = P(u_0)$ on the curve and let:

$$s(u) := \int_{u_0}^{u} \left\| \frac{dP}{du'} \right\| du' . \tag{2.13}$$

By the fundamental theorem of calculus the map $(a, b) \ni u \mapsto s(u)$, called **arclength**, is C^1 with positive first derivative given by the integrand in (2.13). Because of that it is strictly increasing hence invertible with C^1 inverse. Then s can be used to reparametrise Γ. From the elementary theory of curves [Apo91I] we know $s(u)$ is the *length* of the piece of curve from $P(u_0)$ to $P(u)$. This length comes with a sign: it is positive for points that follow $P(u_0)$ and negative for points that precede $P(u_0)$, where the orientation is determined by the initial parameter u.

If we now change parameter to $v = v(u)$ to describe the same Γ, assuming that $v = v(u)$ is C^1 with strictly positive derivative, then $v = v(u)$ is bijective with C^1 inverse, and the curve Γ can be described by a function $P(v) := P(u(v))$ with $v \in (v(a), v(b))$. Here too we can define the arclength $s = s(v)$ using (2.13). Directly from that, we find

$$s(u) = \int_{u_0}^{u} \left\| \frac{dP}{du'} \right\| du' = \int_{v(u_0)}^{v(u)} \left\| \frac{dP}{du'} \right\| \frac{du'}{dv'} dv' = \int_{v(u_0)}^{v(u)} \left\| \frac{dP}{dv'} \right\| dv' = s(v(u)) .$$

In this sense the arclength is *invariant* under reparametrisations of Γ (given by C^1 maps with strictly positive derivative).

Suppose we parametrise Γ using s, so $P(s) := P(u(s))$ with $s \in (s_a, s_b)$. First of all the curve's tangent vector associated with parameter $s \in (s_a, s_b)$:

$$\mathbf{t}(s) := \frac{dP}{ds} \tag{2.14}$$

has unit length. In fact from (2.13), using s as parameter and applying the fundamental theorem of calculus we obtain

$$1 = \frac{ds}{ds} = \left\| \frac{dP}{ds} \right\| .$$

In the sequel, as there will not be any ambiguity, we will denote with $\cdot$ the inner product in the Euclidean space $E_{\mathscr{R}}$. There is a second unit vector associated with Γ when $\mathbf{t}$ is not constant, the **unit normal**:

$$\mathbf{n}(s) := \rho(s) \frac{d\mathbf{t}}{ds} , \quad \text{with } \rho(s) := \left\| \frac{d\mathbf{t}}{ds} \right\|^{-1} . \tag{2.15}$$

The number $\rho(s)$ is called **curvature radius** at $P(s)$. If $\mathbf{t}$ and $\mathbf{n}$ are both defined, then $\mathbf{t} \perp \mathbf{n}$:

$$\mathbf{t} \cdot \mathbf{n} = \rho \mathbf{t} \cdot \frac{d\mathbf{t}}{ds} = \frac{\rho}{2} \frac{d\mathbf{t} \cdot \mathbf{t}}{ds} = \frac{\rho}{2} \frac{d}{ds} 1 = 0 \ .$$

If $\mathbf{t}$ and $\mathbf{n}$ are defined, there is a third unit vector perpendicular to both, called **binormal**:

$$\mathbf{b}(s) := \mathbf{t}(s) \wedge \mathbf{n}(s) \ . \tag{2.16}$$

The unit vectors $\mathbf{t}(s)$, $\mathbf{n}(s)$, $\mathbf{b}(s)$, when they exist, form a *right-handed orthonormal* trihedron, by construction. It is called **Frenet trihedron** of Γ at the point $P(s)$.

Remarks 2.19

(1) The arclength s is defined up to the choice of origin of Γ (the point where $s = 0$), so if s is an arclength, so is $s'(s) := s + c$, where $c \in \mathbb{R}$ is an arbitrary constant. The unit vectors of a Frenet trihedron, being constructed using the derivatives of s, are not affected by the above freedom to choose s.

(2) For a regular curve Γ we have $\left\| \frac{d\mathbf{t}}{ds} \right\| = 0$ for any $s \in (s_1, s_2)$ if and only if Γ can be reparametrised (using a parameter u with $u = u(s)$ of class C^1 and everywhere non-zero derivative) as a straight segment for $s \in (s_1, s_2)$. In fact, if Γ can be reparametrised as segment from $P(s_1)$ to $P(s_2)$, then $P(s(u)) = P(s_1) + u(P(s_2) - P(s_1))$ between those points. Applying the definition of arclength immediately shows that $s(u) = u\|P(s_2) - P(s_1)\|$ and so, as we intuitively expected,

$$P(s) = P(s_1) + s \frac{P(s_2) - P(s_1)}{\|P(s_2) - P(s_1)\|}$$

and then $\mathbf{t} = \frac{P(s_2) - P(s_1)}{\|P(s_2) - P(s_1)\|}$ is constant and $\left\| \frac{d\mathbf{t}}{ds} \right\| = 0$. If, conversely, $\left\| \frac{d\mathbf{t}}{ds} \right\| = 0$, then $\mathbf{t}$ is constant, and therefore integrating the equation $\frac{dP(s)}{ds} = \mathbf{t}$ in the interval (s_1, s_2) we find the equation of a line segment:

$$P(s) = P(s_1) + s\mathbf{t} \ .$$

■

If a point particle (with world line γ) is constrained to move on the C^2 curve Γ, assumed not moving in the frame $\mathscr{R}$, the position of the point as time goes by is completely determined by the arclength s as function of time $s = s(t)$. Hence one should be able to express velocity and acceleration with respect to $\mathscr{R}$ in terms of s and of other geometric features of the curve. We will suppose Γ cannot be parametrised as a line segment, so that each point admits a Frenet trihedron. If the curve has equation $P = P(s)$, the point particle's equation of motion will be $P(t) =$

$P(s(t))$. Applying the definition of velocity and of tangent vector we immediately obtain:

$$\mathbf{v}(t) = \frac{ds}{dt}\,\mathbf{t}(s(t))\,, \tag{2.17}$$

where we have dropped the reference to the frame $|_{\mathscr{R}}$ for simplicity. Differentiating once more in time gives

$$\mathbf{a}(t) = \frac{d^2s}{dt^2}\,\mathbf{t}(s(t)) + \frac{ds}{dt}\frac{d\mathbf{t}}{ds}\frac{ds}{dt}\,,$$

and then, using the definition of $\mathbf{n}$, we find the expression for the acceleration:

$$\mathbf{a}(t) = \frac{d^2s}{dt^2}\,\mathbf{t}(s(t)) + \frac{1}{\rho(s(t))}\left(\frac{ds}{dt}\right)^2 \mathbf{n}(s(t))\,. \tag{2.18}$$

The acceleration is decomposed into a tangential component and a normal component to the constraint curve, which will be of the utmost importance when developing the dynamics of the constrained point particle.

Example 2.20 Consider the curve (helix) $P = P(u)$, given in orthonormal coordinates by parametric equations $x(u) = \cos u$, $y(u) = \sin u$, $z = u$, with $u \in \mathbb{R}$. Let us write the Frenet trihedron in the unit basis $\mathbf{e}_x$, $\mathbf{e}_y$, $\mathbf{e}_z$. The arclength arises by integrating

$$s(u) = \int_0^u \left\|\frac{dP(u')}{du'}\right\| du' = \int_0^u \sqrt{(-\sin u')^2 + (\cos u')^2 + 1}\,du' = \int_0^u \sqrt{2}du'\,.$$

Hence $s(u) = \sqrt{2}u$ if we choose the origin of s at the point $(1, 0, 0)$. All in all the curve has parametric equations: $x(s) = \cos(s/\sqrt{2})$, $y(u) = \sin(s/\sqrt{2})$, $z = (s/\sqrt{2})$. The direct computation of the Frenet trihedron gives:

$$\mathbf{t}(s) = \frac{1}{\sqrt{2}}\left(-\sin(s/\sqrt{2})\,\mathbf{e}_x + \cos(s/\sqrt{2})\,\mathbf{e}_y + \mathbf{e}_z\right)\,, \tag{2.19}$$

$$\mathbf{n}(s) = -\left(\cos(s/\sqrt{2})\,\mathbf{e}_x + \sin(s/\sqrt{2})\,\mathbf{e}_y\right)\,, \tag{2.20}$$

$$\mathbf{b}(s) = \frac{1}{\sqrt{2}}\left(\sin(s/\sqrt{2})\,\mathbf{e}_x - \cos(s/\sqrt{2})\,\mathbf{e}_y + \mathbf{e}_z\right)\,. \tag{2.21}$$

Since the basis $\mathbf{t}, \mathbf{n}, \mathbf{b}$ is orthonormal, as is $\mathbf{e}_x$, $\mathbf{e}_y$, $\mathbf{e}_z$, the above transformation is immediately invertible by replacing the transformation's matrix by its transpose:

$$\mathbf{e}_x = -\frac{1}{\sqrt{2}}\sin(s/\sqrt{2})\mathbf{t}(s) - \cos(s/\sqrt{2})\mathbf{n}(s) + \frac{1}{\sqrt{2}}\sin(s/\sqrt{2})\mathbf{b}(s)\,, \tag{2.22}$$

$$\mathbf{e}_y = \frac{1}{\sqrt{2}} \cos(s/\sqrt{2})\mathbf{t}(s) - \sin(s/\sqrt{2})\mathbf{n}(s) - \frac{1}{\sqrt{2}} \cos(s/\sqrt{2})\mathbf{b}(s) , \quad (2.23)$$

$$\mathbf{e}_z = \frac{1}{\sqrt{2}}\mathbf{t}(s) + \frac{1}{\sqrt{2}}\mathbf{b}(s) . \quad (2.24)$$

Exercises 2.21

(1) Consider the curve Γ, of class C^1 and regular in $\mathbb{E}^3$, parametrised by its arclength $P = P(s)$, $s \in (a, b)$. Show that if Γ is contained in a plane Π and is not reparametrisable as a line segment in any subinterval of (a, b), then the binormal $\mathbf{b}$ is constant along Γ and perpendicular to Π.

(2) Consider the curve Γ, of class C^1 and regular in $\mathbb{E}^3$, parametrised by arclength $P = P(s)$, $s \in (a, b)$. Prove that if the binormal $\mathbf{b}$ exists, is non-zero and constant on Γ, the curve is entirely contained in a plane orthogonal to $\mathbf{b}$.

(3) For a regular C^1 curve parametrised by arclength $P = P(s)$, $s \in (a, b)$, prove the *Darboux formulas*:

$$\frac{d\mathbf{n}(s)}{ds} = -\frac{\mathbf{t}(s)}{\rho(s)} - \frac{\mathbf{b}(s)}{\tau(s)} = \Omega(s) \wedge \mathbf{n}(s) , \quad (2.25)$$

$$\frac{d\mathbf{b}(s)}{ds} = -\frac{\mathbf{n}(s)}{\tau(s)} = \Omega(s) \wedge \mathbf{b}(s) , \quad (2.26)$$

$$\frac{d\mathbf{t}(s)}{ds} = -\frac{\mathbf{n}(s)}{\rho(s)} = \Omega(s) \wedge \mathbf{t}(s) , \quad (2.27)$$

(assuming the vectors and scalars exist). Above, τ, defined in (2.26), is called *torsion* and

$$\Omega(s) := \rho(s)^{-1}\mathbf{b}(s) - \tau(s)^{-1}\mathbf{t}(s)$$

is the *Darboux vector* at the point $P(s)$.

Let us pass to consider a point particle constrained to a surface that is fixed in a frame $\mathscr{R}$. In other words, in rest orthonormal coordinates on $\mathscr{R}$ the equations that determine S do not contain time explicitly. We assume that the surface S is described, in the Euclidean space $\mathbb{E}^3 = E_{\mathscr{R}}$, by parametric equations $P = P(u, v)$, with $(u, v) \in D$, an open (connected) subset of $\mathbb{R}^2$. We also suppose the surface is regular, i.e. $P = P(u, v)$ is at least of class C^1 and:

$$\frac{\partial P}{\partial u} \wedge \frac{\partial P}{\partial v} \neq 0 \quad \text{for any } (u, v) \in D .$$

In these hypotheses the three vectors $\frac{\partial P}{\partial u}, \frac{\partial P}{\partial v}, \frac{\partial P}{\partial u} \wedge \frac{\partial P}{\partial v}$ form, at each point on S, a basis. Note that the third vector is by construction always perpendicular to S. To study the kinematics, and even more so the dynamics, of a point particle constrained to S, it is convenient to parametrise the point's motion by coordinates

$(u, v) = (u(t), v(t))$ and then express the velocity and the acceleration using the above basis or the corresponding basis of *unit vectors*. Within certain limits it is possible to choose other parameters u, v to describe S so to simplify the computations. In particular, when possible, it is very convenient to choose u, v so that the above basis is orthonormal.

Example 2.22

(1) Consider the portion of cone S (Fig. 2.3) given by $0 < \kappa z = \sqrt{x^2 + y^2}$ ($\kappa > 0$ fixed) in orthonormal coordinates $\mathbf{x} = (x, y, z)$ in $\mathbb{E}^3 = E_{\mathscr{R}}$. We may parametrise S by $u = x$ and $v = y$ and the parametric equations become $x = u, y = v, z = \kappa^{-1}\sqrt{u^2 + v^2}$, where the coordinates' domain is $D = \{(u, v) \in \mathbb{R}^2 \mid 0 < u^2 + v^2\}$. It is straightforward to verify that the regularity condition in Cartesian coordinates

$$\frac{\partial \mathbf{x}}{\partial u} \wedge \frac{\partial \mathbf{x}}{\partial v} \neq 0 \quad \text{for any } (u, v) \in D$$

holds. Define $\zeta := \sqrt{(x^2 + y^2) + z(x, y)^2} = \sqrt{(1 + \kappa^{-2})(x^2 + y^2)}$, whose geometric meaning is clear (the distance of a point of coordinates $\mathbf{x} \in C$ to the cone's vertex O), and call ϕ the usual angular polar coordinate in the xy-plane. If we reparametrise the surface S using coordinates $\zeta \in (0, +\infty)$ and $\phi \in (-\pi, \pi)$ (these are not global because the half-line $\phi = \pm\pi$ is not part of the range), the basis of vectors associated with the coordinates can be taken orthonormal.

A point on the cone is described by

$$x = (\sin \alpha)\zeta \cos \phi, \quad y = (\sin \alpha)\zeta \sin \phi, \quad z = (\cos \alpha)\zeta,$$

where $\kappa = \tan \alpha$. The unit vectors $\mathbf{e}_\zeta$ and $\mathbf{e}_\phi$ tangent to the coordinate lines ζ and ϕ (oriented in the increasing direction), and their cross product $\mathbf{n} := \mathbf{e}_\zeta \wedge \mathbf{e}_\phi$,

Fig. 2.3 Example 2.22.1

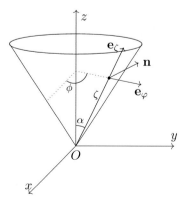

normal to S, are, in the Cartesian basis:

$$\mathbf{e}_\zeta = \sin\alpha \, \cos\phi \, \mathbf{e}_x + \sin\alpha \, \sin\phi \, \mathbf{e}_y + \cos\alpha \, \mathbf{e}_z \,, \tag{2.28}$$

$$\mathbf{e}_\varphi = -\sin\phi \, \mathbf{e}_x + \cos\phi \, \mathbf{e}_y \,, \tag{2.29}$$

$$\mathbf{n} = -\cos\alpha \, \cos\phi \, \mathbf{e}_x - \cos\alpha \, \sin\phi \, \mathbf{e}_y + \sin\alpha \, \mathbf{e}_z \,. \tag{2.30}$$

These relations can be inverted (taking the transformation's transpose matrix) to

$$\mathbf{e}_x = \sin\alpha \, \cos\phi \, \mathbf{e}_\zeta - \sin\phi \, \mathbf{e}_\varphi - \cos\alpha \, \cos\phi \, \mathbf{n} \,, \tag{2.31}$$

$$\mathbf{e}_y = \sin\alpha \, \sin\phi \, \mathbf{e}_\zeta + \cos\phi \, \mathbf{e}_\varphi - \cos\alpha \, \sin\phi \, \mathbf{n} \,, \tag{2.32}$$

$$\mathbf{e}_z = \cos\alpha \, \mathbf{e}_\zeta + \sin\alpha \, \mathbf{n} \,. \tag{2.33}$$

Suppose $\zeta = \zeta(t)$, $\phi = \phi(t)$ describe the evolution of the point $P(t)$ on S. The position of $P(t)$ can be defined by the position vector: $P(t) - O = \zeta(t)\mathbf{e}_\zeta(t)$. The time derivative defines the point's velocity in the frame $\mathcal{R}$ in which the cone S is fixed. One has to bear in mind that $\mathbf{e}_\zeta$ moves with time. Using the dot for the time derivative, we have:

$$\mathbf{v}(t) = \dot{\zeta}(t)\mathbf{e}_\zeta(t) + \zeta(t)\dot{\mathbf{e}}_\zeta(t) \,.$$

The time derivative of the unit vectors can be found by differentiating the right-hand side of (2.28)–(2.30) (assuming $\zeta = \zeta(t)$, $\phi = \phi(t)$) and then using (2.31)–(2.33) in the result. The calculation gives:

$$\dot{\mathbf{e}}_\zeta = \dot{\phi}\sin\alpha \, \mathbf{e}_\varphi \,, \tag{2.34}$$

$$\dot{\mathbf{e}}_\varphi = -\dot{\phi}\sin\alpha \, \mathbf{e}_\zeta + \dot{\phi}\cos\alpha \, \mathbf{n} \,, \tag{2.35}$$

$$\dot{\mathbf{n}} = -\dot{\phi}\cos\alpha \, \mathbf{e}_\varphi \,, \tag{2.36}$$

and so

$$\mathbf{v}(t) = \dot{\zeta}(t)\mathbf{e}_\zeta(t) + \dot{\phi}(t)\zeta(t)\sin\alpha \, \mathbf{e}_\varphi(t) \,. \tag{2.37}$$

We have expressed the velocity in the basis of the surface to which the point is constrained. Differentiating (2.37) in time

$$\mathbf{a}(t) = \ddot{\zeta}(t)\mathbf{e}_\zeta(t) + \dot{\zeta}(t)\dot{\mathbf{e}}_\zeta(t) + \dot{\phi}(t)\dot{\zeta}(t)\sin\alpha \, \mathbf{e}_\varphi(t) + \ddot{\phi}(t)\zeta(t)\sin\alpha \, \mathbf{e}_\varphi$$
$$+ \dot{\phi}(t)\zeta(t)\sin\alpha \, \dot{\mathbf{e}}_\varphi(t) \,,$$

and proceeding as above (using the expressions obtained for the derivatives of the unit vectors), we arrive at the acceleration decomposed in normal and tangent components on the conical surface:

$$\mathbf{a}(t) = \left(\ddot{\zeta}(t) - \dot{\phi}(t)^2 \zeta(t) \sin^2 \alpha\right) \mathbf{e}_\zeta(t) + \left(\ddot{\phi}(t)\zeta(t) + 2\dot{\phi}(t)\dot{\zeta}(t)\right) \sin \alpha \mathbf{e}_\varphi(t)$$

$$+ \frac{\dot{\phi}(t)^2 \zeta(t) \sin 2\alpha}{2} \mathbf{n}(t). \tag{2.38}$$

(2) The formulas of Exercises 2.18.**2** and 2.18.**3** allow to write the velocity and the acceleration of a point particle constrained to, respectively, a spherical surface and a cylindrical surface. Describing the former in spherical coordinates r, ϕ, θ and the surface as $r = R$, the coordinates ϕ and θ turn out to be admissible coordinates on the surface, with $\mathbf{e}_\varphi \perp \mathbf{e}_\theta$. The unit normal $\mathbf{n}$ coincides with $\mathbf{e}_r$. The expression of the velocity and the acceleration using this orthonormal basis then is:

$$\mathbf{v} = R\dot{\theta}\, \mathbf{e}_\theta + R\dot{\phi} \sin\theta\, \mathbf{e}_\varphi , \tag{2.39}$$

$$\mathbf{a} = (R\ddot{\theta} - R\dot{\phi}^2 \sin\theta \cos\theta)\, \mathbf{e}_\theta + (R\ddot{\phi} \sin\theta + 2R\dot{\phi}\dot{\theta} \cos\theta)\, \mathbf{e}_\varphi$$

$$-(R\dot{\theta}^2 + R\dot{\phi}^2 \sin^2\theta)\mathbf{n} . \tag{2.40}$$

For the cylindrical surface described by $r = R$ in cylindrical coordinates r, ϕ, z, the coordinates ϕ, z are admissible on the surface, so $\mathbf{e}_\varphi \perp \mathbf{e}_z$ and the unit normal is still $\mathbf{e}_r$. In the end:

$$\mathbf{v} = R\dot{\phi}\, \mathbf{e}_\varphi + \dot{z}\, \mathbf{e}_z , \tag{2.41}$$

$$\mathbf{a} = R\ddot{\phi}\, \mathbf{e}_\varphi + \ddot{z}\, \mathbf{e}_z - R\dot{\phi}^2 \mathbf{n} . \tag{2.42}$$

2.4 Relative Point-Particle Kinematics

In this section we shall deal with certain relations occurring among the various descriptions of the motion of a single point particle in two (or more) frames. These relations are the mathematical content of **relative Kinematics**.

Consider a frame $\mathscr{R}$. Choose a point $P \in E_{\mathscr{R}}$ (as purely geometric entity) and a vector $\mathbf{v} \in V_{\mathscr{R}}$. These two mathematical objects live in the frame's rest space, but they can be identified *bijectively* with analogous mathematical objects, respectively $P(t)$ and $\mathbf{v}(t)$, in a given absolute space Σ_t once we fix a time t. It is just a matter of defining $P(t) := (\Pi_{\mathscr{R}}|_{\Sigma_t})^{-1}(P)$ and $\mathbf{v}(t) := d(\Pi_{\mathscr{R}}|_{\Sigma_t})^{-1}(\mathbf{v})$. Vice versa, the mathematical objects $P(t)$ and $\mathbf{v}(t)$ may be viewed in some given absolute space as existing in the Euclidean rest space of a frame $\mathscr{R}'$ by inverting the above relationships (and using a different $\mathscr{R}'$ from $\mathscr{R}$ if necessary): $P' := \Pi_{\mathscr{R}'}|_{\Sigma_t}(P(t))$ and $\mathbf{v}' := d(\Pi_{\mathscr{R}'}|_{\Sigma_t})(\mathbf{v}(t))$. All these identifications arise through isomorphisms (affine isometries and vector space isomorphisms respectively) under which *the structure of the admissible operations between the entities considered is preserved.* For instance, the difference of two points in Σ_t becomes the difference of the corresponding points in the space $E_{\mathscr{R}'}$.

If γ is a world line of a point particle that, in the frame $\mathscr{R}$ and at time t, has velocity $\mathbf{v}|_{\mathscr{R}}(t)$, the latter can be thought of as in the absolute space Σ_t rather than in the Euclidean rest space of $\mathscr{R}$. Or it can be viewed in the Euclidean rest space of *another* frame $\mathscr{R}'$. In this way we can compare the velocities of *the same* point particle, but referred to *two different* frames. *For this reason, in absence of ambiguity, from now on we will indicate with a unique symbol two points in distinct Euclidean spaces (example: $E_{\mathscr{R}}$, Σ_t), but identified under $\Pi_{\mathscr{R}}$. The same convention will apply to the vectors in the corresponding spaces of displacements. Similarly, we shall denote inner products only with $\cdot$ (dropping the dependence on time, too), except in cases where the different metric structures are compared.*

2.4.1 The ω Vector and the Poisson Formulas

Consider a function $\mathbb{R} \ni t \mapsto \mathbf{u}(t) \in V_t$ mapping an instant t to a vector in the corresponding absolute space Σ_t (more precisely, in the space of displacements V_t). We will say that such function is *differentiable*, if the components of $t \mapsto d(\Pi_{\mathscr{R}}|_{\Sigma_t})(\mathbf{u}(t))$ are differentiable functions of time in an arbitrary orthonormal basis of $V_{\mathscr{R}}$.

Remarks 2.23 It is immediate to see, identifying vectors and points in rest spaces and so using the transformations (2.2) between rest Cartesian coordinates of different frames, that the differentiability of a curve $\mathbb{R} \ni t \mapsto \mathbf{u}(t) \in V_t$ does not depend on the chosen frame. ∎

Given a differentiable function $\mathbb{R} \ni t \mapsto \mathbf{u}(t) \in V_t$, consider orthonormal bases $\mathbf{e}_1, \mathbf{e}_2, \mathbf{e}_3 \in V_{\mathscr{R}}$ and $\mathbf{e}'_1, \mathbf{e}'_2, \mathbf{e}'_3 \in V_{\mathscr{R}'}$ in the spaces of displacements of *distinct* frames $\mathscr{R}$ and $\mathscr{R}'$, respectively. In every Σ_t:

$$\mathbf{u}(t) = \sum_{i=1}^{3} u^i(t)\mathbf{e}_i(t) = \sum_{j=1}^{3} u'^{j}(t)\mathbf{e}'_j(t) .$$

Similarly, in $\mathscr{R}$, where each $\mathbf{e}_i$ is a constant vector in time:

$$\mathbf{u}(t) = \sum_{i=1}^{3} u^i(t)\mathbf{e}_i = \sum_{j=1}^{3} u'^{j}(t)\mathbf{e}'_j(t) .$$

In this formula $\mathbf{u}(t)$ is viewed as a vector of $V_{\mathscr{R}}$, so it should actually be written $d(\Pi_{\mathscr{R}}|_{\Sigma_t})(\mathbf{u}(t))$, but we will use a lighter notation for heuristic reasons. At last, in $\mathscr{R}'$, where each $\mathbf{e}'_j$ is a constant vector in time:

$$\mathbf{u}(t) = \sum_{i=1}^{3} u^i(t)\mathbf{e}_i(t) = \sum_{j=1}^{3} u'^j(t)\mathbf{e}'_j .$$

In $\mathscr{R}$ we can differentiate $\mathbf{u}(t)$ with respect to time simply by differentiating the components of $\mathbf{u}(t)$ in the basis $\mathbf{e}_i$.

$$\frac{d}{dt}\bigg|_{\mathscr{R}} \mathbf{u}(t) := \sum_{i=1}^{3} \frac{d}{dt} u^i(t)\mathbf{e}_i .$$

In $\mathscr{R}'$ we can differentiate $\mathbf{u}(t)$ with respect to time, simply by differentiating the components of $\mathbf{u}(t)$ in the basis $\mathbf{e}'_j$:

$$\frac{d}{dt}\bigg|_{\mathscr{R}'} \mathbf{u}(t) := \sum_{j=1}^{3} \frac{d}{dt} u'^j(t)\mathbf{e}'_j .$$

Hence we can transport these vectors in spacetime, more precisely in each Σ_t, respectively obtaining

$$\frac{d}{dt}\bigg|_{\mathscr{R}} \mathbf{u}(t) := \sum_{i=1}^{3} \frac{du^i(t)}{dt}\mathbf{e}_i(t) \quad \text{and} \quad \frac{d}{dt}\bigg|_{\mathscr{R}'} \mathbf{u}(t) := \sum_{j=1}^{3} \frac{du'^j(t)}{dt}\mathbf{e}'_j(t) .$$

Note that now the unit vectors *depend* on time, but the derivatives acts *only* on the components, *when these are expressed in the basis with the frame in which the derivative is computed.* Let us state all of this in a formal definition.

Definition 2.24 (Time Derivative with Respect to a Frame) Consider a curve $(a, b) \ni t \mapsto \mathbf{u}(t) \in V_t$.

(1) We will say the curve is **differentiable** if the components of $(a, b) \ni t \mapsto d(\Pi_{\mathscr{R}}|_{\Sigma_t})(\mathbf{u}(t)) \in V_{\mathscr{R}}$ in the orthonormal basis of the rest space of a frame $\mathscr{R}$ are differentiable maps.[3]
(2) In that case, for any $t \in (a, b)$, the **derivative of u with respect to the frame** $\mathscr{R}$ is the vector in V_t

$$\frac{d}{dt}\bigg|_{\mathscr{R}} \mathbf{u}(t) := \sum_{i=1}^{3} \frac{du^i(t)}{dt}\mathbf{e}_i(t) ,$$

---

[3] As noted before, if this condition holds in one frame $\mathscr{R}$ it will hold in all other frames $\mathscr{R}'$.

where $\mathbf{e}_1, \mathbf{e}_2, \mathbf{e}_3$ form an orthonormal basis of $V_{\mathscr{R}}$ in which $\mathbf{u} = \sum_{i=1}^{3} u^i \mathbf{e}_i$, and we use the convention $\mathbf{e}_i(t) := d(\Pi_{\mathscr{R}}|_{\Sigma_t})^{-1}(\mathbf{e}_i)$ for $i = 1, 2, 3$.   ◇

Remarks 2.25

(1) By construction, the derivative with respect to the frame $\mathscr{R}$ does not depend on the basis chosen in $V_{\mathscr{R}}$ nor, more generally, on the frame's rest orthonormal coordinates. The proof, if we reduce to the rest space of $\mathscr{R}$ by $d(\Pi_{\mathscr{R}}|_{\Sigma_t})$, is the same we have given in Sect. 2.3.1. There, we showed that the derivative of a curve, or a vector-valued curve, in a Euclidean space, does not depend on the Cartesian coordinates in the former case, and on the basis in the latter. The aforementioned independence becomes manifest if we observe that by construction we have the identity:

$$\left.\frac{d}{dt}\right|_{\mathscr{R}} = d(\Pi_{\mathscr{R}}|_{\Sigma_t})^{-1} \frac{d}{dt} d(\Pi_{\mathscr{R}}|_{\Sigma_t}) .$$

From this expression it is also clear that $\left.\frac{d}{dt}\right|_{\mathscr{R}}$ satisfies the usual properties of the differentiation operator, for instance the familiar properties of linearity on linear combinations (with constant coefficients) of vector-valued functions, as well as:

$$\left.\frac{d}{dt}\right|_{\mathscr{R}} f(t)\mathbf{u}(t) = \frac{df}{dt}\mathbf{u}(t) + f(t) \left.\frac{d}{dt}\right|_{\mathscr{R}} \mathbf{u}(t) , \tag{2.43}$$

if $f : (a, b) \to \mathbb{R}$ is a differentiable map and $\mathbf{u}$ is as above.

(2) If γ is a world line in $\mathbb{V}^4$, by Definition 2.17 the velocity of γ with respect to $\mathscr{R}$ can be computed differentiating $\Pi_{\mathscr{R}}(\gamma(t))$ in orthogonal Cartesian coordinates with origin $O_{\mathscr{R}}$ and axes $\mathbf{e}_1, \mathbf{e}_2, \mathbf{e}_3$ in $E_{\mathscr{R}}$. Equivalently, that velocity arises differentiating the vector-valued curve in V_t, $t \mapsto \Pi_{\mathscr{R}}(\gamma(t)) - O_{\mathscr{R}}$. With this approach we may use the definition of derivative with respect to a frame and view the velocity as a vector in Σ_t given by:

$$\mathbf{v}|_{\mathscr{R}}(t) = \left.\frac{d}{dt}\right|_{\mathscr{R}} (\gamma(t) - O_{\mathscr{R}}(t)) , \tag{2.44}$$

where $O_{\mathscr{R}}(t) = (\Pi_{\mathscr{R}}|_{\Sigma_t})^{-1}(O_{\mathscr{R}})$ and $O_{\mathscr{R}} \in E_{\mathscr{R}}$.   ∎

The issue we want to study now is the relationship between the two operators $\left.\frac{d}{dt}\right|_{\mathscr{R}}$ and $\left.\frac{d}{dt}\right|_{\mathscr{R}'}$ when $\mathscr{R}$ is different from $\mathscr{R}'$. The matter is of great physical interest because it allows to construct relative Kinematics: it permits us to write the velocity and the acceleration of a point particle in a frame when we know them in another frame and the relative motion of the two frames is given. The following famous theorem due to Poisson holds.

Theorem 2.26 (Poisson Formulas)   *Let $\mathscr{R}$ and $\mathscr{R}'$ be two frames and $\{\mathbf{e}'_1, \mathbf{e}'_2, \mathbf{e}'_3\} \in V_{\mathscr{R}'}$ a right-handed orthonormal basis in the space of displacements of $\mathscr{R}'$. Setting*

$$\boldsymbol{\omega}_{\mathscr{R}'|\mathscr{R}}(t) := \frac{1}{2} \sum_{j=1}^{3} \mathbf{e}'_j(t) \wedge \left.\frac{d}{dt}\right|_{\mathscr{R}} \mathbf{e}'_j(t) \,, \tag{2.45}$$

we then have:

$$\left.\frac{d}{dt}\right|_{\mathscr{R}} = \left.\frac{d}{dt}\right|_{\mathscr{R}'} + \boldsymbol{\omega}_{\mathscr{R}'|\mathscr{R}}(t) \wedge \,. \tag{2.46}$$

Proof Let us represent all vectors in the space $V_{\mathscr{R}}$. There, the vectors $\mathbf{e}_i$ do not depend on time, whereas the vectors $\mathbf{e}'_j$ do. For a generic differentiable curve (of class C^1) $t \mapsto \mathbf{u}(t)$, if we keep (2.43) into account we have:

$$\left.\frac{d}{dt}\right|_{\mathscr{R}} \mathbf{u}(t) = \left.\frac{d}{dt}\right|_{\mathscr{R}} \left(\sum_{j=1}^{3} u'^j(t) \mathbf{e}'_j(t) \right) = \sum_{j=1}^{3} \frac{du'^j}{dt} \mathbf{e}'_j(t) + \sum_{j=1}^{3} u'^j(t) \left.\frac{d}{dt}\right|_{\mathscr{R}} \mathbf{e}'_j(t) \,.$$

In other words,

$$\left.\frac{d}{dt}\right|_{\mathscr{R}} \mathbf{u}(t) = \left.\frac{d}{dt}\right|_{\mathscr{R}'} \mathbf{u}(t) + \sum_{j=1}^{3} u'^j(t) \left.\frac{d}{dt}\right|_{\mathscr{R}} \mathbf{e}'_j(t) \,.$$

To end the proof, remembering that the cross product is linear in the second argument, it is sufficient to show that

$$\left.\frac{d}{dt}\right|_{\mathscr{R}} \mathbf{e}'_j(t) = \boldsymbol{\omega}_{\mathscr{R}'|\mathscr{R}}(t) \wedge \mathbf{e}'_j(t) \,,$$

i.e., considering how $\boldsymbol{\omega}_{\mathscr{R}'|\mathscr{R}}$ was defined,

$$\left.\frac{d}{dt}\right|_{\mathscr{R}} \mathbf{e}'_j(t) = \left(\frac{1}{2} \sum_{k=1}^{3} \mathbf{e}'_k(t) \wedge \left.\frac{d}{dt}\right|_{\mathscr{R}} \mathbf{e}'_k(t) \right) \wedge \mathbf{e}'_j(t) \,. \tag{2.47}$$

Let us demonstrate (2.47) and finish the proof. Using the known general relation:

$$(\mathbf{a} \wedge \mathbf{b}) \wedge \mathbf{c} = (\mathbf{a} \cdot \mathbf{c})\mathbf{b} - (\mathbf{b} \cdot \mathbf{c})\mathbf{a}$$

and that $\mathbf{e}'_r(t) \cdot \mathbf{e}'_s(t) = \delta_{rs}$, the right side of the claimed identity reads

$$\frac{1}{2}\sum_{k=1}^{3}\left(\mathbf{e}'_k(t) \wedge \frac{d}{dt}\bigg|_{\mathscr{R}}\mathbf{e}'_k(t)\right) \wedge \mathbf{e}'_j(t)$$

$$=\frac{1}{2}\sum_{k=1}^{3}\left[\delta_{kj}\frac{d}{dt}\bigg|_{\mathscr{R}}\mathbf{e}'_k(t) - \left(\left(\frac{d}{dt}\bigg|_{\mathscr{R}}\mathbf{e}'_k(t)\right)\cdot\mathbf{e}'_j(t)\right)\mathbf{e}'_k(t)\right] .$$

Now, on the right we have:

$$\sum_{k=1}^{3}\delta_{kj}\frac{d}{dt}\bigg|_{\mathscr{R}}\mathbf{e}'_k(t) = \frac{d}{dt}\bigg|_{\mathscr{R}}\mathbf{e}'_j(t) ,$$

so:

$$\frac{1}{2}\sum_{k=1}^{3}\left(\mathbf{e}'_k(t) \wedge \frac{d}{dt}\bigg|_{\mathscr{R}}\mathbf{e}'_k(t)\right) \wedge \mathbf{e}'_j(t)$$

$$=\frac{1}{2}\frac{d}{dt}\bigg|_{\mathscr{R}}\mathbf{e}'_j(t) - \frac{1}{2}\sum_{k=1}^{3}\left(\left(\frac{d}{dt}\bigg|_{\mathscr{R}}\mathbf{e}'_k(t)\right)\cdot\mathbf{e}'_j(t)\right)\mathbf{e}'_k(t) .$$

Moreover:

$$\left(\frac{d}{dt}\bigg|_{\mathscr{R}}\mathbf{e}'_k(t)\right)\cdot\mathbf{e}'_j(t) = \frac{d}{dt}(\mathbf{e}'_k(t)\cdot\mathbf{e}'_j(t)) - \mathbf{e}'_k(t)\cdot\frac{d}{dt}\bigg|_{\mathscr{R}}\mathbf{e}'_j(t) = \mathbf{0} - \mathbf{e}'_k(t)\cdot\frac{d}{dt}\bigg|_{\mathscr{R}}\mathbf{e}'_j(t) ,$$

where we used that $\mathbf{e}'_k(t)\cdot\mathbf{e}'_j(t) = \delta_{kj}$ *does not* depend on time, so its time derivative vanishes, and then

$$\frac{1}{2}\sum_{k=1}^{3}\left(\mathbf{e}'_k(t) \wedge \frac{d}{dt}\bigg|_{\mathscr{R}}\mathbf{e}'_k(t)\right)\wedge\mathbf{e}'_j(t) = \frac{1}{2}\frac{d}{dt}\bigg|_{\mathscr{R}}\mathbf{e}'_j(t) + \frac{1}{2}\sum_{k=1}^{3}\left(\mathbf{e}'_k(t) \cdot \frac{d}{dt}\bigg|_{\mathscr{R}}\mathbf{e}'_j(t)\right)\mathbf{e}'_k(t).$$

If, in the last sum, we use the formula:

$$\mathbf{s} = \sum_{k=1}^{n}(\mathbf{f}_k \cdot \mathbf{s})\mathbf{f}_k$$

for decomposing a vector $\mathbf{s}$ in an orthonormal basis $\mathbf{f}_1, \ldots, \mathbf{f}_n$ of an n-dimensional real vector space with inner product $\cdot$, the identity we have found becomes:

$$\frac{1}{2} \sum_{k=1}^{3} \left(\mathbf{e}'_k(t) \wedge \frac{d}{dt}\bigg|_{\mathscr{R}} \mathbf{e}'_k(t) \right) \wedge \mathbf{e}'_j(t) = \frac{1}{2} \frac{d}{dt}\bigg|_{\mathscr{R}} \mathbf{e}'_j(t) + \frac{1}{2} \frac{d}{dt}\bigg|_{\mathscr{R}} \mathbf{e}'_j(t) = \frac{d}{dt}\bigg|_{\mathscr{R}} \mathbf{e}'_j(t) ,$$

which is precisely (2.47). This ends the proof.   □

Remarks 2.27

(1) Actually, the definition of $\boldsymbol{\omega}_{\mathscr{R}'|\mathscr{R}}$ does not depend on the bases of the rest spaces of the frames $\mathscr{R}$ and $\mathscr{R}'$, but only on the frames themselves. In fact the basis chosen in $\mathscr{R}$ does not play any role in (2.45) (remember (1) in Remarks 2.25). As for the basis $\mathbf{e}'_j$ in the space of displacements $V_{\mathscr{R}'}$ of the rest space $E_{\mathscr{R}}$ of $\mathscr{R}$, note the following. If $\mathbf{f}'_j = \sum_k R^k{}_j \mathbf{e}'_k$, with $j = 1, 2, 3$, is another orthonormal basis in the same space, then (2.45) gives the same vector $\boldsymbol{\omega}_{\mathscr{R}'|\mathscr{R}}$ because:

$$\sum_{j=1}^{3} \mathbf{f}'_j(t) \wedge \frac{d}{dt}\bigg|_{\mathscr{R}} \mathbf{f}'_j(t) = \sum_{i,j,k=1}^{3} R^k{}_j R^i{}_j \mathbf{e}'_k(t) \wedge \frac{d}{dt}\bigg|_{\mathscr{R}} \mathbf{e}'_i(t)$$

$$= \sum_{i,k=1}^{3} \delta_{ki} \mathbf{e}'_k(t) \wedge \frac{d}{dt}\bigg|_{\mathscr{R}} \mathbf{e}'_i(t)$$

$$= \sum_{k=1}^{3} \mathbf{e}'_k(t) \wedge \frac{d}{dt}\bigg|_{\mathscr{R}} \mathbf{e}'_k(t) ,$$

where we have used the orthonormality relationships $\sum_{j=1}^{3} R^k{}_j R^i{}_j = \delta_{ki}$ (compactly, $RR^t = I$) and the time independence of the coefficients $R^p{}_q$.

(2) As usual, the vectors $\boldsymbol{\omega}_{\mathscr{R}'|\mathscr{R}}(t)$ can be understood as belonging to the space of displacements of $\mathscr{R}$, $\mathscr{R}'$, or any other frame $\mathscr{R}''$, even to the space of displacements of Σ_t. This possibility should not be forgotten in what follows.   ∎

We have the following proposition, which in particular establishes the *composition rule* for $\boldsymbol{\omega}$ vectors.

Proposition 2.28 (Composition Rule for $\boldsymbol{\omega}$ Vectors) *Let $\mathscr{R}$, $\mathscr{R}'$ and $\mathscr{R}''$ be frames in spacetime $\mathbb{V}^4$. At any given time $t \in \mathbb{R}$ the following hold.*

(1) **Composition rule***:*

$$\boldsymbol{\omega}_{\mathscr{R}''|\mathscr{R}}(t) = \boldsymbol{\omega}_{\mathscr{R}''|\mathscr{R}'}(t) + \boldsymbol{\omega}_{\mathscr{R}'|\mathscr{R}}(t) . \tag{2.48}$$

(2) **Inversion rule***:*

$$\boldsymbol{\omega}_{\mathscr{R}'}|_{\mathscr{R}}(t) = -\boldsymbol{\omega}_{\mathscr{R}}|_{\mathscr{R}'}(t) . \tag{2.49}$$

(3) **Absoluteness of the derivative of** ω *(with respect to $\mathscr{R}$ and $\mathscr{R}'$):*

$$\frac{d}{dt}\bigg|_{\mathscr{R}} \boldsymbol{\omega}_{\mathscr{R}'}|_{\mathscr{R}}(t) = \frac{d}{dt}\bigg|_{\mathscr{R}'} \boldsymbol{\omega}_{\mathscr{R}'}|_{\mathscr{R}}(t) . \tag{2.50}$$

$\diamond$

Proof

(1) The following formulas hold simultaneously:

$$\frac{d}{dt}\bigg|_{\mathscr{R}} = \frac{d}{dt}\bigg|_{\mathscr{R}'} + \boldsymbol{\omega}_{\mathscr{R}'}|_{\mathscr{R}}(t)\wedge ,$$

$$\frac{d}{dt}\bigg|_{\mathscr{R}'} = \frac{d}{dt}\bigg|_{\mathscr{R}''} + \boldsymbol{\omega}_{\mathscr{R}''}|_{\mathscr{R}'}(t)\wedge ,$$

$$\frac{d}{dt}\bigg|_{\mathscr{R}} = \frac{d}{dt}\bigg|_{\mathscr{R}''} + \boldsymbol{\omega}_{\mathscr{R}''}|_{\mathscr{R}}(t)\wedge . \tag{2.51}$$

Using the second one in the first we find:

$$\frac{d}{dt}\bigg|_{\mathscr{R}} = \frac{d}{dt}\bigg|_{\mathscr{R}''} + \big(\boldsymbol{\omega}_{\mathscr{R}''}|_{\mathscr{R}'}(t) + \boldsymbol{\omega}_{\mathscr{R}'}|_{\mathscr{R}}(t)\big)\wedge ,$$

which compared with (2.51) gives, for any $t \in \mathbb{R}$:

$$\big(\boldsymbol{\omega}_{\mathscr{R}''}|_{\mathscr{R}'}(t) + \boldsymbol{\omega}_{\mathscr{R}'}|_{\mathscr{R}}(t) - \boldsymbol{\omega}_{\mathscr{R}''}|_{\mathscr{R}}(t)\big)\wedge = 0 .$$

The zero on the right should be understood as the *zero operator*. In other words, the above formula is interpreted as follows. *For any curve C^1 $(a, b) \ni t \mapsto$ $\mathbf{v}(t) \in V_t$ and any $t \in (a, b)$:*

$$\big(\boldsymbol{\omega}_{\mathscr{R}''}|_{\mathscr{R}'}(t) + \boldsymbol{\omega}_{\mathscr{R}'}|_{\mathscr{R}}(t) - \boldsymbol{\omega}_{\mathscr{R}''}|_{\mathscr{R}}(t)\big)\wedge \mathbf{v}(t) = \mathbf{0} . \tag{2.52}$$

If $t_0 \in (a, b)$, let $\mathbf{u}_0 := \big(\boldsymbol{\omega}_{\mathscr{R}''}|_{\mathscr{R}'}(t_0) + \boldsymbol{\omega}_{\mathscr{R}'}|_{\mathscr{R}}(t_0) - \boldsymbol{\omega}_{\mathscr{R}''}|_{\mathscr{R}}(t_0)\big)$. We can easily construct a curve C^1 $\mathbb{R} \ni t \mapsto \mathbf{v}(t) \in V_t$ (it is enough to work in the rest orthonormal coordinates of some frame) such that, exactly at $t = t_0$, has $||\mathbf{v}(t_0)|| = c > 0$ and $\mathbf{v}(t_0) \perp \mathbf{u}_0$. Then (2.52) implies for $t = t_0$ that

$$c\,||\,\boldsymbol{\omega}_{\mathscr{R}''}|_{\mathscr{R}'}(t_0) + \boldsymbol{\omega}_{\mathscr{R}'}|_{\mathscr{R}}(t_0) - \boldsymbol{\omega}_{\mathscr{R}''}|_{\mathscr{R}}(t_0)|| = 0 .$$

As $c > 0$, we must conclude

$$\boldsymbol{\omega}_{\mathscr{R}''}|_{\mathscr{R}'}(t_0) + \boldsymbol{\omega}_{\mathscr{R}'}|_{\mathscr{R}}(t_0) - \boldsymbol{\omega}_{\mathscr{R}''}|_{\mathscr{R}}(t_0) = \mathbf{0}.$$

Since this holds for any $t_0 \in \mathbb{R}$, (1) is true.

(2) follows directly from (1) if we choose $\mathscr{R}'' = \mathscr{R}$, because, as is immediate by Definition (2.45): $\boldsymbol{\omega}_{\mathscr{R}}|_{\mathscr{R}} = \mathbf{0}$.

(3) From (2.46) we obtain:

$$\frac{d}{dt}\bigg|_{\mathscr{R}} \boldsymbol{\omega}_{\mathscr{R}'}|_{\mathscr{R}}(t) = \frac{d}{dt}\bigg|_{\mathscr{R}'} \boldsymbol{\omega}_{\mathscr{R}'}|_{\mathscr{R}}(t) + \boldsymbol{\omega}_{\mathscr{R}'}|_{\mathscr{R}}(t) \wedge \boldsymbol{\omega}_{\mathscr{R}'}|_{\mathscr{R}}(t) = \frac{d}{dt}\bigg|_{\mathscr{R}'} \boldsymbol{\omega}_{\mathscr{R}'}|_{\mathscr{R}}(t).$$

$\square$

Example 2.29

(1) Consider moving frames $\mathscr{R}$ and $\mathscr{R}'$ with respective rest orthonormal coordinates t, x, y, z, origin $O \in E_{\mathscr{R}}$, axes $\mathbf{e}_1, \mathbf{e}_2, \mathbf{e}_3 \in V_{\mathscr{R}}$, and t', x', y', z', origin $O' \in E_{\mathscr{R}}$ and axes $\mathbf{e}'_1, \mathbf{e}'_2, \mathbf{e}'_3 \in V_{\mathscr{R}'}$. Suppose the frames are related by:

$$x = x' \cos\phi(t) - y' \sin\phi(t) + X(t), \tag{2.53}$$

$$y = x' \sin\phi(t) + y' \cos\phi(t) + Y(t), \tag{2.54}$$

$$z = z' + Z(t), \tag{2.55}$$

where $\phi = \phi(t)$, $X = x(t)$, $Y = Y(t)$ and $Z = Z(t)$ are given differentiable maps. Since the difference of the coordinates of a point define the coordinates of vectors due to the affine structure of rest spaces, we immediately find that the transformation rule for the components of a vector $u = \sum_i u^i \mathbf{e}_i = \sum_j u'^j \mathbf{e}'_j \in V_t$ is

$$u^1 = u'^1 \cos\phi(t) - u'^2 \sin\phi(t), \tag{2.56}$$

$$u^2 = u'^1 \sin\phi(t) + u'^2 \cos\phi(t), \tag{2.57}$$

$$u^3 = u'^3, \tag{2.58}$$

Then we can find the transformation rule for unit vectors, noting that $\mathbf{e}'_k = \sum_j \delta^j_k \mathbf{e}'_j$, so

$$\mathbf{e}'_1 = \cos\phi(t)\,\mathbf{e}_1 + \sin\phi(t)\,\mathbf{e}_2, \tag{2.59}$$

$$\mathbf{e}'_2 = -\sin\phi(t)\,\mathbf{e}_1 + \cos\phi(t)\,\mathbf{e}_2, \tag{2.60}$$

$$\mathbf{e}'_3 = \mathbf{e}_3. \tag{2.61}$$

At this point we can compute $\omega_{\mathscr{R}'|\mathscr{R}}$ directly from (2.45), obtaining:

$$\omega_{\mathscr{R}'|\mathscr{R}}(t) = \frac{d\phi(t)}{dt} \, \mathbf{e}_3 \,. \tag{2.62}$$

If we suppose both axes' trihedra are right-handed, when $\phi(t) = vt$ with $v > 0$ the motion of the primed triple with respect to the unprimed one is a uniform counter-clockwise rotation (viewed from above) and $\omega = v\mathbf{e}_3$, pointing upwards, coincides with the angular velocity vector of elementary texts.

(2) Consider a generic coordinate transformation between rest orthogonal Cartesian coordinates on $\mathscr{R}$ and $\mathscr{R}'$ respectively. We know from Exercise 2.11.2 that the transformation rule is:

$$\begin{cases} t = t' + c \,, \\ x^i = c^i(t) + \sum_{j=1}^{3} R^i{}_j(t)\, x'^j \,, \quad i = 1, 2, 3 \,, \end{cases} \tag{2.63}$$

where $c \in \mathbb{R}$ is a constant, the maps $\mathbb{R} \ni t \mapsto c^i(t)$ and $\mathbb{R} \ni t \mapsto R^i{}_j(t)$ are C^∞ and, for any $t \in \mathbb{R}$, the coefficient matrix $R^i{}_j(t)$ is real 3×3 and orthogonal.

The corresponding transformation rule for the unit vectors is, with the obvious notation,

$$\mathbf{e}'_j = \sum_{i=1}^{3} R^i{}_j(t)\mathbf{e}_i \,.$$

Computing $\omega_{\mathscr{R}'|\mathscr{R}}$ with (2.45) gives:

$$\omega_{\mathscr{R}'|\mathscr{R}}(t) = \frac{1}{2} \sum_{i,j,k=1}^{3} R^i{}_j(t)\frac{d R^k{}_j(t)}{dt} \, \mathbf{e}_i \wedge \mathbf{e}_k \,.$$

Assuming the orthonormal basis $\mathbf{e}_1, \mathbf{e}_2, \mathbf{e}_3$ is right-handed,

$$\mathbf{e}_i \wedge \mathbf{e}_k = \sum_{h=1}^{3} \epsilon_{ikh} \, \mathbf{e}_h \,, \tag{2.64}$$

where we have introduced the **Ricci symbol** ϵ_{ijk}, equal to: 1 if ijk is a cyclic permutation of 123, -1 if ijk is a non-cyclic permutation of 123, and 0 otherwise. In summary, if the transformation rules (2.63) hold, the vector

$\omega_{\mathscr{R}'|\mathscr{R}}(t)$ can be written as:

$$\omega_{\mathscr{R}'|\mathscr{R}}(t) = \frac{1}{2} \sum_{i,j,h,k=1}^{3} \epsilon_{ikh} R^i{}_j(t) \frac{d R^k{}_j(t)}{dt} \, \mathbf{e}_h \,. \tag{2.65}$$

(3) A rigid disc is constrained to a vertical plane Π. In Π the disc rotates around its axis uniformly with constant angular velocity $d\theta/dt = \nu$, where θ is the angular polar coordinate in Π with origin O at the disc's centre. The plane Π rotates, with respect to the frame $\mathscr{R}$, around the vertical line $z \in \Pi$, which does not intersect the disc, under the law $\phi(t) = \sin(\lambda t)$, where ϕ is the usual polar angle in the normal plane to z. Take a moving frame $\mathscr{R}'$ on the disc (i.e. in which the disc is always at rest) and find $\omega_{\mathscr{R}'|\mathscr{R}}$.

The problem is solved by composing the ω vectors. In a moving frame $\mathscr{R}''$ on the plane Π, the disc's frame $\mathscr{R}'$ has ω vector given, from Example (1), by: $\omega_{\mathscr{R}'|\mathscr{R}''} = \nu \mathbf{n}_\Pi$, where $\mathbf{n}_\Pi$ is the plane's unit normal *oriented so that the disc's rotation is counter-clockwise when we observe the disc from the tip of* $\mathbf{n}_\Pi$. Usually this orientation of the rotation axis is called *positive* with respect to the direction in which the angle increases. The frame $\mathscr{R}''$ rotates around the z-axis with respect to $\mathscr{R}$. Using Example (1) again, $\omega_{\mathscr{R}''|\mathscr{R}} = \frac{d\phi}{dt} \mathbf{e}_z = \lambda \cos(\lambda t) \mathbf{e}_z$, where $\mathbf{e}_z$ is the unit vector parallel to z and oriented positively with ϕ. The composition rule for ω vectors finally gives:

$$\omega_{\mathscr{R}'|\mathscr{R}}(t) = \lambda \cos(\lambda t) \mathbf{e}_z + \nu \mathbf{n}_\Pi(t) \,.$$

If necessary one can make the time dependence of $\mathbf{n}_\Pi$ explicit, recalling that it rotates with the plane Π. For that let us consider two perpendicular unit vectors $\mathbf{e}_x$, $\mathbf{e}_y$, forming with $\mathbf{e}_z$ (in that order) a rest right-handed orthonormal trihedron for $\mathscr{R}$. In this case, under our assumptions: $\mathbf{n}_\Pi(t) = \cos(\phi(t)+\alpha)\mathbf{e}_x + \sin(\phi(t) + \alpha) \mathbf{e}_y$ for some constant α. Hence:

$$\omega_{\mathscr{R}'|\mathscr{R}}(t) = \nu \cos(\sin(\lambda t) + \alpha) \, \mathbf{e}_x + \nu \sin(\sin(\lambda t) + \alpha) \, \mathbf{e}_y + \lambda \cos(\lambda t) \mathbf{e}_z \,.$$

Exercises 2.30

(1) Consider two rest right-handed orthonormal coordinates systems for the frames $\mathscr{R}$ and $\hat{\mathscr{R}}$ respectively, with axes $\mathbf{e}_1$, $\mathbf{e}_2$, $\mathbf{e}_3$ and $\hat{\mathbf{e}}_1$, $\hat{\mathbf{e}}_2$, $\hat{\mathbf{e}}_3$ and origins O and $\hat{O}$. Suppose $O = \hat{O}$ (in every Σ_t) at each instant. Let us introduce the so-called **Euler angles**: $\psi \in [0, 2\pi)$, $\theta \in [0, \pi]$, $\phi \in [0, 2\pi)$ that determine the trihedron of $\mathscr{R}$ from that of $\hat{\mathscr{R}}$ with the following procedure.

  (i) Rotate the trihedron of $\hat{\mathscr{R}}$ around $\hat{\mathbf{e}}_3$ by ψ in the positive direction with respect to $\hat{\mathbf{e}}_3$. After the rotation the axis $\hat{\mathbf{e}}_1$ has moved to the so-called **line of nodes**, whose unit vector $\mathbf{N}$ is normal to $\hat{\mathbf{e}}_3$ and $\mathbf{e}_3$.
  (ii) The trihedron obtained in (i) is rotated around the line of nodes by a positive angle θ. Thus the $\hat{\mathbf{e}}_3$-axis moves to another axis called $\mathbf{e}_3$.

(iii) The trihedron obtained in (ii) is rotated by a positive ϕ around $\mathbf{e}_3$. In this way $\mathbf{N}$ moves to an axis called $\mathbf{e}_1$.

The final trihedron is the triple $\mathbf{e}_1$, $\mathbf{e}_2$, $\mathbf{e}_3$ associated with $\mathscr{R}$; $\mathbf{e}_3$ and $\mathbf{e}_1$ are the axes obtained in (ii) and (iii) respectively, and $\mathbf{e}_2$ is the position reached by the initial trihedron's $\hat{\mathbf{e}}_2$-axis after the three rotations. By construction $\mathbf{e}_2 = \mathbf{e}_3 \wedge \mathbf{e}_1$.

Write the trihedron associated with $\mathscr{R}$ in terms of the trihedron of $\hat{\mathscr{R}}$ using the Euler angles. In other words prove that:

$$\mathbf{e}_1 = (\cos\phi\cos\psi - \sin\phi\sin\psi\cos\theta)\,\hat{\mathbf{e}}_1 + (\cos\phi\sin\psi$$
$$+ \sin\phi\cos\theta\cos\psi)\,\hat{\mathbf{e}}_2 + \sin\phi\sin\theta\,\hat{\mathbf{e}}_3 \,,$$
$$\mathbf{e}_2 = -(\sin\phi\cos\psi + \cos\phi\sin\psi\cos\theta)\,\hat{\mathbf{e}}_1 + (-\sin\phi\sin\psi$$
$$+ \cos\phi\cos\theta\cos\psi)\,\hat{\mathbf{e}}_2 + \cos\phi\sin\theta\,\hat{\mathbf{e}}_3 \,,$$
$$\mathbf{e}_3 = \sin\theta\,\sin\psi\,\hat{\mathbf{e}}_1 - \sin\theta\,\cos\psi\,\hat{\mathbf{e}}_2 + \cos\theta\,\hat{\mathbf{e}}_3 \,.$$

(2) Referring to the previous exercise, show that:

$$\hat{\mathbf{e}}_1 = (\cos\phi\cos\psi - \sin\phi\sin\psi\cos\theta)\,\mathbf{e}_1 - (\sin\phi\cos\psi$$
$$+ \cos\phi\sin\psi\cos\theta)\,\hat{\mathbf{e}}_2 + \sin\theta\,\sin\psi\,\mathbf{e}_3 \,,$$
$$\hat{\mathbf{e}}_2 = (\cos\phi\sin\psi + \sin\phi\cos\theta\cos\psi)\,\mathbf{e}_1 + (-\sin\phi\sin\psi$$
$$+ \cos\phi\cos\theta\cos\psi)\,\mathbf{e}_2 - \sin\theta\,\cos\psi\,\mathbf{e}_3 \,,$$
$$\hat{\mathbf{e}}_3 = \sin\phi\sin\theta\,\mathbf{e}_1 + \cos\phi\sin\theta\,\mathbf{e}_2 + \cos\theta\,\mathbf{e}_3 \,.$$

(3) Referring to the previous two exercises, let us suppose the trihedron of $\mathscr{R}$ moves, with respect to trihedron of $\hat{\mathscr{R}}$, under the functions $\phi = \phi(t), \psi = \psi(t), \theta = \theta(t)$. Express $\omega_{\hat{\mathscr{R}}|\mathscr{R}}$ in function of the time derivatives of the Euler angles and of the Euler angles themselves in the basis of $\mathscr{R}$. Check that (Fig. 2.4):

$$\omega_{\mathscr{R}|\hat{\mathscr{R}}} = (\dot{\psi}\sin\theta\sin\phi + \dot{\theta}\cos\phi)\mathbf{e}_1 + (\dot{\psi}\sin\theta\cos\phi - \dot{\theta}\sin\phi)\mathbf{e}_2 + (\dot{\phi} + \dot{\psi}\cos\theta)\mathbf{e}_3$$

where the dot is the time derivative of the Euler angles.

(4) $\boxed{\text{AC}}$ Consider $R \in SO(3)$ as an operator acting on vectors of $\mathbb{E}^3$. For any R there always exists a unit vector $\mathbf{u}$ such that $R(\mathbf{u}) = \mathbf{u}$, called **axis of rotation of** R. (In orthonormal coordinates R is given by a real matrix. We can always view this matrix as a unitary matrix with real coefficients. The spectral theorem says that a unitary matrix is diagonalisable and its eigenvalues are of the form e^{ic} for some $c \in \mathbb{R}$. As the matrix' characteristic polynomial is real, the eigenvalues must be: $\lambda_1 = e^{ib} \in \mathbb{R}$, i.e. $\lambda_1 = \pm 1$, and two complex-conjugate eigenvalues: $\lambda_2 = e^{ia}$ and $\lambda_3 = e^{-ia}$. The determinant equals the product $\lambda_1\lambda_2\lambda_3$, and we

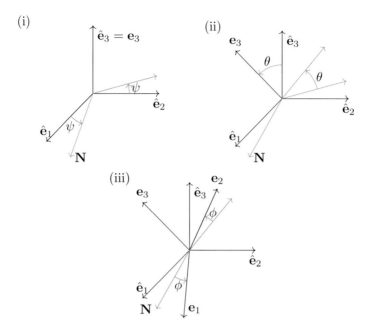

Fig. 2.4 Steps (i), (ii) and (iii) in the procedure of Exercise 2.30.**1**

know that for $R \in SO(3)$ the determinant is 1, so considering all possibilities we easily conclude that $\lambda_1 = 1$.)

Suppose the relationship between the frames' unit vectors is, at time t, $\mathbf{e}'_k(t) = R_t(\mathbf{e}_k)$ for $k = 1, 2, 3$. Show that the following remarkable fact hods:

$$R_t(\boldsymbol{\omega}_{\mathscr{R}'}|_{\mathscr{R}}(t)) = \boldsymbol{\omega}_{\mathscr{R}'}|_{\mathscr{R}}(t) .$$

In other words, the vector $\boldsymbol{\omega}$ is an **instantaneous axis of rotation**.

2.4.2 Velocity and Acceleration as the Frame Varies

We will now use the Poisson formulas, and the vector $\boldsymbol{\omega}$, to write down the formulas relating the velocities and accelerations of a point particle computed in two different frames.

Consider moving frames $\mathscr{R}$ and $\mathscr{R}'$ with rest orthonormal coordinates of origin O and axes $\mathbf{e}_1$, $\mathbf{e}_2$, $\mathbf{e}_3$, and O' and $\mathbf{e}'_1, \mathbf{e}'_2, \mathbf{e}'_3$ respectively. Let γ be a world line describing the motion of a point particle in spacetime. In the sequel, as usual, $P(t) := \gamma(t)$. Then trivially:

$$P - O = P - O' + O' - O . \tag{2.66}$$

Using the Poisson identity

$$\left.\frac{d}{dt}\right|_{\mathscr{R}} = \left.\frac{d}{dt}\right|_{\mathscr{R}'} + \boldsymbol{\omega}_{\mathscr{R}'}|_{\mathscr{R}} \wedge\,,$$

and keeping in account expression (2.44) for the velocity of P with respect to $\mathscr{R}$ and $\mathscr{R}'$, differentiating (2.66) in time we find the relation (we omit indicating the time t explicitly so the notation does not become too heavy):

$$\mathbf{v}_P|_{\mathscr{R}} = \mathbf{v}_P|_{\mathscr{R}'} + \boldsymbol{\omega}_{\mathscr{R}'}|_{\mathscr{R}} \wedge (P - O') + \left.\frac{d}{dt}\right|_{\mathscr{R}} (O' - O)\,,$$

i.e.

$$\mathbf{v}_P|_{\mathscr{R}} = \mathbf{v}_P|_{\mathscr{R}'} + \boldsymbol{\omega}_{\mathscr{R}'}|_{\mathscr{R}} \wedge (P - O') + \mathbf{v}_{O'}|_{\mathscr{R}}\,. \tag{2.67}$$

It is convenient to rewrite the above equation as follows:

$$\begin{cases} \mathbf{v}_P|_{\mathscr{R}} = \mathbf{v}_P|_{\mathscr{R}'} + \mathbf{v}_P^{(tr)}\,, \\ \mathbf{v}_P^{(tr)} := \mathbf{v}_{O'}|_{\mathscr{R}} + \boldsymbol{\omega}_{\mathscr{R}'}|_{\mathscr{R}} \wedge (P - O')\,, \end{cases} \tag{2.68}$$

where $\mathbf{v}_P^{(tr)}$ is the **relative velocity** of $\mathscr{R}'$ with respect to $\mathscr{R}$ at P. This is the velocity $\mathscr{R}$ assigns to the point P when thought of as if it was not moving in $\mathscr{R}'$. Differentiating (2.67) again with respect to $\mathscr{R}$:

$$\mathbf{a}_P|_{\mathscr{R}} = \left.\frac{d}{dt}\right|_{\mathscr{R}} \mathbf{v}_P|_{\mathscr{R}'} + \left.\frac{d}{dt}\right|_{\mathscr{R}} \left(\boldsymbol{\omega}_{\mathscr{R}'}|_{\mathscr{R}} \wedge (P - O')\right) + \left.\frac{d}{dt}\right|_{\mathscr{R}} \mathbf{v}_{O'}|_{\mathscr{R}}\,.$$

Applying (2.46) to both summands on the right, $\mathbf{a}_P|_{\mathscr{R}}$ equals:

$$\mathbf{a}_P|_{\mathscr{R}'} + \boldsymbol{\omega}_{\mathscr{R}'}|_{\mathscr{R}} \wedge \mathbf{v}_P|_{\mathscr{R}'} + \dot{\boldsymbol{\omega}}_{\mathscr{R}'}|_{\mathscr{R}} \wedge (P - O') + \boldsymbol{\omega}_{\mathscr{R}'}|_{\mathscr{R}} \wedge \mathbf{v}_P|_{\mathscr{R}'} + \boldsymbol{\omega}_{\mathscr{R}'}|_{\mathscr{R}} \wedge$$

$$\left(\boldsymbol{\omega}_{\mathscr{R}'}|_{\mathscr{R}} \wedge (P - O')\right) + \mathbf{a}_{O'}|_{\mathscr{R}}\,,$$

where $\dot{\boldsymbol{\omega}}_{\mathscr{R}'}|_{\mathscr{R}} := \left.\frac{d\boldsymbol{\omega}_{\mathscr{R}'}|_{\mathscr{R}}}{dt}\right|_{\mathscr{R}} = \left.\frac{d\boldsymbol{\omega}_{\mathscr{R}'}|_{\mathscr{R}}}{dt}\right|_{\mathscr{R}'}$. Rearranging terms:

$$\begin{cases} \mathbf{a}_P|_{\mathscr{R}} = \mathbf{a}_P|_{\mathscr{R}'} + \mathbf{a}_P^{(tr)} + \mathbf{a}_P^{(C)}\,, \\ \mathbf{a}_P^{(tr)} := \mathbf{a}_{O'}|_{\mathscr{R}} + \boldsymbol{\omega}_{\mathscr{R}'}|_{\mathscr{R}} \wedge (\boldsymbol{\omega}_{\mathscr{R}'}|_{\mathscr{R}} \wedge (P - O')) + \dot{\boldsymbol{\omega}}_{\mathscr{R}'}|_{\mathscr{R}} \wedge (P - O')\,, \\ \mathbf{a}_P^{(C)} := 2\boldsymbol{\omega}_{\mathscr{R}'}|_{\mathscr{R}} \wedge \mathbf{v}_P|_{\mathscr{R}'}\,, \end{cases}$$

$$\tag{2.69}$$

where we have dropped subscripts like $_{\mathscr{R}'|\mathscr{R}}$ in $\mathbf{a}_P^{(tr)}$ and $\mathbf{a}_P^{(C)}$ to keep the notation simple. Above:

(a) $\mathbf{a}_P^{(tr)}$ is the **relative acceleration** of $\mathscr{R}'$ with respect to $\mathscr{R}$ at P. This is the acceleration that $\mathscr{R}$ assigns to the point P when the latter is considered at rest (zero acceleration) in $\mathscr{R}'$.

(b) $\mathbf{a}_P^{(C)}$ is the well-known **Coriolis acceleration**.

Remarks 2.31

(1) We point out that one can prove that $\mathbf{v}_P^{(tr)}$, $\mathbf{a}_P^{(tr)}$ and $\mathbf{a}_P^{(C)}$ *do not* depend on the choice of O'. For $\mathbf{a}_P^{(C)}$ there is actually nothing to prove! Let us then prove the independence from O' of the relative velocity. If $O_1' \neq O'$ is at rest in $\mathscr{R}'$, putting $P = O_1'$ in (2.67) gives the identity: $\mathbf{v}_{O_1'}|_{\mathscr{R}} = \mathbf{0} + \boldsymbol{\omega}_{\mathscr{R}'}|_{\mathscr{R}} \wedge (O_1' - O') + \mathbf{v}_{O'}|_{\mathscr{R}}$, i.e.:

$$\mathbf{v}_{O_1'}|_{\mathscr{R}} - \mathbf{v}_{O'}|_{\mathscr{R}} + \boldsymbol{\omega}_{\mathscr{R}'}|_{\mathscr{R}} \wedge (O_1' - O') = \mathbf{0}$$

Using the second relation in (2.68), and evaluating the relative velocity $\mathbf{v}_P^{(tr1)}$ of P, but referred to O_1' rather than O', we immediately find:

$$\mathbf{v}_P^{(tr1)} - \mathbf{v}_P^{(tr)} = \mathbf{v}_{O_1'}|_{\mathscr{R}} - \mathbf{v}_{O'}|_{\mathscr{R}} + \boldsymbol{\omega}_{\mathscr{R}'}|_{\mathscr{R}} \wedge (O_1' - O') = \mathbf{0} \,,$$

hence $\mathbf{v}_P^{(tr1)} = \mathbf{v}_P^{(tr)}$.

More quickly, note that $\mathbf{v}_P|_{\mathscr{R}}$ and $\mathbf{v}_P|_{\mathscr{R}'}$ do not depend on O' (and O), and so $\mathbf{v}_P^{(tr)}$ in (2.68) must be independent. For the relative acceleration one can proceed similarly.

(2) Note, with obvious notation, that from the identity: $\mathbf{v}_P|_{\mathscr{R}} = \mathbf{v}_P|_{\mathscr{R}'} + \mathbf{v}_{P\mathscr{R}'}^{(tr)}|_{\mathscr{R}}$ and $\mathbf{v}_P|_{\mathscr{R}'} = \mathbf{v}_P|_{\mathscr{R}} + \mathbf{v}_{P\mathscr{R}}^{(tr)}|_{\mathscr{R}'}$ we deduce: $\mathbf{v}_{P\mathscr{R}'}^{(tr)}|_{\mathscr{R}} = -\mathbf{v}_{P\mathscr{R}}^{(tr)}|_{\mathscr{R}'}$.

(3) If, in a three-dimensional Euclidean space, a point Q rotates by $\Delta\theta$ around an axis through O parallel to the unit vector $\mathbf{n}$ (assuming a positive rotation about the axis), and reaches a position Q', then

$$Q' = Q + \Delta\theta \mathbf{n} \wedge (Q - O) + \Delta\theta \mathbf{O}(\Delta\theta) \,,$$

where $\mathbf{O}(\Delta\theta) \to \mathbf{0}$ as $\Delta\theta \to 0$. We leave the reader to prove that fact (it suffices to represent the rotation in orthonormal coordinates with origin $\mathbf{O}$ and axis $\mathbf{n} = \mathbf{e}_3$, then expand in Taylor series the trigonometric functions appearing the rotation matrix).

If P is a point particle at rest in $\mathscr{R}'$ (with origin O') whose world line is differentiable, the Taylor expansion in time of (2.67) gives the position of P in the rest space of $\mathscr{R}$ at time $t + \Delta t$ in function of its position at time t:

$$P(t + \Delta t) = P(t) + \mathbf{v}_{O'}|_{\mathscr{R}}(t)\Delta t + \Delta t \boldsymbol{\omega}_{\mathscr{R}'}|_{\mathscr{R}}(t) \wedge (P(t) - O'(t)) + \Delta t \mathbf{O}(\Delta t) \,, \tag{2.70}$$

where $\mathbf{O}(\Delta t)$ is an infinitesimal function as $\Delta t \to 0$. Formula (2.70) shows that, *to first order in* Δt, the motion of P in $\mathscr{R}$ is a translation with velocity $\mathbf{v}_{O'}|_{\mathscr{R}}(t)$, together with a positive rotation around $O'(t)$, with respect to the axis $\boldsymbol{\omega}_{\mathscr{R}'}|_{\mathscr{R}}(t)$, by $|\boldsymbol{\omega}_{\mathscr{R}'}|_{\mathscr{R}}(t)|\Delta t$. In this way we may view $\boldsymbol{\omega}_{\mathscr{R}'}|_{\mathscr{R}}$ as an *instantaneous axis of rotation* (Exercise 2.30.4 will give a second, more mathematical, interpretation of this). Note that the origin O' at rest with $\mathscr{R}'$ can be fixed arbitrarily. The decomposition of the motion of $\mathscr{R}'$ with respect to $\mathscr{R}'$ in terms of a displacement and a rotation (in the "infinitesimal" neighbourhood of a time instant) can be recovered in several ways depending on the choice of O'.   ∎

Exercises 2.32

(1) Consider rest right-handed orthonormal coordinates x, y, z of $\mathscr{R}$ centred at O, and two rigid discs D and D' constrained to move on the plane $z = 0$ of the rest space of $\mathscr{R}$. They have respective radii $R > 0$ and $r > 0$, centres O and O' along the x-axis, and are always tangent to one another since $||O-O'|| = r+R$. The discs' angular positions are given by angles Θ and θ, respectively formed by $S-O$ and $S'-O'$ and the x-axis, where $S \in D$ and $S' \in D'$ are given points, and the angles are positively oriented with respect to the $\mathbf{e}_z$-axis. Determine the relationship between θ and Θ under the assumption that the discs *rotate without slipping* at their geometric contact point Q. The absence of sliding, by definition, means that the point particles $P \in D$ and $P' \in D'$ that correspond to Q at each time always have the same velocity in $\mathscr{R}$ (hence in any frame!) (Fig. 2.5).

(2) Consider rest right-handed orthonormal coordinates x, y, z on $\mathscr{R}$ and centred at O, and two rigid discs D and D' constrained to move on the plane $z = 0$ of the rest space of $\mathscr{R}$. They have respective radii $R > 0$ and $r > 0$ and centres O and O'. The second point O' moves along a circle with centre O and radius $R + r$,

Fig. 2.5 Exercise 2.32.1

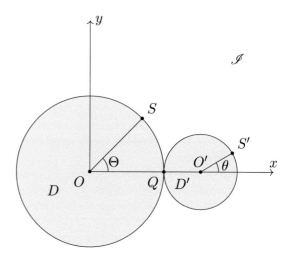

in such a way that the two discs are always tangent to each other. We define
the position of the first disc by the polar angle Θ of a given point measured
with respect to the x-axis oriented positively with respect to $\mathbf{e}_z$. The position
of the second disc around O' is similarly given by the polar angle θ of a given
point of D', measured with respect to axes x' and y' parallel to x and y and
passing through O'. Finally, the position of O' is determined by the polar angle
ϕ that $O' - O$ forms with the x-axis, oriented positively with respect to $\mathbf{e}_z$. Find
the relationship between θ, Θ and ϕ, with the assumption that the discs *rotate
without slipping* at their geometric contact point Q. By definition, this means
that the point particles $P \in D$ and $P' \in D'$ that correspond to Q at each time
always have the same velocity in $\mathscr{R}$ (hence in any frame!).

Chapter 3
Newtonian Dynamics: A Conceptual Critical Review

Classical Mechanics goes beyond Kinematics in that it deals with **Dynamics**, that is, the problem of determining the motion of a body when we assign the "initial conditions" (the system state of motion at a given time) and the "interactions" with other bodies, understood as the causes of motion itself. In Classical Physics interactions are described by the concept of *force*. We shall present this notion in its modern take, as was critically reformulated by E. Mach in the nineteenth century, although it still relies on the original postulates stated by Newton at the end of the seventeenth century, also based on previous results by Galilei. Mathematically, a force at time t is a function of the system's motion at time t (positions and velocities at time t). Finding the motion boils down to solving a *system of differential equations* involving functions that describe the forces and special physical constants associated with the physical system's point particles, called the *masses* of the particles.

3.1 Newton's First Law of Motion

To introduce Dynamics we must select a privileged class of frames called *inertial*. Let us begin by understanding how this class arises physically. The core idea is to study the motion of bodies when they are placed far apart from the other bodies present in the universe. Let us recall that our idealisation of a body is that of a *point particle*, i.e. a physical body whose inner structure is irrelevant and whose spacetime evolution can be described by a world line.

V. Moretti, *Analytical Mechanics*, La Matematica per il 3+2 150,
https://doi.org/10.1007/978-3-031-27612-5_3

3.1.1 Inertial Frames

If we have only one point particle in the universe and we have not chosen any special frame, any proposition about the particle's state of motion makes no physical sense: the state of motion depends on the choice of a frame, and the motion can be fixed arbitrarily by choosing the frame suitably. But if we consider a collection of several point particles, in general it is not possible to find a frame in which *each* point particle has a given state of motion *simultaneously*. In this respect the **Principle of inertia** or **Newton's first law of motion** establishes the state of motion of an arbitrary number of bodies when they are at enormous distances from one another and from all other bodies in the universe. Point particles satisfying this request of "apartness" (in a portion of their world line) are called **isolated** (in said portion of their world line). Exactly because we cannot in general fix the state of motion of several point particles at the same time by choosing a suitable frame, the Principle of inertia has a highly non-trivial physical significance, since it expresses a remarkable (ideally at least) experimental fact.

C1. Principle of Inertia
There exists a class of frames in each of which every particle inside an arbitrary collection of isolated point particles moves **in a straight line with constant velocity** *in modulus, direction and orientation, depending on the point particle. Such reference frames are called* **inertial frames**.

Remarks

(1) We spoke of a collection of inertial frames and not a single one because, as we will show shortly, if $\mathscr{R}$ is an inertial frame and $\mathscr{R}$ and $\mathscr{R}'$ move in a straight line with constant velocity relative on each other, then $\mathscr{R}'$ too is inertial.

(2) It is clear that in practice it is very difficult to tell with certainty whether a frame is inertial or not. At any rate inertial frames can be determined a posteriori after one develops the theory to more accessible experimental levels. For example, once we develop Dynamics and introduce the concept of mass, one can prove that for a set of bodies sufficiently apart from others but not isolated among themselves, the system's centre of mass must be at rest in an inertial frame. If one of the bodies C in the set has much larger mass than the rest, the motion of its centre of mass will be close to the motion of the set's centre of mass, so there is an inertial frame where the centre of mass of C is approximately at rest. This situation occurs in the solar system, where C is the sun: an inertial frame can be taken to be the one where the sun's centre of mass is at rest and where the *fixed stars* are seen as not moving. To confirm this, note that in such a frame the planets roughly obey *Kepler's laws*, which are consequences of Newton's laws of motion valid in inertial frames.

(3) The existence of inertial frames is as a matter of fact disproved by successive developments in Physics, namely the theory of General Relativity. However, in that context as well one sees that almost-inertial frames can be defined in regions of the universe that are cosmologically speaking relatively small, but

still bigger than the solar system, and containing a quantity of mass of smaller order of magnitude that the universe's total mass. For this we shall assume we can select in the universe a spatial region R as above, in which we will develop Classical Mechanics; and when speaking of **sufficiently distant** or **isolated** bodies we will mean bodies in R with mutual distances at least equal to the region's radius.

(4) When stating Newton's first law of motion we have carefully avoided to mention *forces*, so to not fall into an acritical vicious circle, in a sense "traditional", that alas is found in many textbooks and that certainly does not help understand of the Principle of inertia.

3.1.2 Galilean Transformations

To characterise such a motion it is convenient to discuss first when a frame *moves in a straight line and uniformly with respect to another frame.* We begin with a crucial definition.

Definition 3.1 Consider two frames $\mathscr{R}$ and $\mathscr{R}'$. We say $\mathscr{R}$ is in **uniform linear motion with respect to** $\mathscr{R}'$ if the relative velocity $\mathbf{v}_P^{(tr)}$ of $P \in E_{\mathscr{R}}$ with respect to $\mathscr{R}'$:

(1) is independent of $P \in E_{\mathscr{R}}$,
(2) is constant in time.

This relative velocity, indicated by $\mathbf{v}_{\mathscr{R}|\mathscr{R}'}$, is called **relative velocity of** $\mathscr{R}$ **with respect to** $\mathscr{R}'$. $\diamond$

We have the following useful proposition to characterise the uniform linear motion of two frames.

Proposition 3.2 *Let $\mathscr{R}$ and $\mathscr{R}'$ be frames in $\mathbb{V}^4$. The following facts hold.*

*(1) Let $\mathscr{R}$ be in uniform linear motion with respect to $\mathscr{R}'$, and t, x^1, x^2, x^3, t', x'^1, x'^2, x'^3 right-handed orthonormal coordinates on $\mathscr{R}$ and $\mathscr{R}'$ with origins $O \in E_{\mathscr{R}}$, $O' \in E_{\mathscr{R}'}$ and axes $\mathbf{e}_1, \mathbf{e}_2, \mathbf{e}_3 \in V_{\mathscr{R}}$, $\mathbf{e}'_1, \mathbf{e}'_2, \mathbf{e}'_3 \in V_{\mathscr{R}'}$ respectively. The transformation rule between the frames is a **Galilean transformation**, i.e. a transformation from $\mathbb{R}^4$ to $\mathbb{R}^4$ of the form:*

$$\begin{cases} t' = t + c \,, \\ x'^i = c^i + t v^i + \sum_{j=1}^{3} R^i{}_j x^j \,, & i = 1, 2, 3. \end{cases} \tag{3.1}$$

where $c \in \mathbb{R}$, $c^i \in \mathbb{R}$ and $v^i \in \mathbb{R}$ are constants, and the constants $R^i{}_j$ define a real 3×3 orthogonal matrix with positive determinant. Furthermore

$$\mathbf{v}_{\mathscr{R}|\mathscr{R}'} := \sum_{i=1}^{3} v^i \mathbf{e}'_i \, . \tag{3.2}$$

(2) *If, conversely, with respect to rest orthonormal coordinates on $\mathscr{R}$ and $\mathscr{R}'$, Eq. (3.1) hold, then $\mathscr{R}$ is in uniform linear motion with respect to $\mathscr{R}'$ with relative velocity (3.2).*

(3) *If the frame $\mathscr{R}$ is in uniform linear motion with respect to $\mathscr{R}'$, then $\mathscr{R}'$ is in uniform linear motion with respect to $\mathscr{R}$ and $\mathbf{v}_{\mathscr{R}|\mathscr{R}'} = -\mathbf{v}_{\mathscr{R}'|\mathscr{R}}$.*

(4) *If $\mathscr{R}$ is in uniform linear motion with respect to $\mathscr{R}'$ and $\mathscr{R}'$ is in uniform linear motion with respect to $\mathscr{R}''$, then $\mathscr{R}$ is in uniform linear motion with respect to $\mathscr{R}''$.*

(5) *If $\mathscr{R}$ is in uniform linear motion with respect to $\mathscr{R}'$ then $\boldsymbol{\omega}_{\mathscr{R}'|\mathscr{R}}$, $\boldsymbol{\omega}_{\mathscr{R}|\mathscr{R}'}$, the relative acceleration and the Coriolis acceleration of $\mathscr{R}'$ with respect to $\mathscr{R}$, and the similar accelerations of $\mathscr{R}$ with respect to $\mathscr{R}'$ are all zero at every time.*

Proof

(1) In the general case, the coordinate transformation between the frames $\mathscr{R}$ and $\mathscr{R}'$ must have the form (see Exercise 2.11.2):

$$\begin{cases} t' = t + c \, , \\ x'^i = c^i(t) + \displaystyle\sum_{j=1}^{3} R^i{}_j(t)\, x^j \, , \quad i = 1, 2, 3 \, , \end{cases} \tag{3.3}$$

where the functions are all C^k ($k = 2, 3, \ldots, \infty$ being the chosen regularity of $\mathbb{V}^4$) and the coefficients $R^i{}_j(t)$, for any $t \in \mathbb{R}$ define an orthogonal matrix. The world line of the point $O \in E_{\mathscr{R}}$, corresponding to $(0, 0, 0)$ for $\mathscr{R}$, has equation $x'^i(t') = c^i(t' - c)$ for $\mathscr{R}'$. Differentiating with respect to time, by definition we must find the components of the relative velocity of O with respect to $\mathscr{R}'$ expressed in the basis $\mathbf{e}'_1, \mathbf{e}'_2, \mathbf{e}'_3 \in V_{\mathscr{R}'}$. By assumption these components should not depend on time. We conclude that, calling v^i the constants $dc^i(t' - c)/dt'$, (3.3) reads:

$$\begin{cases} t' = t + c \, , \\ x'^i = c^i + v^i t + \displaystyle\sum_{j=1}^{3} R^i{}_j(t)\, x^j \, , \quad i = 1, 2, 3 \, , \end{cases} \tag{3.4}$$

where the c^i are further constants. The relative velocity of the generic point $P \in E_{\mathscr{R}}$ of coordinates (x^1, x^2, x^3) for $\mathscr{R}$ will have, from (3.4), components $v^i + \sum_j x^j dR^i{}_j(t)/dt$ with respect to the basis $\mathbf{e}'_1, \mathbf{e}'_2, \mathbf{e}'_3$. By hypothesis these components cannot depend on the point, i.e. on the coordinates x^i, so differentiating with respect to the generic coordinates x^k we find that $dR^i{}_k(t)/dt = 0$ for any i and k and for any time t. We conclude that (3.1) holds, and hence so does (3.2).

(2) Consider a world line $t \mapsto P(t) \in \mathbb{V}^4$ (with $P(t) \in \Sigma_t$) describing the evolution of a point P in the rest space of $\mathscr{R}$ given, in the coordinates of $\mathscr{R}$, by $(y^1, y^2, y^3) \in \mathbb{R}^3$. In said rest coordinates (on $\mathscr{R}$) the world line will therefore be described by $x^i(t) = y^i$ constant and $t = t$. In the frame $\mathscr{R}'$ the same world line will read

$$x'^i(t') = c^i + (t' - c)v^i + \sum_j R^i{}_j y^j .$$

The coordinates c^i determine a point Q in $E_{\mathscr{R}'}$ at $t' = c$. Differentiating in time, with respect to $\mathscr{R}'$, the vector

$$P(t) - Q := \sum_i (x'^i(t') - c^i) \mathbf{e}'_i$$

we obtain the velocity of P with respect to $\mathscr{R}'$. The result is trivially the vector: $\sum_{i=1}^3 v^i \mathbf{e}'_i$ constant in time and independent of $P \in E_{\mathscr{R}}$, which coincides with $\mathbf{v}_{\mathscr{R}|\mathscr{R}'}$ by construction.

(3) Under the hypotheses we made, (3.1) holds in some coordinates of $\mathscr{R}'$ and $\mathscr{R}$ respectively. Let us invert (3.1) to obtain

$$\begin{cases} t = t' - c, \\ x^j = \sum_{i=1}^3 (R^{-1})^j{}_i \left(cv^i - c^i \right) - t' \sum_{i=1}^3 (R^{-1})^j{}_i v^i \\ \qquad + \sum_{i=1}^3 (R^{-1})^j{}_i x'^i, \quad i = 1, 2, 3. \end{cases} \qquad (3.5)$$

By part (2), $\mathscr{R}'$ is in uniform linear motion with respect to $\mathscr{R}$ and

$$\mathbf{v}_{\mathscr{R}'|\mathscr{R}} = - \sum_{i,j=1}^3 (R^{-1})^j{}_i v^i \mathbf{e}_j .$$

Since by construction $\sum_{j=1}^3 (R^{-1})^j{}_i \mathbf{e}_j = \mathbf{e}'_i$, we also have $\mathbf{v}_{\mathscr{R}|\mathscr{R}'} = -\mathbf{v}_{\mathscr{R}'|\mathscr{R}}$.

(4) This is immediate if we choose right-handed orthonormal coordinate systems in the three frames. The transformation between the coordinates associated with $\mathscr{R}$ and $\mathscr{R}'$ and the transformation between the coordinates associated with $\mathscr{R}'$ and $\mathscr{R}''$ must be of the form (3.1) by (1). The transformation between the coordinates associated with $\mathscr{R}$ and $\mathscr{R}''$ then is the composite of the previous

two. By direct computation the transformation thus found is still of type (3.1), so (2) implies that $\mathscr{R}$ is in uniform linear motion with respect to $\mathscr{R}''$.

(5) This follows directly from (3.1) and (3).

$\square$

Transformation (3.1) is determined by the quadruple $(c, \vec{c}, \vec{v}, R) \in \mathbb{R} \times \mathbb{R}^3 \times \mathbb{R}^3 \times SO(3)$, where $\vec{c} := (c^1, c^2, c^3)$, $\vec{v} := (v^1, v^2, v^3)$ and R is the matrix in the group $SO(3)$ of real 3×3 orthogonal matrices of positive determinant (i.e. 1) and coefficients $R^i{}_j$. The quadruple corresponds to 10 real parameters (remember that an $SO(3)$ matrix is determined by 3 real numbers). Since map composition is associative, the composite of Galilean transformations is still a Galilean transformation, the inverse of a Galilean transformation is Galilean, and finally, the Galilean transformation with parameters $(0, \vec{0}, \vec{0}, I)$ is the identity transformation, we conclude that the set of transformations (3.1), seen as maps from $\mathbb{R}^4$ to $\mathbb{R}^4$, as $(c, \vec{c}, \vec{v}, R) \in \mathbb{R} \times \mathbb{R}^3 \times \mathbb{R}^3 \times SO(3)$ varies, has the algebraic structure of a group under composition. This group is called the **Galilean group**.

Exercises 3.3

(1) Consider three frames $\mathscr{R}$, $\mathscr{R}'$ and $\mathscr{R}''$ such that, in right-handed orthonormal coordinates for each, the coordinate transformation from $\mathscr{R}'$ to $\mathscr{R}$ is Galilean and determined by the quadruple $(c_1, \vec{c}_1, \vec{v}_1, R_1)$, and the coordinate transformation from $\mathscr{R}''$ to $\mathscr{R}'$ is Galilean and determined by the quadruple $(c_2, \vec{c}_2, \vec{v}_2, R_2)$. Prove that the coordinate transformation from $\mathscr{R}''$ to $\mathscr{R}$ obtained by composing the previous two is the Galilean transformation determined by the quadruple:

$$(c_1 + c_2, \ R_2\vec{c}_1 + c_1\vec{v}_2 + \vec{c}_2, \ R_2\vec{v}_1 + \vec{v}_2, \ R_2R_1) \ .$$

(2) Show that $\mathbb{R} \times \mathbb{R}^3 \times \mathbb{R}^3 \times SO(3)$, with composition rule

$$(c_2, \vec{c}_2, \vec{v}_2, R_2) \circ (c_1, \vec{c}_1, \vec{v}_1, R_1) := (c_1 + c_2, \ R_2\vec{c}_1 + c_1\vec{v}_2 + \vec{c}_2, \ R_2\vec{v}_1$$
$$+\vec{v}_2, \ R_2R_1) \ ,$$

is a group where, in particular, the neutral element is $(0, \vec{0}, \vec{0}, I)$ and the inverse is

$$(c, \vec{c}, \vec{v}, R)^{-1} = \left(-c, \ -R^{-1}(\vec{c} - c\vec{v}), \ -R^{-1}\vec{v}, \ R^{-1}\right) \ .$$

(3) $\boxed{\text{AC}}$ Show that the Galilean group is isomorphic to a subgroup of the group $GL(5, \mathbb{R})$ of non-singular, real 5×5 matrices.

3.1.3 Relative Motion of Inertial Frames

What we wish to prove now is the following fundamental result about the relative motion of inertial frames, which uses the notions and results just established.

Theorem 3.4 *Let us assume that, for any frame $\mathscr{R}$ and any event $p \in V^4$, there is a point particle whose world line crosses p with velocity (with respect to $\mathscr{R}$) of magnitude and direction that can be chosen arbitrarily, and that the particle is isolated for some finite time interval around p. Then the following facts hold.*

(1) *If $\mathscr{R}$ is an inertial frame, another frame $\mathscr{R}'$ is inertial if and only if it moves linearly and uniformly with respect to $\mathscr{R}$.*
(2) *The transformations between the rest orthonormal coordinates on inertial frames are the Galilean transformations.*
(3) *If $\mathscr{R}, \mathscr{R}'$ are both inertial, $\boldsymbol{\omega}_{\mathscr{R}'|\mathscr{R}}$, the relative acceleration and the Coriolis acceleration of $\mathscr{R}'$ with respect to $\mathscr{R}$ are always zero.*

Proof

(1) Suppose $\mathscr{R}$ is inertial and $\mathscr{R}'$ moves linearly and uniformly with respect to $\mathscr{R}$. In this case, referring to orthonormal coordinates in $\mathscr{R}$ and $\mathscr{R}'$ respectively, we have the transformation rule

$$\begin{cases} t' = t + c\,, \\ x'^i = c^i + tv^i + \displaystyle\sum_{j=1}^{3} R^i{}_j x^j\,, \quad i = 1, 2, 3. \end{cases} \tag{3.6}$$

If γ is the world line of an isolated particle in $\mathscr{R}$, it must generate a curve $P = P(t)$ with constant velocity $\mathbf{v}_P|_{\mathscr{R}} = \sum_{j=1}^{3} u^j \mathbf{e}_j$. In coordinates x^j in $\mathscr{R}$, the law of motion will then have the form $x^j(t) = tu^j + x_0^j$, for $j = 1, 2, 3$ and $t \in \mathbb{R}$, where the x_0^j are constants. By (3.6), the law of motion of γ in the coordinates x'^i in $\mathscr{R}'$ will be:

$$x'^i(t') = c^i + (t' - c)v^i + \sum_{j=1}^{3} R^i{}_j \left((t' - c)u^j + x_0^j\right).$$

Computing the velocity immediately gives the constant vector

$$\mathbf{v}_P|_{\mathscr{R}'} = \sum_{i=1}^{3} \left(v^i + \sum_{j=1}^{3} R^i{}_j u^j\right) \mathbf{e}'_i\,.$$

Hence *every isolated point particle moves, for $\mathscr{R}'$, uniformly in a straight line*: $\mathscr{R}'$ is inertial.

Suppose conversely that $\mathscr{R}$ and $\mathscr{R}'$ are both inertial. We want to study the motion, in either frame, of an isolated point particle P, keeping into account relationships (2.69):

$$\begin{cases} \mathbf{a}_P|_{\mathscr{R}} = \mathbf{a}_P|_{\mathscr{R}'} + \mathbf{a}_P^{(tr)} + \mathbf{a}_P^{(C)}, \\ \mathbf{a}_P^{(tr)} := \mathbf{a}_{O'}|_{\mathscr{R}} + \boldsymbol{\omega}_{\mathscr{R}'}|_{\mathscr{R}} \wedge (\boldsymbol{\omega}_{\mathscr{R}'}|_{\mathscr{R}} \wedge (P - O')) + \dot{\boldsymbol{\omega}}_{\mathscr{R}'}|_{\mathscr{R}} \wedge (P - O'), \\ \mathbf{a}_P^{(C)} := 2\boldsymbol{\omega}_{\mathscr{R}'}|_{\mathscr{R}} \wedge \mathbf{v}_P|_{\mathscr{R}'}. \end{cases}$$

O' is an arbitrary point at rest for $\mathscr{R}'$. In the case under exam, as P is isolated and both frames are inertial, we must have $\mathbf{a}_P|_{\mathscr{R}} = \mathbf{a}_P|_{\mathscr{R}'} = \mathbf{0}$ at every time (in the time interval where the particle is isolated). Substituting in the first formula above we find that at each instant and for any isolated point particle P:

$$\mathbf{a}_P^{(tr)} + \mathbf{a}_P^{(C)} = \mathbf{0}.$$

i.e.

$$\mathbf{a}_{O'}|_{\mathscr{R}} + \boldsymbol{\omega}_{\mathscr{R}'}|_{\mathscr{R}} \wedge (\boldsymbol{\omega}_{\mathscr{R}'}|_{\mathscr{R}} \wedge (P - O')) + \dot{\boldsymbol{\omega}}_{\mathscr{R}'}|_{\mathscr{R}} \wedge (P - O') + 2\boldsymbol{\omega}_{\mathscr{R}'}|_{\mathscr{R}} \wedge \mathbf{v}_P|_{\mathscr{R}'} = \mathbf{0}$$

Choosing the point particle to be isolated with zero velocity in $\mathscr{R}'$ and coinciding with O', we conclude that at the time considered: $\mathbf{a}_{O'}|_{\mathscr{R}} = \mathbf{0}$. Because we can argue as above for any instant, the result must hold always. To sum up:

$$\boldsymbol{\omega}_{\mathscr{R}'}|_{\mathscr{R}} \wedge (\boldsymbol{\omega}_{\mathscr{R}'}|_{\mathscr{R}} \wedge (P - O')) + \dot{\boldsymbol{\omega}}_{\mathscr{R}'}|_{\mathscr{R}} \wedge (P - O') + 2\boldsymbol{\omega}_{\mathscr{R}'}|_{\mathscr{R}} \wedge \mathbf{v}_P|_{\mathscr{R}'} = \mathbf{0},$$
(3.7)

at every instant and for any isolated point particle P. Choosing again P coinciding with O' and with velocity $\mathbf{v}_P|_{\mathscr{R}'} \neq \mathbf{0}$ orthogonal to the value of $\boldsymbol{\omega}_{\mathscr{R}'}|_{\mathscr{R}}$ at that time, from (3.7) we deduce: $||\boldsymbol{\omega}_{\mathscr{R}'}|_{\mathscr{R}}|| \, ||\mathbf{v}_P|_{\mathscr{R}'}|| = 0$ whence $\boldsymbol{\omega}_{\mathscr{R}'}|_{\mathscr{R}} = \mathbf{0}$ at the given instant, and therefore at any instant. Inserting that in (2.68): $\mathbf{v}_P|_{\mathscr{R}} = \mathbf{v}_P|_{\mathscr{R}'} + \boldsymbol{\omega}_{\mathscr{R}'}|_{\mathscr{R}} \wedge (P - O') + \mathbf{v}_{O'}|_{\mathscr{R}}$, which shows that any point P at rest in $\mathscr{R}'$ has velocity with respect to $\mathscr{R}$ equal to $\mathbf{v}_{O'}|_{\mathscr{R}}$, irrespective of P. On the other hand, since we have found that $\frac{d\mathbf{v}_{O'}|_{\mathscr{R}}}{dt}|_{\mathscr{R}}(= \mathbf{a}_{O'}|_{\mathscr{R}}) = \mathbf{0}$ for any instant, we conclude that the velocity of particles at rest in $\mathscr{R}'$ is *constant in time and does not depend on the particle*. By definition, $\mathscr{R}'$ is therefore in uniform linear motion with respect to $\mathscr{R}$.

(2) The proof is an immediate consequence of this theorem's part (1) and the first two items of Proposition 3.2.

(3) Immediate from (2).

$\square$

Exercises 3.5 Consider the *giant wheel* of a *carnival fair*. Suppose that the frame $\mathscr{R}$ of the ground is approximatively inertial and that a second frame $\hat{\mathscr{R}}$ moves with one of the seats (which is free to stay in a horizontal position as the wheel goes around). What is the vector $\omega_{\hat{\mathscr{R}}}|_{\mathscr{R}}$? Is the frame $\hat{\mathscr{R}}$ inertial?

3.1.4 $\boxed{\text{AC}}$ *The Affine Galilean Structure of* $\mathbb{V}^4$

By means of the class of *inertial* frames we can equip the spacetime of Classical Physics with a privileged *affine structure*. This is defined by the fact that rest orthonormal coordinate on *inertial* frames are Cartesian coordinates for the given affine structure (Definition 1.3). It is crucial to stress that such an affine structure is determined by the Newtonian *dynamics* and not by the kinematics. There is no such affine structure on $\mathbb{V}^4$ that satisfies the aforementioned condition for the entire collection of rest coordinate systems on *all* possible frames. This was already discussed in (1) Remark 2.12.

Theorem 3.6 *There exists a 4-dimensional affine structure* $(\mathbb{V}^4, V, -)$ *on* $\mathbb{V}^4$ *such that every system of rest orthonormal coordinates*

$$\sigma_{\mathscr{R}} : \mathbb{V}^4 \ni e \mapsto (t(e), x^1(e), x^2(e), x^3(e)) \in \mathbb{M}^4$$

on every inertial frame $\mathscr{R}$ *is also a system of Cartesian coordinates on* $(\mathbb{V}^4, V, -)$. *This structure is unique up to affine isomorphisms of* $\mathbb{V}^4$ *that restrict to the identity on events. It is called* **affine Galilean structure** *of* $\mathbb{V}^4$.

Proof Let us recall first that $\mathbb{R}^4$ (though the reasoning trivially extends to $\mathbb{R}^n$) is naturally a 4-dimensional affine space, as follows.

(1) The vector space of displacements is $\mathbb{R}^4$ itself, now seen as real vector space;
(2) the vector difference of two points (a, b, c, d) and (a', b', c', d') in the *affine* space $\mathbb{R}^4$ is defined by

$$(a, b, c, d) - (a', b', c', d') := (a - a', b - b', c - c', d - d'),$$

where the quadruple on the right is an element of the *vector* space $\mathbb{R}^4$.

Consider now an inertial frame $\mathscr{R}$ and a system of rest orthonormal coordinates:

$$\sigma_{\mathscr{R}} : \mathbb{V}^4 \ni p \mapsto (t(p), x^1(p), x^2(p), x^3(p)) \in \mathbb{R}^4 .$$

This system induces an affine structure $(\mathbb{V}^4, V, -)$ on $\mathbb{V}^4$ simply by requesting $V := \mathbb{R}^4$ and $e' - e := \sigma_{\mathscr{R}}(e') - \sigma_{\mathscr{R}}(e)$ for any $e, e' \in \mathbb{V}^4$. Consider now a second inertial frame $\mathscr{R}'$ and rest orthonormal coordinates on it:

$$\sigma_{\mathscr{R}'} : \mathbb{V}^4 \ni p \mapsto (t'(p), x'^1(p), x'^2(p), x'^3(p)) \in \mathbb{R}^4 .$$

Let $\psi : \mathbb{R}^4 \to \mathbb{R}^4$ be the Galilean transformation relating the coordinates of the first system to the coordinates of the second $\psi(t, x^1, x^2, x^3) = (t', x'^1, x'^2, x'^3)$. By construction

$$\sigma_{\mathscr{R}'} = \psi \circ \sigma_{\mathscr{R}} : V^4 \to \mathbb{R}^4 .$$

It is easy to prove that the composite $\sigma' := \psi \circ \sigma$ of an affine bijection from $\mathbb{R}^n$ to $\mathbb{R}^n$

$$\psi : \mathbb{R}^n \ni (x^1, \ldots, x^n) \mapsto (x'^1, \ldots, x'^n) \in \mathbb{R}^n$$

with a Cartesian system $\sigma : \mathbb{A}^n \to \mathbb{R}^n$ on the affine space $\mathbb{A}^n$ is again a Cartesian system σ' on $\mathbb{A}^n$. The new system $\sigma' : \mathbb{A}^n \to \mathbb{R}^n$ is obtained by suitably changing the origin and the axes of the initial system σ on $\mathbb{A}^n$. Back to our concrete case, since every Galilean transformation ψ is an affine map from $\mathbb{R}^4$ to $\mathbb{R}^4$, $\sigma_{\mathscr{R}'} = \psi \circ \sigma_{\mathscr{R}}$ still is a system of Cartesian coordinates of the affine structure $(V^4, V, -)$ described earlier. The affine structure $(V^4, V, -)$ therefore admits any system of rest orthonormal coordinates on any inertial frame as an admissible system of Cartesian coordinates, as required by the first part of the claim.

Let us pass to the uniqueness up to affine isomorphisms of V^4. Consider a generic affine structure $(V^4, V_1, -_1)$ with the same property of $(V^4, V, -)$ with respect to inertial frames: any system of rest orthonormal coordinates on any inertial frame is a Cartesian system for $(V^4, V_1, -_1)$. Let us show this structure is necessarily isomorphic to the $(V^4, V, -)$ we have defined using a specific inertial frame $\mathscr{R}$. The isomorphism between $(V^4, V_1, -_1)$ and $(V^4, V, -)$ is trivially given by composing the above chart $\sigma_{\mathscr{R}}$, representing a system of rest orthonormal coordinates on the inertial frame $\mathscr{R}$, with its inverse. The chart $\sigma_{\mathscr{R}} : V^4 \to \mathbb{R}^4$, here thinking V^4 as referring to the structure $(V^4, V, -)$, is by construction an affine isomorphism from V^4 to $\mathbb{R}^4$. On the other hand, viewing V^4 relative to $(V^4, V_1, -_1)$, the map $\sigma_{\mathscr{R}} : V^4 \to \mathbb{R}^4$ is by hypothesis a system of Cartesian coordinates *also* for this structure, hence still an affine isomorphism. The map:

$$\phi := (\sigma_{\mathscr{R}})^{-1} \circ (\sigma_{\mathscr{R}}) : V^4_{\text{with structure } (V^4, V, -)} \ni p \mapsto p \in V^4_{\text{with structure } (V^4, V_1, -_1)}$$

is therefore an isomorphism between the above two affine structures on V^4 (as composite of affine isomorphisms). Clearly this map fixes the points of V^4. Despite that, $d\phi : V \to V_1$ is not the identity because the source and target are distinct vector spaces. $\square$

Remarks 3.7 $\boxed{\text{AC}}$ Extending the formalism and introducing the *affine connection* on V^4 induced by the affine Galilean structure, one can show that the *geodesics* of the affine connection that satisfy the additional requisite $\langle V, dT \rangle \neq 0$ everywhere, where V is the tangent vector to the geodesic at each event, are nothing but the *inertial motions of point particles*. Absolute time is an affine parameter for such geodesics. In this sense Classical Mechanics and the theory of General Relativity are closer than what one would expect at first sight. The proximity to the theory of Special Relativity will become manifest from the introduction we will give to it in Chap. 10.   ■

3.2   General Formulation of the Classical Dynamics of Systems of Point Particles

Let us explain how to build up the dynamics once we assume the existence of the class of inertial frames with the features we said. Considering *isolated* point particles, i.e. sufficiently far from one another, we know that their motion, described by Newton's first law of motion, is in a straight line with constant velocity in every inertial frame. The problem is then to study and clarify what happens to the motion, always described in an inertial frame, when the particles are *no longer* isolated. Experience shows that their motion becomes accelerated.

3.2.1   Masses, Impulses and Forces

If we have a system of point particles, we will say it is **isolated** when its particles are far from the other bodies in the universe not in the set (the particles in the set may instead be arbitrarily close). More precisely, the distance of the system's particles to the remaining bodies in the universe should be such that if we removed all point particles except for one, the remaining particle would move uniformly in inertial frames. To proceed with the formulation of Dynamics, one assumes that point particles have an associated physical quantity called **mass** that is additive and is preserved. Using the mass we can state the conservation law of the *impulse* or *linear momentum*.

C2. Impulse and Mass
Assume that for any point particle Q there exists a strictly positive constant m called **mass** *such that the following facts hold.*

(1) **Conservation principle of the impulse** *(for pairs of point particles). In every inertial frame $\mathscr{R}$ and for any pair of point particles Q, Q' of respective masses m, m' forming an isolated system,*

$$m\mathbf{v}_Q|_{\mathscr{R}} + m'\mathbf{v}_{Q'}|_{\mathscr{R}}$$

is a constant vector in time, even if $\mathbf{v}_Q|_{\mathscr{R}}$ and $\mathbf{v}_{Q'}|_{\mathscr{R}}$ are not separately constant. The vector $\mathbf{p} := m\mathbf{v}_Q|_{\mathscr{R}}$ is called **impulse** *or* **linear momentum** *of the point particle Q with respect to $\mathscr{R}$.*

(2) Conservation principle/additivity of the mass. *In processes where a point particle decomposes into several point particles (or many point particles coalesce into a single point particle), the initial (final) mass of the only point particle equals the sum of the masses of the final (initial) constituents.*

Remarks 3.8

(1) Masses are independent of the frame, whereas impulses are not and transform when changing frames. Recalling (2.68):

$$\mathbf{p}_{\mathscr{R}} = \mathbf{p}_{\mathscr{R}'} + m\mathbf{v}_Q^{(tr)} \,. \tag{3.8}$$

However, if both $\mathscr{R}$ and $\mathscr{R}'$ are inertial, due to Theorem 3.4 (which sanctions in particular that the relative velocity is constant in time, independent of the point and equal to $\mathbf{v}_{\mathscr{R}'|\mathscr{R}}$), the above formula simplifies to

$$\mathbf{p}_{\mathscr{R}} = \mathbf{p}_{\mathscr{R}'} + m\mathbf{v}_{\mathscr{R}'|\mathscr{R}} \,. \tag{3.9}$$

(2) We may assume that the mass of a reference point particle P_1 is 1, and using the conservation of the impulse, measure the mass of another point particle P_2 by letting it interact with P_1, supposing all other bodies are far away. m_{P_2} is the unique number satisfying, for arbitrary instants $t \neq t'$: $m_{P_2}(\mathbf{v}_{P_2}|_{\mathscr{R}}(t') - \mathbf{v}_{P_2}|_{\mathscr{R}}(t)) = 1(\mathbf{v}_{P_1}|_{\mathscr{R}}(t') - \mathbf{v}_{P_2}|_{\mathscr{R}}(t))$, when the particles do not have constant velocity in time (due to the mutual interaction). ∎

Let us now introduce the concept of *force* to handle the interactions among nearby particles.

C3. Newton's Second Law of Motion (First Part)
Consider two point particles Q and Q' of respective mass m and m', forming an isolated system. Assume that for any inertial frame $\mathscr{R}$ there exist two differentiable functions called **forces** *acting on the point particles,*

$$\mathbf{F}_{\mathscr{R}} : E_{\mathscr{R}} \times E_{\mathscr{R}} \times V_{\mathscr{R}} \times V_{\mathscr{R}} \to V_{\mathscr{R}} \,, \quad \mathbf{F}'_{\mathscr{R}} : E_{\mathscr{R}} \times E_{\mathscr{R}} \times V_{\mathscr{R}} \times V_{\mathscr{R}} \to V_{\mathscr{R}}$$

such that, for any $t \in \mathbb{R}$, **Newton's second law of motion**

$$\begin{cases} m\mathbf{a}_Q|_{\mathscr{R}}(t) = \mathbf{F}_{\mathscr{R}}\left(Q(t), Q'(t), \mathbf{v}_Q|_{\mathscr{R}}(t), \mathbf{v}_{Q'}|_{\mathscr{R}}(t)\right) , \\ m'\mathbf{a}_{Q'}|_{\mathscr{R}}(t) = \mathbf{F}'_{\mathscr{R}}\left(Q'(t), Q(t), \mathbf{v}_{Q'}|_{\mathscr{R}}(t), \mathbf{v}_Q|_{\mathscr{R}}(t)\right) \end{cases} \tag{3.10}$$

holds. More precisely, $\mathbf{F}_{\mathscr{R}}$ is called the force that **acts on Q due to Q'** *and $\mathbf{F}'_{\mathscr{R}}$ the force that* **acts on Q' due to Q**. *$\mathbf{F}_{\mathscr{R}}$ is called* **reaction** *associated with $\mathbf{F}'_{\mathscr{R}}$, and $\mathbf{F}'_{\mathscr{R}}$* **reaction** *associated with $\mathbf{F}_{\mathscr{R}}$.*

As the inertial frame varies (using the identification of $E_{\mathscr{R}}$ and $E_{\mathscr{R}'}$ with Σ_t) the forces' value is **invariant**, *i.e.:*

$$\mathbf{F}_{\mathscr{R}}\left(P_1, P_2, \mathbf{u}_1, \mathbf{u}_2\right) = \mathbf{F}_{\mathscr{R}'}\left(P_1, P_2, \mathbf{u}_1', \mathbf{u}_2'\right) , \tag{3.11}$$

for $P_1, P_2 \in \Sigma_t$ and $\mathbf{u}_1, \mathbf{u}_2 \in V_t$, and where we put $\mathbf{u}' := \mathbf{u} + \mathbf{u}_{\mathscr{R}|\mathscr{R}'}$ for any $\mathbf{u} \in V_t$.

As we will say more generally shortly, in rest orthonormal coordinates x^1, x^2, x^3 on $\mathscr{R}$, Eq. (3.10) become a *system of differential equations of order two* that *determines completely* the particle's laws of motion once we know their positions and velocities at an arbitrary instant t_0.

Physically, once Newton's second law has been formulated, one should determine the various *force laws*, i.e. the explicit form of the functions

$$\mathbf{F}_{\mathscr{R}} = \mathbf{F}_{\mathscr{R}}\left(P_1, P_2, \mathbf{u}_1, \mathbf{u}_2\right) ,$$

which exist in nature and depend on further characteristics of the point particles. One example is the electric charge of interacting point particles, associated with the *Coulomb force*.

Remarks 3.9

(1) In order to not violate the Newton's first law, every force that makes physical sense must decrease and eventually vanish as the distance between the particles increases, so to give zero accelerations for pairs of bodies "sufficiently far from one another", possibly at infinite distance.

(2) The invariance request (3.11) follows from imposing (3.10) in every inertial frame. In fact, we know that if $\hat{\mathscr{R}}$ is another inertial frame, it will be related to the original one by a Galilean transformation, so $\boldsymbol{\omega}_{\hat{\mathscr{R}}|\mathscr{R}} = \boldsymbol{\omega}_{\mathscr{R}|\hat{\mathscr{R}}} = 0$ and the relative acceleration of any particle at rest in one frame is null when referred to the other frame. Using (2.69) we discover that *the acceleration of a particle P at a given time instant does not change when changing frames, passing from an inertial frame $\mathscr{R}$ to another inertial frame $\hat{\mathscr{R}}$*: $\mathbf{a}_{Q}|_{\hat{\mathscr{R}}} = \mathbf{a}_{Q}|_{\mathscr{R}}$ and $\mathbf{a}_{Q'}|_{\hat{\mathscr{R}}} = \mathbf{a}_{Q'}|_{\mathscr{R}}$. Consequently (3.10) can be rewritten so that the left-hand sides refer to $\hat{\mathscr{R}}$ and the right-hand sides to $\mathscr{R}$:

$$\begin{cases} m\mathbf{a}_{Q}|_{\hat{\mathscr{R}}}(t) = \mathbf{F}_{\mathscr{R}}\left(Q(t), Q'(t), \mathbf{v}_{Q}|_{\mathscr{R}}(t), \mathbf{v}_{Q'}|_{\mathscr{R}}(t)\right) , \\ m'\mathbf{a}_{Q'}|_{\hat{\mathscr{R}}}(t) = \mathbf{F}'_{\mathscr{R}}\left(Q'(t), Q(t), \mathbf{v}_{Q'}|_{\mathscr{R}}(t), \mathbf{v}_{Q}|_{\mathscr{R}}(t)\right) . \end{cases}$$

But since the second law of motion holds in $\hat{\mathscr{R}}$, too:

$$\begin{cases} m\mathbf{a}_{Q}|_{\hat{\mathscr{R}}}(t) = \mathbf{F}_{\hat{\mathscr{R}}}\left(Q(t), Q'(t), \mathbf{v}_{Q}|_{\hat{\mathscr{R}}}(t), \mathbf{v}_{Q'}|_{\hat{\mathscr{R}}}(t)\right) , \\ m'\mathbf{a}_{Q'}|_{\hat{\mathscr{R}}}(t) = \mathbf{F}'_{\hat{\mathscr{R}}}\left(Q'(t), Q(t), \mathbf{v}_{Q'}|_{\hat{\mathscr{R}}}(t), \mathbf{v}_{Q}|_{\hat{\mathscr{R}}}(t)\right) . \end{cases}$$

As m, m' do not depend on the frame, we conclude that (3.11) must hold, if we assume that point particles can assume, at a given instant, any possible position and velocity.

(3) By virtue of (3.11), from now on we will drop the subscript $\mathscr{R}$ to specify the inertial frame in which the forces are given. In the same way we will avoid writing $|_\mathscr{R}$ in the accelerations when it is understood they are evaluated with respect to an inertial frame (it does not matter which one, because of what was said before!). Which inertial frame is in use will be specified by the velocities appearing in the forces. Observe that the forces might still *depend on different positions and velocities* as the inertial frame varies. We will tackle this problem in Sect. 3.4.1 and explain that the dependence is actually the same, due to the *Principle of Galilean invariance*. ∎

The conservation of the impulse together with Newton's second law immediately imply what, in traditional presentations, goes under the name of **Newton's third law of motion** for a system of two point particles,[1] or **Action-reaction principle**. It relates the force acting on Q with the force acting on Q':

$$\mathbf{F}\left(Q(t), Q'(t), \mathbf{v}_Q|_{\mathscr{R}}(t), \mathbf{v}_{Q'}|_{\mathscr{R}}(t)\right) = -\mathbf{F}'\left(Q'(t), Q(t), \mathbf{v}_{Q'}|_{\mathscr{R}}(t), \mathbf{v}_Q|_{\mathscr{R}}(t)\right) .$$

The Action-reaction principle **in strong form** demands that the forces are applied along $Q(t) - Q'(t)$; this request will be useful to obtain, as a theorem, the conservation of the angular momentum for systems of particles:

C3. Newton's Second Law of Motion (Second Part)] *For any* $t \in \mathbb{R}$, *if* $Q(t) - Q'(t)$ *is not the zero vector,*

$$\mathbf{F}\left(Q(t), Q'(t), \mathbf{v}_Q|_{\mathscr{R}}(t), \mathbf{v}_{Q'}|_{\mathscr{R}}(t)\right) \quad \text{and} \quad \mathbf{F}'\left(Q'(t), Q(t), \mathbf{v}_{Q'}|_{\mathscr{R}}(t), \mathbf{v}_Q|_{\mathscr{R}}(t)\right)$$

are collinear with the vector (and oppositely oriented).

3.2.2 Superposition of Forces

If a point particle, Q, interacts simultaneously with several particles we assume that the total force applied on Q is the sum of the forces between all possible pairs that include Q.

---

[1] For us, a simple corollary of our principles.

C4. Principle of Superposition of Forces
Given a system of N *point particles* $Q_1, \ldots, Q_N$, *for any particle* Q_k *and in any inertial frame* $\mathcal{R}$ **Newton's second law of motion in complete form** *holds:*

$$m_k \mathbf{a}_{Q_k}(t) = \sum_{k \neq i=1}^{i=N} \mathbf{F}_{ki} \left(Q_k(t), Q_i(t), \mathbf{v}_{Q_k} | \mathcal{R}(t), \mathbf{v}_{Q_i} | \mathcal{R}(t) \right) ,$$

where the function $\mathbf{F}_{ki}$ *is the force acting on* Q_k *due to* Q_i, *obtained placing the system's other particles sufficiently far apart.*

Remarks 3.10 It is well known that the Action-reaction principle is already false when we consider electromagnetic forces in presence of moving charges. Less known is that the superposition principle, too, is not valid for certain physical systems, even ones that are not that sophisticated. It happens for example when we consider three or more polarisable molecules. These molecules change their structure and consequently also the electric force they exert on the other molecules due to the electric field which permeates them, i.e. the electric forces to which they are subjected.   ∎

If $\mathbf{x}_k$ denotes the triple of coordinates determining Q_k with respect to rest orthonormal coordinates on a frame $\mathcal{R}$, the set of equations

$$\frac{d^2 \mathbf{x}_k(t)}{dt^2} = m_k^{-1} \mathbf{F}_i \left(\mathbf{x}_1(t), \ldots, \mathbf{x}_k(t), \frac{d\mathbf{x}_1(t)}{dt}, \ldots, \frac{d\mathbf{x}_N(t)}{dt} \right) , \quad k = 1, 2, \ldots, N ,$$

(3.12)

where, for $i = 1, \ldots, N$,

$$\mathbf{F}_i \left(\mathbf{x}_1(t), \ldots, \mathbf{x}_k(t), \frac{d\mathbf{x}_1(t)}{dt}, \ldots, \frac{d\mathbf{x}_N(t)}{dt} \right)$$

$$:= \sum_{k \neq i=1}^{i=N} \mathbf{F}_{ki} \left(\mathbf{x}_k(t), \mathbf{x}_i(t), \frac{d\mathbf{x}_k(t)}{dt}, \frac{d\mathbf{x}_i(t)}{dt} \right) ,$$

(3.13)

forms an ODE system, to which the Existence and uniqueness theorems discussed in the next section apply.

3.2.3   The Determinism of Classical Mechanics

The problem of determining the **motion of a system**, i.e. the *law of motion* of a system's point particles, when we are given the forces acting on each particle and the *initial conditions*, that is, the position and velocity of each particle of the system at a certain instant with respect to a given inertial frame, is called the **fundamental**

problem of Dynamics. All in all, with a few very important generalisations we will talk about soon, the problem reduces to solving, in rest orthonormal coordinates on an inertial frame $\mathscr{R}$, the differential system (3.12) when the initial conditions $x_k(t_0)$ and $\frac{dx_k(t)}{dt}|_{t=t_0}$ for $k = 1, 2, \ldots, N$ are assigned.

The whole theory is based on the following theorem, which we will discuss, weakening the hypotheses and improving the thesis, and prove in the final Complement on differential equations:

Theorem 3.11 *Let $D, D' \subset \mathbb{R}^n$ be non-empty open sets, $I \subset \mathbb{R}$ a non-empty open interval. Consider the differential equation*

$$\frac{d^2\mathbf{x}}{dt^2} = \mathbf{F}\left(t, \mathbf{x}, \frac{d\mathbf{x}}{dt}\right). \tag{3.14}$$

where $\mathbf{F} : I \times D \times D' \to \mathbb{R}^n$ is a function with $\mathbf{F} \in C^1(I \times D \times D'; \mathbb{R}^n)$. Then for any $(t_0, \mathbf{x}_0, \dot{\mathbf{x}}_0) \in \mathbb{R} \times D \times D'$ there exists an open interval containing t_0 on which there exists a unique C^2 solution $\mathbf{x} = \mathbf{x}(t)$ to (3.14) that satisfies the initial conditions $\mathbf{x}(t_0) = \mathbf{x}_0$ and $\frac{d\mathbf{x}}{dt}|_{t=t_0} = \dot{\mathbf{x}}_0$.

Assuming suitable regularity conditions on the force functions, at least locally, Newton's second law determines the *future, but also past,* motion once some initial conditions are prescribed. This mathematical result goes under the name of **determinism** and has massive importance in Classical Physics.

Remarks 3.12

(1) Heuristically, Theorem 3.11 is plausible if we further strengthen the assumptions, assuming $\mathbf{F}$ of class C^∞ or stronger still, real analytic. In fact, inserting the initial conditions $\mathbf{x}(t_0) = \mathbf{x}_0$ and $\frac{d\mathbf{x}}{dt}|_{t=t_0} = \dot{\mathbf{x}}_0$ in the right-hand side of (3.14), the left-hand side gives the second derivative of the unknown solution $\mathbf{x}(t)$ at t_0. With this information, repeating the procedure we obtain the third derivative of $\mathbf{x}(t)$ at t_0. Iterating infinitely many times we find all derivatives of $\mathbf{x}(t)$ at t_0, $\frac{d^n\mathbf{x}}{dt^n}|_{t_0}$. At this point we expect that if the series:

$$\sum_{n=0}^{+\infty} \frac{d^n\mathbf{x}}{dt^n}\bigg|_{t_0} \frac{(t - t_0)^n}{n!}$$

converges around t_0, the function of t thus obtained will solve (3.14) with the given initial conditions. Clearly, in the class of analytic solutions, if non-empty, that one will be the only local solution with the given initial conditions.

(2) Observe that, for the ODE at hand, if it had not been possible to isolate the second derivative on the left and leave lower-order terms on the right, the above heuristic procedure might not have worked. We will prove Theorem 3.11 in the mathematical complement to Chap. 14, making use of a procedure that requires much weaker assumptions.

(3) As explained in the Complement on ODEs at the end of the book, in reality there is a stronger form of the above theorem. This version says that, under the same hypotheses, there is an open interval $J \ni t_0$ on which the differential equation with the given initial conditions has a solution. The latter, called *maximal solution* of the problem, is moreover characterised by the following two properties: (a) it is not the restriction of a solution defined on a larger interval $J' \supsetneq J$, and (b) any other solution for the same initial data is a restriction of it to some subinterval $J' \subsetneq J$. ∎

Examples 3.13

(1) Three point particles P_1, P_2, P_3 are pairwise joint by springs of zero length at rest and elastic constant $\kappa > 0$. We want to describe the forces of each and the system of equations that determine their motion.

The force of a spring between P and Q of zero length at rest is simply

$$\mathbf{F}_P(P, Q) = -\kappa(P - Q), \tag{3.15}$$

where $\mathbf{F}_P$ is the force the spring exerts on P. This type of law is known as **elastic force**. Hence, for $i \neq k$:

$$\mathbf{F}_{ki}(P_k, P_i) = -\kappa(P_k - P_i).$$

Note that the Action-reaction principle is automatic. The system of equations that determines the three particles' motion then is:

$$m_k \mathbf{a}_{P_k}(t) = \sum_{k \neq i=1}^{i=3} -\kappa(P_k - P_i) \quad k = 1, 2, 3. \tag{3.16}$$

(2) Consider again three particles P_1, P_2, P_3 subject to the **electric force** or **Coulomb force**. We want to write the system of equations that determine their motion.

The electric force between two point particles P, Q has the expression:

$$\mathbf{F}_P(P, Q) = -k\frac{q_P q_Q}{||P - Q||^3}(P - Q), \tag{3.17}$$

known as Coulomb law. Above, $k > 0$ is the electric constant, which depends on the units of measurement, and $q_P, q_Q \in \mathbb{R}$ are the *electric charges* of P and Q respectively.

The system of equations describing the motion for three particles is:

$$m_k \mathbf{a}_{P_k}(t) = \sum_{k \neq i=1}^{i=3} -k\frac{q_k q_i}{||P_k - P_i||^3}(P_k - P_i) \quad k = 1, 2, 3. \tag{3.18}$$

(3) Take three particles P_1, P_2, P_3 subject to the **gravitational force**. We want to write the system of equations that determine their motion.

   The gravitational force between two point particles P, Q of mass m_P, m_Q respectively has the form discovered by Newton

$$\mathbf{F}_P(P, Q) = -G \frac{m_P m_Q}{\|P - Q\|^3}(P - Q), \tag{3.19}$$

where $G > 0$ is the gravitational constant.

   Hence the system of equations describing the motion for three particles is:

$$m_k \mathbf{a}_{P_k}(t) = \sum_{k \neq i=1}^{i=3} -G \frac{m_k m_i}{\|P_k - P_i\|^3}(P_k - P_i) \quad k = 1, 2, 3. \tag{3.20}$$

It is interesting to note how in this formula the masses m_k play two different roles at the same time: that of the constants appearing in the second law of motion $F = ma$, which in that context is called *inertial mass*, and that of "gravitational charge", better known as *gravitational mass*,[2] analogous to the electric charge.

(4) With the exclusion of so-called *inertial forces* discussed below, concrete cases of forces *between two point particles* that depend (also) on the velocities are much more delicate to handle. These laws appear for instance in Electrodynamics. They have however proved to be problematic, and have eventually shown that Newton's formulation, as we have presented it, is inadequate. We will mention this issue in Sect. 3.4.2.

3.3 More General Dynamical Situations

The Newtonian scheme we explained in the previous sections cannot always be fully implemented in the form discussed. In particular, the fundamental problem of Dynamics must, sometimes, be set up differently. Let us address three cases of major interest.

---

[2] The coincidence of the values of the gravitational and inertial masses has been verified very accurately and has led Einstein to formulate what we call the *equivalence principle*. It is one of the conceptual pillars on which the theory of General Relativity is built.

3.3.1   Case 1: Prescribed Motion of a Subsystem and Time-Dependent Forces

The first interesting situation is a system of N interacting point particles $P_1, \ldots, P_N$, some of which, say $P_{k+1}, \ldots, P_N$, have a prescribed motion of some sort.[3] The remaining point particles $P_1, P_2, \ldots, P_k$ will have forces of the form, for $i = 1, \ldots, k$,

$$\mathbf{F}_i = \mathbf{F}_i(t, P_1, \ldots, P_k, \mathbf{v}_{P_1}, \ldots, \mathbf{v}_{P_k}) .$$

The explicit time dependency accounts for the prescribed motion of the other particles $P_{k+1}, \ldots, P_N$. The second law of motion, for the system of particles P_i, $i = 1, \ldots, k$, may be extended to the more general form

$$m_i \mathbf{a}_{P_i} = \mathbf{F}_i(t, P_1, \ldots, P_k, \mathbf{v}_{P_1}, \ldots, \mathbf{v}_{P_k}), \quad \text{for } i = 1, \ldots, k. \tag{3.21}$$

Mathematically, passing to the orthonormal coordinates of an inertial frame, the system of differential equations arising from system (3.21) is covered by the same theory of the case of Theorem 3.11. If the force functions are regular enough, system

$$\frac{d^2 \mathbf{x}_i(t)}{dt^2} = m_i^{-1} \mathbf{F}_i\left(t, \mathbf{x}_1(t), \ldots, \mathbf{x}_k(t), \frac{d\mathbf{x}_1(t)}{dt}, \ldots, \frac{d\mathbf{x}_k(t)}{dt}\right), \quad i = 1, \ldots, k, \tag{3.22}$$

admits a unique solution once the particles' positions and velocities are assigned at an instant. Note that we have a system of $3k$ scalar equations and there are exactly $3k$ unknown functions. This setup comprises situations where a point particle interacts with a more complex system, for example a fluid with given motion in a known frame (supposing the motion is not significantly altered by the presence of the point particle). In that case the dynamical action of the external ambient on the particle is still represented by a force that, in general, explicitly depends on time.

Examples 3.14

(1) The first example, completely trivial, is a point particle P of mass m subject to the force of a spring of zero rest length and elastic constant κ, whose one end, O, is constrained to stay still in an inertial frame $\mathscr{R}$. (From the practical point of view this is realised fixing the end to a body of mass $M \gg m$.) In this case

---

[3] This case is typical for instance when the point particles form an isolated system and one mass is enormously larger than the others. The approximation gets increasingly better the more this mass prevails over the rest, and this body moves in a straight line with constant velocity in an(y) inertial frame.

the equation determining the motion of P is just

$$m\mathbf{a}_P = -\kappa(P - O).$$ (3.23)

(2) Consider a point particle P of mass m immersed in a liquid at rest in the frame $\mathscr{R}'$. The force that acts on it is caused by the liquid's **viscous friction**, and is expressed, in certain velocity regimes, by the function

$$\mathbf{F}(t, P, \mathbf{v}_P|_\mathscr{R}) = -\gamma \mathbf{v}_P|_{\mathscr{R}'},$$ (3.24)

valid in the inertial frame $\mathscr{R}$, where $\gamma > 0$ is a coefficient describing the friction between particle and liquid. If $\mathscr{R}' \neq \mathscr{R}$ and $\mathscr{R}'$ is inertial, we can directly work in $\mathscr{R}'$ and the above expression simplifies, giving the equation of motion (assuming no other forces act on the particle)

$$m\mathbf{a}_P = -\gamma \mathbf{v}_P|_{\mathscr{R}'}.$$ (3.25)

This differential equation can be solved by standard methods once written in Cartesian coordinates of the frame $\mathscr{R}'$.

Let us assume now $\mathscr{R}'$ is *not* inertial, but that its motion is given with respect to the inertial frame $\mathscr{R}$. In this case we cannot reduce to working in $\mathscr{R}'$, since there the axioms of Dynamics do not hold (we will discuss in Sect. 3.3.3 what happens if we insist on reformulating Mechanics in non-inertial frames). Let us stay in $\mathscr{R}$ and write $\mathbf{v}_P|_{\mathscr{R}'}$ in function of $\mathbf{v}_P|_\mathscr{R}$. By Eq. (2.68) we have $\mathbf{v}_P|_{\mathscr{R}'} = \mathbf{v}_P|_\mathscr{R} + \boldsymbol{\omega}_\mathscr{R}|_{\mathscr{R}'} \wedge (P - O) + \mathbf{v}_O|_{\mathscr{R}'}$ where $O \in E_\mathscr{R}$ is a point at rest in $\mathscr{R}$ (typically the origin of the orthonormal coordinates of $\mathscr{R}$). The *functions of time* $\boldsymbol{\omega}_\mathscr{R}|_{\mathscr{R}'} = \boldsymbol{\omega}_\mathscr{R}|_{\mathscr{R}'}(t)$ and $\mathbf{v}_O|_{\mathscr{R}'} = \mathbf{v}_O|_{\mathscr{R}'}(t)$ are known because the relative motion of $\mathscr{R}$ and $\mathscr{R}'$ is given. All in all, the equations of motion for P (assuming on the particle there are no further acting forces) read

$$m\mathbf{a}_P = -\gamma \mathbf{v}_P|_\mathscr{R}(t) - \gamma \boldsymbol{\omega}_\mathscr{R}|_{\mathscr{R}'} \wedge (P - O) - \gamma \mathbf{v}_O|_{\mathscr{R}'}(t).$$ (3.26)

3.3.2 Case 2: Geometric Constraints and Constraint Forces

Another case, of great mathematical interest, of the situation where we cannot use in full the dynamical approach discussed above, is that in which point particles $P_1, P_2, \ldots, P_k$ must also obey *constraint equations* of geometrical/kinematical type. For example, we may demand that one or more particles subject to some forces must move along a curve or on a surface, or that certain conditions on the velocity of the particle (or particles) must hold, or finally that certain conditions of metric nature should be met, for instance the motion of two or more particles, subject to some forces, should preserve the particles' distances in time. In all of these situations we further suppose that there exists, due to the constraint, a force (in addition to the ones already present) acting on the point particle. Such a force, called **constraint reaction** or **constraint force** is denoted by the letter $\boldsymbol{\phi}$. One assumes that it satisfies, by hypothesis, the Action-reaction principle (third law of motion) with respect to the structure of the constraint and that it contributes to the generalised equation for the second law:[4]

$$m_i \mathbf{a}_{P_i} = \mathbf{F}_i(t, P_1, \ldots, P_k, \mathbf{v}_{P_1}, \ldots, \mathbf{v}_{P_k}) + \boldsymbol{\phi}_i, \quad \text{con } i = 1, \ldots, k. \quad (3.27)$$

The force $\boldsymbol{\phi}$ is not prescribed as a function of position and velocity of the particle subject to the constraint. *In principle, though, we assume that this is possible by carefully examining the constraint's physical structure.* In practice the force $\boldsymbol{\phi}$ is an *unknown* for the problem of motion, as is the law of motion itself. In order for the overall problem to be solvable, we must furnish the differential equation (more generally, the system of differential equations) arising from applying Newton's second law with some additional information. First of all the equation of the constraint itself. This might not be enough, and often one must add a **constitutive characterisation of the constraint** that fixes some relationship between the components of the constraint reaction. The use of these additional relations, when possible, serves the purpose of determining a system of differential equations, *not containing the unknown reactions $\boldsymbol{\phi}$*, and satisfying Existence and uniqueness theorems if initial conditions are provided. This system, if it exists, is called **system of pure equations of motion**. The reactions $\boldsymbol{\phi}$ are found afterwards, if possible, once the equation of motion has been found and using Newton's second and third laws.

Let us consider some examples, in the case of a *single* point particle, to clarify what we have said. In the exercises we will treat several point particles. The case of one or more point particles can be treated in complete generality introducing the *Lagrangian formulation* of Mechanics and the postulate of ideal constraints, which we will not treat here.

Examples 3.15

(1) Consider a particle constrained to a regular C^1 curve Γ, different from a straight segment and moving with an inertial frame $\mathscr{R}$, of equation $P = P(u)$. If $\mathbf{F} = \mathbf{F}(t, P, \mathbf{v}_P|_{\mathscr{R}})$ is the total force acting on P as above, excluding the constraint reaction $\boldsymbol{\phi}$, Newton's second law says:

$$m\mathbf{a}_P = \mathbf{F}(t, P, \mathbf{v}_P) + \boldsymbol{\phi}(t),$$

a relation capable of determining the motion even if we assign the equation of Γ as $P = P(u)$. It is convenient, given that the motion happens along Γ, to

---

[4] Some texts distinguish between *active forces*, those whose expression, called by us $\mathbf{F}$, is known, and *reactive forces*, $\boldsymbol{\phi}$, whose explicit form is unknown and are due to the constraints.

describe P using the arclength s parametrised by time: $s = s(t)$. From (2.18), in relationship to the *moving trihedron* of Γ introduced in Sect. 2.3.3, Newton's second law can be rewritten as

$$m\frac{d^2s}{dt^2}\,\mathbf{t}(s(t)) + \frac{m}{\rho(s(t))}\left(\frac{ds}{dt}\right)^2 \mathbf{n}(s(t)) = F^t(t,\, P(s(t)),\, \mathbf{v}_P)\mathbf{t}(s(t))$$

$$+ F^n(t,\, P(s(t)),\, \mathbf{v}_P)\mathbf{n}(s(t)) + F^b(t,\, P(s(t)),\, \mathbf{v}_P)\mathbf{b}(s(t))$$

$$+ \phi^t(t)\mathbf{t}(s(t)) + \phi^n(t)\mathbf{n}(s(t)) + \phi^b(t)\mathbf{b}(s(t))\ .$$

It is convenient to write the velocity too, which appears as argument, in terms of the arclength and the unit tangent vector as in (2.17): $\mathbf{v}(t) = \frac{ds}{dt}\,\mathbf{t}(s(t))$. In other words, Newton's law gives the system of three equations

$$m\frac{d^2s}{dt^2} = F^t\left(t,\, s(t),\, \frac{ds}{dt}\right) + \phi^t(t)\ , \tag{3.28}$$

$$\frac{m}{\rho(s(t))}\left(\frac{ds}{dt}\right)^2 = F^n\left(t,\, s(t),\, \frac{ds}{dt}\right) + \phi^n(t)\ , \tag{3.29}$$

$$0 = F^b\left(t,\, s(t),\, \frac{ds}{dt}\right) + \phi^b(t)\ . \tag{3.30}$$

The simplest constitutive characterisation of the constraint given by the curve Γ is that of a **smooth curve**. This costitutive characterisation is implemented by imposing that the tangent component of $\boldsymbol{\phi}$ to Γ is always zero. In such case the above equations simplify to

$$m\frac{d^2s}{dt^2} = F^t\left(t,\, s(t),\, \frac{ds}{dt}\right)\ , \tag{3.31}$$

$$\frac{m}{\rho(s(t))}\left(\frac{ds}{dt}\right)^2 = F^n\left(t,\, s(t),\, \frac{ds}{dt}\right) + \phi^n(t)\ , \tag{3.32}$$

$$0 = F^b\left(t,\, s(t),\, \frac{ds}{dt}\right) + \phi^b(t)\ . \tag{3.33}$$

Equation (3.31) does not contain the unknown component functions of the constraint force, and it belongs to the class of differential equations covered by Theorem 3.11. Therefore, any assigned initial position and velocity (tangent to Γ!) translate into initial conditions for $s = s(t)$, which is uniquely determined from (3.31). Equation (3.31) is thus a **pure equation of motion** for the system considered.

Once the law of motion $s = s(t)$ is known, substituting it in the left-hand side of (3.32) and (3.33) allows to find the unknown functions ϕ^n and ϕ^b.

(2) Consider once more a particle constrained to a C^1 curve Γ, different from a segment, moving with an inertial frame $\mathscr{R}$ and of equation $P = P(u)$, subject to a given force $\mathbf{F} = \mathbf{F}(t, P, \mathbf{v}_P|_{\mathscr{R}})$, so that we arrive at Eq. (3.30) again. There is a different and more physically realistic constitutive relation for the constraint Γ, namely that it be a **rough curve**. This constitutive characterisation is expressed by asking that the tangential component ϕ^t of the constraint force $\boldsymbol{\phi}$ acting on P because of Γ is related to the normal component $\phi^n \mathbf{n} + \phi^b \mathbf{b}$ by

$$|\phi^t(t)| \leq \mu_s \sqrt{(\phi^n(t))^2 + (\phi^b(t))^2} \,, \tag{3.34}$$

so long as the particle P is still with respect to Γ (i.e. at rest in $\mathscr{R}$). The coefficient $\mu_s > 0$ is called **coefficient of static friction**. When instead the particle is moving we have

$$\phi^t(s(t)) = -\frac{\frac{ds}{dt}}{\left|\frac{ds}{dt}\right|} \mu_d \sqrt{(\phi^n(t))^2 + (\phi^b(t))^2} \,, \tag{3.35}$$

where the coefficient $\mu_d > 0$ is called **coefficient of dynamic friction**, and typically $\mu_d < \mu_s$. The coefficient $\frac{ds}{dt}/\left|\frac{ds}{dt}\right|$ multiplied by $\mathbf{t}$ determines the particle's unit velocity vector, so during the motion $\phi^t \mathbf{t}$ is always collinear with the velocity but with opposite orientation. In our case, assuming the particle initially in motion, the pure equation of motion is found as follows. From (3.29) and (3.30) we obtain

$$\frac{m}{\rho(s(t))} \left(\frac{ds}{dt}\right)^2 - F^n\left(t, s(t), \frac{ds}{dt}\right) = \phi^n(t) \,, \tag{3.36}$$

$$-F^b\left(t, s(t), \frac{ds}{dt}\right) = \phi^b(t) \,. \tag{3.37}$$

From these, by (3.35), we have:

$$\phi^t(s(t)) = -\frac{\mu_d \frac{ds}{dt}}{\left|\frac{ds}{dt}\right|}$$

$$\sqrt{\left[\frac{m}{\rho(s(t))} \left(\frac{ds}{dt}\right)^2 - F^n\left(t, s, \frac{ds}{dt}\right)\right]^2 + F^b\left(t, s(t), \frac{ds}{dt}\right)^2} \,.$$

Inserting the expression of $\phi(s(t))$ in (3.28) produces an equation without unknown constraint reactions, but only containing the unknown function $s =$

$s(t)$, which is therefore the pure equation of motion,

$$\frac{d^2 s}{dt^2} = m^{-1} F^t \left(t, s(t), \frac{ds}{dt} \right) - \frac{\mu_d \frac{ds}{dt}}{m \left| \frac{ds}{dt} \right|}$$

$$\sqrt{\left[\frac{m}{\rho(s(t))} \left(\frac{ds}{dt} \right)^2 - F^n \left(t, s, \frac{ds}{dt} \right) \right]^2 + F^b \left(t, s(t), \frac{ds}{dt} \right)^2}.$$

(3.38)

This is still covered by Theorem 3.11 as long as the active force, and the curve Γ as well, are sufficiently regular, and if the first derivative of the solution $s = s(t)$ does not vanish (notice that by virtue of this, the coefficient $\frac{ds}{dt} / \left| \frac{ds}{dt} \right|$ is constant and equals ± 1 during the motion). Once we have the function $s = s(t)$ for given initial conditions, we may recover the constraint reaction as a function of time, $\phi = \phi(t)$, directly from Eqs. (3.28)–(3.29) by substituting in $s(t)$ the solution to the pure equation of motion. If that solution at $t = t_1$ satisfies $ds/dt|_{t=t_1} = 0$, the particle stops at $s_1 = s(t_1)$, and from that moment onwards its state of motion must be studied using the constitutive relation (3.34), together with Eqs. (3.28)–(3.29) now simplified to

$$- F^t (t, s_1, 0) = \phi^t (t),$$

(3.39)

$$- F^n (t, s_1, 0) = \phi^n (t),$$

(3.40)

$$- F^b (t, s_1, 0) = \phi^b (t).$$

(3.41)

Until the moment the constraint

$$|F^t (t, s_1, 0)| < \mu_s \sqrt{F^n (t, s_1, 0)^2 + F^b (t, s_1, 0)^2}$$

(3.42)

holds, as t varies, the particle is at rest. If for $t = t_2$ the two sides above become equal, the particle starts to move and the motion must be studied using (3.38) with initial conditions $s(t_2) = s_1$ and $ds/dt|_{t=t_2} = 0$. The singular coefficient $\frac{ds}{dt} / \left| \frac{ds}{dt} \right|$ should be replaced by the sign of $F^t (t_2, s_1, 0)$, because we assume that the motion starts in the direction of the force that has prevailed over the static friction.

(3) Let us pass to consider a point particle P of mass m moving on a surface S, which we assume is spherical of radius R and centre O (though what we will say is general), and at rest in the inertial frame $\mathscr{R}$. Let $\mathbf{F} = \mathbf{F}(t, P, \mathbf{v}_P|_{\mathscr{R}})$ be the total force acting on P, a known expression. Excluding the constraint ϕ, Newton's second law gives:

$$m \mathbf{a}_P = \mathbf{F}(t, P, \mathbf{v}_P) + \phi(t).$$

We may parametrise the point on the sphere by spherical coordinates θ and φ and express the latter in function of time $\theta = \theta(t)$, $\varphi = \varphi(t)$. We seek the pure equation of motion in these variables after imposing the constraint's constitutive characterisation. Once again, it is convenient to decompose this equation along the trihedron of vectors associated with the constraint, here given by (2.40). Newton's law, where implicitly $\theta = \theta(t)$, $\varphi = \varphi(t)$ and for conciseness time derivatives are indicated with dots, reads:

$$m(R\ddot{\theta} - R\dot{\varphi}^2 \sin\theta \cos\theta)\, \mathbf{e}_\theta + m(R\ddot{\varphi} \sin\theta + 2R\dot{\varphi}\dot{\theta} \cos\theta)\, \mathbf{e}_\varphi$$

$$- m(R\dot{\theta}^2 + R\dot{\varphi}^2 \sin^2\theta)\mathbf{n}$$

$$= F^n(t, P((\theta, \varphi)), \mathbf{v}_P(\theta, \varphi))\, \mathbf{e}_r + F^\theta(t, P((\theta, \varphi)), \mathbf{v}_P(\theta, \varphi))\, \mathbf{e}_\theta$$

$$+ F^\varphi(t, P(\theta, \varphi), \mathbf{v}_P(\theta, \varphi))\mathbf{e}_\varphi$$

$$+ \phi^n(t)\, \mathbf{e}_r + \phi^\theta(t)\mathbf{n}(s(t))\, \mathbf{e}_\theta + \phi^\varphi(t)\mathbf{e}_\varphi$$

Separating the equations in the three components and using (2.39)

$$\mathbf{v}_P = R\dot{\theta}\, \mathbf{e}_\theta + R\dot{\varphi} \sin\theta\, \mathbf{e}_\varphi \, ,$$

we find the following ODE system

$$m\left(R\frac{d^2\theta}{dt^2} - R\left(\frac{d\varphi}{dt}\right)^2 \sin\theta(t) \cos\theta(t) \right)$$

$$= F^\theta\left(t, \theta(t), \varphi(t), \frac{d\theta}{dt}, \frac{d\varphi}{dt} \right) + \phi^\theta(t) \,, \tag{3.43}$$

$$m\left(R\frac{d^2\varphi}{dt^2} \sin\theta(t) + 2R\frac{d\theta}{dt}\frac{d\varphi}{dt} \cos\theta(t) \right)$$

$$= F^\varphi\left(t, \theta(t), \varphi(t), \frac{d\theta}{dt}, \frac{d\varphi}{dt} \right) + \phi^\varphi(t) \,, \tag{3.44}$$

$$- m\left(R\left(\frac{d\theta}{dt}\right)^2 + R\left(\frac{d\varphi}{dt}\right)^2 \sin^2\theta(t) \right)$$

$$= F^n\left(t, \theta(t), \varphi(t), \frac{d\theta}{dt}, \frac{d\varphi}{dt} \right) + \phi^n(t) \,. \tag{3.45}$$

The constitutive characterisation of a **smooth surface** requires that the components of ϕ tangent to the surface are null. Inserting that in the above system produces the **system of pure equations of motion**

$$m \left(R \frac{d^2\theta}{dt^2} - R \left(\frac{d\varphi}{dt} \right)^2 \sin\theta(t) \, \cos\theta(t) \right)$$
$$= F^\theta \left(t, \theta(t), \varphi(t), \frac{d\theta}{dt}, \frac{d\varphi}{dt} \right), \qquad (3.46)$$

$$m \left(R \frac{d^2\varphi}{dt^2} \sin\theta(t) + 2R \frac{d\theta}{dt} \frac{d\varphi}{dt} \cos\theta(t) \right)$$
$$= F^\varphi \left(t, \theta(t), \varphi(t), \frac{d\theta}{dt}, \frac{d\varphi}{dt} \right), \qquad (3.47)$$

which only contains the functions $\theta = \theta(t)$, $\varphi = \varphi(t)$ giving the particle's motion, and not the unknown reaction force. In fact the above system, under suitable regularity assumptions, admits a unique solution for given initial conditions $\theta(t_0) = \theta_0$ and $\varphi(t_0) = \varphi_0$, since it is covered by Theorem[5] 3.11. We obtain the reaction force, reduced to the sole normal component to the surface, in function of time when we have determined the motion, directly from (3.45). As a final comment, we should stress that the coordinates θ, φ *do not cover the sphere entirely*, so the complete study of the motion of the particle on the sphere requires more than one chart.

3.3.3 Case 3: Dynamics in Non-inertial Frames and the Notion of Inertial Force

What happens if we attempt to reformulate the laws of motion in a non-inertial frame $\mathscr{R}'$ of which we know the motion with respect to an inertial frame $\mathscr{R}$? Is it possible to set up the scheme based on Newton's second law so to obtain an ODE system warranting, under certain hypotheses, determinism?

For simplicity we treat the case of one point particle P. If $\mathscr{R}$ is an inertial frame, as usual the second law of motion (remembering the various extensions described earlier) reads:

$$m_P \mathbf{a}_P|_{\mathscr{R}} = \mathbf{F}(t, P, \mathbf{v}_P|_{\mathscr{R}}) + \boldsymbol{\phi}, \qquad (3.48)$$

We have written $|_{\mathscr{R}}$ explicitly to emphasise that the acceleration is evaluated with respect to the (inertial) frame $\mathscr{R}$. Now we can express the velocity and acceleration

---

[5] If $\sin\theta = 0$ the system cannot be put in normal form, yet $\sin\theta = 0$ defines the points $\theta = 0, \pi$ where the spherical coordinates are not defined: at all other points the system can be written in normal form.

of P with respect to $\mathscr{R}$ in terms of the similar quantities with respect to $\mathscr{R}'$ using Eqs. (2.68)–(2.69). If $O' \in E_{\mathscr{R}'}$ is at rest with $\mathscr{R}'$ we have:

$$m_P \mathbf{a}_P|_{\mathscr{R}'} + m_P \left(\mathbf{a}_{O'}|_{\mathscr{R}} + \boldsymbol{\omega}_{\mathscr{R}'}|_{\mathscr{R}} \wedge (\boldsymbol{\omega}_{\mathscr{R}'}|_{\mathscr{R}} \wedge (P - O')) + \dot{\boldsymbol{\omega}}_{\mathscr{R}'}|_{\mathscr{R}} \wedge (P - O')\right.$$

$$\left. + 2\boldsymbol{\omega}_{\mathscr{R}'}|_{\mathscr{R}} \wedge \mathbf{v}_P|_{\mathscr{R}'}\right)$$

$$= \mathbf{F}\left(t, P, \mathbf{v}_P|_{\mathscr{R}'} + \boldsymbol{\omega}_{\mathscr{R}'}|_{\mathscr{R}} \wedge (P - O') + \mathbf{v}_{O'}|_{\mathscr{R}}\right) + \boldsymbol{\phi}$$

It is convenient to rewrite the above as:

$$m_P \mathbf{a}_P|_{\mathscr{R}'} = \mathbf{F}(t, P, \mathbf{v}_P|_{\mathscr{R}'}) + \boldsymbol{\phi}(t) + \mathbf{F}_{\mathscr{R}'}(t, P, \mathbf{v}_P|_{\mathscr{R}'}), \qquad (3.49)$$

having defined

$$\mathbf{F}_{\mathscr{R}'}(t, P, \mathbf{v}_P|_{\mathscr{R}'}) := -m_P \mathbf{a}_{O'}|_{\mathscr{R}}(t) - m_P \boldsymbol{\omega}_{\mathscr{R}'}|_{\mathscr{R}}(t) \wedge (\boldsymbol{\omega}_{\mathscr{R}'}|_{\mathscr{R}}(t) \wedge (P - O'))$$

$$- 2m_P \boldsymbol{\omega}_{\mathscr{R}'}|_{\mathscr{R}}(t) \wedge \mathbf{v}_P|_{\mathscr{R}'}$$

$$- m_P \dot{\boldsymbol{\omega}}_{\mathscr{R}'}|_{\mathscr{R}} \wedge (P - O'). \qquad (3.50)$$

With improper notation, $\mathbf{F}(t, P, \mathbf{v}_P|_{\mathscr{R}'})$ actually stands for the vector-valued map

$$\mathbf{F}\left(t, P, \mathbf{v}_P|_{\mathscr{R}'} + \boldsymbol{\omega}_{\mathscr{R}'}|_{\mathscr{R}}(t) \wedge (P - O'(t)) + \mathbf{v}_{O'}|_{\mathscr{R}}(t)\right),$$

where *the functions of time:* $\mathbf{v}_{O'}|_{\mathscr{R}} = \mathbf{v}_{O'}|_{\mathscr{R}}(t)$ *and* $\boldsymbol{\omega}_{\mathscr{R}'}|_{\mathscr{R}} = \boldsymbol{\omega}_{\mathscr{R}'}|_{\mathscr{R}}(t)$ *are known, since the motion of* $\mathscr{R}'$ *with respect to* $\mathscr{R}$ *is given.*

From all of that we deduce the following: if we wish to recover the formulation of dynamics in terms of "$F = ma$" in a non-inertial frame $\mathscr{R}'$, compatibly with the mechanics of inertial frames, we are forced to add new forces. The latter, represented by the right-hand side of (3.50), are functions of the particle's position and velocity in the *non-inertial* frame. This warrants the determinism underpinning Theorem 3.11 as long as the functions are regular enough. These functions depend on the motion, which must be known, of the non-inertial frame $\mathscr{R}'$ with respect to an inertial frame ($\mathscr{R}$ in our case). The forces $\mathbf{F}_{\mathscr{R}'}$ are called **fictitious forces** or **inertial forces** acting on P; by contrast, those introduced in the previous sections when working in inertial frames (constraint reactions included) are called **real forces**. Observe that

(a) inertial forces acting on P *do not obey the Action-reaction principle*; it makes no sense to say they are caused by some other point particle P' (another physical system);
(b) inertial forces have values that *depend on the frame*, as opposed to real forces that have the same value in any frame, *inertial and non-inertial*.

Within Classical Mechanics inertial forces do not represent interactions, but just a mathematical artifice. In the theory of General Relativity this point of view will be completely overhauled, as we will show that inertial forces, or rather the corresponding mathematical objects in the new formulation of Dynamics, describe an interaction of the same nature as the gravitational interaction.

There is a finer classification of inertial forces.

$$\mathbf{F}^{(Cent)}(t, P) := -m_P \boldsymbol{\omega}_{\mathscr{R}'}|_{\mathscr{R}}(t) \wedge (\boldsymbol{\omega}_{\mathscr{R}'}|_{\mathscr{R}}(t) \wedge (P - O'))$$

is called **centrifugal force**.

$$\mathbf{F}^{(Coriolis)}(t, \mathbf{v}_P|_{\mathscr{R}'}) := -2m_P \boldsymbol{\omega}_{\mathscr{R}'}|_{\mathscr{R}}(t) \wedge \mathbf{v}_P|_{\mathscr{R}'}$$

is called **Coriolis force**. The latter does not act if the point particle P has zero velocity in $\mathscr{R}'$. The former, instead, thinking $\mathbf{F}^{(Cent)}$ applied to P and $\boldsymbol{\omega}_{\mathscr{R}'}|_{\mathscr{R}}$ outgoing from O', lies on the plane of $P - O'$ and $\boldsymbol{\omega}_{\mathscr{R}'}|_{\mathscr{R}}$, is perpendicular to $\boldsymbol{\omega}_{\mathscr{R}'}|_{\mathscr{R}}$ and oriented outwards from the axis $\boldsymbol{\omega}_{\mathscr{R}'}|_{\mathscr{R}}$. Its norm equals $m||\boldsymbol{\omega}_{\mathscr{R}'}|_{\mathscr{R}}||^2 R$ where $R > 0$ is the *arm* of $P - O'$ with respect to $\boldsymbol{\omega}_{\mathscr{R}'}|_{\mathscr{R}}$ (i.e. $||P - O'|| \, | \sin \alpha|$, and α is the angle between $P - O'$ and $\boldsymbol{\omega}_{\mathscr{R}'}|_{\mathscr{R}}$).

The term

$$- m_P \dot{\boldsymbol{\omega}}_{\mathscr{R}'}|_{\mathscr{R}} \wedge (P - O')$$

is sometimes called **Euler force**.

Examples 3.16

(1) The simplest example is certainly the following. Consider a point particle P of mass m not moving in the inertial frame $\mathscr{R}$. Take another frame $\mathscr{R}'$, non-inertial, whose motion with respect to $\mathscr{R}$ is a pure uniform rotation about the z-axis, with angular speed $\omega > 0$. In other words, with respect to orthogonal coordinates x, y, z and x', y', z' on $\mathscr{R}$ and $\mathscr{R}'$ respectively, we have the transformations $z' = z$, $x = x' \cos(\omega t) - y' \sin(\omega t)$ and $y = x' \sin(\omega t) + y' \cos(\omega t)$. Hence $\boldsymbol{\omega}_{\mathscr{R}'}|_{\mathscr{R}} = \omega \, \mathbf{e}_z$. In $\mathscr{R}'$ the motion of P is described as a uniform rotation around the point $O \equiv O'$ of coordinates $(0, 0, 0)$ in both frames. If $(x_P, y_P, 0)$ are the coordinates of P in $\mathscr{R}'$, then P in $\mathscr{R}$ has law of motion $x'_P(t) = x \cos(\omega t) - y \sin(\omega t)$, $y'_P(t) = x \sin(\omega t) + y \cos(\omega t)$ (and $z'_P(t) = 0$). In particular, then, $\mathbf{v}_{\mathscr{R}'}(t) = \boldsymbol{\omega}_{\mathscr{R}}|_{\mathscr{R}'} \wedge (P(t) - O)$.

From elementary Physics we know that, dynamically, a point particle P rotates uniformly when the total force acting on P is *centripetal*, i.e. it has the same direction and orientation[6] of $-(P(t) - O)$, and modulus $\omega^2 R$, where R is the distance of P to the axis of rotation. The fictitious force $\mathbf{F}_c(t, P)$ is on the contrary *centrifugal*, so by itself it cannot generate the necessary centripetal force. In realty, the direct computation of the entire inertial force appearing in

---

[6] Beware that this is true in the case at hand because $P - O$ is perpendicular to $\boldsymbol{\omega}_{\mathscr{R}'}|_{\mathscr{R}}$.

$\mathscr{R}'$ under (3.50) produces immediately

$$\mathbf{F}_{\mathscr{R}'}(t, P, \mathbf{v}_P|_{\mathscr{R}'}) := -m\mathbf{a}_O|_{\mathscr{R}}(t) - m\boldsymbol{\omega}_{\mathscr{R}'}|_{\mathscr{R}} \wedge (\boldsymbol{\omega}_{\mathscr{R}'}|_{\mathscr{R}} \wedge (P - O))$$

$$- 2m\boldsymbol{\omega}_{\mathscr{R}'}|_{\mathscr{R}} \wedge \mathbf{v}_P|_{\mathscr{R}'}$$

$$- m\dot{\boldsymbol{\omega}}_{\mathscr{R}'}|_{\mathscr{R}} \wedge (P - O)$$

(recall $O \equiv O'$ in our case) where the first and last summands on the right are zero by assumption, while the centrifugal and Coriolis force are respectively equal to (note $\omega \perp (P - O)$ here):

$$\mathbf{F}_c(t, P) = -m\boldsymbol{\omega}_{\mathscr{R}'}|_{\mathscr{R}} \wedge (\boldsymbol{\omega}_{\mathscr{R}'}|_{\mathscr{R}} \wedge (P(t) - O))$$

$$= -m\omega \, \mathbf{e}_z \wedge (\omega \, \mathbf{e}_z \wedge (P(t) - O)) = m\omega^2(P(t) - O),$$

$$\mathbf{F}^{(Coriolis)}(t, \mathbf{v}_P|_{\mathscr{R}'}) = -2m\boldsymbol{\omega}_{\mathscr{R}'}|_{\mathscr{R}} \wedge \mathbf{v}_P|_{\mathscr{R}'}$$

$$= -2m\boldsymbol{\omega}_{\mathscr{R}'}|_{\mathscr{R}} \wedge (\boldsymbol{\omega}_{\mathscr{R}}|_{\mathscr{R}'} \wedge (P(t) - O))$$

$$= -2m\omega^2(P(t) - O).$$

The sum of the two forces produces precisely the necessary centripetal force $-m\omega^2(P(t) - O)$.

(2) Consider a point particle P of mass m constrained to the smooth circle Γ of radius R and centre the origin O', lying on the plane $y' = 0$ in a system of rest orthonormal coordinates x', y', x' on the frame $\mathscr{R}'$. The latter is not inertial, and $\boldsymbol{\omega}_{\mathscr{R}'}|_{\mathscr{R}} = \omega\mathbf{e}_{z'}$, with $\omega > 0$ constant, where $\mathscr{R}$ is an inertial frame of orthonormal coordinates x, y, z with origin $O \equiv O'$ and z-axis always coinciding with the z'-axis of $\mathscr{R}'$. We assume P is subject to the weight $-mg\mathbf{e}_{z'}$, besides the constraint reaction $\boldsymbol{\phi}$ due to Γ. We want to find the pure equation of motion of the particle in $\mathscr{R}'$. It is convenient to use the Frenet trihedron on Γ. We parametrise the circle with the angle φ between $P - O$ and $\mathbf{e}_z = \mathbf{e}_{z'}$, oriented positively with respect to $\mathbf{e}_{y'}$. With that choice for φ, define polar coordinates φ, r on the $z'x'$-plane. The unit tangent vector $\mathbf{t}$ coincides with $\mathbf{e}_\varphi$, the normal vector $\mathbf{n}$ coincides with $-\mathbf{e}_r$ and the binormal coincides with $\mathbf{e}_{y'}$. The curvature radius is just $\rho = R$, constant. A straightforward calculation gives the arclength function $s = s(\varphi)$ as $s = R\varphi$, having chosen as origin the highest point where Γ intercepts the z'-axis. The pure equations of motion are obtained projecting Newton's second law along the unit tangent to Γ, after expressing velocity and acceleration in terms of the arclength. Hence Newton's second law, in the non-inertial frame $\mathscr{R}'$, reads

$$m\mathbf{a}_P|_{\mathscr{R}'} = -mg\mathbf{e}_{z'} + \boldsymbol{\phi} + \mathbf{F}_{\mathscr{R}'},$$

and passing to the arclength coordinate as in (1) Examples 3.15,

$$m\frac{d^2s}{dt^2}\,\mathbf{e}_\varphi(s(t)) - \frac{m}{R}\left(\frac{ds}{dt}\right)^2 \mathbf{e}_r(s(t)) = -mg\mathbf{e}_{z'} + \boldsymbol{\phi}(t) + \mathbf{F}_{\mathscr{R}'}(t, P(t), \mathbf{v}_{\mathscr{R}'}(t)).$$

The pure equation of motion is found by taking the inner product of either side with $\mathbf{e}_\varphi = -\sin\varphi\mathbf{e}_{z'} + \cos\varphi\mathbf{e}_{x'}$. Thus, keeping in account that the reaction $\boldsymbol{\phi}$ is normal to Γ and that $\varphi = s/R$, we have

$$m\frac{d^2s}{dt^2} = mg\sin\left(\frac{s(t)}{R}\right) + [-m\boldsymbol{\omega}_{\mathscr{R}'}|_{\mathscr{R}} \wedge (\boldsymbol{\omega}_{\mathscr{R}'}|_{\mathscr{R}} \wedge (P - O))$$

$$-2m\boldsymbol{\omega}_{\mathscr{R}'}|_{\mathscr{R}} \wedge \mathbf{v}_P|_{\mathscr{R}'}] \cdot \mathbf{e}_\varphi , \qquad (3.51)$$

where we have ignored the first and last summand in the right-hand side of (3.50), both being zero under our assumptions. Moreover:

$$P(t) - O = R\left(\sin\left(\frac{s(t)}{R}\right)\mathbf{e}_{x'} + \cos\left(\frac{s(t)}{R}\right)\mathbf{e}_{z'}\right)$$

from which

$$\mathbf{v}_P|_{\mathscr{R}'}(t) = \frac{ds}{dt}\left(\cos\left(\frac{s(t)}{R}\right)\mathbf{e}_{x'} - \sin\left(\frac{s(t)}{R}\right)\mathbf{e}_{z'}\right).$$

Since $\boldsymbol{\omega}_{\mathscr{R}'}|_{\mathscr{R}} = \omega\mathbf{e}_{z'}$, the inner product on the right in (3.51), after a few passages (in particular the last summand in the inner product in (3.51) has zero contribution), equals

$$m\omega^2 R \sin\left(\frac{s(t)}{R}\right)\cos\left(\frac{s(t)}{R}\right).$$

Finally, the pure equation of motion, passing to the variable $\varphi := s/R$, is (Fig. 3.1):

$$mR\frac{d^2\varphi}{dt^2} - m\omega^2 R\,\sin\varphi\,\cos\varphi - mg\sin\varphi = 0 . \qquad (3.52)$$

Exercises 3.17

(1) Consider two point particles P and Q, both of mass m, moving along the smooth curve Γ of equation $x = \cos u$, $y = \sin u$, $z = u$ with $u \in \mathbb{R}$, where x, y, z are orthonormal coordinates of an inertial frame $\mathscr{R}$. Suppose the particles are subject to, apart from the reaction due to the curve Γ, the weight $-mg\mathbf{e}_z$

Fig. 3.1 Exercise 3.17.1

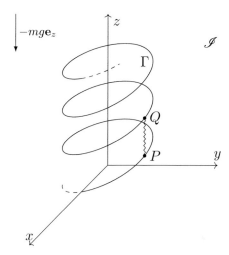

and that they are joint together by an ideal spring of zero rest length and elastic constant κ. Write the system of pure equations of motion of the two particles.

(2) Referring to Example 3.16.2, prove that the quantity:

$$E = \frac{1}{2} m \left(\frac{ds}{dt} \right)^2 + \frac{m\omega^2 R^2}{4} \cos \left(\frac{2s(t)}{R} \right) - mgR \cos \left(\frac{s(t)}{R} \right)$$

is constant during every motion of the system (the constant value depends on the solution considered). What is the physical meaning of E?

3.4   Comments on the General Formulation of Newtonian Dynamics

3.4.1   Galilean Invariance

The laws of Dynamics C2-C4, in the way we have formulated them, have the remarkable property of having the same form when written in the rest orthonormal coordinates of an inertial frame, as the inertial system varies. The proof is immediate considering the various laws separately. Such invariance property extends to the functions expressing the forces, and has a profound physical meaning. Let us see how this principle is enforced and formalised. Let us go back to two point particles, though what follows can trivially be generalised to several point particles using the superposition principle of forces. Now we interpret the action of the Galilean group in *active* sense. Said equivalently, considering the generic Galilean transformation:

$$\begin{cases} t' = t + c \,, \\ x'^i = c^i + t v^i + \sum_{j=1}^{3} R^i{}_j x^j \,, & i = 1, 2, 3, \end{cases} \tag{3.53}$$

we are using a *unique* system of rest right-handed orthogonal coordinates on a *given* inertial frame $\mathscr{R}$. Now the unprimed coordinates refer to an event p, while primed ones (in the same coordinates) refer to an event p' (different from the first) obtained after transforming p. The Galilean group contains a number of subgroups, listed below.[7]

(1) The group of **temporal displacements**, containing the transformations:

$$\begin{cases} t' = t + c \,, \\ x'^i = x^i \,, & i = 1, 2, 3 \end{cases} \tag{3.54}$$

    with $c \in \mathbb{R}$.

(2) the group of **spatial displacements**, corresponding to the transformations

$$\begin{cases} t' = t \,, \\ x'^i = c^i + x^i \,, & i = 1, 2, 3 \end{cases} \tag{3.55}$$

for any triple of numbers $(c^1, c^2, c^3) \in \mathbb{R}^3$. This group is nothing but the group of standard displacements $V_{\mathscr{R}}$ of the Euclidean rest space $E_{\mathscr{R}}$, since (c^1, c^2, c^3) is the representation in components of a displacement.

(3) The group of **rotations**, corresponding to transformations

$$\begin{cases} t' = t \,, \\ x'^i = \sum_{j=1}^{3} R^i{}_j x^j \,, & i = 1, 2, 3 \end{cases} \tag{3.56}$$

    with $R \in SO(3)$.

---

[7] Recall that a subset S in a group G is called a **subgroup** of G if it is closed under the group's composition law and inversion and if it contains the neutral element of G. In that case, clearly, S itself is a group for the composition law of G. Checking that the following sets of transformations are subgroups in the Galilean group is straightforward.

(4) The group of **pure Galilean transformations**, corresponding to transformations

$$\begin{cases} t' = t \,, \\ x'^i = tv^i + x^i \,, \quad i = 1, 2, 3 \end{cases} \tag{3.57}$$

with $(v^1, v^2, v^3) \in \mathbb{R}^3$.

Let us then pass to the conditions associated with each Galilean subgroup when we describe the forces between *two point particles isolated from all the others*.

(a) When assigning the forces' general form we assumed, right from the start, they are explicitly independent of absolute time t. This is equivalent to the forces being invariant under temporal displacements: for any $\Delta t \in \mathbb{R}$,

$$\mathbf{F}\left(t + \Delta t, Q, Q', \mathbf{v}_Q|_{\mathscr{R}}, \mathbf{v}_{Q'}|_{\mathscr{R}}\right) = \mathbf{F}\left(t, Q, Q', \mathbf{v}_Q|_{\mathscr{R}}, \mathbf{v}_{Q'}|_{\mathscr{R}}\right) \,.$$

This invariance request is called postulate of **temporal homogeneity**.

(b) A second postulate is **spatial homogeneity**, which asks that the forces are invariant under the action of the group of spatial displacements. In other words, for any vector $\mathbf{u}$:

$$\mathbf{F}\left(t, Q + \mathbf{u}, Q' + \mathbf{u}, \mathbf{v}_Q|_{\mathscr{R}}, \mathbf{v}_{Q'}|_{\mathscr{R}}\right) = \mathbf{F}\left(t, Q, Q', \mathbf{v}_Q|_{\mathscr{R}}, \mathbf{v}_{Q'}|_{\mathscr{R}}\right) \,.$$

Obviously this should imply that the forces depend *only on the difference vectors* $Q(t) - Q'(t)$, i.e. on the point particles' relative positions at any given time.

(c) A third postulate is **spatial isotropy**, whereby if $g_{O,R}$ is an isometry associated with the orthogonal matrix R and the point O, that rotates the points of Σ_t around $O \in \Sigma_t$ (i.e. $g_{O,R}(P) := O + R(P - O)$), then:[8]

$$\mathbf{F}\left(t, g_{O,R}(Q), g_{O,R}(Q'), R\mathbf{v}_Q|_{\mathscr{R}}, R\mathbf{v}_{Q'}|_{\mathscr{R}}\right) = (R\mathbf{F})\left(t, Q, Q', \mathbf{v}_Q|_{\mathscr{R}}, \mathbf{v}_{Q'}|_{\mathscr{R}}\right).$$

This request corresponds to a sort of "invariance" of the forces under the group of rotations. However, by the above identity, the forces are not *invariant* under rotations as they were under displacements, because the rotation R in the right-hand side acts on the force's value. The previous relationship is called **equivariance** of the forces under the group of rotations. The principle of spatial isotropy may not hold in this elementary version if we assume a more complex notion of point particle, where the particle comes, in its rest system, with some vector bearing physical meaning (electric momentum, magnetic momentum, *spin* etc.)

---

[8] Recall that $\mathbf{F}$ is a vector, so the rotation action affects its direction too. The equation simply says that "rotating the arguments of the function $\mathbf{F}$ is the same as rotating the vector $\mathbf{F}$".

(d) The third postulate concerns **pure Galilean invariance** and posits that the forces are invariant under pure Galilean transformations. For any vector $\mathbf{V}$ we must have:

$$\mathbf{F}\left(t,\, Q + t\mathbf{V},\, Q' + t\mathbf{V},\, \mathbf{v}_Q|_{\mathscr{R}} + \mathbf{V},\, \mathbf{v}_{Q'}|_{\mathscr{R}} + \mathbf{V}\right) = \mathbf{F}\left(t,\, Q,\, Q',\, \mathbf{v}_Q|_{\mathscr{R}},\, \mathbf{v}_{Q'}|_{\mathscr{R}}\right).$$

This principle, together with spatial homogeneity, prescribes that the forces' expressions only contain the *difference of the velocities* of the point particles with respect to $\mathscr{R}$.

Since every element of the Galilean group is built multiplying elements in the aforementioned 4 subgroups, the forces between pairs of point particles (isolated from the rest of the universe) are invariant under the action of *every* element of the Galilean group. Invariance under the Galilean group (henceforth including in the term "invariance" also *equivariance* under rotations, for historical reasons) is considered, in Classical Mechanics, a *general principle* that every law formulated in Mechanics must comply with (not only those prescribing the forces):

C5. Principle of Galilean Relativity
All the laws of Classical Mechanics, in particular those prescribing forces, are invariant under the active action of the Galilean group.

Remarks 3.18

(1) The principle of Galilean relativity has a cardinal importance for the solution of the equations of motion of a system of point particles. Consider an experiment studying n point particles in the rest orthonormal coordinates of an inertial frame $\mathscr{R}$. Fix initial conditions for the particles by assigning the positions and velocities at a generic time t_0 and consider the solution of Newton's equations found with the superposition principle. By Galilean invariance and the fact that Newton's equations admit a unique solution for given initial conditions, the following result follows easily (its proof is left as an exercise). If we let a transformation g of the Galilean group act on a solution to the problem of motion (hence, the same transformation on every point and at any instant), the time evolution of the n particles thus found is just the solution of the problem of motion itself, but with initial conditions transformed by g.

(2) There exists a "dual" formulation of the Principle of Galilean invariance, based on the fact that every Galilean transformation can be interpreted passively: that is, as transformation rule between the coordinates of different inertial frames. It is easy to convince ourselves that from this point of view the above principle can be reformulated in the following way (writing $\mathfrak{I}$ for the class of inertial frames).

C5.' Principle of Galilean Relativity   *All the laws of Classical Mechanics have the same form in any frame's rest orthonormal coordinates. In particular, suppose that in rest orthonormal coordinates for $\mathscr{R} \in \mathfrak{I}$ the force acting on the* *k-th point particle has components*

$$f^i_{Q_k} = \mathscr{F}^i(x^1_{Q_1}, x^2_{Q_1}, x^3_{Q_1}, \ldots x^1_{Q_N}, x^2_{Q_N}, x^3_{Q_N}, v^1_{Q_1}, v^2_{Q_1}, v^3_{Q_1} \ldots,$$
$$v^1_{Q_N}, v^2_{Q_N}, v^3_{Q_N}).$$

Then in the rest orthonormal coordinates of another $\mathscr{R}' \in \mathfrak{I}$ *the force on the k-th particle still has components of the form*

$$f'^i_{Q_k} = \mathscr{F}^i(x'^1_{Q_1}, x'^2_{Q_1}, x'^3_{Q_1}, \ldots x'^1_{Q_N}, x'^2_{Q_N}, x'^3_{Q_N}, v'^1_{Q_1}, v'^2_{Q_1}, v'^3_{Q_1} \ldots,$$
$$v'^1_{Q_N}, v'^2_{Q_N}, v'^3_{Q_N}),$$

where the functions $\mathscr{F}^i$ *are the same as before.*

Remark (1) can be formulated in *passive terms* as follows. Consider an experiment studying *n* point particles (interacting among themselves but isolated from the external world) in rest right-handed orthonormal coordinates of an inertial frame $\mathscr{R}$. Fix initial conditions for the particles by prescribing the positions and velocities at the generic time t_0, and consider the solution of Newton's equations found with the superposition principle. Now pass to another inertial frame $\mathscr{R}' \neq \mathscr{R}$, and assign, for the same system (externally isolated but internally interacting), initial conditions that in components coincide with the initial conditions given in $\mathscr{R}$. Then the system's evolution, read in the coordinates of $\mathscr{R}'$ with the new initial conditions, will be *exactly* the same as in the first case when read in the coordinates of $\mathscr{R}$ and with the old initial conditions. In this sense *there is no way to select an inertial frame over the others via purely mechanical experimental results*. All in all, we will see that the Principle of Galilean invariance certifies that all inertial frames are completely equivalent toward the formulation and application of Classical Mechanics; hence it makes no physical sense to prefer a particular one. For this reason the same principle is often stated by physicists in a mathematically more ambiguous way as follows:

C5". Principle of Galilean Invariance (Physical Formulation) *There is no way to select one particular frame in the class of inertial frames by purely mechanical experimental results.*

(3) In a more advanced formulation of Mechanics, and Physics in general, one can prove there is a deep relationship between the invariance under transformation groups and the existence of preserved physical quantities (in particular: impulse, angular momentum and energy) during the system's evolution. We will discuss this to a great extent, in the part of the book on Noether's theorem in Lagrangian Mechanics, and also in the subsequent Hamiltonian formulation of Mechanics. ∎

3.4.2   The Failure of the Newtonian Programme

Despite everything, nowadays we know that the Newtonian programme of describing interactions in terms of forces between point particles has failed. The problem lies at the root of the entire construction: the structure of spacetime as we know it is not what is purported in the Galilean-Newtonian layout of Classical Physics (even if read with modern spectacles). In reality we know that the structure of spacetime is more coherently described by the relativistic theories, which are heavily removed from the classical formulation when the speeds at stake are large (close to the speed of light, as we will see in Chap. 10) and in regimes with a strong gravitational field. Perhaps there is no structure of spacetime able to describe accurately the physical world, as the non-local phenomena of quantum nature seems to suggest. Moreover, restricting to a notion of force as we have introduced it earlier, the first axiom to fail—already in limit situations for Classical Physics—is most definitely the Action-reaction principle (in strong and weak form), in case we work with forces between electric charges in accelerated relative motion. The problem is due to the speed of propagation of the perturbations of the electromagnetic field, which is responsible for the forces on the charges. An interesting aspect of this matter is that the conservation of the impulse (and the angular momentum, and the energy) continues to hold, even though the Action-reaction principle does not, provided we keep into consideration the impulse (and the angular momentum, and the energy) of the electromagnetic field. In this way the Newtonian programme, despite its clear inadequacy, which is already contained in its theorems, re-elaborated and successively turned into principles some of the fundamental ingredients of the subsequent programmes of Modern Physics.

3.4.3   What Remains Today of "Mach's Principle"?

In Mach's and Einstein's vision the uniform inertial motion of an isolated point particle, i.e. removed from all other bodies in the universe, must in some way or another be *imposed* by the other masses in the universe, albeit very distant, through some sort of interaction. This is in a nutshell the physical content of what we call "Mach's principle" (although it was stated by Einstein). Anyhow, such interaction *cannot be described in terms of forces* because any force describes, by definition, the interactions between "close" bodies. For this reason the Principle of inertia cannot have a dynamical explanation within Classical Mechanics, and must be taken as a principle. From the perspective of Mach's principle, inertial frames are mere "signposts" for faraway masses, and only with respect to these very distant bodies, or with respect to their "average motion", isolated point particles move in a straight line with uniform velocity.[9] For Mach and Einstein, the existence of inertial frames should not make sense if the universe were empty or if it contained only one body (in contrast to Newton, who declared that inertial frames would exist anyway, in the *Principia*'s famous discourse regarding "Newton's bucket"). Einstein speculated at length about the physical meaning of a possible interaction that accounted for inertial motion, believing it was of the same nature as gravity. That idea though cannot be developed within Classical Mechanics precisely because the gravitational interaction is classically described by a force. Also through this type of speculation Einstein arrived at formulating the theory of General Relativity where Mach's principle can, at least partially, be developed.

We would finally like to say that, nowadays, the ideas at the heart of Mach's principle should be reviewed in the light of novel ways to understand the notion of mass and how bodies (the elementary particles) possess a mass. The experimental discovery of the *Higgs boson*, which according to the *Standard Model* of fundamental interactions would give particles (fermions and heavy electroweak mediating bosons) a mass, brings us to a very different picture of what Mach and Einstein has envisioned, where masses were essentially concentrated in relatively few bodies in the universe instead of being "diluted" in quantum fields.

---

[9] Related to this, the experience shows that an optimal inertial frame is one that is simultaneously at rest with the infinitely distant, so-called "fixed stars" and with the sun's centre of mass, or more precisely the solar system's centre of mass.

Chapter 4
Balance Equations and First Integrals in Mechanics

The aim of this chapter is to introduce certain physical quantities relative to point-particle systems entering particular "balance equations" and "conservation laws". These laws are actually theorems, that follow from the principles of Classical Mechanics stated in Chap. 3. They deal in particular with: the linear momentum, the angular momentum and the mechanical energy. We will discuss together the cases of one particle and systems of several particles. Usually (but there are exceptions) these theorems take the general form of an equality between the time derivative of a certain quantity G, which is a function of the positions and velocities of the particles forming a physical system (in a fixed frame), and a second quantity X, normally referring to the world "external" to the system:

$$\frac{dG(P_1(t), \ldots, P_N(t), \mathbf{v}_{P_1}(t), \ldots, \mathbf{v}_{P_N}(t))}{dt}$$
$$= X(t, P_1(t), \ldots, P_N(t), \mathbf{v}_{P_1}(t), \ldots \mathbf{v}_{P_N}(t), \text{"ext. var."}) .$$

Equality occurs when the system evolves under laws C1–C4 of Chap. 3. When the right-hand side is zero (at least for the motion considered), the governing equation becomes a *conservation* law in time for the quantity of the left, *when evaluated on a particular motion of the physical system.* From the perspective of the theory of ODE systems, quantities that are preserved in time are called **first integrals** of the motion (see Definition 14.2).

V. Moretti, *Analytical Mechanics*, La Matematica per il 3+2 150,
https://doi.org/10.1007/978-3-031-27612-5_4

4.1 Governing Equations, Conservation of the Impulse and the Angular Momentum

We now introduce a number of relations, valid for all mechanical systems made of point particles obeying principles C1–C4 (including the generalised versions with constraints and inertial forces) stated in Chap. 3, called *governing equations* of Dynamics.

4.1.1 Total Quantities of Systems of Point Particles

Definition 4.1 Consider a system S of N point particles P_i, $i = 1, \ldots, N$ of respective masses m_i, $i = 1, \ldots, N$. We define the following notions.

(1) The **total mass of the system** is:

$$M := \sum_{k=1}^{N} m_k .$$

(2) The **centre of mass** of system S at time $t \in \mathbb{R}$, $G(t)$, is the point (not necessarily a point particle of the system) in each Σ_t determined by the equation:

$$M(G(t) - O) = \sum_{k=1}^{N} m_k (P_k(t) - O) ,$$

where $O \in \Sigma_t$ is an arbitrary point.

(3) The **(total) impulse** or **(total) momentum** of system S with respect to a frame $\mathscr{R}$ at time t is the vector in V_t:

$$\mathbf{P}|_{\mathscr{R}}(t) := \sum_{k=1}^{N} m_k \mathbf{v}_{P_k}|_{\mathscr{R}}(t) .$$

(4) If $O = O(t)$ is any world line (not necessarily the one in (1)) and $\mathscr{R}$ a frame, the system's **(total) angular momentum** with respect to the pole O and the frame $\mathscr{R}$ at time t is the vector in V_t:

$$\boldsymbol{\Gamma}_O|_{\mathscr{R}}(t) := \sum_{k=1}^{N} m_k (P_k(t) - O(t)) \wedge \mathbf{v}_{P_k}|_{\mathscr{R}}(t) .$$

$\diamondsuit$

The importance of the above quantities is essentially due to the fact that, under certain hypotheses and in a fixed frame, either they are preserved in time or they appear in the defining expressions of quantities that are preserved in time during the system's temporal evolution. In many cases knowing the values of preserved quantities provides relevant information on the system's motion, even when one cannot solve explicitly the equation of motion.

Remarks 4.2

(1) *From now on, when talking of a system of point particles, we shall always assume there is a finite number N of particles.*
(2) Clearly the above definitions hold when $N = 1$, giving back the definitions seen for one particle.
(3) The definition of G is well posed, in the sense that G is uniquely determined, once O is chosen, by:

$$G := O + \frac{1}{M} \sum_{k=1}^{N} m_k (P_k - O) \, ;$$

moreover G *does not depend on* O. In fact, if we define G_O and $G_{O'}$ by:

$$M(G_O - O) = \sum_{k=1}^{N} m_k (P_k - O) \quad \text{and} \quad M(G_{O'} - O') = \sum_{k=1}^{N} m_k (P_k - O') \, ,$$

then:

$$G_O - G_{O'} = (O - O') + \frac{1}{M} \sum_{k=1}^{N} m_k (P_k - O) - \frac{1}{M} \sum_{k=1}^{N} m_k (P_k - O')$$

$$= (O - O') - (O - O') = \mathbf{0} \, .$$

(4) If $\mathbf{v}_G|_{\mathscr{R}}$ is the velocity of the centre of mass in the frame $\mathscr{R}$ for a system of point particles of total mass M, then

$$\mathbf{P}|_{\mathscr{R}} = M \mathbf{v}_G|_{\mathscr{R}} \, . \tag{4.1}$$

In other words: *the system's total impulse is the impulse of a single point particle of mass M concentrated in the system's centre of mass.*

Verifying this is straightforward by choosing a world line $O = O(t)$, differentiating in time the identity $M(G(t) - O(t)) = \sum_{k=1}^{N} m_k (P_k(t) - O(t))$ and remembering that $M = \sum_k m_k$. ∎

Exercises 4.3

(1) Prove that when we switch frames from $\mathcal{R}$ to $\mathcal{R}'$, the momentum (keeping the pole O fixed) transforms under:

$$\mathbf{\Gamma}_O|_{\mathcal{R}}(t) = \mathbf{\Gamma}_O|_{\mathcal{R}'}(t) + \mathbf{I}_{t,O(t)}(\boldsymbol{\omega}_{\mathcal{R}'}|\mathcal{R}) , \qquad (4.2)$$

where we have introduced the **inertia tensor** at time t with respect to the pole O, given by the linear map

$$V_t \ni \mathbf{u} \mapsto \mathbf{I}_{t,O(t)}(\mathbf{u}) := \sum_{k=1}^{N} m_k (P_k(t) - O(t)) \wedge (\mathbf{u} \wedge (P_k(t) - O(t))) . \qquad (4.3)$$

(2) Show that, when changing pole from O to O', but staying in the same frame $\mathcal{R}$, the momentum (at fixed time $t \in \mathbb{R}$) transforms as:

$$\mathbf{\Gamma}_O|_{\mathcal{R}} = \mathbf{\Gamma}_{O'}|_{\mathcal{R}} + (O' - O) \wedge \mathbf{P}|_{\mathcal{R}} . \qquad (4.4)$$

In particular, choosing $O' = G$, we can always write $\mathbf{\Gamma}_O|_{\mathcal{R}}$ as the sum of the total angular momentum in $\mathcal{R}$ with respect to G plus the angular momentum of a unique point particle at G having mass equal to the system's total mass:

$$\mathbf{\Gamma}_O|_{\mathcal{R}} = \mathbf{\Gamma}_G|_{\mathcal{R}} + (G - O) \wedge \mathbf{P}|_{\mathcal{R}} . \qquad (4.5)$$

4.1.2 Governing Equations

We can now prove the so-called *governing equations of Dynamics* and then deduce the corresponding balance/conservation laws. Let us first define the concept of *internal* and *external forces*. If we have a system S of point particles, a force $\mathbf{F}$ acting on $P \in S$ is said to be **internal** if the corresponding reaction acts on a particle $P' \in S$. If not internal, we say the force is **external**. Inertial forces are always considered external.

Theorem 4.4 (Governing Equations of Dynamics for Systems of Point Particles) *Consider a system of N point particles obeying principles C1–C4 (including inertial forces and constraint reactions). If $\mathcal{R}$ is a frame, for any given instant of absolute time t the following relations hold, respectively called* **first and second governing equations of Dynamics for systems of point particles***.*

(1) If at the instant considered $\mathbf{F}_i^{(e)}$ is the sum of the external forces acting on the i-th particle (including, if necessary, inertial forces and constraint reactions), then:

$$\frac{d}{dt}\bigg|_{\mathscr{R}} \mathbf{P}|_{\mathscr{R}} = \sum_{i=1}^{N} \mathbf{F}_i^{(e)} . \tag{4.6}$$

(2) With the same notation, for any choice of pole O:

$$\frac{d}{dt}\bigg|_{\mathscr{R}} \boldsymbol{\Gamma}_O|_{\mathscr{R}} + \mathbf{v}_O|_{\mathscr{R}} \wedge \mathbf{P}|_{\mathscr{R}} = \sum_{i=1}^{N} (P_i - O) \wedge \mathbf{F}_i^{(e)} . \tag{4.7}$$

Proof

(1) Newton's second law for P_i reads:

$$\mathbf{F}_i^{(e)} + \sum_j \mathbf{F}_{ij}^{(i)} = m_i \frac{d^2 P_i}{dt^2}\bigg|_{\mathscr{R}} ,$$

where $\mathbf{F}_{ij}^{(i)}$ is the force acting on P_i due to P_j, while $\mathbf{F}_i^{(e)}$ is the sum of the external forces acting on P_i (including inertial forces if $\mathscr{R}$ is not inertial). From this we find:

$$\sum_i m_i \frac{d^2 P_i}{dt^2}\bigg|_{\mathscr{R}} = \sum_{i,j} \mathbf{F}_{ij}^{(i)} + \sum_i \mathbf{F}_i^{(e)} .$$

Renaming indices:

$$\sum_{i,j} \mathbf{F}_{ij}^{(i)} = \sum_{j,i} \mathbf{F}_{ji}^{(i)} .$$

On the other hand, by the Action-reaction principle: $\mathbf{F}_{ij}^{(i)} = -\mathbf{F}_{ji}^{(i)}$, so:

$$\sum_{i,j} \mathbf{F}_{ij}^{(i)} = -\sum_{j,i} \mathbf{F}_{ij}^{(i)} = -\sum_{i,j} \mathbf{F}_{ij}^{(i)} ,$$

where we have used the fact that the summation order (first i then j, or the other way around) is irrelevant. In conclusion $\sum_{i,j} \mathbf{F}_{ij}^{(i)} = \mathbf{0}$ and so

$$\sum_i m_i \frac{d^2 P_i}{dt^2}\bigg|_{\mathscr{R}} = \sum_i \mathbf{F}_i^{(e)} ,$$

i.e.:

$$\frac{d}{dt}\Bigg|_{\mathscr{R}} \sum_i m_i \mathbf{v}_{P_i}|_{\mathscr{R}} = \sum_i \mathbf{F}_i^{(e)} \, .$$

By the definition of total impulse we immediately find (4.6).

(2) In the same notation, by the definition of $\boldsymbol{\Gamma}_O|_{\mathscr{R}}$ we have:

$$\frac{d}{dt}\Bigg|_{\mathscr{R}} \boldsymbol{\Gamma}_O|_{\mathscr{R}} = \sum_i m_i (\mathbf{v}_{P_i}|_{\mathscr{R}} - \mathbf{v}_O|_{\mathscr{R}}) \wedge \mathbf{v}_{P_i}|_{\mathscr{R}}$$

$$+ \sum_{i,j} (P_i - O) \wedge \mathbf{F}_{ij}^{(i)} + \sum_i (P_i - O) \wedge \mathbf{F}_i^{(e)} \, .$$

In other words:

$$\frac{d}{dt}\Bigg|_{\mathscr{R}} \boldsymbol{\Gamma}_O|_{\mathscr{R}} + \sum_i m_i \mathbf{v}_O|_{\mathscr{R}} \wedge \mathbf{v}_{P_i}|_{\mathscr{R}} = \sum_i (P_i - O) \wedge \mathbf{F}_i^{(e)} + \sum_{i,j} (P_i - O) \wedge \mathbf{F}_{ij}^{(i)} \, .$$

As $\sum_i m_i \mathbf{v}_O|_{\mathscr{R}} \wedge \mathbf{v}_{P_i}|_{\mathscr{R}} = \mathbf{v}_O|_{\mathscr{R}} \wedge \mathbf{P}|_{\mathscr{R}}$, to finish it is enough to show

$$\sum_{i,j} (P_i - O) \wedge \mathbf{F}_{ij}^{(i)} = \mathbf{0} \, .$$

Relabelling indices, that reads:

$$\sum_{i,j} (P_i - O) \wedge \mathbf{F}_{ij}^{(i)} = \frac{1}{2} \left(\sum_{i,j} (P_i - O) \wedge \mathbf{F}_{ij}^{(i)} + \sum_{j,i} (P_j - O) \wedge \mathbf{F}_{ji}^{(i)} \right) \, .$$

Since the summation order is not relevant, we find:

$$\sum_{i,j} (P_i - O) \wedge \mathbf{F}_{ij}^{(i)} = \frac{1}{2} \sum_{i,j} \left((P_i - O) \wedge \mathbf{F}_{ij}^{(i)} + (P_j - O) \wedge \mathbf{F}_{ji}^{(i)} \right) \, .$$

Using the Action-reaction principle:

$$\sum_{i,j} (P_i - O) \wedge \mathbf{F}_{ij}^{(i)} = \frac{1}{2} \sum_{i,j} \left((P_i - O) \wedge \mathbf{F}_{ij}^{(i)} - (P_j - O) \wedge \mathbf{F}_{ij}^{(i)} \right)$$

$$= \frac{1}{2} \sum_{i,j} (P_i - P_j) \wedge \mathbf{F}_{ij}^{(i)} \, .$$

The *third law of motion in strong form* (**C3**) ensures that $\mathbf{F}_{ij}^{(i)}$ is parallel to $P_i - P_j$, so $(P_i - P_j) \wedge \mathbf{F}_{ij}^{(i)} = \mathbf{0}$, ending the proof. □

Remarks 4.5

(1) The vectors:

$$\mathbf{R}^{(e)} := \sum_{i=1}^{N} \mathbf{F}_i^{(e)} \quad \text{and} \quad \mathbf{M}_O^{(e)} := \sum_{i=1}^{N} (P_i - O) \wedge \mathbf{F}_i^{(e)},$$

are called, respectively, **net force** and **net momentum of the external forces** with respect to the pole O. In general, if O is an arbitrary point and $\mathbf{F}$ a force acting on the point particle P, $(P - O) \wedge \mathbf{F}$ is called the **momentum of the force F with respect to the pole** O.

(2) Because of (4.1) the first governing equation can be reformulated equivalently as:

$$M\mathbf{a}_G|_{\mathscr{R}} = \sum_{i=1}^{N} \mathbf{F}_i^{(e)}. \tag{4.8}$$

In this form the equations says that the centre of mass evolves under Newton's second law as if the system were a unique point particle in which the entire mass is concentrated, and it is subject to the sum of all *external* forces acting on the system.

(3) In a similar fashion we define the **net internal force**, $\mathbf{R}^{(i)}$, and the **net momentum of the internal forces**, $\mathbf{M}_O^{(i)}$. The proof of the governing equations shows that

$$\mathbf{R}^{(i)} = \mathbf{0} \quad \text{and} \quad \mathbf{M}_O^{(i)} = \mathbf{0}$$

for any system of point particles obeying principles C1–C4 (including the extended versions) and, regarding the net momentum, also independently of the pole O.

(4) In general, the two governing equations are not capable of determining the system's motion once the initial condition have been given. Nonetheless, if the system satisfies the *rigidity constraint* (the mutual distances between the system's particles are constant irrespective of external forces and initial conditions), the two equations do determine, for assigned initial conditions, the system's motion. ■

4.1.3   Balance/Conservation Laws of Impulse and Angular Momentum

By the first governing equation, if the system is isolated or more weakly if the sum of the external forces is zero (during a time interval), then the total impulse is a *first integral of motion*, that is, it is preserved in time (in that time lapse). Because of the second governing equation, if the system is isolated or, more weakly, the sum of the external momenta with respect to O vanishes (in some time interval), and in either case we chose $O \equiv G$ or O with zero velocity in $\mathscr{R}$, then the total angular momentum is a *first integral of motion*, i.e. it is preserved in time (in that time lapse).

The results mentioned above hold more generally along a fixed direction.

Proposition 4.6 *Under the hypotheses of Theorem 4.4 the following facts hold, with reference to an arbitrary unit vector* $\mathbf{n}$ *constant in time in* $\mathscr{R}$.

(1) If $\mathbf{R}^{(e)} \cdot \mathbf{n} = \mathbf{0}$ *in some time interval, then* $\mathbf{P}|_{\mathscr{R}} \cdot \mathbf{n}$ *is constant in that time interval on every motion of the system (the value depending on the specific motion).*

(2) If $O = G$ *or* O *is at rest in* $\mathscr{R}$, *and* $\mathbf{M}_O^{(e)} \cdot \mathbf{n} = 0$ *in some time interval, then* $\Gamma_O|_{\mathscr{R}} \cdot \mathbf{n}$ *is constant in that time interval on every motion of the system (the value depending on the specific motion).*

(1) and (2) hold in particular in any inertial frame and for any direction $\mathbf{n}$, *provided the system of point particles is* **isolated**, *i.e. not subject to external forces.*

Proof Part (1) is immediate from the first governing equation. The proof of (2) is straightforward from the second governing equation, remembering that if $O = G$ or O is at rest in $\mathscr{R}$ then $\mathbf{v}_O|_{\mathscr{R}} \wedge \mathbf{P}|_{\mathscr{R}} = \mathbf{0}$ (in the first case since $\mathbf{v}_O|_{\mathscr{R}} = \mathbf{v}_G|_{\mathscr{R}}$ and by (4.1)). □

Remarks 4.7

(1) Throughout the history of Physics it has become clear that, extending the class of physical systems examined, it is always possible to define a contribution to the system's total impulse (which might not be totally mechanical, for instance if it contains a contribution due to the electromagnetic field) so that the total impulse of an isolated system remains the same. This principle, namely that the notion of impulse can be extended so to achieve a conservation law, goes under the name of **Principle of conservation of the impulse**. The history of Physics shows that this principle is way more important than the corresponding theorem in Mechanics, that it holds both in relativistic theories and quantum ones, and that it is related to the *homogeneity* of space in inertial frames.

(2) In the same way, we now know that by extending the class of physical systems studied, it is always possible to define a contribution to the system's total angular momentum (which might not be totally mechanical, for instance if it contains a contribution due to the electromagnetic field) so that the total angular momentum of an isolated system does not change. This principle, namely that the notion of angular momentum can be extended so to achieve a conservation law, goes under the name of **Principle of conservation of the angular momentum**. The history of Physics shows that this principle is way more important than the corresponding theorem in Mechanics, that it holds both in relativistic theories and quantum ones, and is related to the *isotropy* properties of space in inertial frames. ∎

4.2 Mechanical Energy

We introduce here the fundamental notions referring to the mechanical energy of single point particles and systems.

Definition 4.8 Let P be a point particle of mass m, subject to the force $\mathbf{F}$ at time t.

(1) The **kinetic energy** of P in the frame $\mathscr{R}$ at time t is the number:

$$\tau|_{\mathscr{R}}(t) := \frac{1}{2}m\mathbf{v}_P^2|_{\mathscr{R}}(t) ,$$

where $\mathbf{u}^2 := \mathbf{u} \cdot \mathbf{u}$ for $\mathbf{u} \in V_t$.

(2) The **power** produced by the force $\mathbf{F}$ at time t is

$$\pi|_{\mathscr{R}}(t) := \mathbf{v}_P|_{\mathscr{R}}(t) \cdot \mathbf{F} .$$

(3) If we have a system S of point particles P_k with masses m_k, $k = 1, 2, \ldots, N$, the number:

$$\mathscr{T}|_{\mathscr{R}}(t) := \sum_{k=1}^{N} \frac{1}{2}m_k\mathbf{v}_{P_k}^2|_{\mathscr{R}}(t)$$

is the **total kinetic energy** of S in the frame $\mathscr{R}$ at time t. ◇

Examples 4.9

(1) There exist forces, called **dissipative** with respect to a frame $\mathscr{R}$, that are characterised by the fact they dissipate strictly negative power with respect to $\mathscr{R}$ when the velocity is non-zero:

$$\mathbf{F}(t, P, \mathbf{v}_P|_{\mathscr{R}}) \cdot \mathbf{v}_P|_{\mathscr{R}} < 0 , \quad \text{for any } \mathbf{v}_P|_{\mathscr{R}} \in V_{\mathscr{R}}, \mathbf{v}_P|_{\mathscr{R}} \neq \mathbf{0}.$$

The typical example is *viscous friction*. A fluid at rest in a frame $\mathscr{R}$ exerts on a point particle P a force of the form

$$\mathbf{F} = -g(||\mathbf{v}_P|_{\mathscr{R}}||)\frac{\mathbf{v}_P|_{\mathscr{R}}}{||\mathbf{v}_P|_{\mathscr{R}}||} .$$

The function g is non-negative. At small speeds (say, up to $2m/s$ in the air) g is constant. At higher speeds the behaviour is more complicated and obeys a polynomial law of the kind $k||\mathbf{v}_P|_{\mathscr{R}}||^n$ with $k > 0$.

(2) **Gyrostatic** forces, with respect to a frame $\mathscr{R}$, are characterised by constantly null dissipated power in $\mathscr{R}$:

$$\mathbf{F}(t, P, \mathbf{v}_P|_{\mathscr{R}}) \cdot \mathbf{v}_P|_{\mathscr{R}} = 0 , \quad \text{for any } \mathbf{v}_P|_{\mathscr{R}} \in V_{\mathscr{R}}.$$

The standard example is the *Lorentz force* on a point particle of charge e when immersed in a magnetic field $\mathbf{B}(t, P)$ in the frame $\mathscr{R}$ (c is the speed of light):

$$\mathbf{F} = \frac{e}{c}\mathbf{v}_P|_{\mathscr{R}} \wedge \mathbf{B}(t, P) .$$

Another example, this time inertial, is the Coriolis force discussed in Chap. 3:

$$\mathbf{F}^{(Coriolis)} = -2m\boldsymbol{\omega}_{\mathscr{R}'}|_{\mathscr{R}} \wedge \mathbf{v}_P|_{\mathscr{R}'} .$$

The kinetic energy of a system of particles can be decomposed in a canonical way into the energy of the centre of mass and the energy "surrounding the centre of mass". Such a splitting is useful in several areas. We introduce it under the form of an elementary but renowned theorem.

Theorem 4.10 (König's Theorem) *For a system of point particles P_k, $k = 1, \ldots, N$, the kinetic energy in the frame $\mathscr{R}$ equals the kinetic energy in the frame $\mathscr{R}_G$, where G is at rest and $\boldsymbol{\omega}_{\mathscr{R}_G}|_{\mathscr{R}} = \mathbf{0}$, plus the kinetic energy of a point particle at G at every instant, with mass equal to the system's total mass M. In formulas (dropping time dependency for simplicity),*

$$\mathscr{T}|_{\mathscr{R}} = \frac{1}{2}M\mathbf{v}_G^2|_{\mathscr{R}} + \mathscr{T}|_{\mathscr{R}_G} . \tag{4.9}$$

Proof With the given hypotheses, using the definition of kinetic energy and (2.67)

$$\mathscr{T}|_{\mathscr{R}} = \sum_k \frac{1}{2}m_k\mathbf{v}_{P_k}|_{\mathscr{R}} \cdot \mathbf{v}_{P_k}|_{\mathscr{R}} = \sum_k \frac{1}{2}m_k \left(\mathbf{v}_{P_k}|_{\mathscr{R}_G} + \mathbf{v}_G|_{\mathscr{R}}\right) \cdot \left(\mathbf{v}_{P_k}|_{\mathscr{R}_G} + \mathbf{v}_G|_{\mathscr{R}}\right)$$

$$= \sum_k \frac{1}{2}m_k\mathbf{v}_{P_k}^2|_{\mathscr{R}_G} + \sum_k \frac{1}{2}m_k\mathbf{v}_G^2|_{\mathscr{R}} + \sum_k m_k\mathbf{v}_{P_k}|_{\mathscr{R}_G} \cdot \mathbf{v}_G|_{\mathscr{R}}$$

$$= \sum_k \frac{1}{2}m_k\mathbf{v}_{P_k}^2|_{\mathscr{R}_G} + \frac{1}{2}M\mathbf{v}_G^2|_{\mathscr{R}} + \sum_k m_k\mathbf{v}_{P_k}|_{\mathscr{R}_G} \cdot \mathbf{v}_G|_{\mathscr{R}}$$

$$= \sum_k \frac{1}{2}m_k\mathbf{v}_{P_k}^2|_{\mathscr{R}_G} + \frac{1}{2}M\mathbf{v}_G^2|_{\mathscr{R}} + M\mathbf{v}_G|_{\mathscr{R}_G} \cdot \mathbf{v}_G|_{\mathscr{R}} .$$

Above we used the definition of centre of mass. The last summand vanishes because in our assumptions $\mathbf{v}_G|_{\mathscr{R}_G} = \mathbf{0}$, so:

$$\mathscr{T}|_{\mathscr{R}} = \frac{1}{2} M v_G^2|_{\mathscr{R}} + \mathscr{T}|_{\mathscr{R}_G} .$$

$\square$

4.2.1 Kinetic Energy Theorem

Next up is the first balance result for the kinetic energy, historically called theorem of 'living forces'.

Theorem 4.11 (Kinetic Energy Theorem) *Consider a system S of point particles* P_k *of masses* m_k, $k = 1, 2, \ldots, N$, *obeying principles C1–C4 (including inertial forces and constraint reactions). In a frame* $\mathscr{R}$, *and for any given instant of absolute time t, the relation:*

$$\Pi^{(e)}|_{\mathscr{R}} + \Pi^{(i)}|_{\mathscr{R}} = \frac{d\mathscr{T}|_{\mathscr{R}}}{dt} \tag{4.10}$$

holds. At the instant considered, $\Pi^{(e)}|_{\mathscr{R}}$ *and* $\Pi^{(i)}|_{\mathscr{R}}$ *are respectively the* **total external power**, *i.e. the sum of all powers dissipated by external forces at each point in the system, and the* **total internal power**, *i.e. the sum of all powers dissipated by internal forces at each point in the system.*

Proof If $\mathbf{F}_k^{(i)}$ and $\mathbf{F}_k^{(e)}$ respectively denote the sum of the internal and the external forces acting on the kth particle, by Newton's second law we have:

$$\mathbf{v}_{P_k}|_{\mathscr{R}} \cdot (\mathbf{F}_k^{(i)} + \mathbf{F}_k^{(e)}) = m_k \mathbf{v}_{P_k}|_{\mathscr{R}} \cdot \frac{d\mathbf{v}_{P_k}|_{\mathscr{R}}}{dt} = \frac{d}{dt} \left(\frac{1}{2} m_k v_{P_k}^2|_{\mathscr{R}} \right) .$$

Summing over k produces (4.11). $\square$

Remarks 4.12 We explicitly wrote the frame in the total power dissipated by the internal forces, although one can prove that this power *does not depend on the frame*. This remarkable result is known as **Principle of mechanical indifference** and has a great impact on the development of Physics, when one passes from Mechanics to Thermodynamics, to define the system's internal thermodynamic energy as a function independent of the frame. Let us prove this independence. Internal forces are "real", so by definition they do not depend on the frame $\mathbf{F}_k^{(i)}|_{\mathscr{R}} = \mathbf{F}_k^{(i)}|_{\mathscr{R}'}$. The relationship between the various velocities of each particle, as the frame varies, is given by (2.67):

$$\mathbf{v}_{P_k}|_{\mathscr{R}} = \mathbf{v}_{P_k}|_{\mathscr{R}'} + \mathbf{v}_{O'}|_{\mathscr{R}} + \boldsymbol{\omega}_{\mathscr{R}'}|_{\mathscr{R}} \wedge (P_k - O') .$$

From that, with the obvious notation, we have:

$$\Pi^{(i)}|_{\mathscr{R}} - \Pi^{(i)}|_{\mathscr{R}'} = \mathbf{v}_{O'}|_{\mathscr{R}} \cdot \sum_k \mathbf{F}_k^{(i)} + \sum_k \boldsymbol{\omega}_{\mathscr{R}'}|_{\mathscr{R}} \wedge (P_k - O') \cdot \mathbf{F}_k^{(i)} \,.$$

The familiar properties of the cross product then give: $\boldsymbol{\omega}_{\mathscr{R}'}|_{\mathscr{R}} \wedge (P_k - O') \cdot \mathbf{F}_k^{(i)} = (P_k - O') \wedge \mathbf{F}_k^{(i)} \cdot \boldsymbol{\omega}_{\mathscr{R}'}|_{\mathscr{R}}$, so:

$$\Pi^{(i)}|_{\mathscr{R}} - \Pi^{(i)}|_{\mathscr{R}'} = \mathbf{v}_{O'}|_{\mathscr{R}} \cdot \sum_k \mathbf{F}_k^{(i)} + \boldsymbol{\omega}_{\mathscr{R}'}|_{\mathscr{R}} \cdot \sum_k (P_k - O') \wedge \mathbf{F}_k^{(i)} \,.$$

In other words

$$\Pi^{(i)}|_{\mathscr{R}} - \Pi_{\mathscr{R}'}^{(i)} = \mathbf{v}_{O'}|_{\mathscr{R}} \cdot \mathbf{R}^{(i)} + \boldsymbol{\omega}_{\mathscr{R}'}|_{\mathscr{R}} \cdot \mathbf{M}_{O'}^{(i)} \,.$$

By (3) in Remarks 4.5 we conclude that $\mathbf{R}^{(i)} = \mathbf{0}$ and $\mathbf{M}_{O'}^{(i)} = \mathbf{0}$, and eventually:

$$\Pi^{(i)}|_{\mathscr{R}} - \Pi^{(i)}|_{\mathscr{R}'} = 0 \,.$$

∎

4.2.2 Conservative Forces

We shall introduce a class of forces of major importance in Physics, that depend on particles' positions.

Definition 4.13 We say a system S of point particles $P_k, k = 1, 2, \ldots, N$, is subject to **position-depending forces** in a frame $\mathscr{R}$, if for any particle P_k there is a force (we exclude *inertial* forces) $\mathbf{F}_k$, that in $\mathscr{R}$ has the form $\mathbf{F}_k = \mathbf{F}_k(P_1, \ldots, P_N)$. Furthermore, we say a system of position-depending forces is **conservative** in $\mathscr{R}$ if there exists a function $\mathscr{U}|_{\mathscr{R}} \in C^1(E_{\mathscr{R}} \times \cdots \times E_{\mathscr{R}})$, called **potential energy** associated with the system of forces in $\mathscr{R}$, such that in $\mathscr{R}$:

$$\mathbf{F}_k(P_1, \ldots, P_N) = -\nabla_{P_k} \mathscr{U}|_{\mathscr{R}}(P_1, \ldots, P_N) \,,$$

$$\text{for any } P_k \in E_{\mathscr{R}} \text{ and any } k = 1, 2, \ldots, N \,. \tag{4.11}$$

where, if $\mathbf{e}_1, \mathbf{e}_2, \mathbf{e}_3$ is an orthonormal basis of the space of displacements of $E_{\mathscr{R}}$, we put

$$\nabla_{P_k} f(P_1, \ldots, P_N) := \sum_{j=1}^{3} \frac{\partial f(P_1, \ldots, P_N)}{\partial x_k^j} \mathbf{e}_j$$

for any differentiable map $f = f(P_1, \ldots, P_N)$ and where $P_j - O = \sum_{k=1}^{3} x_k^j \mathbf{e}_j$ is the position vector of P_k in $E_{\mathscr{R}}$. $\diamondsuit$

Remarks 4.14

(1) The function $-\mathscr{U}|_{\mathscr{R}}$ is called **potential** of the system of forces. Needless to say, both $\mathscr{U}|_{\mathscr{R}}$ and the potential function are defined up to additive constants.

(2) There are systems whose forces can be written as in (4.11), where though $\mathscr{U}|_{\mathscr{R}}$ also depends on time. In such a case $\mathscr{U}|_{\mathscr{R}}$ is *no longer* called potential energy, but $-\mathscr{U}|_{\mathscr{R}}$ is still referred to as the force's potential. This happens when, given a system of conservative forces in a frame $\mathscr{R}$, we pass to a new frame by a Galilean transformation involving a relative velocity.

(3) Special cases of conservative forces, for systems consisting of a unique particle, are the so-called *central forces*: a force $\mathbf{F}$ is **central** in $\mathscr{R}$ with **centre** $O \in E_{\mathscr{R}}$, if it depends only on the position and it satisfies two conditions:

  (i) $\mathbf{F}(P)$ is parallel to $P - O$, for any $P \in E_{\mathscr{R}}$,
  (ii) $\mathbf{F}(P)$ is a function of $||P - O||$ only.

The proof that a central force is conservative is postponed to the exercises. It is instead easy to show that every conservative force $\mathbf{F}$ in a frame $\mathscr{R}$, that satisfies (i) with respect to $O \in E_{\mathscr{R}}$, is central with respect to O. In rest spherical coordinates on $\mathscr{R}$ with origin O, we have

$$\mathbf{F}(P) = -\frac{\partial \mathscr{U}|_{\mathscr{R}}}{\partial r} \mathbf{e}_r - \frac{1}{r} \frac{\partial \mathscr{U}|_{\mathscr{R}}}{\partial \theta} \mathbf{e}_\theta - \frac{1}{r \sin \theta} \frac{\partial \mathscr{U}|_{\mathscr{R}}}{\partial \varphi} \mathbf{e}_\varphi .$$

In order to satisfy constraint (i), the last two derivatives must vanish, and therefore the potential energy U, whence the force itself, cannot depend on θ and φ, but only on $r = ||P - O||$. $\blacksquare$

Using well-known Analysis results [Apo91I] one easily proves the following theorem, characterising conservative forces for systems made of one particle. The result is easy to extend to several point particles, and we leave that generalisation to the reader. Regarding (2) let us recall that in $\mathbb{R}^n$ every non-empty connected open set is smoothly path-connected.

Theorem 4.15 *Consider a continuous position-depending force in a frame $\mathscr{R}$, $\mathbf{F} : \Omega \to \mathbb{R}^3$, with $\Omega \subset E_{\mathscr{R}}$ open and connected.*

(1) The following conditions are equivalent:

  (i) $\mathbf{F}$ is conservative;

  (ii) for any pair of points $P, Q \in E_{\mathscr{R}}$, the integral $\int_{P\,\Gamma}^{Q} \mathbf{F}(\mathbf{x}) \cdot d\mathbf{x}$ does not depend on the regular curve $\Gamma : \mathbf{x} = \mathbf{x}(s)$ between P and Q as long as Γ is entirely contained in Ω;

(iii) for any closed regular curve Γ entirely contained in Ω we have

$$\oint_{\Gamma} \mathbf{F}(\mathbf{x}) \cdot d\mathbf{x} = 0 .$$

(2) If $\mathbf{F}$ is conservative the potential energy $\mathscr{U}$ is of class $C^1(\Omega)$ and can be defined by:

$$\mathscr{U}|_{\mathscr{R}}(P) := -\int_{O\,\Gamma}^{P} \mathbf{F}(\mathbf{x}) \cdot d\mathbf{x} ,$$

where $O \in \Omega$ is an arbitrary point, fixed once and for all, and Γ is any C^1 curve joining O and P all contained in Ω.
*(3) If $\mathbf{F} \in C^1(\Omega)$ is conservative then $\mathbf{F}$ is **irrotational**, i.e.:*

$$\nabla \wedge \mathbf{F}(P) = \mathbf{0} , \quad \text{for any } P \in E_{\mathscr{R}} . \tag{4.12}$$

(4) If $\mathbf{F} \in C^1(\Omega)$, Ω is simply connected and $\mathbf{F}$ is irrotational, then $\mathbf{F}$ is conservative.

Purely mathematically, the above theorem holds if we replace $E_{\mathscr{R}}$ and $\mathbb{R}^3$ with $\mathbb{R}^n$, but the vector field $\mathbf{F}$ will no longer model a force field. In the general case of a field $\mathbf{F}(\mathbf{x}) := \sum_{i=1}^{n} F^i(\mathbf{x})\mathbf{e}_i$ defined on an open set $\Omega \subset \mathbb{R}^n$ with values in $\mathbb{R}^n$, being irrotational at $\mathbf{x}_0$ (which is (4.12) for $n = 3$), becomes:

$$\left.\frac{\partial F^i}{\partial x^j}\right|_{\mathbf{x}_0} = \left.\frac{\partial F^j}{\partial x^i}\right|_{\mathbf{x}_0} .$$

Let us finally point out that the above theorem has an alternative, but completely equivalent, formulation in terms of differential 1-forms: the vector field being irrotational is replaced by the closure of a differential form (see Theorem B.11, and the subsequent comments in Appendix B).

Exercises 4.16

(1) Prove that if $\mathbf{F} : \Omega \to V_{\mathscr{R}}$ is continuous and central with respect to $O \in E_{\mathscr{R}}$, with Ω open, then $\mathbf{F}$ is conservative.
(2) Consider a point particle P subject to the sole central force $\mathbf{F}$ with centre O in the inertial frame $\mathscr{R}$. Show that in $\mathscr{R}$ the particle moves on a plane orthogonal to the angular momentum of $\Gamma_O|_{\mathscr{R}}$, which is constant in time.

Remarks 4.17

(1) One can consider forces that are conservative *irrespective of the frame*. This situation presents itself for systems of *two* point particles P and Q, when they interact under a pair of forces (constituting an *action-reaction pair*) $\mathbf{F}(P, Q)$

and $-\mathbf{F}(P, Q)$ acting on P and Q respectively, for which there is a function $U = U(r)$ of class $C^1(\mathbb{R})$ (or on some open subset of $\mathbb{R}$) such that

$$\mathbf{F}(P, Q) = -\nabla_P \mathscr{U}(P, Q) \tag{4.13}$$

for $\mathscr{U}(P, Q) := U(||P - Q||)$. In this case, $-\nabla_Q U(||P - Q||) = -\mathbf{F}(P, Q)$, so the role of the particles is interchangeable. Asking that U is a function of $||P - Q||$ and not of $P - Q$ immediately implies that $\mathbf{F}$ is directed along the segment joining P to Q, as imposed by the strong Action-reaction principle. In that case, (4.13) holds in every frame (the forces are real and hence frame-invariant) if we assume, as the notation tacitly suggests, that the expression of U does not depend on the frame. This is coherent since the distance $d_t(P, Q) = ||P - Q||$ is absolute and does not depend on the frame. In this case, too, $\mathscr{U}$ is called *potential energy*, and is associated with the *pair* of forces $\mathbf{F}(P, Q)$ and $-\mathbf{F}(P, Q)$. This picture generalises to several point particles whenever, for any pair of particles P_i, P_j with $i \neq j$, the relative action-reaction pair of the system of forces has a potential energy $\mathscr{U}_{ij}(P_i, P_j) = U_{ij}(||P_i - P_j||)$ of the previous type. The total potential energy, for which (4.13) holds, is obtained adding the energies of all possible pairs:

$$\mathscr{U}(P_1, P_2, \ldots, P_N) = \sum_{i<j} \mathscr{U}_{ij}(P_i, P_j) .$$

Note that the pairs P_i, P_j and P_j, P_i contribute one potential energy function $\mathscr{U}_{ij}$.

(2) Returning to two point particles P, Q with potential energy

$$\mathscr{U}(P, Q) = U(||P - Q||) ,$$

if the position of Q is fixed, at rest, in a frame $\mathscr{R}$ via additional forces acting on Q, the force acting on P is actually central, with centre Q, when described in the frame $\mathscr{R}$. ∎

4.2.3 Balance and Conservation of the Mechanical Energy

What we have seen gives us fairly straightforwardly the next fundamental theorem, which prescribes the balance equation for the mechanical energy and its conservation in case all forces are conservative.

Theorem 4.18 (Balance and Conservation of Mechanical Energy) *Consider a system S of point particles P_k, with $k = 1, 2, \ldots, N$, of masses m_k respectively, satisfying principles C1–C4 (also in generalised form). Suppose S is subject to, in addition to possibly non-conservative forces, a system of conservative forces in a frame $\mathscr{R}$ with potential energy $\mathscr{U}|_{\mathscr{R}} = \mathscr{U}|_{\mathscr{R}}(P_1, \ldots, P_N)$.*

Let:

$$\mathscr{E}|_{\mathscr{R}} := \mathscr{T}|_{\mathscr{R}} + \mathscr{U}|_{\mathscr{R}}$$

be the **system's total mechanical energy** *in the frame $\mathscr{R}$. For any instant of absolute time t and on every motion of the system, the balance equation*

$$\Pi|_{\mathscr{R}}^{(noncons.)} = \frac{d\mathscr{E}|_{\mathscr{R}}}{dt}$$

holds, where $\Pi|_{\mathscr{R}}^{(noncons)}$ is the total power of the non-conservative forces in $\mathscr{R}$.

In case all forces acting on the system are conservative, the mechanical energy is a first integral of motion, *i.e. it is preserved during time on the system's motions.*

Proof By Eq. (4.10) it is sufficient to show that the power of the conservative forces $\Pi|_{\mathscr{R}}^{(cons)}$ satisfies

$$\Pi|_{\mathscr{R}}^{(cons)} = -\frac{d\mathscr{U}|_{\mathscr{R}}(P_1(t), \ldots, P_N(t))}{dt},$$

where $P_k = P_k(t)$ solves the equation of motion for particle systems. Indeed, from (4.11) we deduce:

$$\Pi|_{\mathscr{R}}^{(cons)} = \sum_k \pi^{(cons)}|_{\mathscr{R}k} = -\sum_k \mathbf{v}_{P_k} \cdot \nabla_{P_k} \mathscr{U}|_{\mathscr{R}}(P_1(t), \ldots, P_N(t))$$

$$= -\frac{d\mathscr{U}|_{\mathscr{R}}(P_1(t), \ldots, P_N(t))}{dt},$$

as claimed.   □

Remarks 4.19 In general, the *mechanical* energy is not preserved in real isolated physical systems, due to the non-conservative internal forces, which are always present in nature (internal frictions). By enlarging the class of physical systems studied, over the years it was understood that one can always define a contribution to the system's total energy (which may not be entirely mechanical, for example it could contain the electromagnetic field's energy or the internal thermodynamic energy) so that the total energy of an *isolated* system is preserved in time when evaluated in an inertial frame. The general principle stating that we can extend the notion of energy so to eventually achieve a conservation law is called **Principle of conservation of the energy**. The history of Physics tells that this principle is much more important than the above theorem (proved in a mechanical context) and holds in both relativistic theories and quantum ones alike; it is related to the homogeneity of time in inertial frames.   ■

Fig. 4.1 Exercise 4.20.**1**

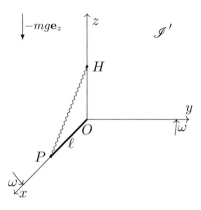

Exercises 4.20

(1) Consider a point particle P (Fig. 4.1) of mass m moving along a smooth horizontal rod ℓ passing through O. The rod is at rest in a frame $\mathscr{R}'$. Define right-handed orthonormal coordinates on $\mathscr{R}'$, centred at O, with x-axis along ℓ, y-axis horizontal and orthogonal to ℓ, and z-axis vertical and pointing toward a point H. P is subject to gravity $-mg\mathbf{e}_z$ and to a force generated by an ideal spring of zero rest length and elastic constant $\kappa > 0$, fixed at one end to P and at the other to the fixed point H at height h above O. The frame $\mathscr{R}'$ rotates about the $\mathbf{e}_z$-axis with constant angular velocity $\boldsymbol{\omega}_{\mathscr{R}'|\mathscr{R}} = \omega\mathbf{e}_z$ in an inertial frame $\mathscr{R}$.

   (i) Write the pure equations of motion for P and solve them in general, depending on the ratio $m\omega^2/\kappa$.
   (ii) In case $m\omega^2/\kappa < 1$, express the constraint reaction $\boldsymbol{\phi}$ in function of time, for the motion with initial conditions $P(0) - O = x_0\mathbf{e}_x$, with $x_0 > 0$, and $\mathbf{v}_P|_{\mathscr{R}'}(0) = \mathbf{0}$.
   (iii) Prove that the component parallel to z of the angular momentum of P with pole O is not preserved in $\mathscr{R}$ in general.
   (iv) Find all initial conditions (at $t = 0$) that produce motions for which the component parallel to z of the angular momentum of P with pole O with respect to $\mathscr{R}$ is preserved during time.

(2) Consider the physical system of the previous exercise.

   (i) Prove that in $\mathscr{R}'$ it makes mathematical sense to define a potential energy of some of the inertial forces, in addition to the spring's, so that the total mechanical energy is preserved in $\mathscr{R}'$.
   (ii) Find all initial conditions (at $t = 0$) producing motions where the mechanical energy is preserved in $\mathscr{R}$.

(3) Consider two point particles P and Q (Fig. 4.2) of mass M and m respectively. They move on the cylindrical surface C of equation $x^2 + y^2 = R^2$ with $R > 0$,

Fig. 4.2 Exercise 4.20.3

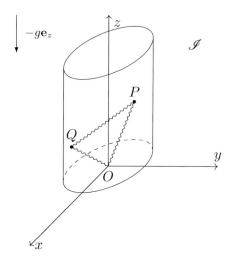

where x, y, z are orthonormal coordinates on the inertial frame $\mathscr{R}$. The particles are subject to the gravitational force given by the acceleration $\mathbf{g} := -g\,\mathbf{e}_z$, they are joint by a spring of elastic constant $\kappa > 0$ and zero rest length, and each is joint to the origin O by a further spring of elastic constant $\gamma > 0$ and zero length at rest.

(i) Write the pure equations of motion in cylindrical coordinates r, φ, z adapted to C, with $\varphi = 0$ along the x-axis.

(ii) Compute the constraint forces acting on the particles at the initial instant for initial conditions $P(0) - O = R\mathbf{e}_x$, $Q(0) - O = \mathbf{e}_y$, $\mathbf{v}_P(0) = \mathbf{v}_Q(0) = \mathbf{0}$.

(iii) Prove that the quantity $M\dot{\varphi}_P + m\dot{\varphi}_Q$ (as usual the dot denotes a time derivative) is preserved during time on every motion of the system.

4.3 Two Conservation Laws Arising from Invariance Properties of the Potential Energy

The use of conservative forces allows to start studying the relationship between invariance properties and the existence of first integrals, one of the core subjects of the book. To conclude we will therefore prove how the potential energy's translational and rotational invariance respectively imply that the total linear momentum and the total angular momentum are preserved, provided one works in inertial frames. All this does not depend on whether one assumes or not the Action-reaction principle (in weak or strong form).

In the sequel, given $O, P \in E_{\mathscr{R}}$ we will use the notation $g_{O,R}(P) := O + R(P - O)$ for a rotation of P about a fixed point O under the rotation R acting on the space of displacements $V_{\mathscr{R}}$. In particular, $R_{\mathbf{n},\theta}\mathbf{v}$ will be the counter-clockwise rotation of the vector $\mathbf{v}$ by θ around the unit vector $\mathbf{n}$.

Proposition 4.21 *Consider a system of N point particles $P_1, \ldots, P_N$ in an inertial frame $\mathscr{R}$ and suppose the only forces acting on the system are conservative and due to a C^1 potential energy $\mathscr{U}|_{\mathscr{R}} = \mathscr{U}|_{\mathscr{R}}(P_1, \ldots, P_N)$. The following facts hold.*

(1) If $\mathscr{U}|_{\mathscr{R}}$ is invariant under rigid displacements of the system along $\mathbf{n}$, i.e.:

$$\mathscr{U}|_{\mathscr{R}}(P_1 + \epsilon \mathbf{n}, \ldots, P_N + \epsilon \mathbf{n}) = \mathscr{U}|_{\mathscr{R}}(P_1, \ldots, P_N) \quad \forall \epsilon \in \mathbb{R},$$

then $\mathbf{P}|_{\mathscr{R}} \cdot \mathbf{n}$ is constant in time on the system's motions.

(2) If $\mathscr{U}|_{\mathscr{R}}$ is invariant under rigid rotations of the system about the axis $\mathbf{n}$ from the origin $O \in E_{\mathscr{R}}$, i.e.:

$$\mathscr{U}|_{\mathscr{R}}(g_{O, R_{\mathbf{n},\theta}}(P_1), \ldots, g_{O, R_{\mathbf{n},\theta}}(P_N)) = \mathscr{U}|_{\mathscr{R}}(P_1, \ldots, P_N) \quad \forall \theta \in \mathbb{R},$$

then $\boldsymbol{\Gamma}_O|_{\mathscr{R}} \cdot \mathbf{n}$ is constant in time on the system's motions.

Proof

(1) The assumptions imply

$$0 = -\left.\frac{\partial}{\partial \epsilon}\right|_{\epsilon=0} \mathscr{U}|_{\mathscr{R}}(P_1 + \epsilon \mathbf{n}, \ldots, P_N + \epsilon \mathbf{n})$$

$$= -\sum_{j=1}^{N} \mathbf{n} \cdot \nabla_{P_j} \mathscr{U}|_{\mathscr{R}}(P_1, \ldots, P_N)$$

$$= \mathbf{n} \cdot \sum_{j=1}^{N} \mathbf{F}_j(P_1, \ldots, P_N) \,.$$

At this point (1) in Proposition 4.6 implies the claim.

(2) The proof is analogous. By assumption,

$$0 = -\left.\frac{\partial}{\partial \epsilon}\right|_{\theta=0} \mathscr{U}|_{\mathscr{R}}(R_{\mathbf{n},\theta} P_1, \ldots, R_{\mathbf{n},\theta} P_N)$$

$$= -\sum_{j=1}^{N} \mathbf{n} \wedge (P_j - O) \cdot \nabla_{P_j} \mathscr{U}|_{\mathscr{R}}(P_1, \ldots, P_N)$$

$$= \sum_{j=1}^{N} \mathbf{n} \wedge (P_j - O) \cdot \mathbf{F}_j(P_1, \ldots, P_N)$$

where we have used (prove it by choosing right-handed orthonormal axes with $\mathbf{e}_z = \mathbf{n}$):

$$\frac{d}{d\theta}\bigg|_{\theta=0} R_{\mathbf{n},\theta}\mathbf{x} = \mathbf{n} \wedge \mathbf{x}.$$

Using the cyclic property of the mixed product:

$$0 = \sum_{j=1}^{N} \mathbf{n} \wedge (P_j - O) \cdot \mathbf{F}_j(P_1, \dots, P_N)$$

$$= \sum_{j=1}^{N} (P_j - O) \wedge \mathbf{F}_j(P_1, \dots, P_N) \cdot \mathbf{n} = \mathbf{n} \cdot \sum_{j=1}^{N} (P_j - O) \wedge \mathbf{F}_j(P_1, \dots, P_N).$$

Then (2) in Proposition 4.6 allows to conclude.   □

4.4   The Necessity of the Description in Terms of Continua and Fields in Classical Mechanics

The description given in basic Classical Mechanics is not sufficient to cover all phenomena known in classical Physics. There are in fact at least two very important situations where the point-particle picture is inadequate to describe the physical reality.

The first case is when we work with bodies called *continua*, in particular those called *fluids*. These are physical systems that appear, macroscopically, as *extended* (usually homogeneous and isotropic) and *continuous*: their spatial configurations, in absolute space and at any instant, are described geometrically by sets homeomorphic to open connected subsets of $\mathbb{R}^3$. In such cases the description given earlier must be modified in order to introduce suitable *density functions* that generalise the various quantities introduced. For example, one has densities of mass and (vector-valued) densities of forces, which integrated over portions of the body (or surface) respectively give back the total mass and total force on that portion. It is fundamental to this setup (for instance if we want to introduce the total force acting on a finite portion of a continuum) that the absolute space at any time admits an affine structure allowing to consider vectors that do not belong to the manifold's tangent space, and also in order to add vectors in different tangent spaces. The generalisation of particle Mechanics to continuous bodies is goes by the name of **Continuum Mechanics**. An important point, with a fundamental impact on the development of Physics, is that Continuum Mechanics also admits a *local* formulation of the laws of Dynamics: instead of referring to finite portions of a continuous body, it deals with points in a continuum. In such a simplified formulation, the integral-differential equations of finite portions of continuum are translated into purely differential equations at each point of the continuum. This formulation is based on the existence (proved by Cauchy) of a particular tensor field called *Cauchy stress tensor*. The relevant point is that once we have the local formulation, the affine structure is no longer needed and the formalism can be naturally adapted to more general situations. In particular, Einstein's theory of General Relativity exploits this possibility to the fullest extend, since it is formulated on a spacetime that cannot have an affine structure by deep physical reasons (the presence of the gravitational field).

The second case in which the formulation in terms of point particles is not sufficient to describe the physical reality is when we have *electromagnetic forces*. Here the forces acting on point particles are *mediated* by *force fields*. The basic idea is that every electrically charged particle (*source*) generates a *force field*, i.e. a pair of vector fields $\mathbf{E} = \mathbf{E}(t, P)$ (**electric field**) and $\mathbf{B} = \mathbf{B}(t, P)$ (**magnetic field**) defined at every point P in space and at every instant t, whose specific expressions depend on the characteristics and state of motion of the source. The other electrically charged particles in space are affected by the forces exclusively through the vector field at the point where they are located. More precisely, for a point particle of mass m and electric charge q in vacuum, that is at $P(t)$ at time t in the rest space of the inertial frame $\mathcal{R}$, we still have

$$m\mathbf{a}_P|_{\mathcal{R}}(t) = \mathbf{F}_L(t, P(t), \mathbf{v}_P|_{\mathcal{R}}(t))$$

where the term on the right is the **Lorentz force**

$$\mathbf{F}_L(t, P(t), \mathbf{v}_P|_{\mathcal{R}}(t)) := q\mathbf{E}(t, P(t)) + \frac{q\mathbf{v}_P|_{\mathcal{R}}(t)}{c} \wedge \mathbf{B}(t, P(t)) \qquad (4.14)$$

and c is a constant equal to the speed of light in vacuum. Observe the explicit presence of t in $\mathbf{F}_L(t, P(t), \mathbf{v}_P|_{\mathcal{R}}(t))$, highlighting that the field source depends on the motion. The *experimental* fact making the idea of field crucial, rather than useless, is that if the source changes its state of motion, the field around it deforms with some delay: the field's perturbations move at finite speed and the forces acting on the other charged particles are affected by the delay. In this situation it is clear that, in presence of moving charges, the Action-reaction principle is no longer valid and therefore *the Principle of conservation of the impulse stops holding, if by impulse we mean the sum of the impulses of the single charged particles.* We can however prove that the Principle of conservation of the impulse continues to be valid for the overall system made of *charges + electromagnetic field*, as long as we define a contribution to the total impulse coming from the field itself. This contribution arises by integrating over space a momentum density associated with the electromagnetic field. *In a similar manner, the conservation laws of energy and angular momentum extend* as well. Finally, it can be proved that the dynamics of the overall system can be formulated with local equations (in the aforementioned sense) equivalent to the initial integral-differential equations. At the end of the day, the electromagnetic field is treated, albeit with important differences, in a similar way to what we said about continua (In particular, one defines a stress tensor for the electromagnetic field, called Maxwell tensor).

Remarks 4.22

(1) It is clear that, in reality, continua do not exist, because all physical bodies, when examined microscopically, reveal a molecular hence discrete structure. It is at this level that the new very important phenomena occur, and the only known description for the latter is the one based on Quantum Mechanics. The electromagnetic field, too, has a quantum nature based on the concept of *photon*, which also shares a relativistic character.

(2) To all effects the elementary electric charges, the electrons, look like point particles, meaning they are dimensionless (their only structure is the one due to the so-called *spin*). In truth, at the classical level (including the non-quantised relativistic study of charges and of the electromagnetic field) the picture of the electric charge as a point and of a non-quantised electromagnetic field produces severe mathematical inconsistencies signalled by the appearance of infinite quantities (eigenforce, eigenenergy, . . .) [Jac98]. These inconsistencies are related to the fact that spatially extended electric charges, born out of the equations of electromagnetism, interact with themselves via the electromagnetic field they produce. This phenomenon is physically evident in the *Bremsstrahlung* occurring when we try to accelerate a charge.[1]   ∎

---

[1] For this reason the *definition* itself of mass of a charged particle is problematic; indeed, it has been often proposed, in earlier classical or relativistic contexts, to define the electron's mass via the decelerating self-interaction with the field that the charge itself emits [Jac98]. Such theories, despite being very interesting, never succeeded to properly convince the community.

Chapter 5
Introduction to Rigid Body Mechanics

The goal of this chapter is to introduce students to the rudiments of the theory of *rigid bodies*. Rigid bodies, sometimes also called *solid bodies*, are those physical systems that due to their inner structure, i.e. the inner constraints, do not change shape (metrically, not only topologically) when subjected to any kind of external force. Such bodies are clearly idealisations, a very useful one in the practice, of several physical bodies that surround us. In reality, perfectly rigid bodies do not exist, since any known physical system deforms when subjected to external forces. The theory describing the relationship between stresses and deformations is studied in *Continuum Mechanics*, which is not an elementary part of Mechanics and therefore will not be treated in this book. It is also important to remember that, physically, strictly rigid bodies cannot exist by first principles, due to the finite velocity of any type of interaction, which ultimately is responsible for the metric deformations of the physical objects.

5.1 The Rigidity Constraint for Discrete and Continuous Systems

There are two classes of rigid bodies: rigid bodies described by systems of finitely many point particles, and continuous systems, also called continuous rigid bodies, which require a generalisation of the definitions gives thus far. In the sequel we shall examine both.

V. Moretti, *Analytical Mechanics*, La Matematica per il 3+2 150,
https://doi.org/10.1007/978-3-031-27612-5_5

5.1.1   Generic Rigid Bodies

Definition 5.1 A system S of finitely many (at least 2) point particles is called, indifferently, **rigid system**, **solid system**, **rigid body** or **solid body** if, due to the internal forces, the distances (measured in every absolute space at time t) between the system's particles are constant in time, for any motion of the system and irrespective of the external forces acting on S. In other words, a rigid body is a system of point particles that satisfy the **rigidity constraint**.   ◇

Practically speaking, the following theorem is hugely important because it allows to set up in a simple way the problem of a rigid body's dynamics.

Proposition 5.2 *A system of point particles S is a rigid body if and only if there exists frame $\mathscr{R}_S$ in which the points of S are always at rest. The frame $\mathscr{R}_S$ is said to be* **moving** *with S and the vector $\boldsymbol{\omega}_{\mathscr{R}_S|\mathscr{R}}$, where $\mathscr{R}$ is any frame, is the $\boldsymbol{\omega}$* **vector of** *S* **with respect to** *$\mathscr{R}$.*

Proof If there is a frame $\mathscr{R}_S$ where all points of S are at rest, the distances between the points of S do not depend on time and hence are constant in time. As distances are absolute, this will happen in any other frame (for the absolute distance on every absolute space at time t), and so S will be a rigid body. Conversely, assume S is a rigid body. If S contains at least three non-collinear points P_1, P_2, P_3 at a certain instant, we can orthonormalise the basis of vectors $P_1 - P_2$, $P_3 - P_2$, $(P_1 - P_2) \wedge (P_3 - P_2)$ (possibly changing their orientations) and choose one point as origin O, so to obtain a frame $\mathscr{R}_S$ in which the above vectors are always at rest. The vector basis and the origin O define rest orthonormal Cartesian coordinates on $\mathscr{R}_S$. By virtue of the rigidity constraint, any other point particle of S must be at rest in $\mathscr{R}_S$, since its coordinates are constant in time in the orthonormal coordinates of $\mathscr{R}_S$. $\mathscr{R}_S$ is therefore a moving frame on S. In case all point particles of S are collinear, at time t, say along the line $\mathbf{u}$, they will stay collinear forever and keep at the same distance from each other, by the rigidity constraint. Taking the axis $\mathbf{u}$, one point in S as origin and any two vectors forming with $\mathbf{u}$ a right-handed orthonormal basis, we obtain a moving frame $\mathscr{R}_S$ on S.   □

Definition 5.3 A point O, evolving along a certain world line, is **at rest with the rigid body** S if it is at rest in a moving frame on S.   ◇

Remarks 5.4

(1) It is clear that if S is made of at least 3 non-collinear points there is a unique moving frame on S. In fact, since the triple of axes built as in the proof must be at rest in any moving frame of S, only changes of Cartesian coordinates that are independent of time are admissible when we change the moving frame. In other words the two frames coincide. Vice versa, if S only contains points along $\mathbf{u}$, there are infinitely many frames moving with S, since once we fix $\mathscr{R}_S$ as in the proof, any other frame that rotates arbitrarily with respect to the former around $\mathbf{u}$ moves with S.

(2) Only when S consists of collinear points, the ω vector of S with respect to a frame $\mathscr{R}$ actually depends on the choice of moving frame $\mathscr{R}_S$. ∎

The following proposition explains the usefulness of the existence of moving frames for rigid bodies.

Proposition 5.5 *Consider a system of point particles S subject to some forces (possibly reactive and/or inertial) with total net force $\mathbf{R}$ and total momentum $\mathbf{M}_O$ with respect to a point O at rest with S. The system's total power, with respect to an arbitrary frame $\mathscr{R}$, satisfies:*

$$\Pi|_{\mathscr{R}} = \omega_{\mathscr{R}_S}|_{\mathscr{R}} \cdot \mathbf{M}_O + \mathbf{v}_O|_{\mathscr{R}} \cdot \mathbf{R} . \tag{5.1}$$

In particular, the total power of the internal forces acting on S is always zero, irrespective of the frame $\mathscr{R}$.

Proof If $\mathbf{f}_i$, $i = 1, \ldots, N$, is the system of forces, where $\mathbf{f}_i$ acts on the i-th particle P_i of S with velocity $\mathbf{v}_i|_{\mathscr{R}}$ in $\mathscr{R}$, by the known kinematical relations of Sect. 2.4.2 we have:

$$\Pi|_{\mathscr{R}} = \sum_{i=1}^{N} \mathbf{f}_i \cdot \mathbf{v}_i|_{\mathscr{R}} = \sum_{i=1}^{N} \mathbf{f}_i \cdot \left(\mathbf{v}_O|_{\mathscr{R}} + \omega_{\mathscr{R}_S}|_{\mathscr{R}} \wedge (P_i - O) \right) ,$$

where we have used that $\mathbf{v}_i|_{\mathscr{R}_S} = \mathbf{0}$ by definition of moving frame on S. The formula immediately leads to (5.1). In case the system of forces is made of the internal forces of S, we know from Sect. 4.1.2 that the total force and total momentum are zero, so the last claim holds. □

Remarks 5.6 *The possible configurations of a rigid body made of 3 non-collinear points at least are described by the points of the manifold $\mathbb{R}^3 \times SO(3)$, where $SO(3)$ is the group of 3×3 real orthogonal matrices. In fact, determining completely the position of a rigid body S with at least three non-collinear points, in the rest space of a given frame at a given instant, is equivalent to determining the position of a right-handed triple of orthonormal axes moving with S. Given right-handed orthonormal coordinates on $\mathscr{R}$, the origin of the triple moving with S is completely determined by a position vector, i.e. a column vector in $\mathbb{R}^3$, while the position of the axes of the moving triple of S is completely determined, with respect to the coordinate axes of $\mathscr{R}$, by an orthogonal matrix of determinant 1, since both triples are right-handed. In this way we have proved that the configurations of S are in 1-1 correspondence with $\mathbb{R}^3 \times SO(3)$. To find the rotation matrix we may use the 3 Euler angles between the triples (see Exercises 2.30). In conclusion, a rigid body with at least three non-collinear points has 6 "degrees of freedom".* ∎

5.1.2   Continuous Rigid Bodies

On every space at absolute time Σ_t there is a natural volume measure dv associated with the affine Euclidean structure. The simplest way to define it is to think of it as induced by any system of orthonormal coordinates on Σ_t, by putting $dv := dx^1 dx^2 dx^3$, where $dx^1 dx^2 dx^3$ is the ordinary Lebesgue measure on $\mathbb{R}^3$. Since the Jacobian matrix of an orthonormal coordinate change is a rotation $R \in O(3)$ with $\det R = 1$, the measure dv is indeed invariant under orthonormal coordinate transformations and therefore well defined. Such a measure, obviously, induces a measure on each rest space $E_{\mathcal{R}}$ of any frame $\mathcal{R}$ in spacetime, under the identification of $E_{\mathcal{R}}$ with Σ_t.

Similarly, we can define a natural measure for immersed surfaces in any Σ_t and for the length of (rectifiable) curves immersed in every Σ_t. These measures coincide with the analogous measures on the frames' rest spaces in spacetime. With these measures it is easy to see that compact sets (segments, planar regions and portions of space) have finite measure.

In the sequel we will consider an elementary generalisation of the notion of rigid body, namely a **continuous system**, or **continuous rigid body**. By this we shall mean a physical system S described, in the rest space $E_{\mathcal{R}_S}$ of a frame $\mathcal{R}_S$, by a connected, closed and bounded set $C \subset E_{\mathcal{R}_S}$. C may be a finite union of segments, of planar regions, of portions of space, or of these three types together. The only portions of the plane or space we will consider are closures of open bounded sets whose boundaries are piecewise C^∞ curves or surfaces, respectively. Each segment I will have a mass $m_I \in [0, +\infty)$, obtained integrating a **linear density of mass** given by a strictly positive continuous function λ_I defined on the segment. Each planar region Σ will have a mass $m_\Sigma \in [0, +\infty)$, obtained integrating a **surface density of mass** given by a strictly positive continuous function σ_Σ defined on the region. Each portion of space V will be equipped with a mass $m_V \in [0, +\infty)$, obtained integrating a **linear density of mass** given by a strictly positive continuous function ρ_V defined on the spatial region. The integrals are obviously taken with respect to the aforementioned measures. In case the continuous rigid body consists of a single segment, a single planar region or a single portion of space, and the corresponding density of mass is constant, the continuous rigid body is called **homogeneous**. We further suppose the external forces acting on the continuous rigid body are finite in number and act on given points in the body. We will therefore not treat *densities of forces*. Discussing internal forces would be much more complex and it would not be possible to avoid force densities. We will not need that since it will suffice, generalising the result for discrete systems and Proposition 5.5, to assume that the total internal force, the total momentum with respect to any point O and frame $\mathcal{R}$, and the total power of internal forces with respect to any frame, are all zero.

Remarks 5.7 What we said in Remarks 5.6 applies immediately to continuous rigid bodies: *the possible configurations of a continuous rigid body made of at least 3 non-collinear points are described by the points of the manifold $\mathbb{R}^3 \times SO(3)$. Note* that S will not possess three non-collinear points only if it is a linear continuous rigid body, i.e. a concrete segment.   ∎

Definition 5.8 Consider a continuous rigid body S determined by the space region V (closure of a non-empty open set) in the rest space of a moving frame on S. Let $\mathscr{R}$ be a second frame such that S is described by the collection of regions $\{V(t) \subset E_{\mathscr{R}} \mid t \in \mathbb{R}\}$ at time varies. Finally, let ρ denote the density of mass of S and $M := \int_V \rho(P)\, dv(P) = \int_{V(t)} \rho(Q)\, dv(Q)$ the total mass of S, where dv is the standard Lebesgue measure on Σ_t induced by that of $\mathbb{R}^3$.

We define the following concepts.

(1) The **centre of mass** of S at time $t \in \mathbb{R}$, $G(t)$, is the point (not necessarily a point particle of the system) on every Σ_t given by the equation:

$$M(G(t) - O) = \int_{V(t)} (P - O)\,\rho(P)\, dv(P),$$

where $O \in \Sigma_t$ is an arbitrary point.

(2) The **(total) impulse** or **(total) momentum** of S with respect to $\mathscr{R}$ at time t is the vector of V_t:

$$\mathbf{P}|_{\mathscr{R}}(t) := \int_{V(t)} \rho(P)\mathbf{v}_P|_{\mathscr{R}}(t)\, dv(P).$$

(3) If $O = O(t)$ is any world line (not necessarily that of (1)) and $\mathscr{R}$ is a frame, the system's **(total) angular momentum** with respect to the pole O and $\mathscr{R}$ at time t is the vector of V_t:

$$\mathbf{\Gamma}_O|_{\mathscr{R}}(t) := \int_{V(t)} \rho(P)(P - O(t)) \wedge \mathbf{v}_P|_{\mathscr{R}}(t)\, dv(P).$$

$\diamond$

From now on we will omit, as usual, to indicate the time dependency unless necessary.

The relevance of the above quantities is essentially due to the fact that, under certain assumptions and for a given mechanical system, either they are preserved in time or they crop up in the definition of other quantities that are preserved in time. In many cases the knowledge of the values of preserved quantities provides important information on the system's motion, even when we cannot solve the equation of motion explicitly.

The above definitions extend easily to continuous rigid systems defined by segments, planar regions or systems made of unions of volumes, surfaces and segments.

Remarks 5.9

(1) G is well defined, in the sense that it is uniquely determined, once O is fixed, by:

$$G := O + \frac{1}{M} \int_V \rho(P)(P - O)\, dv(P)\,,$$

and furthermore G *does not depend on the choice of* O. In fact, if we define G_O and $G_{O'}$ respectively by

$$M(G_O - O) = \int_V \rho(P)(P - O)\, dv(P) \quad \text{and} \quad M(G_{O'} - O')$$

$$= \int_V \rho(P)(P - O')\, dv(P)\,,$$

then:

$$G_O - G_{O'} = (O - O') + \int_V \frac{\rho(P)}{M}(P - O)dv(P)$$

$$- \int_V \frac{\rho(P)}{M}(P - O')\, dv(P)$$

$$= (O - O') - (O - O') = \mathbf{0}\,.$$

(2) If $\mathbf{v}_G|_{\mathscr{R}}$ is the velocity of the centre of mass in $\mathscr{R}$ for a system of point particles of total mass M, then

$$\mathbf{P}|_{\mathscr{R}} = M\mathbf{v}_G|_{\mathscr{R}}\,. \tag{5.2}$$

In other words: *the system's total impulse is the same that a single point particle of mass M concentrated at the centre of mass would have.*
    Checking this is immediate: choose a world line $O = O(t)$, differentiate in time the identity $M(G(t) - O(t)) = \int_V \rho(P)(P - O(t))\, dv(P)$ and remember that $M = \int_V \rho(P)\, dv(P)$.

(3) By the definition of centre of mass, if we split a system into finitely many subsystems, the centre of mass of the overall system coincides with the centre of mass of a system of point particles made by the centres of mass of the single subsystems, each having as mass the total mass of the corresponding subsystem.

(4) From the definition of centre of mass we deduce that the centre of mass of a continuous rigid system lies on any symmetry plane of the system.

(5) Exactly as for systems made by finitely many point particles, also for continuous rigid bodies one can easily prove the following relations, replacing the sums with suitable integrals (see Exercises 4.3 and their solutions). Passing from pole O to O', and staying in the same frame $\mathscr{R}$, the angular momentum (at given time

$t \in \mathbb{R}$) changes by the formula:

$$\mathbf{\Gamma}_O|_{\mathscr{R}} = \mathbf{\Gamma}_{O'}|_{\mathscr{R}} + (O' - O) \wedge \mathbf{P}|_{\mathscr{R}}. \tag{5.3}$$

In particular, choosing $O' = G$, $\mathbf{\Gamma}_O|_{\mathscr{R}}$ can always be written as the sum of the total angular momentum in $\mathscr{R}$ with respect to G plus the angular momentum of one point particle placed at G with mass equal to the system's total mass:

$$\mathbf{\Gamma}_O|_{\mathscr{R}} = \mathbf{\Gamma}_G|_{\mathscr{R}} + (G - O) \wedge \mathbf{P}|_{\mathscr{R}}. \tag{5.4}$$

∎

5.2 The Inertia Tensor and Its Properties

We introduce now an extremely useful mathematical instrument that will allows us to formulate very clearly the relationships between the quantities appearing in the mechanical laws of a rigid body S (continuous or not), in particular the governing equation of Dynamics. The tool in question is a linear operator $\mathbf{I}_O$ of the space of displacements $V_{\mathscr{R}_S}$ of the frame of S, called **inertia tensor**, that depends on a point $O \in E_{\mathscr{R}_S}$.

5.2.1 The Inertia Tensor

To introduce the inertia tensor, we start by finding the expression for the angular momentum $\mathbf{\Gamma}_O|_{\mathscr{R}}$ of a rigid body with respect to a frame $\mathscr{R}$ and pole O, in two situations. In the sequel $\mathscr{R}_S$ will be a frame moving with the rigid body S, which for simplicity we shall assume discrete, made of N point particles P_i of masses m_i; the relations we will find hold for continuous rigid bodies too, provided we replace in the proofs the sums with integrals.

Case 1   *The pole O is at rest with the rigid body.* Here we have

$$\mathbf{\Gamma}_O|_{\mathscr{R}} = \sum_{i=1}^N (P_i - O) \wedge m_i \mathbf{v}_i|_{\mathscr{R}} = \sum_{i=1}^N (P_i - O) \wedge m_i [\mathbf{v}_O|_{\mathscr{R}} + \boldsymbol{\omega}_{\mathscr{R}_S}|_{\mathscr{R}} \wedge (P_i - O)].$$

Hence, if $M = \sum_i m_i$ is the rigid body's total mass,

$$\mathbf{\Gamma}_O|_{\mathscr{R}} = M(G - O) \wedge \mathbf{v}_O|_{\mathscr{R}} + \mathbf{I}_O(\boldsymbol{\omega}_{\mathscr{R}_S|_{\mathscr{R}}}), \tag{5.5}$$

where we have introduced the **inertia tensor** $\mathbf{I}_O : V_{\mathscr{R}_S} \to V_{\mathscr{R}_S}$ of the rigid body S with respect to the point O,

$$\mathbf{I}_O(\mathbf{a}) := \sum_{i=1}^{N} m_i (P_i - O) \wedge [\mathbf{a} \wedge (P_i - O)], \quad \text{for any } \mathbf{a} \in V_{\mathscr{R}_S}. \qquad (5.6)$$

In the continuous case, for instance in 3 dimensions, we have instead

$$\mathbf{I}_O(\mathbf{a}) = \int_{V_S} \rho_V(P)(P-O) \wedge [\mathbf{a} \wedge (P-O)] dv(P), \quad \text{for any } \mathbf{a} \in V_{\mathscr{R}_S}. \qquad (5.7)$$

Case 2 *The pole O is in motion.* In this case

$$\boldsymbol{\Gamma}_O|_{\mathscr{R}} = \sum_{i=1}^{N}(P_i - O) \wedge m_i \mathbf{v}_i|_{\mathscr{R}} = \sum_{i=1}^{N}(P_i - G) \wedge m_i \mathbf{v}_i|_{\mathscr{R}} + \sum_{i=1}^{N}(G - O) \wedge m_i \mathbf{v}_i|_{\mathscr{R}}.$$

We can reduce to the previous case regarding the term $\sum_i (P_i - G) \wedge m_i \mathbf{v}_i|_{\mathscr{R}}$ on the right, because by the rigidity of S the centre of mass G of S will always move with S. Hence:

$$\boldsymbol{\Gamma}_O|_{\mathscr{R}} = M(G - G) \wedge \mathbf{v}_G|_{\mathscr{R}} + I_G(\boldsymbol{\omega}_{\mathscr{R}_S}|_{\mathscr{R}}) + M(G - O) \wedge \mathbf{v}_G|_{\mathscr{R}}.$$

To sum up:

$$\boldsymbol{\Gamma}_O|_{\mathscr{R}} = M(G - O) \wedge \mathbf{v}_G|_{\mathscr{R}} + I_G(\boldsymbol{\omega}_{\mathscr{R}_S}). \qquad (5.8)$$

Let us show that the kinetic energy of S can also be expressed by a formula involving the inertia tensor. We consider a discrete rigid body, though the final formulas will hold in the continuous case, too, as is easy to prove.

Case 1 *Due to the reactive forces, in the moving frame $\mathscr{R}_S$ there is a point O that remains at rest with $\mathscr{R}$.* In this case, using the identity $\mathbf{a} \cdot \mathbf{b} \wedge \mathbf{c} = \mathbf{b} \cdot \mathbf{c} \wedge \mathbf{a}$, we have:

$$\mathscr{T}|_{\mathscr{R}} = \sum_{i=1}^{N} \frac{1}{2} m_i \mathbf{v}_i|_{\mathscr{R}} = \sum_{i=1}^{N} \frac{1}{2} m_i \left[\boldsymbol{\omega}_{\mathscr{R}_S}|_{\mathscr{R}} \wedge (P_i - O) \right]^2$$

$$= \frac{1}{2} \sum_i m_i \left[\boldsymbol{\omega}_{\mathscr{R}_S}|_{\mathscr{R}} \wedge (P_i - O) \right] \cdot \left[\boldsymbol{\omega}_{\mathscr{R}_S}|_{\mathscr{R}} \wedge (P_i - O) \right]$$

$$= \frac{1}{2} \boldsymbol{\omega}_{\mathscr{R}_S}|_{\mathscr{R}} \cdot \sum_i m_i (P_i - O) \wedge \left[\boldsymbol{\omega}_{\mathscr{R}_S}|_{\mathscr{R}} \wedge (P_i - O) \right].$$

Therefore, using Definition (5.6), we have found that

$$\mathcal{T}|_{\mathcal{R}} = \frac{1}{2}\boldsymbol{\omega}_{\mathcal{R}_S}|_{\mathcal{R}} \cdot \mathbf{I}_O(\boldsymbol{\omega}_{\mathcal{R}_S}|_{\mathcal{R}}) \, . \tag{5.9}$$

Case 2   *The rigid system S is in motion.* In this situation we can use König's theorem (Theorem 4.10) and obtain:

$$\mathcal{T}|_{\mathcal{R}} = \frac{1}{2}M\mathbf{v}_G|_{\mathcal{R}}^2 + \mathcal{T}|_{\mathcal{R}_G} \, .$$

Above, the frame $\mathcal{R}_G$ is such that G is at rest in it, and furthermore $\boldsymbol{\omega}_{\mathcal{R}_G}|_{\mathcal{R}} = \mathbf{0}$ by definition. To compute $\mathcal{T}|_{\mathcal{R}_G}$ we can then use the previous result (as G moves with S but is at rest for $\mathcal{R}_G$) and find

$$\mathcal{T}|_{\mathcal{R}_G} = \frac{1}{2}\boldsymbol{\omega}_{\mathcal{R}_S}|_{\mathcal{R}_G} \cdot I_G(\boldsymbol{\omega}_{\mathcal{R}_S}|_{\mathcal{R}_G}) \, .$$

By the composition law for $\boldsymbol{\omega}$ vectors we immediately discover that (see Proposition 2.28) $\boldsymbol{\omega}_{\mathcal{R}_S}|_{\mathcal{R}_G} = \boldsymbol{\omega}_{\mathcal{R}_S}|_{\mathcal{R}}$. The final formula is therefore:

$$\mathcal{T}|_{\mathcal{R}} = \frac{1}{2}M\mathbf{v}_G|_{\mathcal{R}}^2 + \frac{1}{2}\boldsymbol{\omega}_{\mathcal{R}_S}|_{\mathcal{R}} \cdot I_G(\boldsymbol{\omega}_{\mathcal{R}_S}|_{\mathcal{R}}) \, . \tag{5.10}$$

We have seen how the inertia tensor allows to write more compactly the angular momentum and the kinetic energy of a rigid system. In the next section we shall study the properties of the inertia tensor so to make its computation easier in concrete cases. We finish off with the formal definitions of the inertia tensor and the *moment of inertia*.

Definition 5.10   Consider a rigid body S and let O be a point in the space of a frame $\mathcal{R}_S$ moving with S. The **inertia tensor of S with respect to the point** O is the linear operator $\mathbf{I}_O : V_{\mathcal{R}_S} \to V_{\mathcal{R}_S}$ defined by: formula (5.6), if S is a discrete rigid system of N point particles P_i with masses m_i; by (5.7) in case S is a continuous rigid system defined on the portion of space V_S with density of mass ρ_S; by the analogous formulas in more complex situations that involve surfaces and segments.
   If $\mathbf{n} \in V_{\mathcal{R}_S}$ is a unit vector,

$$I_{O,\mathbf{n}} := \mathbf{n} \cdot \mathbf{I}_O(\mathbf{n}) \tag{5.11}$$

is called **moment of inertia of S with respect to the axis through O parallel to n.**   ◇

5.2.2 Principal Triples of Inertia

The inertia tensor possesses a number of properties that simplify its calculation. We will see in a short while that for any given point O moving with a rigid system S, there are three axes originating from O with respect to which the matrix representing the operator $\mathbf{I}_O$ is diagonal, and hence determined by 3 coefficients. Among other things we will show how to determine, via symmetry considerations, these axes.

The following proposition explains the main properties of the inertia tensor.

Proposition 5.11 *The inertia tensor $\mathbf{I}_O : V_{\mathscr{R}_S} \to V_{\mathscr{R}_S}$ of a rigid body S (discrete, continuous, or made of mixed parts) satisfies the following properties.*

(1) Suppose S is the union of two rigid systems S_1 and S_2, assumed disjoint if both discrete, or with $S_1 \cap S_2$ of zero measure if S_1, S_2 are continuous and of the same dimension. In these cases the inertia tensor $\mathbf{I}_O$ is the sum of the inertia tensors S_1 and S_2 with respect to the same point O.

(2) $\mathbf{I}_O$ is a symmetric operator, in other words its matrix in any orthonormal basis in $V_{\mathscr{R}_S}$ is symmetric. The matrix coefficients I_{Oij}, in the right-handed orthonormal basis $\mathbf{e}_1, \mathbf{e}_2, \mathbf{e}_3$, for simplicity thought of at O, have the form

$$I_{Oij} = \mathbf{e}_i \cdot \mathbf{I}_O(\mathbf{e}_j) = \sum_{k=1}^{N} m_k \left(\mathbf{x}_{(k)}^2 \delta_{ij} - x_{(k)i} x_{(k)j} \right) , \tag{5.12}$$

if

$$\mathbf{x}_{(k)} := P_k - O = \sum_{j=1}^{3} x_{(k)j} \mathbf{e}_j \quad for \ k = 1, \dots, N$$

where the N points P_k of masses m_K form S, or

$$I_{Oij} = \mathbf{e}_i \cdot \mathbf{I}_O(\mathbf{e}_j) = \int_{V_S} \rho(\mathbf{x}) \left(\mathbf{x}^2 \delta_{ij} - x_i x_j \right) dv(\mathbf{x}) , \tag{5.13}$$

for a continuous rigid body given by a portion of space V with density of mass ρ. Similar expressions hold for continuous rigid bodies given by parts of surfaces or segments, or even more complex bodies.

(3) If S is not made of collinear points along one axis, then $\mathbf{I}_O$ is positive definite. In other words it is positive semi-definite:

$$\mathbf{a} \cdot \mathbf{I}_O(\mathbf{a}) \geq 0 \quad for \ any \ \mathbf{a} \in V_{\mathscr{R}_S}$$

and furthermore:

$$\mathbf{a} \cdot \mathbf{I}_O(\mathbf{a}) = 0 \quad \textit{implies } \mathbf{a} = \mathbf{0}.$$

In this case every moment of inertia $I_{O,\mathbf{n}}$ is strictly positive.

(4) *If S consists of collinear points along one axis with unit vector $\mathbf{n}$, then $\mathbf{I}_O$ is positive definite, but $\mathbf{a} \cdot \mathbf{I}_O(\mathbf{a}) = 0$ if and only if $\mathbf{a} = \alpha\mathbf{n}$, with $\alpha \in \mathbb{R}$. In particular all moments of inertia are strictly positive, except for $I_{O,\mathbf{n}} = 0$.*

(5) *For any unit vector $\mathbf{u} \in E_{\mathscr{R}_S}$ and point $O \in E_{\mathscr{R}_S}$, the formula*

$$I_{O,\mathbf{u}} = \mathbf{u} \cdot \mathbf{I}_O(\mathbf{u}) = \sum_{k=1}^{N} m_k d_k^2 \tag{5.14}$$

holds, where d_k is the distance of the point particle P_k of S with mass m_k to the axis through O with tangent vector $\mathbf{u}$. In the continuous case the formula is similar,

$$I_{O,\mathbf{u}} = \mathbf{u} \cdot \mathbf{I}_O(\mathbf{u}) = \int_{V_S} \rho_V(P) d(P)^2 \, dv(P) \,, \tag{5.15}$$

and analogously for one- and two-dimensional continuous bodies.

Proof Statement (1) is trivial by the additivity of the sum or the integral, so the proof is immediate.

Let us pass to (2). By definition of inertia tensor (5.6) and (5.7) and the vector identity $\mathbf{a} \wedge (\mathbf{b} \wedge \mathbf{c}) = \mathbf{a} \cdot \mathbf{c}\,\mathbf{b} - \mathbf{a} \cdot \mathbf{b}\,\mathbf{c}$, we obtain (5.12) and (5.13). From the latter expressions it is clear that the matrix of $\mathbf{I}_O$ is symmetric, and so is the operator $\mathbf{I}_O$. As for the definiteness in (3) and (4), for any unit vector $\mathbf{u}$, because of (5.12), we have:

$$\mathbf{u} \cdot \mathbf{I}_O(\mathbf{u}) = \sum_{k=1}^{N} m_k \left(\mathbf{x}_{(k)}^2 \mathbf{u} \cdot \mathbf{u} - (\mathbf{x}_{(k)} \cdot \mathbf{u})^2 \right) = \sum_{k=1}^{N} m_k \left(\mathbf{x}_{(k)}^2 - (\mathbf{x}_{(k)} \cdot \mathbf{u})^2 \right) \,.$$

In other words, if $d_k \geq 0$ is the distance of P_k (of mass $m_k > 0$) to the axis through O parallel to $\mathbf{u}$, then

$$\mathbf{u} \cdot \mathbf{I}_O(\mathbf{u}) = \sum_{k=1}^{N} m_k d_k^2 \geq 0 \,.$$

That, together with (5.14), shows that in all cases $\mathbf{I}_0$ is positive semi-definite. We still do not know whether the tensor is positive when the points of S are not collinear. Since each mass m_k is strictly positive, for the right-hand side to vanish it is necessary, and sufficient, that all P_k lie along an axis parallel to $\mathbf{u}$ and passing through O. In this way all distances $d_k \geq 0$ would vanish. When the points of S are not collinear that cannot happen, and $\mathbf{u} \cdot \mathbf{I}_O(\mathbf{u}) > 0$ however we pick the unit vector $\mathbf{u}$. In this case, since for a generic vector $\mathbf{a} := \alpha\mathbf{u}$ for some $\alpha \in \mathbb{R}$ and some unit vector $\mathbf{u}$, then $\mathbf{a} \cdot \mathbf{I}_O(\mathbf{a}) = 0$ implies $\alpha^2\mathbf{u} \cdot \mathbf{I}_O(\mathbf{u}) = 0$, so $\alpha = 0$ i.e. $\mathbf{a} = 0$. If S is made by points collinear with $\mathbf{u}_0$, the previous reasoning can be repeated for any other unit vector $\mathbf{u} \neq \mathbf{u}_0$ and any $\mathbf{a}$ not parallel to $\mathbf{u}_0$. The proof of the continuous case is totally similar. The proof of (5) in the discrete case was given above, and the continuous case is analogous.   □

That the operator $\mathbf{I}_0$ is symmetric, as shown above, has a very important consequence. Elementary Linear Algebra teaches us that if $A : V \rightarrow V$ is a symmetric linear operator on a finite-dimensional real vector space V with an inner product as in Definition 1.9, there exists an orthonormal basis (orthonormal for the inner product) made of eigenvectors of A. Therefore the symmetric matrix of A is diagonal in that basis and contains on the main diagonal the eigenvalues of A. Consequently, for a rigid body S and a given point $O \in \mathscr{R}_S$, there always exists an *orthonormal basis* of $V_{\mathscr{R}_S}$ (right-handed, without loss of generality) *made of eigenvectors of* $\mathbf{I}_O$. In this basis $\mathbf{I}_O$ is *diagonal*. Observe that if I_k is an eigenvalue of $\mathbf{I}_O$ with eigenvector $\mathbf{e}_k$ in the orthonormal basis, then

$$\mathbf{e}_k \cdot \mathbf{I}_O(\mathbf{e}_k) = \mathbf{e}_k \cdot I_k\mathbf{e}_k = I_k 1 = I_k \ .$$

Hence, still in the orthonormal basis of eigenvectors of $\mathbf{I}_O$, the eigenvalues I_k, i.e. the diagonal elements of the matrix of $\mathbf{I}_O$ in that basis, are the *moments of inertia with respect to axes parallel to the unit basis at* O. We may summarise all of this in a definition.

Definition 5.12 Let S be a rigid body and O a point in the rest space $E_{\mathscr{R}}$ of a moving frame $\mathscr{R}_S$ on S.

(1) A right-handed orthonormal basis of eigenvectors of the inertia tensor $\mathbf{I}_O$ of S with respect to O (which always exists by (2) in Proposition 5.11) is called **principal triple of inertia** of S with respect to O.
(2) The unit vectors of this basis are the **principal axes of inertia** of S with respect to O.
(3) The eigenvalues of this basis, i.e. the diagonal elements of the diagonal matrix of $\mathbf{I}_O$ in the principal basis are called **principal moments of inertia** of S with respect to O.   ◇

Using principal triple of inertia simplifies considerably the expression of the angular momentum and kinetic energy of a rigid body. Suppose $I_{O'k}$, $k = 1, 2, 3$, are the principal moments of inertia of the rigid body S with respect to the generic point $O' \in \mathscr{R}_S$, that $\{\hat{\mathbf{e}}_k\}_{k=1,2,3}$ is the corresponding principal triple of inertia, and set

$$\omega_{\mathscr{R}_S|\mathscr{R}} = \sum_{k=1}^{3} \hat{\omega}_k\hat{\mathbf{e}}_k \ .$$

If M is the body's total mass, we have the following possibilities for the total angular momentum $\boldsymbol{\Gamma}_O|_{\mathscr{R}}$.

Case 1   *The pole O is at rest with the rigid body.* Choosing $O' = O$, (5.5) reduces to:

$$\boldsymbol{\Gamma}_O|_{\mathscr{R}} = M(G - O) \wedge \mathbf{v}_O|_{\mathscr{R}} + \sum_{k=1}^{3} I_{Ok}\hat{\omega}_k\hat{\mathbf{e}}_k \, . \tag{5.16}$$

Case 2   *The pole O is in motion.* If $O' = G$, (5.8) reduces to:

$$\boldsymbol{\Gamma}_O|_{\mathscr{R}} = M(G - O) \wedge \mathbf{v}_G|_{\mathscr{R}} + \sum_{k=1}^{3} I_{Gk}\hat{\omega}_k\hat{\mathbf{e}}_k \, . \tag{5.17}$$

For the total kinetic energy $\mathscr{T}|_{\mathscr{R}}$ we have two cases.

Case 1   *Due to the constraints, in the moving frame $\mathscr{R}_S$ there is a point O that remains at rest with $\mathscr{R}$.* In this case, for $O = O'$, (5.9) reduces to:

$$\mathscr{T}|_{\mathscr{R}} = \frac{1}{2}\sum_{k=1}^{3} I_{Ok}\hat{\omega}_k^2 \, . \tag{5.18}$$

Case 2   *The rigid system S is in motion.* Here, for $O' = G$, (5.10) reduces to:

$$T|_{\mathscr{R}} = \frac{1}{2}M\mathbf{v}_G|_{\mathscr{R}}^2 + \frac{1}{2}\sum_{k=1}^{3} I_{Gk}\hat{\omega}_k^2 \, . \tag{5.19}$$

Examples 5.13 A particularly interesting application of these formulas is the study of the motion of a rigid body S in a reference frame $\mathscr{R}$ where a point O has zero velocity $\mathbf{v}_O|_{\mathscr{R}}$ at the considered instant. When that happens, all points P at rest with S on the axis r parallel to $\boldsymbol{\omega}_{\mathscr{R}_S}|_{\mathscr{R}}$ through O have velocity $\mathbf{v}_P|_{\mathscr{R}} = \mathbf{0}$. This is straightforward from

$$\mathbf{v}_P|_{\mathscr{R}} = \mathbf{v}_O|_{\mathscr{R}} + \boldsymbol{\omega}_{\mathscr{R}_S}|_{\mathscr{R}} \wedge (P - O) = \mathbf{0} + \mathbf{0} \, .$$

We then say that at that instant r is an **instantaneous axis of rotation** of S for $\mathscr{R}$. In certain situations the instantaneous axis of rotation is also a principal axis of inertia. For instance when the rigid body S has a plane of symmetry π and $\boldsymbol{\omega}_{\mathscr{R}_S}|_{\mathscr{R}}$ is perpendicular to π. The proof follows from Remark (3) below. A homogeneous disc, or cylinder, that rolls without slipping on a straight line satisfies the condition when $\mathscr{R}$ is the line's frame. The point O of S (the instantaneous axis is viewed as axis of the point particles of S) on the line changes at each instant. Formula (5.16) for $\boldsymbol{\Gamma}_O|_{\mathscr{R}}$ reduces to, *at the instant considered,*

$$\boldsymbol{\Gamma}_O|_{\mathscr{R}} = \sum_{k=1}^{3} I_{Ok}\hat{\omega}_k \hat{\mathbf{e}}_k \ .$$

As $\boldsymbol{\omega}_{\mathscr{R}_S}|_{\mathscr{R}}$, in the principal triple of inertia, has only one non-zero component, say the first one (the one along the instantaneous axis of rotation through O), the above formula becomes

$$\boldsymbol{\Gamma}_O|_{\mathscr{R}} = I_{O1}\hat{\omega}_1 \hat{\mathbf{e}}_1 \ .$$

Similarly, and in the same hypotheses, case 1 for the kinetic energy $\mathscr{T}|_{\mathscr{R}}$ produces the formula

$$\mathscr{T}|_{\mathscr{R}} = \frac{1}{2} I_{O1}\hat{\omega}_1^2 \ .$$

In other words, the following proposition holds.

Proposition 5.14 *If a rigid body S admits, at the instant considered, an instantaneous axis of rotation r for a frame $\mathscr{R}$ and $O \in r$, then*

$$\boldsymbol{\Gamma}_O|_{\mathscr{R}} = I_r \boldsymbol{\omega}_{\mathscr{R}_S}|_{\mathscr{R}} \ , \tag{5.20}$$

and

$$\mathscr{T}|_{\mathscr{R}} = \frac{1}{2} I_r (\boldsymbol{\omega}_{\mathscr{R}_S}|_{\mathscr{R}})^2 \ , \tag{5.21}$$

where I_r denotes the moment of inertia with respect to the instantaneous axis of rotation.

Remarks 5.15 The following observations are very important in the applications, when one need to find the principal triple of inertia quickly.

(1) Suppose a rigid body S is *symmetric with respect to a plane π moving with S.* In other words, *for any point particle P_k in S of mass m_k there is another point particle $P_{k'}$ in S, symmetric to P_k with respect to π and with the same mass $m_{k'} = m_k$*; in the continuous case we assume the similar property in terms of mass density. Then *if $O \in \pi$, there always exists a principal triple of inertia with respect to O with principal axis of inertia normal to π.*

The proof follows by observing that, by Definition (5.6) or (5.7), the subspace of $V_{\mathscr{R}_S}$ of vectors orthogonal to π and its orthogonal space (the vectors parallel to π) are invariant under $\mathbf{I}_O$ (the proof is an easy direct computation). By restricting the operator to each subspace we still have a symmetric operator, which can then be diagonalised separately on the two subspaces. In this way, by construction, the unit normal to π and the two (normalised) eigenvectors tangent to π form a principal triple of inertia with respect to O.

(2) Suppose a necessarily continuous rigid body S is *symmetric about an axis r moving with S, given by a unit tangent vector* **u** *through O. In other words, the density of mass (one-, two or three-dimensional) is invariant under rotations about r.* In this case, *if $O \in r$, there always exists a principal triple of inertia with respect to O with principal axis of inertia* **u**. More precisely, *every right-handed orthonormal triple with one axis along r (and the others orthogonal to r) is a principal triple of inertia.*

The proof follows by observing that, by Definition (5.7) (or the corresponding ones in one and two dimensions), the subspace of $V_{\mathscr{R}_S}$ of orthogonal vectors to π and its orthogonal space (the vectors parallel to π) are invariant under $\mathbf{I}_O$ (the proof is an easy direct computation). Restricting the operator to each subspace we still have a symmetric operator, that can then be diagonalised separately on the two subspaces. In this way, by construction, the unit normal to π and the two (normalised) eigenvectors tangent to π form a principal triple of inertia with respect to O. On the other hand, due to the axial symmetry, rotating the triple around r cannot change the inertia tensor, and therefore any right-handed orthonormal triple with one axis parallel to r is a principal triple of inertia.

(3) In the situation described in (2), the moments of inertia for the axes perpendicular to r are equal by symmetry. A rigid body with a principal triple of inertia with respect to a point O in which two principal moments of inertia have the same value λ is called **gyroscopic**. The axis through O perpendicular to π and generated by the principal axes with equal principal moments is called **gyroscopic axis**.

It is crucial to observe that any other triple obtained rotating the initial one around the gyroscopic axis is still principal with respect to the same point O, and has the same principal moments of inertia as the initial triple.

This happens because if we restrict $\mathbf{I}_O$ to the subspace $U_{O,\pi}$ of vectors at O lying on π, since the operator is diagonal on that space with only one eigenvalue λ, then $\mathbf{I}_O|_{U_{O,\pi}} = \lambda I$, where I is the identity operator on $U_{O,\pi}$. Consequently, and trivially, *every* orthonormal basis of $U_{O,\pi}$ is made of eigenvectors with eigenvalue λ. Completing such a basis with the unit normal to π we obtain, by definition, a principal triple of inertia with the same principal moments of the beginning.

In presence of two gyroscopic axes in a principal triple of inertia with respect to a point O moving with S, the inertia tensor $\mathbf{I}_O$ is proportional to the identity operator, since the three principal moments must coincide (and therefore all moments of inertia are equal, for any axis passing through O). A body of this type is called **totally gyroscopic**.

(4) Suppose S is a *planar* rigid body and $\{\mathbf{e}_k\}_{k=1,2,3}$ is a principal triple of inertia of S with respect to O, moving with S, such that $\mathbf{e}_1$ *is perpendicular to the plane of S.*

Then the principal moments of inertia satisfy the relationship

$$I_{O1} = I_{O2} + I_{O3} \,. \tag{5.22}$$

The proof is straightforward. Suppose the system discrete, for the continuous case is completely similar. I_{O1} is the moment of inertia with respect to the axis parallel to $\mathbf{e}_1$ through O, hence perpendicular to the system. By (5.14), therefore, if d_i is the distance of P_k to that axis:

$$I_{O1} = \sum_{k=1}^{N} m_k d_i^2 = \sum_{k=1}^{N} m_k x_{2i}^2 + \sum_{k=1}^{N} m_k x_{3i}^2 = I_{O2} + I_{O3} \,,$$

where we have used Pythagoras's theorem for $P_k - O = x_{2k}\mathbf{e}_2 + x_{3k}\mathbf{e}_3$. ∎

Examples 5.16

(1) Consider a cube with edge L, homogeneous and of total mass M. By symmetry the centre of mass G coincides with the intersection of the three planes halving the cube. These are symmetry planes, so their unit normals give a principal triple of inertia with respect to G. Due to the problem's symmetry, the moments of inertia with respect to those axes are equal. However, the rigid body *is not spherically symmetric*. Computing the moments of inertia is elementary:

$$I = \int_{-L/2}^{L/2} dz \left(\int_{-L/2}^{L/2} dx \int_{-L/2}^{L/2} dy \frac{M}{L^3}(x^2 + y^2) \right)$$

$$= \frac{ML}{L^3} \left(\int_{-L/2}^{L/2} dx \int_{-L/2}^{L/2} dy x^2 + \int_{-L/2}^{L/2} dx \int_{-L/2}^{L/2} dy y^2 \right)$$

$$= \frac{ML}{L^3} 2L \frac{1}{3} 2 \left(\frac{L}{2} \right)^3 = \frac{ML^2}{6} \,.$$

The inertia tensor in said basis is given by a matrix with coefficients $I_{Gij} = \frac{ML^2}{6}\delta_{ij}$.

Another principal triple of inertia in G is the one formed by the orthonormal vectors along the lines joining G to the 6 vertices. This is because the planes perpendicular to these lines through G are still symmetry planes. By Remark (3) above, the principal moments of inertia of this triple are again all equal to $\frac{ML^2}{6}$.

By the above Remark (3) a right-handed orthonormal triple centred at G and placed arbitrarily with respect to the cube's faces is still a principal triple of inertia for S with respect to G, and once more $I_{Gij} = \frac{ML^2}{6}\delta_{ij}$.

(2) Consider a plane homogeneous disc of mass M and radius R. The centre coincides with the centre of mass G. Any right-handed orthonormal triple at G, with unit normal $\mathbf{e}_z$ to the disc, must be a principal triple of inertia, since the unit normal determines a gyroscopic axis. Computing the principal moment of inertia for this axis is immediate in polar coordinates:

$$I_{Gz} = \int_0^{2\pi} d\theta \int_0^R r\, dr \frac{M}{\pi R^2} r^2 = 2\pi \frac{M}{\pi R^2} \frac{R^4}{4} = \frac{MR^2}{2}\ .$$

The principal moments of inertia with respect to the axes $\mathbf{e}_x$ and $\mathbf{e}_y$ are equal by symmetry. At the same time $I_{Gx} + I_{Gy} = I_{Gz}$ by the last observation listed above. Hence $I_{Gx} = I_{Gy} = \frac{MR^2}{4}$. In summary, with respect to the aforementioned basis $\mathbf{e}_x, \mathbf{e}_y, \mathbf{e}_z$, the matrix of the inertia tensor $\mathbf{I}_G$ is

$$\frac{MR^2}{4}\,\mathrm{diag}\,(1, 1, 2)\ .$$

(3) Consider a homogeneous rigid square of mass M and edge L with a round hole of radius $R < L/2$ in the middle. The right-handed orthonormal triple of axes $\mathbf{e}_x, \mathbf{e}_y, \mathbf{e}_z$ at the square's centre G, with $\mathbf{e}_z$ normal to the square and $\mathbf{e}_x$ and $\mathbf{e}_y$ perpendicular to the edges, is surely principal with respect to G, exactly as in the previous example. The matrix of $\mathbf{I}_G$ in this basis is easy to find, by (1) in Proposition 5.11, subtracting the inertia matrix of the hollow disc from the inertia matrix of the full square (without hole). The final result is the diagonal matrix of $\mathbf{I}_G$ in the given basis:

$$\frac{M}{12}(L^2 - 3R^2)\mathrm{diag}(1, 1, 2)\ .$$

(4) Consider the same rigid body as the previous example. The right-handed orthonormal triple centred at G obtained rotating the previous example's triple by θ around the z-axis is still a principal triple of inertia, and has the same principal moments of the original triple. This is a consequence of the discussion in the above Remark (3).

5.2.3 Huygens-Steiner Formula

To conclude we present the **Huygens-Steiner formula**, expressing the matrix of the inertia tensor $\mathbf{I}_O$ in terms of the matrix of the inertia tensor I_G, when we use the same orthonormal basis of free vectors:

$$I_{O\,ij} = I_{G\,ij} + M\left((G - O)^2\delta_{ij} - (G - O)_i(G - O)_j\right)\ . \tag{5.23}$$

As usual, M is the rigid system's total mass. The intuitive meaning is rather evident: $I_{O\,ij}$ equals $I_{G\,ij}$ plus the inertia matrix with respect to O of a point particle at G with mass M. The proof follows from (5.12):

$$I_{Oij} = \sum_{k=1}^{N} m_k \left((P_k - O)^2 \delta_{ij} - (P_k - O)_i (P_k - O)_j \right)$$

$$= \sum_{k=1}^{N} m_k \left((P_k - G + G - O)^2 \delta_{ij} - (P_k - G + G - O)_i (P_k - G + G - O)_j \right).$$

Expanding the products and remembering that $\sum_k m_k (P_k - G) = M(G - G) = \mathbf{0}$ and $\sum_k m_k (P_k - G) = M(G - O)$, we find

$$I_{Oij} = M \left((G - O)\delta_{ij} - (G - O)_i (G - O)_j \right)$$

$$+ \sum_{k=1}^{N} m_k \left((P_k - G)^2 \delta_{ij} - (P_k - G)_i (P_k - G)_j \right),$$

which is (5.23). For continuous rigid bodies the argument is completely analogous, with integrals instead of sums.

Applying (5.23) to the component of a unit vector $\mathbf{n}$ and taking the inner product with $\mathbf{n}$ gives the useful relation

$$I_{O,\mathbf{n}} = I_{G,\mathbf{n}} + Md^2, \tag{5.24}$$

where d is the distance between the two axes. The formula expresses the moment of inertia for two parallel axes with tangent vector $\mathbf{n}$, one passing through O and the other through G.

5.3   Rigid Body Dynamics: Introduction to the Theory of Euler Equations

As we explained in Remarks 5.6 (and extended to continuous rigid bodies), a rigid body (with at least three non-collinear points) has 6 degrees of freedom, since its configurations correspond 1–1 to $\mathbb{R}^3 \times SO(3)$. The two governing equations of the dynamics of systems involve 6 unknown functions (and their derivatives) carrying the same information as the 6 freedom degrees. Once we assign the external forces acting on a rigid system in terms of these degrees (and their first derivatives), in principle the governing equations of Dynamics introduced in Sect. 4.1.2 will determine the motion of the rigid body. (In presence of (non-internal) constraints, it will be necessary to prescribe constitutive relations for the constraints in order to obtain pure equations of motion.) Rigid body dynamics is therefore described by the pair of governing equations of the system's dynamics. We will recall them below, adapting them to rigid bodies.

Suppose we have a rigid body S, subject to external forces (the distinction between external and internal forces is the same as for point particles) applied to points $P_1, \ldots, P_N \in S$. Note that these points do no exhaust, in general, the system, which may be a *continuous* rigid body. Furthermore, the P_i should be understood as *geometric points* and not point particles, because we assume they can vary instant by instant, as material particles. Think of an ordinary disc that rolls without slipping along a line: the contact point with the line changes at each instant, but the reaction of the constraint line acts on it. During the evolution, the rigid body S must obey the following two **governing equations of Dynamics**.

C5. Laws of Motion for Continuous Rigid Bodies *Consider a continuous rigid body S, subject to external forces $\mathbf{F}_i^{(e)}$ (either active or reactive) applied at points $P_1, \ldots, P_N \in S$. We assume the rigid body's motion satisfies the following two equations, in any inertial frame $\mathscr{R}$ and for any pole O with given motion in $\mathscr{R}$:*

$$\left.\frac{d}{dt}\right|_{\mathscr{R}} \mathbf{P}|_{\mathscr{R}} = \sum_{i=1}^{N} \mathbf{F}_i^{(e)} \tag{5.25}$$

$$\left.\frac{d}{dt}\right|_{\mathscr{R}} \mathbf{\Gamma}_O|_{\mathscr{R}} + \mathbf{v}_O|_{\mathscr{R}} \wedge \mathbf{P}|_{\mathscr{R}} = \sum_{i=1}^{N} (P_i - O) \wedge \mathbf{F}_i^{(e)} \tag{5.26}$$

(where the second summand on the left is absent in case $O = G$ is the centre of mass of S).

Remarks 5.17

(1) The extension to non-inertial reference frame and inertial forces is completely obvious.

(2) In C5 one considers only forces applied to *single points* of the rigid body. We thus exclude generic situations where, for *continuous* rigid bodies, the forces are described by *densities of forces* to be integrated. This kind of forces are treated by the generalisation of the formalism known as *Continuum Mechanics*, which we will not address. It is though important to realise, always for continuous rigid bodies, that the restriction in C5 on the type of admissible force may still include gravitational forces, in cases where the gravitational acceleration is constant $-g\mathbf{e}_z$, with g independent of the position (but possibly depending on time). In fact, computing the total gravitational force coming from the single particles' gravitational forces and the corresponding total momentum, via an obvious integration, one sees immediately that this kind of force determines: a force $-Mg\mathbf{e}_z$ in the right-hand side of (5.25), and a corresponding momentum in the right-hand side of (5.26), equal to the unique force $-Mg\mathbf{e}_z$ thought of as applied to G.

To sum up, *if we only want to use the two governing equations, the overall action of the constant gravitational force corresponds to a unique force applied to the system's centre of mass and equal to the total force* $-Mg\mathbf{e}_z$ *produced by the various gravitational forces on each point particle of the body.*

(3) Finally, it is important to stress that the above two equations, *in the case of continuous rigid systems*, is here assumed axiomatically, even if it could be deduced from first principles in the general framework of Continuum Mechanics. That these equations determine only one solution for given initial conditions (as we will prove when reformulating the theory in Lagrangian terms, in Sect. 7.3) is warranty that the assumption is reasonable. For discrete rigid systems made of a finite number of point particles, C5 is a consequence of postulates C1–C4 that hold in point-particle Newtonian Mechanics, as we have seen in Chap. 4 with Theorem 4.4.   ∎

Here we will limit ourselves to studying a subcase of the governing equations given by the so-called *Euler equations*. These are differential equations, in the unknown components of the ω vector, that correspond to a rigid body's second governing equation with respect to a pole O, in the simplest possible situation from the point of view of *constraints*: when the rigid body is not subject to constraint forces and O is the centre of mass, or when the system is constrained to the one point O only. In either case other forces—not arising from constraints—are allowed. Together with the first governing equation for systems, the Euler equations determine the dynamics of the rigid body. The special situation we are studying is such that, sometimes, the Euler equations are independent of the first governing equation, and so they can be studied separately. Despite the apparent simplification of the physics of the constraints involved, the list of possible solutions is incredibly variegated and forms an important chapter of Classical Mechanics. We shall only provide a brief introduction to this vast subject.

5.3.1   Euler Equations

Consider a rigid body S, of total mass M, subject to forces such that one for the following two situations hold with respect to a frame $\mathcal{R}$ in which we describe the motion of S.

Case 1   There is a point O moving with S (possibly the centre of mass)—that will be used as pole to state the second governing equation of Dynamics—which remains at rest in $\mathcal{R}$ under constraint forces, while S is free to move around O without other constraints.

Case 2   The pole O coincides with the centre of mass G of S and S is not subject to constraints.

In both cases, we know from Chap. 4 that the second governing equation for S takes the form:

$$\frac{d}{dt}\bigg|_{\mathscr{R}} \boldsymbol{\Gamma}_O|_{\mathscr{R}} = \mathbf{M}_O \tag{5.27}$$

where $\mathbf{M}_O$ is the total momentum of the external forces with respect to O. *The possible constraint acting on O does not contribute to the total momentum since the pole would coincide with the point of application.* As the body is rigid, Eq. (5.27) can be written as

$$\frac{d}{dt}\bigg|_{\mathscr{R}} \mathbf{I}\left(\boldsymbol{\omega}_{\mathscr{R}_S}|_{\mathscr{R}}\right) = \mathbf{M}_O \, , \tag{5.28}$$

where $\mathscr{R}_S$ is the usual moving frame of S. Consider a principal triple of inertia $\hat{\mathbf{e}}_1, \hat{\mathbf{e}}_2, \hat{\mathbf{e}}_3$ referred to O. In the sequel we will use the decomposition

$$\boldsymbol{\omega}_{\mathscr{R}_S}|_{\mathscr{R}} = \sum_{i=1}^{3} \hat{\omega}_i \hat{\mathbf{e}}_i \, , \quad \mathbf{M}_O = \sum_{i=1}^{3} \hat{M}_{Oi} \hat{\mathbf{e}}_i \, ,$$

and call I_{O1}, I_{O2}, I_{O3} the principal moments of inertia for that triple. Expanding (5.28) along the basis $\hat{\mathbf{e}}_1, \hat{\mathbf{e}}_2, \hat{\mathbf{e}}_3$, we obtain the ODE system

$$\begin{cases} I_{O1}\dfrac{d\hat{\omega}_1}{dt} = (I_{O2} - I_{O3})\hat{\omega}_2\hat{\omega}_3 = \hat{M}_{O1} \, , \\[2mm] I_{O2}\dfrac{d\hat{\omega}_2}{dt} = (I_{O3} - I_{O1})\hat{\omega}_3\hat{\omega}_1 = \hat{M}_{O2} \, , \\[2mm] I_{O3}\dfrac{d\hat{\omega}_3}{dt} = (I_{O1} - I_{O2})\hat{\omega}_1\hat{\omega}_2 = \hat{M}_{O3} \, . \end{cases} \tag{5.29}$$

These equations are called **Euler equations**. Together with the first governing equation they determine the motion of S around O in the two cases considered. For that to be possible we need to express the components $\hat{\omega}_k$ in terms of the Euler angles and their first derivatives, which give the principal triple of inertia moving with S in terms of a triple moving with $\mathscr{R}$. The same must be done for the components of the momentum $\mathbf{M}_O$, recalling that if O remains still in $\mathscr{R}$, then $\mathbf{M}_O$ cannot involve terms due to the unknown constraint forces, as noted above. If the Euler angles are not enough to express $\mathbf{M}_O$ as said, but we need further parameters such as positions and velocities of points in S, in particular those of the centre of mass G, the Euler equations must be accompanied by the equations coming from the first governing equation for systems. In case $\mathbf{M}_O$ can be written solely in terms of the Euler angles and their time derivatives, the Euler equations become a normal system of ODEs of order two (in the Euler angles), at least locally. Also observe that after we determine the motion of S around O, the possible unknown constraint force acting on O can be found using the first governing equation for systems:

$$\frac{d}{dt}\bigg|_{\mathscr{R}} M\mathbf{v}_G|_{\mathscr{R}} = \mathbf{R}\,,\qquad(5.30)$$

where $\mathbf{R}$ is the total external force acting on S, that includes any constraint reactions if present. In case $O = G$ and S moves in $\mathscr{R}$ freely, the motion of G is determined by the second governing equation, noting that in this case there are no constraint forces.

Remarks 5.18 There are at least two important cases in which the Euler equations are independent of the first governing equation and can therefore be studied separately. One case is that of a spinning top with tip O on a rough plane π at rest in an inertial frame $\mathscr{R}$. The top is subject to a constant gravitational force normal to π and pointing downwards. In the second case $\mathbf{M}_O$ is identically zero. This is the physical situation, for example, of a rigid body fixed to the point O at rest in the inertial frame $\mathscr{R}$, in absence of other forces. Or a rigid body in free fall in the constant gravitational field, with O at the centre of mass G. In the *non-inertial* frame in free fall with G, the situation appears exactly as in the rigid body fixed to O and without gravity or other forces.   ■

5.3.2 Poinsot Motions

When the right-hand side of (5.29) is identically zero, the motions determined by the Euler equations are called **Poinsot motions**. The corresponding equations then are:

$$\begin{cases} I_{O1}\dfrac{d\hat{\omega}_1}{dt} = (I_{O2} - I_{O3})\hat{\omega}_2\hat{\omega}_3 = 0\,, \\[2mm] I_{O2}\dfrac{d\hat{\omega}_2}{dt} = (I_{O3} - I_{O1})\hat{\omega}_3\hat{\omega}_1 = 0\,, \\[2mm] I_{O3}\dfrac{d\hat{\omega}_3}{dt} = (I_{O1} - I_{O2})\hat{\omega}_1\hat{\omega}_2 = 0\,. \end{cases}\qquad(5.31)$$

We will examine the non-degenerate case where S is not made of collinear points. In that case the inertia tensor is positive definite. If $\boldsymbol{\omega} := \boldsymbol{\omega}_{\mathscr{R}_S}|_{\mathscr{R}}$, we may rewrite the equations in compact form:

$$\mathbf{I}_O\left(\frac{d\boldsymbol{\omega}}{dt}\right) + \boldsymbol{\omega} \wedge \mathbf{I}_O(\boldsymbol{\omega}) = \mathbf{0}\,,$$

and since the operator $\mathbf{I}_O$ is invertible, being positive definite, we can eventually write the Poinsot equations as:

$$\frac{d\boldsymbol{\omega}}{dt} = -\mathbf{I}_O^{-1}\left(\boldsymbol{\omega} \wedge \mathbf{I}_O(\boldsymbol{\omega})\right), \tag{5.32}$$

where the time derivative can be taken with respect to $\mathscr{R}$ or $\mathscr{R}_S$ indistinctly, as is well known from Chap. 2. We can interpret this ODE as a first-order (non-linear) equation in the vector-valued function $\boldsymbol{\omega} = \boldsymbol{\omega}(t)$. It is clear that the equation is in normal form with C^∞ source term, so we can invoke the Existence and uniqueness theorems for the solutions.

5.3.3   Permanent Rotations

We will see in more generality in Chap. 6 that the *singular points* of the first-order system (5.32), i.e. the configurations $\boldsymbol{\omega}_0$ that kill the right-hand side, correspond to the solutions with $\boldsymbol{\omega}(t) = \boldsymbol{\omega}_0$, constant in modulus, direction and orientation, both in $\mathscr{R}$ and in $\mathscr{R}_S$. This sort of solutions are called **permanent rotations**. As $\mathbf{I}_O^{-1}$ is an injective function, the right side of (5.32) vanishes at $\boldsymbol{\omega} = \boldsymbol{\omega}_0$ if and only if:

$$\boldsymbol{\omega}_0 \wedge \mathbf{I}_O(\boldsymbol{\omega}_0) = \mathbf{0}.$$

This means $\boldsymbol{\omega}_0$ and $\mathbf{I}_O(\boldsymbol{\omega}_0)$ are parallel vectors. We have proved that $\boldsymbol{\omega} = \boldsymbol{\omega}(t) = \boldsymbol{\omega}_0$ constant for $t \in \mathbb{R}$ is a permanent rotation if and only if, for some $\lambda \in \mathbb{R}$,

$$\mathbf{I}_O(\boldsymbol{\omega}_0) = \lambda\boldsymbol{\omega}_0.$$

In other words permanent rotations determine eigenvectors of $\mathbf{I}_O$ if they do not correspond to $\omega_0 = \mathbf{0}$, or the trivial rotation $\boldsymbol{\omega}_0 = \mathbf{0}$. Any eigenvector of a symmetric operator on a finite-dimensional real vector space is part of a basis made of eigenvectors of the operator.[1] Hence the unit vector of $\boldsymbol{\omega}_0 \neq \mathbf{0}$ is always part of a principal triple of inertia for S with respect to O. We conclude that the following proposition holds.

Proposition 5.19 *For a rigid body S not made of collinear points along one axis, and whose equations of motion are the Poinsot equations (5.32), we have solutions of **permanent rotation** type:*

$$\boldsymbol{\omega}(t) = \boldsymbol{\omega}_0 \quad \textit{for any } t \in \mathbb{R},$$

---

[1] The proof descends from the recursive construction that proves the aforementioned diagonalisation theorem for symmetric operators.

if and only if the constant vector $\boldsymbol{\omega}_0$ lies along a principal axis of inertia, or when
$\boldsymbol{\omega}_0 = \mathbf{0}$.

Remarks 5.20

(1) The above proposition refers to every principal axis of inertia at O, and not only those in the principal triple chosen for writing the Euler equations. In fact, in presence of some symmetry about O, there may be several principal triples of inertia with respect to the same O. For instance, in a homogeneous cube any axis through the centre (which is the centre of mass) is a principal axis of inertia.

(2) Applying the theory of stability developed in Chap. 6 we can prove that if the three principal moments of inertia of a principal triple for S with respect to O satisfy $I_{O1} < I_{O2} < I_{O3}$, then only the permanent rotations around the first and third axis are stable, together with the trivial permanent rotation with $\boldsymbol{\omega}_0 = \mathbf{0}$. Permanent rotations about the axis with intermediate moment of inertia are unstable, instead.

To expand on these considerations it is important to note that, in the case S is constrained to $\mathscr{R}$ only at O and there are no other forces if not the constraint force at O, we have two simultaneous conservation laws: $\boldsymbol{\Gamma}_O|_{\mathscr{R}} = \text{constant}$, so $\Gamma_O|_{\mathscr{R}}^2 = \text{constant}$ and $\mathscr{T}|_{\mathscr{R}} = \text{constant}$. The former descends from the second governing equation, the latter from the fact that the only acting force does not do any work. The vector $\boldsymbol{\omega}(t) = \boldsymbol{\omega}_{\mathscr{R}_S|\mathscr{R}}(t)$ solving the Poinsot equations must therefore lie along the intersection of two surfaces given by the initial conditions: $(\mathbf{I}_O(\boldsymbol{\omega}(t)))^2 = (\mathbf{I}_O(\boldsymbol{\omega}(t_0)))^2$ and $\boldsymbol{\omega}(t) \cdot \mathbf{I}_O(\boldsymbol{\omega}(t)) = \boldsymbol{\omega}(t_0) \cdot \mathbf{I}_O(\boldsymbol{\omega}(t_0))$. In a principal triple of inertia with respect to O, and defining orthonormal coordinates for this triple, $\hat{x}_i := \frac{I_{Oi}}{\Gamma} \hat{\omega}_i$ where $\Gamma := ||\boldsymbol{\Gamma}_O|_{\mathscr{R}}(t_0)||$ and the $\hat{\omega}_i$ are the components of $\boldsymbol{\omega}$, the previous surfaces have respective equations:

$$\sum_{i=1}^{3} \hat{x}_i^2 = 1 \, , \quad \sum_{i=1}^{3} \frac{\Gamma^2}{2I_{0i}\,\mathscr{T}|_{\mathscr{R}}(t_0)} \hat{x}_i^2 = 1 \, .$$

Hence we are looking at the intersection of a sphere and an ellipsoid. The evolution of $\boldsymbol{\omega}(t)$ (rescaled by constant factors are indicated above) occurs on this intersection. ∎

5.3.4 Poinsot Motions for Gyroscopic Bodies

Let us now study solutions to the Poinsot equations that are less trivial than permanent rotations. We shall examine only systems whose principal moments of inertia are all non-zero (i.e. rigid systems not along a single axis). Consider the situation where $I_{O1} = I_{O2} =: I$ in Eq. (5.32). We are therefore working with a gyroscopic body with gyroscopic axis $\hat{\mathbf{e}}_3$. The Poinsot equations for $\boldsymbol{\omega} := \boldsymbol{\omega}_{\mathscr{R}_S|\mathscr{R}}$ read:

$$\begin{cases} I\dfrac{d\hat{\omega}_1}{dt} = (I - I_{O3})\hat{\omega}_2\hat{\omega}_3 = 0\,, \\[2mm] I\dfrac{d\hat{\omega}_2}{dt} = -(I_{O3} - I)\hat{\omega}_3\hat{\omega}_1 = 0\,, \\[2mm] I_{O3}\dfrac{d\hat{\omega}_3}{dt} = 0\,. \end{cases} \tag{5.33}$$

For a moment we also assume $I \neq I_{O3}$, and at the end we will say what happens in the limiting case.

The first equation has a unique trivial solution ω_3 constant in time. If we put $z := \hat{\omega}_1 + i\hat{\omega}_2$, multiply by i both sides of the second equation and add with the first, the two equations can be combined into

$$I\frac{dz}{dt} = -i(I - I_{O3})\hat{\omega}_3 z\,. \tag{5.34}$$

The (maximal and complete) solution will then be of the form:

$$z(t) = z(0)e^{-i\frac{I-I_{O3}}{I}\hat{\omega}_3 t}\,.$$

Back to real variables, the general solution to (5.33) reads:

$$\boldsymbol{\omega}(t) = \Omega\cos\left(\frac{I - I_{O3}}{I}\hat{\omega}_3 t\right)\hat{\mathbf{e}}_1 - \Omega\sin\left(\frac{I - I_{O3}}{I}\hat{\omega}_3 t\right)\hat{\mathbf{e}}_2 + \hat{\omega}_3\hat{\mathbf{e}}_3\,, \quad \text{for any } t \in \mathbb{R}, \tag{5.35}$$

where $\hat{\omega}_1, \Omega \in \mathbb{R}$ are arbitrary constants.

Let us now study solutions to the Poinsot equations that are less trivial than permanent rotations. We will only consider systems whose principal moments of inertia are all non-rigid systems not collinear with one axis). Consider the situation where in Eq. (5.32) we have $I_{O1} = I_{O2} =: I$. Since the angular momentum $\boldsymbol{\Gamma}_{O|\mathscr{R}}$ is constant in time for $\mathscr{R}$ (it is precisely this that the Poinsot equations are saying!), we choose $\mathbf{e}_3$ so that $\boldsymbol{\Gamma}_{O|\mathscr{R}} = \Gamma\mathbf{e}_3$ with $\Gamma > 0$ (and we omit the limiting case $\boldsymbol{\Gamma}_{O|\mathscr{R}} = \mathbf{0}$). We claim that the angle θ that $\hat{\mathbf{e}}_3$ forms with $\mathbf{e}_3$ is constant in time. From the decomposition of $\boldsymbol{\Gamma}_{O|\mathscr{R}} = \mathbf{I}_O(\boldsymbol{\omega})$ along the principal triple of inertia of S with respect to O:

$$\boldsymbol{\Gamma}_{O|\mathscr{R}} = I\hat{\omega}_1\hat{\mathbf{e}}_1 + I\hat{\omega}_2\hat{\mathbf{e}}_2 + I_{O3}\hat{\omega}_3\hat{\mathbf{e}}_3 \tag{5.36}$$

we have

$$I_3\hat{\omega}_3 = \boldsymbol{\Gamma}_{O|\mathscr{R}} \cdot \hat{\mathbf{e}}_3 = \Gamma\cos\theta\,.$$

On the other hand, as I_3, $\hat{\omega}_3$, Γ are constants, also θ is constant. So we have found that

$$\hat{\omega}_3 = \frac{\Gamma \cos \theta}{I_{O3}} \, . \tag{5.37}$$

Using (5.37) we can eliminate $\hat{\omega}_3$ from (5.35), and reduce it to

$$\boldsymbol{\omega}(t) = \Omega \cos \left(\frac{I - I_{O3}}{I} \hat{\omega}_3 t \right) \hat{\mathbf{e}}_1 - \Omega \sin \left(\frac{I - I_{O3}}{I} \hat{\omega}_3 t \right) \hat{\mathbf{e}}_2$$

$$+ \frac{\Gamma \cos \theta}{I_{O3}} \hat{\mathbf{e}}_3 \, , \quad \text{for any } t \in \mathbb{R}, \tag{5.38}$$

Substituting (5.37) in (5.36) gives:

$$\boldsymbol{\Gamma}_O|_{\mathscr{R}} = I \hat{\omega}_1 \hat{\mathbf{e}}_1 + I \hat{\omega}_2 \hat{\mathbf{e}}_2 + \Gamma \cos \theta \mathbf{e}_3 \, . \tag{5.39}$$

Comparing the above with (5.38) gives the formula:

$$\boldsymbol{\omega}(t) = \frac{\Gamma}{I} \mathbf{e}_3 + \frac{I - I_{O3}}{I} \frac{\Gamma}{I_{O3}} \cos \theta \hat{\mathbf{e}}_3(t) \, . \tag{5.40}$$

In the right-hand side of (5.40) only the unit vector $\hat{\mathbf{e}}_3$ evolves in time in $\mathscr{R}$. In conclusion, the solution (5.40) to the Poinsot equations in the case considered is given by the sum of two terms. One is constant (in $\mathscr{R}$) and aligned with $\mathbf{e}_3$. It is called **precession term**:

$$\boldsymbol{\omega}_{pre} := \frac{\Gamma}{I} \mathbf{e}_3 = \frac{\boldsymbol{\Gamma}_O|_{\mathscr{R}}}{I} \, . \tag{5.41}$$

The remaining part

$$\boldsymbol{\omega}_{rot} := \frac{I - I_{O3}}{I} \frac{\Gamma}{I_{O3}} \cos \theta \hat{\mathbf{e}}_3(t) \tag{5.42}$$

is the **rotation term**. Note that $\boldsymbol{\omega}_{rot}$ has constant modulus and is directed along the gyroscopic axis $\hat{\mathbf{e}}_3$ of S. The axis $\hat{\mathbf{e}}_3$ rotates—in the argot, *precesses*—around $\mathbf{e}_3$ (i.e. $\boldsymbol{\Gamma}_O|_{\mathscr{R}}$) with ω vector given precisely by $\boldsymbol{\omega}_{pre}$, since obviously:

$$\frac{d}{dt}\bigg|_{\mathscr{R}} \hat{\mathbf{e}}_3(t) = \boldsymbol{\omega} \wedge \hat{\mathbf{e}}_3 = \boldsymbol{\omega}_{pre} \wedge \hat{\mathbf{e}}_3 + \mathbf{0} \, .$$

As $\boldsymbol{\omega}_{pre}$ is constant in time, the tip of $\hat{\mathbf{e}}_3(t)$ rotates around $\mathbf{e}_3$ with constant angular speed. From (5.40) we deduce that $\boldsymbol{\omega}$, $\boldsymbol{\Gamma}_O|_{\mathscr{R}}$, $\hat{\mathbf{e}}_3$ are always *coplanar vectors*. In case $I = I_{O3}$ (in particular, for totally gyroscopic bodies), we have $\boldsymbol{\omega}_{rot} = \mathbf{0}$. This means the motion occurs without precession of $\boldsymbol{\omega}$ around $\boldsymbol{\Gamma}_O|_{\mathcal{R}}$, but the two vectors always stay parallel.

Remarks 5.21 Consider the motion of a rigid body S in a frame $\mathcal{R}$ such that there is always a point O moving with S but still for $\mathcal{R}$. We call it **precession motion** when $\boldsymbol{\omega}_{\mathcal{R}_S|\mathcal{R}} = \boldsymbol{\omega}_{pre} + \boldsymbol{\omega}_{rot}$, where the precession term $\boldsymbol{\omega}_{pre}$ has constant unit vector in time in $\mathcal{R}$, whereas the rotation term $\boldsymbol{\omega}_{rot}$ has constant unit vector in the frame $\mathcal{R}_S$ moving with S. The precession is further called **regular** when the precession and rotation terms have constant modulus in time as well. The Poinsot motions we studied above are therefore regular precessions.   ∎

5.3.5   Poinsot Motions for Non-gyroscopic Bodies

Let us take a rigid system whose principal moments of inertia are all non-zero (i.e. rigid systems not lying along one axis), and examine the generic situation where in Eq. (5.32) I_{O1}, I_{O2}, I_{O3} are distinct.

To begin with we introduce the so-called *inertia ellipsoid*. Given a rigid body S, let O move with S and call $\mathbf{I}_O$ the inertia tensor with respect to O. The surface $\mathcal{E}_O$ of points $P \in E_{\mathcal{R}_S}$ such that:

$$(P - O) \cdot \mathbf{I}_O(P - O) = 1 \tag{5.43}$$

is called **inertia ellipsoid** of S with respect to O. It is indeed an ellipsoid: choosing a principal triple of inertia centred at O with axes $\hat{\mathbf{e}}_1, \hat{\mathbf{e}}_2, \hat{\mathbf{e}}_3$, if $P - O = \mathbf{x} = \hat{x}_1\hat{\mathbf{e}}_1 + \hat{x}_2\hat{\mathbf{e}}_2 + \hat{x}_3\hat{\mathbf{e}}_3$ and the principal moments of inertia are I_{O1}, I_{O2}, I_{O3}, the equation of $\mathcal{E}_O$ is

$$\frac{\hat{x}_1^2}{I_{O1}^{-1}} + \frac{\hat{x}_2^2}{I_{O2}^{-1}} + \frac{\hat{x}_3^2}{I_{O3}^{-1}} = 1 \ .$$

$\mathcal{E}_O$ is therefore an ellipsoid of centre O and semi-axes $I_{O1}^{-1/2}, I_{O2}^{-1/2}, I_{O3}^{-1/2}$. Soon it will be useful to recall that the normal vector to the inertia ellipsoid, in the given coordinates, is:

$$N(\hat{x}_1, \hat{x}_2, \hat{x}_3) = 2I_{O1}\hat{x}_1\hat{\mathbf{e}}_1 + 2I_{O2}\hat{x}_2\hat{\mathbf{e}}_2 + 2I_{O3}\hat{x}_3\hat{\mathbf{e}}_3 \ ,$$

i.e., if $\mathbf{x} = P - O$ determines a point on $\mathcal{E}_O$,

$$N(\mathbf{x}) = 2\mathbf{I}_O(\mathbf{x}) \ . \tag{5.44}$$

In the sequel we shall assume the point O is at rest in the non-inertial frame $\mathcal{R}$ and that there are no external forces on S apart from the constraint reaction at O, so that the Poinsot equations (5.31) hold. We know these equations correspond to the second governing equation for systems, and in our case they say $\Gamma_O|_{\mathscr{R}}$ is constant in time in $\mathscr{R}$. Hence it is convenient to choose a system of orthonormal coordinates for $\mathscr{R}$ centred at O and with $\mathbf{e}_3$-axis parallel to $\Gamma_O|_{\mathscr{R}}$. For later convenience we also choose $\mathbf{e}_z$ so that $\Gamma_O|_{\mathscr{R}} = \Gamma\mathbf{e}_z$ where $\Gamma > 0$ is a constant (the trivial case $\Gamma = 0$ corresponds to S being at rest in $\mathscr{R}$). As usual $\boldsymbol{\omega}(t) := \boldsymbol{\omega}_{\mathscr{R}_S}|_{\mathscr{R}}(t)$, and the latter's components along the principal triple of inertia of S with respect to O mentioned above are the functions $\hat{\omega}_i$ appearing in (5.31). We want to study qualitatively the evolution of the vector $\boldsymbol{\omega} = \boldsymbol{\omega}(t)$ solving the Poinsot equations. To that end let us introduce the normalised vector

$$\mathbf{x}(t) := \frac{\boldsymbol{\omega}(t)}{\sqrt{2\mathscr{T}|_{\mathscr{R}}}} \, .$$

Observe that the kinetic energy of S with respect to $\mathscr{R}$: $\mathscr{T}|_{\mathscr{R}} := \frac{1}{2}\boldsymbol{\omega} \cdot \mathbf{I}_O(\boldsymbol{\omega})$ is a constant of motion, because the only external force acting on S is the constrain reaction at O, which is at rest in $\mathscr{R}$, and therefore this force does not do any work. We shall examine the qualitative behaviour, in time, of the normalised vector $\mathbf{x} = \mathbf{x}(t)$, rather than $\boldsymbol{\omega}(t)$.

As $2\mathscr{T}|_{\mathscr{R}} = I_{O1}\hat{\omega}_1^2 + I_{O2}\hat{\omega}_2^2 + I_{O3}\hat{\omega}_3^2$, necessarily

$$\mathbf{x} \cdot \mathbf{I}_O(\mathbf{x}) = 1 \, , \tag{5.45}$$

so $\mathbf{x}(t)$ lies on the inertia ellipsoid at each instant. However, it does not stay still on the surface, but its position on $\mathcal{E}_O$ normally varies in time. Since $\mathcal{E}_O$ moves with S and with $\mathscr{R}_S$, $\mathbf{x}(t)$ will be in motion for both $\mathscr{R}$ and $\mathscr{R}_S$. Another piece of information on $\mathbf{x}$ comes from observing that

$$\mathbf{x}(t) \cdot \mathbf{e}_3 = \Gamma^{-1}\mathbf{x}(t) \cdot \Gamma_O|_{\mathscr{R}} = \Gamma^{-1}\mathbf{x}(t) \cdot \mathbf{I}_O\left(\Gamma\sqrt{2\mathscr{T}|_{\mathscr{R}}}\mathbf{x}(t)\right) = \frac{\sqrt{2\mathscr{T}|_{\mathscr{R}}}}{\Gamma} \tag{5.46}$$

It means that the point $\mathbf{x}(t)$, apart from staying on the inertia ellipsoid, also lies on the plane π of equation

$$(P - O) \cdot \mathbf{e}_3 = \frac{1}{\Gamma\sqrt{2\mathscr{T}|_{\mathscr{R}}}}$$

at rest in $\mathscr{R}$. The latter is called **absolute plane**. Its normal vector is $\mathbf{e}_3$ and passes through the point on the x_3-axis with coordinate $\dfrac{1}{\Gamma\sqrt{2\mathscr{T}|_{\mathscr{R}}}}$.

Using (5.44) we discover something else. From that relation we see that, at the contact point $\mathbf{x}(t)$ between the absolute plane π moving with $\mathscr{R}$ and the inertia ellipsoid $\mathcal{E}_O$ moving with $\mathscr{R}_S$, the ellipsoid's normal vector is always parallel to $\mathbf{e}_3$ (i.e. to $\Gamma_O|_{\mathscr{R}}$). In fact,

$$N(\mathbf{x}) = 2\mathbf{I}_O(\mathbf{x}) = 2\mathbf{I}_O\left(\frac{\boldsymbol{\omega}(t)}{\sqrt{2\mathscr{T}|_{\mathscr{R}}}}\right) = 2\frac{\Gamma\mathbf{e}_3}{\sqrt{2\mathscr{T}|_{\mathscr{R}}}} \, .$$

We can finally compute the velocity in $\mathscr{R}$, instant by instant, of the point $Q(t)$ given by $\mathbf{x}(t)$ *thought of as moving with* S:

$$\mathbf{v}_Q|_{\mathscr{R}} = \mathbf{v}_O|_{\mathscr{R}_S} + \boldsymbol{\omega} \wedge (Q(t) - O) = \mathbf{0} + \boldsymbol{\omega} \wedge \mathbf{x}(t) = \mathbf{0} \,.$$

The contact point between $\mathcal{E}_O$ and π, given by $\mathbf{x}(t)$ instant by instant, always has zero velocity.

Summarising: *the motion of S in $\mathscr{R}$ is such that the inertia ellipsoid moving with S rolls without slipping on the absolute plane π that moves with $\mathscr{R}$, and the contact point is, instant by instant and up to rescaling by a constant, given by the vector* $\omega(t) = \omega_{\mathscr{R}_S}|_{\mathscr{R}}(t)$ *solving the Poinsot equations (5.31)*.

During its time evolution $\mathbf{x}(t)$ traces out a cone around $\mathbf{e}_3$ (i.e. $\Gamma_O|_{\mathscr{R}}$). The tip of $\mathbf{x}(t)$ draws a curve on π called **erpolhode**. The corresponding curve traced by the same vector on $\mathcal{E}_O$ is the **polhode**.

Exercises 5.22

(1) Consider a homogeneous square $ABCD$ (Fig. 5.1) of mass M and edge L. Suppose it hangs from the ceiling (which has inertial frame $\mathscr{R}$) by a rigid rod attached to the square, of mass m and length L, that extends out from the diagonal AC and joins the vertex A to the point O on the ceiling. The rod, thus identified with the segment AO, is free to rotate around O in the vertical plane. At the vertex B is a point particle P of mass m. The overall rigid system S, given by the square, the rod and the particle P, is subject to gravity, given by the constant vertical acceleration $\mathbf{g}$ pointing downwards. Solve the following problems.

 (i) Write the system's equations of motion and the constraint reactions.
(ii) Find a first integral and discuss its physical meaning.

Hint: describe the system using the angle θ between the rod AO and the vertical direction.

Fig. 5.1 Exercise 5.22.**1**

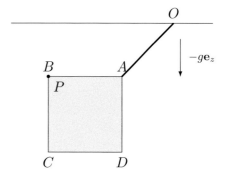

Fig. 5.2 Exercise 5.22.2

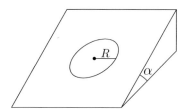

(2) A homogeneous disc of mass m (Fig. 5.2) and radius R lies on a rough ramp, tilted by an angle α. Assume known the coefficients of static friction μ and dynamic friction f, and suppose the disc is not moving at the initial time. Study its motion under the assumption that the ramp is not moving with respect to an inertial frame $\mathscr{R}$.

Chapter 6
Introduction to Stability Theory with Applications to Mechanics

In this chapter we introduce the basics in the theory of Lyapunov stability, which was first formulated at the end of the nineteenth century. We will work with *autonomous* systems of differential equations on $\mathbb{K}^n = \mathbb{R}^n$ or $\mathbb{C}^n$ without distinction, even though with petty adaptations the definitions and results extend to differentiable manifolds. For all notions and definitions regarding systems of differential equations used in this chapter we shall always refer to Complement in Chap. 14.

6.1 Singular Points and Equilibrium Configurations

The fundamental notion we shall study is that of a *singular point*, also known as *critical point*.

Definition 6.1 (Singular Point) Let $D \subset \mathbb{K}^n$ be non-empty and open, and $I \subset \mathbb{R}$ a non-empty open interval. Consider a system of differential equations of order one on $I \times D \subset \mathbb{R} \times \mathbb{K}^n$,

$$\frac{d\mathbf{x}}{dt} = \mathbf{f}(t, \mathbf{x}(t)) , \qquad (6.1)$$

with $\mathbf{f} : I \times D \to \mathbb{K}^n$. The point $\mathbf{x}_0 \in D$ is called a **singular point** (or, equivalently, a **critical point**) of $\mathbf{f}$ if and only if, for any $t_0 \in I$, every solution to the initial value problem given by Eq. (6.1) with initial condition

$$\mathbf{x}(t_0) = \mathbf{x}_0 , \qquad (6.2)$$

is a restriction of the constant map $\mathbf{x}(t) = \mathbf{x}_0$ for any $t \in I$. $\diamondsuit$

Remarks 6.2

(1) If $\mathbf{x}_0$ is a singular point, immediately from the definition the constant function $\mathbf{x}(t) = \mathbf{x}_0$ for any $t \in I$ is the only maximal solution of the initial value problem with Eq. (6.1) and initial condition (6.2), irrespective of $t_0 \in I$.

(2) As the above definition is completely *local* as far as $\mathbf{x}$ is concerned, it can be reformulated for an ODE system given by a vector field on a differentiable manifold, as per Sect. 14.5. We leave this to the reader.

(3) The same definition can be given, slightly altering the wording, for a *second-order* problem. This clearly points to the fundamental problem of Dynamics when $\mathbb{K} = \mathbb{R}$ and $n = 3$ (but also more complicated cases, involving *systems* of point particles). ∎

Here is the definition mentioned in Remark (2) above:

Definition 6.3 (Equilibrium Point) Consider an ODE system of the second order son $I \times D \times D'$, with $D, D' \subset \mathbb{K}^n$ non-empty open sets and $I \subset \mathbb{R}$ a non-empty open interval,

$$\frac{d^2\mathbf{x}}{dt^2} = \mathbf{F}\left(t, \mathbf{x}(t), \frac{d\mathbf{x}}{dt}\right) ,\tag{6.3}$$

where $\mathbf{F} : I \times D \times D' \to \mathbb{K}^n$. The point $\mathbf{x}_0 \in D$ is called an **equilibrium point** (or equivalently **equilibrium configuration**) for $\mathbf{F}$ if and only if, for any $t_0 \in I$, every solution to the Cauchy problem given by (6.3) with initial conditions

$$\begin{cases} \mathbf{x}(t_0) = \mathbf{x}_0, \\ \dfrac{d\mathbf{x}}{dt}\Big|_{t_0} = \mathbf{0}, \end{cases}\tag{6.4}$$

is a restriction of the constant function $\mathbf{x}(t) = \mathbf{x}_0$ for any $t \in I$.   ◇

Remarks 6.4

(1) This definition applies, for example, to a point particle P described by the vector $\mathbf{x} = P - O$, with O at rest in a frame $\mathscr{R}$, when $\mathbf{F}$ is the total force acting on the point, assumed independent of time in $\mathscr{R}$. *One should be well aware that an equilibrium point's definition depends on the chosen frame of reference.*

(2) The definition applies to systems of N particles, when

$$D = D' = \mathbb{R}^3 \times \cdots \times \mathbb{R}^3 = \mathbb{R}^{3N} .$$

In this case the $3N$ components of $\mathbf{x}$ describe, three by three, the components of each of the N points, in given orthonormal coordinates moving with a frame where all forces are time-independent. The $3N$ components of $\mathbf{F}$ describe, in groups of three, the components of the total force acting on each of the N particles, *each one divided by the corresponding particle's mass*. ∎

We have the following obvious proposition, giving us necessary and sufficient conditions for a point to be singular, i.e. an equilibrium, under the hypotheses of the global Existence and uniqueness Theorem 14.23.

Proposition 6.5 *Referring to Definition 6.1, if the function* $\mathbf{f} : I \times D \to \mathbb{K}^n$ *is continuous and locally Lipschitz in* $\mathbf{x}$, *then the point* $\mathbf{x}_0 \in D$ *is singular if and only if*

$$\mathbf{f}(t, \mathbf{x}_0) = \mathbf{0} \quad \text{for any } t \in I . \tag{6.5}$$

Proof That condition (6.5) is necessary is obvious by definition of singular point: for given $t_0 \in I$, the map $\mathbf{x}(t) = \mathbf{x}_0$, $t \in I$, solves the initial value problem with initial condition $\mathbf{x}(t_0) = \mathbf{x}_0$. In particular $0 = \frac{d\mathbf{x}}{dt}|_{t=t_0} = f(t_0, \mathbf{x}_0)$ for any $t_0 \in I$. The sufficiency is straightforward from Theorem 14.23, because if $\mathbf{f}(t, \mathbf{x}_0) = \mathbf{0}$ then $\mathbf{x}(t) = \mathbf{x}_0$ constant for any $t \in I$ is certainly a maximal solution of the Cauchy problem with initial condition $\mathbf{x}(t_0) = \mathbf{x}_0$, $t_0 \in I$ given. It must be the only maximal solution by virtue of the aforementioned theorem, and any other solution is a restriction of it. □

Next is the analogue statement for equilibrium configurations.

Proposition 6.6 *Referring to Definition 6.3, if* $\mathbf{F} = \mathbf{F}(t, \mathbf{x}, \mathbf{v}) \in \mathbb{K}^n$, *with* $(t, \mathbf{x}, \mathbf{v}) \in I \times D \times D'$, *is continuous and locally Lipschitz in* $(\mathbf{x}, \mathbf{v})$ *(or, more strongly, of class* $C^1(I \times D \times D'; \mathbb{K}^n)$)*, then* $\mathbf{x}_0 \in D$ *is an equilibrium configuration if and only if*

$$\mathbf{F}(t, \mathbf{x}_0, \mathbf{0}) = \mathbf{0} \quad \text{for any } t \in I . \tag{6.6}$$

6.1.1 Stable and Unstable Equilibria

From a physical point of view the above notion of equilibrium configuration does not characterise appropriately the real equilibrium configurations of physical systems (and of other concrete systems that are describable by differential equations). One reason, perhaps the most important one, is that the configurations of physical systems do not correspond to *ideal* geometric configurations, because the state of any physical system (the position and shape itself of objects) is continuously subjected to interactions with the surroundings—interactions of thermodynamical nature, mostly—that make it fluctuate, albeit very slightly. For this reason the physical equilibrium configurations we can *observe in the physical reality* are *stable* under small perturbations of the initial state. These stability properties are defined depending on the mathematical model one adopts. In the sequel we will refer to the simplest possible notions, those that can be formulated within the elementary theory of ODE systems. As usual, we will start from first-order systems, although later we shall be interested in applying all of this to order two.

The core idea in the definition of a stable singular point is that the point's orbit remains *close* to the point even when we start from initial conditions that differ a little from the singular point. In the sequel we will restrict to *autonomous* systems, i.e. where the right-hand side of (6.1) does not depend on the variable t. In this setting we can always choose the initial condition as $t_0 = 0$, by redefining the origin of t if necessary.

Definition 6.7 (Stable Singular Point) Consider Definition 6.1 for an *autonomous system*, i.e. when in (6.1) $\mathbf{f} = \mathbf{f}(\mathbf{x})$ does not depend on t. A singular point $\mathbf{y}_0 \in D$ is said to be:

(1) **future-stable** if, for any open neighbourhood $U \ni \mathbf{y}_0$, there exists an open neighbourhood $V \ni \mathbf{y}_0$ such that every solution to (6.1), $\mathbf{x} : J \to D$ (where $J \subset I$ is an open interval containing $t = 0$) with $\mathbf{x}(0) = \mathbf{x}_0 \in V$, satisfies $\mathbf{x}(t) \in U$ for $J \ni t \geq 0$;

(2) **past-stable** if, for any open neighbourhood $U \ni \mathbf{y}_0$, there exists an open neighbourhood $V \ni \mathbf{y}_0$ such that every maximal solution to (6.1), $\mathbf{x} : J \to D$ (with $J \subset I$ an open interval containing $t = 0$) with $\mathbf{x}(0) = \mathbf{x}_0 \in V$, satisfies $\mathbf{x}(t) \in U$ if $J \ni t \leq 0$;

(3) **stable** if it is both future-stable and past-stable;

(4) **(future-** or **past-)unstable** if it is not stable (not future-stable or not past-stable, respectively);

(5) **asymptotically stable in the future** if future-stable and there exists an open neighbourhood $A \ni \mathbf{y}_0$ such that every maximal solution to (6.1) $\mathbf{x}$ with $\mathbf{x}(0) = \mathbf{x}_0 \in A$, is future-complete and satisfies $\mathbf{x}(t) \to \mathbf{y}_0$ as $t \to +\infty$;

(6) **asymptotically stable in the past**, if past-stable and there exists an open neighbourhood $A \ni \mathbf{y}_0$ such that every maximal solution to (6.1) $\mathbf{x}$ with $\mathbf{x}(0) = \mathbf{x}_0 \in A$, is past-complete and satisfies $\mathbf{x}(t) \to \mathbf{y}_0$ as $t \to -\infty$.

In cases (5) and (6), the set A is called a **basin of attraction** of $\mathbf{x}_0$. $\diamondsuit$.

Remarks 6.8

(1) When $\mathbf{f}$ is locally Lipschitz and $\mathbf{y}_0$ is future-stable, by Proposition 14.34 the maximal solutions with initial condition in V are certainly future-complete (we can take U with compact closure). The same happens for past-stability.

(2) The above definitions extend immediately to the system of order two (6.3) where $\mathbf{F}$ does not depend on t. If so, one speaks of **stable equilibrium configurations (in the past** and **in the future)** and **asymptotically stable equilibrium configurations (in the past** and **in the future)**.

(3) When dealing with second-order systems, we must take into account that the neighbourhoods U, V and A are subsets in the space of pairs $(\mathbf{x}, \mathbf{v})$ and not just the variable $\mathbf{x}$. In this sense, when studying the stability of configurations of physical systems, *the notion of stability includes information, and requests, on the velocities and not just the configurations.*   ∎

Examples 6.9

(1) Consider the ODE in $\mathbb{R}$:

$$\frac{dx}{dt} = -kx$$

where $k \neq 0$ is given. The global Existence and uniqueness theorem holds because the source term is C^∞. The only singular point is $x = 0$. As the maximal solutions have the form

$$x(t) = x_0 e^{-kt}, \quad \text{with } t \in \mathbb{R},$$

if $k > 0$ then $|x(t)| < |x_0|$ for $t > 0$. Hence the point is future-stable: for any open neighbourhood $U \ni 0$, the neighbourhood $V \ni 0$ can be any interval $(-\delta, \delta) \subset U$. Furthermore, $x(t) \to 0$ as $t \to +\infty$. We conclude that for $k > 0$ the singular point $x = 0$ is asymptotically stable in the future. On the other hand, still for $k > 0$: $x(t) \to +\infty$ as $t \to -\infty$. Therefore for $k > 0$ the singular point is past-unstable. The picture is the opposite for $k < 0$: the singular point $x = 0$ is asymptotically stable in the future, but unstable in the past.

(2) Consider the ODE system on $\mathbb{R}^2$:

$$\frac{d\mathbf{x}}{dt} = A\mathbf{x},$$

where $\mathbf{x} = (x, y)$ and A is the diagonal matrix $diag(1, -1)$. The global Existence and uniqueness theorem holds since the right-hand-side term is C^∞. The unique singular point is the origin $\mathbf{0}$ of $\mathbb{R}^2$. We easily conclude that the singular point is both past- and future-unstable. In fact, the system decouples into two equations:

$$\frac{dx}{dt} = x, \quad \frac{dy}{dt} = -y.$$

From the previous example we know that every initial condition $\mathbf{x}_0 = (x_0, 0)$ produces a divergent solution (in the x-component only) as $t \to -\infty$. This fact alone is enough to show the singular point $\mathbf{0}$ is past-unstable. Vice versa, again from the previous example, the initial condition $\mathbf{x}_0 = (0, y_0)$ produces a divergent solution (in the y-component only) as $t \to +\infty$. That, by itself, is sufficient to prove $\mathbf{0}$ is future-unstable.

(3) Consider the harmonic oscillator, made of a point particle of mass $m > 0$ attached to a spring of constant $\kappa > 0$, for simplicity of zero length at rest, whose other end is fixed to the point O of an inertial frame. The system's ODE is clearly

$$\frac{d^2\mathbf{x}}{dt^2} = -k^2\mathbf{x},$$

where $k := \sqrt{\kappa/m}$ and $\mathbf{x} = P - O$. We are in the assumptions of the global Existence and uniqueness theorem since the source term is C^∞. Reducing to order one, we have the equivalent system in $\mathbb{R}^6$:

$$\frac{d\mathbf{v}}{dt} = -k^2\mathbf{x}, \quad \frac{d\mathbf{x}}{dt} = \mathbf{v}.$$

In orthonormal coordinates on the inertial frame the system decouples into 3 systems (we will write only one) of the form:

$$\frac{dv}{dt} = -k^2 x, \quad \frac{dx}{dt} = v.$$

Note that the only *equilibrium configuration* is $x = 0$, i.e. $\mathbf{x} = \mathbf{0}$ putting together the components. The system's general integral is found with the methods of Chap. 14, and it is

$$x(t) = A\sin(kt) + B\cos(kt), \quad v(t) = kA\cos(kt) - kB\sin(kt),$$

where $A, B \in \mathbb{R}$ are arbitrary constants. Reassembling the 3 components and fixing the constants in terms of the initial conditions, the general solutions read

$$\mathbf{x}(t) = k^{-1}\sin(kt)\mathbf{v}_0 + \cos(kt)\mathbf{x}_0, \quad \mathbf{v}(t) = \cos(kt)\mathbf{v}_0 - k\sin(kt)\mathbf{x}_0. \quad (6.7)$$

From these, since the functions sine and cosine oscillate in $[-1, 1]$, we deduce:

$$k||\mathbf{x}(t)|| \leq ||\mathbf{v}_0|| + k||\mathbf{x}_0||, \quad ||\mathbf{v}(t)|| \leq ||\mathbf{v}_0|| + k||\mathbf{x}_0||, \quad \text{for any } t \in \mathbb{R}. \quad (6.8)$$

As on $\mathbb{K}^n$ all norms are equivalent (they generate the same topology), we can use on $\mathbb{R}^6$ the norm:

$$||(\mathbf{x}, \mathbf{v})||' := \max(k||\mathbf{x}||, ||\mathbf{v}||).$$

Inequality (6.8) immediately implies

$$||(\mathbf{x}(t), \mathbf{v}(t))||' \leq 2||(\mathbf{x}(0), \mathbf{v}(0))||', \quad \text{for any } t \in \mathbb{R}.$$

This, in turn, guarantees that the configuration $\mathbf{x}_0 = 0$ is a stable equilibrium. In fact, fix an open neighbourhood U of $(\mathbf{x}_0, \mathbf{v}_0) = (\mathbf{0}, \mathbf{0})$, and take an open ball B_δ (for the norm $|| \cdot ||'$) centred at $(\mathbf{0}, \mathbf{0})$ with radius $\delta > 0$ such that $B_\delta \subset U$. By choosing initial conditions in $B_{\delta/2}$ we make sure that the complete maximal solution's orbit will never leave B_δ, so obviously it will not leave U either. Clearly $\mathbf{x}_0 = 0$ is not asymptotically stable in the past nor in the future, since the limits of the right-hand side in (6.7) as $t \to \pm\infty$ do not exist, if we exclude those of the trivial solution.

6.1.2 Introduction to Lyapunov's Methods for Studying Stability

In the previous section's examples we studied the stability of singular points by first computing the ODE's general solution. From a practical viewpoint this is scarcely effective, since solutions are explicitly known in a small number of cases, and even fewer are interesting for the applications. For this reason there exist theorems that give sufficient conditions, for the stability of the critical point considered, that can be checked without having to solve the ODE.

We shall present two theorems due to Lyapunov in this setting, for autonomous systems, that represent the simplest instance of the mathematical apparatus developed by Lyapunov and other mathematicians.

Consider a first-order autonomous system on $D \subset \mathbb{K}^n$ (non-empty and open)

$$\frac{d\mathbf{x}}{dt} = \mathbf{f}(\mathbf{x}(t)) , \tag{6.9}$$

with $\mathbf{f} : D \to \mathbb{K}^n$. Take a differentiable map $W : U \to \mathbb{K}$ with $U \subset D$ open (possibly D itself). If we know a solution $\mathbf{x} = \mathbf{x}(t)$ to (6.9), the derivative of the composite $t \mapsto W(\mathbf{x}(t))$, supposing it is defined, reads:

$$\frac{dW(\mathbf{x}(t))}{dt} = \frac{d\mathbf{x}}{dt} \cdot \nabla W(\mathbf{x})|_{\mathbf{x}(t)} = [\mathbf{f}(\mathbf{x}) \cdot \nabla W(\mathbf{x})]|_{\mathbf{x}=\mathbf{x}(t)} .$$

If we set:

$$\mathcal{W}(\mathbf{x}) := \mathbf{f}(\mathbf{x}) \cdot \nabla W(\mathbf{x}) , \quad \text{for any } \mathbf{x} \in U , \tag{6.10}$$

then for any solution $\mathbf{x} = \mathbf{x}(t)$ to (6.9):

$$\frac{dW(\mathbf{x}(t))}{dt} = \mathcal{W}(\mathbf{x}(t)) \tag{6.11}$$

provided the term on the right is defined.

Remarks 6.10

(1) The function W was defined without solving the ODE explicitly, and yet it retains information on the solutions.

(2) Recall that given $\mathbf{f} : D \to \mathbb{K}^n$, if $W : U \to \mathbb{K}$ (with $U \subset D$ open) is differentiable, we say W is a **first integral** of system (6.9) on U whenever $I \ni t \mapsto W(\mathbf{x}(t))$ is constant along any solution $I \ni t \mapsto \mathbf{x}(t) \subset U$ of (6.9), as $t \in I$ varies.   ∎

We have the following result:

Proposition 6.11 *If $\mathbf{f} : D \to \mathbb{K}^n$ ($D \subset \mathbb{K}^n$ open) is locally Lipschitz, then $W :$ $U \to \mathbb{K}$ ($U \subset D$ open) satisfies:*

$$W(\mathbf{x}) < 0, \ \forall \mathbf{x} \in U, \quad or \quad W(\mathbf{x}) \leq 0, \ \forall \mathbf{x} \in U, \quad or \quad W(\mathbf{x}) = 0, \ \forall \mathbf{x} \in U,$$

if and only if, for any solution $\mathbf{x} : I \to U$ to (6.9), we respectively have:

$$\frac{dW(\mathbf{x}(t))}{dt} < 0, \ \forall t \in I, \ or \ \frac{dW(\mathbf{x}(t))}{dt} \leq 0, \ \forall t \in I, \ or \ \frac{dW(\mathbf{x}(t))}{dt} = 0, \ \forall t \in I \ .$$

A similar statement holds swapping $<$ and $>$.

Proof If $W(\mathbf{x}) < 0$ on U, then (6.11) implies $dW(\mathbf{x}(t))/dt < 0$ for any solution $\mathbf{x} : I \to U$ to system (6.9). Vice versa, if $dW(\mathbf{x}(t))/dt < 0$ for any solution $\mathbf{x} : I \to U$ to (6.9), by (6.11) necessarily $W(\mathbf{x}_0) < 0$ if $\mathbf{x}_0 \in U$ is a point in the orbit of a solution to (6.9). That $\mathbf{f}$ is locally Lipschitz ensures that for any point $\mathbf{x}_0 \in U$ there is a solution to (6.9) with initial condition $\mathbf{x}(0) = \mathbf{x}_0$. Such a solution's orbit passes through $\mathbf{x}_0$ by definition. The other cases are proved analogously.   □

Now that we have defined W for an arbitrary $W : U \to \mathbb{K}$, we can state the first Lyapunov-Barbasin Theorem. From now on, if $F : A \to \mathbb{R}$ is a map on $A \subset \mathbb{K}^n$ open, we will say it has a **strict (local) minimum** at $\mathbf{x}_0 \in A$ if $F(\mathbf{x}) > F(\mathbf{x}_0)$ for $\mathbf{x} \in A_0, \mathbf{x} \neq \mathbf{x}_0$, where $A_0 \subset A$ is an open neighbourhood of $\mathbf{x}_0$.

Theorem 6.12 (Lyapunov-Barbasin) *If $D \subset \mathbb{K}^n$ is open, and $\mathbf{x}_0 \in D$ is a singular point for the locally Lipschitz map $\mathbf{f} : D \to \mathbb{K}^n$, then:*

(1) $\mathbf{x}_0$ is future-stable if on an open neighbourhood $G_0 \subset D$ of $\mathbf{x}_0$ there is a differentiable map[1] $W : G_0 \to \mathbb{R}$ such that:

    (i) W has a strict (local) minimum at $\mathbf{x}_0$,

    (ii) $\overset{\cdot}{W}(\mathbf{x}) \leq 0$ for $\mathbf{x} \in G_0$.

---

[1] The function W is real-valued even when we work in $\mathbb{K}^n = \mathbb{C}^n$.

(2) $\mathbf{x}_0$ *is past-stable if on an open neighbourhood* $G_0 \subset D$ *of* $\mathbf{x}_0$ *there is a differentiable map* $W : G_0 \to \mathbb{R}$ *satisfying:*

  (i) W *has a strict (local) minimum in* $\mathbf{x}_0$,
  (ii) $W(\mathbf{x}) \geq 0$ *for* $\mathbf{x} \in G_0$.

(3) $\mathbf{x}_0$ *is a stable singular point if on an open neighbourhood* $G_0 \subset D$ *of* $\mathbf{x}_0$ *there is a differentiable map* $W : G_0 \to \mathbb{R}$ *such that:*

  (i) W *has a strict (local) minimum at* $\mathbf{x}_0$,
  (ii) W *is a first integral on* G_0 *for the system associated with* $\mathbf{f}$.

Proof The proof of (2) is completely similar to (1). (3) immediately follows from (1), (2) and Proposition 6.11.

Let us prove (1). For simplicity we will work in Cartesian coordinates on $\mathbb{K}^n$ centred at $\mathbf{x}_0$, which therefore becomes the point $\mathbf{0}$. Consider an open neighbourhood $U \ni \mathbf{0}$ and suppose $U \subset G_0$. We must find a second open neighbourhood $V \ni \mathbf{0}$ such that, when we take solutions with initial conditions in V, the orbits do not exit U. We define V as follows. Let $B \subset U$ be an open ball centred at $\mathbf{0}$ with $\overline{B} \subset U$. Since W is continuous on the compact boundary $\mathcal{F}(B)$ of B, we set:

$$\ell := \min \{ W(\mathbf{x}) \mid \mathbf{x} \in \mathcal{F}(B) \} \, ,$$

which exists, is finite, and is the image of some point $\mathbf{x}_m \in \mathcal{F}(B)$ by the Weierstrass Theorem. We can always assume $\ell > W(\mathbf{0})$ by taking the radius of B small enough, since $\mathbf{0}$ is a strict local minimum point for W and $\mathbf{x}_m \neq \mathbf{0}$. Define V to be an open ball centred at $\mathbf{0}$ and with $V \subset B \cap \{\mathbf{x} \in G_0 \mid W(\mathbf{x}) < \ell\}$. Such a set V exists because $\{\mathbf{x} \in B \mid W(\mathbf{x}) < \ell\}$ is certainly open (as continuous pre-image of an open set) and non-empty (it contains $\mathbf{0}$ as $W(\mathbf{0}) < \ell$, as shown above). We claim that every solution with initial condition belonging to V has orbit confined to B for $t > 0$, and therefore contained in U.

Suppose the maximal solution $\mathbf{x} = \mathbf{x}(t)$, with $t \in I$ open interval (containing 0) and $\mathbf{x}(0) \in V$, passes through $\mathbf{x}(t_1) \notin B$ for $t_1 > 0$. If $R > 0$ is the radius of B, the continuous map $I \ni t \mapsto R - ||\mathbf{x}(t)||$ is positive at $t = 0$ and non-positive at $t = t_1$, so it must be zero at some $t_2 \in (0, t_1]$. Hence $||\mathbf{x}(t_2)|| = R$, i.e. $\mathbf{x}(t_2) \in \mathcal{F}(B)$. Observe that there may be several points $t > 0$ where $R - ||\mathbf{x}(t)|| = 0$. The set of such points is bounded below by 0, so it admits greatest lower bound, let us call it $t_2 > 0$. By continuity $R - ||\mathbf{x}(t_2)|| = 0$, and by definition of infimum $R - ||\mathbf{x}(t_2)|| > 0$ if $0 < t < t_2$, so in particular $\mathbf{x}(t) \in B \subset G_0$ if $0 < t < t_2$. But by hypothesis $[0, t_2] \ni t \mapsto W(\mathbf{x}(t))$ is non-increasing because:

$$\frac{dW(\mathbf{x}(t))}{dt} = \dot{W}(\mathbf{x}(t)) \leq 0 \quad \text{if } \mathbf{x}(t) \in G_0.$$

Consequently

$$W(\mathbf{x}(t_2)) \leq W(\mathbf{x}(0)) < \ell. \tag{6.12}$$

On the other hand, by definition of ℓ,

$$W(\mathbf{x}(t_2)) \geq \ell,$$

which contradicts (6.12). This goes to show that $\mathbf{x} = \mathbf{x}(t)$ stays inside $B \subset U$ for any $t \in I$ with $t > 0$.   □

The function W used in the theorem is called a **Lyapunov function**. Lyapunov's method to study stability relies on finding Lyapunov functions adapted to the concrete case at hand. The previous theorem is accompanied by the following result, which handles asymptotic stability.

Theorem 6.13 (Lyapunov-Barbasin 2) *Consider, in the hypotheses of Theorem 6.12, and besides (1) or (2), the stronger assumption: $\dot{W}(\mathbf{x}) < 0$ for $\mathbf{x} \neq \mathbf{x}_0$, or $\dot{W}(\mathbf{x}) > 0$ for $\mathbf{x} \neq \mathbf{x}_0$. Then $\mathbf{x}_0$ is asymptotically stable in the future, or in the past, respectively.*

Proof We shall prove future stability, as past stability is completely similar. Redefining W by adding a constant allows us to assume $W(\mathbf{x}_0) = 0$. Let B be an open ball centred at $\mathbf{x}_0$ with $\overline{B} \subset G_0$. If $\epsilon > 0$, define the non-empty open set (containing x_0):

$$A_\epsilon := \{\mathbf{x} \in B \mid W(x) < \epsilon\}.$$

Consider a maximal solution $\mathbf{x} = \mathbf{x}(t)$ with initial condition in A_1. First, the orbit cannot reach $\mathbf{x}_0$ in finite time T, since the Cauchy problem with initial condition $\mathbf{x}(T) = \mathbf{0}$ has as unique maximal solution (hence, before time T as well) the constant $\mathbf{x}(t) = \mathbf{0}$ for any $t \in \mathbb{R}$. Therefore the orbit stays in a set where $\dot{W} < 0$. The latter implies $dW(\mathbf{x}(t))/dt < 0$, and then $t \mapsto W(\mathbf{x}(t))$ is strictly decreasing. In particular it admits limit ℓ as $t \to +\infty$. If ℓ were positive, from a certain $T > 0$ onwards the orbit would live in the compact set $\overline{B} \setminus A_\epsilon$ for a small enough $\epsilon > 0$. But then, calling $-M = \max\{\dot{W}(\mathbf{x}) \mid \mathbf{x} \in \overline{B} \setminus A_\epsilon\} < 0$, we would have $dW(\mathbf{x}(t))/dt < -M$ for any $t > T$, and the mean value theorem (with $\xi \in (T, t)$) would say:

$$W(\mathbf{x}(t)) - W(\mathbf{x}(T)) = (t - T)\frac{dW(\mathbf{x}(t))}{dt}\Big|_\xi \leq -M(t - T) \to -\infty, \quad \text{as } t \to +\infty.$$

This is impossible because W is continuous on the compact set $\overline{B} \setminus A_n$ and hence bounded. Therefore $W(\mathbf{x}(t)) \to \ell = 0$ as $t \to +\infty$. In other words:

for any $\epsilon > 0$ there exists $T_\epsilon > 0$ such that, if $t > T_\epsilon$, then $\mathbf{x}(t) \in A_\epsilon$. $\qquad$ (6.13)

If B_r is the open ball of radius r and centre x_0, for any $n = 1, 2, \ldots$ there is a small enough $\epsilon > 0$ such that[2] $A_\epsilon \subset B_{1/n}$. Consequently, and keeping Proposition (6.13) in account, we conclude the following: for any $n > 0$ there exists a time T_n after which $\mathbf{x}(t) \in B_{1/n}$. This means that $\mathbf{x}(t) \to \mathbf{x}_0$ as $t \to +\infty$.   □

Examples 6.14

(1) Consider the singular point $x = 0$ of the ODE:

$$\frac{dx}{dt} = -kx$$

on $\mathbb{R}$, where $k \neq 0$ is given. The function $W(x) := x^2$, defined on $\mathbb{R}$, satisfies: $W(x) > 0$ if $x \neq 0$, so it has a strict local minimum at $x = 0$, and $W = -2kx^2 < 0$ if $x \neq 0$. By the first Lyapunov theorem $x = 0$ is a future-stable singular point. By the second Lyapunov theorem we conclude the point is asymptotically stable in the future.

(2) Consider the differential equation on $\mathbb{R}^2$:

$$\frac{d\mathbf{x}}{dt} = -\sin(||\mathbf{x}||^2)A\mathbf{x} \quad \text{where} \quad A := \begin{bmatrix} 2 & 1 \\ 1 & 2 \end{bmatrix}.$$

Since $\det A = 3 > 0$ and $\operatorname{tr} A = 4 > 0$, the eigenvalues of the symmetric matrix A are strictly positive and so $A\mathbf{x} = \mathbf{0}$ if and only if $\mathbf{x} = \mathbf{0}$. The latter is a singular point. We can study its stability even knowing that we cannot solve explicitly. The function $W(\mathbf{x}) = ||\mathbf{x}||^2$ defined on $\mathbb{R}^2$ has a strict local minimum at the origin, and

$$W(\mathbf{x}) = -2\sin(||\mathbf{x}||^2)\mathbf{x} \cdot A\mathbf{x}$$

is strictly negative on a neighbourhood of the origin (origin excluded). In fact $\sin(||\mathbf{x}||^2) > 0$ around $\mathbf{0}$ if $\mathbf{x} \neq \mathbf{0}$, and the quadratic form $\mathbf{x} \cdot A\mathbf{x}$ is positive definite since the symmetric matrix A has positive eigenvalues.

We conclude the origin is stable and asymptotically stable in the future.

(3) Consider again the ODE, now in $\mathbb{C}^2$:

$$\frac{d\mathbf{x}}{dt} = -\sin(||\mathbf{x}||^2)A\mathbf{x} \quad \text{where} \quad A := \begin{bmatrix} 2 & 1+i \\ 1-i & 2 \end{bmatrix}.$$

---

[2] If not, outside a certain ball $B_{1/n*}$ there would be a sequence $\mathbf{x}_1, \mathbf{x}_2, \ldots$ with $W(\mathbf{x}_k) \to 0$ as $k \to +\infty$. Since the $\mathbf{x}_k$ would belong to the compact set $K := \overline{B} \setminus B_{1/n*}$, we could find a subsequence $\{\mathbf{x}_{k_m}\}_m$ with $\mathbf{x}_{k_m} \to \mathbf{x}^* \in K$ as $m \to +\infty$. By continuity $W(\mathbf{x}^*) = 0$. But since $\mathbf{x}^* \neq \mathbf{x}_0$, W could not be strictly minimised by x_0.

We still have $\det A = 2 > 0$ and $\operatorname{tr} A = 4 > 0$. The eigenvalues of the Hermitian matrix A are strictly positive and $A\mathbf{x} = \mathbf{0}$ if and only if $\mathbf{x} = \mathbf{0}$. $\mathbf{x} = \mathbf{0}$ is still singular. If $\mathbf{x}^*$ is the (component-wise) conjugate vector to $\mathbf{x}$, and $\cdot$ denotes the usual matrix product, the real-valued map $W(\mathbf{x}) = \mathbf{x}^* \cdot \mathbf{x} = ||\mathbf{x}||^2$, defined on $\mathbb{C}^2$, has a strict local minimum at the origin. Moreover

$$W(\mathbf{x}) = \sin(||\mathbf{x}||^2)\left(\mathbf{x}^* \cdot A\mathbf{x} + \mathbf{x} \cdot A\mathbf{x}^*\right)$$

is strictly negative around the origin, except at the point. On a neighbourhood of $\mathbf{0}$ in fact, $\sin(||\mathbf{x}||^2) > 0$ if $\mathbf{x} \neq 0$, and the Hermitian form $\mathbf{x}^* \cdot A\mathbf{x}$ is strictly positive since the Hermitian matrix A has positive eigenvalues. Also, if $Re(z)$ denotes the real part of the complex number $z \in \mathbb{C}$:

$$\left(\mathbf{x}^* \cdot A\mathbf{x} + \mathbf{x} \cdot A\mathbf{x}^*\right) = 2Re(\mathbf{x}^* \cdot A\mathbf{x})\,.$$

Hence the origin is both future-stable and asymptotically future-stable.

6.1.3   More on Asymptotic Stability

We shall prove a second result that generalises the previous statement by providing a weaker sufficient condition for asymptotic stability. We also introduce a number of very useful technical concepts.

Definition 6.15 ($\Lambda_\pm(\mathbf{x}_0)$ Sets) Consider a system of autonomous ODEs of order one on a non-empty open set $D \subset \mathbb{K}^n$,

$$\frac{d\mathbf{x}}{dt} = \mathbf{f}(\mathbf{x}(t))\,, \tag{6.14}$$

with $\mathbf{f} : D \to \mathbb{K}^n$ locally Lipschitz. Call $I \ni t \mapsto \mathbf{x}(t|\mathbf{x}_0)$ the maximal solution of (6.14) with initial condition $\mathbf{x}_0 \in D$.

(1) If $I \ni t \mapsto \mathbf{x}(t|\mathbf{x}_0)$ is future-complete, we let

$$\Lambda_+(\mathbf{x}_0) := \{\mathbf{y} \in D \mid \mathbf{x}(t_k|\mathbf{x}_0) \to \mathbf{y}\,,\ k \to +\infty,$$
$$\text{for some } \{t_k\}_{k\in\mathbb{N}} \subset I \text{ with } t_k \to +\infty, k \to +\infty\}\,.$$

(2) If $I \ni t \mapsto \mathbf{x}(t|\mathbf{x}_0)$ is past-complete, we let

$$\Lambda_-(\mathbf{x}_0) := \{\mathbf{y} \in D \mid \mathbf{x}(t_k|\mathbf{x}_0) \to \mathbf{y}\,,\ k \to +\infty,$$
$$\text{for some } \{t_k\}_{k\in\mathbb{N}} \subset I \text{ with } t_k \to -\infty \text{ if } k \to +\infty\}\,.$$

Notice that the sets $\Lambda_{\pm}(\mathbf{x}_0)$ might be empty.

Proposition 6.16 *Referring to Definition 6.15, if $\Lambda_+(\mathbf{x}_0)$ (or $\Lambda_-(\mathbf{x}_0)$) is non-empty then the following hold.*

(1) Every maximal solution with initial condition in $\Lambda_+(\mathbf{x}_0)$ (resp. $\Lambda_-(\mathbf{x}_0)$) is complete.

(2) The orbit of any maximal solution with initial condition in $\Lambda_+(\mathbf{x}_0)$ (resp. $\Lambda_-(\mathbf{x}_0)$) is a subset of $\Lambda_+(\mathbf{x}_0)$ (resp. $\Lambda_-(\mathbf{x}_0)$).

Proof Let us show the result for $\Lambda_+(\mathbf{x}_0)$, the other case being analogous. Consider $\{t_k\}_{k\in\mathbb{N}} \subset I$ with $\mathbf{x}(t_k|\mathbf{x}_0) \to \mathbf{y}_0$ as $k \to +\infty$. Given $t \in \mathbb{R}$, there will be a sufficiently large $k \in \mathbb{N}$ such that $t + t_k > 0$. Then it makes sense to write (note that the uniqueness theorem holds): $\mathbf{x}(t + t_k|\mathbf{x}_0) = \mathbf{x}(t|\mathbf{x}(t_k|\mathbf{x}_0))$. Taking the limit as $k \to +\infty$, the right-hand side tends to $\mathbf{x}(t|\mathbf{y}_0)$ by (2) in Theorem 14.37. Hence the limit of $\mathbf{x}(t + t_k|\mathbf{x}_0)$ as $k \to +\infty$ exists, and it equals:

$$\lim_{k\to+\infty} \mathbf{x}(t + t_k|\mathbf{x}_0) = \mathbf{x}(t|\mathbf{y}_0) \,.$$

In particular, since t is arbitrary in $\mathbb{R}$, the solution $t \mapsto \mathbf{x}(t|\mathbf{y}_0)$ is complete and (1) is thus proved. For (2), defining $t'_k := t + t_k$, we have:

$$\mathbf{x}(t|\mathbf{y}_0) = \lim_{k\to+\infty} \mathbf{x}(t'_k|\mathbf{x}_0) \,;$$

but then, by definition of $\Lambda_+(\mathbf{x}_0)$, $\mathbf{x}(t|\mathbf{y}_0) \in \Lambda_+(\mathbf{x}_0)$ for any $t \in \mathbb{R}$. $\square$

We can then prove a third Lyapunov theorem.

Theorem 6.17 (Lyapunov-Barbasin 3) *In the hypotheses of Theorem 6.12, if, besides (1) or (2) or (3), for some $\epsilon > 0$ the set*

$$A_\epsilon := \{\mathbf{x} \in G_0 \mid 0 \neq ||\mathbf{x} - \mathbf{x}_0|| \leq \epsilon \,, \ W(\mathbf{x}) = 0\}$$

does not fully contain all complete maximal solutions, then $\mathbf{x}_0$ is asymptotically stable, respectively in the future, in the past, or both.

Proof The proposition is equivalent to asserting that: if $\mathbf{x}_0$ is stable, but not asymptotically stable (in the three cases), then every set A_ϵ, for $\epsilon > 0$, contains the orbit of at least one complete maximal solution. So let us prove the latter. We suppose, to fix ideas, $\mathbf{x}_0$ is not asymptotically stable *in the future*, for the other situations are similar.

If there is no asymptotic stability in the future, given a ball B_{ϵ_0} with centre $\mathbf{x}_0$ and radius $\epsilon_0 > 0$ as small as we like, there exists an initial condition $\mathbf{y}_0 \in B_{\epsilon_0}$ for which the associated maximal solution $t \mapsto \mathbf{x}(t|\mathbf{y}_0)$ does not tend to $\mathbf{x}_0$ as $t \to +\infty$. Observe that $t \mapsto \mathbf{x}(t|\mathbf{y}_0)$ is contained in a ball with compact closure B_ϵ, with $\epsilon > \epsilon_0 > 0$, due to the stability. This implies the future completeness of $t \mapsto \mathbf{x}(t|\mathbf{y}_0)$ by (1) in Proposition 14.34. That $\mathbf{x}(t|\mathbf{y}_0) \not\to \mathbf{x}_0$ as $t \to +\infty$ implies that however large we fix t_k, there is a successive time t_{k+1} for which $\mathbf{x}(t_{k+1}|\mathbf{y}_0) \notin B_{\epsilon_0}$. Hence we have a sequence $t_1 < t_2 < \cdots < t_k \to +\infty$ such that every $\mathbf{x}(t_k|\mathbf{y}_0)$ belongs in the compact set $K := \overline{B_\epsilon \setminus B_{\epsilon_0}}$. We can then extract a subsequence $\{\mathbf{x}(t_{k_r}|\mathbf{y}_0)\}_{r \in \mathbb{N}}$ converging to $\mathbf{x}' \in K$. By definition $\mathbf{x}' \in \Lambda_+(\mathbf{y}_0)$. Since $W(\mathbf{x}(t|\mathbf{y}_0))$ is continuous and non-decreasing, necessarily

$$\lim_{t \to +\infty} W(\mathbf{x}(t|\mathbf{y}_0)) = \lim_{r \to +\infty} W(\mathbf{x}(t_{t_r}|\mathbf{y}_0)) = W(\mathbf{x}'),$$

and then:

$$W(\Lambda_+(\mathbf{y}_0)) = W(\mathbf{x}') \quad \text{constant}. \tag{6.15}$$

This fact prevents $\mathbf{x}_0 \in \Lambda_+(\mathbf{y}_0)$, since $W(\mathbf{x}') > W(\mathbf{x}_0)$ and W has a strict local minimum at $\mathbf{x}_0$.

Consider finally $\mathbf{z} \in \Lambda_+(\mathbf{y}_0)$. We know the maximal solution $t \mapsto \mathbf{x}(t|\mathbf{z})$ is complete and $\mathbf{x}(t|\mathbf{z}) \in \Lambda_+(\mathbf{y}_0)$ for any $t \in \mathbb{R}$, by Proposition 6.16. On the other hand, due to (6.15),

$$\frac{dW(\mathbf{x}(t|\mathbf{z}))}{dt} = 0 \quad \text{i.e.} \quad W(\mathbf{x}(t|\mathbf{z})) = 0 \text{, for any } t \in \mathbb{R}.$$

To sum up,

$$A_\epsilon := \{\mathbf{x} \in G_0 \mid 0 \neq ||\mathbf{x} - \mathbf{x}_0|| \leq \epsilon, \quad W(\mathbf{x}) = 0\}$$

entirely contains the orbit of a complete maximal solution: $\mathbf{x} = \mathbf{x}(t|\mathbf{z})$ with $t \in \mathbb{R}$.

$\square$

Remarks 6.18 If the Lyapunov function of Theorem 6.12 satisfies the stronger requirement $W(\mathbf{x}) < 0$ for $\mathbf{x} \neq \mathbf{x}_0$, or $W(\mathbf{x}) > 0$ for $\mathbf{x} \neq \mathbf{x}_0$, all sets A_ϵ are empty, so we trivially fall back to the hypotheses of Theorem 6.17. This shows that Theorem 6.13 is actually a corollary of Theorem 6.17. $\blacksquare$

6.1.4 An Instability Criterion Based on Linearisation

To conclude we provide a simple sufficient criterion, not based on Lyapunov functions, to decide whether a singular point is unstable. The idea is to "linearise" the system around the singular point, study the linear version and then extend the result to the original system.

Consider the usual first-order autonomous system defined on a non-empty open subset D of $\mathbb{K}^n$:

$$\frac{d\mathbf{x}}{dt} = \mathbf{f}(\mathbf{x}). \tag{6.16}$$

Suppose $\mathbf{f}$ is $C^1(D)$ and $\mathbf{x}_0$ is singular. Using the Taylor expansion at $\mathbf{x}_0$ we write:

$$\mathbf{f}(\mathbf{x}) = \mathbf{0} + \nabla \mathbf{f}|_{\mathbf{x}_0} \cdot (\mathbf{x} - \mathbf{x}_0) + ||\mathbf{x} - \mathbf{x}_0|| \mathbf{O}(\mathbf{x} - \mathbf{x}_0) \,,$$

where the function $\mathbf{O}$ satisfies $\mathbf{O}(\mathbf{x}) \to \mathbf{0}$ as $\mathbf{x} \to \mathbf{0}$. Without loss of generality we may always change variables to $\mathbf{y} := \mathbf{x} - \mathbf{x}_0$, so the above expansion reads:

$$\mathbf{f}(\mathbf{x}) = A\mathbf{y} + ||\mathbf{y}|| \mathbf{O}(\mathbf{y}) \,, \tag{6.17}$$

where the linear map $A : \mathbb{K}^n \to \mathbb{K}^n$ is $A\mathbf{y} := \nabla \mathbf{f}|_{\mathbf{x}_0} \cdot \mathbf{y}$. The ODE system:

$$\frac{d\mathbf{y}}{dt} = A\mathbf{y} \,, \tag{6.18}$$

where $A\mathbf{y} := \mathbf{y} \cdot \nabla \mathbf{f}|_{\mathbf{x}_0}$ for any $\mathbf{y} \in \mathbb{K}^n$, is called **linearisation around the singular point** $\mathbf{x}_0$ of system (6.16).

We now have a simple yet important result.

Proposition 6.19 *Consider the ODE system on $\mathbb{K}^n$:*

$$\frac{d\mathbf{y}}{dt} = A\mathbf{y} \,,$$

where A is any $n \times n$ matrix with coefficients in $\mathbb{K}$.

(1) If A has an eigenvalue with positive real part, the singular point $\mathbf{y}_0 = 0$ is unstable in the future.

(2) If A has an eigenvalue with negative real part, the singular point $\mathbf{y}_0 = 0$ is unstable in the past.

Proof Let us show (1), given that (2) is completely similar. There will exist $\mathbf{u} \in \mathbb{C}^n \setminus \{\mathbf{0}\}$ such that $A\mathbf{u} = \lambda \mathbf{u}$ and $\mathrm{Re}\lambda > 0$.

Suppose first $\mathbb{K} = \mathbb{C}$. For $\delta > 0$ small consider the map:

$$\mathbf{x}(t) = \delta e^{\lambda t} \mathbf{u} \,, \quad \text{with } t \in \mathbb{R} \,. \tag{6.19}$$

It satisfies $\frac{d\mathbf{y}}{dt} = A\mathbf{y}$, with initial condition $\mathbf{y}(0) = \delta \mathbf{u}$, as is immediate to see (it is also the only maximal solution with that initial condition, since $\mathbf{y} \mapsto A\mathbf{y}$ is C^∞). On the other hand, as $\mathrm{Re}\lambda > 0$ we also have:

$$||\mathbf{x}(t)||^2 = \delta^2 (\mathbf{u}^* \cdot \mathbf{u}) e^{2(\mathrm{Re}\lambda)t} \to +\infty \quad \text{as } t \to +\infty.$$

Given an open neighbourhood $U \ni \mathbf{0}$, an arbitrarily small open neighbourhood $V \ni \mathbf{0}$ will contain a point $\delta \mathbf{u}$, so the initial condition $\mathbf{y}(0) = \delta \mathbf{u}$ produces a divergent solution as $t \to +\infty$. Hence, a fortiori, its orbit leaves U for t large enough. Therefore the singular point $\mathbf{0}$ cannot be future-stable.

Consider now $\mathbb{K} = \mathbb{R}$. The solution (6.19) cannot be considered since it is $\mathbb{C}^n$-valued and we want to work in $\mathbb{R}^n$. Since A is real, from $A\mathbf{u} = \lambda\mathbf{u}$ ($\mathbf{u} \neq \mathbf{0}$) we deduce $A\mathbf{u}^* = \lambda^*\mathbf{u}^*$. Multiply $\mathbf{u}$ by a complex number, if necessary, so that $\mathbf{u}+\mathbf{u}^* \neq \mathbf{0}$, and consider the $\mathbb{R}^n$-valued map:

$$\mathbf{x}(t) = \delta e^{\lambda t}\mathbf{u} + \delta e^{\lambda^* t}\mathbf{u}^*, \quad \text{with } t \in \mathbb{R}.$$

Clearly this does not solve the ODE with initial condition $\mathbf{x}(0) = \delta(\mathbf{u} + \mathbf{u}^*)$. On the other hand

$$||\mathbf{x}(t)||^2 = \delta^2 \left(e^{\lambda t}\mathbf{u} + e^{\lambda^* t}\mathbf{u}^*\right) \cdot \left(e^{\lambda t}\mathbf{u} + e^{\lambda^* t}\mathbf{u}^*\right)$$

$$= \delta^2 \left[2e^{2(Re\lambda)t}||\mathbf{u}||^2 + 2Re\left(\mathbf{u} \cdot \mathbf{u} e^{2\lambda t}\right)\right]$$

$$= \delta^2 e^{2(Re\lambda)t} \left[2||\mathbf{u}||^2 + 2Re\left(\mathbf{u} \cdot \mathbf{u} e^{2i\,Im(\lambda)t}\right)\right]$$

$$= \delta^2 e^{2(Re\lambda)t} \left[2||\mathbf{u}||^2 + 2\left(Re(\mathbf{u} \cdot \mathbf{u})\cos(2Im(\lambda)t) - Im(\mathbf{u} \cdot \mathbf{u})\sin(2Im(\lambda)t)\right)\right].$$

Therefore:

$$||\mathbf{x}(t)||^2 = 2\delta^2 e^{2(Re\lambda)t} \left[||\mathbf{u}||^2 + |\mathbf{u} \cdot \mathbf{u}|\cos\left(2Im(\lambda)t + \phi\right)\right],$$

where the angle ϕ is defined by: $|\mathbf{u}\cdot\mathbf{u}|\cos\phi = Re(\mathbf{u}\cdot\mathbf{u})$ and $|\mathbf{u}\cdot\mathbf{u}|\sin\phi = Im(\mathbf{u}\cdot\mathbf{u})$. In case $Im\lambda = 0$, the above formulas[3] imply that $||\mathbf{x}(t)||^2 \to +\infty$ as $t \to +\infty$. If, conversely, $Im\lambda \neq 0$, then $||\mathbf{x}(t_k)||^2 \to +\infty$ as $k \to +\infty$, provided we set $t_k := \frac{2\pi k - \phi}{2Im\lambda}$ with $k \in \mathbb{N}$. In either case, given an open neighbourhood $U \ni \mathbf{0}$ any arbitrarily small open neighbourhood $V \ni \mathbf{0}$ will contain a point $\delta(\mathbf{u}+\mathbf{u}^*)$ for which the initial condition $\mathbf{y}(0) = \delta(\mathbf{u} + \mathbf{u}^*)$ produces a solution $\mathbf{x} = \mathbf{x}(t)$ with $\mathbf{x}(t_k) \notin U$ for infinitely many values $t_k \to +\infty$. Hence $\mathbf{0}$ cannot be future-stable. □

The proposition can be generalised to non-linear systems. The instability of a singular point in a linearised system implies the point's instability in the non-linear system. Indeed, the following theorem holds [Mal52]:

Theorem 6.20 *Consider a first-order autonomous system*

$$\frac{d\mathbf{x}}{dt} = \mathbf{f}(\mathbf{x}(t))$$

---

[3] We cannot have $||\mathbf{u}||^2 + |\mathbf{u} \cdot \mathbf{u}|\cos\phi = 0$, i.e. $||\mathbf{u}||^2 + Re(\mathbf{u} \cdot \mathbf{u}) = 0$, since the computation would force $\mathbf{u}^* = -\mathbf{u}$, which we have excluded a priori.

with $\mathbf{f} : D \to \mathbb{K}^n$ *locally Lipschitz and* $D \subset \mathbb{K}$ *non-empty and open. Suppose* $\mathbf{x}_0 \in D$ *is a singular point and:*

(1) $\mathbf{f}(\mathbf{x}) = \nabla \mathbf{f}|_{\mathbf{x}_0} \cdot (\mathbf{x} - \mathbf{x}_0) + \mathbf{R}(\mathbf{x})$ *with* $||\mathbf{R}(\mathbf{x})|| \leq K ||\mathbf{x} - \mathbf{x}_0||^{1+\alpha}$ *around* $\mathbf{x}_0$, *for some* $K \geq 0$ *and* $\alpha > 0$;
(2) the matrix $A = \nabla \mathbf{f}|_{\mathbf{x}_0}$ *has at least one eigenvalue with positive, or negative, real part.*

Then $\mathbf{x}_0$ *is unstable in the future, respectively in the past.*

Remarks 6.21

(1) Condition (1) is automatic if $\mathbf{f} \in C^2(D; \mathbb{K}^n)$. This follows trivially from the second-order Taylor expansion, for in that case every component f^i of $\mathbf{f}$ reads

$$f^i(\mathbf{x}) = f^i(\mathbf{x}_0) + \nabla f^i|_{\mathbf{x}_0} \cdot (\mathbf{x} - \mathbf{x}_0) + ||\mathbf{x} - \mathbf{x}_0||^2 \mathbf{n} \cdot \mathbf{R}_{\mathbf{x}}^i \mathbf{n} ,$$

where $\mathbf{n}$ is $\mathbf{x} - \mathbf{x}_0$ normalised, and the matrix $\mathbf{R}_{\mathbf{x}}^i$ defines a bounded map when $\mathbf{x}$ in a neighbourhood of $\mathbf{x}_0$. Note that just the differentiability of $\mathbf{f}$ on D is not enough to guarantee (1).
(2) There is an analogous criterion for stability rather than instability: it is easy to prove that for the linear system (6.18), if A has *all* eigenvalues with negative real part, the critical point $\mathbf{y}_0 = \mathbf{0}$ is asymptotically stable in the future. Such a result extends to non-linear systems exactly as Theorem 6.20. The critical point $\mathbf{x}_0$ is future-stable if (1) in Theorem 6.20 holds, and moreover all eigenvalues of $A = \nabla \mathbf{f}|_{\mathbf{x}_0}$ have negative real part [Mal52]. A similar theorem holds for past-stability if we ask the eigenvalues of $A = \nabla \mathbf{f}|_{\mathbf{x}_0}$ to have positive real part. ∎

6.2 Applications to Physical Systems in Classical Mechanics

We apply the theory explained in the previous section to particular physical systems. At the end of the section we will apply the theory of stability to the study of permanent rotations of rigid bodies, while below we deal with systems of particles subject to conservative forces, in a given reference frame $\mathscr{R}$, possibly with gyrostatic and dissipative forces. More precisely, we will examine a system of N point particles P_k of masses m_k, $k = 1, \ldots, N$. The system is subject to conservative forces with respect to $\mathscr{R}$ with potential energy $\mathscr{U}|_{\mathscr{R}} = \mathscr{U}|_{\mathscr{R}}(P_1, \ldots, P_N)$. Furthermore, each particle is subject to a force $\mathbf{F}_k = \mathbf{F}_k(P_1, \ldots, P_N, \mathbf{v}_{P_1}|_{\mathscr{R}}, \ldots, \mathbf{v}_{P_N}|_{\mathscr{R}})$ that in $\mathscr{R}$ does not depend on time, so that the following two conditions are met:

(1) $\mathbf{F}_k = \mathbf{F}_k(P_1, \ldots, P_N, \mathbf{0}, \ldots, \mathbf{0}) = \mathbf{0}$ for any $k = 1, \ldots, N$;
(2) $\displaystyle\sum_{k=1}^{N} \mathbf{F}_k(P_1, \ldots, P_N, \mathbf{v}_{P_1}|_{\mathscr{R}}, \ldots, \mathbf{v}_{P_N}|_{\mathscr{R}}) \cdot \mathbf{v}_{P_k}|_{\mathscr{R}} \leq 0$ for all velocities in $\mathscr{R}$:
$\mathbf{v}_{P_1}|_{\mathscr{R}}, \ldots, \mathbf{v}_{P_N}|_{\mathscr{R}} \in V_{\mathscr{R}}$.

Examples 6.22

(1) The simplest example of the situation at hand is when the forces $\mathbf{F}_k$ are all zero and there only are conservative forces. A concrete example is a system of electric charges if we ignore irradiation, or a system of planets interacting under mutual gravitational pull.

(2) A less drastic example where the forces $\mathbf{F}_k$ are not zero, but still satisfy (1) and (2), is the above system of charges together with certain magnetic forces due to an external field. These do not generate power:

$$\sum_{k=1}^{N} \mathbf{F}_k(P_1, \ldots, P_N, \mathbf{v}_{P_1}|_{\mathscr{R}}, \ldots, \mathbf{v}_{P_N}|_{\mathscr{R}}) \cdot \mathbf{v}_{P_k}|_{\mathscr{R}} = 0, \text{ for any choice of}$$

$(\mathbf{v}_{P_1}|_{\mathscr{R}}, \ldots, \mathbf{v}_{P_N}|_{\mathscr{R}})$.

(3) Another example are particles that interact through conservative forces, for instance elastic forces, and the system is immersed in a liquid at rest in a frame $\mathscr{R}$. The power of the conservative forces is non-positive and satisfies the stronger condition

$$\sum_{k=1}^{N} \mathbf{F}_k(P_1, \ldots, P_N, \mathbf{v}_{P_1}|_{\mathscr{R}}, \ldots, \mathbf{v}_{P_N}|_{\mathscr{R}}) \cdot \mathbf{v}_{P_k}|_{\mathscr{R}} < 0 \,,$$

for $(\mathbf{v}_{P_1}|_{\mathscr{R}}, \ldots, \mathbf{v}_{P_N}|_{\mathscr{R}}) \neq (\mathbf{0}, \ldots, \mathbf{0})$.

6.2.1   The Lagrange-Dirichlet Theorem

Fix orthogonal coordinates on $\mathscr{R}$ with origin O. Thus every point P_k is determined by the vector $\mathbf{x}_k$ of $\mathbb{R}^3$ whose components are those of $P_k - O$ in the given coordinates. By Newton's second law the equations of motion, reduced to first order and dropping $|_{\mathscr{R}}$ for simplicity, are:

$$\frac{d\mathbf{v}_k}{dt} = -\frac{1}{m_k}\nabla_{\mathbf{x}_k}\mathscr{U}(\mathbf{x}_1, \ldots, \mathbf{x}_n)$$

$$+\frac{1}{m_k}\mathbf{F}_k(\mathbf{x}_1, \ldots, \mathbf{x}_N, \mathbf{v}_1, \ldots, \mathbf{v}_N), \quad k=1, \ldots, N, \quad (6.20)$$

$$\frac{d\mathbf{x}_k}{dt} = \mathbf{v}_k, \quad k = 1, \ldots, N. \quad (6.21)$$

We will assume $\mathbf{F}_k \in C^1(\mathbb{R}^{6N}; \mathbb{R}^{3N})$, while $\mathscr{U} \in C^2(\mathbb{R}^{3N})$.

The system equilibria are, as explained in Sect. 6.1, the configurations $(\mathbf{x}_{0,1}, \ldots, \mathbf{x}_{0,N}) \in \mathbb{R}^{3N}$ such that the above system's right-hand side vanishes at $(\mathbf{x}_{0,1}, \ldots, \mathbf{x}_{0,N}, \mathbf{0}, \ldots, \mathbf{0}) \in \mathbb{R}^{3N} \times \mathbb{R}^{3N}$, where the N null vectors $\mathbf{0}$ are the velocities $\mathbf{v}_k$. Because of assumption (1) above (in particular, when the non-conservative forces $\mathbf{F}_k$ are absent), the following proposition holds.

Proposition 6.23 *Consider N point particles P_k with masses $m_k > 0$, $k = 1, \ldots, N$. Suppose the laws of motion in $\mathscr{R}$ have the form (where $\mathbf{x}_i = P_i - O$ with O at rest in $\mathscr{R}$):*

$$m_k \frac{d^2 \mathbf{x}_k}{dt^2} = -\nabla_{\mathbf{x}_k} \mathscr{U}|_{\mathscr{R}}(\mathbf{x}_1, \ldots, \mathbf{x}_n) + \mathbf{F}_k \left(\mathbf{x}_1, \ldots, \mathbf{x}_N, \frac{d\mathbf{x}_1}{dt}\Big|_{\mathscr{R}}, \ldots, \frac{d\mathbf{x}_N}{dt}\Big|_{\mathscr{R}} \right),$$

$$(6.22)$$

where the potential energy $\mathscr{U}|_{\mathscr{R}} = \mathscr{U}|_{\mathscr{R}}(\mathbf{x}_1, \ldots, \mathbf{x}_N)$ corresponds to conservative forces in $\mathscr{R}$, and the non-conservative forces $\mathbf{F}_k$ (if present) satisfy:

$$\mathbf{F}_k(\mathbf{x}_1, \ldots, \mathbf{x}_N, \mathbf{0}, \ldots, \mathbf{0}) = \mathbf{0}, \quad \text{for any } k = 1, \ldots, N \text{ and any } (\mathbf{x}_1, \ldots, \mathbf{x}_N) \in V_{\mathscr{R}}^N.$$

The $\mathbf{F}_k$ are assumed C^1 and $\mathscr{U}|_{\mathscr{R}}$ is C^2 on non-empty open subsets in $V_{\mathscr{R}}^N \times V_{\mathscr{R}}^N$ and $V_{\mathscr{R}}^N$ respectively.
Then $(\mathbf{x}_{0,1}, \ldots, \mathbf{x}_{0,N}) \in V_{\mathscr{R}}^N$ is an equilibrium configuration if and only if it is a stationary point for the potential energy, i.e.:

$$\nabla_{\mathbf{x}_k} \mathscr{U}|_{\mathscr{R}}(\mathbf{x}_{0,1}, \ldots, \mathbf{x}_{0,N}) = \mathbf{0}, \quad \text{for any } k = 1, \ldots, N. \tag{6.23}$$

What we now want is to establish when an equilibrium is stable. The Lagrange-Dirichlet Theorem, that we are about to state and prove, is a very useful criterion to that end.

Theorem 6.24 (Lagrange-Dirichlet) *Consider N point particles P_k with masses $m_k > 0$, $k = 1, \ldots, N$. Suppose the laws of motion in $\mathscr{R}$ have the form (with $\mathbf{x}_i = P_i - O$ with O at rest in $\mathscr{R}$):*

$$m_k \frac{d^2 \mathbf{x}_k}{dt^2} = -\nabla_{\mathbf{x}_k} \mathscr{U}|_{\mathscr{R}}(\mathbf{x}_1, \ldots, \mathbf{x}_n) + \mathbf{F}_k \left(\mathbf{x}_1, \ldots, \mathbf{x}_N, \frac{d\mathbf{x}_1}{dt}\Big|_{\mathscr{R}}, \ldots, \frac{d\mathbf{x}_N}{dt}\Big|_{\mathscr{R}} \right),$$

$$(6.24)$$

where the potential energy $\mathscr{U}|_{\mathscr{R}} = \mathscr{U}|_{\mathscr{R}}(\mathbf{x}_1, \ldots, \mathbf{x}_N)$ corresponds to conservative forces in $\mathscr{R}$, and the non-conservative forces $\mathbf{F}_k$ (if present) satisfy:

$$\mathbf{F}_k(\mathbf{x}_1, \ldots, \mathbf{x}_N, \mathbf{0}, \ldots, \mathbf{0}) = \mathbf{0}, \quad \text{for any } k = 1, \ldots, N \text{ and any } (\mathbf{x}_1, \ldots, \mathbf{x}_N) \in V_{\mathscr{R}}^N.$$

The $\mathbf{F}_k$ are C^1 and $\mathscr{U}|_{\mathscr{R}}$ is C^2 on non-empty open subsets in $V_{\mathscr{R}}^N \times V_{\mathscr{R}}^N$ and $V_{\mathscr{R}}^N$ respectively.
An equilibrium configuration $(\mathbf{x}_{0,1}, \ldots, \mathbf{x}_{0,N})$ is:

(1) future-stable if the restriction of $\mathscr{U}|_{\mathscr{R}}$ to an open neighbourhood of $(\mathbf{x}_{0,1}, \ldots, \mathbf{x}_{0,N})$ has a strict local minimum there, and moreover:

$$\sum_{k=1}^{N} \mathbf{F}_k(\mathbf{x}_1, \ldots, \mathbf{x}_N, \mathbf{v}_{P_1}|_{\mathscr{R}}, \ldots, \mathbf{v}_{P_N}|_{\mathscr{R}}) \cdot \mathbf{v}_{P_k}|_{\mathscr{R}} \leq 0 \text{ for all velocities}$$
$$\mathbf{v}_{P_1}|_{\mathscr{R}}, \ldots, \mathbf{v}_{P_N}|_{\mathscr{R}};$$

(2) *past-stable and future-stable if the restriction of $\mathcal{U}|_{\mathcal{R}}$ to an open neighbour-hood of $(\mathbf{x}_{0,1}, \ldots, \mathbf{x}_{0,N})$ has a strict local minimum there, and moreover:*

$$\sum_{k=1}^{N} \mathbf{F}_k(\mathbf{x}_1, \ldots, \mathbf{x}_N, \mathbf{v}_{P_1}|_{\mathcal{R}}, \ldots, \mathbf{v}_{P_N}|_{\mathcal{R}}) \cdot \mathbf{v}_{P_k}|_{\mathcal{R}} = 0 \text{ for all velocities}$$

$\mathbf{v}_{P_1}|_{\mathcal{R}}, \ldots, \mathbf{v}_{P_N}|_{\mathcal{R}}.$

Proof

(1) Fix orthonormal coordinates in $\mathcal{R}$ with origin O, and identify the points P_k with vectors $\mathbf{x}_k \in \mathbb{R}^3$ whose components are those of the $P_k - O$. The velocities $\mathbf{v}_{P_k}|_{\mathcal{R}}$ are completely determined by the corresponding vectors $\mathbf{v}_k \in \mathbb{R}^3$. Consider the function defining the mechanical energy in $\mathcal{R}$:

$$\mathscr{E}(\mathbf{x}_1, \ldots, \mathbf{x}_N, \mathbf{v}_1, \ldots, \mathbf{v}_N) := \sum_{k=1}^{N} \frac{1}{2} m_k \mathbf{v}_k^2 + \mathscr{U}(\mathbf{x}_1, \ldots, \mathbf{x}_N),$$

where we have dropped $|_{\mathcal{R}}$ for simplicity. Let U be an open neighbourhood of $(\mathbf{x}_{0,1}, \ldots, \mathbf{x}_{0,N}) \in \mathbb{R}^{3N}$ where the restriction of $\mathscr{U}$ has a strict local minimum. If $G_0 := U \times \mathbb{R}^{3N}$, the restriction of $\mathscr{E}$: $W := \mathscr{E}|_{G_0} : G_0 \to \mathbb{R}$ evidently has a strict local minimum at $(\mathbf{x}_{0,1}, \ldots, \mathbf{x}_{0,N}, \mathbf{0}, \ldots, \mathbf{0})$. Moreover, by Theorem 4.18, along every solution of the equations of motions (6.20) and (6.21) we have:

$$\frac{d}{dt} W(\mathbf{x}_1(t), \ldots, \mathbf{x}_N(t), \mathbf{v}_1(t), \ldots, \mathbf{v}_N(t))$$

$$= \sum_{k=1}^{n} \mathbf{F}_K(\mathbf{x}_1(t), \ldots, \mathbf{x}_N(t), \mathbf{v}_1(t), \ldots, \mathbf{v}_N(t)) \cdot \mathbf{v}_k \leq 0.$$

In other words, by Proposition 6.11 on G_0 we have:

$$W(\mathbf{x}_1, \ldots, \mathbf{x}_N, \mathbf{v}_1, \ldots, \mathbf{v}_N) \leq 0.$$

We are in the hypotheses of Theorem 6.12 for the Lyapunov function W, so $(\mathbf{x}_{0,1}, \ldots, \mathbf{x}_{0,N}, \mathbf{0}, \ldots, \mathbf{0})$ is a future-stable singular point and hence $(\mathbf{x}_{0,1}, \ldots, \mathbf{x}_{0,N})$ is a future-stable equilibrium configuration.

(2) Exactly as in the proof of case (1), now we have the stronger condition

$$W(\mathbf{x}_1, \ldots, \mathbf{x}_N, \mathbf{v}_1, \ldots, \mathbf{v}_N) = 0.$$

Again by Theorem 6.12, we also gain stability in the past.

$\square$

Remarks 6.25 Requesting a strict local minimum point for the potential energy in correspondence to a stable equilibrium configuration is *not* a necessary condition, even in the absence of non-conservative forces. An elementary example is the system of one point particle on the real line under a conservative force with twice differentiable potential energy: $\mathscr{U}(x) := x^4 \sin(1/x)$ if $x \neq 0$ and $\mathscr{U}(0) := 0$. The configuration $x = 0$ is a stable equilibrium, but it does not correspond to a strict local minimum. ∎

The following result completes the Lagrange-Dirichlet statement, by providing sufficient conditions for asymptotic stability.

Proposition 6.26 *In the hypotheses of Theorem 6.24:*

(3) the equilibrium configuration $(\mathbf{x}_{0,1}, \ldots, \mathbf{x}_{0,N})$ *is both stable and asymptotically stable in the future if the strict local minimum point* $(\mathbf{x}_{0,1}, \ldots, \mathbf{x}_{0,N})$ *is an isolated[4] critical point of the potential energy* $\mathscr{U}_{\mathscr{R}}|$, *and the non-conservative forces satisfy:*

$$\sum_{k=1}^{N} \mathbf{F}_k(\mathbf{x}_1, \ldots, \mathbf{x}_N, \mathbf{v}_{P_1}|_{\mathscr{R}}, \ldots, \mathbf{v}_{P_N}|_{\mathscr{R}}) \cdot \mathbf{v}_{P_k}|_{\mathscr{R}} < 0, \textit{for } (\mathbf{v}_{P_1}|_{\mathscr{R}}, \ldots, \mathbf{v}_{P_N}|_{\mathscr{R}}) \neq (\mathbf{0}, \ldots, \mathbf{0}).$$

Proof The conventions and notations of the previous theorem still apply. Restrict the neighbourhood U (hence $G_0 = U \times \mathbb{R}^{3N}$) of the stationary point so that in U there are no further critical points of the potential energy. We want to use Theorem 6.17, and therefore we first need to find the sets A_ϵ of the statement. In our assumptions, proceeding as in the previous theorem, it is straightforward to see that the set of points where:

$$W(\mathbf{x}_1, \ldots, \mathbf{x}_N, \mathbf{v}_1, \ldots, \mathbf{v}_N) = 0$$

coincides with the locus:

$$\sum_{k=1}^{N} \mathbf{F}_k(\mathbf{x}_1, \ldots, \mathbf{x}_N, \mathbf{v}_1, \ldots, \mathbf{v}_N) \cdot \mathbf{v}_k = 0 \,.$$

As $\sum_{k=1}^{N} \mathbf{F}_k(P_1, \ldots, P_N, \mathbf{v}_{P_1}|_{\mathscr{R}}, \ldots, \mathbf{v}_{P_N}|_{\mathscr{R}}) \cdot \mathbf{v}_{P_k}|_{\mathscr{R}} < 0$ for $(\mathbf{v}_{P_1}|_{\mathscr{R}}, \ldots, \mathbf{v}_{P_N}|_{\mathscr{R}}) \neq (\mathbf{0}, \ldots, \mathbf{0})$, that is possible only if the velocities are all zero. Hence the set A_ϵ contains the points of G_0 such that: $(\mathbf{x}_1, \ldots, \mathbf{x}_N, \mathbf{0}, \ldots, \mathbf{0}) \neq (\mathbf{x}_{0,1}, \ldots, \mathbf{x}_{0,N}, \mathbf{0}, \ldots, \mathbf{0})$ and

$$||(\mathbf{x}_1, \ldots, \mathbf{x}_N, \mathbf{0}, \ldots, \mathbf{0}) - (\mathbf{x}_{0,1}, \ldots, \mathbf{x}_{0,N}, \mathbf{0}, \ldots, \mathbf{0})|| \leq \epsilon \,.$$

---

[4] That $(\mathbf{x}_{0,1}, \ldots, \mathbf{x}_{0,N})$ gives a strict local minimum for the potential energy is not enough to ensure it is an isolated stationary point: consider, for $N = 1$, the map $\mathscr{U}(x) := x^4(2 + \sin(1/x))$ if $x \neq 0$ and $\mathscr{U}(0) := 0$. $\mathscr{U}$ is twice differentiable, $x = 0$ is a strict local minimum but not an isolated stationary point.

If $\mathbb{R} \ni t \mapsto (\mathbf{x}_1(t), \ldots, \mathbf{x}_N(t), \mathbf{v}_1(t), \ldots, \mathbf{v}_N(t))$ is a complete maximal solution belonging to some A_ϵ, necessarily $\mathbf{v}_k(t) = \mathbf{0}$ for any $t \in \mathbb{R}$, and from (6.20) we deduce:

$$\nabla_{\mathbf{x}_k} \mathscr{U}(\mathbf{x}_1(t), \ldots, \mathbf{x}_n(t)) + \mathbf{F}_k(\mathbf{x}_1(t), \ldots, \mathbf{x}_N(t), \mathbf{0}, \ldots, \mathbf{0}) = 0 \,.$$

By condition $\mathbf{F}_k = \mathbf{F}_k(P_1, \ldots, P_N, \mathbf{0}, \ldots, \mathbf{0}) = \mathbf{0}$ for any $k = 1, \ldots, N$, on the aforementioned complete maximal solution:

$$\nabla_{\mathbf{x}_k} \mathscr{U}(\mathbf{x}_1(t), \ldots, \mathbf{x}_n(t)) = \mathbf{0} \,.$$

Therefore all configurations $(\mathbf{x}_1(t), \ldots, \mathbf{x}_n(t))$ (they belong to U by construction) are critical points of the potential energy. But on the open set U this is allowed only if

$$(\mathbf{x}_1(t), \ldots, \mathbf{x}_n(t)) = (\mathbf{x}_{0,1}, \ldots \mathbf{x}_{0,N}) \,,$$

since $(\mathbf{x}_{0,1}, \ldots \mathbf{x}_{0,N})$ is an isolated critical point of $\mathscr{U}$. As $(\mathbf{x}_{0,1}, \ldots \mathbf{x}_{0,N}) \notin A_\epsilon$, we conclude that A_ϵ does not contain orbits of complete maximal solutions, and therefore we have asymptotic stability in the future.   □

Remarks 6.27 The arguments for Proposition 6.23, the Lagrange-Dirichlet Theorem and Proposition 6.26 are uniquely based on the form of Eq. (6.22) and on the fact that the masses m_k are positive. It does not matter what the dimension of the vector space where each $\mathbf{x}_i$ lives is, for it may be different from 3 (though it must be finite). This observation is important since it allows to extend the theorems to constrained systems and when the pure equations of motion have the form (6.22), where $\mathbf{x}_i$ is replaced by the k-th coordinate in some coordinate system, and m_i is a strictly positive constant that may or not have the meaning of a mass. ∎

6.2.2 An Instability Criterion

To conclude we prove a result about systems of point particles under purely conservative forces. It establishes a sufficient condition for instability based on the potential energy's Hessian matrix.

Proposition 6.28 *Consider N point particles P_k of masses $m_k > 0$, $k = 1, \ldots, N$, whose laws of motion in $\mathscr{R}$ have the form ($\mathbf{x}_i = P_i - O$ with O at rest in $\mathscr{R}$):*

$$m_k \frac{d^2 \mathbf{x}_k}{dt^2} = -\nabla_{\mathbf{x}_k} \mathscr{U}|_{\mathscr{R}}(\mathbf{x}_1, \ldots, \mathbf{x}_n) \,, \tag{6.25}$$

where the C^3 potential energy $\mathscr{U}|_{\mathscr{R}} = \mathscr{U}|_{\mathscr{R}}(\mathbf{x}_1, \ldots, \mathbf{x}_N)$ corresponds to conservative forces in $\mathscr{R}$.

If at an equilibrium configuration $(\mathbf{x}_{0,1}, \ldots, \mathbf{x}_{0,N})$ *the potential energy's Hessian matrix has a negative eigenvalue, then* $(\mathbf{x}_{0,1}, \ldots, \mathbf{x}_{0,N})$ *is past- and future-unstable.*

Proof Consider Cartesian coordinates on $\mathscr{R}$ with origin O at the critical point $(\mathbf{x}_{0,1}, \ldots, \mathbf{x}_{0,N}) = (\mathbf{0}, \ldots \mathbf{0})$. In these coordinates the system's equations of motion, written as first-order equations, are:

$$\frac{d\mathbf{v}'_k}{dt} = -\nabla_{\mathbf{x}'_k} \mathscr{U}(\mathbf{x}'_1/\sqrt{m_1}, \ldots, \mathbf{x}'_n/\sqrt{m_N}), \quad k = 1, \ldots, N, \quad (6.26)$$

$$\frac{d\mathbf{x}'_k}{dt} = \mathbf{v}'_k, \quad k = 1, \ldots, N. \quad (6.27)$$

where we have introduced the new variables $\mathbf{x}_i := \sqrt{m_i}\mathbf{x}_i$ and $\mathbf{v}'_i := \sqrt{m_i}\mathbf{v}_i$. Asking $\mathscr{U}$ to be of class C^3 is sufficient to warrant the hypotheses of Theorem 6.20, keeping into account (1) in Remarks 6.21. Linearising the equations at the singular point $(\mathbf{0}, \ldots, \mathbf{0}, \mathbf{0}, \ldots, \mathbf{0})$, we obtain the first-order linear system:

$$\frac{du'_j}{dt} = -\sum_{i=1}^{N} H'_{ji} y'_i, \quad j = 1, \ldots, 3N, \quad (6.28)$$

$$\frac{dy'_j}{dt} = u'_j, \quad j = 1, \ldots, 3N. \quad (6.29)$$

where y'_1, y'_2, y'_3 are the components of $\mathbf{x}'_1$. Similarly, y'_4, y'_5, y'_6 are the components of $\mathbf{x}'_2$ etc, and

$$H'_{ji} := M_j^{-1/2} H_{ji} M_i^{-1/2}.$$

Moreover $M_1 = M_2 = M_3 = m_1$ and $M_4 = M_5 = M_6 = m_2$ and so on, and finally the real symmetric matrix H with coefficients H_{ij} is the Hessian matrix of $\mathscr{U}(\mathbf{x}_1, \ldots, \mathbf{x}_N)$ at the critical configuration $(\mathbf{x}_{01}, \ldots, \mathbf{x}_{0N})$. In our hypotheses, H has an eigenvector:

$$\mathbf{y}_\lambda = (y_{\lambda 1}, \ldots, y_{\lambda 3N})^t,$$

with eigenvalue $\lambda < 0$: that is to say, $H\mathbf{y}_\lambda = \lambda \mathbf{y}_\lambda$. But then H', of coefficients H'_{ji}, also has a non-zero eigenvector:

$$\mathbf{y}'_\lambda := \left(\sqrt{M_1}\, y_{\lambda 1}, \ldots, \sqrt{M_{3N}}\, y_{\lambda 3N}\right)^t$$

with eigenvalue $\lambda < 0$:

$$H'\mathbf{y}'_\lambda = \lambda \mathbf{y}'_\lambda. \quad (6.30)$$

Defining the column vector $\mathbf{Y} = (u'_1, \ldots, u'_{3N}, y'_1, \ldots y'_{3N})^t$ in $\mathbb{R}^{6N}$, the linear system (6.28) and (6.29) can be written as:

$$\frac{d\mathbf{Y}}{dt} = A\mathbf{Y}, \tag{6.31}$$

where the matrix A is:

$$A := \left[\begin{array}{c|c} 0 & -H' \\ \hline I & 0 \end{array}\right]. \tag{6.32}$$

Consider the no-zero vector (note that $-\lambda > 0$ by hypothesis!):

$$\mathbf{U} := \left(y'_{\lambda,1}, \ldots, y'_{\lambda,3N}, (-\lambda)^{-1/2}y'_{\lambda,1}, \ldots, (-\lambda)^{-1/2}y'_\lambda\right)^t.$$

From (6.32) and (6.30), immediately:

$$A\mathbf{U} = \sqrt{-\lambda}\,\mathbf{U}.$$

We have proved that in the linear system (6.31), i.e. (6.28) and (6.29), the matrix on the right has a positive eigenvalue ($\sqrt{-\lambda}$). By Proposition 6.19 the singular point, i.e. the origin of $\mathbb{R}^{6N}$, is future-unstable. Applying Theorem 6.20 we conclude that the singular point $(\mathbf{x}_{0,1}, \ldots, \mathbf{x}_{0,N}, \mathbf{0}, \ldots, \mathbf{0})$ of system (6.26) and (6.27) is future-unstable. This immediately implies the equilibrium configuration $(\mathbf{x}_{0,1}, \ldots, \mathbf{x}_{0,N})$ of the system of point particles is future-unstable. Similarly one shows that

$$\mathbf{U}' := \left(y'_{\lambda,1}, \ldots, y'_{\lambda,3N}, -(-\lambda)^{-1/2}y'_{\lambda,1}, \ldots, -(-\lambda)^{-1/2}y'_\lambda\right)^t$$

satisfies

$$A\mathbf{U}' = -\sqrt{-\lambda}\,\mathbf{U}',$$

and therefore we have instability in the past as well.   □

Remarks 6.29 The proof of Proposition 6.28 relies uniquely on the form of Eq. (6.25) and the fact that the masses m_k are positive. The dimension of the vector space in which the vectors $\mathbf{x}_i$ live is irrelevant, it may as well be different from 3 (but finite). This observation is important in that it permits to extend the theorems to constrained systems where (6.22) holds for the pure equations for motion, and where $\mathbf{x}_i$ is replaced by the k-th component in a certain coordinate system, and m_i is a strictly positive constant that does not necessarily have the meaning of a mass. ■

Examples 6.30

(1) The simplest possible example is a point particle P subject to a conservative force, in a frame $\mathscr{R}$, with potential energy $\mathscr{U}|_{\mathscr{R}} = \mathscr{U}|_{\mathscr{R}}(P)$ (of class C^2). If $P_0 \in E_{\mathscr{R}}$ is a strict local minimum point on a neighbourhood of P_0, it is a stable equilibrium configuration, both in the past and in the future by the Lagrange-Dirichlet theorem. Introducing a viscous force $-\gamma\mathbf{v}_P|_{\mathscr{R}}$ makes the equilibrium point asymptotically stable in the future in view of Proposition 6.26.

(2) Consider a point particle subject to a conservative force, in a frame $\mathscr{R}$, with potential energy $\mathscr{U}|_{\mathscr{R}} = \mathscr{U}|_{\mathscr{R}}(P)$ (of class C^2). (What we are about to say extends trivially to systems of point particles only subject to conservative forces.) In absence of other forces, the stability of a stationary point P_0 of $\mathscr{U}|_{\mathscr{R}}$ can be examined by looking at the Hessian matrix of $\mathscr{U}|_{\mathscr{R}}$ at P_0. More precisely:

    (a) if the Hessian matrix at P_0 has strictly positive eigenvalues only, then we have a strict local minimum at P_0, and therefore the stationary point is a stable equilibrium configuration (past and future).

    (b) If the Hessian matrix at P_0 ha at least one strictly negative eigenvalue, the stationary point is a (future-)unstable equilibrium configuration.

Observe that the presence of a strict local minimum does not mean the eigenvalues must be positive: some might be zero. Consider for example

$$\mathscr{U}|_{\mathscr{R}}(x, y, x) := x^2 + y^2 + z^4 \ .$$

It has a strict local minimum at the origin but the Hessian's eigenvalue of the unit vector $\mathbf{e}_z$ is zero (while the others are positive). Conversely, instability might not be detected just by the presence of a negative Hessian eigenvalue. For instance, a point particle under the conservative force (with respect to the reference frame $\mathscr{R}$):

$$\mathscr{U}|_{\mathscr{R}}(x, y, x) := x^2 + y^2 - z^4$$

has a future-unstable equilibrium configuration at the origin. But the Hessian's eigenvalues at the origin are non-negative. The instability is caused by the z-direction, corresponding to the zero eigenvalue.

6.2.3 Stability of Permanent Rotations for Non-gyroscopic Rigid Bodies

Consider a non-gyroscopic rigid body S with moments of inertia $I_{O1} < I_{O2} < I_{O3}$ with respect to a principal triple of inertia $\mathbf{e}_1, \mathbf{e}_2, \mathbf{e}_3$ moving with the body and centred at the point $O \in S$. The latter is fixed in the inertial frame $\mathscr{R}$ and not subject to any force apart from the constraint reaction at O. What we will say holds, with small alterations in case of non-gyroscopic rigid bodies in free fall in the gravitational field $-g\mathbf{e}_z$ in an inertial frame $\mathscr{R}$, by putting ourselves in a second, non-inertial reference frame, which slides with respect to $\mathscr{R}$ and stays at rest with the body's centre of mass $O = G$.

We want to apply the elementary theory of stability to the solutions of the Euler equations (5.31) given by the *permanent rotations* around the principal axes of inertia from Sect. 5.3.3. We will show that the permanent rotations around $\mathbf{e}_1$ and $\mathbf{e}_3$ are stable, whereas the permanent rotations around $\mathbf{e}_2$, the axis with intermediate moment of inertia, are unstable. The Euler equations' system (5.31) is of order one and autonomous, and the permanent rotations about the axes are singular points, so we can apply directly the theory of first-order systems.

Regarding the stability of permanent rotations around $\mathbf{e}_1$ and $\mathbf{e}_3$, the idea is to construct Lyapunov functions using two first integrals. They are, respectively, the total mechanical energy $\mathscr{E}|_{\mathscr{R}}$, which is preserved since the constraint cannot do any work (its power is zero, as $\mathbf{v}_O|_{\mathscr{R}} = \mathbf{0}$):

$$A(\boldsymbol{\omega}) = \mathscr{E}|_{\mathscr{R}} = \frac{1}{2}\left(I_{O1}\omega_1^2 + I_{O2}\omega_2^2 + I_{O3}\omega_3^2\right) . \qquad (6.33)$$

Above, we kept into account that the mechanical energy reduces in this case to the sole kinetic energy (5.18) and that the axes are principal, so they diagonalise the inertia tensor relative to O. Clearly the ω_i are the components of the $\boldsymbol{\omega}$ vector of the rigid body S in $\mathscr{R}$. The other first integral is the square of the angular momentum of S, in $\mathscr{R}$, calculated with respect to the pole O. By (5.17):

$$B(\boldsymbol{\omega}) = \boldsymbol{\Gamma}_O|_{\mathscr{R}}^2 = I_{O1}^2\omega_1^2 + I_{O2}^2\omega_2^2 + I_{O3}^2\omega_3^2 . \qquad (6.34)$$

This quantity is preserved in time, along the system's motions, due to the second governing equation (5.26), because the reaction, namely the only force acting on the system, has zero momentum with respect to the pole O where it is applied.

Observe first that $A(\boldsymbol{\omega}) > 0$ if $\boldsymbol{\omega} \neq \mathbf{0}$ and $A(\mathbf{0}) = 0$. Since $A(\boldsymbol{\omega})$ is a first integral, so $\dot{A} = 0$, we can apply Lyapunov's Theorem 6.12 to the trivial permanent rotation $\boldsymbol{\omega} = 0$, which is then both past- and future-stable. Consider now a permanent rotation around $\mathbf{e}_1$, given by $\boldsymbol{\omega} = \Omega\mathbf{e}_1$ with $\Omega \neq 0$. To prove it is stable we start by taking the Lyapunov potential function:

$$W_1'(\boldsymbol{\omega}) := I_{O1}^{-1}B(\boldsymbol{\omega}) - 2A(\boldsymbol{\omega}) = \frac{I_{O2}}{I_{O1}}(I_{O2} - I_{O1})\omega_2^2 + \frac{I_{O3}}{I_{O1}}(I_{O3} - I_{O1})\omega_3^2 .$$

This is a first integral and a non-negative map (recall $I_{O1} < I_{O2}$, I_{O3}), and it equals zero at $\omega_1 = \omega_2 = 0$. Therefore it has a minimum also along the permanent rotation. However this is not a strict local minimum on $(\Omega, 0, 0)$, so W_1' is not the Lyapunov function we are interested in. Consider another first integral of the form:

$$W_1''(\boldsymbol{\omega}) := (2A(\boldsymbol{\omega}) - I_{O1}\Omega^2)^2 = \left[I_{O3}\omega_3^2 + I_{O2}\omega_2^2 + I_{O1}(\omega_1^2 - \Omega^2)\right]^2 .$$

This function is still a first integral and satisfies $W_1''(\omega) \geq 0$. Adding the two produces the non-negative first integral

$$W_1(\omega) := W_1'(\omega) + W_1''(\omega) = \left(\sqrt{\frac{I_{02}}{I_{01}}}(I_{02} - I_{01})\omega_2\right)^2 + \left(\sqrt{\frac{I_{03}}{I_{01}}}(I_{03} - I_{01})\omega_3\right)^2$$

$$+ \left[I_{03}\omega_3^2 + I_{02}\omega_2^2 I_{01}(\omega_1^2 - \Omega^2)\right]^2,$$

which vanishes only if all quadratic summands are separately zero, being a sum of squares. This happens only when: $\omega_2 = \omega_2 = 0$ and $\omega_1 = \pm\Omega$. In particular, this first integral has a strict local minimum exactly at $\omega = \Omega e_1$ (with $\Omega \neq 0$). So we can use W_1 as Lyapunov function and conclude that every permanent rotation around e_1 (the one with smallest moment of inertia) is past- and future-stable (since $W_1 = 0$).

We can argue similarly to prove the stability of the permanent rotations $\omega = \Omega e_3$, with $\Omega \neq 0$, around e_3, the axis with largest moment of inertia. In this case the Lyapunov function is:

$$W_3(\omega) := 2A(\omega) - I_{03}^{-1}B(\omega) + (2A(\omega) - I_{03}\Omega^2)^2.$$

To finish we will show that the permanent rotations around e_2, whose moment of inertia is intermediate, are both past- and future-unstable. This follows easily from Theorem 6.20 once we linearise the right-hand side of system (5.31) written in normal form, at the singular point Ωe_2 with $\Omega \neq 0$. The elementary direct computation shows that the matrix obtained on the right after linearising has characteristic equation:

$$\lambda\left(\lambda^2 - \Omega^2 \frac{(I_{02} - I_{03})(I_{01} - I_{02})}{I_{01}I_{03}}\right) = 0,$$

with solutions, apart from $\lambda = 0$:

$$\lambda_{\pm} = \pm|\Omega|\sqrt{\frac{(I_{02} - I_{03})(I_{01} - I_{02})}{I_{01}I_{03}}}.$$

Observe that the radicand is certainly positive since $I_{01} < I_{02} < I_{03}$. As there is a strictly positive eigenvalue and a strictly negative one ($\Omega \neq 0$ by hypotheses), Theorem 6.20 leads us to conclude that the singular point, i.e. the permanent rotation Ωe_2 around e_2, is future- and past-unstable.

Exercises 6.31

(1) Consider a point particle of mass m subject to the conservative force

$$\mathcal{U}|_{\mathcal{R}}(x, y, x) := \kappa(x^2 + y^2 - z^4),$$

Fig. 6.1 Exercise 6.31.2

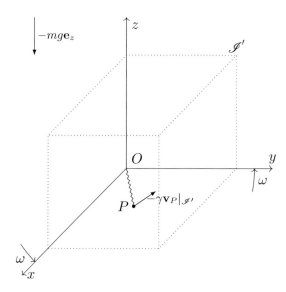

in orthonormal coordinates x, y, z moving with a frame $\mathscr{R}$, and where $\kappa > 0$ is a constant. Prove that the origin determines an unstable equilibrium configuration.

(2) Consider a point particle P (Fig. 6.1) of mass $m > 0$ under the gravitational force $-mg\,\mathbf{e}_z$, attached to the point O by a spring with zero length at rest and elastic constant $\kappa > 0$, and immersed in a liquid. The point O and the liquid are at rest in a non-inertial frame $\mathscr{R}'$ that rotates around the z-axis at O, with respect to the inertial frame $\mathscr{R}$, with angular velocity $\omega_{\mathscr{R}'|\mathscr{R}} = \omega\,\mathbf{e}_z$, $\omega > 0$ constant. Due to the liquid, P is subject to the viscous force $\mathbf{F}(\mathbf{v}_P|_{\mathscr{R}'}) = -\gamma\mathbf{v}_P|_{\mathscr{R}'}$ with $\gamma \geq 0$ constant.

  (i) Write the equations of motion of P in $\mathscr{R}'$.

  (ii) Prove that, in $\mathscr{R}'$, in absence of viscosity ($\gamma = 0$) the point undergoes a gyrostatic force and a remaining conservative force. Write the potential energy of the latter.

  (iii) Determine the equilibrium configurations of P in the frame $\mathscr{R}'$ and discuss their stability, when $k > m\omega^2$.

Chapter 7
Foundations of Lagrangian Mechanics

In this chapter we shall introduce the *Lagrangian formulation* of Classical Mechanics. We remind that the presence of constraint reactions with unknown expression typically makes Newton's equations *non-deterministic*, precisely because these forces appear as additional unknowns. In order to render the system of equations deterministic it is necessary to include information of various kind: geometric/kinematic ("the shape of constraints") and physical (the reactions' constitutive characterisation). *Lagrangian Mechanics*, or the *Lagrangian formulation of Classical Mechanics*, was born in the eighteenth century out of the work of the Italian-French mathematician J.L.Lagrange as a universal procedure to extract the so-called *pure equations of motion*. These are deterministic differential equations in which no constraint reaction appears—as these are unknowns of the problem—which allow to determine the system's motion. All this for systems of constrained point particles, when the constraints are *ideal*. This request generalises, as we shall see, the characterisation of smooth constraints and includes very general situations. The pure equations of motion are called, in this context, *Euler-Lagrange equations*.

Lagrangian Mechanics is applicable, with minor modifications, to continuous rigid systems. Clearly the Lagrangian formulation can be applied to systems of point particles without constraints, in which case the equations are completely equivalent to Newton's deterministic equations. They have though the peculiarity that they can be written, in the same form, in any coordinates system.

Finally, in all cases, Lagrange's formulation of Mechanics makes very evident the tight relationship between a physical system's symmetry under a group of transformations, and the existence of special physical quantities preserved in time during the motion.

Lagrangian Mechanics is an incredibly powerful approach, whose application has gone way beyond classical Newtonian Mechanics and the problem of the unknown constraint reactions. It has survived the relativistic revolution and applies to continuous systems and field theories, with tremendous success. The Lagrangian formulation, in fact, has had in the nineteenth century a massive influence on the

V. Moretti, *Analytical Mechanics*, La Matematica per il 3+2 150,
https://doi.org/10.1007/978-3-031-27612-5_7

development of all of Mathematical Physics, an influence that was recognised as decisive for the birth of Modern Physics in the twentieth century, as well as for its growth in the twenty-first.

7.1 An Introductory Example

As a first example we shall deduce the Euler-Lagrange equations for a particular system. The rest of the chapter will be devoted to generalising this elementary case.

Consider a system $\mathscr{S}$ of 3 point particles P_1, P_2, P_3 with masses m_1, m_2, m_3 respectively, subject to *active forces* (forces of which we know the expression depending on time and on the particles' position and velocities) $\mathbf{F}_1, \mathbf{F}_2, \mathbf{F}_3$. We suppose the forces are not of inertial type (later we will drop this, but for the moment we will work in a relatively simple situation). We also suppose of being in a given inertial frame $\mathscr{R}$ and that the point particles move along curves and surfaces *at rest or in motion* in this frame. We will call $\boldsymbol{\phi}_1, \boldsymbol{\phi}_2, \boldsymbol{\phi}_3$, respectively, the *constraint reactions* on the points, also known as *reactive forces* (forces whose expression is unknown, but must exist due to Newton's second law), caused by the constrained motion. To fix ideas, assume P_1, P_2 move on a surface Σ, while P_3 is constrained to a curve Γ. We also suppose Σ is at rest in $\mathscr{R}$. As we know from Chap. 3, to study the system's dynamics it is convenient to describe the position of the particles in terms of coordinates (in general local) adapted to the constraints. Let, say, q^1, q^2 denote local coordinates for P_1 on Σ, once we have a suitable coordinates system of the surface. For instance, if Σ is a sphere, q^1 and q^2 could be the spherical angles θ and φ of P_1 on Σ. Similarly q^3, q^4 are the coordinates of P_2 on Σ. Finally, q^5 is the coordinate of P_3 along Γ at rest in the frame $\mathscr{R}'$ moving (in a given way) with respect to $\mathscr{R}$. For example q^5 could be the arclength of P_3 on Γ.

If O is an origin point in the rest space $E_{\mathscr{R}}$, and $\mathbf{x}_i := P_i - O$ in $\mathscr{R}$, we then have the known relations:

$$\mathbf{x}_i = \mathbf{x}_i(t, q^1, \dots, q^5), \quad i = 1, 2, 3, \tag{7.1}$$

which allow to find the positions of the P_i in $E_{\mathscr{R}}$ once we know the coordinates $q^1, \dots, q^5$ determining the points on Σ and Γ. Actually, in this particular case, $\mathbf{x}_1$ and $\mathbf{x}_2$ only depend on q^1, q^2 and q^3, q^4 respectively, *but not upon time t*. On the other hand $\mathbf{x}_3$ is a function of q^5 *and of time t*, since Γ is not at rest in $\mathscr{R}$ but its motion is given. That said, we will write $\mathbf{x}_i = \mathbf{x}_i(t, q^1, \dots, q^5)$ to stay general (Fig. 7.1).

Note that $\mathscr{S}$ contains 3 point particles, so a priori we need $3 \times 3 = 9$ Cartesian coordinates to determine their positions in $\mathscr{R}$. Yet the constraint equations reduce to 5 the system's freedom degrees. These 5 *degrees of freedom* are described by the 5 coordinates $q^1, \dots, q^5$ adapted to the constraints. The system's motion will therefore be described by 5 *unknown functions of time* $q^k = q^k(t), k = 1, 2, 3, 4, 5$.

Fig. 7.1 Example of
Sect. 7.1

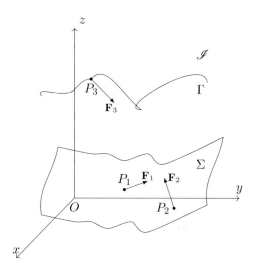

The system's motion should be given by the solution $q^k = q^k(t)$, $k = 1, 2, 3, 4, 5$, of an ODE system of order two, once we have the initial conditions $q^1(t_0), \ldots, q^5(t_0), dq^1/dt(t_0), \ldots, dq^5/dt(t_0)$, and after we have written the forces $\mathbf{F}_i := \mathbf{F}_i(t, \mathbf{x}_1, \mathbf{x}_2, \mathbf{x}_3, \frac{d\mathbf{x}_1}{dt}, \frac{d\mathbf{x}_2}{dt}, \frac{d\mathbf{x}_2}{dt})$ in terms of the $q^1, \ldots, q^5$ and their first time derivatives. The problem, however, is that we do not know the explicit form of the reactive forces $\boldsymbol{\phi}_i$, so Newton's 3 equations:

$$m_i \mathbf{a}_{P_i}|_{\mathscr{R}} = \mathbf{F}_i + \boldsymbol{\phi}_i, \quad i = 1, 2, 3, \tag{7.2}$$

i.e.

$$\frac{d^2\mathbf{x}_i}{dt^2} = \frac{1}{m_i}\mathbf{F}_i\left(t, \mathbf{x}_1(t), \mathbf{x}_2(t), \mathbf{x}_3(t), \frac{d\mathbf{x}_1}{dt}, \frac{d\mathbf{x}_2}{dt}, \frac{d\mathbf{x}_2}{dt}\right) + \frac{1}{m_i}\boldsymbol{\phi}_i, \quad i = 1, 2, 3$$

are not a second-order normal ODE system—not even when written in terms of the 5 effective freedom degrees $q^1, \ldots, q^5$—precisely because of the unknown forces ϕ_i.

Nonetheless, it can be proved that the problem in the unknowns $q^k = q^k(t)$ becomes deterministic if, beside the above equations, we add certain *constitutive characterisations of the constraints*.[1] In the case at hand, we will assume the constraints are smooth. In other words $\boldsymbol{\phi}_1, \boldsymbol{\phi}_2$ will always be normal to Σ, and $\boldsymbol{\phi}_3$ will be normal to Γ. As we saw in Chap. 3, and we will see in this more general case without resorting to those results, by adding constitutive equations we will be able to extract from Newton's equations a subsystem of *pure equations of motion*. These

---

[1] The equations describing Σ and Γ have already been used when, in Newton's equations, we pass from 9 Cartesian coordinates to the 5 effective degrees of freedom $q^1, \ldots, q^5$.

are differential equations for the unknown functions $q^k = q^k(t)$ $(k = 1, 2, 3, 4, 5)$ that are in normal form. Therefore, under rather weak regularity hypotheses and because of the Existence and uniqueness theorem, they define the system's motion once the initial conditions $q^1(t_0), \ldots, q^5(t_0)$ and $dq^1/dt(t_0), \ldots, dq^5/dt(t_0)$ are assigned. In particular, these equations *do not* contain the (unknown) constraint reactions acting on the system's particles. The time-dependent constraint reactions $\boldsymbol{\phi}_i(t)$ are found successively, using Newton's equations, once we know the system's motion.

What we want to introduce in this section is a way leading directly to the pure equations of motion. The discussion should be generalisable to the generic case of N particles constrained in a more general way that what we have now, when the constraints satisfy a broader constitutive characterisation than the present smooth constraints. The procedure in its general form will be studied in the next section.

Having a smooth constraint is *equivalent* to asking that, if $\delta \mathbf{x}_i$ is a generic *tangent* vector to the surface or curve on which the point P_i lies *at the given time* t, the following relations hold: $\boldsymbol{\phi}_i \cdot \delta \mathbf{x}_i = \mathbf{0}$ for $i = 1, 2, 3$ and for *any* vectors $\delta \mathbf{x}_i$. A fortiori we should have:

$$\sum_{i=1}^{3} \boldsymbol{\phi}_i \cdot \delta \mathbf{x}_i = \mathbf{0} \,, \text{ for } \textit{every} \text{ vector } \delta \mathbf{x}_i \text{ of the above type .} \tag{7.3}$$

The second request is a consequence of the first conditions $\boldsymbol{\phi}_i \cdot \delta \mathbf{x}_i = \mathbf{0}$, $i = 1, 2, 3$, but it does not imply them, and as such is less restrictive.

Now we want to write (7.3) in an equivalent, but more useful way, concretely. Consider $P_1 = P_1(q^1, q^2)$. Given the nature of the first two coordinates q^k, that they are coordinates on Σ, at the instant t the tangent vectors to Σ at P_1 have the form:

$$\delta \mathbf{x}_1 = \sum_{k=1}^{2} \frac{\partial \mathbf{x}_1}{\partial q^k} \delta q^k = \sum_{k=1}^{5} \frac{\partial \mathbf{x}_1}{\partial q^k} \delta q^k \,,$$

where the δq^k are arbitrary real numbers. The second identity comes from the fact that $\mathbf{x}_1$ does not depend on q^3, q^4, q^5 and therefore the terms for $i = 3, 4, 5$, added to the sum (the one with $i = 1, 2$) are null. At any given time t, the same expression also holds for the tangent vectors to the corresponding constraints (surface or curve) to which the other two points belong:

$$\delta \mathbf{x}_i = \sum_{k=1}^{5} \frac{\partial \mathbf{x}_i}{\partial q^k} \bigg|_{(t, q^1, \ldots, q^5)} \delta q^k \,, \quad i = 1, 2, 3 \,, \tag{7.4}$$

where the δq^k are arbitrary reals. Equation (7.3) can then be rewritten as

$$\sum_{k=1}^{5}\sum_{i=1}^{3}\boldsymbol{\phi}_i \cdot \frac{\partial \mathbf{x}_i}{\partial q^k}\delta q^k = \mathbf{0}\,.$$

If we remember that the δq^k are arbitrary, we can choose $\delta q^1 = 1$ and the rest zero, then $\delta q^2 = 1$ and the rest zero, and so forth. We conclude that (7.3) is equivalent to the 5 equations:

$$\sum_{i=1}^{3}\boldsymbol{\phi}_i \cdot \frac{\partial \mathbf{x}_i}{\partial q^k} = 0 \quad \text{for } k = 1, \ldots, 5, \tag{7.5}$$

This is the characterisation of so-called *ideal constraints* that we will later extend.

Remarks 7.1 Intuitively speaking, the vectors $\delta \mathbf{x}_i$, in case the δq^k are very small, may be viewed as vectors joining two infinitely close configurations—one given by $(t, q^1, \ldots, q^5)$ and one by $(t, q^1 + \delta q^1, \ldots, q^5 + \delta q^5)$—that are compatible with the system's constraints *at the same time* t. We should stress that the $\delta \mathbf{x}_i$ are not necessarily (infinitesimal) real displacements that the system undergoes during the motion that solves Newton's equations, since real displacements occur during a time interval and the constraints may change expression in that time (this happens to the constraint of P_3, in the case under exam), whereas we are working with fixed t when we define the $\delta \mathbf{x}_i$. For this reason the vectors $\delta \mathbf{x}_i$ in (7.4) were called, in the 1800s language of the "classical" formulation of Lagrangian Mechanics, *virtual displacements of the system* at time t with respect to the configuration $(t, q^1, \ldots, q^5)$, and $\sum_{i=1}^{3} \boldsymbol{\phi}_i \cdot \delta \mathbf{x}_i$ was called the *virtual work* (of the constraint reactions) at time t with respect to the configuration $(t, q^1, \ldots, q^5)$. ∎

Let us finally show how the above characterisation of the ideal constraints allows to write the pure equations of motion. Recalling that

$$m_i \mathbf{a}_{P_i}|_{\mathscr{R}} - \mathbf{F}_i = \boldsymbol{\phi}_i\,, \quad i = 1, 2, 3\,,$$

the set of 5 equations (7.5) reads:

$$\sum_{i=1}^{3} m_i \mathbf{a}_{P_i}|_{\mathscr{R}} \cdot \frac{\partial \mathbf{x}_i}{\partial q^k} = \sum_{i=1}^{3} \mathbf{F}_i(t, q^1, \ldots, q^5, \dot{q}^1, \ldots, \dot{q}^5) \cdot \frac{\partial \mathbf{x}_1}{\partial q^k} \quad \text{for } k = 1, \ldots, 5. \tag{7.6}$$

We want to express these equations in a way that is easier to handle. For starters note that, in our assumptions, for $i = 1, 2, 3$ the velocity $\mathbf{v}_{P_i}|_{\mathscr{R}}(t)$ of P_i in $\mathscr{R}$ as function of the degrees of freedom $q^1, \ldots, q^5$ and their time derivatives, is:

$$\mathbf{v}_{P_i}|_{\mathscr{R}}(t) = \frac{\partial \mathbf{x}_i}{\partial t} + \sum_{k=1}^{5} \frac{\partial \mathbf{x}_i}{\partial q^k} \dot{q}^k, \tag{7.7}$$

where $\dot{q}^k := \frac{dq^k}{dt}$. Keeping (7.7) in account we have the kinetic energy $\mathscr{T}|_{\mathscr{R}} = \mathscr{T}|_{\mathscr{R}}(t, q^1, \ldots, q^5, \dot{q}^1, \ldots, \dot{q}^5)$ of $\mathscr{S}$ in $\mathscr{R}$:

$$\mathscr{T}|_{\mathscr{R}} = \sum_{i=1}^{3} \frac{m_i}{2} \mathbf{v}_{P_i}|_{\mathscr{R}}(t)^2 = \sum_{i=1}^{3} \frac{m_i}{2} \left(\frac{\partial \mathbf{x}_i}{\partial t} + \sum_{k=1}^{5} \frac{\partial \mathbf{x}_i}{\partial q^k} \dot{q}^k \right) \cdot \left(\frac{\partial \mathbf{x}_i}{\partial t} + \sum_{k=1}^{5} \frac{\partial \mathbf{x}_i}{\partial q^k} \dot{q}^k \right).$$

Using the computations that we will perform explicitly in the next section, we then find:

$$\sum_{i=1}^{3} m_i \mathbf{a}_{P_i}|_{\mathscr{R}} \cdot \frac{\partial \mathbf{x}_i}{\partial q^k} = \frac{d}{dt} \frac{\partial \mathscr{T}|_{\mathscr{R}}}{\partial \dot{q}^k} - \frac{\partial \mathscr{T}|_{\mathscr{R}}}{\partial q^k},$$

where the $\dot{q}^k$ should be thought of as variables independent of the q^k when we differentiate $\frac{\partial \mathscr{T}|_{\mathscr{R}}}{\partial \dot{q}^k}$. By this identity we conclude that equations (7.6) can be equivalently written in the freedom degrees $q^1, \ldots, q^5$ only, as:

$$\begin{cases} \frac{d}{dt} \left(\frac{\partial \mathscr{T}|_{\mathscr{R}}}{\partial \dot{q}^k} \Big|_{(t,q(t),\dot{q}(t))} \right) - \frac{\partial \mathscr{T}|_{\mathscr{R}}}{\partial q^k} \Big|_{(t,q(t),\dot{q}(t))} = \mathscr{Q}_k|_{\mathscr{R}}(t, q(t), \dot{q}(t)), \\ \frac{dq^k}{dt} = \dot{q}^k(t) \end{cases}$$

$$\text{for } k = 1, \ldots, 5, \tag{7.8}$$

where:

$$\mathscr{Q}_k|_{\mathscr{R}}(t, q^1, \ldots, q^n, \dot{q}^1, \ldots, \dot{q}^n) := \sum_{i=1}^{3} \mathbf{F}_i(t, q^1, \ldots q^5, \dot{q}^1, \ldots, \dot{q}^5) \cdot \frac{\partial \mathbf{x}_i}{\partial q^k}. \tag{7.9}$$

The $\dot{q}^k$ should be understood as independent of the q^k in the first relation in (7.8) (otherwise we would have a problem in interpreting the time derivative $\dot{q}^k$). Furthermore, the second equation reminds us that the dotted coordinates should be viewed as the time derivatives of the undotted coordinates.

Remarks 7.2 Pay attention to the procedure indicated in (7.8), which we explain in detail below.

(a) Differentiate $\mathscr{T}|_{\mathscr{R}}(t, q, \dot{q})$ with respect to $\dot{q}^k$ as if the latter was independent of t, of the other q^j and of $\dot{q}^j$.

(b) Restrict the derivative to the unknown curves $I \ni t \mapsto (q^1(t), \ldots, q^5(t), \dot{q}^1(t), \ldots, \dot{q}^n(t))$ to obtain a function in one variable t.

(c) Differentiate $\frac{\partial \mathcal{T}|_{\mathcal{R}}}{\partial \dot{q}^k}\Big|_{(t,q(t),\dot{q}(t))}$ in t, the only variable left, remembering *what depends on t*; then substitute this derivative $\left(\frac{d}{dt}\left(\frac{\partial \mathcal{T}|_{\mathcal{R}}}{\partial \dot{q}^k}\Big|_{(t,q(t),\dot{q}(t))}\right)\right)$ in the first relation of (7.8), again restricting $\frac{\partial \mathcal{T}|_{\mathcal{R}}}{\partial q^k}$ and the right-hand side to the curves $I \ni t \mapsto (q^1(t), \ldots, q^5(t), \dot{q}^1(t), \ldots, \dot{q}^n(t))$.

(d) In the second equation of system (7.8), remember that the 'dot coordinates' are the derivatives of the 'undotted' coordinates.

In this way we produce a differential system of order two in the 5 components of the unknown curves $I \ni t \mapsto (q^1(t), \ldots, q^5(t))$. ∎

Equation (7.8) are called *Euler-Lagrange equations* and form a system of 10 differential equations of *order one*, or equivalently, 5 equations of *order two*. A crucial result we will prove is that our system can be put in *normal form*, on an open set in $\mathbb{R}^5$ (if we work with order two). As we saw, these equations are important for finding a curve parametrised by time in terms of the degrees of freedom $q^1 = q^1(t), \ldots, q^5 = q^5(t)$. Exactly because they can be put in normal form, if the $\mathcal{Q}_k = \mathcal{Q}_k(t, q^1, \ldots, q^n, \dot{q}^1, \ldots, \dot{q}^n)$ are regular enough the Euler-Lagrange equations admits a unique solution $q^k = q^k(t)$, $k = 1, 2, 3, 4, 5$, for t in some open interval $I \subset \mathbb{R}$, once we have the initial conditions: the values $q^k(t_0)$ and their time derivatives $\dot{q}^k(t_0)$ at a given instant $t_0 \in I$. This solution, in the abstract coordinates $q^1, \ldots, q^5$, determines the system's motion in the physical space $E_{\mathcal{R}}$:

$$\mathbf{x}_i = \mathbf{x}_i(t, q^1(t), \ldots, q^5(t)) \quad \text{for } i = 1, 2, 3 \text{ and } t \in I,$$

we have used the known functions (7.1).

Let us finally examine the case where, in the frame $\mathcal{R}$, the forces $\mathbf{F}_i$, different from the constraint reactions, acting on P_1, P_2, P_3 arise from a potential $\mathcal{V}|_{\mathcal{R}} = \mathcal{V}|_{\mathcal{R}}(t, \mathbf{x}_1, \mathbf{x}_2, \mathbf{x}_3)$:

$$\mathbf{F}_i(t, \mathbf{x}_1, \mathbf{x}_2, \mathbf{x}_3) = \nabla_{\mathbf{x}_i} \mathcal{V}|_{\mathcal{R}}(t, \mathbf{x}_1, \mathbf{x}_2, \mathbf{x}_3), \quad i = 1, 2, 3.$$

Now expressions $\mathcal{Q}_k|_{\mathcal{R}}$ simplify:

$$\mathcal{Q}_k|_{\mathcal{R}}(t, q^1, \ldots, q^5) = \sum_{i=1}^{3} \mathbf{F}_i \cdot \frac{\partial \mathbf{x}_i}{\partial q^k} = \sum_{i=1}^{3} (\nabla_{\mathbf{x}_i} \mathcal{V}|_{\mathcal{R}}) \cdot \frac{\partial \mathbf{x}_i}{\partial q^k} = \frac{\partial \mathcal{V}|_{\mathcal{R}}}{\partial q^k},$$

where in the last derivative $\mathcal{V}|_{\mathcal{R}}$ indicates, slightly improperly, the composite

$$\mathcal{V}|_{\mathcal{R}} = \mathcal{V}|_{\mathcal{R}}(t, \mathbf{x}_1(q^1, \ldots, q^5), \mathbf{x}_2(q^1, \ldots, q^5), \mathbf{x}_3(q^1, \ldots, q^5))$$

The Euler-Lagrange equations for the unknown curves $I \ni t \mapsto (q^1(t), \ldots, q^5(t))$ read

$$\begin{cases} \dfrac{d}{dt}\left(\left.\dfrac{\partial\mathscr{T}|_{\mathscr{R}}}{\partial\dot{q}^k}\right|_{(t,q(t),\dot{q}(t))}\right) - \left.\dfrac{\partial\mathscr{T}|_{\mathscr{R}}}{\partial q^k}\right|_{(t,q(t),\dot{q}(t))} = \left.\dfrac{\partial\mathscr{V}|_{\mathscr{R}}}{\partial q^k}\right|_{(t,q(t),\dot{q}(t))} , & \text{for } k = 1, \ldots, 5, \\ \dfrac{dq^k}{dt} = \dot{q}^k(t) \end{cases}$$

As $\mathscr{V}|_{\mathscr{R}}$ does not depend on the variables $\dot{q}^k$, by introducing the *Lagrangian* (function) of $\mathscr{S}$ in $\mathscr{R}$:

$$\mathscr{L}|_{\mathscr{R}}(t, q^1, \ldots, q^n, \dot{q}^1, \ldots, \dot{q}^n) := \mathscr{T}|_{\mathscr{R}}(t, q^1, \ldots, q^n, \dot{q}^1, \ldots, \dot{q}^n)$$
$$+ \mathscr{V}|_{\mathscr{R}}(t, q^1, \ldots, q^n),$$

the previous equations become

$$\begin{cases} \dfrac{d}{dt}\left(\left.\dfrac{\partial\mathscr{L}|_{\mathscr{R}}}{\partial\dot{q}^k}\right|_{(t,q(t),\dot{q}(t))}\right) - \left.\dfrac{\partial\mathscr{L}|_{\mathscr{R}}}{\partial q^k}\right|_{(t,q(t),\dot{q}(t))} = 0 , & \text{for } k = 1, \ldots, 5. \quad (7.10) \\ \dfrac{dq^k}{dt} = \dot{q}^k(t) \end{cases}$$

Remarks 7.3

(1) The Euler-Lagrange equations (7.8) or (7.10) are therefore the *pure equations of motion of system* $\mathscr{S}$, the same equations we obtain when projecting Newton's equations onto bases adapted to the surfaces and curves, as discussed in Chap. 3. Apart from the elementary situation considered in this section, the procedure we have explained proves to be a general technique to project Newton's equations along the directions that do not contain constraint reactions, irrespective of the specific form of the constraining surfaces and curves.

(2) We should emphasise that the two ingredients leading up to the deterministic equations (7.8) or (7.10) are: Newton's equations (7.2), and identities (7.3), which hold when all constraints are smooth, but also in more generality as we shall see shortly.   ∎

Examples 7.4 Consider one point particle P of mass $m > 0$ moving on the surface of equation $z = a(x^2 + y^2)$, with $a > 0$ constant, seen as smooth surface. The point is subject to gravity $-mg\mathbf{e}_z$ and to a massless spring of constant $k > 0$, with one end at $O \equiv (0, 0, 0)$ and zero length at rest. x, y, z are orthonormal coordinates on an inertial frame $\mathscr{R}$. We want to calculate the equations of motion of P using as free coordinates $q^1 = x$, $q^2 = y$. The point particle's kinetic energy in $\mathscr{R}$ is:

$$\mathscr{T}|_{\mathscr{R}} = \frac{m}{2}(\dot{x}^2 + \dot{y}^2 + (2ax)^2\dot{x}^2 + (2ay)^2\dot{y}^2 + 8a^2xy\dot{x}\dot{y})$$

$$= \frac{m}{2}((1 + 4a^2x^2)\dot{x}^2 + (1 + 4a^2y^2)\dot{y}^2 + 8a^2xy\dot{x}\dot{y}),$$

where we used $z(t) = a(x(t)^2 + y(t)^2)$ so

$$\frac{dz}{dt} = 2ax\dot{x} + 2ay\dot{y} \,.$$

The active forces are conservative and gravity has potential energy

$$mgz = mga(x^2 + y^2) \,,$$

while the spring has potential energy

$$\frac{k}{2}(x^2 + y^2 + z^2) = \frac{k}{2}(x^2 + y^2 + a^2(x^2 + y^2)^2) \,.$$

Consequently

$$\mathscr{U}|_{\mathscr{R}} = \left(mga + \frac{k}{2}\right)(x^2 + y^2) + \frac{ka^2}{2}(x^2 + y^2)^2 \,.$$

The system's Lagrangian will then be

$$\mathscr{L}|_{\mathscr{R}} = \frac{m}{2}((1 + 4a^2x^2)\dot{x}^2 + (1 + 4a^2y^2)\dot{y}^2 + 8a^2xy\dot{x}\dot{y})$$
$$- \left(mga + \frac{k}{2}\right)(x^2 + y^2) - \frac{ka^2}{2}(x^2 + y^2)^2 \,.$$

We can compute the terms entering the Euler-Lagrange equations for the unknown curves $x = x(t)$, $y = y(t)$. Given the symmetry in x with y we shall only refer to the coordinate x. Then

$$\frac{\partial \mathscr{L}|_{\mathscr{R}}}{\partial \dot{x}} = m(1 + 4a^2x^2)\dot{x} + 4a^2mxy\dot{y} \,,$$

$$\frac{\partial \mathscr{L}|_{\mathscr{R}}}{\partial x} = 4ma^2x\dot{x}^2 + 4ma^2y\dot{x}\dot{y} - 2\left(mga + \frac{k}{2}\right)x - 2ka^2(x^2 + y^2)x \,.$$

We thus obtain the system of Euler-Lagrange equations:

$$\frac{d}{dt}\left(m(1 + 4a^2x(t)^2)\frac{dx}{dt} + 4a^2mx(t)y(t)\frac{dy}{dt}\right)$$
$$= 4ma^2x(t)\left(\frac{dx}{dt}\right)^2 + 4ma^2y(t)\frac{dx}{dt}\frac{dy}{dt}$$
$$-2\left[\left(mga + \frac{k}{2}\right) - ka^2(x(t)^2 + y(t)^2)\right]x(t) \,, \qquad (7.11)$$
$$\frac{d}{dt}\left(m(1 + 4a^2y(t)^2)\frac{dy}{dt} + 4a^2mx(t)y(t)\frac{dx}{dt}\right)$$

$$= 4ma^2 y(t) \left(\frac{dy}{dt}\right)^2 + 4ma^2 x(t) \frac{dy}{dt} \frac{dx}{dt}$$

$$-2 \left[\left(mga + \frac{k}{2}\right) - ka^2 (x(t)^2 + y(t)^2)\right] y(t) \tag{7.12}$$

where we assumed $\dot{x} = \frac{dx}{dt}$ and $\dot{y} = \frac{dy}{dt}$, so we have a second-order ODE system in the unknown curves $x = x(t), y = y(t)$. Even if it is far from evident from the expression, the system obtained from the above (after differentiating in time on the left) can be put in normal form, as we shall prove later on in a completely general way.

7.2 The General Case: Holonomic Systems and Euler-Lagrange Equations

In this section we will introduce a new terminology, and generalise and complete the previous introductory section's discussion, to arrive at the Euler-Lagrange equations for constrained physical systems that obey a constitutive characterisation extending that of smooth constraints, as was announced in the characterisation of the *ideal constraint*.

Consider a system $\mathscr{S}$ of N point particles $P_1, \ldots, P_N$ with masses $m_1, \ldots, m_N$ respectively. For various reasons, in many branches of Classical Physics (in particular *Statistical Mechanics*) it is often useful to study the above system's kinematics and dynamics not in spacetime $\mathbb{V}^4$, but on a manifold $\mathbb{V}^{3N+1}$ of dimension $3N + 1$ called **spacetime of configurations (without constraints)**.[2] Its structure is identical to that of $\mathbb{V}^4$ except that the leaves Σ_t at constant absolute time $T : \mathbb{V}^{3N+1} \to \mathbb{R}$, seen as regular surjective function, are now replaced by the (embedded) submanifolds of dimension $3N$ given by the Cartesian product of N copies of Euclidean space Σ_t, i.e. $\Sigma_t^N = \Sigma_t \times \cdots \times \Sigma_t$. The k-th copy of Σ_t should be viewed as the absolute space of P_k for every $k = 1, \ldots, N$, and is called **space of configurations (in absence of constraints)**. Note Σ_t^N is a Euclidean space, whose space of displacements is the Cartesian product of the spaces of displacements $V_t^N = V_t \times \cdots (N \text{ times}) \cdots \times V_t$. (The latter is a natural vector space, the direct sum of the V_t seen as subspaces.) The inner product on V_t^N, in the sense of Definition 1.9, is defined by:

$$(\mathbf{v}_1, \ldots, \mathbf{v}_N) \cdot (\mathbf{u}_1, \ldots, \mathbf{u}_N) := \sum_{k=1}^{N} \mathbf{v}_k \cdot \mathbf{u}_k .$$

---

[2] In Statistical Mechanics one studies physical systems with an extremely large number N of elements, be they atoms or molecules, of the order of the *Avogadro number*: 6.02×10^{23}.

A point $p \in \mathbb{V}^{3N+1}$ is then given by a number $t := T(p)$ and an N-tuple of points $c_t := (Q_1, \ldots, Q_N) \in \Sigma_t^N$ called **configuration (without constraints) at time** t. Clearly, assigning a configuration c_t at time t is the same as assigning the value of $3N$ coordinates: a triple of numbers for each point P_k. Keeping into account the value t of absolute time at the moment we assign c_t, to find a point in $\mathbb{V}^{3N+1}$ we need to fix $3N + 1$ coordinates. Consequently the spacetime of configurations is a manifold of dimension $3N + 1$, justifying the notation $\mathbb{V}^{3N+1}$.

The evolution of system $\mathscr{S}$ is given by a **world line (without constraints)**, i.e. a C^k curve (at least $k \geq 2$ to write the dynamics) $\Gamma = \Gamma(t) \in \mathbb{V}^{3N+1}$ with $t \in I$ an open interval of $\mathbb{R}$ that cannot meet each Σ_t^N more than once, so to avoid the usual time paradoxes. Mathematically, this restriction is imposed, as in the case of a single particle, by asking $T(\Gamma(t)) = t$ (for a suitable choice of the arbitrary constant in the definition of absolute time) for any $t \in I$.

Identifying the absolute time of $\mathbb{V}^{3N+1}$ with the absolute time of spacetime $\mathbb{V}^4$, for any $k = 1, \ldots, N$, the surjective differentiable map:

$$\Pi_k : \mathbb{V}^{3N+1} \ni (t, (Q_1, \ldots, Q_N)) \mapsto (t, Q_k) \in \mathbb{V}^4$$

extracts the configuration of the single P_k from the system's total configuration, at each instant t of absolute time. $\gamma_k(t) := \Pi_k(\Gamma(t))$, for $t \in I$, defines the world line of the k-th particle, when we have the world line Γ of $\mathscr{S}$. Using the Π_k it makes sense, for example, to define the distance of two points $P_k(t)$ and $P_h(t)$ in the configuration $c_t = \Gamma(t)$ at a given time t by the distance d_t on each absolute space $\Sigma_t \subset \mathbb{V}^4$. To do that it suffices to view $P_k(t)$ and $P_h(t)$ as in $\mathbb{V}^4$ under the maps Π_k and Π_h.

7.2.1 Spacetime of Configurations $\mathbb{V}^{n+1}$ in Presence of Holonomic Constraints

Suppose the N points of $\mathscr{S}$ are subject to positional constraints given by $C < 3N$ ($C \geq 0$) C^k functions $f_j : \mathbb{V}^{3N+1} \to \mathbb{R}$ ($k \geq 1$ for the moment, or at least large enough to justify what comes next):

$$f_j(t, P_1(t), \ldots, P_N(t)) = 0, \quad j = 1, \ldots, C \tag{7.13}$$

that should hold on every motion $\Gamma = \Gamma(t)$, $t \in I$. In this situation, very frequent in reality, the physical system's configurations at time t are not all those represented by the points of Σ_t^N. The constraints, at every instant t, will determine a subset $\mathbb{Q}_t \subset \Sigma_t^N$ of the configurations that are really accessible to the system at that time. For mathematical-physical simplicity, we want these subsets $\mathbb{Q}_t$ to have a structure allowing for the usual analytical techniques based on the differential equations describing the system's evolution laws. An optimal situation (not the only possible one) is that in which the subsets $\mathbb{Q}_t$ are differentiable manifolds of dimension $n \leq 3N$. it is natural to suppose this structure is the one inherited from each Σ_t^N. In other words we will assume that the $\mathbb{Q}_t$ are (embedded) *submanifolds* (see Appendix A) in the manifold Σ_t^N. The dimension n will be fixed by the number of constraint functions f_j: we expect that, imposing C *independent* constraint equations at each time t, these are able to leave us free to pick $n := 3N - C$ coordinates for the N point particles forming the system. Mathematically this situation occurs whenever the constraint equations (7.13) fulfil the following technical assumptions.

(H1) For any $t \in \mathbb{R}$

$$f_j(t, P_1, \ldots, P_N) = 0 , \quad j = 1, \ldots, C \tag{7.14}$$

hold on a non-empty set.

(H2) At every instant $t \in \mathbb{R}$, the f_j are **functionally independent**.

Note *functionally independent* means that around any point $p \in \mathbb{V}^{3N+1}$, if we write the f_j in terms of local coordinates $t, x^1, \ldots, x^{3N}$ on $\mathbb{V}^{3N+1}$, where t is the absolute time and $x^1, \ldots, x^{3N}$ are local coordinates[3] on Σ_t^N, these functions are C^k with $k \geq 1$ and the Jacobian matrix with entries

$$\frac{\partial f_j}{\partial x^i} , \quad i = 1, \ldots, 3N , j = 1, \ldots, C ,$$

has rank C, for any instant t, at the points of $\mathbb{V}^{3N+1}$ for which

$$f_j(t, x^1, \ldots, x^{3N}) = 0 .$$

As explained in Appendix A, this characterisation is independent of the choice of local coordinates.

We will call **system of holonomic constraints** a system of constraints, imposed on a system of N point particles, obeying (H1) and (H2).

We have the following proposition.

Proposition 7.5 *For a system $\mathscr{S}$ of N point particles in $\mathbb{V}^{3N+1}$ subject to constraints (7.14), given by $C \leq 3N$ functions f_j of class C^k ($k \geq 1$), consider the set $\mathbb{V}^{n+1} \subset \mathbb{V}^{3N+1}$ determined by (7.14) with $n := 3N - C$.*
Assuming the constraints are holonomic, the following facts hold.

(1) $\mathbb{V}^{n+1}$ is an (embedded) C^k submanifold of $\mathbb{V}^{3N+1}$ of dimension $n + 1$.
(2) For any given time t, the set $\mathbb{Q}_t := \mathbb{V}^{n+1} \cap \Sigma_t^N$ is an (embedded) C^k submanifold of Σ_t^N of dimension n.

---

[3] Such coordinates always exist (as the surfaces for $T = t$ constant are submanifolds) and can be constructed in the same way we have defined, for $N = 1$, the orthonormal coordinates of the frame.

(3) Around any point $p \in \mathbb{V}^{n+1}$ we can always choose coordinates $t, q^1, \ldots, q^n$
where:

(1) t measures absolute time,
(2) for any given absolute time t, $q^1, \ldots, q^n$ are admissible local coordinates
on $\mathbb{Q}_t$.

In particular $q^1, \ldots, q^n$ can be chosen among the Cartesian coordinates of the
N points in an arbitrary reference frame.

Proof $\boxed{AC}$ Put on $\mathbb{V}^{3N+1}$ Cartesian coordinates $t, x^1, \ldots, x^{3N}$ for a given frame
$\mathscr{R}$. For example, x^4, x^5, x^6 are the Cartesian coordinates in $\mathscr{R}$ of the point $P_2 \in \mathscr{S}$.
Now consider the set $\mathbb{V}^{n+1} \subset \mathbb{V}^{3N+1}$ determined by (7.14), under the assumption
of holonomic constraints. By Theorem A.26 (Regular-value theorem), $\mathbb{V}^{n+1}$ is an
embedded $(n + 1)$-submanifold of $\mathbb{V}^{3N+1}$. This is immediate from the fact that on
$\mathbb{V}^{n+1}$, where the local coordinates are: $y^0 := t$ and $x^i := y^i$ for $i = 1, \ldots, 3N$, the
Jacobian matrix

$$\frac{\partial f_j}{\partial y^i}, \quad i = 0, \ldots, 3N, j = 1, \ldots, C,$$

has rank C, since by (H2) it has C linearly independent columns. (In fact,
suppressing the column of the y^0-coordinate, the resulting Jacobian matrix has
maximal rank by (H2), so the C remaining rows are linearly independent. This
means there are C linearly independent columns. When we append the y^0-column,
the above C columns remain linearly independent so the larger Jacobian matrix has
rank at least C, i.e. maximal rank equal to C because the number of rows C does
not change.) Request (H2) implies, for the same reason, that at any given time t,
$\mathbb{Q}_t := \mathbb{V}^{n+1} \cap \Sigma_t^N$ is an embedded submanifold of Σ_t^N of dimension n. Note that
$\mathbb{Q}_t \neq \varnothing$ due to (H1).
The second part of Theorem A.26 has another consequence. Since the Jacobian
matrix

$$\frac{\partial f_j}{\partial y^i}, \quad i = 0, \ldots, 3N, j = 1, \ldots, C,$$

has rank C, and this is caused by the submatrix

$$\frac{\partial f_j}{\partial y^i}, \quad i = 1, \ldots, 3N, j = 1, \ldots, C,$$

around any $p \in \mathbb{V}^{n+1}$ we can use $n + 1$ coordinates to describe $\mathbb{V}^{n+1}$ by choosing
the coordinates among $y^0, \ldots, y^{3N}$ as follows. The first coordinate is always y^0
(absolute time), and the other $n = 3N - C$ are picked from $y^1, \ldots, y^{3N}$ so that the
$C \times C$ Jacobian submatrix corresponding to the coordinates y^k that were not chosen
is non-singular at p. $\square$

As usual, in the sequel the word *differentiable* on its own will mean C^k with $k \geq 1$, equal to the differentiability order of the manifold/manifolds we are dealing with. The order will be specified by the context, if necessary.

Definition 7.6 (Spacetime of Configurations with Holonomic Constraints) Consider a system $\mathscr{S}$ of N point particles subject to $C < 3N$ differentiable holonomic constraints $f_j(t, P_1, \ldots, P_N) = 0$ with $j = 1, \ldots, C$. Let $T : \mathbb{V}^{3N+1} \to \mathbb{R}$ denote the absolute time function (as usual assumed surjective, differentiable and non-singular).

(1) The submanifold $\mathbb{V}^{n+1}$ of dimension $n := 3N - C$, given by (7.14), is called **spacetime of configurations** for the constrained system $\mathscr{S}$.
    The number $n := 3N - C$ counts the **degrees of freedom** of system $\mathscr{S}$.

(2) For any instant of absolute time $t \in \mathbb{R}$, $\mathbb{Q}_t := \mathbb{V}^{n+1} \cap \Sigma_t^N$ is called the **space of configurations** at time t, for the constrained system $\mathscr{S}$. A **configuration** of the constrained system $\mathscr{S}$ at time t is therefore a point in $\mathbb{Q}_t$.

(3) Any system of local coordinates $\phi : U \ni p \mapsto (t(p), q^1(p), \ldots q^n(p)) \in \mathbb{R}^{n+1}$, where $U \subset \mathbb{V}^{n+1}$ is open and $t = T(p)$—so that $(q^1, \ldots q^n)$ define a local chart on $\mathbb{Q}_t$ for any t—is called **system of natural local coordinates** on $\mathbb{V}^{n+1}$.

    The local coordinates $q^1, \ldots, q^n$ on every $\mathbb{Q}_t$ are called **free coordinates** or **Lagrangian coordinates**.

(4) A system of natural local coordinates $\phi : U \ni p \mapsto (t(p), q^1(p), \ldots q^n(p)) \in \mathbb{R}^{n+1}$ as in (3) is said to **move with the reference frame** $\mathscr{R}$ when, calling $\mathbf{x}_i = P_i - O$ the position vector of P with respect to the origin O of $\mathscr{R}$ (the components of $\mathbf{x}_i$ are then the coordinates of P in orthonormal coordinates on $\mathscr{R}$), the parametric relationship between $\mathbf{x}_i$ and the coordinates does not contain time explicitly: $\mathbf{x}_i = \mathbf{x}_i(q^1, \ldots, q^n)$, for any $i = 1, \ldots, N$.   ◇

Remarks 7.7

(1) Clearly, natural local coordinates form an atlas of $\mathbb{V}^{n+1}$.
    If $t, q^1, \ldots, q^n$ and $t', q'^1, \ldots q'^n$ are distinct natural local coordinates on two domains $U, U' \subset \mathbb{V}^{n+1}$ with $U \cap U' \neq \varnothing$, on the intersection the coordinates are related as follows:

$$q'^k = q'^k(t, q^1, \ldots, q^n), \tag{7.15}$$

$$t' = t + c, \tag{7.16}$$

where c is a constant expressing the known ambiguity in defining absolute time. A fundamental role in the sequel is played by the following observation: *with respect to two natural coordinate systems, the Jacobian determinant of the matrix with coefficients $\frac{\partial q'^k}{\partial q^r}$ is non-zero everywhere on $\mathbb{V}^{n+1}$.*

The proof of this descends from (7.15)–(7.16) and the corresponding equations obtained swapping primed and unprimed coordinates. Composing these transformations, we may write

$$q'^k = q'^k(t' - c, q^1(t', q'^1, \ldots, q'^n), \ldots, q^n(t', q'^1, \ldots, q'^n)) .$$

Differentiate the left-hand side in the q'^l and use the chain rule to find:

$$\delta_l^k = \frac{\partial q'^k}{\partial q'^l} = \sum_{r=1}^n \frac{\partial q'^k}{\partial q^r} \frac{\partial q^r}{\partial q'^l} + \frac{\partial q'^k}{\partial t} \frac{\partial t}{\partial q'^l} .$$

The last summand obviously vanishes since $\frac{\partial t}{\partial q'^l} = 0$, as we deduce from (7.16) after swapping the coordinate systems. All in all:

$$\delta_l^k = \sum_{r=1}^n \frac{\partial q'^k}{\partial q^r} \frac{\partial q^r}{\partial q'^l} ,$$

which can be read as saying that the product of the Jacobian matrices on the right equals the identity on the left. Consequently the Jacobian matrices with coefficients $\frac{\partial q'^k}{\partial q^r}$ and $\frac{\partial q^r}{\partial q'^l}$ are invertible, hence with non-zero determinant.

(2) A space of configurations $\mathbb{Q}_t$ at time t is a **fibre** of $\mathbb{V}^{n+1}$ at time t. Natural local coordinates on $\mathbb{V}^{n+1}$ are also known as natural local coordinates **adapted to the fibres**[4] of $\mathbb{V}^{n+1}$.

(3) Configurations can be seen at the same time as points of $\mathbb{V}^{n+1}$ and of $\mathbb{V}^{3N+1}$, since every $\mathbb{Q}_t$ is a submanifold of Σ_t^N.

(4) If the points $P_i \in \mathbb{V}^4$ of the constrained system are given by position vectors $P_i - O = \mathbf{x}_i$ with respect to some origin O with prescribed evolution, and if $t, q^1, \ldots, q^n$ are natural coordinates around a configuration, we should be able to write, as in Definition 7.6:

$$\mathbf{x}_i = \mathbf{x}_i(t, q^1, \ldots, q^n) ,$$

where time appears explicitly, in general. The system's motion will then be determined, at least locally, by differentiable maps $q^k = q^k(t)$ describing in time the position vectors of the physical system's points, hence their evolution:

$$\mathbf{x}_i = \mathbf{x}_i(t, q^1(t), \ldots, q^n(t)) .$$

∎

---

[4] AC   This slight abuse of language refers to Definition B.4, since there may be coordinates adapted to the fibres of $\mathbb{V}^{n+1}$ where the t coordinate is not absolute time, but a twice differentiable function of t.

Examples 7.8

(1) Consider the elementary case $N = 1$, i.e. one point particle P, subject to the two constraints:

$$x^2 + y^2 = 1 + ct^2 \quad \text{and} \quad z = x, \quad \text{for any } t \in \mathbb{R}.$$

The coordinates x, y, z are orthonormal coordinates on a frame $\mathcal{R}$, t is the usual absolute time and $c > 0$. So we have two constraints that vanish on a non-empty set for any $t \in \mathbb{R}$. The functions are: $f_1(t, x, y, z) = x^2 + y^2 - (1 + ct^2)$ and $f_2(t, x, y, z) = z - x$, clearly C^∞. The constraints' Jacobian matrix J is rectangular

$$J := \begin{pmatrix} \frac{\partial f_1}{\partial x} & \frac{\partial f_1}{\partial y} & \frac{\partial f_1}{\partial z} \\ \frac{\partial f_2}{\partial x} & \frac{\partial f_2}{\partial y} & \frac{\partial f_3}{\partial z} \end{pmatrix} = \begin{pmatrix} 2x & 2y & 0 \\ 1 & 0 & -1 \end{pmatrix}.$$

When $x^2 + y^2 = 1 + ct^2$ the first row is never zero. Moreover it is never linearly dependent on the second (i.e. proportional). Hence (H1) and (H2) are satisfied. For any time t the two constraints define a curve $\mathbb{Q}_t = \Gamma_t$, intersection of the cylinder of radius $\sqrt{1 + ct^2}$ with z-axis and the plane $z = x$. This is an ellipse whose shape varies in time. The point particle then has one freedom degree: $1 = 3 - 2$.

Let us show how to find free coordinates. The constraints' Jacobian matrix consists of the two above row vectors. Looking at J, we spot for example the submatrix obtained suppressing the last column,

$$\begin{pmatrix} 2x & 2y \\ 1 & 0 \end{pmatrix}.$$

It is non-singular when $y \neq 0$. By the Regular-value Theorem A.26 the coordinate $q^1 := z$ (relative to the suppressed column) is a free coordinate for the point P on $\mathbb{Q}_t$ around every configuration on $\mathbb{Q}_t$ where $y \neq 0$. It is clear that these coordinates are just local, since when we describe the projection of Γ_t on the xy-plane in terms of the polar angle $\varphi \in (-\pi, +\pi)$ (from polar coordinates in the xy-plane), the points on Γ_t with angle $\pm\varphi$ have the same z. However, on each branch $(-\pi, 0)$ and $(0, \pi)$ the coordinate z can be taken as free. The boundary points of these branches are those with $y = 0$. Around them we can use $q'^1 := x$ as new free coordinate (because x and y cannot vanish simultaneously, the matrix J without the first column is non-singular). The two local charts associated with q^1 and q'^1 together cover $\mathbb{Q}_t$, so they form an atlas on the manifold.

An often useful choice of free coordinate on $\mathbb{Q}_t$ is $q^1 := \varphi$, where φ is the above polar angle. In this case the configuration corresponding to $\varphi \pm \pi$ is excluded. The Cartesian coordinates of P in $\mathcal{R}$ are then, in function of q^1 and for any instant t:

$$x(t, q^1) = \sqrt{1 + ct^2} \cos q^1 , \tag{7.17}$$

$$y(t, q^1) = \sqrt{1 + ct^2} \sin q^1 , \tag{7.18}$$

$$z(t, q^1) = \sqrt{1 + ct^2} \cos q^1 . \tag{7.19}$$

The reader can easily show that the new coordinate on $\mathbb{Q}_t$ defines an admissible system, by proving that φ is related to q^1, q'^1 in the above charts by a differentiable map with differentiable inverse on the respective domains.

(2) Consider the less elementary case of $N = 2$ point particles P, Q at a fixed distance $d > 0$. The spacetime of configurations has initially $6 + 1 = 7$ dimensions. Using a frame $\mathscr{R}$ we can employ global coordinates t, x, y, z, X, Y, Z on $\mathbb{V}^{6+1}$, where t is the usual absolute time and (x, y, z) and (X, Y, Z) are the Cartesian coordinates of P and Q respectively in a orthonormal system on $\mathscr{R}$ with origin O. The distance constraint can be expressed, by defining the C^∞ map:

$$f(t, x, y, z, X, Y, Z) := (x - X)^2 + (y - Y)^2 + (z - Z)^2 - d^2 ,$$

as:

$$f(x, y, z, X, Y, Z) = 0 .$$

To verify the functional independence we must compute the gradient of f and check it never vanishes on the constraint. The gradient is the row vector:

$$(2(x - X), 2(y - Y), 2(z - Z), 2(X - x), 2(Y - y), 2(Z - z)) .$$

It is clearly impossible that the above is zero when $(x - X)^2 + (y - Y)^2 + (z - Z)^2 = d^2$ for $d > 0$. For any time $t \in \mathbb{R}$ we can solve the constraint equation for X, say, in terms of the other 5 variables:

$$X = x \pm \sqrt{d^2 - (y - Y)^2 - (z - Z)^2}$$

The presence of the $\pm$ indicates that the coordinates $(t, q^1, q^2, q^3, q^4, q^5) := (t, x, y, z, Y, Z)$ are local, and we need more than one coordinate systems to cover, at each instant, the spacetime of configurations $\mathbb{V}^{5+1}$. A more useful system of local coordinates for the applications, and certainly more visual, is the following: q^1, q^2, q^3 are the three Cartesian coordinates, in $\mathscr{R}$, of the midpoint M of P and Q; q^3, q^4 are the polar angles of $P - M$ with respect to orthonormal axes at M that stay parallel to the trihedron at O moving with $\mathscr{R}$. One can prove as exercise that these axes are admissible on $\mathbb{Q}_t$ and that $(t, q^1, \dots, q^5)$ are natural local coordinates on $\mathbb{V}^{5+1}$.

AC Although not strictly necessary, we will make an additional technical assumption on $\mathbb{V}^{n+1}$, which may be disregarded unless one is interested in differential-geometric details.

(H3) For any $t \in \mathbb{R}$, axis of absolute time, there exist a non-empty open interval $J \subset \mathbb{R}$, with $J \ni t$, and a corresponding collection of natural coordinates $\{(U_i, \phi_i)\}_{i \in I}$ with $\phi_i : U_i \ni p \mapsto (t_i(p), q_i^1(p), \ldots q_i^n(p)) \in \mathbb{R}^{n+1}$, such that:

(i) $\bigcup_{i \in I} U_i \supset \mathbb{Q}_\tau$ for any $\tau \in J$;

(ii) $\phi_i(U_i) \supset J \times U_i'$, with $U_i' \subset \mathbb{R}^n$ open, for any $i \in I$;

(iii) for any $i, j \in I$ with $U_i \cap U_j \neq \varnothing$, the functions $\phi_i \circ \phi_j^{-1} : \phi_j(U_i \cap U_j) \to \phi_i(U_i \cap U_j)$ have the form:

$$t_i = t_j, \qquad q_i^k = q_i^k(q_j^1, \ldots, q_j^n).$$

That is to say, Lagrangian coordinate changes do not depend explicitly on time.

It should be rather clear that (H3) holds whenever there is a reference frame $\mathscr{R}$ such that, calling $\mathbf{x}_i$ the position vector of the i-th point particle in the rest space of $\mathscr{R}$, the constraints' equations are independent of time, i.e. they have the form $f_j(\mathbf{x}_1, \mathbf{x}_2, \ldots, \mathbf{x}_N) = 0$, $j = 1, \ldots, c$. It may however happen that there is no frame in which the constraints do not depend on time as above: think of a point moving on a surface that deforms non-isometrically in time. If so, (H3) must be considered an additional hypothesis.

Remarks 7.9

(1) AC As will be made more precise in Sect. 7.5.1, keeping in account Definition 7.6 of $\mathbb{V}^{n+1}$, (H3) is equivalent to asking $\mathbb{V}^{n+1}$ is a *bundle* with *base* $\mathbb{R}$ (the axis of absolute time), *standard fibre* $\mathbb{Q}$ and *canonical projection* given by the absolute time function $T : \mathbb{V}^{n+1} \to \mathbb{R}$. The standard fibre $\mathbb{Q}$ is diffeomorphic to each configuration space at time t, $\mathbb{Q}_t$, so that the latter are all diffeomorphic. This hypothesis is a true restriction on the possible constraints of a physical systems because, for example, it imposes that the topology on the space of configurations cannot vary in time, which is normally not the case. Exactly as spacetime, $\mathbb{V}^{n+1}$ is a trivialisable bundle, albeit not canonically.

(2) There are other types of constraints, different from holonomic ones. See [Bis16, BRSG16] for an exhaustive treatment, also in relationship to the definition of ideal constraint we shall discuss later, and which in the aforementioned references are called *perfect constraints*. The constraints considered in this book are *geometrical*, and force point particles to live on curves or surfaces, possibly abstract ones. A weaker demand would for example be that a point particle remains confined to some region without crossing its boundary. This type of geometric constraint is called *unilateral* (whereas our ones are *bilateral*). Another class of *non-geometric* constraints are those describing restrictions on the configurations and velocities of the system's points, called *kinetic*. *Rolling constraints* for rigid bodies are of this type.   ∎

7.2.2 Tangent Vectors to the Space of Configurations $\mathbb{Q}_t$

Consider the usual system of N point particles $\mathscr{S}$ subject to C holonomic constraints, described in the spacetime of configurations $\mathbb{V}^{n+1}$ as (embedded) submanifold of $\mathbb{V}^{3N+1}$. We want to characterise, in general, *ideal constraints*. For this, in this section we will first need to introduce some notions and geometric notation.

Take two configurations *at the same time* t: $c_t = (P_1, \ldots, P_N) \subset \Sigma_t \times \cdots \times \Sigma_t$ and $c_t' = (P_1', \ldots, P_N') \subset \Sigma_t \times \cdots \times \Sigma_t$. In $\mathbb{V}^{3N+1}$ we define the row of vectors

$$\Delta \mathbf{P} = (\Delta P_1, \ldots, \Delta P_N) \quad \text{where } \Delta P_i = P_i' - P_i \text{ for } i = 1, \ldots, N .$$

Now we use that $\mathbb{Q}_t \subset \Sigma_t^N$. If the points P_i and P_i' are respectively given by position vectors $P_i - O = \mathbf{x}_i$ and $P_i' - O = \mathbf{x}_i'$, in the rest space of a given frame $\mathscr{R}$, and if $t, q^1, \ldots, q^n$ are natural coordinates around c_t, then

$$\mathbf{x}_i = \mathbf{x}_i(t, q^1, \ldots, q^n) \quad \text{and} \quad \mathbf{x}_i' = \mathbf{x}_i(t, q'^1, \ldots, q'^n) .$$

Consequently, the i-th point will move by:

$$\Delta P_i = \mathbf{x}_i(t, q'^1, \ldots, q'^n) - \mathbf{x}_i(t, q^1, \ldots, q^n) = \sum_{k=1}^{n} \frac{\partial \mathbf{x}_i}{\partial q^k}\Big|_{c_t} \delta q^k + O_i((\delta q)^2)$$

where $\delta q^k := q'^k - q^k$ and O_i is an infinitesimal function of order two as $\sqrt{\sum_{k=1}^{n} |\delta q^k|^2} \to 0$. To begin with, we can neglect higher-order terms provided the configurations c_t and c_t' are sufficiently close. Irrespective of that, the vector of V_t^N:

$$\delta \mathbf{P} = \left(\sum_{k=1}^{n} \frac{\partial \mathbf{x}_1}{\partial q^k}\Big|_{c_t} \delta q^k, \cdots, \sum_{k=1}^{n} \frac{\partial \mathbf{x}_N}{\partial q^k}\Big|_{c_t} \delta q^k \right) , \tag{7.20}$$

where the $\delta q^k \in \mathbb{R}$ are arbitrary numbers (possibly large!), represents a generic vector of V_t^N that is tangent to $\mathbb{Q}_t$ at the point c_t. As the coefficients $\delta q^k \in \mathbb{R}$, $k = 1, \ldots, n$, vary, the left-hand side of (7.20) describes all possible tangent vectors to $\mathbb{Q}_t$ at c_t.

Remarks 7.10

(1) The single tangent vector δP_i, *does not* actually depend on the frame on spacetime $\mathbb{V}^4$, nor on the frame's coordinates, since δP_i is defined at constant time (if there were a displacement between two instants we would need a frame to identify the absolute spaces $\Sigma_{t'}$ and Σ_t also in the limit $t \to t'$). Similarly, the expression of a tangent vector to $\mathbb{Q}_t$ at c_t given in (7.20) is also independent of the free coordinates $q^1, \ldots, q^n$ (and of the origin of absolute time), despite the coefficients δq^k may depend on the chosen free coordinates.

In fact, changing free coordinates around a configuration c_t as in (7.15) and (7.16), we find

$$\frac{\partial}{\partial q^k} = \sum_{h=1}^n \frac{\partial q'^h}{\partial q^k} \frac{\partial}{\partial q'^h}$$

where the Jacobian matrix $\frac{\partial q'^h}{\partial q^k}$ is non-singular. Consequently:

$$\delta \mathbf{P} = \left(\sum_{k=1}^n \frac{\partial \mathbf{x}_1}{\partial q^k} |_{c_t} \delta q^k, \cdots, \sum_{k=1}^n \frac{\partial \mathbf{x}_N}{\partial q^k} |_{c_t} \delta q^k \right)$$

$$= \left(\sum_{h=1}^n \frac{\partial \mathbf{x}_1}{\partial q'^h} |_{c_t} \delta q'^h, \cdots, \sum_{h=1}^n \frac{\partial \mathbf{x}_N}{\partial q'^h} |_{c_t} \delta q'^h \right),$$

where

$$\delta q'^h = \sum_{k=1}^n \frac{\partial q'^h}{\partial q^k} |_{c_t} \delta q^k . \tag{7.21}$$

In conclusion, every tangent vector to $\mathbb{Q}_t$, $\delta \mathbf{P}$ can be written indifferently in the coordinates $t, q^1, \ldots, q^n$ or $t', q'^1, \ldots, q'^n$ using (7.20) and possibly changing the components δq^k.

(2) We may rewrite (7.20) as:

$$\delta \mathbf{P} = \sum_{k=1}^n \left(\frac{\partial \mathbf{x}_1}{\partial q^k} |_{c_t}, \cdots, \frac{\partial \mathbf{x}_N}{\partial q^k} |_{c_t} \right) \delta q^k , \tag{7.22}$$

Having functionally independent constraints implies the n vectors

$$\mathbf{b}_k := \left(\frac{\partial \mathbf{x}_1}{\partial q^k} |_{c_t}, \cdots, \frac{\partial \mathbf{x}_N}{\partial q^k} |_{c_t} \right) , \quad k = 1, \ldots, n$$

are linearly independent. *Hence these vectors form a basis of the subspace of* V_t^N *of tangent vectors to* $\mathbb{Q}_t$ *at* c_t. To show the linear independence of the $\mathbf{b}_k$ observe that if we change natural coordinates around c_t

$$\mathbf{b}'_k = \left(\frac{\partial \mathbf{x}_1}{\partial q'^k} |_{c_t}, \cdots, \frac{\partial \mathbf{x}_N}{\partial q'^k} |_{c_t} \right) = \sum_h \left(\frac{\partial \mathbf{x}_1}{\partial q^h} |_{c_t}, \cdots, \frac{\partial \mathbf{x}_N}{\partial q^h} |_{c_t} \right) \frac{\partial q^h}{\partial q'^k} = \sum_h \frac{\partial q^h}{\partial q'^k} \mathbf{b}_h .$$

As the matrix of coefficients $\frac{\partial q^h}{\partial q'^k}$ is invertible, the n vectors $\mathbf{b}_h$ are linearly independent if and only if the n vectors $\mathbf{b}'_k$ are linearly independent. So, we may prove the linear independence of the $\mathbf{b}_k$ using a particular choice of natural coordinates $t, q^1, \ldots, q^n$. We may in particular choose $q^1, \ldots, q^n$ to be among the components of the N vectors $\mathbf{x}_i$. As already observed, this follows precisely from the constraints' functional independence. With such q^k, suppose then $\sum_{k=1}^{n} c_k \mathbf{b}_k = 0$. We wish to show that $c_k = 0$ for any $k = 1, 2, \ldots, n$. But $\sum_{k=1}^{n} c_k \mathbf{b}_k = 0$ can be equivalently written, making the $\mathbf{b}_k$ explicit, as:

$$\sum_{k=1}^{n} c_k \frac{\partial \mathbf{x}_i}{\partial q^k} = 0, \quad \text{for } i = 1, 2, \ldots, N. \tag{7.23}$$

If now q^1 corresponds to the coordinate x_{ab}, i.e. the b-th component of $\mathbf{x}_a$, choosing $i = a$ in (7.23) and thus only looking at the b-th component of the identity obtained, we have

$$\sum_{k=1}^{n} c_k \frac{\partial x_{ab}}{\partial q^k} = 0 \quad \text{i.e.} \quad \sum_{k=1}^{n} c_k \frac{\partial q^1}{\partial q^k} = 0$$

which reduces to

$$\sum_{k=1}^{n} c_k \delta_k^1 = 0 \quad \text{i.e.} \quad c_1 = 0.$$

Proceeding in the same way for the other $n - 1$ coordinates $q^2, q^3, \ldots, q^n$ we end up with $c_k = 0$ for $k = 1, \ldots, n$, as requested. Hence we have shown that the n vectors $\mathbf{b}_k$ are linearly independent. ∎

The next elementary, but fundamental, technical proposition is about vectors *orthogonal* to the space of configurations $\mathbb{Q}_t$.

Proposition 7.11 *Consider a system $\mathscr{S}$ of N point particles subject to $C < 3N$ holonomic constraints. The following are equivalent for a vector $\mathbf{N} = (\mathbf{n}_1, \ldots, \mathbf{n}_N) \in V_t^N$ and free coordinates (U, ψ) with $U \ni p \mapsto (t(p), q^1(p), \ldots, q^n(p))$ defined around a configuration $c_t \in \mathbb{Q}_t$ at a given time $t \in \mathbb{R}$.*

(1) $\mathbf{N}$ *is **normal** to $\mathbb{Q}_t$ at the configuration c_t, i.e.:*

$$\mathbf{N} \cdot \delta \mathbf{P} = 0 \quad \text{for any } \delta \mathbf{P} \in V_t^N \text{ tangent to } \mathbb{Q}_t \text{ at } c_t; \tag{7.24}$$

(2) with respect to the local chart (U, ψ):

$$\sum_{i=1}^{N} \mathbf{n}_i \cdot \frac{\partial \mathbf{x}_i}{\partial q^k}\bigg|_{c_t} = 0 \quad \textit{for } k = 1, \ldots, n. \tag{7.25}$$

Proof As a generic tangent vector to $\mathbb{Q}_t$ is given by (7.20), immediately:

$$\delta \mathbf{P} = \sum_{k=1}^{n} \left(\frac{\partial \mathbf{x}_1}{\partial q^k}\bigg|_{c_t}, \ldots, \frac{\partial \mathbf{x}_N}{\partial q^k}\bigg|_{c_t} \right) \delta q^k .$$

Then it is clear that (2) implies (1). To show that (1) implies (2), by the above identity we recast (1) as:

$$\sum_{k=1}^{n} \sum_{i=1}^{N} \mathbf{n}_i \cdot \frac{\partial \mathbf{x}_i}{\partial q^k}\bigg|_{c_t} \delta q^k = 0 \quad \text{for any } \delta q^k \in \mathbb{R}, k = 1, \ldots, n.$$

As the δq^k are arbitrary, we can chose them all zero except $\delta q^{k_0} = 1$, so in particular:

$$\sum_{i=1}^{N} \mathbf{n}_i \cdot \frac{\partial \mathbf{x}_i}{\partial q^{k_0}}\bigg|_{c_t} = 0 .$$

Since k_0 is arbitrary, (2) follows.   □

Remarks 7.12 $\boxed{\text{AC}}$ We may equivalently view the tangent vector to $\mathbb{Q}_t$, $\delta \mathbf{P}$, as a vector in the *tangent space* $T_{c_t}\mathbb{Q}_t$ at $c_t \in \mathbb{Q}_t$, defined by:

$$\delta \mathbf{P} = \sum_{k=1}^{n} \delta q^k \frac{\partial}{\partial q^k}\big|_{c_t} . \tag{7.26}$$

When changing the natural coordinates' local chart, by general principles the components of $\delta \mathbf{P}$ are related to the components in the initial coordinates by the known relationships:

$$\delta q'^h = \sum_{k=1}^{n} \frac{\partial q'^h}{\partial q^k} \delta q^k,$$

which are exactly the transformation laws (7.21) seen earlier.   ∎

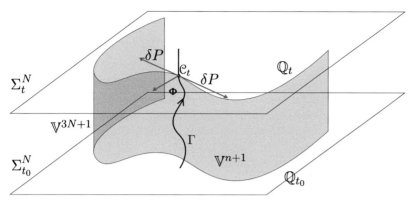

Fig. 7.2 The spacetime of configurations with constraints $\mathbb{V}^{n+1}$

7.2.3   Ideal Constraints

Let us work with a system of N point particles $\mathscr{S}$ subject to $C < 3N$ holonomic constraints. For a given frame $\mathscr{R}$, suppose that on the generic i-th particle there is a force (in general this will be the sum of forces of various type acting on the particle, of known expression):

$$\mathbf{F}_{\mathscr{R}i}(t, P_1, \ldots, P_N, \mathbf{v}_{P_1}|_{\mathscr{R}}, \ldots, \mathbf{v}_{P_N}|_{\mathscr{R}}) .$$

To distinguish it from the reaction due to the constraints we shall call it **active force**. We have written the subscript $\mathscr{R}$ explicitly since we will assume $\mathbf{F}_{\mathscr{R}}$ may include inertial forces when $\mathscr{R}$ is non-inertial. Newton's equations read:

$$m_{P_i}\mathbf{a}_{P_i}|_{\mathscr{R}} = \mathbf{F}_{\mathscr{R}i} + \boldsymbol{\phi}_i , \quad \text{for any } i = 1, \ldots, N. \tag{7.27}$$

As said, physically the constraints are imposed via forces, more precisely **constraint reactions**, or **reactive forces** or **passive forces**, acting on each P_i. We will call $\boldsymbol{\phi}_i$, $i = 1, \ldots, N$ the constraint reaction on the i-th point particle. We can now define *ideal constraints*,[5] which generalise the notion of smooth constraint seen previously.

Definition 7.13 (Ideal Constraints) Consider a system $\mathscr{S}$ of N point particles subject to $C < 3N$ holonomic constraints. The constraint reactions acting on a world line $\Gamma = \Gamma(t)$ of the system are called **ideal** if, for any time t and any configuration $\Gamma(t)$ crossed by the world line at time t, the vector of constraint reactions:

$$\Phi = (\boldsymbol{\phi}_1, \ldots, \boldsymbol{\phi}_N) \in V_t^N \quad \text{is always normal to } \mathbb{Q}_t \text{ at } \Gamma(t).$$

---

[5] See Fig. 7.2.

Constraints are called **ideal** if the constraint reactions are ideal on any world line solving the system's equations of motion (7.27). ◇

Remarks 7.14

(1) In the wording dating back to the nineteenth century, a tangent vector $\delta\mathbf{P}$ to the space of configurations $\mathbb{Q}_t$ at time t was called *virtual displacement of the physical system*. The term "virtual" refers to a displacement measured at a fixed time, whereas the functions describing the constraints change with time, $f_j(t, P_1, \ldots, P_N) \neq f_j(t', P_1, \ldots, P_N)$ for $t \neq t'$. Consequently a virtual displacement does not correspond, in general, to any real displacement of the system during a finite time.

(2) In the nineteenth century tradition *ideal constraints* are also called "constraints that obey the D'Alembert principle" or "constraints that satisfy the principle of virtual work". Indeed, back then the *virtual work* of the constraint reactions $(\boldsymbol{\phi}_1, \ldots, \boldsymbol{\phi}_N)$ during the virtual displacement $\delta\mathbf{P} = (\delta P_1, \ldots, \delta P_N)$ for the configuration c_t referred to the quantity:

$$\delta L := \Phi \cdot \delta\mathbf{P} = \sum_{i=1}^{N} \boldsymbol{\phi}_i \cdot \delta P_i . \tag{7.28}$$

Since virtual displacements are nothing but tangent vectors to $\mathbb{Q}_t$, it is then clear that the 'ideality' of the reactions is the same as:

$$\Phi \cdot \delta\mathbf{P} = 0 , \quad \text{for any virtual displacement } \delta\mathbf{P}. \tag{7.29}$$

Indeed, the above expresses that the constraint reactions' virtual work is zero for any virtual displacement of the system. ∎

Examples 7.15

(1) The simplest example of particle system subject to ideal holonomic constraints is what we discussed in the chapter's introduction: a system of interacting point particles, subject to external forces, each moving on a surface or a smooth curve at rest in some frame. We do not demand the frame is the same for all point particles. In that case, the vectorial components δP_i of the generic vector $\delta\mathbf{P} = (\delta P_1, \ldots \delta P_N) \in V_t^N$ tangent to the space of configurations $\mathbb{Q}_t$ are each tangent to the surface or curve to which the i-th particle is constrained. At any instant, then, each $\boldsymbol{\phi}_i \cdot \delta P_i$, whose sum is the inner product $\Phi \cdot \delta\mathbf{P}$, is zero because the curves and surfaces have no friction. Hence Φ is always normal to $\mathbb{Q}_t$.

(2) Another very important example is the *rigidity constraint* (we assume the strong Action-reaction principle holds): a system of N point particles connected by ideal supports so that they remain at fixed distances during any motion. The admissible configurations (i.e. satisfying the rigidity constraint) at a given time

t will be related by a generic isometry of $\Sigma_t \equiv \mathbb{E}^3$:

$$\mathbf{x}_i = \mathbf{t} + R\mathbf{x}_{0i}\,, \quad \text{for any } i = 1, \ldots, N, \tag{7.30}$$

where $\mathbf{t}$ and R are a generic displacement (a vector in the space of displacements of Σ_t) and a rotation in $O(3)$, both acting actively. $\mathbf{x}_{0i}$ and $\mathbf{x}_i$ are the position vectors of the i-th particle, in two configurations compatible with the rigidity constraint, with respect to an origin $O \in \Sigma_t$ fixed once and for all. The position vectors $\mathbf{x}_{0i}$ define a reference configuration, fixed once and for all, and all other configurations compatible with the constraints will be given by the position vectors $\mathbf{x}_i$. More generally, let us fix a frame $\mathscr{R}$ and orthonormal coordinates at O with axes $\mathbf{e}_1, \mathbf{e}_2, \mathbf{e}_3$. The vectors $\mathbf{x}_{0i}$ and $\mathbf{x}_i$ will be then identified with column vectors in $\mathbb{R}^3$ giving the components in that basis of the above position vectors. From now on we shall assume this. Therefore R is an orthogonal matrix in $SO(3)$. To describe (locally) the class of rigid transformations (7.30) we need exactly 6 real parameters $q^1, \ldots, q^6$:

(1) the coordinates $(q^1, q^2, q^3) \in \mathbb{R}^3$ giving the displacement $\mathbf{t}(q^1, q^2, q^3) = \sum_{j=1}^3 q^j \mathbf{e}_j$,
(2) three angles q^4, q^5, q^6 (for example the *Euler angles* that vary in their domains) giving $R(q^4, q^5, q^6) \in SO(3)$.

Thus the generic configuration $(\mathbf{x}_1(q^1, \ldots, q^6), \ldots, \mathbf{x}_n(q^1, \ldots, q^6))$ compatible with the rigidity constraint will be:

$$\mathbf{x}_i(q^1, \ldots, q^6) = \mathbf{t}(q^1, q^2, q^3) + R(q^4, q^5, q^6)\mathbf{x}_{0i}\,, \quad \text{for any } i = 1, \ldots, N.$$

Expanding the right-hand side to order one around the configuration $(q^1, \ldots, q^6)$, we have:

$$\delta P_i := \delta \mathbf{x}_i = \sum_{k=1}^3 \delta q^k \mathbf{e}_k + \sum_{k=4}^6 \delta q^k \frac{\partial R}{\partial q^k}\mathbf{x}_{0i}\,, \quad \text{for any } i = 1, \ldots, N,$$

where R and its derivatives are computed with respect to $(q^1, \ldots, q^6)$. Since $R^t R = I$,

$$\delta P_i = \sum_{k=1}^3 \delta q^k \mathbf{e}_k + \sum_{k=4}^6 \delta q^k \frac{\partial R}{\partial q^k} R^t R\mathbf{x}_{0i}\,, \quad \text{for any } i = 1, \ldots, N.$$

Furthermore,

$$R\mathbf{x}_{0i} = \mathbf{x}_i - \mathbf{t}\,,$$

so

$$\delta P_i = \sum_{k=1}^{3} \delta q^k \mathbf{e}_k + \sum_{k=4}^{6} \delta q^k \frac{\partial R}{\partial q^k} R^t (\mathbf{x}_i - \mathbf{t}), \quad \text{for any } i = 1, \ldots, N.$$

Each one of the 3 matrices $A_k := \frac{\partial R}{\partial q^k} R^t$ is skew-symmetric. In fact, from $I = RR^t$ we have

$$0 = \frac{\partial RR^t}{\partial q^k} = \frac{\partial R}{\partial q^k} R^t + R \frac{\partial R^t}{\partial q^k} = \frac{\partial R}{\partial q^k} R^t + \left(\frac{\partial R}{\partial q^k} R^t \right)^t = A_k + A_k^t.$$

Only due to the skew-symmetry, the action of A_k on the vectors can be written as $A_k \cdot = \boldsymbol{\omega}_k \wedge \cdot$, where the components of $\boldsymbol{\omega}_k$, signs apart, are the 3 independent elements of A_k. Hence there exist three vectors $\boldsymbol{\omega}_4, \boldsymbol{\omega}_5, \boldsymbol{\omega}_6$ (dependent on the configuration $q^1, \ldots, q^6$) for which $A_k \mathbf{x} = \boldsymbol{\omega}_k \wedge \mathbf{x}$. We conclude that the vectors δP_i, components of the generic $\boldsymbol{\delta P}$ tangent to $\mathbb{Q}_t$ at configuration $(\mathbf{x}_1(q^1, \ldots, q^6), \ldots, \mathbf{x}_n(q^1, \ldots, q^6))$, are:

$$\delta P_i = \sum_{k=1}^{3} \delta q^k \mathbf{e}_k + \sum_{k=4}^{6} \delta q^k \boldsymbol{\omega}_k(q^4, q^5, q^6) \wedge (\mathbf{x}_i(q^1, \ldots, q^6) - \mathbf{t}(q^1, q^2, q^3)),$$

for $i = 1, \ldots, N$,

where $\delta q^1, \ldots, \delta q^6 \in \mathbb{R}$ and

$$\boldsymbol{\omega}_k(q^4, q^5, q^6) \wedge := \frac{\partial R}{\partial q^k}\Big|_{(q^4, q^5, q^6)} R^t(q^4, q^5, q^6)$$

$$\text{and} \quad \mathbf{t}(q^1, q^2, q^3) := \sum_{j=1}^{3} q^j \mathbf{e}_j.$$

If $\boldsymbol{\phi}_{ij}$ is the constraint reaction on the i-th particle given by $\mathbf{x}_i$ caused by the j-th particle ($j \neq i$) given by $\mathbf{x}_j$, the inner product between the reactions' vector and a generic tangent vector $\boldsymbol{\delta P}$ to the space of configurations is

$$\Phi \cdot \boldsymbol{\delta P} = \sum_{i,j \, i \neq j} \sum_{k=1}^{3} \delta q^k \mathbf{e}_k \cdot \boldsymbol{\phi}_{ij} + \sum_{i,j \, i \neq j} \sum_{k=4}^{6} \delta q^k \boldsymbol{\omega}_k \wedge (\mathbf{x}_i - \mathbf{t}) \cdot \boldsymbol{\phi}_{ij}.$$

After simple manipulations (recall $\mathbf{a} \wedge \mathbf{b} \cdot \mathbf{c} = \mathbf{b} \wedge \mathbf{c} \cdot \mathbf{a} = \mathbf{a} \cdot \mathbf{b} \wedge \mathbf{c}$):

$$\delta L = \sum_{k=1}^{3} \delta q^k \mathbf{e}_k \cdot \sum_{i,j\ i\neq j}^{3} \boldsymbol{\phi}_{ij} + \sum_{k=4}^{6} \delta q^k \boldsymbol{\omega}_k \cdot \sum_{i,j\ i\neq j} (\mathbf{x}_i - \mathbf{t}) \wedge \boldsymbol{\phi}_{ij}$$

$$= \sum_{k=1}^{3} \delta q^k \mathbf{e}_k \cdot \sum_{i,j\ i\neq j}^{3} \boldsymbol{\phi}_{ij} + \sum_{k=4}^{6} \delta q^k \boldsymbol{\omega}_k \cdot \sum_{i,j\ i\neq j} \mathbf{x}_i \wedge \boldsymbol{\phi}_{ij} - \sum_{k=4}^{6} \delta q^k \boldsymbol{\omega}_k \cdot \mathbf{t} \wedge \sum_{i,j\ i\neq j} \boldsymbol{\phi}_{ij}.$$

But one easily proves that

$$\sum_{i,j\ i\neq j} \boldsymbol{\phi}_{ij} = \mathbf{0} \quad \text{and} \quad \sum_{i,j\ i\neq j} \mathbf{x}_i \wedge \boldsymbol{\phi}_{ij} = \mathbf{0},$$

since $\boldsymbol{\phi}_{ij} = -\boldsymbol{\phi}_{ji}$, and $\boldsymbol{\phi}_{ij}$ is parallel to $\mathbf{x}_i - \mathbf{x}_j$ (the proof is just like Theorem 4.4 for the internal forces). Therefore the rigidity constraint is an ideal constraint. Intuitively, the result generalises, passing to continuous bodies, to physical systems made of extended rigid bodies. The generalisation, however, is not as obvious as it might seem, since the interaction occurring inside a continuous body are described by *densities of forces* rather than finite forces on discrete points.

(3) Also ideal are the constraints on particle system made by the previous two situations simultaneously. For example a system of three point particles P_1, P_2, P_3, with P_1, P_2 joined to P_3 by massless bars and supposing the angle formed by the bars at P_3 varies arbitrarily (one could also suppose the bars move on a fixed plane) due to ideal hinges with no friction (i.e. no friction momenta). Particle P_3 moves along a given smooth curve at rest in some frame.

A similar case is a system of point particles $\mathscr{S}$ subject to the rigidity constraint and attached at one point particle by some massless rigid rod to a geometric point at rest in a frame. The joints are ideal hinges without friction moments. This pendulum still satisfies the ideal constraint condition. In particular, the constraint reaction due to the rod does not contribute to any product $\Phi \cdot \delta \mathbf{P}$, since every vector $\delta \mathbf{P}$ is normal to the axis while the constraint reaction is parallel to it (it might not be if the rod had a mass).

(4) We can consider point particles on smooth curves or surfaces whose shape changes with time. The product $\Phi \cdot \delta \mathbf{P}$, computed at fixed time, is the same as for the case of smooth fixed curves or surfaces: any product $\Phi \cdot \delta \mathbf{P}$ is always zero.

(5) At last we mention *rolling constraints* by examining the Lagrangian Mechanics of the continuous rigid bodies considered later. We are in presence of a rolling constraint when a physical system $\mathscr{S}$, made by two rigid bodies $\mathscr{S}_1$ and $\mathscr{S}_2$ touching at a geometric point P, corresponding to point particles $Q_1 \in \mathscr{S}_1$ and $Q_2 \in \mathscr{S}_2$, satisfies $\mathbf{v}_{Q_1}|_{\mathscr{R}} = \mathbf{v}_{Q_2}|_{\mathscr{R}}$ at any instant, for a frame $\mathscr{R}$. Note Q_1 and Q_2 vary instant by instant (imagine $\mathscr{S}_1$ and $\mathscr{S}_2$ as wheels in a plane that roll

without slipping against each other, and Q_1, Q_2 are the contact points at any instant). The condition $\mathbf{v}_{Q_1}|_{\mathscr{R}} = \mathbf{v}_{Q_2}|_{\mathscr{R}}$ does not actually depend on the frame, since $Q_1 - Q_2 = \mathbf{0}$ by assumptions, and so:

$$
\mathbf{v}_{Q_1}|_{\mathscr{R}} - \mathbf{v}_{Q_2}|_{\mathscr{R}} = (\mathbf{v}_{Q_1}|_{\mathscr{R}'} + \mathbf{v}_{O'}|_{\mathscr{R}} + \boldsymbol{\omega}_{\mathscr{R}'}|_{\mathscr{R}} \wedge (Q_1 - O'))
$$

$$
- (\mathbf{v}_{Q_2}|_{\mathscr{R}'} + \boldsymbol{\omega}_{\mathscr{R}'}|_{\mathscr{R}} \wedge (Q_2 - O'))
$$

$$
= \mathbf{v}_{Q_1}|_{\mathscr{R}'} - \mathbf{v}_{Q_2}|_{\mathscr{R}'} + \boldsymbol{\omega}_{\mathscr{R}'}|_{\mathscr{R}} \wedge (Q_1 - Q_2) = \mathbf{v}_{Q_1}|_{\mathscr{R}'} - \mathbf{v}_{Q_2}|_{\mathscr{R}'} .
$$

Secondly, the rolling constraint is *not* in general holonomic, because it involves the velocities. Nonetheless, in some cases it can be *integrated* and made holonomic. In such a case virtual displacements are proper displacements during a time interval δt, approximated to order one by the usual Taylor expansion. In particular, for the components of a tangent vector to the space of configurations $\boldsymbol{\delta P} = (\delta P_1, \delta P_2)\, \delta Q_1 - \delta Q_2 = \mathbf{0}$. As the constraint reaction $\boldsymbol{\phi}_1$ at Q_1, due to $\mathscr{S}_2$, coincides with $-\boldsymbol{\phi}_2$, where $\boldsymbol{\phi}_2$ is the constraint reaction at Q_2 due to $\mathscr{S}_1$, we have $\boldsymbol{\Phi} \cdot \boldsymbol{\delta P} = \boldsymbol{\phi}_1 \cdot \delta Q_1 + \boldsymbol{\phi}_2 \cdot \delta Q_2 = 0$, and there is no contribution to the overall virtual work of $\mathscr{S}$.

7.2.4 Kinematic Quantities and Kinetic Energy

To write the Euler-Lagrange equations we still need a technical ingredient to do with Kinematics. Consider the usual system of N point particles $P_1, \ldots, P_N$ subject to $C < 3N$ holonomic constraints described in the spacetime of configurations $\mathbb{V}^{n+1}$. Using natural coordinates $t, q^1, \ldots, q^n$, we can write the points' positions as $P_i = P_i(t, q^1, \ldots, q^n)$. A world line $\Gamma = \Gamma(t) \in \mathbb{V}^{3N+1}$ that obeys the constraints is locally described in natural coordinates by a differentiable curve $q^k = q^k(t)$. Given a frame $\mathscr{R}$, we have $P_i(t) = \mathbf{x}_i(t) + O$, where we have identified every space Σ_t with the frame's rest space $E_{\mathscr{R}}$ where O is an origin and $\mathbf{x}_i$ is the position vector of P_i in $\mathscr{R}$. Then:

$$
\mathbf{v}_{P_i}|_{\mathscr{R}}(t) = \frac{\partial \mathbf{x}_i}{\partial t} + \sum_{k=1}^{n} \frac{\partial \mathbf{x}_i}{\partial q^k} \frac{dq^k}{dt} . \tag{7.31}
$$

Thus we can write the system's kinetic energy in $\mathscr{R}$ using the natural coordinates. As $\mathscr{T}|_{\mathscr{R}} = \sum_i \frac{1}{2} m_{P_i} \mathbf{v}_{P_i}|_{\mathscr{R}}^2$, immediately:

$$
\mathscr{T}|_{\mathscr{R}} = \mathscr{T}_2|_{\mathscr{R}} + \mathscr{T}_1|_{\mathscr{R}} + \mathscr{T}_0|_{\mathscr{R}} \tag{7.32}
$$

where:

$$\mathcal{T}_2|_{\mathscr{R}} := \sum_{h,k=1}^{n} a_{hk}(t, q^1, \ldots, q^n) \frac{dq^h}{dt} \frac{dq^k}{dt}, \qquad \mathcal{T}_1|_{\mathscr{R}} := \sum_{k=1}^{n} b_k(t, q^1, \ldots, q^n) \frac{dq^k}{dt},$$

$$\mathcal{T}_0|_{\mathscr{R}} := c(t, q^1, \ldots, q^n),$$

and by definition:

$$a_{hk}(t, q^1, \ldots, q^n) := \frac{1}{2} \sum_{i=1}^{N} m_i \frac{\partial \mathbf{x}_i}{\partial q^h} \cdot \frac{\partial \mathbf{x}_i}{\partial q^k}, \tag{7.33}$$

$$b_k(t, q^1, \ldots, q^n) := \sum_{i=1}^{N} m_i \frac{\partial \mathbf{x}_i}{\partial q^k} \cdot \frac{\partial \mathbf{x}_i}{\partial t}, \tag{7.34}$$

$$c(t, q^1, \ldots, q^n) := \frac{1}{2} \sum_{i=1}^{N} m_i \frac{\partial \mathbf{x}_i}{\partial t} \cdot \frac{\partial \mathbf{x}_i}{\partial t}. \tag{7.35}$$

7.2.5   Euler-Lagrange Equations for Systems of Point Particles

Consider the usual system of N point particles $P_1, \ldots, P_N$, with masses $m_1, \ldots, m_N$ respectively, subject to $C < 3N$ ideal holonomic constraints and described in the configuration spacetime $\mathbb{V}^{n+1}$ where $n := 3N - C$. To any world line $\Gamma = (\mathbf{x}_1(t), \ldots, \mathbf{x}_N(t)) \in \mathbb{V}^{3N+1}$ obeying the constraints on $\mathscr{S}$ there corresponds 1-1 a curve $\gamma = \gamma(t) \in \mathbb{V}^{n+1}$ written using local free coordinates $q^k = q^k(t)$, $k = 1, \ldots, n$. That is, $\mathbf{x}_i(t) = \mathbf{x}_i(q^1(t), \ldots, q^n(t))$ for any $i = 1, \ldots, N$. Then the following theorem holds.[6]

Theorem 7.16 *Consider a system $\mathscr{S}$ of N point particles $P_1, \ldots, P_N$, with masses $m_1, \ldots, m_N$ respectively, subject to $C < 3N$ ideal holonomic constraints, of class C^2 at least, described on the spacetime of configurations $\mathbb{V}^{n+1}$. Given a frame $\mathscr{R}$, suppose the (total) active force $\mathbf{F}_{\mathscr{R}i}(t, P_1, \ldots, P_N, \mathbf{v}_{P_1}|_{\mathscr{R}}, \ldots, \mathbf{v}_{P_N}|_{\mathscr{R}})$ and the (total) constraint reaction $\boldsymbol{\phi}_i$ act on the generic i-th particle, so that Newton's equation of motion (7.27) holds. If the constraint reactions on a world line $\Gamma = \Gamma(t) \in \mathbb{V}^{3N+1}$ of $\mathscr{S}$ are ideal, then the curve $\gamma = \gamma(t) \in \mathbb{V}^{n+1}$, corresponding*

---

[6] See Fig. 7.2.

*to Γ and described in natural local coordinates by $q^k = q^k(t)$, satisfies the **Euler-Lagrange equations**:*

$$
\begin{cases}
\dfrac{d}{dt}\left(\dfrac{\partial \mathscr{T}|_{\mathscr{R}}}{\partial \dot{q}^k}\bigg|_{(t,q(t),\dot{q}(t))}\right) - \dfrac{\partial \mathscr{T}|_{\mathscr{R}}}{\partial q^k}\bigg|_{(t,q(t),\dot{q}(t))} = \mathscr{Q}_k|_{\mathscr{R}}(t,q(t),\dot{q}(t))\,, \\[2mm]
\dfrac{dq^k}{dt} = \dot{q}^k(t)
\end{cases}
$$

$$\text{for } k = 1,\ldots,n. \tag{7.36}$$

*Above $n := 3N - C$, $\mathscr{T}|_{\mathscr{R}} = \mathscr{T}|_{\mathscr{R}}(t,q^1,\ldots,q^n,\dot{q}^1,\ldots,\dot{q}^n)$ is the kinetic energy of $\mathscr{S}$ in frame $\mathscr{R}$ (7.32) and $\mathscr{Q}_k|_{\mathscr{R}}$ are the **Lagrangian components of the active forces**:*

$$
\mathscr{Q}_k|_{\mathscr{R}}(t,q^1,\ldots,q^n,\dot{q}^1,\ldots,\dot{q}^n) := \sum_{i=1}^{N} \mathbf{F}_{\mathscr{R}i}(t,q^1,\ldots q^n,\dot{q}^1,\ldots,\dot{q}^n) \cdot \frac{\partial \mathbf{x}_i}{\partial q^k} \tag{7.37}
$$

In (7.37), and in the computation of the explicit expression of $\mathscr{T}|_{\mathscr{R}}(t,q^1,\ldots,q^n,\dot{q}^1,\ldots,\dot{q}^n)$, we have used (7.31) to write the velocities of the points of $\mathscr{S}$ in terms of the q^k and their time derivatives $\dot{q}^k$, the latter seen as independent of the q^k.

Proof The proof goes along the lines of the chapter's introductory section. From Newton's equations (7.27), the characterisation of ideal constraints (valid at the configuration c_t reached by the world line):

$$\Phi := (\boldsymbol{\phi}_1,\ldots,\boldsymbol{\phi}_N) \quad \text{is normal to } \mathbb{Q}_t \text{ at } c_t, \tag{7.38}$$

can be equivalently written:

$$(m_{P_1}\mathbf{a}_{P_1}|_{\mathscr{R}} - \mathbf{F}_{\mathscr{R}1},\ldots,m_{P_N}\mathbf{a}_{P_N}|_{\mathscr{R}} - \mathbf{F}_{\mathscr{R}N}) \quad \text{is normal to } \mathbb{Q}_t \text{ at } c_t \,.$$

Set $\mathbf{x}_i := P_i - O$ with O at rest in $\mathscr{R}$, so we have another equation equivalent to (7.38) by Proposition 7.11:

$$\sum_{i=1}^{N}\left(m_{P_i}\mathbf{a}_{P_i}|_{\mathscr{R}} - \mathbf{F}_{\mathscr{R}i}\right) \cdot \frac{\partial \mathbf{x}_i}{\partial q^k} = 0\,, \quad \text{for any } k = 1,\ldots,n\,. \tag{7.39}$$

The functions $\mathbf{x}_i = \mathbf{x}_i(t,q^1,\ldots,q^n)$ are C^2 if the constraints are C^2. To continue we will show that on any world line:

$$\frac{d}{dt}\frac{\partial \mathscr{T}|_{\mathscr{R}}}{\partial \dot{q}^k} - \frac{\partial \mathscr{T}|_{\mathscr{R}}}{\partial q^k} = \sum_{i=1}^{N} m_i \mathbf{a}_{P_i}|_{\mathscr{R}} \cdot \frac{\partial \mathbf{x}_i}{\partial q^k}\,, \tag{7.40}$$

(where we have imposed $\dot{q}^k = \frac{dq^k}{dt}$ only *after* computing the left-hand side). Let us prove (7.40). In the sequel we will just write $\mathbf{v}_i$ instead of $\mathbf{v}_{P_i}|_{\mathscr{R}}$ and $\mathbf{a}_i$ instead of $\mathbf{a}_{P_i}|_{\mathscr{R}}$. Consider the left-hand side of (7.40), i.e. explicitly:

$$\frac{d}{dt}\frac{\partial \mathscr{T}|_{\mathscr{R}}}{\partial \dot{q}^k} - \frac{\partial \mathscr{T}|_{\mathscr{R}}}{\partial q^k} = \sum_i \frac{d}{dt}\frac{\partial}{\partial \dot{q}^k}\left(\frac{1}{2}m_i v_i^2\right) - \sum_i \frac{\partial}{\partial q^k}\left(\frac{1}{2}m_i v_i^2\right) =$$

$$\sum_i \frac{m_i}{2} 2 \frac{d}{dt}\left(\mathbf{v}_i \cdot \frac{\partial \mathbf{v}_i}{\partial \dot{q}^k}\right) - \sum_i \frac{m_i}{2} 2\mathbf{v}_i \cdot \frac{\partial \mathbf{v}_i}{\partial q^k} = \sum_i m_i \frac{d\mathbf{v}_i}{dt}\cdot\frac{\partial \mathbf{v}_i}{\partial \dot{q}^k}$$

$$+ \sum_i m_i \mathbf{v}_i \cdot \frac{d}{dt}\frac{\partial \mathbf{v}_i}{\partial \dot{q}^k} - \sum_i m_i \mathbf{v}_i \cdot \frac{\partial \mathbf{v}_i}{\partial q^k}$$

$$= \sum_i m_i \mathbf{a}_i \cdot \frac{\partial \mathbf{v}_i}{\partial \dot{q}^k} + \sum_i m_i \mathbf{v}_i \cdot \left(\frac{d}{dt}\frac{\partial \mathbf{v}_i}{\partial \dot{q}^k} - \frac{\partial \mathbf{v}_i}{\partial q^k}\right).$$

The velocity's expression, because $dq^k/dt = \dot{q}^k$, is:

$$\mathbf{v}_i = \frac{\partial \mathbf{x}_i}{\partial t} + \sum_h \frac{\partial \mathbf{x}_i}{\partial q^h}\dot{q}^h, \qquad (7.41)$$

and then, as the q^k are independent of the $\dot{q}^k$, we obtain:

$$\frac{\partial \mathbf{v}_i}{\partial \dot{q}^k} = \frac{\partial \mathbf{x}_i}{\partial q^k}.$$

Substituting in the above expression for the left-hand side of (7.40), we conclude:

$$\frac{d}{dt}\frac{\partial \mathscr{T}|_{\mathscr{R}}}{\partial \dot{q}^k} - \frac{\partial \mathscr{T}|_{\mathscr{R}}}{\partial q^k} = \sum_i m_i \mathbf{a}_i \cdot \frac{\partial \mathbf{x}_i}{\partial q^k} + \sum_i m_i \mathbf{v}_i \cdot \left(\frac{d}{dt}\frac{\partial \mathbf{v}_i}{\partial \dot{q}^k} - \frac{\partial \mathbf{v}_i}{\partial q^k}\right).$$

To finish the proof of (7.40) we just need to show:

$$\frac{d}{dt}\frac{\partial \mathbf{v}_i}{\partial \dot{q}^k} - \frac{\partial \mathbf{v}_i}{\partial q^k} = 0.$$

This is immediate from (7.41), once we ask $dq^k/dt = \dot{q}^k$:

$$\frac{d}{dt}\frac{\partial \mathbf{v}_i}{\partial \dot{q}^k} - \frac{\partial \mathbf{v}_i}{\partial q^k} = \frac{d}{dt}\frac{\partial}{\partial \dot{q}^k}\left(\sum_h \frac{\partial \mathbf{x}_i}{\partial q^h}\dot{q}^h + \frac{\partial \mathbf{x}_i}{\partial t}\right) - \frac{\partial}{\partial q^k}\left(\sum_h \frac{\partial \mathbf{x}_i}{\partial q^h}\dot{q}^h + \frac{\partial \mathbf{x}_i}{\partial t}\right)$$

$$= \frac{d}{dt}\left(\frac{\partial \mathbf{x}_i}{\partial q^k}\right) - \left(\sum_h \frac{\partial^2 \mathbf{x}_i}{\partial q^k \partial q^h}\dot{q}^h + \frac{\partial^2 \mathbf{x}_i}{\partial q^k \partial t}\right)$$

$$= \left(\sum_h \frac{\partial^2 \mathbf{x}_i}{\partial q^h \partial q^k}\dot{q}^h + \frac{\partial^2 \mathbf{x}_i}{\partial t \partial q^k}\right) - \left(\sum_h \frac{\partial^2 \mathbf{x}_i}{\partial q^k \partial q^h}\dot{q}^h + \frac{\partial^2 \mathbf{x}_i}{\partial q^k \partial t}\right) = 0\,,$$

by the Schwarz Theorem. Hence (7.40) is proved.

Therefore the Euler-Lagrange equations can be written, recalling Definition (7.37) for the Lagrangian components of the active forces $\mathcal{Q}_k$:

$$\begin{cases} \sum_{i=1}^{N} m_i \mathbf{a}_{P_i}|_{\mathscr{R}} \cdot \dfrac{\partial \mathbf{x}_i}{\partial q^k} = \sum_{i=1}^{N} \mathbf{F}_{\mathscr{R}i}(t, q^1, \ldots q^n, \dot{q}^1, \ldots, \dot{q}^n) \cdot \dfrac{\partial \mathbf{x}_i}{\partial q^k}\,, & \text{for } k = 1, \ldots, n, \\[4mm] \dfrac{dq^k}{dt} = \dot{q}^k(t) \end{cases}$$

i.e.

$$\begin{cases} \sum_{i=1}^{N} \left(m_i \mathbf{a}_{P_i}|_{\mathscr{R}} - \mathbf{F}_{\mathscr{R}i}(t, q^1, \ldots q^n, \dot{q}^1, \ldots, \dot{q}^n)\right) \cdot \dfrac{\partial \mathbf{x}_i}{\partial q^k} = 0\,, \\[4mm] \dfrac{dq^k}{dt} = \dot{q}^k(t) \\[2mm] \qquad\qquad \text{for } k = 1, \ldots, n, \end{cases} \tag{7.42}$$

This result concludes the proof. In fact, suppose $q^k = q^k(t)$ defines a world line $\Gamma = \Gamma(t) \in \mathbb{V}^{3N+1}$ for $\mathscr{S}$ that solves Newton's equations (7.27) with ideal constraints. Then $m_i \mathbf{a}_{P_i}|_{\mathscr{R}} - \mathbf{F}_{\mathscr{R}i} = \boldsymbol{\phi}_i$ is the constraint reaction on the i-th particle and the above system (7.42) holds by virtue of characterisation (7.39) for ideal constraints.

$\square$

Remarks 7.17

(1) We must point out that the Lagrangian components of the active forces are *independent* of the frame $\mathscr{R}$ employed to define the position vectors $\mathbf{x}_i$ when the active forces are *true forces*. Changing frame and using position vectors $\mathbf{x}_i'$ with respect to $\mathscr{R}'$, we have

$$\mathbf{x}_i'(t, q^1, \ldots, q^n) = \mathbf{x}_i(t, q^1, \ldots, q^n) + \mathbf{X}(t) \tag{7.43}$$

where $\mathbf{X}(t) = O'(t) - O(t)$ is the *pre-ordained* function describing the relative motion of the two origins chosen in the rest spaces of $\mathscr{R}$ and $\mathscr{R}'$. To compare position vectors we are identifying the vectors $\mathbf{x}_i$ and $\mathbf{x}'_i$ with corresponding vectors in the rest space of some frame $\mathscr{R}_0$ where all computations are performed. If $\mathbf{e}_{01}, \mathbf{e}_{02}, \mathbf{e}_{03}$ is an orthonormal basis of the rest space of $\mathscr{R}_0$, (7.43) becomes, with obvious notation:

$$\sum_{j=1}^{3} x_i'^{\,j}(t, q^1, \ldots, q^n)\mathbf{e}_{0j} = \sum_{j=1}^{3} x_i^j(t, q^1, \ldots, q^n)\mathbf{e}_{0j} + \sum_{j=1}^{3} o^j(t)\mathbf{e}_{0j}.$$

$$(7.44)$$

Then

$$\mathscr{Q}_k|_{\mathscr{R}'}(t, q^1, \ldots, q^n, \dot{q}^1, \ldots, \dot{q}^n) := \sum_{i=1}^{N} \mathbf{F}_{\mathscr{R}'i}(t, q^1, \ldots q^n, \dot{q}^1, \ldots, \dot{q}^n) \cdot \frac{\partial \mathbf{x}'_i}{\partial q^k}$$

$$= \sum_{i=1}^{N} \mathbf{F}_{\mathscr{R}i}(t, q^1, \ldots q^n, \dot{q}^1, \ldots, \dot{q}^n) \cdot \frac{\partial \mathbf{x}_i}{\partial q^k} = \mathscr{Q}_k|_{\mathscr{R}}(t, q^1, \ldots, q^n, \dot{q}^1, \ldots, \dot{q}^n),$$

where we used that true forces do not depend on the frame, and also:

$$\frac{\partial \mathbf{x}_i}{\partial q^k} = \frac{\partial \mathbf{x}'_i}{\partial q^k}$$

because of (7.44). The above result does not depend on $\mathscr{R}_0$, as is easy to see using the usual isometric identifications between the frames' rest spaces.

    If we work in a non-inertial frame, part of the active forces is inertial and hence *frame-dependent*. This affects the Lagrangian components of the active forces.

(2) It is extremely important to stress that the variables $\dot{q}^k$ and q^k are independent, and become dependent only when we impose the Euler-Lagrange system, which is of order one. If we do not assume the $\dot{q}^k$ independent of the q^k, the first line in (7.36) would be hard to interpret.

(3) The superiority of the Euler-Lagrange equations over Newton's equations is mainly based on the fact the former are *pure equations of motion* automatically, for they do not involve constraint reactions. Moreover, as we will see shortly, they can be put in normal form, so they abide by the main assumption of the Existence and uniqueness theorem. Practically, the Euler-Lagrange equations can be written in any local coordinates $t, q^1, \ldots, q^n$. This simplifies enormously the problem's maths. Later we shall return to this procedure and make it more precise from a differential-geometric point of view.

(4) As seen in the proof, having ideal constraints (for the configurations reached by a world line of the physical system) is equivalent to (7.39). The latter can be

written as:

$$\sum_{i=1}^{N} \boldsymbol{\phi}_i \cdot \frac{\partial \mathbf{x}_i}{\partial q^k} = 0 \,,$$

which we may understand as saying that *the Lagrangian components of ideal constraint reactions are zero (for configurations reached by a world line of the physical system)*.

(5) Theorem 7.16 holds for more complex physical systems, given by finitely many point particles. In fact it is possible to extend the previous result (we will do just that in the next section) to physical systems made of finitely many subsystems, which may be point particles and/or continuous rigid bodies, subject to ideal constraints (both external or internal to the system itself, for example constraints due to hinges between two rigid bodies). ∎

7.3 Extension to Systems of Continuous Rigid Bodies and Point Particles

In this section we will show how the Euler-Lagrange equations holds for more complicated mechanical systems, made of point particles and continuous rigid bodies, when the constraint reactions are ideal.

This is the stage where manuals usually invoke some non-entirely transparent procedure to extend the formalism to physical systems with an infinite number of point particles, and then to continuous, albeit rigid, physical systems. In the author's opinion such a procedure might be valid only within Continuous Mechanics, after one extends Newtonian Mechanics. But it is impossible to justify it only by relying on the treatment of the dynamics of discrete systems we have given in previous chapters. For this reason we shall follow a different, and certainly more rigorous path: we will start from the physical hypothesis C5 introduced in Sect. 5.3, whereby the dynamics of continuous rigid bodies must obey the governing equations (5.25)–(5.26). As a matter of fact we will prove that these two equations, together with the assumption that constraints are ideal, is equivalent to the Euler-Lagrange equations for continuous rigid bodies. Even more, we will extend that to systems made of rigid bodies and point particles, called *articulated systems*.

7.3.1 Articulated Systems

Consider a system $\mathscr{S}$ made of N subsystems $S_1, S_2, \ldots, S_N$, of respective masses $M_1, \ldots, M_N$. Some (or all) subsystems S_k, say the first N', are continuous rigid bodies with centre of mass G_k, while the other N'' are point particles (in which case G_k coincides with the point itself).

The configuration spacetime of $\mathscr{S}$, without constraints, will then be a differentiable manifold $\mathbb{V}^{6N'+3N''+1}$ similar to $\mathbb{V}^{3N+1}$, whose spatial leaves Σ_t^N (isomorphic to $\mathbb{R}^{3N}$) are replaced by (embedded) submanifolds isomorphic to $(\mathbb{R}^3 \times SO(3))^{N'} \times \mathbb{R}^{3N''}$. To have that, we have used that the configurations of a rigid body (with at least three non-collinear points—the only situation we shall consider in the sequel) are given by the points of a differentiable manifold isomorphic to $\mathbb{R}^3 \times SO(3)$ (see Remarks 5.6). The system's evolution will be described in that space by a (differentiable) world line $\Gamma = \Gamma(t) \in \mathbb{V}^{6N'+3N''+1}$.

Suppose the system is subject to $C < 6N' + 3N''$ holonomic constraints, i.e., as for the point particle case, the evolution passes through configurations that kill C functions $f_j : \mathbb{V}^{6N'+3N''+1} \to \mathbb{R}$ and verify the usual two conditions:

(H1) For any $t \in \mathbb{R}$: $f_j(t, X) = 0$, $\quad j = 1, \ldots, C$ on a non-empty set.
(H2) At any instant $t \in \mathbb{R}$, the functions f_j are functionally independent.

Exactly as for constrained systems of point particles, the freedom degrees of $\mathscr{S}$ reduce to $n := 6N' + 3N'' - C$; in other words the **configuration spacetime (with constraints)** is a manifold $\mathbb{V}^{n+1}$ of dimension n, given by the disjoint union of n-dimensional submanifolds $\mathbb{Q}_t$ labelled by absolute time t and defining the system's **spaces of configurations at time** t. The points c_t of any $\mathbb{Q}_t$ are called **configurations** of system $\mathscr{S}$ at time t. The configurations at time t of $\mathscr{S}$ will be locally determined by n **free** or **Lagrangian coordinates** $q^1, \ldots, q^n$ that define local coordinates on $\mathbb{Q}_t$. Together with absolute time t, these local coordinates define **systems of natural local coordinates** on $\mathbb{V}^{n+1}$, exactly as for point particle systems. Every evolution $\Gamma = \Gamma(t) \in \mathbb{V}^{6N'+3N''+1}$ will be equivalent to a (differentiable) curve $\gamma = \gamma(t) \in \mathbb{V}^{n+1}$ given by

$$t = t, \quad q^j = q^j(t), \quad j = 1, \ldots, n$$

in natural local coordinates.

From the point of view of Dynamics, a subsystem S_k is acted upon by active *external* forces[7] $\mathbf{f}_1^{(k)}, \ldots, \mathbf{f}_{L_k}^{(k)}$ applied, respectively, to the points $Q_1^{(k)}, \ldots, Q_{N_k}^{(k)} \in S_k$, and reactive *external* forces $\boldsymbol{\phi}_1^{(k)}, \ldots, \boldsymbol{\phi}_{L_k}^{(k)}$ applied, respectively, to $P_1^{(k)}, \ldots, P_{L_k}^{(k)} \in S_k$ (obviously, if S_k is a point, say G_k, all forces are applied to G_k). We will obviously assume all external forces at play, both active and reactive, satisfy the Action-reaction principle. That is to say, they are assigned on pairs of vectors that at any given instant have same modulus, are directed along the line joining the points of application and are oppositely oriented. As already pointed out when discussing the Euler equations (see Sect. 5.3), the points $P_i^{(k)}$ and $Q_j^{(k)}$ should be understood as

---

[7] These forces may be inertial and hence frame-dependent, but we will not write the subscript $\mathscr{R}$ to keep the notation simpler.

geometric points and not point particles of S_k in case of continuous rigid bodies, since we allow them to vary instant by instant, as particles. Think of the routine situation where S_k is a disc rolling on a guide: the point particle touching the guide changes instant by instant, but the guide's constraint reaction is acting on it. The evolution of any subsystem S_k will satisfy the governing equations (7.47)–(7.48) (or just the first if S_k is a single point particle), where the right-hand side is the total contribution of the active and reactive external forces acting on S_k. We will call such a complex system of physical bodies an **articulated system**.

$\boxed{\text{AC}}$ To wrap up the geometrical aspects we will assume, here as well, condition (H3).

(H3) For any given $t \in \mathbb{R}$, the axis of absolute time, there exist a non-empty open interval $J \subset \mathbb{R}$, with $J \ni t$, and a corresponding collection of natural coordinates $\{(U_i, \phi_i)\}_{i \in I}$ with $\phi_i : U_i \ni p \mapsto (t_i(p), q_i^1(p), \ldots q_i^n(p)) \in \mathbb{R}^{n+1}$, satisfying the following three requests:

(i) $\cup_{i \in I} U_i \supset \mathbb{Q}_\tau$ for any $\tau \in J$;

(ii) $\phi_i(U_i) \supset J \times U_i'$, with $U_i' \subset \mathbb{R}^n$ open, for any $i \in I$;

(iii) for any $i, j \in I$ such that $U_i \cap U_j \neq \varnothing$, the functions $\phi_i \circ \phi_j^{-1} : \phi_j(U_i \cap U_j) \to \phi_i(U_i \cap U_j)$ have the form:

$$t_i = t_j, \quad q_i^k = q_i^k(q_j^1, \ldots, q_j^n)$$

i.e. the transformation rule between free coordinates does not depend explicitly on time.

Remarks 7.18 We are supposing the active external forces work on single points in the rigid bodies S_k making up the system. As observed in Chap. 5 (see (2) Remarks 5.17) it is important to realise that this restriction may include gravitational forces, when the gravitational acceleration is constant $-g\mathbf{e}_z$, with g independent of the position (but possibly dependent on time). *Solely for the purpose of using the two governing equations, the overall action of the constant gravitational force on S_k corresponds to a unique force applied to the system's centre of mass G_k, and equal to the resultant $-M_k g\mathbf{e}_z$ of all gravitational forces on the single particles in the rigid body.* ∎

As in the case of systems of point particles, we can give the following definition. As usual we define:

$$\delta P_i^{(k)} := \sum_{j=1}^n \frac{\partial P_i^{(k)}}{\partial q^j}\bigg|_{c_t} \delta q^j \quad \text{for } \delta q^j \in \mathbb{R} \text{ and } j = 1, \ldots, n. \tag{7.45}$$

Thus $\{\delta P_i^{(k)}\}_{k=1,\ldots,N, i=1,\ldots,L_k}$ determines the generic tangent vector $\delta \mathbf{P}$ to $\mathbb{Q}_t$ at c_t.

Definition 7.19 For an articulated system $\mathscr{S}$, the constraint reactions acting on a world line $\Gamma = \Gamma(t)$ are said **ideal** if, for any time t and any configuration $c_t \in \mathbb{Q}_t$

reached by the world line at time t, the vector formed by the constraint reactions (*external* to the subsystems) is normal to $\mathbb{Q}_t$ at c_t, i.e.:

$$\sum_{k=1}^{N} \sum_{i=1}^{L_k} \boldsymbol{\phi}_i^{(k)} \cdot \delta P_i^{(k)} = 0 \quad \text{for any tangent vector to } \mathbb{Q}_t \text{ at } c_t , \tag{7.46}$$

where, for any k, we only considered constraint reactions $\boldsymbol{\phi}_i^{(k)}$ *external* to subsystem S_k. Constraints are called **ideal** when the constraint reactions are ideal on any world line solving Newton's equations of motion (7.47)–(7.48) (see Sect. 5.3). ◇

Later we will show that under said assumptions, if the system's overall motion satisfies Newton's equations of motion in $\mathscr{R}$, i.e.:

$$M_k \mathbf{a}_{G_k}|_{\mathscr{R}} = \sum_{j=1}^{N_k} \mathbf{f}_j^{(k)} + \sum_{i=1}^{L_k} \boldsymbol{\phi}_i^{(k)} \quad \text{for } k = 1, \ldots, N \tag{7.47}$$

and in case S_k is a continuous rigid body, also:

$$\frac{d}{dt}\bigg|_{\mathscr{R}} \boldsymbol{\Gamma}_{G_k}|_{\mathscr{R}} = \sum_{j=1}^{N_k} (Q_j^{(k)} - G_k) \wedge \mathbf{f}_j^{(k)} + \sum_{i=1}^{L_k} (P_j^{(k)} - G_k) \wedge \boldsymbol{\phi}_i^{(k)} , \quad k = 1, \ldots, N, \tag{7.48}$$

then the same evolution, in terms of Lagrangian coordinates $q^j = q^j(t)$, $j = 1, \ldots, n$, satisfies the Euler-Lagrange equations:

$$\begin{cases} \dfrac{d}{dt} \dfrac{\partial \mathscr{T}|_{\mathscr{R}}}{\partial \dot{q}^k} - \dfrac{\partial \mathscr{T}|_{\mathscr{R}}}{\partial q^k} = \mathcal{Q}_k|_{\mathscr{R}} , \\[2mm] \dfrac{dq^k}{dt} = \dot{q}^k(t) \end{cases} \quad \text{for } k = 1, \ldots, n. \tag{7.49}$$

Above, $\mathscr{T}|_{\mathscr{R}} = \mathscr{T}|_{\mathscr{R}}(t, q^1, \ldots, q^n, \dot{q}^1, \ldots, \dot{q}^n)$ is the kinetic energy of $\mathscr{S}$ in $\mathscr{R}$, obtained adding the kinetic energies of the subsystems:

$$\mathscr{T}|_{\mathscr{R}} = \sum_{k=1}^{N} \mathscr{T}_k|_{\mathscr{R}} , \tag{7.50}$$

and $\mathscr{Q}_k|_{\mathscr{R}}$ are the Lagrangian components of the active forces:

$$\mathscr{Q}_k|_{\mathscr{R}}(t,q^1,\ldots,q^n,\dot{q}^1,\ldots,\dot{q}^n) := \sum_{k=1}^{N}\sum_{i=1}^{N_k}\mathbf{f}_i^{(k)}(t,q^1,\ldots q^n,\dot{q}^1,\ldots,\dot{q}^n)\cdot\frac{\partial\mathbf{x}_i^{(k)}}{\partial q^k}$$

$$(7.51)$$

where $\mathbf{x}_i^{(k)} := Q_i^{(k)} - O$ and O is a point at rest in $\mathscr{R}$.

After that, we will prove how the Euler-Lagrange equations (7.49), together with the hypothesis (7.46) that the constraints are ideal, force (7.47)–(7.48).

To proceed with the proof we need an explicit expression for the vectors $\delta P_i^{(k)}$ and for the kinetic energy $\mathscr{T}_k|_{\mathscr{R}}$ in case S_k is a continuous rigid body. We will also need a couple of identities, that for continuous rigid bodies correspond to the generalisation of the remarkable relationship (7.40).

7.3.2   Computing the Tangent Vectors $\delta P_i^{(k)}$ and the Kinetic Energy of Rigid Bodies

Let S_k be a continuous rigid body. Under this assumption we consider a triple of axes $\mathbf{e}_1^{(k)}, \mathbf{e}_2^{(k)}, \mathbf{e}_3^{(k)}$ moving with S_k and centred at the centre of mass G_k. If $\mathbf{x}'_{P^{(k)}} := P^{(k)} - G_k$ is the position vector of a given point particle $P^{(k)} \in S_k$, we write $\mathbf{x}_{P^{(k)}} := P^{(k)} - O = (P^{(k)} - G_k) + (G_k - O) = \mathbf{x}'_{P^{(k)}} + \mathbf{x}_{G_k}$ for its position vector with respect to the origin O in the rest space of $\mathscr{R}$. The possible values of $\mathbf{x}_{P^{(k)}}$ compatible with the constraints at time t are given by the free coordinates $q^1,\ldots,q^n$ in terms of the function ("solving the constraint equations") $\mathbf{x}_{P^{(k)}} = \mathbf{x}_{P^{(k)}}(t,q^1,\ldots,q^n)$ and similarly $\mathbf{x}_{G_k} = \mathbf{x}_{G_k}(t,q^1,\ldots,q^n)$. Hence:

$$\delta P^{(k)} = \sum_{j=1}^{n}\frac{\partial\mathbf{x}_{G_k}}{\partial q^j}\delta q^j + \sum_{j=1}^{n}\frac{\partial\mathbf{x}'_{P^{(k)}}}{\partial q^j}\delta q^j\,.$$

The second summand on the right can be computed explicitly. Fix Cartesian coordinates on $\mathscr{R}$, write the unit vectors $\mathbf{e}_s^{(k)}$ of S_k at G_k in terms of the "fixed" unit vectors $\mathbf{e}_1, \mathbf{e}_2, \mathbf{e}_3$ at O moving with $\mathscr{R}$. As the free coordinates and time vary, the unit vectors $\mathbf{e}_s^{(k)}$ will change orientation with respect to the "fixed" vectors $\mathbf{e}_1, \mathbf{e}_2, \mathbf{e}_3$ according to functions $\mathbf{e}_s^{(k)} = \mathbf{e}_s^{(k)}(t,q^1,\ldots,q^n)$. (All functions in $t, q^1,\ldots,q^n$ will be at least C^2). Proceeding as in the proof of the Poisson formulas (see the proof of Theorem 2.26), replacing in that proof the derivative $\frac{d}{dt}\big|_{\mathscr{R}}$ with $\frac{\partial}{\partial q^k}$ we obtain:

$$\frac{\partial\mathbf{x}'_{P^{(k)}}}{\partial q^j} = \mathbf{\Omega}_j^{(k)} \wedge (P^{(k)} - G_k)\,,\qquad(7.52)$$

where:

$$\mathbf{\Omega}_j^{(k)}(t, q^1, \ldots, q^n) := \frac{1}{2} \sum_{s=1}^{3} \mathbf{e}_s^{(k)} \wedge \frac{\partial \mathbf{e}_s^{(k)}}{\partial q^j} . \tag{7.53}$$

Putting:

$$\mathbf{T}_j^{(k)}(t, q^1, \ldots, q^n) := \frac{\partial \mathbf{x}_{G_k}}{\partial q^j} , \tag{7.54}$$

we conclude that the vectors $\delta P_i^{(k)}$ satisfy the explicit formula:

$$\delta P^{(k)} = \sum_{j=1}^{n} \mathbf{T}_j^{(k)} \delta q^j + \sum_{j=1}^{n} \mathbf{\Omega}_j^{(k)} \wedge (P^{(k)} - G_k) \delta q^j . \tag{7.55}$$

Note that if S_k is a point particle (coinciding with G_k), the second term on the right vanishes and we recover the usual expression for the vector $\delta P^{(k)}$ of the point particle G_k. It will be useful later to observe that the same computation giving (7.55) also shows that for any point particle $P \in S_k$, the position vector $\mathbf{x}_{P^{(k)}}$ with respect to the origin O of $\mathcal{R}$ satisfies:

$$\frac{\partial \mathbf{x}_{P^{(k)}}}{\partial q^j} = \mathbf{T}_j^{(k)} + \mathbf{\Omega}_j^{(k)} \wedge (P^{(k)} - G_k) . \tag{7.56}$$

The kinetic energy $\mathcal{T}_k|_{\mathcal{R}}$ will be given by (5.10):

$$\mathcal{T}_k|_{\mathcal{R}} = \frac{1}{2} M_k \mathbf{v}_{G_k}|_{\mathcal{R}}^2 + \frac{1}{2} \boldsymbol{\omega}_{\mathcal{R}_{S_k}}|_{\mathcal{R}} \cdot I_{G_k}(\boldsymbol{\omega}_{\mathcal{R}_{S_k}}|_{\mathcal{R}}) \tag{7.57}$$

where I_{G_k} is the inertia tensor of the rigid body S_k and its ω vector in $\mathcal{R}$ will be given (in terms of free coordinates and of the above axes) by:

$$\boldsymbol{\omega}_{\mathcal{R}_{S_k}}|_{\mathcal{R}}(t, q^1, \ldots, q^n, \dot{q}^1, \ldots, \dot{q}^n) = \frac{1}{2} \sum_{i=1}^{3} \mathbf{e}_i^{(k)} \wedge \left(\sum_{j=1}^{n} \frac{\partial \mathbf{e}_i^{(k)}}{\partial q^j} \dot{q}^j + \frac{\partial \mathbf{e}_i^{(k)}}{\partial t} \right) . \tag{7.58}$$

7.3.3 Generalisation of Identity (7.40) to Continuous Rigid Bodies

Consider a continuous rigid body S_k and define, relatively to (7.57):

$$\mathscr{T}_k'|_{\mathscr{R}} := \frac{1}{2} M_k \mathbf{v}_{G_k}|_{\mathscr{R}}^2 , \qquad \mathscr{T}_k''|_{\mathscr{R}} := \frac{1}{2} \boldsymbol{\omega}_{\mathscr{R}_{S_k}}|_{\mathscr{R}} \cdot I_{G_k}(\boldsymbol{\omega}_{\mathscr{R}_{S_k}}|_{\mathscr{R}}) . \tag{7.59}$$

We want to show that, with these definitions and using (7.53) and (7.54), we have:

$$M_k \mathbf{a}_{G_k}|_{\mathscr{R}} \cdot \mathbf{T}_j^{(k)} = \frac{d}{dt}\left(\frac{\partial \mathscr{T}_k'|_{\mathscr{R}}}{\partial \dot{q}^j}\right) - \frac{\partial \mathscr{T}_k'|_{\mathscr{R}}}{\partial q^j} \quad \text{if} \quad \frac{dq^j}{dt} = \dot{q}^j \quad j = 1, \dots, n , \tag{7.60}$$

$$\frac{d\boldsymbol{\Gamma}_{G_k}|_{\mathscr{R}}}{dt}\bigg|_{\mathscr{R}} \cdot \boldsymbol{\Omega}_j^{(k)} = \frac{d}{dt}\left(\frac{\partial \mathscr{T}_k''|_{\mathscr{R}}}{\partial \dot{q}^j}\right) - \frac{\partial \mathscr{T}_k''|_{\mathscr{R}}}{\partial q^j} \quad \text{if} \quad \frac{dq^j}{dt} = \dot{q}^j \quad j = 1, \dots, n . \tag{7.61}$$

If S_k is a single point particle (coinciding with G_k) the second relation is meaningless, while the first one reduces to (7.40). The proof of (7.60) is identical to that of (7.40) so we omit it, and we pass to (7.61). First of all let us choose the basis $\mathbf{e}_i^{(k)}$ to be a principal triple of inertia for S_k, with moments of inertia I_1, I_2, I_3. Henceforth, to simplify the notation, we will drop the labels $^{(k)}$, $_{G_k}$, $_{\mathscr{R}}$, so for instance $\mathbf{e}_i$ will stand for $\mathbf{e}_i^{(k)}$. To prove (7.61) observe that (7.58) implies:

$$\frac{\partial \boldsymbol{\omega}}{\partial \dot{q}^j} = \frac{1}{2}\sum_{i=1}^{3} \mathbf{e}_i \wedge \frac{d\mathbf{e}_i}{dt} = \boldsymbol{\Omega}_j . \tag{7.62}$$

Looking at the right-hand side of (7.61) we have:

$$\frac{d\boldsymbol{\Gamma}}{dt}\cdot\boldsymbol{\Omega}_j = \frac{d\boldsymbol{\Gamma}}{dt}\cdot\frac{\partial\boldsymbol{\omega}}{\partial\dot{q}^j} = \frac{dI(\boldsymbol{\omega})}{dt}\cdot\frac{\partial\boldsymbol{\omega}}{\partial\dot{q}^j} = \frac{d}{dt}\left(I(\boldsymbol{\omega})\cdot\frac{\partial\boldsymbol{\omega}}{\partial\dot{q}^j}\right) - \left(\frac{d}{dt}\frac{\partial\boldsymbol{\omega}}{\partial\dot{q}^j}\right)\cdot I(\boldsymbol{\omega}) .$$

It is immediate to see, taking the components of the diagonal inertia operator I, that:

$$I(\boldsymbol{\omega})\cdot\frac{\partial\boldsymbol{\omega}}{\partial\dot{q}^j} = \sum_{i=1}^{3} I_i \omega^i \frac{\partial\omega^i}{\partial\dot{q}^j} = \frac{\partial}{\partial\dot{q}^j}\left(\frac{1}{2}\sum_{i=1}^{3}\omega^i I_i \omega^i\right) = \frac{\partial}{\partial\dot{q}^j}\left(\frac{1}{2}\boldsymbol{\omega}\cdot I(\boldsymbol{\omega})\right) ,$$

and then:

$$\frac{d\boldsymbol{\Gamma}}{dt}\cdot\boldsymbol{\Omega}_j = \frac{d}{dt}\frac{\partial}{\partial\dot{q}^j}\left(\frac{1}{2}\boldsymbol{\omega}\cdot I(\boldsymbol{\omega})\right) - \left(\frac{d}{dt}\frac{\partial\boldsymbol{\omega}}{\partial\dot{q}^j}\right)\cdot I(\boldsymbol{\omega}) .$$

The right side is precisely what appears in (7.61) provided that:

$$\left(\frac{d}{dt}\frac{\partial \boldsymbol{\omega}}{\partial \dot{q}^j}\right) \cdot I(\boldsymbol{\omega}) = \frac{\partial}{\partial q^j}\left(\frac{1}{2}\boldsymbol{\omega} \cdot I(\boldsymbol{\omega})\right) . \tag{7.63}$$

The latter immediately follows from the remarkable identity:

$$\frac{d}{dt}\frac{\partial \boldsymbol{\omega}}{\partial \dot{q}^j} = \sum_{k=1}^{3} \frac{\partial \omega^k}{\partial q^j}\mathbf{e}_k . \tag{7.64}$$

In fact, using the above identity on the left of (7.63) and the fact that the orthonormal $\mathbf{e}_i$ are a principal triple of inertia:

$$\left(\frac{d}{dt}\frac{\partial \boldsymbol{\omega}}{\partial \dot{q}^j}\right) \cdot I(\boldsymbol{\omega}) = \sum_{i=1}^{3} \frac{\partial \omega^i}{\partial q^j}\mathbf{e}_i \cdot \sum_{k=1}^{3} I_k \omega_k \mathbf{e}_k = \sum_{i=1}^{3} \frac{\partial \omega^i}{\partial q^j}I_i \omega_i = \frac{\partial}{\partial q^j}\sum_{i=1}^{3}\frac{1}{2}\omega_i I_i \omega_i$$

$$= \frac{\partial}{\partial q^j}\left(\frac{1}{2}\boldsymbol{\omega} \cdot I(\boldsymbol{\omega})\right) ,$$

i.e. exactly (7.63). To finish the proof let us show (7.64). The left-hand side equals:

$$\frac{d}{dt}\frac{\partial \boldsymbol{\omega}}{\partial \dot{q}^j} = \frac{d}{dt}\left(\frac{1}{2}\sum_{i=1}^{3}\mathbf{e}_i \wedge \frac{\partial \mathbf{e}_i}{\partial q^j}\right) = \frac{1}{2}\sum_{i=1}^{3}\frac{d\mathbf{e}_i}{dt}\wedge\frac{\partial \mathbf{e}_i}{\partial q^j} + \frac{1}{2}\sum_{i=1}^{3}\mathbf{e}_i \wedge \frac{\partial}{\partial q^j}\frac{d\mathbf{e}_i}{dt}$$

$$= \frac{1}{2}\sum_{i=1}^{3}\frac{d\mathbf{e}_i}{dt}\wedge\frac{\partial \mathbf{e}_i}{\partial q^j} + \frac{\partial}{\partial q^j}\left(\frac{1}{2}\sum_{i=1}^{3}\mathbf{e}_i \wedge \frac{d\mathbf{e}_i}{dt}\right) - \frac{1}{2}\sum_{i=1}^{3}\frac{\partial \mathbf{e}_i}{\partial q^j}\wedge\frac{d\mathbf{e}_i}{dt} .$$

The first and last terms are equal, so:

$$\frac{d}{dt}\frac{\partial \boldsymbol{\omega}}{\partial \dot{q}^j} = \sum_{i=1}^{3}\frac{d\mathbf{e}_i}{dt}\wedge\frac{\partial \mathbf{e}_i}{\partial q^j} + \frac{\partial \boldsymbol{\omega}}{\partial q^j} . \tag{7.65}$$

Furthermore:

$$\sum_{i=1}^{3}\frac{d\mathbf{e}_i}{dt}\wedge\frac{\partial \mathbf{e}_i}{\partial q^j} = \sum_{i=1}^{3}(\boldsymbol{\omega}\wedge\mathbf{e}_i)\wedge\frac{\partial \mathbf{e}_i}{\partial q^j} = -\sum_{i=1}^{3}\left(\mathbf{e}_i \cdot \frac{\partial \mathbf{e}_i}{\partial q^j}\right)\boldsymbol{\omega} + \sum_{i=1}^{3}\left(\frac{\partial \mathbf{e}_i}{\partial q^j}\cdot\boldsymbol{\omega}\right)\mathbf{e}_i .$$

Since $\mathbf{e}_i \cdot \frac{\partial \mathbf{e}_i}{\partial q^j} = \frac{1}{2} \frac{\partial}{\partial q^j} \mathbf{e}_i \cdot \mathbf{e}_i = 0$, we have:

$$\sum_{i=1}^{3} \frac{d\mathbf{e}_i}{dt} \wedge \frac{\partial \mathbf{e}_i}{\partial q^j} = \sum_{i=1}^{3} \left(\frac{\partial \mathbf{e}_i}{\partial q^j} \cdot \boldsymbol{\omega} \right) \mathbf{e}_i = \sum_{i=1}^{3} \left[\frac{\partial}{\partial q^j} (\mathbf{e}_i \cdot \boldsymbol{\omega}) \right] \mathbf{e}_i - \sum_{i=1}^{3} \left(\mathbf{e}_i \cdot \frac{\partial \boldsymbol{\omega}}{\partial q^j} \right) \mathbf{e}_i ,$$

i.e.

$$\sum_{i=1}^{3} \frac{d\mathbf{e}_i}{dt} \wedge \frac{\partial \mathbf{e}_i}{\partial q^j} = \sum_{i=1}^{3} \frac{\partial \omega^i}{\partial q^j} \mathbf{e}_i - \frac{\partial \boldsymbol{\omega}}{\partial q^j} .$$

Inserting that in the right-hand side of (7.65) gives (7.64).   □

7.3.4 Euler-Lagrange Equations for Articulated Systems

To close the section, we can finally prove the following theorem, which relates the Newtonian and Lagrangian Mechanics of articulated systems subject to ideal constraints.

Theorem 7.20 *Let $\mathscr{S}$ be an articulated system, as in Sect. 7.3.1, subject to $C <$ $6N'+3N''$ holonomic constraints of class C^2 at least, described in the configuration spacetime $\mathbb{V}^{n+1}$. If a world line $\Gamma = \Gamma(t) \in \mathbb{V}^{6N'+3N''+1}$ of $\mathscr{S}$ satisfies Newton's equations of motion (7.47)–(7.48) with ideal constraint reactions, then the curve $\gamma = \gamma(t) \in \mathbb{V}^{n+1}$, corresponding to Γ and given by $q^k = q^k(t)$ in natural local coordinates, satisfies the Euler-Lagrange equations (7.49).*

Proof For starters, by (7.55) we can express the fact that (7.46) are ideal for the configurations reached by Γ as follows:

$$\sum_{j=1}^{n} \left(\sum_{k=1}^{N} \sum_{i=1}^{L_k} \boldsymbol{\phi}_i^{(k)} \cdot \mathbf{T}_j^{(k)} + \sum_{k=1}^{N} \sum_{i=1}^{L_k} \boldsymbol{\Omega}_j^{(k)} \wedge (P_i^{(k)} - G_k) \cdot \boldsymbol{\phi}_i^{(k)} \right) \delta q^j = 0 ,$$

$$\forall \delta q^j \in \mathbb{R}, \, j = 1, \ldots, n ,$$

for any configuration on γ. As usual, since the δq^j are arbitrary, the above are equivalent to the n conditions:

$$\sum_{k=1}^{N} \sum_{i=1}^{L_k} \boldsymbol{\phi}_i^{(k)} \cdot \mathbf{T}_j^{(k)} + \sum_{k=1}^{N} \sum_{i=1}^{L_k} (P_i^{(k)} - G_k) \wedge \boldsymbol{\phi}_i^{(k)} \cdot \boldsymbol{\Omega}_j^{(k)} = 0, \, j = 1, \ldots, n \quad (7.66)$$

where we have used the elementary relation $\mathbf{a} \wedge \mathbf{b} \cdot \mathbf{c} = \mathbf{b} \wedge \mathbf{c} \cdot \mathbf{a}$. As the above hold on any motion satisfying the governing equations (7.47) and (7.48), for $j = 1, \ldots, n$

we will also have:

$$\sum_{k=1}^{N} \left(M_k \mathbf{a}_{G_k}|_{\mathscr{R}} - \sum_{r=1}^{N_k} \mathbf{f}_r^{(k)} \right) \cdot \mathbf{T}_j^{(k)}$$

$$+ \sum_{k=1}^{N} \left(\frac{d}{dt}\bigg|_{\mathscr{R}} \Gamma_{G_k}|_{\mathscr{R}} - \sum_{j=1}^{N_k} (Q_j^{(k)} - G_k) \wedge \mathbf{f}_j^{(k)} \right) \cdot \mathbf{\Omega}_j^{(k)} = 0 \,.$$

Remembering (7.60) and (7.61) and that $\mathscr{T}|_{\mathscr{R}} = \sum_k (\mathscr{T}_k'|_{\mathscr{R}} + \mathscr{T}_k''|_{\mathscr{R}})$, the previous equations can be written without $|_\gamma$ to unburden the notation:

$$\frac{d}{dt}\left(\frac{\partial \mathscr{T}|_{\mathscr{R}}}{\partial \dot{q}^j} \right) - \frac{\partial \mathscr{T}|_{\mathscr{R}}}{\partial q^j} = \sum_{k=1}^{N} \sum_{r=1}^{N_k} \mathbf{f}_r^{(k)} \cdot \mathbf{T}_j^{(k)} + \sum_{r=1}^{N_k} (Q_j^{(k)} - G_k) \wedge \mathbf{f}_r^{(k)} \cdot \mathbf{\Omega}_j^{(k)}$$

where we used the constraint $\frac{dq^j}{dt} = \dot{q}^j$ holding during motion. Using $\mathbf{a} \wedge \mathbf{b} \cdot \mathbf{c} = \mathbf{b} \wedge \mathbf{c} \cdot \mathbf{a}$ again, we can rearrange like this:

$$\frac{d}{dt}\left(\frac{\partial \mathscr{T}|_{\mathscr{R}}}{\partial \dot{q}^j} \right) - \frac{\partial \mathscr{T}|_{\mathscr{R}}}{\partial q^j} = \sum_{k=1}^{N} \sum_{r=1}^{N_k} \mathbf{f}_r^{(k)} \cdot \mathbf{T}_j^{(k)} + \sum_{r=1}^{N_k} \mathbf{\Omega}_j^{(k)} \wedge (Q_r^{(k)} - G_k) \cdot \mathbf{f}_j^{(k)} \,.$$

From (7.56), we can then rewrite as, for $j = 1, \ldots, n$:

$$\frac{d}{dt}\left(\frac{\partial \mathscr{T}|_{\mathscr{R}}}{\partial \dot{q}^j} \right) - \frac{\partial \mathscr{T}|_{\mathscr{R}}}{\partial q^j} = \sum_{k=1}^{N} \sum_{r=1}^{N_k} \mathbf{f}_i^{(k)} \cdot \frac{\partial \mathbf{x}_{Q_i^{(k)}}}{\partial q^j} \,.$$

By definition (7.51), where $\mathbf{x}_{Q_i^{(k)}}$ is called $\mathbf{x}_i^{(k)}$, we have proved that the time evolution of the articulated system $\mathscr{S}$, under the given assumptions, solves the Euler-Lagrange equations (7.49):

$$\frac{d}{dt}\left(\frac{\partial \mathscr{T}|_{\mathscr{R}}}{\partial \dot{q}^j}\bigg|_{\gamma(t)} \right) - \frac{\partial \mathscr{T}|_{\mathscr{R}}}{\partial q^j}\bigg|_{\gamma(t)} = \mathscr{Q}_j(\gamma(t)), \quad \frac{dq^j}{dt} = \dot{q}^j, \quad \text{for} \quad j = 1, \ldots, n.$$

$\square$

Remarks 7.21

(1) The internal constraint reactions of any rigid body in $\mathscr{S}$ are difficult to handle in the continuous case, because they should be described by means of force densities rather than true forces. For rigid bodies made of finitely many particles, internal reactions are ideal irrespective of the other constraint reactions of the large system, as clarified in (3) Examples 7.15. In the continuous case it would be necessary to involve densities of forces, and use the notion of *stress tensor* that we will not address here. Moreover, without any further information on the constitutive characterisation of the rigid body, it is impossible to compute explicitly the constraint reactions on any internal point of a rigid body made by sufficiently many particles, even if the system's motion is known. That being said, it is important to stress that in our treatment, to arrive at the Euler-Lagrange equations for constrained systems we have assumed (7.46) *only* for the *non-internal* constraint reactions in each body of the system, which are caused by the rigidity constraint, and for the mutual constraint reactions between different bodies, or between bodies and external guides.

(2) We must say that the model of articulated body is still rather "abstract" and does not allow to treat physically relevant situations such as the collisions between the (rigid) bodies in the system, since in these cases there appear impulsive reactive forces (instantaneous, in theory) which were ignored in our model. Furthermore, in the same way, friction phenomena (constraint reactions due to friction) are usually impossible to handle in this simplified model, apart from special cases like two rolling bodies (rolling against one another or onto an external guide), for this is subsumed by the notion of holonomic constraint if the rolling constraint is integrable.

7.4 Elementary Properties of the Euler-Lagrange Equations

In this section we examine the most relevant among the elementary properties of the Euler-Lagrange equations. For simplicity we discuss physical systems made by a finite number of point particles, although almost everything carries over to systems made of subsystems consisting of point particles and/or extended rigid bodies, with suitable adaptations.   ∎

7.4.1 Normality of Euler-Lagrange Equations and Existence and Uniqueness Theorem

We will show that the Euler-Lagrange equations can always be put in normal form. We will do so for systems of point particles. The normal form, as we know, ensures that the equations admit *a unique solution for given initial conditions*, under suitable regularity assumptions on the known functions appearing in the Euler-Lagrange equations.

Recalling the kinetic energy formula (7.32), the Euler-Lagrange equations have the explicit form (note $a_{hk} = a_{kh}$):

$$\sum_{h=1}^{n} 2a_{kh}(t, q^1(t), \ldots, q^n(t))\frac{d^2q^h}{dt^2} = G_k\left(t, q^1(t), \ldots, q^n(t), \frac{dq^1}{dt}, \ldots, \frac{dq^n}{dt}\right),$$

$$k = 1, \ldots, n,\tag{7.67}$$

where the G_k arise by adding the explicit form of the $\mathcal{Q}_k$, $-2\sum_h \dfrac{da_{kh}}{dt}\dfrac{dq^h}{dt}$ and the derivatives of $\mathcal{T}_1|_{\mathcal{R}}$, $\mathcal{T}_0|_{\mathcal{R}}$. If the constraints are at least of class C^3 and the active forces at least C^1, then the coefficients $a_{kh} = a_{kh}(t, q^1, \ldots, q^n)$ are C^2 and the right-hand side of (7.67) are C^1. If we show the square matrix $a(t, q^1, \ldots, q^n)$ of the coefficients $a_{kh}(t, q^1, \ldots, q^n)$ is invertible for any $(t, q^1, \ldots, q^n)$, automatically the Euler-Lagrange equations can be put in normal form with C^1 right-hand-side term, and this in turn guarantees the corresponding Cauchy problem has exactly one solution. More precisely, by Cramer's rule for the inverse, any $(a^{-1})_{kh} = (a^{-1})_{kh}(t, q^1, \ldots, q^n)$ is as differentiable as the $a_{rs}(1, q^1, \ldots, q^n)$, i.e. C^1, being a quotient of polynomials in the variables $a_{rs}(1, q^1, \ldots, q^n)$ divided by a polynomial in the same variables, namely the (non-zero) determinant of a. We could, as said, write (7.67) in normal form with C^2 source in the variables t, q^k and $\dot{q}^k (= \frac{dq^k}{dt})$, the latter independent of the former:

$$\frac{d^2q^h}{dt^2} = \sum_{k=1}^{n} 2(a^{-1})_{hk}(t, q^1(t), \ldots, q^n(t))G_k$$

$$\times \left(t, q^1(t), \ldots, q^n(t), \frac{dq^1}{dt}, \ldots, \frac{dq^n}{dt}\right), \quad h = 1, \ldots, n,\tag{7.68}$$

Note that if we change free coordinates $q'^h = q'^h(t, q^1, \ldots, q^n)$ (and $t' = t+$ constant), the coefficients of the new square matrix a', relative to the new coordinates, obey: $a'_{ij} = \sum_{h,k} \frac{\partial q^k}{\partial q'^i}\frac{\partial q^h}{\partial q'^j}a_{kh}$. As the Jacobian matrix of coefficients $\frac{\partial q^r}{\partial q'^s}$ is non-singular, a is invertible if and only if a' is invertible. Consider then the *kernel* of a, for some Lagrangian coordinates $t, q^1, \ldots, q^n$ which we will specify shortly. The claim is that the kernel only contains the zero vector, making a bijective.

Now, $(c^1, \ldots, c^n) \in \mathbb{R}^m$ belongs to $\ker(a)$ if and only if $\sum_{h=1}^{n} a_{kh}c^h = 0$ for any

$k = 1, \ldots, n$, in which case: $\displaystyle\sum_{h,k=1}^{n} c^k c^h a_{kh} = 0$. From (7.33) for the matrix a, the above is equivalent to:

$$\sum_{i=1}^{N} m_i \left(\sum_h c^h \frac{\partial \mathbf{x}_i}{\partial q^h} \right) \cdot \left(\sum_k c^k \frac{\partial \mathbf{x}_i}{\partial q^k} \right) = 0, \quad \text{that is,} \quad \sum_{i=1}^{N} m_i \left\| \sum_h c^h \frac{\partial \mathbf{x}_i}{\partial q^h} \right\|^2 = 0.$$

As the masses m_i are strictly positive, necessarily

$$\sum_{h=1}^{n} c^h \frac{\partial x_i^j}{\partial q^h} = \mathbf{0}, \quad \text{for any } i = 1, \ldots N \text{ and } j = 1, 2, 3. \tag{7.69}$$

Now we will show that (7.69), applied to a for special Lagrangian coordinates, implies $c^1 = c^2 = \cdots = c^n = 0$, i.e. the kernel of a is trivial. By Proposition 7.5 the coordinates $t, q^1, \ldots, q^n$ can be chosen, locally, to be the absolute time t, and n among the x_i^j. Without loss of generality (it suffices to rename the x_i^j) we can assume the free coordinates are $q^1 := x_1^1, q^2 := x_1^2$ and so forth up to q^n. Choosing $x_i^j = x_1^1$, (7.69) reduces to $c^1 = 0$; for $x_i^j = x_1^2$, (7.69) reduces to $c^2 = 0$ et cetera, up to $c^n = 0$. Hence the square matrix a is non-singular. Since:

$$\sum_{h,k=1}^{n} c^k c^h a_{kh} = \sum_{i=1}^{N} m_i \left\| \sum_h c^h \frac{\partial \mathbf{x}_i}{\partial q^h} \right\|^2 \geq 0,$$

we have also shown that the matrix of coefficients a_{kh} is *positive definite*.

Remarks 7.22 $\boxed{\text{AC}}$ A quicker way to obtain $c^1 = c^2 = \cdots c^n = 0$, which involves tangent vectors to manifolds, goes as follows. As the constraints are functionally independent, $q^1, \ldots, q^n$ are coordinates on the submanifold $\mathbb{Q}_t$. Using the Cartesian coordinates $x_1^1, x_1^2, x_1^3, \ldots, x_N^1, x_N^2, x_N^3$ on Σ_t^N, identity (7.69) reads:

$$\sum_{h=1}^{n} c^h \sum_{j,i} \frac{\partial x_i^j}{\partial q^h} \frac{\partial}{\partial x_i^j} = \mathbf{0}, \quad \text{which we may recast using (B.8) as:} \quad \sum_{h=1}^{n} c^h \frac{\partial}{\partial q^h} = \mathbf{0}.$$

This is possible only when all coefficients c^h vanish, since the $\frac{\partial}{\partial q^h}$ are a basis of the tangent space $T_{c_t}\mathbb{Q}_t$, hence linearly independent. ∎

We have then proved the following result, which keeps Theorem 14.16 and Proposition 14.15 into account.

Theorem 7.23 (Existence and Uniqueness for Euler-Lagrange Equations) *Under the assumptions of Theorem 7.16, in any natural local coordinates $(t, q^1, \ldots, q^n)$ on $\mathbb{V}^{n+1}$ the $n \times n$ matrix (7.33) with coefficients $a_{kh}(t, q^1, \ldots, q^n)$ is positive definite for any $(t, q^1, \ldots, q^n)$. Consequently the Euler-Lagrange* *equations (7.36) can always be put in normal form:*

$$\frac{d^2 q^k}{dt^2} = z^k \left(t, q^1(t), \ldots, q^n(t), \frac{dq^1}{dt}(t), \ldots, \frac{dq^n}{dt}(t) \right) \quad \text{for any } k = 1, \ldots, n.$$

$$(7.70)$$

The right-hand side is a C^1 function in all its variables (seen as independent) when $\mathscr{T}|_{\mathscr{R}}$ is of class[8] C^2 and the Lagrangian components $\mathscr{Q}_k$ of the active forces are at least C^1 in the coordinates[9] $(t, q^1, \ldots, q^n, \dot{q}^1, \ldots \dot{q}^n)$. If so, in the given chart there is global existence and uniqueness (Theorem 14.23) for the Euler-Lagrange system (7.36) with given Cauchy data $q^k(t_0) = q_0^k$, $\frac{dq^k}{dt}(t_0) = \dot{q}_0^k$, $k = 1, \ldots, n$.

Remarks 7.24

(1) The above result trivially extends to physical system made by point particles and/or continuous rigid bodies when the inertia tensor of the centre of mass of any continuous rigid body is positive definite, so that the matrix of coefficients a_{hk} is positive definite and therefore invertible. Consider a physical system made of a single continuous rigid body *not subject to constraints* and with positive-definite barycentric inertia tensor. It was shown earlier that the Euler-Lagrange equations descend from Newton's equations of motion—i.e. the two governing equations of the Dynamics of systems (see Sect. 5.3)—and from having ideal constraints. Hence the previous existence and uniqueness result for a continuous rigid body without constraints implies a similar result for the ODE system made by the governing equations of Dynamics. All of this, naturally, if the forces are sufficiently regular.

(2) In the concrete situation of a physical system of N point particles (articulated systems are similar) subject to ideal holonomic constraints, the initial conditions in the physical space are:

$$\mathbf{x}_i(t_0) = \mathbf{x}_{i0}, \quad \mathbf{v}_i(t_0) = \mathbf{v}_{i0}, \quad \text{for } i = 1, \ldots, N.$$

The position vectors and the velocities refer to some frame $\mathscr{R}$ that we will not mention explicitly, just to simplify the notation. Given natural local coordinates (U, ψ) on $\mathbb{V}^{n+1}$, i.e. $\psi : U \ni c \mapsto (t(c), q^1(c), \ldots, q^n(c)) \in \psi(U) \subset \mathbb{R}^{n+1}$, Theorem 7.23 wants initial conditions in those coordinates:

$$q^k(t_0) = q_0^k, \quad \frac{dq^k}{dt}(t_0) = \dot{q}_0^k, \quad \text{for } k = 1, \ldots, n.$$

How do we get this initial data from the previous data?

---

[8] This happens in particular if the constraints are at least C^3 on $\mathbb{V}^{N+1}$.

[9] This occurs in particular if the active forces $\mathbf{F}_{\mathscr{R}i}$ are at least C^1 with respect to time and to the particles' positions and velocities.

First, using $\mathbf{x}_i = \mathbf{x}_i(t, q^1, \ldots, q^n)$ we recover from the vectors $\mathbf{x}_{i0}$ the numbers $\dot{q}_0^k$, because the map given by the N functions $\psi(U) \ni (t, q^1, \ldots, q^n) \mapsto \mathbf{x}_i(t, q^1, \ldots, q^n) \in \mathbb{V}^4$, $i = 1, \ldots, n$, is 1-1 by definition of local chart on $\mathbb{V}^{n+1} \subset \mathbb{V}^{3N+1}$. Secondly,

$$
\mathbf{v}_{i0} = \frac{\partial \mathbf{x}_i}{\partial t}\Big|_{(t_0, q_0^1, \ldots, q_0^n)} + \sum_{k=1}^n \frac{\partial \mathbf{x}_i}{\partial q^k}\Big|_{(t_0, q_0^1, \ldots, q_0^n)} \dot{q}_0^k,
$$

for some $(\dot{q}_0^1, \ldots, \dot{q}_0^n) \in \mathbb{R}^n$. The only issue is whether this n-tuple is unique. If the identity still holds when we insert $(\dot{q}_1^1, \ldots, \dot{q}_1^n)$ in the above identity in place of $(\dot{q}_0^1, \ldots, \dot{q}_0^n)$, then putting $c^k := \dot{q}_1^k - \dot{q}_0^k$ we must have

$$
\sum_{k=1}^n c^k \frac{\partial \mathbf{x}_i}{\partial q^k}\Big|_{(t_0, q_0^1, \ldots, q_0^n)} = 0 \quad \text{for } i = 1, \ldots, N.
$$

The same argument used to prove Theorem 7.23 immediately gives $c^k = 0$ for any $k = 1, \ldots, n$ when the local coordinates $(t, q^1, \ldots, q^n)$ are chosen suitably around $(t_0, \mathbf{x}_{01}, \ldots, \mathbf{x}_{0N}) \in \mathbb{Q}_{t_0}$. In reality the argument is completely general, because for different natural coordinates around $(t_0, \mathbf{x}_{01}, \ldots, \mathbf{x}_{0N})$ the new coefficients c^k will come from the initial ones via a non-singular Jacobian matrix, so they will still be zero. All in all $(\dot{q}_0^1, \ldots, \dot{q}_0^n)$ is completely determined by the initial velocities $\mathbf{v}_i(t_0) = \mathbf{v}_{i0}$ for $i = 1, \ldots, N$.

∎

7.4.2 Spacetime of Kinetic States $A(\mathbb{V}^{n+1})$

When we wrote the Euler-Lagrange equations (7.36) on $\mathbb{V}^{n+1}$ we treated the coordinates $\dot{q}^k$ as independent of the q^k, and their dependency was fixed precisely by the n "trivial" equations in the second line of (7.36). Here we would like to discuss the nature of the coordinates $\dot{q}^k$ once and for all. Consider two opens sets $U, V \subset \mathbb{V}^{n+1}$ with natural coordinates $t, q^1, \ldots, q^n$ and $t', q'^1, \ldots, q'^n$ respectively, and $U \cap V \neq \varnothing$. The coordinate sets are related on $U \cap V$ by the transformation rule (7.15)–(7.16). Let us examine the ensuing relationship between the dotted coordinates. As we said above the dotted coordinates are *independent* of the undotted ones, but equal the latter's derivatives along any solution of the Euler-Lagrange equations: if $q^k = q^k(t)$ solve the Euler-Lagrange equations, then at any instant

$$
\dot{q}'^k = \frac{dq'^k}{dt'} = \frac{\partial q'^k}{\partial t} + \sum_{h=1}^n \frac{\partial q'^k}{\partial q^h} \frac{dq^h}{dt} = \frac{\partial q'^k}{\partial t} + \sum_{h=1}^n \frac{\partial q'^k}{\partial q^h} \dot{q}^h.
$$

We may think that the left-most and right-most sides are equal *even if no motion is given*, since these identities would anyway reduce on any motion to the correct conditions for the equations of motion. So we shall assume that the dotted coordinates are independent of the undotted ones, and also that, when changing coordinates:

$$t' = t + c \,, \tag{7.71}$$

$$q'^k = q'^k(t, q^1, \ldots, q^n) \,, \tag{7.72}$$

$$\dot{q}'^k = \frac{\partial q'^k}{\partial t} + \sum_{h=1}^{n} \frac{\partial q'^k}{\partial q^h} \dot{q}^h \,. \tag{7.73}$$

This discussion leads us to the definition of *spacetime of kinetic states*, where the relations among the coordinates are *assumed by definition* as the transformation rules in a privileged atlas. This space is therefore the natural ambient to make sense of the Euler-Lagrange equations when we suppose the dotted coordinates are true coordinates, independent of the undotted ones. The construction and meaning of such space within the theory of fibre bundles is provided by Sect. 7.5.

Definition 7.25 (Spacetime of Kinetic States) The **spacetime of kinetic states** $A(\mathbb{V}^{n+1})$, built on the configuration spacetime $\mathbb{V}^{n+1}$, is a C^k manifold of dimension $2n+1$, with $k \geq 2$. It admits a privileged atlas whose local charts are called **systems of natural local coordinates of** $A(\mathbb{V}^{n+1})$. Calling $t, q^1, \ldots, q^n, \dot{q}^1, \ldots, \dot{q}^n$ the coordinates in a generic natural local chart, the following conditions hold.

(1) The coordinates $t, q^1, \ldots, q^n$ are identified with natural local coordinates on $\mathbb{V}^{n+1}$: in particular, t is absolute time up to an additive constant, seen as differentiable and surjective (non-singular) function $T : A(\mathbb{V}^{n+1}) \to \mathbb{R}$, and the remaining $\dot{q}^1, \ldots, \dot{q}^n$, for any $t, q^1, \ldots, q^n$, take values in the *entire* $\mathbb{R}^n$. The above coordinates on $\mathbb{V}^{n+1}$ are uniquely determined by the coordinates on $A(\mathbb{V}^{n+1})$, and any system of natural local coordinates on $\mathbb{V}^{n+1}$ extends to give a system on $A(\mathbb{V}^{n+1})$ as above.
(2) Natural local coordinates on $A(\mathbb{V}^{n+1})$ transform under invertible C^k maps with inverses (7.71), (7.72), and (7.73).

The **space of kinetic states at time** t is by definition the (embedded) $2n$-dimensional submanifold $\mathbb{A}_t := T^{-1}(t)$ of $A(\mathbb{V}^{n+1})$ obtained fixing an absolute time $t \in \mathbb{R}$.[10] A **kinetic state at time** t is by definition a point in $\mathbb{A}_t$. ◇

Remarks 7.26

(1) The $(n+1)$-tuples $(t, q^1, \ldots, q^n)$ vary in open subsets of $\mathbb{R}^{n+1}$ whose shape differs case by case; vice versa the n-tuples $(\dot{q}^1, \ldots, \dot{q}^n)$ always cover the *whole* $\mathbb{R}^n$.

---

[10] $A(\mathbb{V}^{n+1})$ is therefore the disjoint union of the various $\mathbb{A}_t$, $t \in \mathbb{R}$.

(2) Every kinetic-state space at time t, $\mathbb{A}_t$, is the **fibre of** $A(\mathbb{V}^{n+1})$ **at time** t, so
the natural local coordinates on $A(\mathbb{V}^{n+1})$ are local coordinates **adapted to the
fibres** of $A(\mathbb{V}^{n+1})$ in the sense of Definition B.4 (though they are not the only
such). In the assumptions made, it is easy to prove (the reader should do the
exercise!) that for any configuration c_t at time t, considering two sets of natural
local coordinates, the matrix of coefficients $\frac{\partial q'^k}{\partial q^h}|_{c_t}$ is non-singular (it has non-
zero determinant).   ■

The Euler-Lagrange equations (7.36) are now viewed as differential equations
on $A(\mathbb{V}^{n+1})$, for the moment defined only locally, that determine C^1curves $I \ni
t \mapsto \gamma(t) \in A(\mathbb{V}^{n+1})$ parametrised by the special coordinate of absolute time, up
to constants: $T(\gamma(t)) = t + c$. In the natural local chart (U, ψ) with coordinates
$(t, q^1, \ldots, q^n, \dot{q}^1, \ldots, \dot{q}^n)$, the solution γ reads:

$$I \ni t \mapsto (t, q^1(t), \ldots, q^n(t), \dot{q}^1(t), \ldots, \dot{q}^n(t)) \quad \text{with} \quad \dot{q}^k(t) = \frac{dq^k}{dt},$$

$$, k = 1, \ldots, n .$$

Curves of this sort, in terms of the geometry of $A(\mathbb{V}^{n+1})$, are **local sections** of the
bundle $A(\mathbb{V}^{n+1})$ over the base $\mathbb{R}$ of absolute time (Sect. 7.5) (Fig. 7.3)

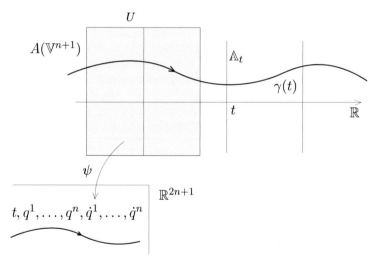

Fig. 7.3 Spacetime of kinetic states $A(\mathbb{V}^{n+1})$

7.4.3 Non-dependency of the Euler-Lagrange Solutions on Coordinates

We pass to examine the independence of the Euler-Lagrange equations' solutions of the choice of local natural coordinates on $A(\mathbb{V}^{n+1})$. We begin with a few elementary technical facts. From (7.71)–(7.73) we immediately obtain:

$$\frac{\partial \dot{q}'^k}{\partial \dot{q}^h} = \frac{\partial q'^k}{\partial q^h}, \tag{7.74}$$

Furthermore, if on the curve $q'^k = q'^k(t)$ we identify dotted variables with the time derivatives of undotted variables, (7.71)–(7.73) imply:

$$\left(\frac{d}{dt} \frac{\partial q^h}{\partial q'^k} \right) = \frac{\partial \dot{q}^h}{\partial q'^k}. \tag{7.75}$$

Assuming transformations (7.71)–(7.73), we find the useful identities:

$$\frac{\partial}{\partial t'} = \frac{\partial}{\partial t} + \sum_{h=1}^{n} \frac{\partial q^h}{\partial t'} \frac{\partial}{\partial q^h} + \sum_{h=1}^{n} \frac{\partial \dot{q}^h}{\partial t'} \frac{\partial}{\partial \dot{q}^h}, \tag{7.76}$$

$$\frac{\partial}{\partial \dot{q}'^k} = \sum_{h=1}^{n} \frac{\partial \dot{q}^h}{\partial \dot{q}'^k} \frac{\partial}{\partial \dot{q}^h} = \sum_{h=1}^{n} \frac{\partial q^h}{\partial q'^k} \frac{\partial}{\partial \dot{q}^h}, \tag{7.77}$$

$$\frac{\partial}{\partial q'^k} = \sum_{h=1}^{n} \frac{\partial q^h}{\partial q'^k} \frac{\partial}{\partial q^h} + \sum_{h=1}^{n} \frac{\partial \dot{q}^h}{\partial q'^k} \frac{\partial}{\partial \dot{q}^h}. \tag{7.78}$$

With these results in place we can now state and prove the first important result.

Proposition 7.27 *Under the assumptions of Theorem 7.16, let* $B, B' \subset A(\mathbb{V}^{n+1})$ *be open sets with natural coordinates* $t, q^1, \ldots, q^n, \dot{q}^1, \ldots, \dot{q}^n$ *and* $t', q'^1, \ldots, q'^n, \dot{q}'^1, \ldots, \dot{q}'^n$ *respectively and* $B \cap B' \neq \varnothing$, *so that the transformation rules (7.72),(7.71), (7.73) hold.*

(1) Given a frame $\mathscr{R}$, *with the obvious notation, on* $B \cap B'$ *we have:*

$$\mathscr{T}'|_{\mathscr{R}}(t', q', \dot{q}') = \mathscr{T}|_{\mathscr{R}}(t, q, \dot{q}), \tag{7.79}$$

$$\mathscr{Q}'_k|_{\mathscr{R}}(t', q', \dot{q}') = \sum_{h=1}^{n} \frac{\partial q^h}{\partial q'^k} \mathscr{Q}_h|_{\mathscr{R}}(t, q, \dot{q}). \tag{7.80}$$

(2) *Consider a curve γ intersecting $B \cap B'$, in coordinates $q^k = q^k(t)$, $\dot{q}^k = \dot{q}^k(t)$ and $q'^k = q'^k(t)$, $\dot{q}'^k = \dot{q}'^k(t)$ (for $k = 1, \ldots, n$), and not necessarily satisfying the Euler-Lagrange equations. Then at any instant:*

$$\frac{d}{dt'} \left(\frac{\partial \mathscr{T}'|_{\mathscr{R}}}{\partial \dot{q}'^k} \Big|_\gamma \right) - \frac{\partial \mathscr{T}'|_{\mathscr{R}}}{\partial q'^k} \Big|_\gamma = \sum_{h=1}^{n} \frac{\partial q^h}{\partial q'^k} \Big|_\gamma \left[\frac{d}{dt} \left(\frac{\partial \mathscr{T}|_{\mathscr{R}}}{\partial \dot{q}^h} \Big|_\gamma \right) - \frac{\partial \mathscr{T}|_{\mathscr{R}}}{\partial q^h} \Big|_\gamma \right],$$

$$(7.81)$$

whenever $\dot{q}^h = \frac{dq^h}{dt}$, and necessarily also $\dot{q}'^k = \frac{dq'^k}{dt}$ according to (7.73).

Proof

(1) The first two identities are proved by direct computation using the definitions of the quantities involved, and pose no difficulty.

(2) Identity (7.81) involves complicated calculations when not done properly. We shall prove it and suppress the subscript $|_\gamma$ for simplicity. The right-hand side splits into two parts:

$$-\sum_{h=1}^{n} \frac{\partial q^h}{\partial q'^k} \frac{\partial \mathscr{T}|_{\mathscr{R}}}{\partial q^h} \qquad\qquad (7.82)$$

and

$$\sum_{h=1}^{n} \frac{\partial q^h}{\partial q'^k} \frac{d}{dt} \frac{\partial \mathscr{T}|_{\mathscr{R}}}{\partial \dot{q}^h} = \sum_{h=1}^{n} \frac{d}{dt} \left(\frac{\partial q^h}{\partial q'^k} \frac{\partial \mathscr{T}|_{\mathscr{R}}}{\partial \dot{q}^h} \right) - \sum_{h=1}^{n} \left(\frac{d}{dt} \frac{\partial q^h}{\partial q'^k} \right) \frac{\partial \mathscr{T}|_{\mathscr{R}}}{\partial \dot{q}^h} .$$

Using (7.74) we have:

$$\sum_{h=1}^{n} \frac{\partial q^h}{\partial q'^k} \frac{d}{dt} \frac{\partial \mathscr{T}|_{\mathscr{R}}}{\partial \dot{q}^h} = \sum_{h=1}^{n} \frac{d}{dt} \left(\frac{\partial \dot{q}^h}{\partial \dot{q}'^k} \frac{\partial \mathscr{T}|_{\mathscr{R}}}{\partial \dot{q}^h} \right) - \sum_{h=1}^{n} \left(\frac{d}{dt} \frac{\partial q^h}{\partial q'^k} \right) \frac{\partial \mathscr{T}|_{\mathscr{R}}}{\partial \dot{q}^h} .$$

From (7.77), we may also write:

$$\sum_{h=1}^{n} \frac{\partial q^h}{\partial q'^k} \frac{d}{dt} \frac{\partial \mathscr{T}|_{\mathscr{R}}}{\partial \dot{q}^h} = \frac{d}{dt} \frac{\partial \mathscr{T}|_{\mathscr{R}}}{\partial \dot{q}'^k} - \sum_{h=1}^{n} \left(\frac{d}{dt} \frac{\partial q^h}{\partial q'^k} \right) \frac{\partial \mathscr{T}|_{\mathscr{R}}}{\partial \dot{q}^h} .$$

Adding this to (7.82) we conclude that the right-hand side of (7.81) equals:

$$\frac{d}{dt} \frac{\partial \mathscr{T}|_{\mathscr{R}}}{\partial \dot{q}'^k} - \sum_{h=1}^{n} \left[\left(\frac{d}{dt} \frac{\partial q^h}{\partial q'^k} \right) \frac{\partial \mathscr{T}|_{\mathscr{R}}}{\partial \dot{q}^h} + \frac{\partial q^h}{\partial q'^k} \frac{\partial \mathscr{T}|_{\mathscr{R}}}{\partial q^h} \right] .$$

On any curve $q^h = q^h(t)$, where the dotted variables are the time derivatives of the undotted variables, we can use (7.75) to write the right-hand side of (7.81) as:

$$\frac{d}{dt}\frac{\partial \mathscr{T}|_{\mathscr{R}}}{\partial \dot{q}'^k} - \sum_{h=1}^{n}\left[\frac{\partial \dot{q}^h}{\partial q'^k}\frac{\partial \mathscr{T}|_{\mathscr{R}}}{\partial \dot{q}^h} + \frac{\partial q^h}{\partial q'^k}\frac{\partial \mathscr{T}|_{\mathscr{R}}}{\partial q^h}\right].$$

Eventually, from (7.78), the right-hand side of (7.81) is:

$$\frac{d}{dt'}\frac{\partial \mathscr{T}'|_{\mathscr{R}}}{\partial \dot{q}'^k} - \frac{\partial \mathscr{T}'|_{\mathscr{R}}}{\partial q'^k},$$

where differentiating in t is the same as differentiating in t' because the two variables differ by an additive constant. This proves (7.81).

□

The next theorem, consequence of the previous proposition, answers the question regarding the independence of the solutions of the Euler-Lagrange equations of local coordinates.

Theorem 7.28 *Under the assumptions of Theorem 7.16, consider the C^1 curve $I \ni t \mapsto \gamma(t) \in A(\mathbb{V}^{n+1})$:*

$$I \ni t \mapsto (t, q^1(t), \ldots, q^n(t), \dot{q}^1(t), \ldots, \dot{q}^n(t)) \in \psi(U)$$

in a natural local chart (U, ψ) (where $I \subset \mathbb{R}$ is an open interval) that solves the Euler-Lagrange equations in a local chart of $A(\mathbb{V}^{n+1})$ for the functions $\mathscr{T}|_{\mathscr{R}}, \mathscr{Q}|_{\mathscr{R}}$. The same curve, in the coordinates $t', q'^1, \ldots, q'^n, \dot{q}'^1, \ldots, \dot{q}'^n$ of another natural local chart containing the curve, will solve the Euler-Lagrange equations for the above functions $\mathscr{T}'|_{\mathscr{R}}, \mathscr{Q}'|_{\mathscr{R}}$.

7.4.4   Maximal Solutions of the Euler-Lagrange Equations Defined Globally on $A(\mathbb{V}^{n+1})$

Theorem 7.28 has an important consequence regarding the hypotheses that $\mathscr{T}|_{\mathscr{R}}$ is of class C^2 and the $\mathscr{Q}_k|_{\mathscr{R}}$ are C^1 in any natural local chart, which then guarantees the Existence and uniqueness theorem (Theorem 14.23) for the Euler-Lagrange equations in any such chart.

So far we have built the theory of the Euler-Lagrange equations using an arbitrary system of (natural) local coordinates on the differentiable manifold $A(\mathbb{V}^{n+1})$. In general however $A(\mathbb{V}^{n+1})$ cannot be covered by one local chart, and when we look at a solution of the Euler-Lagrange equations in a local chart (B, ψ), we expect this solution to reach the boundary $\mathscr{F}(B)$ of the chart. From Physics' point of view a solution obtained in a local chart for a certain initial datum $a \in B \subset A(\mathbb{V}^{n+1})$ is expected to be extendable beyond the chart, since the particular chart has no special physical meaning. We can try to use different coordinate systems by glueing the various solutions obtained, but a priori we do not know that the underpinning theory will be coherent. The proposition we have proved says that the various ingredients used in formulating the Euler-Lagrange equations behave consistently when we change coordinates. If in fact (B', ψ') is another local chart with $B \cap B' \neq \varnothing$, and the solution γ_B reaches a point $\gamma_B(t_1) \in B \cap B'$, we can reset the Euler-Lagrange initial-value problem now *written in coordinates*[11] on B', using as initial conditions $\gamma_B(t_1)$ and $\dot{\gamma}_B(t_1)$ at time t_1. The unique solution $\gamma_{B'}$ we obtain will, on one hand, latch on to the solution existing on B by the Uniqueness Theorem on B, and on the other hand it will extend on V *also outside of* B due to the Existence Theorem on B'. Clearly, in this way, using the same argument of the global Existence and uniqueness theorem (Theorem 14.23), we eventually find the unique *maximal solution* of the Euler-Lagrange equations with initial datum a taking values *anywhere* in $A(\mathbb{V}^{n+1})$. In particular, any other solution with the same initial condition will be a restriction of the maximal solution.

7.4.5 The Notion of Lagrangian

Suppose the active forces acting on the system of point particles $\mathscr{S}$ (with ideal holonomic constraints) are all conservative in frame $\mathscr{R}$. Then there will be a potential energy:

$$\mathscr{U}|_{\mathscr{R}} = \mathscr{U}|_{\mathscr{R}}(P_1, \ldots, P_N) \quad \text{such that} \quad \mathbf{F}_{i\mathscr{R}} = -\nabla_{P_i} \mathscr{U}(P_1, \ldots, P_N) .$$

We can weaken that, simply assuming there exists a function, called **force potential**, $\mathscr{V}|_{\mathscr{R}} = \mathscr{V}|_{\mathscr{R}}(t, P_1, \ldots, P_N)$, in general depending on time (so the conservation of the mechanical energy is not warranted), for which

$$\mathbf{F}_{i\mathscr{R}} = \nabla_{P_i} \mathscr{V}|_{\mathscr{R}}(t, P_1, \ldots, P_N) .$$

In natural coordinates $t, q^1, \ldots, q^n$ on $\mathbb{V}^{n+1}$ with origin O moving with $\mathscr{R}$, so that $P_i = \mathbf{x}_i + O$, the active forces' Lagrangian components will be (where we omit $|_{\mathscr{R}}$ for the sake of simplicity)

$$\mathscr{Q}_k(t, q^1, \ldots, q^n) = \sum_{i=1}^{N} \frac{\partial \mathbf{x}_i}{\partial q^k} \nabla_{\mathbf{x}_i} \mathscr{V}(t, \mathbf{x}_1, \ldots, \mathbf{x}_N) = \frac{\partial}{\partial q^k} \mathscr{V}(t, q^1, \ldots, q^n) .$$

---

[11] If necessary, we can move the origin of time in B' so that it matches the time coordinate in B.

Above $\mathscr{V}(t, q^1, \ldots, q^n)$ actually denotes the composite

$$\mathscr{V}(\mathbf{x}_1(t, q^1, \ldots, q^n), \ldots, \mathbf{x}_N(t, q^1, \ldots, q^n)) \, .$$

Indicating the above function with $\mathscr{V}(t, q^1, \ldots, q^n)$ is not very rigorous but quite effective, and we will take advantage of this notation very often in the rest of book. Inserting the expression for $\mathscr{Q}_k$ in the Euler-Lagrange equations, we are immediately able to rewrite them as:

$$\begin{cases} \dfrac{d}{dt} \dfrac{\partial(\mathscr{T}|_{\mathscr{R}} + \mathscr{V}|_{\mathscr{R}})}{\partial \dot{q}^k} - \dfrac{\partial(\mathscr{T}|_{\mathscr{R}} + \mathscr{V}|_{\mathscr{R}})}{\partial q^k} = 0 \, , \\[2mm] \dfrac{dq^k}{dt} = \dot{q}^k(t) \end{cases} \quad \text{for } k = 1, \ldots, n,$$

where we used that $\mathscr{V}|_{\mathscr{R}}$ does *not* depend on the coordinates $\dot{q}^k$. Defining the **system's Lagrangian**:

$$\mathscr{L}|_{\mathscr{R}}(t, q, \dot{q}) := \mathscr{T}|_{\mathscr{R}}(t, q, \dot{q}) + \mathscr{V}|_{\mathscr{R}}(t, q) \, , \tag{7.83}$$

the Euler-Lagrange equations take the classical form, for $\mathscr{L} := \mathscr{L}|_{\mathscr{R}}$

$$\begin{cases} \dfrac{d}{dt} \left(\left. \dfrac{\partial \mathscr{L}}{\partial \dot{q}^k} \right|_{(t, q(t), \dot{q}(t))} \right) - \left. \dfrac{\partial \mathscr{L}}{\partial q^k} \right|_{(t, q(t), \dot{q}(t))} = 0 \, , \\[2mm] \dfrac{dq^k}{dt} = \dot{q}^k(t) \end{cases} \quad \text{for } k = 1, \ldots, n. \tag{7.84}$$

We will say a system $\mathscr{S}$ of point particles subject to ideal holonomic constraints **admits a Lagrangian** whenever the Euler-Lagrange equations can be written as (7.84) for *some* Lagrangian $\mathscr{L}$. This happens in particular when, in a frame $\mathscr{R}$, *all* active forces admits a potential $\mathscr{V}|_{\mathscr{R}}$, and in that case $\mathscr{L} = \mathscr{T}|_{\mathscr{R}} + \mathscr{V}|_{\mathscr{R}}$. More generally, some forces will not be expressible by a potential, but will still have Lagrangians components $\mathscr{Q}_k|_{\mathscr{R}}$ (where $|_{\mathscr{R}}$ remind us that these components may depend on the reference frame, just as for inertial forces). Then the Euler-Lagrange equations take a mixed form:

$$\begin{cases} \dfrac{d}{dt} \left(\left. \dfrac{\partial \mathscr{L}|_{\mathscr{R}}}{\partial \dot{q}^k} \right|_{(t, q(t), \dot{q}(t))} \right) - \left. \dfrac{\partial \mathscr{L}|_{\mathscr{R}}}{\partial q^k} \right|_{(t, q(t), \dot{q}(t))} = \mathscr{Q}_k|_{\mathscr{R}}(t, q(t), \dot{q}(t)) \, , \\[2mm] \dfrac{dq^k}{dt} = \dot{q}^k(t) \\[2mm] \quad \text{for } k = 1, \ldots, n, \end{cases} \tag{7.85}$$

where the Lagrangian only accounts for the forces with potential, and the $\mathscr{Q}_k|_{\mathscr{R}}$ are the remaining active forces (which depend on the choice of the reference frame $\mathscr{R}$ only if they are referred to inertial forces).

When the physical system is made of subsystems given by point particles and/or continuous rigid systems, the definition of Lagrangian is the same given above, if necessary viewing the forces $\mathbf{F}_{i\mathscr{R}}$ as applied to given points P_i in the continuous rigid bodies.

Examples 7.29

(1) Here is an elementary case. Consider a system of N point particles, $P_1, \ldots, P_N$ of masses $m_1, \ldots, m_N$ respectively, *not subject to constraints* but interacting under forces coming from a total potential energy $\mathscr{U}|_{\mathscr{R}}$. $\mathscr{R}$ is an inertial frame where the points are given by position vectors $\mathbf{x}_1 := P_1 - O, \ldots, \mathbf{x}_n := P_N - O$. O is the origin of orthonormal coordinates on $\mathscr{R}$ and the components of any $\mathbf{x}_i$ are the components of $P - O$ in these coordinates. Using the components of all vectors $\mathbf{x}_i$ as free coordinates, with obvious notation the system's Lagrangian is:

$$\mathscr{L}(\mathbf{x}_1, \ldots, \mathbf{x}_N, \dot{\mathbf{x}}_1, \ldots, \dot{\mathbf{x}}_N) = \sum_{i=1}^{N} \frac{m_i}{2} \dot{\mathbf{x}}_i^2 - \mathscr{U}|_{\mathscr{R}}(\mathbf{x}_1, \ldots, \mathbf{x}_N).$$

The Euler-Lagrange equations are simply Newton's equations for the system of point particles in the inertial frame $\mathscr{R}$. In fact, if $\mathbf{x}_i = \sum_{k=1}^{3} x_i^k \mathbf{e}_k$, then:

$$\left(\frac{d}{dt} \frac{\partial}{\partial \dot{x}_i^k} - \frac{\partial}{\partial x_i^k} \right) \left(\sum_{i=1}^{N} \frac{m_i}{2} \dot{\mathbf{x}}_i^2 - \mathscr{U}|_{\mathscr{R}}(\mathbf{x}_1, \ldots, \mathbf{x}_N) \right)$$

$$= \frac{d}{dt} \frac{\partial \frac{m_i}{2} \dot{\mathbf{x}}_i^2}{\partial \dot{x}_i^k} + \frac{\partial}{\partial x_i^k} \mathscr{U}|_{\mathscr{R}}(\mathbf{x}_1, \ldots, \mathbf{x}_N)$$

$$= m_i \frac{d\dot{x}_i^k}{dt} + \frac{\partial}{\partial x_i^k} \mathscr{U}|_{\mathscr{R}}(\mathbf{x}_1, \ldots, \mathbf{x}_N).$$

The Euler-Lagrange equations, written as second-order equations (using $\dot{x}_i^k = dx_i^k/dt$), become Newton's equations:

$$m_i \frac{d^2 x_i^k}{dt^2} = -\frac{\partial}{\partial x_i^k} \mathscr{U}|_{\mathscr{R}}(\mathbf{x}_1, \ldots, \mathbf{x}_N), \quad \text{for } i = 1, \ldots, N \text{ and } k = 1, 2, 3.$$

(2) Consider a physical system of three point particles P, P_1, P_2 of masses m, m_1, m_2 respectively. P_1 and P_2 are joined by a weightless ideal rigid rod of length $d > 0$. The point P is joined to the rod through a spring of elastic constant $K > 0$ attached to the rod's centre of mass G (accounting for the two masses). We suppose the spring has zero length at rest, and that there are no other forces on the points except the obvious constraint reactions. We want to write the equations of motion (the pure equations of motion) in an inertial frame $\mathscr{R}$.

Fix orthonormal coordinates on $\mathscr{R}$ with origin O and axes $\mathbf{e}_1, \mathbf{e}_2, \mathbf{e}_3$. The following 8 coordinates can be the free coordinates: the three components x^1, x^2, x^3, with respect to the axes, of the position vector $P - O$; the three components X^1, X^2, X^3, with respect to the axes, of the position vector $G - O$; the two polar angles θ, ϕ of the position vector $P_2 - G$ in Cartesian coordinates at G with axes $\mathbf{e}_1, \mathbf{e}_2, \mathbf{e}_3$ (those of $\mathscr{R}$) forming the frame of the centre of mass $\mathscr{R}_G$.

As the only active force, the spring's, is conservative, the Euler-Lagrange equations can be written using the Lagrangian $\mathscr{L}|_{\mathscr{R}} = \mathscr{T}|_{\mathscr{R}} - \mathscr{U}|_{\mathscr{R}}$. Using König's theorem (Theorem 4.10) to compute the kinetic energy in $\mathscr{R}$ of P_1, P_2, we have:

$$\mathscr{T}|_{\mathscr{R}} = \frac{1}{2}\sum_{j=1}^{3} m(\dot{x}^j)^2 + \frac{1}{2}\sum_{j=1}^{3} M(\dot{X}^j)^2 + \mathscr{T}|_{\mathscr{R}_G} \,,$$

where $M := m_1 + m_2$. The kinetic energy of the centre of mass in $\mathscr{R}_G$ is just the sum of the kinetic energies in $\mathscr{R}_G$. Call d_1 and d_2 the distances of P_1 and P_2 to G, determined by $m_1 d_1 = m_2 d_2$ and $d_1 + d_2 = d$. Now look at P_2. Its position vector is $P_2 - G = d_2\, \mathbf{e}_r$ where $\mathbf{e}_r$ is the unit radial vector of the polar coordinates at G with respect to $\mathbf{e}_1, \mathbf{e}_2, \mathbf{e}_3$. Consequently, using Exercise 2 in Exercises 2.18:

$$\mathbf{v}_{P_2}|_{\mathscr{R}_g} = d_1\, \dot{\mathbf{e}}_r = d_1(\dot{\theta}\mathbf{e}_\theta + \dot{\phi}\sin\theta\mathbf{e}_\phi)\,.$$

As $P_1 - G = -\frac{d_1}{d_2}(P_2 - G)$ we also have

$$\mathbf{v}_{P_2}|_{\mathscr{R}_g} = -d_2\, \dot{\mathbf{e}}_r = -d_2(\dot{\theta}\mathbf{e}_\theta + \dot{\phi}\sin\theta\mathbf{e}_\phi)\,.$$

Since the unit vectors $\mathbf{e}_\theta$ and $\mathbf{e}_\phi$ are orthogonal, squaring the above expressions gives

$$(\mathbf{v}_{P_2}|_{\mathscr{R}_g})^2 = d_2^2(\dot{\theta}^2 + \dot{\phi}^2\sin^2\theta) \quad \text{and} \quad (\mathbf{v}_{P_1}|_{\mathscr{R}_g})^2 = d_1^2(\dot{\theta}^2 + \dot{\phi}^2\sin^2\theta)\,.$$

Hence:

$$\mathscr{T}|_{\mathscr{R}} = \frac{m}{2}\sum_{j=1}^{3}(\dot{x}^j)^2 + \frac{M}{2}\sum_{j=1}^{3}(\dot{X}^j)^2 + \frac{1}{2}(m_1^2 d_1^2 + m_2 d_2^2)(\dot{\theta}^2 + \dot{\phi}^2\sin^2\theta)\,.$$

The springs' potential energy (frame-independent, in this case), equals

$$\mathscr{U} = \frac{K}{2}\sum_{j=1}^{3}(X^j - x^j)^2\,.$$

Therefore the Lagrangian of the three point particles in the inertial frame $\mathscr{R}$ is:

$$\mathscr{L}|_{\mathscr{R}} = \frac{m}{2}\sum_{j=1}^{3}(\dot{x}^j)^2 + \frac{M}{2}\sum_{j=1}^{3}(\dot{X}^j)^2 + \frac{1}{2}(m_1^2 d_1^2 + m_2 d_2^2)(\dot{\theta}^2 + \dot{\phi}^2\sin^2\theta)$$

$$- \frac{K}{2}\sum_{j=1}^{3}(X^j - x^j)^2\,.$$

We thus have a system of Euler-Lagrange equations consisting of 8 equations, obtained from (7.84) for q^k respectively equal to x^k, X^k with $k = 1, 2, 3$ and θ, ϕ. The equations are as follows:

$$\begin{cases} m\dfrac{d^2x^k}{dt^2} = -K(x^k - X^k)\,, \\[2mm] M\dfrac{d^2X^k}{dt^2} = -K(X^k - x^k)\,, \\[2mm] \dfrac{d^2\theta}{dt^2} = 2\left(\dfrac{d\phi}{dt}\right)^2\sin\theta\cos\theta\,, \\[2mm] \dfrac{d}{dt}\left(\dfrac{d\phi}{dt}\sin^2\theta\right) = 0\,. \end{cases} \qquad \text{for } k = 1, 2, 3,$$

The system has order two and can be put in normal form if $\sin\theta \neq 0$: when that happens we are outside the domain of the polar coordinates. As the system's right-hand side in normal form is C^∞, the equations admit a unique solution, the motion, for given initial conditions $x^i(t_0)$, $X^i(t_0)$, $\theta(t_0)$ and $\phi(t_0)$ and corresponding derivatives $(dx^i/dt)(t_0)$, $(dX^i/dt)(t_0)$, $(d\theta/dt)(t_0)$ and $(d\phi/dt)(t_0)$.

(3) When studying general physical systems $\mathscr{S}$ made of two (or more) subsystems $\mathscr{S}_1, \mathscr{S}_2$, the system's Lagrangian (in a frame $\mathscr{R}$ we will take as implicit) often has the form:

$$\mathscr{L} = \mathscr{L}_1 + \mathscr{L}_2 + \mathscr{L}_I\,,$$

where $\mathscr{L}_1$ and $\mathscr{L}_2$ are the Lagrangians of the two systems seen as non-interacting between themselves—each Lagrangians contains *only* the coordinates of its subsystem, while $\mathscr{L}_I$ is the *interaction Lagrangian*, which contains the coordinates of *both* subsystems. The simplest example is that of two particles of masses m_1, m_2, with natural coordinates $(\mathbf{x}_1, \dot{\mathbf{x}}_1)$ and $(\mathbf{x}_2, \dot{\mathbf{x}}_2)$ using the conventions of Example (1) above (in the inertial frame $\mathscr{R}$). The free Lagrangians are, for example:

$$\mathscr{L}_1 := \frac{m\dot{\mathbf{x}}_1^2}{2}\,, \quad \mathscr{L}_2 := \frac{m\dot{\mathbf{x}}_2^2}{2}\,,$$

while a possible interaction Lagrangian is given by a conservative force with potential energy that depends on the particles's positions:

$$\mathscr{L}_I := -\mathscr{U}(\|\mathbf{x}_1 - \mathbf{x}_2\|) .$$

When we consider systems made of several subsystems, the interaction Lagrangian can often be decomposed into so-called "two-body" Lagrangians, "three-body" Lagrangians etc. With 3 particles, for instance, we could have:

$$\mathscr{L}_I = -\mathscr{U}^{(2)}(\mathbf{x}_1, \mathbf{x}_2) - \mathscr{U}^{(2)}(\mathbf{x}_2, \mathbf{x}_3) - \mathscr{U}^{(2)}(\mathbf{x}_1, \mathbf{x}_3) - \mathscr{U}^{(3)}(\mathbf{x}_1, \mathbf{x}_2, \mathbf{x}_3) ,$$

where $\mathscr{U}^{(2)}$ represents the interaction (conservative in this case) of two bodies and $\mathscr{U}^{(3)}$ that of three bodies. In concrete applications, like Statistical Mechanics, one often uses numerical methods; these neglect the interaction Lagrangian or part of it (say, the three-body one) at the first approximation, and account for it only in successive ones.

(4) Consider an articulated mechanical system made of two homogeneous discs S_1 and S_2, (Fig. 7.4) of radius r and equal mass M, and a point particle P of mass m, organised as follows. S_1 is fixed at its centre O on a vertical plane in the inertial frame $\mathscr{R}$, while S_2 rolls on S_1 on the same plane. The point P is fixed to the boundary of S_2. Apart from the ideal constraint reactions, the forces acting on the system are the weight with gravitational acceleration $-g\mathbf{e}_z$ and an ideal spring of no mass, zero length at rest, elastic constant k, attached to O and P. We want to write the system's Lagrangian. If we ignored the rolling, the system would have 3 degrees of freedom. The rolling constraint, as we shall see, reduces the degrees to 2. First, we find the configurations of S_1 using the angle θ_1 between a fixed radius of S_1 and the vertical half-line from O, measured

Fig. 7.4 Example 7.29.4

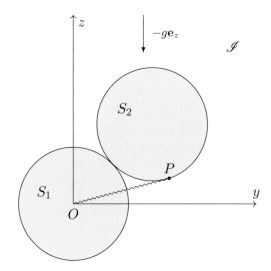

counter-clockwise. We call θ_2 the similar angle for S_2 between the vertical half-line at the centre O' of S_2 and the radius through P. Finally, let ϕ be the angle between the vertical half-line from O and the segment $O' - O$. It is easy to see that the rolling constraint —i.e. that the contact points have zero relative velocity—forces: $r\dot{\theta}_1 = 2r\dot{\phi} - r\dot{\theta}_2$. Consequently we can find $\phi = \frac{\theta_1+\theta_2}{2}$ if we assume that for $\theta_1 = \theta_2 = 0$ the point P is on the vertical line through O and O', which is the situation we will be working in (the generic case is completely similar). The total kinetic energy will be the sum of the 3 kinetic energies of the constituents S_1, S_2 and P, in this order:

$$\mathscr{T} = \frac{1}{2}Mr^2\dot{\theta}_1^2 + \left(\frac{1}{2}Mr^2\dot{\theta}_2^2 + \frac{1}{2}M(2r)^2\dot{\phi}^2\right) + \frac{1}{2}mr^2[\dot{\phi}^2 + \dot{\theta}_1^2 + 2\dot{\phi}\dot{\theta}_2\cos(\phi - \theta_2)].$$

Using the angles' relationship coming from the rolling constraint:

$$\mathscr{T} = \frac{1}{2}Mr^2\dot{\theta}_1^2 + \left(\frac{1}{2}Mr^2\dot{\theta}_2^2 + \frac{1}{2}Mr^2(\dot{\theta}_1 + \dot{\theta}_2)^2\right)$$
$$+ \frac{1}{2}mr^2\left[\frac{1}{4}(\dot{\theta}_1 + \dot{\theta}_2)^2 + \dot{\theta}_1^2 + (\dot{\theta}_1 + \dot{\theta}_2)\dot{\theta}_2\cos\left(\frac{\theta_1 - \theta_2}{2}\right)\right].$$

Hence:

$$\mathscr{T} = Mr^2(\dot{\theta}_1^2 + \dot{\theta}_2^2 + \dot{\theta}_1\dot{\theta}_2) + \frac{mr^2}{8}\left[\dot{\theta}_1^2 + \dot{\theta}_2^2\left(1 + 4\cos\left(\frac{\theta_1 - \theta_2}{2}\right)\right)\right.$$
$$\left. + 2\dot{\theta}_1\dot{\theta}_2\left(4 + \cos\left(\frac{\theta_1 - \theta_2}{2}\right)\right)\right],$$

which we rewrite as:

$$\mathscr{T} = \dot{\theta}_1^2\left(Mr^2 + \frac{mr^2}{8}\right) + \dot{\theta}_2^2\left[Mr^2 + \frac{mr^2}{8}\left(1 + 4\cos\left(\frac{\theta_1 - \theta_2}{2}\right)\right)\right]$$
$$+ \dot{\theta}_1\dot{\theta}_2\left[Mr^2 + \frac{mr^2}{4}\left(4 + \cos\left(\frac{\theta_1 - \theta_2}{2}\right)\right)\right].$$

The potential energy is the sum of the three gravitational potential energies (we can assume the one of S_1 is zero) plus the spring's potential energy:

$$\mathscr{U} = Mg\cos\phi + mg(\cos\phi + \cos\theta_2) + kr\cos(\phi - \theta_2),$$

where we have dropped the additive constant in the spring's potential energy. Using the angles' relation:

$$\mathscr{U} = Mg\cos\left(\frac{\theta_1 + \theta_2}{2}\right) + mg\left(\cos\left(\frac{\theta_1 + \theta_2}{2}\right) + \cos\theta_2\right) + kr\cos\left(\frac{\theta_1 - \theta_2}{2}\right)$$

and therefore

$$\mathscr{U} = (Mg + mg)\cos\left(\frac{\theta_1 + \theta_2}{2}\right) + mg\cos\theta_2 + kr\cos\left(\frac{\theta_1 - \theta_2}{2}\right).$$

Eventually, the system's Lagrangian reads:

$$\mathscr{L}(\theta_1, \theta_2, \dot{\theta}_1, \dot{\theta}_2) = \dot{\theta}_1^2\left(Mr^2 + \frac{mr^2}{8}\right)$$

$$+ \dot{\theta}_2^2\left[Mr^2 + \frac{mr^2}{8}\left(1 + 4\cos\left(\frac{\theta_1 - \theta_2}{2}\right)\right)\right]$$

$$+ \dot{\theta}_1\dot{\theta}_2\left[Mr^2 + \frac{mr^2}{4}\left(4 + \cos\left(\frac{\theta_1 - \theta_2}{2}\right)\right)\right]$$

$$- (Mg + mg)\cos\left(\frac{\theta_1 + \theta_2}{2}\right) - mg\cos\theta_2$$

$$- kr\cos\left(\frac{\theta_1 - \theta_2}{2}\right).$$

(5) We might imagine that a physical system subject to friction will not admit a Lagrangian description. If *viscous friction* is at play, this is false: it is possible to describe the dynamics using a Lagrangian. For example, take the elementary case of a point particle P of mass m, with position vector $\mathbf{x} = P - O$ in the inertial frame $\mathscr{R}$, not subject to constraints but subject to a conservative force in $\mathscr{R}$ with potential energy $\mathscr{U}$ and a viscous force $-\gamma\mathbf{v}_P|_{\mathscr{R}}$. The equations of motion in $\mathscr{R}$ are:

$$m\frac{d^2\mathbf{x}}{dt^2} = -\nabla\mathscr{U}(\mathbf{x}) - \gamma\frac{d\mathbf{x}}{dt}.$$

We leave it to the reader to check (it is straightforward) that the same equations arise from the Lagrangian:

$$\mathscr{L}(t, \mathbf{x}, \dot{\mathbf{x}}) := e^{\frac{\gamma t}{m}}\left(\frac{1}{2}m\dot{\mathbf{x}}^2 - \mathscr{U}(\mathbf{x})\right). \tag{7.86}$$

7.4.6 Regularity of Lagrangians in Standard Form

Consider an atlas $\mathcal{A}$ of $A(\mathbb{V}^{n+1})$ of natural local charts, and in any chart with coordinates $t, q^1, \ldots, q^n, \dot{q}^1, \ldots, \dot{q}^n$ let us define the Lagrangian in standard form:

$$\mathcal{L}(t, q, \dot{q}) := \sum_{k,h=1}^{n} a_{kh}(t, q)\dot{q}^k\dot{q}^h + \sum_{k=1}^{n} \beta_k(t, q)\dot{q}^k + \gamma(t, q) \tag{7.87}$$

where the matrix of coefficients $a_{hk}(t, q^1, \ldots, q^n)$ is symmetric and all functions a_{ak}, β_k and γ are of class C^k with $k \geq 1$. By the transformation laws (7.71)–(7.73) between coordinates we have the following elementary proposition.

Proposition 7.30 *The functions in (7.87), for any (local natural) chart in the atlas $\mathcal{A}$, define a unique C^k function $\mathcal{L} : A(\mathbb{V}^{n+1}) \to \mathbb{R}$ whenever, under coordinate change, they obey:*

$$a'_{rs}(t', q') = \sum_{h,k=1}^{n} \frac{\partial q^h}{\partial q'^r}\frac{\partial q^k}{\partial q'^s} a_{hk}(t, q), \tag{7.88}$$

$$\beta'_r(t', q') = \sum_{h=1}^{n} \frac{\partial q^h}{\partial q'^r}\beta_h(t, q) + 2\sum_{h=1}^{n} \frac{\partial q^h}{\partial q'^r}\frac{\partial q^k}{\partial t'} a_{hk}(t, q), \tag{7.89}$$

$$\gamma'(t', q') = \gamma(t, q) + \frac{\partial q^h}{\partial t'}\beta_h(t, q) + \sum_{h,k=1}^{n} \frac{\partial q^h}{\partial t'}\frac{\partial q^k}{\partial t'} a_{hk}(t, q) \tag{7.90}$$

with the obvious notation. In any natural local chart of $A(\mathbb{V}^{n+1})$, possibly not belonging in $\mathcal{A}$, the Lagrangian $\mathcal{L}$ still has form (7.87).

Proof The proof is immediate by direct computation. ☐

Notice that in any natural chart the Jacobian matrix of $\mathcal{L}$ in the dotted coordinates coincides with the matrix of coefficients a_{hk}. All Lagrangians arising from a system's mechanics have local structure (7.87). Following the proof of Theorem 7.23, the fact we can write the Euler-Lagrange equations in normal form, and hence the Existence and uniqueness theorem holds if $k \geq 2$, is guaranteed once the Jacobian matrix of $\mathcal{L}$ in the dotted coordinates is non-singular on the charts of atlas $\mathcal{A}$ (and therefore on any natural local chart of $A(\mathbb{V}^{n+1})$). When the Lagrangian is of the type seen thus far, $\mathcal{L}|_{\mathscr{R}} + \mathscr{V}|_{\mathscr{R}}$, the non-singularity is in turn a consequence of the fact the matrix a_{hk} is positive definite, as we know very well.

Relatively to the above standard Lagrangian we have the following elementary fact, which relates the Lagrangian's regularity to the regularity of its known constituents. We shall use it often in the sequel.

Proposition 7.31 *The right side of (7.87) is jointly of class C^k ($k \geq 1$) in the variables $(t, q^1, \ldots, q^n, \dot{q}^1, \ldots, \dot{q}^n)$ if and only if all functions a_{hk}, β_h, γ are jointly of class C^k in the variables $(t, q^1, \ldots, q^n)$.*

Proof The proof of the non-trivial implication is based on the following observation. If the right-hand side of (7.87) is C^k in $(t, q^1, \ldots, q^n, \dot{q}^1, \ldots, \dot{q}^n)$ we can in particular compute any mixed derivative up to order k with respect to the first $n + 1$ variables $t, q^1, \ldots, q^n$, thus obtaining a continuous function in all variables, dotted ones included. Putting $\dot{q}^r = 0$ in this expression, for $r = 1, \ldots, n$, we conclude γ is of class C^k on its domain in the variables $t, q^1, \ldots, q^n$. Consequently also the right-hand side of (7.87) *without* the term γ is C^k *for any given value of* $(\dot{q}^1, \ldots, \dot{q}^n) \in \mathbb{R}^n$. This happens for $(\dot{q}^1, \ldots, \dot{q}^n)$ but also for $(-\dot{q}^1, \ldots, -\dot{q}^n)$. Comparing the derivatives in the variables $t, q^1, \ldots, q^n$ of the right-hand side of (7.87) (without γ) evaluated for these choices of dotted coordinates, we immediately obtain that the summands $\sum_{k,h=1}^n a_{kh}(t, q^1, \ldots, q^n)\dot{q}^k\dot{q}^h$ and $\sum_{k=1}^n \beta_k(t, q^1, \ldots, q^n)\dot{q}^k$ must be C^k in each $t, q^1, \ldots, q^n$ for any choice of $(\dot{q}^1, \ldots, \dot{q}^n) \in \mathbb{R}^n$. At this point it is not hard to show that for any choice of h and k the functions a_{hk} and β_k are C^k provided we choose $(\dot{q}^1, \ldots, \dot{q}^n)$ suitably (in particular, remembering that $a_{kh} = a_{hk}$). $\quad\square$

7.4.7 Change of Inertial Frame and Lagrangian Non-uniqueness

Consider a physical system $\mathscr{S}$ of N point particles with n degrees of freedom, described by the Lagrangian components of the active forces $\mathscr{Q}_k|_{\mathscr{R}}$ in a frame $\mathscr{R}$ and with kinetic energy $\mathscr{T}|_{\mathscr{R}}$. We make no assumption on the nature of the coordinates $q^1, \ldots, q^n$, which may or not move with the frame. By construction the kinetic energy $\mathscr{T}|_{\mathscr{R}} = \mathscr{T}|_{\mathscr{R}}(t, q, \dot{q})$, the potential $\mathscr{V}|_{\mathscr{R}} = \mathscr{V}|_{\mathscr{R}}(t, q)$ and the Lagrangian $\mathscr{L}|_{\mathscr{R}} = \mathscr{L}|_{\mathscr{R}}(t, q, \dot{q})$ are *scalar fields* on $A(\mathbb{V}^{n+1})$ when the reference frame $\mathscr{R}$ is given but the free coordinates are allowed to change. Fixing $\mathscr{R}$, but changing the initial natural coordinates $(t, q^1, \ldots, q^n, \dot{q}^1, \ldots \dot{q}^n)$ to new coordinates (obviously local and natural), the new Lagrangian will be simply given, supposing for simplicity $n = 1$, in the obvious form:

$$\mathscr{L}|_{\mathscr{R}}(t(t'), q(t', q'), \dot{q}(t', q', \dot{q}')) .$$

Similar results hold for the other scalar fields. Let us see what happens when we fix the free coordinates and instead change frame from $\mathscr{R}$ to $\mathscr{R}'$, for the moment assuming both inertial. We expect the new Euler-Lagrange equations to be the same as those in $\mathscr{R}$, since there is no physical reason to prefer one frame over the other, and the natural coordinates $t, q^1, \ldots, q^n$ bear no particular relation to either frame. The Lagrangian components of the active forces will stay the same, as noted in (1) Remarks 7.17, whereas the kinetic energy in $\mathscr{R}'$, $\mathscr{T}|_{\mathscr{R}'}$, will differ from $\mathscr{T}|_{\mathscr{R}}$ which is in $\mathscr{R}$. Since for any P_i of $\mathscr{S}$:

$$\mathbf{v}_{P_i}|_{\mathscr{R}} = \mathbf{v}_{P_i}|_{\mathscr{R}'} + \mathbf{v}_{\mathscr{R}'}|_{\mathscr{R}} \,,$$

where $\mathbf{v}_{\mathscr{R}'}|_{\mathscr{R}}$ is constant in time and space (the frames are in uniform linear motion) we immediately find:

$$\Delta \mathscr{T} = \mathscr{T}|_{\mathscr{R}} - \mathscr{T}|_{\mathscr{R}'} = \frac{M}{2}(\mathbf{v}_{\mathscr{R}'}|_{\mathscr{R}})^2 + \mathbf{v}_{\mathscr{R}'}|_{\mathscr{R}} \cdot \sum_{i=1}^{N} m_i \mathbf{v}_{P_i}|_{\mathscr{R}'} \,,$$

where $M := \sum_{i=1}^{N} m_i$. If $\mathbf{x}_i' = \mathbf{x}_i'(t, q^1, \ldots, q^n)$ is the position vector of P_i in $\mathscr{R}'$, we have:

$$\mathbf{v}_{P_i}|_{\mathscr{R}'} = \frac{\partial \mathbf{x}_i'}{\partial t} + \sum_{k=1}^{n} \frac{\partial \mathbf{x}_i'}{\partial q^k} \frac{dq^k}{dt} \,,$$

so, *along the motion*:

$$\Delta \mathscr{T} = \frac{d}{dt}\left(\frac{M}{2}(\mathbf{v}_{\mathscr{R}'}|_{\mathscr{R}})^2 t + \mathbf{v}_{\mathscr{R}'}|_{\mathscr{R}} \cdot \sum_{i=1}^{N} m_i \mathbf{x}_i'(t, q^1(t), \ldots, q^n(t)) \right) \,.$$

Working on $A(\mathbb{V}^{n+1})$, and not thinking the $\dot{q}^k$ as derivatives of the q^k, we can write

$$\Delta \mathscr{T} := \frac{\partial g}{\partial t} + \sum_{k=1}^{n} \frac{\partial g}{\partial q^k} \dot{q}^k \,, \tag{7.91}$$

where $g = g(t, q^1, \ldots, q^n)$ is the function on $\mathbb{V}^{n+1}$ in brackets in the right-hand side of the previous identity:

$$g(t, q^1, \ldots, q^n) := \frac{M}{2}(\mathbf{v}_{\mathscr{R}'}|_{\mathscr{R}})^2 t + \mathbf{v}_{\mathscr{R}'}|_{\mathscr{R}} \cdot \sum_{i=1}^{N} m_i \mathbf{x}_i'(t, q^1, \ldots, q^n) \,.$$

Immediately, for $\Delta \mathscr{T}$ of the form (7.91):

$$\frac{d}{dt} \frac{\partial \Delta \mathscr{T}}{\partial \dot{q}^k} - \frac{\partial \Delta \mathscr{T}}{\partial q^k} = 0 \,.$$

The immediate consequence of this is that the term $\Delta \mathscr{T}$ does not contribute to the Euler-Lagrange equations, and since the Lagrangian components of the (real) active forces do not depend on the frame, we conclude that: *the Euler-Lagrange equations in the inertial frame $\mathscr{R}$ coincide with the Euler-Lagrange equations in the inertial frame $\mathscr{R}'$*.

This result has as straightforward corollary that if a physical system admits a Lagrangian, the latter is not unique. Suppose in fact the Lagrangian components of the active forces (necessarily real, as we are in an inertial frame) are given by a potential $\mathscr{V}|_{\mathscr{R}} = \mathscr{V}|_{\mathscr{R}}(t, q, \dot{q})$ in the natural local coordinates. Hence the physical system will have Lagrangian in the inertial frame $\mathscr{R}$:

$$\mathscr{L}|_{\mathscr{R}} = \mathscr{T}|_{\mathscr{R}}(t, q^1, \ldots, q^n, \dot{q}^1, \ldots, \dot{q}^n) + \mathscr{V}|_{\mathscr{R}}(t, q^1, \ldots, q^n) \, .$$

Passing to $\mathscr{R}'$ (even non-inertial) *without changing the natural local coordinates*, the Lagrangian components $\mathscr{Q}_k|_{\mathscr{R}'}$ of the real active forces will still have a potential $\mathscr{V}|_{\mathscr{R}'}(t, q^1, \ldots, q^n)$, which we can choose to be exactly $\mathscr{V}|_{\mathscr{R}}(t, q^1, \ldots, q^n)$. This follows immediately from the fact that the Lagrangian components of real forces in given natural local coordinates satisfy:

$$\mathscr{Q}_k|_{\mathscr{R}}(t, q, \dot{q}) = \mathscr{Q}_k|_{\mathscr{R}'}(t, q, \dot{q}) \, ,$$

as clarified in (1) of Remark 7.17. Consequently:

$$\mathscr{Q}_k|_{\mathscr{R}'}(t, q, \dot{q}) = \mathscr{Q}_k|_{\mathscr{R}}(t, q, \dot{q}) = \left. \frac{\partial \mathscr{V}|_{\mathscr{R}}}{\partial q^k} \right|_{(t, q, \dot{q})} \, .$$

As $\mathscr{V}|_{\mathscr{R}} = \mathscr{V}|_{\mathscr{R}'}$, the only difference between the Lagrangians in $\mathscr{L}|_{\mathscr{R}}$ and $\mathscr{L}|_{\mathscr{R}'}$ regards the kinetic energy:

$$\Delta \mathscr{L} = \mathscr{L}|_{\mathscr{R}} - \mathscr{L}|_{\mathscr{R}'} = \Delta \mathscr{T} \, .$$

As mentioned above, if $\mathscr{R}'$ is inertial as well, the term $\Delta \mathscr{T}$ does not contribute to the Euler-Lagrange equations, which thus stay the same.

We have shown that for a given physical system there are at least two Lagrangians giving the same Euler-Lagrange equations. They have a precise physical meaning, as they refer to distinct inertial frames. In any case the result is completely general: if we add to a Lagrangian $\mathscr{L}|_{\mathscr{R}}$ a function on $A(\mathbb{V}^{n+1})$ that, in natural local coordinates, has local form:

$$\Delta \mathscr{L} := \frac{\partial g}{\partial t} + \sum_{k=1}^{n} \frac{\partial g}{\partial q^k} \dot{q}^k \, , \tag{7.92}$$

where $g = g(t, q^1, \ldots, q^n)$ is an arbitrary function on $\mathbb{V}^{n+1}$, then le equations given by $\mathscr{L}' := \mathscr{L}|_{\mathscr{R}} + \Delta \mathscr{L}$ when $\mathscr{L}'$ is inserted in (7.84) are the same equations given by $\mathscr{L}_{\mathscr{R}}$. The proof is the same as before, and is based on the fact that:

$$\frac{d}{dt} \frac{\partial \Delta \mathscr{L}}{\partial \dot{q}^k} - \frac{\partial \Delta \mathscr{L}}{\partial q^k} = 0 \, ,$$

which is clear from the definitions we have given.

We have examined the case where both $\mathscr{R}$, $\mathscr{R}'$ are inertial. The situation is more complicated if $\mathscr{R}$ is inertial and $\mathscr{R}'$ is not. In that case $\mathscr{T}|_{\mathscr{R}} - \mathscr{T}|_{\mathscr{R}'}$ is not a total formal derivative as (7.92), so in $\mathscr{R}'$ there will appear new Lagrangian components, obviously due to the inertial forces appearing in $\mathscr{R}'$. We postpone the discussion to Sect. 9.2.4.

Remarks 7.32

(1) If $g : \mathbb{V}^{n+1} \to \mathbb{R}$ is a scalar field (C^1 at least) there exists a unique scalar field $G : A(\mathbb{V}^{n+1}) \to \mathbb{R}$ (C^0 at least) such that in any natural coordinates $t, q^1, \ldots, q^n, \dot{q}^1, \ldots, \dot{q}^n$ on $A(\mathbb{V}^{n+1})$ we have

$$G(t, q^1, \ldots, q^n, \dot{q}^1, \ldots, \dot{q}^n) = \frac{\partial g}{\partial t} + \sum_{k=1}^{n} \frac{\partial g}{\partial q^k} \dot{q}^k . \tag{7.93}$$

The proof is elementary, for it suffices to show that when changing natural coordinates on $A(\mathbb{V}^{n+1})$ to $t', q'^1, \ldots, q'^n, \dot{q}'^1, \ldots, \dot{q}'^n$, and writing the scalar field g in the new coordinates, in the intersection of the two local charts we have

$$\frac{\partial g}{\partial t} + \sum_{k=1}^{n} \frac{\partial g}{\partial q^k} \dot{q}^k = \frac{\partial g}{\partial t'} + \sum_{h=1}^{n} \frac{\partial g}{\partial q'^h} \dot{q}'^h$$

with obvious notation. We leave it to the reader to check this, using (7.76)–(7.78). The new scalar field $G : A(\mathbb{V}^{n+1}) \to \mathbb{R}$ built from $g : \mathbb{V}^{n+1} \to \mathbb{R}$ is called **formal total derivative** of g.

(2) Very often one finds in the literature that the right-hand side of (7.92) is called a *total derivative*. In other words:

$$\frac{\partial g}{\partial t} + \sum_{k=1}^{n} \frac{\partial g}{\partial q^k} \dot{q}^k = \frac{dg}{dt} .$$

If taken literally, this is wrong. In fact *it is not true*, when we write (7.91), that the $\dot{q}^k$ are the derivatives of the q^k. That holds when we have a solution of the Euler-Lagrange equations and we work with it. For this reason we used the word *formal*, as explained by the previous comment. ∎

Exercises 7.33 Figures 7.5, 7.6, 7.7 and 7.8.

(1) Consider two point particles P and Q, both of mass m, on the smooth curve Γ of equations $x = R \cos \phi$, $y = R \sin \phi$, $z = R\phi$ ($R > 0$ constant) with $\phi \in \mathbb{R}$, where x, y, z are orthonormal coordinates on an inertial frame $\mathscr{R}$. Suppose the points undergo, apart from the constraint reaction of Γ, the weight $-mg\mathbf{e}_z$ and that they are joined by an ideal spring of zero length at rest and elastic constant $\kappa > 0$. Finally, suppose there is a viscous friction $-\gamma \mathbf{v}_P|_{\mathscr{R}}$, $\gamma \geq 0$ constant, acting on P.

Fig. 7.5 Exercise 7.33.2

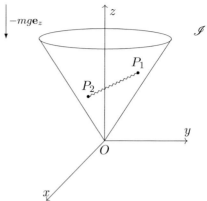

Fig. 7.6 Projection on the xz-plane of Exercise 7.33.3 in frame $\mathscr{R}$

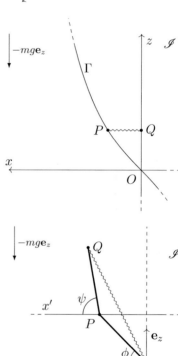

Fig. 7.7 Exercise 7.33.4

(a) Show that if $\gamma = 0$ the system admits a Lagrangian $\mathscr{L}|_{\mathscr{R}}$. Find its explicit form. Write the system of pure equations of motion when $\gamma = 0$ and when $\gamma > 0$. Use as free coordinates the angles ϕ_P and ϕ_Q.

Fig. 7.8 Exercise 7.33.1

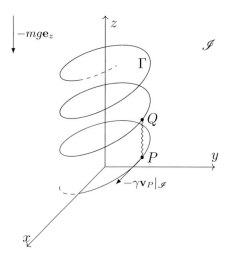

(b) Writing the equations of motion in the new free variables $\theta := \phi_P - \phi_Q$ and $\tau := \phi_P + \phi_Q$, and assuming $\gamma = 0$ and $g = 0$, prove that the quantity:

$$I := \frac{\partial \mathcal{L}|_{\mathcal{R}}}{\partial \dot{\tau}}$$

is conserved during motion. Explain its physical meaning.

(2) Consider two point particles P_1 and P_2 of mass m moving on the smooth conical surface C of equation $z = \sqrt{x^2 + y^2}$ in orthogonal coordinates of origin O of some inertial frame $\mathcal{R}$. The points are joined by an ideal spring of zero length at rest and elastic constant $\kappa > 0$, and subject to gravity $-mg\, \mathbf{e}_z$.

   (a) Write the Euler-Lagrange equations of the system. Describe the points by polar coordinates (r_1, φ_1) and (r_2, φ_2) obtained projecting P_1 and P_2 on the plane $z = 0$.
   (b) Write the equations of motion in the new free variables $\Phi := \varphi_1 - \varphi_2$ and $\Theta := \varphi_1 + \varphi_2$. Prove that the quantity:

$$I := \frac{\partial \mathcal{L}|_{\mathcal{R}}}{\partial \dot{\Theta}}$$

is conserved during motion. Explain the physical meaning of such quantity.

(3) Consider an inertial frame $\hat{\mathcal{R}}$ with orthonormal coordinates $\hat{x}, \hat{y}, \hat{z}$ of origin $\hat{O}$. A second, non-inertial, frame $\mathcal{R}$ is given by orthonormal coordinates moving with $\mathcal{R}$, with origin $O \equiv \hat{O}$ and axes x, y, z. The z-axis coincides with $\hat{z}$ instant by instant, while x and y rotate on the plane $z = 0$ so that $\boldsymbol{\omega}_{\mathcal{R}}|_{\hat{\mathcal{R}}} = \Omega\, \mathbf{e}_z$ with $\Omega > 0$ constant.
A point particle P of mass $m > 0$ travels along the smooth curve Γ, fixed in $\mathcal{R}$, of equation $z = \sinh x$. Beside the constraint reaction $\boldsymbol{\phi}$, P is subject to gravity

$-mg\,\mathbf{e}_z$ and to an ideal spring (zero length at rest) of elastic constant κ joining P to the point Q on the z-axis that is always at the same height as P.

(a) Using as free coordinate for P the x-coordinate, write the equation of motion of P. Work with the Lagrangian of P in an inertial frame $\hat{\mathscr{R}}$.

(b) Show that the quantity:

$$\mathscr{H}(x,\dot{x}) := \dot{x}\,\frac{\partial\mathscr{L}|_{\hat{\mathscr{R}}}}{\partial\dot{x}} - \mathscr{L}|_{\hat{\mathscr{R}}}$$

is a first integral and explain its physical meaning.

(c) Write the x-component of the constraint reaction $\boldsymbol{\phi}$ in terms of x and $\dot{x}$.

(d) Considering the motion with initial conditions $x(0)=0$ and $\dot{x}(0)=v>0$, find along it the x-component of the constraint reaction $\boldsymbol{\phi}$ in terms of x only.

(4) In a frame $\mathscr{R}$ define orthonormal coordinates with origin O and axes $\mathbf{e}_x, \mathbf{e}_y, \mathbf{e}_z$. Consider the system of two point particles P and Q of mass $m > 0$: P is joined to O by a massless rigid rod of length $L > 0$, and Q is joined to P by another massless rigid rod of length L. At the junctures we have ideal constraints allowing the rods to rotate freely in the plane $y = 0$, around O and P respectively. An ideal spring (zero length at rest) of elastic constant $k > 0$ connects Q to O. The frame $\mathscr{R}$ rotates about the $\mathbf{e}_z$-axis with respect to an inertial frame $\hat{\mathscr{R}}$ with $\boldsymbol{\omega}$ vector $\boldsymbol{\omega}_{\mathscr{R}}|_{\hat{\mathscr{R}}} = \Omega\mathbf{e}_z$, $\Omega > 0$ constant.

The point particles are subject to the constraint reactions, the inertial forces and the weight $-mg\mathbf{e}_z$.

Describe the configurations using these angles: ϕ between the rod $P - O$ and the x-axis, and ψ between $Q - P$ and the x'-axis parallel to x through P (fix the counter-clockwise direction with respect to $-\mathbf{e}_y$ as the positive orientation of both angles).

(a) Using the Lagrangian $\mathscr{L} = \mathscr{L}|_{\hat{\mathscr{R}}}$ with respect to $\hat{\mathscr{R}}$, write the Euler-Lagrange equations of motion of the system of points.

(b) Write the equations explicitly in normal form.

(c) If $q^1 = \phi$ and $q^2 = \psi$, show that the Hamiltonian

$$\mathscr{H} := \sum_{k=1,2} \dot{q}^k \frac{\partial\mathscr{L}}{\partial\dot{q}^k} - \mathscr{L}$$

associated with $\mathscr{L}$ and with the free coordinates is a first integral, and discuss its physical meaning.

(d) Prove that the 4 configurations $(\phi_\pm, \psi_\pm) = (\pm\pi/2, \pm\pi/2)$ are *equilibrium configurations* in $\mathscr{R}$: the only solution to the equations of motion that has either one of the 4 configurations as initial condition $(\phi(0), \psi(0))$, together with $(d\phi/dt(0), d\psi/dt(0)) = (0,0)$, is the rest configuration in $\mathscr{R}$: $(\phi(t), \psi(t)) = (\phi(0), \psi(0))$ for any $t \in \mathbb{R}$.

7.5 |AC| Global Differential-Geometric Formulation of the Euler-Lagrange Equations

In this section we will se that we can make precise sense of $\mathbb{V}^{n+1}$ and $A(\mathbb{V}^{n+1})$ within the theory of *fibre bundles* (see Sect. B.3), and also of how the Euler-Lagrange equations should be interpreted in this framework. We will freely use the theory of differentiable manifolds and vector fields (Sect. A.6.5 and ff.).

7.5.1 The Bundle Structures of $\mathbb{V}^{n+1}$ and $A(\mathbb{V}^{n+1})$

Consider Definition 7.6 with $\mathbb{V}^{n+1}$ also satisfying (H3). In this situation $\mathbb{V}^{n+1}$ is a *bundle* (cf. Definition B.4) with *base* given by the axis $\mathbb{R}$ (or equivalently, the Euclidean space $\mathbb{E}^1$) defining absolute time. The fibres at the various base points t are the spaces $\mathbb{Q}_t$, and the *canonical projection* is obviously absolute time T : $\mathbb{V}^{n+1} \to \mathbb{R}$.

Proposition 7.34 *Assuming $\mathbb{V}^{n+1}$ in Definition 7.6 satisfies (H3), then $\mathbb{V}^{n+1}$ is a bundle over $\mathbb{R}$ with canonical projection given by absolute time $T : \mathbb{V}^{n+1} \to \mathbb{R}$ and fibres given by the configuration spaces $\mathbb{Q}_t$ at time $t \in \mathbb{R}$.*
The natural coordinates in the sense of Definition 7.6 are local coordinates adapted to the fibres as per Definition B.4.

Proof The last claim is obvious from Definition 7.6. To prove the first part we only need to check $\mathbb{V}^{n+1}$ is locally diffeomorphic to the product $\mathbb{R} \times \mathbb{Q}$, $\mathbb{Q}$ being a given n-dimensional manifold, the standard fibre, diffeomorphic to any $\mathbb{Q}_t$. More precisely, we have to show that for any $t \in \mathbb{R}$ there exist an open interval $J \ni t$ and a diffeomorphism (a *local trivialisation*) $f : J \times \mathbb{Q} \to T^{-1}(J)$ such that $T(f(\tau, p)) = \tau$ for any $(\tau, p) \in J \times \mathbb{Q}$.
Condition (H3) allows to construct f by choosing $\mathbb{Q} := \mathbb{Q}_t$. If (U_i, ϕ_i) with ϕ_i : $U_i \ni p \mapsto (t(p), q^1(p), \ldots, q^n(p))$ is a local chart as prescribed in (H3), we can associate with it a corresponding local chart (V_i, ψ_i) on $\mathbb{Q}_t$ where $V_i := U_i \cap \mathbb{Q}_t$ and $\psi_i(p) := (q^1(p), \ldots, q^n(p))$. It is then easy to see there is a unique map h : $T^{-1}(J) \to J \times \mathbb{Q}_t$ such that $h|_{U_i} : U_i \ni p \mapsto (t, \psi_i^{-1}(q_i^1(p), \ldots q_i^n(p))) \in J \times \mathbb{Q}_t$, and that it is a diffeomorphism. Putting $f := h^{-1}$ we see that $T(f(\tau, p)) = \tau$ for any $(\tau, p) \in J \times \mathbb{Q}_t$. In this way every fibre $\mathbb{Q}_{t'}$ with $t' \in I$ is diffeomorphic to $\mathbb{Q}_t$, where the diffeomorphism is $g : \mathbb{Q}_t \ni p \mapsto f(t', p) \in \mathbb{Q}_{t'}$. To finish we just have to show all fibres $\mathbb{Q}_t$, for $t \in \mathbb{R}$, are diffeomorphic (so that any one can be taken as $\mathbb{Q}$). This goes as follows: take two instants $t_2 > t_1$, so on the neighbourhood J of any $t \in [t_1, t_2]$ we can define the corresponding diffeomorphism f. In particular that shows $\mathbb{Q}_t$ is diffeomorphic to any $\mathbb{Q}_\tau$ with $\tau \in J$. As $[t_1, t_2]$ is compact, we extract a finite subcover made of intervals J. After finitely many passages, composing the various fibre diffeomorphisms g, $\mathbb{Q}_{t_1}$ becomes diffeomorphic to $\mathbb{Q}_{t_2}$. $\square$

Remarks 7.35

(1) Let us forget for the moment the construction leading to Definition 7.6 from the notion of constraint, and the subsequent hypothesis (H3). We could define, right from the start, a configuration spacetime as a bundle $\mathbb{V}^{n+1}$ over $\mathbb{R}$ with canonical projection $T : \mathbb{V}^{n+1} \to \mathbb{R}$—that at this juncture we view as the physical absolute time—and standard fibre $\mathbb{Q}$ of dimension n. Then $(H3)$ would automatically hold. If $t \in \mathbb{R}$, the local coordinates $\{(U_i, \phi_i)\}_{i \in I}$ defined around $\mathbb{Q}_t := T^{-1}(t)$ that satisfy (H3) are exactly the local coordinates on the product $J \times \mathbb{Q}$ diffeomorphic under f to $T^{-1}(J)$, where $t \in J$ and $T(f(t, p)) = t$. Such coordinates are constructed in the obvious way, from local coordinates on $\mathbb{Q}$ and adding T as first coordinate. The existence of the open interval J and of the diffeomorphism f is warranted by $\mathbb{V}^{n+1}$ being a bundle over $\mathbb{R}$.

(2) On the other hand, one can build systems of constraints where assumption (H3) is *not* satisfied. This is equivalent to saying that the configuration spacetime $\mathbb{V}^{n+1}$ would not be locally diffeomorphic to a product $J \times \mathbb{Q}$ for some given standard fibre $\mathbb{Q}$. For example let us consider a point particle moving on the intersection between a cone, with axis parallel to $\mathbf{e}_z$ moving with a frame $\mathscr{R}$, and a plane whose slope in the direction $\mathbf{e}_z$ varies with time. The *conic sections* to which the particle is constrained at any instant vary, in time, from ellipses to parabolas to hyperbolas. While the first type of curves is compact, the other two are not, so they certainly cannot be diffeomorphic. Thus we would have non-diffeomorphic configuration spaces $\mathbb{Q}_t$ for (some) distinct values t. In this case it is impossible for the configuration spacetime, defined by "solving the constraints" in $\mathbb{V}^{3N+1}$, to be a bundle over the absolute time axis $\mathbb{R}$.

(3) Let us think $\mathbb{V}^{n+1}$, from now on a bundle over the absolute time axis, as immersed in spacetime $\mathbb{V}^{3N+1}$, itself a bundle over $\mathbb{R}$ via absolute time. The former bundle structure determines the latter. In particular, $\mathbb{V}^{n+1}$ is a trivialisable bundle, as is $\mathbb{V}^{3N+1}$, if there exists a frame where the constraints used to build $\mathbb{V}^{n+1}$ do not depend on time. As the base $\mathbb{R}$ is contractible, $\mathbb{V}^{n+1}$ is trivialisable by general Differential Geometry results, irrespective of its physical nature and of the existence privileged frames. Such a trivialisation, however, can be defined in several ways and is very much non-canonical. ∎

To construct $A(\mathbb{V}^{n+1})$ the following elementary notion is important. A **local section** of the bundle $\mathbb{V}^{n+1}$ is by definition a differentiable map (up to the needed order) $I \ni u \mapsto \gamma(u) \in \mathbb{V}^{n+1}$, where $I \subset \mathbb{R}$ is an open interval, such that $T(\gamma(u)) = u$. We can extend this by asking $T(\gamma(u)) = u + c$ for the present purposes, where the constant c reminds us that absolute time T is defined up to additive constants. (The section is called **global** if $I = \mathbb{R}$.) In other words a section is a differentiable curve, of the required class, in $\mathbb{V}^{n+1}$ that may be parametrised by absolute time t. Hence it may have the meaning of a motion of the system.

The *spacetime of kinetic states* $A(\mathbb{V}^{n+1})$ is defined as follows, recalling that $\mathbb{V}^{n+1}$ has a preferred function $T : \mathbb{V}^{n+1} \to \mathbb{R}$ given by absolute time, coinciding with the bundle's projection. First we take the *tangent bundle* (Sect. B.3): $T\mathbb{V}^{n+1}$. The differentiable absolute time map $T : \mathbb{V}^{n+1} \to \mathbb{R}$ extends to $T\mathbb{V}^{n+1}$ to a map

$$\tilde{T} : T\mathbb{V}^{n+1} \ni (p, v_p) \mapsto T(p) .$$

Henceforth $\tilde{T}$ will be denoted by T. Natural local coordinates $(t, q^1, \ldots, q^n)$ on $\mathbb{V}^{n+1}$ define canonical local natural coordinates on $T\mathbb{V}^{n+1}$: $t, q^1, \ldots, q^n, \dot{t}, \dot{q}^1, \ldots,$ $\dot{q}^n$ where $\dot{t}, \dot{q}^1, \ldots, \dot{q}^n$ are the components $\partial/\partial t, \partial/\partial q^1, \ldots \partial/\partial q^n$ of a vector of $T_{(t, q^1, \ldots, q^n)} \mathbb{V}^{n+1}$ on the natural base (associated with the coordinates on $T\mathbb{V}^{n+1}$). A possible motion, i.e. a *local section* $I \ni t \mapsto \gamma(t) \in \mathbb{V}^{n+1}$ of class C^2, parametrised by absolute time $t \in I$ open interval of $\mathbb{R}$, defines a local C^1 section on $T\mathbb{V}^{n+1}$, called its **lift**: $I \ni t \mapsto (\gamma(t), \dot{\gamma}(t))$. In natural local coordinates on $\mathbb{V}^{n+1}$ and $T\mathbb{V}^{n+1}$ we then have, explicitly:

$$I \ni t \mapsto (t, q^1(t), \ldots, q^n(t))$$

$$\text{and} \quad I \ni t \mapsto \left(t, q^1(t), \ldots, q^n(t), 1, \frac{dq^1}{dt}, \ldots, \frac{dq^n}{dt} \right)$$

for the curves. Note that the t-component of the tangent vector $\dot{\gamma}$ is always 1 by construction (even if we change the constant of absolute time). This happens in particular for any motion. We can write this condition more intrinsically using Appendix A:

$$\langle \dot{\gamma}(t), dT_{\gamma(t)} \rangle = 1 .$$

If we consider only points (p, v_p) in $T\mathbb{V}^{n+1}$ that satisfy $\langle v_p, dT_p \rangle - 1 = 0$, by the Regular-value theorem we obtain an embedded submanifold of dimension $2n + 2 - 1 = 2n + 1$, called $A(\mathbb{V}^{n+1})$. This manifold contains all pairs (p, v_p) where p is reached by a motion with tangent vector v_p. $A(\mathbb{V}^{n+1})$ can be seen as a bundle, in two distinct ways:

(1) over $\mathbb{V}^{n+1}$, with the subsets of any tangent space $T_p \mathbb{V}^{n+1}$:

$$\left\{ (p, v_p) \in T_p \mathbb{V}^{n+1} \mid \langle v_p, dT_p \rangle = 1 \right\}$$

as fibres;

(2) over the time axis $\mathbb{R}$, with the *spaces of kinetic states* $\mathbb{A}_t$, for any time $t \in \mathbb{R}$,

$$\mathbb{A}_t = \left\{ (c, v_{(t,c)}) \in T_{(t,c)} \mathbb{V}^{n+1} \mid c \in \mathbb{Q}_t , \quad \langle v_{(t,c)}, dT v_{(t,c)} \rangle = 1 \right\}$$

as fibres.

For any system of natural local coordinates $(t, q^1, \ldots, q^n)$ on the configuration spacetime $\mathbb{V}^{n+1}$, the local coordinates $(t, q^1, \ldots, q^n, \dot{q}^1, \ldots, \dot{q}^n)$ are then local coordinates on the spacetime of kinetic states $A(\mathbb{V}^{n+1})$ adapted to the fibres of both bundles, in the obvious sense (Definition B.4): on the first bundle, $(t, q^1, \ldots, q^n)$ are coordinates on the base and the other $(\dot{q}^1, \ldots, \dot{q}^n)$ are fibre coordinates; on the second bundle, t is a coordinate on the base and the $(q^1, \ldots q^n, \dot{q}^1, \ldots, \dot{q}^n)$ are coordinates on each fibre. Either collection of adapted local coordinates form an atlas by construction. The transformation rules for coordinates of this type are precisely (7.71), (7.72), and (7.73).

That $A(\mathbb{V}^{n+1})$ is a bundle in the sense of (2) above means it has local trivialisations over $\mathbb{R}$ with absolute time projection and canonical fibre $\mathbb{A}$ diffeomorphic to any $\mathbb{A}_t$ as per Definition B.4. All this is proved exactly as for $\mathbb{V}^{n+1}$, see the proof of Proposition 7.34. For $\mathbb{V}^{n+1}$ the proof was based on the existence of natural local charts satisfying (H3). With the above construction of $A(\mathbb{V}^{n+1})$, the same charts give natural coordinates and satisfy similar conditions: for any absolute time $t \in \mathbb{R}$ there is an open, non-empty interval $J \subset \mathbb{R}$ containing t and also natural coordinates $\{(U_i, \phi_i)\}_{i \in I}$ with $\phi_i : U_i \ni a \mapsto (t_i(a), q_i^1(a), \ldots q_i^n(a), \dot{q}^1(a), \ldots, \dot{q}^n(a)) \in \mathbb{R}^{2n+1}$, satisfying the following three requirements:

(i) $\cup_{i \in I} U_i \supset \mathbb{A}_\tau$ for any $\tau \in J$;
(ii) $\phi_i(U_i) \supset J \times U_i'$, with $U_i' \subset \mathbb{R}^{2n}$ open, for any $i \in I$;
(iii) for any $i, j \in I$ with $U_i \cap U_j \neq \varnothing$ the functions $\phi_i \circ \phi_j^{-1} : \phi_j(U_i \cap U_j) \to \phi_i(U_i \cap U_j)$ have the form:

$$t_i = t_j, \quad q_i^k = q_i^k(q_j, \ldots, q_j^n), \quad \dot{q}_i^k = \sum_h \frac{\partial q_i^k}{\partial q_j^h} \dot{q}_j^h$$

where, i.e., the transformation rule for the Lagrangian coordinates does not depend explicitly on time.

At this point, the same glueing procedure of Proposition 7.34 shows that for the aforementioned J around the given $t \in \mathbb{R}$, there is a diffeomorphism (a local trivialisation) $f : J \times \mathbb{A}_t \to T^{-1}(A(\mathbb{V}^{n+1}))$ such that $T(f(\tau, a)) = \tau$ if $\tau \in J$ and $a \in \mathbb{A}_t$.

Remarks 7.36

(1) Also $A(\mathbb{V}^{n+1})$ is trivialisable, since it fibres over $\mathbb{R}$. Some trivialisations are inherited from $\mathbb{V}^{n+1}$, for example by the choice of a frame in $\mathbb{V}^4$. There are however other trivialisations that do not come from a frame.

(2) $A(\mathbb{V}^{n+1})$ is a canonical object in the theory of *jet bundles*, which we shall not address but only touch upon in what follows. $A(\mathbb{V}^{n+1}) = j^1(\mathbb{V}^{n+1}) := \cup_{t \in \mathbb{R}} \mathbb{A}_t$ is the first (hence the superscript [1]) jet bundle of $\mathbb{V}^{n+1} \to \mathbb{R}$. Within the framework of jet bundles the definition of $j^1(\mathbb{V}^{n+1}) = A(\mathbb{V}^{n+1})$ is different but equivalent to the above one for $A(\mathbb{V}^{n+1})$. One starts from $\mathbb{V}^{n+1} \to \mathbb{R}$, and given an open interval $I \subset \mathbb{R}$ containing t_0, one considers *local sections* $I \ni t \mapsto \gamma(t) \in \mathbb{V}^{n+1}$ through $(t_0, q_0^1, \ldots, q_0^n)$, in natural coordinates. Two such sections γ, γ' are equivalent if their tangent vectors at $(t_0, q_0^1, \ldots, q_0^n)$ are the same. Clearly, given natural coordinates $(t, q^1, \ldots, q^n)$, the space of equivalence classes of $(t_0, q_0^1, \ldots, q_0^n)$ is parametrised by coordinates $(\dot{q}^1, \ldots, \dot{q}^n)$ giving all possible tangent vectors to the sections of the equivalence class of $(t_0, q_0^1, \ldots, q_0^n)$. If we vary the configuration, we obtain a local parametrisation depending on $(t, q^1, \ldots, q^n, \dot{q}^1, \ldots, \dot{q}^n)$. By construction, the transformation rules for coordinates of this type are still (7.71), (7.72), and (7.73). In this manner we recover the local structure of $A(\mathbb{V}^{n+1})$ described earlier.   ∎

7.5.2   The Dynamic Vector Field Associated with the Euler-Lagrange Equations

Theorem 7.23, together with Theorem 7.28 and the discussion of Sect. 7.4.4, prove that if the Lagrangian has the usual form $\mathscr{T}|_{\mathscr{R}} + \mathscr{V}|_{\mathscr{R}}$, maximal solutions of the Euler-Lagrange equations (7.36) on $A(\mathbb{V}^{n+1})$ exist, are unique for every initial kinetic state and are independent of the local coordinates, even if the equations are by nature local. The result can be generalised immediately to C^2 Lagrangians of the form (7.87) when the matrix a_{hk} is non-singular in any natural local chart.

Assuming the manifold $A(\mathbb{V}^{n+1})$ is at least of class C^3, let us consider a generic function $\mathscr{L} : A(\mathbb{V}^{n+1}) \to \mathbb{R}$ of class C^3. By (7.77) it follows immediately that for any pair of natural local coordinate systems and any given kinetic state $a \in A(\mathbb{V}^{n+1})$, with obvious notation

$$\left.\frac{\partial^2 \mathscr{L}}{\partial \dot{q}'^r \partial \dot{q}'^s}\right|_a = \sum_{k,h=1}^{n} \left.\frac{\partial q^k}{\partial q'^r}\right|_a \left.\frac{\partial q^h}{\partial q'^s}\right|_a \left.\frac{\partial^2 \mathscr{L}}{\partial \dot{q}^k \partial \dot{q}^h}\right|_a .$$

As the Jacobian matrix $\frac{\partial q^k}{\partial q'^r}$ is always non-singular, the identity shows that there is no contradiction in supposing that the Lagrangian of a generic function has Hessian matrix, with respect to the dotted coordinates, *everywhere* non-singular on $A(\mathbb{V}^{n+1})$ in *every* natural local chart. Even more is true: to guarantee that condition, it suffices for it to hold in a possibly non-maximal atlas of natural local charts. In case the Lagrangian has form (7.87) the Hessian matrix will coincide with the matrix of coefficients a_{kh}, and will only depend on time and on the undotted coordinates. In general, the Hessian matrix will still depend on the dotted coordinates. In any case the proof of Theorems 7.23 and 7.28 carry through in essentially the same way under the more general assumptions of a C^3 Lagrangian with Hessian, in the dotted coordinates, everywhere non-singular on an atlas of natural local charts.[12] Repeating the argument of Sect. 7.4.4 we find the next result.

---

[12] The purpose of demanding class C^3 is so that the Hessian is C^1 in the general case. If $\mathscr{L}$ is of form (7.87), C^2 is sufficient in what follows.

Theorem 7.37 *Let $\mathscr{L} : A(\mathbb{V}^{n+1}) \to \mathbb{R}$ be of class C^3 with Hessian matrix $\frac{\partial^2 \mathscr{L}}{\partial \dot{q}^k \partial \dot{q}^h}$ everywhere non-singular on an atlas of natural local charts. For any $a \in A(\mathbb{V}^{n+1})$ there exists a unique section $I \ni t \mapsto \gamma(t) \in A(\mathbb{V}^{n+1})$ of class C^1 that:*

(1) satisfies the Euler-Lagrange equations (7.84) in any natural local chart it crosses,
(2) passes through a,
(3) is maximal among sections respecting (1) and (2).

Any other C^1 section satisfying (1) and (2) is a restriction of γ.

With this technical premise in place, which we shall use a lot henceforth, we pass to a completely different approach for interpreting the Euler-Lagrange equations. Consider first the case where the active forces on the physical system are given by Lagrangian components $\mathscr{Q}_k|_{\mathscr{R}}$ and the Euler-Lagrange equations are in form (7.36) after we fix a frame $\mathscr{R}$ in which we define the kinetic energy $\mathscr{T}|_{\mathscr{R}}$. If, in natural coordinates, $I \ni t \mapsto (t, q^1(t), \ldots, q^n(t), \dot{q}^1(t), \ldots, \dot{q}^n(t))$ solves the Euler-Lagrange equations, from (7.36) its tangent vector is:

$$\frac{\partial}{\partial t} + \sum_{k=1}^{n} \dot{q}^k(t) \frac{\partial}{\partial q^k} + \sum_{k=1}^{n} z^k(t, q^1(t), \ldots, q^n(t), \dot{q}^1(t), \ldots, dq^n(t)) \frac{\partial}{\partial \dot{q}^k} .$$

Suppose the assumptions guaranteeing the existence and uniqueness for the maximal solution of the Euler-Lagrange equations through a given kinetic state $a \in A(\mathbb{V}^{n+1})$ do hold. As established in Theorem 7.23 and from the discussion of Sect. 7.4.4, it is enough for $\mathscr{T}|_{\mathscr{R}}$ to be C^2 and the $\mathscr{Q}_k|_{\mathscr{R}}$ to be C^1. For systems described by a C^3 Lagrangian $\mathscr{L} : A(\mathbb{V}^{n+1}) \to \mathbb{R}$ without obvious physical meaning, we will assume that on an atlas of natural coordinates the Hessian matrix $\frac{\partial^2 \mathscr{L}}{\partial \dot{q}^k \partial \dot{q}^h}$ is everywhere non-singular, so that Theorem 7.37 holds. Then we can define a vector field on $A(\mathbb{V}^{n+1})$ using the tangent vector to the only solution passing through any a. We denote this field with Z and call it **dynamic vector field**. In natural local coordinates Z is then given by

$$Z(t, q^1, \ldots, q^n, \dot{q}^1, \ldots, \dot{q}^n) = \frac{\partial}{\partial t} + \sum_{k=1}^{n} \dot{q}^k \frac{\partial}{\partial q^k}$$

$$+ \sum_{k=1}^{n} z^k(t, q^1, \ldots, q^n, \dot{q}^1, \ldots, \dot{q}^n) \frac{\partial}{\partial \dot{q}^k} . \tag{7.94}$$

If γ solves the Euler-Lagrange equations, then:

$$\frac{d\gamma}{dt} = Z(\gamma(t)) , \tag{7.95}$$

and so

$$Z(f(\gamma(t))) = \frac{d}{dt} f(\gamma(t)) \tag{7.96}$$

will hold for any differentiable map $f : A(\mathbb{V}^{n+1}) \to \mathbb{R}$, even if defined locally. Keeping (7.96) in account, equations (7.36) and (7.84) read, in natural local coordinates:

$$Z\left(\frac{\partial \mathscr{T}|_{\mathscr{R}}}{\partial \dot{q}^k}\right) - \frac{\partial \mathscr{T}|_{\mathscr{R}}}{\partial q^k} = \mathscr{Q}_k|_{\mathscr{R}}, \quad \text{for any } k = 1, \ldots, n, \tag{7.97}$$

and

$$Z\left(\frac{\partial \mathscr{L}}{\partial \dot{q}^k}\right) - \frac{\partial \mathscr{L}}{\partial q^k} = 0, \quad \text{for any } k = 1, \ldots, n \tag{7.98}$$

respectively. This changes the perspective regarding the meaning of the Euler-Lagrange equations: we may think them as *algebraic* equations in the unknown components z^k of the vector Z appearing in (7.94) in terms of: the source terms $\mathscr{Q}_k|_{\mathscr{R}}$ and the first and second derivatives of $\mathscr{T}|_{\mathscr{R}}$ in case (7.97), or the Lagrangian $\mathscr{L}$ and its first and second derivatives in case (7.98). In fact the explicit expression of the n functions $z^k(t, q^1, \ldots, q^n, \dot{q}^1, \ldots, \dot{q}^n)$ is found by writing in normal form the Euler-Lagrange equations, as proved in Theorem 7.23: inverting the matrix a_{hk} obtained from the kinetic energy, that is, from the Hessian of $\mathscr{L}$ in the dotted coordinates. Once we have built the field Z of the form (7.94) on $A(\mathbb{V}^{n+1})$ and found the components z^k from the *algebraic* equations (7.97) or (7.98), the curves describing the system's motion with given initial conditions are just the integral curves of Z with those initial conditions. Under the assumptions we made, in particular, Z is C^1 and therefore ODE (7.95) itself ensures (Theorem 14.40) existence and uniqueness for the maximal solution with the given initial conditions.

Remarks 7.38 It has to be pointed out that the dynamic vector field is unique, since it is directly associated with the solutions to the Euler-Lagrange equations, provided the latter obey the Existence and uniqueness theorem. This is in contrast to the Lagrangian, which apart from not being unique, may not exist at all if the Lagrangian components of the active forces do not have a suitable form. ∎

The notion of dynamic vector field allows us to define rigorously the total formal derivative of $g : \mathbb{V}^{n+1} \to \mathbb{R}$. First, any such function extends naturally to $A(\mathbb{V}^{n+1})$ simply by composing $g \circ P$, where $P : A(\mathbb{V}^{n+1}) \to \mathbb{V}^{n+1}$ is the canonical projection $(t, q, \dot{q}) \mapsto (t, q)$ in any natural coordinates.

Definition 7.39 Given $g : \mathbb{V}^{n+1} \to \mathbb{R}$ of class C^1, the **total formal derivative** of g is:

$$Z(g \circ P),$$

where Z is an arbitrary dynamic vector field defined on $A(\mathbb{V}^{n+1})$. ◇

It is clear the definition does not depend on Z, since in any natural coordinates $Z(g \circ P)$ has the form appearing in the right-hand side of (7.93), and that is independent of the components with respect to the base vectors $\frac{\partial}{\partial \dot{q}^k}$.

7.5.3 Contact Forms, Poincaré-Cartan Form and Intrinsic Formulation of the Euler-Lagrange Equations Induced by a Lagrangian

Introducing a few additional theoretical notions and using the *Lie derivative* (see Sects. B.1.2 and B.4.3) we can finally present an even more intrinsic version of Eq. (7.97) to determine the dynamic field Z given by a Lagrangian $\mathscr{L}$ that completely describes the system. Consider the atlas of natural local charts on $A(\mathbb{V}^{n+1})$ and fix a generic local chart (U, ψ) with coordinates $t, q^1, \ldots, q^n, \dot{q}^1, \ldots, \dot{q}^n$. On a such chart we define n **contact 1-forms**:

$$\omega^k := dq^k - \dot{q}^k dt \, . \tag{7.99}$$

Beware that the superscript ^{k} in ω^k *does not* indicate a component but one of the n contact forms on U in the given natural coordinates. There is a contact form for any coordinates q^k, and each such lives on $A(\mathbb{V}^{n+1})$ (we could add the trivial term $0 d\dot{q}^k$ on the right in (7.99)). Hence it defines, for any kinetic state $a \in U$, a cotangent vector at a with respect to the basis of 1-forms $dt|_a, dq^1|_a, \ldots dq^n|_a, d\dot{q}^1|_a, \ldots, d\dot{q}^n|_a$ of $T_a^* A(\mathbb{V}^{n+1})$. Contact forms enjoy a remarkable property. If (U', ψ') is another natural chart with coordinates $t', q'^1, \ldots, q'^n, \dot{q}'^1, \ldots, dq'^n$ such that $U \cap U' \neq \varnothing$, then for $a \in U \cap U'$

$$\omega'^k|_a = \sum_{h=1}^{n} \frac{\partial q'^k}{\partial q^h}\Big|_a \omega^h|_a \, . \tag{7.100}$$

The proof is obvious, it suffices to decompose the basis of $T_a^* A(\mathbb{V}^{n+1})$ in the second coordinates with respect to the natural basis of the first coordinates.

If $T : A(\mathbb{V}^{n+1}) \to \mathbb{R}$ is *absolute time* on the spacetime of kinetic states, extended from $\mathbb{V}^{n+1}$ in the obvious way, the structure (7.94) of a generic vector field Z on $A(\mathbb{V}^{n+1})$ is patently equivalent to the following pair of conditions, which justify the introduction of contact forms:

$$\langle Z, dT \rangle = 1 \, , \tag{7.101}$$

$$\langle Z, \omega^k \rangle = 0 \quad \text{for any contact form } \omega^k \text{ in any natural coordinates.} \tag{7.102}$$

The first relation is equivalent to asking that in any natural coordinates on $A(\mathbb{V}^{n+1})$, the component of Z along $\frac{\partial}{\partial t}$ equals 1. The second relation is equivalent to asking that the components of Z along $\frac{\partial}{\partial q^k}$ are equal to $\dot{q}^k$. In other words the second condition implies n "trivial Euler-Lagrange equations", namely the second row of (7.36).

The following abstract definition prescinds from having a Lagrangian or any Euler-Lagrange equations.

Definition 7.40 A vector field Z on $A(\mathbb{V}^{n+1})$ is called **dynamic vector field**, or just **dynamic vector**, whenever (7.101) and (7.102) hold. $\diamond$

Given a dynamic vector field Z (of class C^1), using (B.29) we immediately verify the relationships:

$$\mathcal{L}_Z(dt) = \mathcal{L}_Z(dT) = d\langle Z, dT \rangle = 0, \tag{7.103}$$

$$\mathcal{L}_Z(dq^k) = d\dot{q}^k, \tag{7.104}$$

$$\mathcal{L}_Z(\omega^k) = d\dot{q}^k - z^k dt . \tag{7.105}$$

The last ingredient is the **Poincaré-Cartan 1-form** $\Omega_{\mathscr{L}}$ on $A(\mathbb{V}^{n+1})$ associated with a function $\mathscr{L} : A(\mathbb{V}^{n+1}) \to \mathbb{R}$, of class C^1 at least and viewed as abstract Lagrangian (though not necessarily of the form $\mathscr{T}|_{\mathscr{R}} + \mathscr{V}|_{\mathscr{R}}$). Such 1-form is completely determined by the fact that for any natural coordinates on $A(\mathbb{V}^{n+1})$ it looks like

$$\Omega_{\mathscr{L}} = \sum_{k=1}^{n} \frac{\partial \mathscr{L}}{\partial \dot{q}^k} \omega^k + \mathscr{L} dT . \tag{7.106}$$

Using (7.100) and (7.77) it is easy to check that on natural charts' overlaps the right-hand sides of (7.106) coincide. If we compute the Lie derivative of $\Omega_{\mathscr{L}}$ along a vector Z of the form (7.94) we find:

$$\mathcal{L}_Z(\Omega_{\mathscr{L}}) = \sum_k Z\left(\frac{\partial \mathscr{L}}{\partial \dot{q}^k}\right) \omega^k + \sum_k \frac{\partial \mathscr{L}}{\partial \dot{q}^k} \mathcal{L}_Z(\omega^k) + Z(\mathscr{L})dT + \mathscr{L}\mathcal{L}_Z(dT)$$

$$= \sum_k Z\left(\frac{\partial \mathscr{L}}{\partial \dot{q}^k}\right)(dq^k - \dot{q}^k dt) + \sum_k \frac{\partial \mathscr{L}}{\partial \dot{q}^k} d\dot{q}^k - \sum_k \frac{\partial \mathscr{L}}{\partial \dot{q}^k} z^k dt + Z(\mathscr{L})dt$$

$$= \sum_k Z\left(\frac{\partial \mathscr{L}}{\partial \dot{q}^k}\right) dq^k + \sum_k \frac{\partial \mathscr{L}}{\partial \dot{q}^k} d\dot{q}^k + \left(Z(\mathscr{L}) - Z\left(\frac{\partial \mathscr{L}}{\partial \dot{q}^k}\right)\dot{q}^k - \sum_k \frac{\partial \mathscr{L}}{\partial \dot{q}^k} z^k\right) dt$$

Suppose Z arises from $\mathscr{L}$ as per (7.97) when the Lagrangian is quadratic in the dotted coordinates in a natural atlas on $A(\mathbb{V}^{n+1})$ where the (dotted) Hessian of $\mathscr{L}$ is everywhere non-singular. In other words Z comes from (in the known components

z^k of Z)

$$Z\left(\frac{\partial \mathscr{L}}{\partial \dot{q}^k}\right) - \frac{\partial \mathscr{L}}{\partial q^k} = 0, \quad \text{for any } k = 1, \ldots, n.$$ (7.107)

With the assumptions made, the above can always be solved, as we discussed in Sect. 7.5.2. Then the above expression for $\mathcal{L}_Z(\Omega_{\mathscr{L}})$ simplifies to

$$\mathcal{L}_Z(\Omega_{\mathscr{L}}) = \sum_k \frac{\partial \mathscr{L}}{\partial q^k} dq^k + \sum_k \frac{\partial \mathscr{L}}{\partial \dot{q}^k} d\dot{q}^k + \left(Z(\mathscr{L}) - \frac{\partial \mathscr{L}}{\partial q^k} \dot{q}^k - \sum_k \frac{\partial \mathscr{L}}{\partial \dot{q}^k} z^k\right) dt.$$

Using (7.94) this finally gives

$$\mathcal{L}_Z(\Omega_{\mathscr{L}}) = d\mathscr{L}.$$ (7.108)

We conclude that if the physical system is completely described by a Lagrangian $\mathscr{L} : A(\mathbb{V}^{n+1}) \to \mathbb{R}$ and Z is the tangent vector field to the Euler-Lagrange solutions, (7.108) holds. The argument can be reversed: if Z is a dynamic vector that satisfies (7.108) for a Lagrangian $\mathscr{L}$, then it must satisfy (7.107) in any natural coordinates. To complete the discussion, the following theorem provides an alternative and equivalent form of (7.108) in purely algebraic terms, by using the *inner product* of vectors and forms (B.17).

Theorem 7.41 *Consider a Lagrangian $\mathscr{L} : A(\mathbb{V}^{n+1}) \to \mathbb{R}$ of class*[13] *C^3 such that the Hessian $\frac{\partial^2 \mathscr{L}}{\partial \dot{q}^k \partial \dot{q}^h}$ is everywhere non-singular in an atlas of natural local charts. Let Z be a C^1 dynamic vector field. The following facts are equivalent.*

(1) The integral curves of Z satisfy the Euler-Lagrange equations (7.84) with respect to the Lagrangian $\mathscr{L}$ in any system of natural coordinates on $A(\mathbb{V}^{n+1})$;

(2) Z satisfies conditions (7.107):

$$Z\left(\frac{\partial \mathscr{L}}{\partial \dot{q}^k}\right) - \frac{\partial \mathscr{L}}{\partial q^k} = 0, \quad \text{for any } k = 1, \ldots, n$$

in any system of natural coordinates on $A(\mathbb{V}^{n+1})$;

(3) Z satisfies relation (7.108) everywhere on $A(\mathbb{V}^{n+1})$:

$$\mathcal{L}_Z(\Omega_{\mathscr{L}}) = d\mathscr{L};$$

(4) Z satisfies the algebraic equation:

$$Z \lrcorner d\Omega_{\mathscr{L}} = 0$$ (7.109)

---

[13] If $\mathscr{L}$ is of form (7.87) C^2 is enough.

everywhere on $A(\mathbb{V}^{n+1})$.

Furthermore:

(5) *there exists a unique dynamic field Z of class C^1 satisfying (1)–(4);*

(6) *through $a \in A(\mathbb{V}^{n+1})$ there is one, and only one, maximal integral curve of such Z; this curve is the unique maximal solution to the Euler-Lagrange equations (7.84) for $\mathscr{L}$ with initial condition a.*

Proof We have already settled the fact that (1) and (2) are equivalent if Z satisfies (7.101)–(7.102). Let us show that (3) implies (2). If Z is as in (7.94), the same manipulations produce:

$$\mathcal{L}_Z(\Omega_{\mathscr{L}}) - d\mathscr{L} = \sum_k \left(Z\left(\frac{\partial \mathscr{L}}{\partial \dot{q}^k} \right) - \frac{\partial \mathscr{L}}{\partial q^k} \right) dq^k - \sum_k \left(Z\left(\frac{\partial \mathscr{L}}{\partial \dot{q}^k} \right) - \frac{\partial \mathscr{L}}{\partial q^k} \right) \dot{q}^k dt ,$$

valid in any natural chart of $A(\mathbb{V}^{n+1})$. As the $n + 1$ one-forms dq^k and dt are linearly independent, Z must satisfy (7.107). Hence (3) implies (2). The converse was proved earlier and so (2) and (3) are equivalent. To show that (3) and (4) are equivalent we observe that *Cartan's magic formula* (B.30) gives $\mathcal{L}_Z(\Omega_{\mathscr{L}}) = Z \lrcorner\, d\Theta_{\mathscr{L}} + d(Z \lrcorner \Theta_{\mathscr{L}})$. But (7.101) and (7.102) immediately give $Z \lrcorner \Theta_{\mathscr{L}} = \mathscr{L}$. Hence the left-hand side of (7.108) reads $Z \lrcorner\, d\Theta_{\mathscr{L}} + d\mathscr{L}$, from which (7.108) is equivalent to (7.109), that is, (3) $\Longleftrightarrow$ (4). We have proved that (1), (2), (3), and (4) are equivalent under the assumptions on Z. As for (5) and (6), we start from (2) and write the algebraic equation for the coefficients z^k of Z in (7.94). This equation is solvable when the Hessian $\frac{\partial^2 \mathscr{L}}{\partial \dot{q}^h \partial \dot{q}^k}$ is everywhere non-singular in an atlas of natural local coordinates, and the z^k are C^1 functions in the natural coordinates. The Existence and uniqueness theorem 14.40 applied to the integral equation (7.95), together with Theorem 7.37, prove the final claim. $\qquad\square$

Equations (7.108) and (7.109), in view of (7.101) and (7.102), respectively represent the *most intrinsic version of the Euler-Lagrange equations* in purely *differential* or *algebraic* terms.

Definition 7.42 Given a Lagrangian $\mathscr{L}$ satisfying Theorem 7.41, the unique dynamic vector field Z fulfilling (7.108) or the equivalent (7.109) is said to be **associated** with $\mathscr{L}$. $\diamond$

7.5.4 Dynamic Vector Field on $A(\mathbb{V}^{n+1})$ Without Global Lagrangian

The introduction of the globally defined dynamic vector field Z also allows us to deal with another matter of theoretical interest. Until this moment we have worked with physical systems that may admit a Lagrangian or not, but in either case the vector Z could be defined. There is a third situation. In dealing with physical systems that admit Lagrangian we have always assumed the Lagrangian is defined on the entire spacetime of kinetic states: $\mathscr{L} : A(\mathbb{V}^{n+1}) \to \mathbb{R}$. The dynamics is thus determined by the field Z built from $\mathscr{L}$. Yet there are physical systems that, albeit admitting a field Z, only have a local Lagrangian, instead of a global one. Consider one of the simplest cases, $\mathbb{V}^{2+1} = \mathbb{R} \times (\mathbb{R}^2 \backslash \{(0, 0)\})$. As usual $\mathbb{R}$ is the absolute time axis and $\mathbb{R}^2 \backslash \{(0, 0)\}$ is any configuration space at time t. Suppose the equations of motion are, in canonical coordinates on $\mathbb{R}^2$:

$$m \frac{d^2 x}{dt^2} = -\frac{ky}{x^2 + y^2} , \quad m \frac{d^2 x}{dt^2} = \frac{kx}{x^2 + y^2} .$$

If we consider the global field on $A(\mathbb{V}^{n+1})$:

$$Z = \frac{\partial}{\partial t} + \dot{x} \frac{\partial}{\partial x} + \dot{y} \frac{\partial}{\partial y} - \frac{k}{m} \frac{y}{x^2 + y^2} \frac{\partial}{\partial \dot{x}} + \frac{k}{m} \frac{x}{x^2 + y^2} \frac{\partial}{\partial \dot{y}} ,$$

the integral curves correspond to the above equations of motion. Indeed, the Euler-Lagrange equations can always be written as (7.97), where:

$$\mathscr{Q}_x = -ky(x^2 + y^2)^{-1} , \quad \mathscr{Q}_y = kx(x^2 + y^2)^{-1} .$$

Since the above are C^1 (actually C^∞), and by direct computation they define an irrotational field because:

$$\frac{\partial \mathscr{Q}_x}{\partial y} = \frac{\partial \mathscr{Q}_y}{\partial x} ,$$

at least on open and simply connected subsets of $\mathbb{R}^2$ there will exist a function $\mathscr{U} = \mathscr{U}(x, y)$ of class C^2 (actually C^∞) for which

$$\mathscr{Q}_x = -\frac{\partial \mathscr{U}}{\partial x} , \quad \mathscr{Q}_y = -\frac{\partial \mathscr{U}}{\partial y} .$$

If $\mathscr{U}$ were defined on $\mathbb{R}^2$ the system would be described by a standard Lagrangian $\mathscr{L} = \frac{m}{2}(\dot{x}^2 + \dot{y}^2) - \mathscr{U}(x, y)$. However there is no global function $\mathscr{U}$ on $\mathbb{R}^2 \backslash \{(0, 0)\}$ whose gradient produces $\mathscr{Q}_x$ and $\mathscr{Q}_y$. If there were, the integral of the vector field with those components along any closed path in the domain would have to vanish; but it does not. We can for example integrate along a counter-clockwise circle of radius R centred at the origin: the integral equals $2\pi k$. Hence there cannot be a global Lagrangian, if we want it of the form $\mathscr{T} - \mathscr{U}$. Nevertheless the system is described, globally, by the above Z.

Exercises 7.43 Cut $\mathbb{R}^2 \setminus \{(0, 0)\}$ along a half-line s from the origin and consider the simply connected domain obtained by removing s from the plane. Using polar coordinates, show that any function $\mathcal{U}$ giving the above Lagrangian components must necessarily be $\mathcal{U}(r, \theta) = -k\theta$, where θ is the counter-clockwise angle measured from s.

Chapter 8
Symmetries and Conservation Laws in Lagrangian Mechanics

In this chapter we consider results in Lagrangian Mechanics that had key conse-quences on the successive developments of Theoretical and Mathematical Physics, even in contexts far removed from Classical Mechanics. We will introduce theorems relating the symmetries of the Lagrangian to the existence of quantities that remain *constant during the system's dynamical evolution*, known as *first integrals*. We will, in particular, prove Noether's Theorem and Jacobi's Theorem. In the final section we shall reformulate Noether's Theorem in a very general form using the suitable language of Differential Geometry, and show that Jacobi's Theorem is a special case of Noether's.

8.1 The Relationship Between Symmetry and Conservation Laws: Cyclic Coordinates

In this section we begin the study of the relationship between the symmetries of a system that admits a Lagrangian and some of its first integrals, i.e. quantities conserved along the system's motions. We will concentrate on the concept of *cyclic coordinate* and the associated conjugate momentum. We shall exemplify the results by showing how, under suitable assumptions on the physical system, the invariance under displacements has to do with the conservation of the impulse, while the invariance under rotations is related to the conservation of the angular momentum.

V. Moretti, *Analytical Mechanics*, La Matematica per il 3+2 150,
https://doi.org/10.1007/978-3-031-27612-5_8

8.1.1 Cyclic Coordinates and Constancy of Conjugate Momenta on the Motion

Consider a physical system of point particles $\mathscr{S}$ described on the spacetime of kinetic states $A(\mathbb{V}^{n+1})$ by a Lagrangian $\mathscr{L}$. The latter might be associated with a frame (inertial or not), but for the moment we shall stay general. Suppose we have chosen natural local coordinates

$$(t, q, \dot{q}) := (t, q^1, \ldots, q^n, \dot{q}^1, \ldots, \dot{q}^n) \in I \times U \times \mathbb{R}^n$$

and that, in the spacetime of kinetic states with the above coordinates, relatively to a given coordinate q^j:

$$\frac{\partial \mathscr{L}(t, q, \dot{q})}{\partial q^j} = 0, \quad \text{for any choice of } (t, q^1, \ldots, q^n, \dot{q}^1, \ldots, \dot{q}^n).$$

If so we will say the coordinates q^j is **cyclic** or **ignorable**. Cyclicity with respect to q^j is completely *equivalent* to the Lagrangian's *invariance* under transformations that only alter q^j and leave the others unchanged:

$$(q^1, \ldots, q^{j-1}, q^j, q^{j+1}, \ldots, q^n) \mapsto (q^1, \ldots, q^{j-1}, q^j + \Delta q^j, q^{j+1}, \ldots, q^n).$$

To say that (8.2) holds is indeed clearly the same as:

$$
\begin{cases}
\mathscr{L}(t, q^1, \ldots, q^{j-1}, q^j + \Delta q^j, q^{j+1}, \ldots, q^n, \dot{q}^1, \ldots, \dot{q}^n) \\
\quad = \mathscr{L}(t, q^1, \ldots, q^{j-1}, q^j, q^{j+1}, \ldots, q^n, \dot{q}^1, \ldots, \dot{q}^n), \\
\\
\quad\quad \text{for any choice of } (t, q^1, \ldots, q^n, \dot{q}^1, \ldots, \dot{q}^n).
\end{cases}
$$

Now the Euler-Lagrange equations (7.84) immediately give the identity:

$$\frac{d}{dt}\left(\left.\frac{\partial \mathscr{L}}{\partial \dot{q}^j}\right|_{(t,q(t),\dot{q}(t))}\right) = \left.\frac{\partial \mathscr{L}}{\partial q^j}\right|_{(t,q(t),\dot{q}(t))}$$

where $I \ni t \mapsto (t, q(t), \dot{q}(t))$ is a generic solution of the Euler-Lagrange equations in the above local coordinates. The hypothesis that q^j is cyclic kills the partial derivative on the right, so:

$$\frac{d}{dt}\left(\left.\frac{\partial \mathscr{L}}{\partial \dot{q}^j}\right|_{(t,q(t),\dot{q}(t))}\right) = 0.$$

This equation says that the **conjugate momentum** to q^j, i.e. the map in the local coordinates on $A(\mathbb{V}^{n+1})$:

$$p_j(t, q, \dot{q}) := \frac{\partial \mathscr{L}(t, q^1, \ldots, q, \dot{q})}{\partial \dot{q}^j} \tag{8.1}$$

is constant in time *along any motion of the system*. In the language of Mechanics the function p_j is called a *first integral*.

More generally, a **first integral** of the motion is a differentiable map I : $A(\mathbb{V}^{n+1}) \to \mathbb{R}$ (possibly defined on one natural chart only) such that $J \ni t \mapsto I(\gamma(t)) \in \mathbb{R}$ is *constant* for every solution $J \ni t \mapsto \gamma(t) \in A(\mathbb{V}^{n+1})$ of the Euler-Lagrange equations, and the constant typically depends on γ.

We have thus proved the following useful proposition.

Proposition 8.1 *Consider a physical system of point particles $\mathscr{S}$ described on the spacetime of kinetic states $A(\mathbb{V}^{n+1})$ by a Lagrangian $\mathscr{L}$ of class C^2. Suppose we have picked natural local coordinates $t, q, \dot{q}$ in which*

$$\frac{\partial \mathscr{L}(t, q, \dot{q})}{\partial q^j} = 0\,, \quad \text{for any } (t, q, q) \text{ in the chosen local chart,} \tag{8.2}$$

i.e., equivalently, that the Lagrangian $\mathscr{L}$ is invariant under the transformations

$$(q^1, \ldots, q^{j-1}, q^j, q^{j+1}, \ldots, q^n) \mapsto (q^1, \ldots, q^{j-1}, q^j + \Delta q^j, q^{j+1}, \ldots, q^n)\,, \tag{8.3}$$

for any $(q^1, \ldots, q^n, \dot{q}^1, \ldots, \dot{q}^n)$ and any Δq^j (compatibly with the chart). Then the conjugate momentum $p_j = p_j(t, q, \dot{q})$ of q^j defined in (8.1) is a first integral of the motion (relatively to the restriction of the solution to the Euler-Lagrange equations on the same chart).

The constant's value will depend on the solution considered. The physical meaning of p_j depends on the situation. We will shortly see two important examples where the Lagrangian has the standard form "kinetic energy plus potential energy" in a given reference frame $\mathscr{R}$.

Consider a system $\mathscr{S}$ of N point particles $P_1, \ldots, P_N$ of masses $m_1, \ldots, m_N$ respectively, subject to ideal holonomic constraints and to forces with potential $\mathscr{V}|_{\mathscr{R}}$ with respect to $\mathscr{R}$.

$$\mathscr{L}|_{\mathscr{R}}(t, q, \dot{q}) = \mathscr{T}|_{\mathscr{R}}(t, q, \dot{q}) + \mathscr{V}|_{\mathscr{R}}(t, q)\,.$$

If $\mathbf{x}_i := P_i - O$ is the position vector of the i-th point at the origin O of orthonormal coordinates for $\mathscr{R}$, then

$$\mathbf{x}_i = \mathbf{x}_i(t, q^1, \ldots, q^n)\,. \tag{8.4}$$

In the case under exam the kinetic energy can be written as :

$$\mathscr{T}|_{\mathscr{R}}(t, q, \dot{q}) = \sum_{i=1}^{N} \frac{m_i}{2} \sum_{h,k=1}^{n} \frac{\partial \mathbf{x}_i}{\partial q^k} \cdot \frac{\partial \mathbf{x}_i}{\partial q^h} \dot{q}^k \dot{q}^h + \sum_{i=1}^{N} m_i \sum_{h,k=1}^{n} \frac{\partial \mathbf{x}_i}{\partial q^k} \cdot \frac{\partial \mathbf{x}_i}{\partial t} \dot{q}^k$$

$$+ \sum_{i=1}^{N} \frac{m_i}{2} \sum_{h,k=1}^{n} \frac{\partial \mathbf{x}_i}{\partial t} \cdot \frac{\partial \mathbf{x}_i}{\partial t} \, .$$

Easy calculations give:

$$p_j = \sum_{i=1}^{N} \sum_{k=1}^{n} m_i \left(\frac{\partial \mathbf{x}_i}{\partial q^k} \dot{q}^k + \frac{\partial \mathbf{x}_i}{\partial t} \right) \cdot \frac{\partial \mathbf{x}_i}{\partial q^j} \tag{8.5}$$

that is to say, using (8.4):

$$p_j = \sum_{i=1}^{N} m_i \mathbf{v}_i|_{\mathscr{R}} \cdot \frac{\partial \mathbf{x}_i}{\partial q^j} \tag{8.6}$$

on the system's motions.

8.1.2 Translation-Invariance and Conservation of the Impulse

Consider again a system $\mathscr{S}$ of N point particles $P_1, \ldots, P_N$ of masses $m_1, \ldots, m_N$ respectively, subject to ideal holonomic constraints and forces with potential $\mathscr{V}|_{\mathscr{R}}$ in a frame $\mathscr{R}$. We will have a Lagrangian:

$$\mathscr{L}|_{\mathscr{R}}(t, q, \dot{q}) = \mathscr{T}|_{\mathscr{R}}(t, q, \dot{q}) + \mathscr{V}|_{\mathscr{R}}(t, q)$$

in some natural local chart. We can then apply formulas (8.5) and (8.6) to find the conjugate momenta.

Now suppose in particular that the coordinate q^j describes *rigid displacements of the system along a direction* $\mathbf{n}$, that is to say, the coordinate q^j is **translational** along $\mathbf{n}$ in $\mathscr{R}$. This means that for any $\Delta q^j \in \mathbb{R}$, at any instant t we have:

$$\begin{cases} \mathbf{x}_i(t, q^1, \ldots, q^{j-1}, q^j + \Delta q^j, q^{j+1}, \ldots, q^n) \\ = \mathbf{x}_i(t, q^1, \ldots, q^{j-1}, q^j, q^{j+1}, \ldots, q^n) + \Delta q^j \mathbf{n} \, , \\ \\ \qquad\qquad \text{for any } i = 1, \ldots, N, \end{cases} \tag{8.7}$$

compatibly with the extension of the local chart in use.

Note that *all* points are shifted, as q^j varies, by the *same* quantity $\Delta q^j \mathbf{n}$. In this sense the displacement is *rigid*. We also assume the coordinate q^j is cyclic for the Lagrangian $\mathscr{L}|_{\mathscr{R}}$. Then, using (8.6), the conjugate momentum p_j assumes the form

$$p_j = \sum_{i=1}^{N} m_i \mathbf{v}_i|_{\mathscr{R}} \cdot \frac{\partial \mathbf{x}_i}{\partial q^j} = \sum_{i=1}^{N} m_i \mathbf{v}_i|_{\mathscr{R}} \cdot \mathbf{n},$$

on the motion, i.e.

$$p_j = \mathbf{P}|_{\mathscr{R}} \cdot \mathbf{n}. \tag{8.8}$$

So we have the following proposition.

Proposition 8.2 *Consider a system of point particles $\mathscr{S}$ subject to ideal holonomic constraints and described in $\mathscr{R}$, where the active forces are given by a potential function $\mathscr{V}|_{\mathscr{R}}$ and where we use free coordinates $q^1, \ldots, q^n$. The following hold.*

(1) If the coordinate q^j describes rigid displacements in $\mathscr{R}$ of $\mathscr{S}$ along $\mathbf{n}$, the conjugate momentum p_j coincides with the component along $\mathbf{n}$ of the total impulse (with respect to $\mathscr{R}$) on any motion of $\mathscr{S}$.

(2) If, under the assumptions of (1), the Lagrangian $\mathscr{L}|_{\mathscr{R}} = \mathscr{T}|_{\mathscr{R}} + \mathscr{V}|_{\mathscr{R}}$ (of class C^2) is invariant under rigid displacements along $\mathbf{n}$, i.e. q^j is cyclic, the total impulse of $\mathscr{S}$ (with respect to $\mathscr{R}$) along $\mathbf{n}$ is conserved for any solution of the Euler-Lagrange equations (restricted to the chart considered).

Examples 8.3

(1) The simplest example is a system $\mathscr{S}$ of N point particles $P_1, \cdots, P_N$ with masses $m_1, \ldots, m_N$, *not* subject to constraints and where the forces (all active) are internal and conservative. The potential energy only depends on the differences $P_i - P_j$ between the particles' positions, $i, j = 1, \ldots, N$ and $i \neq j$. We can describe $\mathscr{S}$ in an inertial frame $\mathscr{R}$, taking as free coordinates the $n = 3N$ Cartesian coordinates of the points of $\mathscr{S}$ in an orthonormal system in $\mathscr{R}$ with axes $\mathbf{e}_1, \mathbf{e}_2, \mathbf{e}_3$ and origin O. Using a linear map we change coordinates so that the first 3 free coordinates q^1, q^2, q^3 describe rigid displacements of $\mathscr{S}$ along $\mathbf{e}_1, \mathbf{e}_2, \mathbf{e}_3$ respectively. These 3, which we denote by q_G^1, q_G^2, q_G^3, can be taken to be the components of the centre of mass G. The remaining $3N - 3$ free coordinates may, for instance, be the triples of components of the $N - 1$ vectors $P_i - P_1$, $1 < i \leq N$. We will call the latter q_i^1, q_i^2, q_i^3, $i = 2, 3 \ldots, N$. Choosing P_1 as origin and putting $M = \sum_{i=1}^{N} m_i$, by the definition of centre of mass we find:

$$G - P_1 = \sum_{i=1}^{N} \frac{m_i}{M} (P_i - P_1) = \sum_{i=1}^{N} \frac{m_i q_i^k}{M} \mathbf{e}_k.$$

Using $P_j = P_1 + \sum_{k=1}^{3} q_j^k \mathbf{e}_k$, $j = 2, 1 \ldots, N$, we have:

$$P_1 = O + \sum_{k=1}^{3} q_G^k \mathbf{e}_k - \sum_{i=1}^{N} \sum_{k=1}^{3} \frac{m_i q_i^k \mathbf{e}_k}{M} \quad \text{and}$$

$$P_j = O + \sum_{k=1}^{3} q_G^k \mathbf{e}_k - \sum_{i=1}^{N} \sum_{k=1}^{3} \frac{m_i q_i^k \mathbf{e}_k}{M} + \sum_{k=1}^{3} q_j^k \mathbf{e}_k .$$

Evidently the three coordinates q_G^k are translational coordinates along the $\mathbf{e}_k$, $k = 1, 2, 3$. The kinetic energy is only function of the dotted coordinates, not of the undotted ones: this is true for the initial Cartesian coordinates and for the Lagrangian coordinates as well, since the latter are obtained from the former under a linear transformation (prove this fact as exercise). The potential energy depends on the differences $P_i - P_j$, which are invariant under rigid displacements of the entire system. In particular the above formulas show that these differences do not depend on q_G^1, q_G^2, q_G^3. Hence the Lagrangian is invariant under displacements along $\mathbf{e}_1, \mathbf{e}_2, \mathbf{e}_3$ i.e. q_G^1, q_G^2, q_G^3 are cyclic. From the previous proposition we then deduce that the components of the total impulse are conserved.

(2) In the previous example we can take an alternative coordinate system, called **Jacobi coordinates**. The configuration of N point particles without constraints, in orthonormal coordinates for $\mathscr{R}$ with axes $\mathbf{e}_1, \mathbf{e}_2, \mathbf{e}_3$ and origin O, is defined by the following $3N$ coordinates: the 3 components of the position vector $\mathbf{X} := G - O$ of the centre of mass, and for $i = 1, \ldots, N - 1$ the 3 components of

$$\boldsymbol{\xi}_i := \mathbf{x}_i - \frac{\sum_{j=i+1}^{N} m_j \mathbf{x}_j}{\sum_{j=i+1}^{N} m_j} ,$$

where $\mathbf{x}_j := P_j - O$ and m_j is the mass of the j-th particle. $\boldsymbol{\xi}_i$ is the position vector of the i-th particle with respect to the centre of mass of the subsystem $P_{i+1}, P_{i+2} \ldots, P_N$. Let us introduce the **reduced masses**, for $i = 1, \ldots, N-1$:

$$\mu_i := m_i \frac{\sum_{j=i+1}^{N} m_j}{m_i + \sum_{j=i+1}^{N} m_j}$$

and the total mass $M := \sum_{i=1}^{N} m_i$ of the system. Then it is easy to show by induction that:

$$\mathscr{T}|_{\mathscr{R}} = \frac{1}{2} M \dot{\mathbf{X}}^2 + \frac{1}{2} \sum_{i=1}^{N-1} \mu_i \dot{\boldsymbol{\xi}}_i^2$$

and so:

$$\mathscr{L}|_{\mathscr{R}} = \frac{1}{2} M \dot{\mathbf{X}}^2 + \frac{1}{2} \sum_{i=1}^{N-1} \mu_i \dot{\boldsymbol{\xi}}_i^2 + \mathscr{V}(\mathbf{x}_1 - \mathbf{x}_2, \dots, \mathbf{x}_{N-1} - \mathbf{x}_N) .$$

The differences $\mathbf{x}_i - \mathbf{x}_j$ can be expresses in terms of the coordinates $\boldsymbol{\xi}_i$ only, because if the position of the centre of mass changes the coordinates of the vectors $\mathbf{x}_i - \mathbf{x}_j$ and $\boldsymbol{\xi}_i$ do not. Clearly then the Lagrangian is invariant under displacements along $\mathbf{e}_1$, $\mathbf{e}_2$, $\mathbf{e}_3$. By the previous proposition and the Lagrangian's form, we immediately obtain that the three components of the total impulse are conserved.

Even if it is not so relevant at this stage, let us point out that the total angular momentum $\boldsymbol{\Gamma}_O|_{\mathscr{R}}$ decomposes along the Jacobi coordinates as follows:

$$\boldsymbol{\Gamma}_O|_{\mathscr{R}} = \mathbf{X} \wedge M\dot{\mathbf{X}} + \sum_{i=1}^{N-1} \boldsymbol{\xi}_i \wedge \mu_i \dot{\boldsymbol{\xi}}_i . \tag{8.9}$$

(3) Consider a system $\mathscr{S}$ of two point particles P_1 and P_2 of masses m_1 and m_2, moving along a smooth straight line r at rest in an inertial frame $\mathscr{R}$. The line r has unit direction $\mathbf{n}$ and passes through the origin O of orthonormal coordinates for $\mathscr{R}$. The particles interact with each other under a pair of conservative forces of potential energy $\mathscr{U} = \mathscr{U}(\|P_1 - P_2\|)$. In contrast to the previous example, here we have (ideal) reactions due to the constraint (4 functionally independent equations) and the number of freedom degrees reduces to 2. Using free coordinates $q^1 := (s_1 + s_2)/2$ and $q^2 := (s_1 - s_2)/2$, where s_i is the coordinate of P_i on r measured from O and oriented as $\mathbf{n}$, we see that q^1 generates rigid displacements. This is straightforward from the relations: $s_1 = q^1 + q^2$ and $s_2 = q^1 - q^2$. The Lagrangian $\mathscr{L}|_{\mathscr{R}} = \mathscr{T}|_{\mathscr{R}} - \mathscr{U}|_{\mathscr{R}}$ will not depend, by construction, on q^1 (prove it!). We conclude, by the previous proposition, that the total impulse along $\mathbf{n}$ is conserved.

8.1.3   Rotation-Invariance and Conservation of the Angular Momentum

Consider again a system $\mathscr{S}$ of N point particles $P_1, \dots, P_N$ of masses $m_1, \dots, m_N$ respectively, subject to ideal holonomic constraints and to forces with potential $\mathscr{V}|_{\mathscr{R}}$ in a frame $\mathscr{R}$. The Lagrangian, in natural local coordinates, is:

$$\mathscr{L}|_{\mathscr{R}}(t, q, \dot{q}) = \mathscr{T}|_{\mathscr{R}}(t, q, \dot{q}) + \mathscr{V}|_{\mathscr{R}}(t, q) .$$

We can then apply formulas (8.5) and (8.6) to compute the conjugate momenta.

Suppose in particular that q^j describes *rigid rotations around an axis* **n** *through a point* O, i.e. q^j is a **rotational coordinate around n through** O **in the frame** $\mathcal{R}$. Put otherwise, for any $\Delta q^j \in \mathbb{R}$ at any instant t:

$$
\begin{cases}
\mathbf{x}_i(t, q^1, \ldots, q^{j-1}, q^j + \Delta q^j, q^{j+1}, \ldots, q^n) \\
= R_{\mathbf{n}, \Delta q^j} \mathbf{x}_i(t, q^1, \ldots, q^{j-1}, q^j, q^{j+1}, \ldots, q^n), \\
\qquad\qquad \text{for any } i = 1, \ldots, N.
\end{cases}
\tag{8.10}
$$

$R_{\mathbf{n}, \theta} : V^3 \to V^3$ is the operator rotating vectors at O by a positive angle θ around the unit vector **n** at O. Note that *all* position vectors $\mathbf{x}_i = P_i - O$ rotate, as q^j varies, under the *same* rotation $R_{\mathbf{n}, \Delta q^1}$. Hence the rotation is *rigid*. By the above relation,

$$
\frac{\mathbf{x}_i(t, q^1, \ldots, q^{j-1}, q^j + \Delta q^j, q^{j+1}, \ldots, q^n)}{-\mathbf{x}_i(t, q^1, \ldots, q^{j-1}, q^j, q^{j+1}, \ldots, q^n)}{\Delta q^j}
$$

$$
= \frac{R_{\mathbf{n}, \Delta q^j} - I}{\Delta q^j} \mathbf{x}_i(t, q^1, \ldots, q^{j-1}, q^j, q^{j+1}, \ldots, q^n)
$$

$$
= \frac{R_{\mathbf{n}, 0 + \Delta q^j} - R_{\mathbf{n}, 0}}{\Delta q^j} \mathbf{x}_i(t, q^1, \ldots, q^{j-1}, q^j, q^{j+1}, \ldots, q^n).
$$

As is well known (prove it as exercise, choosing right-handed orthonormal axes with $\mathbf{e}_z = \mathbf{n}$):

$$
\left. \frac{d}{d\theta} \right|_{\theta=0} R_{\mathbf{n}, \theta} \mathbf{x} = \mathbf{n} \wedge \mathbf{x}.
$$

Along the motion, therefore, the conjugate momentum p_j becomes, by (8.6),

$$
p_j = \sum_{i=1}^N m_i \mathbf{v}_i|_{\mathcal{R}} \cdot \frac{\partial \mathbf{x}_i}{\partial q^j} = \sum_{i=1}^N m_i \mathbf{v}_i|_{\mathcal{R}} \cdot \left. \frac{d}{d\theta} \right|_{\theta=0} R_{\mathbf{n}, \theta} \mathbf{x}_i = \sum_{i=1}^N m_i \mathbf{v}_i|_{\mathcal{R}} \cdot \mathbf{n} \wedge \mathbf{x}_i.
$$

Since $\mathbf{v}_i|_{\mathcal{R}} \cdot \mathbf{n} \wedge \mathbf{x}_i = \mathbf{n} \cdot \mathbf{x}_i \wedge m_i \mathbf{v}_i$, we can rewrite p_j as:

$$
p_j = \mathbf{n} \cdot \sum_{i=1}^N \mathbf{x}_i \wedge m_i \mathbf{v}_i = \Gamma_O|_{\mathcal{R}} \cdot \mathbf{n}.
\tag{8.11}
$$

where $\Gamma_O|_{\mathcal{R}}$ is the total angular momentum of $\mathscr{S}$ with respect to the pole O in $\mathcal{R}$. Then the following proposition holds.

Proposition 8.4 *Consider a system of point particles $\mathscr{S}$ described in a frame $\mathscr{R}$ where the active forces are given by a potential $\mathscr{V}|_{\mathscr{R}}$, and use free coordinates $q^1, \ldots, q^n$.*

(1) If q^j describes rigid rotations in $\mathscr{R}$ about $\mathbf{n}$ with respect to O, the conjugate momentum p_j coincides with the $\mathbf{n}$-component of the total angular momentum of $\mathscr{S}$ with respect to the pole O in $\mathscr{R}$.

(2) Suppose, under the assumptions of (1), that the Lagrangian $\mathscr{L}|_{\mathscr{R}} = \mathscr{T}|_{\mathscr{R}} + \mathscr{V}|_{\mathscr{R}}$ (of class C^2) is invariant under rigid rotations about $\mathbf{n}$ with respect to O, i.e. the variable q^j is cyclic. Then the $\mathbf{n}$-component of the total angular momentum with respect to the pole O in $\mathscr{R}$ is conserved on every solution to the Euler-Lagrange equations (restricted to the local chart considered).

.

Examples 8.5

(1) The easiest example is a system $\mathscr{S}$ of N point particles $P_1, \cdots, P_N$, with masses $m_1, \ldots, m_n$, *not* subject to constraints and whose forces (all active) are internal and conservative. Thus the potential energy depends only on the differences $P_i - P_j$, $i, j = 1, \ldots, N$, $i \neq j$, of the particles' positions. Then $\mathscr{S}$ can be described in an inertial frame $\mathscr{R}$, with free coordinates the $3N$ cylindrical coordinates φ_i, r_i, z_i of the N points, with respect to a common system of orthogonal axes of origin O for $\mathscr{R}$. Using a linear transformation on the angles, we can make q^1 the angle coordinate of the centre of mass, and the other angular coordinates are the differences $\varphi_i - \varphi_1$, $1 < i \leq N$. The remaining free coordinates (those of type r_i and z_i) are unaffected. Thus q^1 describes rigid rotations of $\mathscr{S}$ around the z-axis. As the kinetic energy is independent of the undotted coordinates and the potential energy depends on the norms of the differences $P_i - P_j$, which are invariant by rigid rotations of the entire system, the Lagrangian is invariant under rotations about z with respect to O. By the previous proposition, the z-component of the total angular momentum, with respect to the pole O and in $\mathscr{R}$, is preserved. Actually this holds *for every axis and any origin O* (both at rest in $\mathscr{R}$), since the initial choice of where to place z is completely arbitrary.

(2) Consider a system $\mathscr{S}$ of two point particles P_1 and P_2 of masses m_1 and m_2, travelling on a smooth circle C. The curve, at rest in an inertial frame $\mathscr{R}$, has centre at the origin O of orthonormal coordinates for $\mathscr{R}$ and is perpendicular to the z-axis. The particles interact under a pair of conservative forces of potential energy $\mathscr{U} = \mathscr{U}(\|P_1 - P_2\|)$. In contrast to the previous example, now we have (ideal) reactions due to the constraint (4 independent equations) and the system's freedom degrees reduce to 2. Using free coordinates $q^1 := (\varphi_1 + \varphi_2)/2$ and $q^2 := (\varphi_1 - \varphi_2)/2$, where φ_i is the polar angular coordinate of P_i on the plane $z = 0$, we see that q^1 generates rigid displacements. In fact: $\varphi_1 = q^1 + q^2$ and $\varphi_2 = q^1 - q^2$. The Lagrangian $\mathscr{L}|_{\mathscr{R}} = \mathscr{T}|_{\mathscr{R}} - \mathscr{U}|_{\mathscr{R}}$ will not, by construction, depend on q^1 (prove it!). Hence, by the previous proposition, along any motion of $\mathscr{S}$ the $\mathbf{e}_z$-component of the total angular momentum with respect to the pole O in $\mathscr{R}$ is conserved.

8.2   The Relationship Between Symmetries and Conservation Laws: Emmy Noether's Theorem

In this section we state and prove Emmy Noether's Theorem (for a mechanical system), considered one of the most important results in Mathematical and Theoretical Physics. The theorem was formulated[1] at the start of the twentieth century and explains the tight relationship between the symmetry of a Lagrangian system under a group of transformations and the existence of conserved quantities along the system's motions. The result can be formulated for general physical systems, including continuous systems and fields (relativistic as well) as long as they are described by a Lagrangian, and has had a profound influence over the Theoretical Physics of the twentieth and twentyfirst centuries.

8.2.1   Transformations on $A(\mathbb{V}^{n+1})$

Consider a mechanical system $\mathscr{S}$ described on the spacetime of configurations $\mathbb{V}^{n+1}$, and let us fix natural local coordinates $(t, q^1, \ldots, q^n) \in (a, b) \times \mathcal{V}$ defined on an open set in $\mathbb{V}^{n+1}$. In the following we shall assume all maps on the natural local chart of $A(\mathbb{V}^{n+1})$ are at least C^2 in the coordinates. Recall that the $q^1, \ldots, q^n$ are local coordinates on each configuration space $\mathbb{Q}_t$ at any time t. Consider an active transformation in $(t_1, t_2) \times \mathcal{U} \subset (a, b) \times \mathcal{V}$, which defines a diffeomorphism between points of coordinates $(t, q^1, \ldots, q^n)$ and points of coordinates $(t', q'^1, \ldots, q'^n) \in (t'_1, t'_1) \times \mathcal{U}' \subset (a, b) \times \mathcal{V}$:

$$\begin{cases} t' = t\,, \\[2mm] q'^k = q'^k(t, q^1, \ldots, q^n) \quad \text{for } k = 1, \ldots, n. \end{cases} \qquad (8.12)$$

In other words, for any instant t, *the points in $\mathbb{Q}_t$ move, but are sent bijectively and bi-differentiably to the same configuration space $\mathbb{Q}_t$ at time t* (in particular, $t_1 = t'_1$ and $t_2 = t'_2$).

Since the configuration space at time t is a subset, defined by the constraints, in the absolute space at time t (precisely, in the product of the absolute spaces Σ_t

---

[1] The general statement, valid for continuous systems including fields, first appeared in "Invariante Variationsprobleme" Nachr.d. König. Gesellsch. d. Wiss. zu Göttingen, Math-phys. Klasse, 235–257 (1918). For an English translation see http://arxiv.org/pdf/physics/0503066.

at time t of the point particles making $\mathscr{S}$), these transformations are nothing but local differentiable maps with differentiable inverse on the configurations of $\mathscr{S}$ that: (1) respect the system's constraints, and (2) act on absolute space ad each instant, sending points of the absolute space at time t to points of the absolute space at the same instant t. What these transformations do *not* do is change absolute space: they do not map Σ_t to $\Sigma_{t'}$ if $t \neq t'$.

A local transformation on $\mathbb{V}^{n+1}$ of that type is said to **preserve the fibres** of $\mathbb{V}^{n+1}$. We recall that the fibres of $\mathbb{V}^{n+1}$ are precisely the spaces $\mathbb{Q}_t$ for any t.

Examples 8.6

(1) If $\mathscr{S}$ is a single particle and q^1, q^2, q^3 are its Cartesian coordinates in a frame $\mathscr{R}$, a transformation of the above sort is one that at each instant rotates the particle's position around a given axis through a certain point (both independent of time) by an angle ϵ.

(2) The previous case generalises immediately to two points P, Q connected by a rigid rod of fixed length, described by free coordinates $q^1, \ldots, q^5$, where q^1, q^2, q^3 are the coordinates of P in an orthonormal system with axes $\mathbf{e}_x, \mathbf{e}_y, \mathbf{e}_z$ and origin O in a frame $\mathscr{R}$, while q^4, q^5 are spherical angles defining Q with respect to a orthonormal axes at P parallel to $\mathbf{e}_x, \mathbf{e}_y, \mathbf{e}_z$. The rotation around an axis $\mathbf{n}$ through O (both time-independent) by ϵ that *preserves the distance of the two points* also preserves the fibres of $\mathbb{V}^{5+1}$.

A fibre-preserving map extends immediately to the dotted coordinates, i.e. from the spacetime of configurations $\mathbb{V}^{n+1}$ to the spacetime of kinetic states $A(\mathbb{V}^{n+1})$, in the following manner:

$$
\begin{cases}
t' = t, \\[2mm]
q'^k = q'^k(t, q^1, \ldots, q^n) \quad \text{for } k = 1, \ldots, n. \\[2mm]
\dot{q}'^k = \dfrac{\partial q'^k}{\partial t}(t, q^1, \ldots, q^n) + \sum_{j=1}^{n} \dfrac{\partial q'^k}{\partial q^j}(t, q^1, \ldots, q^n)\dot{q}^j \quad \text{for } k = 1, \ldots, n,
\end{cases}
\tag{8.13}
$$

where $(t, q^1, \ldots, q^n, \dot{q}^1, \ldots, \dot{q}^n) \in (t_1, t_2) \times \mathcal{U} \times \mathbb{R}^n$.

In writing the last line we have heuristically assumed the $\dot{q}^k$ are the time derivatives of the q^k along a given curve in $\mathbb{V}^{n+1}$ parametrised by time. In reality such a curve does not exist, since the $\dot{q}^k$ are *independent* of the q^k. The interpretation is nonetheless correct for any curve in $\mathbb{V}^{n+1}$ parametrised by time as long as we understand the $\dot{q}^k$ as time derivatives of the q^k. This happens in particular along the solutions of the Euler-Lagrange equations. Transformation (8.13) is said to be **induced** by the fibre-preserving map (8.12).

8.2.2 Noether's Theorem in Elementary Local Form

Consider a physical system of point particles $\mathscr{S}$ described in $A(\mathbb{V}^{n+1})$. Choose natural local coordinates $(t, q^1, \ldots, q^n) \in (a, b) \times \mathcal{V}$ on an open set of $\mathbb{V}^{n+1}$ and suppose there is a *family* of local diffeomorphisms ϕ, that preserve the fibres of $\mathbb{V}^{n+1}$, parametrised by $\epsilon \in (-\alpha, \alpha)$, $\alpha > 0$ (including $\alpha = +\infty$):

$$\phi_\epsilon : \begin{cases} t' = t, \\ q'^k = q'^k(\epsilon, t, q^1, \ldots, q^n) & \text{for } k = 1, \ldots, n. \end{cases} \tag{8.14}$$

Assume the right-hand side maps are C^2 in all variables (ϵ included), that for $\epsilon = 0$ the local diffeomorphism is the identity:

$$q'^k(0, t, q^1, \ldots, q^n) = q^k \quad \text{for } k = 1, \ldots, n, \tag{8.15}$$

and finally that, for $k = 1, \ldots, n$:

$$q'^k\left(\epsilon', t, q'(\epsilon, t, q)\right) = q'^k(\epsilon' + \epsilon, t, q) \tag{8.16}$$

whenever both sides are defined. Then ϕ is called a **one-parameter group of fibre-preserving local diffeomorphisms of** $\mathbb{V}^{n+1}$. (See Sect. 14.5.3 for the general theory of 1-parameter groups of local diffeomorphisms.)

Examples 8.7 Relatively to Example (2) in 8.6, consider the family obtained as ϵ varies in $(-\pi, \pi)$. When $\epsilon = 0$ we have the trivial rotation so (8.15) holds. Moreover, the composite of two rotations, first by ϵ and then by ϵ', coincides with a single rotation by $\epsilon + \epsilon'$. Hence (8.16) is true.

Definition 8.8 (Lagrangian System Invariant Under a Transformation Group)
We say a system of point particles $\mathscr{S}$ described on $\mathbb{V}^{n+1}$ is **invariant** under the 1-parameter group ϕ of C^2 local diffeomorphisms preserving the fibres of $\mathbb{V}^{n+1}$, given by (8.14), if the following hold.

(1) There is a C^2 Lagrangian $\mathscr{L} : A(\mathbb{V}^{n+1}) \to \mathbb{R}$ (for example, of the usual form $\mathscr{T}|_{\mathscr{R}} + \mathscr{V}|_{\mathscr{R}}$) that describes the dynamics of $\mathscr{S}$.
(2) In the natural coordinates used to define ϕ:

$$\left.\frac{\partial \mathscr{L}(t', q', \dot{q}')}{\partial \epsilon}\right|_{\epsilon=0} = 0, \quad \text{for any } t, q^1, \ldots, q^n, \dot{q}^1, \ldots, \dot{q}^n, \tag{8.17}$$

where:

$$\dot{q}'^k = \frac{\partial q'^k}{\partial t}(\epsilon, t, q) + \sum_{j=1}^{n} \frac{\partial q'^k}{\partial q^j}(\epsilon, t, q)\dot{q}^j \quad \text{for } k = 1, \ldots, n. \qquad (8.18)$$

Then ϕ is called a **symmetry group** (group of symmetry transformations) of $\mathscr{S}$.

We shall say $\mathscr{S}$, described on $\mathbb{V}^{n+1}$, is **weakly invariant** under ϕ if there is a Lagrangian $\mathscr{L} : A(\mathbb{V}^{n+1}) \to \mathbb{R}$ (of class C^2) describing the dynamics of $\mathscr{S}$ for which

$$\frac{\partial \mathscr{L}(t', q', \dot{q}')}{\partial \epsilon}\bigg|_{\epsilon=0} = G(t, q, \dot{q}), \quad \text{for any } t, q^1, \ldots, q^n, \dot{q}^1, \ldots, \dot{q}^n \qquad (8.19)$$

where: $t, q^1, \ldots, q^n, \dot{q}^1, \ldots, \dot{q}^n$ are local coordinates on $A(\mathbb{V}^{n+1})$ used to define ϕ, Eq. (8.18) hold, and finally G is a *formal total derivative*:

$$G(t, q, \dot{q}) = \frac{\partial g}{\partial t}(t, q) + \sum_{k=1}^{n} \frac{\partial g}{\partial q^k}(t, q)\dot{q}^k \qquad (8.20)$$

for some $g = g(t, q)$ of class C^1 defined at least locally on $\mathbb{V}^{n+1}$.

In this case ϕ is a **weak symmetry group** of $\mathscr{S}$. $\qquad \diamond$

Remarks 8.9 Here is an important point. We might be led to think that (8.19) is a weaker demand than the similar condition on all values of ϵ, not just $\epsilon = 0$:

$$\frac{\partial \mathscr{L}(t', q', \dot{q}')}{\partial \epsilon} = G(t, q', \dot{q}'), \quad \text{for any } \epsilon \text{ and } t, q^1, \ldots, q^n, \dot{q}^1, \ldots, \dot{q}^n \qquad (8.21)$$

Yet, the two are equivalent, since the natural local coordinates' domain is *invariant* under the symmetry group, which therefore maps the domain to itself for any $\epsilon \in \mathbb{R}$. Since the kinetic state $(q^1, \ldots, q^n, \dot{q}^1, \ldots, \dot{q}^n)$ (at time t) is arbitrary in (8.19), we can always choose it to be $(q'^1, \ldots, q'^n, \dot{q}'^1, \ldots, \dot{q}'^n)$, i.e. pick a generic ϵ in $(q^1, \ldots, q^n, \dot{q}^1, \ldots, \dot{q}^n)$. Differentiating in ϵ the Lagrangian $\mathscr{L}(t, q', \dot{q}')$ at $\epsilon = \epsilon_0$ is the same as differentiating $\mathscr{L}(t, q', \dot{q}')$ in ϵ' for $\epsilon' = 0$, taking the kinetic state (at time t) $(q'^1, \ldots, q'^n, \dot{q}'^1, \ldots, \dot{q}'^n)$ coming from $(q^1, \ldots, q^n, \dot{q}^1, \ldots, \dot{q}^n)$ with $\epsilon = \epsilon_0 + \epsilon'$. Thus (8.19) implies (8.21). This is true in particular when $G \equiv 0$. Then (8.17) is equivalent to the apparently stronger:

$$\frac{\partial \mathscr{L}(t', q', \dot{q}')}{\partial \epsilon} = 0, \quad \text{for any } \epsilon \text{ and } t, q^1, \ldots, q^n, \dot{q}^1, \ldots, \dot{q}^n. \qquad (8.22)$$

Even though the domain is not invariant under the group, the above identity is valid for a given kinetic state $t, q^1, \ldots, q^n, \dot{q}^1, \ldots, \dot{q}^n$ provided ϵ is taken in a small enough interval containing $\epsilon = 0$. ∎

We are ready to state and prove *Noether's Theorem* in its elementary and local form.

Theorem 8.10 (Noether's Theorem in Elementary Local Form) *Consider a system of point particles $\mathscr{S}$ described on $\mathbb{V}^{n+1}$ and invariant, or weakly invariant, under the 1-parameter group ϕ of local diffeomorphisms that preserve the fibres of $\mathbb{V}^{n+1}$, as in (8.14).*

(1) If the Lagrangian $\mathscr{L}$ of $\mathscr{S}$ satisfies (8.17), on the system's motions (until they stay within the domain of the coordinates used to describe the local ϕ-action on $A(\mathbb{V}^{n+1})$) the function:

$$I_\phi(t, q, \dot{q}) := \sum_{k=1}^{n} \frac{\partial \mathscr{L}}{\partial \dot{q}^k} \frac{\partial q'^k}{\partial \epsilon}\bigg|_{\epsilon=0} (t, q, \dot{q}) \tag{8.23}$$

is constant in time when evaluated on a solution of the Euler-Lagrange equations.

(2) If the maps $\mathscr{L}$ and g satisfy (8.19) and (8.20), then on the system's motions (until they stay in the domain of the coordinates used to describe the local action of ϕ on $A(\mathbb{V}^{n+1})$) the map:

$$I_{\phi,g}(t, q, \dot{q}) := -g(t, q) + \sum_{k=1}^{n} \frac{\partial \mathscr{L}}{\partial \dot{q}^k} \frac{\partial q'^k}{\partial \epsilon}\bigg|_{\epsilon=0} (t, q, \dot{q}) \tag{8.24}$$

is constant in time along a solution of the Euler-Lagrange equations.

I_ϕ and $I_{\phi,g}$ are called **Noether's first integrals**.

Proof Let us write explicitly the left-hand side of (8.17):

$$0 = \frac{\partial \mathscr{L}(t', q', \dot{q}')}{\partial \epsilon}\bigg|_{\epsilon=0} = \sum_k \frac{\partial \mathscr{L}(t', q', \dot{q}')}{\partial q'^k} \frac{\partial q'^k}{\partial \epsilon}\bigg|_{\epsilon=0}$$
$$+ \sum_k \frac{\partial \mathscr{L}(t', q', \dot{q}')}{\partial \dot{q}'^k} \frac{\partial \dot{q}'^k}{\partial \epsilon}\bigg|_{\epsilon=0}.$$

Using (8.15), this reads:

$$0 = \frac{\partial \mathscr{L}(t', q', \dot{q}')}{\partial \epsilon}\bigg|_{\epsilon=0} = \sum_k \frac{\partial \mathscr{L}(t, q, \dot{q})}{\partial q^k} \frac{\partial q'^k}{\partial \epsilon}\bigg|_{\epsilon=0} + \sum_k \frac{\partial \mathscr{L}(t, q, \dot{q})}{\partial \dot{q}^k} \frac{\partial \dot{q}'^k}{\partial \epsilon}\bigg|_{\epsilon=0}.$$
$$\tag{8.25}$$

Now consider a motion $q^k = q^k(t)$, $\dot{q}^k(t) = dq^k/dt(t)$. From (8.18), expanding the two sides and comparing:

$$\frac{\partial \dot{q}'^k(\epsilon, t, q(t))}{\partial \epsilon} = \frac{d}{dt} \frac{\partial q'^k(\epsilon, t, q(t))}{\partial \epsilon}. \tag{8.26}$$

(Here is the proof. The left-hand side of the above is

$$\frac{\partial \dot{q}'^k}{\partial \epsilon} = \frac{\partial}{\partial \epsilon} \left(\frac{\partial q'^k}{\partial t}(\epsilon, t, q) + \sum_{j=1}^{n} \frac{\partial q'^k}{\partial q^j}(\epsilon, t, q)\dot{q}^j \right)$$

$$= \frac{\partial^2 q'^k}{\partial \epsilon \partial t}(\epsilon, t, q) + \sum_{j=1}^{n} \frac{\partial^2 q'^k}{\partial \epsilon \partial q^j}(\epsilon, t, q)\dot{q}^j$$

$$= \frac{\partial^2 q'^k}{\partial t \partial \epsilon}(\epsilon, t, q) + \sum_{j=1}^{n} \frac{\partial^2 q'^k}{\partial q^j \partial \epsilon}(\epsilon, t, q)\dot{q}^j.$$

Replace q with the solution $q = q(t)$, so $\dot{q}^k = \frac{dq^k}{dt}$, and the last term above coincides with

$$\frac{d}{dt} \frac{\partial q'^k(\epsilon, t, q(t))}{\partial \epsilon},$$

which proves (8.26).) Inserting the right-hand side of (8.26) in the corresponding term in the right side of (8.25), where now *we assume explicitly that the Lagrangian's variables are evaluated on the motion*, we find:

$$0 = \frac{\partial}{\partial \epsilon}\bigg|_{\epsilon=0} \mathscr{L}(t', q', \dot{q}') = \sum_{k} \frac{\partial \mathscr{L}(t, q, \dot{q})}{\partial q^k} \frac{\partial q'^k}{\partial \epsilon}\bigg|_{\epsilon=0}$$

$$+ \sum_{k} \frac{\partial \mathscr{L}(t, q, \dot{q})}{\partial \dot{q}^k} \frac{d}{dt} \frac{\partial q'^k(\epsilon, t, q(t))}{\partial \epsilon}\bigg|_{\epsilon=0}.$$

All in all:

$$\sum_{k} \frac{\partial \mathscr{L}}{\partial q^k} \frac{\partial q'^k}{\partial \epsilon}\bigg|_{\epsilon=0} + \sum_{k} \frac{\partial \mathscr{L}}{\partial \dot{q}^k} \frac{d}{dt} \frac{\partial q'^k}{\partial \epsilon}\bigg|_{\epsilon=0} = 0.$$

Since everything was evaluated on a solution to the Euler-Lagrange equations, we may rewrite the above as:

$$\sum_k \left(\frac{d}{dt} \frac{\partial \mathscr{L}}{\partial \dot{q}^k} \right) \frac{\partial q'^k}{\partial \epsilon} \bigg|_{\epsilon=0} + \sum_k \frac{\partial \mathscr{L}}{\partial \dot{q}^k} \frac{d}{dt} \frac{\partial q'^k}{\partial \epsilon} \bigg|_{\epsilon=0} = 0 .$$

Further, if $\gamma : t \mapsto (q(t), \dot{q}(t))$ is the Euler-Lagrange solution, we can write

$$\frac{d}{dt} \left(\sum_{k=1}^{n} \frac{\partial \mathscr{L}}{\partial \dot{q}^k} \bigg|_{\gamma(t)} \frac{\partial q'^k}{\partial \epsilon} \bigg|_{\gamma(t),\epsilon=0} \right) = 0 ,$$

where the left-hand side, prior to differentiating in time, *is evaluated on any solution of the Euler-Lagrange equations*. This is the claim in the first case. As for the second case, the argument is essentially the same. Weak invariance (8.19) can be written, on any solution of the Euler-Lagrange equations, as:

$$\frac{d}{dt} \left(\sum_{k=1}^{n} \frac{\partial \mathscr{L}}{\partial \dot{q}^k} \bigg|_{\gamma(t)} \frac{\partial q'^k}{\partial \epsilon} \bigg|_{\gamma(t),\epsilon=0} \right) = \frac{d}{dt} g(t, \gamma(t)) ,$$

and the claim is immediate.   □

Examples 8.11

(1) The case of a Lagrangian with cyclic coordinates is a subcase of Noether's Theorem: Noether's first integral is the preserved conjugate momentum. In fact if $\mathscr{L} = \mathscr{L}(t, q, \dot{q})$ is such that $\partial \mathscr{L}/\partial q^j = 0$, consider the local one-parameter group ϕ that preserves the fibres of $\mathbb{V}^{n+1}$ given by: $t' = t$, $q'^j = q^j + \epsilon$ and $q'^k = q^k$ for $k \neq j$. Now (8.18) immediately gives:

$$\dot{q}'^k = \dot{q}^k \quad \text{for } k = 1, \ldots, n. \tag{8.27}$$

Trivially, the Lagrangian satisfies (8.17) under ϕ, as it does not depend on q^j explicitly. The first integral from Noether's Theorem is then:

$$I_\phi = \sum_k \frac{\partial \mathscr{L}}{\partial \dot{q}^k} \frac{\partial q'^k}{\partial \epsilon} \bigg|_{\epsilon=0} = \sum_k \frac{\partial \mathscr{L}}{\partial \dot{q}^k} \delta_{kj} = \frac{\partial \mathscr{L}}{\partial \dot{q}^j} = p_j .$$

Hence, for systems that are invariant under rigid displacements or rigid rotations with Lagrangian $\mathscr{T} + \mathscr{V}$, the conserved first integrals of Noether's Theorem are the components of the total impulse and the total angular momentum, as we saw in the previous section's Examples 8.6 and 8.7.

(2) A more interesting example is this. Consider a system $\mathscr{S}$ of N point particles $P_1, \cdots, P_N$, with masses $m_1, \ldots, m_N$, *not* subject to constraints and with internal and conservative forces (all active). The potential energy depends only on the position differences $P_i - P_j$ with $i, j = 1, \ldots, N$ and $i \neq j$. We can describe $\mathscr{S}$ in an inertial frame $\mathscr{R}$, using as free coordinates the $3N$ coordinates of the P_i in an orthonormal system for $\mathscr{R}$ of origin O and axes

$\mathbf{e}_x$, $\mathbf{e}_y$, $\mathbf{e}_z$. Consider the action of a 1-parameter subgroup ϕ of the *subgroup of pure Galilean transformations* on the system's particles: for $i = 1, \ldots, N$, if $\mathbf{x}_i := P_i - O$,

$$\mathbf{x}_i \to \mathbf{x}_i' = \mathbf{x}_i + \epsilon t \mathbf{n}, \quad \text{with } \epsilon \in \mathbb{R}. \tag{8.28}$$

The unit vector $\mathbf{n}$ is fixed once and for all. The transformation depends parametrically on time. The action on the dotted coordinates is:

$$\dot{\mathbf{x}}_i \to \dot{\mathbf{x}}_i' = \dot{\mathbf{x}}_i + \epsilon \mathbf{n}, \quad \text{with } \epsilon \in \mathbb{R}. \tag{8.29}$$

Under these maps the potential energy is unchanged as it depends on the positions' differences. Instead the kinetic energy transforms as:

$$\mathscr{T} \to \mathscr{T}' = \mathscr{T} + \epsilon \sum_{i=1}^{N} m_i \dot{\mathbf{x}}_i \cdot \mathbf{n} + \frac{\epsilon^2}{2} \sum_{i=1}^{N} m_i.$$

Therefore:

$$\mathscr{L}' = \mathscr{L} + \epsilon \sum_{i=1}^{N} m_i \dot{\mathbf{x}}_i \cdot \mathbf{n} + \frac{\epsilon^2}{2} \sum_{i=1}^{N} m_i$$

and so:

$$\left. \frac{\partial \mathscr{L}'}{\partial \epsilon} \right|_{\epsilon=0} = \mathbf{n} \cdot \sum_{i=1}^{N} m_i \dot{\mathbf{x}}_i.$$

The right-hand side is the formal total derivative of:

$$g(t, \mathbf{x}) := \mathbf{n} \cdot \sum_{i=1}^{N} m_i \mathbf{x}_i.$$

We have shown that the system is weakly invariant under ϕ. By Noether's Theorem we have a first integral depending explicitly on time, called (classical) **boost**:

$$I_\phi(t, \mathbf{x}_1, \ldots, \mathbf{x}_N, \dot{\mathbf{x}}_1 \ldots, \dot{\mathbf{x}}_N) := \mathbf{n} \cdot t \sum_{i=1}^{N} m_i \dot{\mathbf{x}}_i - \mathbf{n} \cdot \sum_{i=1}^{N} m_i \mathbf{x}_i. \tag{8.30}$$

On any motion the above first integral can be written:

$$I_\phi(t, \mathbf{x}_1, \ldots, \mathbf{x}_N, \dot{\mathbf{x}}_1 \ldots, \dot{\mathbf{x}}_N) := \mathbf{n} \cdot \left(t\mathbf{P}_{\mathscr{R}} - \sum_{i=1}^{N} m_i \mathbf{x}_i \right).$$

The conservation law $dI_\phi/dt = 0$, since the total impulse $\mathbf{P}_{\mathscr{R}}$ is conserved in our assumptions, gives:

$$\mathbf{n} \cdot \left(\mathbf{P}|_{\mathscr{R}} - \frac{d}{dt} \sum_{i=1}^{N} m_i \mathbf{x}_i(t) \right) = 0,$$

which holds for any direction $\mathbf{n}$ because the initial unit vector $\mathbf{n}$ is arbitrary. To sum up, using the notation od Sect. 4.1, we have:

$$\mathbf{P}|_{\mathscr{R}} = M\mathbf{v}_G|_{\mathscr{R}},$$

This was already proved in Sect. 4.1, there expressed by (4.1).

8.2.3 Noether's First Integral's Independence of the Coordinate System

We now wish to put the stress on how the first integral associated with a one-parameter group of local diffeomorphisms ϕ *does not* depend on the natural coordinates representing ϕ. Consider a one-parameter group ϕ. To describe it we use a local chart (U, ψ) on $\mathbb{V}^{n+1}$ where $\psi : U \ni p \mapsto (t, q^1, \ldots, q^n)$ and $U \subset \mathbb{V}^{n+1}$ is open. In these coordinates ϕ is described by the active transformations: $\phi_\epsilon : (t, q^1, \ldots, q^n) \mapsto (t', q'^1, \ldots, q'^n)$, depending on the parameter $\epsilon \in (-\delta, \delta)$. They are expressed in coordinates of (U, ψ) by:

$$\phi_\epsilon : \begin{cases} t' = t, \\ \\ q'^k = q'^k(\epsilon, t, q^1, \ldots, q^n) & \text{for } k = 1, \ldots, n. \end{cases} \tag{8.31}$$

Let us pass to a local chart (V, ϕ) where $\eta : V \ni q \mapsto (\hat{t}, \hat{q}^1, \ldots, \hat{q}^n)$ and $V \subset \mathbb{V}^{n+1}$ open such that $V \cap U \neq \varnothing$. The coordinate change, as the two systems are adapted to the fibres of $\mathbb{V}^{n+1}$, has the usual form: $\hat{t} = t + c$ and $\hat{q}^k = \hat{q}^k(t, q^1, \ldots, q^n)$ for $k = 1, \ldots, n$. The group ϕ on $V \cap U$ can be written as:

$$\hat{T}_\epsilon := \psi \circ (\eta^{-1} \circ \phi_\epsilon \circ \eta) \circ \psi^{-1}$$

in the new coordinates $\hat{t}, \hat{q}^1, \ldots, \hat{q}^n$. They correspond to

$$
\begin{cases}
\hat{t}' = \hat{t}, \\
\hat{q}'^k = \hat{q}'^k \left(\epsilon, \hat{t}, q \left(\epsilon, t(\hat{t}, \hat{q}) \right) \right) & \text{for } k = 1, \ldots, n.
\end{cases}
\tag{8.32}
$$

Suppose the physical system described by the Lagrangian $\mathscr{L}$ is weakly invariant under ϕ:

$$
\frac{\mathscr{L}(t', q', \dot{q}')}{\partial \epsilon} \bigg|_{\epsilon=0} = G(t, q, \dot{q}),
\tag{8.33}
$$

where G is the usual formal total derivative of g. G and $\mathscr{L}$ are *thought of as scalar functions* on $A(\mathbb{V}^{n+1})$. Clearly, condition (8.33) does not depend on the coordinates: when we pass to $\hat{t}, \hat{q}^1, \ldots, \hat{q}^n$ it continues to hold, as a direct computation shows. In the same way (8.32) implies (we leave the computation as exercise):

$$
I_{\phi,g} = \sum_{k=1}^n \frac{\partial \mathscr{L}}{\partial \dot{q}^k} \frac{\partial q'^k}{\partial \epsilon} \bigg|_{\epsilon=0} - g(t, q) = \sum_{k=1}^n \frac{\partial \mathscr{L}}{\partial \dot{\hat{q}}^k} \frac{\partial \hat{q}'^k}{\partial \epsilon} \bigg|_{\epsilon=0} - g(\hat{t}, \hat{q}).
\tag{8.34}
$$

8.2.4 Action of the (Weak) Symmetries on the Solutions of the Euler-Lagrange Equations

In this section we want to show the dual aspect of a physical system's invariance under a transformation group. We will prove that if a physical system is *weakly invariant* (in particular, invariant) under a one-parameter group ϕ of local diffeomorphisms that preserve the time fibres, the action of ϕ maps solutions of the Euler-Lagrange equations to other solutions.

Theorem 8.12 *Referring to Theorem 8.10, consider a system of point particles $\mathscr{S}$ described on $\mathbb{V}^{n+1}$ and weakly invariant under the 1-parameter group of local diffeomorphisms ϕ given by (8.14) that preserve the fibres of $\mathbb{V}^{n+1}$. Let $\mathscr{L}$ and g be the functions in (8.19) and (8.20).*

Consider a solution to the Euler-Lagrange equations for $\mathscr{L}$:

$$
q^k = q^k(t), \quad \dot{q}^k = \dot{q}^k(t), \quad \text{for } t \in I
$$

and the curve obtained when ϕ acts on $q^k = q^k(t)$, $\dot{q}^k = \dot{q}^k(t)$ for $t \in I$:

$$\begin{cases} q'^k(t) = q'^k(\epsilon_1, t, q(t)) \quad \text{for } k = 1, \dots, n. \\[2mm] \dot{q}'^k(t) = \dfrac{\partial q'^k}{\partial t}(\epsilon_1, t, q(t)) + \displaystyle\sum_{j=1}^{n} \dfrac{\partial q'^k}{\partial q^j}(\epsilon_1, t, q(t))\dot{q}^j(t) \quad \text{for } k = 1, \dots, n. \end{cases} \tag{8.35}$$

for some chosen value $\epsilon = \epsilon_1$ of the parameter of ϕ. The curve $q'^k = q'^k(t)$, $\dot{q}'^k = \dot{q}'^k(t)$ for $t \in I$, defined locally on $A(\mathbb{V}^{n+1})$, is a solution of the Euler-Lagrange equations for the same $\mathscr{L}$, provided $\epsilon_1 \in \mathbb{R}$ is small enough (so that the curves $q'^k = q'^k(t)$, $\dot{q}'^k = \dot{q}'^k(t)$, for $t \in I$, are contained in the domain of the local chart used for ϕ, for any $\epsilon \in [0, \epsilon_1]$ when $\epsilon_1 > 0$ or $\epsilon \in [\epsilon_1, 0]$ when $\epsilon_1 < 0$).

Proof For $|\epsilon| \leq |\epsilon_1|$ let:

$$\begin{cases} q_\epsilon^k(t) = q'^k(\epsilon, t, q^1(t), \dots, q^n(t)) \quad \text{for } k = 1, \dots, n. \\[2mm] \dot{q}_\epsilon^k(t) = \dfrac{\partial q'^k}{\partial t}(\epsilon, t, q^1(t), \dots, q^n(t)) \\[2mm] \quad\quad + \displaystyle\sum_{j=1}^{n} \dfrac{\partial q'^k}{\partial q^j}(\epsilon, t, q^1, \dots, q^n)\dot{q}^j \quad \text{for } k = 1, \dots, n. \end{cases} \tag{8.36}$$

To simplicity let us assume there is $n = 1$ freedom degree; the general case $n > 1$ is straightforward. Since (8.19) implies (8.21), integrating the latter produces:

$$\mathscr{L}(t, q'(t), \dot{q}'(t)) = \mathscr{L}(t, q(t), \dot{q}(t)) + \int_0^{\epsilon_1} G(t, q_\epsilon(t), \dot{q}_\epsilon(t))d\epsilon . \tag{8.37}$$

Let us apply to both sides the differential operator:

$$\dfrac{d}{dt}\bigg|_{(t,q(t),\dot{q}(t))} \left(\dfrac{\partial}{\partial \dot{q}}\right) - \dfrac{\partial}{\partial q}\bigg|_{(t,q(t),\dot{q}(t))} . \tag{8.38}$$

The first summand on the right in (8.37) will vanish since the curve $q = q(t)$, $\dot{q} = \dot{q}(t)$ solves the Euler-Lagrange equations. The second summand can be analysed by swapping derivatives and integral (because the maps are continuously differentiable in all arguments, and $[-\epsilon_1, \epsilon_1]$ is compact). Therefore we find:

$$\int_0^{\epsilon_1} \left[\dfrac{d}{dt}\bigg|_{(t,q(t),\dot{q}(t))} \left(\dfrac{\partial}{\partial \dot{q}}\right) - \dfrac{\partial}{\partial q}\bigg|_{(t,q(t),\dot{q}(t))} \right] G(t, q_\epsilon(t), \dot{q}_\epsilon(t))d\epsilon .$$

The operator (8.38) on the left-hand side of (8.37) instead gives:

$$\frac{d}{dt}\bigg|_{(t,q(t),\dot{q}(t))}\left(\frac{\partial\mathscr{L}}{\partial\dot{q}'}\frac{\partial\dot{q}'}{\partial\dot{q}}\right) - \frac{\partial\mathscr{L}}{\partial q'}\frac{\partial q'}{\partial q}\bigg|_{(t,q(t),\dot{q}(t))} - \frac{\partial\mathscr{L}}{\partial\dot{q}'}\frac{\partial\dot{q}'}{\partial q}\bigg|_{(t,q(t),\dot{q}(t))}.$$

Computing the various derivatives and using (8.36) we arrive at:

$$\left[\frac{d}{dt}\left(\frac{\partial\mathscr{L}}{\partial\dot{q}}\right) - \frac{\partial\mathscr{L}}{\partial q}\right]\bigg|_{(t,q'(t),\dot{q}'(t))} \frac{\partial q'}{\partial q}\bigg|_{(t,q(t),\dot{q}(t))}$$

$$= \int_0^{\epsilon_1}\left[\frac{d}{dt}\left(\frac{\partial G}{\partial\dot{q}}\right) - \frac{\partial G}{\partial q}\right]\bigg|_{(t,q_\epsilon(t),\dot{q}_\epsilon(t))} \frac{\partial q_\epsilon}{\partial q}\bigg|_{(t,q(t),\dot{q}(t))} d\epsilon.$$

The function of t in the integral vanishes identically since $G(t,q,\dot{q})$ is a "total derivative". In conclusion, for any instant $t \in I$:

$$\left[\frac{d}{dt}\bigg|_{(t,q'(t),\dot{q}'(t))}\left(\frac{\partial\mathscr{L}}{\partial\dot{q}}\right) - \frac{\partial\mathscr{L}}{\partial q}\bigg|_{(t,q'(t),\dot{q}'(t))}\right]\frac{\partial q'}{\partial q}\bigg|_{(t,q(t),\dot{q}(t))} = 0.$$

Since the transformation sending the q to the q' is differentiable, invertible and with differentiable inverse, the Jacobian matrix must be everywhere invertible. We can then drop the second factor on the left by multiplying by the inverse Jacobian, so eventually:

$$\frac{d}{dt}\bigg|_{(t,q'(t),\dot{q}'(t))}\left(\frac{\partial\mathscr{L}}{\partial\dot{q}}\right) - \frac{\partial\mathscr{L}}{\partial q}\bigg|_{(t,q'(t),\dot{q}'(t))} = 0,$$

for any $t \in I$. This is what we wanted to prove. □

8.3 Jacobi's First Integral, Invariance Under "Temporal Displacements" and Conservation of the Mechanical Energy

In this section we will discuss a first integral, already known in the nineteenth century—well before Noether's Theorem—called *Jacobi's first integral*. We shall state the corresponding theorem straightaway.

Theorem 8.13 (Jacobi's Theorem) *Let $\mathscr{S}$ be a system of N point particles subject to $C = 3N - n \geq 0$ ideal holonomic constraints and completely described by the Lagrangian $\mathscr{L}$ on $A(\mathbb{V}^{n+1})$ (not necessarily of the form $\mathscr{T} - \mathscr{U}$ for some potential energy $\mathscr{U}$) of class C^2. Suppose that in natural coordinates*

$(t, q^1, \ldots, q^n, \dot{q}^1, \ldots, \dot{q}^n)$ *on the open subset* U *of the spacetime of kinetic states* $A(\mathbb{V}^{n+1})$ *we have:*

$$\frac{\partial \mathscr{L}}{\partial t} = 0 , \quad \textit{everywhere on } U. \tag{8.39}$$

Then the following hold.

(1) The function, defined on U *and called the* **system's Hamiltonian,**

$$\mathscr{H}(t, q, \dot{q}) := \sum_{k=1}^{n} \dot{q}^k \frac{\partial \mathscr{L}}{\partial \dot{q}^k} - \mathscr{L}(t, q, \dot{q}) , \tag{8.40}$$

is conserved in time on the system's motions (the portions in U *), in which case it is called* **Jacobi's first integral**. *In other terms*

$$\mathscr{H}(t, q(t), \dot{q}(t)) = constant$$

if $t \mapsto (t, q^1(t), \ldots, q^n(t), \dot{q}^1(t), \ldots, \dot{q}^n(t))$ *is a solution of the Euler-Lagrange equations. The constant depends on the solution. More generally, even without assuming (8.39), for any instant* t:

$$-\frac{\partial \mathscr{L}}{\partial t}\bigg|_{(t, q(t), \dot{q}(t)))} = \frac{d}{dt} \mathscr{H}(t, q(t), \dot{q}(t)) \tag{8.41}$$

on any solution of the Euler-Lagrange equations.

(2) Suppose the following hypotheses hold:

  (i) the active forces of $\mathscr{S}$ *are conservative in the frame* $\mathscr{R}$ *for which the Lagrangian is* $\mathscr{T}|_{\mathscr{R}} - \mathscr{U}|_{\mathscr{R}}$,
  (ii) the constraint functions, in orthonormal coordinates for $\mathscr{R}$, *do not depend on time explicitly,*
  (iii) the free coordinates $(t, q^1, \ldots, q^n)$ *move with* $\mathscr{R}$.

Then $\mathscr{H}$ *is the total mechanical energy of* $\mathscr{S}$ *in* $\mathscr{R}$:

$$\mathscr{H}(q, \dot{q}) := \mathscr{T}|_{\mathscr{R}}(q, \dot{q}) + \mathscr{U}|_{\mathscr{R}}(q) . \tag{8.42}$$

Remarks 8.14 As the proof will show, (8.42) holds even if the Lagrangian $\mathscr{L}|_{\mathscr{R}} = \mathscr{T}|_{\mathscr{R}} - \mathscr{U}|_{\mathscr{R}}$ does *not* describe $\mathscr{S}$ completely, but in the Euler-Lagrange equations there *also* appear source Lagrangian components of *non*-conservative active forces. In this case (8.42) is valid as long as $\mathscr{U}|_{\mathscr{R}}$ describes all conservative active forces acting on $\mathscr{S}$. If so, $\mathscr{H}$ will not be a first integral in general, as discussed in (1), Remarks 8.15.   ∎

Proof

(1) Let $t \mapsto (t, q^1(t), \ldots, q^n(t), \dot{q}^1(t), \ldots, \dot{q}^n(t))$ be a solution of the Euler-Lagrange equations. By the definition of $\mathscr{H}$ we have:

$$\frac{d}{dt}\mathscr{H}(t, q(t), \dot{q}(t))$$

$$= \sum_k \frac{d\dot{q}^k}{dt}\frac{\partial\mathscr{L}}{\partial\dot{q}^k} + \sum_k \dot{q}^k \frac{d}{dt}\frac{\partial\mathscr{L}}{\partial\dot{q}^k} - \frac{d}{dt}\mathscr{L}(t, q(t), \dot{q}(t)).$$

Using the Euler-Lagrange equations we can rewrite the above relation as:

$$\frac{d}{dt}\mathscr{H}(t, q(t), \dot{q}(t)) = \sum_k \frac{d\dot{q}^k}{dt}\frac{\partial\mathscr{L}}{\partial\dot{q}^k} + \sum_k \dot{q}^k \frac{\partial\mathscr{L}}{\partial q^k} - \frac{d}{dt}\mathscr{L}(t, q(t), \dot{q}(t)).$$

Adding and subtracting $\partial\mathscr{L}/\partial t$ on the right gives:

$$\frac{d}{dt}\mathscr{H} = \sum_k \frac{d\dot{q}^k}{dt}\frac{\partial\mathscr{L}}{\partial\dot{q}^k} + \sum_k \dot{q}^k \frac{\partial\mathscr{L}}{\partial q^k} + \frac{\partial\mathscr{L}}{\partial t} - \frac{d}{dt}\mathscr{L} - \frac{\partial\mathscr{L}}{\partial t}.$$

On the other hand, as $\mathscr{L} = \mathscr{L}(t, q(t), \dot{q}(t))$, we must also have:

$$\frac{d}{dt}\mathscr{L}(t, q(t), \dot{q}(t)) = \sum_k \frac{d\dot{q}^k}{dt}\frac{\partial\mathscr{L}}{\partial\dot{q}^k} + \sum_k \dot{q}^k \frac{\partial\mathscr{L}}{\partial q^k} + \frac{\partial\mathscr{L}}{\partial t}.$$

Inserting the above in the previous expression, on any solution of the Euler-Lagrange equations (8.41) holds:

$$\frac{d}{dt}\mathscr{H}(t, q(t), \dot{q}(t)) = -\frac{\partial}{\partial t}\mathscr{L}(t, q(t), \dot{q}(t)).$$

Then from (8.39) we deduce:

$$\frac{d}{dt}\mathscr{H}(t, q(t), \dot{q}(t)) = -\frac{\partial}{\partial t}\mathscr{L}(t, q(t), \dot{q}(t)) = 0,$$

so on any solution of the Euler-Lagrange equations $\mathscr{H}(t, q(t), \dot{q}(t))$ is constant in time.

(2) With our assumptions, using Sect. 7.2.4:

$$\mathscr{T}|_{\mathscr{R}}(q, \dot{q}) = \sum_{h,k=1}^n a_{hk}(q)\dot{q}^h\dot{q}^k \quad \text{with} \quad a_{hk}(q) = \sum_{i=1}^N \frac{m_i}{2}\frac{\partial\mathbf{x}_i}{\partial q^h} \cdot \frac{\partial\mathbf{x}_i}{\partial q^k},$$

and $\mathbf{x}_i = P_i(q^1, \ldots, q^n) - O$ is the position vector in $\mathscr{R}$ of the i-th particle (O is the origin of orthonormal coordinates for $\mathscr{R}$). Therefore:

$$\mathscr{L}|_{\mathscr{R}}(t, q, \dot{q}) = \sum_{h,k=1}^{n} a_{hk}(q)\dot{q}^h\dot{q}^k - \mathscr{U}(q) .$$

A direct calculation using (8.40) produces:

$$\mathscr{H}(t, q, \dot{q}) = \mathscr{U}|_{\mathscr{R}}(q) + \sum_{h,k=1}^{n} a_{hk}(q)\dot{q}^h\dot{q}^k \quad \text{i.e.} \quad \mathscr{H}(t, q, \dot{q}) = \mathscr{T}|_{\mathscr{R}} + \mathscr{U}|_{\mathscr{R}} .$$

This ends the proof.

$\square$

Remarks 8.15

(1) What happens if the system is also subject to non-conservative forces $\mathbf{F}_i$ with Lagrangian components $\mathscr{Q}_k$? (Note that such non-conservative forces might be described by a time-dependent potential, or a generalised potential). Then the following facts hold.

   (a) Adapting in the obvious way the first part of Jacobi's Theorem we immediately see that if $\mathscr{L}|_{\mathscr{R}} = \mathscr{T}|_{\mathscr{R}} - \mathscr{U}|_{\mathscr{R}}$ is the Lagrangian *relative to the sole conservative active forces in* $\mathscr{R}$, then on any solution of the Euler-Lagrange equations:

$$\frac{d\mathscr{H}}{dt} = \sum_{k=1}^{n} \frac{dq^k}{dt}\mathscr{Q}_k , \quad \text{provided} \quad \frac{\partial\mathscr{L}|_{\mathscr{R}}}{\partial t} = 0 \text{ on } U.$$

   (b) When the Lagrangian coordinates $q^1, \ldots, q^n$ move with $\mathscr{R}$, it is easy to show (do it as exercise):

$$\sum_{k=1}^{n} \frac{dq^k}{dt}\mathscr{Q}_k = \sum_{i=1}^{N} \mathbf{F}_i \cdot \mathbf{v}_{P_i}|_{\mathscr{R}}$$

   (irrespective of whether $\frac{\partial\mathscr{L}|_{\mathscr{R}}}{\partial t} = 0$ on U). The right-hand side is the total power in $\mathscr{R}$ dissipated by the forces that do not admit a potential.

   (c) If the Lagrangian coordinates move with $\mathscr{R}$, and besides conservative forces there are non-conservative forces, $\mathscr{H}$ is still the total mechanical energy $\mathscr{E}|_{\mathscr{R}} = \mathscr{T}|_{\mathscr{R}} + \mathscr{U}|_{\mathscr{R}}$, where the potential energy accounts only for the conservative forces (again, all this irrespective of $\frac{\partial\mathscr{L}|_{\mathscr{R}}}{\partial t} = 0$).

(2) It would seem that Jacobi's Theorem is logically unrelated to Noether's Theorem, especially since the transformations of Jacobi's Theorem alter the time coordinates (see the next remark) and hence do not preserve the fibres of $A(\mathbb{V}^{n+1})$. In reality this is due to the fact that we have stated Noether's Theorem in too elementary a way to subsume Jacobi's Theorem. In Sect. 8.5 we will show that Jacobi's Theorem is actually a subcase of a more general version of Noether's result.

(3) Condition (8.39):

$$\frac{\partial \mathscr{L}}{\partial t} = 0, \quad \text{everywhere on } U,$$

can be interpreted as *invariance of the physical system under temporal displacements*. In other words (8.39) establishes that the physical system has a Lagrangian satisfying, for any possible t, τ and $q^1, \dots, q^n, \dot{q}^1, \dots, \dot{q}^n$ in the chart of the case,

$$\mathscr{L}(t + \tau, q, \dot{q}) = \mathscr{L}(t, q, \dot{q}).$$

In this sense, under (2), Jacobi's Theorem says that the mechanical energy is conserved when the Lagrangian is invariant under temporal displacements. We should make it crystal clear that (8.39) depends *in an essential way* on the free coordinates $(t, q^1, \dots, q^n)$. Condition (8.39) in other free coordinates $(t', q'^1, \dots, q'^m)$:

$$\frac{\partial \mathscr{L}(t' + c, q(t', q'), \dot{q}(t', q', \dot{q}'))}{\partial t'} = 0$$

(for simplicity with $n = 1$) might be false due to the possible explicit t'-dependency of the relationship between the coordinate sets:

$$t = t' + c, \quad q^k = q^k(t', q'^1, \dots, q'^m).$$

Similarly, the Hamiltonian $\mathscr{H}$ depends essentially on the free coordinates: in contrast to $\mathscr{L}$, it is *not* a scalar field on $A(\mathbb{V}^{n+1})$. In general, moreover, $\mathscr{H}$ is *not* globally defined on the spacetime of kinetic states, but only on every open set U with fibre-adapted natural coordinates.

(4) Suppose $\mathscr{L}$, given by $\mathscr{T}|_{\mathscr{R}} - \mathscr{U}|_{\mathscr{R}}$, does not depend explicitly on time, as required by (8.39), but the free coordinates are not adapted to $\mathscr{R}$. The Jacobi first integral still exists, but it does not represent the total mechanical energy in $\mathscr{R}$. It may be the total mechanical energy in another reference frame. A concrete example of this situation is described in Exercise 7.33.**3**.

(5) Condition $\frac{\partial \mathscr{L}}{\partial t} = 0$ (holding everywhere on the domain of the natural coordinates we are using) implies that the Euler-Lagrange equations are *autonomous*. That is to say, when we write the Euler-Lagrange equations in normal form, in the right-hand side the variabile t does not show up explicitly.

The straightforward, well-known consequence is that if

$$I \ni t \mapsto (q^1(t), \ldots, q^n(t), \dot{q}^1(t), \ldots, \dot{q}^n(t))$$

solves the equations, so does

$$I_\tau \ni t \mapsto (q^1(t+\tau), \ldots, q^n(t+\tau), \dot{q}^1(t+\tau), \ldots, \dot{q}^n(t+\tau))$$

for any constant τ, where $I = (a, b)$ and $I_\tau = (a-\tau, b-\tau)$. This is yet another, perhaps more physical, way to understand the invariance under temporal displacements: looking at the set of solutions rather than the Lagrangian. Notice that $\frac{\partial \mathscr{L}}{\partial t} = 0$ implies that the equations of motion are autonomous, but not vice versa. If we add to a Lagrangian satisfying $\frac{\partial \mathscr{L}}{\partial t} = 0$ a *formal total derivative* (at least C^1)

$$f(t, q, \dot{q}) = \frac{\partial g}{\partial t} + \sum_{k=1}^{n} \frac{\partial g}{\partial q^k} \dot{q}^k,$$

that depends explicitly on time, the equations of motion do not change and therefore stay autonomous, but the new Lagrangian $\mathscr{L}' = \mathscr{L} + f$ depends explicitly on time.   ∎

8.4 Comments on the Relationship Between Symmetries and Constant of Motion

We finish with a few remarks on the Galilean invariance of Noether's and Jacobi's Theorems and on the importance of the latter in the subsequent development of Theoretical Physics.

8.4.1 Galilean Invariance in Classical Lagrangian Mechanics

We can now state the postulate of invariance of the theory under the Galilean group, referring to Sect. 3.4.1 and adapting it to the physical description in terms of Lagrangians. The requirement is simple. *In a given inertial frame, every system of interacting point particles, not subject to constraints, externally isolated and admitting a Lagrangian description, must be such that the Lagrangian or an equivalent (i.e. producing the same Euler-Lagrange equations) is invariant or weakly invariant under the 1-parameter subgroups of the Galilean group: displacements along a given (arbitrary) direction, rotations about a given (arbitrary) axis, temporal displacements, pure Galilean transformations in a given (arbitrary) direction.*

In other words we have imposed that for the class of inertial frames the description of Mechanics should satisfy, in Lagrangian language: *spatial homogeneity*, *spatial isotropy*, *temporal homogeneity* and *pure Galilean invariance*. By virtue of Noether's Theorem (which *subsumes* Jacobi's Theorem in the next section's advanced formulation), to each such symmetry property there corresponds a conserved physical quantity.

Let us summarise the chapter's results regarding isolated systems described in an inertial frame $\mathscr{R}$ by a Lagrangian of the form[2] $\mathscr{T}|_{\mathscr{R}} - \mathscr{U}|_{\mathscr{R}}$, with potential energy $\mathscr{U}$ depending only on the point particles' mutual distances. Galilean invariance is verified through calculations, and the Lagrangian is (weakly) invariant under the Galilean group and its 10 subgroups. Noether's Theorem then has the following consequences:

(1) *spatial homogeneity implies the conservation of the total impulse,*
(2) *spatial isotropy implies the conservation of the total angular momentum,*
(3) *temporal homogeneity implies the conservation of the total mechanical energy,*
(4) *pure Galilean invariance generates a conserved quantity whose conservation law corresponds to (4.1) (see 2 in Examples 8.11 for details).*

8.4.2 The Noether and Jacobi Theorems Beyond Classical Mechanics

To finish off last section's summary it would be necessary to discuss what happens for systems admitting a Lagrangian different from $\mathscr{T}|_{\mathscr{R}} - \mathscr{U}|_{\mathscr{R}}$. Such a discourse would be pure speculation since the physical world *is not* invariant under the Galilean group; rather, it is invariant (at least away from strong gravitational fields) under the Poincaré group. The discussion would require extending the Lagrangian formalism and the theorems of Noether and Jacobi to relativistic theories (we shall say something about it in Chap. 10), including field theories, and merge them into a single result, still called Noether's Theorem. Here it will suffice to say that such an extension exists, is very natural and has proved to be extremely powerful in the formulations of Modern Physics (especially the theory of fundamental interactions and elementary particles), where the relationship symmetries-conservation laws has played and still plays a decisive role. To mention one example, think of how complicated it is, in Electrodynamics, to define the impulse of the electromagnetic field (or its angular momentum). Extending the Lagrangian formalism and Noether's Theorem to field theories allows to *define* the electromagnetic impulse as the quantity that is conserved due to space homogeneity (invariance of the electromagnetic field's Lagrangian under spatial displacements).

---

[2] There might be Lagrangians with different structure, in which case one should check the Galilean invariance separately.

The evidence confirms that this definition is the *physically* correct one, also because, in particular, it consents to implement the Principle of conservation of the impulse. With that definition one proves that in an isolated system, made of interacting particles and electromagnetic fields, the total impulse is conserved in time. One proceeds similarly for all conserved quantities known from Classical Physics and defined in Modern Physics. When we pass to the quantum formulations and the relativistic quantum formulations born in the twentieth century (which have more sophisticated versions of Noether's Theorem) some theories are built precisely on the premise that they support some symmetry, or that a certain symmetry is broken.

8.5 $\boxed{\text{AC}}$: General and Global Formulation of Noether's Theorem

There are two ways to make Noether's Theorem *global*, i.e. state it for quantities defined on the entire $A(\mathbb{V}^{n+1})$ rather than a local chart. One way is to assume that the one-parameter group of fibre-preserving local diffeomorphisms that leave the Lagrangian invariant is global. This means (a) it is defined on the entire $A(\mathbb{V}^{n+1})$ and (b) it is defined for any value of the parameter. This is certainly too strong an assumption, and hard to check in several concrete cases. The second possibility is to use the notion of *infinitesimal* symmetry and refer, rather than to the group itself, to the *vector field* X (Sect. A.6.5) whose integral curves define the *flow* of X, i.e. the *associated one-parameter group of local diffeomorphisms* (Sect. 14.5.3). This alternative still produces a *global* version of Noether's Theorem when the symmetry-generating field is defined globally. In this section we shall provide a *general global formulation of Noether's Theorem* using the language of vector fields (Theorem 14.46, Proposition 14.51 and Definition 14.52) and consider a larger class of symmetries than the one treated so far. We will show that this setup is general enough to include Jacobi's Theorem as a subcase of Noether's Theorem.

Henceforth the manifold $A(\mathbb{V}^{n+1})$ will always be (at least) C^3, since the Lagrangians we will take on it will be C^3 (or C^2 if of the usual special form).

8.5.1 Symmetries and First Integrals in Terms of Vector Fields on $A(\mathbb{V}^{n+1})$

With every one-parameter group of local diffeomorphisms ϕ that preserve the fibres of $\mathbb{V}^{n+1}$, as in (8.14), we can associate a unique a vector field X. Consider in fact the vector field:

$$X(t,q) := \sum_{k=1}^{n} \frac{\partial q'^{k}(\epsilon, t, q)}{\partial \epsilon}\bigg|_{\epsilon=0} \frac{\partial}{\partial q^k}\bigg|_{(t,q)} + 0 \frac{\partial}{\partial t}\bigg|_{(t,q)}$$

on the natural local chart $\psi : U \ni c \mapsto (t(c), q^1(c), \ldots, q^n(c)) \in \mathbb{R}^{n+1}$ of $\mathbb{V}^{n+1}$. Using (8.15) and (8.16) it is easy to see that, for a given configuration with coordinates $t, q^1, \ldots, q^n$, the curve $\epsilon \mapsto (t'(\epsilon, t, q)), q'(\epsilon, t, q)) \in \psi(U)$, obtained letting ϕ act on the configuration, satisfies:

$$\frac{dt'}{d\epsilon} = 0, \quad \frac{dq'^k}{d\epsilon} = X^k(t, q'(\epsilon, q)).$$

In other words the one-parameter group's action can be read off the *integral curves* of X: the configuration $(t'(\epsilon, t, q), q'(\epsilon, t, q))$ resulting from ϕ acting on the initial configuration is precisely the configuration that we find, when the parameter equals ϵ, on the unique integral curve of X in $\mathbb{V}^{n+1}$ starting at the initial configuration $(t, q^1, \ldots, q^n)$ for $\epsilon = 0$. In this sense X is the *generating vector field of the group*.

Consider now the action of ϕ extended locally to $A(\mathbb{V}^{n+1})$. Easily, the set of transformations induced by ϕ on $A(\mathbb{V}^{n+1})$, which *locally* are

$$\Phi_\epsilon^{(X)} : \begin{cases} t' = t, \\[2ex] q'^k = q'^k(\epsilon, t, q) \quad \text{for } k = 1, \ldots, n. \\[2ex] \dot{q}'^k = \dfrac{\partial q'^k}{\partial t}(\epsilon, t, q^1, \ldots, q^n) \\[1ex] \quad + \displaystyle\sum_{j=1}^n \dfrac{\partial q'^k}{\partial q^j}(\epsilon, t, q^1, \ldots, q^n)\dot{q}^j \quad \text{for } k = 1, \ldots, n. \end{cases} \tag{8.43}$$

can be obtained directly from the integral curves of the locally defined field on $A(\mathbb{V}^{n+1})$ (still called X):

$$X(t, q, \dot{q}) := \sum_{k=1}^n \left. \frac{\partial q'^k(\epsilon, t, q)}{\partial \epsilon} \right|_{\epsilon=0} \left. \frac{\partial}{\partial q^k} \right|_{(t,q,\dot{q})} + 0 \left. \frac{\partial}{\partial t} \right|_{(t,q,\dot{q})}$$

$$+ \sum_{k=1}^n \left. \frac{\partial \dot{q}'^k(\epsilon, t, \dot{q})}{\partial \epsilon} \right|_{\epsilon=0} \left. \frac{\partial}{\partial \dot{q}^k} \right|_{(t,q,\dot{q})}. \tag{8.44}$$

Working with the integral curves (8.44) of this field (though the result is general, see Sect. 14.5.3) one can prove that the analogue properties to (8.15) and (8.16) hold:

$$\Phi_0^{(X)} = id_{A(\mathbb{V}^{n+1})}, \quad \Phi_s^{(X)} \circ \Phi_u^{(X)} = \Phi_{s+u}^{(X)}, \tag{8.45}$$

where the second formula holds on states $a \in A(\mathbb{V}^{n+1})$ for which the left-hand side (acting on a) is well defined. When X is globally defined on $A(\mathbb{V}^{n+1})$, the above relations are valid everywhere on $A(\mathbb{V}^{n+1})$; the only restrictions are on the range of the parameter ϵ in $\Phi_\epsilon^{(X)}(a)$, which may depend on a.

In full generality, and in agreement with the theory of 1-parameter groups of local diffeomorphisms of Sect. 14.5.3, we may say the following. When X is defined everywhere (in the concrete situation we are studying it has the form (8.44) in every natural local chart, but the present remark is totally general) the domain $\Gamma_X \subset \mathbb{R} \times A(\mathbb{V}^{n+1})$ of $\Phi^{(X)} : \Gamma_X \ni (\epsilon, a) \mapsto \Phi_\epsilon^{(X)}(a) \in A(\mathbb{V}^{n+1})$, also known as the *flow* of X, is an open set of the form:

$$\Gamma_X = \bigcup_{a \in A(\mathbb{V}^{n+1})} I_a \times \{a\}, \tag{8.46}$$

where $I_a \subset \mathbb{R}$ is an open interval containing 0 corresponding to the domain of the maximal integral curve $I_a \ni \epsilon \mapsto \gamma(\epsilon)$ of X with initial condition $\gamma(0) = a$. By definition:

$$\Phi_\epsilon^{(X)}(a) := \gamma(\epsilon), \quad \epsilon \in I_a. \tag{8.47}$$

In the concrete case at hand Eq. (8.43) merely represent the action of $\Phi^{(X)}$ in any natural local coordinate system on $A(\mathbb{V}^{n+1})$.

In this formalism the Lagrangian's invariance condition (8.17) can be written in coordinate-free form, viewing X as a differential operator:

$$X(\mathscr{L}) = 0 \quad \text{everywhere on } A(\mathbb{V}^{n+1}). \tag{8.48}$$

Proposition 8.16 *If X is C^1 on $A(\mathbb{V}^{n+1})$, conditions (8.48) and (8.17) are equivalent on any natural local chart of $A(\mathbb{V}^{n+1})$.*

Proof Equation (8.48) implies in particular that along any integral curve (8.44) of X, $I_a \ni \epsilon \mapsto \gamma(\epsilon) \in A(\mathbb{V}^{n+1})$, we have:

$$\frac{d}{d\epsilon}\bigg|_{\epsilon=0} \mathscr{L}(\gamma(\epsilon)) = X(\mathscr{L}(a)) = 0.$$

Therefore (8.17) holds true if we restrict to a natural local chart. If, vice versa, we have (8.17), as there is an integral curve of X departing from any $a \in A(\mathbb{V}^{n+1})$ by the Existence Theorem (we are assuming X is C^1), then (8.17) implies (8.48) on any local chart where (8.17) holds. □

As a matter of fact (8.48) is equivalent to another condition that seems much stronger at first, as we now show.

Proposition 8.17 *Suppose X is a C^1 vector field on $A(\mathbb{V}^{n+1})$ (not necessarily as in (8.44) in natural local coordinates) and $\mathscr{L} : A(\mathbb{V}^{n+1}) \to \mathbb{R}$ is also C^1. Then condition (8.48) is equivalent to*

$$\mathscr{L} \circ \Phi_\epsilon^{(X)}(a) = \mathscr{L}(a), \quad \text{for any } a \in A(\mathbb{V}^{n+1}) \text{ and } \epsilon \in I_a. \tag{8.49}$$

Proof Using (8.45) and (8.48) we have:

$$\frac{d}{d\epsilon} \mathscr{L}(\gamma(\epsilon)) = \lim_{h \to 0} \frac{1}{h} \left(\mathscr{L} \circ \Phi_{\epsilon+h}^{(X)}(a) - \mathscr{L} \circ \Phi_{\epsilon}^{(X)}(a) \right)$$

$$= \lim_{h \to 0} \frac{1}{h} \left[\mathscr{L} \left(\Phi_h^{(X)} \left(\Phi_{\epsilon}^{(X)}(a) \right) \right) - \mathscr{L} \left(\Phi_{\epsilon}^{(X)}(a) \right) \right] = X \left(\mathscr{L}(\Phi_{\epsilon}^{(X)}(a)) \right) = 0$$

so (8.49) follows immediately. Vice versa, (8.49) implies (8.48) if we differentiate it at $\epsilon = 0$. □

Summarising: conditions (8.48) and (8.49) are *completely equivalent*, even though the former might seem weaker. On any natural chart, moreover, they are equivalent to (8.17).

In the sequel we will state a generalised version of Noether's Theorem using the identification between local 1-parameter groups of diffeomorphisms and vector fields: instead of starting from 1-parameter groups of transformations $\Phi^{(X)}$ we will speak of vector fields X. The main hypothesis in Noether's Theorem, i.e. that the Lagrangian is invariant under a certain transformation group, will be expressed in terms of vector fields as in (8.48), and we will extend that request by formalising, in the language of vector fields, the notion of weak invariance. As already observed, since vector fields are definable globally on $A(\mathbb{V}^{n+1})$ without too much fuss, what we gain is that the theorem becomes completely independent of any choice of coordinates, and thus it turns global.

Let us then translate, keeping the above added benefits, the conclusion of Noether's Theorem using vector fields on the spacetime of kinetic states. Let us recall that by Theorem 7.41 the maximal solutions to the Euler-Lagrange equations are global on $A(\mathbb{V}^{n+1})$, and they are nothing but the integral curves of the globally defined *dynamic vector field* Z (Definition 7.40) as per (7.94), described in natural local coordinates by:

$$Z(t, q, \dot{q}) = \frac{\partial}{\partial t} + \sum_{k=1}^{n} \dot{q}^k \frac{\partial}{\partial q^k} + \sum_{k=1}^{n} z^k(t, q, \dot{q}) \frac{\partial}{\partial \dot{q}^k}. \tag{8.50}$$

The evolution in time under the dynamics generated by Z will then be given in $A(\mathbb{V}^{n+1})$ by a corresponding *flow* $\Phi^{(Z)}$ on $A(\mathbb{V}^{n+1})$. Let $I_a \ni t \mapsto \gamma(t) \in A(\mathbb{V}^{n+1})$ denote the maximal solution of the Euler-Lagrange equations associated with Z with initial condition $\gamma(t_a) = a$ if $t_a := T(a)$, where we have parametrised the solution by absolute time because by definition it is a section of the bundle $A(\mathbb{V}^{n+1})$ over absolute time. Then the flow $\Phi^{(Z)}$ is completely defined by:[3]

$$\Phi_s^{(Z)}(a) = \gamma(t_a + s), \quad s \in I_a - t_a, \tag{8.51}$$

and its domain has the form (8.46), clearly adapted.

From now on we will refer to Sects. 7.5.2 and 7.5.3 for the properties of Z and the intrinsic treatment of the Euler-Lagrange equations seen in Theorem 7.41. Using Z we have the following proposition:

Proposition 8.18 *Consider a dynamic vector field Z on $A(\mathbb{V}^{n+1})$ of class C^1. The condition that a C^1 function $f : A(\mathbb{V}^{n+1}) \to \mathbb{R}$ is a first integral—i.e. constant along any solution of the Euler-Lagrange equations associated with Z—is equivalent to:*

$$Z(f) = 0, \quad everywhere \ on \ A(\mathbb{V}^{n+1}). \tag{8.52}$$

Proof Since the solutions of the Euler-Lagrange equations are precisely the integral curves of Z, the proof is the same as Proposition 8.16, replacing X with Z and $\mathscr{L}$ with f.   □

In the thesis of Noether's Theorem the conservation of the first integral I along the solutions of the Euler-Lagrange equations will therefore just be stated as: $Z(I) = 0$ everywhere on the spacetime of kinetic states.

8.5.2 Noether's Theorem in General Global Form

Now we will state Noether's Theorem using vector fields. We begin with a discussion in natural coordinates on $A(\mathbb{V}^{n+1})$, and then pass to a more intrinsic approach. The field X will be defined globally on $A(\mathbb{V}^{n+1})$ and viewed as generating a one-parameter group of local diffeomorphisms that fix the Lagrangian of a physical system (at least weakly). In natural coordinates of $A(\mathbb{V}^{n+1})$ every field X reads:

$$X = X^t \frac{\partial}{\partial t} + \sum_{k=1}^{n} X^k \frac{\partial}{\partial q^k} + \sum_{k=1}^{n} \dot{X}^k \frac{\partial}{\partial \dot{q}^k}, \tag{8.53}$$

where X^t, X^k and $\dot{X}^k$ are differentiable in $(t, q^1, \ldots, q^n, \dot{q}^1, \ldots, \dot{q}^n)$. Under the elementary form of Noether's Theorem, thinking Z as differential operator acting

---

[3] For given a we can always redefine the absolute time function by adding an additive so that $t_a = 0$ and $s = t$.

on differentiable maps on $A(\mathbb{V}^{n+1})$, we assumed that

$$\dot{X}^k = Z\left(X^k\right), \quad k = 1, \ldots, n .$$

In fact, this is true for the generator of any local group of form (8.44):

$$Z\left(\frac{\partial q'^k(\epsilon, t, q)}{\partial \epsilon}\Big|_{\epsilon=0}\right) = \frac{\partial}{\partial t}\frac{\partial q'(\epsilon, t, q)}{\partial \epsilon}\Big|_{\epsilon=0} + \sum_k \dot{q}^k \frac{\partial q'(\epsilon, t, q)}{\partial \epsilon}\Big|_{\epsilon=0}$$

$$= \frac{\partial \dot{q}'(\epsilon, t, q)}{\partial \epsilon}\Big|_{\epsilon=0} ,$$

where in the final passage we used the last equation of (8.43). This hypothesis looks essential from a physical point of view: it tells the transformation group that the world of velocities (that of the coordinates $\dot{q}^k$) and the world of positions (the coordinates q^k) are close relatives. In the sequel we will still keep this, even when we assume, in greater generality than for the elementary Noether Theorem, that the functions X^k depend on the $\dot{q}^k$ as well. The ensuing one-parameter groups' action on the spacetime of configurations depends not just on the configurations themselves, but also on the velocities the system assumes in those configurations. We speak in this case of **dynamic symmetries** (where X^k, and not just $\dot{X}^k$, are functions of all the coordinates $(t, q^1, \ldots, q^n, \dot{q}^1, \ldots, \dot{q}^n)$) rather than **geometric symmetries** (where the X^k only depend on $(t, q^1, \ldots, q^n)$).

Assuming $\dot{X}^k = Z\left(X^k\right)$, as is natural, in *every system of natural coordinates on* $A(\mathbb{V}^{n+1})$, imposes restrictions on the field X. Namely, the following proposition holds.

Proposition 8.19 *If a C^1 field X on $A(\mathbb{V}^{n+1})$ satisfies*

$$\dot{X}^k = Z\left(X^k\right) ,$$

for $k = 1, \ldots, n$ in every system of natural coordinates on $A(\mathbb{V}^{n+1})$, (8.54)

for the C^1 dynamic vector field Z in (8.50) on $A(\mathbb{V}^{n+1})$, then

$$Z(\langle X, dT \rangle) = 0 . \tag{8.55}$$

Conversely if (8.55) holds, and we have (8.54) in an atlas of natural local coordinates of $A(\mathbb{V}^{n+1})$, then (8.54) holds in any natural local coordinates of $A(\mathbb{V}^{n+1})$.

Proof Locally, (8.55) reads $Z(X^t) = 0$ in any natural coordinates. As the components X^t transform as scalars, when we change natural local chart by:

$$X''^{t'} = \frac{\partial t'}{\partial t} X^t = X^t ,$$

to show the first claim it suffices to prove that there exist natural coordinates around any point where $Z(X^t) = 0$. Suppose (8.54) holds in any natural local coordinates adapted to the fibres of $A(\mathbb{V}^{n+1})$. If $(t, q^1, \ldots, q^n, \dot{q}^1, \ldots, \dot{q}^n)$ and $(t', q'^1, \ldots, q'^n, \dot{q}'^1, \ldots, \dot{q}'^n)$ are such coordinates on a common open set $U \subset A(\mathbb{V}^{n+1})$, from (7.71), (7.72) and (7.73) we deduce, for $j = 1, \ldots, n$:

$$\dot{X}'^j = \sum_{k=1}^{n} \frac{\partial \dot{q}'^j}{\partial \dot{q}^k} \dot{X}^k + \sum_{k=1}^{n} \frac{\partial \dot{q}'^j}{\partial q^k} X^k + \frac{\partial \dot{q}'^k}{\partial t} X^t , \tag{8.56}$$

$$X'^j = \sum_{k=1}^{n} \frac{\partial q'^j}{\partial q^k} X^k + \frac{\partial q'^j}{\partial t} X^t , \tag{8.57}$$

$$X'^{t'} = X^t .$$

But (8.54) holds in both systems, so:

$$Z\left(\sum_{k=1}^{n} \frac{\partial q'^j}{\partial q^k} X^k + \frac{\partial q'^j}{\partial t} X^t \right) = \sum_{k=1}^{n} \frac{\partial \dot{q}'^j}{\partial \dot{q}^k} Z(X^k) + \sum_{k=1}^{n} \frac{\partial \dot{q}'^j}{\partial q^k} X^k + \frac{\partial \dot{q}'^j}{\partial t} X^t . \tag{8.58}$$

From (7.73), then:

$$\frac{\partial \dot{q}'^j}{\partial \dot{q}^k} = \frac{\partial q'^j}{\partial q^k} ,$$

and we can recast the above as:

$$Z\left(\sum_{k=1}^{n} \frac{\partial q'^j}{\partial q^k} X^k + \frac{\partial q'^j}{\partial t} X^t \right) = \sum_{k=1}^{n} \frac{\partial q'^j}{\partial q^k} Z(X^k) + \sum_{k=1}^{n} \frac{\partial \dot{q}'^j}{\partial q^k} X^k + \frac{\partial \dot{q}'^j}{\partial t} X^t ,$$

or explicitly:

$$\sum_{k=1}^{n} Z\left(\frac{\partial q'^j}{\partial q^k} \right) X^k + \sum_{k=1}^{n} \frac{\partial q'^j}{\partial q^k} Z\left(X^k \right) + \sum_{k=1}^{n} Z\left(\frac{\partial q'^j}{\partial t} \right) X^t + \frac{\partial q'^j}{\partial t} Z\left(X^t \right)$$

$$= \sum_{k=1}^{n} \frac{\partial q'^j}{\partial q^k} Z(X^k) + \sum_{k=1}^{n} \frac{\partial \dot{q}'^j}{\partial q^k} X^k + \frac{\partial \dot{q}'^j}{\partial t} X^t . \tag{8.59}$$

Now, as:

$$Z\left(\frac{\partial q'^j}{\partial q^k} \right) = \frac{\partial \dot{q}'^j}{\partial q^k} \quad \text{and} \quad Z\left(\frac{\partial q'^j}{\partial t} \right) = \frac{\partial \dot{q}'^j}{\partial t} ,$$

(8.59) reduces to:

$$0 = \frac{\partial q'^j}{\partial t} Z\left(X^t\right) \qquad \text{for any natural local coordinate systems on } A(\mathbb{V}^{n+1}) \,.$$
(8.60)

We claim the above implies the thesis. Around any point $p \in A(\mathbb{V}^{n+1})$, given natural coordinates $(t, q^1, \ldots, q^n, \dot{q}^1, \ldots, \dot{q}^n)$ with origin at p, we can define coordinates $(t', q'^1, \ldots, q'^n, \dot{q}'^1, \ldots, \dot{q}'^n)$ so that $q'^j = q^j$ for $j > 1$ and $q'^1 = q^1 + ct$, where $c \neq 0$ (these are admissible around p since the Jacobian of the coordinate change has non-zero determinant at p). Then (8.60) at p implies $0 = cZ\left(X^t\right)|_p$, so $Z\left(X^t\right)|_p = 0$. This holds for any $p \in A(\mathbb{V}^{n+1})$ in natural coordinates around p, as we wanted.

The second part of the theorem is now obvious. In fact, suppose that under (8.55), in an atlas of (unprimed) natural coordinates we have (8.54). Then for any (primed) natural local coordinates (8.58) must hold. Using (8.54) for the (unprimed) coordinates of the atlas, we conclude from (8.56)–(8.57) that $Z(X'^j) = \dot{X}'^j$, i.e. (8.54) for any natural local coordinates.   □

Remarks 8.20

(1) In the applications it will be good to remember that (8.55) reads:

$$Z(X^t) = 0$$
(8.61)

in natural local coordinates and relatively to decomposition (8.53).

(2) We can produce vector fields X on $A(\mathbb{V}^{n+1})$ satisfying (8.54) by *lifting* to $A(\mathbb{V}^{n+1})$ any vector field $\tilde{X}$ on $\mathbb{V}^{n+1}$ *by means of* Z, provided certain assumptions hold. In this case one speaks of *geometric symmetries* induced by the lift X, as already said. If, in natural local coordinates on $\mathbb{V}^{n+1}$:

$$\tilde{X} = \tilde{X}^t \frac{\partial}{\partial t} + \tilde{X}^k \frac{\partial}{\partial q^k} \,,$$

where $X^k = X^k(t, q^1, \ldots, q^n)$, the lift X to $A(\mathbb{V}^{n+1})$ is defined by:

$$X = \tilde{X}^t \frac{\partial}{\partial t} + \tilde{X}^k \frac{\partial}{\partial q^k} + Z(\tilde{X}^k) \frac{\partial}{\partial \dot{q}^k} \,.$$

This expression does not depend on the local coordinates, so it defines a global field on $A(\mathbb{V}^{n+1})$ whenever $Z(\tilde{X}^t) = 0$, as we know from the previous proposition. Since $\tilde{X}^t$ can only depend on t and q^k, not on the dotted coordinates, the latter condition reduces to:

$$\frac{\partial \tilde{X}^t}{\partial t} + \sum_{k=1}^{n} \frac{\partial \tilde{X}^t}{\partial q^k} \dot{q}^k = 0 \,,$$

which must hold everywhere. From that we deduce that the partial derivatives of $\tilde{X}^t$ in t and q^k must vanish identically. Hence the only possibility is that $\tilde{X}^t$ = constant. We have proved, in greatest generality under $X^t \neq 0$, that a *geometric symmetry* acts trivially on time coordinates: integrating the 1-parameter group $\Phi_\epsilon^{(X)}$ generated by X on $A(\mathbb{V}^{n+1})$, it transforms the kinetic state $(t, q^1, \ldots, q^n, \dot{q}^1, \cdots, \dot{q}^n)$ into the kinetic state $(t'_\epsilon, q'^1_\epsilon, \ldots, q'^n_\epsilon, \dot{q}'^1_\epsilon, \cdots, \dot{q}'^n_\epsilon)$, where in particular

$$t'_\epsilon = t + \epsilon X^t, \quad X^t \text{ constant.}$$

The *preservation of the temporal fibres* we imposed in the elementary version of Noether's Theorem is valid only if the constant X^t is null.

(3) Let us consider again the geometric symmetries induced by the field X. Even if the constant X^t is not zero, the finite local action of the local group generated by X maps local sections of the bundle $A(\mathbb{V}^{n+1})$ (curves $I \ni t \mapsto \gamma(t) \in A(\mathbb{V}^{n+1})$ with $T(\gamma(t)) = t + c$ for some constant c) to local sections. This is immediate because $t'_\epsilon = t + \epsilon X^t$ (in fact $T(\phi_\epsilon^{(X)}(\gamma(t))) = t + c + \epsilon X^t$). In other words the symmetry's action maps potential evolutions of the system to potential evolutions. ∎

Let us now translate (8.54) and (8.55) in more intrinsic terms. We will use the *contact forms*:

$$\omega^k = dq^k - \dot{q}^k dt, \quad k = 1, 2, \ldots, n$$

associated with any natural coordinates ((7.99) in Sect. 7.5.3) and the Lie derivative of differential forms $\mathcal{L}_X(\omega)$ (Sects. B.1.2 and B.4.3).

Proposition 8.21 *Consider a C^1 vector field X on $A(\mathbb{V}^{n+1})$ satisfying condition (8.55):*

$$Z(\langle X, dT \rangle) = 0.$$

Then X satisfies (8.54) in any natural coordinates on $A(\mathbb{V}^{n+1})$ for the dynamic field Z if and only if

$$\langle \mathcal{L}_Z(X), \omega^k \rangle = 0, \tag{8.62}$$

for any contact form ω^k in any natural local coordinates.

Proof Condition (8.62) can be equivalently written as: $Z(\langle X, \omega^k \rangle) - \langle X, \mathcal{L}_Z(\omega^k) \rangle = 0$. By (7.105), passing to components:

$$0 = Z(X^k) - Z(X^t \dot{q}^k) - \dot{X}^k + X^t z^k = Z(X^k) - \dot{X}^k - Z(X^t \dot{q}^k) + X^t z^k$$
$$= Z(X^k) - \dot{X}^k - X^t z^k + X^t z^k,$$

i.e. $Z(X^k) - \dot{X}^k = 0$. For this we have used the hypothesis $Z(\langle X, dT \rangle) = 0$, which is $Z(\dot{q}^k) = z^k$ and $Z(X^t) = 0$ in coordinates. $\qquad\qquad\square$

We can now state and prove Noether's Theorem in general.

Theorem 8.22 (Noether's Theorem in General Global Form) *Consider a Lagrangian $\mathscr{L}$ of class[4] C^3 such that the Hessian matrix $\frac{\partial^2 \mathscr{L}}{\partial \dot{q}^k \partial \dot{q}^h}$ is everywhere non-singular on an atlas of natural local charts of $A(\mathbb{V}^{n+1})$, so that there is a dynamic field Z associated with $\mathscr{L}$ in the sense of Definition 7.42.*

Let X be a C^1 vector field on $A(\mathbb{V}^{n+1})$ satisfying:

(1) Equations (8.55) and (8.62):

$$Z(\langle X, dT \rangle) = 0, \quad \langle \mathcal{L}_Z(X), \omega^k \rangle = 0$$

for any contact form ω^k on any natural local coordinate system;

*(2) **weak invariance** of $\mathscr{L}$ under the action of X for some C^1 function f : $A(\mathbb{V}^{n+1}) \to \mathbb{R}$, possibly identically zero:*

$$X(\mathscr{L}) = Z(f) . \tag{8.63}$$

Then if $\Omega_{\mathscr{L}}$ is the Poincaré-Cartan form (7.106), the function on $A(\mathbb{V}^{n+1})$

$$I_{X,f} := \langle X, \Omega_{\mathscr{L}} \rangle - f \tag{8.64}$$

is a first integral:

$$Z(I_{X,f}) = 0 .$$

*$I_{X,f}$ is called **Noether's first integral** and in natural local coordinates on $A(\mathbb{V}^{n+1})$:*

$$I_{X,f} := \sum_{k=1}^{n} \frac{\partial \mathscr{L}}{\partial \dot{q}^k} (X^k - X^t \dot{q}^k) + X^t \mathscr{L} - f , \tag{8.65}$$

where X is (8.53).

---

[4] If $\mathscr{L}$ has form (7.87) it suffices to assume C^2.

Proof That $I_{X,f}$ is a well defined scalar field on the entire $A(\mathbb{V}^{n+1})$ is immediate because the Poincaré -Cartan 1-form is global and f is a scalar field. The direct calculation gives:

$$Z(I_{X,f}) = \langle \mathcal{L}_Z(X), \Omega_{\mathscr{L}} \rangle + \langle X, \mathcal{L}_Z(\Omega_{\mathscr{L}}) \rangle - Z(f)$$

$$= \sum_{k=1}^{n} \frac{\partial \mathscr{L}}{\partial \dot{q}^k} \langle \mathcal{L}_Z(X), \omega^k \rangle + \mathscr{L} \langle \mathcal{L}_Z(X), dT \rangle + \langle X, d\mathscr{L} \rangle - Z(f),$$

where we used (7.108). Observe that $\langle \mathcal{L}_Z(X), \omega^k \rangle = 0$ by (8.62), and $\langle \mathcal{L}_Z(X), dT \rangle = Z(\langle X, dT \rangle) - \langle X, \mathcal{L}_Z(dT) \rangle = 0$, using (8.55) and (7.103). All in all, $Z(I_{X,f}) = \langle X, d\mathscr{L} \rangle - Z(f) = X(\mathscr{L}) - Z(f) = Z(f) - Z(f) = 0$. $\square$

Remarks 8.23

(1) This version of Noether's Theorem is more sophisticated than Theorem 8.10 because here the first integral is defined globally. It is also more general since the f in the weak symmetry condition (8.63) can depend, locally, on the coordinates $\dot{q}^k$, in contrast to the elementary case of Theorem 8.10. Finally, (8.55), which is automatic with our assumptions on X, generalises and contains as special case the demand that the one-parameter group generated by X preserves the temporal fibres.

(2) On the other hand, the above version of Noether's uses the dynamic vector field Z which, in order to be defined, requires that the Euler-Lagrange equations have unique solutions for any initial condition. The elementary version of Theorem 8.10, instead, only needs one solution of said equations, and does not mention uniqueness at all. When we study relativistic Lagrangians in Chap. 10 (in particular Sect. 10.5.4) this fact will prove to be crucial, because we will deal with a Lagrangian, very popular in Relativistic Physics, whose Euler-Lagrange solutions exist but are not unique.   ∎

8.5.3   Properties of Vector Fields X Generating Symmetries

As mentioned before we will call $\Phi^{(Z)}$ the one-parameter group of local diffeomorphisms on $A(\mathbb{V}^{n+1})$ generated by the dynamic vector field Z (Sect. 14.5.3). $\Phi_\tau^{(Z)}(a)$ then corresponds to the time evolution, under the Euler-Lagrange equations associated with Z, from the initial condition $a \in \mathbb{A}_t$ to state $a_\tau \in \mathbb{A}_{t+\tau}$. In natural local coordinates: from $(t, q(t), \dot{q}(t))$ to $(t + \tau, q(t + \tau), \dot{q}(t + \tau))$.

If we abandon the demand that the one-parameter group of X preserves the fibres of $A(\mathbb{V}^{n+1})$ there is no guarantee the group action on a section of $A(\mathbb{V}^{n+1})$ will produce a section of $A(\mathbb{V}^{n+1})$, except for the case of geometric symmetries (see (3), Remarks 8.20). In other words a curve $I \ni u \mapsto \gamma(u) \in A(\mathbb{V}^{n+1})$, which can be represented as function of time because it is a section, i.e. $T(\gamma(u)) = u + c$ for some constant c, and is thus understandable as a system's motion, may be transformed into a curve $I \ni u \mapsto \gamma_\epsilon(u) = \Phi_\epsilon^{(X)}(\gamma(u))$ that is no longer a section: there is no c such that $T(\gamma_\epsilon(u)) \neq u + c$ for $u \in I$. Then it would not be possible to interpret γ_ϵ as a system's motion any longer. Physically, it would be problematic to assign a physical meaning to such symmetries of the Lagrangian $\mathscr{L}$, even though they might still be associated with a first integral by Noether's Theorem. We will show that for any symmetry of $\mathscr{L}$ (including weak ones) induced by a vector field X on $A(\mathbb{V}^{n+1})$ and respecting Theorem 8.22, there always is a second symmetry $\tilde{X}$ (in general *not* geometric) that obeys Theorem 8.22 and induces the same first integral as X. But $\tilde{X}$ *additionally satisfies* $\langle \tilde{X}, dT \rangle = 0$ and therefore its one-parameter group of local diffeomorphisms maps sections of $A(\mathbb{V}^{n+1})$ to sections of $A(\mathbb{V}^{n+1})$, because it preserves the temporal fibres.

Proposition 8.24 *Let $\mathscr{L}$ be as in Theorem 8.22 and Z the associated dynamic field. Suppose X is a vector field on $A(\mathbb{V}^{n+1})$ that satisfies (8.55) and (8.62) (equivalent to (8.54)) and represents a symmetry of the Lagrangian $\mathscr{L}$, in the sense that it satisfies (8.63) as well. Then the vector field on $A(\mathbb{V}^{n+1})$:*

$$\tilde{X} := X - \langle X, dT \rangle Z$$

(1) satisfies (8.55) in strong sense, namely it preserves the fibres of $A(\mathbb{V}^{n+1})$ over the time axis: $\langle \tilde{X}, dT \rangle = 0$ everywhere on $A(\mathbb{V}^{n+1})$;

(2) satisfies (8.62) and is still a symmetry of $\mathscr{L}$, since it satisfies (8.63) with f replaced by $\tilde{f} := f - \langle X, dT \rangle \mathscr{L}$;

(3) produces the same Noether first integral as X.

Proof (1) and (2). Properties $\langle \tilde{X}, dT \rangle = 0$ and (8.55) are clear by construction. Now, (8.63) holds since: $\tilde{X}(\mathscr{L}) = X(\mathscr{L}) - \langle X, dT \rangle Z(\mathscr{L}) = Z(f) - Z(\langle X, dT \rangle \mathscr{L}) = Z(f - \langle X, dT \rangle \mathscr{L})$, where we used the fact that $Z(\langle X, dT \rangle) = 0$. The above, plus $\mathcal{L}_Z Z = [Z, Z] = 0$, imply $\tilde{X}$ satisfies (8.62) since X does. (3) From $\langle Z, \omega_k \rangle = 0$, finally: $\langle X, \Omega_\mathscr{L} \rangle - f = \langle \tilde{X}, \Omega_\mathscr{L} \rangle - f + \langle X, dT \rangle \mathscr{L} = \langle \tilde{X}, \Omega_\mathscr{L} \rangle - \tilde{f}$. □

Another interesting property of symmetry-generating vector fields X on $A(\mathbb{V}^{n+1})$ is that the symmetries map Euler-Lagrange solutions to Euler-Lagrange solutions. We saw this in Theorem 8.12 for geometric symmetries. In the sequel we shall formulate the infinitesimal version of that fact: because the solutions to the Euler-Lagrange equations are integral curves of the dynamic vector Z, that they are preserved corresponds to the fact Z and X commute.

Theorem 8.25 *Let $\mathscr{L}$ be as in Theorem 8.22, Z the associated dynamic field, and X a C^2 vector field on $A(\mathbb{V}^{n+1})$ generating a geometric symmetry of $\mathscr{L}$. More precisely, assume X satisfies:*

(1) the strong form of (8.55), whereby the fibres of $A(\mathbb{V}^{n+1})$ over the time axis are preserved: $\langle X, dT \rangle = 0$;

(2) (8.54) in an atlas of natural charts where the X^k depend on t and $q^1, \ldots, q^n$ but not on the dotted coordinates, i.e. X is the lift to $A(\mathbb{V}^{n+1})$ under Z of a vector field on $\mathbb{V}^{n+1}$;

(3) $X(\mathscr{L}) = Z(f)$, where $f : \mathbb{V}^{n+1} \to \mathbb{R}$ is the trivial[5] C^2 lift to $A(\mathbb{V}^{n+1})$.

Then:

$$[X, Z] = 0 \quad \text{everywhere on } A(\mathbb{V}^{n+1}).$$

Proof If X is C^2 it generates a 1-parameter group of local diffeomorphisms $\Phi_\epsilon^{(X)}$ sending C^2 curves to C^2 curves. Theorem 8.12 can be phrased by saying that for given $a \in A(\mathbb{V}^{n+1})$ and if ϵ and τ are small enough, $I \ni \tau \mapsto \Phi_\epsilon^{(X)}(\Phi_\tau^{(Z)}(a))$ is still a solution of the Euler-Lagrange equations for $\mathscr{L}$. (Note we are covered by the Existence and uniqueness theorem for the equations and for the integral curves of Z, due to (1) and (2) in Theorem 7.41). In other words, considering the flows of X and Z: $\Phi_\epsilon^{(X)}(\Phi_\tau^{(Z)}(a)) = \Phi_\tau^{(Z)}(a_\epsilon)$ for some $a_\epsilon \in A(\mathbb{V}^{n+1})$. At $\tau = 0$ we find $\Phi_\epsilon^{(X)}(a) = a_\epsilon$. Hence for small τ and ϵ, $\Phi_\epsilon^{(X)}(\Phi_\tau^{(Z)}(a)) = \Phi_\tau^{(Z)}(\Phi_\epsilon^{(X)}(a))$ for any $a \in A(\mathbb{V}^{n+1})$. The discussion at the beginning of Sect. 14.5.4 implies the claim. $\square$

As a last comment, consider a field X generating a *geometric symmetry* that does not necessarily preserve the temporal fibres. As we saw in (3) Remark 8.20, such symmetries (locally) map sections of $A(\mathbb{V}^{n+1})$ to sections of the same bundle over the time axis. Condition (8.54) also ensures that if a section $I \ni t \mapsto (t, q(t), \dot{q}(t))$ satisfies $\frac{dq^k}{dt} = \dot{q}^k(t)$ (we are not asking it solves the Euler-Lagrange equations for some Lagrangian), then the transformed section under the flow $\Phi^{(X)}$ also satisfies that condition. Given a section as above, if we fix ϵ we may define another section, locally, parametrised by t'_ϵ, as follows:

$$(t'_\epsilon, q'_\epsilon(t_\epsilon), \dot{q}'_\epsilon(t_\epsilon)) := \Phi_\epsilon^{(X)}(t, q(t), \dot{q}(t)) = \Phi_\epsilon^{(X)}(t'_\epsilon - \epsilon X^t, q(t'_\epsilon - \epsilon X^t), \dot{q}(t'_\epsilon - \epsilon X^t)),$$

where we used the trivial relation $t'_\epsilon = t + \epsilon X^t$ because X^t is constant, as seen earlier (for geometric symmetries!). If $X(\mathscr{L}) = 0$, integrating X locally we arrive, as seen, at $\mathscr{L} \circ \Phi_\epsilon^{(X)} = \mathscr{L}$ (provided the left-hand side is meaningful). Evaluating on a generic section of $\mathbb{V}^{n+1}$ lifted to $A(\mathbb{V}^{n+1})$, imposing $\frac{dq^k}{dt} = \dot{q}^k(t)$ and then integrating over time, we finally obtain:

$$\int_{t_1}^{t_2} \mathscr{L}(t, q(t), \dot{q}(t)) dt = \int_{t'_{1\epsilon}}^{t'_{2\epsilon}} \mathscr{L}(t'_\epsilon, q'_\epsilon(t'_\epsilon), \dot{q}'_\epsilon(t'_\epsilon)) dt'_\epsilon . \tag{8.66}$$

Above we used $dt'_\epsilon/dt = 1$ when changing variables from t to t'_ϵ. Condition (8.66), valid when both sides are defined ($\epsilon > 0$ and $t_2 - t_1 > 0$ small enough), expresses the invariance of the *action functional* in local coordinates (addressed in Sect. 9.1) under

---

[5] In other words, on every natural local chart of $A(\mathbb{V}^{n+1})$, f is *not* a function of the $\dot{q}^k$.

the Lagrangian's *geometric symmetries*. In classical Lagrangian Mechanics, restricting ourselves to *geometric symmetries*, the Lagrangian's invariance is equivalent to the invariance of the associated action. Whereas in advanced theories (field theory) it is convenient to express symmetries using the action, in Classical Mechanics it is better to use the Lagrangian directly. We point out that formulation 8.22 of Noether's Theorem includes a broader class of symmetries than mere geometric ones, and assumes as hypothesis the invariance of the Lagrangian rather than the action's invariance.

8.5.4 Jacobi's First Integral as Consequence of Noether's Theorem

The existence of Jacobi's first integral is most certainly not a consequence of Noether's Theorem in its elementary form 8.10, since it cannot be associated with any one-parameter group of local diffeomorphisms that preserve the fibres of $A(\mathbb{V}^{n+1})$. Next we show that we can actually deduce the existence of Jacobi's first integral from the general version 8.22 of Noether's Theorem. Let us put ourselves in the hypotheses of Jacobi's Theorem 8.13, and let Z be the dynamic vector associated with the Lagrangian $\mathscr{L}$. Consider natural coordinates $(t, q^1, \ldots, q^n, \dot{q}^1, \ldots, \dot{q}^n)$ on the open subset U in the spacetime of kinetic states $A(\mathbb{V}^{n+1})$ where Jacobi's Theorem's assumptions hold for the vector field on $A(\mathbb{V}^{n+1})$:

$$Y := \frac{\partial}{\partial t} .$$

The above satisfies Theorem 8.22 (in particular $\langle Y, dT \rangle = 1$ everywhere) by construction. Moreover, under Jacobi's hypotheses:

$$Y(\mathscr{L}) = 0 ,$$

so the f in Theorem 8.22 is zero. Then we know there must exist a first integral $\mathscr{H}_Y := -I_{Y,0}$ that, *in the coordinates considered*, has form (8.65). For convenience we change the sign:

$$\mathscr{H}_Y := \sum_k \frac{\partial \mathscr{L}}{\partial \dot{q}^k}(\dot{q}^k - Y^k) - \mathscr{L} = \sum_k \frac{\partial \mathscr{L}}{\partial \dot{q}^k}\dot{q}^k - \mathscr{L} ,$$

since in the given coordinates $Y^k = 0$ identically. We have thus recovered Jacobi's first integral $\mathscr{H} = \mathscr{H}_Y$.

8.5.5 Jacobi's Global First Integral from the Global Noether Theorem

Observe that in Jacobi's Theorem the field $Y := \frac{\partial}{\partial t}$ is just defined on an open subset of $A(\mathbb{V}^{n+1})$, so the first integral is only local. However, with the approach based on Noether's Theorem just seen, the procedure generalises globally when the field Y is global. Physically, the choice of a reference frame $\mathscr{R}$ (if existent) in which the possible holonomic constraints on the Lagrangian system are time-independent, determines such a global vector field $Y_{\mathscr{R}}$. In fact, such a frame $\mathscr{R}$ defines a privileged atlas $\mathcal{A}_{0\mathscr{R}}$ on $\mathbb{V}^{n+1}$. This is the atlas of Lagrangian coordinates *moving* with $\mathscr{R}$. In turn, the atlas induces a preferred atlas on $A(\mathbb{V}^{n+1})$, which we shall indicate by $\mathcal{A}_{\mathscr{R}}$, given by natural coordinate systems built by putting together the coordinates $\dot{q}^k$ with the coordinates of $\mathcal{A}_{0\mathscr{R}}$. It is immediate to see that if (U, ψ) and (U', ψ') are local charts of $\mathcal{A}_{\mathscr{R}}$ with $U \cap U' \neq \varnothing$, $\psi : U \ni a \mapsto (t(a), q^1(a), \ldots, q^n(a), \dot{q}^1(a), \ldots, \dot{q}^n(a))$ and $\psi' : U' \ni a \mapsto (t'(a), q'^1(a), \ldots, q'^n(a), \dot{q}'^1(a), \ldots, \dot{q}'^n(a))$, then:

$$\frac{\partial}{\partial t} = \frac{\partial}{\partial t'}$$

on $U \cap U'$, precisely because the coordinate change *does not depend on time* (cf. (2) in Remark 9.11):

$$t' = t + c, \quad q'^k = q'^k(q^1, \ldots, q^n), \quad \dot{q}'^k = \sum_l \frac{\partial q'^k}{\partial q^l} \dot{q}^l, \tag{8.67}$$

since both coordinate sets work for the frame $\mathscr{R}$. Consequently we define a vector field $Y_{\mathscr{R}}$, on the whole $A(\mathbb{V}^{n+1})$, satisfying $\langle Y_{\mathscr{R}}, dT \rangle = 1$, when we impose that in any local coordinates of $\mathcal{A}_{\mathscr{R}}$:

$$Y_{\mathscr{R}} := \frac{\partial}{\partial t}.$$

Now consider Noether's first integral (8.64) associated with $Y_{\mathscr{R}}$; as we know, it is globally defined on $A(\mathbb{V}^{n+1})$. In any coordinates of $\mathcal{A}_{\mathscr{R}}$ it reads:

$$\mathscr{H}_{Y_{\mathscr{R}}} := \sum_k \frac{\partial \mathscr{L}}{\partial \dot{q}^k} \dot{q}^k - \mathscr{L}. \tag{8.68}$$

Now change coordinates, always in $\mathcal{A}_{\mathscr{R}}$. The right-hand side does not change since $\mathscr{L}$ is a scalar map on $A(\mathbb{V}^{n+1})$, while by (8.67):

$$\sum_k \frac{\partial \mathscr{L}}{\partial \dot{q}^k} \dot{q}^k = \sum_{k,l,r} \frac{\partial \mathscr{L}}{\partial \dot{q}'^l} \frac{\partial q'^l}{\partial q^k} \dot{q}'^r \frac{\partial q'^r}{\partial q^k} = \sum_r \frac{\partial \mathscr{L}}{\partial \dot{q}'^r} \dot{q}'^r.$$

Therefore, confirming what we already knew, (8.68) defines $\mathscr{H}_{Y_{\mathscr{R}}}$ globally on $A(\mathbb{V}^{n+1})$.

Remarks 8.26 With the aforementioned hypotheses on Y, the function $\mathscr{H}_{Y_{\mathscr{R}}} : A(\mathbb{V}^{n+1}) \to \mathbb{R}$ is well defined everywhere by (8.68), even in case $Y_{\mathscr{R}}(\mathscr{L}) \neq 0$, so even when $\mathscr{H}_{Y_{\mathscr{R}}}$ is not conserved on the solutions of the Euler-Lagrange equations of $\mathscr{L}$. ∎

8.5.6 The Runge-Lenz Vector from Noether's Theorem

We wish to present an interesting *dynamic symmetry* and the corresponding first integral coming from the above generalisation of Noether's Theorem. Consider, in an inertial frame $\mathscr{R}$, a particle P of mass m and position vector $\mathbf{x} = P - O$ in $\mathscr{R}$, subject to a central Coulomb force with centre the origin O of orthonormal coordinates of $\mathscr{R}$. We have a potential energy and an associated Lagrangian:

$$\mathscr{U}(\mathbf{x}) = -\frac{mg}{||\mathbf{x}||}, \quad \mathscr{L}(\mathbf{x}, \dot{\mathbf{x}}) = \frac{1}{2}m\dot{\mathbf{x}}^2 + \frac{mg}{||\mathbf{x}||},$$

where $g \neq 0$ is a constant and we used the Cartesian components of $\mathbf{x}$ as free coordinates. The equations of motion, trivially, are:

$$m\frac{d^2 x^i}{dt^2} = \frac{mg}{||\mathbf{x}||^3}x^i, \quad \text{for i=1,2,3.}$$

As the Lagrangian is patently invariant under rotations about any axis $\mathbf{n}$ through O, by Noether's Theorem (see Example 8.11.**1**) the angular momentum $\mathbf{\Gamma}_{O}|_{\mathscr{R}} = \mathbf{x}(t) \wedge m\mathbf{v}(t)$ will be conserved in time. Hence if $\mathbf{\Gamma}_{O}|_{\mathscr{R}} \neq \mathbf{0}$, the particle will move on the plane through O orthogonal to $\mathbf{\Gamma}_{O}|_{\mathscr{R}}$. Supposing $\mathbf{\Gamma}_{O}|_{\mathscr{R}} \neq \mathbf{0}$, we rotate the axes so that $\mathbf{e}_3$ is aligned with $\mathbf{\Gamma}_{O}|_{\mathscr{R}}$. The above Lagrangian, restricted to the plane $z = 0$, becomes:

$$\mathscr{L}(\mathbf{x}, \dot{\mathbf{x}}) = \frac{1}{2}m\dot{\mathbf{x}}^2 + \frac{mg}{||\mathbf{x}||},$$

where now $\mathbf{x} = x^1\mathbf{e}_1 + x^2\mathbf{e}_2$ and the equations of motion are:

$$m\frac{d^2 x^i}{dt^2} = \frac{mg}{||\mathbf{x}||^3}x^i, \quad \text{for i=1,2.}$$

In other words the Lagrangian restricted to $x^3 = 0$ produces the initial Lagrangian's motions that are confined to $x^3 = 0$. In polar coordinates, on $x^3 = 0$ the new Lagrangian can be written as:

$$\mathscr{L} = \frac{m}{2}\left(r^2\dot{\varphi}^2 + \dot{r}^2\right) + \frac{mg}{r}. \tag{8.69}$$

We will use from now on the reduced Lagrangian (8.69) and limit ourselves to study the motions of P on the plane $z = 0$ (with non-zero initial angular momentum). This Lagrangian does not seem to have any evident symmetry capable of producing further first integrals, besides the one we already have (from the full Lagrangian) and Jacobi's first integral (the system's mechanical energy). We want to prove that there exists a *dynamic symmetry* that gives a novel first integral, called *Runge-Lenz vector*. The latter is well known in classical Dynamics, but its existence cannot be easily explained using the general principles of elementary formulations of Mechanics.

For a given $\alpha \in \mathbb{R}$, consider the vector field **X** on $A(\mathbb{V}^{3+1})$:

$$\mathbf{X} = r^2\dot{\varphi}\sin(\varphi + \alpha)\frac{\partial}{\partial r} + r\dot{\varphi}\cos(\varphi + \alpha)\frac{\partial}{\partial\varphi} + Z(X^r)\frac{\partial}{\partial\dot{r}} + Z(X^\varphi)\frac{\partial}{\partial\dot{\varphi}}, \tag{8.70}$$

where we used the notation of the previous section:

$$X^r := r^2\dot{\varphi}\sin(\varphi + \alpha), \quad \text{and} \quad X^\varphi := r\dot{\varphi}\cos(\varphi + \alpha),$$

and where the dynamic vector Z associated with the Lagrangian is, after easy calculations:

$$Z = \frac{\partial}{\partial t} + \dot{r}\frac{\partial}{\partial r} + \dot{\varphi}\frac{\partial}{\partial\varphi} + \left(r\dot{\varphi}^2 - \frac{mg}{r}\right)\frac{\partial}{\partial\dot{r}} - \frac{2\dot{r}\dot{\varphi}}{r}\frac{\partial}{\partial\dot{\varphi}} \tag{8.71}$$

The explicit computation gives:

$$\mathbf{X}(\mathscr{L}) = Z(mg\cos(\varphi + \alpha)). \tag{8.72}$$

We are in the assumptions of Theorem 8.22, more precisely $\langle \mathbf{X}, dT \rangle = 0$, so there will be a first integral $I_{\mathbf{X}}$ associated with the weak symmetry induced by **X**. Note the one-parameter group of local diffeomorphisms generated by **X** is not of the type treated in the elementary versions of Noether's Theorem (Theorem 8.10), since the components X^r and X^φ *also* depend on the dotted coordinates. We are therefore dealing with a genuine dynamic symmetry. The direct calculations furnishes:

$$I = (\sin\alpha)\left[m\dot{r}\dot{\varphi}r^2\cos\varphi - (mr^3\dot{\varphi}^2 - mg)\sin\varphi\right]$$
$$+ (\cos\alpha)\left[m\dot{r}\dot{\varphi}r^2\sin\varphi + (mr^3\dot{\varphi}^2 - mg)\cos\varphi\right].$$

Note there is a first integral for every $\alpha \in [0, 2\pi]$. The full family of first integrals arises by taking linear combinations of the above for $\alpha = 0$ and $\alpha = \pi/2$. In particular, the following vector, called *Runge-Lenz vector*, carries the same information as the entire family of first integrals:

$$\mathbf{K} := \left[m\dot{r}\dot{\varphi}r^2 \sin\varphi + (mr^3\dot{\varphi}^2 - mg) \cos\varphi \right] \mathbf{e}_1$$

$$+ \left[-m\dot{r}\dot{\varphi}r^2 \cos\varphi + (mr^3\dot{\varphi}^2 - mg) \sin\varphi \right] \mathbf{e}_2 .$$

Recall that in our hypotheses $m\dot{\varphi}r^2\mathbf{e}_3 = \mathbf{\Gamma}_O|_{\mathscr{R}}$. Then easily:

$$\mathbf{K} = \frac{\mathbf{p}|_{\mathscr{R}} \wedge \mathbf{\Gamma}_O|_{\mathscr{R}}}{m} - mg\frac{\mathbf{x}}{||\mathbf{x}||} ,$$

where $\mathbf{p}|_{\mathscr{R}} = m\mathbf{v}|_{\mathscr{R}}$ is the particle's impulse (changing instant by instant). The vector $\mathbf{K}$ is a constant of motion since its components are (they are first integrals). In case the particle's trajectories are closed, hence ellipses (the Coulomb force is attractive, so $g > 0$), when the particle crosses the ellipse's major axis then $\mathbf{p}|_{\mathscr{R}} \wedge \mathbf{\Gamma}_O|_{\mathscr{R}}$ and $\mathbf{x}$ are both aligned with the axis. As $\mathbf{K}$ is constant in time, it will always be aligned with the major axis. In case the motion traces out a circle, it is easy to see that $\mathbf{K} = 0$.

Chapter 9
Advanced Topics in Lagrangian Mechanics

In this chapter we consider more advanced topics in Lagrangian Mechanics that have had a tremendous impact on the later development of Mathematical and Theoretical Physics, even in faraway contexts from Classical Mechanics. We will introduce the variational formulation of the Euler-Lagrange equations, the notion of generalised potential and some definitions and results on stability theory.

9.1 The Stationary-Action Principle for Systems that Admit a Lagrangian

We wish to show how the Euler-Lagrange equations can be obtained from a *variational principle*. In other words the solutions of the Euler-Lagrange equations, for systems that admit a Lagrangian, are *stationary points* of a certain functional called *action*. This result can be generalised in several directions, from Continuum Mechanics to Classical and Relativistic Field Theories. The variational approach has also had an important development within theoretical Quantum Physics, and has led to the formulation of Quantum Mechanics (and Quantum Field Theory) based on the so-called *Feynman integral*.

9.1.1 Rudiments of Calculus of Variations

Let us start from some general concepts. If $f : \Omega \to \mathbb{R}$ is a function defined on an open set $\Omega \subset \mathbb{R}^n$, a point $p \in \Omega$ is called *stationary* or *critical* if f is differentiable at p and the gradient of f at p vanishes. Every minimum point, maximum point or saddle point of f is critical, so being stationary is a necessary condition for extrema (internal to the domain). We want to extend this notion to *functionals*, i.e. real-valued

© The Author(s), under exclusive license to Springer Nature Switzerland AG 2023
V. Moretti, *Analytical Mechanics*, La Matematica per il 3+2 150,
https://doi.org/10.1007/978-3-031-27612-5_9

maps whose domain is a set of functions. Let us look at a few examples. Consider a map $F : D \to \mathbb{R}$ where D is a set of functions defined on an interval $I = [a, b]$ and with values in $\mathbb{R}^n$. We shall call it a **functional on D**.

Examples 9.1 Here is a rather obvious example of a functional:

$$F[\gamma] := \gamma(a), \quad \text{for any } \gamma \in D.$$

Supposing curves in D are differentiable, if $c \in (a, b)$ is a given point, another easy example is:

$$F[\gamma] := \frac{d\gamma}{dt}\Big|_c, \quad \text{for any } \gamma \in D.$$

Less trivially, for $n = 1$ and for continuous curves in D with continuous derivative:

$$F[\gamma] := \int_a^b \left(\gamma(u) + \frac{d\gamma}{du} \right) du, \quad \text{for any } \gamma \in D.$$

Let us go beck to the notion of critical points for functionals. The definition based on the gradient's vanishing is too complex, because it requires a notion of gradient in an *infinite-dimensional* space. This can be captured by the *Fréchet derivative*. Yet to introduce it we would have to define new topological notions that are not relevant to our elementary context. There is a simpler way that uses directional derivatives, hence derivatives in one variable (in the proper setting this would be a *weak Gateaux derivative*, technically). To arrive at such object we observe that an equivalent way of phrasing that p is stationary for a differentiable f on an open domain $\Omega \ni p$ is to say that all directional derivatives vanish:

$$\nabla f|_p \cdot \mathbf{u} = 0, \quad \text{for any vector } \mathbf{u} \in \mathbb{R}^n.$$

The latter can be recast in terms of one-variable functions, rather than using a gradient in several variables, as:

$$\frac{d}{d\alpha}\Big|_{\alpha=0} f(p + \alpha\mathbf{u}) = 0, \quad \text{for any vector } \mathbf{u} \in \mathbb{R}^n.$$

Returning to the functional $F : D \to \mathbb{R}$, where D is a set of $\mathbb{R}^n$-valued functions defined on $I = [a, b]$, we will say $\gamma_0 \in D$ is a stationary point of F if:

$$\begin{cases} \dfrac{d}{d\alpha}\Big|_{\alpha=0} F[\gamma_0 + \alpha\eta] = 0, \\[2mm] \text{for any } \eta : I \to \mathbb{R}^n \text{ such that } \gamma_0 + \alpha\eta \in D \text{ for } \alpha \in (-\delta_\eta, \delta_\eta). \end{cases} \tag{9.1}$$

Above, $\delta_\eta > 0$ in general depends on η. Clearly if F assumes maximum or minimum at γ_0, then (9.1) is true for any η, provided $\alpha \mapsto F[\gamma_0 + \alpha\eta]$ is well defined around $\alpha = 0$ and differentiable at the point. If for example γ_0 is a minimum point then $\alpha^{-1}(F[\gamma_0 + \alpha\eta] - F[\gamma_0]) \geq 0$ for $\alpha > 0$ and $\alpha^{-1}(F[\gamma_0 + \alpha\eta] - F[\gamma_0]) \leq 0$ for $\alpha < 0$, so the limit as $\alpha \to 0$ must be zero. A similar argument holds for maximum points.

What we want to show now is that *the solutions of the Euler-Lagrange equations are critical points of a certain functional defined on a suitable space of curves.*

9.1.2   Hamilton's Stationary-Action Principle

Fix natural coordinates on $\mathbb{V}^{n+1}$: $t, q^1, \ldots, q^n$ and assume, without loss of generality, they vary in $I \times \Omega$, where $I = [a, b]$, $a < b$ and $\Omega \subset \mathbb{R}^n$ is open (and non-empty). Consider the Lagrangian $\mathscr{L} : I \times \Omega \times \mathbb{R}^n \to \mathbb{R}$ of class C^2 in all variables (in general defined with respect to a frame $\mathscr{R}$ that will be implicit). Fix two points $Q_a, Q_b \in \Omega$ and consider the functional, called **action**:

$$I_{Q_a, Q_b}^{(\mathscr{L})}[\gamma] := \int_a^b \mathscr{L}\left(t, q^1(t), \ldots, q^n(t), \frac{dq^1}{dt}(t), \ldots, \frac{dq^n}{dt}(t)\right) dt, \qquad (9.2)$$

where the curves $\gamma : I \ni t \mapsto (q^1(t), \ldots q^n(t)) \in \Omega$ belong to the domain:

$$D_{Q_a, Q_b} := \left\{\gamma \in C^2(I) \;\middle|\; \gamma(t) \in \Omega \text{ for any } t \in I \text{ with } \gamma(a) = Q_a \text{ and } \gamma(b) = Q_b\right\}.$$

Fix $\gamma \in D_{Q_a, Q_b}$. As γ is continuous and hence $\gamma([a, b])$ is compact (continuous image of a compact set), we obtain that γ is contained in a finite union of open balls centred at points on the curve and all contained in Ω. With this open neighbourhood of γ it is easy to see that every C^2 curve $\eta : I \to \mathbb{R}^n$ with $\eta(a) = \eta(b) = 0$ satisfies $\gamma + \alpha\eta \in D_{Q_a, Q_b}$ for $\alpha \in (-\delta_\eta, \delta_\eta)$ if $\delta_\eta > 0$ is small enough (this follows from Proposition A.5, which holds under weaker hypotheses). Therefore it makes sense to use the definition of stationary point for the action functional, thus obtaining the following remarkable theorem of Hamilton's.

Theorem 9.2 (Hamilton's Principle of Stationary Action)   *With the above definitions, a curve $\gamma : I \to (q^1(t), \ldots, q^n(t)) \in \Omega$ belonging to D_{Q_a, Q_b} (so $\gamma(a) = Q_a$, $\gamma(b) = Q_b$) satisfies the Euler-Lagrange equations (7.84), for a C^2 Lagrangian $\mathscr{L} : I \times \Omega \times \mathbb{R}^n \to \mathbb{R}$ in all its variables, if and only if γ is stationary for the action functional $I_{Q_a, Q_b}^{(\mathscr{L})}$ in (9.2).*   ◇

Proof By construction, and with obvious notations:

$$I_{Q_a, Q_b}^{(\mathscr{L})}[\gamma + \alpha\eta] := \int_a^b \mathscr{L}\left(t, \gamma(t) + \alpha\eta(t), \frac{d\gamma}{dt}(t) + \alpha\frac{d\eta}{dt}(t)\right) dt.$$

After we pass the derivative in α inside the integral using Theorem A.2, in the obvious notation:

$$\frac{d}{d\alpha} I^{(\mathscr{L})}_{Q_a, Q_b}[\gamma + \alpha\eta] = \int_a^b \{\eta \cdot \nabla_\gamma \mathscr{L}(t, \gamma + \alpha\eta, \dot{\gamma} + \alpha\dot{\eta})$$

$$+ \dot{\eta} \cdot \nabla_{\dot{\gamma}} \mathscr{L}(t, \gamma + \alpha\eta, \dot{\gamma} + \alpha\dot{\eta})\} \, dt \, ,$$

where we have *defined* $\dot{\eta} := d\eta/dt$. For $\alpha = 0$ we find:

$$\frac{d}{d\alpha}\bigg|_{\alpha=0} I^{(\mathscr{L})}_{Q_a, Q_b}[\gamma + \alpha\eta] = \int_a^b \{\eta \cdot \nabla_\gamma \mathscr{L}(t, \gamma, \dot{\gamma}) + \dot{\eta} \cdot \nabla_{\dot{\gamma}} \mathscr{L}(t, \gamma, \dot{\gamma})\} \, dt \, .$$

Integrating by parts:

$$\frac{d}{d\alpha}\bigg|_{\alpha=0} I^{(\mathscr{L})}_{Q_a, Q_b}[\gamma + \alpha\eta] = \int_a^b \eta \cdot \left\{\nabla_\gamma \mathscr{L}(t, \gamma, \dot{\gamma}) - \frac{d}{dt}\nabla_{\dot{\gamma}} \mathscr{L}(t, \gamma, \dot{\gamma})\right\} \, dt \, .$$

We have omitted a term in the formula for integration by parts that does not contribute, namely

$$\int_a^b \frac{d}{dt}\left(\eta \cdot \nabla_\gamma \mathscr{L}\right) dt = \eta(b) \cdot \nabla_\gamma \mathscr{L} - \eta(a) \cdot \nabla_\gamma \mathscr{L} \, . \tag{9.3}$$

Since we are looking at variations $\gamma + \alpha\eta$ of the curve γ while keeping the ends fixed, necessarily $\gamma(a) + \alpha\eta(a) = Q_a$ and $\gamma(b) + \alpha\eta(b) = Q_b$ for any α near 0. Therefore $\eta(a) = \eta(b) = 0$, and the left-hand side of (9.3) vanishes.

All in all, in components:

$$\frac{d}{d\alpha}\bigg|_{\alpha=0} I^{(\mathscr{L})}_{Q_a, Q_b}[\gamma + \alpha\eta] = \int_a^b \sum_{k=1}^n \eta^k \left[\frac{\partial \mathscr{L}}{\partial q^k} - \frac{d}{dt}\left(\frac{\partial \mathscr{L}}{\partial \dot{q}^k}\right)\right] dt \, , \tag{9.4}$$

where, implicitly, $\dot{q}^k = dq^k/dt$. The above relationship immediately implies that if $\gamma : t \mapsto (q^1(t), \ldots, q^n(t))$ solves the Euler-Lagrange equations then γ is a stationary point of $I^{(\mathscr{L})}_{Q_a, Q_b}$, since by (9.4) it satisfies:

$$\frac{d}{d\alpha}\bigg|_{\alpha=0} I^{(\mathscr{L})}_{Q_a, Q_b}[\gamma + \alpha\eta] = 0 \, , \tag{9.5}$$

for any $\eta : I \to \mathbb{R}^n$ such that $\gamma + \alpha\eta \in D_{Q_a, Q_b}$, and α in some interval $(-\delta_\eta, \delta_\eta)$.

Conversely, let us assume (9.5) for any $\eta : I \to \mathbb{R}^n$ as above. By (9.4):

$$\int_a^b \sum_{k=1}^n \eta^k \left[\frac{\partial \mathscr{L}}{\partial q^k} - \frac{d}{dt}\left(\frac{\partial \mathscr{L}}{\partial \dot{q}^k}\right)\right] dt = 0 \, , \tag{9.6}$$

for any $\eta : I \to \mathbb{R}^n$ such that $\gamma + \alpha \eta \in D_{Q_a, Q_b}$ for α in some interval $(-\delta_\eta, \delta_\eta)$. We claim that forces γ, as the η^k are arbitrary, to satisfy the Euler-Lagrange equations. By contradiction, suppose there is a point $t_0 \in I$ where, for some k, *the Euler-Lagrange equations do not hold*. In other words:

$$\left[\frac{\partial \mathscr{L}}{\partial q^k} - \frac{d}{dt} \left(\frac{\partial \mathscr{L}}{\partial \dot{q}^k} \right) \right] \Bigg|_{(t_0, \gamma(t_0), \dot{\gamma}(t_0))} \neq 0$$

Without loss of generality we take $k = 1$ and $t_0 \in (a, b)$ (by continuity, the following reasoning will give the cases $t_0 = a$ and $t_0 = b$). As the map

$$I \ni t \mapsto \left[\frac{\partial \mathscr{L}}{\partial q^k} - \frac{d}{dt} \left(\frac{\partial \mathscr{L}}{\partial \dot{q}^k} \right) \right]$$

is continuous in t_0, it will have a certain sign on an open neighbourhood $I' = (c, d)$ of t_0, say $+$ (if $-$, the argument is the similar). Shrinking I' we can always make the map positive on $[c, d]$. Hence

$$\left[\frac{\partial \mathscr{L}}{\partial q^1} - \frac{d}{dt} \left(\frac{\partial \mathscr{L}}{\partial \dot{q}^1} \right) \right] \Bigg|_{(t, \gamma(t), \dot{\gamma}(t))} \geq C > 0$$

for any $t \in [c, d]$ and for some constant C. To conclude we can pick an element η whose components η^k are zero except η^1, which is non-negative on $[c, d]$ and zero on the rest of I, and with integral equal to $L > 0$ (it suffices to "smoothen it out" to make it a C^2 map whose graph is a column of height $L/(d - c)$ over $[c, d]$). Therefore

$$\int_a^b \sum_{k=1}^n \eta^k \left[\frac{\partial \mathscr{L}}{\partial q^k} - \frac{d}{dt} \left(\frac{\partial \mathscr{L}}{\partial \dot{q}^k} \right) \right] dt = \int_c^d \eta^1 \left[\frac{\partial \mathscr{L}}{\partial q^1} - \frac{d}{dt} \left(\frac{\partial \mathscr{L}}{\partial \dot{q}^1} \right) \right] dt$$

$$\geq C \int_c^d \eta^1 dt \geq LC > 0 \,,$$

which contradicts (9.6). We conclude that if γ is critical for I_{Q_a, Q_b} then, for any $t \in I$:

$$\left[\frac{\partial \mathscr{L}}{\partial q^k} - \frac{d}{dt} \left(\frac{\partial \mathscr{L}}{\partial \dot{q}^k} \right) \right] \Bigg|_{(t, \gamma(t), \dot{\gamma}(t))} = 0$$

and so γ also solves the Euler-Lagrange equations. $\qquad \square$

Remarks 9.3

(1) We should emphasise *there may not exist a solution of the Euler-Lagrange equations* $\gamma : I \ni t \mapsto (q^1(t), \ldots, q^n(t)) \in \Omega$ *when the ends* $\gamma(a) = Q_a$, $\gamma(b) = Q_b$ *as assigned*. And if such a solution does exist, *it may not be unique*. The Existence Theorem warrants the existence of a solution *in presence of (Cauchy) initial conditions* and not *boundary conditions*, i.e. the endpoints of the required curve. For a boundary-value problem of this type there are no global existence and uniqueness results, in general (we will address local results in Sect. 12.4.6 and the one preceding it). In this sense Hamilton's variational principle should be seen and a procedure for finding the equations of motion rather than their solutions.

(2) If we consider two inertial frames $\mathscr{R}$ and $\mathscr{R}'$, the Lagrangians associated with one physical system differ by a formal total derivative (see Sect. 7.4.7)

$$\Delta\mathscr{L} := \sum_{k=1}^{n} \frac{\partial g}{\partial q^k}\dot{q}^k + \frac{\partial g}{\partial t} . \tag{9.7}$$

Passing to the variational approach where we consider curves in the variables q^k, and the $\dot{q}^k$ are understood as derivatives of the q^k, the aforementioned term truly is a total derivative with respect to time. Each term of this type provides a *constant* contribution to the action functional (9.2), since:

$$\int_a^b \Delta\mathscr{L}dt = \int_a^b \left(\sum_{k=1}^{n} \frac{\partial g}{\partial q^k}\frac{dq^k}{dt} + \frac{\partial g}{\partial t} \right) dt$$

$$= g(b, q^1(b), \ldots, q^n(b)) - g(a, q^1(a), \ldots, q^n(a)) .$$

As the curves $\gamma : [a, b] \ni t \mapsto (q^1(1), \ldots, q^n(t))$ vary in the domain D_{Q_a, Q_b},

$$g(b, q^1(b), \ldots, q^n(b)) - g(a, q^1(a), \ldots, q^n(a))$$

is constant by construction, because all curves in the domain have the same endpoints $(q^1(a), \ldots, q^n(a))$ and $(q^1(b), \ldots, q^n(b))$ by definition. Clearly, then, the two action functionals, the one in the inertial frame $\mathscr{R}$ and the one in $\mathscr{R}'$, have the same critical points since they differ by a constant. In other words they determine the same Euler-Lagrange equations, as we observed in Sect. 7.4.7 even before tackling variational calculus.

(3) $\boxed{\text{AC}}$ We formulated the theory in $\mathbb{R}^n$, or more precisely in a local chart of $\mathbb{V}^{n+1}$. One could reformulate Hamilton's variational principle in a more general ambient, involving several charts when the curves are not contained in a single chart, by using *partitions of unity*. ∎

9.2 Generalised Potentials

Analytical Mechanics is the place where the notion of potential naturally gives rise to an extension, which covers potentials not depending on the "dotted coordinates" used to compute a force's Lagrangian components.

9.2.1 The Case of the Lorentz Force

Before we start with the two definitions let us consider a situation of major physical relevance. Consider a particle P of mass m without constraints, with electric charge e and subject in the inertial frame $\mathscr{R}$ to an electric field $\mathbf{E}(t, \mathbf{x})$ and a magnetic field $\mathbf{B}(t, \mathbf{x})$. In the sequel $\mathbf{x} := P - O$ is the particle's position vector in $\mathscr{R}$ with respect to the origin O, and the orthonormal components of $\mathbf{x}$ in $\mathscr{R}$ will be used as free coordinates for the particle; $\mathbf{v}$ will denote the velocity with respect to $\mathscr{R}$. The Lorentz force

$$\mathbf{F}_L(t, \mathbf{x}, \mathbf{v}) := e\mathbf{E}(t, \mathbf{x}) + \frac{e}{c}\mathbf{v} \wedge \mathbf{B}(t, \mathbf{x}) \tag{9.8}$$

acts on the particle, c being the speed of light. To describe the electric and magnetic fields we introduce the *scalar potential* $\varphi(t, \mathbf{x})$ and the *vector potential* $\mathbf{A}(t, \mathbf{x})$ so that, if $\nabla_\mathbf{x}$ is the gradient evaluated at $\mathbf{x}$:

$$\mathbf{E}(t, \mathbf{x}) = -\frac{1}{c}\frac{\partial \mathbf{A}}{\partial t}(t, \mathbf{x}) - \nabla_\mathbf{x}\varphi(t, \mathbf{x}), \tag{9.9}$$

$$\mathbf{B}(t, \mathbf{x}) = \nabla_\mathbf{x} \wedge \mathbf{A}(t, \mathbf{x}). \tag{9.10}$$

In terms of the two potentials, (9.8) reads:

$$\mathbf{F}_L(t, \mathbf{x}, \mathbf{v}) = -\frac{e}{c}\frac{\partial \mathbf{A}}{\partial t}(t, \mathbf{x}) - e\nabla_x\varphi(t, \mathbf{x}) + \frac{e}{c}\mathbf{v} \wedge (\nabla_\mathbf{x} \wedge \mathbf{A}(t, \mathbf{x})). \tag{9.11}$$

Computing the last cross product we may finally write:[1]

$$\mathbf{F}_L(t, \mathbf{x}, \mathbf{v}) = -\frac{e}{c}\frac{\partial \mathbf{A}}{\partial t}(t, \mathbf{x}) - e\nabla_x\varphi(t, \mathbf{x}) - \frac{e}{c}(\mathbf{v} \cdot \nabla_\mathbf{x})\mathbf{A}(t, \mathbf{x}) + \frac{e}{c}\nabla_\mathbf{x}(\mathbf{v} \cdot \mathbf{A}(t, \mathbf{x})).$$
$$\tag{9.12}$$

---

[1] If $\mathbf{e}$ is a constant vector and $\mathbf{B}$ a C^1 vector field on $\mathbb{R}^3$: $\mathbf{e} \wedge (\nabla_\mathbf{x} \wedge \mathbf{B}) = \nabla_\mathbf{x}(\mathbf{e} \cdot \mathbf{B}) - (\mathbf{e} \cdot \nabla_\mathbf{x})\mathbf{B}$.

We would like to show that the Lorentz force's expression (9.12) can be written, using $\mathbf{v}$ for the three coordinates $\dot{\mathbf{x}}$,

$$\mathbf{F}_L(t, \mathbf{x}, \mathbf{v}) = \left(\nabla_{\mathbf{x}} - \frac{d}{dt} \nabla_{\dot{\mathbf{x}}} \right) \mathscr{V}(t, \mathbf{x}, \dot{\mathbf{x}}) \tag{9.13}$$

where we have introduced the generalised electromagnetic potential:

$$\mathscr{V}(t, \mathbf{x}, \dot{\mathbf{x}}) := -e\varphi(t, \mathbf{x}) + \frac{e}{c}\mathbf{A}(t, \mathbf{x}) \cdot \dot{\mathbf{x}} . \tag{9.14}$$

Indeed, by direct calculation:

$$\left(\nabla_{\mathbf{x}} - \frac{d}{dt} \nabla_{\dot{\mathbf{x}}} \right) \mathscr{V}(t, \mathbf{x}, \dot{\mathbf{x}}) = -e\nabla_{\mathbf{x}}\varphi(t, \mathbf{x}) + \frac{e}{c}\nabla_{\mathbf{x}}(\dot{\mathbf{x}} \cdot \mathbf{A}(t, \mathbf{x})) - \frac{d}{dt}\left(\frac{e}{c}\mathbf{A}(t, \mathbf{x}) \right)$$

$$= -e\nabla_x\varphi(t, \mathbf{x}) + \frac{e}{c}\nabla_{\mathbf{x}}(\dot{\mathbf{x}} \cdot \mathbf{A}(t, \mathbf{x})) - \frac{e}{c}\frac{\partial \mathbf{A}}{\partial t}(t, \mathbf{x}) - \frac{e}{c}\dot{\mathbf{x}} \cdot \nabla_{\mathbf{x}}\mathbf{A}(t, \mathbf{x})$$

$$= -\frac{e}{c}\frac{\partial \mathbf{A}}{\partial t}(t, \mathbf{x}) - e\nabla_{\mathbf{x}}\varphi(t, \mathbf{x}) - \frac{e}{c}(\mathbf{v} \cdot \nabla_{\mathbf{x}})\mathbf{A}(t, \mathbf{x}) + \frac{e}{c}\nabla_{\mathbf{x}}(\mathbf{v} \cdot \mathbf{A}(t, \mathbf{x})) = \mathbf{F}_L(t, \mathbf{x}, \mathbf{v}).$$

This immediately implies that the charge's equations of motion take the form of Euler-Lagrange equations as long as we use the Lagrangian

$$\mathscr{L}(t, \mathbf{x}, \dot{\mathbf{x}}) = \frac{m}{2}\dot{\mathbf{x}}^2 + \mathscr{V}(t, \mathbf{x}, \dot{\mathbf{x}}) \tag{9.15}$$

where the function $\mathscr{V} : A(\mathbb{V}^{3+1}) \to \mathbb{R}$ was defined in (9.14). In fact, the Euler-Lagrange equations are:

$$\left(\frac{d}{dt}\nabla_{\dot{\mathbf{x}}} - \nabla_{\mathbf{x}} \right) \frac{m}{2}\dot{\mathbf{x}}^2 = \left(\nabla_{\mathbf{x}} - \frac{d}{dt}\nabla_{\dot{\mathbf{x}}} \right) \mathscr{V}(t, \mathbf{x}, \dot{\mathbf{x}}) , \quad \dot{\mathbf{x}} = \frac{d\mathbf{x}}{dt} .$$

Keeping (9.13) in account and working out the left-hand side explicitly:

$$m\frac{d\dot{\mathbf{x}}}{dt} = \mathbf{F}_L(t, \mathbf{x}, \dot{\mathbf{x}}) , \quad \dot{\mathbf{x}} = \frac{d\mathbf{x}}{dt} ,$$

so eventually:

$$m\frac{d^2\mathbf{x}}{dt^2} = \mathbf{F}_L(t, \mathbf{x}, \dot{\mathbf{x}}) .$$

Therefore the Lagrangian of a particle with charge e under a given electromagnetic field in an inertial frame $\mathscr{R}$, in terms of the electric potential φ and the magnetic potential $\mathbf{A}$, is:

$$\mathscr{L}(t, \mathbf{x}, \dot{\mathbf{x}}) = \frac{m}{2}\dot{\mathbf{x}}^2 - e\varphi(t, \mathbf{x}) + \frac{e}{c}\mathbf{A}(t, \mathbf{x}) \cdot \dot{\mathbf{x}}. \tag{9.16}$$

Remarks 9.4 The fields $\mathbf{E}$ and $\mathbf{B}$ in (9.9) and (9.10) remain unaltered if we change potentials by a *gauge transformation*:

$$\varphi(t, \mathbf{x}) \to \varphi'(t, \mathbf{x}) := \varphi(t, \mathbf{x}) - \frac{1}{c}\frac{\partial \chi}{\partial t}(t, \mathbf{x}),$$

$$\mathbf{A}(t, \mathbf{x}) \to \mathbf{A}'(t, \mathbf{x}) := \mathbf{A}(t, \mathbf{x}) + \nabla_{\mathbf{x}}\chi(t, \mathbf{x}), \tag{9.17}$$

where χ is any C^2 function. Under gauge transformations the Lagrangian (9.16) is *not* invariant. Nonetheless, the explicit calculation shows immediately that under gauge transformations:

$$\mathscr{L}(t, \mathbf{x}, \dot{\mathbf{x}}) \to \mathscr{L}'(t, \mathbf{x}, \dot{\mathbf{x}}) := \mathscr{L}(t, \mathbf{x}, \dot{\mathbf{x}}) + \frac{d}{dt}\frac{e}{c}\chi(t, \mathbf{x}),$$

so the equations of motion do not change, since the new Lagrangian differs from the old one by a formal total derivative only. ∎

9.2.2 Generalisation of the Notion of Potential

The previous example shows that in certain cases the Lagrangian components $\mathscr{Q}_k$ of a force can be found by applying the Euler-Lagrange operator $\frac{\partial}{\partial q^k} - \frac{d}{dt}\frac{\partial}{\partial \dot{q}^k}$ to a function $\mathscr{V} : A(\mathbb{V}^{n+1}) \to \mathbb{R}$ that depends both on the coordinates q^k (plus time) and *the coordinates* $\dot{q}^k$. The case where the Lagrangian components come from a potential (a function of t and q^k only) is just a special situation. For the above Lorentz force, the function $\mathscr{V}$ depended *linearly* on the dotted coordinates. Thus the Lagrangian components of the forces associated with $\mathscr{V}$ depended on the *first time derivatives* of the coordinates q^k (i.e. the $\dot{q}^k$) and not on the *second time derivatives* of the q^k (the first derivatives of the $\dot{q}^k$). The dependency of q^k on the second time derivatives is entirely due, in the Euler-Lagrange equations, to the term associated with the system's kinetic energy. Because of this dependency, we know the Euler-Lagrange equations can be put in normal form, as shown by Theorem 7.23. This is a very important point, both physically and mathematically, on which we should take a moment. Suppose the forces' Lagrangian components $\mathscr{Q}_k$ arise by applying the Euler-Lagrange operator $\frac{\partial}{\partial q^k} - \frac{d}{dt}\frac{\partial}{\partial \dot{q}^k}$ to some $\mathscr{V} : A(\mathbb{V}^{n+1}) \to \mathbb{R}$ depending both on the q^k (and time) and *the coordinates* $\dot{q}^k$. It is reasonable to assume $\mathscr{V}$ depends on the $\dot{q}^k$ in a *linear* fashion. If $\mathscr{V}$ depended on the $\dot{q}^k$ in an arbitrary way, we could end up with Lagrangian components for the forces, remembering that $dq^k = dq^k/dt$, containing the *second time derivatives* of the q^k. We would then find ourselves beyond the reach of Theorem 7.23, and it would no longer be guaranteed that the Euler-Lagrange equations can be put in normal form. We would have to relinquish the solutions' existence and uniqueness. We would, in general, have to forego the *determinism* of Physics. Because of that we give the following definition.

Definition 9.5 (Generalised Potential) Take a system $\mathscr{S}$ of N point particles $P_1, \ldots, P_N$ under ideal holonomic constraints and described on $A(\mathbb{V}^{n+1})$. Consider a distribution of forces $\mathbf{F}_i|_{\mathscr{R}}$, $i = 1, \ldots, N$, with $\mathbf{F}_i|_{\mathscr{R}}$ acting on P_i (depending on $\mathscr{R}$ only in case the forces are inertial). This distribution has the scalar field $\mathscr{V} : A(\mathbb{V}^{n+1}) \to \mathbb{R}$ as **generalised potential** if, for any natural local coordinates on $A(\mathbb{V}^{n+1})$, the Lagrangian components $\mathscr{Q}_k|_{\mathscr{R}}$ are:

$$\mathscr{Q}_k|_{\mathscr{R}}(t, q, \dot{q}) = \frac{\partial \mathscr{V}|_{\mathscr{R}}}{\partial q^k} - \frac{d}{dt} \frac{\partial \mathscr{V}|_{\mathscr{R}}}{\partial \dot{q}^k} , \qquad (9.18)$$

where the function $\mathscr{V}|_{\mathscr{R}}$ (more precisely $\mathscr{V}|_{\mathscr{R}} \circ \psi^{-1}$ in any natural local chart (U, ψ) on $A(\mathbb{V}^{n+1})$), depends at most linearly on the $\dot{q}^k$, and the symbol d/dt, in an imprecise but coherent way, should be understood as the *formal total derivative*:

$$\frac{d}{dt} f(t, q) := \frac{\partial f}{\partial t} + \sum_{k=1}^{n} \dot{q}^k \frac{\partial f}{\partial q^k} \qquad (9.19)$$

for any function f of the local coordinates $(t, q^1, \ldots, q^n)$.   ◇

Remarks 9.6

(1) As we have already discussed elsewhere, since in (9.18) we did not assume any relationship between the q^k and the $\dot{q}^k$—in particular we did not fix any curve parametrised by t—it is necessary to define d/dt in formula (9.18). Such definition is then the standard one when (9.18) is deployed on the solutions to the Euler-Lagrange equations, along which we explicitly assume $\dot{q}^k = dq^k/dt$.

(2) If we assume $\mathscr{V}|_{\mathscr{R}}$ is a scalar field on $A(\mathbb{V}^{n+1})$, the same procedure as in Proposition 7.27 shows that:

$$\frac{\partial \mathscr{V}|_{\mathscr{R}}}{\partial q^k} - \frac{d}{dt} \frac{\partial \mathscr{V}|_{\mathscr{R}}}{\partial \dot{q}^k} = \sum_{j=1}^{n} \frac{\partial q'^j}{\partial q^k} \left(\frac{\partial \mathscr{V}|_{\mathscr{R}}}{\partial q'^j} - \frac{d}{dt} \frac{\partial \mathscr{V}|_{\mathscr{R}}}{\partial \dot{q}'^j} \right)$$

when changing natural coordinates. The relations are compatible with those for the Lagrangian components under coordinate change:

$$\mathscr{Q}_k|_{\mathscr{R}}(t, q, \dot{q}) = \sum_{j=1}^{n} \frac{\partial q'^j}{\partial q^k} \mathscr{Q}'_j|_{\mathscr{R}}(t', q', \dot{q}') .$$

(3) Because of the linearity in the dotted coordinates, in given natural coordinates the explicit form of $\mathscr{V}$ can only be

$$\mathscr{V}(t, q, \dot{q}^n) = \sum_{k=1}^{n} A_h(t, q)\dot{q}^h + B(t, q) . \tag{9.20}$$

It is straightforward to see that a coordinate change preserves the above form, and remains linear in the dotted coordinates. Under coordinate changes on $A(\mathbb{V}^{n+1})$, the functions A and B change as follows, with the obvious notations:

$$A'_k(t', q') = \sum_{h=1}^{n} \frac{\partial q^h}{\partial q'_k} A_h(t, q) , \quad B'(t', q') = B(t, q) + \sum_{h=1}^{n} \frac{\partial q^h}{\partial t'} A_h(t, q) .$$

In any natural local coordinates on $A(\mathbb{V}^{n+1})$ the computation of $\mathscr{Q}_k|_{\mathscr{R}}$, using (9.20) and (9.18), immediately gives:

$$\mathscr{Q}_k|_{\mathscr{R}} = \sum_{h=1}^{n} \left(\frac{\partial A_h}{\partial q^k} - \frac{\partial A_k}{\partial q^h} \right) \dot{q}^h - \frac{\partial A_k}{\partial t} + \frac{\partial B}{\partial q^k} . \tag{9.21}$$

It then become evident that the force's Lagrangian components as well, and not just the generalised potential generating them, must be *linear* in the dotted coordinates $\dot{q}^k$.

(4) Under (9.20), even in presence of a generalised potential $\mathscr{V}|_{\mathscr{R}}$, a Lagrangian of the form $\mathscr{L}|_{\mathscr{R}} = \mathscr{T}|_{\mathscr{R}} + \mathscr{V}|_{\mathscr{R}}$ must have the standard structure (7.87), since the generalised potential is at most linear in the dotted coordinates for any natural coordinates on $A(\mathbb{V}^{n+1})$. ∎

9.2.3 Conditions for the Existence of the Generalised Potential

If we have a system of forces with Lagrangian components:

$$\mathscr{Q}_k|_{\mathscr{R}} = \mathscr{Q}_k|_{\mathscr{R}}(t, q, \dot{q})$$

it is no easy task to decide whether they can be given by a generalised potential defined on the domain of the natural local chart employed. There exist, nevertheless, necessary and locally sufficient conditions that we will now study. The situation is similar to when one asks if a certain force is the gradient of a scalar field. (In that case, if the force is a C^2 field only depending on positions, Theorem 4.15 says that Eq. (4.12) are necessary and also locally sufficient). In the sequel we will not indicate the dependency $|_{\mathscr{R}}$ for simplicity. We seek relationships involving only the $\mathscr{Q}_k$, but neither $\mathscr{V}$ nor the functions A_k and B, that should hold whenever the Lagrangian components $\mathscr{Q}_k$ have a generalised potential satisfying (9.20). From (9.21), immediately:

$$\frac{\partial \mathscr{Q}_k}{\partial \dot{q}^h} = \frac{\partial A_h}{\partial q^k} - \frac{\partial A_k}{\partial q^h}$$

and therefore:

$$\frac{\partial \mathscr{Q}_k}{\partial \dot{q}^h} + \frac{\partial \mathscr{Q}_h}{\partial \dot{q}^k} = 0 , \quad \text{everywhere and for any } k, h = 1, \ldots, n.$$

In the same way, starting from (9.21) it easy to see, by comparing the two sides:

$$\frac{\partial \mathscr{Q}_h}{\partial q^k} - \frac{\partial \mathscr{Q}_k}{\partial q^h} = \frac{d}{dt}\left(\frac{\partial \mathscr{Q}_h}{\partial \dot{q}^k}\right) , \quad \text{everywhere and for any } k, h = 1, \ldots, n.$$

The following remarkable result holds.

Theorem 9.7 *Relatively to Definition 9.5 the following hold.*

(1) If a system of forces admits a generalised potential, the associated Lagrangian components $\mathscr{Q}_k$, seen as C^2 maps on $A(\mathbb{V}^{n+1})$ in arbitrary natural local coordinates on $A(\mathbb{V}^{n+1})$, satisfy:

$$\frac{\partial \mathscr{Q}_k}{\partial \dot{q}^h} + \frac{\partial \mathscr{Q}_h}{\partial \dot{q}^k} = 0 , \quad \textit{everywhere and for any } k, h = 1, \ldots, n. \tag{9.22}$$

$$\frac{\partial \mathscr{Q}_k}{\partial q^h} - \frac{\partial \mathscr{Q}_h}{\partial q^k} = \frac{d}{dt}\left(\frac{\partial \mathscr{Q}_k}{\partial \dot{q}^h}\right) , \quad \textit{everywhere and for any } k, h = 1, \ldots, n. \tag{9.23}$$

(2) Vice versa, the Lagrangian components $\mathscr{Q}_k$, seen as C^2 maps on $A(\mathbb{V}^{n+1})$ and linear in the $\dot{q}^k$, admit a generalised potential in a sufficiently enough neighbourhood of any kinetic state $(t_0, q_0^1, \ldots, q_0^n, \dot{q}_0^1, \ldots \dot{q}_0^n)$ if they satisfy (9.22) and (9.23). In that case the generalised potential is:

$$\mathscr{V}(t, q, \dot{q}) =$$

$$\int_0^1 \sum_{k=1}^n (q^k - q_0^k) \mathscr{Q}_k(t, q_0^1 + s(q^1 - q_0^1), \ldots, q_0^n$$

$$+ s(q^n - q_0^n), \dot{q}_0^1 + s(\dot{q}^1 - \dot{q}_0^1), \ldots, \dot{q}^n + s(\dot{q}^n - \dot{q}_0^n))ds . \tag{9.24}$$

This relation holds whenever the right-hand side is defined in the considered local chart.

Proof The proof of (1) was shown above, so let us pass to (2). Consider a point $a \in A(\mathbb{V}^{n+1})$ of coordinates $(t_0, q_0^1, \ldots, q_0^n, \dot{q}_0^1, \ldots, \dot{q}_0^n) = (t_0, 0, \ldots, 0, 0, \ldots, 0)$ in the local chart $\psi : U \subset A(\mathbb{V}^{n+1}) \to \mathbb{R}^{2n+1}$. This is no restriction since we can always move the coordinates' origin starting from a local chart around the kinetic state a. Restrict then to an open neighbourhood $V \subset \psi(U) \subset \mathbb{R}^{2n+1}$ of $(t_0, 0, \ldots, 0, 0, \ldots, 0)$ where $(t, sq^1, \ldots, sq^n, s\dot{q}^1, \ldots, s\dot{q}^n)$ for $s \in [0, 1]$ correspond to points of U when $(t, q^1, \ldots, q^n, \dot{q}^1, \ldots, \dot{q}^n) \in V$. In other words, if $(q^1, \ldots, q^n, \dot{q}^1, \ldots, \dot{q}^n) \in V$, for any given time t, the points on the segment parametrised by $s \in [0, 1]$ and joining the origin of $\mathbb{R}^{2n}$ with $(q^1, \ldots, q^n, \dot{q}^1, \ldots, \dot{q}^n)$, fall in $\psi(U)$ for $s \in [0, 1]$. There are no "holes" in V that contain segments not in $\psi(U)$. As we will integrate along such segments certain quantities defined on $A(\mathbb{V}^{n+1})$, the above shape of V is necessary to carry out the proof. We leave the reader to prove such a neighbourhood V exists, using the fact that Cartesian products of open time intervals and small enough open balls for $q^1, \ldots, q^n, \dot{q}^1, \ldots, \dot{q}^n$ form a basis of the topology of $\mathbb{R}^{2n+1}$ (and induce an analogous basis for the topology of $A(\mathbb{V}^{n+1})$ under the charts' local homeomorphisms). The goal is the prove that, working in V, from (9.24) we deduce

$$\frac{\partial \mathcal{V}}{\partial q^k} - \frac{d}{dt}\frac{\partial \mathcal{V}}{\partial \dot{q}^k} = \mathcal{Q}_k(t, q, \dot{q}) \,.$$

Working in V, expression (9.24) is well defined and gives the following two relations, with obvious notation:

$$\frac{\partial \mathcal{V}}{\partial q^k} = \int_0^1 ds\, \mathcal{Q}_k(t, sq, s\dot{q}) + \int_0^1 ds \sum_h sq^h \frac{\partial \mathcal{Q}_h(t, sq, s\dot{q})}{\partial q^k} \,,$$

$$\frac{\partial \mathcal{V}}{\partial \dot{q}^k} = \int_0^1 ds \sum_h sq^h \frac{\partial \mathcal{Q}_h(t, sq, s\dot{q})}{\partial \dot{q}^k} \,.$$

In the sequel one should remember that by assumption $\partial \mathcal{Q}_h / \partial \dot{q}^k$ does not depend on the $\dot{q}$. From the previous identities

$$\frac{\partial \mathcal{V}}{\partial q^k} - \frac{d}{dt}\frac{\partial \mathcal{V}}{\partial \dot{q}^k} = \int_0^1 ds\, \mathcal{Q}_k(t, sq, s\dot{q}) + \int_0^1 ds \sum_h sq^h \frac{\partial \mathcal{Q}_h(t, sq, s\dot{q})}{\partial q^k}$$

$$- \int_0^1 ds \sum_h s\dot{q}^h \frac{\partial \mathcal{Q}_h(t, sq, s\dot{q})}{\partial \dot{q}^k} - \int_0^1 ds \sum_h sq^h \frac{d}{dt}\frac{\partial \mathcal{Q}_h(t, sq, s\dot{q})}{\partial \dot{q}^k}$$

Hence rearranging the right-hand side:

$$\frac{\partial \mathcal{V}}{\partial q^k} - \frac{d}{dt}\frac{\partial \mathcal{V}}{\partial \dot{q}^k} = \int_0^1 ds\, \mathcal{Q}_k(t, sq, s\dot{q}) - \int_0^1 ds \sum_h s\dot{q}^h \frac{\partial \mathcal{Q}_h(t, sq, s\dot{q})}{\partial \dot{q}^k}$$

$$- \int_0^1 ds \sum_h sq^h \left(\frac{d}{dt}\frac{\partial \mathcal{Q}_h(t, sq, s\dot{q})}{\partial \dot{q}^k} - \frac{\partial \mathcal{Q}_h(t, sq, s\dot{q})}{\partial q^k} \right)$$

Now, using (9.22) and (9.23) on the right we obtain:

$$\frac{\partial \mathcal{V}}{\partial q^k} - \frac{d}{dt}\frac{\partial \mathcal{V}}{\partial \dot{q}^k} = \int_0^1 ds\, \mathcal{Q}_k(t, sq, s\dot{q}) + \int_0^1 ds \sum_h sq^h \frac{\partial \mathcal{Q}_k(t, sq, s\dot{q})}{\partial q^h}$$

$$+ \int_0^1 ds \sum_h s\dot{q}^h \frac{\partial \mathcal{Q}_k(t, sq, s\dot{q})}{\partial \dot{q}^h}$$

i.e. :

$$\frac{\partial \mathcal{V}}{\partial q^k} - \frac{d}{dt}\frac{\partial \mathcal{V}}{\partial \dot{q}^k} = \int_0^1 ds \frac{d}{ds} s\, \mathcal{Q}_k(t, sq, s\dot{q}) \,,$$

and finally:

$$\frac{\partial \mathcal{V}}{\partial q^k} - \frac{d}{dt}\frac{\partial \mathcal{V}}{\partial \dot{q}^k} = \mathcal{Q}_k(t, q, \dot{q})$$

as claimed. $\square$

Remarks 9.8 The theorem's conditions guarantee the local existence of a generalised potential of a collection of Lagrangian components. It is not obvious though that the various potentials thus obtained on each neighbourhood agree to define a global structure. This issue is addressed on a case-by-case basis.   ■

9.2.4   Generalised Potentials of Inertial Forces

Consider the usual system $\mathscr{S}$ of N point particles P_i of masses m_i. Suppose the system undergoes $c = 3N - n$ ideal holonomic constraints, so we can describe it on $A(\mathbb{V}^{n+1})$ by generic Euler-Lagrange equations:

$$\frac{d}{dt}\frac{\partial \mathcal{T}|_{\mathscr{R}}}{\partial \dot{q}^k} - \frac{\partial \mathcal{T}|_{\mathscr{R}}}{\partial q^k} = \mathcal{Q}_k \,, \quad \text{for } k = 1, \ldots, n, \tag{9.25}$$

$$\frac{dq^k}{dt} = \dot{q}^k \,, \quad \text{for } k = 1, \ldots, n, \tag{9.26}$$

where $\mathscr{R}$ is an inertial frame, the $\mathcal{Q}_k$ are the Lagrangian components of the (real) active forces, and we use natural local coordinates on $A(\mathbb{V}^{n+1})$. We want to compare the above to the equations for a non-inertial frame $\mathscr{R}'$ with given motion with respect to $\mathscr{R}$. We will continue to use the same local coordinates $(t, q^1, \ldots, q^n, \dot{q}^1, \ldots, \dot{q}^n)$.

$$\frac{d}{dt} \frac{\partial \mathscr{T}|_{\mathscr{R}'}}{\partial \dot{q}^k} - \frac{\partial \mathscr{T}|_{\mathscr{R}'}}{\partial q^k} = \mathcal{Q}_k + \mathcal{Q}_k|_{\mathscr{R}'}, \quad \text{for } k = 1, \ldots, n, \tag{9.27}$$

$$\frac{dq^k}{dt} = \dot{q}^k, \quad \text{for } k = 1, \ldots, n, \tag{9.28}$$

where the new Lagrangian components $\mathcal{Q}_k|_{\mathscr{R}'}$ refer to the inertial forces present in the non-inertial frame $\mathscr{R}'$ and depend on the frame, in contrast to the components $\mathcal{Q}_k$.

For the above sets of Euler-Lagrange equations (assuming the Existence Theorem holds) any solution of the former must solve the latter and conversely. Indeed, consider a solution $\gamma : I \ni t \mapsto (t, q(t), \dot{q}(t))$ to (9.25)–(9.26). Remembering how we found the Euler-Lagrange equations in Theorem 7.16, we conclude that the curve satisfies:

$$\sum_{i=1}^N \left. \left(m_i \frac{d\mathbf{v}_i}{dt} \Big|_{\mathscr{R}} - \mathbf{f}_i \right) \right|_{\gamma(t)} \cdot \left. \frac{\partial \mathbf{x}_i}{\partial q^k} \right|_{\gamma(t)} = 0, \quad \dot{q}^k(t) = \frac{dq}{dt}, \quad k = 1, \ldots, N,$$

where we used the position vectors $\mathbf{x}_i(\gamma(t))$ in $\mathscr{R}$ and the velocities $\mathbf{v}_i = \mathbf{v}_i|_{\mathscr{R}}(\gamma(t))$. In particular the forces $\mathbf{f}_j$ have these vectors ($i = 1, \ldots, N$) as arguments, together with time t, so they can be defined along γ. Invoking the usual frame-change formulas, the equations read

$$\sum_{i=1}^N \left. \left(m_i \frac{d\mathbf{v}_i'}{dt} \Big|_{\mathscr{R}'} - \mathbf{f}_i - \mathbf{f}_i|_{\mathscr{R}'} \right) \right|_{\gamma(t)} \cdot \left. \frac{\partial \mathbf{x}_i}{\partial q^k} \right|_{\gamma(t)} = 0, \quad \dot{q}^k(t) = \frac{dq}{dt}, \quad k = 1, \ldots, N,$$

where we used the position vectors $\mathbf{x}_i'(\gamma(t))$ in $\mathscr{R}$ and the velocities $\mathbf{v}_i' = \mathbf{v}_i|_{\mathscr{R}'}(\gamma(t))$, and we introduced inertial forces $\mathbf{f}_i|_{\mathscr{R}'}$. At the same time we also know ((1) Remarks 7.17):

$$\left. \frac{\partial \mathbf{x}_i}{\partial q^k} \right|_{\gamma(t)} = \left. \frac{\partial \mathbf{x}_i'}{\partial q^k} \right|_{\gamma(t)},$$

so γ satisfies:

$$\sum_{i=1}^N \left. \left(m_i \frac{d\mathbf{v}_i'}{dt} \Big|_{\mathscr{R}'} - \mathbf{f}_i - \mathbf{f}_i|_{\mathscr{R}'} \right) \right|_{\gamma(t)} \cdot \left. \frac{\partial \mathbf{x}_i'}{\partial q^k} \right|_{\gamma(t)} = 0, \quad \dot{q}^k(t) = \frac{dq}{dt}, \quad k = 1, \ldots, N.$$

These relations are equivalent to (9.27) and (9.28). Swapping the roles of the two reference frames we see that any solution to (9.27) and (9.28) solves (9.25)–(9.26). Hence we have the following fact.

Proposition 9.9 *Consider a system $\mathscr{S}$ of N point particles described on $A(\mathbb{V}^{n+1})$ and let us work in a natural local chart with coordinates $(t, q^1, \ldots, q^n, \dot{q}^1, \ldots, \dot{q}^n)$. Suppose that:*

(1) the position vectors $\mathbf{x}_i := P_i - O$ of each particle in $\mathscr{S}$ in the inertial frame $\mathscr{R}$ are jointly C^3 in the above coordinates and in time;

(2) given right-handed orthonormal coordinates on the two frames, the motion of $\mathscr{R}'$ relative to $\mathscr{R}$ is given by C^3 functions of time (so that $t \mapsto O'(t) - O$ is C^3 and $t \mapsto \boldsymbol{\omega}_{\mathscr{R}'}|_{\mathscr{R}}(t)$ is C^2).

Then the Lagrangian components $\mathscr{Q}_k|_{\mathscr{R}'}$ of the inertial forces in $\mathscr{R}'$ are given by a generalised potential $\mathscr{V}|_{\mathscr{R}'}$ such that:

$$\mathscr{V}|_{\mathscr{R}'}(t, q, \dot{q}) = \mathscr{T}|_{\mathscr{R}}(t, q, \dot{q}) - \mathscr{T}|_{\mathscr{R}'}(t, q, \dot{q}) . \qquad (9.29)$$

In particular $\mathscr{T}|_{\mathscr{R}} - \mathscr{T}|_{\mathscr{R}'}$ is always at most linear in the $\dot{q}^k$ in the given chart.

Proof The kinetic energy $\mathscr{T}|_{\mathscr{R}'}$ will be different from $\mathscr{T}|_{\mathscr{R}}$. In the sequel we will indicate by $\mathbf{x}_i(t, q^1, \ldots, q^n) = P_i - O$ the position vector of P_i with respect to $\mathscr{R}$, and by $\mathbf{x}'_i(t, q^1, \ldots, q^n) := P_i - O'$ the position vector of the same particle but in $\mathscr{R}'$. These functions are C^2 by hypothesis. Similarly, let $\mathbf{v}_i(t, q^1, \ldots, q^n, \dot{q}^1, \ldots, \dot{q}^n)$ and $\mathbf{v}'_i(t, q^1, \ldots, q^n, \dot{q}^1, \ldots, \dot{q}^n)$ be the corresponding velocities, which as well known are *linear* in the $\dot{q}^k$ because:

$$\mathbf{v}_i(t, q^1, \ldots, q^n, \dot{q}^1, \ldots, \dot{q}^n) = \frac{\partial \mathbf{x}_i}{\partial t}(t, q^1, \ldots, q^n) + \sum_{k=1}^{n} \dot{q}^k \frac{\partial \mathbf{x}_i}{\partial q^k}(t, q^1, \ldots, q^n) .$$

Analogously:

$$\mathbf{v}'_i(t, q^1, \ldots, q^n, \dot{q}^1, \ldots, \dot{q}^n) = \frac{\partial \mathbf{x}'_i}{\partial t}(t, q^1, \ldots, q^n) + \sum_{k=1}^{n} \dot{q}^k \frac{\partial \mathbf{x}'_i}{\partial q^k}(t, q^1, \ldots, q^n) .$$

The transformation law for the velocities reads:

$$\mathbf{v}_i = \mathbf{v}'_i + \mathbf{v}_{O'}|_{\mathscr{R}}(t) + \boldsymbol{\omega}_{\mathscr{R}'}|_{\mathscr{R}}(t) \wedge \mathbf{x}'_i$$

where $\mathbf{v}_{O'}|_{\mathscr{R}}(t)$ and $\boldsymbol{\omega}_{\mathscr{R}'}|_{\mathscr{R}}(t)$ are given and C^2. The expressions of the kinetic energy in $\mathscr{R}$ and $\mathscr{R}'$:

$$\mathscr{T}|_{\mathscr{R}} = \sum_{i=1} \frac{1}{2} m_i \mathbf{v}_i^2 , \qquad \mathscr{T}|_{\mathscr{R}'} = \sum_{i=1} \frac{1}{2} m_i \mathbf{v}'^2_i$$

are both C^1 on $A(\mathbb{V}^{n+1})$. In the same way, the Lagrangian components of the inertial forces

$$F_i|_{\mathscr{R}'} = -m_i \mathbf{a}_{O'}|_{\mathscr{R}}(t) - m_i \boldsymbol{\omega}_{\mathscr{R}'}|_{\mathscr{R}}(t) \wedge (\boldsymbol{\omega}_{\mathscr{R}'}|_{\mathscr{R}}(t) \wedge \mathbf{x}'_i) - 2m_i \boldsymbol{\omega}_{\mathscr{R}'}|_{\mathscr{R}}(t) \wedge \mathbf{v}'_i$$
$$- m_P \dot{\boldsymbol{\omega}}_{\mathscr{R}'}|_{\mathscr{R}} \wedge \mathbf{x}'_i ,$$

are C^1 on $A(\mathbb{V}^{n+1})$. Comparing the two kinetic energies, in $\mathscr{R}$ and in $\mathscr{R}'$, we find that:

$$\mathscr{V}|_{\mathscr{R}'}(t, q, \dot{q}) := \mathscr{T}|_{\mathscr{R}}(t, q, \dot{q}) - \mathscr{T}|_{\mathscr{R}'}(t, q, \dot{q})$$

is C^2, and moreover:

$$\mathscr{V}|_{\mathscr{R}'}(t, q^1, \ldots, q^n, \dot{q}^1, \ldots, \dot{q}^n) := \sum_{i=1}^{N} \frac{1}{2} m_i \left(\boldsymbol{\omega}_{\mathscr{R}'}|_{\mathscr{R}}(t) \wedge \mathbf{x}'_i(t, q^1, \ldots, q^n) \right)^2$$

$$+ \frac{M}{2} \mathbf{v}_{O'}|_{\mathscr{R}}(t)^2 + \mathbf{v}_{O'}|_{\mathscr{R}}(t) \cdot \sum_{i=1}^{N} m_i \mathbf{v}'_i(t, q^1, \ldots, q^n, \dot{q}^1, \ldots, \dot{q}^n)$$

$$+ \boldsymbol{\omega}_{\mathscr{R}'}|_{\mathscr{R}}(t) \cdot \sum_{i=1}^{N} m_i \mathbf{x}'_i \wedge \left(\mathbf{v}_{O'}|_{\mathscr{R}}(t) + \mathbf{v}'_i(t, q^1, \ldots, q^n, \dot{q}^1, \ldots, \dot{q}^n) \right) ,$$

where $M := \sum_{i=1}^{N} m_i$. By construction, $\mathscr{V}(t, q^1, \ldots, q^n, \dot{q}^1, \ldots, \dot{q}^n)$ is *linear* in the $\dot{q}^k$, as the $\dot{q}^k$ appear, linearly, when we solve for $\mathbf{v}'_i$, and the latter's components appear linearly in the above $\mathscr{V}$. Hence $\mathscr{V}$ fulfils the most important necessary condition to be a *generalised potential*; in particular the derivative d/dt can be interpreted as in (9.19). There remains to show that applying (9.18) to our $\mathscr{V}$, the left side defines the Lagrangian components $\mathscr{Q}_k|_{\mathscr{R}'}$ in (9.28) when evaluated on kinetic states $(t, q^1, \ldots, q^n, \dot{q}^1, \ldots, \dot{q}^n)$ of the relevant chart. We may compare (9.25) and (9.27), where we are allowed to choose the Lagrangian components $\mathscr{Q}_k$ to be identically zero. As noticed above, since the first system's solutions solve the second and, conversely, we have sufficient regularity for the Existence and uniqueness theorem to hold, immediately

$$\frac{\partial \mathscr{V}|_{\mathscr{R}'}}{\partial q^k} - \frac{d}{dt} \frac{\partial \mathscr{V}|_{\mathscr{R}'}}{\partial \dot{q}^k} = \mathscr{Q}_k|_{\mathscr{R}'}(t, q, \dot{q}) , \qquad (9.30)$$

on any kinetic state $(t, q^1, \ldots, q^n, \dot{q}^1, \ldots, \dot{q}^n)$ *reached by a solution of the Euler-Lagrange equations* (9.25)–(9.26) and (9.27)–(9.28) with $\mathscr{Q}_k$ identically null. The Existence Theorem holds, so any kinetic state $(t, q^1, \ldots, q^n, \dot{q}^1, \ldots, \dot{q}^n)$ is reached by a solution: the one with such kinetic state as initial condition. Hence (9.30) holds for any kinetic state, and the theorem is proved. $\qquad \square$

9.3 Equilibrium and Stability in the Lagrangian Formulation

In this section we will specialise some definitions and results from Chap. 6 to physical systems of point particles admitting a Lagrangian description. The first concept we need is *equilibrium configuration*. This will be a special case of Definition 6.3 for the Lagrangian setting.

9.3.1 Equilibrium Configurations with Respect to a Frame

We start from the most important definition.

Definition 9.10 (Equilibrium Configuration with Respect to a Reference Frame) Consider a system $\mathscr{S}$ of N point particles under $c = 3N - n \geq 0$ ideal holonomic constraints, described on $A(\mathbb{V}^{n+1})$ and C^k ($k \geq 2$). Fix a reference frame $\mathscr{R}$ in the physical space. Then take a natural local chart on $A(\mathbb{V}^{n+1})$,

$$\psi : U \ni a \mapsto (t, q^1, \ldots, q^n, \dot{q}^1, \ldots, \dot{q}^n) \in \mathbb{R}^{2n+1}$$

where $\psi(U) = \mathbb{R} \times V \times \mathbb{R}^n$, i.e.:

(1) t varies on the *entire* $\mathbb{R}$,
(2) $(q^1, \ldots, q^n)$ vary in the whole open set $V \subset \mathbb{R}^n$,
(3) $(\dot{q}^1, \ldots, \dot{q}^n)$ vary in the *whole* $\mathbb{R}^n$.

We further impose that:

(4) the coordinates $(t, q^1, \ldots, q^n)$ move (see Definition 7.6) with $\mathscr{R}$.

Suppose the Lagrangian components $\mathscr{Q}_k$ of all active forces do not depend explicitly on time in the Lagrangian coordinates: $\mathscr{Q}_k = \mathscr{Q}_k(q^1, \ldots, q^n, \dot{q}^1, \ldots, \dot{q}^n)$.

Then one says the n-tuple $(q_0^1, \ldots, q_0^n) \in V$ defines an **equilibrium configuration in frame** $\mathscr{R}$, if every solution in $\mathbb{R} \times V \times \mathbb{R}^n$ of the Euler-Lagrange equations (7.36) with initial conditions

$$(q^1(0), \ldots, q^n(0), \dot{q}^1(0), \ldots, \dot{q}^n(0)) = (q_0^1, \ldots, q_0^n, 0, \ldots, 0) .$$

is a restriction of the curve:

$$(q^1(t), \ldots, q^n(t), \dot{q}^1(t), \ldots, \dot{q}^n(t)) = (q_0^1, \ldots, q_0^n, 0, \ldots, 0) , \quad \text{for any } t \in \mathbb{R} .$$

$\diamond$

Remarks 9.11

(1) It is important to note that the n-tuple $(q_0^1, \ldots, q_0^n) \in V$ *does not* determine a configuration in $\mathbb{V}^{n+1}$, but a *family of configurations*, more precisely *a solution* *to the Euler-Lagrange equations.* That curve looks like it is describing a point *at rest in $\mathscr{R}$.* The definition of "equilibrium configuration" depends thus on the chosen reference frame $\mathscr{R}$.

(2) However, the definition *does not* depend on the natural coordinates, as long as the latter move with $\mathscr{R}$, and assuming the constraints are C^k. The proof is immediate from the following. If the two natural coordinate systems defined on the same open subset in $\mathbb{V}^{n+1}$: $(t, q^1, \ldots, q^n)$ and $(t', q'^1, \ldots, q'^n)$, adapted to the fibres of $\mathbb{V}^{n+1}$, both move with $\mathscr{R}$, the coordinate change $q^k = q^k(t', q'^1, \ldots, q'^n)$, $k = 1, \ldots, n$ (and the inverse) cannot depend on time explicitly. In fact, if $\mathbf{x}_i$ is the position vector in $\mathscr{R}$ of $P_i \in \mathscr{S}$, $\mathbf{x}_i$ cannot depend explicitly on time when parametrised by primed or unprimed coordinates, by Definition 7.6. On the other hand the parametric dependency of $\mathbf{x}_i$ on primed coordinates arises by writing (a) $\mathbf{x}_i$ as (C^k) function of the unprimed coordinates and then (b) writing the latter as (C^k) functions of the primed ones:

$$\mathbf{x}_i = \mathbf{x}_i(q^1(t', q'^1, \ldots, q'^n) \ldots, q^n(t', q'^1, \ldots, q'^n))$$

For any $(t', q'^1, \ldots, q'^n)$ in the considered neighbourhood, therefore:

$$\frac{\partial}{\partial t'}\mathbf{x}_i(q^1(t', q'^1, \ldots, q'^n) \ldots, q^n(t', q'^1, \ldots, q'^n)) = 0.$$

In other words:

$$\sum_{k=1}^{n} \frac{\partial \mathbf{x}_i}{\partial q^k} \frac{\partial q^k}{\partial t'} = \mathbf{0}.$$

Multiplying each side by the mass m_i, adding over i and squaring gives

$$\sum_{k,h=1}^{n} \left(\sum_{i=1}^{N} m_i \frac{\partial \mathbf{x}_i}{\partial q^h} \cdot \frac{\partial \mathbf{x}_i}{\partial q^k} \right) \frac{\partial q^h}{\partial t'} \frac{\partial q^k}{\partial t'} = \sum_{k,h=1}^{n} 2a_{hk} \frac{\partial q^h}{\partial t'} \frac{\partial q^k}{\partial t'} = 0.$$

The coefficient matrix a_{hk} is that of Theorem 7.23. There, we proved this matrix is positive definite. We then conclude that for any $h = 1, \ldots, n$ and any configuration given by n-tuples $(q^1, \ldots, q^n)$ in V,

$$\frac{\partial q^h}{\partial t'} = 0.$$

∎

The following proposition provides necessary and sufficient conditions for a configuration to be an equilibrium.

Proposition 9.12 *Relatively to Definition 9.10 consider natural local coordinates on $A(\mathbb{V}^{n+1})$, $(t, q^1, \ldots, q^n, \dot{q}^1, \ldots, \dot{q}^n)$ in the open set $\mathbb{R} \times V \times \mathbb{R}^n$ and moving with the frame $\mathscr{R}$. In such coordinates consider the Euler-Lagrange equations:*

$$
\begin{cases}
\dfrac{d}{dt} \dfrac{\partial \mathscr{T}|_{\mathscr{R}}}{\partial \dot{q}^k} - \dfrac{\partial \mathscr{T}|_{\mathscr{R}}}{\partial q^k} = \mathcal{Q}_k|_{\mathscr{R}} , & \\
\dfrac{dq^k}{dt} = \dot{q}^k(t) & \quad \text{for } k = 1, \ldots, n, .
\end{cases}
$$

Suppose the function $\mathscr{T}|_{\mathscr{R}}(q, \dot{q})$ (that does not depend on t since the natural local coordinates move with $\mathscr{R}$) is C^2. Then the following hold.

(1) If the $\mathcal{Q}_k|_{\mathscr{R}} = \mathcal{Q}_k|_{\hat{\mathscr{R}}}(t, q, \dot{q})$, for $k = 1, \ldots, n$ are C^1 and do not depend explicitly on time, $(q_0^1, \ldots, q_0^n) \in V$ is an equilibrium configuration in $\mathscr{R}$ if and only if

$$
\mathcal{Q}_k|_{\mathscr{R}}(q_0^1, \ldots, q_0^n, 0, \ldots, 0) = 0, \text{ for } k = 1, \ldots, n. \tag{9.31}
$$

(2) If the $\mathcal{Q}_k|_{\mathscr{R}}$ are all described by a potential energy $\mathscr{U}|_{\mathscr{R}}$ and so $\mathscr{S}$ admits a Lagrangian $\mathscr{L}|_{\mathscr{R}} = \mathscr{T}|_{\mathscr{R}} - \mathscr{U}|_{\mathscr{R}}$ with $\mathscr{U}|_{\hat{\mathscr{R}}} = \mathscr{U}|_{\hat{\mathscr{R}}}(q^1, \ldots, q^n)$ of class C^2 on V, $(q) \in V$ is an equilibrium configuration in $\mathscr{R}$ if and only if $(q_0^1, \ldots, q_0^n)$ is a stationary point of $\mathscr{U}|_{\mathscr{R}}$:

$$
\frac{\partial}{\partial q^k} \mathscr{U}|_{\mathscr{R}}(q^1, \ldots, q^n) \bigg|_{(q_0^1, \ldots, q_0^n)} = 0 , \quad \text{for } k = 1, \ldots, n. \tag{9.32}
$$

Proof

(1) As the $t, q^1, \ldots, q^n$ move with $\mathscr{R}$, the functions of the position vectors $\mathbf{x}_i$ of $P_i \in \mathscr{S}$ in $\mathscr{R}$ will not depend explicitly on time. Hence the kinetic energy $\mathscr{T}|_{\mathscr{R}}$ will not depend explicitly on time and will take the form (see Sect. 7.2.4):

$$
\mathscr{T}|_{\mathscr{R}}(q, \dot{q}) = \sum_{h,k=1}^{n} a_{hk}(q) \dot{q}^h \dot{q}^k ,
$$

where the coefficients $a_{hk}(q)$ are C^2 in the above coordinates. The Euler-Lagrange equations, since the a_{hk} do not depend explicitly on time, will be:

$$
2 \sum_{k} a_{hk} \frac{d^2 q^k}{dt^2} + 2 \sum_{k,r} \frac{\partial a_{hk}}{\partial q^r} \frac{dq^k}{dt} \frac{dq^r}{dt} - \sum_{r,k} \frac{\partial a_{rk}}{\partial q^h} \frac{dq^r}{dt} \frac{dq^k}{dt} = \mathcal{Q}_h|_{\mathscr{R}}(q, \dot{q}) . \tag{9.33}
$$

(By construction, the left-hand side vanishes along a constant curve $q^k(t) = q_0^k$ for any $k = 1, \ldots, n, t \in \mathbb{R}$; this might have been false if $\mathscr{T}|_{\mathscr{R}}$ had depended on time explicitly (why?)). Such equations become normal if we multiply by the inverse of the matrix a_{hk} and with other trivial manipulations. Apart from the $\mathscr{Q}^k$ then, the remaining functions in the $q^1, \ldots, q^n$ in the above explicit Euler-Lagrange equations are C^1. The coefficients of the inverse of $[a_{hk}]$ are C^2 in $q^1, \ldots, q^n$ because they are rational maps (with non-zero denominators) in the a_{hk}. Since, excluding the $\mathscr{Q}^k$, the dependency on the derivatives of the q^k is polynomial and the $\mathscr{Q}^k$ are C^1, we conclude that *the right-hand side in the normal Euler-Lagrange equations must be jointly C^1 (at least) in the variables q^k and their derivatives.* This falls under the global (Existence and) uniqueness Theorem 14.23. To prove the claim it then suffices to verify that if the constant solution:

$$(q^1(t), \ldots, q^n(t), \dot{q}^1(t), \ldots, \dot{q}^n(t)) = (q_0^1, \ldots, q_0^n, 0, \ldots, 0), \quad \text{for any } t \in \mathbb{R}$$
$$(9.34)$$

solves the Euler-Lagrange equations, then (9.31) holds; and if (9.31) holds then

$$(q^1(t), \ldots, q^n(t), \dot{q}^1(t), \ldots, \dot{q}^n(t)) = (q_0^1, \ldots, q_0^n, 0, \ldots, 0), \quad \text{for any } t \in \mathbb{R}$$

solves the Euler-Lagrange equations.

If the *constant* solution (9.34) solves the Euler-Lagrange equations (9.33), the left side vanishes by construction. Hence we have (9.31). If on the other hand (9.31) holds, the constant solution (9.34) solves the Euler-Lagrange equations (9.33) because it kills the left-hand side while the right-hand side is already zero by (9.31).

(2) The proof is straightforward from the previous one, because:

$$\mathscr{Q}_k|_{\mathscr{R}} = -\frac{\partial}{\partial q^k} \mathscr{U}|_{\mathscr{R}}(q).$$

□

Remarks 9.13 The $\mathscr{Q}_k|_{\mathscr{R}}$ depend on $\mathscr{R}$ only when we consider frames $\mathscr{R}$ that are non-inertial: the inertial forces depend on the frame. The real forces, instead, do not depend on the frame. ∎

Proposition can be tweaked to give a useful extension.

Proposition 9.14 *Relatively to Definition 9.10 consider natural local coordinates on $A(\mathbb{V}^{n+1})$, $(t, q^1, \ldots, q^n, \dot{q}^1, \ldots, \dot{q}^n)$ in the open set $\mathbb{R} \times V \times \mathbb{R}^n$ moving with $\mathscr{R}$. In such coordinates consider the Euler-Lagrange equations:*

$$\begin{cases} \dfrac{d}{dt} \dfrac{\partial \mathscr{T}|_{\hat{\mathscr{R}}}}{\partial \dot{q}^k} - \dfrac{\partial \mathscr{T}|_{\hat{\mathscr{R}}}}{\partial q^k} = \mathscr{Q}_k|_{\hat{\mathscr{R}}}, \\ \dfrac{dq^k}{dt} = \dot{q}^k(t) \end{cases} \quad \text{for } k = 1, \ldots, n, .$$

where $\hat{\mathscr{R}} \neq \mathscr{R}$. Suppose the function $\mathscr{T}|_{\hat{\mathscr{R}}}(t, q, \dot{q})$ is C^2 and does not depend explicitly on time. Then the following facts hold.

(1) Consider the standard expansion of $\mathscr{T}|_{\hat{\mathscr{R}}}$ (see Sect. 7.2.4):

$$\mathscr{T}|_{\hat{\mathscr{R}}} = \sum_{h,k} a_{hk} \dot{q}^h \dot{q}^k + \sum_k b_k \dot{q}^k + c. \tag{9.35}$$

If the $\mathscr{Q}_k|_{\hat{\mathscr{R}}} = \mathscr{Q}_k|_{\hat{\mathscr{R}}}(t, q, \dot{q})$, for $k = 1, \ldots, n$ are C^1 and do not depend explicitly on time, $(q_0^1, \ldots, q_0^n) \in V$ is an equilibrium configuration in $\mathscr{R}$ if and only if:

$$\mathscr{Q}_k|_{\hat{\mathscr{R}}}(q_0^1, \ldots, q_0^n, 0, \ldots, 0) = -\frac{\partial c(q_0^1, \ldots, q_0^n, 0, \ldots, 0)}{\partial q^k}, \quad \text{for } k = 1, \ldots, n. \tag{9.36}$$

(2) If system $\mathscr{S}$ has a Lagrangian of the form $\mathscr{L}|_{\hat{\mathscr{R}}} = \mathscr{T}|_{\hat{\mathscr{R}}} - \mathscr{U}|_{\hat{\mathscr{R}}}$ with $\mathscr{U}|_{\hat{\mathscr{R}}} = \mathscr{U}|_{\hat{\mathscr{R}}}(q^1, \ldots, q^n)$ of class C^2 on V, $(q_0^1, \ldots, q_0^n) \in V$ is an equilibrium configuration in $\mathscr{R}$ if and only if $(q_0^1, \ldots, q_0^n)$ is a stationary point of $\mathscr{U}' := \mathscr{U}|_{\hat{\mathscr{R}}} - c$:

$$\frac{\partial}{\partial q^k} \mathscr{U}'(q) \Big|_{(q_0^1, \ldots, q_0^n)} = 0, \quad \text{for } k = 1, \ldots, n. \tag{9.37}$$

Proof We have:

$$\mathscr{T}|_{\hat{\mathscr{R}}}(q, \dot{q}) = \sum_{h,k=1}^{n} a_{hk}(t, q) \dot{q}^h \dot{q}^k + \sum_{k=1}^{n} b_k(t, q) \dot{q}^k + c(t, q),$$

where on the left we did not write the variabile t because by hypothesis $\mathscr{T}|_{\hat{\mathscr{R}}}$ does not depend explicitly on t. Differentiating in t:

$$0 = \sum_{h,k=1}^{n} \frac{\partial a_{hk}}{\partial t} \dot{q}^h \dot{q}^k + \sum_{k=1}^{n} \frac{\partial b_k}{\partial t} \dot{q}^k + \frac{\partial c}{\partial t}.$$

For any $(t, q^1, \ldots, q^n)$, the quadratic polynomial in the $\dot{q}^k$ (independent of the other variables) on the right is zero. A polynomial is identically zero if and only if its coefficients are zero. Hence the partial derivatives in t of a_{hk}, b_k and c are identically null. Therefore:

$$\mathscr{T}|_{\hat{\mathscr{R}}}(q, \dot{q}) = \sum_{h,k=1}^{n} a_{hk}(q) \dot{q}^h \dot{q}^k + \sum_{k=1}^{n} b_k(q) \dot{q}^k + c(q),$$

and the proof proceeds as in the previous case, remembering that now the Euler-Lagrange equations read:

$$2\sum_k a_{hk}\frac{d^2q^k}{dt^2} + 2\sum_{k,r}\frac{\partial a_{hk}}{\partial q^r}\frac{dq^k}{dt}\frac{dq^r}{dt} + \sum_r\frac{\partial b_h}{\partial q^r}\frac{dq^r}{dt}$$

$$-\sum_{r,k}\frac{\partial a_{rk}}{\partial q^h}\frac{dq^r}{dt}\frac{dq^k}{dt} + \sum_k^n\frac{\partial b_k}{\partial q^h}\frac{dq^k}{dt}$$

$$=\frac{\partial c(q_0^1,\ldots,q_0^n,0,\ldots,0)}{\partial q^k} + \mathcal{Q}_k|_{\hat{\mathcal{R}}}(q^1,\ldots,q^n,\dot{q}^1,\ldots,\dot{q}^n).$$

□

Remarks 9.15

(1) As in the previous case, the $\mathcal{Q}_k|_{\hat{\mathcal{R}}}$ depend on $\hat{\mathcal{R}}$ only when we consider non-inertial frames $\hat{\mathcal{R}}$: inertial forces depend on the frame. Real forces do not.

(2) We may ask whether there are cases in which $\mathcal{R} \neq \hat{\mathcal{R}}$, the free coordinates move with $\mathcal{R}$, but the kinetic energy with respect to $\hat{\mathcal{R}}$ does not depend explicitly on time. The answer is yes, one simple example being that where $\hat{\mathcal{R}}$ is inertial while $\mathcal{R}$ is not, but $\mathcal{R}$ rotates without sliding with respect to $\hat{\mathcal{R}}$ due to the constant vector $\boldsymbol{\omega}_{\mathcal{R}}|_{\hat{\mathcal{R}}} \neq \mathbf{0}$. ∎

Examples 9.16 Fix right-handed orthogonal axes $\mathbf{e}_x, \mathbf{e}_y, \mathbf{e}_z$ on the frame $\mathcal{R}$ with origin O. Consider a point particle P of mass m, moving on the (smooth) x-axis, joined to O by a massless spring of elastic constant $k > 0$ and zero length at rest. The frame $\mathcal{R}$ rotates with respect to the inertial frame $\hat{\mathcal{R}}$ with $\boldsymbol{\omega}_{\mathcal{R}}|_{\hat{\mathcal{R}}} = \Omega \mathbf{e}_z$, where $\Omega > 0$ is constant. On P, besides the elastic force and the constraint reaction, we have a viscous force $-\gamma \mathbf{v}_P|_{\mathcal{R}}$ with $\gamma \geq 0$ constant.

Using the coordinate x of P as free coordinate, we have natural coordinates (t, x) satisfying all the requirements of Definition 9.10, with respect to $\mathcal{R}$. Immediately:

$$\mathcal{T}|_{\hat{\mathcal{R}}} = \frac{m}{2}\dot{x}^2 + \frac{m}{2}\Omega^2 x^2.$$

The function c is then

$$c(x) = \frac{m}{2}\Omega^2 x^2.$$

Regarding the forces (we will drop the subscript $|_{\hat{\mathcal{R}}}$ as $\hat{\mathcal{R}}$ is inertial and the forces are real):

$$\mathcal{Q} = -kx - \gamma\dot{x}.$$

The equilibrium configurations x_0 in the non-inertial frame $\hat{\mathscr{R}}$ are found, by Proposition 9.14, imposing:

$$\frac{dc(x)}{dx}\bigg|_{x_0} = \mathscr{Q}(x_0, 0) \,.$$

In other words $m\Omega^2 x_0 = k x_0$, so $x_0 = 0$, where we have assumed $m\Omega^2/k \neq 1$.

We can study the equilibrium problem using Proposition 9.12 directly. In this case we only need $\mathscr{R}$, in which

$$\mathscr{T}|_{\mathscr{R}} = \frac{m}{2}\dot{x}^2 \,.$$

Notice that in $\mathscr{R}$ we also have the inertial Coriolis force $-2m\boldsymbol{\omega}_{\mathscr{R}}|_{\hat{\mathscr{R}}} \wedge \mathbf{v}_P|_{\mathscr{R}}$, and the centrifugal force $-m\boldsymbol{\omega}_{\mathscr{R}}|_{\hat{\mathscr{R}}} \wedge (\boldsymbol{\omega}_{\mathscr{R}}|_{\hat{\mathscr{R}}} \wedge \mathbf{x})$, where $\mathbf{x} := x\mathbf{e}_x$. Their sum equals:

$$-2m\Omega\dot{x}\mathbf{e}_y + m\Omega^2 x\mathbf{e}_x \,.$$

As for the forces we reinstate $|_{\mathscr{R}}$ since $\mathscr{R}$ *is not* inertial (and part of the active forces has non-inertial nature):

$$\mathscr{Q}|_{\mathscr{R}} = -kx - \gamma\dot{x} + m\Omega^2 x \,.$$

The Coriolis force does not give Lagrangian components since:

$$-2m\Omega\dot{x}\mathbf{e}_y \cdot \frac{\partial \mathbf{x}}{\partial x} = -2m\Omega\dot{x}\mathbf{e}_y \cdot \mathbf{e}_x = 0 \,.$$

The equilibrium configurations x_0 in the non-inertial frame $\hat{\mathscr{R}}$ are found, by Proposition 9.12, imposing:

$$\mathscr{Q}|_{\mathscr{R}}(x_0, 0) = 0 \,.$$

In other words $-kx_0 + m\Omega^2 0 = 0$, so exactly as before: $x_0 = 0$, where we assumed $m\Omega^2/k \neq 1$.

9.3.2 Stability and the Lagrange-Dirichlet Theorem

We shall repeat for Lagrangian systems some of the considerations made in general in Chap. 6. In Physics the above notion of equilibrium configuration does not characterise properly real equilibrium configurations of physical systems (and other systems described by differential equations). One reason, perhaps the most important one, is that the configurations of physical systems do not correspond to *ideal* geometric configurations, because the state of every physical system (the position and shape of objects) interacts constantly with the surrounding ambient (mostly through thermodynamical interactions), and that produces fluctuating states, even if the fluctuations are very small. For this reason the physical equilibrium configurations that are *observed in the physical reality* also display *stability* under small perturbations of the initial state. These stability features are defined according to the mathematical model one adopts. In the sequel we shall refer to the simplest notions and specialise to the Lagrangian case the concepts of Chap. 6. The core idea in the definition of stable equilibrium configuration (in a frame $\mathscr{R}$) is that every solution stays *close* to the configuration (with respect to $\mathscr{R}$) even when we start from small variations of the initial conditions.

Definition 9.17 (Stable Equilibrium Configuration with Respect to a Frame)
We shall refer to Definition 9.10 and its notation (in particular, the Lagrangian coordinates *move* with the frame $\mathscr{R}$). Suppose the physical system is described by the Lagrangian components of all active forces $\mathscr{Q}_k|_{\mathscr{R}} = \mathscr{Q}_k|_{\mathscr{R}}(q^1, \ldots, q^n, \dot{q}^1, \ldots, \dot{q}^n)$ *with no explicit dependency on time*, so that the resulting Euler-Lagrange equations are *autonomous*.

 If the n-tuple $Q_0 := (q_0^1, \ldots, q_0^n) \in V$ defines an equilibrium configuration in $\mathscr{R}$, we set $A_0 := (q_0^1, \ldots, q_0^n, 0, \ldots, 0) \in V \times \mathbb{R}^n$. Q_0 is said to be:

(1) **future stable (with respect to $\mathscr{R}$)**, if for any open neighbourhood $B \ni A_0$ there is an open neighbourhood $E \ni A_0$ such that every solution of the Euler-Lagrange equations (7.36) in $V \times \mathbb{R}^n$, $\gamma : J \to V \times \mathbb{R}^n$ (where J is an open interval in $\mathbb{R}$ containing $t = 0$) with $\gamma(0) = \gamma_0 \in E$, satisfies $\gamma(t) \in B$ if $J \ni t \geq 0$ (Fig. 9.1);
(2) **past stable (with respect to $\mathscr{R}$)**, if for any open neighbourhood $B \ni A_0$ there is an open neighbourhood $E \ni A_0$ such that every solution to (7.36) in $V \times \mathbb{R}^n$, $\gamma : J \to V \times \mathbb{R}^n$ (where J is an open interval in $\mathbb{R}$ containing $t = 0$) with $\gamma(0) = \gamma_0 \in E$, satisfies $\gamma(t) \in B$ if $J \ni t \leq 0$;
(3) **stable (with respect to $\mathscr{R}$)**, if it is future stable and past stable;
(4) **unstable (with respect to $\mathscr{R}$) (in the future** or **in the past)**, if it is not stable (respectively in the future or the past);

Fig. 9.1 Future stability for $n = 1$

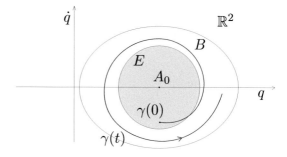

(5) **asymptotically future stable (with respect to $\mathscr{R}$)**, if it is future stable and there is an open neighbourhood $F \ni A_0$ such that any maximal solution to (7.36) in $V \times \mathbb{R}^n$, γ with $\gamma(0) = \gamma_0 \in F$, is future-complete and $\gamma(t) \to A_0$ as $t \to +\infty$;
(6) **asymptotically past stable (with respect to $\mathscr{R}$)**, if it is past stable and there is an open neighbourhood $P \ni A_0$ such that any maximal solution of (7.36) in $V \times \mathbb{R}^n$, γ with $\gamma(0) = \gamma_0 \in P$, is past-complete and $\gamma(t) \to A_0$ as $t \to -\infty$.
$\diamondsuit$

Remarks 9.18

(1) In case the $\mathscr{Q}_k|_{\mathscr{R}}$ are C^1 and the $\mathscr{T}|_{\mathscr{R}}$ are C^2, if $Q_0 := (q_0^1, \ldots, q_0^n) \in V$ is an equilibrium configuration for $\mathscr{R}$ that is future stable, by Proposition 14.34 the Euler-Lagrange equations' maximal solutions with initial condition in E are certainly future-complete (we may pick B to be relatively compact). The same holds for past stability.
(2) It must be crystal clear, when talking about the stability of the configurations of a physical systems, in particular Lagrangian systems, that *the notion of stability contains information and requirements regarding the velocities as well, not just the configurations.* ∎

We can now state the Lagrangian analogue to Theorem 6.24, including the analogue of Theorem 6.26.

Theorem 9.19 (Lagrange-Dirichlet) *Consider a system $\mathscr{S}$ of N point particles $P_1, \ldots, P_N$ under $c = 3N - n \geq 0$ ideal holonomic constraints and described on $A(\mathbb{V}^{n+1})$. Consider natural coordinates $(t, q^1, \ldots, q^n)$ on $\mathbb{R} \times V$ ($V \subset \mathbb{R}^n$ open) moving with $\mathscr{R}$, and suppose the active forces decompose into a conservative part for $\mathscr{R}$ with total potential energy $\mathscr{U}|_{\mathscr{R}}$ (C^2 in $(q^1, \ldots, q^n)$) and a non-conservative part coming from forces $\mathbf{F}_i|_{\mathscr{R}} = \mathbf{F}_i(\mathbf{x}_1, \ldots, \mathbf{x}_N, \mathbf{v}_{P_1}|_{\mathscr{R}}, \ldots, \mathbf{v}_{P_N}|_{\mathscr{R}})$, $i = 1, \ldots, N$, not depending on time explicitly (whose Lagrangian components are C^1) and possibly null, where $\mathbf{x}_i := P - O$ is the position vector of P_i in $\mathscr{R}$.*

Assume the kinetic energy $\mathscr{T}|_{\mathscr{R}} = \mathscr{T}|_{\mathscr{R}}(q, \dot{q})$ is C^2. Then an equilibrium configuration $Q_0 := (q_0^1, \ldots, q_0^n) \in V$ for $\mathscr{R}$ is:

(1) *future stable if the restriction of $\mathscr{U}|_{\mathscr{R}}$ to V has a strict local minimum at that point, and for $(\mathbf{x}_1, \ldots, \mathbf{x}_N)$ corresponding to any configuration in V:*

$$\sum_{i=1}^{N} \mathbf{F}_i(\mathbf{x}_1, \ldots, \mathbf{x}_N, \mathbf{v}_{P_1}|_{\mathscr{R}}, \ldots, \mathbf{v}_{P_N}|_{\mathscr{R}}) \cdot \mathbf{v}_{P_i}|_{\mathscr{R}} \leq 0 \text{ for any vectors}$$

$\mathbf{v}_{P_1}|_{\mathscr{R}}, \ldots, \mathbf{v}_{P_N}|_{\mathscr{R}}$;

(2) *stable (in the past and in the future) if the restriction of $\mathscr{U}|_{\mathscr{R}}$ to V has a strict local minimum at Q_0, and for $(\mathbf{x}_1, \ldots, \mathbf{x}_N)$ corresponding to any configuration in V:*

$$\sum_{i=1}^{N} \mathbf{F}_i(\mathbf{x}_1, \ldots, \mathbf{x}_N, \mathbf{v}_{P_1}|_{\mathscr{R}}, \ldots, \mathbf{v}_{P_N}|_{\mathscr{R}}) \cdot \mathbf{v}_{P_i}|_{\mathscr{R}} = 0 \text{ for any vectors}$$

$\mathbf{v}_{P_1}|_{\mathscr{R}}, \ldots, \mathbf{v}_{P_N}|_{\mathscr{R}}$; *this happens in particular if all forces $\mathbf{F}_i$ are zero;*

(3) future stable and asymptotically future stable if the strict local minimum point Q_0 is an isolated critical point[2] of the potential energy $\mathcal{U}|_{\mathcal{R}}$, and for $(\mathbf{x}_1, \ldots, \mathbf{x}_N)$ corresponding to any configuration in V:

$$\sum_{i=1}^{N} \mathbf{F}_k(\mathbf{x}_1, \ldots, \mathbf{x}_N, \mathbf{v}_{P_1}|_{\mathcal{R}}, \ldots, \mathbf{v}_{P_N}|_{\mathcal{R}}) \cdot \mathbf{v}_{P_i}|_{\mathcal{R}} < 0, \text{for } (\mathbf{v}_{P_1}|_{\mathcal{R}}, \ldots, \mathbf{v}_{P_N}|_{\mathcal{R}})$$
$$\neq (\mathbf{0}, \ldots, \mathbf{0}).$$

Proof (1) and (2). The proof (essentially identical to that of Theorem 6.24) relies on a suitable Lyapunov function that coincides with the mechanical energy of system $\mathscr{S}$ in $\mathscr{R}$:

$$\mathscr{H}(q^1, \ldots, q^n, \dot{q}^1, \ldots, \dot{q}^n) := \mathscr{T}|_{\mathscr{R}}(q^1, \ldots, q^n, \dot{q}^1, \ldots \dot{q}^n) + \mathscr{U}|_{\mathscr{R}}(q^1, \ldots, q^n),$$

where

$$\mathscr{T}|_{\mathscr{R}}(q^1, \ldots, q^n, \dot{q}^1, \ldots \dot{q}^n) := \sum_{h,k=1}^{n} a_{hk}(q^1, \ldots, q^n)\dot{q}^h\dot{q}^k.$$

First, for any n-tuple $(q^1, \ldots, q^n)$, the quadratic polynomial in the $\dot{q}^k$: $\sum_{h,k=1}^{n} a_{hk}(q^1, \ldots, q^n)\dot{q}^h\dot{q}^k$ has a strict local minimum at $\dot{q}^k = 0$, $k = 1, \ldots, n$, because the coefficient matrix a_{hk} is positive definite by Theorem 7.23. Hence in particular

$$\mathscr{T}_{\mathscr{R}}(q^1, \ldots, q^n, \dot{q}^1, \ldots \dot{q}^n) > \mathscr{T}|_{\mathscr{R}}(q_0^1, \ldots, q_0^n, 0, \ldots, 0) = 0$$

if $(\dot{q}^1, \ldots, \dot{q}^n) \neq (0, \ldots, 0)$. If the potential energy also has a strict local minimum at $(q_0^1, \ldots, q_0^n)$, then:

$$\mathscr{U}|_{\mathscr{R}}(q^1, \ldots, q^n) > \mathscr{U}|_{\mathscr{R}}(q_0^1, \ldots, q_0^n)$$

if $(q^1, \ldots, q^n) \neq (q_0^1, \ldots, q_0^n)$ around the point. Adding the inequalities immediately tells, as said above, that the Hamiltonian has a strict local minimum in $(q_0^1, \ldots, q_0^n, 0, \ldots, 0)$ since:

$$\mathscr{H}(q^1, \ldots, q^n, \dot{q}^1, \ldots, \dot{q}^n) > \mathscr{H}(q_0^1, \ldots, q_0^n, 0, \ldots, 0)$$

if $(q^1, \ldots, q^n, \dot{q}^1, \ldots, \dot{q}^n) \neq (q_0^1, \ldots, q_0^n, 0, \ldots, 0)$ near the point. Let us show that also the other condition for $\mathscr{H}$ to be Lyapunov holds. It concerns $\mathscr{H}$. In our

---

[2] That $(q_0^1, \ldots, q_0^n)$ gives a strict local minimum for the potential energy is not enough for it to be an isolated critical point: consider, for $N = 1$, the function $\mathscr{U}(q) := q^4(2 + \sin(1/q))$ if $q \neq 0$ and $\mathscr{U}(0) := 0$. $\mathscr{U}$ is twice differentiable and has a strict local minimum at $q = 0$, which is not an isolated stationary point.

case the Euler-Lagrange equations are:

$$2 \sum_k a_{hk} \frac{d^2 q^k}{dt^2} + 2 \sum_{k,r} \frac{\partial a_{hk}}{\partial q^r} \frac{dq^k}{dt} \frac{dq^r}{dt} - \sum_{r,k} \frac{\partial a_{rk}}{\partial q^h} \frac{dq^r}{dt} \frac{dq^k}{dt} + \frac{\partial \mathcal{U}|_{\mathcal{R}}}{\partial q^h}$$

$$= \mathcal{Q}_h|_{\mathcal{R}}(q^1, \dots, q^n, \dot{q}^1, \dots, \dot{q}^n),$$

where the $\mathcal{Q}_k$ are the Lagrangian components of the non-conservative forces $\{\mathbf{F}_i|_{\mathcal{R}}\}_{i=1,\dots,N}$. Multiplying by $\dot{q}^h$, adding over h and using $\dot{q}^k = dq^k/dt$ (valid along the motions), we easily find that along the solutions of the Euler-Lagrange equations:

$$\frac{d}{dt} \left(\sum_{h,k=1}^n a_{hk}(q^1(t), \dots, q^n(t)) \dot{q}^h(t) \dot{q}^k(t) + \mathcal{U}|_{\mathcal{R}}(q^1(t), \dots, q^n(t)) \right)$$

$$= \sum_{h=1}^n \dot{q}^h \mathcal{Q}_h|_{\mathcal{R}}(q^1(t), \dots, q^n(t), \dot{q}^1(t), \dots, \dot{q}^n(t)).$$

The function inside the total time derivative is nothing but the Hamiltonian evaluated along the solution of the Euler-Lagrange equations we are using. By (9.38) therefore:

$$\frac{d}{dt} \mathcal{H}(q^1(t), \dots, q^n(t), \dot{q}^1(t), \dots, \dot{q}^n(t))$$

$$= \sum_{k=1}^N \mathbf{F}_k(\mathbf{x}_1, \dots, \mathbf{x}_N, \mathbf{v}_{P_1}|_{\mathcal{R}}, \dots, \mathbf{v}_{P_N}|_{\mathcal{R}}) \cdot \mathbf{v}_{P_k}|_{\mathcal{R}},$$

where $\mathbf{x}_i = \mathbf{x}_i(q^1(t), \dots, q^n(t))$ and $\mathbf{v}_{P_k}|_{\mathcal{R}} = \mathbf{v}_{P_k}|_{\mathcal{R}}(q^1(t), \dots, q^n(t), \dot{q}^1(t), \dots, \dot{q}^n(t))$ are evaluated along the solution. As we know from Proposition 6.11, by the Euler-Lagrange equations' Existence Theorem this fact is equivalent to saying (viewing the equations as a first-order system) that:

$$\mathcal{H}(q^1, \dots, q^n, \dot{q}^1, \dots, \dot{q}^n) = \sum_{k=1}^N \mathbf{F}_k(q^1, \dots, q^n, \dot{q}^1, \dots, \dot{q}^n)$$

$$\cdot \mathbf{v}_{P_k}|_{\mathcal{R}}(q^1, \dots, q^n, \dot{q}^1, \dots, \dot{q}^n).$$

Under our assumptions on the right-hand side, (1) and (2) are now immediate from Lyapunov's theorem (Theorem 6.12) using $\mathcal{H}$ (restricted to a neighbourhood of $(q_0^1, \dots, q_0^n, 0, \dots, 0)$) as Lyapunov function W. (3) The proof is essentially identical to Theorem 6.26 by the above remarks. $\square$

Remarks 9.20

(1) If $\mathscr{Q}_k$ are the Lagrangian components of the *non-conservative* force $\mathbf{F}_i$, we have the relationship, employed in the proof:

$$\sum_{k=1}^{n} \mathscr{Q}_k(q^1, \ldots, q^n, \dot{q}^1, \ldots, \dot{q}^n) \dot{q}^k = \sum_{i=1}^{N} \mathbf{F}_k(\mathbf{x}_1, \ldots, \mathbf{x}_N, \mathbf{v}_{P_1}|_{\mathscr{R}}, \ldots, \mathbf{v}_{P_N}|_{\mathscr{R}})$$

$$\cdot \mathbf{v}_{P_i}|_{\mathscr{R}}. \tag{9.38}$$

It expresses the *total power* of the forces $\{\mathbf{F}_i\}_{i=1,\ldots,N}$ with respect to $\mathscr{R}$ in natural coordinates on $\mathscr{R}$, and allows to recast the above theorem alternatively, without mentioning any forces but only using their Lagrangian components.

(2) Using coordinates on $\mathscr{R}$, case (2) in the theorem holds in particular when the active forces are all given by a generalised potential:

$$\mathscr{V}|_{\mathscr{R}}(q, \dot{q}) = \sum_{k=1}^{n} A_k(q) \dot{q}^k + B(q)$$

not depending explicitly on time. If B, the part independent of $\dot{q}$, has a strict local maximum at some configuration Q_0, the latter is a past-stable and future-stable equilibrium. In fact, if so, the total mechanical energy in $\mathscr{R}$ (coinciding with the Hamiltonian) is just $\mathscr{E}|_{\mathscr{R}}(q, \dot{q}) = \mathscr{T}|_{\mathscr{R}}(q, \dot{q}) - B(q)$ and the non-conservative forces, associated only with $\sum_{k=1}^{n} A_k(q) \dot{q}^k$, will obey (2) in the theorem, as implied by (9.21):

$$\sum_{i=1}^{N} \mathbf{F}_i \cdot \mathbf{v}_i = \sum_{k=1}^{n} \mathscr{Q}_k \dot{q}^k = \sum_{h,k=1}^{n} \left(\frac{\partial A_h}{\partial q^k} - \frac{\partial A_k}{\partial q^h} \right) \dot{q}^h \dot{q}^k = 0.$$

The end result is 0 since $\dot{q}^h \dot{q}^k$ is symmetric if we swap k and h while $\left(\frac{\partial A_h}{\partial q^k} - \frac{\partial A_k}{\partial q^h} \right)$ is skew-symmetric. ∎

To finish, let us state the proposition similar to Proposition 6.28, an sufficient condition for instability based on the Hessian of the potential energy, for systems with either purely conservative forces or such that the non-conservative active forces have no Lagrangian components. This happens in certain cases with gyrostatic forces such as the Coriolis force. The proof is left as exercise, as it resembles Proposition 6.28.

Proposition 9.21 *Under the hypotheses of Theorem 9.19, suppose the active forces are purely conservative (i.e. there are no $\mathbf{F}_i$) or the possible non-conservative active forces have no Lagrangian components. An equilibrium configuration for $\mathscr{R}$ given by $(q_0^1, \ldots, q_0^n) \in V$ is past unstable and future unstable if the Hessian matrix of the potential energy has a negative eigenvalue at $(q_0^1, \ldots, q_0^n)$.*

9.4 Introduction to the Theory of Small Vibrations and Normal Coordinates

Consider a physical system with n degrees of freedom, described in free coordinates $q^1, \ldots, q^n$ on a frame $\mathscr{R}$ where the constraints do not depend on time explicitly. The physical system's dynamics is governed by a Lagrangian $\mathscr{L}|_{\mathscr{R}} = \mathscr{T}|_{\mathscr{R}} - \mathscr{U}|_{\mathscr{R}}$, viewed as C^2 in the coordinates of $A(\mathbb{V}^{n+1})$ so to warrant the existence and uniqueness of the solutions to the Euler-Lagrange equations. In the sequel we will not write $|_{\mathscr{R}}$ for simplicity. Now assume, possibly redefining the coordinates' origin, that the configuration $(q_0^1, \ldots, q_0^n) = (0, \ldots, 0)$ is a both future and past stable equilibrium for $\mathscr{R}$. More precisely, suppose the potential energy $\mathscr{U}$, for which the configuration is stationary by hypothesis, has positive-definite Hessian matrix H of coefficients

$$H_{hk} := \frac{\partial^2 \mathscr{U}}{\partial q^h \partial q^k}\Big|_{(0,\ldots,0)}. \tag{9.39}$$

The Lagrange-Dirichlet theorem guarantees stability. In other words the eigenvalues of the symmetric matrix H are all positive. Finally assume $\mathscr{U}(0, \ldots, 0) = 0$, using an immaterial redefinition of the additive constant in $\mathscr{U}$ if necessary. Approximating $\mathscr{U}$ with its quadratic Taylor expansion centred at the origin (order two is the first significant polynomial), near the equilibrium configuration we can write:

$$\mathscr{L} \simeq \sum_{k,h=1}^n a_{hk}(q^1, \ldots, q^n)\dot{q}^h\dot{q}^k - \frac{1}{2}\sum_{k,h=1}^n H_{hk}q^h q^k \, .$$

Approximating the kinetic energy's symmetric matrix by the value at the equilibrium:

$$A_{hk} := 2a_{hk}(0, \ldots, 0) \, , \tag{9.40}$$

the rough approximation around the equilibrium is:

$$\mathscr{L} \simeq \mathscr{L}_0 := \frac{1}{2}\sum_{k,h=1}^n \dot{q}^h A_{hk}\dot{q}^k - \frac{1}{2}\sum_{k,h=1}^n q^h H_{hk}q^k \, . \tag{9.41}$$

That this is a good approximation of the real value of $\mathscr{L}$ along a solution of the Euler-Lagrange equations with initial conditions near the equilibrium is a consequence of the fact the motion stays near the configuration, precisely by definition of stable equilibrium and because $\mathscr{L}$ is continuous (actually C^2) in the q^k and the $\dot{q}^k$ by assumption. At this juncture a natural question arises (one that often we do not perceive, thus making a dangerous logical mistake). If the initial conditions do not change, are *the solutions* to the Euler-Lagrange equations of the approximated Lagrangian $\mathscr{L}_0$ good approximations of the *solutions* of the analogous equations for the exact Lagrangian $\mathscr{L}$ (at least when such solutions stay close to the equilibrium and have small velocities)?

The answer is yes and this kind of analysis can be carried out starting from Theorem 14.35. We shall not do that, but rather work directly with the Euler-Lagrange equations coming from $\mathscr{L}_0$, assuming that such solutions are indeed good approximations of the system's true motion, associated with the complete Lagrangian $\mathscr{L}$, around the equilibrium configuration at small velocities.

9.4.1 Linearised and Decoupled Equations: Normal Coordinates

The equations descending from the approximate Lagrangian $\mathscr{L}_0$ in (9.41) are *linear*, so they are called **linearised equations**, and the aforementioned procedure for passing from $\mathscr{L}$ to $\mathscr{L}_0$ is called **linearisation of the equations** when we speak of the corresponding solutions.

The linearised equations obtained as Euler-Lagrange equations from $\mathscr{L}_0$ in (9.41) thus have the form:

$$\sum_{k=1}^{2} A_{hk} \frac{d^2 q^k}{dt^2} = -\sum_{k=1}^{2} H_{hk} q^k .$$

To simplify the notation we will use the column vector $Q := (q^1, \ldots, q^n)^t$ and write the above equations as:

$$A \frac{d^2 Q}{dt^2} = -HQ .$$

As A is invertible, as proved in Theorem 7.23 whereby the Euler-Lagrange equations can be put in normal form, we can write the above differential system in normal form:

$$\frac{d^2 Q}{dt^2} = -A^{-1} HQ . \tag{9.42}$$

We want to show that it is always possible to change free coordinates $Q := (q^1, \ldots, q^n)^t$ to $\widetilde{Q} := (\widetilde{q}^1, \ldots, \widetilde{q}^n)^t$ by a simple linear transformation:

$$\widetilde{Q} = DQ , \quad \text{with } D \text{ real, } n \times n, \text{ non-singular}, \tag{9.43}$$

so that the final equations decouple:

$$\frac{d^2\widetilde{Q}}{dt^2} = -\mathrm{diag}(\omega_1^2, \ldots, \omega_n^2)\, \widetilde{Q}\,, \tag{9.44}$$

where $diag(\omega_1^2, \ldots, \omega_n^2)$ is the diagonal matrix with eigenvalues $\omega_k^2 > 0$, the n solutions ω^2 (all positive) of the **characteristic equation**:

$$\det\left(\omega^2 A - H\right) = 0\,. \tag{9.45}$$

In this way (9.44) are n equations independent from one other:

$$\frac{d^2\widetilde{q}^k}{dt^2} = -\omega_k^2 \widetilde{q}^k\,, \quad k = 1, 2, \ldots, n\,. \tag{9.46}$$

The new coordinates $\widetilde{Q} := (\widetilde{q}^1, \ldots, \widetilde{q}^n)^t$, in which the linearised equations are decoupled as in (6.8), are called **normal coordinates**. They are defined by the system and by the stable equilibrium configuration at which the Euler-Lagrange equations are linearised.

We proceed heuristically to prove this fact. Once we define the new coordinates by (9.43), Eq. (9.42) read, multiplying by D on the left:

$$\frac{d^2 DQ}{dt^2} = -DA^{-1}HQ\,.$$

As D is invertible $Q = D^{-1}\widetilde{Q}$, so:

$$\frac{d^2\widetilde{Q}}{dt^2} = -DA^{-1}HD^{-1}\widetilde{Q}\,.$$

By construction these equations are *equivalent* to the ones in (9.42). That these equations can be written as (9.44) is immediate from the following proposition, which shows that in out setup $A^{-1}H$ is always diagonalisable. In this case in particular, H is positive definite by assumption and hence the characteristic equations' solutions are positive, so they can be written ω_k^2 with $\omega_k > 0$.

Proposition 9.22 *If A and H are $n \times n$ real symmetric matrices and A is positive definite, there exists an $n \times n$ real non-singular matrix D such that:*

$$DA^{-1}HD^{-1} = diag(\lambda_1, \ldots, \lambda_n)\,. \tag{9.47}$$

Furthermore, the following hold.

(1) The n numbers λ_k do not depend on D, being precisely the solutions to

$$det\,(\lambda A - H) = 0\,. \tag{9.48}$$

(2) $\lambda_k \geq 0$, $k = 1, \ldots, n$, if and only if H is positive semi-definite.
(3) $\lambda_k > 0$, $k = 1, \ldots, n$, if and only if H is positive definite.

Proof We begin from (1). Suppose there is a non-singular matrix D such that (9.47) holds, i.e. a matrix diagonalising $A^{-1}H$, and let us prove the λ_k are exactly the solutions to (9.48). By elementary linear algebra we know that if B is diagonalisable over $\mathbb{K} = \mathbb{R}$ or $\mathbb{C}$, the diagonalised matrix contains the eigenvalues of B. The latter are the solutions λ to $\det(\lambda I - B) = 0$ over $\mathbb{K}$. In our case the eigenvalues—the numbers in the matrix diagonalising $A^{-1}H$—are the real solutions of:

$$\det\left(\lambda I - A^{-1}H\right) = 0\,. \tag{9.49}$$

We claim (9.48) is equivalent to such equation, thus proving (1). As the real symmetric matrix A is strictly positive it is invertible, and so is A^{-1}: $\det(\lambda A - H) = 0$ if and only if $\det(A^{-1})\det(\lambda A - H) = 0$. But $\det(A^{-1})\det(\lambda A - H) = \det\left[A^{-1}(\lambda A - H)\right] = \det\left(\lambda I - A^{-1}H\right)$.

Let us pass to the first assertion in the theorem. We must exhibit a non-singular D satisfying (9.47). Take the product $A^{-1}H$. As A is symmetric, A^{-1} is symmetric (in general $(A^{-1})^t = (A^t)^{-1}$, so the symmetry of A implies $(A^{-1})^t = (A^t)^{-1} = A^{-1}$). Further, A is positive definite and so is A^{-1}. In fact A^{-1} is invertible by construction and positive definite, so for a generic column vector $X \in \mathbb{R}^n$, if $X' := A^{-1}X$:

$$X^t A^{-1} X = (AX')^t A^{-1} AX' = X'^t A^t A^{-1} AX' = X'^t A^t X'$$
$$= (X'^t A^t X')^t = X'^t AX' \geq 0\,.$$

Being real symmetric, A^{-1} is diagonalisable by some $R \in O(n)$, and because it is positive definite, on the main diagonal there will appear strictly positive numbers: $RAR^{-1} = \text{diag}(a_1, \ldots, a_n)$, where $a_k > 0$. Hence:

$$R\,A^{-1}H\,R^{-1} = RA^{-1}R^{-1}\,RHR^{-1} = \text{diag}(a_1, \ldots, a_n)RHR^{-1}\,.$$

If $S := \text{diag}(a_1^{-1/2}, \ldots, a_n^{-1/2})$ (well defined since a_k is strictly positive):

$$SR\,A^{-1}H\,R^{-1}S^{-1} = \text{diag}(a_1^{1/2}, \ldots, a_n^{1/2})RHR^{-1}\text{diag}(a_1^{1/2}, \ldots, a_n^{1/2})$$
$$= S^{-1}RHR^t S^{-1t}\,,$$

i.e.

$$SR\,A^{-1}H\,(SR)^{-1} = S^{-1}RH(S^{-1}R)^t \tag{9.50}$$

where we used $R^{-1} = R^t$ as $R \in O(n)$ and that by construction $S^{-1} = S^{-1t}$. To finish, the matrix $S^{-1}RH(S^{-1}R)^t$ in the right-hand side of (9.50) is symmetric by construction:

$$\left(S^{-1}RH(S^{-1}R)^t\right)^t = S^{-1}RH^t(S^{-1}R)^t = S^{-1}RH(S^{-1}R)^t \,,$$

so there will be $R' \in O(n)$ such that

$$R'S^{-1}RH(S^{-1}R)^t R'^{-1} = \text{diag}(\lambda_1, \ldots, \lambda_n) \,. \tag{9.51}$$

Putting everything together, applying $R' - R'^{-1}$ to (9.50), we have found:

$$R'SR\,A^{-1}H\,(SR)^{-1}R'^{-1} = \text{diag}(\lambda_1, \ldots, \lambda_n)$$

$$\text{i.e.} \quad R'SR\,A^{-1}H\,(R'SR)^{-1} = \text{diag}(\lambda_1, \ldots, \lambda_n) \,.$$

Now set $D := R'SR$ to obtain (9.47). To conclude the proof, let us establish (2) and (3). By (9.51), the λ_k are the eigenvalues of the symmetric matrix:

$$H' := C^tHC \,.$$

where $C := (S^{-1}R)^t$. Note C is non-singular in our hypotheses, since $\det(S^{-1}R)^t = \det(S^{-1}R) = \det(S^{-1})\det R \neq 0$ because S^{-1} is invertible and $\det R = \pm 1$. Let us consider the positivity of H':

$$X^t H' X = X^t C^t HCX = (CX)^t H(CX) = Y^t HY \geq 0$$

where $Y = CX$ and H is positive definite. As C is non-singular and hence bijective, the positivity of H is equivalent to the positivity of H'. Therefore the eigenvalues of H', i.e. the λ_k whose sign we are examining, are non-negative if and only if H is positive definite. Similarly, to say H is positive definite is the same as saying H is positive definite and $\det H \neq 0$. For the same reason as above, this is in turn equivalent to H' being positive definite and $\det H' \neq 0$, so its eigenvalues λ_k are strictly positive.   □

Remarks 9.23

(1) We leave it as exercise to show that the matrix D satisfying (9.47) is not uniquely determined by A and H. Nonetheless, the λ_k are uniquely determined by A and H (to be precise, by $A^{-1}H$).

(2) In practice, to find the D diagonalising $A^{-1}H$ it is not necessary to follow in the above footsteps. That proof's argument was only meant to show that such a matrix exists. Once we know that, it suffices to go about as follows, using the general recipe for diagonalising a matrix. If $E_{(k)}$ is the k-th canonical vector of $\mathbb{R}^n$, the identity:

$$DA^{-1}HD^{-1} = \text{diag}(\omega_1^k, \ldots, \omega_n^2)$$

(true for some non-singular D), immediately implies:

$$DA^{-1}HD^{-1}E_{(k)} = \omega_k^2 E_{(k)} \quad \text{and so} \quad A^{-1}HD^{-1}E_{(k)} = \omega_k^2 D^{-1}E_{(k)} \ .$$

In other words:

$$N_{(k)} := D^{-1}E_{(k)}$$

is an eigenvector of $A^{-1}H$ with eigenvalue ω_k^2. The inner product between $N_{(k)}$ and $E_{(h)}$ gives the h-th component of $N_{(k)}$, and the above identity says:

$$\left(D^{-1}\right)_{hk} = E_{(h)}^t D^{-1} E_{(k)} = E_{(h)}^t N_{(k)} = \left(N_{(k)}\right)_h \ .$$

Hence, to find D^{-1} and therefore D the steps are: (a) seek the constants ω_k^2 that solve the characteristic equation (9.45), (b) for each of the k values ω_k^2 find a column vector $N_{(k)} \in \mathbb{R}^n$ solving the eigenvector equation:

$$A^{-1}HN_{(k)} = \omega_k N_{(k)} \ .$$

We know there are n linearly independent vectors (not unique) solving the equation since $A^{-1}H$ is diagonalisable. (c) The matrix D^{-1} will eventually be given by:

$$(D^{-1})_{hk} = (N_{(k)})_h \ , \quad \text{for } h, k = 1, \ldots, n, \tag{9.52}$$

where $(N_{(k)})_h$ is the h-th component of the column vector $N_{(k)}$ in the canonical basis of $\mathbb{R}^n$. Since $Q = D^{-1}\widetilde{Q}$ we then have:

$$q^k = \sum_{h=1}^n (N_{(k)})_h \widetilde{q}^h \ . \tag{9.53}$$

∎

9.4.2 Natural Frequencies (or Eigenfrequencies) and Normal Vibration Modes

Let us return to the linearised equation in normal coordinates (9.46). The general solution is now immediate, since the numbers ω_k^2 are strictly positive, and conventionally chosen so that the ω_k are strictly positive, too. The general solution has the form, for arbitrary but given numbers $A_k, B_k \in \mathbb{R}$:

$$\widetilde{q}^k(t) = A_k \cos(\omega_k t) + B_k \sin(\omega_k t) \ , \quad k = 1, 2, \ldots, n \ . \tag{9.54}$$

The frequencies ω_k, obtained from the characteristic equation, are called **natural frequencies** or **eigenfrequencies** (around the equilibrium configuration under exam). Every solution arising when the coefficients A_k and B_k are all zero, except for A_{k_0} and/or B_{k_0}, is known as k_0-**th normal vibration mode**. Physically, a paramount observation (for the consequences in quantum applications, crystallography and other areas), is that the normal vibration modes are the ways in which the system moves *collectively*, or more precisely, how it *vibrates collectively*. This is due to the fact that the normal coordinates are *collective coordinates* in the following sense. Usually, the starting coordinates $q^1, \ldots, q^n$ are chosen so they identify, separately or in groups, certain parts of the system. For example, the first two coordinates might be for a point particle in the constrained system, the next 3 might define another point and so on. The coordinate change to normal coordinates, mediated by a linear map D, shuffles the initial affiliation between coordinates and point particles. This is manifest when we write solution (9.54) in the initial coordinates and then we look at the system's motion, which in normal coordinates is given by a precise normal mode, thus turning off all others. With this procedure:

$$q^h(t) = \sum_{k=1}^{n} (D^{-1})_{hk} \left(A_k \cos (\omega_k t) + B_k \sin (\omega_k t) \right), \quad h = 1, 2, \ldots, n. \quad (9.55)$$

Even when all normal modes are off except for mode k_0, the solution in the initial coordinates takes the form:

$$q^h(t) = (D^{-1})_{hk_0} \left(A_{k_0} \cos \left(\omega_{k_0} t \right) + B_{k_0} \sin \left(\omega_{k_0} t \right) \right), \quad h = 1, 2, \ldots, n. \quad (9.56)$$

We should emphasise that even when *only* mode k_0 is active, in the initial coordinates, in general *all* coordinates q^h will vibrate—producing a collective motion—due to the presence of the coefficients $(D^{-1})_{hk_0}$, $h = 1, 2, \ldots, n$, in (9.56).

Examples 9.24 Consider a horizontal segment of length $3l > 0$ on which two point particles of mass $m > 0$ glide without friction. The first, P_1, is determined by the coordinate x, measured from the rod's left end, while the second, P_2, has coordinate y measured from the same position. Particle P_1 is attached to a spring of elastic constant $k > 0$ whose other end is fixed to the rod's left end. The spring has length l at rest and negligible mass. P_2 is attached to a spring with the same elastic constant $k > 0$, whose second end is fixed to the rod's right end. Also this spring has length l at rest and negligible mass. A third spring of the same type as the other two joins P_1 and P_2. The rod is at rest in an inertial frame. The system's Lagrangian is then:

$$\mathcal{L} = \frac{m}{2} \left(\dot{x}^2 + \dot{y}^2 \right) - \frac{k}{2} (l - x)^2 - \frac{k}{2} (2l - y)^2 - \frac{k}{2} (y - x - l)^2.$$

The equilibrium configurations, as usual, come from setting equal to zero the gradient of the potential energy:

$$\mathscr{U}(x, y) = \frac{k}{2}(l - x)^2 + \frac{k}{2}(2l - y)^2 + \frac{k}{2}(y - x - l)^2 ,$$

which we used in the Lagrangian. We have

$$\frac{\partial \mathscr{U}}{\partial x} = -k(l - x) - k(y - x - l) = 0 , \qquad \frac{\partial \mathscr{U}}{\partial y} = -k(2l - y) + k(y - x - l) = 0 ,$$

determining a unique configuration (as one would intuitively expect):

$$(x_0, y_0) = (l, 2l) .$$

It si convenient to redefine the free coordinates so that the equilibrium configuration coincides with the origin:

$$q^1 := x - l , \qquad q^2 := y - 2l .$$

The configuration is a stable equilibrium, since the potential energy's Hessian matrix H, evaluated at the equilibrium, is:

$$H = k \begin{bmatrix} 2 & -1 \\ -1 & 2 \end{bmatrix} ,$$

and has two positive eigenvalues: 1 and 3. Therefore:

$$A^{-1}H = \frac{k}{m} \begin{bmatrix} 2 & -1 \\ -1 & 2 \end{bmatrix} .$$

The eigenfrequencies then are:

$$\omega_1 = \sqrt{\frac{k}{m}} , \qquad \omega_1 = \sqrt{3\frac{k}{m}} .$$

Diagonalising $A^{-1}H$ and following the steps in (2), Remarks 9.23, then applying (9.53), we immediately find a pair of normal coordinates:

$$q^1 = \tilde{q}^1 + \tilde{q}^2 , \quad q^2 = \tilde{q}^1 - \tilde{q}^2 , \qquad \text{equivalent to} \quad \tilde{q}^1 := \frac{q^1 + q^2}{2} , \quad \tilde{q}^2 := \frac{q^1 - q^2}{2} .$$

As regards normal vibration modes, we have the solutions:

$$\widetilde{q}^{\,1}(t) = A_1 \cos\left(\sqrt{\frac{k}{m}}\, t\right) + B_1 \sin\left(\sqrt{\frac{k}{m}}\, t\right),$$

$$\widetilde{q}^{\,2}(t) = A_2 \cos\left(\sqrt{3\frac{k}{m}}\, t\right) + B_2 \sin\left(\sqrt{3\frac{k}{m}}\, t\right).$$

Evidently, when we turn on only the first mode, in the initial free coordinates we have a collective motion in which the points vibrate in phase:

$$q^1(t) = A_1 \cos\left(\sqrt{\frac{k}{m}}\, t\right) + B_1 \sin\left(\sqrt{\frac{k}{m}}\, t\right),$$

$$q^2(t) = A_1 \cos\left(\sqrt{\frac{k}{m}}\, t\right) + B_1 \sin\left(\sqrt{\frac{k}{m}}\, t\right).$$

Vice versa, if we turn on only the second normal mode, in the initial coordinates we witness a collective motion of two points in anti-phase, with different frequency from before:

$$q^1(t) = A_2 \cos\left(\sqrt{3\frac{k}{m}}\, t\right) + B_2 \sin\left(\sqrt{3\frac{k}{m}}\, t\right),$$

$$q^2(t) = -A_2 \cos\left(3\sqrt{\frac{k}{m}}\, t\right) - B_2 \sin\left(\sqrt{3\frac{k}{m}}\, t\right).$$

Remarks 9.25

(1) We briefly mentioned earlier that the concept of normal mode has had a massive success in the theory of the structure of matter. There, to account for the vibrational degrees of freedom, crystalline structures can be described by a sequence of variously coupled oscillators (atoms or molecules) with a strong symmetry. To study the dynamics of such systems it is convenient to use normal coordinates, under which the motion decouples. The subsequent quantisation procedure can start from the classical model described in normal coordinates. But we have to say that, contrary to what happens in Classical Physics, the quantum description is not, in general, invariant under arbitrary choices of the coordinates employed to described the system. The evidence has shown that the quantisation of normal modes theoretically accounts for well-known experimental features of physical systems (for example the behaviour of the specific heat capacity of crystals in function of the temperature) that would otherwise be theoretically inexplicable. The notion of *phonon* as quantum (quasi-)particle associated with classical normal vibration modes arises in this context, and is one of the conceptual tools leading, for example, to the first models that explain *superconductivity*.

(2) The decomposition of a motion in normal modes generalises to continuous physical systems and force fields. This extension was one of the most fecund ideas in the entire history of 20th century Physics, when one passes from a classical description to a quantum one: the quantisation of the vibrational modes of continuous systems, in particular fields, is called *second quantisation* or *Quantum Field Theory*. It is in this context that one studies the notion of *photon*.   ■

Chapter 10
$\boxed{\text{AC}}$ Mathematical Introduction to Special Relativity and the Relativistic Lagrangian Formulation

In this chapter we will discuss the mathematical principles underpinning the theory of Special Relativity from a geometrical and axiomatic point of view. The motivation for the axioms, which are based on crucial experimental evidence and the ensuing physical postulates due to Einstein, will be discussed in Complement in Chap. 15.

It must be said that the *theory of Special Relativity* is one of the theories that has received the largest number of experimental confirmations in the history of science, and is without a doubt "more true" than Classical Mechanics. The word "theory" does not mean it is something still awaiting corroboration, but refers to an organic construction of logically coherent propositions.

We shall assume readers are familiar with the physical principles, also in their historical development, at the heart of the theory of Special Relativity [Rin06]; here we shall limit ourselves to illustrate those principles as pure consequences of the geometric axioms adopted. This approach, albeit certainly elegant and economical, is conceptually *dangerous* if not supported by physical justifications, in that it tends to sweep under the rug conceptual issues regarding the physical nature of the theoretical hypotheses. For example, in the present chapter we shall not deal at all with the problem of choosing the synchronisation procedure, nor with the partial conventionality of the common method known as synchronisation *à la Einstein*. We will not make explicit use of the elementary tensor formalism, apart a very brief mention, even though all formulas written in components naturally lend themselves to it. The emphasis will be on the intrinsic geometrical aspect of the physical-mathematical objects. The entire physical discussion will be postponed to Chap. 15.

In the last part of the chapter we will introduce the Lagrangian formulation of Dynamics only for the case of point particles (without constraints!) subject to external interactions of electromagnetic type.

V. Moretti, *Analytical Mechanics*, La Matematica per il 3+2 150,
https://doi.org/10.1007/978-3-031-27612-5_10

10.1   Linear Algebra Preliminaries

We need to recall elementary geometric structures of a certain physical importance in Special Relativity. The reader might skip this part and come back to it later.

10.1.1   The Dual of a Finite-Dimensional Real Vector Space

If V is a real vector space, the (algebraic) **dual space** V^* is the set of linear maps $f :$ $V \to \mathbb{R}$. It is canonically a real vector space under pointwise linear combinations of linear maps:

$$(af + bg)(u) := af(u) + bg(u) \quad \text{for } a, b \in \mathbb{R}, f, g \in V^* \text{ and } u \in V.$$

Proposition 10.1 *If* $\dim(V) < +\infty$ *then* V^* *has the same dimension as* V.

Proof Let $n := \dim(V)$. Given a basis $\{e_1, \ldots, e_n\} \subset V$, the n elements $e^{*1}, \ldots, e^{*n}$ of V^* defined by $e^{*k}(e_h) = \delta_h^k$ form a basis of V^*. In fact they are linearly independent: if $\sum_{k=1}^n c_k e^{*k} = 0$ (the zero linear functional), then $c^h = \sum_{k=1}^n c_k e^{*k}(e_h) = 0$, for all $h = 1, 2, \ldots, n$. Moreover these elements generate V^*. In fact if $f \in V^*$ then $f(e_h) = \sum_{k=1}^n f(e_k) e^{*k}(e_h)$ for any $h = 1, \ldots, n$. By linearity $f(u) = \left(\sum_{k=1}^n f(e_k) e^{*k} \right)(u)$ for any $u \in V$.   □

Vectors in V are often called **contravariant vectors**, while those in V^* are **covariant vectors**. Examining the above proof we can state the following definition.

Definition 10.2 Let V be a real vector space of dimension $n = 1, 2, \ldots$. The basis $\{e^{*k}\}_{k=1,\ldots,n} \subset V^*$ associated with a basis $\{e_k\}_{k=1,\ldots,n} \subset V$ and uniquely defined by

$$e^{*k}(e_h) = \delta_h^k, \quad h, k := 1, 2, \ldots, n,$$

is called the **dual basis** of $\{e_k\}_{k=1,\ldots,n} \subset V$.   ◇

The components of vectors in V^*, when written using a dual basis, customarily have lower indices. In particular, for $f \in V^*$:

$$f = \sum_{k=1}^n c_k e^{*k}, \quad \text{where } c_k = f(e_k) \text{ for } k = 1, 2, \ldots, n. \tag{10.1}$$

The proof that $c_k = f(e_k)$ is contained in Proposition 10.1.

10.1.2 Indefinite Inner Products, Covariant and Contravariant Components

Let us recall **Sylvester's Theorem** [Nor86] also know as **Sylvester's law of inertia**.

Theorem 10.3 *Let V be a real vector space of finite dimension* $n = 1, 2, \ldots$ *and consider a symmetric bilinear map* $g : V \times V \to \mathbb{R}$. *The following hold.*

(1) There exist a basis $\mathbf{e}_1, \ldots, \mathbf{e}_n \in V$, *called **pseudo-orthonormal**, such that*

$$g(\mathbf{e}_i, \mathbf{e}_j) = h_i \delta_{ij} \quad with \quad h_i \in \{-1, 0, 1\} \, .$$

*(We call it simply **orthonormal** when* $m = z = 0$, *where* m, z *are defined below.)*
(2) The number m of elements $\mathbf{e}_i$ *with* $h_i = -1$, *the number z of elements* $\mathbf{e}_i$ *with* $h_i = 0$ *and the number p of elements* $\mathbf{e}_i$ *with* $h_i = +1$ *does not depend on the chosen pseudo-orthonormal basis.*

The triple (m, z, p) *is the **signature** of the quadratic form (note* $m + z + p = n$).

Next to *positive-definite* inner products (Definition 1.9) there is another concept of fundamental importance in Relativity. Positive-definite inner products are *non-degenerate*: $g(u, v) = 0$ for any $u \in V$ forces $v = 0$. In fact, choosing $u = v$, the condition $g(v, v) = 0$ for a positive-definite inner product implies $v = 0$. If we relax the positivity, but keep the non-degeneracy, we obtain the following definition, which mimicks the previous one.

Definition 10.4 An **indefinite inner product** on a finite-dimensional real vector space V is a bilinear symmetric map $g : V \times V \to \mathbb{R}$ such that:

(1) g is **non-degenerate**: $g(u, v) = 0$ for any $u \in V$ implies $v = 0$;
(2) the signature (m, z, p) of g satisfies $m \cdot p \neq 0$.   ◇

Non-degeneracy is *equivalent* to asking $z = 0$, as shown below. Hence the signature of an indefinite inner product reduces to the pair (m, p).

Proposition 10.5 *Let* $g : V \times V \to \mathbb{R}$ *be a bilinear symmetric form with signature* (m, z, p) *on a real vector space V of dimension* $n = 1, 2, \ldots$. *Then g is non-degenerate if and only if* $z = 0$. *In such a case:*

(1) if $m \cdot p \neq 0$ *then g is indefinite;*
(2) if $m = 0$, *then g is positive definite;*
(3) if $p = 0$, *then* $-g$ *is positive definite.*

Proof Take $V = \mathbb{R}^n$ $(n \geq 1)$ and $g(u, v) = u^t S v$ with $S = \text{diag}(-1, \ldots, -1, 0, \ldots, 0, 1, \ldots, 1)$, where $-1, 0$ and 1 appear m, z, p times respectively. The claim follows by direct computation, as we will show. If $z \neq 0$, any $u \in \mathbb{R}^n$ with non-zero components corresponding to the z eigenvectors with zero eigenvalue, trivially satisfies $g(u, v) = 0$ for any $v \in \mathbb{R}^n$. If, conversely, $z = 0$ and $g(u, v) = 0$ for any $v \in \mathbb{R}^n$, we are free to pick $v \in \mathbb{R}^n$ with zero components corresponding to the

p eigenvectors with eigenvalue 1 in S, and with the remaining m components equal to the corresponding components of u. Then $g(u, v) = \sum_{k=1}^{m} u^k u^k = 0$ forces the m components of u to vanish. The other p components of u must similarly vanish. Hence $u = 0$ and the condition $z = 0$ implies g is non-degenerate. When $V \neq \mathbb{R}^n$ we reduce to $V = \mathbb{R}^n$ by working in components with respect to some pseudo-orthonormal basis. Assuming $z = 0$, then (1) holds by definition while (2) and (3) are immediate using a (pseudo-)orthonormal basis.   □

The presence of an indefinite, or positive-definite, inner product implies there is a canonical isomorphism between V and V^* whenever the dimension of V is finite.

Proposition 10.6 *If $g : V \times V \to \mathbb{R}$ is an inner product (indefinite or positive definite) on the real vector space V of finite dimension $n = 1, 2, \ldots$, the map:*

$$S : V \ni v \mapsto g(v,\ ) \in V^* \tag{10.2}$$

is an isomorphism of vector spaces.

Proof It suffices to show that the linear map S is 1-1, for then the surjectivity follows from V and V^* having the same finite dimension n. Injectivity is straightforward from non-degeneracy: if $S(u) = 0$ then $g(v, u) = g(u, v) = 0$ for any $u \in V$ hence $u = 0$. If s is a positive-definite inner product as per Definition 1.9, the non-degeneracy is implicit in the definition: $g(u, v) = 0$ for any $u \in V$ implies in particular $g(u, u) = 0$, and so $u = 0$ by Definition 1.9.   □

Let V be a finite-dimensional real vector space with indefinite (or positive definite) inner product $g : V \times V \to \mathbb{R}$, and $\{e_1, \ldots, e_n\} \subset V$ a basis, not necessarily pseudo-orthonormal (orthonormal), with dual basis $\{e^{*1}, \ldots e^{*n}\} \subset V^*$. The canonical isomorphism S is explicit, since decomposing the functional $S(v) \in V^*$ under (10.1):

$$S(v) = \sum_{h=1}^{3}(S(v))(e_h)e^{*h} = \sum_{h}\left(S\left(\sum_k v^k e_k\right)\right)(e_h)e^{*h}$$

$$= \sum_{k,h=1}^{n} v^k g(e_k, e_h)e^{*h} = \sum_{k,h=1}^{n} v^k s_{kh}e^{*h},$$

where we have introduced the symmetric matrix of elements $g_{hk} := g(e_k, e_h)$. All in all, if $v = \sum_k v^k e_k$, the dual vector $S(v) := \sum_h v_h e^{*h}$ has components:

$$v_h := \sum_{k=1}^{n} g_{hk} v^k = \sum_{k=1}^{n} g_{kh} v^k . \tag{10.3}$$

The numbers v_h are the **covariant components** of the vector $v \in V$ in the dual basis $\{e^{*h}\}_{h=1,\ldots,n} \subset V^*$ of $\{e_h\}_{h=1,\ldots,n} \subset V$. Note that (10.3) must be invertible since

S is bijective. Using the linearity, the contravariant components of $v \in V$ can be expressed in terms of the covariant ones:

$$v^k := \sum_{h=1}^{n} g^{kh} v_h = \sum_{h=1}^{n} g^{hk} v_h , \qquad (10.4)$$

where the symmetric matrix of elements g^{hk} is the inverse of the matrix with elements g_{hr}.

10.1.3   Applied Vectors

When, in the sequel, we speak of **applied vectors** at $p \in \mathbb{A}^n$ or **vectors at** $p \in \mathbb{A}^n$, where $\mathbb{A}^n$ is an affine space with space of displacements V^n, we mean pairs $(p, v) \in \mathbb{A}^n \times V^n$. The set of applied vectors at p is a vector space, written $T_p \mathbb{A}^n$, isomorphic to V^n, and hence equipped with the topology of V^n. The set of pairs (p, v) will be indicated by $T \mathbb{A}^n$.

Remarks 10.7

(1) We discussed at great lengths in Sect. A.6.7 that $T_p \mathbb{A}^n$ coincides with the *tangent space* (see Sect. A.6.4) of $\mathbb{A}^n$ at $p \in \mathbb{A}^n$, when we view $\mathbb{A}^n$ as differentiable n-manifold with its natural C^∞ structure. (Occasionally we shall make use of the C^∞ structure induced by the atlas of Cartesian coordinates of an affine space $\mathbb{A}^n$ when speaking of continuous or differentiable vector fields and differentiable curves on $\mathbb{A}^n$.)

(2) Clearly $T \mathbb{A}^n$ coincides with the *tangent bundle* (see Sect. B.3) when, as before, $\mathbb{A}^n$ is viewed as differentiable manifold.

■

10.2   The Geometry of Special Relativity

We will now introduce the fundamental geometric structure of Special Relativity, known as *Minkowski spacetime*. It is a four-dimensional affine space similar to the *affine Galilean structure* of classical Dynamics seen in Sect. 3.1.4. The analogy, however, ends here: in the relativistic case, in contrast to the classical situation, spacetime is not foliated by a collection of three-dimensional Euclidean absolute spaces parametrised by absolute time. The metric properties corresponding to measuring instruments of space and time available in the various reference frames are now absorbed into a single indefinite inner product of Lorentzian signature. Mathematically this new metric structure of spacetime underpins all physical principles at the basis of Special Relativity, such as the invariance of the speed of light (see Chap. 15).

10.2.1 Minkowski Spacetime, Light Cone and Time Orientation

Exactly as in Classical Mechanics, in *Relativistic Theories* everything that exists (in the past, present and future of any observer) is described by minimal spacetime determinations called *events*. These are determined locally by 4 coordinates that can be fixed in several equivalent ways, thus producing a 4-dimensional differentiable manifold called *spacetime*, provided we impose a few rather natural physical-mathematical assumptions.

In Special Relativity, spacetime is referred to as *Minkowski spacetime*. This differentiable manifold, due to the further physical hypotheses related to the invariance of the speed of light in inertial frames, turns out to carry additional geometric features that we will introduce *a priori*, in an axiomatic manner, and that will be justified only *subsequently*.

Definition 10.8 The spacetime of Special Relativity, called **Minkowski spacetime**, is a 4-dimensional affine space $\mathbb{M}^4$ whose points are called **events** and whose space of displacements V^4 has an indefinite inner product g, called **Minkowski metric**, with **Lorentzian** signature: $(m, p) = (1, 3)$. ◇

The above general definition allows to introduce certain elementary *causal structures*.

Definition 10.9 At any event $p \in \mathbb{M}^4$ there is an open cone of applied vectors:

$$\mathcal{V}_p := \left\{ (p, v) \in T_p\mathbb{M}^4 \mid v \in V^4 , \ g(v, v) < 0 \right\} ,$$

called **light cone** at p. The elements of $\overline{\mathcal{V}_p} \setminus \{0\}$ (the bar denotes the topological closure) are called **causal vectors** at p. A causal vector v is called:

(a) **lightlike**, or **null**,[1] if $g(v, v) = 0$;
(b) **timelike** if $g(v, v) < 0$.

Vectors at p that do not belong in $\overline{\mathcal{V}_p} \setminus \{0\}$ are called **spacelike**.

If $I \subset \mathbb{R}$ is a non-trivial interval, a C^1 map $I \ni s \mapsto \gamma(s) \in \mathbb{M}^4$ is called **causal curve** (**lightlike**, **timelike**, **spacelike**) if its tangent vectors $\dot{\gamma}(s)$, $s \in I$, are causal (resp. lightlike, timelike, spacelike). ◇

Notation From now on, unless strictly necessary for the understanding, we will just write u instead of (p, u) for applied vectors of $T_p\mathbb{M}^4$. Further, using pseudo-

---

[1] Note that $v \neq 0$ by definition!

orthonormal bases, Greek indices α, β, μ, ν etc. will vary between 0 and 3 while Roman ones a, b, h, k etc. will vary from 1 to 3.

The light cone $\mathcal{V}_p$ is the disjoint union of two open convex subsets, the two half-cones called **future and past light cones**. Choose in fact a pseudo-orthonormal basis e_0, e_1, e_2, e_3 of V^4, e_0 being the unique timelike vector, such that:

$$g(e_\mu, e_\nu) = g_{\mu\nu}, \quad \text{where} \quad [g_{\alpha\beta}]_{\alpha,\beta=0,1,2,3} = \text{diag}(-1, 1, 1, 1).$$

Considering the basis of $T_p\mathbb{M}^4$, we may decompose the light cone into the two above halves:

$$\mathcal{V}_p = \left\{ \sum_{\alpha=0}^{3} v^\alpha e_\alpha \ \middle| \ (v^0)^2 > \sum_{a=1}^{3} (v^a)^2, \ v^0 > 0 \right\}$$

$$\bigcup \left\{ \sum_{\alpha=0}^{3} v^\alpha e_\alpha \ \middle| \ (v^0)^2 > \sum_{a=1}^{3} (v^a)^2, \ v^0 < 0 \right\}.$$

This splitting does not depend on the pseudo-orthonormal basis. In fact the two half cones are open, convex (hence connected) and disjoint with open union $\mathcal{V}_p$. Hence the above is just $\mathcal{V}_p$ decomposed in *connected components*. Since connected components are unique, the decomposition coming from another pseudo-orthonormal basis would produce the same two halves, possibly swapped.

Definition 10.10 Two causal vectors u, v in $p \in \mathbb{M}^4$ are said to have **the same time orientation** if u and v belong to the closure of the same half of $\mathcal{V}_p$. $\diamond$

Evidently *having the same time orientation* defines an equivalence relation on causal vectors of $T_p\mathbb{M}^4$.

Proposition 10.11 *If the vectors $u, v \in T_p\mathbb{M}^4$ are timelike then:*

(1) $g(u, v) \neq 0$;
(2) u and v have the same time orientation if and only if $g(u, v) < 0$.

Proof

(1) Using a pseudo-orthonormal basis, the Schwarz inequality produces:

$$\left(\sum_{a=1}^{3} u^a v^a \right)^2 \leq \sum_{a=1}^{3} (u^a)^2 \sum_{b=1}^{3} (v^a)^2 < (u^0)^2 (v^0)^2,$$

where the last inequality is a consequence of the fact that the vectors are timelike so

$$(u^0)^2 > \sum_{a=1}^{3} (u^a)^2 \quad \text{and} \quad (v^0)^2 > \sum_{a=1}^{3} (v^a)^2.$$

Since:

$$\left| \sum_{a=1}^{3} u^a v^a \right| < |u^0 v^0| , \tag{10.5}$$

it follows that:

$$g(u, v) = -u^0 v^0 + \sum_{a=1}^{3} u^a v^a \neq 0 . \tag{10.6}$$

(2) In a pseudo-orthonormal basis, the decomposition of $\mathcal{V}_p$ implies that u and v belong to the same half cone if and only if u^0 and v^0 have equal sign, i.e. $u^0 v^0 > 0$. Inequality (10.5) and (10.6) prove the claim.

The spacetime $\mathbb{M}^4$ is assumed to have a **time orientation**, i.e. a *continuous* choice of one half cone in any $\mathcal{V}_p$, indicated by $\mathcal{V}_p^+$. We say *continuous* in the sense that it passes through a C^0 vector field X on $\mathbb{M}^4$ (C^0 for the natural C^∞ structure of the affine space $\mathbb{M}^4$) that is everywhere timelike. $\mathcal{V}_p^+$ is *defined* to be the half cone containing X_p for any $p \in \mathbb{M}^4$. □

Definition 10.12 If $\mathbb{M}^4$ is time-oriented and $p \in \mathbb{M}^4$,

(1) $\mathcal{V}_p^+$ is called **future light cone** at p, and the other component $\mathcal{V}_p^-$ is the **past light cone** at p.
(2) Causal vectors in $\overline{\mathcal{V}_p^+}$ ($\overline{\mathcal{V}_p^-}$) are **future-directed** (resp. **past-directed**) or **future-like** (resp. **past-like**).
(3) A causal (time, light) curve $\gamma : I \to \mathbb{M}^4$ of class C^1 is **of future causal type (time, light)** or **future-directed** if its tangent vector $\dot{\gamma}(s)$ is in $\overline{\mathcal{V}_p^+}$ for any $s \in I$. There are similar names in $\overline{\mathcal{V}_p^-}$ for curves of **past causal type (time, light)** or **past-directed**. ◇

Proposition 10.13 *There exist only two time orientations on $\mathbb{M}^4$.*

Proof There is certainly one time orientation given by the vector field $X(p) = (p, \mathbf{e}_0)$, where $\mathbf{e}_0$ is a timelike vector in V^4 with $g(\mathbf{e}_0, \mathbf{e}_0) = -1$. Such vector field is C^0 since its components are constant in any Cartesian coordinates on $\mathbb{M}^4$. If X and X' are continuous timelike vector fields on $\mathbb{M}^4$, the function $\mathbb{M}^4 \ni p \mapsto g(X(p), X'(p))$ is continuous and never zero by Proposition 10.11, so it cannot change sign because $\mathbb{M}^4$ is connected. By Proposition 10.11 $X(p)$ and $X'(p)$ are either in the same half cone for any $p \in \mathbb{M}^4$, or in opposite halves for any $p \in \mathbb{M}^4$. Hence there are only two ways to give $\mathbb{M}^4$ a time orientation. □

In the sequel we shall always assume $\mathbb{M}^4$ is time-oriented.

10.2.2 Physical Correspondences: Proper Time, Four-Velocity and Causality

Exactly as in Classical Physics, the **histories** or **world lines** of point particles are described by curves in $\mathbb{M}^4$ at least of class C^1. These are now chosen to be *future-directed*. Timelike histories describe the evolution of particles travelling at the speed of light (in any reference frame, as we shall shortly see), which we will always denote by c.

Definition 10.14 If $\gamma : I \ni s \mapsto \gamma(s) \in \mathbb{M}^4$ is a C^1 future-directed timelike curve, where s is some parameter and $s_0 \in I$, the $1/c$-rescaled arc length:

$$\tau(s) = \frac{1}{c} \int_{s_0}^{s} \sqrt{-g(\dot{\gamma}(\xi), \dot{\gamma}(\xi))} d\xi , \quad s \in I \tag{10.7}$$

is called **proper time**.   ◇

The function $\tau = \tau(s)$ is C^1 and satisfies $\frac{d\tau}{ds} = c^{-1}\sqrt{-g(\dot{\gamma}(s), \dot{\gamma}(s))} > 0$ by definition, so its inverse is C^1 as well. Therefore we may always reparametrise any C^1 future-directed timelike curve by its proper time; this leads to the following definition.

Definition 10.15 If a C^1 future timelike curve γ is (re-)parametrised by proper time τ, the tangent vector is written $V(\tau) = \dot{\gamma}(\tau)$ and called **four-velocity** of γ.   ◇

The length:

$$\ell(\gamma) := \int_{s_0}^{s_1} \sqrt{-g(\dot{\gamma}(s), \dot{\gamma}(s))} ds ,$$

of a C^1 timelike curve $\gamma : [s_0, s_1] \ni s \mapsto \gamma(s) \in \mathbb{M}^4$ is *invariant under reparametrisations* if the new parameter s' is related to the old one by a C^1 map with $\frac{ds'}{ds} > 0$ for $s \in [s_0, s_1]$. The proof is immediate. This allows us to prove an elementary but important fact.

Proposition 10.16 *Consider a C^1 future timelike curve γ. Then*

$$g(V(\tau), V(\tau)) = -c^2 , \tag{10.8}$$

and $V(\tau)$ is future timelike.

Proof Reparametrise γ by proper time. As arc length (i.e. proper time) is invariant under reparametrisations, using τ in place of s in (10.7) we find

$$1 = \frac{d\tau}{d\tau} = c^{-1}\sqrt{-g(\dot{\gamma}(\tau), \dot{\gamma}(\tau))} = c^{-1}\sqrt{-g(V(\tau), V(\tau))} .$$

Hence V satisfies (10.8). Finally, $V(\tau)$ is timelike and future-directed just like $\dot{\gamma}(s)$, because the two differ by a positive factor and hence have the same time orientation by Proposition 10.11.   □

From a physical point of view proper time is supposed to correspond to the time measured by an *ideal clock* (operationally, a clock defined as in Classical Physics) at rest with the particle. This physical request goes by the name of **Clock postulate**. As we can construct two timelike curves γ and γ' of different length joining events p and q, perhaps the most important consequence of the Clock postulate is that ideal clocks do not stay synchronised over time. If two ideal clocks at rest with point particles with world lines γ and γ' mark the same time at the first instant they meet at p, they might mark different times at the next meeting at q. This phenomenon was observed experimentally with great accuracy using atomic clocks (in particular in the uber-famous experiment of *Haefele and Keating*). In summary, in contrast to what happens in Classical Physics, in Special Relativity one can no longer define an absolute time that is marked in the same way by every ideal clock.

Staying within the physical perspective, the interactions are transported by causal curves of future type. Therefore two spacetime regions A and B are said to be **causally connected** if there exists a future causal curve joining an event in A to an event in B or vice versa. If there is no such curve, A and B are **causally separated**. For that reason one introduces the following physically relevant sets related to $A \subset \mathbb{M}^4$. They are the **causal future** of A and the **causal past** of A, respectively:[2]

$$J^+(A) := A \cup \{p \in \mathbb{M}^4 \mid \exists \gamma : [a, b] \to \mathbb{M}^4, \quad C^1,$$

$$\text{future causal and such that} \quad \gamma(a) \in A, \ \gamma(b) = p\},$$

$$J^-(A) := A \cup \{p \in \mathbb{M}^4 \mid \exists \gamma : [a, b] \to \mathbb{M}^4, \quad C^1,$$

$$\text{past causal and such that} \quad \gamma(b) \in A, \ \gamma(a) = p\},$$

where the numbers $a < b$ depend on γ.

All in all, A and B are causally separated if and only if $B \subset \mathbb{M}^4 \setminus (J^+(A) \cap J^-(A))$ (equivalently, $A \subset \mathbb{M}^4 \setminus (J^+(B) \cap J^-(B))$).

Besides the sets $J^\pm(A)$ one defined the **chronological future** of A and its **chronological past**, respectively by:

$$I^+(A) := \{p \in \mathbb{M}^4 \mid \exists \gamma : [a, b] \to \mathbb{M}^4, \quad C^1,$$

$$\text{future causal and such that} \quad \gamma(a) \in A, \ \gamma(b) = p\},$$

---

[2] The curves' differentiability order could be changed without affecting the definition, but we will not concern ourselves with that.

$$I^-(A) := \{p \in \mathbb{M}^4 \mid \exists \gamma : [a, b] \to \mathbb{M}^4, \quad C^1,$$

$$\text{past causal and such that} \quad \gamma(b) \in A, \ \gamma(a) = p\}.$$

These four types of sets are defined in a similar way in General Relativity, and are the stepping stone of the modern theory of *Lorentzian causality* [ONe83, Min19].

Remarks 10.17 In the physical reality, when we consider quantum phenomena, there might be statistical correlations between physical facts occurring in causally separated regions ("EPR correlations"). This is one of the surprises of Quantum Theory. We will not get into such matters (but see [Mor19, Chapter 5]). ∎

10.2.3 Minkowski Coordinates and Minkowski Frames

We can now introduce a class of natural coordinates on Minkowski spacetime which account for both the affine structure and the indefinite inner product.

If we were working in a Euclidean space the corresponding coordinates would be the right-handed orthonormal ones. In the spacetime of Classical Physics V^4, instead, these coordinates are a relativistic version of what we would have called Cartesian coordinates moving with an (inertial) reference frame.

Definition 10.18 Cartesian coordinates on $\mathbb{M}^4$

$$\mathbb{M}^4 \ni p \mapsto (x^0, x^1, x^2, x^3) \in \mathbb{R}^4$$

with origin $O \in \mathbb{M}^4$ and axes $\mathbf{e}_0, \mathbf{e}_1, \mathbf{e}_2, \mathbf{e}_3 \in V^4$ are called **Minkowski coordinates** if:

(i) the basis $\mathbf{e}_0, \mathbf{e}_1, \mathbf{e}_2, \mathbf{e}_3 \in V^4$ is pseudo-orthonormal:

$$g(\mathbf{e}_\mu, \mathbf{e}_\nu) = g_{\mu\nu}, \quad \text{where} \quad [g_{\alpha\beta}]_{\alpha,\beta=0,1,2,3} = \text{diag}(-1, 1, 1, 1) \,; \quad (10.9)$$

(ii) the only timelike basis vector $\mathbf{e}_0$ is future-like. ◇

Physically, a set of Minkowski coordinates $\psi : \mathbb{M}^4 \ni p \mapsto (x^0, x^1, x^2, x^3) \in \mathbb{R}^4$ is thought of as Cartesian coordinates moving with a *Minkowski frame* $\mathscr{R}$, in the following sense.

Definition 10.19 A **Minkowski frame** on $\mathbb{M}^4$ is a future-directed timelike vector $\mathscr{R} \in V^4$ such that $g(\mathscr{R}, \mathscr{R}) = -1$. We say the Minkowski coordinates $\psi : \mathbb{M}^4 \ni p \mapsto (x^0, x^1, x^2, x^3) \in \mathbb{R}^4$ **move with** $\mathscr{R}$ if they are built from an origin $O \in \mathbb{M}^4$ and a pseudo-orthonormal basis $\mathbf{e}_0, \mathbf{e}_1, \mathbf{e}_2, \mathbf{e}_3$ such that $\mathbf{e}_0 = \mathscr{R}$. ◇

Remarks 10.20

(1) The above definition is completely geometrical. For a number of reasons one would expect Minkowski frames to be the inertial ones. This is true, as we will show later. In the physical formulation of Special Relativity the construction of the formalism is turned upside down, and one supposes that in inertial frames (defined by the same property of the motion of isolated bodies, as in Classical Mechanics) certain kinematical facts regarding the invariance of the speed of light occur. From the latter one constructs the indefinite inner product g, and the Minkowski frames are defined to be the inertial frames. Then they are automatically equipped with coordinates, which we called Minkowski coordinates. Under our axiomatic-geometric procedure, that goes the other way around, the Principle of inertia will be a consequence of the formalism of Sect. 10.3.3.

(2) Any set of Minkowski coordinates move with a unique Minkowski frame $\mathscr{R}$. The latter is entirely determined by $\mathscr{R} = \mathbf{e}_0$. This vector also coincides with the tangent vector to the x^0-coordinate curve passing through any point of $\mathbb{M}^4$.

(3) Two sets of Minkowski coordinates therefore move with the same Minkowski frame $\mathscr{R}$ if and only if they define the same vector $\mathbf{e}_0 = \mathscr{R}$.

(4) Given $\mathscr{R}$, the only arbitrary choices in defining Minkowski coordinates are the origin $O \in \mathbb{M}^4$ and the other basis vectors $\mathbf{e}_1, \mathbf{e}_2, \mathbf{e}_3$, which at any rate must obey the above pseudo-orthonormality with respect to $\mathbf{e}_0$.

∎

10.2.4 Physical and Kinematic Properties of Minkowski Coordinates and Minkowski Frames

Given Minkowski coordinates $\psi : \mathbb{M}^4 \ni p \mapsto (x^0, x^1, x^2, x^3) \in \mathbb{R}^4$ on $\mathscr{R}$, we have a number of mathematical concepts of great physical importance.

(a) $t = x^0/c$ corresponds to a **global time coordinate** of $\mathscr{R}$ that determines when an event occurs with respect to the frame. All other Minkowski coordinates on $\mathscr{R}$ simply change that global coordinate by an additive constant;

(b) (x^1, x^2, x^3) describe orthonormal Cartesian coordinates in the **rest spaces** of $\mathscr{R}$

$$E_{\mathscr{R}, t_0} := \{ p \in \mathbb{M}^4 \mid x^0(p) = ct_0 \} .$$

They are three-planes in $\mathbb{M}^4$. Any other Minkowski coordinates on $\mathscr{R}$ transform the above three-dimensional Cartesian coordinates by an arbitrary three-dimensional roto-translation (not depending on x^0).

(c) (i) On each three-dimensional rest space $E_{\mathcal{R},x^0}$ of $\mathcal{R}$ the indefinite inner product g induces a standard inner product, *i.e. positive definite*:

$$\mathbf{v} \cdot \mathbf{u} = g(\mathbf{u}, \mathbf{v}) \quad \text{if } \mathbf{u}, \mathbf{v} \in V^3_{\mathcal{R},x^0}, \text{ the space of displacements of } E_{\mathcal{R},x^0}.$$

(ii) The temporal distance in $\mathcal{R}$ measured by $t = x^0/c$ coincides with the temporal distance measured by proper time along any x^0-curve.

In this way it is clear that the indefinite inner product g simultaneously induces:

(a) the Euclidean metric structure employed to measure objects in the three-dimensional physical space of any frame $\mathcal{R}$, and that corresponds to the frame's ideal rigid rulers;

(b) the metric structure on the time axis of $\mathcal{R}$ described by the frame's ideal clocks.

The crucial difference with Classical Physics is that now there no longer is a three-dimensional absolute space nor a time axis common to all reference frames.

In the light of the above physical meaning of Minkowski frames we can pass to the physical interpretation of the properties of world lines in Minkowski coordinates.

We begin with an elementary technical lemma, of fundamental physical importance because it shows we can always adopt global time to parametrise the histories of particles.

Lemma 10.21 *Let $\psi : \mathbb{M}^4 \ni p \mapsto (x^0, x^1, x^2, x^3) \in \mathbb{R}^4$ be arbitrary Minkowski coordinates on $\mathbb{M}^4$. Any C^1 future causal curve $\gamma : I \to \mathbb{M}^4$ can be reparametrised, preserving the time orientation, using the temporal coordinate $t = x^0/c$.*

Proof As $g(\dot{\gamma}(s), \dot{\gamma}(s)) = -\left(\frac{dx^0}{ds}\right)^2 + \sum_{a=1}^{3}\left(\frac{dx^a}{ds}\right)^2 \leq 0$ and $\dot{\gamma}(s) \neq 0$ by assumption, then $\frac{dx^0}{ds} \neq 0$. More precisely $\frac{dx^0}{ds} > 0$, since the curve is of future type and $\dot{\gamma}(s)$ must belong in the same half cone as $\mathbf{e}_0$. $\square$

Definition 10.22 Let $\psi : \mathbb{M}^4 \ni p \mapsto (x^0, x^1, x^2, x^3) \in \mathbb{R}^4$ be Minkowski coordinates on the Minkowski frame $\mathcal{R}$ and γ a C^1 future causal curve reparametrised by $t = x^0/c$ as $I \ni t \mapsto (ct, x^1(t), x^2(t), x^3(t))$. The **velocity** of γ in $\mathcal{R}$ at time t is

$$\mathbf{v}_\gamma|_{\mathcal{R}}(t) := \sum_{a=1}^{3} v^a(t)\mathbf{e}_a \in E_{\mathcal{R},t}, \tag{10.10}$$

where $v^a(t) := \frac{dx^a}{dt}(t)$. $\diamond$

Remarks 10.23 The notion of velocity of a point particle introduced here depends on the frame and is defined in its rest space. *The notion of four-velocity, on the contrary, is absolute and does not depend on any reference frame.*   ∎

Suppose now the curve γ is *future-like* and parametrised by the time t of Minkowski coordinates as $I \ni t \mapsto (ct, x^1(t), x^2(t), x^3(t))$. Since γ is future timelike, we can find the four-velocity $V(\tau) := \dot{\gamma}(\tau)$:

$$V = \sum_{\alpha=0}^{3} V^\alpha \mathbf{e}_\alpha = \sum_{\alpha=0}^{3} \frac{dx^\alpha(t(\tau))}{d\tau} \mathbf{e}_\alpha , \tag{10.11}$$

where τ is the proper time of (10.7) with t replacing s. The following result's proof is elementary and left as exercise.

Proposition 10.24 *If $I \ni \tau \mapsto (ct, x^1(t), x^2(t), x^3(t))$ is the representation in Minkowski coordinates on $\mathscr{R}$ of a C^1 future-directed timelike curve, the components of the four-velocity (10.11) take the following form in terms of the velocity in $\mathscr{R}$:*

$$V^0(t) = \frac{c}{\sqrt{1 - \frac{v(t)^2}{c^2}}} , \quad V^a(t) = \frac{v^a(t)}{\sqrt{1 - \frac{v(t)^2}{c^2}}} , \quad a = 1, 2, 3 . \tag{10.12}$$

The inverse relations are:

$$v^a(t) = c \frac{V^a(t)}{V^0(t)} , \quad v(t) = c \sqrt{1 - \frac{c^2}{(V^0(t))^2}} . \tag{10.13}$$

Above we have used the notation of Definition 10.22 and set

$$v(t) := ||\mathbf{v}_\gamma|_{\mathscr{R}}(t)|| = \sqrt{\sum_{a=1}^{3} (v^a(t))^2} ,$$

the norm referring to the Euclidean inner product on $E_{\mathscr{R},t}$.

Remarks 10.25

(1) From expression (10.12) for V^0 and V^a we easily see that the condition that $V(t)$ is timelike:

$$-(V^0)^2 + \sum_{a=1}^{3} (V^a)^2 < 0$$

is equivalent to $v(t) < c$. Thus the formalism encompasses the physical demand that *the highest-possible velocity for point particles equals c in every Minkowski frame (later we will see the latter are the inertial ones)*.

(2) Let $I \ni s \mapsto (ct(s), x^1(s), x^2(s), x^3(s))$ represent the history γ in Minkowski coordinates on $\mathscr{R}$. Being lightlike, $g(\dot{\gamma}(s), \dot{\gamma}(s)) = 0$ (recall $\dot{\gamma}(s) \neq 0$) is trivially equivalent to:

$$c^2 = \frac{\sum_{a=1}^{3}\left(\frac{dx^a}{ds}\right)^2}{\left(\frac{dt}{ds}\right)^2} = \sum_{a=1}^{3}\left(\frac{dx^a}{dt}\right)^2 = ||\mathbf{v}_\gamma|_{\mathscr{R}}(t)||^2, \tag{10.14}$$

where we have reparametrised the curve using t. It then become clear that lightlike curves describe histories of particles moving at the speed of light c in *every* Minkowski frame. Furthermore, if the particle's velocity in a Minkowski frame equals c, then its history is lightlike, and so the velocity will equal c in any other Minkowski frame. In this sense the geometric structure we have introduced incorporates the **Postulate of invariance of the speed of light**.

∎

Examples 10.26 Consider a particle created at the event $o \in \mathbb{M}^4$ that evolves with constant velocity $v > 0$ in the Minkowski frame $\mathscr{R}$ along $\mathbf{e}_1$ (a spacelike unit vector perpendicular to $\mathscr{R}$) for an interval of proper time $\Delta\tau$, and then decays. With respect to the global time of $\mathscr{R}$, how much time Δt has passed from the particle's birth at o to its death at o'?

The calculation is easy. Decompose $o' - o = \Delta t \mathbf{e}_0 + \Delta x \mathbf{e}_1$ along the pseudo-orthonormal basis on $\mathscr{R}$, with Minkowski coordinates, obtained completing $\mathbf{e}_0, \mathbf{e}_1$ with some vectors $\mathbf{e}_2, \mathbf{e}_3$. We know that

$$\Delta x = v\Delta t.$$

Hence

$$-c^2\Delta\tau^2 = g(o - o', o - o') = -c^2\Delta t^2 + (v\Delta t)^2,$$

from which we obtain the famous **time dilation** formula:

$$\Delta t = \frac{\Delta\tau}{\sqrt{1 - \frac{v^2}{c^2}}}.$$

Even if the particle has a rather short lifespan $\Delta\tau$ in its reference frame, it could still travel much farther than what is permitted in Classical Physics. In fact $\Delta t > \Delta\tau$, and the more v approaches the limit value c, the larger the time interval becomes. This phenomenon is well known to scholars studying cosmic rays: *muons*, particles created in the atmosphere's outer layers, beyond 15 *km* above the earth's surface, could never be detected by surface-bound instruments if Classical Physics were at play, by virtue of their short lifespan when measured at rest. Keeping their velocity into account, their average lifetime would allow them to penetrate the atmosphere by 500 m only. Yet muons are detected on the surface, because of the time dilation described above.

10.3   Introduction to Relativistic Dynamics

We set out to briefly describe the simplest structure of the dynamics of a point particle in Special Relativity. It is a four-dimensional and intrinsic version (no need to choose frames) of Newtonian Dynamics.

10.3.1   Mass, Four-Momentum and Their Elementary Properties

In Special Relativity, when introducing Dynamics, one defines the *mass m* of a particle as in Classical Physics. The notion of impulse generalises to the *four-momentum*. We have the following principle, whose name is explained by the ensuing crucial definition.

Principle of Conservation of the Four-Momentum *Consider a process in which N histories of point particles, described by line segments, collide in a unique event, and this event initiates M histories of point particles described by other line segments. There exist numbers $m_1, \ldots, m_N > 0, m'_1, \ldots, m'_N > 0$, called **masses**, one for each point particle and constant along the corresponding world line, for which the **Principle of conservation of the four-momentum** holds:*

$$\sum_{i=1,\ldots,N} m_i V_i = \sum_{j=1,\ldots,M} m'_j V'_j ,$$

where the left-hand side is evaluated before the collision and the right-hand side after.

Remarks 10.27 Physically, the value of $m > 0$ can be thought of as *classical*, i.e. measured in an inertial frame where the particle is seen as at rest, at the considered instant, if such a frame exists. The idea behind the operational definition is that nearby a time instant where the particle has zero velocity in an inertial frame, we may assume Classical Physics to hold and therefore the mass can be defined (and measured) by classical methods.   ∎

Definition 10.28 If $V(\tau)$ is the four-velocity of a point particle with future timelike world line $\gamma = \gamma(\tau), \tau \in I$ (not necessarily a line segment) and $m > 0$ is the particle's mass, the vector

$$P(\tau) := m V(\tau) \qquad (10.15)$$

tangent to γ is the particle's **four-momentum** at proper time τ.   ◇

As immediate consequence of the definition we have:

$$g(P, P) = -m^2 c^2 . \tag{10.16}$$

To complement what we said let us say that Special Relativity also takes into account particles of mass $m = 0$, which do not exist in Classical Physics. If the mass of a particle satisfies $m = 0$ and the history is generically future causal, one assumes the four-momentum is still defined and corresponds to a tangent vector $P(s)$ in some privileged future-directed parametrisation s of its history. One further assumes the four-momentum satisfies: (a) the obvious extension of the aforementioned conservation principle, and (b) formula (10.16).

Remarks 10.29

(1) As P is always tangent to the particle's history, from the general validity of (10.16) we know that for a certain value of the parameter:

   (i) particles with zero mass are precisely those with future lightlike histories, i.e. particles whose velocity equals c in any Minkowski frame.
   (ii) particles with positive mass are precisely those with future timelike histories.

(2) As straightforward consequence of (10.16) we have the useful formula:

$$P^0 = \sqrt{\sum_{\alpha=1}^{3} (P^\alpha)^2 + m^2 c^2} \tag{10.17}$$

   where we kept into account that $P^0 > 0$ since P is future-directed.

(3) In a Minkowski frame $\mathscr{R}$ where the *spatial* components of P are zero[3] (at some instant) the time component of P carries the information regarding the mass $P^0 = mc$ due to (10.17). By construction a similar Minkowski frame is **at rest** (at the instant of proper time considered) with the particle, because the spatial components of the four-velocity will be zero as well, and so the velocity.

(4) There exists a Minkowski frame at rest with a particle if and only if the particle has strictly positive mass: if the mass is zero also P^0 vanishes and P would not be causal. If, on the other hand, P is timelike (i.e. $m > 0$), by choosing $\mathscr{R}$ parallel to P the spatial components of P are automatically zero in $\mathscr{R}$.

■

---

[3] This condition is independent of the spatial orthonormal vectors $\mathbf{e}_1, \mathbf{e}_2, \mathbf{e}_3$ in which the components are taken and that complete $\mathscr{R} = \mathbf{e}_0$ to a pseudo-orthonormal basis.

10.3.2   The So-Called Mass-Energy Equivalence Principle

The conservation law of the four-momentum, as opposed to the conservation law of the impulse in Classical Physics, does not require an (inertial) frame in order to be stated: it is an intrinsic law of geometrical nature. It is however important to understand what it implies in case we do choose a Minkowski frame $\mathscr{R}$ (the generalisation of a classical inertial frame, as we will see shortly) and write the principle in its coordinates.

The conservation of the four-momentum's spatial components in $\mathscr{R}$

$$\sum_{i=1,\dots,N} \frac{m_i v_i^a}{\sqrt{1 - \frac{v_i^2}{c^2}}} = \sum_{j=1,\dots,M} \frac{m_j' v_j'^a}{\sqrt{1 - \frac{v_j'^2}{c^2}}} \quad a = 1,2,3$$

is a direct generalisation of the conservation law of the classical impulse in the rest spaces of $\mathscr{R}$. When $v, v' \ll c$, approximating

$$\sqrt{1 - \frac{v^2}{c^2}} \approx 1, \quad \sqrt{1 - \frac{v'^2}{c^2}} \approx 1$$

recovers the classical conservation law.

The conservation of the time component P^0 is instead much more interesting to examine. Consider $N = 1$, when a unique particle decays into $M = 2$ particles. The four-momentum's conservation assumes then an interesting form. Let us suppose the initial particle moves (before decaying) with constant velocity in a Minkowski frame $\mathscr{R}$. We change to a frame $\mathscr{R}_0$ in which the velocity is zero. In the latter reference frame the four-momentum P of the initial particle, say of mass m, only has time component, equal to mc. The conservation law of the four-momentum in space and time components in $\mathscr{R}_0$ provides (we shall not write v' but simply v):

$$mc = \frac{m_1 c}{\sqrt{1 - \frac{v_1^2}{c^2}}} + \frac{m_2 c}{\sqrt{1 - \frac{v_2^2}{c^2}}}, \quad 0 = \frac{m_1 v_1^a}{\sqrt{1 - \frac{v_1^2}{c^2}}} + \frac{m_2 v_2^a}{\sqrt{1 - \frac{v_2^2}{c^2}}} = 0, \quad a = 1,2,3.$$

The second relation, for v small compared to c, recovers the classical conservation of the impulse, as already said. The first, dividing by c, says:

$$m = \frac{m_1}{\sqrt{1 - \frac{v_1^2}{c^2}}} + \frac{m_2}{\sqrt{1 - \frac{v_2^2}{c^2}}}$$

$$= m_1 + m_2 + \mathscr{T}_1|_{\mathscr{R}_0}/c^2 + \mathscr{T}_2|_{\mathscr{R}_0}/c^2 + O(v_1^4/c^4) + O(v_2^2/c^4),$$

where we used the expansion:

$$\frac{m_i}{\sqrt{1 - \frac{v_i^2}{c^2}}} = \frac{1}{2}m_i v_i^2/c^2 + O(v_i^4/c^4) \,.$$

The first conclusion is that the mass *does not* satisfy a conservation law. It is conserved only up to contributions proportional to the classical kinetic energy (in the frame we are using) by the factor $1/c^2$, and all this up to corrections $O(v^4/c^4)$. This is one of the starting points to arrive at the **Mass-energy equivalence principle**, by which we assume that:

(i) the mass of an isolated system is not conserved, but in the balance equations it can gain (or shed) contributions from various forms of energy under the relation $\Delta E = \Delta mc^2$;

(ii) conversely, the total energy of an isolated system is always conserved, even if part of the energy appears in the form of mass under the above relationship.

The conservation law of the four-momentum expresses the conservation of the total energy when we account for its time component.

With (i) and (ii) in place we can say the following. In Special Relativity a particle four-momentum's time component multiplied by c

$$cP^0 = \frac{mc^2}{\sqrt{1 - \frac{v^2}{c^2}}}$$

represents the particle's *total mechanical energy content* in the Minkowski frame $\mathscr{R}$ with respect to which P is decomposed in Minkowski components (in particular, P^0 is a feature of $\mathscr{R}$ and not of the chosen Minkowski coordinates on $\mathscr{R}$, as $P^0 = -g(P, \mathscr{R})$). More precisely

$$cP^0 = mc^2 + T|_{\mathscr{R}} \quad \text{where} \quad T|_{\mathscr{R}} := \frac{mc^2}{\sqrt{1 - \frac{v^2}{c^2}}} - mc^2 \,, \tag{10.18}$$

is, by definition, the **relativistic kinetic energy**, while mc^2 is the energy contribution due to the pure mass: in contrast to the kinetic energy part, it does not depend on the frame of reference. As we have seen earlier, the relativistic kinetic energy reduces to the classical kinetic energy:

$$T|_{\mathscr{R}} = \frac{1}{2}mv^2 + O(v^4/c^4) \,.$$

in the approximation where speeds are small compared to the speed of light.

Remarks 10.30 The name "Mass-energy equivalence principle" is not entirely appropriate and should be understood as a relic of history, given that mass and energy *are definitely not equivalent precisely by virtue of the principle itself!* The mass is not conserved, whereas the energy is. To any quantity of mass (measured at rest with a physical system) there corresponds a quantity of energy that adds to the other forms of energy the systems possesses. But a quantity of energy does not automatically have a corresponding quantity of mass associated with it, except for rather specific situations (see for instance the second example below).   ∎

Examples 10.31

(1) The *photon* is an example of a massless relativistic particle with an energy, whose value depends on the reference frame. Truth be told, the photon is a quantum concept and here we will discuss a semiclassical version in which the *angular frequency* is assumed to have a precise value. We shall not address the problem of defining operationally the photon's history in spacetime, and we will only refer to the four-momentum.[4] From a semiclassical (not quantum) point of view, in any Minkowski coordinates:

$$P^0 = \hbar \frac{\omega}{c}, \quad P^a = \hbar k^a, \quad a = 1, 2, 3,$$

where $\omega > 0$ is the *angular frequency* of the particle's light wave and k^a are the three components of the *wave vector* in the frame. Above $\hbar = \frac{h}{2\pi}$, where h is *Planck's constant* (6.626×10^{-27}erg sec). As

$$\sum_{a=1}^{3} (k^a)^2 = \left(\frac{\omega}{c}\right)^2$$

from electromagnetism, in this case we must have $g(P, P) = 0$ and so the photon is massless and its world lines are lightlike. Clearly there is no Minkowski frame at rest with a photon, because imposing $k^a = 0$ for $a = 1, 2, 3$ would force $\omega = 0$ and the four-momentum would not be causal. Sometimes one reads that photons are "pure energy". Utterances of this sort are pure nonsense. Energy is a *property* of the physical systems, whose value depends on the frame of reference as well. Photons are physical systems with a number of properties, one of which is energy, equal to $cP^0 = \hbar \omega$ if the photon has angular frequency ω in the Minkowski frame we are working in. On the other hand a photon (in the classical limit, as a quantum particle) does not only have a four-momentum but also another measurable feature called *polarisation*, which at present will not concern us.

---

[4] A deeper analysis would discuss *paraxial* "wave trains" of photons whose wave vector and angular frequency are only approximate; one can still find a small region in space where the packet is defined and generates a spacetime history, again in an approximate way.

(2) Consider a physical system of N point particles of masses $m_i > 0$, with $i = 1, \ldots, N$, that interact only among themselves. Suppose there is a Minkowski frame $\mathscr{R}_0$, i.e. inertial as we will shortly see, where all velocities are small compared to c and where Classical Physics is expected to hold with best-possible approximation. Finally, assume that in $\mathscr{R}_0$ the interaction forces are conservative and described by a potential energy $\mathscr{U}$ depending on the distances $||\mathbf{x}_i - \mathbf{x}_j||$, $i \neq j$, where $\mathbf{x}_i$ is the spatial position vector of the i-th particle in the rest space of $\mathscr{R}_0$. As Classical Physics holds, there is an inertial frame at rest with the system's centre of mass where the total momentum is zero. In this frame the velocities are small relatively to the speed of light. Henceforth we assume $\mathscr{R}_0$ is such a reference frame at rest with the (classical) centre of mass. We also assume the system is bound, namely that the orbits cluster around the centre of mass so that we can view the latter as a point particle. The centre of mass will evolve and define a curve γ in spacetime. The classical total mechanical energy in $\mathscr{R}_0$, conserved in time, is:

$$cP_0^0 = \sum_{i=1}^{N} m_i c^2 + \sum_{i=1}^{N} T_i|_{\mathscr{R}_0} + \mathscr{U}|_{\mathscr{R}_0} \,.$$

We added the additive constant $\sum_{i=1}^{N} m_i c^2$ which, although classically irrelevant, matches the Mass-energy equivalence principle. Note that we could add any other constant, but our choice agrees with the idea that when the particles' mutual distances increase, $\mathscr{U}|_{\mathscr{R}_0}$ turns off and the system's energy equals the sum of the kinetic energy plus the term due to the masses. Under these hypotheses on the potential energy, and since we are supposing the N orbits are bound to remain "close" to the centre of mass, the typical situation is that

$$\mathscr{U}|_{\mathscr{R}_0}(||\mathbf{x}_1(t) - \mathbf{x}_2(t)||, ||\mathbf{x}_1(t) - \mathbf{x}_3(t)||, \ldots) < 0$$

during the entire evolution of the N particles, even though the proof of this well-known fact (think of the solar system) would require a detailed discussion. In the frame $\mathscr{R}_0$, by assumption

$$P_0^a = 0 \,, \quad a = 1, 2, 3$$

because the momenta are classical (we neglect the terms v_i^2/c^2) and we are in the centre of mass's frame of reference. We can define a four-vector P of components P_0^0 and P_0^a in $\mathscr{R}_0$. By construction P is timelike and (approximately) tangent to γ. If we change Minkowski frame the four-momentum P will acquire spatial components and the system will be described as a single particle of mass

$$M = \sqrt{-\frac{1}{c^2}g(P, P)} = \frac{1}{c^2}P_0^0 = \sum_{i=1}^{N} m_i + \sum_{i=1}^{N} \frac{T_i|_{\mathscr{R}_0}}{c^2} + \frac{1}{c^2}\mathscr{U}|_{\mathscr{R}_0}$$

$$< \sum_{i=1}^{N} m_i + \sum_{i=1}^{N} \frac{T_i|_{\mathscr{R}_0}}{c^2},$$

since the potential energy is assumed negative. In concrete cases, say molecules, the terms $\frac{T_i|_{\mathscr{R}_0}}{c^2}$ are extremely small compared to m_i, so that

$$M < \sum_{i=1}^{N} m_i .$$

This inequality is observed experimentally in bound physical systems such as molecules, atoms and nuclei. The *mass defect* $\Delta M := M - \sum_{i=1}^{N} m_i > 0$ corresponds to the energy ΔMc^2 the system needs in order to break the bounds tying up the masses in the system. Conversely, when a bound system is created it "releases" the energy ΔMc^2. This is how hydrogen bombs work, or thermonuclear fusion reactors, or even stars. For example, the fusion of two protons and two neutrons produces a Helium nucleus whose total mass is less than the sum of the four masses. In fact, simultaneous to the creation of the Helium nucleus we witness particle emission (typically gamma photons) whose total energy is exactly ΔMc^2. These kinds of processes, however, are based on Quantum Mechanics and the *strong interaction*, which this book does not cover. At any rate, the principle of equivalence between mass and energy is upheld in these contexts too.

10.3.3 Relativistic Equation of Motion and Identification Between Minkowski and Inertial Frames

Apart from the pointwise interactions, those localised at single events, in the conservation principle of the four-momentum, the interactions of point particles are described in Special Relativity by the concept of *four-force*. We shall deal with the case of a single particle of mass $m > 0$, and only at the end we will briefly discuss multi-particle systems with strictly positive masses.

In the sequel we will use the open subset in $T\mathbb{M}^4$:

$$\mathscr{V}^+\mathbb{M}^4 = \{(p, v) \in T\mathbb{M}^4 \mid (p, v) \in \mathscr{V}_p^+\}. \tag{10.19}$$

We will define four-forces only on this open set since we are interested in future-directed timelike curves. Here is the definition.

Definition 10.32 A **four-force** is a C^k function, $k \geq 1$,

$$F : \mathcal{V}^+\mathbb{M}^4 \ni (p, v) \to F(p, v) \in T\mathbb{M}^4 ,$$

where $F(p, v) \in T_p\mathbb{M}^4$ for any $(p, u) \in \mathcal{V}^+\mathbb{M}^4$, satisfying

$$g(F(p, v), v) = 0 \quad \text{if } (p, v) \in \Omega , \qquad (10.20)$$

where $\Omega \subset \mathcal{V}^+\mathbb{M}^4$ is an open neighbourhood of the set of vectors $(p, v) \in \mathcal{V}^+\mathbb{M}^4$ with $g(v, v) = -c^2$. ∎

The justification of the definition is contained in the dynamical meaning of four-forces.

In the particle's relativistic dynamics world lines $I \ni \tau \mapsto \gamma(\tau) \in \mathbb{M}^4$, for a point particle of mass $m > 0$ and parametrised by proper time τ, are supposed to satisfy the **relativistic equation of motion**:

$$\frac{dP}{d\tau} = F(\gamma(\tau), V(\tau)) , \qquad (10.21)$$

where the four-force on the right is a known function representing all particle's interactions.

Consider now the constraint (10.16) for the four-momentum $P = mV$, written as:

$$g(V, P) = -mc^2 .$$

If we differentiate with respect to τ *assuming m is constant* and use (10.21), then condition (10.20) holds for $p = \gamma(\tau)$ and $v = V(\tau)$. As we expect that through each pair (p, u) there is a solution if u is future timelike and $g(u, u) = -c^2$, that fact justifies (10.20) at least for vectors u such that $g(u, u) = -c^2$. If $F(p, u)$ is linear in u, the explanation is complete, because even if $g(u, u) = -c^2$ did not hold we could rescale u to reduce to the present case. Imposing (10.20) on an arbitrarily small open set around points $(p, v) \in \mathcal{V}^+\mathbb{M}^4$ where $g(u, u) = -c^2$ is at any rate useful, because (10.20), written as above, is employed in the Existence and uniqueness theorem proved below.

Remarks 10.33 The justification of (10.20) crucially depends on $m > 0$ being constant during the whole evolution of the particle. The orthogonality of F and V:

$$g(F, V) = 0 ,$$

is equivalent to saying $m > 0$ does not change in time. There are more sophisticated formulations of the relativistic dynamics of a particle in which m varies. This possibility is already included in the structure of equation (10.21) when we *also* allow m to vary. In fact, if m varies, there are four scalar quantities to be determined (not only the three independent components of the four-velocity) and we have four equations in (10.21).   ∎

Mathematically, the problem of motion based on the definition of four-force and on the fact equation (10.21) is well posed. The following theorem explains why.

Theorem 10.34 *Consider a point particle of constant mass $m > 0$ whose history $\gamma = \gamma(s) \in \mathbb{M}^4$, seen as future-directed timelike and at least C^1, satisfies:*

$$m\frac{d\dot\gamma}{ds} = F(\gamma(s), \dot\gamma(s)),$$

where $F : \mathscr{V}^+\mathbb{M}^4 \to T\mathbb{M}^4$ is a four-force of class C^1. For any choice of initial conditions

$$\gamma(s_0) = p, \ \dot\gamma(s_0) = u \quad \text{with } g(u, u) = -c^2 \text{ and } u \text{ of future type,}$$

there exists exactly one maximal solution: $I \ni s \mapsto \gamma(s) \in \mathbb{M}^4$ for some open interval $I \ni s_0$. (In particular any other solution with the same initial conditions is a restriction of the former.) Such curve is C^3 and future-like, the parameter s is the proper time τ of the world line, and $\dot\gamma(s)$ coincides with the four-velocity $V(\tau)$.

Proof Fix Minkowski coordinates to work in. We then have the Cauchy problem in normal form:

$$\frac{d^2 x^\mu}{ds^2} = \frac{1}{m}F^\mu\left(x^0(s), x^1(s), x^2(s), x^3(s), \frac{dx^0}{ds}, \frac{dx^1}{ds}, \frac{dx^2}{ds}, \frac{dx^3}{ds}\right),$$

$$\mu = 0, 1, 2, 3$$

with initial data:

$$x^\mu(s_0) = x_0^\mu, \quad \frac{d^3 x^\mu}{ds} = u^\mu, \quad \mu = 0, 1, 2, 3,$$

where x_0^μ is the μ-coordinate of p. Under the given regularity hypotheses on the four-force's components we have the Existence and uniqueness theorem, whereby there is a unique maximal solution $x^\mu = x^\mu(s)$, $\mu = 0, 1, 2, 3$, on an open interval $I \ni s_0$. This solution is C^3 by construction, as is easy to see from the differential equation and the regularity of the F^μ. Let us suppose $g(\dot\gamma(s_0), \dot\gamma(s_0)) = -c^2$ and that $\dot\gamma(s_0)$ is of future type. By continuity there exists an open interval $(s_0-\delta, s_0+\delta')$ where the tangent vector to γ stays future timelike and remains inside Ω. With an eye to (10.20), we then have:

$$\frac{d}{ds}g(\dot\gamma(s), \dot\gamma(s)) = \frac{1}{m}g(F(\gamma(s), \dot\gamma(s)), \dot\gamma(s))) = 0$$

if $s \in (s_0 - \delta, s_0 + \delta')$. Therefore $g(\dot{\gamma}(s), \dot{\gamma}(s)) = g(\dot{\gamma}(s_0), \dot{\gamma}(s_0)) = -c^2$ if $t \in (s_0 - \delta, s_0 + \delta')$. We have proved that if $s \in (s_0 - \delta, s_0 + \delta')$ then $\gamma = \gamma(s)$ is future timelike and s is its proper time. Define $d \leq +\infty$ to be the supremum over $r \in I$, $r > s_0$, for which $\gamma = \gamma(s)$ is future timelike and s is its proper time in $[s_0, r)$. The set of such s is non-empty because $s_0 + \delta$ is one. If $d < \sup I$ we would reach a contradiction: by the above argument, we could show $\gamma = \gamma(s)$ is future timelike and s its proper time in $[s_0, d + \epsilon)$ for some $\epsilon > 0$. Using the same idea for times smaller than s_0, we conclude $\gamma = \gamma(s)$ is future timelike and s its proper time on the entire I. $\square$

Consider the special case of a particle with $m > 0$ that is *isolated*, i.e. with zero four-force and not subject to interactions with other point particles in the proper time interval of concern. The relativistic equation of motion implies P stays constant along the history γ of the particle. More precisely, the history will be a straight segment in the affine space $\mathbb{M}^4$ parametrised by proper time:

$$\gamma(\tau) = \gamma(s_0) + (\tau - \tau_0)V(\tau_0) . \tag{10.22}$$

The same type of inhomogeneous linear equation holds if we parametrise the curve by the global time t of a Minkowski frame:

$$\gamma(\tau(t)) = \gamma(s_0) + (t - t(\tau_0))\sqrt{1 - \left(\frac{v}{c}\right)^2}V(\tau_0) .$$

Above we used (10.13) from Proposition 10.24, from which in particular the *velocity* v of our particle, which has no four-forces acting on it, is constant in any Minkowski frame. In this sense:

(1) Minkowski frames are identified with inertial frames;
(2) world lines of the form (10.22) describe the **inertial motions** of massive particles.

A massless particle evolving with constant four-momentum has similar properties to massive particles. $P(s)$ anyway corresponds to the tangent vector for some parametrisation s. Having $P(s) \neq 0$ constant implies

$$\gamma(s) = \gamma(s_0) + (s - s_0)P(s_0) .$$

Physically, this world line describes the **inertial motion** of a massless particle. Using Minkowski coordinates on any inertial frame, the reasoning behind (10.17) implies the particle's velocity is constantly equal to c (along a constant spatial direction, since its history is a segment).

Note *In the sequel we will treat the terms 'inertial frame' and 'Minkowski frame' as synonyms.*

Remarks 10.35 The notion of four-force and the relativistic equation of motion (10.21) have as consequence a formula generalising the *Kinetic energy theorem* (Theorem 4.11), even though for one particle. Decomposing (10.21) in Minkowski coordinates on $\mathcal{R}$, the time component produces:

$$m\frac{dV^0}{d\tau} = F^0 .$$

On the other hand we know that:

$$0 = g(F, V) = -F^0 V^0 + \sum_{a=1}^{3} F^a V^a .$$

The two together give:

$$\sum_{a=1}^{3} F^a V^a = V^0 \frac{dm V^0}{d\tau} .$$

Using expressions (10.12) for the four-velocity in terms of the velocities in $\mathcal{R}$ where we have the Minkowski coordinates, we find:

$$\frac{d}{d\tau} \frac{mc^2}{\sqrt{1 - \frac{v^2}{c^2}}} = \sum_{a=1}^{3} F^a v^a .$$

The differentiated term can be modified by adding a constant, whose derivative would anyhow vanish. So we can write, correctly:

$$\frac{d}{d\tau} \left(\frac{mc^2}{\sqrt{1 - \frac{v^2}{c^2}}} - mc^2 \right) = \sum_{a=1}^{3} F^a v^a .$$

By the definition of *relativistic kinetic energy* in the inertial frame $\mathcal{R}$ appearing in (10.18), the above equation extends the kinetic energy theorem and is often called **Relativistic kinetic energy theorem**:

$$\frac{d}{d\tau} T|_\mathcal{R} = F|_\mathcal{R} \cdot \mathbf{v}_\gamma|_\mathcal{R} .$$

We have introduced, in the rest space $E_{\mathcal{R},\tau}$ of $\mathcal{R}$, a vector that accounts for the sole spatial components of the four-force:

$$F|_\mathcal{R}(\gamma(\tau), \dot{\gamma}(\tau)) := \sum_{a=1}^{3} F^a(\gamma(\tau), \dot{\gamma}(\tau))\mathbf{e}_a .$$

The above relationship corroborates the idea of defining the relativistic kinetic energy by (10.18). ∎

To conclude we will comment on the case of N interacting point particles, when the interactions are not particle-particle and are restricted to single events, as in the situation of the Postulate of conservation of the four-momentum. In principle we would expect the dynamics to be described by a set of equations of the type:

$$\frac{dP_i}{d\tau_i} = F_i(\gamma(\tau_1), V_1(\tau_1), \ldots, \gamma(\tau_N), V_N(\tau_N)), \quad i = 1, \ldots, N. \tag{10.23}$$

where $P_i = m_i V_i$ and the masses $m_1, \ldots, m_N$ may or not be constant (in agreement with Remark 10.33).

It is immediate to see that the above system of equations is problematic. It does not have the usual normal form: each particle appears with its own proper time, since there is no way to identify proper times on distinct world lines.

An alternative (in most cases a viable one) is not to use proper time to parametrise world lines, but rather opt for the common time coordinate t of some frame for any world line. However, this formulation would be based on arbitrary choices and might not seem natural, given that the equations we have encountered so far only contain intrinsic geometric quantities such as the four-velocity, and not the velocity in some frame.

As a matter of fact there is a more profound physical reason for rejecting (10.23) as a general case (although that system ma be used in particular types of interactions, in almost classical velocity regimes where each τ_i is well approximated by a global time t of an inertial frame). In the description given thus far there is a prominent actor missing: the *interaction field*. Particles are not just subject to interactions; they *generate* them. Each particle given by a world line $I \ni s \mapsto \gamma(s) \in \mathbb{M}^4$ generates an interaction field (e.g., electromagnetic) typically described by *vector fields or tensor fields* on $\mathbb{M}^4$, whose support is in $J^+(\gamma(I))$ to respect causality. Any other particle whose world line crosses $J^+(\gamma(I))$ undergoes a four-force due to the presence of the interaction field produced by the first particle.

A physically sensible description of the dynamics of N relativistic point particles is still based on the relativistic equation of motion (10.21) for any particle in the system. The four-force on the right is produced by the remaining $N - 1$ particles. Or better: it is due to interaction fields generated by those particles. To assign the four-force in terms of the interaction fields, therefore, we must add to the N Eq. (10.21) three further equations describing how point particles generate the interaction fields (for example Maxwell's equations in relativistic form, where the sources are the world lines of the field-generating charges).

The final system of differential equations contains at the same time the equations of the particles' evolution, *as if the interaction fields were assigned*, and the fields' evolution equations, *as if the sources (the world lines) were assigned*. In reality neither is truly assigned, but we are given initial conditions for both particles and fields.

Another important point is that the fields carry four-momenta (and other conserved quantities) exactly as point particles. Therefore, in presence of interaction fields, the conservation law of the four-momentum needs to account for the field contribution as well. (The explicit form we have used for it does not require possible interaction fields, which may always be present, due to the locality of the interaction, which occurs at a single event. But this is a very special situation.) The procedure to account for conserved quantities obtained integrating densities produced by interaction fields involves the so-called *energy-momentum tensor* (which is important even for continuous physical systems like relativistic fluids, and in Cosmology alike) and the extension of *Noether's Theorem* to field theories [Rin06, Wal84]. This is an obligatory route, especially when passing from Special to General Relativity, where the gravitational interaction is 'geometrised'.

We cannot delve further in these matters given the introductory character of the chapter.

10.3.4   The Geometry of the So-Called Twin Paradox

We use this section to examine a geometric property of the inertial motions (10.22) that is of physical interest. The inertial motion described by segments of temporal curves possesses in Special Relativity a very interesting physical/geometric peculiarity. Consider two events o and o' joined by the histories of two particles $\gamma^{(inertial)}$ and γ, of class C^1 and future-like, each propagating in $\mathbb{M}^4$ from o to o'. Suppose $\gamma^{(inertial)}$ describes the inertial motion of a massive physical body, in other words $\gamma^{(inertial)}$ is the only (future-like) line segment between o and o', while γ is a generic future-directed causal curve from o to o'. If $\Delta\tau^{(inertial)}$ and $\Delta\tau$ are the proper time intervals measured along $\gamma^{(inert)}$ and γ between o and o', we have:

$$\Delta\tau^{(inertial)} > \Delta\tau . \tag{10.24}$$

That is to say, if the curves represent the histories of two twins that go their own way at o, at their next meeting at o' the twin that has moved inertially will be older. This phenomenon goes by the name of **twin paradox**. The allegedly paradoxical aspect is caused by the relative motion, which leads to conclude naively that the twins' status is interchangeable ("because motion is relative"), whereas the two outcomes predicted by Relativity can definitely not be swapped. *In reality there is no paradox at all, since the situation is completely asymmetrical: only one twin moves inertially from o to o', because there is a unique line segment between two events in an affine space.* Let us prove (10.24).

Proposition 10.36 *Consider a future-directed timelike segment*

$$\gamma : [0, 1] \ni s \mapsto \gamma(s) = \gamma(0) + s(\gamma(1) - \gamma(0)) \in \mathbb{M}^4$$

joining the events $o = \gamma(0)$ and $o' = \gamma(1)$. Then

(i) γ has maximal length among all C^1 future causal curves from o to o'.
(ii) γ is the unique curve in the above class that maximises length.

Proof Consider Minkowski coordinates x^0, x^1, x^2, x^3 where $\mathbf{e}_0$ is parallel to V, and fix the origin O so that $o = O$. In coordinates, by Lemma 10.21, we can parametrise by $t = x^0/c$ any C^1 future causal curve γ' from o to o':

$$[0, \Delta t] \ni t \mapsto (ct, x^1(t), x^2(t), x^3(t)) \in \mathbb{R}^4$$

where $c\Delta t$ is just the length of γ. The length of γ' will be less than that of γ, thus proving (i), because:

$$c\Delta\tau = c\int_0^{\Delta t} \sqrt{1 - \frac{1}{c^2}\sum_{a=1}^{3}\left(\frac{dx^a}{dt}\right)^2}\, dt$$

$$\leq c\Delta t \max_{u \in [0, \Delta t]} \sqrt{1 - \frac{1}{c^2}\sum_{a=1}^{3}\left(\frac{dx^a}{du}\right)^2} \leq c\Delta t. \tag{10.25}$$

The final inequality comes from the fact that γ is timelike if and only if:

$$g(\dot\gamma(t), \dot\gamma(t)) = -c^2\left(\frac{dt}{dt}\right)^2 + \sum_{a=1}^{3}\left(\frac{dx^a}{dt}\right)^2 < 0.$$

To prove (ii), if we have $\Delta\tau = \Delta t$, from (10.25) we deduce $\frac{dx^a}{dt} = 0$ on $[0, \Delta t]$ for $a = 1, 2, 3$. Hence $(ct, x^1(t), x^2(t), x^3(t)) = (ct, x^1(0), x^2(0), x^3(0)) = (ct, 0, 0, 0)$ for any t. This is precisely γ in our coordinates, and therefore $\gamma = \gamma'$.

$\square$

Remarks 10.37

(1) The fact that the length of the only timelike segment between o and o' is *greater* than the length of any other (causal) curve between the same points strongly disagrees with Euclidean geometry, where segments *minimise* the length of curves! But the geometry of $\mathbb{M}^4$ relies on the indefinite inner product g, and not on the positive-definite inner product of $\mathbb{E}^n$.
(2) In Special Relativity inertial motions are therefore described by affine segments parametrised by arc length. They are extremal as regards the length of timelike curves between two events. Curves of this kind are called *geodesics* (according to one of several characterisations of such curves). This observation represents an entrance door to the theory of General Relativity [Rin06].

■

10.4   The Lorentz and Poincaré Groups

We introduce the transformation groups between inertial frames that replace the Galilean transformations (3.1) when we pass from Classical Physics to Special Relativity.

10.4.1   The Lorentz and Poincaré Groups and Their Orthochronous Subgroups

Given two sets of Minkowski coordinates:

$$\psi : \mathbb{M}^4 \ni p \mapsto (x^0, x^1, x^2, x^3) \in \mathbb{R}^4$$

$$\text{and } \psi' : \mathbb{M}^4 \ni p \mapsto (x'^0, x'^1, x'^2, x'^3) \in \mathbb{R}^4 \,,$$

asking the corresponding bases $\mathbf{e}_0, \mathbf{e}_1, \mathbf{e}_2, \mathbf{e}_3$ and $\mathbf{e}'_0, \mathbf{e}'_1, \mathbf{e}'_2, \mathbf{e}'_3$ to be pseudo-orthonormal with future-directed time vectors translates into:

$$g_{\mu\nu} = g(\mathbf{e}_\mu, \mathbf{e}_\nu) = \sum_{\alpha,\beta=0}^{3} \Lambda^\alpha{}_\mu \Lambda^\beta{}_\nu g(\mathbf{e}'_\alpha, \mathbf{e}'_\beta) = \sum_{\alpha,\beta=0}^{3} \Lambda^\alpha{}_\mu \Lambda^\beta{}_\nu g_{\alpha\beta} \qquad (10.26)$$

and

$$0 > g(\mathbf{e}_0, \mathbf{e}'_0) = \sum_{\alpha=0}^{3} g(\Lambda^\alpha{}_0 \mathbf{e}'_\alpha, \mathbf{e}'_0) = \Lambda^0{}_0 g_{00} = -\Lambda^0{}_0 \,, \qquad (10.27)$$

where we used the coefficients $g_{\alpha\beta}$ from (10.9), plus the decomposition

$$\mathbf{e}_\mu = \sum_{\alpha=0}^{3} \Lambda^\alpha{}_\mu \mathbf{e}'_\alpha \,.$$

In conclusion, if

$$\psi : \mathbb{M}^4 \ni p \mapsto (x^0, x^1, x^2, x^3) \in \mathbb{R}^4$$

$$\text{and } \psi' : \mathbb{M}^4 \ni p \mapsto (x'^0, x'^1, x'^2, x'^3) \in \mathbb{R}^4$$

are Minkowski coordinates, in view of Exercises 1.7.**1**,

$$\psi \circ \psi'^{-1} \quad \text{is represented by} \quad x^\alpha = c^\alpha + \sum_{\mu=0}^{3} \Lambda^\alpha{}_\mu x'^\mu, \quad \alpha = 0, 1, 2, 3 \qquad (10.28)$$

where $c^0, c^1, c^2, c^3 \in \mathbb{R}$ and the matrix $\Lambda := [\Lambda^{\alpha}{}_{\mu}]_{\alpha,\mu=0,1,2,3}$ satisfies

$$\Lambda^t \eta \Lambda = \eta, \quad \Lambda^0{}_0 > 0, \tag{10.29}$$

by (10.26) and (10.27), with

$$\eta := \operatorname{diag}(-1, 1, 1, 1) = [g_{\alpha\beta}]_{\alpha,\beta=0,1,2,3}. \tag{10.30}$$

Conversely, it is evident that if ψ' defines Minkowski coordinates and $\psi : M^4 \ni p \mapsto (x^0, x^1, x^2, x^3) \in \mathbb{R}^4$ another set of global coordinates on M^4, and if we assume (10.28) with (10.29), then ψ too defines Minkowski coordinates (prove it as an exercise).

We have the following definitions.

Definition 10.38 Consider transformations from $\mathbb{R}^4$ to $\mathbb{R}^4$ of the form

$$x^{\alpha} = c^{\alpha} + \sum_{\mu=0}^{3} \Lambda^{\alpha}{}_{\mu} x'^{\mu}, \quad \alpha = 0, 1, 2, 3 \tag{10.31}$$

(1) The set of maps (10.31) with $\Lambda \in M(4, \mathbb{R})$ satisfying:

$$\Lambda^t \eta \Lambda = \eta \tag{10.32}$$

and where $(c^0, c^1, c^2, c^3) \in \mathbb{R}^4$ are arbitrary, is called **Poincaré group** and written $IO(1, 3)$.
(2) The set of matrices $\Lambda \in M(4, \mathbb{R})$ obeying (10.32) is called **Lorentz group** $O(1, 3)$.
(3) The set of transformations (10.31) with $\Lambda \in M(4, \mathbb{R})$ satisfying (10.29):

$$\Lambda^t \eta \Lambda = \eta \quad \text{and} \quad \Lambda^0{}_0 > 0$$

and where $(c^0, c^1, c^2, c^3) \in \mathbb{R}^4$ are arbitrary is the **orthochronous Poincaré group** $IO(1, 3)_+$.
(4) The set of matrices $\Lambda \in M(4, \mathbb{R})$ satisfying (10.29) is the **orthochronous Lorentz group** $O(1, 3)_+$. $\diamond$

Based on the previous discussion and (10.31), it is clear that $O(1, 3)_+$ is the set of possible Poincaré transformations between Minkowski coordinates with the same spacetime origin.

We have used the word "group" because the above sets of transformations and matrices have a natural composition law making them groups, as we show next.

Theorem 10.39 *The following facts hold.*

(1) $IO(1, 3)$ and $IO(1, 3)_+$ are groups under the composition of maps $\mathbb{R}^4 \to \mathbb{R}^4$, while $O(1, 3)$ and $O(1, 3)_+$ are groups for the matrix product on $M(4, \mathbb{R})$.

(2) $O(1, 3) \subset IO(1, 3)$, $O(1, 3)_+ \subset IO(1, 3)_+$, $IO(1, 3)_+ \subset IO(1, 3)$, $O(1, 3)_+ \subset O(1, 3)$ as transformation subgroups.

(3) If $\Lambda \in O(1, 3)$, then

    (i) $\Lambda^t \in O(1, 3)$,
    (ii) $\Lambda^{-1} = \eta \Lambda^t \eta$.

(4) If $\Lambda \in O(1, 3)_+$, then

    (i) $\Lambda^t \in O(1, 3)_+$,
    (ii) $\Lambda^{-1} = \eta \Lambda^t \eta$.

(5) $O(1, 3) = \{\Lambda, T\Lambda \,|\, \Lambda \in O(1, 3)_+\}$, and similarly for $IO(1, 3)$ and $IO(1, 3)_+$. $T := \eta \in O(1, 3)$ is the so-called **time reversal** operator, which satisfies $TT = I$.

Proof (1), (3) and (4). It suffices to show the claims for the Lorentz groups. Once we know these are indeed groups, the argument for the Poincaré groups boils down to verifying that the composite of transformations like (10.31) is of the same type, and the inverse of such as well, using the property of Lorentz groups. We leave these elementary checks to the reader, and concentrate on showing $O(1, 3)$ and $O(1, 3)_+$ are closed under inversion and matrix product. (That is enough since it is clear that $I \in O(1, 3)$ and $I \in O(1, 3)_+$, directly by definition. The product's associativity is guaranteed by the associativity of matrix multiplication.)

(a) We claim that if $\Lambda \in O(1, 3)_+$ then Λ^{-1} is defined and belongs to $O(1, 3)_+$. The same proof, simplified (dropping $\Lambda^0{}_0 > 0$), also shows that if $\Lambda \in O(1, 3)$ then Λ^{-1} exists and lives in $O(1, 3)$. As for existence, if $\Lambda \in O(1, 3)_+$, or even if $\Lambda \in O(1, 3)$: $\det(\Lambda^t \eta \Lambda) = \det \eta = -1$. Therefore $(\det \Lambda)(\det \eta) \det \Lambda = 1$, i.e. $\det(\Lambda)^2 = 1$. Hence Λ^{-1} exists if $\Lambda \in O(1, 3)_+$ or $\Lambda \in O(1, 3)$. The direct calculation shows $\Lambda^{-1} = \eta \Lambda^t \eta$. On the other hand $(\eta \Lambda^t \eta)^0{}_0 = \Lambda^0{}_0 > 0$ in case $\Lambda \in IO(1, 3)_+$. So, it is enough to prove $(\Lambda^t)^t \eta \Lambda^t = \eta$ if $\Lambda^t \eta \Lambda = \eta$. Starting from $\Lambda^{-1} = \eta \Lambda^t \eta$ we have $I = \Lambda \eta \Lambda^t \eta$. Multiplying by η on the right gives $\eta = \Lambda \eta \Lambda^t$, which we write as $(\Lambda^t)^t \eta \Lambda^t = \eta$, as required. In this way we have also proved (i) and (ii) in (3) and (4).

(b) Let us prove that if $\Lambda, \Lambda' \in O(1, 3)_+$ (or $O(1, 3)$) then $\Lambda \Lambda' \in O(1, 3)_+$ (respectively, $O(1, 3)$). Now, $(\Lambda \Lambda')^t \eta \Lambda \Lambda' = \eta$ is obvious since $(\Lambda \Lambda')^t = \Lambda'^t \Lambda^t$, $\Lambda^t \eta \Lambda = \eta$ and $\Lambda'^t \eta \Lambda' = \eta$, concluding the claim for $O(1, 3)$. What is less obvious concerns $O(1, 3)_+$. Namely, that $\Lambda^0_0 > 0$ and $\Lambda'^0_0 > 0$ imply $(\Lambda \Lambda')^0_0 > 0$. So let us show that.

We will use repeatedly the following identity. From the sole $\Lambda^t \eta \Lambda = \eta$ we have:

$$-1 = \eta_{00} = -(\Lambda^0{}_0)^2 + \sum_{a=1}^{3} \Lambda^a{}_0 \Lambda^a{}_0. \tag{10.33}$$

Let us take:

$$(\Lambda\Lambda')^0{}_0 = \Lambda^0{}_0\Lambda'^0{}_0 + \sum_{a=1}^{3} \Lambda^0{}_a\Lambda'^a{}_0$$

and write it as:

$$(\Lambda\Lambda')^0{}_0 = (\Lambda^t)^0{}_0\Lambda'^0{}_0 + \sum_{a=1}^{3}(\Lambda^t)^a{}_0\Lambda'^a{}_0.$$

Now it suffices to prove

$$\left|\sum_{a=1}^{3}(\Lambda^t)^a{}_0\Lambda'^a{}_0\right| \le |(\Lambda^t)^0{}_0\Lambda'^0{}_0|, \tag{10.34}$$

so that the sign of $(\Lambda\Lambda')^0{}_0$, never zero by (10.33), coincides with the sign of $\Lambda^0{}_0\Lambda'^0{}_0$, whence positive. Let us prove (10.34) with the Cauchy-Schwarz inequality

$$\left|\sum_{a=1}^{3}(\Lambda^t)^a{}_0\Lambda'^a{}_0\right|^2 \le \left(\sum_{a=1}^{3}(\Lambda^t)^a{}_0(\Lambda^t)^a{}_0\right)\left(\sum_{b=1}^{3}\Lambda'^b{}_0\Lambda'^b{}_0\right).$$

Applying (10.33) to Λ and Λ'' (recall Λ'' satisfies $(\Lambda'')^t\eta\Lambda'' = \eta$, as seen) (10.33) immediately implies:

$$\sum_{a=1}^{3}\Lambda^a{}_0\Lambda^a{}_0 \le (\Lambda^0{}_0)^2 \quad \text{and} \quad \sum_{b=1}^{3}(\Lambda'')^b{}_0(\Lambda'')^b{}_0 \le ((\Lambda'')^0{}_0)^2,$$

and these two imply (10.34).

(2) $O(1,3) \subset IO(1,3)$ since $O(1,3)$ contains the elements of $IO(1,3)_+$ with $c^0 = c^1 = c^2 = c^3 = 0$ in (10.31). Immediately, these transformations form a subgroup. In the same way one shows the subgroup inclusion $O(1,3)_+ \subset IO(1,3)_+$. The latter two inclusions are obvious because the groups on the left additionally have $\Lambda^0{}_0$.

(5) From (10.33), if $\Lambda \in O(1,3)$ then $\Lambda^0{}_0 > 0$ or $\Lambda^0{}_0 < 0$. Consider a matrix $\Lambda \in O(1,3)$ satisfying the second case. We wish to prove we can write it as $T\Lambda'$ for some $\Lambda' \in O(1,3)_+$, thus ending the proof. If $\Lambda^0{}_0 < 0$ then $T\Lambda \in O(1,3)$ since $T = \eta \in O(1,3)$, but also $(T\Lambda)^0{}_0 > 0$, i.e. $T\Lambda \in O(1,3)_+$, $(T\Lambda)^0{}_0 = -\Lambda^0{}_0$ by computation. Finally, since $TT = I$, we have $\Lambda = T\Lambda'$ where $\Lambda' := T\Lambda \in O(1,3)_+$.   □

Remarks 10.40

(1) The group $IO(1, 3)$ (and similarly $IO(1, 3)_+$) can be viewed abstractly as the set of pairs $\{(\Lambda, C) \mid \Lambda \in O(1, 3), C \in \mathbb{R}^4\}$ under the group law[5]

$$(\Lambda, C) \circ (\Lambda', C') = (\Lambda \Lambda', \Lambda C' + C)$$

as is immediate composing the corresponding transformations (10.31).

(2) The Poincaré group $IO(1, 3)$ and the Lorentz group $O(1, 3)$ seem physically slightly less interesting than their orthochronous subgroups because they may invert the temporal order and thus not act between inertial frames. In reality this is not true because they do play a crucial role in Quantum Field Theories, for example.

■

10.4.2 Special and Special Orthochronous Subgroups, Discrete Transformations

From the previous proof we deduce that $\det \Lambda = \pm 1$ if $\Lambda \in O(1, 3)$. This allows to define two further subgroups of $IO(1, 3)$ and $O(1, 3)$ called **proper** Poincaré and Lorentz groups respectively. They are indicated by $ISO(1, 3)$ and $SO(1, 3)$ and defined by:

$$\Lambda^t \eta \Lambda = \eta \quad \text{and} \quad \det \Lambda = 1 . \tag{10.35}$$

Easily, in analogy to the orthochronous case:

$$O(1, 3) = \{\Lambda, P\Lambda \mid \Lambda \in SO(1, 3)\} \tag{10.36}$$

where $P = -\eta \in O(1, 3)$ is called **parity inversion**. It satisfies $PP = I$.

We finally define two further subgroups of $IO(1, 3)$ and $O(1, 3)$, respectively called **proper orthochronous Poincaré group** $ISO(1, 3)_+$ and **proper orthochronous Lorentz group** $SO(1, 3)_+$, by imposing the conditions:

$$\Lambda^t \eta \Lambda = \eta , \quad \Lambda^0{}_0 > 0 , \quad \det \Lambda = 1 . \tag{10.37}$$

---

[5] This is a special case of a *semi-direct product* of two groups, here $O(1, 3)$ and $\mathbb{R}^4$.

The group $O(1, 3)$ is reconstructed from $SO(1, 3)_+$ using time reversal and parity inversion, since the following elementary fact holds:

$$O(1, 3) = \{\Lambda\, , P\Lambda\, , T\Lambda\, , PT\Lambda \mid \Lambda \in SO(1, 3)_+\}\, . \tag{10.38}$$

Note that the **composite inversion** PT satisfies $PT = -I$ and $(PT)(PT) = I$.

The **subgroup of discrete Lorentz transformations** is generated by $\{I, P, T, PT\}$, which easily form a group. This sort of discrete operations (together with the so-called *charge conjugation*) have a fundamental importance in the theory of elementary particles. We may recast (10.38) as:

$$O(1, 3) = SO(1, 3)_+ \cup TSO(1, 3)_+ \cup PSO(1, 3)_+ \cup PTSO(1, 3)_+ \tag{10.39}$$

in which the 4 subsets on the right are pairwise disjoint because they differ in the signs of Λ_0^0, $\det\Lambda$ or both. The Lorentz group $O(1, 3)$ and its subgroups $O(1, 3)_+$, $SO(1, 3)$, $SO(1, 3)_+$ are subgroups of $GL(4, \mathbb{R})$, the groups of 4×4 real invertible matrices. The latter is a *topological group*, i.e. also a topological space for the induced topology of $\mathbb{R}^{4\times 4}$ with continuous group operations (product and inversion). The groups $O(1, 3)$, $O(1, 3)_+$, $SO(1, 3)$, $SO(1, 3)_+$ are closed subsets of $GL(4, \mathbb{R})$ and inherit the topological group structure. It can be proved that $SO(1, 3)_+$ is path-connected hence connected (for example with Theorem 10.51, see below), and the four discrete transformations are continuous with continuous inverses. Then it is easy to conclude that (10.39) is precisely the decomposition of $O(1, 3)$ into its connected components. Among the four sets only $SO(1, 3)_+$ is a subgroup of $O(1, 3)$: the others do not contain the identity matrix.

By virtue of (1) Remarks 10.40, a similar discourse can be made regarding the Poincaré group and its subgroups, viewing the pair $(\Lambda, C) \in IO(1, 3)$ as a 5×5 real matrix in $GL(5, \mathbb{R})$:

$$\begin{bmatrix} \Lambda & C \\ 0^t & 1 \end{bmatrix}\, ,$$

and noting that the matrix product of such corresponds to the composite of the transformations of $IO(1, 3)$:

$$\begin{bmatrix} \Lambda & C \\ 0^t & 1 \end{bmatrix} \begin{bmatrix} \Lambda' & C' \\ 0^t & 1 \end{bmatrix} = \begin{bmatrix} \Lambda\Lambda' & \Lambda C' + C \\ 0^t & 1 \end{bmatrix}\, .$$

The final observation is that all groups introduced so far are *Lie groups*, i.e. real-analytic manifolds with real-analytic group operations. These aspects go beyond the elementary presentation of this book.

10.4.3   Elementary Properties of $O(1, 3)_+$ and $IO(1, 3)_+$

Let us examine the elementary structure of the group $O(1, 3)_+$. The first thing is that $O(3) \subset O(1, 3)_+$ as a subgroup. In fact the transformations

$$\Omega_R := \begin{bmatrix} 1 & 0 & 0 & 0 \\ \hline 0 & & & \\ 0 & & R & \\ 0 & & & \end{bmatrix}, \tag{10.40}$$

where $R \in O(3)$ are elements of $O(1, 3)_+$, as is immediate by direct computation. In particular $O(3) \ni R \mapsto \Omega_R \in O(3)_+$ is an injective group homomorphism.

Together with **spacetime displacements**:

$$x^\alpha = c^\alpha + x'^\alpha, \qquad \alpha = 0, 1, 2, 3 \tag{10.41}$$

the rotations form the group of **internal transformations** of a given inertial frame. These are the spatial roto-translations and temporal displacements:

$$x^\alpha = c^\alpha + \sum_{\mu=0}^{3} (\Omega_R)^\alpha{}_\mu x'^\mu, \qquad \alpha = 0, 1, 2, 3 \tag{10.42}$$

where $(c^0, c^1, c^2, c^3) \in \mathbb{R}^4$ and $R \in O(3)$. This subgroup is also a subgroup of the Galilean group. The group of internal transformations can be split into $1 + 3 + 3 = 7$ continuous 1-parameter subgroups representing, respectively,

  (i) temporal displacements along $\mathbf{e}_0$ (with parameter $c^0 \in \mathbb{R}$),
  (ii) the 3 spatial displacements along $\mathbf{e}_1, \mathbf{e}_2, \mathbf{e}_3$ respectively (with parameters $c^1, c^2, c^3 \in \mathbb{R}$ respectively),
  (iii) the 3 spatial rotations about $\mathbf{e}_1, \mathbf{e}_2, \mathbf{e}_3$ (with angle in $\mathbb{R}$ as parameter).

Each subgroup is *additive* with respect to its parameters.

We finally have three more transformations, not internal, called **special Lorentz transformations**, respectively along $\mathbf{e}_1, \mathbf{e}_2$ and $\mathbf{e}_3$:

$$\Lambda_1(v) := \begin{bmatrix} \gamma & \gamma v/c & 0 & 0 \\ \gamma v/c & \gamma & 0 & 0 \\ 0 & 0 & 1 & 0 \\ 0 & 0 & 0 & 1 \end{bmatrix}, \qquad \Lambda_2(v) := \begin{bmatrix} \gamma & 0 & \gamma v/c & 0 \\ 0 & 1 & 0 & 0 \\ \gamma v/c & 0 & \gamma & 0 \\ 0 & 0 & 0 & 1 \end{bmatrix},$$

$$\Lambda_3(v) := \begin{bmatrix} \gamma & 0 & 0 & \gamma v/c \\ 0 & 1 & 0 & 0 \\ 0 & 0 & 1 & 0 \\ \gamma v/c & 0 & 0 & \gamma \end{bmatrix} \qquad \text{where } \gamma(v) := 1/\sqrt{1 - v^2/c^2} \text{ and } v \in (-c, c).$$

As $v \in (-c, c)$ varies each set of matrices forms a continuous one-parameter subgroup of $O(1, 3)_+$, hence of $IO(1, 3)_+$. It is also easy to see that, in contrast to the previous 7, the parameter v is not additive:

$$\Lambda_3(v)\Lambda_3(v') \neq \Lambda_3(v + v')$$

in general, and the right-hand side is not even always defined, since we may have $v + v' \notin (-c, c)$. We can achieve an additive reparametrisation in terms of the so-called **rapidity** $\chi \in \mathbb{R}$, related to v by the smooth diffeomorphism from $\mathbb{R}$ to $(-c, c)$:

$$v(\chi) = c \sinh \chi \, , \quad \chi \in \mathbb{R} \, . \tag{10.43}$$

With that choice:

$$\Lambda_3(v(\chi)) := \begin{bmatrix} \cosh \chi & 0 & 0 & \sinh \chi \\ 0 & 1 & 0 & 0 \\ 0 & 0 & 1 & 0 \\ \sinh \chi & 0 & 0 & \cosh \chi \end{bmatrix} \tag{10.44}$$

and a direct computation re-instates additivity:

$$\Lambda_k(v(\chi))\Lambda_k(v(\chi')) = \Lambda_k(v(\chi + \chi')) \, , \quad \chi, \chi' \in \mathbb{R} \, , \quad k = 1, 2, 3 \, .$$

The explicit action of a Lorentz transformation $\Lambda_3(v)$ is the following.

$$\Lambda_3(v) : \begin{cases} t' = \gamma(v)(t + \frac{v}{c^2}x^3) \, , \\ x'^1 = x^1 \, , \\ x'^2 = x^2 \, , \\ x'^3 = \gamma(v)(x^3 + vt) \, , \end{cases}$$

where (ct, x^1, x^2, x^3) and (ct', x'^1, x'^2, x'^3) are the coordinates of one event in Minkowski coordinates of two frames $\mathscr{R}$ and $\mathscr{R}'$ respectively. Let us consider the physical meaning of the parameter v. A point at rest in $\mathscr{R}$ has equations of motion in $\mathscr{R}'$:

$$x'^1(t') = x^1 \, , \quad x'^2(t') = x^2 \, , \quad x'^3(t') = \gamma(x^3 + v(t'/\gamma + vx^3/c))$$

Hence each point at rest with $\mathscr{R}$ seems, for $\mathscr{R}'$, to move with constant velocity in space and time equal to $v e'_3$, which is thus the relative velocity of $\mathscr{R}$ with respect to $\mathscr{R}'$.

Remarks 10.41

(1) Special Lorentz transformations, for example $\Lambda_3(v)$, show how *the simultaneity of events is relative to the frame of reference*: two events o and o' with the same time coordinate t in $\mathscr{R}$ and occurring at different places (different x^3-coordinate) appear in $\mathscr{R}'$ to have distinct time coordinate t'.

(2) Using Λ_3 it is easy to see that if $o - o'$ is spacelike, we can find a frame $\mathscr{R}$ for which $t'(o) > t(o')$ and another frame $\mathscr{R}'$ where $t'(o) < t'(o')$. The temporal ordering between causally separated events depends on the chosen reference frame. The temporal order is instead fixed if $o - o'$ is causal, even if we employ all transformations of $IO(1, 3)_+$, because orthochronous Poincaré transformations do not allow one to leave the half light cone in which $o - o'$ lives by definition.

(3) If we neglect all quadratic and higher-order terms in c^2 (including those appearing in $\gamma(v)$), the transformations $\Lambda_3(v)$ reduce to pure Galilean transformations along $\mathbf{e}'_3$:

$$\begin{cases} t' = t' \,, \\ x'^1 = x^1 \,, \\ x'^2 = x^2 \,, \\ x'^3 = x^3 + vt \,, \end{cases}$$

which goes to show that Minkowski frames correspond to inertial frames.

∎.

Examples 10.42 The structure of special transformations Λ_3 prompts us to address an important relativistic phenomenon known as the **contraction of lengths**. Together with the *dilation of time intervals* discussed in Example 10.26, the phenomenon proves how time intervals and spatial distances, even if measured with ideal instruments, stop being absolute in Special Relativity. Consider an ideal rigid ruler at rest in an inertial frame $\mathscr{R}$ along the spatial axis $\mathbf{e}_3$ with ends x_0^3 and $x_0^3 + \ell$, where the ruler's length in $\mathscr{R}$ equals $\ell > 0$. Consider another inertial frame $\mathscr{R}'$ and suppose that, under a suitable choice of Minkowski coordinates, $\mathscr{R}'$ is related to the Minkowski coordinates of $\mathscr{R}$ by a special transformation $\Lambda_3(v)$. We want to assess the length ℓ' assigned to the ruler in frame $\mathscr{R}'$, which now moves along $\mathbf{e}'_3$ with velocity v. The computation is as follows. The histories γ_1 and γ_2 of the first and second end of the ruler are, in coordinates of $\mathscr{R}$:

$$\gamma_1 : \mathbb{R} \ni x^0 \mapsto (x^0, 0, 0, x_0^3) \,, \quad \gamma_2 : \mathbb{R} \ni x^0 \mapsto (x^0, 0, 0, x_0^3 + \ell) \,.$$

These world lines can be described in $\mathscr{R}'$ using expression (10.44) for $\Lambda_3(\chi)$ with $\chi = \sinh^{-1} \frac{v}{c}$. In this way:

$$\gamma_1 : \mathbb{R} \ni x^0 \mapsto \left(x^0 \cosh \chi + x_0^3 \sinh \chi, 0, 0, x^0 \sinh \chi + x_0^3 \cosh \chi \right)$$

and

$$\gamma_2 : \mathbb{R} \ni t \mapsto \left(x^0 \cosh \chi + (x_0^3 + \ell) \sinh \chi, 0, 0, x^0 \sinh \chi + (x_0^3 + \ell) \cosh \chi \right) .$$

(10.45)

Let us reparametrise the curves with the global time of $\mathscr{R}'$, which we read from the first components

$$x'^0 = x^0 \cosh \chi + x_0^3 \sinh \chi , \quad x'^0 = x^0 \cosh \chi + (x_0^3 + \ell) \sinh \chi ;$$

we then find:

$$x^0 = \frac{x'^0 - x_0^3 \sinh \chi}{\cosh \chi} , \quad x^0 = \frac{x'^0 - (x_0^3 + \ell) \sinh \chi}{\cosh \chi} .$$

Inserting these in (10.42) and (10.45), and recalling that $\cosh^2 \chi - \sinh^2 \chi = 1$, we obtain the world lines describing the histories of the rod's ends parametrised by the time coordinate x'^0 of $\mathscr{R}'$ (Fig. 10.1).

$$\gamma_1 : \mathbb{R} \ni x'^0 \mapsto \left(\frac{x'^0 + x_0^3 (\sinh \chi)(\cosh \chi - 1)}{\cosh \chi}, 0, 0, \frac{x'^0 - x_0^3}{\cosh \chi} \right)$$

(10.46)

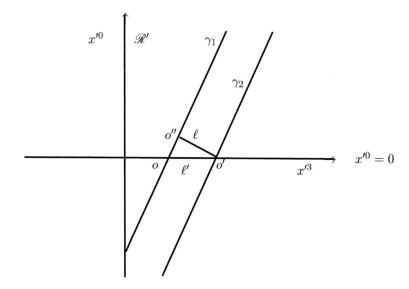

Fig. 10.1 Length contraction: in contrast to Eq. (10.48), it appears that $\ell' > \ell$. This is due to our habit of unconsciously using Euclidean geometry on the plane, whereas now the geometry is Lorentzian

and

$$\gamma_2 : \mathbb{R} \ni t \mapsto \left(\frac{x'^0 + (x_0^3 + \ell)(\sinh \chi)(\cosh \chi - 1)}{\cosh \chi}, 0, 0, \frac{x'^0 - (x_0^3 + \ell)}{\cosh \chi} \right).$$
(10.47)

The ruler's length ℓ' measured in $\mathscr{R}'$ is then the difference between the last two coordinates corresponding to the intersections of the world lines of the ruler's ends in the rest spaces $E_{\mathscr{R}',x'^0}$ of $\mathscr{R}'$. We have found that $\ell' = \frac{\ell}{\cosh \chi}$. In terms of the velocity v of $\mathscr{R}$ relative to $\mathscr{R}'$ we immediately recover the celebrated *law of contraction of lengths*:

$$\ell' = \ell \sqrt{1 - \frac{v^2}{c^2}}.$$
(10.48)

Note that indeed $\ell' < \ell$ if $v > 0$. It is paramount, to fully grasp the phenomenon, to understand the following fact. The measurement of the ruler's length at rest in $\mathscr{R}$ is done in $\mathscr{R}'$ using a pair of events, simultaneous in $\mathscr{R}'$, that *are not simultaneous in $\mathscr{R}$* albeit lying on the world lines of the rod's ends. In $\mathscr{R}'$ the ruler appear as if it were moving. To measure its length in $\mathscr{R}'$ we must compare it at some time instant $t' = x'^0/c$ of $\mathscr{R}'$ to an ideal ruler at rest in $\mathscr{R}'$. This selects a pair of points on the world lines of the moving ruler's ends. These two points, say o and o' at time $t' = 0$ in the figure, are simultaneous events in $\mathscr{R}'$ with distinct spatial coordinate x^3. As observed in (1) Remarks 10.41, o and o' are not simultaneous in $\mathscr{R}$. The length ℓ in $\mathscr{R}$ is instead measured using two events, simultaneous in $\mathscr{R}$, along γ_1 and γ_2. The figure shows one pair of such events o', o'.

10.4.4   Relevance of Pure Lorentz Transformations

Let us pass to describe a type of Lorentz transformation that generalises special ones. By the definition it is evident that any frame $\mathscr{R}$ has an associated *four-velocity* $V_{\mathscr{R}} := c\mathscr{R}$. Given a second frame $\mathscr{R}'$, the relative velocity $\mathbf{v}_{\mathscr{R}|\mathscr{R}'} = \sum_{a=1}^3 v^a \mathbf{e}'_a$ of $\mathscr{R}$ with respect to $\mathscr{R}'$ is defined by the components of the four-velocity $V_{\mathscr{R}} = \sum_{\alpha=0}^3 V^\alpha \mathbf{e}_\alpha$ in Minkowski coordinates of $\mathscr{R}'$:

$$v^a = c \frac{V^a}{V^0}.$$
(10.49)

The components of $\mathbf{v}_{\mathscr{R}|\mathscr{R}'}$ and $V_{\mathscr{R}}$ in $\mathscr{R}'$ are constant, as they should be. We can recover these components directly from the Poincaré transformation of the Minkowski coordinates of $\mathscr{R}$ and $\mathscr{R}'$:

$$x'^\alpha = \sum_{\beta=0}^3 \Lambda^\alpha{}_\beta x^\beta + b^\alpha.$$
(10.50)

We then have:

$$\mathbf{e}_\alpha = \sum_{\beta=0}^{3} \Lambda^\alpha{}_\beta \mathbf{e}'_\beta \,,$$

and so:

$$V_{\mathcal{R}} = c\mathbf{e}_0 = \sum_{\mu=0}^{3} c\Lambda^\mu{}_0 \mathbf{e}_\mu \,. \tag{10.51}$$

Hence:

$$\mathbf{v}_{\mathcal{R}|\mathcal{R}'} = \sum_{a=1}^{3} c\frac{\Lambda^a{}_0}{\Lambda^0{}_0} \mathbf{e}'_a \,. \tag{10.52}$$

Now the natural question is whether we can reconstruct the matrix $\Lambda \in O(1,3)_+$ in (10.50) starting from the components of the relative velocity of $\mathcal{R}$ as seen by $\mathcal{R}'$. The answer is yes. Fix the Minkowski frame $\mathcal{R}$ of components $c^{-1}V^\alpha$ in some other frame $\mathcal{R}'$, and define the column vector $\vec{V} \in \mathbb{R}^3$ whose components are the spatial components of $V_{\mathcal{R}}$ in Minkowski coordinates x'^0, x'^1, x'^2, x'^3 of $\mathcal{R}'$. The matrix:

$$\Lambda_{V_{\mathcal{R}}|\mathcal{R}'} := \left[\begin{array}{c|c} V^0/c & \vec{V}^t/c \\ \hline \vec{V}/c & I + \frac{\vec{V}\vec{V}^t}{c^2(1+V^0/c)} \end{array} \right] \tag{10.53}$$

belongs to $SO(1,3)_+$. Moreover, there are Minkowski coordinates on $\mathcal{R}$, defined up to 4 constants $b^0, b^1, b^2, b^3 \in \mathbb{R}$, such that:

$$x'^\alpha = \sum_{\beta=0}^{3} \left(\Lambda_{V_{\mathcal{R}}|\mathcal{R}'}\right)^\alpha{}_\beta x^\beta + b^\alpha \,. \tag{10.54}$$

In fact, taking advantage of conditions (10.55) below, which trivially hold here, we have

(i) $\left(\Lambda_{V_{\mathcal{R}}|\mathcal{R}'}\right)^t \eta \Lambda_{V_{\mathcal{R}}|\mathcal{R}'} = \eta$,

(ii) $(\Lambda_{V_{\mathcal{R}}|\mathcal{R}'})^0{}_0 \geq 0$,

(iii) $\mathbf{e}_0 = \mathcal{R}$, where $\mathbf{e}_0$ is the unit vector of the x^0-axis in the Cartesian coordinates on $\mathbb{M}^4$ defined implicitly by (10.54). These coordinates are therefore Minkowski coordinates on $\mathcal{R}$.

The numbers b^α do not play any role and can be chosen all zero by changing the origin in $\mathcal{R}$ or in $\mathcal{R}'$.

Definition 10.43 The matrices $\Lambda \in M(4, \mathbb{R})$ of the form (10.53), for any quadruple $(V^0, V^1, V^2, V^3) \in \mathbb{R}^4$ such that

$$V^0 > 0, \quad -(V^0)^2 + \sum_{a=1}^{3}(V^a)^2 = -c^2, \quad (10.55)$$

(so that (i) and (ii) are valid) are called **pure Lorentz transformations**. If we allow non-zero constants b^α in (10.50) and Λ is pure we obtain a **pure Poincaré transformation**.   ◇

The above discussion can be summarised in the following proposition.

Proposition 10.44 *Given inertial frames $\mathscr{R}$ and $\mathscr{R}'$ it is always possible to fix the axes and the origin of time of Minkowski coordinates on the frames so that $\mathscr{R}$ and $\mathscr{R}'$ are related by a pure Lorentz transformation (10.54) with $b^0 = b^1 = b^2 = b^3 = 0$. The quadruple of constants $(V^0, V^1, V^2, V^3) \in \mathbb{R}^4$ in the pure transformation is given by the components of the four-velocity $V_\mathscr{R}$ of $\mathscr{R}$ written in Minkowski coordinates of $\mathscr{R}'$.*

Remarks 10.45

(1) Pure Lorentz transformations include special transformations. A special transformation $\Lambda_a(v)$ along $\mathbf{e}'_a$ is nothing but a pure transformation where the components of $V_\mathscr{R}$ in $\mathscr{R}'$ are all zero except the one of $\mathbf{e}'_a$. In that case $v = cV^a/V^0 \in (-c, c)$ due to (10.55).

(2) The reader can prove as exercise that pure transformations do not form a subgroup of $O(1, 3)_+$. In fact, excluding the case of two special transformations along the same axis, the product of pure transformations is not pure; in some sense it contains an extra rotation. This mathematical fact has important consequences in Physics, such as the so-called *Thomas precession*.

■

10.4.5   Two Decomposition Results for the Lorentz Group

Pure Lorentz transformations may be characterised in a completely abstract manner as follows.

Proposition 10.46 $\Lambda \in O(1, 3)_+$ *is pure if and only if it is symmetric and positive semi-definite. Moreover, every pure Lorentz transformation is proper (det $\Lambda > 0$) and positive definite.*

Dimostrazione A pure Lorentz matrix is certainly symmetric and positive semi-definite because of (10.53). But because it is invertible, it cannot have zero eigenvalues, and therefore it is positive definite. Then the determinant is positive, so it must be proper as well. To finish let us check that any symmetric and positive semi-definite $\Lambda \in O(1, 3)_+$ can be put in form (10.53) with (10.55), in other words a pure Lorentz matrix. By assumption

$$\Lambda = \begin{bmatrix} \gamma & \vec{S}^t \\ \vec{S} & A \end{bmatrix}$$

where $\gamma > 0$ to respect orthochronicity, $\vec{S} \in \mathbb{R}^3$ and $A \in M(2, \mathbb{R})$ satisfies $A = A^t$ because Λ is symmetric. Λ is symmetric, positive semi-definite and invertible by definition, as noted above, hence positive definite. This implies A must be positive definite. Asking $\Lambda \in O(1, 3)$ is equivalent to

$$\vec{S}^t \vec{S} = \gamma^2 - 1 , \tag{10.56}$$

$$A\vec{S} = \gamma \vec{S} , \tag{10.57}$$

$$A^2 = I + \vec{S}\vec{S}^t . \tag{10.58}$$

Set $V^a := cS^a$ and $V^0 := c\gamma$. Then Λ takes a form similar to (10.53) in which (10.55) holds by virtue of (10.56)

$$\Lambda = \begin{bmatrix} V^0/c & \vec{V}^t/c \\ \vec{V}/c & A \end{bmatrix}$$

where we have $A^2 = I + \vec{V}\vec{V}^t/c^2$ and $A\vec{V} = V^0\vec{V}/c$. To conclude it suffices to show that the unique solution to the equations of A is $A = I + \frac{\vec{V}\vec{V}^t}{c^2(1+V^0/c)}$. Assume $\vec{V} \neq 0$, and remember that A is symmetric so it has an orthonormal basis of eigenvectors. Equation $A\vec{V} = V^0\vec{V}/c$ fixes A in the eigenspace spanned by $\vec{V}$ in $\mathbb{R}^3$, so we should find the symmetric operator A on the two-dimensional invariant subspace N orthogonal to $\vec{V}$. Note that N is an eigenspace of A with eigenvalues λ_1, λ_2, possibly equal. From $A^2 = I + \vec{V}\vec{V}^t/c^2$ it follows that A^2 is the identity on N. As λ_1^2, λ_2^2 are the eigenvalues of A^2 on N, necessarily $\lambda_1 = \pm 1$ and $\lambda_2 = \pm 1$. But A is positive definite, so $\lambda_1 = \lambda_2 = 1$. In other words A itself is the identity on N. Keeping into account that $\frac{\vec{V}\vec{V}^t}{\vec{V}^t\vec{V}}$ is the orthogonal projection operator onto the eigenspace generated by $\vec{V}$, whence $P_N = I - \frac{\vec{V}\vec{V}^t}{\vec{V}^t\vec{V}} = I - \frac{\vec{V}\vec{V}^t}{(V^0)^2 - c^2}$ is the orthogonal projection onto N, we deduce

$$A = 1\left(I - \frac{\vec{V}\vec{V}^t}{(V^0)^2 - c^2}\right) + \frac{V^0}{c}\frac{\vec{V}\vec{V}^t}{(V^0)^2 - c^2} = I + \frac{\vec{V}\vec{V}^t}{c^2(1 + V^0/c)} ,$$

ending the proof in case $\vec{V} \neq 0$. In the other case, equation $A^2 = I + \frac{\vec{V}\vec{V}^t}{c^2} = I$ forces $A = I$ because A is symmetric (hence diagonalisable) and positive definite. Here, too, Λ takes form (10.53) with $\vec{V} = 0$. $\qquad\square$

We pass to state and prove[6] a first classical decomposition, whereby any $\Lambda \in O(1, 3)_+$ can always be written uniquely as product of a pure transformation and a space rotation. In this sense we can describe the entire orthochronous Lorentz group (and hence the orthochronous Poincaré group) only using pure transformations and spatial rotations.

Theorem 10.47 *For any* $\Lambda \in O(1, 3)_+$ *there exist a unique* $R \in O(3)$ *and a unique pure Lorentz transformation* Λ_P *such that:*

$$\Lambda = \Omega_R \Lambda_P , \qquad (10.59)$$

where Ω_R *is defined in (10.40). The above is the polar decomposition (see below) of* Λ.

Remarks 10.48 | AC | We shall make use of elementary techniques regarding *spectral decompositions*. The real vector space V is always *finite dimensional* and endowed with a positive-definite inner product $\mathbf{u} \cdot \mathbf{v}$. A^t denotes the **adjoint operator** to the linear operator $A : V \to V$, defined by the relation $A^t \mathbf{u} \cdot \mathbf{v} = \mathbf{u} \cdot A\mathbf{v}$ for any $\mathbf{u}, \mathbf{v} \in V$. It coincides with the transpose matrix when $V = \mathbb{R}^n$ with the standard Euclidean inner product. A is **symmetric** if $A = A^t$.

(1) By the **finite-dimensional spectral theorem** (Theorem A.11) Any symmetric linear operator $A : V \to V$ on a real vector space V with positive-definite inner product decomposes uniquely as:

$$A = \sum_{\lambda \in \sigma(A)} \lambda P_\lambda , \qquad (10.60)$$

where $\sigma(A) \subset \mathbb{R}$ is a finite set and the linear maps $P_\lambda : V \to V$, with $P_\lambda \neq 0$, are **orthogonal projectors** ($P_\lambda^2 = P_\lambda$ and $P_\lambda^t = P_\lambda$) onto subspaces of V that satisfy $P_\lambda P_\mu = 0$ for $\lambda \neq \mu$ and $\sum_{\lambda \in \sigma(A)} P_\lambda = I$. Furthermore, the **spectrum** $\sigma(A)$ of the symmetric operator A is the collection of eigenvalues of A and every P_λ is the orthogonal projector of the eigenspace associated with $\lambda \in \sigma(A)$. We call (10.60) the **spectral decomposition** of A.

(2) Given an arbitrary map $f : \mathbb{R} \to \mathbb{R}$, one defines $f(A) := \sum_{\lambda \in \sigma(A)} f(\lambda) P_\lambda$. When f is a polynomial, this expression evidently coincides with the standard definition for which A^n is the n-fold composite of A with itself and $A^0 = I$. More generally,

$$f(A)g(A) = (f \cdot g)(A)$$

---

[6] See for example: R.U. Sexl and H.K. Urbanke, *Relativity, Groups Particles*. Springer (2000) and V. Moretti, *The interplay of the polar decomposition theorem and the Lorentz group*, Lecture Notes, Seminario Interdisciplinare di Matematica 5-153 (2006).

where $(f \cdot g)(\lambda) := f(\lambda)g(\lambda)$. Indeed,

$$f(A)g(A) = \sum_{\lambda \in \sigma(A)} f(\lambda)P_\lambda \sum_{\lambda' \in \sigma(A)} g(\lambda')P_{\lambda'} = \sum_{\lambda, \lambda' \in \sigma(A)} f(\lambda)g(\lambda')\delta_{\lambda\lambda'}P_\lambda^2$$

$$= \sum_{\lambda \in \sigma(A)} f(\lambda)g(\lambda)P_\lambda = (f \cdot g)(A) .$$

(3) The uniqueness of (10.60) easily implies the following:

Proposition 10.49 *If the symmetric operators A, B are positive semi-definite (so $\sigma(A), \sigma(B) \in [0, +\infty)$), then $A^2 = B^2$ implies $A = B$. In particular, for every symmetric positive semi-definite operator $C : V \to V$ there is a unique symmetric positive semi-definite operator $\sqrt{C} : V \to V$ such that $(\sqrt{C})^2 = C$.*

Proof $A^2 = B^2$ translates into

$$A^2 = \sum_{\lambda \in \sigma(A)} \lambda^2 P_\lambda^{(A)} = \sum_{\lambda \in \sigma(B)} \lambda^2 P_\lambda^{(B)} = B^2$$

i.e.

$$A^2 = \sum_{\mu \in \sigma(A)^2} \mu P_{\sqrt{\mu}}^{(A)} = \sum_{\mu \in \sigma(B)^2} \mu P_{\sqrt{\mu}}^{(B)} = B^2 .$$

As the map $\mathbb{R}_+ \ni x \mapsto \sqrt{x} \in \mathbb{R}_+$ is bijective, by the spectral decomposition's uniqueness we conclude $\sigma(A^2) = \sigma(A)^2$, $\sigma(B^2) = \sigma(B)^2$ and $P_{\sqrt{\mu}}^{(A)} = P_\mu^{(A^2)} = P_\mu^{(B^2)} = P_{\sqrt{\mu}}^{(B)}$ for any $\sqrt{\mu} \in \sigma(A) = \sigma(B)$, with clear notation. Because $P_\lambda^{(A)} = P_\lambda^{(B)}$ when $\lambda \in \sigma(A) = \sigma(B)$, from the spectral decomposition of A, B we deduce $A = B$. The last sentence immediately follows form the first statement defining $\sqrt{C} = \sum_{\lambda \in \sigma(C)} \sqrt{\lambda} P_\lambda^{(C)}$ where $C = \sum_{\lambda \in \sigma(C)} \lambda P_\lambda^{(C)}$ is the spectral decomposition of C. $\square$

(4) The well-known *polar decomposition* holds.

Theorem 10.50 *If $A : V \to V$ is symmetric and non-singular ($A\mathbf{v} = \mathbf{0}$ implies $\mathbf{v} = \mathbf{0}$), then A admits **polar decomposition***

$$A = UP ,$$

where $P : V \to V$ is symmetric, positive-definite operator and $U : V \to V$ is and orthogonal operator: $U^t U = I$. The pair U, P is uniquely defined by A.

Proof *(Existence)* The matrix $A^t A$ is symmetric and positive definite: if $A^t A\mathbf{v} = \mathbf{0}$ then $A\mathbf{v} \cdot A\mathbf{v} = 0$ so $A\mathbf{v} = \mathbf{0}$, which forces $\mathbf{v} = \mathbf{0}$. Define the symmetric operator $P := \sqrt{A^t A}$. It is non-singular: $P\mathbf{v} = \mathbf{0}$ implies $\sqrt{A^t A}\sqrt{A^t A}\mathbf{v} = \mathbf{0}$, hence

$A^t A \mathbf{v} = \mathbf{0}$ and then $\mathbf{v} = \mathbf{0}$. Therefore P^{-1} exists. Now write $A = AP^{-1}P$ and set $U := AP^{-1}$. Then $U^t U = (P^{-1})^t A^t AP^{-1} = P^{-1}P^2 P^{-1} = I$, as required. (Uniqueness) If $A = U'P'$ with U', P' as above, then $P'P' = P'U'^t U'P' = A^t A = PU^t UP = PP$ and so $P = P'$, by Remark (3). Hence $U'P' = U'P = A = UP$ implies $U'P = UP$, and applying P^{-1} to either side we find $U' = U$.
$\qquad\qquad\qquad\qquad\qquad\qquad\qquad\qquad\qquad\qquad\qquad\qquad\square\blacksquare$

AC   **Proof of Theorem 10.47.** Consider the polar decomposition $\Lambda = UP$ of $\Lambda \in O(1,3)_+$ and let us show P is a pure Lorentz matrix. First, we know that $\Lambda^t \in O(1,3)$ by Theorem 10.39. Hence $P^2 = \Lambda^t \Lambda \in O(1,3)$. Because $E \in O(1,3)$, i.e. $E^t \eta E = \eta$, if $E = E^t$ then $\eta E \eta = E^{-1}$. Hence P^2 satisfies $\eta PP\eta = (P^{-1})^2 P^{-1}$, that is to say $(\eta P\eta)(\eta P\eta) = (P^{-1})^2$. By construction P^{-1} and $\eta P\eta$ are symmetric and positive definite, (the eigenvalues of P^{-1} are the reciprocals of the eigenvalues of P, which are positive, while $\mathbf{v}\cdot\eta P\eta\mathbf{v} = \eta^t \mathbf{v}\cdot P\eta\mathbf{v} = \eta\mathbf{v}\cdot P\eta\mathbf{v} \geq 0$ and $\det(\eta P\eta) = \det(P) \neq 0$). Hence $\eta P\eta = \sqrt{(\eta P\eta)(\eta P\eta)} = \sqrt{(P^{-1})^2} = P^{-1}$ as per (3) in Remarks 10.48. We have shown that $\eta P\eta = P^{-1}$, and so $P\eta P = \eta$. But $P = P^t$, so $P \in O(1,3)$. Since P is positive definite its component P_0^0 must be positive and therefore $P \in O(1,3)_+$. Proposition 10.46 now tells that P is a pure Lorentz transformation. As $UP = \Lambda \in O(1,3)_+$, and P belongs to the same group, we conclude $U = \Lambda P^{-1} \in O(1,3)_+$. Condition $U \in O(4) \cap O(1,3)_+$ immediately shows that $U_0^0 = 1$, $U_0^i = U_i^0 = 0$, and therefore $U = \Omega_R$ as in (10.40). That $R^t R = I$ is a straightforward consequence of the form of $\Omega_R = U$ and of $U^t U = I$. We have then proved the existence the claimed decomposition, provided we set $\Lambda_P := P$.

The uniqueness of $\Lambda = \Omega_R \Lambda_P$ follows from the polar decomposition's uniqueness. In fact, Λ_P is symmetric positive definite by Proposition 10.46 and Ω_R is orthogonal on $\mathbb{R}^4$, so $\Lambda = \Omega_R \Lambda_P$ is the polar decomposition of Λ. $\square$

We can finally ask, given frames $\mathscr{R}$ and $\mathscr{R}'$ related by a generic Poincaré transformation, whether one can modify the respective Minkowski coordinates *by internal transformations* so that the Poincaré transformation is actually special Lorentz, for example along $\mathbf{e}_3'$. Clearly, by translating the spacetime origins in the two rest spaces we can always make the coordinate change an element in $O(1,3)_+$. There remains to see whether we can rotate the spatial axes so that the residual Lorentz transformation becomes special along $\mathbf{e}_3'$. The answers is still yes, and is based on a second decomposition (itself a consequence of the first decomposition examined earlier).

Theorem 10.51 *For any $\Lambda \in O(1,3)_+$ there exist $R_1, R_2 \in O(3)$ and $v \in (-c, c)$ such that:*

$$\Lambda = \Omega_{R_1}\Lambda_3(v)\Omega_{R_2}.$$

Proof If $R \in O(3)$ and Λ_P is a pure Lorentz matrix as in (10.53), by computation we have:

$$\Omega_R \Lambda_P \Omega_R^t := \left[\begin{array}{c|c} V^0/c & (R\vec{V})^t/c \\ \hline R\vec{V}/c & I + \frac{R\vec{V}(R\vec{V})^t}{c^2(1+V^0/c)} \end{array} \right], \tag{10.61}$$

So we can find a rotation $R_0 \in O(3)$ making $R\vec{V}$ parallel to $\mathbf{e}_3'$, and then $\Omega_{R_0} \Lambda_P \Omega_{R_0}^t = \Lambda_3(v)$ is a special Lorentz transformation along $\mathbf{e}_3'$ (for some $v \in (-c, c)$). Let now $\Lambda \in O(1, 3)_+$ be generic. By Theorem 10.47 we can write $\Lambda = \Omega_R \Lambda_P$ for a unique $R \in O(3)$ and a unique pure transformation Λ_P. Putting the two facts together:

$$\Lambda = \Omega_R \Lambda_P = \Omega_R \Omega_{R_0}^{-1} \Lambda_3(v) \Omega_{R_0} = \Omega_R \Omega_{R_0}^t \Lambda_3(v) \Omega_{R_0}$$

$$= \Omega_R \Omega_{R_0^t} \Lambda_3(v) \Omega_{R_0} = \Omega_{RR_0^t} \Lambda_3(v) \Omega_{R_0} .$$

The claim is then true for $R_1 = R R_0^t = R R_0^{-1}$ and $R_2 = R_0$. $\qquad\square$

By that, if the Lorentz transformation between the initial coordinates is Λ, it suffices to rotate the spatial axes in $\mathscr{R}$ and $\mathscr{R}'$ using *internal transformations* in order to eliminate the rotations Ω_{R_1} and Ω_{R_2}. The Minkowski coordinates chosen in this way are related by a special transformation along $\mathbf{e}_3'$.

10.5 Introduction to the Lagrangian Formalism in Special Relativity

We shall study in this section certain Lagrangians for the massive and charged relativistic particle in an electromagnetic field, and the corresponding equations of motion in terms of the four-force, making use of the previous section's results. We will examine three Lagrangians: the first is quadratic and seldom used in Physics; a second one that is non-quadratic and mathematically rather pathological, but very much employed in Physics. Either two do not depend on reference frames: they are 'covariant', or put in other terms, defined in an intrinsic geometric way. This separates them from the third Lagrangian we will discuss, which does depend on frames. The action functional of the third Lagrangian is independent of the frame, though. In all three cases, and nearly always in absence of an electromagnetic field, we will study the first integrals borne by Noether's (and Jacobi's) Theorem, with an eye to the 1-parameter subgroups of the Poincaré group.

So, consider a point particle of mass $m > 0$ in Minkowski spacetime $\mathbb{M}^4$. We will assume it carries an electric charge $e \in \mathbb{R}$ and is immersed in an external electromagnetic field (which might be zero) described in any Minkowski coordinate system by the **electromagnetic four-potential field**:

$$A(p) = \sum_{\alpha=0}^{3} A^\alpha \mathbf{e}_\alpha := \varphi(p)\mathbf{e}_0 + \sum_{a=1}^{3} A^\alpha(p)\mathbf{e}_\alpha \in T_p\mathbb{M}^4 , \quad p \in \mathbb{M}^4$$

where $p \equiv (x^0, x^1, x^2, x^3)$ in the chosen coordinates.

Remarks 10.52 Since A is a vector in the space of displacements at any point, when changing Minkowski coordinates its components must change under the familiar Lorentz transformations:

$$A'^\mu(p) = \sum_{\nu=0}^{3} \Lambda^\mu{}_\nu A^\nu(p) .$$

This fact is pretty much non-obvious, but true when we consider the electromagnetic field. ∎

10.5.1 The Covariant Quadratic Lagrangian for the Charged Relativistic Particle

A rather natural Lagrangian for a particle in an electromagnetic field is:

$$\mathscr{L}(\tau, \gamma, \dot{\gamma}) = \frac{m}{2} g(\dot{\gamma}, \dot{\gamma}) + \frac{e}{c} g(\dot{\gamma}, A(\gamma)) . \tag{10.62}$$

The expression is entirely geometrical and does not depend on coordinates, so we did *not* indicate any reference frame in the formula. If we had fixed Minkowski coordinates the expression for $\mathscr{L}$ (see (10.64) below) would have been the same in any system of Minkowski coordinates.

This last remark can be understood in a slightly more elaborate way, which we now discuss. Fix Minkowski coordinates, write the Lagrangian and then let the group $IO(1, 2)_+$ act on each term: the Lagrangian's final expression will not vary, because the procedure corresponds, in passive sense, to a Minkowski coordinate change. The property is often called **covariance**, and in the jargon it means invariance, or equivariance, under $IO(1, 3)_+$ (or even $IO(1, 3)$) and its *tensor* and *spinor* representations. In this sense the quadratic Lagrangian (10.62) is covariant.

The Euler-Lagrange equations for the above Lagrangian must detect curves $\gamma = \gamma(s)$ representing particles' histories described by a parameter $s \in I$ with the dimensions of a time. We expect s to be proper time eventually, but for the moment s is any old parameter and the nature of the curves it parametrises is not necessarily causal.

The natural ambient space to describe said Lagrangian is the spacetime of kinetic states $A(\mathbb{R} \times \mathbb{M}^4)$, the first jet space of the spacetime configurations given by the trivial bundle $\mathbb{R} \times \mathbb{M}^4$ over $\mathbb{R} \ni s$. The latter was described in (2) in Remarks 7.36 and alternatively, but equivalently, in Sect. 7.5.1. Actually in this case we clearly have:

$$A(\mathbb{R} \times \mathbb{M}^4) = \mathbb{R} \times T\mathbb{M}^4 .$$

The vectors $\dot{\gamma}$ should be thought of as belonging to the fibres of the first jet space of the bundle $\mathbb{R} \times \mathbb{M}^4$ over $\mathbb{R}$. They are, in other words, vectors of $T_p\mathbb{M}^4$. The natural atlas of $A(\mathbb{R} \times \mathbb{M}^4)$ contains every Minkowski coordinate system on $\mathbb{M}^4$ extended by the usual set of 4 *dotted coordinates* plus the s coordinate for the base: $(s, x^0, x^1, x^2, x^3, \dot{x}^0, \dot{x}^1, \dot{x}^2, \dot{x}^3)$.

In components:

$$\mathscr{L}(s, x, \dot{x}) = \sum_{\mu,\nu=0}^{3} \frac{m}{2} g_{\mu\nu} \dot{x}^{\mu} \dot{x}^{\nu} + \sum_{\mu,\nu=0}^{3} \frac{e}{c} g_{\mu\nu} \dot{x}^{\mu} A^{\nu}(x) , \tag{10.63}$$

or even more explicitly:

$$\mathscr{L}(s, x, \dot{x}) = \frac{m}{2} \left(-(\dot{x}^0)^2 + \sum_{a=1}^{3} (\dot{x}^a)^2 \right) - \frac{e}{c} \dot{x}^0 \varphi(\gamma) + \frac{e}{c} \sum_{a=1}^{3} \dot{x}^a A^a(x) . \tag{10.64}$$

The symbol $\dot{x}^\alpha$ denotes an independent coordinate, *not* a component of the tangent vector (unlike $\dot{\gamma}$). It will become a tangent component only along solutions of the Euler-Lagrange equations.

Remarks 10.53 Here is a non-rigorous yet physically important consideration. Supposing s is proper time, at small speeds ($v \ll c$) t and s become indistinguishable and $\dot{x}^0 \approx c$. The Lagrangian then becomes:

$$\mathscr{L}(s, x(t), \dot{x}(t)) \approx \frac{m}{2} \sum_{a=1}^{3} \left(\frac{dx^a}{dt} \right)^2 - e\varphi(x(t)) + \frac{e}{c} \sum_{a=1}^{3} \frac{dx^a}{dt} A^a(x(t)) - \frac{mc^2}{2} \tag{10.65}$$

where we have spelt out the derivatives explicitly to emphasise the change in parameter from s to t. This is the Lagrangian in (9.16) up to an inessential additive constant (that could be subtracted already in the initial definition of the Lagrangian). ∎

Now we are interested in future-directed timelike curves $I \ni s \mapsto \gamma(s) \in \mathbb{M}^4$ whose associated local sections of $\mathbb{R} \times T\mathbb{M}^4$ (over $\mathbb{R}$):

$$I \ni s \mapsto (s, \gamma(s), \dot{\gamma}(s)) \in \mathbb{R} \times T\mathbb{M}^4$$

solve the Euler-Lagrange equations generated by $\mathscr{L}$. The equations obtained from (10.62) are, to second order:

$$m \sum_{\alpha=0}^{3} g_{\alpha\beta} \frac{d}{ds}\left(\frac{dx^{\alpha}}{ds} + \frac{e}{c} A^{\alpha}(x(s))\right) = \frac{e}{c} \sum_{\alpha,\delta=0}^{3} g_{\alpha\delta} \frac{dx^{\alpha}}{ds} \frac{\partial A^{\delta}}{\partial x^{\beta}}\bigg|_{x(s)}.$$

Trivial calculations produce the *relativistic equations of motion of a charged particle in an external electromagnetic field*:

$$\sum_{\alpha=0}^{3} g_{\alpha\beta} m \frac{d^2 x^{\alpha}}{ds^2} = \sum_{\mu=0}^{3} \frac{e}{c} \frac{dx^{\mu}}{ds} \left(\sum_{\nu=0}^{3} g_{\mu\nu} \frac{\partial A^{\nu}}{\partial x^{\beta}} - \frac{\partial A^{\alpha}}{\partial x^{\mu}}\right)\bigg|_{x(s)}, \quad \alpha = 0, 1, 2, 3,$$

or in equivalent form:

$$\sum_{\alpha=0}^{3} g_{\alpha\beta} m \frac{d^2 x^{\alpha}}{ds^2} = \sum_{\mu=0}^{3} \frac{e}{c} \frac{dx^{\mu}}{ds} F_{\mu\beta}\big|_{x(s)}, \quad \alpha = 0, 1, 2, 3, \quad (10.66)$$

where we have introduced the **components of the electromagnetic tensor**

$$F_{\mu\nu} := \frac{\partial A_{\mu}}{\partial x^{\nu}} - \frac{\partial A_{\nu}}{\partial x^{\mu}}, \quad \mu, \nu = 0, 1, 2, 3, \quad (10.67)$$

involving the *covariant components* (see Sect. 10.1.2) of the four-potential

$$A_{\mu} := \sum_{\nu=0}^{3} g_{\mu\nu} A^{\nu}, \quad \mu = 0, 1, 2, 3.$$

Remarks 10.54 Relative to the electric and magnetic fields obtained from $F_{\mu\nu}$ in Minkowski coordinates moving with an inertial frame $\mathscr{R}$ we have:

$$[F_{ab}]_{a,b=0,1,2,3} = \begin{bmatrix} 0 & -E_x/c & -E_y/c & -E_z/c \\ E_x/c & 0 & B_z/c & -B_y/c \\ E_y/c & -B_z/c & 0 & B_x/c \\ E_z/c & B_y/c & -B_x/c & 0 \end{bmatrix} \quad (10.68)$$

where E^a and B^a, for $a = 1, 2, 3$, are the components of the electric and magnetic field respectively. ∎

Equations (10.66) can be put in normal form using the matrix $[g^{\alpha\beta}]_{\alpha,\beta=0,1,2,3} = \eta$ satisfying:

$$\sum_{\beta=0}^{3} g^{\alpha\beta} g_{\beta\sigma} = \delta^{\alpha}_{\sigma}, \quad \alpha, \sigma = 0, 1, 2, 3,$$

which merely say $[g_{\alpha\beta}]_{\alpha,\beta=0,1,2,3}$ is the inverse of $[g_{\alpha\beta}]_{\alpha,\beta=0,1,2,3}$. Applying the inverse to the differential equations we eventually find the relativistic equations of motion of a charged particle in an external electromagnetic field, written in normal form:

$$m\frac{d^2 x^\alpha}{ds^2} = \frac{e}{c} \sum_{\beta,\mu=0}^{3} g^{\alpha\beta} \frac{dx^\mu}{ds} F_{\mu\beta}\big|_{x(s)}, \quad \alpha = 0, 1, 2, 3. \tag{10.69}$$

These, too, are *covariant* in the sense explained earlier (more precisely, they are $IO(1,3)$-equivariant). They can be written more economically by involving the *electromagnetic tensor* and making use of a little tensor calculus [Rin06]; this, though, would require we introduce more mathematical technology and push the treatise beyond the intended elementary introduction. At any rate we can write the equations as:

$$m\frac{d\dot{\gamma}}{ds} = F(\gamma(s), \dot{\gamma}(s)). \tag{10.70}$$

The right side contains the Lorentz **electromagnetic four-force**, which we can view as defined on the entire $T\mathbb{M}^4$ and not just $\mathcal{V}^+\mathbb{M}^4$:

$$F(p, v) = \frac{e}{c} \sum_{\alpha,\mu=0}^{3} g^{\beta\alpha} v^\mu F_{\mu\alpha}\big|_p \, \mathbf{e}_\beta. \tag{10.71}$$

In any inertial frame $\mathscr{R}$, in terms of the electromagnetic field:

$$F^0 = \frac{q\mathbf{E}\cdot\mathbf{v}|_{\mathscr{R}}}{\sqrt{1-\frac{v^2}{c^2}}}, \qquad F^a = \frac{eE^a}{\sqrt{1-\frac{v^2}{c^2}}} + \frac{e}{c}\left(\frac{\mathbf{v}|_{\mathscr{R}}}{\sqrt{1-\frac{v^2}{c^2}}} \wedge \mathbf{B}\right)^a \quad a = 1, 2, 3.$$

A calculation shows that

$$g(F(p, v), v) = 0 \quad \text{if } (p, v) \in T_p\mathbb{M}^4. \tag{10.72}$$

In fact:

$$g(F(p, v), v) = \frac{e}{c} \sum_{\alpha,\beta,\delta,\mu=0}^{3} v^\delta g_{\delta\beta} g^{\beta\alpha} v^\mu F_{\mu\alpha} = \frac{e}{c} \sum_{\alpha,\mu=0}^{3} v^\alpha v^\mu F_{\mu\alpha} = 0,$$

because $v^\alpha v^\beta$ is symmetric under an index swap while $F_{\alpha\beta}$ is skew-symmetric. With (10.72) in place we can finally invoke Theorem 10.34 and obtain the following result.

Theorem 10.55 *If the electromagnetic four-potential A is C^2, equation (10.69) admits a unique future-like maximal solution if the initial data are future-like. (In particular, any other solution with the same initial data is a restriction of it.) Finally, if the initial tangent vector satisfies $g(\dot{\gamma}(s_0), \dot{\gamma}(s_0)) = -c^2$, then the parameter s is proper time and the tangent vector to the maximal solution is the four-velocity.*

10.5.2 First Integrals of the Covariant Quadratic Lagrangian

Since the Lagrangian in (10.62) does not depend explicitly on s and we can therefore apply Jacobi's Theorem, the Hamiltonian:

$$\mathcal{H}(s, \gamma, \dot{\gamma}) = \frac{m}{2} g(\dot{\gamma}, \dot{\gamma}) \tag{10.73}$$

is conserved, as s varies, along the solutions of the Euler-Lagrange equations. (The curves with the physical meaning of massive particles' histories are evidently those for which the above right-hand side is negative.) The result confirms what we already knew from Proposition 10.34: curves with tangent timelike vector at the initial time stay timelike. Now we know more: the length of the tangent vector to a solution of the Euler-Lagrange equations remains constant as s varies (though even this was known from the proof of Proposition 10.34).

Let us move on to the constants of motion produced by the Poincaré invariance of $\mathcal{L}$ in absence of an electromagnetic field (also when present, actually, but with additional symmetry assumptions on the electromagnetic four-potential). We will list the 10 one-parameter additive subgroups of $IO(1, 3)_+$ seen earlier, and write their action on events in $\mathbb{M}^4$ in given Minkowski coordinates on $\mathcal{R}$.

(1) **Temporal displacements**: $\begin{cases} x'^0_\epsilon = x^0 + \epsilon \\ x'^a_\epsilon = x^a, \quad a = 1, 2, 3 \end{cases}$ where $\epsilon \in \mathbb{R}$;

(2) **spatial displacements along e_1**: $\begin{cases} x'^1_\epsilon = x^1 + \epsilon \\ x'^\beta_\epsilon = x^\beta, \quad \beta = 0, 2, 3 \end{cases}$ where $\epsilon \in \mathbb{R}$,

plus the other two analogue spatial displacements along e_2 and e_3;

(3) **spatial rotations around e_3**: $\begin{cases} x'^1_\epsilon = x^1 \cos \epsilon - x^2 \sin \epsilon \\ x'^2_\epsilon = x^1 \sin \epsilon + x^2 \cos \epsilon \\ x'^\beta_\epsilon = x^\beta, \quad \beta = 0, 3 \end{cases}$ where $\epsilon \in \mathbb{R}$, plus

the other two analogue spatial rotations around e_1 and e_2;

(4) **special Lorentz transformations along e_3**: $\begin{cases} x'^0_\epsilon = x^0 \cosh \epsilon + x^3 \sinh \epsilon \\ x'^3_\epsilon = x^0 \sinh \epsilon + x^3 \cosh \epsilon \\ x'^a_\epsilon = x^a, \quad a = 1, 2 \end{cases}$

$\epsilon \in \mathbb{R}$ plus the other two similar special Lorentz transformations along e_2 and e_1.

The extension of the 10 subgroups' action on dotted coordinates is obtained as prescribed in Sect. 8.2.1, with the crucial interpretative difference that now the bundle's base is no longer absolute time but the s-axis, and the fibres are copies of Minkowski spacetime. Hence we can extend the above to $A(\mathbb{R} \times \mathbb{M}^4) = \mathbb{R} \times T\mathbb{M}^4$ simply by replacing in the formulas all coordinates with the corresponding dotted coordinates, except for spacetime displacements where trivially:

$$\dot{x}_\epsilon'^\mu = \dot{x}^\mu, \quad \mu = 0, 1, 2, 3.$$

The Lagrangian

$$\mathcal{L}(s, x, \dot{x}) = \frac{m}{2} \sum_{\mu,\nu=0}^{3} g_{\mu\nu} \dot{x}^\mu \dot{x}^\nu,$$

satisfies $\mathcal{L} \circ \phi_\epsilon = \mathcal{L}$ for any $\epsilon \in \mathbb{R}$, where $\{\phi_\epsilon\}_{\epsilon \in \mathbb{R}}$ is any one of the previous 1-parameter subgroups of $O(1,3)_+$ (with action extended to $A(\mathbb{R} \times \mathbb{M}^4)$). Correspondingly, in agreement with Theorem 8.10, we have a bunch of Noether first integrals of the form:

$$I(s, x, \dot{x}) := \sum_{\mu=0}^{3} \frac{\partial \mathcal{L}}{\partial \dot{x}^\mu} \frac{\partial x_\epsilon'^\mu}{\partial \epsilon}\bigg|_{\epsilon=0}.$$

(1) For temporal displacements Noether's first integral is

$$m\dot{x}^0 = P^0;$$

namely, the 0-th component in $\mathcal{R}$ of the four-momentum along the motion (provided s is proper time, as is the case under the usual hypotheses on the initial conditions).

(2) For spatial displacements along $\mathbf{e}_1$ we have Noether first integral

$$m\dot{x}^1 = P^1,$$

the 1st component in $\mathcal{R}$ of the four-momentum along the motion (provided s is proper time). Spatial displacements along $\mathbf{e}_2$ and $\mathbf{e}_3$ produce analogue first integrals.

(3) For spatial rotations around $\mathbf{e}_3$ we have Noether first integral:

$$x^1 P^2 - x^2 P^1 = L^3,$$

the 3rd component in $\mathcal{R}$ of the angular momentum along the motion with respect to the spatial origin of the axes (provided s is proper time). Spatial rotations about $\mathbf{e}_1$ and $\mathbf{e}_2$ give similar first integrals.

(4) For special Lorentz transformations along $\mathbf{e}_3$ we have Noether first integral:

$$x^0 P^3 - x^3 P^0 = K^3 \,,$$

the 3rd component in $\mathscr{R}$ of the **boost vector** (provided s is proper time). Special transformations along $\mathbf{e}_1$ and $\mathbf{e}_2$ produce analogue first integrals.

In presence of an electromagnetic field whose Lagrangian is still (10.63) and, say, does not explicitly depend on time $t = x^0/c$ in frame $\mathscr{R}$, we have a first integral accounting for the external fields. In the example at hand:

$$I = \left. \frac{\partial \mathscr{L}}{\partial \dot{x}^0} \frac{\partial x'^0_\epsilon}{\partial \epsilon} \right|_{\epsilon = 0} = m\dot{x}^0 + \frac{e}{c}\varphi(x^1, x^2, x^3) \,.$$

Multiplying by c we find that along the system's motion the following quantity, of the dimensions of an energy, is conserved in (proper) time:

$$cI = \frac{mc^2}{\sqrt{1 - \frac{v^2}{c^2}}} + e\varphi(x^1, x^2, x^3) \,.$$

The first summand is the relativistic total mechanical energy of the particle, as we already know; the second is due to the electromagnetic field. At small speeds v compared to c, the above first integral reads

$$I \approx mc^2 + \frac{1}{2}mv^2 + e\varphi(x^1, x^2, x^3) \,.$$

On the right, apart from the constant mc^2, we recognise the total energy in $\mathscr{R}$ given by the sum of the classical kinetic energy plus the electric potential energy.

Similar electromagnetic contributions come from the other first integrals when, in some inertial frame $\mathscr{R}$, the electromagnetic field's configuration is invariant under the 1-parameter group of Poincaré symmetries considered.

10.5.3 The Non-quadratic Covariant Lagrangian for the Charged Relativistic Particle

Another Lagrangian, even more frequently employed in the literature than (10.62), is the following:

$$\mathscr{L}_2(r, \gamma, \dot{\gamma}) = -mc\sqrt{-g(\dot{\gamma}, \dot{\gamma})} + \frac{e}{c}g(\dot{\gamma}, A(\gamma)) \,. \tag{10.74}$$

The ambient space is always $A(\mathbb{R} \times \mathbb{M}^4) = \mathbb{R} \times T\mathbb{M}^4$. In principle we seek future-directed timelike curves $I \ni r \mapsto \gamma(r) \in \mathbb{M}^4$ such that the associated local sections of $\mathbb{R} \times T\mathbb{M}^4$ (over $\mathbb{R}$):

$$I \ni r \mapsto (r, \gamma(r), \dot\gamma(r)) \in \mathbb{R} \times T\mathbb{M}^4$$

solve the Euler-Lagrange equations generated by $\mathscr{L}_2$. The condition that γ is timelike ensures that $\sqrt{-g(\dot\gamma, \dot\gamma)} > 0$.

In reality we will see that the solution method is very involved, and the problem is technically *ill posed* since solutions are not unique, despite the fact that uniqueness is somehow re-instated when we look at the graphs of solutions.

For dimensional reasons, r must be a time in order for $\mathscr{L}_2$ to have the dimensions of an energy. The hope (eventually unheeded) is that r coincides with proper time along solutions, or even only that we can reparametrise every solution by proper time τ (this actually will be possible).

Remarks 10.56 A heuristic argument of the sort used in Remarks 10.53 shows that at small velocities (now computed in the parameter r viewed as time) in Minkowski coordinates, renaming r by t, if we assume that $r \approx t$ in classical regimes then:

$$\mathscr{L}_2(s, x(t), \dot x(t)) \approx \frac{m}{2} \sum_{a=1}^{3} \left(\frac{dx^a}{dt} \right)^2 - e\varphi(x(t)) + \frac{e}{c} \sum_{a=1}^{3} \frac{dx^a}{dt} A^a(x(t)), \qquad (10.75)$$

which is the corresponding classical Lagrangian . ∎

In Minkowski coordinates where γ is represented by $I \ni s \mapsto (x^0(r), x^1(r), x^2(r), x^3(r)) \in \mathbb{R}^4$, the order-two Euler-Lagrange equations deduced from $\mathscr{L}_2$ to find γ have the form:

$$mc \sum_{\alpha=1}^{3} g_{\alpha\beta} \frac{d}{dr} \left(\frac{1}{\sqrt{-\sum_{\mu,\nu} g_{\mu\nu} \frac{dx^\mu}{dr} \frac{dx^\nu}{dr}}} \frac{dx^\alpha}{dr} \right)$$

$$= \sum_{\mu=0}^{3} \frac{e}{c} \frac{dx^\nu}{dr} F_{\nu\beta}\Big|_{x(r)}, \qquad \alpha = 0, 1, 2, 3 . \qquad (10.76)$$

where the given vector field A is C^2. The Cauchy problem is set up by adding initial conditions for $r_0 \in I$:

$$(x^0(r_0), x^1(r_0), x^2(r_0), x^3(r_0)) = (x_0^0, x_0^1, x_0^2, x_0^3) , \qquad (10.77)$$

$$\left(\frac{dx^0}{dr}\Big|_{r_0}, \frac{dx^1}{dr}\Big|_{r_0}, \frac{dx^2}{dr}\Big|_{r_0}, \frac{dx^3}{dr}\Big|_{r_0} \right) = (\dot x_0^0, \dot x_0^1, \dot x_0^2, \dot x_0^3) , \qquad (10.78)$$

so that

$$\sum_{\alpha=0}^{3} \dot{x}_0^\alpha \mathbf{e}_\alpha \text{ is a future timelike vector.} \tag{10.79}$$

The system is properly defined only if the γ we are seeking satisfies $\sqrt{-g(\dot{\gamma}, \dot{\gamma})} > 0$. Hence we include this constraint in the problem:

$$\sum_{\alpha=1}^{3} x^\alpha(r) \mathbf{e}_\alpha \text{ is a future timelike vector for } r \in I. \tag{10.80}$$

The general strategy to solve the problem goes as follows. We multiply by $(\sqrt{-\sum_{\mu,\nu=0}^{3} g_{\mu\nu} \frac{dx^\mu}{d\tau} \frac{dx^\nu}{d\tau}})^{-1}$ and introduce the new parameter τ that satisfies:

$$\frac{d}{d\tau} = \frac{c}{\sqrt{-\sum_{\mu,\nu} g_{\mu\nu} \frac{dx^\mu}{dr} \frac{dx^\nu}{dr}}} \frac{d}{dr}.$$

We end up with the same Eq. (10.69) in normal form:

$$m \frac{d^2 x^\alpha}{d\tau^2} = \frac{e}{c} \sum_{\beta,\mu=0}^{3} g^{\alpha\beta} \frac{dx^\mu}{d\tau} F_{\mu\beta}\big|_{x(\tau)}, \quad \alpha = 0, 1, 2, 3. \tag{10.81}$$

Let $\tau_0 := s(r_0)$. As already seen, Eq. (10.81) together with the initial conditions

$$(x^0(\tau_0), x^1(\tau_0), x^2(\tau_0), x^3(\tau_0)) = (x_0^0, x_0^1, x_0^2, x_0^3), \tag{10.82}$$

$$\left(\frac{dx^0}{d\tau}\bigg|_{\tau_0}, \frac{dx^1}{d\tau}\bigg|_{\tau_0}, \frac{dx^2}{d\tau}\bigg|_{\tau_0}, \frac{dx^3}{d\tau}\bigg|_{\tau_0} \right) = \frac{c}{\sqrt{-\sum_{\mu,\nu} g_{\mu\nu} \dot{x}_0^\mu \dot{x}_0^\nu}} (\dot{x}_0^0, \dot{x}_0^1, \dot{x}_0^2, \dot{x}_0^3),$$

$$\tag{10.83}$$

determine a future-directed timelike curve (if the initial conditions are future-like, as in our case). Moreover, as the initial tangent vector satisfies $g(\dot{\gamma}(\tau_0), \dot{\gamma}(\tau_0)) = -c^2$, Theorem 10.34 implies that the parameter τ is proper time, as the notation was trying to imply, and the tangent vector to the maximal solution is the four-velocity.

Despite there is existence and uniqueness when we pass to the parameter τ, the initial problem in r is *ill posed*: it admits infinitely many solutions on every interval $I \ni r_0$. Physically, this is not a major issue, because we still have a sort of uniqueness: the graphs of all these solutions are parts of a *single* graph in $\mathbb{M}^4$, and they are associated with a *unique* curve parametrised by proper time. Let us prove all this.

Theorem 10.57 *Consider the Cauchy problem defined by (10.76), with A of class C^2, and initial conditions (10.77) and (10.78) satisfying (10.79) and also (10.80). This problem admits infinitely many distinct solutions on any interval $I \ni r_0$, for any initial data obeying (10.79).*

Every solution (even non-maximal) comes from the maximal solution

$$J \ni s \mapsto (x^0(\tau), x^1(\tau), x^2(\tau), x^3(\tau)) \in \mathbb{R}^4$$

to problem (10.81) with initial conditions (10.82) and (10.83) (assuming (10.79)) as follows:

$$I \ni r \mapsto (x^0(\tau(r)), x^1(\tau(r)), x^2(\tau(r)), x^3(\tau(r))) \in \mathbb{R}^4 , \tag{10.84}$$

where $I \ni r_0$ is an arbitrary open interval, and $\tau : I \to J$ is any C^2 function with $\frac{d\tau}{dr} > 0$ on I, $\tau(r_0) = \tau_0$ and $\frac{d\tau}{dr}|_{r_0} = c^{-1}\sqrt{-\sum_{\mu,\nu} g_{\mu\nu}\dot{x}_0^\mu \dot{x}_0^\nu}$.

Proof Look at the maximal solution $J \ni s \mapsto (x^0(\tau), x^1(\tau), x^2(\tau), x^3(\tau))$ to problem (10.81) with initial conditions (10.82) and (10.83) (assuming (10.79)). The parameter τ is by construction proper time. Hence if we reparametrise the curve as in (10.84) we may write:

$$0 < \frac{d\tau}{dr} = \sqrt{-\sum_\alpha g_{\alpha\beta} \frac{dx^\alpha(\tau(r))}{dr} \frac{dx^\beta(\tau(r))}{dr}} .$$

As the maps under square root are C^1, the function $\tau = \tau(r)$ is C^2. Therefore system (10.81), for the curve parametrised by τ, can be equivalently written as (10.76) for the curve reparametrised by r. Eventually, any function (10.84) solves the Cauchy problem (10.76)–(10.80).

On a single interval $I \ni r_0$ we can find infinitely many C^2 maps $\tau : I \to J$ with $\frac{d\tau}{dr} > 0$ on I, $\tau(r_0) = \tau_0$ and $\frac{d\tau}{dr}|_{r_0} = c^{-1}\sqrt{-\sum_{\mu,\nu} g_{\mu\nu}\dot{x}_0^\mu \dot{x}_0^\nu}$. Therefore there exist infinitely many solutions to the given problem.

Next we show that problem (10.76)–(10.80) can be put in the above form. If

$$I \ni r \mapsto (z^0(r), z^1(r), z^2(r), z^3(r)) \in \mathbb{R}^4$$

is a solution (hence C^2), we can reparametrise it by arc length $\tau = \tau(r)$ (a C^2 map by construction, with C^2 inverse since $\frac{d\tau}{dr} > 0$ at present). The new function

$$\tau(I) \ni \tau \mapsto (z^0(r(\tau)), z^1(r(\tau)), z^2(r(\tau)), z^3(r(\tau))) \in \mathbb{R}^4$$

is necessarily a restriction of the maximal solution $J \ni \tau \mapsto (x^0(\tau), x^1(\tau), x^2(\tau), x^3(\tau))$ to (10.81)–(10.79). The initial solution then reads:

$$I \ni r \mapsto (z^0(r), z^1(r), z^2(r), z^3(r))$$
$$= (x^0(\tau(r)), x^1(\tau(r)), x^2(\tau(r)), x^3(\tau(r))) \in \mathbb{R}^4 .$$

By construction: $\frac{d\tau}{dr} > 0$ on I, $\tau(r_0) = \tau_0$ and $\frac{d\tau}{dr}|_{r_0} = c^{-1} \sqrt{-\sum_{\mu,\nu} g_{\mu\nu} \dot{x}_0^\mu \dot{x}_0^\nu}$.   □

Now, an interesting point is what might be the obstruction to applying Theorem 14.23 to a system of type (10.76) reduced to first order by the usual technique. That there must be some obstruction is evident a posteriori, because otherwise the theorem would also give us unique solutions, which we know is not true! Indeed, the problem is that the system *cannot be put in normal form*. The computation shows that, under $\sqrt{-g(\dot{\gamma}(r), \dot{\gamma}(r))} > 0$, system (10.76) is equivalent to:

$$mc \sum_{\alpha=0}^{3} \left(\delta_\alpha^\beta - \frac{\sum_\delta g_{\alpha\delta} \frac{dx^\delta}{dr} \frac{dx^\beta}{dr}}{\sum_{\sigma,\tau} g_{\sigma\tau} \frac{dx^\sigma}{dr} \frac{dx^\tau}{dr}} \right) \frac{d^2 x^\alpha}{dr^2}$$

$$= \sum_{\mu=0}^{3} \frac{e}{c} \frac{dx^\nu}{dr} F_{\nu\beta}\Big|_{x(r)} , \qquad \alpha = 0, 1, 2, 3 . \tag{10.85}$$

But since

$$\det \left(I - v u^t \right) = 1 - v^t u \quad \text{for column vectors } u, v \text{ in } \mathbb{R}^n ,$$

we have

$$\det \left[\delta_\alpha^\beta - \frac{\sum_\delta g_{\alpha\delta} \frac{dx^\delta}{dr} \frac{dx^\beta}{dr}}{\sum_{\sigma,\tau} g_{\sigma\tau} \frac{dx^\sigma}{dr} \frac{dx^\tau}{dr}} \right]_{\alpha,\beta=0,1,2,3} = 1 - \frac{\sum_{\delta,\alpha} g_{\alpha\delta} \frac{dx^\delta}{dr} \frac{dx^\alpha}{dr}}{\sum_{\sigma,\tau} g_{\sigma\tau} \frac{dx^\sigma}{dr} \frac{dx^\tau}{dr}} = 0 .$$

Therefore, we cannot write system (10.76) in normal form multiplying both sides of (10.85) by the inverse of:

$$\left[\delta_\alpha^\beta - \frac{\sum_{\delta=0}^{3} g_{\alpha\delta} \frac{dx^\delta}{dr} \frac{dx^\beta}{dr}}{\sum_{\sigma,\tau=0}^{3} g_{\sigma\tau} \frac{dx^\sigma}{dr} \frac{dx^\tau}{dr}} \right]_{\alpha,\beta=0,1,2,3} ,$$

simply because the inverse does not exist. At any rate, we have manufactured an independent existence argument, and checked directly the lack of uniqueness in the proof of Theorem 10.57 .

10.5.4   First Integrals of the Non-quadratic Covariant Lagrangian

The Lagrangian $\mathscr{L}_2$ in (10.74) does not depend on the parameter r explicitly and so we can invoke Jacobi's Theorem, obtaining that the Hamiltonian is conserved along the solutions of the Euler-Lagrange equations. Notwithstanding, the direct computation shows that:

$$\mathscr{H}(r, \gamma, \dot{\gamma}) = 0, \tag{10.86}$$

where zero means the *zero function*. This fact has a number of serious consequences in *Hamilton's formulation*, which is presented in the next chapter. The Lagrangian (10.74), in contrast to (10.62), *does not* admit a Hamiltonian formulation. (It can be shown that this pathology is related to the property of the *action functional* built from (10.74) of being invariant under reparametrisations.)

Let us pass to study the constants of motion generated by the Poincaré invariance of $\mathscr{L}_2$ in absence of electromagnetic field (or in presence of it, but with additional symmetry assumption on the electromagnetic four-potential).

Remarks 10.58 It is immaterial that we are working with a Lagrangian whose Euler-Lagrange equations do not have unique solutions. The version of Noether's Theorem we shall use (Theorem 8.10) only assumes that there exist solutions, since it does not mention the dynamic vector field Z on $A(\mathbb{R} \times \mathbb{R}^4)$. This field does not exist for the Lagrangian $\mathscr{L}_2$, as opposed to $\mathscr{L}$, precisely because to define Z one needs the Existence and uniqueness theorem.   ∎

We have the familiar 10 one-parameter (additive) subgroups of $IO(1, 3)_+$ already seen as regards $\mathscr{L}$.

(1) **Temporal displacements**: $\begin{cases} x_\epsilon'^0 = x^0 + \epsilon \\ x_\epsilon'^a = x^a, \quad a = 1, 2, 3 \end{cases}$ where $\epsilon \in \mathbb{R}$;

(2) **spatial displacements along $\mathbf{e}_1$**: $\begin{cases} x_\epsilon'^1 = x^1 + \epsilon \\ x_\epsilon'^\beta = x^\beta, \quad \beta = 0, 2, 3 \end{cases}$ where $\epsilon \in \mathbb{R}$,

plus the two, similar, spatial displacements along $\mathbf{e}_2$ and $\mathbf{e}_3$;

(3) **spatial rotations around $\mathbf{e}_3$**: $\begin{cases} x_\epsilon'^1 = x^1 \cos \epsilon - x^2 \sin \epsilon \\ x_\epsilon'^2 = x^1 \sin \epsilon + x^2 \cos \epsilon \\ x_\epsilon'^\beta = x^\beta, \quad \beta = 0, 3 \end{cases}$ where $\epsilon \in \mathbb{R}$, plus

the two analogous spatial rotations around $\mathbf{e}_1$ and $\mathbf{e}_2$;

(4) **special Lorentz transformations along $\mathbf{e}_3$**: $\begin{cases} x_\epsilon'^0 = x^0 \cosh \epsilon + x^3 \sinh \epsilon \\ x_\epsilon'^3 = x^0 \sinh \epsilon + x^3 \cosh \epsilon \\ x_\epsilon'^a = x^a, \quad a = 1, 2 \end{cases}$

$\epsilon \in \mathbb{R}$ plus the two other special Lorentz transformations along $\mathbf{e}_1$ and $\mathbf{e}_2$.

The extended action of these 10 subgroups to the dotted coordinates is defined as explained in Sect. 8.2.1, with the usual crucial interpretative difference: now the bundle's base is no longer absolute time but the r-axis, and the fibres are copies of Minkowski spacetime. The above groups act on the dotted coordinates by the same formulas where all coordinates become dotted coordinates, except for spacetime displacements where:

$$\dot{x}_\epsilon'^\mu = \dot{x}^\mu , \quad \mu = 0, 1, 2, 3 .$$

The Lagrangian:

$$\mathcal{L}_2(r, x, \dot{x}) = -mc\sqrt{-\sum_{\mu,\nu=0}^{3} g_{\mu\nu}\dot{x}^\mu \dot{x}^\nu} ,$$

satisfies $\mathcal{L}_2 \circ \phi_\epsilon = \mathcal{L}_2$ for any $\epsilon \in \mathbb{R}$, where $\{\phi_\epsilon\}_{\epsilon \in \mathbb{R}}$ is any one of the previous one-parameter subgroups of $O(1, 3)_+$ extended to $A(\mathbb{R} \times \mathbb{M}^4)$. We therefore have corresponding Noether first integrals:

$$I(s, x, \dot{x}) := \sum_{\mu=0}^{3} \frac{\partial \mathcal{L}_2}{\partial \dot{x}^\mu} \frac{\partial x_\epsilon'^\mu}{\partial \epsilon}\bigg|_{\epsilon=0}$$

in agreement with Theorem 8.10.

(1) For temporal displacements Noether's first integral is

$$mc\frac{\dot{x}^0}{\sqrt{-\sum_{\mu,\nu} g_{\mu\nu}\dot{x}^\mu \dot{x}^\nu}} = P^0 ,$$

corresponding, along the motion, to the 0th component in $\mathcal{R}$ of the four-momentum, since the parameter τ that satisfies:

$$\frac{d}{d\tau} = \frac{c}{\sqrt{-\sum_{\mu,\nu} g_{\mu\nu}\frac{dx^\mu}{dr}\frac{dx^\nu}{dr}}}\frac{d}{dr} ,$$

is proper time.

(2) For spatial displacements along $\mathbf{e}_1$ we have Noether first integral

$$mc\frac{\dot{x}^1}{\sqrt{-\sum_{\mu,\nu} g_{\mu\nu}\dot{x}^\mu \dot{x}^\nu}} = P^1 ,$$

i.e. the 1st component in $\mathcal{R}$ of the four-momentum along the motion, when using proper time τ. The other spatial displacements along $\mathbf{e}_2$ and $\mathbf{e}_3$ give similar first integrals.

(3) In the case of spatial rotations around $\mathbf{e}_3$ we have Noether first integral

$$x^1 P^2 - x^2 P^1 = L^3 \,,$$

corresponding along the motion to the 3rd component in $\mathscr{R}$ of the angular momentum in $\mathscr{R}$ with respect to the axes' spatial origin, when using proper time τ. The spatial rotations around $\mathbf{e}_1$, $\mathbf{e}_2$ give similar first integrals.

(4) For special Lorentz transformations along $\mathbf{e}_3$ we obtain Noether first integral (up to a sign)

$$x^0 P^3 - x^3 P^0 = K^3 \,,$$

corresponding along the motion to the 3rd component in $\mathscr{R}$ of the **boost vector**, using proper time τ. The remaining two special transformations, along $\mathbf{e}_1$ and $\mathbf{e}_2$, give similar first integrals.

In presence of an electromagnetic field whose Lagrangian is still (10.63) and not, for example, explicitly dependent on time $t = x^0/c$ in the frame $\mathscr{R}$, we have a first integral that accounts for the external fields. In the case at hand:

$$I = \left.\frac{\partial \mathscr{L}_2}{\partial \dot{x}^0}\frac{\partial x_\epsilon'^0}{\partial \epsilon}\right|_{\epsilon=0} = \frac{mc\dot{x}^0}{\sqrt{-\sum_{\mu,\nu}g_{\mu\nu}\frac{dx^\mu}{dr}\frac{dx^\nu}{dr}}} + \frac{e}{c}\varphi(x^1, x^2, x^3) \,.$$

Using proper time along the motion we find, as for $\mathscr{L}$:

$$cI = \frac{mc^2}{\sqrt{1 - \frac{v^2}{c^2}}} + e\varphi(x^1, x^2, x^3) \,.$$

In a non-relativistic regime this recovers the energy conservation of a charge in an electrostatic field (apart from the customary, and classically superfluous, additive constant mc^2):

$$cI \approx mc^2 + \frac{1}{2}mv^2 + e\varphi(x^1, x^2, x^3) \,.$$

Similar electromagnetic contributions are present for the other first integrals whenever, in some inertial frame $\mathscr{R}$, the electromagnetic field's configuration is invariant under the 1-parameter group of Poincaré symmetries of concern.

10.5.5   The Non-quadratic and Non-covariant Lagrangian for the Charged Relativistic Particle

Finally, let us examine a Lagrangian not in intrinsic form, that is to say a Lagrangian depending on a chosen Minkowski coordinate system $\psi : \mathbb{M}^4 \ni p \mapsto (x^0 = ct, x^1, x^2, x^3) \in \mathbb{R}^4$.

The spacetime of configurations is now the trivial bundle $\mathbb{R} \times \mathbb{R}^3$ where the base, the factor $\mathbb{R}$, is the axis of global time t and the configuration space $\mathbb{R}^3$ is where the spatial coordinates x^1, x^2, x^3 live. The Lagrangian

$$\mathscr{L}_3(t, x, \dot{x}) := -mc^2 \sqrt{1 - \sum_{a=1}^{3} \left(\frac{\dot{x}^a}{c}\right)^2} - e\varphi(t, x) + \frac{e}{c} \sum_{a=1}^{3} \dot{x}^a A^a(t, x) \qquad (10.87)$$

is defined on the spacetime of kinetic states $A(\mathbb{R} \times \mathbb{R}^3) = \mathbb{R} \times \mathbb{R}^3 \times \mathbb{R}^3$. The third term $\mathbb{R}^3$ is where the dotted coordinates $\dot{x}^1, \dot{x}^2, \dot{x}^3$ live. Assuming $\sqrt{1 - \sum_{a=1}^{3} \left(\frac{\dot{x}^a}{c}\right)^2} > 0$, the equations of motion of order two read:

$$m \frac{d}{dt}\left(\frac{1}{\sqrt{1 - \sum_b \frac{1}{c^2}\left(\frac{dx^b}{dt}\right)^2}} \frac{dx^a}{dt}\right) = -e\frac{\partial\varphi}{\partial x^a} + \frac{e}{c}\sum_{b=1}^{3}\frac{dx^b}{dt}F_{ab}, \ a = 1, 2, 3,$$

$$(10.88)$$

with initial conditions:

$$(x^1(t_0), x^2(t_0), x^3(t_0)) = (x_0^1, x_0^2, x_0^3), \qquad (10.89)$$

$$\left(\frac{dx^1}{dt}\bigg|_{t_0}, \frac{dx^2}{dt}\bigg|_{t_0}, \frac{dx^3}{dt}\bigg|_{t_0}\right) = (\dot{x}_0^1, \dot{x}_0^2, \dot{x}_0^3), \qquad (10.90)$$

with the additional request that the initial four-vector is future-directed and timelike, meaning:

$$\sum_{a=1}^{3}(\dot{x}^a)^2 < c^2. \qquad (10.91)$$

The following result holds.

Theorem 10.59 *If φ and A^1, A^2, A^3 are C^2 maps, there is a unique maximal solution to Cauchy problem (10.88) with initial data (10.89) and (10.90) satisfying (10.91). (In particular any other solution with that initial data is a restriction.)*

Reparametrising the maximal solution by proper time, the new curve is the only maximal solution (10.69) with initial conditions corresponding to (10.89) and (10.90).

Proof Suppose there exists a future-directed timelike solution to (10.88). Multiplying (10.88) by $1/\sqrt{1 - \sum_b \frac{1}{c^2}\left(\frac{dx^b}{dt}\right)^2}$ we can write the equation using proper time τ, and thus recover (10.69) with τ replacing s in the 3 spatial components of the four-velocity:

$$m\frac{dV^a}{d\tau} = \frac{e}{c}\sum_{\mu,\beta=0}^{3} V^\mu g^{a\beta}\, F_{\mu\beta}\big|_{x(\tau)}\,, \qquad a = 1,2,3\,.$$

The time component V^0 satisfies the same system due to a general fact ((8) in Exercises 10.61). Let us provide an argument in any case. From $g(V,V) = -c^2$ and $V^0 > 0$:

$$m\frac{dV^0}{d\tau} = m\frac{d\sqrt{c^2 + \sum_{b=1}^{3}(V^b)^2}}{d\tau} = m\frac{1}{V^0}\sum_{a=1}^{3} V^a\frac{dV^a}{d\tau} = \frac{e}{c}\sum_{\mu=0}^{3}\sum_{a=1}^{3}\frac{V^a}{V^0}V^\mu\, F_{\mu a}\big|_{x(\tau)}$$

$$= \frac{e}{c}\sum_{a,b=1}^{3}\frac{V^a V^b}{V^0}\, F_{ba}\big|_{x(\tau)} + \frac{e}{c}\sum_{a=1}^{3}\frac{V^a V^0}{V^0}\, F_{0a}\big|_{x(\tau)}\,.$$

On the last line the first summand vanishes since F_{ab} is skew-symmetric. Hence:

$$m\frac{dV^0}{d\tau} = \frac{e}{c}\sum_{a=1}^{3} V^a\, F_{0a}\big|_{x(\tau)} = \frac{e}{c}\sum_{\mu=0}^{3} V^\mu\, F_{0\mu}\big|_{x(\tau)}$$

$$= -\frac{e}{c}\sum_{\mu,\beta=0}^{3} V^\mu\, g^{0\beta} F_{\beta\mu}\big|_{x(\tau)} = \frac{e}{c}\sum_{\mu,\beta=0}^{3} V^\mu\, g^{0\beta} F_{\mu\beta}\big|_{x(\tau)}$$

where we used $F_{00} = 0$, $g^{0\beta} = -\delta^{0\beta}$ and $F_{\beta\mu} = -F_{\mu\beta}$. We have shown that the time component of the four-velocity, too, satisfies system (10.88). All in all, every solution of (10.88) written in terms of four-velocity satisfies (10.69) with s being proper time. The procedure also goes the other way around: starting from a future-directed timelike solution to (10.69) and proceeding backwards we find a solution to (10.88) when we read the spatial components of the four-velocity in terms of the velocity. More precisely: every solution to problem (10.88) with initial conditions subject to (10.91) solves problem (10.69) with corresponding future-directed timelike initial data. And conversely. Since, for φ and A^1, A^2, A^3 of class C^2, problem (10.69) with future-directed timelike initial data satisfies the Existence and uniqueness theorem for maximal solutions, the proof is concluded. □

Remarks 10.60

(1) Despite $\mathscr{L}_3$ depends on the Minkowski coordinates, the *action* functional does not, and coincides with the action functional built from the Lagrangian $\mathscr{L}_2$. For simplicity let us examine the case $e = 0$, as the general situation is not dissimilar:

$$I[\gamma] = -mc^2 \int_{t_1}^{t_2} \sqrt{1 - \sum_{a=1}^{3} \frac{1}{c^2} \left(\frac{dx^a}{dt} \right)^2 } \, dt$$

$$= -mc \int_{t_1}^{t_2} \sqrt{ \left(\frac{dx^0}{dt} \right)^2 - \sum_{a=1}^{3} \left(\frac{dx^a}{dt} \right)^2 } \, dt$$

$$= -mc \int_{s_1}^{s_2} \sqrt{-g(\dot{\gamma}(s), \dot{\gamma}(s))} \, ds \; .$$

The functional I works on future-like curves $\gamma = \gamma(t)$ connecting two given events $p_1 \equiv (ct_1, x_1^1, x_1^2, x_1^3)$ and $p_2 \equiv (ct_2, x_1^2, x_2^2, x_2^3)$, and $s = s(t)$ is any reparametrisation (C^2, with $\frac{ds}{dt} > 0$) such that $s_1 := s(t_1)$, $s_2 = s(t_2)$.

(2) We would find the same action functional from the previous relation (even when $e \neq 0$) using other Minkowski coordinates. Remembering Theorem 9.2, the solutions of the Euler-Lagrange equations generated by $\mathscr{L}_3$ in some coordinates are solutions for the analogous Lagrangian associated to different Minkowski coordinates, whenever the first solutions are written in the second coordinate set. We leave the details of the proof as exercise.

(3) As usual, when speeds are small compared to c, $\mathscr{L}_3$ approximates the classical Lagrangian up to the additive constant mc^2. The proof is left to the reader.

(4) The Lagrangian we are treating, neither quadratic nor covariant, is so to speak the "physically correct" one in terms of the Hamiltonian formulation (addressed in Chap. 11) and the quantum theory of elementary particles. Within the quantum formulation the notion of proper time for elementary particles loses its meaning, and the temporal evolution is described with respect to the global time of the Minkowski frame in use. Working with the non-quadratic and non-covariant Lagrangian (in absence of electromagnetic field) the Hamiltonian (see Chap. 11 and Sect. 13.2.3) coincides with the time component of the four-momentum. This fact is required, after the quantisation process, to describe elementary particles *à la Wigner*. ∎

10.5.6 First Integrals of the Non-quadratic, Non-covariant Lagrangian

The Lagrangian (10.87) admits a non-trivial Hamiltonian function:

$$\mathscr{H}(t, x, \dot{x}) = \frac{mc^2}{\sqrt{1 - \sum_{a=1}^{3}\left(\frac{\dot{x}^a}{c^2}\right)^2}} + e\varphi(t, x^1, x^2, x^3) \,. \tag{10.92}$$

If φ and the three components A^a are not functions of t, in particular if they are zero, the Lagrangian is invariant under the continuous group of *temporal displacements*. We can apply Jacobi's Theorem, so the Hamiltonian is conserved, as t varies, along the solutions of the Euler-Lagrange equations. It was shown several times that the above coincides, at small velocities, with the total mechanical + electrostatic energy, reflecting the conservation law we have in Classical Physics. Finally observe that in absence of external fields:

$$\mathscr{H}(t, x, \dot{x}) = cP^0 \,,$$

i.e. the Hamiltonian coincides with the time component of the four-momentum (times a factor c) along the system's motion.

Let us study the constants of motion produced by the invariance of:

$$\mathscr{L}_3(t, x, \dot{x}) = -mc^2\sqrt{1 - \sum_{a=1}^{3}\left(\frac{\dot{x}^a}{c}\right)^2} \,, \tag{10.93}$$

under the Poincaré group, hence in absence of electromagnetic field.

We already have accounted for the group of temporal displacements, so we concentrate on the other 9 (additive) subgroups of $IO(1, 3)_+$.

Let us start from the groups of *internal transformations* of the frame $\mathscr{R}$ on which the Minkowski coordinates we are using move. Such groups preserve the temporal fibres of the trivial bundle $\mathbb{R} \times \mathbb{R}^3$, where $\mathbb{R}$ is the time axis. The extension to the dotted coordinates proceeds in the usual way known from Classical Mechanics as per Sect. 8.2.1. Let us write the lifted action to $A(\mathbb{R} \times \mathbb{R}^3) = \mathbb{R} \times \mathbb{R} \times \mathbb{R}^3$. First of all we have the group of **spatial displacements along** x^3: $\begin{cases} x'^3_\epsilon = x^3 + \epsilon \\ x'^b_\epsilon = x^b \,, & b = 1, 2 \\ \dot{x}'^a_\epsilon = \dot{x}^a \,, & \text{a=1,2,3,} \end{cases}$

where $\epsilon \in \mathbb{R}$, and the analogues along x^1 and x^2. Evidently $\mathscr{L}_3$ is invariant under such 1-parameter subgroups. For example, invariance under spatial displacements along $\mathbf{e}_1$, by Theorem 8.10, gives Noether first integral:

$$\sum_{a=1}^{3} \frac{\partial \mathscr{L}_3}{\partial \dot{x}^a} \frac{\partial x_\epsilon'^a}{\partial \epsilon}\bigg|_{\epsilon=0} = \frac{m\dot{x}^1}{\sqrt{1 - \sum_{a=1}^{3} \left(\frac{\dot{x}^a}{c^2}\right)^2}} = P^1,$$

corresponding along the motion to the component P^1 in $\mathscr{R}$ of the four-momentum. The other two spatial displacements along $\mathbf{e}_2$ and $\mathbf{e}_3$ produce similar first integrals. Now to **spatial rotations around $\mathbf{e}_3$**:

$$\begin{cases} x_\epsilon'^1 = x^1 \cos \epsilon - x^2 \sin \epsilon \\ x_\epsilon'^2 = x^1 \sin \epsilon + x^2 \cos \epsilon \\ x_\epsilon'^b = x^b, \quad b = 0, 3 \\ \dot{x}_\epsilon'^1 = \dot{x}^1 \cos \epsilon - \dot{x}^2 \sin \epsilon \\ \dot{x}_\epsilon'^2 = \dot{x}^1 \sin \epsilon + \dot{x}^2 \cos \epsilon \\ \dot{x}_\epsilon'^b = \dot{x}^b, \quad b = 0, 3 \end{cases}$$

where $\epsilon \in \mathbb{R}$, remembering we have spatial rotations around $\mathbf{e}_1$ and $\mathbf{e}_2$ as well. Again, $\mathscr{L}_3$ is invariant under such 1-parameter subgroups. For rotations around $\mathbf{e}_3$, Theorem 8.10 gives us Noether first integral

$$\sum_{a=1}^{3} \frac{\partial \mathscr{L}_3}{\partial \dot{x}^a} \frac{\partial x_\epsilon'^a}{\partial \epsilon}\bigg|_{\epsilon=0} = x^1 P^2 - x^2 P^1 = L^3,$$

i.e. the component L^3 in $\mathscr{R}$ of the angular momentum in $\mathscr{R}$ with respect to the spatial origin of the axes, along the motion. Spatial rotations around $\mathbf{e}_1$ and $\mathbf{e}_2$ give analogous first integrals.

Way more difficult is the study of the group of **special Lorentz transformations along $\mathbf{e}_3$**:

$$\begin{cases} x_\epsilon'^0 = x^0 \cosh \epsilon + x^3 \sinh \epsilon \\ x_\epsilon'^3 = x^0 \sinh \epsilon + x^3 \cosh \epsilon \\ x_\epsilon'^b = x^b, \quad b = 1, 2 \end{cases}$$

where $\epsilon \in \mathbb{R}$. (Pure transformations along $\mathbf{e}_1$ and $\mathbf{e}_2$ are similar.) We have only written the action on $\mathbb{R} \times \mathbb{R}^3$. The induced action on $A(\mathbb{R} \times \mathbb{R}^3)$ is not immediate, since the transformations do not preserve the temporal fibres of the bundle $\mathbb{R} \times \mathbb{R}^3$ over the time line, parametrised by $\mathbb{R} \ni t$. In order to understand the action on the dotted coordinates we would better remember that the dotted coordinates are identified with tangent vectors to local sections of $\mathbb{R} \times \mathbb{R}^3$ over $\mathbb{R}$. The above special transformations can be used to map local sections $x^a = x^a(t)$ to local sections

$x'^a_\epsilon = x'^a_\epsilon(t'_\epsilon)$ as follows:

$$\begin{cases} ct'_\epsilon = ct \cosh\epsilon + x^3(t)\sinh\epsilon \\ x'^3_\epsilon(t'_\epsilon) = ct \sinh\epsilon + x^3(t)\cosh\epsilon \\ x'^b_\epsilon(t'_\epsilon) = x^b(t)\,, \quad b = 1,2 \end{cases}$$

Therefore:

$$\frac{dx'^3_\epsilon}{dt'_\epsilon} = \frac{dx'^3_\epsilon}{dt}\left(\frac{dt'_\epsilon}{dt}\right)^{-1} = \frac{c\sinh\epsilon + \frac{dx^3}{dt}\cosh\epsilon}{\cosh\epsilon + \frac{1}{c}\frac{dx^3}{dt}\sinh\epsilon}$$

and so:

$$\begin{cases} \dot{x}'^3_\epsilon(t'_\epsilon, x'_\epsilon) = \frac{c\sinh\epsilon + \dot{x}^3(t,x)\cosh\epsilon}{\cosh\epsilon + \frac{1}{c}\dot{x}^3(t,x)\sinh\epsilon} \\ \dot{x}'^b_\epsilon(t'_\epsilon, x'_\epsilon) = \dot{x}^b(t,x)\,, \quad b = 1,2 \end{cases}$$

We would obtain the same if we let the transformation $\Lambda_3(\epsilon)$ act on the components of the four-velocity. Given the complexity, it is better to work with the vector fields on $A(\mathbb{R}\times\mathbb{R}^3)$ that generate the extended 1-parameter group, as we extensively explained in Sect. 8.5. Directly (computing the derivatives in ϵ at $\epsilon = 0$ of t'_ϵ, x'^a_ϵ, $\dot{x}'^a_\epsilon$):

$$X = \frac{x^3}{c}\frac{\partial}{\partial t} + ct\frac{\partial}{\partial x^3} + c\left(1 - \left(\frac{\dot{x}^3}{c}\right)^2\right)\frac{\partial}{\partial \dot{x}^3}. \tag{10.94}$$

It is here that a number of problems crop up in order for us to apply Noether's Theorem. For starters, the dynamic vector field of the Lagrangian (10.93) is clearly:

$$Z = \frac{\partial}{\partial t} + \sum_{a=1}^{3}\dot{x}^a\frac{\partial}{\partial x^a}\,, \tag{10.95}$$

since the solutions of the Euler-Lagrange equations of such Lagrangian are the curves $x^a(t) = c^a + b^a t$ for any possibile constant $c^a, b^a \in \mathbb{R}$, $a = 1, 2, 3$. Compared to the general form (8.53), now the field X in (10.94) violates both (8.54) (i.e. (8.62)) and (8.55), and these conditions are crucial for Noether's Theorem in the formulation of Theorem 8.22. At last, the Lagrangian does not satisfy $X(\mathcal{L}_3) = 0$. That said, we next show that an *ad hoc* adaptation of Noether's proof allows us to deduce the conservation of the boost's K^3 component under the 1-parameter group generated by X. First of all, we notice that using (10.94) and the explicit expressions of Z and $\mathcal{L}_3$, trivial computations produce

$$X(\mathcal{L}_3) = Z\left(\frac{mcx^3\left(1 - \frac{(\dot{x}^3)^2}{c^3}\right)}{\sqrt{1 - \sum_{a=1}^3 \left(\frac{\dot{x}^a}{c}\right)^2}}\right).$$

We now need to re-write X as:

$$X = \frac{x^3}{c}\left(\frac{\partial}{\partial t} + \dot{x}^3 \frac{\partial}{\partial x^3}\right) + c\left(t - \frac{x^3\dot{x}^3}{c^2}\right)\frac{\partial}{\partial x^3} + c\left(1 - \frac{(\dot{x}^3)^2}{c^2}\right)\frac{\partial}{\partial \dot{x}^3}$$

In other terms:

$$X = \frac{x^3}{c}\left(\frac{\partial}{\partial t} + \dot{x}^3 \frac{\partial}{\partial x^3}\right) + c\left(t - \frac{x^3\dot{x}^3}{c^2}\right)\frac{\partial}{\partial x^3} + Z\left(c\left(t - \frac{x^3\dot{x}^3}{c^2}\right)\right)\frac{\partial}{\partial \dot{x}^3}.$$

$$(10.96)$$

Since $\mathcal{L}_3$ does not explicitly depend on t and x^3,

$$X(\mathcal{L}_3) = 0 + c\left(t - \frac{x^3\dot{x}^3}{c^2}\right)\frac{\partial}{\partial x^3}\mathcal{L}_3 + Z\left(c\left(t - \frac{x^3\dot{x}^3}{c^2}\right)\right)\frac{\partial}{\partial \dot{x}^3}\mathcal{L}_3.$$

As $Z\left(\frac{\partial}{\partial \dot{x}^i}\mathcal{L}_3\right) = \frac{\partial}{\partial \dot{x}^i}\mathcal{L}_3$, we can re-arrange the above as

$$X(\mathcal{L}_3) = c\left(t - \frac{x^3\dot{x}^3}{c^2}\right)Z\left(\frac{\partial}{\partial \dot{x}^3}\mathcal{L}_3\right) + Z\left(c\left(t - \frac{x^3\dot{x}^3}{c^2}\right)\right)\frac{\partial}{\partial \dot{x}^3}\mathcal{L}_3$$

$$= Z\left(c\left(t - \frac{x^3\dot{x}^3}{c^2}\right)\frac{\partial\mathcal{L}_3}{\partial \dot{x}^3}\right).$$

Comparing the two expressions of $X(\mathcal{L}_3)$ we find:

$$Z\left(c\left(t - \frac{x^3\dot{x}^3}{c^2}\right)\frac{\partial\mathcal{L}_3}{\partial \dot{x}^3} - \frac{mcx^3\left(1 - \frac{(\dot{x}^3)^2}{c^2}\right)}{\sqrt{1 - \sum_{a=1}^3 \left(\frac{\dot{x}^a}{c}\right)^2}}\right) = 0.$$

Hence the argument of Z, let us call it I, is a first integral. Its expression is:

$$
I = c\left(t - \frac{x^3\dot{x}^3}{c^2}\right)\frac{\partial\mathscr{L}_3}{\partial\dot{x}^3} - \frac{mcx^3\left(1 - \frac{(\dot{x}^3)^2}{c^2}\right)}{\sqrt{1 - \sum_{a=1}^{3}\left(\frac{\dot{x}^a}{c}\right)^2}}
$$

$$
= \frac{mc\left(t - \frac{x^3\dot{x}^3}{c^2}\right)\dot{x}^3}{\sqrt{1 - \sum_{a=1}^{3}\left(\frac{\dot{x}^a}{c}\right)^2}} - \frac{mcx^3\left(1 - \frac{(\dot{x}^3)^2}{c^2}\right)}{\sqrt{1 - \sum_{a=1}^{3}\left(\frac{\dot{x}^a}{c}\right)^2}}
$$

$$
= \frac{mct\dot{x}^3 - mcx^3}{\sqrt{1 - \sum_{a=1}^{3}\left(\frac{\dot{x}^a}{c}\right)^2}} = K^3 .
$$

10.5.7 Extension of the Formalism to N Point-Particles

In contrast to the previous two Lagrangian formulations, the use of the frame's global time coordinate $t = x^0/c$ as evolution parameter allows us to extend the third Lagrangian formulation immediately, to the case of N non-interacting particles, or particles inside an external electromagnetic field. Physically, at least one portion of this field would be generated by the particles themselves, so we would need to complement the particles' equations of motion with Maxwell's equations,[7] whose role is to describe how the particles generate the electromagnetic field (see the discussion relative to equation (10.23)).

The Lagrangian of the system just introduced coincides with the sum of the single Lagrangians, when using the frame's time coordinate $t = x^0/c$ as common parameter for the evolution:

$$
\mathscr{L}_3(t, x_1, \ldots, x_N, \dot{x}_1, \ldots \dot{x}_N) := \sum_{i=1}^{N}\left(-m_ic^2\sqrt{1 - \sum_{a=1}^{3}\left(\frac{\dot{x}_i^a}{c^2}\right)^2}\right.
$$

$$
\left. -e_i\varphi(t, x_i) + \frac{e_i}{c}\sum_{a=1}^{3}\dot{x}^a A^a(t, x_i)\right) . \tag{10.97}
$$

Clearly such a Lagrangian produces the correct Euler-Lagrange equations for each single particle.

---

[7] As is well known [Jac98], the ensuing theory produces mathematical inconsistencies caused by the *infinite eigenforce* each particles exerts on itself, and there is no (non-quantum, and mathematically consistent) theory for charged *point-particles* that are affected by and participate in creating the electromagnetic field.

Now we shall focus on the case where there is *no electromagnetic field* and make some comments on the first integrals coming from Noether's (and Jacobi's) theorem as a consequence of the Lagrangian's invariance under the Poincaré group, with the only exception of special Lorentz transformations. As for the one-particle case, the first-order action of special transformations (along the spatial axis x^a) on the Lagrangian is generated by the vector field sum of the single particles' analogous vector fields:

$$X_{(a)} = \sum_{i=1}^{N} \left(\frac{x_i^a}{c} \frac{\partial}{\partial t} + ct \frac{\partial}{\partial x_i^a} + c \left(1 - \left(\frac{\dot{x}_i^a}{c} \right)^2 \right) \frac{\partial}{\partial \dot{x}_i^a} \right),$$

as the reader can easily check. Furthermore:

$$X_{(a)}(\mathscr{L}_3) = Z \left(\sum_{i=1}^{N} \frac{m_i c x_i^a \left(1 - \frac{(\dot{x}_i^a)^2}{c^3} \right)}{\sqrt{1 - \sum_{b=1}^{3} \left(\frac{\dot{x}_i^b}{c} \right)^2}} \right),$$

where the dynamic vector filed of the N non-interacting particles reads:

$$Z = \frac{\partial}{\partial t} + \sum_{i=1}^{N} \sum_{b=1}^{3} \dot{x}_i^b \frac{\partial}{\partial x_i^b}.$$

In this situation, the same procedure employed for one particle allows to adapt ad hoc Noether's theorem, as we used it above, and shows that the system's total boost

$$K^a(t, x_1, \ldots, x_N, \dot{x}_1, \ldots \dot{x}_N) := \sum_{i=1}^{N} \frac{m_i ct\dot{x}_i^a - m_i c x_i^a}{\sqrt{1 - \sum_{b=1}^{3} \left(\frac{\dot{x}_i^b}{c} \right)^2}},$$

understood as parametrically (or explicitly) depending on time, is preserved along the motion; that is it is preserved when we replace in the above formula each x_i by the curve $\mathbb{R} \ni t \mapsto x_i(t)$ describing the evolution in time of the i-th particle.

The action of *each* one-parameter subgroup of the Poincaré group corresponds to the trivial action on each one of the particles separately. For each one-parameter subgroup, the conserved quantities are just sums of the first integrals of each particle alone. In particular, The Lagrangian invariance under spatial displacements along the axis x^a will give the conservation of the *total relativistic impulse*:

$$P^a(t, x_1, \ldots, x_N, \dot{x}_1, \ldots \dot{x}_N) := \sum_{i=1}^{N} P_i^a = \sum_{i=1}^{N} \frac{m_i \dot{x}_i^a}{\sqrt{1 - \sum_{a=1}^{3} \left(\frac{\dot{x}_i^a}{c^2} \right)^2}}$$

on the system's motions. The invariance under time displacements, by Jacobi's theorem, forces the conservation along the motions of the *total relativistic energy*:

$$\mathscr{H}(t, x_1, \ldots, x_N, \dot{x}_1, \ldots \dot{x}_N) := \sum_{i=1}^{N} \mathscr{H}_i = \sum_{i=1}^{N} \frac{m_i c^2}{\sqrt{1 - \sum_{a=1}^{3} \left(\frac{\dot{x}_i^a}{c^2}\right)^2}} .$$

Keeping in account that P^a is conserved, we may recast the conservation of K^a as:

$$P^a|_{\text{E-L solution}} = \frac{d}{dt}\bigg|_{\text{E-L solution}} \sum_{i=1}^{N} \frac{\mathscr{H}_i}{c^2} x_i^a . \tag{10.98}$$

At this point we can then examine the relativistic equivalent of (2) in Examples 8.11 in the case of non-interacting particles. If we want to interpret equation (10.98) as a *relativistic total impulse theorem*, that is, a relativistic version of (4.1), we had better define the centre of mass' position in terms of the total energy content $\mathscr{H}_i = c P_i^0$ of the particles (in the current inertial frame), and not just the energy of the single masses m_i. The position of the relativistic centre of mass must therefore be defined to be

$$x_G^a := \frac{1}{\mathscr{H}} \sum_{i=1}^{N} \mathscr{H}_i x_i^a .$$

In this way the relativistic total impulse theorem becomes, transcribing (10.98) in the new notation:

$$P^a|_{\text{E-L solution}} = \frac{d}{dt}\bigg|_{\text{E-L solution}} \frac{\mathscr{H}}{c^2} x_G^a .$$

This relation essentially states that the system of N non-interacting point-particles corresponds to a unique point particle of mass equal to the *total energy times* $1/c^2$. Remembering (10.18), we find in particular that *the kinetic energy's contribution abides by Einstein's equation $E = mc^2$.*

Problems 5, 6, 7 in Exercises 10.61 take up this discussion and generalise some of the above results to the case of particles with localised interactions, which cannot be handled within the Lagrangian formalism.

Exercises 10.61

(1) Consider two events $o, o' \in \mathbb{M}^4$ and show that

(1) o and o' are causally separated, i.e. $o \notin J^+(o') \cup J^-(o')$, if and only if there is a spacelike segment joining o to o';

(2) $o \in J^+(o') \cup J^-(o')$ if and only if there is a causal segment between the events;

(3) $o \in I^+(o') \cup I^-(o')$ if and only if there is a timelike segment joining the points.

(2) Consider a pointwise electromagnetic source rotating with constant angular speed Ω around the spatial origin of an inertial frame, at a distance r from it. If the source emits a light radiation at frequency ν_0, measured at rest with the source, what frequency ν will a motionless observer at the frame's origin measure?

(3) A vehicle of length L travels along the x-axis of an inertial frame $\mathscr{R}$. Travelling at (constant) relativistic speed it manages to remain trapped *for one instant* in a garage of length $L/2$ positioned along the x-axis, and then immediately exits from the other side. The front and back doors of the garage open and close instantaneously (as the problem is one-dimensional, the opening and closing processes are single events). Compute the vehicle's velocity v in $\mathscr{R}$ for which the above phenomenon is possible. Show that in the eyes of the driver, who is another inertial system $\mathscr{R}'$, the garage has length $L/4$. Explain the alleged paradox.

(4) Let o and o' be events in $\mathbb{M}^4$ joint by a spacelike segment. Prove that the greatest lower bound of the lengths of all C^1 spacelike curves between the two events is zero.

(5) Consider a system of N point particles with four-momenta P_i and masses $m_i > 0$ for $i = 1, \ldots, N$. Suppose each particle is isolated. Prove that there exists an inertial frame $\mathscr{R}_G$, called *barycentric*, in which the total four-momentum P has only the time component $P^0_{\mathscr{R}_G}$. Define the *system's total mass*

$$M := \frac{1}{c} P^0_{\mathscr{R}_G} \, .$$

Prove that, in any Minkowski coordinates, the centre of mass has equation of motion:

$$P^a = \frac{M}{\sqrt{1 - \left(\frac{v_G}{c}\right)^2}} \frac{dx^a_G}{dt} \, , \qquad a = 1, 2, 3 \, , \tag{10.99}$$

where t is the global time coordinate of the Minkowski system, and $x^a_G(t)$, $v^a_G(t)$ are the coordinates and velocity components of the *centre of mass* in the Minkowski coordinates, at time t. By definition:

$$v^a_G(t) = \frac{dx^a_G}{dt} \, , \qquad \frac{M x^a_G(t)}{\sqrt{1 - \left(\frac{v_G}{c}\right)^2}} := \sum_{i=1}^{N} \frac{m_i x^a_i(t)}{\sqrt{1 - \left(\frac{v_i}{c}\right)^2}} \, , \qquad a = 1, 2, 3 \tag{10.100}$$

Above, x^a_i is the a-th spatial coordinate of the i-th particle, and v_i the velocity, both in the Minkowski coordinates we are using. Prove that (10.99) can be written as

$$P^a = M \frac{dx_G^a}{d\tau_G}, \quad a = 1, 2, 3,$$

where τ_G is the barycentric time coordinate.

(6) Consider a system of N point particles of four-momenta P_i and masses $m_i > 0$ for $i = 1, \ldots, N$. Assume each particle is isolated, except for one event o where all world lines meet and from which N' point particles exit, each with four-momentum P'_j and mass $m'_j > 0$, $j = 1, \ldots, N'$. Let Σ denote the rest space of an arbitrary inertial frame $\mathscr{R}$ at any time t in the past of o. Define:

$$P := \sum_{i=1}^{N} P_i \quad \text{where the } P_i \text{ are evaluated on } \Sigma.$$

Show that P does not depend on the choice of Σ, $\mathscr{R}$ and t. Then prove that if Σ' is the rest space of any inertial frame $\mathscr{R}'$ at any time t' in the future of o, then

$$P' := \sum_{j=1}^{N'} P'_j \quad \text{where the } P'_j \text{ are evaluated on } \Sigma',$$

does not depend on Σ', $\mathscr{R}'$ and t', and furthermore

$$P = P'.$$

(7) Extend Exercise (5) to the case of multiple interaction events $o_1, o_2, \ldots, o_M$ on subsets of the point particles P_i, assuming that each world line represent an isolated point between two interaction events joining pairs of events o_i. Conclude that in this context both the barycentric frame $\mathscr{R}_G$ and the total four-momentum P are well defined, even if the particles interact or if the number of particles during the process and at the end of it is different from the initial number, in general. Define the total mass:

$$M := \frac{1}{c}\sqrt{-g(P, P)}.$$

Show that the results of Exercise (4) are still valid. (Note that the number of particles N in (10.100) can now vary in time.)

(8) Consider a four-force F and a world line γ for a particle of mass $m > 0$. Fix Minkowski coordinates in which:

$$m \frac{dV^a}{d\tau} = F^a, \quad a = 1, 2, 3.$$

Show that we must have:

$$m\frac{dV^0}{d\tau} = F^0 .$$

(9) Consider two frames $\mathscr{R}$ and $\mathscr{R}'$ for which the coordinate change from $\mathscr{R}$ to $\mathscr{R}'$ is a pure Lorentz transformation Λ_3 with parameter $v \in (-c, c)$ (the relative velocity of $\mathscr{R}$ with respect to $\mathscr{R}'$). Suppose a point particle has, at a certain event, velocity $u\mathbf{e}_3$ in $\mathscr{R}$ and $u'\mathbf{e}'_3$ in $\mathscr{R}'$. Prove the **relativistic velocity-addition formula**:

$$u' = \frac{v + u}{1 + \frac{vu}{c^2}} .$$

Chapter 11
Fundamentals of Hamiltonian Mechanics

This chapter is devoted to *Hamilton's formulation* of Classical Mechanics. The term refers to a reworking of Lagrangian Mechanics that was started in the nineteenth century by W.R.Hamilton and then boosted by several other mathematical physicists until the present day. Apart from the indisputable importance within classical Mathematical Physics, Hamiltonian Mechanics has had a deep influence on the theoretical development of many areas of Physics such as modern *Statistical Mechanics* and *Quantum Mechanics* at the start of the twentieth century.

In Hamilton's formulation the key mathematical object, playing a role not dissimilar to the function $\mathscr{L}$ in the Lagrangian formulation and carrying all the information regarding the dynamics of the physical system under exam, is the Hamiltonian function $\mathscr{H}$, which was introduced together with Jacobi's Theorem. As we know, $\mathscr{H}$ is related to the mechanical energy, and coincides with it in certain cases. In contrast to the Lagrangian, therefore, the Hamiltonian is, under certain hypotheses, overtly related to experience and participates *directly* in some physical phenomena.[1]

In Hamilton's formulation the equations of motion, equivalent to the Euler-Lagrange equations in the Lagrangian variables $q^k, \dot{q}^k$, are written in terms of *Hamiltonian variables*, also known as *canonical variables*: q^k, p_k. The equations of motion, which retain the theory's entire structure, have a much more symmetrical form than the Euler-Lagrange equations and the ensuing theory when we swap the roles of the two types of variables. This point of view was decisive in the passage from Classical to Quantum Mechanics, which in its elementary formulation inherited from Hamiltonian Mechanics most of its language and initial setup.

---

[1] This fact had a certain importance in the development of classical Statistical Mechanics, where macroscopic quantities in thermodynamic equilibrium can only be functions of the system's first integrals. The most important first integral is the mechanical energy. Distributions in Statistical Mechanics are described by functions of the system's Hamiltonian. The result also holds in Quantum Statistical Mechanics.

V. Moretti, *Analytical Mechanics*, La Matematica per il 3+2 150, https://doi.org/10.1007/978-3-031-27612-5_11

11.1   The Phase Spacetime and Hamilton's Equations

The ambient space in which Hamilton's formulation is set up is *phase spacetime*. Let us show heuristically how to define it. In Sect. 8.1.1 we introduced, for a system described on $A(\mathbb{V}^{n+1})$ by the Lagrangian $\mathscr{L}$, the *conjugate momentum* of the coordinates q^k:

$$p_k(t, q, \dot{q}) := \frac{\partial \mathscr{L}(t, q, \dot{q})}{\partial \dot{q}^k} .$$

The idea at the centre of Hamilton's formulation is to use the coordinates $(t, q^1, \ldots, q^n, p_1, \ldots, p_n)$ to write the equations of motion instead of the Lagrangian coordinates $(t, q^1, \ldots, q^n, \dot{q}^1, \ldots, \dot{q}^n)$. In this sense the coordinates $(p_1, \ldots, p_n)$ are seen as *independent* of the q^k and employed *instead of* the $\dot{q}^k$. The space (a $(2n + 1)$-dimensional manifold) on which $(t, q^1, \ldots, q^n, p_1, \ldots, p_n)$ are natural local coordinates is precisely the *phase spacetime*, which plays a symmetrical role to the spacetime of kinetic states.

If we choose new coordinates on $A(\mathbb{V}^{n+1})$ so that the coordinate changes are the ones we know well:

$$t' = t + c , \tag{11.1}$$

$$q'^k = q'^k(t, q^1, \ldots, q^n) , \tag{11.2}$$

$$\dot{q}'^k = \sum_{h=1}^{n} \frac{\partial q'^k}{\partial q^h} \dot{q}^h + \frac{\partial q'^k}{\partial t} , \tag{11.3}$$

the Lagrangian will satisfy the following property, being a scalar field:

$$\frac{\partial \mathscr{L}(t', q', \dot{q}')}{\partial \dot{q}'^h} = \sum_{r=1}^{n} \frac{\partial \dot{q}^r}{\partial \dot{q}'^h} \frac{\partial \mathscr{L}(t, q, \dot{q}'')}{\partial \dot{q}^r} .$$

As usual, by (11.3) we can rewrite the latter as:

$$\frac{\partial \mathscr{L}(t', q', \dot{q}')}{\partial \dot{q}'^h} = \sum_{r=1}^{n} \frac{\partial q^r}{\partial q'^h} \frac{\partial \mathscr{L}(t, q, \dot{q})}{\partial \dot{q}^r} . \tag{11.4}$$

If we identify p_k with the partial derivative of $\mathscr{L}$ in $\dot{q}^k$, we conclude that performing the natural coordinate change on $A(\mathbb{V}^{n+1})$ through Eqs. (11.1), (11.2) and (11.3) is equivalent to changing $(t, q^1, \ldots, q^n, p_1, \ldots, p_n)$ to:

$$t' = t + c,$$
$$q'^h = q'^h(t, q^1, \ldots, q^n),$$
$$p'_h = \sum_{r=1}^{n} \frac{\partial q^r}{\partial q'^h} p_r.$$

This explains heuristically the transformation laws of the natural coordinates on phase spacetime, which we shall assume from now on. Regardless, it should be *very clear* that the coordinates p_k are identified with the conjugate momenta only *after* we have fixed a Lagrangian. The phase spacetime, instead, is an abstract entity, *independent* of the choice of Lagrangian for the system.

11.1.1   The Phase Spacetime $F(\mathbb{V}^{n+1})$

To describe the space of states of a physical system $\mathscr{S}$ whose dynamics we want to study in Hamilton's formulation, we give the following definition.

Definition 11.1 (Phase Spacetime) The **phase spacetime** $F(\mathbb{V}^{n+1})$ is a $(2n + 1)$-dimensional C^k manifold,[2] $k \geq 2$, built on the configuration spacetime $\mathbb{V}^{n+1}$, that admits a privileged atlas whose local charts are called **natural local coordinate systems on** $F(\mathbb{V}^{n+1})$. Calling $t, q^1, \ldots, q^n, p_1, \ldots, p_n$ the coordinates of a generic natural local chart, the following conditions hold.

(1) The coordinates $t, q^1, \ldots, q^n$ are identified with the coordinates of a system of natural local coordinates on $\mathbb{V}^{n+1}$; in particular t is identified up to additive constants with absolute time, seen as (non-singular) differentiable surjective map $T : F(\mathbb{V}^{n+1}) \to \mathbb{R}$. Furthermore, the remaining coordinates $p_1, \ldots, p_n$, for any $t, q^1, \ldots, q^n$, take values on the *entire* $\mathbb{R}^n$. The above system on $\mathbb{V}^{n+1}$ is uniquely determined by the system on $F(\mathbb{V}^{n+1})$, and every set of natural local coordinates on $\mathbb{V}^{n+1}$ can be completed to a system of natural coordinates on $F(\mathbb{V}^{n+1})$.

---

[2] Unless specified, we shall assume $A(\mathbb{V}^{n+1})$ and $F(\mathbb{V}^{n+1})$ are of class C^k with $k \geq 2$.

(2) Natural local coordinates on $F(\mathbb{V}^{n+1})$ transform under C^k maps with C^k inverses

$$t' = t + c \, , \tag{11.5}$$

$$q'^k = q'^k(t, q^1, \ldots, q^n) \, , \tag{11.6}$$

$$p'_k = \sum_{h=1}^{n} \frac{\partial q^h}{\partial q'^k} p_h \, . \tag{11.7}$$

The points in $F(\mathbb{V}^{n+1})$ are called (Hamiltonian) **states** or **representative points**. We call **phase space at time** t every $2n$-dimensional (embedded) submanifold $\mathbb{F}_t :=$ $T^{-1}(t)$ of $F(\mathbb{V}^{n+1})$ obtained fixing an absolute time value $t \in \mathbb{R}$, and any point in $\mathbb{F}_t$ is a **representative point at time** t. $F(\mathbb{V}^{n+1})$ is therefore the disjoint union of the spaces $\mathbb{F}_t$ as $t \in \mathbb{R}$ varies. $\diamond$

Remarks 11.2

(1) A *representative point or Hamiltonian state at time* t is then given, in the aforementioned natural coordinates, by a $(2n + 1)$-tuple of numbers $(t, q^1, \ldots, q^n, p_1, \ldots, p_n)$. The latter also defines the system's configuration at time t, in $\mathbb{Q}_t$, through the row vector $(q^1, \ldots, q^n)$.

(2) Each phase space at time t, $\mathbb{F}_t$, is also called **fibre of** $F(\mathbb{V}^{n+1})$ **at time** t, so the natural local coordinates on $F(\mathbb{V}^{n+1})$ are local coordinates **adapted to the fibres** of $F(\mathbb{V}^{n+1})$ in the sense of the general Definition B.4. But they are not the only local coordinates with this property. For instance t may be replaced by a differentiable function of t with differentiable inverse, and the coordinates of "type" q and p may be "mixed" by some local diffeomorphism that is not sanctioned by the transformation rules of natural coordinates.

(3) Asking that (11.5)–(11.7) give a differentiable transformation with differentiable inverse (differentiable means C^k with $k \geq 1$ large enough to allow for the computations to work) still of form (11.5)–(11.7) with primed and unprimed coordinates swapped, implies in particular that the Jacobian matrix J of coefficients $\frac{\partial q^h}{\partial q'^k}$ and the Jacobian J' of coefficients $\frac{\partial q'^k}{\partial q^j}$ have non-zero determinant (and are inverses). In fact, from (11.5)–(11.7):

$$\delta^i_j = \frac{\partial q^i}{\partial q^j} = \frac{\partial q^i}{\partial t'} \frac{\partial t'}{\partial q^j} + \sum_k \frac{\partial q^i}{\partial q'^k} \frac{\partial q'^k}{\partial q^j} + \sum_k \frac{\partial q^i}{\partial p'_k} \frac{\partial p'_k}{\partial q^j}$$

$$= 0 + \sum_k \frac{\partial q^i}{\partial q'^k} \frac{\partial q'^k}{\partial q^j} + 0 = (J J')^i{}_j \, ,$$

so $1 = \det I = \det(J J') = \det J \det J'$ and therefore $\det J, \det J' \neq 0$. Hence, with t' and q'^k fixed, (11.7) is a bijective linear map on $\mathbb{R}^n$ in the n coordinates p_h.   ∎

11.1.2 The Legendre Transform

Let us deal with a correspondence mediated by a given Lagrangian $\mathscr{L}$, between the natural local coordinates on phase spacetime and those on the spacetime of kinetic states. Let us fix a natural local chart on the open set $U \subset \mathbb{V}^{n+1}$, with coordinates

$$\psi : U \ni c \mapsto (t, q^1, \ldots, q^n) \in \mathbb{R}^{n+1} .$$

This chart can be extended in two distinct ways: either as natural local chart of $A(\mathbb{V}^{n+1})$, passing to coordinates $(t, q^1, \ldots, q^n, \dot{q}^1, \ldots, \dot{q}^n)$ where $(t, q^1, \ldots, q^n) \in \psi(U)$ is an open set in $\mathbb{R}^{n+1}$ and $(\dot{q}^1, \ldots, \dot{q}^n) \in \mathbb{R}^n$; or as natural local chart on $F(\mathbb{V}^{n+1})$, passing to coordinates $(t, q^1, \ldots, q^n, p_1, \ldots, p_n)$ where $(t, q^1, \ldots, q^n) \in \psi(U)$ is open in $\mathbb{R}^{n+1}$ (the same as before) and $(p_1, \ldots, p_n) \in \mathbb{R}^n$. Now consider the transformation of the two coordinate systems on their domains:

$$t = t , \tag{11.8}$$

$$q^k = q^k , \quad \text{for } k = 1, \ldots, n , \tag{11.9}$$

$$p_k = \frac{\partial \mathscr{L}(t, q, \dot{q})}{\partial \dot{q}^k} , \quad \text{for } k = 1, \ldots, n . \tag{11.10}$$

This locally defined map from the spacetime of kinetic states to the phase spacetime is called **Legendre transformation**, or more precisely, *representation in natural coordinates* of the Legendre transformation. The following three properties of the Legendre transformation are fundamental to the theory.

(a) The transformation (11.8)–(11.10) is *bijective* on the above domains for all the Lagrangians we met, besides being differentiable with differentiable inverse. The Lagrangians we have encountered so far in the Lagrangian formulation of Mechanics, including those containing generalised potentials (see Sect. 9.2.2), are quadratic in the $\dot{q}^k$ and have the form (7.87), where the symmetric matrix of coefficients $a_{hk}(t, q^1, \ldots, q^n)$ is *non-singular* (for Lagrangians deduced from mechanical systems, a_{hk} is even *positive definite*) and all functions in the $(t, q^1, \ldots, q^n)$ are C^k with $k \geq 1$, or $k \geq 2$ to guarantee existence and uniqueness for the solutions of the Euler-Lagrange equations. In this case the non-trivial equation in the Legendre transformation reduces to:

$$p_k = \sum_{h=1}^{n} 2a_{kh}(t, q)\dot{q}^h + \beta_k(t, q) . \tag{11.11}$$

As the symmetric matrix a_{kh} is invertible, the above Legendre transformation, seen as function mapping the $\dot{q}^k$ to the p_k for any given $t, q^1, \ldots, q^n$, reduces to a mere bijective transformation from $\mathbb{R}^n$ to $\mathbb{R}^n$. Hence the coordinate transformation (11.8)–(11.10) is bijective (on the aforementioned coordinate domains) and of class C^k together with its inverse.

(b) The Legendre transformation in coordinates allows to export the Hamiltonian from one natural coordinate system on $A(\mathbb{V}^{n+1})$ to the corresponding system on $F(\mathbb{V}^{n+1})$. In fact as the Legendre transformation is bijective with regular inverse, by inverting it we can write on the given charts the coordinates $\dot{q}$ in terms of (t, q, p). Thus we can redefine the Hamiltonian associated to $\mathscr{L}$ on the natural chart of $A(\mathbb{V}^{n+1})$ as a C^k function on the corresponding chart of $F(\mathbb{V}^{n+1})$:

$$\mathscr{H}(t, q, p) = \sum_{k=1}^{n} p_k \dot{q}^k(t, q, p) - \mathscr{L}(t, q, \dot{q}(t, q, p)) \,.$$

In this situation we say that the Hamiltonian, *written in terms of the Hamiltonian variables that arise by Legendre transforming the Lagrangian variables*, is the **Legendre transform** of the Lagrangian.

(c) The Legendre transformation is at the moment only defined locally; but it has *global* nature. Consider two natural charts (U, ψ) and (U', ψ') on $\mathbb{V}^{n+1}$ and extend them to $A(\mathbb{V}^{n+1})$ over domains V and V' with coordinates $t, q^1, \ldots, q^n, \dot{q}^1, \ldots, \dot{q}^n$ and $t', q'^1, \ldots, q'^n, \dot{q}'^1, \ldots, \dot{q}'^n$ respectively. Let us also extend the two initial charts to natural charts on $F(\mathbb{V}^{n+1})$ with domains W, W' and coordinates $t, q^1, \ldots, q^n, p_1, \ldots, p_n$ and $t', q'^1, \ldots, q'^n, p'_1, \ldots, p'_n$ respectively. Suppose V and V' intersect, and $a \in V \cap V'$. Given a Lagrangian $\mathscr{L} : A(\mathbb{V}^{n+1}) \to \mathbb{R}$, we can act on the kinetic state a by *two distinct* coordinate representations of the Legendre transformation: one that maps V to W and one that sends V' to W'. Can we conclude that the two Legendre transformations define *the same state in $F(\mathbb{V}^{n+1})$ as image of* a? The answer is yes, we can (Proposition 11.22 addresses the matter in general): the two Legendre transformations on $V \cap V'$ act in the same manner on $a \in V \cap V'$, thus giving the same image. In fact, comparing (11.7) and (11.4) we find:

$$p_k(a) = \left. \frac{\partial \mathscr{L}(t, q, \dot{q})}{\partial \dot{q}^k} \right|_a \,, \quad k = 1, \ldots, n \,,$$

$$\Leftrightarrow \quad p'_h(a) = \left. \frac{\partial \mathscr{L}(t', q', \dot{q}')}{\partial \dot{q}'^h} \right|_a \,, \quad h = 1, \ldots, n \,.$$

Due to the nature of the atlases of natural coordinates, we conclude that there is a unique bijection

$$\mathbb{L}_{\mathscr{L}} : A(\mathbb{V}^{n+1}) \to F(\mathbb{V}^{n+1})$$

that reads as (11.8)–(11.10) in coordinates, for any pair of natural local coordinate systems on $A(\mathbb{V}^{n+1})$ and $F(\mathbb{V}^{n+1})$ obtained extending the same natural local chart on $\mathbb{V}^{n+1}$. Such a bijection is differentiable with differentiable inverse whenever written for corresponding charts in the natural atlases of $A(\mathbb{V}^{n+1})$ and $F(\mathbb{V}^{n+1})$. By standard Differential Geometry results, $\mathbb{L}_{\mathscr{L}}$ is therefore a *diffeomorphism* from $A(\mathbb{V}^{n+1})$ to $F(\mathbb{V}^{n+1})$. It is the global **Legendre transform** induced by the Lagrangian $\mathscr{L}$.

Examples 11.3

(1) Consider a system of N point particles $P_1, \ldots, P_N$ of respective masses $m_1, \ldots, m_N$, not subject to constraints but interacting under forces with total potential $\mathscr{V}|_{\mathscr{R}}$. $\mathscr{R}$ is a inertial frame in which the particles are determined by their position vectors $\mathbf{x}_1, \ldots, \mathbf{x}_n$. The system's Lagrangian, using as free coordinates the components of the $\mathbf{x}_i$, will be:

$$\mathscr{L}(t, \mathbf{x}_1, \ldots, \mathbf{x}_N, \dot{\mathbf{x}}_1, \ldots, \dot{\mathbf{x}}_n) = \sum_{i=1}^{N} \frac{m_i}{2} \dot{\mathbf{x}}_i^2 + \mathscr{V}|_{\mathscr{R}}(t, \mathbf{x}_1, \ldots, \mathbf{x}_N) \, .$$

Grouping the p_k coordinates in threes, as in the position vectors, we have:

$$\mathbf{p}_i = \nabla_{\dot{\mathbf{x}}_i} \mathscr{L}(t, \mathbf{x}_1, \ldots, \mathbf{x}_N, \dot{\mathbf{x}}_1, \ldots, \dot{\mathbf{x}}_N) = m_i \dot{\mathbf{x}}_i \, .$$

Hence for physical systems of point particles without constraints, where the forces are either conservative or come from a *non-generalised* potential, if we use as free coordinates the particles' positional components then the associated coordinates p_k are the components of the momenta on the system's motions.

(2) A less trivial case is that of a system of forces that is induced by a generalised potential. Consider a charged particle of mass m, charge e, and immersed in an electromagnetic field with electromagnetic potentials $\mathbf{A}$ and φ. From Sect. 9.2.1 the Lagrangian is, in an inertial frame and using the position vector's components as free coordinates:

$$\mathscr{L}(t, \mathbf{x}, \dot{\mathbf{x}}) = \frac{m}{2} \dot{\mathbf{x}}^2 - e\varphi(t, \mathbf{x}) + \frac{e}{c} \mathbf{A}(t, \mathbf{x}) \cdot \dot{\mathbf{x}} \, .$$

Immediately:

$$\mathbf{p} = m\dot{\mathbf{x}} + \frac{e}{c} \mathbf{A}(t, \mathbf{x}) \, ,$$

so the conjugate momenta associated with the three spatial coordinates are no longer (on the motion) the momentum's components.

(3) Consider a particle of mass m moving on a smooth spherical surface of radius $R > 0$ and centre O that is at rest in an inertial frame $\mathscr{R}$, and subject to a force with potential $\mathscr{V}|_{\mathscr{R}}$. Let us use spherical coordinates θ and ϕ (on $\mathscr{R}$) to locate

P on the sphere. We know (Chap. 2) that in spherical coordinates the particle's velocity in $\mathscr{R}$ is:

$$\mathbf{v} = \dot{r}\,\mathbf{e}_r + r\dot{\theta}\,\mathbf{e}_\theta + r\dot{\phi}\sin\theta\,\mathbf{e}_\varphi\ .$$

In our case $r = R$ is constant in the above formula, so the particle's Lagrangian is:

$$\mathscr{L}(t,\theta,\phi,\dot{\theta},\dot{\phi}) = \frac{mR^2}{2}\left(\dot{\theta}^2 + \dot{\phi}^2\sin^2\theta\right) + \mathscr{V}|_{\mathscr{R}}(t,\theta,\phi)\ .$$

Immediately, the conjugate momenta p_ϕ and ϕ_θ are:

$$p_\phi = mR^2\dot{\phi}\sin^2\theta\ ,\qquad p_\theta = mR^2\dot{\theta}\ .$$

Note the physical meaning of p_ϕ: along the motion, p_ϕ is the *component along z of the angular momentum* (in $\mathscr{R}$ with respect to the pole O). In fact

$$\boldsymbol{\Gamma}_O|_{\mathscr{R}} = m(P - O) \wedge \mathbf{v} = mR\,\mathbf{e}_r \wedge \left(R\dot{\theta}\,\mathbf{e}_\theta + R\dot{\phi}\sin\theta\,\mathbf{e}_\varphi\right)$$

$$= mR^2\left(\dot{\theta}\,\mathbf{e}_\varphi - \dot{\phi}\sin\theta\,\mathbf{e}_\theta\right)\ ,$$

and so:

$$\boldsymbol{\Gamma}_O|_{\mathscr{R}} \cdot \mathbf{e}_z = mR^2\left(\dot{\theta}\,\mathbf{e}_\varphi \cdot \mathbf{e}_z - \dot{\phi}\sin\theta\,\mathbf{e}_\theta \cdot \mathbf{e}_z\right) = 0 + mR^2\dot{\phi}\sin^2\theta = p_\phi\ .$$

The same interpretation is found when studying a particle without constraints (but subject to a potential), using the three spherical parameters in an inertial frame as free coordinates.

11.1.3   Hamilton's Equations and the Local Uniqueness of the Hamiltonian

Now we wish to find which relations on phase spacetime correspond to the Euler-Lagrange equations. The following fundamental proposition gives the answer.

Proposition 11.4 *Consider a C^2 Lagrangian $\mathscr{L} : A(\mathbb{V}^{n+1}) \to \mathbb{R}$ as in (7.87) in an atlas of natural coordinates of $A(\mathbb{V}^{n+1})$ where the matrix a_{hk} is everywhere non-singular. We use natural local coordinates: $(t, q^1, \ldots, q^n, \dot{q}^1, \ldots, \dot{q}^n)$ on*

$A(\mathbb{V}^{n+1})$ and $(t, q^1, \ldots, q^n, p_1, \ldots, p_n)$ on $F(\mathbb{V}^{n+1})$ related by the Legendre transformation:

$$t = t, \tag{11.12}$$

$$q^k = q^k, \quad \text{for } k = 1, \ldots, n, \tag{11.13}$$

$$p_k = \frac{\partial \mathscr{L}(t, q, \dot{q})}{\partial \dot{q}^k}, \quad \text{for } k = 1, \ldots, n. \tag{11.14}$$

Then, setting $(t, q, p) := (t, q^1, \ldots, q^n, p_1, \ldots, p_n)$, we have:

$$\frac{\partial \mathscr{H}}{\partial t} = -\left. \frac{\partial \mathscr{L}}{\partial t} \right|_{(t, q, \dot{q}(t, q, p))}, \tag{11.15}$$

$$\frac{\partial \mathscr{H}}{\partial p_k} = \dot{q}^k(t, q, p), \quad \text{for } k = 1, \ldots, n \tag{11.16}$$

$$\frac{\partial \mathscr{H}}{\partial q^k} = -\left. \frac{\partial \mathscr{L}}{\partial q^k} \right|_{(t, q, \dot{q}(t, q, p))}, \quad \text{for } k = 1, \ldots, n, \tag{11.17}$$

where the Hamiltonian $\mathscr{H}$ associated to $\mathscr{L}$ in the above coordinates reads:

$$\mathscr{H}(t, q, p) = \sum_{k=1}^{n} \left. \frac{\partial \mathscr{L}}{\partial \dot{q}^k} \right|_{(t, q, \dot{q}(t, q, p))} \dot{q}^k(t, q, p) - \mathscr{L}(t, q, \dot{q}(t, q, p)). \tag{11.18}$$

in Hamiltonian variables.

Proof First, by (11.10) we can trivially write equation (11.18) as:

$$\mathscr{H}(t, q, p) = \sum_{k=1}^{n} p_k \dot{q}^k(t, q, p) - \mathscr{L}(t, q, \dot{q}(t, q, p)). \tag{11.19}$$

Consider a C^1 curve in $F(\mathbb{V}^{n+1})$ parametrised by absolute time, $t \mapsto (t, q(t), p(t))$. It will correspond to an analogous curve in $A(\mathbb{V}^{n+1})$: $t \mapsto (t, q(t), \dot{q}(t, q(t), p(t)))$. Note that we are not supposing the second curve solves the Euler-Lagrange equations. Differentiating $\mathscr{H}(t, q(t), p(t))$ in t using (11.19) we find:

$$\frac{\partial \mathscr{H}}{\partial t} + \sum_k \frac{\partial \mathscr{H}}{\partial q^k} \frac{dq^k}{dt} + \sum_k \frac{\partial \mathscr{H}}{\partial p_k} \frac{dp_k}{dt} = \sum_{k=1}^{n} \frac{dp_k}{dt} \dot{q}^k + \sum_{k=1}^{n} \frac{d\dot{q}^k}{dt} p_k$$

$$- \sum_k \frac{\partial \mathscr{L}}{\partial q^k} \frac{dq^k}{dt} - \sum_k \frac{\partial \mathscr{L}}{\partial \dot{q}^k} \frac{d\dot{q}^k}{dt} - \frac{\partial \mathscr{L}}{\partial t}.$$

The terms:

$$\sum_{k=1}^{n} \frac{d\dot{q}^k}{dt} p_k - \sum_k \frac{\partial \mathscr{L}}{\partial \dot{q}^k} \frac{d\dot{q}^k}{dt}$$

could be further expanded to render the derivatives $d\dot{q}^k/dt$ explicit, but this is not convenient since they cancel out under (11.14). Hence we can write:

$$\frac{\partial \mathscr{H}}{\partial t} + \frac{\partial \mathscr{L}}{\partial t} + \sum_k \left(\frac{\partial \mathscr{H}}{\partial q^k} + \frac{\partial \mathscr{L}}{\partial q^k} \right) \frac{dq^k}{dt} + \sum_k \left(\frac{\partial \mathscr{H}}{\partial p_k} - \dot{q}^k \right) \frac{dp_k}{dt} = 0. \quad (11.20)$$

As for every point (t_0, q, p) we can define a C^1 curve, say $t \mapsto (t, q(t), p(t))$, through the points at $t = t_0$. We can also choose $\frac{dp^k}{dt}|_{t=t_0}$ and $\frac{dq^k}{dt}|_{t=t_0}$ arbitrarily, let us for example take them all zero. Then (11.20) becomes:

$$\left. \frac{\partial \mathscr{H}}{\partial t} \right|_{(t_0, q, p)} + \left. \frac{\partial \mathscr{L}}{\partial t} \right|_{(t_0, q, p)} = 0, \quad \text{for every state } (t_0, q, p).$$

But then (11.20), in the region of $F(\mathbb{V}^{n+1})$ we are examining, is:

$$\sum_k \left(\frac{\partial \mathscr{H}}{\partial q^k} + \frac{\partial \mathscr{L}}{\partial q^k} \right) \frac{dq^k}{dt} + \sum_k \left(\frac{\partial \mathscr{H}}{\partial p_k} - \dot{q}^k \right) \frac{dp_k}{dt} = 0,$$

for any C^1 curve $t \mapsto (t, q(t), p(t))$ passing through the region.

Changing the curve through (t_0, q, p) at $t = t_0$ and choosing all terms $\frac{dp_k}{dt}|_{t=t_0}$ and $\frac{dq^k}{dt}|_{t=t_0}$ equal to zero except $\frac{dp_{k_0}}{dt}|_{t=t_0} = 1$, for some $k_0 = 1, \ldots, n$, we conclude:

$$\left. \frac{\partial \mathscr{H}}{\partial p_{k_0}} \right|_{(t_0, q, p)} - \dot{q}^{k_0}|_{(t_0, q, p)} = 0, \quad \text{for every state } (t_0, q, p).$$

But k_0 is arbitrary, so the initial identity reduces to:

$$\sum_k \left(\frac{\partial \mathscr{H}}{\partial q^k} + \frac{\partial \mathscr{L}}{\partial q^k} \right) \frac{dq^k}{dt} = 0.$$

Iterating the procedure, since the state (t_0, q, p) is arbitrary, we arrive at (11.15)–(11.17).   □

Formula (11.16) together with (11.18) immediately proves the following corollary, showing in particular how the coordinate representation of a Legendre transformation is *involutive* (applied twice it gives the identity map).

Corollary 11.5 *In the hypotheses of Proposition 11.4, the Legendre transform's representation in natural local coordinates has inverse:*

$$t = t \,, \tag{11.21}$$

$$q^k = q^k \,, \quad for \ k = 1, \ldots, n \,, \tag{11.22}$$

$$\dot{q}^k = \frac{\partial \mathscr{H}(t, q, p)}{\partial p_k} \,, \quad for \ k = 1, \ldots, n \,. \tag{11.23}$$

The Lagrangian is obtained from the Hamiltonian as follows:

$$\mathscr{L}(t, q, \dot{q}) = \sum_{k=1}^{n} \frac{\partial \mathscr{H}}{\partial p_k}\bigg|_{(t,q,p(t,q,\dot{q}))} p_k(t, q, \dot{q}) - \mathscr{H}(t, q, p(t, q, \dot{q})) \,. \tag{11.24}$$

An immediate consequence of Proposition 11.4 is the fundamental theorem establishing Hamilton's equations.

Theorem 11.6 (Hamilton's Equations) *Consider a C^2 Lagrangian $\mathscr{L}$: $A(\mathbb{V}^{n+1}) \to \mathbb{R}$ as in (7.87) in arbitrary natural coordinates of $A(\mathbb{V}^{n+1})$ where the coefficient matrix a_{hk} is everywhere non-singular.*

Use natural local coordinates $(t, q^1, \ldots, q^n, \dot{q}^1, \ldots, \dot{q}^n)$ on $A(\mathbb{V}^{n+1})$ and $(t, q^1, \ldots, q^n, p_1, \ldots, p_n)$ on $F(\mathbb{V}^{n+1})$, related by the Legendre transformation:

$$t = t \,,$$

$$q^k = q^k \,, \quad for \ k = 1, \ldots, n \,,$$

$$p_k = \frac{\partial \mathscr{L}(t, q, \dot{q})}{\partial \dot{q}^k} \,, \quad for \ k = 1, \ldots, n \,.$$

The C^1 curve $I \ni t \mapsto (t, q(t), \dot{q}(t))$ solves the Euler-Lagrange equations (7.84) (on some open interval $I \subset \mathbb{R}$):

$$\begin{cases} \dfrac{d}{dt}\left(\dfrac{\partial \mathscr{L}}{\partial \dot{q}^k}\bigg|_{(t,q(t),\dot{q}(t))} \right) = \dfrac{\partial \mathscr{L}}{\partial q^k}\bigg|_{(t,q(t),\dot{q}(t))} \,, & for \ k = 1, \ldots, n, \\ \dfrac{dq^k}{dt} = \dot{q}^k(t) & \end{cases}$$

if and only if the transformed curve $I \ni t \mapsto (t, q(t), p(t))$ under (11.12)–(11.14) satisfies **Hamilton's equations:**

$$\begin{cases} \dfrac{dp_k}{dt} = -\dfrac{\partial \mathscr{H}}{\partial q^k}\bigg|_{(t,q(t),p(t))} \,, & for \ k = 1, \ldots, n, \\ \dfrac{dq^k}{dt} = \dfrac{\partial \mathscr{H}}{\partial p_k}\bigg|_{(t,q(t),p(t))} \,, & \end{cases} \tag{11.25}$$

where the Hamiltonian $\mathscr{H}$ associated to $\mathscr{L}$ in the coordinates is written in the Hamiltonian variables (11.24).

Finally, if $I \ni t \mapsto (t, q(t), p(t))$ is a solution to Hamilton's equations, then:

$$\frac{d\mathscr{H}(t, q(t), p(t))}{dt} = \left.\frac{\partial \mathscr{H}}{\partial t}\right|_{(t,q(t),p(t))}, \tag{11.26}$$

implying the Hamiltonian is a first integral when not depending explicitly on time.   ◇

Proof Using $\frac{\partial \mathscr{L}}{\partial \dot{q}^k} = p_k$ and (11.17), the first Euler-Lagrange equation reads, in Hamiltonian coordinates:

$$\frac{dp_k}{dt} = -\frac{\partial \mathscr{H}}{\partial q^k}.$$

The second Euler-Lagrange equation, using (11.16), is:

$$\frac{dq^k}{dt} = \frac{\partial \mathscr{H}}{\partial p_k}$$

in Hamiltonian coordinates. By construction, then, a curve $I \ni t \mapsto (t, q(t), \dot{q}(t))$ solves the Euler-Lagrange equations if and only if $I \ni t \mapsto (t, q(t), p(t, q(t), \dot{q}(t))$ solves Hamilton's equations. This ends the first part.

The last claim is a direct computation involving Hamilton's equations:

$$\frac{d\mathscr{H}}{dt} = \frac{\partial \mathscr{H}}{\partial t} + \sum_k \frac{\partial \mathscr{H}}{\partial q^k}\frac{dq^k}{dt} + \sum_k \frac{\partial \mathscr{H}}{\partial p_k}\frac{dp_k}{dt}$$

$$= \frac{\partial \mathscr{H}}{\partial t} + \sum_k \left(\frac{\partial \mathscr{H}}{\partial q^k}\frac{\partial \mathscr{H}}{\partial p_k} - \frac{\partial \mathscr{H}}{\partial p_k}\frac{\partial \mathscr{H}}{\partial q^k} \right) = \frac{\partial \mathscr{H}}{\partial t}.$$

□

We have seen that the Lagrangian of a physical system is not unique, but can be redefined by adding formal time derivative. Now we want to discover how unique a Hamiltonian on a natural local chart of $F(\mathbb{V}^{n+1})$ might be. The following result is independent of the fact the Hamiltonian comes from a Lagrangian.

Proposition 11.7 *Consider a natural local chart (U, ψ) on $F(\mathbb{V}^{n+1})$, with*

$$\psi : U \ni a \mapsto (t(a), \mathbf{x}(a)) \in \mathbb{R}^{2n+1},$$

such that $T^{-1}(t) \cap U$ is connected for any $t \in T(U)$. Suppose there is a C^2 Hamiltonian in the coordinates of U giving Eq. (11.25). Another C^2 function of the same coordinates $\mathscr{H}' : U \to \mathbb{R}$ gives the same Hamiltonian equations if and only if the C^2 function $\mathscr{H} - \mathscr{H}'$ depends only on t.

Proof If $\mathscr{H} - \mathscr{H}'$ is a C^2 map in the sole coordinate t, Hamilton's equations are evidently the same. Suppose conversely that $\mathscr{H}'$ is C^2 and gives Eq. (11.25). Since the latter are in normal form and the right-hand side is C^1 by construction, the Existence and uniqueness theorem holds. In particular, for any state $a \in U$ there is a solution with that initial condition, since the solutions to Hamilton's equations generated by $\mathscr{H}$ and $\mathscr{H}'$ coincide (the equations are the same). Then the right-hand sides of Hamilton's equations will coincide when evaluated at a. Using coordinates on U, call $\mathbf{x}$ the collection of $(q^1, \ldots, q^n, p_1, \ldots, p_n)$ for which $a \equiv (t, \mathbf{x})$. Then $\nabla_{\mathbf{x}}(\mathscr{H}(t, \mathbf{x}) - \mathscr{H}'(t, \mathbf{x})) = 0$ and hence $\nabla_{\mathbf{x}}(\mathscr{H}(t, \mathbf{x}) - \mathscr{H}'(t, \mathbf{x})) = 0$ for any $(t, \mathbf{x})$ in the given chart. Define $f(t, \mathbf{x}) := \mathscr{H}(t, \mathbf{x}) - \mathscr{H}'(t, \mathbf{x})$, so that $\nabla_{\mathbf{x}} f(t, \mathbf{x}) = 0$. For fixed t, $\mathbf{x}$ lives in a connected open set $A_t := \psi(T^{-1}(t) \cap U)$, whence path-connected. Consider a C^1 curve $\mathbf{x} = \mathbf{x}(s)$, $s \in [0, 1]$, from $\mathbf{x}_1$ to $\mathbf{x}_2$ that always stays in A_t. The composite $[0, 1] \ni s \mapsto f(t, \mathbf{x}(s))$ is $C^1([0, 1])$ by construction, and the Mean-value Theorem says there is some $s_0 \in (0, 1)$ such that: $f(t, \mathbf{x}_2) - f(t, \mathbf{x}_1) = f(t, \mathbf{x}(1)) - f(t, \mathbf{x}(0)) = 1 \frac{d\mathbf{x}}{ds}|_{s_0} \cdot \nabla_{\mathbf{x}} f(t, \mathbf{x})|_{\mathbf{x}(s_0)} = 0$. Therefore $f(t, \mathbf{x}_2) = f(t, \mathbf{x}_1)$ and $f = \mathscr{H} - \mathscr{H}'$ can only depend on t over U (or perhaps not even on that). It will be C^2 as difference of C^2 maps. $\qquad\square$

Remarks 11.8

(1) As noted in the previous proposition's proof, Hamilton's equations are in normal form, so the Existence and uniqueness theorem holds provided the Hamiltonian function is regular enough, for instance if C^2.

(2) Identity (11.26) implies, with (11.17), the first claim in Jacobi's Theorem 8.13.

(3) The form of Hamilton equations accounts for the Hamiltonian version of the relationship between *cyclic variables and conserved quantities*. Clearly, in fact, if the Hamiltonian does not depend explicitly on q^k the conjugate momentum p_k is a constant of motion.

(4) The next property does not have a correspondent in the Lagrangian formulation. If the Hamiltonian does not depend explicitly on the conjugate momentum p_k, the associated q_k is a constant of motion. This result cannot be used in Hamiltonian systems obtained by Legendre transforming pre-existent Lagrangian systems, as we now show. If we assumed that $\mathscr{H}$ did not depend on p_{k_0}, (11.23) would imply:

$$\frac{\partial \dot{q}^{k_0}}{\partial p_k} = \frac{\partial^2 \mathscr{H}}{\partial p_k \partial p_{k_0}} = \frac{\partial 0}{\partial p_k} = 0 \quad k = 1, 2 \ldots, n,$$

hence the matrix of derivatives $\frac{\partial \dot{q}^h}{\partial p_k}$ would have a null row and therefore zero determinant. This is impossible because the inverse Legendre transform is a diffeomorphism (as the Legendre transform itself) and its Jacobian determinant in any local chart cannot vanish. Nonetheless, in natural coordinates, that determinant equals $\det[\frac{\partial \dot{q}^h}{\partial p_k}]_{h,j=1,\ldots,n}$, as is easy to prove as exercise. So assuming the Legendre transform is well defined, if we work in natural coordinates there is no way to cook up Hamiltonians that do not depend on some conjugate momentum explicitly. By extending the formalism it would be possible to have Hamiltonians independent of certain p_k (or even independent of *all* coordinates), even for systems built by Legendre transforming a Lagrangian, but we would need to use *more general coordinates than the natural ones* where Hamilton's equations still hold. We shall introduce such coordinates in Chap. 12, and the issue (Hamiltonian function independent of certain coordinates) will become crucial in the theory of *Hamilton-Jacobi*, Sect. 12.4.

(5) There exist physical systems, also in more sophisticated theories than Classical Mechanics such as field theories, that by nature are described in Hamiltonian formulation, and do not admit a Lagrangian formulation related to the former by a Legendre transform.

■

Examples 11.9

(1) Consider N point particles, $P_1, \ldots, P_N$ of respective masses $m_1, \ldots, m_N$, not subject to constraints but interacting through forces with total potential energy $\mathscr{U}|_{\mathscr{R}}$. $\mathscr{R}$ is an inertial frame in which the particles are defined by their position vectors $\mathbf{x}_1, \ldots, \mathbf{x}_n$. The system's Lagrangian, using the components of the $\mathbf{x}_i$ as free coordinates, is:

$$\mathscr{L}(\mathbf{x}_1, \ldots, \mathbf{x}_N, \dot{\mathbf{x}}_1, \ldots, \dot{\mathbf{x}}_N) = \sum_{i=1}^{N} \frac{m_i}{2} \dot{\mathbf{x}}_i^2 - \mathscr{U}|_{\mathscr{R}}(\mathbf{x}_1, \ldots, \mathbf{x}_N) \,.$$

The Hamiltonian reads:

$$\mathscr{H}(\mathbf{x}_1, \ldots, \mathbf{x}_N, \dot{\mathbf{x}}_1, \ldots, \dot{\mathbf{x}}_N) = \sum_{i=1}^{N} \frac{m_i}{2} \dot{\mathbf{x}}_i^2 + \mathscr{U}|_{\mathscr{R}}(\mathbf{x}_1, \ldots, \mathbf{x}_N) \,.$$

We also know that:

$$\mathbf{p}_i = \nabla_{\dot{\mathbf{x}}_i} \mathscr{L}(t, \mathbf{x}_1, \ldots, \mathbf{x}_N, \dot{\mathbf{x}}_1, \ldots, \dot{\mathbf{x}}_N) = m_i \dot{\mathbf{x}}_i$$

and therefore:

$$\dot{\mathbf{x}}_i = \frac{\mathbf{p}_i}{m_i} \,.$$

We can then write the Hamiltonian function:

$$\mathscr{H}(\mathbf{x}_1, \ldots, \mathbf{x}_N, \mathbf{p}_1, \ldots, \mathbf{p}_N) = \sum_{i=1}^{N} \frac{\mathbf{p}_i^2}{2m_i} + \mathscr{U}|_{\mathscr{R}}(\mathbf{x}_1, \ldots, \mathbf{x}_N)$$

using the Hamiltonian variables. In our case the Hamiltonian does not depend explicitly on time, so it is conserved along Hamilton's equations of motion (indeed, the Hamiltonian, on the motion, is the total mechanical energy). Hamilton's equations are, for $i = 1, \ldots, N$:

$$\frac{d\mathbf{p}_i}{dt} = \left(-\nabla_{\mathbf{x}_i} \mathscr{H} =\right) - \nabla_{\mathbf{x}_i} \mathscr{U}|_{\mathscr{R}}(\mathbf{x}_1, \ldots, \mathbf{x}_N),$$

$$\frac{d\mathbf{x}_i}{dt} = \left(\nabla_{\mathbf{p}_i} \mathscr{H} =\right) \frac{\mathbf{p}_i}{m_i}.$$

The two together give the N Newtonian equations ($i = 1, \ldots, N$):

$$m_i \frac{d^2 \mathbf{x}_i}{dt^2} = -\nabla_{\mathbf{x}_i} \mathscr{U}|_{\mathscr{R}}(\mathbf{x}_1, \ldots, \mathbf{x}_N).$$

(2) Consider a charged particle of mass m and charge e immersed in an electromagnetic field with electromagnetic potentials $\mathbf{A}$ and φ. From Sect. 9.2.1, in an inertial frame and using the position vector's components as free coordinates, the Lagrangian is:

$$\mathscr{L}(t, \mathbf{x}, \dot{\mathbf{x}}) = \frac{m}{2}\dot{\mathbf{x}}^2 - e\varphi(t, \mathbf{x}) + \frac{e}{c}\mathbf{A}(t, \mathbf{x}) \cdot \dot{\mathbf{x}}.$$

Directly, the Hamiltonian is:

$$\mathscr{H}(t, \mathbf{x}, \dot{\mathbf{x}}) = \frac{m}{2}\dot{\mathbf{x}}^2 + e\varphi(t, \mathbf{x}).$$

Notice that the magnetic potential has now disappeared: in Lagrangian formulation the Hamiltonian *does not* see the presence of the magnetic field. We do however know that:

$$\mathbf{p} = m\dot{\mathbf{x}} + \frac{e}{c}\mathbf{A}(t, \mathbf{x}),$$

so:

$$\dot{\mathbf{x}} := \frac{\mathbf{p}}{m} - \frac{e}{mc}\mathbf{A}(t, \mathbf{x}).$$

When we express the Hamiltonian in terms of Hamiltonian variables the information on the magnetic field resurfaces because the conjugate momenta are not simply the impulse's components:

$$\mathscr{H}(t, \mathbf{x}, \mathbf{p}) = \frac{\left(\mathbf{p} - \frac{e}{c}\mathbf{A}(t, \mathbf{x})\right)^2}{2m} + e\varphi(t, \mathbf{x}).$$

Hamilton's equations are:

$$\frac{d\mathbf{p}}{dt} = (-\nabla_{\mathbf{x}}\mathscr{H} =) \ \frac{1}{m}\left(\mathbf{p} - \frac{e}{c}\mathbf{A}\right) \cdot \frac{e}{c}\nabla_{\mathbf{x}}\mathbf{A} - e\nabla\varphi \,,$$

$$\frac{d\mathbf{x}}{dt} = (\ \nabla_{\mathbf{p}}\mathscr{H} =) \ \frac{\mathbf{p}}{m} - \frac{e}{mc}\mathbf{A} \,,$$

which imply:

$$m\frac{d^2\mathbf{x}}{dt^2} = -\frac{e}{c}\frac{\partial\mathbf{A}}{\partial t} - e\nabla_x\varphi - \frac{e}{c}\left(\mathbf{v}\cdot\nabla_{\mathbf{x}}\right)\mathbf{A}(t,\mathbf{x}) + \frac{e}{c}\nabla_{\mathbf{x}}\left(\mathbf{v}\cdot\mathbf{A}\right) \,.$$

The right-hand side is just the Lorentz force, as seen in Sect. 9.2.1. We have eventually recovered the correct equations of motion of a charged particle in an electromagnetic field.

(3) Consider a relativistic particle described on $A(\mathbb{R} \times \mathbb{R}^3)$, as in Sect. 10.5.5, by Minkowski coordinates $(x^0 = ct, x^1, x^2, x^3)$ on Minkowski spacetime $\mathbb{M}^4$. The reference Lagrangian will be (10.87): this is not the standard form, and neither is it defined on the entire $A(\mathbb{R} \times \mathbb{R}^3) = \mathbb{R} \times \mathbb{R}^3 \times \mathbb{R}^3$, only on the open region $\sum_{a=1}^{3}(\dot{x}^a)^2 < c^2$. The Legendre transform is anyhow well defined, and C^∞ with smooth inverse. Hence:

$$\mathbf{p} = \frac{m\dot{\mathbf{x}}}{\sqrt{1 - \left(\frac{\dot{\mathbf{x}}}{c}\right)^2}} + \frac{e}{c}\mathbf{A}(t,\mathbf{x}) \,, \quad a = 1, 2, 3 \tag{11.27}$$

and the Hamiltonian:

$$\mathscr{H}(\mathbf{x},\mathbf{p}) = c\sqrt{m^2c^2 + \left(\mathbf{p} - \frac{e}{c}\mathbf{A}(t,\mathbf{x})\right)^2} + e\varphi(t,\mathbf{x}) \tag{11.28}$$

is defined on $F(\mathbb{R} \times \mathbb{R}^3)$. It is easy to prove that if $m^2c^2 \gg \left(\mathbf{p} - \frac{e}{c}\mathbf{A}\right)^2$ then $\mathscr{H}$ is well approximated by the Hamiltonian of Example (2) up to the mathematically inessential additive constant mc^2. Hamilton's equations are easily found from $\mathscr{H}$, and give equations of order two for the relativistic particle in an electromagnetic field of the form (10.88), as it should be. Again, at small speeds compared to c, these equations are approximated by those in Example (2).

(4) A particle of mass m is constrained on a smooth sphere of radius $R > 0$ and centre O at rest in an inertial frame $\mathscr{R}$. It is subject to a force of potential energy $\mathscr{U}|_{\mathscr{R}}$. In spherical coordinates θ and ϕ the Lagrangian is:

$$\mathscr{L}(\theta,\phi,\dot{\theta},\dot{\phi}) = \frac{mR^2}{2}\left(\dot{\theta}^2 + \dot{\phi}^2\sin^2\theta\right) - \mathscr{U}|_{\mathscr{R}}(\theta,\phi) \,,$$

so the Hamiltonian is:

$$\mathscr{H}(\theta, \phi, \dot{\theta}, \dot{\phi}) = \frac{mR^2}{2}\left(\dot{\theta}^2 + \dot{\phi}^2 \sin^2 \theta\right) + \mathscr{U}|_{\mathscr{R}}(\theta, \phi).$$

Since the two conjugate momenta p_ϕ and ϕ_θ are:

$$p_\phi = mR^2 \dot{\phi} \sin^2 \theta, \qquad p_\theta = mR^2 \dot{\theta},$$

the Hamiltonian reads:

$$\mathscr{H}(\theta, \phi, p_\theta, p_\phi) = \frac{p_\theta^2}{2mR^2} + \frac{p_\phi^2}{2mR^2 \sin^2 \theta} + \mathscr{U}|_{\mathscr{R}}(t, \theta, \phi)$$

in Hamiltonian variables. We leave it to the reader to write down Hamilton's equations.

(5) When studying general physical systems $\mathscr{S}$ made of two (or more) subsets $\mathscr{S}_1, \mathscr{S}_2$, the Hamiltonian often has the form:

$$\mathscr{H} = \mathscr{H}_1 + \mathscr{H}_2 + \mathscr{H}_I$$

in natural local coordinates of $F(\mathbb{V}^{n+1})$, where $\mathscr{H}_1$ and $\mathscr{H}_2$ are the Hamiltonians of the subsystems as if they were not interacting, i.e. each Hamiltonian contains *only* the coordinates of its own subsystem; the third term $\mathscr{H}_I$ is the *interaction Hamiltonian*, involving the coordinates of *both* subsystems. The simples example is two particles of mass m_1 and m_2 respectively, with Hamiltonian coordinates $(\mathbf{x}_1, \mathbf{p}_1)$ and $(\mathbf{x}_2, \mathbf{p}_2)$ in an inertial frame $\mathscr{R}$ (using the same conventions as Example (1)). The free Hamiltonians are, for instance:

$$\mathscr{H}_1 := \frac{\mathbf{p}_1^2}{2m_1}, \qquad \mathscr{H}_2 := \frac{\mathbf{p}_2^2}{2m_2},$$

while the interaction Hamiltonian may be given by a conservative force with potential energy depending on the positions:

$$\mathscr{H}_I := \mathscr{U}(||\mathbf{x}_1 - \mathbf{x}_2||).$$

For systems made by several subsystems the interaction Hamiltonian can be decomposed into so-called: "two-body" Hamiltonians, "three-body" Hamiltonians and so forth. For 3 particles for example, we could have:

$$\mathscr{H}_I = \mathscr{U}^{(2)}(\mathbf{x}_1, \mathbf{x}_2) + \mathscr{U}^{(2)}(\mathbf{x}_2, \mathbf{x}_3) + \mathscr{U}^{(2)}(\mathbf{x}_1, \mathbf{x}_3) + \mathscr{U}^{(3)}(\mathbf{x}_1, \mathbf{x}_2, \mathbf{x}_3),$$

where $\mathscr{U}^{(2)}$ is the interaction (conservative in this case) of two bodies and $\mathscr{U}^{(3)}$ the interaction of three. In practical applications, e.g. Statistical Mechanics, sometimes one uses approximate techniques, which as first step ignore the interaction Hamiltonian, or only part of it (for example, the three-body part), and bring it back in the picture in the successive steps.

Exercises 11.10

(1) Consider a point particle P (Fig. 11.1) of mass m on the smooth surface of a torus $\mathbb{T}^2$ at rest in an inertial frame $\mathscr{R}$. Given orthonormal coordinates on $\mathscr{R}$ with axes $\mathbf{e}_x, \mathbf{e}_y, \mathbf{e}_z$ and origin O, the torus $\mathbb{T}^2$ has equation $P(\theta, \phi) = O + (R + r \cos\theta) \cos\phi \, \mathbf{e}_x + (R + r \cos\theta) \sin\phi \, \mathbf{e}_y + r \sin\theta \, \mathbf{e}_z$, where $R > r > 0$ are constants. Apart from the constraint reaction $\boldsymbol{\phi}$, P is acted upon by an ideal spring (no mass and zero length at rest) of elastic constant $k > 0$ with one end fixed to O, and also a viscous force $-\gamma \mathbf{v}_P$ where $\mathbf{v}_P$ is the velocity of P in $\mathscr{R}$ and $\gamma \geq 0$ is a constant. Answer the following questions using the coordinates $\phi \in (-\pi, \pi), \theta \in (-\pi, \pi)$ to describe the motion of P.

(a) Write the Euler-Lagrange equations of P.
(b) Prove that if $\gamma = 0$, on any motion the z-component of the angular momentum with respect to the pole O in $\mathscr{R}$ and the total mechanical energy in $\mathscr{R}$ are conserved; write explicitly these first integrals in terms of free coordinates.
(c) Take $\gamma = 0$ and pass to Hamilton's formulation: write Hamilton's equations and the Hamiltonian function in Hamiltonian coordinates.
(d) In case $\gamma > 0$, use the Euler-Lagrange equations to solve the equations of motion with initial conditions $\theta(0) = 0$, $\dot{\theta}(0) = 0$, $\phi(0) = 0$, $\dot{\phi}(0) = v > 0$.

Fig. 11.1 Exercise 11.10.1

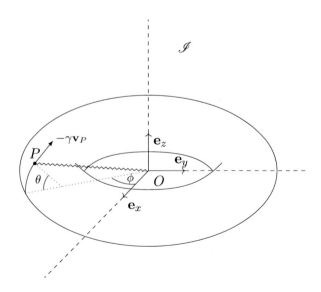

(2) Consider orthonormal coordinates on the non-inertial frame $\mathscr{R}$ of origin O and axes $\mathbf{e}_x, \mathbf{e}_y, \mathbf{e}_z$. A particle P of mass m travels along the smooth curve $z = \cosh x$. A second particle Q of mass m slides along an ideal smooth rigid rod attached to $O + x\mathbf{e}_x$ (the projection of P along the x-axis) and rotating around that point on the plane perpendicular to $\mathbf{e}_x$. The particles are connected by an ideal spring of elastic constant k, and are subject to the gravitational force with constant acceleration $-g\mathbf{e}_z$. The frame $\mathscr{R}$ rotates around the z-axis with respect to an inertial frame $\hat{\mathscr{R}}$ with axes $\hat{\mathbf{e}}_x, \hat{\mathbf{e}}_y, \hat{\mathbf{e}}_z \equiv \mathbf{e}_z$ and origin $\hat{O} \equiv O$, where $\boldsymbol{\omega}_{\mathscr{R}}|_{\hat{\mathscr{R}}} = \Omega\mathbf{e}_z$ and $\Omega > 0$ is constant. Answer the following questions using as free coordinates the following x, θ, r: the coordinate x is the first component in $\mathscr{R}$ of the position vector $P - O$; θ is the angle between the rod $Q - P$ and the y-axis (shifted to P) measured counterclockwise from $\mathbf{e}_x$; r is the distance of Q to $O + x\mathbf{e}_x$.

(a) Write the system's Euler-Lagrange equations.
(b) Prove that the Hamiltonian associated with the Lagrangian in frame $\mathscr{L}|_{\hat{\mathscr{R}}}$, with the above free coordinates, is conserved. Show that this Hamiltonian, along the motion, coincides with the total mechanical energy in the non-inertial frame $\mathscr{R}$.
(c) Pass to Hamilton's formulation and find the conjugate momenta in terms of Lagrangian coordinates and the inverse transformation. Then write the previous Hamiltonian in terms of Hamiltonian variables, and also Hamilton's equations.

11.1.4 The Hamiltonian's Dependency on the Local Chart

The Lagrangian $\mathscr{L} : A(\mathbb{V}^{n+1}) \to \mathbb{R}$ is a scalar field, i.e. a map defined on the manifold $A(\mathbb{V}^{n+1})$ without any system of coordinates (local or global), as attested by the notation we have employed. This means that if we choose two sets of natural coordinates on the spacetime of kinetic states:

$$\psi_1 : U_1 \ni a \mapsto (t(a), q^1(a), \ldots, q^n(a), \dot{q}^1(a), \ldots, \dot{q}^n(a)) \in \psi_1(U_1) \subset \mathbb{R}^{2n+1}$$

and

$$\psi_2 : U_2 \ni a \mapsto (t'(a), q'^1(a), \ldots, q'^n(a), \dot{q}'^1(a), \ldots, \dot{q}'^n(a)) \in \psi_2(U_2) \subset \mathbb{R}^{2n+1}$$

and work with the images of the intersection $U_1 \cap U_2$ of the charts' domains in the respective spaces $\mathbb{R}^{2n+1}$, with obvious notation we have:

$$\mathscr{L}_2(t'(t), q'(t,q), \dot{q}'(t,q,\dot{q})) = \mathscr{L}_1(t,q,\dot{q}),$$

where $\mathscr{L}_i := \mathscr{L} \circ \psi_i^{-1}$ if $i = 1, 2$. For this reason, we have always indicated the Lagrangian function by $\mathscr{L}$, slightly improperly, even when it is written in different natural coordinates.

On the other hand, the Hamiltonian associated with a given Lagrangian *is not a scalar field on* $F(\mathbb{V}^{n+1})$ *since it depends on the natural coordinates chosen*, so writing $\mathscr{H} : F(\mathbb{V}^{n+1}) \to \mathbb{R}$ *is incorrect*. Let us show this.

Suppose, for one Lagrangian $\mathscr{L} : A(\mathbb{V}^{n+1}) \to \mathbb{R}$, that the Hamiltonians $\mathscr{H}$ and $\mathscr{H}'$ on two natural local charts with non-empty intersection are associated with natural local coordinates $(t, q^1, \ldots, q^n, p_1, \ldots, p_n)$ and $(t', q'^1, \ldots, q'^n, p'_1, \ldots, p'_n)$ such that:

$$t' = t + c,$$

$$q'^k = q'^k(t, q^1, \ldots, q^n),$$

$$p'_k = \sum_{h=1}^{n} \frac{\partial q^h}{\partial q'^k} p_h,$$

on the charts' intersection. Then we have the equivalent identities:

$$\mathscr{H}'(t'(t), q'(t, q), p'(t, q, p)) = \mathscr{H}(t, q, p) + \sum_{k=1}^{n} \frac{\partial q'^k}{\partial t}\bigg|_{(t,q,p)} p'_k(t, q, p). \qquad (11.29)$$

and

$$\mathscr{H}'(t'(t), q'(t, q), p'(t, q, p)) = \mathscr{H}(t, q, p) - \sum_{h=1}^{n} p_h \frac{\partial q^h}{\partial t'}\bigg|_{(t'(t),q'(t,q),p'(t,q,p))}. \qquad (11.30)$$

The second one follows easily from the definition of Hamiltonian arising in natural coordinates from a Lagrangian, and from the coordinate transformation from $F(\mathbb{V}^{n+1})$ to $A(\mathbb{V}^{n+1})$:

$$\mathscr{H}'(t', q', p') = \sum_{k} p'_k \dot{q}'^k - \mathscr{L} = \sum_{k,h} p_h \frac{\partial q^h}{\partial q'^k} \dot{q}'^k - \mathscr{L}$$

$$= \sum_{k,h} p_h \left(\frac{\partial q^h}{\partial q'^k} \dot{q}'^k + \frac{\partial q^h}{\partial t'} \right) - \mathscr{L} - \sum_{h} p_h \frac{\partial q^h}{\partial t'} = \sum_{h} p_h \dot{q}^h - \mathscr{L} - \sum_{h} p_h \frac{\partial q^h}{\partial t'}$$

$$= \mathscr{H}(t, q, p) - \sum_{h} p_h \frac{\partial q^h}{\partial t'}.$$

These prove the second identity above. The first one can be proved similarly using the transformation rule for $\dot{q}'^k$ in terms of $\dot{q}^k$ in $\mathscr{H}'$. Alternatively, the first relation follows from the second one by the general formula:

$$\sum_h p_h \frac{\partial q^h}{\partial t'} = -\sum_k p'_k \frac{\partial q'^k}{\partial t} \ . \tag{11.31}$$

The latter is itself a consequence of the transformation rule between natural coordinates on phase spacetime:

$$p'_k = \sum_h \frac{\partial q^h}{\partial q'^k} p_h \ , \quad 0 = \frac{\partial q^h}{\partial t} = \sum_k \frac{\partial q^h}{\partial q'^k} \frac{\partial q'^k}{\partial t} + \frac{\partial q^h}{\partial t'} \frac{\partial t'}{\partial t} = \sum_k \frac{\partial q^h}{\partial q'^k} \frac{\partial q'^k}{\partial t} + \frac{\partial q^h}{\partial t'}$$

so:

$$-\sum_k p'_k \frac{\partial q'^k}{\partial t} = \sum_{h,k} \frac{\partial q^h}{\partial q'^k} \frac{\partial q'^k}{\partial t} p_h = \sum_h p_h \frac{\partial q^h}{\partial t'} \ .$$

Remarks 11.11 The Hamiltonian depends on the coordinates, yet it is the same for a fixed *reference frame* of the physical space, in the following sense. If we have a frame $\mathscr{R}$, *and in it the constraints do not depend on time*, from (2) in Remarks 9.11 the natural coordinates *moving with* $\mathscr{R}$ obey transformation rules that do not depend explicitly on time. In other words if (U, ψ) and (U', ψ') are natural local charts on $\mathbb{V}^{n+1}$ with $U \cap U' \neq \varnothing$ and both coordinates move with $\mathscr{R}$, their relationship on $U \cap U'$ is

$$t' = t + c \quad \text{and} \quad q'^k = q'^k(q^1, \ldots, q^n) \quad \text{for} \quad k = 1, \ldots, n \ .$$

with obvious notation. Then (11.29) says

$$\mathscr{H}'(t'(t), q'(t, q), p'(t, q, p)) = \mathscr{H}(t, q, p) \ . \tag{11.32}$$

Therefore, *using natural local coordinates on* $F(\mathbb{V}^{n+1})$ *moving with* $\mathscr{R}$, the Hamiltonian can be understood as a scalar field. Such coordinates form an atlas on $F(\mathbb{V}^{n+1})$ (this is easy if we remember how $\mathbb{V}^{n+1}$ is constructed, noting that any frame admits global coordinates on spacetime given by Cartesian coordinates moving with it). Therefore a Hamiltonian $\mathscr{H}|_{\mathscr{R}} : F(\mathbb{V}^{n+1}) \to \mathbb{R}$ is defined on *the entire* phase spacetime whenever we have a frame $\mathscr{R}$ in which the constraints do not depend on time (assuming such a frame exists at all) and when we have a Lagrangian $\mathscr{L}|_{\mathscr{R}}$ built in that frame. That is because the various Hamiltonians built from $\mathscr{L}|_{\mathscr{R}}$ with (11.18) "are glued together" over the atlas of $\mathscr{R}$, as prescribed by (11.32). A frame change produces a new globally defined Hamiltonian $\mathscr{H}|_{\mathscr{R}'}$. But if we change natural local charts on $F(\mathbb{V}^{n+1})$, possibly not moving with either frame, the Hamiltonians in local charts are normally related by (11.29). ∎

11.1.5 Independence of the Solutions to Hamilton's Equations of Local Charts

Despite the Hamiltonian is not a scalar field on $F(\mathbb{V}^{n+1})$, the following important result shows that Hamilton's equations are invariant under coordinate changes in the atlas of natural local charts; that is, the solutions to Hamilton's equations are independent of natural local coordinates provided the Hamiltonian functions are related by (11.29) (or equivalently (11.30)) when we change natural coordinates on $F(\mathbb{V}^{n+1})$. This happens automatically when the Hamiltonians come from one Lagrangian $\mathscr{L} : A(\mathbb{V}^{n+1}) \to \mathbb{R}$ under Legendre transform in arbitrary natural coordinates on $F(\mathbb{V}^{n+1})$. Regardless, one could consider (as one does, in certain cases of physical interest) situations that assign Hamiltonians on every natural chart of $F(\mathbb{V}^{n+1})$ satisfying (11.29), without referring to a common Lagrangian that generates them all.

Theorem 11.12 *Relatively to the phase spacetime $F(\mathbb{V}^{n+1})$, consider a C^1 curve parametrised by absolute time $\gamma : I \ni t \mapsto F(\mathbb{V}^{n+1})$ that solves Hamilton's equations on the open set $V \subset F(\mathbb{V}^{n+1})$ with natural coordinates $(t, q^1, \ldots, q^n, p_1, \ldots, p_n)$ for the C^1 Hamiltonian function $\mathscr{H} = \mathscr{H}(t, q, p)$. Such curve will satisfy Hamilton's equations in any other natural coordinates $(t', q'^1, \ldots, q'^n, p'_1, \ldots, p'_n)$ defined on V with respect to $\mathscr{H}' = \mathscr{H}'(t', q', p')$, provided (11.29) holds, or equivalently (11.30).*

Proof The curve γ in the first system of coordinates is described by $q^k = q^k(t)$ and $p_k = p_k(t)$. In the new coordinates, the same curve in phase space is given by $q'^h(t') = q^k(t(t'), q(t(t')))$ and $p'_h(t') = p'_k(t(t'), q(t(t')), p(t(t')))$, where we used the natural coordinate transformations (11.6) and (11.7). In the sequel we will assume $c = 0$, so $t' = t$, without any loss in generality due to the trivial relationship between t' and t. As γ solves Hamilton's equations for $\mathscr{H}$ in the unprimed coordinates, we have:

$$\frac{dq'^k}{dt} = \frac{\partial q'^k}{\partial t} + \sum_s \frac{\partial q'^k}{\partial q^s} \frac{dq^s}{dt} = \frac{\partial q'^k}{\partial t} + \sum_s \frac{\partial q'^k}{\partial q^s} \frac{\partial \mathscr{H}}{\partial p_s}.$$

The third relation in (11.6) and (11.7) gives:

$$\frac{\partial q'^k}{\partial q^s} = \frac{\partial p_s}{\partial p'_k},$$

and since $\frac{\partial}{\partial p'_k} = \sum_s \frac{\partial p_s}{\partial p'_k} \frac{\partial}{\partial p_s}$ from the inverses of (11.6) and (11.7):

$$\frac{dq'^k}{dt} = \frac{\partial q'^k}{\partial t} + \sum_s \frac{\partial p_s}{\partial p'_k} \frac{\partial \mathscr{H}}{\partial p_s} = \frac{\partial q'^k}{\partial t} + \frac{\partial \mathscr{H}}{\partial p'_k}.$$

We have found

$$\frac{dq'^k}{dt} = \frac{\partial q'^k}{\partial t} + \frac{\partial \mathcal{H}}{\partial p'_k} \ . \tag{11.33}$$

Arguing similarly for the second components we immediately obtain:

$$\frac{dp'_k}{dt} = \frac{\partial p'_k}{\partial t} - \sum_s \frac{\partial q^s}{\partial q'^k} \frac{\partial \mathcal{H}}{\partial q^s} + \sum_s \frac{\partial p'_k}{\partial q^s} \frac{\partial \mathcal{H}}{\partial p_s} \ .$$

On the right we can replace $\frac{\partial p'_k}{\partial q^s}$ by $-\frac{\partial p_s}{\partial q'^k}$ because $\frac{\partial p'_k}{\partial q^s} + \frac{\partial p_s}{\partial q'^k} = 0$, given that:

$$0 = \frac{\partial p_s}{\partial q^k} = \sum_r \frac{\partial p_s}{\partial q'^r} \frac{\partial q'^r}{\partial q^k} + \sum_r \frac{\partial p_s}{\partial p'_r} \frac{\partial p'_r}{\partial q^k} = \sum_r \frac{\partial p_s}{\partial q'^r} \frac{\partial q'^r}{\partial q^k} + \sum_r \frac{\partial q'^r}{\partial q^k} \frac{\partial p'_r}{\partial q^k}$$

$$= \sum_r \frac{\partial q'^r}{\partial q^k} \left(\frac{\partial p_s}{\partial q'^r} + \frac{\partial p'_r}{\partial q^k} \right) \ ,$$

and the coefficient matrix $\frac{\partial q'^r}{\partial q^k}$ is invertible by hypothesis. Hence:

$$\frac{dp'_k}{dt} = \frac{\partial p'_k}{\partial t} - \sum_s \frac{\partial q^s}{\partial q'^k} \frac{\partial \mathcal{H}}{\partial q^s} - \sum_s \frac{\partial p_s}{\partial q'^k} \frac{\partial \mathcal{H}}{\partial p_s} \ .$$

The inverses to (11.6) and (11.7) give $\frac{\partial}{\partial q'^k} = \sum_s \frac{\partial q^s}{\partial q'^k} \frac{\partial}{\partial q^s} + \sum_s \frac{\partial p_s}{\partial q'^k} \frac{\partial}{\partial p_s}$, so we conclude:

$$\frac{dp'_k}{dt} = \frac{\partial p'_k}{\partial t} - \frac{\partial \mathcal{H}}{\partial q'^k} \ . \tag{11.34}$$

Finally:

$$\frac{\partial q'^k}{\partial t} = \frac{\partial}{\partial p'_k} \sum_s p'_s \frac{\partial q'^s}{\partial t}$$

remembering that $\frac{\partial}{\partial p'_k} \frac{\partial q'^s}{\partial t} = \sum_s \frac{\partial p_s}{\partial p'_k} \frac{\partial}{\partial p_s} \frac{\partial q'^s}{\partial t} = 0$. Consequently, (11.33) can be written, using (11.29):

$$\frac{dq'^k}{dt} = \frac{\partial \mathcal{H}'}{\partial p'_k} \ .$$

Therefore the curve γ, in primed coordinates, satisfies the first set of Hamilton's equations for $\mathcal{H}'$. Analogously:

$$-\frac{\partial p'_k}{\partial t} = \frac{\partial}{\partial q'^k} \sum_s p'_s \frac{\partial q'^s}{\partial t} \tag{11.35}$$

which follows from (11.7) as shown below. Therefore γ, in primed coordinates, satisfies the second set of Hamilton's equations for $\mathcal{H}'$, since (11.34) can be written as, using (11.29):

$$\frac{dp'_k}{dt} = -\frac{\partial \mathcal{H}'}{\partial q'^k}.$$

To finish let us prove (11.35). The right-hand side reads:

$$\sum_{h,s} p'_s \frac{\partial q^h}{\partial q'^k} \frac{\partial^2 q'^s}{\partial q^h \partial t} = \sum_{h,s} p'_s \frac{\partial q^h}{\partial q'^k} \frac{\partial}{\partial t} \frac{\partial q'^s}{\partial q^h}$$

$$= \sum_s p'_s \frac{\partial}{\partial t} \left(\sum_h \frac{\partial q^h}{\partial q'^k} \frac{\partial q'^s}{\partial q^h} \right) - \sum_{h,s} p'_s \frac{\partial q'^s}{\partial q^h} \frac{\partial}{\partial t} \frac{\partial q^h}{\partial q'^k}$$

$$= \sum_s p'_s \frac{\partial}{\partial t} \left(\delta^s_k \right) - \sum_{h,s} p'_s \frac{\partial q'^s}{\partial q^h} \frac{\partial}{\partial t} \frac{\partial q^h}{\partial q'^k} = 0 - \sum_{h,s} p'_s \frac{\partial q'^s}{\partial q^h} \frac{\partial}{\partial t} \frac{\partial q^h}{\partial q'^k}$$

$$= -\sum_h p_h \frac{\partial}{\partial t} \frac{\partial q^h}{\partial q'^k} = -\frac{\partial}{\partial t} \sum_h p_h \frac{\partial q^h}{\partial q'^k}$$

$$= -\frac{\partial p'_k}{\partial t}.$$

$\square$

11.1.6 Global Solutions to Hamilton's Equations on $F(\mathbb{V}^{n+1})$

Theorem 11.12 has an important consequence when the Hamiltonians are C^2 in any natural local chart of $F(\mathbb{V}^{n+1})$ so to guarantee the existence and uniqueness of the maximal solution to Hamilton's equations in any such chart.

In general $F(\mathbb{V}^{n+1})$ cannot be covered by one local chart, and looking at a solution to Hamilton's equations on a local chart (U, ψ) with initial datum $p \in U$, we expect it to reach the chart's boundary $\mathcal{F}(U)$. From a physical point of view the solution is expected to continue existing outside the chart, since the latter has no special physical meaning in general. We could try to use different coordinates and patch up the solutions, but a priori we cannot know the overall outcome will be coherent. The proposition just proved tells us that the various ingredients in Hamilton's equations, when we change coordinates, behave compatibly. If, in fact, (V, ϕ) is another local chart with $V \cap U \neq \varnothing$ and the above solution γ_U reaches a point $\gamma_U(t_1) \in U \cap V$, we can reset the initial value problem of Hamilton's equations *written in coordinates*[3] *on V* using $\gamma_U(t_1)$ and $\dot{\gamma}_U(t_1)$ as initial conditions at time t_1. The unique solution γ_V will in general extend to V *also outside of U*. By construction the local sections γ_U, γ_V will overlap differentiably on $U \cap V$, satisfying both sets of Hamiltonian equations in the respective coordinates. Clearly, in this way, using the same argument of the global Existence and uniqueness theorem (Theorem 14.23), we eventually find the unique *maximal solution* to Hamilton's equations with initial state p, taking values in *the entire* $F(\mathbb{V}^{n+1})$. Every other solution with the same initial datum is a restriction of this one.

11.2 Hamilton's Equations from a Variational Principle

We will show how to derive Hamilton's equations from a variational principle, in analogy to what we did for the Euler-Lagrange equations in Theorem 9.2. Here, too, the functional will be the action, i.e. the Lagrangian's integral written in Hamiltonian formalism with the Hamiltonian variables. Furthermore, in contrast to the Lagrangian setting, now the variables p_k will be completely independent of the q^k. (In the Lagrangian case, the variational principle displayed only the q^k, while the $\dot{q}^k$ were immediately the time derivatives of the q^k).

Consider a Hamiltonian system on $F(\mathbb{V}^{n+1})$ and fix natural coordinates $t, q^1, \ldots, q^n, p_1, \ldots, p_n$. Without loss of generality we can view them on $I \times \Omega \times \Omega'$ where $I = [a, b]$, $a < b$, and $\Omega, \Omega' \subset \mathbb{R}^n$ are open (and non-empty). Suppose the Hamiltonian system is described by the Hamiltonian $\mathscr{H} : I \times \Omega \times \Omega' \to \mathbb{R}$, jointly C^1 in all variables. Fix configurations $Q_a, Q_b \in \Omega$ and consider the functional:

$$I_{Q_a, Q_b}^{(\mathscr{H})}[\gamma] := \int_a^b \left[\sum_{k=1}^n p_k \frac{dq^k}{dt} - \mathscr{H}\left(t, q^1(t), \ldots, q^n(t), p_1(t), \ldots, p_n(t)\right) \right] dt ,$$

$$(11.36)$$

---

[3] If necessary, translating the origin of time in V to match the time coordinate in U.

called **action** as in the Lagrangian case. The curves $\gamma \;:\; I \ni t \mapsto$ $(\gamma^1(t), \ldots \gamma^n(t), \gamma_1(t), \ldots, \gamma_n(t)) \;=\; (q^1(t), \ldots q^n(t), p_1(t), \ldots, p_n(t))$ belong to the domain:

$$D_{Q_a, Q_b} := \left\{ \gamma \in C^1(I) \;\middle|\; \gamma(t) \in \Omega \times \Omega' \text{ if } t \in I \right.$$

$$\left. \text{and } \gamma^k(a) = Q_a^k, \gamma^k(b) = Q_b^k \text{ for } k = 1, \ldots, n \right\} \;.$$

Remarks 11.13

(1) In the integrand of (11.36) the Lagrangian appears in terms of the Hamiltonian:

$$\mathscr{L} = \sum_k p_k \dot{q}^k - \mathscr{H} \;.$$

Hence the functional $I_{Q_a, Q_b}^{(\mathscr{H})}$ becomes the one of Theorem 9.2 when we revert to the Lagrangian formulation.

(2) The conditions satisfied by the curves of D_{Q_a, Q_b} at the boundary of $[a, b]$ refer to the first n components only, those of type q^k, and *not* the p_k. ∎

As $\Omega \times \Omega'$ is open, easily every curve $\gamma \in D_{Q_a, Q_b}$ admits an open neighbourhood of $\gamma(I)$ contained in $\Omega \times \Omega'$; with it, we can define a curve $\eta : I \to \Omega \times \Omega'$ such that $\gamma + \alpha \eta \in D_{Q_a, Q_b}$ for $\alpha \in (-\delta_\eta, \delta_\eta)$ and $\delta_\eta > 0$ small enough. It then makes sense to speak of stationary points for the action functional, as per Sect. 9.1.1. This leads to the following remarkable result, still called *Stationary-action principle* (in Hamilton's formulation), as in the Lagrangian case.

Theorem 11.14 (Hamiltonian Stationary-Action Principle) *With the above definitions, a curve $\gamma : I \to (q^1(t), \ldots, q^n(t), p_1(t), \ldots, p_n(t))$ in D_{Q_a, Q_b} satisfies Hamilton's equations (11.25) with respect to a jointly C^1 Hamiltonian $\mathscr{H} :$ $I \times \Omega \times \Omega' \to \mathbb{R}$ if and only if γ is a critical point of the action functional $I_{Q_a, Q_b}^{(\mathscr{H})}$ in (11.36).*

Proof Henceforth $\eta = (\eta^1, \ldots, \eta^n, \eta_1, \ldots, \eta_n)$. Proceeding as for Theorem 9.2 we immediately have:

$$\frac{d}{d\alpha}\bigg|_{\alpha=0} I_{Q_a, Q_b}^{(\mathscr{H})}[\gamma + \alpha\eta] = \int_a^b \sum_k \frac{\partial}{\partial\alpha}\bigg|_{\alpha=0} (p_k + \alpha\eta_k)\left(\frac{dq^k}{dt} + \alpha\frac{d\eta^k}{dt}\right) dt$$

$$- \int_a^b \frac{\partial}{\partial\alpha}\bigg|_{\alpha=0} \mathscr{H}(t, q^1 + \alpha\eta^1, \ldots, q^n + \alpha\eta^n, p_1 + \alpha\eta_1, \ldots, p_n + \alpha\eta_n) dt \;.$$

Differentiating,

$$\frac{d}{d\alpha}\bigg|_{\alpha=0} I^{(\mathscr{H})}_{Q_a,Q_b}[\gamma + \alpha\eta] = \int_a^b \sum_k \left(\eta_k \frac{dq^k}{dt} + p_k \frac{d\eta^k}{dt} - \frac{\partial\mathscr{H}}{\partial q^k}\eta^k - \frac{\partial\mathscr{H}}{\partial p_k}\eta_k \right) dt .$$

Integrating by parts and using that $\eta^k(a) = \eta^k(b) = 0$, because $\gamma + \alpha\eta$ passes through the configurations Q_a and Q_b,

$$\int_a^b \sum_k p_k \frac{d\eta^k}{dt} dt = -\int_a^b \sum_k \eta^k \frac{dp_k}{dt} dt .$$

We conclude:

$$\frac{d}{d\alpha}\bigg|_{\alpha=0} I^{(\mathscr{H})}_{Q_a,Q_b}[\gamma + \alpha\eta] = \int_a^b \sum_{k=1}^n \eta^k(t) \left(-\frac{\partial\mathscr{H}}{\partial q^k}\bigg|_{(t,\gamma(t))} - \frac{dp_k}{dt}\bigg|_{(t,\gamma(t))} \right) dt$$

$$+ \int_a^b \sum_{k=1}^n \eta_k(t) \left(-\frac{\partial\mathscr{H}}{\partial p_k}\bigg|_{(t,\gamma(t))} + \frac{dq^k}{dt}\bigg|_{(t,\gamma(t))} \right) dt .$$

The procedure used for Theorem 9.2, plus the fact both the curve γ and the Hamiltonian are C^1, shows that the vanishing of the above relation's left-hand side (on the set of maps considered) for any η is equivalent to asking, along γ:

$$\frac{\partial\mathscr{H}}{\partial q^k} + \frac{dp_k}{dt} = 0 , \qquad \frac{\partial\mathscr{H}}{\partial p_k} - \frac{dq^k}{dt} = 0 .$$

In other words the curve γ is a critical point of the action functional if and only if it solves Hamilton's equations. □

Remarks 11.15 $\boxed{AC}$ We have set up the theory in $\mathbb{R}^{2n}$, more precisely on a local chart of $F(\mathbb{V}^{n+1})$. We could reformulate the variational principle in a larger ambient, using more charts when the curves do not fall within a single chart, simply employing a *partition of unity*. ∎

11.3 Hamiltonian Formulation on $\mathbb{R} \times \mathbb{R}^{2n}$

This section is devoted to elementary but important properties of Hamiltonian systems on $\mathbb{R} \times \mathbb{R}^{2n}$; in particular, we will see a first version of *Liouville's theorem*.

11.3.1   Hamiltonian Systems on $\mathbb{R} \times \mathbb{R}^{2n}$ and the Symplectic Matrix S

Consider a physical system that admits Hamiltonian description on the phase space $F(\mathbb{V}^{n+1})$. Fix a natural local chart

$$\psi : U \ni a \mapsto (t, q^1, \ldots, q^n, p_1, \ldots, p_n) \in \mathbb{R} \times \mathbb{R}^{2n}$$

on phase space and identify the latter, at least the open region covered by the coordinates, with an open subset of $\mathbb{R} \times \mathbb{R}^{2n}$, where the first factor is the time axis. We will put ourselves in the simpler situation where the coordinates $(t, q^1, \ldots, q^n, p_1, \ldots, p_n)$ vary on *the whole* $\mathbb{R} \times \Omega$, where Ω is an open set in $\mathbb{R}^{2n}$ identified with every fibre $\mathbb{F}_t$. Hence we can view $F(\mathbb{V}^{n+1})$ as the Cartesian product $\mathbb{R} \times \Omega$. Hamilton's equations become:

$$\frac{d\mathbf{x}}{dt} = S\nabla_{\mathbf{x}}\mathscr{H}(t, \mathbf{x}) , \tag{11.37}$$

where: $\mathbf{x}$ is a column vector, $(q^1, \ldots, q^n, p_1, \ldots, p_n)^t$ is the generic point in $\mathbb{R}^{2n}$ and $\mathscr{H} : \mathbb{R} \times \Omega \to \mathbb{R}$ the Hamiltonian function. We will assume $\nabla_{\mathbf{x}}\mathscr{H}$ is $C^1(\mathbb{R} \times \Omega)$ to ensure (this is more than enough) the solutions' existence and uniqueness.[4] The matrix $S \in M(2n, \mathbb{R})$, which we call **symplectic matrix**, has the form:

$$S := \left[\begin{array}{c|c} 0 & I \\ \hline -I & 0 \end{array} \right] , \tag{11.38}$$

where I is the $n \times n$ identity. Finally, $\nabla_{\mathbf{x}} f$ is the column vector

$$\nabla_{\mathbf{x}} f := \left(\frac{\partial f}{\partial q^1}, \ldots \frac{\partial f}{\partial q^n}, \frac{\partial f}{\partial p_1}, \ldots \frac{\partial f}{\partial p_n} \right)^t .$$

Note that

$$S^t = -S = S^{-1} . \tag{11.39}$$

Before continuing let us fix the notation of the rest of the book. In components, (11.38) is:

$$S = [S_{ij}]_{ij=1,\ldots,2n} . \tag{11.40}$$

---

[4] From Complement 14, it would be enough to ask $\nabla_{\mathbf{x}}\mathscr{H} \in C^0(\mathbb{R} \times \Omega)$ be locally Lipschitz in $\mathbf{x}$, for any $t \in \mathbb{R}$.

It is convenient to use coefficients S^{rs} with raised indices, $r, s = 1, 2 \ldots, 2n$, with the property that

$$\sum_{j=1}^{2n} S^{ij} S_{jk} = -\delta_k^i \quad i, k = 1, \ldots, 2n .$$ (11.41)

This can be understood as a matrix product, and since the inverse is unique and $(-S)S = I$ (from (11.39)), we have

$$S^{ij} = S_{ij} \quad i, j = 1, \ldots, 2n .$$ (11.42)

In agreement with tensorial conventions, Hamilton's equations may be written as

$$\frac{dx^k}{dt} = \sum_{j=1}^{2n} S^{kj} \frac{\partial \mathcal{H}}{\partial x^j} \quad k = 1, \ldots, 2n ,$$ (11.43)

so for example, using (11.41),

$$\sum_{k=1}^{2n} S_{jk} \frac{dx^k}{dt} = -\frac{\partial \mathcal{H}}{\partial x^j} \quad j = 1, \ldots, 2n .$$ (11.44)

Let us return to Hamilton's equations. Expression (11.37) has important consequences on the physical-mathematical properties of solutions: amongst others, the cornerstone *Liouville theorem*, the *Liouville equation* and *Poincaré's "recurrence" theorem*. We shall discuss these in full in another chapter, while here we shall merely present a simplified version of the first result. An important issue is to establish mathematical criteria to decide whether a given system of first-order equations:

$$\frac{d\mathbf{x}}{dt} = \mathbf{F}(t, \mathbf{x})$$

on some open set $\mathbb{R} \times \Omega$, $\Omega \subset \mathbb{R}^{2n}$, can, at least locally, reduce to (11.37) once we have a suitable Hamiltonian function.

11.3.2 Hamiltonian Matrices and Hamiltonian Dynamical Systems

To address the problem raised at the end of the previous section let us introduce a family of matrices built from the symplectic matrix S.

Definition 11.16 The real vector space of **Hamiltonian matrices** of order n is

$$sp(n, \mathbb{R}) := \left\{ E \in M(2n, \mathbb{R}) \mid E^t S + SE = 0 \right\} . \tag{11.45}$$

$\diamond$

The set $sp(n, \mathbb{R})$ is manifestly closed under linear combinations and so it is a real vector space.

We can state and prove the result providing necessary and sufficient conditions for a system of first-order differential equations on $\mathbb{R} \times \mathbb{R}^{2n}$ to be expressible in Hamiltonian form.

Theorem 11.17 *For a given $n = 1, 2, \ldots$, consider the ODE system:*

$$\frac{d\mathbf{x}}{dt} = \mathbf{F}(t, \mathbf{x}) \tag{11.46}$$

on $\mathbb{R} \times \Omega$, where Ω is (non-empty and) open in $\mathbb{R}^{2n}$ and $\mathbf{F} \in C^1(\mathbb{R} \times \Omega)$. The following facts hold.

(1) If the system can be written in Hamiltonian form (11.37) for some Hamiltonian function, the matrices $D_{(t,\mathbf{x})}$ whose coefficients are the derivatives:

$$\left. \frac{\partial F^i}{\partial x^j} \right|_{(t,\mathbf{x})}$$

belong to $sp(n, \mathbb{R})$ for any $(t, \mathbf{x}) \in \mathbb{R} \times \Omega$.

(2) Vice versa, if $D_{(t,\mathbf{x})} \in sp(n, \mathbb{R})$ for any $(t, \mathbf{x}) \in \mathbb{R} \times \Omega$, then for any connected and simply connected open subset $B \subset \Omega$, system (11.46) can be put in Hamiltonian form on $\mathbb{R} \times B$.

Proof

(1) The system is Hamiltonian if and only if we are able to define a map (whose gradient will be C^1 since $\mathbf{F}$ is, by hypothesis) $(t, \mathbf{x}) \mapsto \mathscr{H}(t, \mathbf{x})$ such that $\mathbf{F}(t, \mathbf{x}) = S\nabla_{\mathbf{x}} \mathscr{H}(t, \mathbf{x})$. Equivalently, multiplying by S, $\mathbf{G} := -S\mathbf{F}$ is a gradient map for any t. As $\mathbf{G}$ is C^1, being a gradient implies it is irrotational:

$$\frac{\partial G^k}{\partial x^h} = \frac{\partial G^h}{\partial x^k} . \tag{11.47}$$

That is, in components,

$$-\frac{\partial \sum_r S_{kr} F^r}{\partial x^h} = -\frac{\partial \sum_r S_{hr} F^r}{\partial x^k} ,$$

which we recast as:

$$-\sum_r S_{kr} \frac{\partial F^r}{\partial x^h} + \sum_r S_{hr} \frac{\partial F^r}{\partial x^k} = 0 \,,$$

and using $S_{kr} = -S_{rk}$:

$$\sum_r \frac{\partial F^r}{\partial x^h} S_{rk} + \sum_r S_{hr} \frac{\partial F^r}{\partial x^k} = 0 \,.$$

Written more compactly, we have $D^t_{(t,\mathbf{x})} S + S D_{(t,\mathbf{x})} = 0$, in other words $D_{(t,\mathbf{x})} \in sp(n, \mathbb{R})$.

(2) For the converse let us fix a connected and simply connected open set $B \subset \Omega$ and $t \in \mathbb{R}$. Reading the proof of (1) backwards shows that $D_{(t,\mathbf{x})} \in sp(n, \mathbb{R})$ is the same as (11.47), i.e. the C^1 field $B \ni \mathbf{x} \mapsto -S\mathbf{F}(t, \mathbf{x})$ is irrotational. By Theorem 4.15 there exists a potential

$$\mathcal{H}(t, \mathbf{x}) := \int_{\mathbf{x}_0}^{\mathbf{x}} (-S\mathbf{F}(t, \mathbf{x})) \cdot d\mathbf{x} \tag{11.48}$$

on B with gradient $-S\mathbf{F}(t, \mathbf{x})$. The integral is computed along any C^1 path (independent of t) from $\mathbf{x} \in B$ to an arbitrary point $\mathbf{x}_0 \in B$, and is all contained in B. By construction, exactly at time t on B:

$$\mathbf{F}(t, \mathbf{x}) = S\nabla_{\mathbf{x}} \mathcal{H}(t, \mathbf{x}) \,.$$

The function $\mathcal{H}$ depends parametrically on time since the right-hand side of (11.48) depends on time. By construction the right side of (11.48) determines, at every $t \in \mathbb{R}$, a map with gradient $-S\mathbf{F}$ on B, so it is the required Hamiltonian function on $\mathbb{R} \times B$.

$\square$

Remarks 11.18

(1) The theorem proves a more general fact: a vector-valued map $\mathbf{F} := \mathbf{F}(t, \mathbf{x}) \in C^1((a, b) \times \Omega)$, with Ω connected and simply connected, can be written as $\mathbf{F}(t, \mathbf{x}) = S\nabla_{\mathbf{x}} \mathcal{H}(t, \mathbf{x})$ in terms of a scalar function $\mathcal{H}$ defined on the same domain if and only if the matrix with coefficients

$$\frac{\partial F^i}{\partial x^j}\bigg|_{(t,\mathbf{x})}$$

belongs in $sp(n, \mathbb{R})$ for every $(t, \mathbf{x}) \in (a, b) \times \Omega$.

(2) Even if a system of differential equations can be written in Hamiltonian form, in general the Hamiltonian cannot be constructed out of a Lagrangian through a Legendre transformation. An elementary example on $\mathbb{R} \times \mathbb{R}^2$ is

$$\frac{dx}{dt} = a, \quad \frac{dy}{dt} = b,$$

where $a, b \in \mathbb{R} \setminus \{0\}$ are constants. This system is Hamiltonian if we take

$$\mathscr{H}(t, x, p) := ap - bx,$$

where $p := y$. However, it easy to see (exercise) that there is no Lagrangian $\mathscr{L}(t, x, \dot{x})$ on $\mathbb{R} \times \mathbb{R}^2$ such that $\mathscr{H}(t, x, p) = \frac{\partial \mathscr{L}}{\partial \dot{x}} \dot{x}(t, x, p) - \mathscr{L}$ where $t = t$, $x = x$, $p = \frac{\partial \mathscr{L}}{\partial \dot{x}}(t, x, \dot{x})$ is a diffeomorphism (at least local) from $\mathbb{R} \times \mathbb{R}^2$ to $\mathbb{R} \times \mathbb{R}^2$. *The fact that a Hamiltonian function cannot arise from a Lagrangian does not prevent one from exploiting the powerful results of the Hamiltonian formalism, one being the Liouville theorem.*   ∎

11.3.3   Liouville's Theorem in $\mathbb{R} \times \mathbb{R}^{2n}$

The *theorem of Liouville* is definitely one of the most important results in Hamiltonian Mechanics, in *Statistical Mechanics* and in all of Mathematical Physics. In a nutshell the theorem states the following, intuitively speaking. Suppose a collection of representative points of a physical system is described in Hamiltonian fashion in phase *spacetime*, and the cardinality is so large that the points form a continuous set. Then the system evolves so that to conserve in time the points' volume on each fibre (the phase space at time t, $\mathbb{F}_t$) crossed by some history. The measure used to compute the volume in phase space $\mathbb{F}_t$ is the one given by the natural coordinates. For the moment we shall not address the general problem of defining the volume; we shall still work with Hamiltonian systems on $\mathbb{R} \times \Omega$ in given natural coordinates $(t, \mathbf{x})$, and use the standard measure on $\mathbb{R}^{2n}$ on the fibres $\mathbb{F}_t \equiv \Omega$ at constant time.

   Consider the Hamiltonian system on $\mathbb{R} \times \Omega$ described by (11.37) with $\mathscr{H} \in C^k(\mathbb{R} \times \Omega)$, where $k \geq 2$ in order for global Hamiltonian equations to admit unique solutions. The phase spacetime is now $\mathbb{R} \times \Omega$. We want to interpret Hamilton's equations as an *autonomous* system and then use the theory of autonomous differential equations, in particular the description of solutions as one-parameter groups of local diffeomorphisms. To do this, think t as coordinate on phase spacetime and introduce an obvious equation reminding us that t is actually the parameter of the curve.[5] The abstract parameter describing the evolution will be s,

---

[5] $\boxed{\text{AC}}$ They are sections of the bundle $F(\mathbb{V}^{n+1})$ over the time line $\mathbb{R}$.

living in an open interval $I_{(t_0, \mathbf{x}_0)}$ that depends on the initial conditions $(t_0, \mathbf{x}_0)$:

$$
\begin{cases}
\dfrac{d\mathbf{x}}{ds} = S\nabla_{\mathbf{x}}\mathscr{H}(t(s), \mathbf{x}(s)), \\[2mm]
\dfrac{dt}{ds} = 1.
\end{cases}
\tag{11.49}
$$

This differential system on $\mathbb{R} \times \Omega$ describes Hamilton's equations and is autonomous, since on the right the solutions' parameter s does not show up. The maximal solution will then be vector-valued with $1 + 2n$ components: $t = t_0 + s$, $\mathbf{x} = \mathbf{x}(s|t_0, \mathbf{x}_0)$. To be precise, call

$$ I_{(t_0, \mathbf{x}_0)} \ni s \mapsto (t_0 + s, \mathbf{x}(s|t_0, \mathbf{x}_0)) \in \mathbb{R} \times \Omega $$

the maximal solution of (11.49), with maximal domain the open interval $I_{(t_0, \mathbf{x}_0)} \subset \mathbb{R}$ containing 0, and initial conditions $\mathbf{x}(0) = \mathbf{x}_0$ and $t(0) = t_0$. The obvious expression of $t = t(s)$ descends from equation (11.49). Thus we define a map associating with a solution's initial conditions $(t_0, \mathbf{x}_0)$ the value of the solution at s:

$$ \Phi_s(t_0, \mathbf{x}_0) := (t_0 + s, \mathbf{x}(s|t_0, \mathbf{x}_0)) \in \mathbb{R} \times \Omega. \tag{11.50} $$

Φ is called **Hamiltonian flow** on phase spacetime. Section 14.5.3 shows that the map $(s, t_0, \mathbf{x}_0) \mapsto \Phi_s(t_0, \mathbf{x}_0) = (t_0 + s, \mathbf{x}(s|t_0, \mathbf{x}_0))$ is well defined and C^{k-1}, if $\mathscr{H}$ is C^k, on a non-empty open subset of $\Gamma \subset \mathbb{R} \times \mathbb{R} \times \Omega$ (Proposition 14.48) of the form:

$$ \Gamma = \bigcup_{(t_0, \mathbf{x}_0) \in \mathbb{R} \times \Omega} I_{(t_0, \mathbf{x}_0)} \times \{(t_0, \mathbf{x}_0)\}. $$

The Uniqueness Theorem (see Sect. 14.5.3) eventually proves that the flow is a 1-*parameter group of local diffeomorphisms*:

$$ \Phi_0(t, \mathbf{x}) = (t, \mathbf{x}), \qquad \Phi_s \circ \Phi_u(t, \mathbf{x}) = \Phi_{s+u}(t, \mathbf{x}). \tag{11.51} $$

The second relation holds whenever the left-hand side is defined (Proposition 14.57). Since (Proposition 14.50) Γ is open in a small neighbourhood of each $(t_0, \mathbf{x}_0)$ containing $(t, \mathbf{x})$, if s and u belong in a small enough neighbourhood of 0, the second relation holds for sure. At last, if $A \subset \mathbb{R} \times \Omega$ is open and sufficiently small, $\Phi_s(A)$ is open and $\Phi_s : A \to \Phi_s(A)$ is a diffeomorphism.

Remarks 11.19 If every solution to (11.49) is complete, i.e. $I_{(t_0, \mathbf{x}_0)} = (-\infty, +\infty)$ for any t_0 and $\mathbf{x}_0$, then Φ_s is defined globally on $\mathbb{R} \times \Omega$ for any $s \in \mathbb{R}$, and $\{\Phi_s\}_{s \in \mathbb{R}}$ is a 1-parameter group (parameter s) of diffeomorphisms defined and surjective on

$\mathbb{R} \times \Omega$ (see Sect. 14.5.3). If so:

$$\Phi_0 = id_{\mathbb{R} \times \Omega}, \quad \Phi_s \circ \Phi_u = \Phi_{s+u} \quad \forall s, u \in \mathbb{R},$$

without limitations on the domains.   ∎

Fix an instant t_0 and an open set with compact closure $D_{t_0} \subset \mathbb{F}_{t_0}$. We let D_{t_0} evolve under the local group Φ until time $t = t_0 + s$. If D_{t_0} is small enough $\{t\} \times D_t := \Phi_s(\{t_0\} \times D_{t_0})$ is well defined for $D_t \subset \mathbb{F}_t$. By construction D_t contains the representative point at time t of every solution to Hamilton's equations passing through D_{t_0} at time t_0.

Theorem 11.20 (Liouville's Theorem on $\mathbb{R} \times \mathbb{R}^{2n}$) *Relatively to the Hamiltonian system (11.37) on $\mathbb{R} \times \Omega$, with $\Omega \subset \mathbb{R}^{2n}$ open, non-empty and[6] $\mathscr{H} \in C^3(\mathbb{R} \times \Omega)$, the Hamiltonian flow Φ preserves the volume of the fibres of phase spacetime. In other words, if $D_{t_0} \subset \mathbb{F}_{t_0}$ is a small enough open set with compact closure and $t > t_0$ is sufficiently close to t_0 so that $\{t\} \times D_t := \Phi_s(\{t_0\} \times D_{t_0})$ is defined, then:*

$$\int_{D_{t_0}} d^{2n}\mathbf{x} = \int_{D_t} d^{2n}\mathbf{x}. \tag{11.52}$$

where $d^{2n}\mathbf{x}$ is the standard Lebesgue measure on the fibres $\mathbb{F}_t$, i.e. $d^{2n}\mathbf{x} = dq^1 \cdots dq^n dp_1 \cdots dp_n$ in coordinates

$$\mathbf{x} = (q^1, \ldots, q^n, p_1, \ldots, p_n).$$

If all solutions to (11.37) are complete, the above formula holds for any open $D_{t_0} \subset \mathbb{F}_{t_0}$ with compact closure and for any $t \in \mathbb{R}$.

Proof First of all (Fig. 11.2), D_t is open: for $\delta > 0$ small enough, the open set $(t_0 - \delta, t_0 + \delta) \times D_{t_0}$ is mapped to an open set $A \subset \mathbb{R} \times \Omega$ whose projection onto the absolute time axis is $(t - \delta, t + \delta)$. The set $D_t \subset \Omega$ made of the points of A at time t is open in Ω, as one shows using the product topology and the fact that the projection $\mathbb{R} \times \Omega \to \Omega$ is open. Hence D_t is measurable. Changing variable in the second integral we write:

$$\int_{D_t} d^{2n}\mathbf{x} = \int_{D_{t_0}} |\det J_s| d^{2n}\mathbf{x}$$

where $\det J_t$ is the Jacobian determinant of $\mathbf{x}_0 \mapsto \mathbf{x}(s|t_0, \mathbf{x}_0)$, which is jointly C^2 given the assumptions on $\mathscr{H}$. We can differentiate with respect to s inside the integral since $(s, \mathbf{x}) \mapsto J_s$ has continuous first derivatives and $\overline{D_{t_0}}$ is compact. To

---

[6] We could relax this hypothesis by distinguishing the regularity of the various derivatives.

Fig. 11.2 Hamiltonian evolution of D_{t_0}

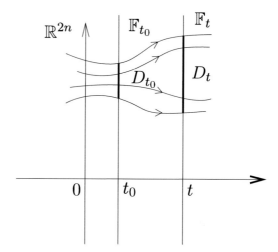

prove the claim it suffices to show:

$$\frac{\partial |\det J_s|}{\partial s} = 0, \quad \forall s \in I_{(t_0, \mathbf{x}_0)}. \tag{11.53}$$

The absolute value, which is not regular at the origin, is actually harmless here. In fact, when $s = 0$, $\det J_0 = \det I = 1 > 0$ and the sign of $\det J_s$ stays positive, by continuity, in every interval of s containing $s = 0$: if the sign changed it would vanish somewhere in the interval. But it can never vanish because $(t_0, \mathbf{x}_0) \mapsto (t_0 + s, \mathbf{x}(s|t_0, \mathbf{x}_0))$ is a diffeomorphism (so it has non-zero Jacobian determinant) and the Jacobian determinant of such transformation equals the Jacobian determinant of $\mathbf{x}_0 \mapsto \mathbf{x}(s|t_0, \mathbf{x}_0)$. Hence the claim is:

$$\frac{\partial \det J_s}{\partial s} = 0, \quad \forall s \in I_{(t_0, \mathbf{x}_0)}.$$

Let us use a simpler notation. For $s > s'$, and making the dependency on t implicit, we will write:

$$\mathbf{x}_{s'}(\mathbf{x}_s) := \mathbf{x}(s'|t(s), \mathbf{x}(s)).$$

Then:

$$\frac{\partial \det J_s}{\partial s} = \lim_{h \to 0} \frac{1}{h} \left(\det \left[\frac{\partial \mathbf{x}_{s+h}(\mathbf{x}_0)}{\partial \mathbf{x}_0} \right] - \det \left[\frac{\partial \mathbf{x}_s(\mathbf{x}_0)}{\partial \mathbf{x}_0} \right] \right)$$

$$= \lim_{h \to 0} \frac{1}{h} \left(\det \left[\frac{\partial \mathbf{x}_{s+h}(\mathbf{x}_s)}{\partial \mathbf{x}_s} \right] \det \left[\frac{\partial \mathbf{x}_s(\mathbf{x}_0)}{\partial \mathbf{x}_0} \right] - \det \left[\frac{\partial \mathbf{x}_s(\mathbf{x}_0)}{\partial \mathbf{x}_0} \right] \right) =$$

$$= \lim_{h \to 0} \frac{1}{h} \left(\det \left[\frac{\partial \mathbf{x}_{s+h}(\mathbf{x}_s)}{\partial \mathbf{x}_s} \right] - 1 \right) \det \left[\frac{\partial \mathbf{x}_s(\mathbf{x}_0)}{\partial \mathbf{x}_0} \right]$$

$$= (\det J_s) \lim_{h \to 0} \frac{1}{h} \left(\det \left[\frac{\partial \mathbf{x}_{s+h}(\mathbf{x}_s)}{\partial \mathbf{x}_s} \right] - 1 \right) .$$

Let us expand the function in square brackets in Taylor series with respect to h at $h = 0$. If $\mathbf{x}$ denotes the generic $\mathbf{x}_s = \mathbf{x}(t)$ ($t = t_0 + s$), and for coherence $\mathbf{x}_{s+h}$ becomes $\mathbf{x}_h$ ($\mathbf{x}_{s+h}$ is the evolution of $\mathbf{x}_s$ under the Hamiltonian flow until h) we have

$$\frac{\partial x^i_{h=0}(\mathbf{x})}{\partial x^j} = \frac{\partial x^i}{\partial x^j} = \delta^i_j .$$

The Hamiltonian flow is C^2 because $\nabla \mathscr{H}$ is, so by the Schwarz theorem:

$$\frac{\partial}{\partial h}\Big|_{h=0} \frac{\partial x^i_h(\mathbf{x})}{\partial x^j} = \frac{\partial}{\partial x^j} \frac{\partial x^i_h}{\partial h}\Big|_{h=0} = \frac{\partial}{\partial x^j} (S\nabla_{\mathbf{x}}\mathscr{H}(t, \mathbf{x}))^i ,$$

where by definition the curve $h \mapsto \mathbf{x}_h(\mathbf{x})$ satisfies Hamilton's equations:

$$\frac{d\mathbf{x}_h}{dh} = S\nabla_{\mathbf{x}}\mathscr{H}(t + h, \mathbf{x}_h)$$

with initial condition $\mathbf{x}_{h=0} = \mathbf{x} = \mathbf{x}(t)$ at $h = 0$. Therefore

$$\frac{\partial x^i_h(\mathbf{x})}{\partial x^j} = \delta^i_j + h \frac{\partial}{\partial x^j} (S\nabla_{\mathbf{x}}\mathscr{H}(t, \mathbf{x}))^i + h O_{t,\mathbf{x}}(h)^i_j$$

where $O_{t,\mathbf{x}}(h)^i_j \to 0$ as $h \to 0$. The determinant of $A(h) = [\frac{\partial x^i_h}{\partial x^j}]_{i,j=1,\ldots,2n}$, whose coefficients are the right-hand side of the expansion, is found using the following technical lemma.   □

Lemma 11.21 *Take matrices $A(h), B, O(h) \in M(m, \mathbb{R})$. If $A(h) = I + hB + hO(h)$ as $h \in (-\delta, \delta)$ and $O(h) \to 0$ as $h \to 0$, then*

$$\det A(h) = 1 + h \, \mathrm{tr}\, B + h\Omega(h) , \tag{11.54}$$

where $\Omega : (-\delta, \delta) \to \mathbb{R}$ satisfies $\Omega(h) \to 0$ as $h \to 0$.   ◇

Proof The proof is direct, using the formula

$$\det C = \sum_{i_1,i_2,\ldots,i_m=1}^{m} \epsilon_{i_1 i_2 \ldots i_n} C_{1i_1} C_{2i_2} \cdots C_{m i_m} ,$$

valid for any $C \in M(m, \mathbb{R})$, where the numbers $\epsilon_{i_1 i_2 \dots i_m}$ are defined as: $\epsilon_{123 \dots m-1m} := 1$ and

$$\epsilon_{i_1 \dots i_{h-1} i_h i_{h+1} \dots i_{k-1} i_k i_{k+1} \dots i_n} = -\epsilon_{i_1 \dots i_{h-1} i_k i_{h+1} \dots i_{k-1} i_h i_{k+1} \dots i_n} \quad \text{for any } h \neq k.$$

That is, the $\epsilon_{i_1 i_2 \dots i_m}$ are skew-symmetric whenever we swap two indices. In particular, if (at least) two indices in $i_1 i_2 \dots i_n$ are equal, $\epsilon_{i_1 i_2 \dots i_n} = 0$. In our case:

$$\det C = \sum_{i_1, i_2, \dots, i_m=1}^{m} \epsilon_{i_1 i_2 \dots i_n} \left(\delta_{i_1}^1 + h B_{i_1}^1 + h O_{i_1}^1(h) \right) \cdots \left(\delta_{i_m}^m + h B_{i_m}^m + h O_{i_m}^m(h) \right)$$

and rearranging by powers of h we immediately see

$$\det C = \epsilon_{123 \dots m-1m} + h \sum_{a=1}^{m} B_a^a + h\Omega(h)$$

with $\Omega(h) \to 0$ as $h \to 0$. In compact form, this is precisely (11.54). From the lemma we have:

$$\det \left[\frac{\partial \mathbf{x}_h(\mathbf{x})}{\partial \mathbf{x}} \right] = 1 + h \sum_{i=1}^{2n} \frac{\partial}{\partial x^i} (S\nabla_{\mathbf{x}} \mathcal{H}(t, \mathbf{x}))^i + h\Omega_{t,\mathbf{x}}(h),$$

and therefore:

$$\lim_{h \to 0} \frac{1}{h} \left(\det \left[\frac{\partial \mathbf{x}_h(\mathbf{x})}{\partial \mathbf{x}} \right] - 1 \right) = \lim_{h \to 0} \left\{ \sum_{i=1}^{2n} \frac{\partial}{\partial x^i} (S\nabla_{\mathbf{x}} \mathcal{H}(t, \mathbf{x}))^i + \Omega_{t,\mathbf{x}}(h) \right\}$$

$$= \sum_{i=1}^{2n} \frac{\partial}{\partial x^i} (S\nabla_{\mathbf{x}} \mathcal{H}(t, \mathbf{x}))^i + 0.$$

So we have found:

$$\frac{\partial |\det J_s|}{\partial s} = J_s \sum_{i=1}^{2n} \frac{\partial}{\partial x^i} (S\nabla_{\mathbf{x}} \mathcal{H}(t, \mathbf{x}))^i,$$

or more explicitly:

$$\sum_{i=1}^{2n} \frac{\partial}{\partial x^i} (S\nabla \mathcal{H}(t, \mathbf{x}))^i = \sum_{i=1}^{2n} \frac{\partial}{\partial x^i} \sum_{j=1}^{2n} S^{ij} \frac{\partial}{\partial x^j} \mathcal{H}(t, \mathbf{x}) = \sum_{i,j=1}^{2n} S^{ij} \frac{\partial^2 \mathcal{H}(t, \mathbf{x})}{\partial x^i \partial x^j}.$$

Now, Schwarz's theorem implies:

$$\sum_{i,j=1}^{2n} S^{ij} \frac{\partial^2 \mathcal{H}(t,\mathbf{x})}{\partial x^i \partial x^j} = \sum_{i,j=1}^{2n} S^{ij} \frac{\partial^2 \mathcal{H}(t,\mathbf{x})}{\partial x^j \partial x^i} \,,$$

and renaming the indices on the right, from $S^{ij} = -S^{ji}$ we deduce:

$$\sum_{i,j=1}^{2n} S^{ij} \frac{\partial^2 \mathcal{H}(t,\mathbf{x})}{\partial x^i \partial x^j} = \sum_{i,j=1}^{2n} S^{ji} \frac{\partial^2 \mathcal{H}(t,\mathbf{x})}{\partial x^i \partial x^j} = -\sum_{i,j=1}^{2n} S^{ij} \frac{\partial^2 \mathcal{H}(t,\mathbf{x})}{\partial x^i \partial x^j} \,.$$

Eventually:

$$\sum_{i,j=1}^{2n} S^{ij} \frac{\partial^2 \mathcal{H}(t,\mathbf{x})}{\partial x^i \partial x^j} = 0 \,,$$

which implies (11.53). This proves the first claim. The other is an immediate consequence, since if the solutions to (11.37) are complete, so are the solutions to (11.49). Then $\{\Phi_s\}_{s \in \mathbb{R}}$ is a 1-parameter group of diffeomorphisms, as explained in Remark 11.19, and it acts globally on $\mathbb{R} \times \Omega$ for any $s \in \mathbb{R}$.   □

11.4 $\boxed{\text{AC}}$ The Bundle $F(\mathbb{V}^{n+1})$ and Hamilton's Equations as Global Equations

In this section we will study in depth the differential-geometric structure of phase space, focussing on the global nature of the Legendre transformation and Hamilton's equations. We will make use of Definition B.4.

11.4.1 The Bundle $F(\mathbb{V}^{n+1})$

The space $F(\mathbb{V}^{n+1})$ can be *defined* as a bundle over $\mathbb{V}^{n+1}$ in the following way. As we did for $A(\mathbb{V}^{n+1})$, we start from a standard bundle. In this case it is convenient to take the cotangent bundle $T^*\mathbb{V}^{n+1}$ rather than the tangent bundle.

For any point $p \in \mathbb{V}^{n+1}$ consider the subspace $K_p \subset T_p^*\mathbb{V}^{n+1}$ of 1-forms generated by dT_p, where $T : \mathbb{V}^{n+1} \to \mathbb{R}$, up to additive constants (which are immaterial for dT_p), is the usual absolute time. Now take the quotient

$$F_p(\mathbb{V}^{n+1}) := T_p^*\mathbb{V}^{n+1}/\sim$$

by the equivalence relation defined by the subspace K_p:

$$\omega \sim \omega', \quad \text{if and only if } \omega - \omega' \in K_p.$$

It is easy to see that the quotient inherits the structure of real vector space of dimension n from $T^*_p\mathbb{V}^{n+1}$. In particular, if $[\omega] \in F_p(\mathbb{V}^{n+1})$ is the equivalence class of $\omega \in T^*\mathbb{V}^{n+1}$, for any $\omega, \omega' \in T^*_p\mathbb{V}^{n+1}$ and $a, b \in \mathbb{R}$:

$$[a\omega + b\omega'] = a[\omega] + b[\omega'], \tag{11.55}$$

while

$$[dt|_p] = 0. \tag{11.56}$$

Let $(t, q^1, \ldots, q^n)$ be natural local coordinates defined around $p \in \mathbb{V}^{n+1}$ with corresponding basis $dt|_p, dq^1|_p, \ldots dq^n|_p \in T^*_p\mathbb{V}^{n+1}$. Then two 1-forms

$$\omega = p_t dt|_p + \sum_{k=1}^n p_k dq^k|_p \quad \text{and} \quad \omega' = p'_t dt|_p + \sum_{k=1}^n p'_k dq^k|_p$$

define the same element in $F_p(\mathbb{V}^{n+1})$ if and only if they differ a most by their time components p_t and p'_t. Clearly, then, if we change basis $[dq^1|_p], \ldots [dq^n|_p] \in F_p(\mathbb{V}^{n+1})$ (the new basis is still induced by natural coordinates on $F_p(\mathbb{V}^{n+1})$):

$$[dq'^k|_p] = \sum_{h=1}^n \left.\frac{\partial q'^k}{\partial q^h}\right|_p [dq^h|_p].$$

This is straightforward from (11.55) and (11.56) and the relations:

$$dq'^k|_p = \sum_{h=1}^n \left.\frac{\partial q'^k}{\partial q^h}\right|_p dq^h|_p + \left.\frac{\partial q'^k}{\partial t}\right|_p dt|_p,$$

recalling that the matrix $\frac{\partial q'^k}{\partial q^h}$ has non-zero determinant.

The manifold $F(\mathbb{V}^{n+1})$ is a bundle in two different ways:

(a) over $\mathbb{V}^{n+1}$, with the vector spaces $F_p(\mathbb{V}^{n+1})$ as fibres;
(b) over the time axis $\mathbb{R}$ (or $\mathbb{E}^1$), with the **phase spaces at time** t:

$$\mathbb{F}_t = \left\{ (c_t, p) \mid c_t \in \mathbb{Q}_t, \quad p \in F_{c_t}(\mathbb{V}^{n+1}) \right\}$$

as fibres, for any $t \in \mathbb{R}$. The bundle projection $T : F(\mathbb{V}^{n+1}) \to \mathbb{R}$ is just *absolute time*.

The local coordinates mapping $(c_t, p) \in F(\mathbb{V}^{n+1})$ to the $(2n + 1)$-tuple $(t, q^1, \ldots, q^n, p_1, \ldots, p_n)$, where $(q^1, \ldots, q^n)$ are the configuration's coordinates at time t, c_t and $p = \sum_{k=1}^n p_k [dq^k|_{c_t}]$ are natural local coordinates adapted to both bundles in the sense of Definition B.4. In fact they induce the bundle's local splitting as base times canonical fibre.

The coordinates' transformation rule is indeed:

$$t' = t + c, \tag{11.57}$$

$$q'^k = q'^k(t, q^1, \ldots, q^n), \tag{11.58}$$

$$p'_k = \sum_{h=1}^n \frac{\partial q^h}{\partial q'^k} p_h. \tag{11.59}$$

Let us see in detail how the above $F(\mathbb{V}^{n+1})$ is a bundle in sense (b): with base $\mathbb{R}$, standard fibre $\mathbb{F}$ and projection T. In other words $F(\mathbb{V}^{n+1})$ is locally diffeomorphic to the product $\mathbb{R} \times \mathbb{F}$, where $\mathbb{F}$ (the **phase space at given time**) is in turn diffeomorphic to each $\mathbb{F}_t$. The proof is the same as for $\mathbb{V}^{n+1}$, see Proposition 7.34. Let us go over it. For $\mathbb{V}^{n+1}$ the argument was based on the existence of natural local charts satisfying condition (H3). With the above construction of $F(\mathbb{V}^{n+1})$, the same charts endow the space with natural coordinates that obey similar conditions: for any absolute time $t \in \mathbb{R}$ there is an open non-empty interval $J \subset \mathbb{R}$ containing t and a corresponding family of natural coordinates $\{(U_i, \phi_i)\}_{i \in I}$ on $F(\mathbb{V}^{n+1})$ with $\phi_i : U_i \ni a \mapsto (t_i(a), q_i^1(a), \ldots q_i^n(a), p_{i1}(a), \ldots, p_{in}(a)) \in \mathbb{R}^{2n+1}$, satisfying the following:

(i) $\bigcup_{i \in I} U_i \supset \mathbb{F}_\tau$ for any $\tau \in J$;

(ii) $\phi_i(U_i) \supset J \times U_i'$, with $U_i \subset \mathbb{R}^{2n}$ open, for any $i \in I$;

(iii) for any $i, j \in I$ such that $U_i \cap U_j \neq \varnothing$ the functions $\phi_i \circ \phi_j^{-1} : \phi_j(U_i \cap U_j) \to \phi_i(U_i \cap U_j)$ have the form:

$$t_i = t_j, \quad q_i^k = q_i^k(q_j^1, \ldots, q_j^n), \quad p_{ik} = \sum_h \frac{\partial q_j^h}{\partial q_i^k} p_{jh};$$

that is, the transformation rule between Lagrangian coordinates does not depend explicitly on time.

Now the same glueing procedure used for Proposition 7.34 shows that for the neighbourhood J of the given point $t \in \mathbb{R}$, there is a diffeomorphism $f : J \times \mathbb{F}_t \to T^{-1}(J)$ such that $T(f(\tau, a)) = \tau$ if $\tau \in J$ and $a \in \mathbb{F}_t$, i.e. a *local trivialisation*. Note that $f^{(\tau)} : a \in \mathbb{F}_t \mapsto f(\tau, a) \in \mathbb{F}_\tau$ is a diffeomorphism from $\mathbb{F}_t$ to $\mathbb{F}_\tau$ for any $\tau \in J$. If $t' \notin J$, say $t' < t$, we can iterate the construction and define a diffeomorphism $f' : J' \times \mathbb{F}_{t'} \to T^{-1}(J')$ such that $T(f'(\tau, a)) = \tau$ if $\tau \in J'$ and $a \in \mathbb{F}_{t'}$. To finish we still need to prove $\mathbb{F}_{t'}$ is diffeomorphic to $\mathbb{F}_t$, so that we can take the latter as standard fibre. Using the compactness of $[t', t]$ it is easy to see that there are finitely many intervals $J_1 = J', J_2, \ldots, J_N = J$ of said type such that $J_k \cap J_{k+1} \neq \varnothing$. Composing a finite number of diffeomorphisms analogous to $f^{(\tau)}$, we end up with a diffeomorphism $g_{t'} : \mathbb{F}_t \to \mathbb{F}_{t'}$. Thus all fibres are diffeomorphic to $\mathbb{F} := \mathbb{F}_t$. The local trivialisations on every $J' \ni t'$ are the diffeomorphisms $g : J' \times \mathbb{F} \ni (\tau, a) \mapsto f'(\tau, g_{t'}(a)) \in T^{-1}(J')$, for which $T(g(\tau, a)) = \tau$ if $\tau \in J'$ and $a \in \mathbb{F}_{t'}$, trivially.

11.4.2 Global Legendre Transformation as Diffeomorphism from $A(\mathbb{V}^{n+1})$ to $F(\mathbb{V}^{n+1})$

Suppose we are assigned a Lagrangian $\mathscr{L} : A(\mathbb{V}^{n+1}) \to \mathbb{R}$ which we will assume of the usual local standard form (7.87), where the matrix a_{hk} is everywhere non-singular or even positive definite. We want to define Legendre transformations in a global way. To this end the following fundamental proposition summarises and extends the discussion of Sect. 11.1.2.

Proposition 11.22 *Consider a Lagrangian $\mathscr{L} : A(\mathbb{V}^{n+1}) \to \mathbb{R}$ of class C^k, $k \geq 2$. The following facts hold.*

(1) *If $\mathscr{L}$ has form (7.87) in natural local coordinates where the coefficient matrix a_{hk} is everywhere non-singular, there exists a unique C^k diffeomorphism, called* **Legendre transform**:

$$\mathbb{L}_{\mathscr{L}} : A(\mathbb{V}^{n+1}) \to F(\mathbb{V}^{n+1}) \tag{11.60}$$

such that, for any natural local charts

$$\psi : V \ni a \mapsto (t, q^1, \ldots, q^n, \dot{q}^1, \ldots, \dot{q}^n) \in \mathbb{R}^{2n+1}$$

on $A(\mathbb{V}^{n+1})$ and

$$\phi : W \ni a \mapsto (t, q^1, \ldots, q^n, p_1, \ldots, p_n) \in \mathbb{R}^{2n+1}$$

on $F(\mathbb{V}^{n+1})$ extending the same natural local chart on $\mathbb{V}^{n+1}$, the coordinate representation of $\phi \circ \mathbb{L}_{\mathscr{L}} \circ \psi^{-1}$ has the form

$$t = t, \tag{11.61}$$

$$q^k = q^k \quad for\ k = 1, \ldots, n, \tag{11.62}$$

$$p_k = \left.\frac{\partial \mathscr{L}}{\partial \dot{q}^k}\right|_{(t, q^1, \ldots, q^n, \dot{q}^1, \ldots, \dot{q}^n)} \quad for\ k = 1, \ldots, n. \tag{11.63}$$

(2) *More weakly, if $\mathscr{L}$ is such that, for any natural coordinates (V, ψ) in $A(\mathbb{V}^{n+1})$*
and (W, ϕ) in $F(\mathbb{V}^{n+1})$ extending the same natural local chart on $\mathbb{V}^{n+1}$, the
transformation (11.61)–(11.63) is bijective with:

$$det\left[\frac{\partial^2 \mathscr{L}}{\partial \dot{q}^h \partial \dot{q}^k}\right]_{h,k=1,\ldots,n} \neq 0 \quad \text{everywhere on } V, \tag{11.64}$$

then there exists a unique C^{k-1} diffeomorphism (11.60) whose representation
in natural local coordinates is (11.61)–(11.63).

Proof

(1) By construction the coordinate transformation (11.61)–(11.63) is C^k with
C^k inverse and maps $\psi(V)$ to $\phi(W)$ bijectively. In fact, if $\mathscr{L}$ is C^k and
of form (7.87), the coefficients $a_{hk}(t, q)$ (non-singular), $\beta_k(t, q)$ and the
coefficients of the inverse $(a^{-1})_{hk}(t, q)$ are C^k; (11.61)–(11.63) and the inverse
transformation read:

$$t = t ,$$

$$q^k = q^k \quad \text{for } k = 1, \ldots, n ,$$

$$p_k = \beta_k(t, q) + 2\sum_{h=1}^{n} a_{kh}(t, q)\dot{q}^h \quad \text{for } k = 1, \ldots, n .$$

and

$$t = t ,$$

$$q^k = q^k \quad \text{for } k = 1, \ldots, n ,$$

$$\dot{q}^k = \frac{1}{2}\sum_{h=1}^{n} a_{kh}^{-1}(t, q)(p_h - \beta_h(t, q)) \quad \text{for } k = 1, \ldots, n .$$

Consider now two further local charts as in the claim:

$$\psi' : V' \ni a \mapsto (t', q'^1, \ldots, q'^n, \dot{q}'^1, \ldots, \dot{q}'^n) \in \mathbb{R}^{2n+1} ,$$

and

$$\phi' : W' \ni a \mapsto (t', q'^1, \ldots, q'^n, p'_1, \ldots, p'_n) \in \mathbb{R}^{2n+1} ,$$

built to extend the same local chart on $\mathbb{V}^{n+1}$ and so that $V \cap V' \neq \varnothing$, whence $W \cap W' \neq \varnothing$. With obvious notation, we look at the transformation

$$t' = t', \tag{11.65}$$

$$q'^k = q'^k \quad \text{for } k = 1, \ldots, n, \tag{11.66}$$

$$p'_k = \frac{\partial \mathscr{L}}{\partial \dot{q}'^k}\bigg|_{(t', q'^1, \ldots, q'^n, \dot{q}'^1, \ldots, \dot{q}'^n)} \quad \text{for } k = 1, \ldots, n \tag{11.67}$$

and prove that it coincides with (11.61)–(11.63) on $a \in V \cap V'$. The proof is immediate if we remember that:

$$p'_k = \sum_j \frac{\partial q^j}{\partial q'^k} p_j, \qquad \frac{\partial \mathscr{L}}{\partial \dot{q}'^k} = \sum_j \frac{\partial q^j}{\partial q'^k} \frac{\partial \mathscr{L}}{\partial \dot{q}^j}.$$

As the natural charts of type V and W form atlases on $A(\mathbb{V}^{n+1})$ and $F(\mathbb{V}^{n+1})$ respectively, what was proved above immediately implies that there exists a C^k diffeomorphism from $A(\mathbb{V}^{n+1})$ to $F(\mathbb{V}^{n+1})$, which on the local charts V, W is (11.61)–(11.63).

(2) The proof is the same, except that we no longer have the C^k regularity guaranteed by $\mathscr{L}$. Now the (11.61)–(11.63) are certainly C^{k-1}. Condition (11.64), plus the trivial (11.61) and (11.62), imply that these transformations are a local diffeomorphism of class C^{k-1} around any $a \in V$. But the map is bijective, so the inverse from $\phi(W)$ to $\psi(V)$ will also be C^{k-1}. Hence (11.61)–(11.63) define a C^{k-1} diffeomorphism mapping $\psi(V)$ to $\phi(W)$. The proof ends by the same argument used in case (1).

$\square$

The previous proposition assumes that the transformation is invertible. In some situations this hypothesis is automatic. Without going into details we will only cite a general result regarding the global invertibility of maps over convex subsets of $\mathbb{R}^n$. We remind that a subset $C \subset \mathbb{A}$ of a real affine space is called **convex** whenever the line segment $[0, 1] \ni s \mapsto p + s(q - p) \in \mathbb{A}$ between any two points $p, q \in C$ is all contained in C.

Proposition 11.23 *Let $F : \Omega \to \mathbb{R}^n$ be a C^k map, $k \geq 1$, with $\Omega \subset \mathbb{R}^n$ open and convex. If the quadratic form Q_{J_p} of the Jacobian matrix of F at any $p \in \Omega$ does not have **isotropic vectors**, i.e. vectors $\mathbf{u} \in \mathbb{R}^n \setminus \{\mathbf{0}\}$ such that $Q_{J_p}(\mathbf{u}) = 0$, then:*

(1) $F(\Omega) \subset \mathbb{R}^n$ is open,

(2) $F : \Omega \to F(\Omega)$ is a C^k diffeomorphism (F is bijective and its inverse is C^k).

Proof We claim that the Jacobian matrix J_p of F at any p has non-vanishing determinant. If the determinant were zero there would be a (non-zero) eigenvector $\mathbf{u}$ with zero eigenvalue: $J_p \mathbf{u} = \mathbf{0}$. Then $Q_{J_p}(\mathbf{u}) := \mathbf{u} \cdot J_p \mathbf{u} = 0$, meaning that the quadratic form would have an isotropic vector. By the Implicit Function Theorem

[KrPa03], F is a local C^k diffeomorphism around any point of Ω. In other words, for any $p \in \Omega$, there is an open neighbourhood $U \ni p$ with $U \subset \Omega$ such that (i) $F(U)$ is open and contained in $F(\Omega)$, and (ii) $F|_U : U \to F(U)$ is a C^k diffeomorphism. In particular, $F(\Omega)$ is open as union of open sets. To conclude we just need to show F is 1-1 on Ω. Suppose there exist $p, q \in \Omega$ with $p \neq q$ and $F(p) = F(q)$. Let $\mathbf{n}$ be the unit vector of $q - p$. If $x(t) = p + t(q - p)$, the map $f(t) := \mathbf{n} \cdot (F(x(t)) - F(p))$, $t \in [0, 1]$ (well defined since Ω is convex!), must have zero derivative at some point $t_0 \in (0, 1)$ by Rolle's Theorem:

$$f'(t_0) = \sum_{i,j=1}^{n} n^i \left. \frac{\partial F^i}{\partial x^j} \right|_{x(t_0)} n^j \, ||p - q|| = ||p - q|| Q_{J_{x(t_0)}}(\mathbf{n}) = 0 \,.$$

Therefore $Q_{J_{x(t_0)}}(\mathbf{n}) = 0$ on the non-zero vector $\mathbf{n}$. But as we said this is impossible, so there are no such p and q. $\square$

Remarks 11.24 Saying Q_{J_p} does not have isotropic vectors is the same as asking that at any $p \in \Omega$ the bilinear form Q_{J_p} is positive definite or negative definite, in view of Sylvester's law of inertia.   ■

11.4.3   Global Intrinsic Hamiltonian Formulation via the Field Z and the Emancipation of the Lagrangian Formulation

When a Lagrangian has been assigned on $A(\mathbb{V}^{n+1})$, we also have a Hamiltonian function $\mathscr{H}$ in natural local coordinates, given by the Lagrangian's *Legendre transform*:

$$\mathscr{H}(t, q, p) := \sum_{k=1}^{n} p_k \dot{q}^k(t, q, p) - \mathscr{L}(t, q, \dot{q}(t, q, p)) \,.$$

A crucial observation we have already made is that, in contrast to the Lagrangian, the Hamiltonian function $\mathscr{H}$ does not have, in general, a scalar behaviour when we change natural coordinates. Directly from the definition, in fact, when we pass to natural coordinates (t', q', p') on phase spacetime, on the intersections of the latter coordinates' domains the new Hamiltonian is related to the old one by:

$$\mathscr{H}'(t'(t, q, p), q'(t, q, p), p'(t, q, p)) = \mathscr{H}(t, q, p) + \sum_{k=1}^{n} \left. \frac{\partial q'^k}{\partial t} \right|_{(t,q,p)} p'_k(t, q, p) \,.$$

$$(11.68)$$

Despite this fact, Hamilton's equations have a global valence on $F(\mathbb{V}^{n+1})$. Namely, they determine sections of the bundle $F(\mathbb{V}^{n+1})$ over absolute time. These sections are the integral curves of the vector field Z on $F(\mathbb{V}^{n+1})$ called *Hamiltonian dynamic vector field* (or simply *dynamic vector*). It is given in natural local coordinates by

$$Z(t, q, p) = \frac{\partial}{\partial t} + \sum_{k=1}^{n} \frac{\partial \mathcal{H}}{\partial p_k} \frac{\partial}{\partial q^k} - \sum_{k=1}^{n} \frac{\partial \mathcal{H}}{\partial q^k} \frac{\partial}{\partial p_k}. \tag{11.69}$$

In fact, it is evident that the integral curves of Z are precisely the solutions to Hamilton's equations (11.25). In the sequel we will assume that the various Hamiltonians are at least C^2, so that Z is C^1 at least.

A computation shows that Z is *defined globally* on $F(\mathbb{V}^{n+1})$, i.e. independent of the choice of local coordinates: indeed one shows that a coordinate change on $\mathbb{V}^{n+1}$ and the corresponding changes of natural coordinates on $F(\mathbb{V}^{n+1})$ produce:

$$\frac{\partial}{\partial t} + \sum_{k=1}^{n} \frac{\partial \mathcal{H}}{\partial p_k} \frac{\partial}{\partial q^k} - \sum_{k=1}^{n} \frac{\partial \mathcal{H}}{\partial q^k} \frac{\partial}{\partial p_k} = \frac{\partial}{\partial t'} + \sum_{k=1}^{n} \frac{\partial \mathcal{H}'}{\partial p'_k} \frac{\partial}{\partial q'^k} - \sum_{k=1}^{n} \frac{\partial \mathcal{H}'}{\partial q'^k} \frac{\partial}{\partial p'_k},$$

on the intersections of the domains of natural coordinates on $F(\mathbb{V}^{n+1})$, *when we assume the Hamiltonians are related by (11.68)*. Another way to see that is from Theorem 11.12: given a Hamiltonian $\mathcal{H}$ in natural local coordinates around a point a in phase spacetime, the vector Z associated with $\mathcal{H}$ is tangent to the unique solution of Hamilton's equations for $\mathcal{H}$, with initial condition a. In view of the Uniqueness theorem for first-order differential equations, Theorem 11.12 ensures that the resulting curve is the same curve we would find in other coordinates around a when replacing $\mathcal{H}$ with the corresponding Hamiltonian under (11.68), provided the initial conditions are the same. But the curve does not depend on the coordinates, so neither will the tangent vector $Z(a)$ to the curve at a. Therefore Z is well defined on the space $F(\mathbb{V}^{n+1})$ and it makes sense to speak of the solution to the equations of motion extended to the entire phase spacetime, not just on a local chart, despite every $\mathcal{H}$ is only defined locally on $F(\mathbb{V}^{n+1})$, not globally in general.

It is fundamental to note that the global existence of the dynamic field Z has nothing to do with the fact the Hamiltonians are Legendre transforms of a given Lagrangian on $A(\mathbb{V}^{n+1})$. It is sufficient to work directly on $F(\mathbb{V}^{n+1})$ and prescribe a function $\mathcal{H}$ in every natural local coordinates, with the proviso that a coordinate change affects the Hamiltonian by (11.68) on the local charts' common domains. *In this way the Hamiltonian formulation turns out to be totally independent of the Lagrangian formulation, and may be stated directly on* $F(\mathbb{V}^{n+1})$ *without mentioning neither* $A(\mathbb{V}^{n+1})$ *nor the Lagrangian on it*. Here is a definition.

Definition 11.25 (Hamiltonian Dynamic Vector Field)   A C^1 vector field Z on $F(\mathbb{V}^{n+1})$ is called **Hamiltonian dynamic vector field** (or **dynamic vector** for short) if it is given in natural local coordinates by

$$Z(t, q, p) = \frac{\partial}{\partial t} + \sum_{k=1}^{n} \frac{\partial \mathscr{H}}{\partial p_k} \frac{\partial}{\partial q^k} - \sum_{k=1}^{n} \frac{\partial \mathscr{H}}{\partial q^k} \frac{\partial}{\partial p_k}, \qquad (11.70)$$

where the C^2 functions $\mathscr{H}$ satisfy relations (11.68) on the intersections of the natural local charts' domains. $\diamond$

Remarks 11.26

(1) In a general setting it is far from obvious that one can assign functions $\mathscr{H}$ in every natural chart of $F(\mathbb{V}^{n+1})$ so to satisfy (11.68) (think of three charts with double intersections but no triple intersection). Constraints (11.68) are instead automatic when the various $\mathscr{H}$ are Legendre transforms of a unique Lagrangian. That said, one can construct physically interesting systems that satisfy evolution equations of Hamiltonian type where the Hamiltonian functions are not generated by a Legendre transforms from $A(\mathbb{V}^{n+1})$ to $F(\mathbb{V}^{n+1})$, because the latter would be non-invertible. This occurs in advanced physical formulations such as Field Theory.

(2) One can, conversely, define Hamiltonian systems whose dynamic field Z is globally defined but locally induced by Hamiltonians, and at the same time there is no global Hamiltonian, even if we use a specific atlas of natural charts with time-independent transition functions, or even if there is a single global chart covering phase spacetime. This follows from the discussion of Sect. 7.5.4. Take $\mathbb{V}^{2+1} = \mathbb{R} \times (\mathbb{R}^2 \setminus \{(0, 0)\})$. As usual $\mathbb{R}$ is absolute time and $\mathbb{R}^2 \setminus \{(0, 0)\}$ is identified with each configuration space at time t. Suppose the equations of motion in standard coordinates on $\mathbb{R}^2$ are:

$$m \frac{d^2 x}{dt^2} = -\frac{ky}{x^2 + y^2}, \qquad m \frac{d^2 x}{dt^2} = \frac{kx}{x^2 + y^2}.$$

Consider, in *global* natural coordinates associated with the standard coordinates of $\mathbb{R}^2$, the globally defined field on $F(\mathbb{V}^{n+1})$:

$$Z = \frac{\partial}{\partial t} + \frac{p_x}{m} \frac{\partial}{\partial x} + \frac{p_y}{m} \frac{\partial}{\partial y} - \frac{k}{m} \frac{y}{x^2 + y^2} \frac{\partial}{\partial p_x} + \frac{k}{m} \frac{x}{x^2 + y^2} \frac{\partial}{\partial p_y}.$$

The associated Hamiltonian equations are clearly equivalent to the above ones to second order. However, there is no way to define a global C^1 Hamiltonian on $F(\mathbb{R} \times (\mathbb{R}^2 \setminus \{(0, 0)\}))$ giving Z. In fact, if there were, it would have to look like

$$\mathscr{H}(x, y, p_x, p_y) = \frac{p_x^2}{2m} + \frac{p_y^2}{2m} + \mathscr{U}(x, y)$$

(up to time-additive maps), where the gradient of $\mathscr{U}$ reproduces the vector field in Newton's equations' right-hand side. Such a $\mathscr{U}$ cannot exist if we want it defined on the entire $\mathbb{R}^2 \setminus \{(0,0)\}$. The reason was explained in Sect. 7.5.4: the force field on the right is irrotational but not conservative, since its path integral along a loop around the origin does not vanish. Using polar coordinates on the plane *minus the negative x-axis*, for example, the field Z is given by the following Hamiltonian:

$$\mathscr{H}(r, \theta, p_r, p_\theta) = \frac{p_r^2}{2m} + \frac{p_\theta^2}{2mr^2} + k\theta .$$

Evidently this function *cannot* be extended, not even continuously, to the entire phase space because it would become multi-valued.

■

11.5 Symplectic Vector Spaces, the Symplectic Group and the Lie Algebra of Hamiltonian Matrices

The present section has a fully mathematical character, and we shall better examine the relationship between the *symplectic matrix S* (11.38)—used to build the theory of Hamilton's equations on $\mathbb{R} \times \mathbb{R}^{2n}$—and *Hamiltonian matrices* (11.45). We will introduce the important notion of *symplectic vector space* and an associated object, the *symplectic group*. For this we shall also revise a few elementary facts regarding the *exponential map* for matrices.

11.5.1 Symplectic Vector Spaces

We will define a structure reminiscent of a vector space equipped with an inner product.

Definition 11.27 (Symplectic Form and Symplectic Vector Space)   Let V be a real vector space. A map $\langle \cdot, \cdot \rangle : V \times V \to \mathbb{R}$ sending pairs (u, v) to a real number $\langle u, v \rangle$ is called a **symplectic form** on V if it enjoys the following properties:

(1) **skew-symmetry**: $\langle u, v \rangle = -\langle v, u \rangle$ for any $u, v \in V$,
(2) **bilinearity**: $\langle \cdot, \cdot \rangle$ is linear in each argument,
(3) **non-degeneracy**: if $\langle u, v \rangle = 0$ for any $v \in V$ then $u = 0$.

The pair $(V, \langle, \rangle)$ is a (real) **symplectic vector space**.   ◇

The importance of the concepts in Hamiltonian Mechanics comes from the fact S is the matrix of a symplectic form.

Proposition 11.28 *The map:*

$$\langle \cdot, \cdot \rangle_S : \mathbb{R}^{2n} \times \mathbb{R}^{2n} \to \mathbb{R} \quad \text{defined by} \quad \mathbb{R}^{2n} \times \mathbb{R}^{2n} \ni (u, v) \mapsto u^t S v \in \mathbb{R} \tag{11.71}$$

induces a symplectic form on $\mathbb{R}^{2n}$.

Proof Bilinearity is obvious, and (11.39) implies skew-symmetry $S^t = -S$. Hence:

$$\langle u, v \rangle_S = u^t S v = (u^t S v)^t = v^t S^t u = -v^t S u = -\langle v, u \rangle_S .$$

Non-degeneracy: by definition of S we have $SS = -I$. If $u^t S v = 0$ for any $v \in \mathbb{R}^{2n}$, we choose $v = Su$ and then $u^t S S u = 0$, i.e. $-u^t u = 0$, so $u^t u = 0$ and finally $u = 0$. □

It can be proved that *every* symplectic form on an even-dimensional real vector space V reduces to (11.71) in suitable Cartesian coordinates. This is a theorem due to Darboux, which we leave to the reader to prove.

Theorem 11.29 (First Darboux Theorem) *Any $2n$-dimensional symplectic space $(V, \langle \cdot, \cdot \rangle)$, $n = 1, 2, \ldots$, admits a so-called **symplectic** basis $\{\mathbf{s}_i\}_{i=1,\ldots,2n}$, in which the symplectic form has matrix S as of (11.38). In other words, if u and v are the column vectors of the components of $\mathbf{u}$ and $\mathbf{v}$ in V, respectively, in the basis $\{\mathbf{s}_i\}_{i=1,\ldots,2n}$:*

$$\langle \mathbf{u}, \mathbf{v} \rangle = u^t S v . \tag{11.72}$$

Remarks 11.30

(1) The basis $\{s_i\}_{i=1,\ldots,2n}$ is not the only one for which the symplectic form reads as in (11.71). Symplectic bases are abundant. Suppose $\mathbf{u} \in V$ is represented by the column vector $u' \in \mathbb{R}^{2n}$ in an arbitrary basis $\{s_i'\}_{i=1,\ldots,2n}$, and by the column vector u in the symplectic basis $\{s_i\}_{i=1,\ldots,2n}$. Then $u = Au'$, where the matrix of change of basis A belongs to the group of $2n \times 2n$ real invertible matrices $GL(2n, \mathbb{R})$. Taking $\mathbf{u}, \mathbf{v} \in V$ and their components, we have:

$$\langle \mathbf{u}, \mathbf{v} \rangle = u^t S v = (Au')^t A v' = u'^t (A^t S A) v' .$$

Hence *the basis $\{s_i'\}_{i=1,\ldots,2n} \subset V$ is symplectic if and only if $A^t S A = S$.*

(2) On a $2n$-dimensional real symplectic space $(V, \langle, \rangle)$ we can represent vectors in a symplectic basis: $\mathbf{u} = \sum_{k=1}^{2n} u^k \mathbf{s}_k$. In this representation the symplectic form reads

$$\langle \mathbf{u}, \mathbf{v} \rangle = \sum_{i,j=1}^{2n} u^i S_{ij} v^j , \tag{11.73}$$

From the point of view of tensor calculus the coefficients S_{ij} are the components of the linear isomorphism:

$$T_S : V \ni \mathbf{u} \mapsto \langle \cdot, \mathbf{u} \rangle \in V^*$$

written *with respect to a symplectic basis of V and the dual basis of V^**. This linear map is an isomorphism (canonical, for finite-dimensional symplectic vector spaces) because the symplectic form is non-degenerate and the spaces are equidimensional. The numbers $-S^{ij}$ in (11.42) that satisfy equation (11.41) are the coefficients of the inverse isomorphism $T_S^{-1} : V^* \to V$, for the same bases (but swapped). The transformation T_S^{-1}, as a skew-symmetric bilinear form on V is the well-known **Poisson bivector**.

(3) $SS = -I$ implies $\det S = \pm 1$. In fact: $(\det S)^2 = (\det S^2) = (-1)^{2n} = 1$, so $\det S = \pm 1$. As a matter of fact one has $\det S = 1$, because exchanging the first n rows with the last n changes S into

$$L := \left[\begin{array}{c|c} -I & 0 \\ \hline 0 & I \end{array} \right] ,$$

Having swapped n row pairs, elementary properties of the determinant say that $\det S = (-1)^n \det L$. But L is diagonal and its determinant is the product of the diagonal elements, so $\det L = (-1)^n (1)^n$. Eventually $\det S = (-1)^{2n} = 1$. ∎

11.5.2   The Symplectic Group

We will show that the matrices appearing in **(1)** Remarks (11.30), which obey $A^t S A = S$ since they transform symplectic bases into symplectic bases, form a subgroup of $GL(2n, \mathbb{R})$.

Proposition 11.31 *For any $n = 1, 2, \ldots$ the set of matrices*

$$Sp(n, \mathbb{R}) := \left\{ A \in M(2n, \mathbb{R}) \mid A^t S A = S \right\}$$

is a subgroup of $GL(n, \mathbb{R})$. Moreover, $A \in Sp(n, \mathbb{R})$ if and only if $A^t \in Sp(n, \mathbb{R})$.

Proof It is enough to show that these matrices (i) form a subset of $GL(2n, \mathbb{R})$, (ii) this subset contains the identity matrix, (iii) the subset is closed under matrix multiplication, and (iv) the subset is closed under inversion.[7] Let us proceed with order.

---

[7] That the set contains the identity matrix actually follows from (iii) and (iv).

(i) If $A^t S A = S$, computing the determinant and using Binet's rule, we find $(\det A^t) \det S \det A = \det S$. Since $\det S \neq 0$ and $\det A^t = \det A$, we have $(\det A)^2 = 1$ and so $\det A \neq 0$. Therefore the matrices satisfying $A^t S A = S$ belong to $GL(2n, \mathbb{R})$.

(ii) I satisfies $I^t S I = S$.

(iii) If $A^t S A = S$ and $B^t S B = S$ then $(AB)^t S(AB) = B^t (A^t S A) B = B^t S B = S$, so $(AB)^t S(AB) = S$.

(iv) Let us start from showing that the set satisfying $A^t S A = S$ is *closed under transposition*. If $A^t S A = S$ then $S A^t S A = -I$, where we used $S S = -I$ (so $S^{-1} = -S$). Multiplying by A on the left and $A^{-1} S$ on the right we find $A S A^t = S$. Hence if A is symplectic so is A^t. We conclude that $A^t S A = S$ implies $A S A^t = S$, so inverting each side: $A^{t-1}(-S) A^{-1} = -S$, and therefore $A^{t-1} S A^{-1} = S$ if $A^t S A = S$.

$\square$

Definition 11.32 (Symplectic Group)   The subgroup of $GL(2n, \mathbb{R})$:

$$Sp(n, \mathbb{R}) := \left\{ A \in M(2n, \mathbb{R}) \ \big| A^t S A = S \right\} \tag{11.74}$$

is called **real symplectic group of order** n. $\diamond$

11.5.3   Intermezzo: Matrix Exponential

To study the properties of the symplectic group it is convenient to introduce general notions and facts about the exponential map of matrices.

If $E \in M(m, \mathbb{R})$ is a given matrix, we define the map (where $E^0 := I$):

$$\mathbb{R} \ni s \mapsto e^{sE} := \sum_{k=0}^{+\infty} \frac{s^k}{k!} E^k \in M(m, \mathbb{R}) \equiv \mathbb{R}^{m^2}. \tag{11.75}$$

The meaning of this is sanctioned by the next proposition.

Proposition 11.33 *For any* $E \in M(m, \mathbb{R})$ *and* $s \in \mathbb{R}$, *the* $\mathbb{R}^{m^2}$*-valued series (11.75) converges in the topology of* $M(m, \mathbb{R}) \equiv \mathbb{R}^{m^2}$ *induced by the standard Euclidean norm of* $\mathbb{R}^{m^2}$:

$$||A|| := \sqrt{\sum_{h,k=1}^{m} A_{hk}^2}.$$

Furthermore, the following facts hold.

(1) *For any $E \in M(m, \mathbb{R})$ the $\mathbb{R}^{m^2}$-valued series of functions (11.75) converges uniformly in the variable s when $s \in [-M, M]$, for any $M > 0$.*

(2) *For all $s \in \mathbb{R}$ the $\mathbb{R}^{m^2}$-valued series of functions*

$$M(m, \mathbb{R}) \ni E \mapsto e^{sE} := \sum_{k=0}^{+\infty} \frac{s^k}{k!} E^k \in M(m, \mathbb{R}) \equiv \mathbb{R}^{m^2} \tag{11.76}$$

converges uniformly in the variable E on the set

$$\{E \in M(m, \mathbb{R}) \mid ||E|| \leq M\}$$

for any $M > 0$.

Proof For the given norm, the Cauchy-Schwarz inequality and the definition of matrix product imply:

$$||AB|| \leq ||A|| \, ||B|| . \tag{11.77}$$

In fact:

$$||AB||^2 = \sum_{a,b,c} |A_{ab} B_{bc}|^2 \leq \sum_{a,c} \left(\sum_b A_{ab}^2 \right) \left(\sum_d B_{bd}^2 \right)$$

$$= \left(\sum_{a,b} A_{ab}^2 \right) \left(\sum_{b,d} B_{bd}^2 \right) = ||A||^2 ||B||^2 .$$

By (11.77):

$$\left|\left| \sum_{k=0}^N \frac{s^k}{k!} E^k - \sum_{k=0}^M \frac{s^k}{k!} E^k \right|\right| = \left|\left| \sum_{k=M+1}^N \frac{s^k}{k!} E^k \right|\right| \leq \sum_{k=M+1}^N \frac{|s|^k}{k!} ||E||^k$$

$$= \sum_{k=0}^N \frac{|s|^k}{k!} ||E||^k - \sum_{k=0}^M \frac{|s|^k}{k!} ||E||^k$$

where the last side is the difference of the partial sums of

$$\mathbb{R} \ni r \mapsto e^{r \, ||E||}$$

which is known to converge on $\mathbb{R}$. Hence the sequence of partial sums is a *Cauchy sequence*:

$$\left| \sum_{k=0}^N \frac{|s|^k}{k!} ||E||^k - \sum_{k=0}^M \frac{|s|^k}{k!} ||E||^k \right| < \epsilon$$

for any $\epsilon > 0$, if $N, M > K_\epsilon$, where K_ϵ is large enough. That shows the series of e^{sE} is Cauchy, so it converges for any $s \in \mathbb{R}$ because $\mathbb{R}^{m^2}$ is complete for the given norm. Furthermore, the series' tail satisfies:

$$\left\| \sum_{k=N+1}^{+\infty} \frac{s^k}{k!} E^k \right\| \leq \sum_{k=N+1}^{+\infty} \frac{|s|^k}{k!} ||E||^k ,$$

where, for given E, the right-hand side tends to zero uniformly in $|s|$ on every interval $|s| \in [0, M]$ (and so does the other side) because it is a power series, namely the Taylor expansion of the exponential map

$$[0, M] \ni r \mapsto e^{r\,||E||} .$$

Hence (11.75) converges *uniformly* on any interval $[-M, M] \ni s$. If we fix $s \in \mathbb{R}$ and vary E, the same argument gives uniform convergence of the series in the variable E (11.76) on the sets $||E|| \leq M$, for any $M > 0$.   □

Based on the elementary facts of Sect. A.3 we can actually say a little more.

Proposition 11.34 *For any $E \in M(m, \mathbb{R})$, the series of vector-valued functions (11.75) of variable s enjoys the following properties:*

(1) $\mathbb{R} \ni s \mapsto e^{sE}$ is C^∞ (actually real analytic);
(2) all derivatives of e^{sE} with respect to s are computed by differentiating term by term.

Proof If we consider the derivative series (the series of the first derivatives in s) of (11.75):

$$\sum_{k=1}^{+\infty} \frac{k s^{k-1}}{k!} E^k ,$$

we immediately see that it is made by continuous functions and it converges uniformly in s around any $s_0 \in \mathbb{R}$. This is proved in the same way as the previous proposition, using (11.77) to reduce to the numerical exponential map. The same thing holds for p-th derivatives in s. As is well known, a series' convergence together with the uniform convergence of the series of derivatives up to order p (assuming all maps continuous) imply that the sum can be differentiated up to order p term by term (see Sect. A.3). As p is arbitrary, the claim is proved.   □

Remarks 11.35 The proposition implies, in particular:

$$\frac{d}{ds} e^{sE} = \sum_{k=1}^{+\infty} \frac{s^{k-1}}{(k-1)!} E^k = E e^{sE} = e^{sE} E . \tag{11.78}$$

We will need this later on.   ■

The following fact regarding $GL(m, \mathbb{R})$ is a known fact:

$$e^{sE} \in GL(m, \mathbb{R}) \quad \text{for all } s \in \mathbb{R} \text{ and } E \in M(m, \mathbb{R}) . \tag{11.79}$$

It is a general property in the theory of *Lie groups*, implied by the next result.

Proposition 11.36 *If* $F \in M(m, \mathbb{R})$ *then* $e^F \in GL(m, \mathbb{R})$.

Proof The proof is straightforward from the known identity (see (2) Exercises 11.40)

$$\det e^F = e^{tr F} \quad \text{for all } F \in M(m, \mathbb{R}) \tag{11.80}$$

As $e^{tr F} \neq 0$, we deduce that $\det e^F \neq 0$ and hence $e^F \in GL(m, \mathbb{R})$. $\square$

The result in (11.79) can be sort of inverted, in the following sense. Think $M(m, \mathbb{R})$ as $\mathbb{R}^{m^2}$, so $GL(m, \mathbb{R})$ is the pre-image of $\mathbb{R} \setminus \{0\}$ under the determinant. As this is continuous, $GL(m, \mathbb{R})$ is an open subset in $\mathbb{R}^{m^2}$ containing the identity $I = e^0$. We can then check that exp is non-singular around the zero matrix and use the Inverse function theorem. One can indeed prove the following.

Proposition 11.37 *The function* $\mathbb{R}^{m^2} \ni F \mapsto e^F \in \mathbb{R}^{m^2}$ *is* C^1, *so in particular its Jacobian matrix at* $F = 0$ *in the natural coordinates of* $\mathbb{R}^{m^2}$ *is non-singular. Hence the exponential mapping defines a local* C^1 *diffeomorphism*[8] *from an open neighbourhood of the zero matrix onto an open neighbourhood of the identity matrix in* $M(m, \mathbb{R})$.

Proof We know that the series

$$e^F = \sum_{n=0}^{+\infty} \frac{F^k}{k!} \tag{11.81}$$

converges for any $F \in M(m, \mathbb{R})$. Let us show it defines a C^1 map. For that it suffices to prove that, around any $F_0 \in M(m, \mathbb{R})$, the derivative series is made of continuous functions and that it converges uniformly. Consequently the uniform limit will be continuous (see Sect. A.3). The series of derivatives with respect to the generic component F_{ab} is:

$$\sum_{k=0}^{+\infty} \frac{\partial}{\partial F_{ab}} \frac{1}{k!} F^k = \sum_{k=0}^{+\infty} \frac{1}{k!} \left(\frac{\partial F}{\partial F_{ab}} F^{k-1} + F \frac{\partial F}{\partial F_{ab}} F^{k-2} + \cdots + F^{k-1} \frac{\partial F}{\partial F_{ab}} \right),$$

$$\tag{11.82}$$

---

[8] The diffeomorphism is easily C^∞.

where we defined

$$\frac{\partial F}{\partial F_{ab}} := \left[\frac{\partial F_{ij}}{\partial F_{ab}} \right]_{i,j=1,\ldots,m} = \left[\delta_{ia} \delta_{jb} \right]_{i,j=1,\ldots,m} ,$$

so in particular:

$$\left\| \frac{\partial F}{\partial F_{ab}} \right\| = 1 .$$

Consequently the generic term of (11.82) satisfies:

$$\left\| \frac{\partial}{\partial F_{ab}} \frac{1}{k!} F^k \right\| \leq \frac{k}{k!} ||F||^{k-1} 1 \leq \frac{||F||^{k-1}}{(k-1)!} .$$

If F belongs to a neighbourhood of F_0 where $||F|| < M$ for some constant $M > 0$, the series:

$$\sum_{k=0}^{+\infty} \frac{\partial}{\partial F_{ab}} \frac{1}{k!} F^k$$

is upper bounded, term by term, by the (convergent) positive-term series:

$$\sum_{k=1}^{\infty} \frac{M^{k-1}}{(k-1)!} = e^M .$$

The same argument of Proposition 11.33 tells us that (11.82) converges uniformly to a function of the variable F. Such function must be continuous, as sum of continuous functions that converge uniformly. Now, the result on differentiating series (Sect. A.3) implies $F \mapsto e^F$ is C^1, and its partial derivatives can be computed term by term from (11.81). In particular

$$\frac{\partial e^F}{\partial F_{ab}} \bigg|_{F=0} = \sum_{k=0}^{+\infty} \frac{\partial}{\partial F_{ab}} \bigg|_{F=0} \frac{1}{k!} F^k = \frac{\partial F}{\partial F_{ab}} = \left[\delta_{ia} \delta_{jb} \right]_{i,j=1,\ldots,m} .$$

In other words $\mathbb{R}^{m^2} \ni F \mapsto e^F \in \mathbb{R}^{m^2}$ has Jacobian matrix equal to the identity at $F = 0$, and therefore is non-singular.   □

To sum up, on a small enough open neighbourhood $\mathcal{O}_I$ of the identity in $GL(m, \mathbb{R})$, every $A \in GL(m, \mathbb{R})$ can be written as e^F, where $F := \ln A$ is uniquely determined by A and belongs in a corresponding small neighbourhood $\mathcal{O}_0$ of 0.

11.5.4 The Symplectic Group and the Lie Algebra of Hamiltonian Matrices

Now that $Sp(n, \mathbb{R})$ is a subgroup of $GL(2n, \mathbb{R})$, we may ask whether there is a subset of matrices (close to the zero matrix, to be specific) whose exponentials are elements of $Sp(n, \mathbb{R})$. The problem fits into the general picture of Lie algebras, and the well-known local correspondence between Lie groups and Lie algebras. We prefer to keep on a very elementary level and hence will not go into this matter. The next proposition shows that such a subset exists (and contains elements that are far from the zero matrix). It is precisely the space $sp(n, \mathbb{R})$ of Hamiltonian matrices of Definition 11.16.

Proposition 11.38 *For any $n = 1, 2, \ldots$:*

(1) $e^{sE} \in Sp(n, \mathbb{R})$ for every $s \in \mathbb{R}$ if and only if $E \in sp(n, \mathbb{R})$.
(2) If $E, E' \in sp(n, \mathbb{R})$ then $EE' - E'E \in sp(n, \mathbb{R})$.

Proof

(1) The expansion of e^{sE} easily implies:

$$(e^{sE})^t = e^{sE^t}$$

and differentiating term by term, with (11.78) we have:

$$\frac{de^{sE}}{ds} = Ee^{sE} = e^{sE}E .$$

If $(e^{sE})^t S e^{sE} = S$, differentiating at $s = 0$ gives $E^t S + SE = 0$. Hence $e^{sE} \in Sp(n, \mathbb{R})$ for any $s \in \mathbb{R}$ implies $E \in sp(n, \mathbb{R})$. Let now $E \in sp(n, \mathbb{R})$, and we will show $e^{sE} \in Sp(n, \mathbb{R})$ for any $s \in \mathbb{R}$. Consider the function with values in $M(2n, \mathbb{R}) \equiv \mathbb{R}^{(2n)^2}$, for $s \in \mathbb{R}$, $f(s) := (e^{sE})^t S e^{sE}$. Computing its derivative in s we see f solves the autonomous linear Cauchy problem on $\mathbb{R}^{(2n)^2}$:

$$\frac{df}{ds} = E^t f(s) + f(s)E \quad \text{with } f(0) = S .$$

The right-hand side's linearity grants global existence and uniqueness because the linear maps are C^∞. The *constant* function $\mathbb{R} \ni s \mapsto f(s) := S$ solves the problem, as $E^t S + SE = 0$ by hypothesis. This is then the only solution (maximal and complete) to the Cauchy problem. In other words:

$$f(s) = (e^{sE})^t S e^{sE} = S \quad \forall s \in \mathbb{R} ,$$

i.e. $e^{sE} \in Sp(n, \mathbb{R})$ for any $s \in \mathbb{R}$ if $E \in sp(n, \mathbb{R})$.

(2) As $Sp(n, \mathbb{R})$ is a group, (1) implies:

$$e^{te^{sE} E' e^{-sE}} = e^{sE} e^{tE'} e^{-sE} \in Sp(n, \mathbb{R}) \quad \forall s, t \in \mathbb{R}.$$

Differentiating at $t = 0$ still by (1) we have:

$$e^{sE} E' e^{-sE} \in sp(n, \mathbb{R}).$$

By definition (11.45) the set $sp(n, \mathbb{R})$ is closed in $M(2n, \mathbb{R})$ since it contains its limit points (if $E_n \to E$ as $n \to +\infty$ and $E_n^t S + S E_n = 0$, by the product's continuity $E^t S + SE = 0$). Since $sp(n, \mathbb{R})$ is a linear subspace:[9]

$$\lim_{h \to 0} \frac{1}{h} \left(e^{hE} E' e^{-hE} - E' \right) \in sp(n, \mathbb{R}).$$

Expanding in series the exponentials (order one is enough) shows that the derivative is precisely $EE' - E'E$, which therefore belongs to $sp(n, \mathbb{R})$. This ends the proof.

$\square$

Remarks 11.39 $GL(m, \mathbb{R})$ and $Sp(n, \mathbb{R})$ are Lie groups, i.e. they have a differentiable structure (induced by $\mathbb{R}^{m^2}$ and $\mathbb{R}^{(2n)^2}$ respectively), and compatible group operations (meaning they are differentiable). $M(m, \mathbb{R})$ is the *Lie algebra* of $GL(m, \mathbb{R})$, that is, the tangent space at the identity point $I \in GL(m, \mathbb{R})$, with Lie bracket given by the **commutator** (as is customary for matrix groups): $[A, B] := AB - BA$ for $A, B \in M(m, \mathbb{R})$. By definition of Lie algebra, the Lie bracket of two elements in a Lie algebra is an element of the Lie algebra.

Similarly, $sp(n, \mathbb{R})$ is the *Lie algebra of* $Sp(n, \mathbb{R})$. In particular $sp(n, \mathbb{R})$ is a subspace of $M(2n, \mathbb{R})$ for which:

$$[E, E'] \in sp(n, \mathbb{R}) \quad \text{if } E, E' \in sp(n, \mathbb{R})$$

as seen above.   ∎

Exercises 11.40

(1) Prove that:

$$e^{A+B} = e^A e^B \quad \text{if } AB = BA \text{ for } A, B \in M(m, \mathbb{R}).$$

This holds in particular if $A = tF$ and $B = t'F$ for some $F \in M(m, \mathbb{R})$ and $t, t' \in \mathbb{R}$.

---

[9] That $sp(n, \mathbb{R})$ is finite-dimensional—since the ambient space $M(2n, \mathbb{R})$ is—implies on its own that $sp(n, \mathbb{R})$ is a subspace.

(2) Prove the identity

$$\det e^{tF} = e^{t\,\mathrm{tr}\,F} \quad \text{for all } F \in M(m, \mathbb{R}) \text{ and } t \in \mathbb{R}.$$

(3) Prove that $A^t = -A \in M(m, \mathbb{R})$ implies $e^A \in SO(m)$.

(4) Prove that $e^{tA} \in SO(m)$ for any $t \in \mathbb{R}$ if and only if $A^t = -A$.

Chapter 12
Canonical Hamiltonian Theory, Hamiltonian Symmetries and Hamilton-Jacobi Theory

In this chapter we shall discuss more advanced topics in Hamiltonian Mechanics. We will introduce the theory of canonical transformations and the Poisson bracket to study the relationship between symmetries and conservation laws in Hamilton's formulation. Together with the canonical transformations of coordinates we will introduce a special atlas on phase spacetime that extends the one of natural coordinates. Using that, we shall reformulate Liouville's theorem and deduce the Poincaré "recurrence" theorem. In the last part we will return to canonical transformations from a novel point of view which will allows us to introduce the Hamilton-Jacobi theory.

With the exclusion of the first section, where we use the *symplectic group* $Sp(n, \mathbb{R})$ introduced in Sect. 11.5.2, see (11.74), the majority of the other sections make use of vector fields (Sect. A.6.5), one-parameter groups of local diffeomorphisms (Sect. 14.5.3) and the general theory of differentiable forms on manifolds, including the Lie derivative (Appendix B).

12.1 Canonical Hamiltonian Theory

We shall address a pivotal topic in Hamiltonian Mechanics. In Chap. 11 we saw that Hamilton's equations have the form:

$$\begin{cases} \dfrac{dp_k}{dt} = -\dfrac{\partial \mathscr{H}}{\partial q^k}, \\ \dfrac{dq^k}{dt} = \dfrac{\partial \mathscr{H}}{\partial p_k}, \end{cases} \quad \text{for } k = 1, \ldots, n, \tag{12.1}$$

© The Author(s), under exclusive license to Springer Nature Switzerland AG 2023
V. Moretti, *Analytical Mechanics*, La Matematica per il 3+2 150,
https://doi.org/10.1007/978-3-031-27612-5_12

for any natural local coordinate system

$$\psi : U \ni a \mapsto (t, q^1, \ldots, q^n, p_1, \ldots, p_n) \in \mathbb{R} \times \mathbb{R}^{2n}$$

on phase spacetime $F(\mathbb{V}^{n+1})$. More precisely, the transformation rules between natural local coordinates are, by definition:

$$t' = t + c, \tag{12.2}$$

$$q'^k = q'^k(t, q^1, \ldots, q^n), \tag{12.3}$$

$$p'_k = \sum_{h=1}^{n} \frac{\partial q^h}{\partial q'^k} p_h. \tag{12.4}$$

As said, Theorem 11.12 says that in the new natural coordinates

$$\psi' : U' \ni a \mapsto (t', q'^1, \ldots, q'^n, p'_1, \ldots, p'_n) \in \mathbb{R} \times \mathbb{R}^{2n},$$

Hamilton's equations have the same form as in the old coordinates:

$$\begin{cases} \dfrac{dp'_k}{dt'} = -\dfrac{\partial \mathscr{H}'}{\partial q'^k}, \\[2mm] \dfrac{dq'^k}{dt'} = \dfrac{\partial \mathscr{H}'}{\partial p'_k}, \end{cases} \quad \text{for } k = 1, \ldots, n,$$

whenever the new Hamiltonian is related to the old one by:

$$\mathscr{H}'(t', q', p') = \mathscr{H}(t(t'), q(t', q', p'), p(t', q', p'))$$

$$- \sum_{k=1}^{n} p_k(t', q', p') \frac{\partial q^k}{\partial t}\bigg|_{(t', q', p')}, \tag{12.5}$$

on $U \cap U'$.

We would like to know if there exist a larger class of local coordinates, containing natural coordinates, for which the equations of motion on $F(\mathbb{V}^{n+1})$ preserve form (12.1), provided we define suitable Hamiltonians. The answer is yes: the coordinate transformations that generalise (12.2)–(12.4) are the so-called *canonical transformations* (there are some topological details that will be addressed in the sequel). In this way we define on $F(\mathbb{V}^{n+1})$ an atlas that includes some of the old natural coordinate systems plus the new ones. On such an atlas Hamilton's equations hold on every local chart. The atlas's coordinates are called *canonical coordinates*.

12.1.1 Canonical Transformations and Canonical Coordinates

We will assume $F(\mathbb{V}^{n+1})$ is a differentiable manifold, at least C^2, and all coordinate transformations will be at least C^2 as well.

By Definition B.4, applied to the bundle $F(\mathbb{V}^{n+1})$ over absolute time T, a local chart

$$\psi : U \ni a \mapsto (s(a), x^1(a), \dots, x^{2n}(a)) \in \mathbb{R}^{2n+1}$$

in $F(\mathbb{V}^{n+1})$ is **adapted to the fibres** of $F(\mathbb{V}^{n+1})$ when the first coordinate is a local coordinate on the subset $T(U) \subset \mathbb{R}$ of the real line, and the other $2n$ define local coordinates on $T^{-1}(t) \cap U$ for any $t \in T(U)$. In particular, s may be taken to be absolute time.

Definition 12.1 (Canonical Transformations and Canonical Coordinates) Consider two local charts (U, ψ) and (U', ψ') on $F(\mathbb{V}^{n+1})$ with $U \cap U' \neq \varnothing$,

$$\psi : U \ni a \mapsto (t(a), x^1(a), \dots, x^{2n}(a)) \in \mathbb{R}^{2n+1} ,$$

$$\psi' : U' \ni a \mapsto (t'(a), x'^1(a), \dots, x'^{2n}(a)) \in \mathbb{R}^{2n+1}$$

adapted to the fibres, where $t(a)$ and $t'(a)$ equal absolute time $T(a)$ up to additive constants. Thus for any t and t', $(x^{\cdot} \dots, x^{2n})$ and $(x', \dots, x'^{2n})$ are coordinates on the corresponding phase spaces $\mathbb{F}_\tau$. Consider the transition function $\psi' \circ \psi$: $\psi^{-1}(U \cap U') \to \psi'(U \cap U')$:

$$t' = t + c , \tag{12.6}$$

$$x'^k = x'^k(t, x^1, \dots, x^{2n}) . \tag{12.7}$$

(1) We call (12.6)–(12.7) **canonical transformation** if the Jacobian matrix of the *non-temporal coordinates*, of coefficients

$$\left. \frac{\partial x'^k}{\partial x^i} \right|_a , \quad \text{for } i, k = 1, \dots, 2n,$$

belongs to the *symplectic group* $Sp(n, \mathbb{R})$ (see (11.74)) for any state on $a \in U \cap U'$.

(2) The coordinate change (12.6)–(12.7) is said to be **completely canonical** if it is canonical, $c = 0$ and t is not appearing explicitly in the right-hand side of (12.7).

(3) The local system (U, ψ) is called a **system of (local) canonical coordinates** if it fulfils the following requisites.

(i) The connected open set $U \subset F(\mathbb{V}^{n+1})$ is fibre-wise **locally simply connected**. This means that $T^{-1}(t) \cap U$ is connected and simply connected for any t in the open interval[1] $I_U := T(U) \subset \mathbb{R}$.

(ii) There exists a natural local coordinate system (V, ϕ) on $F(\mathbb{V}^{n+1})$ where $U = V$ and $\psi \circ \phi^{-1}$ is a canonical transformation. $\diamond$

Remarks 12.2 Let $U \subset F(\mathbb{V}^{n+1})$ be open, connected and locally simply connected. Take a local chart $\psi : U \to \mathbb{R}^{2n+1}$ with coordinates $(t, \mathbf{x})$ adapted to the fibres, where t is absolute time: $T(\psi^{-1}(t, \mathbf{x})) = t$ (we could add a constant on the right and nothing would change) for $(t, \mathbf{x}) \in \psi(U)$. For example, natural local charts on $F(\mathbb{V}^{n+1})$ satisfy the request. The representation $\psi(U)$ of U in coordinates is clearly of the type:

$$\psi(U) = \bigcup_{t \in I} \{t\} \times V_t \, ,$$

where:

(i) $V_t \subset \mathbb{R}^{2n}$ is open, connected and simply connected
(ii) $I = T(U) \subset \mathbb{R}$ is an open interval.

This is consequence of the fact the coordinates $\mathbf{x} \in V_t$ live on $T^{-1}(t) \cap U$ for any t in the open interval $T(U)$ and define a diffeomorphism between the open set V_t and $T^{-1}(t) \cap U$.   ∎

Next we show that canonical coordinates do exist.

Proposition 12.3 *On phase space $F(\mathbb{V}^{n+1})$ the following facts hold.*

(1) Every local system of natural coordinates (U, ψ):

$$\psi : U \ni a \mapsto (t(a), q^1(a), \ldots, q^n(a), p_1(a), \ldots p_n(a)) \in \mathbb{R}^{2n+1}$$

is canonical if and only if the open set U is connected and locally simply connected.

(2) For any $a \in F(\mathbb{V}^{n+1})$ there exists a natural local chart (U, ψ) such that $a \in U$ and U is connected and locally simply connected, so that (U, ψ) is a canonical coordinate system.

(3) Any transformation between natural coordinates is canonical.

---

[1] U is open and connected and $T : F(\mathbb{V}^{n+1}) \to \mathbb{R}$ is continuous and open, so $T(U)$ is open and connected in $\mathbb{R}$, i.e. an open interval.

Proof

(1) It suffices to notice that the identity map

$$(t, q^1, \ldots, q^n, p_1, \ldots p_n) \mapsto (t, q^1, \ldots, q^n, p_1, \ldots p_n)$$

is canonical.

(2) If (V, ψ) defines natural coordinates around $a \in F(\mathbb{V}^{n+1})$ (they exist by definition of phase spacetime), as $\psi(V) = (t(a), \mathbf{x}(a)) \subset \mathbb{R} \times \mathbb{R}^{2n}$ is open we can take a cylinder $I \times D \subset \psi(V)$ such that $I \subset \mathbb{R}$ is an open interval containing $t(a)$ and $D \subset \mathbb{R}^{2n}$ an open ball containing $\mathbf{x}(a)$. By construction $U := \psi^{-1}(I \times D)$ is open, connected and locally simply connected.

(3) Consider the transformation (12.2)–(12.4) representing a generic natural coordinate change. Its Jacobian matrix, only taking derivatives in non-temporal variables, is:

$$J := \left[\begin{array}{c|c} Q & 0 \\ \hline P & Q^{-1t} \end{array} \right] , \tag{12.8}$$

Above, Q is the Jacobian matrix of coefficients

$$Q^k{}_i := \frac{\partial q'^k}{\partial q^i}$$

while P has coefficients

$$P_{kr} := \sum_{l,s=1}^{n} p_s \frac{\partial^2 q^s}{\partial q'^k \partial q'^l} \frac{\partial q'^l}{\partial q^r} .$$

From Schwarz's theorem $Q^t P = P^t Q$, so the form of J implies:

$$J^t S J = S .$$

This proves the claim, in view of Definition (11.74).

$\square$

Definition 12.4 We will call **canonical atlas** of $F(\mathbb{V}^{n+1})$ the collection of local canonical coordinate systems. $\diamond$

Remarks 12.5

(1) Composing canonical transformations produces canonical transformations, since the composite's Jacobian matrix is the product of the Jacobians, and symplectic matrices form a group (closed under multiplication).

(2) It is also clear that the inverse of a canonical transformation is canonical: the Jacobian matrix of the inverse transformation, computed with respect to

the non-temporal variables, is the inverse of the initial Jacobian matrix (also computed with respect to non-temporal variables). Symplectic matrices are closed under inversion, as they form a group, so the inverse of a canonical transformation is canonical.

(3) The condition that $J := [\frac{\partial x'^k}{\partial x^i}]_{k,i=1,\dots,2n}$ belongs to $Sp(n, \mathbb{R})$, namely $J^t S J = S$, reads

$$\sum_{h,k=1}^{2n} \frac{\partial x'^h}{\partial x^i} S_{hk} \frac{\partial x'^k}{\partial x^j} = S_{ij}, \quad i, j \in 1, \dots, 2n \tag{12.9}$$

in components. As $Sp(n, \mathbb{R})$ is closed under transposition, we have $J S J^t = S$. Hence the previous condition is equivalent to:

$$\sum_{i,j=1}^{2n} \frac{\partial x'^h}{\partial x^i} S^{ij} \frac{\partial x'^k}{\partial x^j} = S^{hk}, \quad h, k \in 1, \dots, 2n. \tag{12.10}$$

∎

Proposition 12.6 *If (U, ψ) and (U', ψ') are systems of local canonical coordinates such that $U \cap U' \neq \varnothing$, the transition functions*

$$\psi \circ \psi'^{-1} : \psi'(U \cap U') \to \psi(U \cap U') \quad and \quad \psi' \circ \psi^{-1} : \psi(U \cap U') \to \psi'(U \cap U')$$

define canonical transformations.

Proof The claim is straightforward from (1) and (2) in Remarks 12.5 and from the definition of canonical coordinates. □

Remarks 12.7 There exist canonical coordinates that *are not* natural on phase space. Consider the C^∞ transformation (with C^∞ inverse) sending natural coordinates $(t, q^1, \dots, q^n, p_1, \dots, p_n)$ (varying in a set with the known topological features) to local coordinates $(t, Q^1, \dots, Q^n, P_1, \dots, P_n)$:

$$t' = t,$$
$$Q^k = p_k, \quad k = 1, \dots, n$$
$$P_k = -q^k, \quad k = 1, \dots, n.$$

The Jacobian matrix of non-temporal coordinates is S, and moreover $S^t S S = -SSS = -S(-I) = S$. We cannot think $(t, Q^1, \dots, Q^n, P_1, \dots, P_n)$ as natural coordinates on phase space, and hence related to natural coordinates $(t, Q^1, \dots, Q^n, \dot{Q}^1, \dots, \dot{Q}^n)$ on the spacetime of kinetic states, because the transformation rule between natural coordinates does not allow for the *second* equation above: $Q^k = p_k$. Expression (12.3), in fact, prescribes that on the right there can only be a function of $(t, q^1, \dots, q^n)$. ∎

Examples 12.8 Consider the phase spacetime of a point particle moving on a straight line at rest in some frame. We can identify $F(\mathbb{V}^3)$ with $\mathbb{R} \times \mathbb{R}^2$ with natural coordinates $(t, q, p) \in \mathbb{R}^3$, where q covers the line, p is the conjugate momentum and t is absolute time. The coordinates are therefore defined globally on phase spacetime $U = F(\mathbb{V}^{1+1})$. We define the global chart (U, ϕ) where $\phi : V \ni a \mapsto (t(a), q(a), p(a)) \in \mathbb{R}^3$ with $V = F(\mathbb{V}^3)$. By construction ϕ is surjective and the image covers $\mathbb{R} \times \mathbb{R}^3$. We emphasise that (U, ϕ) defines, by assumption, natural coordinates.

Consider now another chart (U, ψ) where $\psi : U \ni a \mapsto (t'(a), q'(a), p'(a)) \in \mathbb{R}^3$ with $U = F(\mathbb{V}^3)$. Suppose the global coordinate change is:

$$t = t', \quad p = p' \cos \Omega t' - q' \sin \Omega t', \quad q = q' \cos \Omega t' + p' \sin \Omega t' \quad (12.11)$$

for some constant $\Omega > 0$. Easily this map from $\mathbb{R}^3$ onto $\mathbb{R}^3$ is C^∞, invertible, with C^∞ inverse

$$t' = t, \quad p' = p \cos \Omega t + q \sin \Omega t, \quad q' = q \cos \Omega t - p \sin \Omega t. \quad (12.12)$$

In particular $\psi(U)$ covers $\mathbb{R} \times \mathbb{R}^2$. It is not hard to see that the Jacobian matrix J of the second transformation's non-temporal part:

$$J = \begin{bmatrix} \frac{\partial q'}{\partial q} & \frac{\partial q'}{\partial p} \\ \frac{\partial p''}{\partial q} & \frac{\partial p'}{\partial p} \end{bmatrix} = \begin{bmatrix} \cos \Omega t & -\sin \Omega t \\ \sin \Omega t & \cos \Omega t \end{bmatrix} \quad (12.13)$$

satisfies $J^t S J = S$. Therefore both transformations are canonical, and the t', q', p' are canonical since $U = V$ is diffeomorphic to $\mathbb{R} \times \mathbb{R}^2$ and $\mathbb{R}^2$ is simply connected. We may change V by restricting it: $\phi(V) = (a, b) \times D$, where $D := \{(q, p) \in \mathbb{R}^2 \mid q^2 + p^2 < R\}$ for some $R > 0$. As the open set $\mathbb{R} \times D$ is connected, D is simply connected and ϕ is a homeomorphism. The domain V thus defined is connected and locally simply connected, and in coordinates t', q', p' it has the form $(a, b) \times D$, as one verifies using the above coordinate change. Restricting ψ as well to $U := V$ we find a chart (U, ψ), now local, that is canonical but not natural.

12.1.2 Conservation of Hamilton's Equations

The canonical atlas has the remarkable property of preserving the form of Hamilton's equations. More precisely, if we have assigned (coherently) a Hamiltonian function for any natural local coordinate system, the solution curves to Hamilton's equations on $F(\mathbb{V}^{n+1})$ satisfy equations of the form (12.1) even in local canonical coordinates, for a suitable choice of the Hamiltonian, where we conventionally called $q^1, \ldots, q^n$ the first n canonical coordinates other than t, and $p_1, \ldots, p_n$ the other n. The choice of notation is rather arbitrary though, since the canonical coordinates are not natural in general, and we cannot view the first n non-temporal coordinates q^k as coordinates on the configuration space. Anyhow, an equivalent notation for Hamilton's equations is

$$\frac{d\mathbf{x}}{dt} = S\nabla_{\mathbf{x}}\mathscr{H}(t, \mathbf{x}) \,. \tag{12.14}$$

where $(t, \mathbf{x}) \equiv (t, x^1, \dots, x^{2n})$ are our canonical coordinates.

This result is an immediate consequence of the next theorem and of the definition of canonical coordinates.

Theorem 12.9 Let (U, ψ) and (U', ψ') define C^2 local coordinates adapted to the fibres of $F(\mathbb{V}^{n+1})$ such that, calling $(t, q^1, \dots, q^n, p_1, \dots, p_n)$ and $(t', q'^1, \dots, q'^n, p'_1, \dots, p'_n)$ the respective coordinates, t and t' identify with absolute time up to additive constants. Let $\gamma : I \to U$, with $I \subset \mathbb{R}$ open interval, be a solution to

$$\begin{cases} \dfrac{dp_k}{dt} = -\dfrac{\partial\mathscr{H}}{\partial q^k}\,, \\[2mm] \dfrac{dq^k}{dt} = \dfrac{\partial\mathscr{H}}{\partial p_k}\,, \end{cases} \quad \text{for } k = 1, \dots, n, \tag{12.15}$$

for a given (C^1) Hamiltonian $\mathscr{H}$ defined on U.

(1) If the coordinate transformation is canonical and $U \cap U'$ is connected and locally simply connected, then the curve γ satisfies Hamilton's equations on $U \cap U'$ when written in the coordinates of U':

$$\begin{cases} \dfrac{dp'_k}{dt} = -\dfrac{\partial\mathscr{H}'}{\partial q'^k}\,, \\[2mm] \dfrac{dq'^k}{dt} = \dfrac{\partial\mathscr{H}'}{\partial p'_k}\,, \end{cases} \quad \text{for } k = 1, \dots, n, \tag{12.16}$$

where the new Hamiltonian on $U \cap U'$ is:

$$\mathscr{H}'(t', q', p') = \mathscr{H}(t(t'), q(t', q', p'), p(t', q', p')) + \mathscr{K}(t', q', p') \,. \tag{12.17}$$

The C^1 map $\mathscr{K}$ is defined by the canonical transformation:

$$\begin{cases} \dfrac{\partial q'_k}{\partial t} = \dfrac{\partial\mathscr{K}}{\partial p'_k}\,, \\[2mm] \dfrac{\partial p'_k}{\partial t} = -\dfrac{\partial\mathscr{K}}{\partial q'^k}\,, \end{cases} \quad \text{for } k = 1, \dots, n. \tag{12.18}$$

(so, it does not depend on $\mathscr{H}$) up to additive function in the variable t'.

(2) If the coordinate change is completely canonical we may choose:

$$\mathscr{H}'(t', q', p') = \mathscr{H}(t(t'), q(t', q', p'), p(t', q', p')),$$

and in that case (1) holds even when $U \cap U'$ is not connected, or not locally simply connected.

(3) If the coordinates on $F(\mathbb{V}^{n+1})$ are natural we may choose in (12.17):

$$\mathscr{H}(t', q', p') = -\sum_{k=1}^{n} p_k(t', q', p') \left. \frac{\partial q^k}{\partial t'} \right|_{(t', q', p')}$$

and then (1) holds also if $U \cap U'$ is not connected or not locally simply connected.

Proof First, observe that (3) was proved by (11.30), so we shall demonstrate the other claims. For simplicity we shall take $t' = t$, since the general case $t' = t + c$, c constant, is not more complicated. Consider a C^1 curve $\mathbf{x} = \mathbf{x}(t)$ in U (for the moment we shall not assume it solves Hamilton's equations (12.15)). Let us pass to primed coordinates on $U \cap U'$ and obtain the curve $\mathbf{x}'(t) = \mathbf{x}'(t, \mathbf{x}(t))$. Then:

$$\frac{d\mathbf{x}'}{dt} = \frac{\partial \mathbf{x}'}{\partial t} + J \frac{d\mathbf{x}}{dt},$$

where $J(t, \mathbf{x})$ is the Jacobian matrix of the derivatives $\partial x'^i / \partial x^k$ evaluated on $\mathbf{x} = \mathbf{x}(t)$. This curve satisfies Hamilton's equations in the initial coordinates if and only if

$$\frac{d\mathbf{x}}{dt} = S \nabla_{\mathbf{x}} \mathscr{H}(t, \mathbf{x}(t)).$$

Hence $\mathbf{x}' = \mathbf{x}'(t)$ satisfies Hamilton's equations in unprimed coordinates if and only if

$$\frac{d\mathbf{x}'}{dt} = \frac{\partial \mathbf{x}'}{\partial t} + J S \nabla_{\mathbf{x}} \mathscr{H}(t, \mathbf{x}(t)).$$

Since

$$\frac{\partial}{\partial x^k} = \sum_{i=1}^{2n} \frac{\partial x'^i}{\partial x^k} \frac{\partial}{\partial x'^i},$$

we can rewrite the condition as:

$$\frac{d\mathbf{x}'}{dt} = \frac{\partial \mathbf{x}'}{\partial t} + J S J^t \nabla_{\mathbf{x}'} \mathscr{H}(t, \mathbf{x}(t, \mathbf{x}'(t))).$$

But $JSJ^t = S$ from Proposition 11.31 (since $J \in Sp(n, \mathbb{R})$ by assumption, so $J^t \in Sp(n, \mathbb{R})$), and then $\mathbf{x}' = \mathbf{x}'(t)$ solves Hamilton's equations in the initial coordinates if and only if, changing coordinates:

$$\frac{d\mathbf{x}'}{dt} = \frac{\partial \mathbf{x}'}{\partial t} + S\nabla_{\mathbf{x}'}\mathscr{H}(t, \mathbf{x}(t, \mathbf{x}'(t))) .$$

In case the transformation $\mathbf{x}' = \mathbf{x}'(t, \mathbf{x})$ does not depend on time the proof ends, as we have (2), since for the latter we did not make any topological assumption on $U \cap U'$. In that case the proof of (2) ends by putting:

$$\mathscr{H}'(t, \mathbf{x}') := \mathscr{H}(t, \mathbf{x}(t, \mathbf{x}')) .$$

For the general case it is enough to check that the map:

$$\mathbf{F}(t, \mathbf{x}') := \frac{\partial \mathbf{x}'(t, \mathbf{x})}{\partial t}\Big|_{\mathbf{x}=\mathbf{x}(t,\mathbf{x}')}$$

can be written as, under our hypotheses:

$$\mathbf{F}(t, \mathbf{x}') = S\nabla_{\mathbf{x}'}\mathscr{K}(t, \mathbf{x}') \tag{12.19}$$

for some scalar map $\mathscr{K} = \mathscr{K}(t, \mathbf{x}')$. In fact if we return to (t, q, p), identity (12.19) is precisely (12.18). Since $-SS = I$, we need to show that for any t, the vector field on $U \cap U'$:

$$\mathbf{Y}(t, \mathbf{x}') := -S\mathbf{F}(t, \mathbf{x}')$$

is the gradient of a scalar field $\mathscr{K}(t, \mathbf{x}')$ parametrised by time, so that the map $(t, \mathbf{x}) \mapsto \mathscr{K}(t, \mathbf{x}')$ is C^1. $U \cap U'$ is open, connected and locally simply connected, so in coordinates $t, \mathbf{x}'$ we can write it as $\cup_{t \in I} V_t$ where $V_t := \{t\} \times V'_t$ with $V'_t \subset \mathbb{R}^{2n}$ is open, connected and simply connected (see Remarks 12.2). If, for a given t, the C^1 field $V'_t \ni \mathbf{x}' \mapsto \mathbf{Y}(t, \mathbf{x}')$ is irrotational

$$\frac{\partial Y^i}{\partial x'^j} = \frac{\partial Y^j}{\partial x'^i} , \tag{12.20}$$

then we know that it is a gradient:

$$\mathbf{Y}(t, \mathbf{x}') = \nabla_{\mathbf{x}'}\mathscr{K}(t, \mathbf{x}') ,$$

where

$$\mathscr{K}(t, \mathbf{x}') := \int_{\mathbf{z}'(t)}^{\mathbf{x}'} \mathbf{Y}(t, \mathbf{y}) \cdot d\mathbf{y} . \tag{12.21}$$

At given $t \in I$ the line integral is computed along any differentiable path *contained* in V_t' from $\mathbf{z}'(t)$ to $\mathbf{x}'$, and where $\mathbf{z}'(t)$ is an arbitrary point in V_t' (the integral's value does not depend on the path). As t varies in I, (12.21) defines a map of t and $\mathbf{x}'$, defined everywhere on $\cup_{t \in I} V_t$ (i.e. $U \cap U'$ when seen in $F(\mathbb{V}^{n+1})$), whose gradient with respect to $\mathbf{x}'$ is exactly $\mathbf{Y}(t, \mathbf{x}')$, as required. If we pick a smooth curve $I \ni t \mapsto (t, \mathbf{z}'(t)) \in \{t\} \times V_t'$, then

$$\cup_{t \in I} V_t \ni (t, \mathbf{x}') \mapsto \mathscr{K}(t, \mathbf{x}') := \int_{\mathbf{z}'(t)}^{\mathbf{x}'} \mathbf{Y}(t, \mathbf{y}) \cdot d\mathbf{y}$$

is definitely C^1. The only thing to check is that the derivative in t:

$$\frac{\partial}{\partial t} \int_{\mathbf{z}'(t)}^{\mathbf{x}'} \mathbf{Y}(t, \mathbf{y}) \cdot d\mathbf{y} = \int_{\frac{d\mathbf{z}'(t)}{dt}}^{\mathbf{x}'} \mathbf{Y}(t, \mathbf{y}) \cdot d\mathbf{y} + \int_{\mathbf{z}'(t)}^{\mathbf{x}'} \frac{\partial \mathbf{Y}(t, \mathbf{y})}{\partial t} \cdot d\mathbf{y}$$

is jointly continuous in t and $\mathbf{x}'$. We leave this as exercise, with a hint: recall how the integral depends on the endpoints of integration and when it is possible to swap derivative and integral. In order to prove the existence of $\mathscr{K}$ it is then enough to show (12.20) holds for any t. Using (11.40), that condition reads

$$\sum_{k=1}^{2n} \frac{\partial}{\partial x'^j} \left(S_{ik} \left. \frac{\partial x'^k}{\partial t} \right|_{(t, \mathbf{x}(t, \mathbf{x}'))} \right) = \sum_{k=1}^{2n} \frac{\partial}{\partial x'^i} \left(S_{jk} \left. \frac{\partial x'^k}{\partial t} \right|_{(t, \mathbf{x}(t, \mathbf{x}'))} \right).$$

Equivalently:

$$\sum_{r,k} \frac{\partial x^r}{\partial x'^j} S_{ik} \frac{\partial^2 x'^k}{\partial x^r \partial t} = \sum_{r,k} \frac{\partial x^r}{\partial x'^i} S_{jk} \frac{\partial^2 x'^k}{\partial x^r \partial t},$$

and again, by Schwarz's theorem (coordinate changes are C^2)

$$\sum_{r,k} \frac{\partial x^r}{\partial x'^j} S_{ik} \frac{\partial^2 x'^k}{\partial t \partial x^r} = \sum_{r,k} \frac{\partial x^r}{\partial x'^i} S_{jk} \frac{\partial^2 x'^k}{\partial t \partial x^r}.$$

We recast the above as:

$$\sum_{r,k} S_{ik} \frac{\partial^2 x'^k}{\partial t \partial x^r} \frac{\partial x^r}{\partial x'^j} = \sum_{r,k} S_{jk} \frac{\partial^2 x'^k}{\partial t \partial x^r} \frac{\partial x^r}{\partial x'^i}.$$

Call J the Jacobian matrix of coefficients $J^k{}_r = \frac{\partial x'^k}{\partial x^r}$ and $\dot{J}$ the matrix of coefficients $\frac{\partial}{\partial t} J^k{}_r$. What we must prove (the last formula above) is:

$$(S \dot{J} J^{-1})_{ij} = (S \dot{J} J^{-1})_{ji}, \quad \text{in other words} \quad S \dot{J} J^{-1} = (S \dot{J} J^{-1})^t.$$

Computing the product's transpose and recalling $S^t = -S$, we end up with having to show:

$$S \dot{J} J^{-1} + J^{-1t} \dot{J}^t S = 0.$$

Note that transposition and inversion commute. Applying J^t to the left and J to the right we finally obtain:

$$J^t S \dot{J} + \dot{J}^t S J = 0. \tag{12.22}$$

But this is true because the coordinate transformation is canonical, so $J^t S J = S$, and differentiating in t both sides produces just (12.22). Therefore (12.20), i.e. (12.22), does hold, and $\mathcal{K}$ is build as we claimed above. At last, note $\mathcal{K}$ is independent of $\mathcal{H}$ by construction ($\mathcal{H}$ was never used to obtain it) and also defined up to additive constants at any instant, as we now set out to prove. Suppose $\mathcal{K}$ and $\mathcal{K}'$ satisfy (12.18) and let $f(t', \mathbf{x}') := \mathcal{K}(t', \mathbf{x}') - \mathcal{K}'(t', \mathbf{x}')$, so that $\nabla_{\mathbf{x}'} f(t', \mathbf{x}') = 0$. At a given t', $\mathbf{x}'$ varies in an open and connected set $A_{t'} \subset \mathbb{R}^{2n}$, hence path connected (by differentiable paths). Take a C^1 curve $\mathbf{x}' = \mathbf{x}'(s)$ with $s \in [0, 1]$ from $\mathbf{x}'_1$ to $\mathbf{x}'_2$, all contained in $A_{t'}$. The composite $[0, 1] \ni s \mapsto f(t', \mathbf{x}'(s))$ is $C^1([0, 1])$ by construction, so the Mean Value Theorem guarantees there is an $s_0 \in (0, 1)$ such that: $f(t', \mathbf{x}'_2) - f(t', \mathbf{x}'_1) = f(t', \mathbf{x}'(1)) - f(t', \mathbf{x}'(0)) = 1 \frac{d\mathbf{x}'}{ds}|_{s_0} \cdot \nabla_{\mathbf{x}'} f(t'\mathbf{x}')|_{\mathbf{x}'(s_0)} = 0$. Therefore $f(t', \mathbf{x}'_2) = f(t', \mathbf{x}'_1)$, and f can only depend upon t' on $U \cap U'$.   □

We have a very relevant and immediate corollary. In agreement with Sect. 11.1.6, we asked C^2 only to have existence and uniqueness for Hamilton's equations, and thus speak consistently of global solutions.

Proposition 12.10 *Suppose there is on $F(\mathbb{V}^{n+1})$ a C^2 Hamiltonian in any natural coordinates respecting (12.5),[2] so to define global solutions to Hamilton's equations on $F(\mathbb{V}^{n+1})$. Then we can define a C^2 Hamiltonian in any canonical coordinates so that the aforementioned global solutions solve Hamilton's equations on the charts of the canonical atlas for the given Hamiltonians.*

Proof If (U', ψ') is canonical chart, by definition there is a canonical transformation mapping it to a natural chart (U, ϕ) with $U = U'$ connected and locally simply connected. Given a Hamiltonian in any natural local coordinates we have a Hamiltonian in any canonical coordinates by the previous theorem, and in such a way that a curve solves Hamilton's equations on the first chart if and only if it solves them on the second. We know that if Hamiltonians abide by (12.5), for instance if they all come from a single Lagrangian, Hamilton's equations are global on $F(\mathbb{V}^{n+1})$ (see Sect. 11.1.6).   □

---

[2] In particular, when we start from a Lagrangian on $A(\mathbb{V}^{n+1})$.

Remarks 12.11

(1) The dynamic vector field Z of Sect. 11.4.3 assumes the standard form (11.70) in natural coordinates, and is properly defined, when the Hamiltonians respect (12.5). Its integral curves are precisely the solutions to Hamilton's equations. By Theorem 12.9 and Proposition 12.10, the dynamic vector field will always be (11.70), in any local canonical coordinates:

$$Z = \frac{\partial}{\partial t} + \sum_{k=1}^{n} \frac{\partial \mathcal{H}}{\partial p_k} \frac{\partial}{\partial q^k} - \sum_{k=1}^{n} \frac{\partial \mathcal{H}}{\partial q^k} \frac{\partial}{\partial p_k} . \tag{12.23}$$

(2) We may pose the question as to whether the canonical transformations are the most general coordinate transformations that preserve the form of Hamilton's equations of motion. The answer is no. For the sake of an example, the transformation

$$t' = t ,$$
$$q'^k = \lambda q^k , \quad k = 1, \ldots, n,$$
$$p'_k = \mu p_k , \quad k = 1, \ldots, n ,$$

where $\lambda, \mu > 0$ are constants, clearly preserves the equations of motion if we set:

$$\mathcal{H}'(t', q', p') = \lambda \mu \mathcal{H}(t(t'), q(t', q', p'), p(t', q', p')) .$$

However, $J' S J = \mu \lambda S$, so we do not have a canonical transformation in case $\mu \lambda \neq 1$. As a matter of fact one could prove that a differentiable transformation:

$$t' = t + c , \tag{12.24}$$
$$x'^k = x'^k(t, x^1, \ldots, x^{2n}) , \tag{12.25}$$

with differentiable inverse (and both at least C^2) preserves the Hamiltonian form of the equations of motion if and only if the Jacobian matrix of the non-temporal coordinates satisfies $J' S J = a S$ for any $(t, \mathbf{x})$ and for some constant $a \neq 0$.

(3) The fact the Hamiltonian *does not* behave as a scalar under canonical coordinate changes, but acquires a term $\mathcal{K}$ as in (12.17), seems to suggest, for a given $\mathcal{H}$, one should seek canonical coordinates where $\mathcal{H}'$ is identically zero. This would be the same as solving Hamilton's equations for any initial condition, since in the new canonical coordinates the equations are trivial: $q'^k = q'^k_0$ constant, $p'_k = p'_{0k}$ constant. This observation prompted Jacobi to develop a technique that led to the celebrated *Hamilton-Jacobi equation*. ∎

Examples 12.12 Let us use the canonical coordinates t', q', p' of Example 12.8, where U is the open set $(a, b) \times D$ in coordinates. The goal is to find $\mathscr{K}$. We know such a function exists, because the assumptions of Theorem 12.9 hold. Let $\mathbf{x}' := (q', p')^t$, and since $t = t'$ we must have:

$$- S \frac{\partial \mathbf{x}'}{\partial t} \bigg|_{(t, \mathbf{x}(t, \mathbf{x}'))} = \nabla_{\mathbf{x}} \mathscr{K}(t, \mathbf{x}'),$$

translating into the two equations

$$\frac{\partial q'}{\partial t} = \frac{\partial \mathscr{K}}{\partial p'}, \quad \frac{\partial p'}{\partial t} = -\frac{\partial \mathscr{K}}{\partial q'}.$$

Keeping (12.12) into account, we have:

$$- q \sin \Omega t - p \cos \Omega t = \Omega^{-1} \frac{\partial \mathscr{K}}{\partial p'}, \quad -p \sin \Omega t + q \cos \Omega t = -\Omega^{-1} \frac{\partial \mathscr{K}}{\partial q'}.$$

We still need to express unprimed coordinates in terms primed ones. Using (12.11) the equations become:

$$- \Omega p' = \frac{\partial \mathscr{K}}{\partial p'}, \quad \Omega q' = -\frac{\partial \mathscr{K}}{\partial q'}.$$

Integrating the former, and ignoring the domains:

$$\mathscr{K}(t, q', p') = -\frac{\Omega}{2} p'^2 + f(t, q).$$

From this, the second equation gives:

$$\frac{\partial f}{\partial q'} = -\Omega q'.$$

In summary:

$$\mathscr{K}(t, q', p') = -\frac{\Omega}{2} \left(p'^2 + q'^2 \right) + g(t),$$

where g is arbitrary. By Theorem 12.9 this is the only solution $\mathscr{K}$ we are looking for (recall g is an arbitrary function).

We finish with one result on the Hamiltonian's uniqueness in given canonical coordinates when we enhance the regularity to C^2. The next proposition is the version of Proposition 11.7 in canonical coordinates.

Proposition 12.13 *Consider a local canonical chart* (U, ψ) *on* $F(\mathbb{V}^{n+1})$ *(or a natural chart in which every* $\mathbb{F}_t \cap U$ *is connected for all* $t \in T(U)$*) where*

$$\psi : U \ni a \mapsto (t(a), \mathbf{x}(a)) \in \mathbb{R}^{2n+1} .$$

Suppose there exists a C^2 *Hamiltonian* $\mathscr{H}$ *in the coordinates of* U *that produces Hamilton's equations (11.25). Another* C^2 *function in the same coordinates* $\mathscr{H}'$: $U \to \mathbb{R}$ *produces the same Hamiltonian equations if and only if* $\mathscr{H} - \mathscr{H}'$ *is a* C^2 *function that only depends on the coordinate* t.

Proof The proof is the same as that of Proposition 11.7, now using canonical, instead of natural, charts.   □

12.2   Liouville's Theorem in Global Form and Poincaré's "Recurrence" Theorem

The formalism introduced in the previous section allows to state the global version of Liouville's theorem on phase spacetime $F(\mathbb{V}^{n+1})$ without any restriction. One corollary is Poincaré's "recurrence" theorem, which will be examined immediately after.

12.2.1   Liouville Theorem and Liouville Equation

Another fundamental property of canonical transformations is that they preserve the volume of every phase space. Fix a fibre $\mathbb{F}_{\tau_0}$ at time τ_0 and consider local canonical (or natural) coordinates $(t, \mathbf{x}) \equiv (t, q^1, \ldots, q^n, p_1, \ldots, p_n)$ on the neighbourhood $\mathcal{O}$ of a point $p \in \mathbb{F}_{\tau_0}$. Let us pass to another system of canonical (or natural) coordinates on a neighbourhood $\mathcal{O}'$ of the same point: $(t', \mathbf{x}') \equiv (t', q'^1, \ldots, q'^n, p'_1, \ldots, p'_n)$. If $D \subset \mathbb{F}_{\tau_0}$ is a measurable set contained in $\mathcal{O} \cap \mathcal{O}'$, we can associate with it two volumes:

$$\int_{\mathbf{x}(D)} dq^1 \cdots dq^n dp_1 \cdots dp_n \quad \text{and} \quad \int_{\mathbf{x}'(D)} dq'^1 \cdots dq'^n dp'_1 \cdots dp'_n .$$

Since coordinate changes are canonical, the two coincide. In fact:

$$\int_{\mathbf{x}'(D)} dq'^1 \cdots dq'^n dp'_1 \cdots dp'_n$$

$$= \int_{\mathbf{x}(D)} \left| \det J(t, q^1, \ldots, q^n, p_1, \ldots, p_n) \right| dq^1 \cdots dq^n dp_1 \cdots dp_n .$$

J is the Jacobian matrix of the coordinate change, which has absolute value 1 since $J^t S J = S$ implies $(\det J)^2 = 1$, as already seen. All in all:

$$\int_{\mathbf{x}(D)} dq^1 \cdots dq^n dp_1 \cdots dp_n = \int_{\mathbf{x}'(D)} dq'^1 \cdots dq'^n dp'_1 \cdots dp'_n .$$

Canonical coordinates defined on the same open set V in a phase space at time τ_0 define a unique natural measure, which is the Lebesgue measure $dq_1 \cdots dq_n dp_1 \cdots dp_n$ in coordinates. If (U, ψ) is a canonical or local coordinate system around $p \in \mathbb{F}_{\tau_0}$ and $V := U \cap \mathbb{F}_{\tau_0}$, we indicate by ν_{τ_0} the above measure on V. Standard measure-theoretical techniques on differentiable manifolds (involving *partitions of unity*) allow to define a global **canonical measure** μ_{τ_0} on every fibre $\mathbb{F}_{\tau_0}$ of phase space, completely determined by the canonical atlas of $F(\mathbb{V}^{n+1})$ by "glueing" the various measures ν_{τ_0}. Such a measure (positive, Borel and regular [Rud78]) is by definition the only one reducing to $dq^1 \cdots dq^n dp_1 \cdots dp_n$ in any natural or canonical local coordinates. Using this canonical measure plus Liouville's theorem on $\mathbb{R} \times \mathbb{R}^{2n}$, see Chap. 11, one manages to prove Liouville's theorem in the general form below. We shall prove it later by integrating differential forms and relating to Sect. B.4.

To state Liouville's theorem consider a Hamiltonian dynamic vector field Z of class C^2 of the form (12.23), obtained by assigning compatible Hamiltonians in natural local charts. The vector field generates a one-parameter group of local diffeomorphisms $\Phi^{(Z)}$, called **Hamiltonian flow** (in phase spacetime). In Sect. 11.3.3, when referring to the elementary phase spacetime $\mathbb{R} \times \mathbb{R}^{2n}$, it was called Φ.

Theorem 12.14 (Liouville's Theorem) *Consider a physical system described on the phase spacetime $F(\mathbb{V}^{n+1})$ with fibres $\mathbb{F}_\tau$ (phase spaces at time τ) and C^2 Hamiltonian dynamic vector field Z generating the one-parameter group of local diffeomorphisms $\Phi^{(Z)}$ (the Hamiltonian flow on phase spacetime). Take two fibres $\mathbb{F}_{\tau_1}$ and $\mathbb{F}_{\tau_2}$ and a Borel set*[3] *$D_{\tau_1} \subset \mathbb{F}_{\tau_1}$ with $\mu_{\tau_1}(D_{\tau_1}) < +\infty$ so that*

$$D_t := \Phi^{(Z)}_{t-\tau_1}(D_{\tau_1}) \in \mathbb{F}_\tau$$

is defined for $t \in [\tau_1, \tau_2]$.[4] *Then:*

$$\int_{D_{\tau_2}} d\mu_{\tau_2} = \int_{D_{\tau_1}} d\mu_{\tau_1} .$$

Proof See the proof of (6) in Theorem 13.21.   □

---

[3] For example, an open set with compact closure.
[4] That is: for any $a \in D_{\tau_1}$, the maximal integral curve of Z with initial condition a at time τ_1 reaches $\mathbb{F}_t$, so that $D_t \subset \mathbb{F}_t$ when $t \in [\tau_1, \tau_2]$.

Remarks 12.15

(1) The physical meaning of the measure μ_t can be illustrated as follows. It is trivially clear, by construction, that the number of curves describing Hamiltonian evolutions between D_{τ_1} and D_{τ_2} is constant in the tube obtained by applying $\Phi^{(Z)}$ to D_{τ_1}: this tube in fact *consists of these curves*. The measure μ_t "counts", in the technical sense of Measure Theory, how many curves there are. In other words μ_t "counts" the points in the cross sections D_t of the tube, and each point defines exactly one of the curves the tube is made of.

(2) The canonical measure μ_{τ_0} can always be **completed** [Rud78] by: (a) enlarging the class of measurable sets by adding all subsets of zero-measure sets; (b) declaring that the latter have zero measure. The completed measure, too, is invariant under Hamiltonian evolution, as is immediate from the completion process, since the Hamiltonian flow is a diffeomorphism for any value of the parameter and as such it preserves the inclusion. ∎

The presence of a natural evolution-invariant measure on phase space allows to make sense of probability distributions, a notion with big impact on physical applications. The above theorem is in fact the starting point of so-called *Statistical Mechanics*. This is the mathematical theory that studies statistical ensembles of identical physical systems, hence subject to the same Hamiltonian dynamics, but described by distinct representative points instant by instant due to different initial conditions. Such a collection of physical systems may comprise the situation where we do not exactly know the state of the system at each instant. The approach is also employed to provide microscopic descriptions of the thermodynamics of systems in equilibrium when it is assumed that the macroscopic state at equilibrium corresponds to an average of the microscopic states over long periods of time, and that this temporal averaging is the same as the assignment at each instant of a distribution of microscopic states (representative points), instead of a single representative point (ergodic hypothesis).

In Statistical Mechanics one assigns a (C^1) probability density $\rho = \rho(t, q, p) \geq 0$ on phase spacetime; its integral *with respect to the canonical measure* μ, over the measurable set $D_t \subset \mathbb{F}_t$ at time t, defines the probability $P_t(D_t)$ that the physical system's representative point belongs to D_t at time t:

$$P_t(D_t) := \int_{D_t} \rho(t, q, p) \, d\mu_t(q, p) \,. \tag{12.26}$$

The logical condition to be imposed on ρ is the following. Take a measurable set $D_{t_0} \subset \mathbb{F}_{t_0}$ of representative points with probability $P_{t_0}(D_{t_0})$ at time t_0. For $t \neq t_0$ let D_t be obtained from D_{t_0} under Hamiltonian evolution. Then the probability $P_t(D_t)$ satisfies:

$$P_t(D_t) = P_{t_0}(D_{t_0}) \,. \tag{12.27}$$

In particular, for $D_t = \mathbb{F}_t$ the probability must be always 1.

Now, (12.27) can be written as:

$$\frac{d}{dt} \int_{D_t} \rho(t, q, p) \, d\mu_t(q, p) = 0 \, .$$

The latter, restricting ourselves to open sets D_t with compact closure covered by canonical coordinates, equivalently reads:

$$\frac{d}{dt} \int_{D_{t_0}} \rho(t, q(t, q_0, p_0), p(t, q_0, p_0)) \, |\det J_t| \, d\mu_{t_0}(q_0, p_0) = 0 \, ,$$

where $q(t, q_0, p_0), p(t, q_0, p_0)$ are the position in phase spacetime at time t of the representative point that evolves, under Hamilton's equations, from the initial condition (q_0, p_0) at time t_0. As $|\det J_t|$ is constant, in the above identity we can swap derivative and integral to obtain:

$$\int_{D_{t_0}} \frac{d}{dt} (\rho(t, q(t, q_0, p_0), p(t, q_0, p_0))) \, |\det J_t| \, d\mu_{t_0}(q_0, p_0) = 0 \, ,$$

But D_{t_0} is arbitrary (it could be a small standard ball) and the integrand is continuous, so another way of casting the above is:[5]

$$\frac{d}{dt} \rho(t, q(t, q_0, p_0), p(t, q_0, p_0)) = 0 \, .$$

This equation says that along every world line that solves Hamilton's equations, the value of ρ must keep constant (but vary with the curve). This fact, by itself, guarantees that if, at initial time t_0, $\rho \geq 0$ everywhere on $\mathbb{F}_{t_0}$, then ρ always fulfils the condition. In canonical or natural coordinates the above evolution equations reads

$$\frac{\partial \rho}{\partial t} + \sum_{k=1}^{n} \left(\frac{\partial \rho}{\partial p_k} \frac{dp_k}{dt} + \frac{\partial \rho}{\partial q^k} \frac{dq^k}{dt} \right) = 0 \quad \text{at every point in phase space.}$$

Eventually, using Hamilton's equations we discover the celebrated **Liouville equation**:

$$\frac{\partial \rho}{\partial t} = \sum_{k=1}^{n} \left(\frac{\partial \rho}{\partial p_k} \frac{\partial \mathcal{H}}{\partial q^k} - \frac{\partial \rho}{\partial q^k} \frac{\partial \mathcal{H}}{\partial p_k} \right) \quad \text{at every point in phase space .} \tag{12.28}$$

---

[5] If, for a continuous map $g : D \to \mathbb{R}$ on $D \subset \mathbb{R}^n$ open, $\int_A g(x) d^n x = 0$ for any measurable set $A \subset D$, then $g(x) = 0$ everywhere on D. If we had $g(x_0) > 0$ at some x_0, then $0 < \epsilon_0 < g(x_0) - \epsilon < g(x) < g(x_0) + \epsilon$ on an open neighbourhood A_{x_0} of x_0. There, we would have $\int_{A_{x_0}} g(x) d^n x \geq \epsilon_0 \int_{A_{x_0}} d^n x > 0$, against the hypothesis. If $g(x_0) < 0$, using $-g$ we would reach the same contradiction.

Going backwards (this involves a partition of unity), it can be shown that the Liouville equation is a sufficient condition to have (12.27).

12.2.2 Poincaré's "Recurrence" Theorem

The present section uses elementary facts form the theory of (Borel) measures. The text of reference is [Rud78].

Consider a Hamiltonian dynamical system described in phase spacetime $F(\mathbb{V}^{n+1})$ (arising for instance from a Lagrangian theory on $A(\mathbb{V}^{n+1})$) decomposed canonically into the Cartesian product $\mathbb{R} \times \mathbb{F}$. We shall assume the following stronger conditions.

(R1) *$F(\mathbb{V}^{n+1})$ is covered by an atlas $\mathcal{A} = \{(U_i, \psi_i)\}_{i \in I}$ of local canonical, or natural, coordinate systems*

$$\psi_i : U_i \ni a \mapsto (t_i(a), x_i(a), \dots, x_i^{2n}(a)) \in \mathbb{R}^{2n+1}$$

such that $t_i(F(\mathbb{V}^{n+1})) = \mathbb{R}$ for any $i \in I$ and the transformations on $\mathcal{A}$ are completely canonical.[6] In particular, then, $t_i(a) = t_j(a) =: t(a)$ for any $a \in U_i \cap U_j$.

Then we may identify $F(\mathbb{V}^{n+1})$ with $\mathbb{R} \times \mathbb{F}$. $\mathbb{F}$, called **phase space**, is any fibre $\mathbb{F}_{t_0}$ seen as differentiable manifold with atlas $\mathcal{A}' = \{(V_i, \psi_i')\}_{i \in I}$, where $V_i := U_i \cap \mathbb{F}_{t_0}$, $\psi_i' : V_i \ni r \mapsto (x_i(r), \dots, x_i^{2n}(r))$ and the x_i^k are the non-temporal coordinates of the chart (U_i, ψ_i). The diffeomorphism

$$f : \mathbb{R} \times \mathbb{F}_{t_0} \to F(\mathbb{V}^{n+1})$$

between the two manifolds is found by asking, with respect to $\mathcal{A}$ and $\mathcal{A}'$:

$$\psi_i \circ f(t, \psi_i'^{-1}(x_i^1, \dots, x_i^n)) = (t, x_i^1, \dots, x_i^n) \in \mathbb{R}^{2n+1} \quad \text{for any } i \in I .$$

(R2) *The system's Hamiltonians—for the moment seen as C^2 maps that transform under Theorem 12.9 when changing local charts—are independent of time in each such local chart.*

Up to additive constants, therefore, the various Hamiltonians can be "glued" to give one scalar field that is time-independent on the entire phase space(time). Hamilton's equations are at present an *autonomous* system on the phase space manifold $M := \mathbb{F}$ in the sense of Sect. 14.5.

---

[6] There is no need for the coordinate domains to be simply connected when using canonical coordinates. Indeed, (R1) is automatic when the Hamiltonian dynamical formulation comes from a Lagrangian formulation with Lagrangian independent of time in a frame $\mathscr{R}$ for which we define an atlas of $A(\mathbb{V}^{n+1})$ of natural coordinates.

(R3) *Hamilton's equations' maximal solutions are complete, i.e. defined for any* $t \in \mathbb{R}$.

With all of that, consider the flow $\Phi^{(Z)}$ on $\mathbb{R} \times \mathbb{F}$. By definition

$$I_{(t,x)} \ni s \mapsto \Phi_s^{(Z)}(t, x) =: (s, \varphi_s(t, x)) \in \mathbb{R} \times \mathbb{F}$$

is the maximal solution to Hamilton's equations with initial condition $(t, x) \in \mathbb{R} \times \mathbb{F}$ at $s = 0$ (note: the s parametrising the solution is not absolute time, which is $t + s$). Hamilton's equations are autonomous by (R2), so on $\mathbb{F}$ we have (Proposition 14.41):

$$\varphi_{s+\tau}(t, x) = \varphi_s(t + \tau, x)$$

and therefore:

$$\varphi_s(t, x) = \varphi_{s+t}(0, x) =: \varphi_{s+t}(x) \,,$$

implying:

$$\Phi_s^{(Z)}(t, x) = (s + t, \varphi_{s+t}(x)) \,. \tag{12.29}$$

We can then forget the factor $\mathbb{R}$ and directly work on the phase space $\mathbb{F}$, where the maximal solutions to Hamilton's equations give a one-parameter group of local diffeomorphisms φ called the **Hamiltonian flow** on phase space. In fact, the analogous properties of $\Phi^{(Z)}$ produce, under (12.29):

$$\varphi_0 = id_{\mathbb{F}} \,, \qquad \varphi_s \circ \varphi_t = \varphi_{s+t} \,, \tag{12.30}$$

where, as usual, the second relation holds at $x \in \mathbb{F}$ and for values $s, t \in \mathbb{R}$ for which the left-hand side exists. As explained in Sect. 14.5.3 (in particular, Proposition 14.55) condition (R3) implies that $\Phi^{(Z)}$ (hence φ) is a *global* flow, i.e. $I_{(t,x)} = \mathbb{R}$ for any $(t, x) \in \mathbb{R} \times \mathbb{F}$. Then φ is made of proper diffeomorphisms acting on the whole phase space: $\varphi_s : \mathbb{F} \to \mathbb{F}$ for any $s \in \mathbb{R}$. In particular, the second equation in (12.30) holds without domain restrictions for every $s \in \mathbb{R}$ and every $x \in \mathbb{F}$.

At last, observe that with the special atlas $\mathcal{A}$, Liouville's Theorem 12.14 can be recast in terms of φ instead of $\Phi^{(Z)}$: if the Hamiltonian is at least C^3 (i.e. the dynamic field Z is C^2 at least) and

$$D_t := \varphi_t^{(Z)}(D_0) \in \mathbb{F}$$

for some measurable set $D_0 \subset \mathbb{F}$ of finite canonical measure and $t \in \mathbb{R}$, then:

$$\int_{D_t} d\mu = \int_{D_0} d\mu \,.$$

We point out that now there are no restrictions on t and D_0 because the one-parameter group is global. Moreover, the canonical measure μ is the same for any time, and defined on $\mathbb{F}$ rather than on distinct phase spaces at different times. In local or canonical coordinates of $\mathcal{A}$:

$$d\mu = dq^1 \cdots dq^n dp_1 \cdots dp_n .$$

The famous *"recurrence" theorem of Poincaré's* holds.[7]

Theorem 12.16 (Poincaré's "Recurrence Theorem") *Consider a Hamiltonian system with C^3 Hamiltonian function $\mathcal{H}$ satisfying (R1), (R2), (R3), and generating the Hamiltonian flow $\{\varphi_t\}_{t \in \mathbb{R}}$ on phase space. Let μ denote the canonical (positive Borel) measure on $\mathbb{F}$. Assume $\mathcal{R} \subset \mathbb{F}$ fulfils:*

(i) $\varphi_t(\mathcal{R}) \subset \mathcal{R}$ for any $t \in \mathbb{R}$;
(ii) $\mathcal{R}$ is measurable with finite measure: $\mu(\mathcal{R}) < +\infty$.

Take any sufficiently small open neighbourhood $\mathcal{G}_a \subset \mathcal{R}$ of a point $a \in \mathcal{R}$. The following facts hold.

(1) $\mathcal{G}_a$ "returns infinitely often" to itself: that is, there exists a sequence of instants $0 < \tau_1 < \tau_2 < \cdots \to +\infty$ such that $\mathcal{G}_a \cap \varphi_{\tau_n}(\mathcal{G}_g) \neq \varnothing$ where $\mu\left(\mathcal{G}_a \cap \varphi_{\tau_n}(\mathcal{G}_a)\right) > 0$ for any $n \in \mathbb{N}$.
(2) (Assuming μ complete) the points of $\mathcal{G}_a$ that "never return to $\mathcal{G}_a$ after an arbitrary instant $T > 0$" form a zero-measure subset:

$$\mu\left(\{p \in \mathcal{G}_a \mid \varphi_t(p) \notin \mathcal{G}_a , \quad t > T\}\right) = 0 .$$

Proof We start with a lemma. $\qquad\square$

Lemma 12.17 *If $A \subset \mathcal{R}$ is measurable with $\mu(A) > 0$ and $t > 0$ is any constant, every set*

$$A_n := \varphi_{nt}(A) := (\varphi_t)^n(A) , \quad n \in \mathbb{N},$$

intersects A, and $\mu(A \cap A_{k_t}) > 0$ for some $k_t \in \mathbb{N} \setminus \{0\}$.

Proof of Lemma The A_m are measurable as continuous pre-images of measurable sets (continuous maps are measurable because μ is Borel). For $n \in \mathbb{N} \setminus \{0\}$ set $B_n := A_n \setminus (\cup_{m \neq n} A_m)$, so that by construction $B_m \cap B_n = \varnothing$ when $n \neq m$. Suppose *by contradiction* that $\mu(A_n \cap A_m) = 0$ for any $n \neq m$, and let us show this leads to an absurd. If $\mu(A_n \cap A_m) = 0$ then $\mu\left(A_n \cap (\cup_{m \neq n} A_m)\right) = 0$,[8] so splitting

---

[7] Stated by Poincaré but formally proved by Carathéodory, actually.

[8] In fact: $0 \leq \mu\left(A_n \cap (\cup_{m \neq n} A_m)\right) = \mu\left(\cup_{m \neq n}(A_n \cap A_m)\right) \leq \sum_{m \neq n} \mu(A_n \cap A_m) = \sum_{m \neq m} 0 = 0.$

A_n as disjoint union of $A_n \setminus (\cup_{m \neq n} A_m)$ and $A_n \cap (\cup_{m \neq n} A_m)$ we obtain:

$$\mu(A_n) = \mu\left(A_n \setminus (\cup_{m \neq n} A_m)\right) + \mu\left(A_n \cap (\cup_{m \neq n} A_m)\right)$$
$$= \mu\left(A_n \setminus (\cup_{m \neq n} A_m)\right) = \mu(B_n).$$

As $B_m \cap B_n = \varnothing$ for $n \neq m$, we have:

$$\mu\left(\bigcup_{m \in \mathbb{N}} B_n\right) = \sum_{m \in \mathbb{N}} \mu(B_m) = \sum_{m \in \mathbb{N}} \mu(A_m) = \sum_{m \in \mathbb{N}} \mu\left((\varphi_t)^n(A)\right)$$
$$= \sum_{m \in \mathbb{N}} \mu(A) = +\infty.$$

This cannot be, since $\bigcup_{m \in \mathbb{N}} B_n \subset \mathcal{R}$ and hence:

$$\mu\left(\bigcup_{m \in \mathbb{N}} B_n\right) \leq \mu(\mathcal{R}) < +\infty.$$

Therefore necessarily $\mu(A_n \cap A_m) > 0$ for some m, n with $m \neq n$. Supposing n smaller than m, we apply $(\varphi_{nt})^{-1}$ and recall that the transformation is volume-preserving, to obtain

$$0 < \mu(A_n \cap A_m) = \mu\left((\varphi_{nt})^{-1}(A_n \cap A_m)\right) = \mu\left((\varphi_{nt})^{-1}(A_n) \cap (\varphi_{nt})^{-1}(A_m)\right)$$
$$= \mu\left(A \cap (\varphi_{nt})^{-1}(A_m)\right).$$

Let $k_t := m - n > 0$, so that $\varphi_{nt}^{-1}(A_m) = \varphi_{nt}^{-1}(\varphi_{mt} A) = \varphi_{(m-n)t}(A) = A_{k_t}$. Eventually, $\mu(A \cap A_{k_t}) > 0$.   □

This lemma proves claim (1) in the recurrence theorem if we choose A to be the non-empty open set $\mathcal{G}_a$. The latter has non-zero measure because its Lebesgue measure is non-zero (locally, μ is the Lebesgue measure on $\mathbb{F}$ in canonical/natural coordinates). The sequence $\tau_1 < \tau_1 < \cdots$ arises by applying the lemma at times $t_1 < t_2 < t_3, \ldots \to +\infty$, chosen at each step so that $t_2 > k_{t_1} t_1$, $t_3 > k_{t_2} t_2$ etc, and setting $\tau_1 := k_{t_1} t_1$, $\tau_2 := k_{t_2} t_2$, or $\tau_n := k_{t_n} t_n$ in general. By construction $0 < \tau_1 < \tau_2 < \cdots \to +\infty$.

Now let us deal with claim (2). Consider the subset $E \subset \mathcal{G}_a$ of 'non-returning' points $p \in \mathcal{G}_a$ under the discrete evolution of the multiples of a certain time $T' > T$: $\varphi_{nT'}(p) \notin \mathcal{G}_a$ for some $T' > 0$ and any $n = 1, 2, \ldots$. The set E is measurable as countable intersection of measurable sets:

$$E = \{p \in \mathcal{G}_a \mid \varphi_{nT'}(p) \notin \mathcal{G}_a \quad n \in \mathbb{N} \setminus \{0\}\} = \mathcal{G}_a \bigcap_{n \in \mathbb{N} \setminus \{0\}} (\varphi_{nT'})^{-1}(\mathcal{R} \setminus \mathcal{G}_a).$$

If $\mu(E) > 0$, applying the lemma with $A = E$ would give $E \cap \varphi_{kT'}(E) \neq \varnothing$ for some $k \in \mathbb{N} \setminus \{0\}$. This is impossible by definition of E. Therefore $\mu(E) = 0$. The subset

$$E_0 := \{p \in \mathcal{G}_a \mid \varphi_t(p) \notin \mathcal{G}_a, \quad t > T\},$$

made of points of $\mathcal{G}_a$ that never return to $\mathcal{G}_a$ starting from time T, clearly satisfies $E_0 \subset E$. But μ is complete, so E_0 is measurable and then $\mu(E) = \mu(E_0) = 0$.

Remarks 12.18 As a matter of fact, assumptions (R1)–(R3) are only required to state the result in a Hamiltonian setting. It is evident from the proof that the only mathematical hypotheses needed are:

(a) having a 1-parameter group $\{\varphi_t\}_{t \in \mathbb{R}}$ of measurable bijections $\varphi_t : \mathbb{F} \to \mathbb{F}$ on the Borel σ-algebra of a topological space $\mathbb{F}$;
(b) the existence of a (completed) φ-invariant measure μ on the Borel σ-algebra of $\mathbb{F}$;
(c) the existence of a region $\mathcal{R} \subset \mathbb{F}$ satisfying (i) and (ii) in Theorem 12.16.

The neighbourhood $\mathcal{G}_a$ must be assumed to have non-zero measure explicitly. One could more simply work on a measurable space $\mathbb{F}$, even not a topological space, provided $\mathcal{G}_a$ is a measurable set containing a with non-zero measure. ∎

Examples 12.19

(1) Hamiltonian systems that satisfy Poincaré's theorem are rather common. Take a physical system with time-independent forces in an inertial frame, and suppose it admits Lagrangian description, hence a Hamiltonian description too. Working in coordinates moving with the inertial frame, (R1) and (R2) hold. Suppose furthermore the physical system admits a first integral $H : \mathbb{F} \to \mathbb{R}$, typically the energy, whose level surfaces in phase space, obtained by fixing a special value of the first integral:

$$\Sigma_h := \{\mathbf{x} \in \mathbb{F} \mid H(\mathbf{x}) = h\}$$

are compact. Since any solution to Hamilton's equations must stay inside the compact set Σ_h (h being the value of H at the initial instant), every maximal solution will be complete by Proposition 14.43. Therefore also (R3) holds. Finally, if the volume of:

$$\mathcal{R} := \{\mathbf{x} \in \mathbb{F} \mid h_1 < H(\mathbf{x}) < h_2\}$$

has finite measure—and for that it suffices that the closure of $\mathcal{R}$ be compact—then Poincaré's theorem holds, since $\varphi_t(\mathcal{R}) \subset \mathcal{R}$. In fact, if $p \in \mathcal{R}$ then $H(p) = h_p \in (h_1, h_2)$, and as H is a first integral $H(\varphi_t(\mathcal{R})) = h_p \in (h_1, h_2)$ so in the end $\varphi_t(p) \in \mathcal{R}$ for any $t \in \mathbb{R}$.

(2) The previous situation is that of a physical system of N particles with Hamiltonian

$$\mathcal{H} = \sum_{i=1}^{N} \frac{\mathbf{p}_i^2}{2m_i} + \mathcal{U}(\mathbf{x}_1, \ldots, \mathbf{x}_N)$$

where the potential energy $\mathcal{U}$ (C^3) is *bounded from below*:

$$\mathcal{U}(\mathbf{x}_1, \ldots, \mathbf{x}_N) \geq C \quad \text{for any } (\mathbf{x}_1, \ldots, \mathbf{x}_N) \in \mathbb{R}^{3N} \text{ and some constant } C \in \mathbb{R}$$

and such that the sets

$$K_u := \{(\mathbf{x}_1, \ldots, \mathbf{x}_N) \in \mathbb{R}^{3N} \mid \mathcal{U}(\mathbf{x}_1, \ldots, \mathbf{x}_N) \leq u\}$$

are compact for any u in some interval. (This last condition is true if $\mathcal{U}$ is the sum of potential energies describing potential sinks). In this case, if $\mathcal{H}$ is constant and equal to h, and if M is the largest mass m_i for which $\sum_{i=1}^{N} \mathbf{p}_i^2/(2M) \leq \sum_{i=1}^{N} \mathbf{p}_i^2/(2m_i)$, the values of $\sqrt{\sum_{i=1}^{N} \mathbf{p}_i^2}$ cannot exceed $\sqrt{2M(h-C)}$, so the vector $(\mathbf{p}_1, \ldots, \mathbf{p}_N)$ belongs to the closed ball $B_h \subset \mathbb{R}^{3N}$ at the origin with radius $\sqrt{2M(h-C)}$. Therefore the closed surface $\Sigma_h \subset \mathbb{R}^{6N}$

$$\Sigma_h := \{(\mathbf{x}_1, \ldots, \mathbf{x}_N, \mathbf{p}_1, \ldots, \mathbf{p}_N) \in \mathbb{R}^{6N} \mid \mathcal{H}(\mathbf{x}_1, \ldots, \mathbf{x}_N, \mathbf{p}_1, \ldots, \mathbf{p}_N) = h\}$$

is contained in the compact set $K_h \times B_h \subset \mathbb{R}^{6N}$, so it is compact.
As h varies in (h_1, h_2), the compact set K_{h_2} contains every K_h by construction. Consequently

$$\mathcal{R} := \{(\mathbf{x}_1, \ldots, \mathbf{x}_N, \mathbf{p}_1, \ldots, \mathbf{p}_N) \in \mathbb{R}^{6N} \mid \mathcal{H}(\mathbf{x}_1, \ldots, \mathbf{x}_N, \mathbf{p}_1, \ldots, \mathbf{p}_N) \in (h_1, h_2)\}$$

has finite volume, being contained in the compact set $K_{h_2} \times B_{h_2}$, and is by construction invariant under the dynamical flow. A one-dimensional example of this is a particle undergoing a force whose potential energy, around the equilibrium point, describes a lower-bounded potential sink. The simplest concrete case is the potential sink of the harmonic oscillator.

(3) A particle with attractive Coulomb potential may not satisfy the hypotheses, because the impulse's possible values may be arbitrarily large, at the expense of a correspondingly arbitrarily large (but negative) potential energy, in order to maintain constant total energy. Constant-energy surfaces in phase space may be unbounded, hence not compact. Yet, if we consider the angular momentum's first integral, it is possible to define, for any energy value and any *non-zero* value of the angular momentum,[9] compact sets containing the solutions of motion, whose union is an invariant set $\mathcal{R}$ with finite canonical measure.

Remarks 12.20 A physically interesting case where the "recurrence" theorem applies is a gas confined to a lower-bounded potential sink (a gas bottle, for instance) and made of non-interacting particles (an ideal gas) that are kept isolated from the outside so that the system's energy is conserved. A macroscopic state of the gas, given in terms of its macroscopic thermodynamical quantities, should correspond to an enormous number of microscopic states in $6N$-dimensional phase space ($N \sim 6.02 \times 10^{23}$), because a variation of the kinetic state of each single particle (moving from state a to a nearby representative point) will not alter the macroscopic state. Consider a microscopic state a that macroscopically, when looked at in space, appears like the gas is confined to half of the bottle. If we move in an open neighbourhood $\mathcal{G}_a$ of a in $\mathbb{F}$ the macroscopic state will remain indistinguishable, so the microscopic state corresponding to the initial macroscopic state might be another $a' \in \mathcal{G}_a$, different from a. Starting from one of these microscopic states (macroscopically all equal) and waiting a sufficiently long time t while the system evolves under Hamiltonian dynamics, the state might return to $\mathcal{G}_a$. From a macroscopic point of view we would witness, at time t, the gas back into the half bottle where it was when the experiment started. The probability of this occurring is non-zero, and is represented by (or proportional to) the measure of the intersection $\mu(\mathcal{G}_a \cap \varphi_{\tau_n}(\mathcal{G}_a)) > 0$ (depending on the mathematical model adopted to describe macrostates in terms of microstates). The probability that this process does not happen at all, even if we wait an arbitrarily long time t is **zero**, as per the theorem's second part. This situation is clearly at odds with the second law of Thermodynamics when interpreted in the most direct way, which postulates that the macroscopic state always stays the one of largest entropy. This would be the gas filling the entire bottle. Nonetheless, it is possible to estimate the times of "recurrence" for physical systems like the one described, and one discovers that they have order of magnitude comparable with the age of the solar system [Arn92]. Based on that, one should consider the thermodynamics of equilibria an approximate description, valid on rather short time scales. At last, it is important to mention that there is a quantum version of the "recurrence" theorem for Hamiltonian systems with "discrete energy spectrum" (P. Bocchieri and A. Loinger, Phys. Rev. 107, 337–338 (1957)). For the ideal quantum gas inside a bottle, but also for one or several particles under attractive Coulomb forces, the energy spectrum is indeed discrete (for attractive Coulomb interaction potential this happens at negative energies). Also in a quantum setting, therefore, the recurrence theorem's effects regarding the thermodynamics of equilibria continue to hold.   ∎

---

[9] Using radial coordinates, when the angular momentum is non-zero, the theory is described by a potential energy that accounts for the so-called "centrifugal potential", besides the Coulomb potential energy. Due to that repulsive term the actual potential energy is bounded below when the angular momentum is non-zero.

12.3 $\boxed{\text{AC}}$ Symmetries and Conservation Laws in Hamiltonian Mechanics

We will introduce a theoretical and practical useful tool: the *Poisson bracket*. It is a bilinear functional acting on functions defined on phase spacetime. This mathematical instrument's importance is dual. On one hand it allows us to easily study the relationship between symmetries and conserved quantities in the Hamiltonian formulation. On the other, many results of Hamiltonian Mechanics have quantum correspondents when we replace the Poisson bracket with quantum commutators (this is the so-called *Dirac correspondence principle*).

In this section we will use the Poisson bracket formalism to reformulate in Hamiltonian language the relationship of symmetries to conservation laws, which we saw in Lagrangian formulation by means of Noether's theorem.

12.3.1 *Hamiltonian Vector Fields and Poisson Bracket*

Consider a function (a scalar field) $f : F(\mathbb{V}^{n+1}) \to \mathbb{R}$ of class C^1. We will work in canonical local coordinates $(t, q^1, \ldots, q^n, p_1, \ldots, p_n) \equiv (t, \mathbf{x})$. With f we associate the vector field X_f on $F(\mathbb{V}^{n+1})$ defined locally by

$$X_f := \sum_{i,j=1}^{2n} S^{ij} \frac{\partial f}{\partial x^i} \frac{\partial}{\partial x^j} = \sum_{k=1}^{n} \left(\frac{\partial f}{\partial q^k} \frac{\partial}{\partial p_k} - \frac{\partial f}{\partial p_k} \frac{\partial}{\partial q^k} \right), \qquad (12.31)$$

where we used (11.42). This field is actually global, meaning that if for the same f we define the vector field in other canonical coordinates, on intersections of chart domains (if non-empty) the two fields coincide. As canonical local coordinates form an atlas, we obtain a well-defined vector field on $F(\mathbb{V}^{n+1})$ that we shall still call X_f. X_f is therefore the vector field on $F(\mathbb{V}^{n+1})$ with form (12.31) in any canonical local coordinates. To show that fields associated with the same f coincide on the intersections of the domains of canonical coordinate systems, let us suppose $(t', \mathbf{x}')$ are the canonical coordinates on some domain in phase space that intersects the previous. On the intersection we have the following equalities, proving the claim:

$$\sum_{i,j=1}^{2n} S^{ij} \frac{\partial f}{\partial x^i} \frac{\partial}{\partial x^j} = \sum_{i,j,k,r=1}^{2n} S^{ij} \frac{\partial x'^k}{\partial x^i} \frac{\partial x'^r}{\partial x^j} \frac{\partial f}{\partial x'^k} \frac{\partial}{\partial x'^r}$$

$$= \sum_{k,r=1}^{2n} (JSJ^t)^{kr} \frac{\partial f}{\partial x'^k} \frac{\partial}{\partial x'^r} = \sum_{k,r=1}^{2n} S^{kr} \frac{\partial f}{\partial x'^k} \frac{\partial}{\partial x'^r} ,$$

where we used that transformations of canonical coordinates are canonical, so the Jacobian matrix J (and the transpose) of the non-temporal coordinates belongs to the group $Sp(n, \mathbb{R})$, i.e. $JSJ^t = S$.

Remarks 12.21

(1) We obtain the same result and the same global vector field X_f using the atlas of natural coordinates, or its union with the canonical atlas. Indeed, coordinate changes between natural local charts, or from natural to canonical, are canonical transformations.

(2) The Hamiltonian dynamic vector field Z (Definition 11.25), whose integral curves are the solutions to Hamilton's equations, can be *locally* written as:

$$Z = \frac{\partial}{\partial t} - X_{\mathscr{H}}.$$

$\mathscr{H}$ is the Hamiltonian associated with Z in natural/canonical local coordinates $t, q^1, \ldots, q^n, p_1, \ldots, p_n$. Note that the above expression holds in any natural/canonical coordinates, but $\mathscr{H}$ is not a scalar function on phase spacetime, in contrast to the f in (12.31). When changing coordinates, the new $\mathscr{H}'$ compensates for the different vector $\frac{\partial}{\partial t'}$, since the latter changes too as we change the local chart:

$$\frac{\partial}{\partial t'} = \frac{\partial}{\partial t} + \sum_{k=1}^{n} \frac{\partial q'^k}{\partial t} \frac{\partial}{\partial q'^k} + \sum_{k=1}^{n} \frac{\partial p'_k}{\partial t} \frac{\partial}{\partial p'_k}.$$

The sum of the two contributions remains constant and gives the coordinate-independent vector field:

$$Z = \frac{\partial}{\partial t} - X_{\mathscr{H}} = \frac{\partial}{\partial t'} - X_{\mathscr{H}'}.$$

So one should pay attention when writing $X_{\mathscr{H}}$: the symbol makes sense in the natural/canonical coordinates where $\mathscr{H}$ is defined as well. ∎

Consider two C^1 functions $f, g : F(\mathbb{V}^{n+1}) \to \mathbb{R}$. It is useful to introduce a third scalar field, locally defined in canonical coordinates by:

$$\{f, g\} := X_f(g) = -X_g(f) := \sum_{i,j=1}^{2n} S^{ij} \frac{\partial f}{\partial x^i} \frac{\partial g}{\partial x^j} = \sum_{k=1}^{n} \left(\frac{\partial f}{\partial q^k} \frac{\partial g}{\partial p_k} - \frac{\partial f}{\partial p_k} \frac{\partial g}{\partial q^k} \right).$$

(12.32)

It will turn out to be well defined *on the entire phase spacetime*, for the same reason X_f is.

Definition 12.22 Consider $f, g : F(\mathbb{V}^{n+1}) \to \mathbb{R}$ of class C^1.

(1) The **Hamiltonian vector field associated** with f is the vector field X_f on $F(\mathbb{V}^{n+1})$ given by (12.31) in any canonical or natural local coordinates.

(2) The **Poisson bracket of** f **and** g is the scalar field $\{f, g\}$ on $F(\mathbb{V}^{n+1})$ defined by (12.32) in any canonical or natural local coordinates. ◇

Let us now suppose the phase space at a given time is connected, by which we mean each fibre $\mathbb{F}_{t_0} := T^{-1}(t_0)$ is connected in the bundle $T : F(\mathbb{V}^{n+1}) \to \mathbb{R}$, where T is the usual absolute time. The map associating functions to Hamiltonian fields has an important property, stated in part two of the next proposition.

Proposition 12.23 *Let* $T : F(\mathbb{V}^{n+1}) \to \mathbb{R}$ *be the absolute time of phase space seen as projection of the bundle over the absolute time axis, and* $f, f' \in C^1(\mathbb{V}^{n+1})$.

(1) If $f - f'$ *is a function of absolute time* only:

$$h \circ T \quad where\ h : \mathbb{R} \to \mathbb{R}\ (with\ h \in C^1(\mathbb{R}))$$

then

$$X_f = X_{f'}.$$

(2) If $X_f = X_{f'}$*, and the phase space at a given time is a connected manifold, then* $f - f'$ *is a* (C^1) *function of absolute time solely.*

Proof

(1) If $f - f'$ depends only on the time variable, trivially $X_f = X_{f'}$ in canonical coordinates.

(2) Set $g = f - f'$, and suppose $X_f = X_{f'}$, i.e. $X_g = 0$ everywhere. Consider two kinetic states a and a' at time t_0 and let us show $g(a) = g(a')$, so that g can only depend on time. Since every phase space at a given time is connected, there is a C^1 curve $\gamma : [0, 1] \to F(\mathbb{V}^{n+1})$ contained in phase space at time t_0 joining a to a'. The range of γ is compact as continuous image of the compact interval $[0, 1]$. By definition of compact set, $\gamma([0, 1])$ can be covered by finitely many natural or canonical charts (U_i, ψ_i), $i = 1, \ldots, N$. Hence the chain of N charts whose union contains $\gamma([0, 1])$ goes from a to a'. If necessary we may restrict the domains U_i so that U_k meets U_{k-1} and U_{k+1} only (except for $k = 1$ and $k = N$). If the portion of γ in U_i is $t \mapsto (t_0, \mathbf{x}(t))$ in coordinates, then $\frac{d}{dt} g(\gamma(t)) = -\frac{d\mathbf{x}}{dt} \cdot S\nabla_{\mathbf{x}} g + \frac{\partial g}{\partial t} 0 = 0$, so $g \circ \gamma$ is constant on U_i. On the adjacent domains U_{i-1} and U_{i+1}, $g \circ \gamma$ will assume the same constant value it has on U_i. Considering the full chain we will have $g(a) = g(\gamma(0)) = g(\gamma(1)) = g(a')$. Therefore g is constant on every phase space, and can thus only be a function of time. □

Remarks 12.24 Consequently, the vector field X_f carries *less* information than the scalar field f, since $X_f = X_{f+g}$ where g is a C^1 function of time. ∎

Immediately, if (t, q, p) are canonical (or natural) coordinates, on the domain of such chart:

$$\{q^i, q^j\} = \{p_i, p_j\} = 0 \quad \text{and} \quad \{q^i, p_j\} = \delta^i_j . \quad \text{for all } i, j = 1, \ldots, n .$$

Functions $f^i, g_j, i, j = 1, \ldots, n$, defined either on phase spacetime or just locally, that satisfy the above conditions are called **canonically conjugate**. This condition (which will play an important part in Quantum Mechanics, being related to *Heisenberg's uncertainty principle* [Mor18]) is actually *sufficient* to guarantee the functions define canonical coordinates, once the usual topological assumptions on the domains hold. Let us clarify this point.

Proposition 12.25 *Consider two local coordinate systems adapted to the fibres of the C^1 manifold $F(\mathbb{V}^{n+1})$*

$$\psi : U \ni a \mapsto (t(a), q(a), p(a)) \quad \text{and} \quad \psi' : U \ni a \mapsto (t'(a), q'(a), p'(a))$$

defined on a common domain U where $t' = t + c$ is absolute time. Suppose (U, ψ) is a local canonical coordinate system. The following are equivalent.

(1) (U, ψ') is a canonical local chart;
(2) for any $f, g : F(\mathbb{V}^{n+1}) \to \mathbb{R}$ of class C^1:

$$\{f, g\} = \sum_{k=1}^{n} \left(\frac{\partial f}{\partial q'^k} \frac{\partial g}{\partial p'_k} - \frac{\partial f}{\partial p'_k} \frac{\partial g}{\partial q'^k} \right) ;$$

(3) We have:

$$\{q'^i, q'^j\} = \{p'_i, p'_j\} = 0 \quad \text{and} \quad \{q'^i, p'_j\} = \delta^i_j , \quad \text{for all } i, j = 1, \ldots, n .$$
$$(12.33)$$

Proof (1) implies (2) because X_f looks the same in all canonical coordinates, as shown earlier. (2) implies (3) by computation, as remarked after Definition 12.22 when we wrote the Poisson bracket in primed canonical coordinates. Let us show that (3) implies (1). The domain of primed and unprimed coordinates is the same, so connected and locally simply connected. Then we just need to prove that primed coordinates are canonical transforms of natural coordinates. As the composite of canonical transformations is canonical, and unprimed coordinates are canonical from the outset, it suffices to show that the change from primed to unprimed is canonical. We can put all relations in (12.33) together in one as:

$$\sum_{k,h=1}^{2n} S^{kh} \frac{\partial x'^i}{\partial x^k} \frac{\partial x'^j}{\partial x^h} = S^{ij} ,$$

or more compactly: $J^t S J = S$. But this means the coordinate transformation is canonical, so the (t', q', p') are canonical coordinates. □

The next proposition summarises further features of the Poisson bracket.

Proposition 12.26 *The Poisson bracket enjoys the following properties.*

(1) $f, g \mapsto \{f, g\}$ *is a* bilinear skew-symmetric *function mapping* $C^k(F(\mathbb{V}^{n+1}))$ *scalar fields to* $C^{k-1}(F(\mathbb{V}^{n+1}))$ *scalar fields* $(k > 0)$.
(2) *The* **Leibniz rule** *holds:*

$$\{f \cdot g, h\} = f \cdot \{g, h\} + g \cdot \{f, h\}$$

where $\cdot$ *is the pointwise multiplication and* f, g, h *are* C^1 *scalar fields on on* $F(\mathbb{V}^{n+1})$.
(3) *The* **Jacobi identity** *holds:*

$$\{\{f, g\}, h\} + \{\{h, f\}, g\} + \{\{g, h\}, f\} = 0$$

for any C^2 *scalar fields* f, g, h *on* $F(\mathbb{V}^{n+1})$.
(4) *If* $[\cdot, \cdot]$ *is the Lie bracket of vector fields on* $F(\mathbb{V}^{n+1})$ *and* f, g *are* C^2 *scalar fields on* $F(\mathbb{V}^{n+1})$:

$$[X_f, X_g] = X_{\{f,g\}}.$$

Remarks 12.27

(1) The set $C^\infty(F(\mathbb{V}^{n+1}))$ of smooth real-valued functions defined on phase spacetime is a real vector space. The structure is induced by pointwise linear combinations: if $a, b \in \mathbb{R}$ and $f, g \in C^\infty(F(\mathbb{V}^{n+1}))$ then $(af + bg)(p) = af(p) + bg(p)$ for any $p \in F(\mathbb{V}^{n+1})$. Adding the Poisson bracket endows $C^\infty(F(\mathbb{V}^{n+1}))$ with a *Lie algebra* structure (see Definition B.2), since the Poisson bracket is bilinear, skew-symmetric on $C^\infty(F(\mathbb{V}^{n+1}))$, and satisfies the Jacobi identity.
(2) If M is a C^∞ manifold and $\{\cdot, \cdot\} : C^\infty(M) \times C^\infty(M) \to C^\infty(M)$ a bilinear, skew-symmetric map satisfying the Leibniz rule and the Jacobi identity, we say $(M, \{\cdot, \cdot\})$ is a **Poisson manifold**. This type of structure is more general than those born within the Hamiltonian formulation. ∎

Proof of Proposition 12.26 Properties (1) and (2) are obvious, so let us prove (3). We have:

$$\{\{f, g\}, h\} = \sum_{ijkr} S^{kr} S^{ij} \frac{\partial^2 f}{\partial x^k \partial x^i} \frac{\partial g}{\partial x^j} \frac{\partial h}{\partial x^r} + \sum_{ijkr} S^{kr} S^{ij} \frac{\partial f}{\partial x^i} \frac{\partial^2 g}{\partial x^k \partial x^j} \frac{\partial h}{\partial x^r}.$$

The other two summands in $\{\{f, g\}, h\} + \{\{h, f\}, g\} + \{\{g, h\}, f\} = 0$ are obtained from the above by permuting f, g, h cyclically. It suffices to show that, in the sum $\{\{f, g\}, h\} + \{\{h, f\}, g\} + \{\{g, h\}, f\}$, the terms with the second derivatives of f cancel each other out, so the same will happen by symmetry to the second-order derivatives of g and h. The first term with second derivatives of f appears above in $\{\{f, g\}, h\}$. There is another one in $\{\{h, f\}, g\}$ coming from the second summand in the expansion of $\{\{f, g\}, h\}$ under the permutation $f, g, h \to h, f, g$. So, we need to check that:

$$\sum_{ijkr} S^{kr} S^{ij} \frac{\partial^2 f}{\partial x^k \partial x^i} \frac{\partial g}{\partial x^j} \frac{\partial h}{\partial x^r} + \sum_{ijkr} S^{kr} S^{ij} \frac{\partial h}{\partial x^i} \frac{\partial^2 f}{\partial x^k \partial x^j} \frac{\partial g}{\partial x^r} = 0. \tag{12.34}$$

The second term on the left of (12.34), after relabelling indices ($k \leftrightarrow i$ then $i \leftrightarrow j$) reads:

$$\sum_{ijkr} S^{kj} S^{ri} \frac{\partial h}{\partial x^r} \frac{\partial^2 f}{\partial x^k \partial x^i} \frac{\partial g}{\partial x^j} = \sum_{ijkr} S^{kj} S^{ri} \frac{\partial h}{\partial x^r} \frac{\partial^2 f}{\partial x^i \partial x^k} \frac{\partial g}{\partial x^j},$$

where at the end we used Schwarz's theorem (that is why we needed C^2 and not just twice differentiable). Swapping i and k the second summand in (12.34) is:

$$\sum_{ijkr} S^{ij} S^{rk} \frac{\partial h}{\partial x^r} \frac{\partial^2 f}{\partial x^k \partial x^i} \frac{\partial g}{\partial x^j} = -\sum_{ijkr} S^{ij} S^{kr} \frac{\partial h}{\partial x^r} \frac{\partial^2 f}{\partial x^k \partial x^i} \frac{\partial g}{\partial x^j},$$

where in the last passage we used the skew-symmetry of S. Hence:

$$\sum_{ijkr} S^{kr} S^{ij} \frac{\partial h}{\partial x^i} \frac{\partial^2 f}{\partial x^k \partial x^j} \frac{\partial g}{\partial x^r} = -\sum_{ijkr} S^{ij} S^{kr} \frac{\partial h}{\partial x^r} \frac{\partial^2 f}{\partial x^k \partial x^i} \frac{\partial g}{\partial x^j}.$$

Substituting in (12.34), we obtain (12.34).

To finish we will prove (4). If h is a scalar field on $F(\mathbb{V}^{n+1})$ of class C^2 at least:

$$[X_f, X_g]h = X_f X_g h - X_g X_f h = X_f \{g, h\} - X_g \{f, h\} = \{f, \{g, h\}\} - \{g, \{f, h\}\}$$

$$= -\{\{g, h\}, f\} - \{\{h, f\}, g\} = \{\{f, g\}, h\} = X_{\{f,g\}} h .$$

But h is arbitrary, so: $[X_f, X_g] = X_{\{f,g\}}$. $\qquad\square$

Recalling Definition B.2 of Lie algebra and Lie algebra homomorphism, we deduce the following important proposition. In the sequel $V(F(\mathbb{V}^{n+1}))$ will indicate the real vector space of C^∞ vector fields on $F(\mathbb{V}^{n+1})$ with linear structure $(aX + bY)(p) = aX(p) + bY(p)$ for any $p \in F(\mathbb{V}^{n+1})$, $a, b \in \mathbb{R}$ and $X, Y \in V(F(\mathbb{V}^{n+1}))$.

Proposition 12.28 *Consider phase spacetime* $F(\mathbb{V}^{n+1})$ *as* C^∞ *manifold, and take on it:*

(i) *the Lie algebra* $V(F(\mathbb{V}^{n+1}))$ *of* C^∞ *vector fields with commutator as Lie bracket,*

(ii) *the Lie algebra* $C^\infty(F(\mathbb{V}^{n+1}))$ *of real* C^∞ *functions with the Poisson bracket.*

The following hold.

(1) The map

$$L : C^\infty(F(\mathbb{V}^{n+1})) \ni f \mapsto X_f \in V(F(\mathbb{V}^{n+1}))$$

is a homomorphism of Lie algebras.

(2) If the bundle's projection $T : F(\mathbb{V}^{n+1}) \to \mathbb{R}$ *is absolute time, then*

$$Ker(L) \supset \{g \circ T \mid g \in C^\infty(\mathbb{R})\} ,$$

(3) If the generic phase space (the standard fibre of $T : F(\mathbb{V}^{n+1}) \to \mathbb{R})$ *is connected, then:*

$$Ker(L) = \{g \circ T \mid g \in C^\infty(\mathbb{R})\} .$$

(4) If $Z_{C^\infty(F(\mathbb{V}^{n+1}))}$ *denotes the* **centre** *of the Lie algebra* $C^\infty(F(\mathbb{V}^{n+1}))$:

$$Z_{C^\infty(F(\mathbb{V}^{n+1}))} := \{f \in C^\infty(F(\mathbb{V}^{n+1})) \mid \{f, g\} = 0 \quad \forall g \in C^\infty(F(\mathbb{V}^{n+1}))\}$$

then

$$Z_{C^\infty(F(\mathbb{V}^{n+1}))} = Ker(L) .$$

Proof (1)–(3) are immediate from the previous proposition and Proposition 12.23. As for (4), $X_f = 0$ means, by definition of vector field, $X_f(g) = 0$ for any $g \in C^\infty(F(\mathbb{V}^{n+1}))$. By definition of Poisson bracket, $X_f(g) = 0$ is the same as $\{f, g\} = 0$ for any $g \in C^\infty(F(\mathbb{V}^{n+1}))$. ∎

Remarks 12.29

(1) If the phase space at a given time is connected, $Ker(L)$ consists exactly of the functions on $F(\mathbb{V}^{n+1})$ that only depend on absolute time t.

(2) Properties (2), (3) and (4) hold in more generality when $L : f \mapsto X_f$ maps $C^1(F(\mathbb{V}^{n+1}))$ functions to C^0 vector fields on $F(\mathbb{V}^{n+1})$, because the proof of (2)–(4) only needs that regularity. Asking smoothness is only necessary to define the Lie algebra structure coherently: the Poisson bracket $\{f, g\}$ is required to be as differentiable as f and g. ∎

12.3.2　Local One-Parameter Groups of Active Canonical Transformations

We will show how to associate with every scalar field $f : F(\mathbb{V}^{n+1}) \rightarrow \mathbb{R}$ a local one-parameter group of *active* canonical transformations. Active means the transformations are not coordinate changes, but transformations that displace the representative points of the physical system. Consider $f \in C^2(F(\mathbb{V}^{n+1}))$ and the associated field X_f. We want to find its integral curves and use them to construct the 1-parameter group of local diffeomorphisms $\Phi^{(X_f)}$ associated with X_f. By definition (see Sect. 14.5.3), since X_f is defined everywhere on $F(\mathbb{V}^{n+1})$, the flow $\Phi^{(X_f)}$ acts on the *entire* phase spacetime, irrespective of particular local coordinates. That said, in the sequel it will be convenient to study $\Phi^{(X_f)}$ in a given system of local coordinates. So, given one such local system on an open set U in phase spacetime $U \ni a \mapsto (t(a), \mathbf{x}(a)) \in \mathbb{R}^{2n+1}$, first of all we have to solve the first-order differential equation, in coordinates:

$$X_f(t(u), \mathbf{x}(u)) = \frac{dx^k}{du}\frac{\partial}{\partial x^k} + \frac{dt}{du}\frac{\partial}{\partial t}$$

for the curve $(t, \mathbf{x}) = (t(u), \mathbf{x}(u))$, $u \in \mathbb{R}$. Explicitly:

$$\sum_{k=1}^{n} S^{ik}\frac{\partial f}{\partial x^i}\bigg|_{(t(u), \mathbf{x}(u))}\frac{\partial}{\partial x^k} = \frac{dx^k}{du}\frac{\partial}{\partial x^k} + \frac{dt}{du}\frac{\partial}{\partial t}.$$

The function $t = t(u)$ must be constant. Hence the initial differential equation, *instant by instant*, becomes:

$$S^t \nabla_{\mathbf{x}} f|_{(t, \mathbf{x}(u))} = \frac{d\mathbf{x}}{du}. \tag{12.35}$$

The solution curves $(t, \mathbf{x}) = (t(u), \mathbf{x}(u))$ in phase spacetime remain in the same fibre where they started from. In other words, the *flow* $\Phi^{(X_f)}$ of X_f preserves the *fibres* of phase spacetime. Fix an instant t once and for all and consider the local one-parameter group $\varphi^{(t, X_f)}$ on $\mathbb{F}_t$ associated with the autonomous equation (t is now given, and is *not* the curve's parameter!). By definition

$$u \mapsto \varphi_u^{(t, X_f)}(\mathbf{x})$$

is the only local solution of (12.35) passing through $\mathbf{x}$ at $u = 0$.

Remarks 12.30 The function $(u, \mathbf{x}) \mapsto \varphi_u^{(t, X_f)}(\mathbf{x})$ is jointly C^k in its variables, and C^{k+1} in u, where k is the regularity degree of X_f as per Sect. 14.5.3. To avoid discussing regularity further we could assume in the sequel that all maps are C^∞. ∎

We aim to show that for any given value of u the locally defined transformation

$$(t, \mathbf{x}) \mapsto (t', \mathbf{x}') := \Phi_u^{(X_f)}(t, \mathbf{x}) = (t, \varphi_u^{(t, X_f)}(\mathbf{x}))$$

(depending on the value of u) *is canonical, meaning its Jacobian matrix in non-temporal coordinates belongs to* $Sp(n, \mathbb{R})$. For t given define the Jacobian matrix J, dependent on $\mathbf{x}$ and u, of coefficients:

$$J^i{}_k := \frac{\partial x'^i}{\partial x^k}.$$

Consider the only solution $\mathbf{x}' = \mathbf{x}'(u)$ to (12.35) joining $\mathbf{x}$, for $u = 0$, and $\mathbf{x}'_1$, for $u = u_1$. Thus $\mathbf{x}' = \varphi_{u_1}^{(t, X_f)}(\mathbf{x})$. We want to find the derivative in u of

$$u \mapsto (J^t S J)_{\mathbf{x}'(u)}$$

along the curve, and show it vanishes. This will imply $J^t S J$ is constant along the curve and therefore the Jacobian of $\mathbf{x}' = \varphi_{u_1}^{(t, X_f)}(\mathbf{x})$ satisfies $J^t S J = S$, since that happens for $u = 0$. Note that

$$\frac{d}{du}\left(J^t S J\right) = \left(\frac{dJ^t}{du} S J\right) + J^t S \frac{dJ}{du}. \tag{12.36}$$

As

$$\frac{d}{du}(J_{\mathbf{x}'(u)})^i{}_k = \frac{d}{du}\frac{\partial x'^i}{\partial x^k}\bigg|_{\mathbf{x}'(u)} = \frac{\partial^2 x'^i}{\partial u \partial x^k} = \frac{\partial}{\partial x^k}\frac{dx'^i}{du} = \frac{\partial X_f^i}{\partial x^k},$$

writing the right-hand side of (12.36) in components:

$$\sum_{i,k=1}^n \frac{\partial X_f^i}{\partial x^j} S_{ik} \frac{\partial x'^k}{\partial x^l} + \sum_{i,k=1}^n \frac{\partial x'^i}{\partial x^j} S_{ik} \frac{\partial X_f^k}{\partial x^l} = \sum_{i,k,p=1}^n S^{pi} \frac{\partial^2 f}{\partial x^j \partial x'^p} S_{ik} \frac{\partial x'^k}{\partial x^l}$$

$$+ \sum_{i,k,s=1}^n \frac{\partial x'^i}{\partial x^j} S_{ik} S^{sk} \frac{\partial^2 f}{\partial x'^l \partial x'^s}.$$

From (11.41) this simplifies to:

$$-\sum_{k,p=1}^{n} \delta_k^p \frac{\partial^2 f}{\partial x^j \partial x'^p} \frac{\partial x'^k}{\partial x^l} + \sum_{i,s=1}^{n} \frac{\partial x'^i}{\partial x^j} \delta_i^s \frac{\partial^2 f}{\partial x^l \partial x'^s}$$

$$= -\sum_{k=1}^{n} \frac{\partial^2 f}{\partial x^j \partial x'^k} \frac{\partial x'^k}{\partial x^l} + \sum_{s=1}^{n} \frac{\partial x'^s}{\partial x^j} \frac{\partial^2 f}{\partial x^l \partial x'^s}$$

$$= -\frac{\partial^2 f}{\partial x^j \partial x^l} + \frac{\partial^2 f}{\partial x^l \partial x^j} = 0.$$

But for $u = 0$ we have $J = I$, so $J^t S J = S$. We conclude that $J^t S J = S$ for any u, as claimed. That holds in particular for $u = u_1$. Eventually, the transformation

$$(t, \mathbf{x}) \mapsto (t', \mathbf{x}') = \left(t, \varphi_u^{(t, X_f)}(\mathbf{x}) \right)$$

is canonical everywhere it is defined. We have then proved the following proposition.

Proposition 12.31 *The flow* $\Phi^{(X_f)}$ *of the vector field* X_f *associated with a scalar field* $f \in C^2(F(\mathbb{V}^{n+1}))$ *consists of* **active canonical transformations** *in the following sense. If the open set* $U \subset F(\mathbb{V}^{n+1})$ *is the domain of a local canonical chart with coordinates* $(t, \mathbf{x})$, *and* $\Phi_u^{(X_f)}(U') \subset U$ *for some open set* $U' \subset U$ *and some* $u \in \mathbb{R}$, *then*

$$\Phi_u^{(X_f)} : U' \ni (t, \mathbf{x}) \mapsto (t', \mathbf{x}') \in U$$

is such that:

(1) $t' = t$
(2) the Jacobian matrix of elements $\frac{\partial x'^a}{\partial x^b}$ *belongs to* $Sp(n, \mathbb{R})$.

Definition 12.32 The one-parameter group of local diffeomorphisms $\Phi^{(X_f)}$ generated by the Hamiltonian field X_f associated with f is called **local one-parameter group of active canonical transformations generated by** f, and f is its **generating function.** ◇

Examples 12.33 Consider a particle of mass m, moving on a plane with polar coordinates (r, θ). Consider $f(r, \theta, p_r, p_\theta) := p_\theta$ and the associated one-parameter group of canonical transformations. This is obtained integrating the differential equation:

$$S^t \nabla_{\mathbf{x}} f = \frac{d\mathbf{x}}{du}$$

where $\mathbf{x} := (r, \theta, p_r, p_\theta)$. The computation reduces the equation to:

$$\frac{dr}{du} = 0, \quad \frac{d\theta}{du} = -1, \quad \frac{dp_r}{du} = 0, \quad \frac{dp_\theta}{du} = 0.$$

All in all the group $\varphi^{X_{p_\theta}}$ acts by:

$$\varphi_u^{(X_{p_\theta})}(r, \theta, p_r, p_\theta) = (r, \theta - u, p_r, p_\theta) =: (r', \theta', p_r', p_\theta'), \tag{12.37}$$

so on the configuration space it corresponds to rotations by $-u$ around the origin.

If for a physical system we interpret *passively* the canonical transformation induced by f, we may ask what might be the Hamiltonian $\mathscr{H}'(t', \mathbf{x}')$ associated with the coordinates $(t', \mathbf{x}')$ when the Hamiltonian $\mathscr{H}(t, \mathbf{x})$ in the initial coordinates $(t, \mathbf{x})$ is known. As the coordinate transformation is active, i.e. the coordinates stay the same but the representative points in phase spacetime are transformed, the new $\mathscr{H}'$ should be understood as Hamiltonian of a *new* physical system: the system obtained through the active transformation generated by f (for instance, the physical system's rotation by an angle u around the z-axis, induced by the map $f(r, \theta, p_r, p_\theta) := p_\theta$ of the previous example). The point is that the new system admits a Hamiltonian description because the coordinate change is a canonical map. Finally, note that in general the Hamiltonian $\mathscr{H}'$ will not be of the same form as $\mathscr{H}$. In this sense the new physical system will be *different* from the initial one.

What we will do now is find $\mathscr{H}'$ up to second-order infinitesimals of u, and so the finite expression. This form will be useful to link Hamiltonian symmetries and conserved quantities.

Proposition 12.34 *Consider a C^2 Hamiltonian $\mathscr{H} = \mathscr{H}(t, \mathbf{x})$ in canonical/natural coordinates on $F(\mathbb{V}^{n+1})$,*

$$\psi : U \ni a \mapsto (t(a), \mathbf{x}(a)) \in \mathbb{R}^{2n+1},$$

and suppose the Hamiltonian system is acted upon by a one-parameter group of canonical transformations generated by the C^3 map $f : U \to \mathbb{R}$, so that in coordinates

$$\psi(U) \ni (t, \mathbf{x}) \mapsto (t_u', \mathbf{x}_u'(t, \mathbf{x})), \quad \text{where } t_u' = t,$$

whenever the right-hand side belongs to $\psi(U)$.

For any u, associate to the system a new C^2 Hamiltonian $\mathscr{H}_u'$ so that, interpreting the canonical transformation passively, $\mathscr{H}_u'(t_u', \mathbf{x}_u')$ is related to $\mathscr{H}(t, \mathbf{x})$ by (1) in Theorem 12.9.

The following facts hold.

(1) The Hamiltonian $\mathscr{H}'_u$ can be locally chosen so that:

$$\mathscr{H}'_u(t'_u, \mathbf{x}'_u(t, \mathbf{x})) = \mathscr{H}(t, \mathbf{x}) - u \frac{\partial f}{\partial t}(t, \mathbf{x}) + u O_{(t,\mathbf{x})}(u) , \tag{12.38}$$

where $(u, t, \mathbf{x}) \mapsto u O_{(t,\mathbf{x})}(u)$ is a C^2 function such that $O_{(t,\mathbf{x})}(u) \to 0$ as $u \to 0$ for any given $(t, \mathbf{x})$.

(2) With that choice the Hamiltonian of the transformed system (locally) satisfies:

$$\mathscr{H}'_u(t, \mathbf{y}) = \mathscr{H}(t, \mathbf{x}'_{-u}(t, \mathbf{y})) - \int_0^u \frac{\partial}{\partial t} f(t, \mathbf{x}'_{s-u}(t, \mathbf{y})) ds , \tag{12.39}$$

where the partial derivative is taken with respect to the first variable t in f. In particular, the right-most term in (12.39) vanishes when f does not depend explicitly on t.

Proof

(1) We will set $t'_u = t$ in the sequel because the canonical transformation does not affect time. We will also drop the subscript u whenever not necessary. From Theorem 12.9 we have:

$$\mathscr{H}'(t, \mathbf{x}') = \mathscr{H}(t, \mathbf{x}(t, \mathbf{x}')) + \mathscr{K}(t, \mathbf{x}')$$

coming from

$$\frac{\partial \mathbf{x}'}{\partial t} = S \nabla_{\mathbf{x}'} \mathscr{K}(t, \mathbf{x}') ,$$

i.e.

$$\nabla_{\mathbf{x}'} \mathscr{K}(t, \mathbf{x}') = -S \frac{\partial \mathbf{x}'}{\partial t} .$$

Proposition 12.13 says that the above $\mathscr{H}'(t, \mathbf{x}')$ is determined up to additive functions in t. Let us expand with Taylor in the variable u the function $\partial \mathbf{x}'/\partial t$, where $\mathbf{x}' = \mathbf{x}'_u(t, \mathbf{x})$. Note that $\mathbf{x}'_0(t, \mathbf{x}) = \mathbf{x}$ at $u = 0$, so the partial derivative in t is zero. Then:

$$\frac{\partial \mathbf{x}'}{\partial t} = u \left. \frac{\partial}{\partial u} \frac{\partial \mathbf{x}'}{\partial t} \right|_{u=0} + u O_{(t,\mathbf{x})}(u) ,$$

where $(u, t, \mathbf{x}) \mapsto O_{(t,\mathbf{x})}(u)$ is a continuous function defined on an open set such that, for any given $(t, \mathbf{x})$, $O_{(t,\mathbf{x})}(u) \to 0$ as $u \to 0$. In other words, swapping the derivatives and using the definition of X_f:

$$\frac{\partial \mathbf{x}'}{\partial t} = u\frac{\partial}{\partial t} S^t \nabla_{\mathbf{x}} f + u O_{(t,\mathbf{x})}(u) \,.$$

Applying $-S$, recalling $SS^t = I$ and observing that $SO_{(t,\mathbf{x})}(u)$ is infinitesimal as $u \to 0$ (in the sequel $-SO$ will become O), we find:

$$\nabla_{\mathbf{x}'} \mathscr{K}(t, \mathbf{x}') = -u\frac{\partial}{\partial t} \nabla_{\mathbf{x}} f(t, \mathbf{x}) + u O_{(t,\mathbf{x})}(u) \,.$$

On the right we can replace $\nabla_{\mathbf{x}} f(t, \mathbf{x})$ with $\nabla_{\mathbf{x}'} f(t, \mathbf{x}')$ up to errors in u of order higher than one, as is easy to see.[10] To sum up:

$$\nabla_{\mathbf{x}'} \mathscr{K}(t, \mathbf{x}') = -\nabla_{\mathbf{x}'} u \frac{\partial f(t, \mathbf{x}')}{\partial t} + u O_{(t,\mathbf{x}')}(u) \,,$$

where, abusing notation, the above O is the O on the previous line where we wrote the infinitesimal in primed coordinates without changing the behaviour as $u \to 0$. If we work in a simply connected domain (at given t) we may choose:

$$\mathscr{K}(t, \mathbf{x}) = -u\frac{\partial f(t, \mathbf{x})}{\partial t} + u \int_{\mathbf{x}_0}^{\mathbf{x}} O_{(t,\mathbf{y})}(u) \cdot d\mathbf{y}$$

where the integral is computed along any differentiable curve between some $\mathbf{x}_0$ and $\mathbf{x}$ inside the domain. The map $(u, t, \mathbf{x}) \mapsto O_{(t,\mathbf{y})}(u)$ is continuous, hence bounded on compact sets. As $O_{(t,\mathbf{y})}(u) \to 0$ when $u \to 0$ at given $(t, \mathbf{y})$, the Dominated convergence theorem implies:

$$\int_{\mathbf{x}_0}^{\mathbf{x}} O_{(t,\mathbf{y})}(u) \cdot d\mathbf{y} \to 0 \quad \text{as } u \to 0 \,.$$

Redefining once more $O_{(t,\mathbf{x})}(u)$, we obtain (12.38). Observe that $(u, t, \mathbf{x}) \mapsto u O_{(t,\mathbf{x})}(u)$ in (12.38) is certainly C^2 as difference of C^2 maps.

(2) Now we can find the finite expression for (12.38). This formula (recalling that $t' = t$, viewing $(t, \mathbf{x})$ fixed and reinstating the subscript u) can be written as:

$$\frac{d}{du} \mathscr{H}'_u(t, \mathbf{x}'_u(t, \mathbf{x})) = -\frac{\partial f}{\partial t}(t, \mathbf{x}'_u(t, \mathbf{x})) \,,$$

---

[10] The Taylor expansion at $u = 0$ is $u\nabla_{\mathbf{x}} f(t, \mathbf{x}') = u\left(\nabla_{\mathbf{x}} f(t, \mathbf{x}) + u H(t, \mathbf{x})\right) = u\nabla_{\mathbf{x}} f(t, \mathbf{x}) + u O(u)$. The same happens replacing $\nabla_{\mathbf{x}}$ with $\nabla_{\mathbf{x}'}$.

so

$$\mathcal{H}'_u(t, \mathbf{x}'_u(t, \mathbf{x})) = \mathcal{H}(t, \mathbf{x}) - \int_0^u \frac{\partial f}{\partial t}(t, \mathbf{x}'_s(t, \mathbf{x})) ds \ .$$

This implies (12.39) if we put $\mathbf{x} := \mathbf{x}'_{-u}(t, \mathbf{y})$.

□

Remarks 12.35 Without further assumptions, by Proposition 12.13 the transformed Hamiltonian $\mathcal{H}'_u$ can only be redefined locally, with respect to (12.38), by adding a function of u and t (with the proper regularity). At any rate, formula (12.44) will furnish the true, most general formula when we add other reasonable hypotheses. ∎

Next we introduce Hamiltonian systems that are invariant under a 1-parameter group of canonical transformations.

Definition 12.36 Assume the previous proposition's hypotheses and that $\mathcal{H}'_u$ is subject to (12.38). Suppose

$$\mathcal{H}'_u(t, \mathbf{y}) = \mathcal{H}(t, \mathbf{y}) + u O_{(t, \mathbf{y})}(u) \tag{12.40}$$

for any $(t, \mathbf{y}) \in U$ where the two sides are both defined, and that $O_{(t, \mathbf{y})}(u) \to 0$ as $u \to 0$ for any $(t, \mathbf{y})$. Then we say the Hamiltonian $\mathcal{H}$ is **formally invariant to order one** under the local one-parameter group of canonical transformations generated by f. The one-parameter group itself is called a **Hamiltonian symmetry**. ◇.

The meaning should be physically obvious: if the Hamiltonian is *formally invariant*, the Hamiltonian $\mathcal{H}'(t, \mathbf{y})$ arising *from* the (infinitesimal) active action of the canonical transformation on the physical system has the same form $\mathcal{H}(t, \mathbf{y})$ that is has *before* the canonical transformation's action (up to infinitesimals of order higher than one). The active canonical transformation is in this sense a *Hamiltonian symmetry* of the physical system. We will see in the next section that in such a situation the presence of symmetries is very tightly related to the existence of conserved quantities (first integrals).

Remarks 12.37

(1) First-order formal invariance can be written as:

$$\left. \frac{\partial \mathcal{H}'_u(t, \mathbf{x})}{\partial u} \right|_{u=0} = 0 \ . \tag{12.41}$$

Exactly as for the 1-parameter groups in Noether's theorem, if the domain U of canonical coordinates is invariant under the local group of active canonical transformations considered, *the Hamiltonian's formal invariance to first order is equivalent to formal invariance to any order:*

$$\mathcal{H}'_u(t, \mathbf{x}) = \mathcal{H}(t, \mathbf{x}), \quad \text{for any } u \in \mathbb{R} \text{ and any } (t, \mathbf{x}) \in U.$$

The proof is based on a calculation performed on (12.39), from which:

$$(\mathcal{H}'_u)'_v(t, \mathbf{y}) = \mathcal{H}'_{u+v}(t, \mathbf{y}); \qquad (12.42)$$

In other words, letting the one-parameter group of canonical transformations act on the system twice consecutively, is equivalent to letting it act once with parameter value equal to the sum of the parameters, at least as far as the Hamiltonian is concerned. Then (12.41) implies that, for any u in an interval containing $u = 0$ that might depend on $(t, \mathbf{x})$:

$$\frac{\partial \mathcal{H}'_u(t, \mathbf{x})}{\partial u} = 0.$$

and therefore

$$\mathcal{H}'_u(t, \mathbf{x}) = \mathcal{H}'_0(t, \mathbf{x}) = \mathcal{H}(t, \mathbf{x})$$

whenever the two sides exist simultaneously. This is true in particular for any $u \in \mathbb{R}$ and on the entire U when the latter domain is invariant under the symmetry group.

(2) We have already noticed that if the generating function f of the active canonical transformations does not depend explicitly on time, the term containing the time derivative of f in (12.38) is zero. Therefore, the first-order formal invariance of $\mathcal{H}$ coincides with the invariance under such group up to order one. That can be written as

$$X_f(\mathcal{H}) = 0$$

on the chart where $\mathcal{H}$ is defined. Finally, formula (12.39) shows that when $\frac{\partial f}{\partial t} = 0$, first-order formal invariance coincides with true invariance as long as we work with small parameters and on sufficiently small domains. ∎

In Remarks 12.35 we explained that the only way to extend formula (12.43) is

$$\mathcal{H}'_u(t, \mathbf{y}) = \mathcal{H}(t, \mathbf{x}'_{-u}(t, \mathbf{y})) - \int_0^u \frac{\partial}{\partial t} f(t, \mathbf{x}'_{s-u}(t, \mathbf{y}))ds + h(u, t), \qquad (12.43)$$

where $h(u, t)$ is any C^2 map in t, if we assume, as is the case, that the Hamiltonians are at least C^2. But if we impose (12.42) the only possibility, supposing the dependency on u is differentiable, is

$$\mathscr{H}_u'(t, \mathbf{y}) = \mathscr{H}(t, \mathbf{x}_{-u}'(t, \mathbf{y})) - \int_0^u \frac{\partial}{\partial t} f(t, \mathbf{x}_{s-u}'(t, \mathbf{y})) ds + uh(t) , \qquad (12.44)$$

where h is C^2. The proof is easy and left to the reader as exercise. The possible additional term $uh(t)$ can be absorbed into f, by redefining the latter:

$$f(t, \mathbf{x}) \to f_1(t, \mathbf{x}) := f(t, \mathbf{x}) + \int_{t_0}^t h(\tau) d\tau .$$

Since f and f_1 differ by a function of time, the Hamiltonian vector fields generated by f and f_1 coincide, and therefore generate the same local one-parameter group of canonical transformations.

In the sequel we shall assume (12.39), or its infinitesimal version (12.38), for the Hamiltonian transformed under a one-parameter group of canonical transformations.

12.3.3 Symmetries and Conservation Laws: The Hamiltonian Noether Theorem

Suppose we are given Hamiltonian functions $\mathscr{H} = \mathscr{H}(t, q, p)$ on $F(\mathbb{V}^{n+1})$, in every coordinate system of the natural atlas, that transform as (11.29). The system's dynamics is then global and the evolution curves are precisely the integral curves of the Hamiltonian dynamic vector field Z, by Definition 11.25. The vector Z, in turn, induces Hamiltonians on all canonical charts, so (11.29) holds in each one of these.

Let us write, in natural or canonical coordinates, the condition that $f : F(\mathbb{V}^{n+1}) \to \mathbb{R}$ is a first integral of the Hamiltonian system with the prescribed Hamiltonians:

$$\frac{d}{dt} f(t, \mathbf{x}(t)) = 0 , \qquad (12.45)$$

where $\mathbf{x} = \mathbf{x}(t)$ is any solution to Hamilton's equations:

$$\frac{d\mathbf{x}}{dt} = S\nabla_{\mathbf{x}} \mathscr{H}(t, \mathbf{x}(t)) .$$

Expanding the right-hand side of (12.45), we write:

$$\left. \frac{\partial f}{\partial t} \right|_{(t,\mathbf{x}(t))} + \frac{d\mathbf{x}}{dt} \cdot \nabla f \bigg|_{(t,\mathbf{x}(t))} = 0 ,$$

and using Hamilton's equations:

$$\frac{\partial f}{\partial t}\bigg|_{(t,\mathbf{x}(t))} + \sum_{i,j=1}^{2n} S^{ij}\frac{\partial \mathscr{H}}{\partial x^j}\frac{\partial f}{\partial x^i}\bigg|_{(t,\mathbf{x}(t))} = 0.$$

We discovered that f is a first integral if and only if:

$$\left[\frac{\partial f}{\partial t} + \{f, \mathscr{H}\}\right]\bigg|_{(t,\mathbf{x}(t))} = 0$$

on every solution $\mathbf{x} = \mathbf{x}(t)$ to Hamilton's equations. On the other hand, if the Hamiltonian is C^2 and so Hamilton's equations have (unique) solutions, we conclude that for any event $(t, \mathbf{x})$ in phase space there certainly is a solution starting at that event. This means that in order for f to be a first integral it is necessary, and sufficient, that:

$$\frac{\partial f}{\partial t} + \{f, \mathscr{H}\} = 0$$

everywhere on phase spacetime (in any local system of canonical coordinates).

Remarks 12.38 In the above formula the solutions of the equations of motion have disappeared. By this relation, therefore, there is no need to solve the equations of motion to decide whether a function is a first integral or not. ∎

Let us frame this result and add another important result.

Theorem 12.39 *Consider the Hamiltonian dynamic vector field Z on $F(\mathbb{V}^{n+1})$ inducing Hamiltonians $\mathscr{H} = \mathscr{H}(t, q, p)$, as per (11.29), in any coordinates of the natural (or equivalently, canonical) atlas. Suppose the Hamiltonians are C^2. Then the following hold.*

(1) A C^1 function $f : F(\mathbb{V}^{n+1}) \to \mathbb{R}$ is a first integral if and only if:

$$\frac{\partial f}{\partial t} + \{f, \mathscr{H}\} = 0 \tag{12.46}$$

everywhere on $F(\mathbb{V}^{n+1})$.

(2) If f is C^2 and $g : F(\mathbb{V}^{n+1}) \to \mathbb{R}$ a C^2 first integral, the scalar field on $F(\mathbb{V}^{n+1})$ given by the Poisson bracket $\{f, g\}$ is another first integral.

Proof The first claim was proved earlier. Regarding the second:

$$\frac{\partial \{f, g\}}{\partial t} + \{\{f, g\}, \mathscr{H}\} = \left\{\frac{\partial f}{\partial t}, g\right\} + \left\{f, \frac{\partial g}{\partial t}\right\} - \{\{\mathscr{H}, f\}, g\} - \{\{g, \mathscr{H}\}, f\},$$

where we used the Jacobi identity at the end. Since the Poisson bracket is bilinear and skew:

$$\frac{\partial\{f, g\}}{\partial t} + \{\{f, g\}, \mathscr{H}\} = \left\{\frac{\partial f}{\partial t} + \{f, \mathscr{H}\}, g\right\} + \left\{f, \frac{\partial g}{\partial t} + \{g, \mathscr{H}\}\right\} = 0 .$$

But f and g are first integrals, so:

$$\frac{\partial g}{\partial t} + \{g, \mathscr{H}\} = 0 \quad \text{and} \quad \frac{\partial f}{\partial t} + \{f, \mathscr{H}\} = 0 .$$

By linearity $\{f, g\}$ satisfies (12.46), so it, too, is a first integral.   □

Remarks 12.40

(1) The theorem's first claim implies immediately that the Hamiltonian itself is a first integral if it does not depend on time in the canonical coordinates employed, as clearly $\{\mathscr{H}, \mathscr{H}\} = 0$ by skew-symmetry. In contrast to the general case, however, this fact holds only locally, because the Hamiltonian does not define a scalar field as we change charts on phase spacetime.

(2) The maximum number of functionally independent first integrals on $F(\mathbb{V}^{n+1})$ is $2n$. Indeed, functionally independent real maps can be used as local coordinates on $F(\mathbb{V}^{n+1})$—or on embedded submanifolds—by virtue of their independency and because $F(\mathbb{V}^{n+1})$ has precisely dimension $2n + 1$ and one dimension is the time parametrisation (see also Theorem 14.63). So when we repeatedly apply Poisson brackets on first integrals, at some point we will obtain again the same first integrals, or functions of them.

(3) The Liouville equation for a probability density ρ on phase spacetime can be expressed as:

$$\frac{\partial\rho}{\partial t} + \{\rho, \mathscr{H}\} = 0 .$$

     ■

Finally we can present the Hamiltonian theorem that links the presence of symmetries in a Hamiltonian system to the existence of first integrals, thus playing the same role as Noether's theorem in Lagrangian Mechanics. The statement has a global part (1)–(2) and a local one (3).

Theorem 12.41 (Hamiltonian Noether's Theorem) *Let Z be a Hamiltonian dynamic vector field on $F(\mathbb{V}^{n+1})$ with given C^2 Hamiltonians $\mathscr{H}$ on natural/canonical local charts as per (12.23).*

Let $\Phi^{(X_f)}$ denote the local one-parameter group of active canonical transformations associated with some $f \in C^2(F(\mathbb{V}^{n+1}))$. The following hold.

(1) If f is a first integral:

$$Z(f) = 0$$

then:

$$[X_f, Z] = 0 . \tag{12.47}$$

Equivalently, $\Phi^{(X_f)}$ maps solutions to Hamilton's equations (the integral curves of Z) to solutions.

(2) *Conversely, if (12.47) holds and the phase space at a given time is connected, there exists $f_1 \in C^2(F(\mathbb{V}^{n+1}))$ (which might be f itself) such that:*

(i) $\Phi^{(X_f)} = \Phi^{(X_{f_1})}$,

(ii) f_1 *is a first integral.*

If f_1 satisfies (i) and (ii) then $f_1 - f$ is a function of absolute time t only, and is unique up to additive constants.

(3) *Consider natural or canonical local coordinates on $F(\mathbb{V}^{n+1})$ with Hamiltonian $\mathcal{H}$ associated with Z. The following facts are equivalent.*

(i) *f restricted to the charts domains is a first integral.*

(ii) *Everywhere in the above coordinates:*

$$\{f, \mathcal{H}\} + \frac{\partial f}{\partial t} = 0 .$$

(iii) *$\Phi^{(X_f)}$ is a Hamiltonian symmetry: on the aforementioned domains $\mathcal{H}$ is formally invariant to order one under $\Phi^{(X_f)}$, in the sense of Proposition 12.34.*

Remarks 12.42 The statement is actually stronger that the 'Lagrangian' Noether theorem: in a Hamiltonian setup we have two extra facts that are absent, in this generality, in the Lagrangian formulation:

(a) the first integral associated with a symmetry (up to functions of time) is exactly the *symmetry's generating function*, in the sense of local 1-parameter groups of canonical transformations;

(b) we see from (1) and (3) that the Hamiltonian's formal invariance under a certain group of canonical transformations *is equivalent* to Hamilton's solutions staying solutions under that transformation group.

The latter remark is paramount in Physics. In contrast to the Lagrangian version of Noether's theorem, at present we have proved the equivalence of two notions of *invariance of a system under a transformation group*: one of theoretical essence, and sensitive to the mathematical setup (invariance of the Lagrangian, the Hamiltonian etc. under the transformation), and another one of physical nature (the system's motions are transformed into motions). ∎

Proof of Theorem 12.41

(1) We can argue in a generic canonical local chart since they form an atlas on $F(\mathbb{V}^{n+1})$. In coordinates, the dynamic vector Z reads:

$$Z = \frac{\partial}{\partial t} - X_{\mathscr{H}} \, ,$$

so using $\left[X_f, \frac{\partial}{\partial t} \right] = X_{-\frac{\partial f}{\partial t}}$, and in particular (4) in Proposition 12.26:

$$[X_f, Z] = \left[X_f, \frac{\partial}{\partial t} - X_{\mathscr{H}} \right] = \left[X_f, \frac{\partial}{\partial t} \right] + [X_f, X_{-\mathscr{H}}]$$

$$= X_{-\frac{\partial f}{\partial t}} + X_{-\{f,\mathscr{H}\}} = -X_{\frac{\partial f}{\partial t} + \{f,\mathscr{H}\}} = 0 \, .$$

This proves the first claim. The last assertion is immediate from Theorem 14.59.

(2) Assuming (12.47) and going backwards in the proof (1):

$$X_{\frac{\partial f}{\partial t} + \{f,\mathscr{H}\}} = 0 \, .$$

Applying the definition of X_g, for $i = 1, \dots, 2n$:

$$\frac{\partial}{\partial x^i} \left(\frac{\partial f}{\partial t} + \{f, \mathscr{H}\} \right) = \frac{\partial}{\partial x^i} Z(f) = 0$$

so in every natural coordinate domain $h = Z(f)$, C^1 by construction, is a function of time only. The procedure of Proposition 12.23 also says that the connectedness of phase space at a given time implies that such result is global on $F(\mathbb{V}^{n+1})$. Set $g(t) := -\int_{t_0}^{t} h(u)du$ so $f_1 := f + g$. Then f_1 satisfies (ii), since locally:

$$\frac{\partial f_1}{\partial t} + \{f_1, \mathscr{H}\} = 0 \, .$$

But f and f_1 differ by a function of t, so $X_f = X_{f_1}$ and the one-parameter group of canonical transformations they generate is the same. This proves (i). Adding to g a constant does not affect the validity of what we found above. Vice versa, another (C^2) function $g' : F(\mathbb{V}^{n+1}) \to \mathbb{R}$ making $f_1' := f + g'$ satisfy (i) will give $X_f = X_{f_1'}$. By Proposition 12.23 g is a function of absolute time only. Assuming (ii) for f_1' we have $Z(f_1') = Z(f_1) = 0$, which in coordinates means $\frac{d(g-g')}{dt} = 0$ on $\mathbb{R}$. Therefore g' differs from g by a constant.

(3) The equivalence of (i) and (ii) was proved in (1) Theorem 12.39. Let us show they are equivalent to (iii). Formal invariance to order one reads (recall $t = t'$):

$$\mathscr{H}'(t, \mathbf{x}') = \mathscr{H}(t, \mathbf{x}') + u O'_{(t,\mathbf{x}')}(u) \, . \tag{12.48}$$

By Proposition 12.34:

$$\mathscr{H}'(t, \mathbf{x}') = \mathscr{H}(t, \mathbf{x}(t, \mathbf{x}')) - u\frac{\partial f}{\partial t} + u\,O_{(t',\mathbf{x}')}(u)\,, \tag{12.49}$$

so the previous formula gives:

$$\mathscr{H}(t, \mathbf{x}') = \mathscr{H}(t, \mathbf{x}) - u\frac{\partial f}{\partial t} + u\,O''_{(t',\mathbf{x}')}(u)\,. \tag{12.50}$$

Fixing the other variables, dividing by u and taking the limit as $u \to 0$:

$$\frac{d}{du}\bigg|_{u=0} \mathscr{H}(t, \mathbf{x}') + \frac{\partial f}{\partial t} = 0\,.$$

Explicitly:

$$\sum_{k=1}^{n} \frac{\partial \mathscr{H}}{\partial x^k}\frac{dx'^k}{du}\bigg|_{u=0} + \frac{\partial f}{\partial t} = 0 \quad \text{i.e.} \quad \sum_{k,i=1}^{n} \frac{\partial \mathscr{H}}{\partial x^k}S^{ik}\frac{\partial f}{\partial x^i} + \frac{\partial f}{\partial t} = 0\,,$$

which can be put in form (2):

$$\{f, \mathscr{H}\} + \frac{\partial f}{\partial t} = 0\,.$$

Hence, (iii) implies (ii). Going backwards, the above implies (12.50). The latter, due to (12.49), gives (12.48) i.e. (i). To sum up, we have shown (i), (ii) and (iii) are equivalent.

□

Examples 12.43

(1) Consider once again the particle of mass m, moving on a plane with polar coordinates (r, θ), subject to a conservative force with central potential energy $\mathscr{U} = \mathscr{U}(r)$. The Hamiltonian then is:

$$\mathscr{H}(r, \theta, p_r, p_\theta) = \frac{p_r^2}{2m} + \frac{p_\theta^2}{2mr^2} + \mathscr{U}(r)\,. \tag{12.51}$$

Consider the function $f(r, \theta, p_r, p_\theta) := p_\theta$ and the associated one-parameter group of canonical transformations. As we saw earlier, this group acts by:

$$\varphi_u^{(X_{p_\theta})}(r, \theta, p_r, p_\theta) = (r, \theta - u, p_r, p_\theta) =: (r', \theta', p_r', p_\theta')\,, \tag{12.52}$$

so on the configuration space it corresponds to rotations by $-u$ around the origin. Since the canonical transformation $\varphi_u^{(X_{p_\theta})}$ does not depend on time and is thus completely canonical, we set $\mathcal{K} \equiv 0$ and the Hamiltonian $\mathcal{H}'(\mathbf{x}')$ can be defined as $\mathcal{H}'(\mathbf{x}') = \mathcal{H}(\mathbf{x}(\mathbf{x}'))$, i.e. from (12.52):

$$\mathcal{H}'(r', \theta', p'_r, p'_\theta) = \mathcal{H}(r', \theta' + u, p'_r, p'_\theta) .$$

Hence:

$$\mathcal{H}'(r', \theta', p'_r, p'_\theta) = \frac{p'^2_r}{2m} + \frac{p'^2_\theta}{2mr'^2} + \mathcal{U}(r') .$$

Comparing with (12.51) we have:

$$\mathcal{H}'(r', \theta', p'_r, p'_\theta) = \mathcal{H}(r', \theta', p'_r, p'_\theta) ,$$

and the Hamiltonian is formally invariant (not just to order one) under $\varphi^{X_{p_\theta}}$. Consequently p_θ is conserved on the motions. This quantity is the angular momentum of the point particle.
Equivalently, computing $\frac{\partial f}{\partial t} + \{f, \mathcal{H}\}$ shows that:

$$\frac{\partial f}{\partial t} = 0 \quad \text{while} \quad \{f, \mathcal{H}\} = 1\frac{\partial \mathcal{H}}{\partial \theta} = 0 ,$$

since $\mathcal{H}$ does not depend explicitly on time. Therefore $\frac{\partial f}{\partial t} + \{f, \mathcal{H}\} = 0$ and we rediscover, by the above theorem, that f is a first integral.
The theorem's last claim says that if $t \mapsto \mathbf{x}(t)$ is the particle's motion, also $t \mapsto \varphi_u^{(X_{p_\theta})}(\mathbf{x}(t))$ describes a motion, for any given value of the angle u.

(2) Consider a system of N particles of masses m_i, $i = 1, \ldots, N$, described in orthonormal coordinates of some inertial frame. The position vector of each point will be $\mathbf{x}_i$. The conjugate momenta to the position vectors' components are the components of each impulse $\mathbf{p}_i$. Assume the particles interact under conservative forces with C^2 potential energy $\mathcal{U}(\mathbf{x}_i, \ldots, \mathbf{x}_N)$ that only depends on the mutual distances $||\mathbf{x}_i - \mathbf{x}_j||$, $i \neq j$. In the canonical chosen frame the Hamiltonian is:

$$\mathcal{H}(\mathbf{x}_1, \ldots, \mathbf{x}_N, \mathbf{p}_1, \ldots, \mathbf{p}_N) = \sum_{i=1}^{N} \frac{\mathbf{p}_i^2}{2m_i} + \mathcal{U}(\mathbf{x}_1, \ldots, \mathbf{x}_N) .$$

Fix a unit vector $\mathbf{n}$ and the map

$$f(t, \mathbf{x}_1, \ldots, \mathbf{x}_N, \mathbf{p}_1, \ldots \mathbf{p}_N) := \sum_{i=1}^{N} m_i \mathbf{x}_i \cdot \mathbf{n} - t\mathbf{p}_i \cdot \mathbf{n} .$$

A computation shows that

$$\frac{\partial f}{\partial t} + \{f, \mathscr{H}\} = -\sum_{i=1}^{2} \mathbf{p}_i \cdot \mathbf{n} + \sum_{i=1}^{N} \mathbf{p}_i \cdot \mathbf{n} - t\mathbf{n} \cdot \sum_{i=1}^{N} \nabla_{\mathbf{x}_i} \mathscr{U}(\mathbf{x}_1, \ldots, \mathbf{x}_N) = 0,$$

where, by construction,

$$\nabla_{\mathbf{x}_i} \mathscr{U}(\mathbf{x}_1, \ldots, \mathbf{x}_N) = -\nabla_{\mathbf{x}_j} \mathscr{U}(\mathbf{x}_1, \ldots, \mathbf{x}_N) \quad \text{if } i \neq j.$$

By the previous theorem f is a first integral, and the Hamiltonian is formally invariant to order one under the one-parameter group of canonical transformations generated by f. The conservation law of f simply says that on the system's motions (the solutions to Hamilton's equations) we have the well-known formula:

$$\sum_{i=1}^{N} m_i \frac{\mathbf{x}_i(t) - \mathbf{x}_i(0)}{t} \cdot \mathbf{n} = \sum_{i=1}^{N} \mathbf{p}_i \cdot \mathbf{n}.$$

This follows from the conservation of the total impulse and Newton's third law. It states that the system's centre of mass evolves as a point particle of mass equal to the total mass and impulse equal to the total impulse of the system. What can we say about the one-parameter group of canonical transformations generated by f? Its equations are:

$$\frac{d\mathbf{p}_i'}{du} = \nabla_{\mathbf{x}_i'} f(t, \mathbf{x}_1', \mathbf{x}_2', \mathbf{p}_1', \mathbf{p}_2') = m\mathbf{n}, \qquad \frac{d\mathbf{x}_i'}{du} = -\nabla_{\mathbf{p}_i'} f(t, \mathbf{x}_1', \mathbf{x}_2', \mathbf{p}_1', \mathbf{p}_2') = t\mathbf{n}.$$

They can be integrated completely, to give:

$$\mathbf{p}_i' = \mathbf{p}_i + mu\mathbf{n}, \qquad \mathbf{x}_i' = \mathbf{x}_i + tu\mathbf{n}. \tag{12.53}$$

We emphasise that for any given u, the above transformation is a *pure Galilean transformation* (see Chap. 3) along $\mathbf{n}$ and with velocity $u\mathbf{n}$.

The Hamiltonian's formal invariance under such transformation group expresses the fact that the corresponding inertial frames are physically equivalent for the purposes of studying the dynamics.

Let us check the formal invariance directly. In our case (12.39) produces:

$$\mathscr{H}'(t', \mathbf{x}_1', \ldots, \mathbf{x}_N', \mathbf{p}_1', \ldots, \mathbf{p}_N')$$

$$= \mathscr{H}(t, \mathbf{x}_1, \ldots, \mathbf{x}_N, \mathbf{p}_1, \ldots, \mathbf{p}_N) + \int_0^u \sum_{i=1}^{N} (\mathbf{p}_i' \cdot \mathbf{n}) du$$

$$= \sum_{i=1}^{N} \frac{\mathbf{p}_i^2}{2m_i} + \mathscr{U}(\mathbf{x}_1, \ldots, \mathbf{x}_N) + \int_0^u \sum_{i=1}^{N} (\mathbf{p}_i + um_i\mathbf{n}) \cdot \mathbf{n}\, du$$

$$= \sum_{i=1}^{N} \frac{\mathbf{p}_i^2}{2m_i} + \mathscr{U}(\mathbf{x}_1', \ldots, \mathbf{x}_N') + u \sum_{i=1}^{N} \mathbf{p}_i + \frac{u^2}{2} \sum_{i=1}^{N} m_i,$$

where

$$\mathscr{U}(\mathbf{x}_1', \ldots, \mathbf{x}_N') = \mathscr{U}(\mathbf{x}_1, \ldots, \mathbf{x}_N)$$

due to the translational invariance of the potential energy and the second formula in (12.53). Recalling (12.53) and that $t' = t$, the computation gives:

$$\mathscr{H}'(t', \mathbf{x}_1', \ldots, \mathbf{x}_N', \mathbf{p}_1', \ldots, \mathbf{p}_N') = \sum_{i=1}^{N} \frac{\mathbf{p}_i'^2}{2m_i} + \mathscr{U}(\mathbf{x}_1', \ldots, \mathbf{x}_N')$$

$$= \mathscr{H}(t', \mathbf{x}_1', \ldots, \mathbf{x}_N', \mathbf{p}_1', \ldots, \mathbf{p}_N').$$
$$(12.54)$$

The Hamiltonian's invariance is self-evident.

(3) Let us focus on systems of N isolated particles interacting under internal conservative forces (translation- and rotation-invariant). We will see in detail in Sect. 13.2.2 that, if we describe them in orthonormal coordinates on an inertial frame, the Hamiltonian is formally invariant to order one (actually, invariant) under each one of the 9 one-parameter Galilean subgroups, seen as canonical transformations. This excludes temporal displacements, which do not preserve the fibres of phase spacetime and must be discussed separately (we will do that in Sect. 13.2.2). In the sense of the previous theorem (modulo signs), it is clear that:

(1) axial rotations are generated by the total angular momentum component along the rotation axis,
(2) displacements in some direction are generated by the total impulse component in that direction,
(3) pure Galilean transformations in some direction ($\mathbf{n}$) are generated by the total *boost* component $\mathbf{n} \cdot \sum_{i=1}^{N} (t\mathbf{p}_i - m_i\mathbf{x}_i)$ in that direction.

These generating functions are first integrals.

12.4 $\boxed{\text{AC}}$ Poincaré-Cartan Form and Introduction to Hamilton-Jacobi Theory

In this part we go back to the notions of canonical transformation and generating function, taking things from a different perspective. The main goal is to formulate the *Hamilton-Jacobi theory*. This is the theory, founded on the namesake equation, that accentuates a class of generating functions of special canonical transformations: in the new canonical coordinates the system's evolution, read in the first n coordinates $q^1, \ldots, q^n$, is at rest. From the applicative point of view finding this canonical transformation is equivalent to solving the equation of motion. But seen theoretically, the novel perspective brought fresh and very important developments in Hamiltonian Mechanics. In particular, these notions laid the ground for Schrödinger to formulate Quantum Mechanics in terms of the equation bearing his name, using a formal analogy between Geometric Optics and Hamilton-Jacobi theory (see [Gol50] for a critical discussion).

12.4.1 *Lie's Condition and Canonical Transformations*

We seek to prove a characterisation of canonical transformations that will be relevant later. The theorem, due to Lie, involves the **Poincaré-Cartan form**:

$$\Theta_{\mathscr{H}} := \sum_{k=1}^{n} p_k dq^k - \mathscr{H} dt . \tag{12.55}$$

Above it is assumed we are in local coordinates $t, q^1, \ldots, q^n, p_1, \ldots, p_n$ on phase spacetime, defined around a state and adapted to the fibres of $F(\mathbb{V}^{n+1})$. More precisely, the coordinate t is the absolute time of the bundle projection $T : F(\mathbb{V}^{n+1}) \to \mathbb{R}$ up to the usual additive constant, and $q^1, \ldots, q^n, p_1, \ldots, p_n$ are local coordinates on the fibre $\mathbb{F}_t$. We *do not* ask the coordinates to be canonical or natural. The real differentiable map $\mathscr{H} = \mathscr{H}(t, q, p)$ has no special meaning, at least for the time being. The differential form $\Theta_{\mathscr{H}}$ is defined on the domain of the local coordinates and depends explicitly both on the coordinates and on $\mathscr{H}$.

In the sequel we will view $F(\mathbb{V}^{n+1})$ as a C^2 manifold.

Remarks 12.44 If $t, q^1, \ldots, q^n, p_1, \ldots, p_n$ are natural coordinates on phase spacetime and $\mathscr{H}$ is the Hamiltonian of some Lagrangian $\mathscr{L}$, the Poincaré-Cartan form $\Theta_{\mathscr{H}}$ is formally obtained (under Legendre transform) from the Lagrangian Poincaré-Cartan form $\Omega_{\mathscr{L}}$ defined in (7.106), as is easy to see. However, $\Omega_{\mathscr{L}}$ is uniquely defined, globally, on $A(\mathbb{V}^{n+1})$, whereas $\Theta_{\mathscr{H}}$ a priori depends on the natural coordinates. ∎

We will show how Hamilton's equations can be expressed in terms of the Poincaré-Cartan form once we view $\mathscr{H}$ as Hamiltonian of the physical system. Let us begin with an important technical lemma.

Lemma 12.45 *Given $p \in F(\mathbb{V}^{n+1})$, the linear map $d\Theta_{\mathscr{H}}|_p : T_p F(\mathbb{V}^{n+1}) \to T_p^* F(\mathbb{V}^{n+1})$ has 1-dimensional kernel.*

Proof We have:

$$d\Theta_{\mathscr{H}} = \sum_{k=1}^{n} dp_k \wedge dq^k - d\mathscr{H} \wedge dt .$$

Let us view $d\Theta_{\mathscr{H}}|_p$ as $(2n+1) \times (2n+1)$ skew-symmetric matrix acting on column vectors

$$Z = Z^1 \frac{\partial}{\partial q^1} + \cdots + Z^n \frac{\partial}{\partial q^n} + Z_1 \frac{\partial}{\partial p_1} + \cdots + Z_n \frac{\partial}{\partial p_n} + Z' \frac{\partial}{\partial t} . \qquad (12.56)$$

The top $2n \times 2n$ minor is the matrix S, which satisfies $SS = -I$ and has non-zero determinant, $(\det S)^2 = 1$. Hence the matrix of $d\Theta_p$ has rank $\geq 2n$ and the kernel must be 1-dimensional tops. At the same time the matrix is skew-symmetric, hence diagonalisable over $\mathbb{C}$, and the $2n + 1$ eigenvectors are zero or purely imaginary. Their product equals the determinant, which is real since the matrix is real and of odd order, implying one eigenvalue must be zero. We conclude that $\dim Ker\, d\Theta_{\mathscr{H}}|_p = 1$. □

As announced, the Poincaré-Cartan form provides an alternative means to write Hamilton's equations, or rather the equations determining the dynamic vector field Z (12.23) whose integral curves are the solutions to Hamilton's equations. The following useful proposition holds.

Proposition 12.46 *Take a Hamiltonian system described on phase space $F(\mathbb{V}^{n+1})$ by the Hamiltonian dynamic vector field Z as per Definition 11.25. Consider*

(a) a canonical or natural local coordinate system (U, ψ) on $F(\mathbb{V}^{n+1})$:

$$U \ni a \mapsto \psi(a) = (t, q^1, \ldots, q^n, p_1, \ldots, p_n) \in \mathbb{R}^{2n+1} ;$$

(b) a Hamiltonian $\mathscr{H}$ obeying (12.23) with respect to Z in the above coordinates;
(c) the corresponding Poincaré-Cartan form $\Theta_{\mathscr{H}}$ on (U, ψ).

The dynamic vector field Z is completely determined on (U, ψ) by requesting:

$$\langle Z, dT \rangle = 1 , \qquad (12.57)$$

$$(d\Theta_{\mathscr{H}})(Z) = 0 , \qquad (12.58)$$

where $T : F(\mathbb{V}^{n+1}) \to \mathbb{R}$ is absolute time.

Proof Substituting (12.23) in (12.57) and (12.58) we see that they hold. Any other vector satisfying (12.58) must then be a multiple of Z by Lemma 12.45, and the proportionality factor must be 1 if Z is to satisfy (12.57). ☐

Remarks 12.47 Using the *contraction* (B.17) of vectors and forms, (12.58) reads

$$Z \lrcorner d\Theta_{\mathscr{H}} = 0 , \tag{12.59}$$

which is known in the Lagrangian formulation as (7.109), together with (12.57). ∎

We are then ready to state Lie's important characterisation of canonical transformations in terms of Poincaré-Cartan forms.

Theorem 12.48 (Lie)   *Consider two local charts (U, ψ), (U', ψ'):*

$$\psi : U \ni p \mapsto (t, q^1, \dots, q^n, p_1, \dots, p_n) \in \mathbb{R}^{2n+1},$$

$$\psi' : U' \ni p \mapsto (t', q'^1, \dots, q'^n, p'_1, \dots, p'_n) \in \mathbb{R}^{2n+1}$$

on phase spacetime $F(\mathbb{V}^{n+1})$. Suppose $U \cap U' \neq \varnothing$ and that the coordinates are adapted to the fibres as in Definition B.4: $q^1, \dots, q^n, p_1, \dots, p_n$ and $q'^1, \dots, q'^n, p'_1, \dots, p'_n$ are local coordinates on the fibre $\mathbb{F}_\tau$, and even more, t and t' are absolute time up to additive constants, so that on $U \cap U'$:

$$t' = t + c , \tag{12.60}$$

$$q'^k = q'^k(t, q, p) , \quad k = 1, \dots, n , \tag{12.61}$$

$$p'_k = p'_k(t, q, p) , \quad k = 1, \dots, n . \tag{12.62}$$

The following hold.

*(1) If the coordinate transformation (12.60)–(12.62) is canonical, for any C^1 function $\mathscr{H} = \mathscr{H}(t, q, p)$ and any $a \in U \cap U'$ there exist a neighbourhood V_a of a and a pair of C^1 real functions $\mathscr{H}' = \mathscr{H}'(t', q', p')$ and f defined on V_a satisfying **Lie's condition**:*

$$\Theta_{\mathscr{H}} - \Theta'_{\mathscr{H}'} = df . \tag{12.63}$$

$\Theta_{\mathscr{H}}$ is the Poincaré-Cartan form (12.55) and $\Theta'_{\mathscr{H}'}$ is the analogue on (U', ψ') for $\mathscr{H}'$ on V_a.

(2) If, for any given $\mathscr{H} = \mathscr{H}(t, q, p)$ of class C^1 on U, there exists a C^1 map $\mathscr{H}' = \mathscr{H}'(t', q', p')$ on U' satisfying (12.63) on $U \cap U'$ for some C^2 function f on $U \cap U'$, then (12.60)–(12.62) is a canonical transformation.

(3) If the local chart (U, ψ) is canonical or natural and $\mathscr{H} = \mathscr{H}(t, q, p)$ (C^1) is the Hamiltonian of a physical system on that chart, then every (C^1) map $\mathscr{H}' = \mathscr{H}'(t', q', p')$, defined on (U', ψ') and satisfying (12.63) for some f, produces the same Hamiltonian equations as $\mathscr{H}$ on $U \cap U'$.

(4) If t does not appear in the right-hand side of (12.61) and (12.62) (possibly renaming t' = t), then the coordinate transformation is completely canonical if and only if

$$\sum_{k=1}^{n} dp_k \wedge dq^k = \sum_{k=1}^{n} dp'_k \wedge dq'^k .$$

(12.64)

If so, letting $\mathcal{H} = \mathcal{H}(t, q, p)$ be the Hamiltonian on (U, ψ), the Hamiltonian on (U', ψ') giving the same Hamiltonian equations can be chosen to be:

$$\mathcal{H}'(t', q', p') = \mathcal{H}(t, q(q', p'), p(q', p')) .$$

Proof

(1) By the *Poincaré lemma* (Theorem B.11), the claim is equivalent to showing that if the transformation is canonical, then for any $\mathcal{H}$ we can find $\mathcal{H}'$ such that the differential form $\Theta_{\mathcal{H}} - \Theta'_{\mathcal{H}'}$ is closed, $d\Theta_{\mathcal{H}} - d\Theta'_{\mathcal{H}'} = 0$. By Theorem B.9 that means:

$$\sum_{k=1}^{n} dp_k \wedge dq^k - \sum_{k=1}^{n} dp'_k \wedge dq'^k - (d\mathcal{H} - d\mathcal{H}') \wedge dt = 0 ,$$

(12.65)

where $dt = dt'$. Now we want to prove that for any given $\mathcal{H}$, there is an $\mathcal{H}'$ satisfying (12.65) around any point.

Using the notation of Definition 12.1, easily:

$$\sum_{k=1}^{n} dp_k \wedge dq^k - d\mathcal{H} \wedge dt = -\frac{1}{2} \sum_{i,j=1}^{2n} S_{ij} dx^i \wedge dx^j - d\mathcal{H} \wedge dt .$$

Passing to coordinates $(t', x') \equiv (t', q', p')$, we have:

$$\frac{1}{2} \sum_{i,j=1}^{2n} S_{ij} dx^i \wedge dx^j + d\mathcal{H} \wedge dt = \frac{1}{2} \sum_{i,j,k,l} S_{ij} \frac{\partial x^i}{\partial x'^k} \frac{\partial x^j}{\partial x'^l} dx'^k \wedge dx'^l$$

$$+ \left(\sum_{i,j,l} S_{ij} \frac{\partial x^i}{\partial t'} \frac{\partial x^j}{\partial x'^l} dx'^l + d\mathcal{H} \right) \wedge dt' .$$

By assumption the transformation $t' = t + c$, $x' = x'(t, x)$ is canonical, so:

$$\frac{1}{2} \sum_{i,j,k,l} S_{ij} \frac{\partial x^i}{\partial x'^k} \frac{\partial x^j}{\partial x'^l} dx'^k \wedge dx'^l = \frac{1}{2} \sum_{i,j} S_{ij} dx'^i \wedge dx'^j .$$

Hence the left-hand side of (12.65) reads:

$$\frac{1}{2}\sum_{i,j} S_{ij}dx'^i \wedge dx'^j - \frac{1}{2}\sum_{i,j} S_{ij}dx'^i \wedge dx'^j$$

$$+ \left(\sum_{i,j,l} S_{ij}\frac{\partial x^i}{\partial t'}\frac{\partial x^j}{\partial x'^l}dx'^l + d\mathcal{H}\right) \wedge dt - d\mathcal{H}'.$$

That is to say:

$$\left(\sum_{i,j,l} S_{ij}\frac{\partial x^i}{\partial t'}\frac{\partial x^j}{\partial x'^l}dx'^l + d\mathcal{H}\right) \wedge dt - d\mathcal{H}' \wedge dt.$$

To prove (12.65), it then suffices to show that:

$$\sum_{i,j,l} S_{ij}\frac{\partial x^i}{\partial t'}\frac{\partial x^j}{\partial x'^l}dx'^l$$

is closed (differentiating only in the x') *for any given t'*. If that case, in fact, the Poincaré lemma implies that, fixing $p \in \mathbb{F}_{t'_0}$ of coordinates (t'_0, x'_0) and varying t' near t'_0:

$$\sum_{i,j,l} S_{ij}\frac{\partial x^i}{\partial t'}\frac{\partial x^j}{\partial x'^l}dx'^l = dg_{t'}$$

on a ball $B \subset \mathbb{F}_{t'}$ centred at x'_0 of sufficiently small radius independent of t'. Above, the derivative d on the right acts on the x' only. We could then conclude that, as t' varies near t'_0, on a small tubular neighbourhood around p we can define a function g in the $n+1$ coordinates t' and x' such that:[11]

$$dg = \sum_{i,j,l} S_{ij}\frac{\partial x^i}{\partial t'}\frac{\partial x^j}{\partial x'^l}dx'^l + \frac{\partial g}{\partial t'}dt'.$$

---

[11] The function g $=$ $g(t', x')$ is certainly C^1, since we can explicitly integrate $\sum_{i,j,l} S_{ij}\frac{\partial x^i}{\partial t'}\frac{\partial x^j}{\partial x'^l}dx'^l$ along the straight segment from (t', x'_0) to (t', x), see (5) in Remarks B.12.

If we defined $\mathscr{H}' := \mathscr{H} - g$, (12.65) would hold immediately. So to finish we just have to show that, for any given t', $\sum_{i,j,l} S_{ij} \frac{\partial x^i}{\partial t'} \frac{\partial x^j}{\partial x'^l} dx'^l$ is a closed form:

$$\sum_{i,j} S_{ij} \frac{\partial}{\partial x'^k} \frac{\partial x^i}{\partial t'} \frac{\partial x^j}{\partial x'^l} = \sum_{i,j} S_{ij} \frac{\partial}{\partial x'^l} \frac{\partial x^i}{\partial t'} \frac{\partial x^j}{\partial x'^k}$$

i.e.

$$\sum_{i,j} S_{ij} \frac{\partial^2 x^i}{\partial x'^k \partial t'} \frac{\partial x^j}{\partial x'^l} + \sum_{i,j} S_{ij} \frac{\partial x^i}{\partial t'} \frac{\partial^2 x^j}{\partial x'^k \partial x'^l} = \sum_{i,j} S_{ij} \frac{\partial^2 x^i}{\partial x'^l \partial t'} \frac{\partial x^j}{\partial x'^k}$$

$$+ \sum_{i,j} S_{ij} \frac{\partial x^i}{\partial t'} \frac{\partial^2 x^j}{\partial x'^l \partial x'^k} \ .$$

By Schwarz's theorem, this means:

$$\sum_{i,j} S_{ij} \frac{\partial^2 x^i}{\partial x'^k \partial t'} \frac{\partial x^j}{\partial x'^l} = \sum_{i,j} S_{ij} \frac{\partial^2 x^i}{\partial x'^l \partial t'} \frac{\partial x^j}{\partial x'^k}$$

or, by the skew-symmetry of S:

$$\sum_{i,j} S_{ij} \frac{\partial^2 x^i}{\partial x'^k \partial t'} \frac{\partial x^j}{\partial x'^l} + \sum_{i,j} S_{ij} \frac{\partial x^i}{\partial x'^k} \frac{\partial^2 x^j}{\partial x'^l \partial t'} = 0 \ . \tag{12.66}$$

Eventually, (1) holds when (12.66) holds. This identity is valid in our hypotheses because the transformation is canonical, so:

$$\sum_{i,j} S_{ij} \frac{\partial x^i}{\partial x'^k} \frac{\partial x^j}{\partial x'^l} = S_{kl} \ .$$

Differentiating in t', since the coefficients S_{rs} are constants and by Schwarz's theorem we obtain (12.66). This ends (1).

(2) The assumption imply, for given $\mathscr{H}$, (12.65) for some function $\mathscr{H}'$:

$$\frac{1}{2} \sum_{i,j,k,l} S_{ij} \frac{\partial x^i}{\partial x'^k} \frac{\partial x^j}{\partial x'^l} dx'^k \wedge dx'^l + \left(\sum_{i,j} S_{ij} \frac{\partial x^i}{\partial t'} \frac{\partial x^j}{\partial x'^l} dx'^l + d\mathscr{H} \right) \wedge dt'$$

$$- \frac{1}{2} \sum_{k,l} S_{ij} dx'^k \wedge dx'^l - d\mathscr{H}' \wedge dt' = 0$$

that is:

$$\frac{1}{2}\sum_{k,l}\left(\sum_{i,j}S_{ij}\frac{\partial x^i}{\partial x'^k}\frac{\partial x^j}{\partial x'^l}-S_{kl}\right)dx'^k \wedge dx'^l$$

$$+\left(\sum_{i,j}S_{ij}\frac{\partial x^i}{\partial t'}\frac{\partial x^j}{\partial x'^l}dx'^l-d(\mathscr{H}-\mathscr{H}')\right)\wedge dt'=0.$$

Computing the second summand, we find

$$\frac{1}{2}\sum_{k,l}\left(\sum_{i,j}S_{ij}\frac{\partial x^i}{\partial x'^k}\frac{\partial x^j}{\partial x'^l}-S_{kl}\right)dx'^k \wedge dx'^l+\sum_k A_k(t',x')dx'^k \wedge dt'=0$$

for some functions A_k. As $dx'^l \wedge dx'^l$ and $dx'^k \wedge dt'$ are independent, in particular:

$$\frac{1}{2}\sum_{k,l}\left(\sum_{i,j}S_{ij}\frac{\partial x^i}{\partial x'^k}\frac{\partial x^j}{\partial x'^l}-S_{kl}\right)dx'^k \wedge dx'^l=0,$$

which means:

$$\sum_{i,j}S_{ij}\frac{\partial x^i}{\partial x'^k}\frac{\partial x^j}{\partial x'^l}=S_{kl}.\qquad(12.67)$$

By definition, the latter says the transformation (12.60)–(12.62) is canonical.

(3) This claim is immediate from Proposition 12.46 because (12.63) implies $d\Theta_{\mathscr{H}}=d\Theta_{\mathscr{H}'}$, so the dynamic vectors Z and Z' relative to $\mathscr{H}$ and $\mathscr{H}'$ under (12.23) must be equal on $U \cap U'$.

(4) Saying the transformation is canonical means precisely (12.67), or that:

$$\frac{1}{2}\sum_{k,l}\left(\sum_{i,j}S_{ij}\frac{\partial x^i}{\partial x'^k}\frac{\partial x^j}{\partial x'^l}-S_{kl}\right)dx'^k \wedge dx'^l=0.\qquad(12.68)$$

Assume now $\frac{\partial x^i}{\partial t'}=0$. In view of (12.60)–(12.62), these relations are $\frac{\partial x'^k}{\partial t}=0$. Then (12.60)–(12.62) is canonical if and only if it is completely canonical. But $\frac{\partial x^i}{\partial t'}=0$, so (12.68) reads:

$$\frac{1}{2}\sum_{k,l}\left(\sum_{i,j}S_{ij}\frac{\partial x^i}{\partial x'^k}\frac{\partial x^j}{\partial x'^l}-S_{kl}\right)dx'^k\wedge dx'^l+\left(\sum_{i,j}S_{ij}\frac{\partial x^i}{\partial t'}\frac{\partial x^j}{\partial x'^l}dx'^l\right)\wedge dt'=0.$$

Now, the above is just (12.64). Hence, when $\frac{\partial x'^k}{\partial t} = 0$ formula (12.64) is equivalent to system (12.60)–(12.62) being completely canonical. The last claim is true by the final part of Theorem 12.9.

$\square$

12.4.2 Generating Functions of Canonical Transformations

Theorem 12.48 permits us to introduce a very useful notion, that of *generating function* of a canonical transformation. (This should not be confused with the earlier *generating function of a one-parameter group of active canonical transformations*.)

In the sequel all arguments will be local, so we will not pay much attention to the topology of the domains of canonical coordinates: one can always restrict to connected and locally simply connected domains around any point in phase spacetime $F(\mathbb{V}^{n+1})$.

Consider, on $F(\mathbb{V}^{n+1})$, a canonical transformation between canonical coordinates t, q, p and T, Q, P for a physical system with (C^1) Hamiltonian $\mathscr{H}$ in the t, q, p. We know from Theorem 12.48 that, locally, Lie's condition holds:

$$\sum_k P_k dQ^k - \mathscr{K} dT = \sum_k p_k dq^k - \mathscr{H} dt + dF_1$$

for a pair of maps $\mathscr{K}$ and F_1, and that $\mathscr{K}$ can be thought of as the Hamiltonian[12] in the new canonical coordinates T, Q, P. As $dt = dT$, we can rephrase as follows:

$$\sum_k P_k dQ^k - \sum_k p_k dq^k - (\mathscr{K} - \mathscr{H}) dt = dF_1 . \tag{12.69}$$

Hence, if we view F_1 depending on t, q, Q—*assuming this is possible*—we must have:

$$p_k = -\frac{\partial F_1}{\partial q^k} , \quad P_k = \frac{\partial F_1}{\partial Q^k} , \quad \mathscr{K} = \frac{\partial F_1}{\partial t} + \mathscr{H} .$$

These relations can be used in the opposite direction to *define* a canonical transformation! In fact, suppose we have a canonical coordinate system (U, ψ) with C^2 coordinates t, q, p, and there is a C^3 map F_1 only depending on t, q and on local parameters $Q^1, \ldots, Q^n$ defined on an open set $V \subset \mathbb{R}^n$. For simplicity let us redefine U around a given state a so that $\psi(U) = I \times A \times B$, with $A, B \subset \mathbb{R}^n$ open

---

[12] Beware that, in Theorem 12.9, $\mathscr{K}$ denotes the difference between the new and old Hamiltonians!

and $I \subset \mathbb{R}$ open interval. Finally, suppose that *for any* $(t, q) \in I \times A$, the map:

$$p_k = -\frac{\partial F_1}{\partial q^k}(t, q, Q) , \tag{12.70}$$

sending $(Q^1, \ldots, Q^n) \in V$ to $(p_1, \ldots, p_n) \in B$, is bijective with C^2 inverse (in all variables). Now set:

$$P_k := \frac{\partial F_1}{\partial Q^k} , \tag{12.71}$$

$$\mathcal{K} := \frac{\partial F_1}{\partial t} + \mathcal{H} , \tag{12.72}$$

where $\mathcal{H}$ is any C^1 map in t, q, p. Then (12.69) is automatic. Set $W := \frac{\partial F_1}{\partial Q^k}(I \times A \times V) \subset \mathbb{R}^n$, assumed open, and consider the C^2 transformation:

$$I \times B \times C \ni (t, q, p) \mapsto (T, Q, P) \in I \times V \times W$$

given by

$$T = t , \tag{12.73}$$

$$Q^k = Q^k(t, q, p) , \tag{12.74}$$

$$P_k = P_k(t, q, p) . \tag{12.75}$$

It is well defined since the Q^k can be expressed in terms of the p_k with t, Q fixed. This transformation will be canonical by (2) in Theorem 12.48 provided it is bijective with C^2 inverse. In practice, for F_1 to *generate* a canonical transformation we must check that (12.70) is C^2, invertible for fixed (t, q), and also that the transformation (12.73)–(12.75) is bijective with C^2 inverse.

A global *necessary* condition for (12.70) to be invertible (with the required regularity), and a *locally sufficient* one—i.e. warranting a C^2 inverse on some open neighbourhood of any (t_0, q_0, Q_0)—is that:

$$\det \left[\frac{\partial^2 F_1}{\partial q^k \partial Q^h} \right]_{k,h=1,\ldots,n} \Bigg|_{(t,q,Q)} \neq 0 \quad \forall (t, q, Q) \in I \times A \times V . \tag{12.76}$$

This is straightforward from the Inverse function theorem 1.25, whereby W is open. Actually, (12.76) also ensures that (12.73)–(12.75) is bijective with C^2 inverse but *locally*, on some open neighbourhood of (t_0, q_0, p_0), where $p_{0k} := -\frac{\partial F_1}{\partial q^k}|_{(q_0, Q_0)} \in B$. That neighbourhood is diffeomorphic to an open neighbourhood of the point with (new) coordinates (T_0, P_0, Q_0) defined by the right-hand sides of (12.73)–(12.75). In fact the map (12.75) sending q to P with fixed $(t, p) \in I \times B$ can be locally inverted, regularly, by Theorem 1.25 whenever (12.76) holds, since the latter means:

$$\det \left[\frac{\partial P_h}{\partial q^k} \right]_{k,h=1,\ldots,n} \Bigg|_{(t,q,Q)} \neq 0 \quad \forall (t, q, Q) \in I \times A \times V . \tag{12.77}$$

The above relationship defines q in terms of P with Q and $T = t$ fixed in some neighbourhood of (t_0, Q_0, P_0). Together with (12.70), it determines the inverses to (12.73)–(12.75):

$$t = T , \tag{12.78}$$

$$q^k = q^k(T, Q, P) , \tag{12.79}$$

$$p_k = p_k(T, Q, P) . \tag{12.80}$$

By construction these are defined on the open neighbourhood of $a \in U$ with initial coordinates: $t(a) = t_0, q(a) = q_0, p(a) = -\frac{\partial F_1}{\partial q^k}|_{(q_0, Q_0)}$.

Furthermore, by (3) in Theorem 12.48, if $\mathscr{H}$ is the Hamiltonian in canonical coordinates t, q, p, then $\mathscr{K}$ (written in coordinates T, Q, P) is the Hamiltonian in the new canonical coordinates.

Altogether we have proved the following proposition.

Proposition 12.49 *On the C^2 manifold $F(\mathbb{V}^{n+1})$ consider a canonical coordinate system:*

$$\psi : U \ni a \mapsto (t(a), q(a), p(a)) \in \psi(U) = I \times A \times B$$

with $I \subset \mathbb{R}$ interval and $A, B \subset \mathbb{R}^n$ open, and a C^3 function

$$F_1 : I \times A \times V \ni (t, q, Q) \to F_1(t, q, Q) \in \mathbb{R}$$

with $V \subset \mathbb{R}^n$ open. If (12.76) holds:

$$\det \left[\frac{\partial^2 F_1}{\partial q^k \partial Q^h} \right]_{k,h=1,\ldots,n} \Bigg|_{(t,q,Q)} \neq 0 \quad \forall (t, q, Q) \in I \times A \times V ,$$

then:

(1) on an open neighbourhood U_{a_0} of every state $a_0 \in U$ of coordinates (t_0, q_0, p_0), where

$$p_{0k} := -\frac{\partial F_1}{\partial q^k}\Bigg|_{(t_0, q_0, Q_0)} \quad \text{for } t_0, q_0, Q_0 \text{ given and } k = 1, \ldots, n$$

there exists another system of canonical coordinates

$$\phi : U_{a_0} \ni a \mapsto (T(a), Q(a), P(a)) \in \phi(U_{a_0}) \subset \mathbb{R}^{2n+1}$$

where:

$$T = t, \quad P_k = \frac{\partial F_1}{\partial Q^k}(t, q, Q), \quad p_k = -\frac{\partial F_1}{\partial q^k}(t, q, Q)$$

for a generic state $a \in U_{a_0}$ of coordinates (t, q, p) and (T, Q, P);
(2) the Hamiltonian $\mathcal{K}$ in the new canonical coordinates on U_{a_0} (possibly restrict-
ing U_{a_0}) is related to the Hamiltonian $\mathcal{H}$ in the initial coordinates by (12.72):

$$\mathcal{K} = \frac{\partial F_1}{\partial t} + \mathcal{H} .$$

The function $F_1 = F_1(t, q, Q)$ is called **generating function** *of the canonical transformation (12.73)–(12.75).*

Remarks 12.50 It should be clear that asking $\psi(U) = I \times A \times B$ is not essential, given the proposition's local nature: starting from an open $\psi(U)$ we can always find $I \times A \times B$ around any given point in U and then redefine U so that its image is that set, since $\psi : U \rightarrow \psi(U)$ is a diffeomorphism. The same goes for the domain $I \times A \times V$ of F_1.   ∎

We can then proceed similarly for other generating functions:

$$F_2 = F_2(t, q, P), \quad F_3 = F_3(t, Q, p), \quad F_4 = F_4(t, P, p),$$

by asking, respectively:

$$\det \left[\frac{\partial^2 F_2}{\partial q^k \partial P_h} \right]_{k,h=1,...,n} \neq 0, \quad \det \left[\frac{\partial^2 F_3}{\partial Q^k \partial p_h} \right]_{k,h=1,...,n} \neq 0, \quad \det \left[\frac{\partial^2 F_4}{\partial P_k \partial p_h} \right]_{k,h=1,...,n} \neq 0.$$

$$(12.81)$$

The corresponding canonical transformations and the new Hamiltonian are, respectively:

$$Q^k = -\frac{\partial F_2}{\partial P_k}, \quad p_k = -\frac{\partial F_2}{\partial q^k}, \quad \mathcal{K} = \frac{\partial F_2}{\partial t} + \mathcal{H},$$

$$P_k = \frac{\partial F_3}{\partial Q^k}, \quad q^k = \frac{\partial F_3}{\partial p_k}, \quad \mathcal{K} = \frac{\partial F_3}{\partial t} + \mathcal{H},$$

and finally:

$$Q^k = -\frac{\partial F_4}{\partial P_k}, \quad q^k = \frac{\partial F_4}{\partial p_k}, \quad \mathcal{K} = \frac{\partial F_4}{\partial t} + \mathcal{H}.$$

In the three cases, Lie's condition

$$\sum_k P_k dQ^k - \sum_k p_k dq^k - (\mathcal{K} - \mathcal{H})dt = df,$$

which guarantees the transformations are canonical, does hold if we set:

$$f := F_2 + \sum_k Q^k P_k, \quad f := F_2 - \sum_k q^k p_k \quad f := F_2 - \sum_k q^k p_k + \sum_k Q^k P_k.$$

The proof is straightforward.

Remarks 12.51

(1) The various generating functions F_k are related among themselves by Legendre transformations. For example:

$$F_2(t, q, P) = F_1(t, q, Q) - \sum_k P_k Q^k, \tag{12.82}$$

$$-Q^k = \frac{\partial F_2}{\partial P_k}. \tag{12.83}$$

(2) Every canonical transformation is generated by a function of type F_k, $k = 1, 2, 3, 4$, composed with some transformation that replaces the momenta components p_k with coordinate components q^k and flips some signs. ∎

Examples 12.52

(1) The simplest generating function is:

$$F_2(q, P) := -\sum_k q^k P_k.$$

The canonical transformation generated by it is the identity:

$$T = t, \quad Q^k = q^k, \quad P_k = p_k, \quad k = 1, \dots n.$$

(2) Less trivially, consider the canonical transformation:

$$T = t, \quad Q^k = -p_k, \quad P_k = q^k, \quad k = 1, \dots n$$

that swaps the variables q and p up to sign (this is necessary for it to be canonical). Immediately, its generating function is of type F_4:

$$F_4(t, p, P) := -\sum_k p_k P_k .$$

(3) Consider the coordinate transformation on $\mathbb{R}^2$:

$$P = p/\sqrt{1 + (qp)^2} , \quad Q = q\sqrt{1 + (qp)^2} , \tag{12.84}$$

with q, p being the natural coordinates of $\mathbb{R}^2$. This is C^∞ from $\mathbb{R}^2$ to $\mathbb{R}^2$. It is bijective, since by construction $QP = qp$, so the inverse exists and is C^∞, namely:

$$p = P\sqrt{1 + (QP)^2} , \quad q = Q/\sqrt{1 + (QP)^2} . \tag{12.85}$$

Transformation (12.84) (plus the obvious $T = t$) also defines a completely canonical transformation, since time does not appear explicitly and $dP \wedge dQ = dp \wedge dq$ as per (d) in Theorem 12.48. Let us prove this. We have:

$$dQ = \sqrt{1 + (qp)^2}dq + \frac{q(qp^2 dq + q^2 pdp)}{\sqrt{1 + (qp)^2}}$$

and so:

$$PdQ = pdq + \frac{pq}{1 + (qp)^2}(qp^2 dq + q^2 pdp) ,$$

i.e.:

$$PdQ = pdq + \frac{(qp)^2}{1 + (qp)^2}d(qp) = pdq + d(qp - \arctan(qp)) , \tag{12.86}$$

and computing the exterior derivative:

$$dP \wedge dQ = dp \wedge dq .$$

Now, (12.86) implies that irrespective of whether $\mathscr{H}$ is viewed in terms of t, q, p or T, Q, P, Lie's condition:

$$PdQ - pdq - (\mathscr{K} - \mathscr{H})dt = d(qp - \arctan(qp))$$

holds if $\mathscr{K} := \mathscr{H}$. This fact suggests that, locally, $f(q, p) := qp - \arctan(qp)$ can be seen as generating function of the canonical transformation (12.84). To understand things this way we must first express this function in q, Q. From the second equation in (12.85):

$$p = \pm \frac{1}{q}\sqrt{\left(\frac{Q}{q}\right)^2 - 1}$$

if $(Q/q)^2 > 1$ and the right-hand-side's sign is the sign of pq. Condition $(Q/q)^2 > 1$ is equivalent to $qp \neq 0$, from the second relation in (12.84). Therefore, for instance in the quadrant $q, p > 0$ of $\mathbb{R}^2$ (by (12.84) this corresponds to $Q, P > 0$):

$$p = \frac{1}{q}\sqrt{\left(\frac{Q}{q}\right)^2 - 1}, \quad P = p/\sqrt{1 + (qp)^2}.$$

Using the first relation we may rewrite as:

$$p = \frac{1}{q}\sqrt{\left(\frac{Q}{q}\right)^2 - 1}, \quad P = \frac{1}{Q}\sqrt{\left(\frac{Q}{q}\right)^2 - 1}.$$

Now let us go recast $f(q, p) = qp - \arctan(qp)$ in terms of q, Q for $qp > 0$:

$$F_1(q, Q) := \sqrt{\left(\frac{Q}{q}\right)^2 - 1} - \arctan\left(\sqrt{\left(\frac{Q}{q}\right)^2 - 1}\right).$$

By construction, or computing directly:

$$p = \frac{1}{q}\sqrt{\left(\frac{Q}{q}\right)^2 - 1} = -\frac{\partial F_1}{\partial q}, \quad P = \frac{1}{Q}\sqrt{\left(\frac{Q}{q}\right)^2 - 1} = \frac{\partial F_1}{\partial Q}.$$

Besides, as F_1 does not depend explicitly on t:

$$\mathscr{K} = \mathscr{H} + \frac{\partial F_1}{\partial t}.$$

Finally, as is required for F_1 to generate a canonical transformation, in the quadrant $q, p > 0$ (or $Q, P > 0$) we can invert

$$p = -\frac{\partial F_2}{\partial q} = \frac{1}{q}\sqrt{\left(\frac{Q}{q}\right)^2 - 1}$$

for fixed q since:

$$Q = \sqrt{q^2(1 + (qp)^2)}\,.$$

Similar discussions hold in other regions of the plane $\mathbb{R}^2$ that do not contain the origin.

12.4.3 Introduction to Hamilton-Jacobi Theory

Consider a Hamiltonian system on $F(\mathbb{V}^{n+1})$. Referring to a local canonical chart (U, ψ) with

$$U \ni a \mapsto \psi(a) = (t, q, p) \in \mathbb{R}^{2n+1}\,,$$

call $\mathscr{H} = \mathscr{H}(t, q, p)$ the system's Hamiltonian (assumed C^1, or even C^2 so to have unique solutions to Hamilton's equations). A rather drastic way to solve the equations of motion is to seek a canonical transformation (supposing there is one such):

$$t = t\,, \tag{12.87}$$

$$Q^k = Q^k(t, q, p)\,, \qquad k = 1, \ldots, n\,, \tag{12.88}$$

$$P_k = P_k(t, q, p)\,, \qquad k = 1, \ldots, n\,, \tag{12.89}$$

in which the new Hamiltonian $\mathscr{K} = \mathscr{K}(t, Q, P)$ is *identically zero*. If we manage to find such a canonical transformation, the solution to Hamilton's equations in the new canonical coordinates would have trivial form as t varies:

$$Q^k(t) = Q^k(t_0, q_0, p_0) =: Q_0^k\,, \quad P_k(t) = P_k(t_0, q_0, p_0) =: P_{0k}\,, \quad k = 1, \ldots, n, \tag{12.90}$$

where q_0, p_0 are the initial conditions at time t_0 in the initial coordinates. The solution to Hamilton's equations in initial coordinates would then read, inverting the canonical transformation:

$$q^k(t) = q^k(t, Q_0, P_0)\,, \quad p_k(t) = P_k(t, Q_0, P_0)\,, \quad k = 1, \ldots, n\,. \tag{12.91}$$

At least locally, therefore, we can attempt to construct a similar canonical transformation by finding its generating function, for instance of type F_2, i.e.

$S = S(t, q, P)$. As discussed previously, we would require that S is at least C^3 and that:

$$\det \left[\frac{\partial^2 S}{\partial q^k \partial P_h} \right]_{k,h=1,\ldots,n} \neq 0$$

everywhere and for any t in its domain. Lie's condition reads (as explained above):

$$\sum_k p_k dq^k - \mathscr{H} dt = -\sum_k Q^k dP_k + dS ,$$

where the Hamiltonian $\mathscr{H} = \mathscr{H}(t, Q, P)$ is the zero map. From these relations we know that:

$$p_k = \frac{\partial S}{\partial q^k} , \quad Q^k = \frac{\partial S}{\partial P_k} , \qquad k = 1, \ldots, n ,$$

and differentiating in t:

$$\frac{\partial S}{\partial t} + \mathscr{H} = 0 .$$

Using this argument backwards, we arrive at the following remarkable result, known as **Jacobi Theorem**.

Theorem 12.53 *Consider a Hamiltonian system on $F(\mathbb{V}^{n+1})$ with Hamiltonian $\mathscr{H} = \mathscr{H}(t, q, p)$ at least of class C^1 in a local canonical chart (U, ψ) with*

$$U \ni a \mapsto \psi(a) = (t, q, p) \in \mathbb{R}^{2n+1} .$$

Suppose there exists a family of maps $S = S(t, q, P)$, parametrised by $P_1, \ldots, P_n$, that is jointly C^3 in all variables (parameters included), satisfies

$$\det \left[\frac{\partial^2 S}{\partial q^k \partial P_h} \right]_{k,h=1,\ldots,n} \neq 0 \tag{12.92}$$

*everywhere and for any t in its domain, and also solves the **Hamilton-Jacobi equation**:*

$$\frac{\partial S}{\partial t} + \mathscr{H}(t, q, \nabla_q S) = 0 . \tag{12.93}$$

Then, in the sense of Proposition 12.49,

$$p_k = \frac{\partial S}{\partial q^k} , \quad Q^k = \frac{\partial S}{\partial P_k} , \qquad k = 1, \ldots, n , \tag{12.94}$$

locally define a canonical transformation such that in the new canonical coordinates t, Q, P *the Hamiltonian is identically zero. Therefore, the equations of motion are solved in the new chart by solutions (12.90)–(12.91) that are constant in* Q, P.

The following is an important definition.

Definition 12.54 Consider a collection $S = S(t, q, P)$ of local solutions to the Hamilton-Jacobi equation on phase spacetime

$$\frac{\partial S}{\partial t} + \mathscr{H}(t, q, \nabla_q S) = 0 ,$$

parametrised by $(P_1, \ldots, P_n)$ varying in an open set of $\mathbb{R}^n$, and jointly C^3 in (t, q) and in the parameters P. If (12.92) holds where the family is defined, we call S a **complete integral** of the equation. $\diamond$

12.4.4 Local Existence of Complete Integrals: Hamilton's Principal Function

Next we will show that there always exists, locally, a family of solutions of the Hamilton-Jacobi equation giving a complete integral.

If we have a solution to the Hamilton-Jacobi equations $S = S(t, q)$, when we evaluate it on a solution $q = q(t)$, $p = p(t)$ of Hamilton's equations in canonical coordinates, from the definition we find:

$$\frac{d}{dt} S(t, q(t)) = \frac{\partial S}{\partial t} + \sum_k \frac{\partial S}{\partial q^k} \frac{dq^k}{dt} = -\mathscr{H}(t, q(t), p(t)) + \sum_k p_k(t) \frac{dq^k}{dt}$$

$$= \mathscr{L}(t, q(t), p(t)) .$$

Hence the total time derivative along the motion (i.e. a solution to Hamilton's equations) of a solution to the Hamilton-Jacobi equations is nothing else than the system's Lagrangian evaluated on the motion. The idea is then to seek solutions of the Hamilton-Jacobi equation that are *integrals with respect to time* of the Lagrangian, evaluated on solutions to Hamilton's equations or equivalently to the Euler-Lagrange equations.

In a natural local chart of $\mathbb{V}^{n+1}$, the integral curves will be solutions starting at an initial configuration (T, Q) in the configuration spacetime and reaching a final configuration (t, q). The idea is that (t, q) also are the variables of $S = S(t, q)$. The arbitrary parameters needed to have a complete integral will be found by varying the initial configuration.

In order to implement all of this we need to tackle a technically challenging problem. In a natural local chart of $\mathbb{V}^{n+1}$, consider $\mathscr{L} = \mathscr{L}(t, q, \dot{q})$ on $I \times \Omega \times \mathbb{R}^n$, where $I \subset \mathbb{R}$ is an open interval and $\Omega \subset \mathbb{R}^n$ is open. Let $T \in I$ and $Q \in \Omega$

and let us try to find solutions to the Euler-Lagrange equations departing from Q at time T and arriving, at a given $t \in I$, to the configuration $q \in \Omega$. Note that this is not an initial value problem, but a *boundary value problem*. Instead of Cauchy initial conditions, we have the two configurations that the solution connects during the time interval considered. This kind of problem is very difficult and typically ill posed.

We can nonetheless prove the following. In the sequel $B_r(x)$ will be the open ball of radius $r > 0$ centred at $x \in \mathbb{R}^n$, so in particular $B_\delta(r) = (x - \delta, x + \delta)$ if we are on $\mathbb{R}^1$.

Proposition 12.55 *Let $I \subset \mathbb{R}$ be an open interval, $\Omega \subset \mathbb{R}^n$ an open set and consider a Lagrangian $\mathscr{L} \in C^4(I \times \Omega \times \mathbb{R}^n)$ with everywhere non-singular Hessian $\frac{\partial^2 \mathscr{L}}{\partial \dot{q}^k \partial \dot{q}^h}$ and C^2 Lagrangian components[13] $\mathscr{Q}_k(t, q, \dot{q})$. Let finally*

$$\gamma_1 : [T_0, t_1] \to \Omega \quad \text{where } T, t_1 \in I \text{ with } T_0 < t_1 \text{ and } \gamma_1(T_0) =: Q_0, \gamma_1(t_1) =: q_1$$

be the restriction of a solution to the Euler-Lagrange equations (as curve on the configuration spacetime $\mathbb{R} \times \Omega$):

$$\frac{d}{dt}\left(\frac{\partial \mathscr{L}}{\partial \dot{q}^k}\bigg|_{(t,q(t),\dot{q}(t))} \right) - \frac{\partial \mathscr{L}}{\partial q^k}\bigg|_{(t,q(t),\dot{q}(t))}$$

$$= \mathscr{Q}_k(t, q(t), \dot{q}(t)), \quad \frac{dq^k}{dt} = \dot{q}^k(t), \quad k = 1, \ldots, n.$$

If T_0 and t_1 are close enough, there exist a neighbourhood of (T_0, Q_0), a neighbourhood of (t, q) and a family of solutions to the Euler-Lagrange equations, given by differentiable deformations of γ_1, joining any pair of points chosen in the two previous neighbourhoods. In this way any pair of points is connected by exactly one solution in the family.

More precisely, if T_0 and t_1 are sufficiently cose, there exist $r > 0$, $\delta > 0$ and a C^2 map:

$$\bigcup_{(T,t,Q,q) \in B_\delta(T_0) \times B_\delta(t_1) \times B_r(Q_0) \times B_r(q_1)} I_{(T,t,Q,q)} \times \{(T, t, Q, q)\} \ni (\tau, T, Q, t, q)$$

$$\mapsto q'_{\gamma_1, T, t, Q, q}(\tau) \in \Omega$$

---

[13] Physically, these describe forces that cannot be put in the Lagrangian's expression via a potential, even generalised.

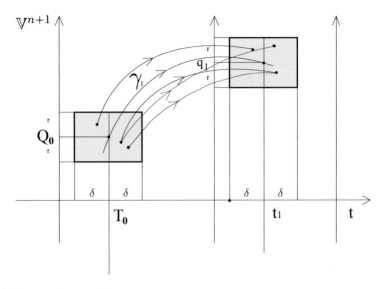

Fig. 12.1 Proposition 12.55

where:

(1) any $I_{(T,t,Q,q)} \subset \mathbb{R}$ is an open interval containing $[T, t]$,

(2) the domain

$$\Gamma := \bigcup_{(T,t,Q,q)\in B_\delta(T_0)\times B_\delta(t_1)\times B_r(Q_0)\times B_r(q_1)} I_{(T,t,Q,q)} \times \{(T, t, Q, q)\} \subset \mathbb{R} \times \mathbb{R} \times \mathbb{R} \times \mathbb{R}^n \times \mathbb{R}^n$$

(12.95)

is open,

(3) $I_{(T,t,Q,q)} \ni \tau \mapsto q'_{\gamma_1,T,t,Q,q}(\tau) \in \Omega$ solves the Euler-Lagrange equations,

(4) such a solution joins (T, Q) to (t, q), i.e.: $q'_{\gamma_1,T,t,Q,q}(T) = Q$ and $q'_{\gamma_1,T,t,Q,q}(t) = q$,

(5) $q'_{\gamma_1,T,t,Q,q_1}|_{[T_0,t_1]} = \gamma_1$ (Fig. 12.1).

Proof As γ_1 is fixed, throughout the proof we will write

$$q'_{T,t,Q,q} := q'_{\gamma_1,T,t,Q,q} \, .$$

By Theorem 7.37 we have existence and uniqueness of the solutions to the Euler-Lagrange equations with given initial data. The dynamic vector field Z associated with the Lagrangian and the Lagrangian components is C^2 and therefore the 1-parameter group of local diffeomorphisms:

$$\Phi_s^{(Z)}(T, Q, \dot{Q}) = (T + s, q(T + s, T, Q, \dot{Q}), \dot{q}(T + s, T, Q, \dot{Q})) , \qquad (12.96)$$

giving the integral curves, is jointly C^2 in its variables $(s, T, Q, \dot{Q}) \in A_0$ where $A_0 \subset \mathbb{R} \times I \times \Omega \times \mathbb{R}^n$ is open. Let us change variables with the diffeomorphism

$$A_0 \ni (s, T, Q, \dot{Q}) \mapsto (T, t := T + s, Q, \dot{Q}) \in A,$$

noting that $t \in I$ by construction and so $A \subset I \times I \times \Omega \times \mathbb{R}^n$. Consider the map, for $t > T$,

$$A \ni (T, t, Q, \dot{Q}) \mapsto (T, t, Q, q(t, T, Q, \dot{Q})), \tag{12.97}$$

where $q(t, T, Q, \dot{Q})$ appears in (12.96). To say $(T, t, Q, \dot{Q}) \in A$ means in particular that we can evolve under $\Phi^{(Z)}$ the initial datum $(T, Q, \dot{Q})$ from time T to time t, so that $q(T, \tau, Q, \dot{Q})$ is defined whenever τ is in an open interval containing $[T, t]$. By the assumptions on the solution γ_1 joining (T_0, Q_0) and (t_1, q_1), we deduce there exists $\dot{Q}_1 \in \mathbb{R}^n$ such that

$$q(T_0, t_1, Q_0, \dot{Q}_1) = q_1.$$

We claim that if $t_1 - T_0 > 0$ is small, then (12.97) is non-singular at $(T_0, t_1, Q_0, \dot{Q}_1)$, and hence it is a local C^2 diffeomorphism from a neighbourhood of $(T_0, t_1, Q_0, \dot{Q}_1)$ to a neighbourhood of (T_0, t_1, Q_0, q_1). The latter point's neighbourhood can be chosen to be $B_\delta(T_0) \times B_\delta(t_1) \times B_r(Q_0) \times B_r(q_1)$. If $(T, t, Q, q) \mapsto (T, t, Q, \dot{L}(T, t, Q, q))$ is the inverse diffeomorphism, then $q'_{T, t, Q, q}(\tau) := q(T, \tau, Q, \dot{L}(T, t, Q, q))$ satisfies the claim. In particular, as the domains of $(T, t, Q, \dot{Q}) \to q(T, t, Q, \dot{Q})$ and $(T, t, Q, q) \to \dot{L}(T, t, Q, q)$ are open and the second map is continuous, immediately $(\tau, T, t, Q, q) \mapsto q'_{T, t, Q, q}(\tau)$ is defined on an open set Γ. Moreover, Γ inherits from the domain of (12.97) the property that at $t > T$ fixed, τ varies in an open interval containing $[T, t]$. Necessarily its form is (12.95). To finish we will prove (12.97) is non-singular at $(T_0, t_1, Q_0, \dot{Q}_1)$ for $t_1 - T_0 > 0$ small. As the first three components of the range and the domain of (12.97) are the same, it will suffice to show that

$$\det \left[\frac{\partial q^k}{\partial \dot{Q}^h} \right] \Bigg|_{(T_0, t_1, Q_0, \dot{Q}_1)} \neq 0.$$

By construction, as we are working with solutions from Q_0 at time T_0, whose initial tangent vector is $\dot{Q}_1$,

$$\frac{\partial q^k}{\partial \dot{Q}^h} \Bigg|_{(T_0, T_0, Q_0, \dot{Q}_1)} = 0 \quad \text{and}$$

$$\frac{\partial^2 q^k}{\partial t \partial \dot{Q}^h} \Bigg|_{(T_0, T_0, Q_0, \dot{Q}_1)} = \frac{\partial^2 q^k}{\partial \dot{Q}^h \partial t} \Bigg|_{(T_0, T_0, Q_0, \dot{Q}_1)} = \frac{\partial \dot{Q}^k}{\partial \dot{Q}^h} \Bigg|_{(T_0, T_0, Q_0, \dot{Q}_1)} = \delta^k_h,$$

where we used Schwarz's theorem, because the function is C^2. Therefore:

$$\left.\frac{\partial q^k}{\partial \dot{Q}^h}\right|_{(T_0,t_1,Q_0,\dot{Q}_1)} = \delta_h^k(t_1 - T_0) + O_{h,T_0,Q_0,\dot{Q}_1}^k(t_1 - T_0),$$

where $|O_{h,T_0,Q_0,\dot{Q}_1}^k(t_1 - T_0)||t_1 - T_0|^{-1} \to 0$ if $t_1 - T_0 \to 0$ and with velocity dependent on T_0, Q_0, $\dot{Q}_1$. Computing the determinant we then have

$$\det\left.\left[\frac{\partial q^k}{\partial \dot{Q}^h}\right]\right|_{(T_0,t_1,Q_0,\dot{Q}_1)} = n(t_1 - T_0) + O_{T_0,Q_0,\dot{Q}_1}(t_1 - T_0) \qquad (12.98)$$

where again $|O_{T_0,Q_0,\dot{Q}_1}(t_1 - T_0)||t_1 - T_0|^{-1} \to 0$ as $t_1 - T_0 \to 0$ and the velocity depends on T_0, Q_0, $\dot{Q}_1$. If those values are fixed and $t_1 - T_0$ is small enough but non-zero, the sign of the right-hand side of (12.98) is decided by the first summand, so strictly positive since $t_1 > T_0$. ☐

Remarks 12.56 The proposition does not prove the existence and uniqueness (on a local chart) for the Euler-Lagrange equations with boundary conditions, rather than initial data. It warrants local existence in the case of time-independent Lagrangians, as we see from Exercise 12.64. In particular:

(a) there might exist several solutions between the initial point Q at time T and a final point q at time t, despite the family contains only one such solution;
(b) there might exist points q that are not reached at time t by any solution departing from Q at time T if t and T are far enough from one another.

Furthermore, the starting hypothesis is that there exists, a priori, a solution γ_1. ∎

Examples 12.57 The restriction on $t_1 - T_0$ is crucial, as the following elementary example clarifies. Consider the Lagrangian on $\mathbb{R}^2$:

$$\mathscr{L}(q,\dot{q}) = \frac{m}{2}\dot{q}^2 - \frac{k}{2}q^2,$$

with $m, k > 0$ given. We do not add Lagrangian components. Referring to the previous construction, we can fix the reference solution γ_1 to be the constant curve $q(\tau) = 0$, for any $\tau \in \mathbb{R}$. Thus we have $Q_0 = q_1 = 0$ irrespective of T_0 and t_1, and the initial tangent vector is $\dot{Q}_1 = 0$. The first step to build the family of solutions $q'_{\gamma_1,T,t,Q,q}$ is to invert the function (with initial configuration Q near Q_0, at time T close to T_0) that gives q at the final time t (close to t_1) in terms of the corresponding initial tangent vector $\dot{Q}$ at T. This is possible because $\frac{dq}{d\dot{Q}}|_{(T_0,t_1,Q_0,\dot{Q}_1)} \neq 0$. Choosing $T_0 = 0$ and t_1 suitably, the map $\dot{Q} \mapsto q(t)$ is pathological and might not satisfy $\frac{dq}{d\dot{Q}}|_{(T_0,t_1,Q_0,\dot{Q}_1)} \neq 0$. Indeed, the solutions to the

Euler-Lagrange equations with the given Lagrangian are:

$$q(t) = Q \cos\left(\sqrt{\frac{k}{m}}t\right) + \sqrt{\frac{m}{k}}\dot{Q} \sin\left(\sqrt{\frac{k}{m}}t\right)$$

and so

$$\frac{dq}{d\dot{Q}} = \sin\left(\sqrt{\frac{k}{m}}t\right).$$

Choosing

$$t = t_1 := \pi\sqrt{\frac{m}{k}}$$

annihilates the above derivative. The pathology is evident in Fig. 12.2. All possible initial velocities $\dot{Q}_1$ (with $Q_0 = 0$ at $T_0 = 0$) produce solutions ending up at the same configuration $q_1 = 0$ for $t = t_1$. The solutions will not all converge to a common configuration if we change t_1 to $t_1 - \epsilon > 0$ ($\epsilon > 0$) and making it closer to $T_0 = 0$. If so, the construction can be continued.

Let us now return to the starting idea and Proposition 12.55, staying in the natural local chart of $\mathbb{V}^{n+1}$ we have given. Henceforth we shall only look at physical systems whose Lagrangian does not have the Lagrangian components of other forces.

Fix the first endpoint $(T, Q) \in (T_0 - \delta, T_0 + \delta) \times B_r(Q_0)$. From the solutions to the Euler-Lagrange equations of the above family we can define the following

Fig. 12.2 Concurrent solutions for $\mathcal{L}(x, \dot{x}) = \frac{m}{2}\dot{x}^2 - \frac{k}{2}x^2$

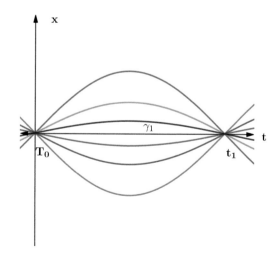

function, called **Hamilton's principal function**:

$$S_{\gamma_1,T,Q}(t,q) := \int_T^t \mathscr{L}\left(\tau, q'_{\gamma_1,T,t,Q,q}(\tau), \frac{\partial q'_{\gamma_1,T,t,Q,q}(\tau)}{\partial \tau}\right) d\tau, \qquad (12.99)$$

where

$$(t,q) \in (t_1 - \delta, t_1 + \delta) \times B_r(q_1).$$

This is well defined once we have a configuration (t_1, q_1) along a solution γ_1 from (T, Q) with t_1 close to T. In particular, $S_{\gamma_1,T,Q}$ depends on the choice of such solution γ_1, not only on the endpoints $(T, Q) \in (T_0 - \delta, T_0 + \delta) \times B_r(Q_0)$ and $(t,q) \in (t_1 - \delta, t_1 + \delta) \times B_r(q_1)$ of the solutions to the Euler-Lagrange equations appearing in the Lagrangian.

This function has a remarkable property that confirms the idea discussed at the beginning.

Proposition 12.58 *Hamilton's principal function $S_{\gamma_1,T,Q}$, defined in (12.99), satisfies the Hamilton-Jacobi equation (12.93) and the fist equation in (12.94).*

Proof Since γ_1, T and Q are fixed, we will write $q'_{t,q}$ instead of $q'_{\gamma_1,T,t,Q,q}$, and sometimes just q', even dropping the t and the q. Similarly, S will stand for $S_{T,Q}$.

By Definition (12.99) we have:

$$\frac{\partial S}{\partial t} = \mathscr{L}\left(t, q'_{t,q}(t), \frac{\partial q'_{t,q}(\tau)}{\partial \tau}\Big|_{\tau=t}\right) + \int_T^t \sum_k \frac{\partial \mathscr{L}}{\partial q'^k}\frac{\partial q'^k}{\partial t}d\tau + \int_T^t \sum_k \frac{\partial \mathscr{L}}{\partial \dot{q}'^k}\frac{\partial^2 q'^k}{\partial t \partial \tau}d\tau.$$

The last integral, after we swap the derivatives in t and τ by Schwarz's theorem, can be computed by parts:

$$-\int_T^t \sum_k \frac{\partial}{\partial \tau}\left(\frac{\partial \mathscr{L}}{\partial \dot{q}'^k}\right)\frac{\partial q'^k}{\partial t}d\tau + \int_T^t \frac{\partial}{\partial \tau}\sum_k \frac{\partial \mathscr{L}}{\partial \dot{q}'^k}\frac{\partial q'^k}{\partial t}d\tau.$$

Substituting the expression of $\frac{\partial S}{\partial t}$ we find:

$$\frac{\partial S}{\partial t} = \mathscr{L}\left(t, q'_{t,q}(t), \frac{\partial q'_{t,q}(\tau)}{\partial \tau}\Big|_{\tau=t}\right) + \int_T^t \sum_k \left(\frac{\partial \mathscr{L}}{\partial q'^k} - \frac{\partial}{\partial \tau}\left(\frac{\partial \mathscr{L}}{\partial \dot{q}'^k}\right)\right)\frac{\partial q'^k}{\partial t}d\tau$$

$$+ \sum_k \frac{\partial \mathscr{L}}{\partial \dot{q}'^k}\frac{\partial q'^k}{\partial t}\Big|_{\tau=t} - \sum_k \frac{\partial \mathscr{L}}{\partial \dot{q}'^k}\frac{\partial q'^k}{\partial t}\Big|_{\tau=T}.$$

That is, recalling that the curve q' solves the Euler-Lagrange equations:

$$\frac{\partial S}{\partial t} = \mathcal{L}\left(t, q'_{t,q}(t), \frac{\partial q'_{t,q}(\tau)}{\partial \tau}\bigg|_{\tau=t}\right) + \sum_k \frac{\partial \mathcal{L}}{\partial \dot{q}'^k}\frac{\partial q'^k}{\partial t}\bigg|_{\tau=t} - \sum_k \frac{\partial \mathcal{L}}{\partial \dot{q}'^k}\frac{\partial q'^k}{\partial t}\bigg|_{\tau=T}.$$

The last summand vanishes because $q'_{t,q}(T) = Q$ is constant in t. Similarly, recalling $q'_{t,q}(t) = q$ and differentiating in t, we obtain:

$$\frac{\partial q'^k}{\partial \tau}\bigg|_{\tau=t} + \frac{\partial q'^k}{\partial t}\bigg|_{\tau=t} = 0.$$

Substituting above we end up with:

$$\frac{\partial S}{\partial t} = \mathcal{L}\left(t, q'_{t,q}(t), \frac{\partial q'_{t,q}(\tau)}{\partial \tau}\bigg|_{\tau=t}\right) - \sum_k \frac{\partial \mathcal{L}}{\partial \dot{q}'^k}\frac{\partial q'^k}{\partial \tau}\bigg|_{\tau=t}.$$

In other words, if $\mathcal{H}$ is the Hamiltonian associated with $\mathcal{L}$:

$$\frac{\partial S}{\partial t}\bigg|_{(t,q)} = -\mathcal{H}(t, q, \dot{q}),$$

where $\dot{q} = \frac{\partial q'_{t,q}(\tau)}{\partial \tau}\big|_{\tau=t}$ is determined by the special family of solutions $q'_{t,q} = q'_{\gamma_1, T, t, Q, q}$ we are using. Passing to the Hamiltonian formulation, if p_k is the conjugate momentum of q^k, we can write:

$$\frac{\partial S(t, q)}{\partial t} + \mathcal{H}(t, q, p) = 0.$$

To finish it is then enough to show:

$$\frac{\partial S(t, q)}{\partial q^h} = p_h(t, q, \dot{q}),$$

where $\dot{q}$ is as above. Let us start from (12.99) and this time differentiate in q^h. Similar calculations involving the Euler-Lagrange equations produce:

$$\frac{\partial S}{\partial q^h} = \sum_k \frac{\partial \mathcal{L}}{\partial \dot{q}'^k}\frac{\partial q'^k}{\partial q^h}\bigg|_{\tau=t} - \sum_k \frac{\partial \mathcal{L}}{\partial \dot{q}'^k}\frac{\partial q'^k}{\partial q^h}\bigg|_{\tau=T}.$$

The last term is zero since $q'_{t,q}(T) = Q$ is constant in q. Analogously, differentiating $q'_{t,q} = q$ in q^h gives:

$$\frac{\partial q'^k}{\partial q^h}\bigg|_{\tau=t} = \delta_h^k.$$

Substituting, we find:

$$\left.\frac{\partial S}{\partial q^h}\right|_{(t,q)} = \left.\frac{\partial \mathscr{L}}{\partial \dot{q}^h}\right|_{(t,q,\dot{q})} = p_h(t,q,\dot{q}) \,, \tag{12.100}$$

where again $\dot{q} = \left.\frac{\partial q'_{t,q}(\tau)}{\partial \tau}\right|_{\tau=t}$ is determined by the special family of solutions $q'_{t,q} = q'_{\gamma_1,T,t,Q,q}$. This ends the proof. $\qquad\qquad\qquad\qquad\qquad\qquad\qquad\qquad\square$

At last we can prove the announced existence result.[14]

Theorem 12.59 *Consider a C^4 Hamiltonian $\mathscr{H}$ defined on a local canonical chart of $F(\mathbb{V}^{n+1})$*

$$\psi : U \ni a \mapsto (t, q^1(a), \ldots, q^n(a), p_1(a), \ldots, p_n(a)) \in \mathbb{R} \times \mathbb{R}^{2n} \,.$$

Suppose that for $a_1 \in U$:

$$det \left.\left[\frac{\partial^2 \mathscr{H}}{\partial p_h \partial p_k}\right]_{h,h=1,\ldots,n}\right|_{a_1} \neq 0 \,. \tag{12.101}$$

Then there exists a complete, jointly C^3 solution to the Hamilton-Jacobi

$$S = S(t, q^1, \ldots, q^n, P_1, \ldots, P_n)$$

where the $(2n+1)$-tuples $(t, q^1, \ldots, q^n, P_1, \ldots, P_n)$ live in an open neighbourhood $O \subset \mathbb{R}^{2n+1}$ of $(t(a_1), q(a_1), P_0)$ for some $P_0 \in \mathbb{R}^n$ such that

$$U_{a_1} := \left\{ \left(t, q^1, \ldots, q^n, \left.\frac{\partial S}{\partial q^1}\right|_{(q,P)}, \ldots, \left.\frac{\partial S}{\partial q^n}\right|_{(t,q,P)}\right) \,\middle|\, (t,q,P) \in O \right\} \tag{12.102}$$

is an open neighbourhood of a_1.

Proof Let us work in coordinates and consider an open neighbourhood of $a_1 \in F(\mathbb{V}^{n+1})$. Condition (12.101) implies, possibly restricting ourselves closer to state a_1, that the inverse Legendre transformation $\mathbb{L}^{-1}$

$$t = t \,, \quad q^k = q^k \,, \quad \dot{q}^k = \frac{\partial \mathscr{H}}{\partial p_k} \quad \text{for } k = 1, \ldots, n,$$

---

[14] Strangely enough, in [FaMa02, p. 417] it is said that "it can be proved that the Hamilton-Jacobi equation does not always admit a complete integral, for instance around an equilibrium point." If taken literally this is false, unless we understand the term *complete integral* in a different way, or think of pathological Hamiltonians.

is a diffeomorphism from the neighbourhood to some corresponding neighbourhood of the image $\mathbb{L}^{-1}(a_1) \in A(\mathbb{V}^{n+1})$. On the latter we define the C^4 Lagrangian:

$$\mathcal{L}(t, q, \dot{q}) := \sum_{k=1}^{n} p_k(t, q, \dot{q})\dot{q}^k - \mathcal{H}(t, q, p(t, q, \dot{q})) .$$

Clearly $\frac{\partial^2 \mathcal{L}}{\partial \dot{q}^k \partial \dot{q}^h}$ is everywhere non-singular because its Legendre transformation inverts $\mathbb{L}$ and is therefore defined near $\mathbb{L}^{-1}(a)$. Consider then a solution $[T_0, t_1] \ni \tau \mapsto \gamma_1(\tau) = (\tau, q^1(\tau), \ldots, q^n(\tau))$ of the Euler-Lagrange equations, in coordinates, such that, putting $t_1 := t(a_1)$, we have $\gamma_1(t_1) = q(a_1)$. Consider an instant $T_0 < t_1$ close to t_1, and set $Q_0 := \gamma_1(T_0)$ so to have the family $q_{\gamma_1, T, t, Q, q}$ of solutions of Proposition 12.55.

Finally, define Hamilton's characteristic function $S_{\gamma_1, T, Q} = S_{\gamma_1, T, Q}(t, q)$ with Proposition 12.58, fixing $T = T_0$. For any Q around Q_0, the functions $S_{\gamma_1, T_0, Q}$ satisfy the Hamilton-Jacobi equation. To finish it suffices to prove that

$$\det \left[\frac{\partial^2 S_{\gamma_1, T, S, Q}}{\partial q^h \partial Q^k} \right]_{h, k = 1, \ldots, n} \bigg|_{(T_0, t_1, Q_0, q_1)} \neq 0 . \tag{12.103}$$

In fact, we can rename the Q^k by P_k, so the complete solution of the Hamilton-Jacobi equation is:

$$S = S(t, q, P) := S_{\gamma_1, T_0, P}(t, q)$$

on a neighbourhood of $(t(a_1), q(a_1), Q_0)$ (the function in (12.103) will not vanish around the point, by continuity). In particular, the set U_{a_1} in (12.102) contains the point describing the state a_1 in canonical initial coordinates, since (12.100) forces in particular $(P_0 := Q_0)$

$$\frac{\partial S_{\gamma_1, T_0, P_0}}{\partial q^k}\bigg|_{(t_1, q_1)} = p_k(a_1) .$$

That U_{a_1} can be taken open follows from Proposition 12.49, where U_{a_1} corresponds to the open set U_{a_0} and where a_0 is our a_1.

Finally, there remains to prove (12.103). An essentially identical computation to the one producing (12.100) shows that:

$$\frac{\partial S_{\gamma_1, T, Q}}{\partial Q^h}\bigg|_{(T, Q, t, q)} = -\frac{\partial \mathcal{L}}{\partial \dot{q}^h}\bigg|_{(T, Q, \dot{Q})} .$$

Hamilton's principal function is here viewed as function of the Lagrangian coordinates (t, q) and the parameters (T, Q), and we have set $\dot{Q} := \frac{\partial q'_{\gamma_1, T, t, Q, q}(\tau)}{\partial \tau}\big|_{\tau = T}$. As the above right-hand side can only depend on q^k via $\dot{Q}^h(T, Q, t, q)$, we have:

$$\frac{\partial^2 S_{\gamma_1, T, Q}}{\partial q^k \partial Q^h}\bigg|_{(T_0, Q_0, t_1, q_1)} = -\sum_r \frac{\partial^2 \mathscr{L}}{\partial \dot{q}^r \partial \dot{q}^h}\bigg|_{(T_0, Q_0, \dot{Q}(T_0, Q_0, t_1, q_1))} \frac{\partial \dot{Q}^r}{\partial q^k}\bigg|_{(T_0, Q_0, t_1, q_1)},$$
$$(12.104)$$

where $\dot{Q}(T, Q, t, q)$ is the tangent vector at time $\tau = T$ of the solution from (T, Q) to (t, q). At the beginning of the proof of Proposition 12.55 we showed that the inverse matrix of $\left[\frac{\partial \dot{Q}^r}{\partial q^k}\right]$ exists at (T_0, Q_0, t_1, q_1) (there, $\dot{Q}(T_0, Q_0, t_1, q_1)$ was called $\dot{Q}_1$). Hence the matrix

$$\frac{\partial \dot{Q}^r}{\partial q^k}\bigg|_{(T_0, Q_0, t_1, q_1)}$$

has non-zero determinant. As we have seen, the Hessian

$$\frac{\partial^2 \mathscr{L}}{\partial \dot{q}^r \partial \dot{q}^h}\bigg|_{(T_0, Q_0, \dot{Q}(T_0, Q_0, t_1, q_1))}$$

has non-zero determinant, so (12.104) implies (12.103), and the proof is concluded.
□

Remarks 12.60 In view of (12.100), condition (12.103) means that the momenta p at the final time t on the family of solutions are diffeomorphisms of the initial configurations Q at time T (always on small enough neighbourhoods of (T_0, Q_0) and (t_1, q_1)). ∎

Examples 12.61 Consider the simple one-dimensional harmonic oscillator with Lagrangian

$$\mathscr{L}(q, \dot{q}) = \frac{1}{2}\dot{q}^2 - \frac{1}{2}q^2$$

on the spacetime of kinetic states $A(\mathbb{R}^{1+1})$. Its Hamiltonian version on $F(\mathbb{R}^{1+1})$ corresponds to the everywhere defined Hamiltonian

$$\mathscr{H}(q, p) = \frac{1}{2}p^2 + \frac{1}{2}q^2.$$

We want to find Hamilton's principal function $S_{T, Q}(t, q)$ for generic T, Q, t, q. The solutions to the equation of motion all have the form

$$q(\tau) = A \cos t + B \sin t.$$

Imposing $q(T) = Q$ and $q(t) = q$ we obtain the general expression:

$$q(\tau) = \frac{q \sin(\tau - T) + Q \sin(t - \tau)}{\sin(t - T)}, \qquad \tau \in \mathbb{R}.$$

Hence:

$$\dot{q}(\tau) = \frac{q \cos(\tau - T) - Q \cos(t - \tau)}{\sin(t - T)}, \qquad \tau \in \mathbb{R}.$$

From the second one:

$$\frac{\partial q}{\partial \dot{Q}} = \sin(t - T),$$

where $\dot{Q}$ is the velocity at initial time T of the solution that starts at Q and reaches q at time $t > T$. This agrees perfectly with the general result (12.98), and goes to show that the local construction of Hamilton's principal function is feasible. To finish, we also have:

$$\mathscr{L}(q(\tau), \dot{q}(\tau)) = \frac{q^2 \cos(2(\tau - T)) + Q^2 \cos(2(t - \tau)) - 2q Q \cos(\tau - (t + T))}{2 \sin^2(t - T)}.$$

At this point we may integrate from T to t to find Hamilton's principal function:

$$S_{T,Q}(t, q) = \frac{(q^2 + Q^2) \cos(t - T) - 2q Q}{2 \sin(t - T)}.$$

In particular, (12.92) holds around any (t, q) by suitably choosing $T = T_0$, because

$$\frac{\partial^2 S_{T,Q}}{\partial q \partial Q} = -\frac{1}{\sin(t - T)},$$

which becomes singular only at $t - T = k\pi$ and $k \in \mathbb{Z}$. Hence, around any $(t, q) \in \mathbb{R}^2$ we can always find a complete solution of the Hamilton-Jacobi equation, choosing the reference time $T = T_0$ in a suitable way.

12.4.5 Time-Independent Hamilton-Jacobi Equation

An interesting case is that where the Hamiltonian in canonical coordinates,

$$\mathscr{H} = \mathscr{H}(q, p)$$

does not depend explicitly on time, as in the two cases examined above. Except for special situations, if we are interested in truly solving the Hamilton-Jacobi equation without knowing the solution set to Hamilton's equations (which we did in the theoretical situations of Examples 12.61), it is better to seek solutions parametrised by $P \equiv (P_1, \ldots, P_n)$ of the form:

$$S = W(q, P) - t E(P), \tag{12.105}$$

where E is a *given* C^3 function, while W is an *unknown* C^3 function called **Hamilton-Jacobi characteristic function**. At present the Hamilton-Jacobi equation (12.93) becomes:

$$E(P) = \mathscr{H}(q, \nabla_q W), \tag{12.106}$$

and (12.92) reads:

$$\left(\det \left[\frac{\partial^2 S}{\partial q^k \partial P_h} \right]_{k,h=1,\ldots,n} = \right) \det \left[\frac{\partial^2 W}{\partial q^k \partial P_h} \right]_{k,h=1,\ldots,n} \neq 0. \tag{12.107}$$

If, for a time-independent Hamiltonian system:

$$\mathscr{H}(q^1, \ldots, q^n, p_1, \ldots, p_n) = h(q^1, p_1) + \mathscr{H}'(q^2, \ldots, q^n, p_2, \ldots, p_n),$$

the variable q^1 is called **separable** if there exists a characteristic function W of the form:

$$W(q^1, \ldots, q^n, P_1, \ldots, P_n) = w(q^1, P_1) + W'(q^2, \ldots, q^n, P_2, \ldots, P_n).$$

Going back to the general setting, once we have found a characteristic function W for a given function $E = E(P)$, we may use two methods to find, locally, the general solution to Hamilton's equations. The first is to consider Hamilton's principal function:

$$S(q, P) := W(q, P) - t E(P) \tag{12.108}$$

and proceed exactly as in the previously discussed general case. The second way consists in defining, locally, a *completely canonical* transformation by imposing:

$$t = t, \tag{12.109}$$

$$p_k = \frac{\partial W}{\partial q^k}, \quad k = 1, \ldots, n \tag{12.110}$$

$$Q^k = \frac{\partial W}{\partial P_k}, \quad k = 1, \ldots, n. \tag{12.111}$$

Now, (12.107) guarantees that, locally, the above relations can be understood as a C^2 bijection with C^2 inverse $Q = Q(q)$. Further, such transformation is completely canonical by (d) in Theorem 12.48, since:

$$dW = \sum_k \frac{\partial W}{\partial q^k} dq^k + \sum_k \frac{\partial W}{\partial P_k} dP_k = \sum_k p_k dq^k + \sum_k Q^k dP_k ,$$

and computing the differential:

$$0 = \sum_k dp_k \wedge dq^k + \sum_k dQ^k \wedge dP_k ;$$

i.e.:

$$\sum_k dq^k \wedge dp_k = \sum_k dQ^k \wedge dP_k .$$

Equation (12.106), plus (4) of Theorem 12.48, says that the Hamiltonian $\mathscr{H}$ in the new coordinates Q, P is:

$$\mathscr{H}(Q, P) = E(P)$$

and therefore Hamilton's equations in the new variables trivially give:

$$P_k(t) = P_k(0) \quad k = 1, \ldots, n , \tag{12.112}$$

$$Q^k(t) = \frac{\partial E}{\partial P_k}|_{P(0)} t + Q^k(0) , \quad k = 1, \ldots, n . \tag{12.113}$$

Examples 12.62

(1) Let us begin with the harmonic oscillator on $F(\mathbb{R}^{1+1})$, with Hamiltonian:

$$\mathscr{H}(q, p) = \frac{p^2}{2m} + \frac{k}{2} q^2 = \frac{1}{2m} \left(p^2 + m^2 \omega^2 q^2 \right) ,$$

where $m, k > 0$ are given and $\omega := \sqrt{\frac{k}{m}}$. We have already found a principal function. Equation (12.106) now reads

$$E(P) = \frac{1}{2m} \left[\left(\frac{\partial W}{\partial q} \right)^2 + m^2 \omega^2 q^2 \right] .$$

Since we only need one parameter P to obtain a complete integral we can directly set

$$P := E .$$

Evidently $E \geq 0$, representing the total mechanical energy, and the above can be written:

$$\frac{\partial W}{\partial q} = \pm\sqrt{2mE}\sqrt{1 - \frac{m\omega^2 q^2}{2E}},$$

with general solution

$$W(q, E) = \pm\frac{1}{2}\sqrt{2mE}\left[q\sqrt{1 - \frac{m\omega^2 q^2}{2E}} + \sqrt{\frac{2E}{m\omega^2}}\arcsin\left(\sqrt{\frac{m\omega^2}{2E}}q\right)\right] + C$$

Furthermore:

$$\frac{\partial^2 W}{\partial q \partial E} = \pm\frac{m}{\sqrt{2mE - m^2\omega^2 q^2}}.$$

At given mechanical energy value E, the region accessible to the motion is $|q| \leq \sqrt{\frac{2E}{m\omega^2}}$; the case $E = 0$ is therefore not treatable by our procedure. In the other cases the mixed derivative is non-zero, but singular for $|q| \neq \sqrt{\frac{2E}{m\omega^2}}$. We have then a pair of complete solutions to the Hamilton-Jacobi equation of the form (12.108) if $E > 0$ and $q \neq \pm\sqrt{\frac{2E}{m\omega^2}}$. The constant C does not play any part since it vanishes when we differentiate. Proceeding with the canonical transformation we have:

$$Q = \frac{\partial W}{\partial E} = \frac{1}{\omega}\arcsin\left(\sqrt{\frac{m\omega^2}{2E}}q\right),$$

$$p = \frac{\partial W}{\partial q} = \pm\sqrt{2mE}\sqrt{1 - \frac{m\omega^2 q^2}{2E}} = \sqrt{2mE}\cos(\omega Q).$$

In view of (12.112) and (12.113), the general solution to Hamilton's equations in the initial coordinates is:

$$q(t) = \sqrt{\frac{2E}{m\omega^2}}\sin(\omega t + Q(0)), \quad p(t) = \sqrt{2mE}\cos(\omega t + Q(0)).$$

At this point the constants E and $Q(0)$ can be determined from the initial conditions.

(2) Consider the Lagrangian of a particle of mass $m > 0$ described in the spherical coordinates r, θ, ϕ of a frame $\mathscr{R}$ (which will not be indicated to lighten up the notation):

$$\mathscr{L}(r, \dot{r}, \theta, \phi, \dot{\theta}, \dot{\phi}) = \frac{m\dot{r}^2}{2} + \frac{mr^2}{2}\left(\dot{\theta}^2 + \dot{\phi}^2\sin^2\theta\right) - \mathscr{U}(r, \theta, \phi).$$

The Legendre transform immediately leads to the Hamiltonian, which coincides with the total mechanical energy in $\mathscr{R}$,

$$\mathscr{H} = \frac{p_r^2}{2m} + \frac{p_\theta^2}{2mr^2} + \frac{p_\phi^2}{2mr^2 \sin^2 \theta} + \mathscr{U}(r, \theta, \phi), \tag{12.114}$$

where:

$$p_r = m\dot{r}, \quad p_\theta = mr^2\dot{\theta}, \quad p_\phi = mr^2 \sin^2 \theta \dot{\phi}.$$

We shall finally supoose:

$$\mathscr{U}(r, \theta, \phi) = U_1(r) + \frac{U_2(\theta)}{r^2} + \frac{U_3(\phi)}{r^2 \sin^2 \theta}, \tag{12.115}$$

and we claim that in this way the three coordinates are separable. This decomposition includes the case, perhaps the most important of all, of a central conservative force where $U_2 = U_3 = 0$. We want to find S of the form (12.105) by using a characteristic function $W(r, \theta, \phi)$ that will consist of three terms:

$$S(r, \theta, \phi; P_1, P_2, P_3) = W_1(r, P_1, P_2, P_3) + W_2(\theta, P_1, P_2, P_3)$$
$$+ W_3(\phi, P_1, P_2, P_3) - tE(P_1, P_2, P_3).$$

Equation (12.106) can be expressed as:

$$E = \frac{1}{2m} \left[\left(\frac{dW_1}{dr} \right)^2 + 2mU_1(r) \right] + \frac{1}{2mr^2} \left[\left(\frac{dW_2}{d\theta} \right)^2 + 2mU_2(\theta) \right]$$
$$+ \frac{1}{2mr^2 \sin^2 \theta} \left[\left(\frac{dW_3}{d\phi} \right)^2 + 2mU_3(\phi) \right]$$

As the functions depend on distinct variables, the three square brackets must be constants. Thus we obtain 3 ODEs parametrised by three constants of motion, namely P_1, P_2, P_3 with $P_1 := E$. First of all:

$$\left(\frac{dW_3(\phi, P_3)}{d\phi} \right)^2 + 2mU_3(\phi) = P_3,$$

so:

$$W_3(\phi, P_3) = \pm \int \sqrt{P_3 - 2mU_3(\phi)} d\phi.$$

The above equation for W becomes:

$$E = \frac{1}{2m} \left[\left(\frac{dW_1}{dr} \right)^2 + 2mU_1(r) \right] + \frac{1}{2mr^2} \left[\left(\frac{dW_2}{d\theta} \right)^2 + 2mU_2(\theta) + \frac{P_3}{\sin^2 \theta} \right].$$

The second ODE together with the second constant of motion P_2 will be:

$$\left(\frac{dW_2(\theta, P_2, P_3)}{d\theta} \right)^2 + 2mU_2(\theta) + \frac{P_3}{\sin^2 \theta} = P_2,$$

and so:

$$W_2(\theta, P_2, P_3) = \pm \int \sqrt{P_2 - 2mU_2(\theta) - \frac{P_3}{\sin^2 \theta}} d\theta.$$

The equation for W now is, with the third constant of motion $P_3 := E$:

$$E = \frac{1}{2m} \left[\left(\frac{dW_1(r, P_2, E)}{dr} \right)^2 + 2mU_1(r) + \frac{P_2}{r^2} \right],$$

and then:

$$W_1(r, P_2, E) = \pm \int \sqrt{2mE - 2mU_1(r) - \frac{P_2}{r^2}} dr.$$

Summing up:

$$S(r, \theta, \phi, E, P_2, P_3) = \pm \int \sqrt{2mE - 2mU_1(r) - \frac{P_2}{r^2}} dr$$

$$\pm \int \sqrt{P_2 - 2mU_2(\theta) - \frac{P_3}{\sin^2 \theta}} d\theta \pm \int \sqrt{P_3 - 2mU_3(\phi)} d\phi - tE.$$

12.4.6   Local Existence of Solutions for Boundary Value Problems of Order Two on Manifolds

The argument employed to prove Proposition 12.55 also allows to state a fairly general theorem for second-order *boundary value* problems on manifolds. Similarly, part (2) of the statement allows to generalise the *exponential map* [KoNo63], which was originally defined for manifolds with an affine (usually metric) connection. We will extend that setup to second-order normal equations different from the geodesic flow equations.

Let M be a C^k manifold of dimension n, with $k \geq 2$, and $\pi : TM \to M$ the canonical projection $TM \ni (p, u) \mapsto p \in M$. Consider the boundary value problem on TM:

$$\begin{cases} \dfrac{d\gamma}{dt} = Y(\gamma(t)) \,, \\ \pi(\gamma(0)) = p_0 \,, \quad \pi(\gamma(t_1)) = p_1 \,, \end{cases} \tag{12.116}$$

where Y is the C^k vector field on TM

$$Y(v) = \sum_{a=1}^{n} \dot{x}^a(v) \left. \frac{\partial}{\partial x^a} \right|_v + \sum_{a=1}^{n} Y^a(v) \left. \frac{\partial}{\partial \dot{x}^n} \right|_v \tag{12.117}$$

in any natural local coordinates $\psi : TU \ni v \mapsto (x^1(v), \ldots, x^n(v), \dot{x}^1(v), \ldots, \dot{x}^n(v)) \in \mathbb{R}^n \times \mathbb{R}^n$ on TM. The ODE in (12.116), together with (12.117), is the most natural way to define intrinsically a system of differential equations of *order two* (in normal form) on a differentiable manifold.

By definition (Sect. 14.5.3) the *flow of Y* is $\phi_t^{(Y)}(v) := \gamma_v(t)$, where $\gamma_v(t)$ is the maximal solution of the Cauchy problem:

$$\begin{cases} \dfrac{d\gamma}{dt} = Y(\gamma(t)) \,, \\ \gamma(0) = v \,, \end{cases} \tag{12.118}$$

defined on the maximal open interval $I_v \ni 0$, at $t \in I_v$. We have:

$$\phi^{(Y)} : \Gamma_Y \ni (t, v) \mapsto \phi_t^{(Y)}(v) \in TM \,,$$

where the domain

$$\Gamma_Y = \bigcup_{v \in TM} I_v \times \{v\} \subset \mathbb{R} \times TM$$

is an *open* set. The flow $\phi^{(Y)}$ has the same regularity as Y, at present C^2, at (t, v) (and the same for $\frac{\partial \phi_t^{(Y)}}{\partial t}$).

Let us state the result we mentioned.

Proposition 12.63 *Consider the boundary value problem (12.116) with $Y \in C^2(TM)$ of the form (12.117) in any natural coordinates on TM.*

(1) For any $p_0 \in M$ there exist an instant $t_1 > 0$ and a ball $B_r(p_0)$ centred at p_0 of radius $r > 0$, both small enough so that (12.116) admits a solution for any $p_1 \in B_r(p_0)$.

(2) *The exists an open neighbourhood $V \subset T_{p_0}M$ of the origin such that, with the definitions of (1),*

$$\exp_{p_0}^{(Y,t_1)} : V \ni u \mapsto \phi_{t_1}^{(Y)}(p_0, u) \in B_r(p_0) \tag{12.119}$$

is a diffeomorphism.

Proof As Γ_X is open, in any natural chart of TM

$$T\psi : TU \ni v \mapsto (x^1(v), \ldots, x^n(v), \dot{x}^1(v), \ldots, \dot{x}^n(v)) \in \mathbb{R}^n \times \mathbb{R}^n$$

with $U \ni p_0$, the flow can be locally written as:

$$J \times TU \ni (t, x, \dot{x}) \mapsto (x_t(x, \dot{x}), \dot{x}_t(x, \dot{x})) \in TU,$$

restricting TU, where $J \ni 0$ is a small open interval. Consider, for given $t \in J$, the map:

$$F^{(t)} : TU \ni (x, \dot{x}) \mapsto (x, x_t(x, \dot{x})) \in U \times U$$

in the local chart. We claim that if t is small enough, then $dF^{(t)}_{(x(p_0),0)}$ is non-singular. As the first component of $F^{(t)}$ is the identity, it is enough to show

$$\det \left[\frac{\partial x_t^j}{\partial \dot{x}^i} \bigg|_{(x(p_0),0)} \right]_{i,j=1,\ldots,n} \neq 0. \tag{12.120}$$

Varying t we can expand the map in brackets in Taylor series at $t = 0$:

$$\frac{\partial x_t^j}{\partial \dot{x}^i} \bigg|_{(x(p_0),0)} = \frac{\partial x_0^j}{\partial \dot{x}^i} \bigg|_{(x(p_0),0)} + t \frac{\partial^2 x_t^j}{\partial t \partial \dot{x}^i} \bigg|_{t=0,(x(p_0),0)} + t O_{(x(p_0),0)\,i}^j(t).$$

We have

$$\frac{\partial x_0^j}{\partial \dot{x}^i} \bigg|_{(x(p_0),0)} = 0$$

since $x_0(x, \dot{x}) = x$ at $t = 0$. Moreover, as the function to be differentiated is C^2,

$$\frac{\partial^2 x_t^j}{\partial t \partial \dot{x}^i} \bigg|_{t=0,(x(p_0),0)} = \frac{\partial^2 x_t^j}{\partial \dot{x}^i \partial t} \bigg|_{t=0,(x(p_0),0)} = \frac{\partial \dot{x}^j}{\partial \dot{x}^i} \bigg|_{t=0,(x(p_0),0)} = \delta_i^j.$$

Hence:

$$\frac{\partial x_t^j}{\partial \dot{x}^i}\bigg|_{(x(p_0),0)} = t\delta_i^j + tO_{(x(p_0),0)\,i}^j(t)\,,$$

implying:

$$\det\left[\frac{\partial x_t^j}{\partial \dot{x}^i}\bigg|_{(x(p_0),0)}\right]_{i,j=1,\dots,n} = tn + tO_{(x(p_0),0)}(t)\,.$$

It is then obvious that if $0 < t < T_{p_0}$ for some $T_{p_0} > 0$ then (12.120) is true, and so $dF_{(x(p_0),0)}^{(t)} \neq 0$. Consequently $F^{(t)}$ is a local diffeomorphism from a neighbourhood of $(p_0, 0) \in M \times T_{p_0}M$ to a neighbourhood of $(p_0, p_0) \in M \times M$. We can always choose the image neighbourhood to be $B_r(p_0) \times B_r(p_0)$, and taking $B_r(p_0) \times V$ as the pre-image we have proved (2), provided $\exp_{p_0}^{(Y,t_1)}(u)$ is given in coordinates by $x_{t_1}(x(p_0), \dot{x}(u))$, for some $0 < t_1 < T_{p_0}$.

The proof of (1) is now obvious: if $q \in B_r(p_0)$, then $u_q := (\exp_{p_0}^{(Y,t_1)})^{-1}(q)$ is the initial condition for the initial tangent vector at p_0 of a solution to problem (12.118). The latter's solution, projected onto M, departs from p_0 at $t = 0$ and reaches q at time t_1. Such solution therefore solves the boundary value problem (12.116).   □

Exercises 12.64

(1) Take a C^3 Lagrangian $\mathscr{L} : \Omega \times \mathbb{R}^n \to \mathbb{R}$:

$$\mathscr{L}(q, \dot{q}) = \sum_{i,j=1}^n a_{ij}(q)\dot{q}^i\dot{q}^j - \mathscr{U}(q)\,,$$

where $\Omega \in \mathbb{R}^n$ is open and the matrix $a_{ij}(q)$ is non-singular at $q \in \Omega$. Consider Lagrangian components $\mathscr{Q}_k(q, \dot{q})$ of class C^2. Prove that given $q_0 \in \Omega$, there exist a small enough time $t_1 > 0$ and a sufficiently small ball $B_r(q_0)$ centred at q_0 of radius $r > 0$ such that the boundary value problem

$$\begin{cases} \dfrac{d}{dt}\left(\dfrac{\partial \mathscr{L}}{\partial \dot{q}^k}\bigg|_{(q(t),\dot{q}(t))}\right) - \dfrac{\partial \mathscr{L}}{\partial q^k}\bigg|_{(q(t),\dot{q}(t))} = \mathscr{Q}_k(q(t), \dot{q}(t)), \\[2mm] \dfrac{dq^k}{dt} = \dot{q}^k(t)\,, \hspace{4.5cm} \text{for } k = 1, \dots, n, \\[2mm] q^k(0) = q_0^k\,, \quad q(t_1) = q_1^k \end{cases}$$

$$(12.121)$$

admits a solution for any $q_1 \in B_r(q_0)$.

(2) Consider the *repulsive* harmonic oscillator with Lagrangian

$$\mathscr{L}(q, \dot{q}) = \frac{1}{2}\dot{q}^2 + \frac{1}{2}q^2 ,$$

on the spacetime of kinetic states $A(\mathbb{R}^{1+1})$, and its Hamiltonian version on $F(\mathbb{R}^{1+1})$ corresponding to global Hamiltonian

$$\mathscr{H}(q, p) = \frac{1}{2}p^2 - \frac{1}{2}q^2 .$$

Write down Hamilton's principal function $S_{T, Q}(t, q)$ for generic T, Q, t, q.

Chapter 13
$\boxed{\text{AC}}$ Hamiltonian Symplectic Structures: An Introduction

The final chapter is devoted to formulating Hamiltonian Mechanics on symplectic manifolds and on bundles over symplectic manifolds. We will take the chance to discuss in detail the phase-space action of the Galilean and Poincaré groups in terms of canonical transformations.

This topic requires more advanced mathematics: Hamilton's formulation of Mechanics from Sect. 11.4, vector fields (Sect. A.6.5), their local 1-parameter groups of diffeomorphisms (Sect. 14.5.3), and differential forms on manifolds, including the Lie derivative (Appendix B).

13.1 The Phase Space $\mathbb{F}$ as Symplectic Manifold

We will introduce *symplectic manifolds* and show how they enter Hamiltonian Mechanics when we study **autonomous Hamiltonian systems**, those in which the globally defined Hamiltonian does not depend explicitly on time.

13.1.1 The Symplectic Structure of Autonomous Hamiltonian Systems

To justify, physically, the relevance of a symplectic formulation we start by looking at Hamiltonian systems whose phase spacetime and Hamiltonians satisfy certain additional hypotheses. We have actually already seen these when presenting Poincaré's "recurrence" (Theorem 12.16). Therefore let us consider a Hamiltonian system described on a phase spacetime $F(\mathbb{V}^{n+1})$ with the further requirement that there is an atlas of canonical/natural coordinates $\mathcal{A} = \{(U_i, \psi_i)\}_{i \in I}$ (as always, C^∞) with the following features.

V. Moretti, *Analytical Mechanics*, La Matematica per il 3+2 150,
https://doi.org/10.1007/978-3-031-27612-5_13

(1) For any chart $(U_i, \psi_i) \in \mathcal{A}$, hence $\psi_i : U \ni a \mapsto (t_i(a), x_i^1(a), \ldots, x_i^{2n}(a))$ where t_i is absolute time up to additive constants, we ask $t_i(U) = \mathbb{R}$. In other words the time coordinate covers the whole temporal axis.

(2) The coordinate transformations $\psi_i \circ \psi_j^{-1} : \psi_j(U_i \cap U_j) \to \psi_i(U_i \cap U_j)$ satisfy $t_i(a) = t_j(a)$ for any $a \in U_i \cap U_j$. In other words there is a unique time common to all coordinates.

(3) The coordinate transformations $\psi_i \circ \psi_j^{-1} : \psi_j(U_i \cap U_j) \to \psi_i(U_i \cap U_j)$ are completely canonical. In other words $\psi_i \circ \psi_j^{-1}$ reads, in coordinates, $t = t$ and $\mathbf{x}_i = \mathbf{x}_i(\mathbf{x}_j)$, where time does not appear in the second transformation.

Then we can identify $F(\mathbb{V}^{n+1})$ with $\mathbb{R} \times \mathbb{F}$. $\mathbb{F}$ is the generic fibre $\mathbb{F}_{t_0}$ as differentiable manifold with atlas $\mathcal{A}' = \{(V_i, \psi_i')\}_{i \in I}$, where $V_i := U_i \cap \mathbb{F}_{t_0}$ and $\psi_i' : V_i \ni r \mapsto (x_i(r), \ldots, x_i^{2n}(r))$; the x_i^k are the non-temporal coordinates of (U_i, ψ_i).

The diffeomorphism $f : \mathbb{R} \times \mathbb{F}_{t_0} \to F(\mathbb{V}^{n+1})$ we refer to is completely determined by asking, with reference to the atlases $\mathcal{A}$, $\mathcal{A}'$:

$$\psi_i \circ f(t, \psi_i'^{-1}(x_i^1, \ldots, x_i^n)) = (t, x_i^1, \ldots, x_i^n)$$

for any given $i \in I$. Thus $T(f(t, r)) = t$ for any $(t, r) \in \mathbb{R} \times \mathbb{F}_{t_0}$. The diffeomorphism f is therefore a *global trivialisation* of the bundle $F(\mathbb{V}^{n+1})$ over $\mathbb{R}$. It is important to note that the decomposition of $F(\mathbb{V}^{n+1})$ as Cartesian product $\mathbb{R} \times \mathbb{F}$ depends on the choice of $\mathcal{A}$ with the above requirements. In this situation $\mathbb{F}$ is called **phase space** of the system.

We will also assume the following conditions.

(4) Every coordinate system in $\mathcal{A}$ has a Hamiltonian $\mathscr{H}_i'$ such that:

$$\mathscr{H}_i'(a) = \mathscr{H}_j'(a) \quad \text{if } a \in U_i \cap U_j, \quad \text{in coordinates} \quad \mathscr{H}_i'(t, \mathbf{x}_i(\mathbf{x}_j)) = \mathscr{H}_j'(t, \mathbf{x}_j)$$

(in agreement with (2) in Theorem 12.9). Then, we have on $F(\mathbb{V}^{n+1})$ a unique global Hamiltonian $\mathscr{H} : \mathbb{R} \times \mathbb{F} \to \mathbb{R}$, which restricts to $\mathscr{H}_i'$ on each domain U_i of the local chart $(U_i, \psi_i) \in \mathcal{A}$.

(5) The above function $\mathscr{H}$ satisfies $\mathscr{H}(t, r) = \mathscr{H}(t', r)$ for any $t, t' \in \mathbb{R}$ and $r \in \mathbb{F}$.

Remarks 13.1

(1) From (2) in Theorem 12.9 we know that there is no need for topological restrictions on the domains U_i of $\mathcal{A}$.

(2) The existence of an atlas obeying (1), (2) and (3) (equivalently, the existence of the corresponding global trivialisation), just like the existence of a global Hamiltonian, is rather natural from a physical viewpoint. As we said in Sect. 8.5.5 in a Lagrangian setting, physically one expects that the choice of frame $\mathscr{R}$ on the physical space where the constraints, if present, are time-independent (assuming such an $\mathscr{R}$ exists), also determines the following two mathematical objects for the Hamiltonian theory of a concrete physical system.

(a) An atlas $\mathcal{A}_{\mathcal{R}}$ on $F(\mathbb{V}^{n+1})$ obeying (1), (2) and (3). This is built from the collection of all local coordinate systems, *moving with* $\mathcal{R}$, on the *spacetime of configurations* whose time coordinates cover the entire absolute time axis, extending each such system to natural coordinates on $F(\mathbb{V}^{n+1})$ by including all coordinate systems coming from *completely canonical* transformations of the former. In the end $\mathcal{A}_{\mathcal{R}}$ contains in particular a sub-atlas of natural coordinates on $F(\mathbb{V}^{n+1})$ and a sub-atlas of canonical coordinates on $F(\mathbb{V}^{n+1})$.

(b) A Hamiltonian $\mathcal{H}_{\mathcal{R}}$ satisfying (4), i.e. defined globally on the phase spacetime $F(\mathbb{V}^{n+1}) \simeq \mathbb{R} \times \mathbb{F}$. The function $\mathcal{H}_{\mathcal{R}}$ is global precisely because the coordinate transformations of $\mathcal{A}_{\mathcal{R}}$ are completely canonical. If we work with natural coordinates in $\mathcal{A}_{\mathcal{R}}$, this follows anyway from (11.29). (See Remark 8.26 for the Lagrangian version on this.) In general, however, (5) will not hold. ∎

Under (1)–(3), we have on phase space $\mathbb{F}$ a special, global 2-form ω called a **symplectic form**. It is completely determined by the fact that, in any local system $(V, \phi) \in \mathcal{A}'$ as above:

$$\omega = -\frac{1}{2} \sum_{i,k=1}^{2n} S_{ik} dx^i \wedge dx^k = \sum_{l=1}^{n} dp_l \wedge dq^l \qquad (13.1)$$

where $\psi : U \ni a \mapsto (x^1(a), \ldots, x^{2n}(a)) = (q^1(a), \ldots, q^n(a), p_1(a), \ldots, p_n(a))$ in standard notation.

This 2-form is well defined on $\mathbb{F}$ (i.e. it does not depend on the coordinates of the type considered) by (4) in Theorem 12.48. The second equality in (13.1), which we have used already, is immediate from the definition of $\wedge$, the components of S and the decomposition of $x^1, \ldots, x^{2n}$ into n coordinates q^k and n coordinates p_k. Directly from the definition of $\wedge$, the 2-form ω defines, for $r \in \mathbb{F}$, a map:

$$\omega|_r : T_r\mathbb{F} \times T_r\mathbb{F} \ni (u, v) \mapsto \omega(u, v) \in \mathbb{R}$$

that is bilinear, skew-symmetric and non-degenerate. The last property is because the skew-symmetric matrix S_{rs} has non-zero determinant.[1] There is a further property of ω worth mentioning. From (13.1), and since $dd = 0$, we have $d\omega = 0$,

---

[1] In components: $-\omega(u, v) = \sum_{i,j} u^i S_{ij} v^j$. If there were a vector $v = \sum_i v^i \frac{\partial}{\partial x^i}|_r \in T_r\mathbb{F}$ with $\omega(u, v) = 0$ for any $u \in T_r\mathbb{F}$, choosing u with components $u^i = \sum_j S_{ij} v^j$ we would infer $0 = -\omega(u, v) = \sum_i \left(\sum_k S_{ik} v^k, \sum_j S_{ij} v^j \right)$ and so $\sum_k S_{ik} v^k = 0$. But $\det S \neq 0$, so the only possibility is that all v^k vanish, i.e. $v = 0$.

i.e. ω is closed. (The Poincaré lemma implies that ω is locally exact, and indeed we always have $\omega = d\left(\sum_{k=1}^{n} p_k dq^k\right)$ in the coordinates we are using).

13.1.2 Symplectic Manifolds and Hamiltonian Mechanics

The structure detected by $\mathbb{F}$ and ω is well known in Mathematics, and defined as follows, irrespective of the trail we have followed to get here within the general Hamiltonian theory.

Definition 13.2 A pair (M, ω) consisting of a differentiable manifold M of dimension $2n$, $n = 1, 2, \ldots$, and a closed 2-form ω of class C^k, $k \geq 1$, such that, for any $p \in M$, the bilinear skew-symmetric map:

$$\omega|_p : T_p M \times T_p M \ni (u, v) \mapsto \omega(u, v) \in \mathbb{R}$$

is non-degenerate, is called a **symplectic manifold**, and ω is a **symplectic form** on M. $\diamond$

Examples 13.3

(1) The *cotangent bundle* T^*N of any C^k manifold of dimension n (see Sect. B.3) is a symplectic manifold in a natural way. Consider a natural local chart of T^*N with coordinates $x^1, \ldots, x^n, p_1, \ldots, p_n$, built from the local chart (U, ψ) on N of coordinates $x^1, \ldots, x^n$. Define on such a chart the so-called **tautological 1-form**

$$\sum_{k=1}^{n} p_k dx^k \, .$$

The exterior derivative is

$$\sum_{k=1}^{n} dp_k \wedge dx^k \, .$$

It is easy to see that these 2-forms, each one defined on a natural local chart, are the restrictions of a unique global symplectic form on $M := T^*N$. Thus T^*N becomes a symplectic manifold, with canonical symplectic form (13.1) in each natural chart.

(2) The situation illustrated for the initial physical example is the one we obtain in particular when starting from a Lagrangian formulation, as we hinted at earlier. The hypotheses are to have an atlas on $\mathbb{V}^{n+1}$ made of natural charts whose time coordinate varies in the entire $\mathbb{R}$, and whose transition functions and Lagrangian do not depend explicitly on time. This happens for instance when the possible constraints of the physical system are independent of time in a frame $\mathscr{R}$, the atlas consists of charts moving with $\mathscr{R}$, and finally the Lagrangian does not depend on time explicitly in $\mathscr{R}$. Then $\mathbb{V}^{n+1}$ is explicitly diffeomorphic to $\mathbb{R} \times \mathbb{Q}$ (its global trivialisation), where $\mathbb{Q}$ is the bundle's standard fibre: the local charts are natural for the Cartesian product $\mathbb{R} \times \mathbb{F}$. We can define $\mathbb{F} = T^*\mathbb{Q}$ and obtain that $F(\mathbb{V}^{n+1})$ is explicitly diffeomorphic to $\mathbb{R} \times \mathbb{F} = \mathbb{R} \times T^*\mathbb{Q}$. Conditions (i)–(iii) are trivially valid. We fall back in this way to the picture of (2), Remarks 13.1. The coordinates p_k in natural local charts of $T^*\mathbb{Q}$ are effectively coordinates on phase space, with the same name. The local symplectic charts that will be defined shortly, together with absolute time t, belong to the canonical atlas of $F(\mathbb{V}^{n+1}) = \mathbb{R} \times T^*\mathbb{Q}$.

A crucial result due to Darboux (see for example [FaMa02, Wes78]) shows that every symplectic manifold has an atlas in which the symplectic form has the standard form (13.1).

Theorem 13.4 (Second Darboux Theorem) *If (M, ω) is a C^k symplectic manifold, $k \geq 2$, there exists an atlas of M, called* **symplectic atlas**, *that is compatible with the differentiable structure of M and made of local charts in which ω has standard form (13.1), where $n := \dim M$.*

By the above theorem if (U, ψ) and (V, ϕ) belong to the symplectic atlas of M, the coordinate transformation $\psi \circ \phi^{-1}$ defined on $\phi(U \cap V)$ has Jacobian matrix J satisfying $J^t S J = S$, so it is equivalent to a completely canonical transformation of $\mathbb{R} \times M$. If we suppose the symplectic atlas is maximal, the condition $J^t S J = S$ is sufficient as well for the local chart (V, ϕ) to be in the symplectic atlas, provided we use a cover for V made of charts (U, ψ) in the symplectic atlas.

All of this allows us to translate our initial demands on the Hamiltonian system and phase spacetime into a symplectic language. We can model phase spacetime as the product $\mathbb{R} \times \mathbb{F}$, where now $\mathbb{F}$ is a symplectic manifold. The interesting coordinate systems for the Hamiltonian theory will be those defined on domains $\mathbb{R} \times V$, where (V, ϕ) is in the symplectic atlas of $\mathbb{F}$, and the coordinates will always be of type $\mathbb{R} \times V \ni (t, r) \mapsto (t, \phi(r)) \in \mathbb{R} \times \mathbb{R}^{2n}$. If $\mathscr{H} : \mathbb{F} \to \mathbb{R}$ is C^2, then certainly (by (2) in Theorem 12.9 and (4) in Theorem 12.48) Hamilton's equations will be well defined by using these charts (in the symplectic atlas), irrespective of the particular local chart, provided that on every chart with domain $\mathbb{R} \times V$ the Hamiltonian is just the restriction of $\mathscr{H}$. As $\mathscr{H}$ is not a function of time, Hamiltonian Dynamics can be interpreted as the theory of an *autonomous system of differential equations* on $\mathbb{F}$.

Remarks 13.5

(1) Actually, the symplectic approach, albeit staying within autonomous Hamiltonian systems, is slightly more general than the standard one. It does not suppose the structure of phase spacetime is built on a configuration spacetime. In the latter case, in fact, we would need to posit the existence of special atlases on $\mathbb{F}$ in which the coordinates p_k cover all of $\mathbb{R}$, and the coordinate transformations in the atlas must be special, namely the p_k change as components of 1-forms. The existence of such is not guaranteed only by assuming $\mathbb{F}$ is symplectic.

Nonetheless, for the symplectic structure on a cotangent bundle, as seen in Examples 13.3, the atlas of natural local charts on $M := T^*N (= T^*\mathbb{Q}$ in the second, more physically oriented, example) is automatically contained in the symplectic atlas of M, and the p_k of any local chart in it vary over $\mathbb{R}$ by definition.

(2) A generic symplectic manifold $(\mathbb{F}, \omega)$ has a *volume form* (see Sect. B.4):

$$\mu = \frac{(-1)^{n(n+1)/2}}{n!} \omega \wedge \cdots (n \text{ times}) \cdots \wedge \omega . \tag{13.2}$$

where $2n = \dim(\mathbb{F})$. This is precisely the form in *Liouville's theorem*, once we view the measure as a form of top degree in the sense of Sect. B.4.4. By definition, in fact, we locally have:

$$\mu = dq^1 \wedge \cdots \wedge dq^n \wedge dp_1 \wedge \cdots \wedge dp_n ,$$

where $q^1, \ldots, q^n, p_1, \ldots, p_n$ are coordinates of any local chart in the symplectic atlas. In the sense of (B.32) the volume form μ (Definition B.24) is therefore a Borel measure, namely $dq^1 \cdots dq^n dp_1 \cdots dp_n$ in any symplectic coordinates. ∎

Observation (2) has an important consequence: every symplectic manifold is orientable, and the symplectic atlas induces a preferred orientation.

Proposition 13.6 *Every symplectic manifold is orientable and the symplectic atlas defines an orientation.*

Proof Suppose (U, ψ) and (U', ψ')—assuming $U \ni x \mapsto (q^1(x), \ldots, q^n(x), p_1(x), \ldots, p_n(x)) \in \mathbb{R}^{2n}$ and similarly $U' \ni x \mapsto (q'^1(x), \ldots, q'^n(x), p_1'(x), \ldots, p_n'(x)) \in \mathbb{R}^{2n}$—are local symplectic coordinate systems with $U \cap U' \neq \emptyset$. It suffices to show that the Jacobian matrix J of the coordinate change has positive determinant at any point $x \in U \cap U'$. By remark (2), on $U \cap U'$

$$dq^1 \wedge \cdots \wedge dq^n \wedge dp^1 \wedge \cdots \wedge dp^n = dq'^1 \wedge \cdots \wedge dq'^n \wedge dp'^1 \wedge \cdots \wedge dp'^n .$$

On the other hand we have two top-degree forms, so:

$$dq^1 \wedge \cdots \wedge dq^n \wedge dp^1 \wedge \cdots \wedge dp^n = (\det J) dq'^1 \wedge \cdots \wedge dq'^n \wedge dp'^1 \wedge \cdots \wedge dp'^n .$$

Therefore $\det J = 1$, and in particular $\det J > 0$. □

Physically, the volume form μ, in practice the one we used in Poincaré's "recurrence" theorem, is constant under the Hamiltonian flow. Indeed, we will show below that:

$$\mathcal{L}_W \mu = 0 ,$$

where W, defined on $\mathbb{F}$ (not on $\mathbb{R} \times \mathbb{F}!$), is the symplectic *dynamic vector field*, whose integral curves are the solutions to Hamilton's equations. W furnishes the infinitesimal symplectic version of Liouville's theorem. Let us be more precise in the following result.

Theorem 13.7 *Let $(\mathbb{F}, \omega)$ be a symplectic manifold and $\mathscr{H} : \mathbb{F} \to \mathbb{R}$ a C^2 map.*

(1) A C^1 curve $I \ni t \mapsto \mathbb{F}$, $I \subset \mathbb{R}$ an open interval, satisfies Hamilton's equations with Hamiltonian $\mathscr{H}$ in any coordinate system (V, ϕ) of the symplectic atlas of $\mathbb{F}$ if and only if:

$$\dot{\gamma}(t) \lrcorner \omega|_{\gamma(t)} = -d\mathscr{H}|_{\gamma(t)} \tag{13.3}$$

where $\dot{\gamma}$ is the tangent vector to γ.

*(2) Let W denote the (**symplectic**) **dynamic vector field**, i.e. the only vector field on $\mathbb{F}$ such that*

$$W \lrcorner \omega = -d\mathscr{H} , \tag{13.4}$$

whose integral curves solve Hamilton's equations. Then:

$$\mathcal{L}_W \omega = 0 \tag{13.5}$$

and hence:

$$\mathcal{L}_W \mu = 0 , \tag{13.6}$$

*where μ is the **symplectic volume form** (13.2).*

Proof Writing everything in symplectic coordinates and expressing

$$\gamma : I \ni t \mapsto (q^1(t), \ldots, q^n(t), p_1(t), \ldots, p_n(t)) ,$$

we find the equivalent relation:

$$\left(\sum_{k=1}^{n} \frac{dq^k}{dt} \frac{\partial}{\partial q^k} + \sum_{k=1}^{n} \frac{dp_k}{dt} \frac{\partial}{\partial p_k} \right) \lrcorner \sum_{l=1}^{n} dp_l \wedge dq^l = -\sum_{k=1}^{n} \frac{\partial \mathscr{H}}{\partial q^k} dq^k - \sum_{k=1}^{n} \frac{\partial \mathscr{H}}{\partial p_k} dp_k .$$

Using the following properties of $\lrcorner$:

$$\frac{\partial}{\partial q^k} \lrcorner dp_l \wedge dq^l = \frac{\partial}{\partial q^k} \lrcorner (-dq^l \wedge dp_l) = -\frac{\partial}{\partial q^k} \lrcorner dq^l \wedge dp_l$$

$$= -\delta_k^l dp_l \quad \text{and} \quad \frac{\partial}{\partial p_k} \lrcorner dp_l \wedge dq^l = \delta_l^k dq^l ,$$

we can recast equivalently as:

$$\sum_{k=1}^{n} \frac{dq^k}{dt} dp_k - \sum_{k=1}^{n} \frac{dp_k}{dt} dq^k = \sum_{k=1}^{n} \frac{\partial \mathcal{H}}{\partial q^k} dq^k + \sum_{k=1}^{n} \frac{\partial \mathcal{H}}{\partial p_k} dp_k . \tag{13.7}$$

Gathering alike components and recalling that the $dq^1, \ldots, dq^n, dp_1, \ldots, dp_n$ are linearly independent, the above reduces to Hamilton's equations in $t, q^1, \ldots, q^n, p_1, \ldots, p_n$. The procedure is evidently reversible, so from Hamilton's equations (13.7) we deduce (13.3). Now, W is pointwise determined by (13.4) because ω is non-degenerate. Thus we have a C^1 field, since ω is C^1 and $\mathcal{H}$ is C^2. Equation (13.6) is immediate from (13.5) using the Lie derivative's property (B.27) (see Sect. B.4). Finally, (13.5) descends from (B.30), since ω is closed, and by definition of W it follows that

$$\mathcal{L}_W \omega = d (W \lrcorner \omega) + W \lrcorner d\omega = -dd\mathcal{H} + W \lrcorner 0 = 0 + 0 = 0 .$$

$\square$

Remarks 13.8

(1) By virtue of Proposition B.31 Eq. (13.6) is the infinitesimal version of Liouville's theorem, and implies the usual integral version as the reader can prove.

(2) From the proof we also evince the dynamic vector field is:

$$W = -\sum_{k=1}^{n} \frac{\partial \mathcal{H}}{\partial q^k} \frac{\partial}{\partial p_k} + \sum_{k=1}^{n} \frac{\partial \mathcal{H}}{\partial p_k} \frac{\partial}{\partial q^k} \tag{13.8}$$

in any local symplectic chart. This expression differs a little from (11.70) for of the Hamiltonian dynamic vector field Z, due to the absence of $\frac{\partial}{\partial t}$. The latter is not a tangent vector field to $\mathbb{F}$ in the formalism we are using. Another fundamental difference is that now $\mathcal{H} : \mathbb{F} \to \mathbb{R}$ is a global function on phase space, and it does not change with the chart: coordinate transformations are in fact completely canonical, since there no longer is a time variable.

(3) The flow $\varphi^{(W)}$ of W is the Hamiltonian flow we called φ when proving Poincaré's recurrence theorem in Sect. 12.2.2. There, we assumed the phase flow was complete as well, i.e. a proper one-parameter group of global diffeomorphisms. $\blacksquare$

13.2   Hamiltonian Vector Fields and the Poisson Bracket on Symplectic Manifolds

Suppose $(\mathbb{F}, \omega)$ is a $2n$-dimensional symplectic manifold. Via (12.32), in symplectic coordinates on $\mathbb{F}$ we can define the **Poisson bracket** $\{f, g\} \in C^{k-1}(\mathbb{F})$ associated with ω for $f, g \in C^k(\mathbb{F})$ ($k \geq 1$). With our conventions, the Poisson bracket and the symplectic form satisfy:[2]

$$\omega(X_f, X_g) = -\{f, g\}, \qquad (13.9)$$

where X_f is the **Hamiltonian vector field** associated with f:

$$X_f = \sum_{k=1}^{n} \frac{\partial f}{dq^k} \frac{\partial}{\partial p_k} - \sum_{k=1}^{n} \frac{\partial f}{dp_k} \frac{\partial}{\partial q^k} \qquad (13.10)$$

in local symplectic charts.

There is another way to define the Poisson bracket using the symplectic structure and without invoking Hamiltonian vector fields. We define the field of 2-vectors on $\mathbb{F}$

$$\pi = \sum_{k=1}^{n} \frac{\partial}{\partial q^k} \wedge \frac{\partial}{\partial p_k} \qquad (13.11)$$

in symplectic coordinates, called **Poisson bivector**. That is, using a compact notation for the coordinates q^k and p_k:

$$\pi = \frac{1}{2} \sum_{i,j=1}^{2n} S^{ij} \frac{\partial}{\partial x^i} \wedge \frac{\partial}{\partial x^j}. \qquad (13.12)$$

The definition does not depend on the symplectic coordinates. By construction, furthermore:

$$\{f, g\} = \pi(df, dg),$$

which allows to define the Poisson bracket without Hamiltonian vectors. Hamilton's equations (13.3) read (now $\lrcorner$ is the contraction of 1-forms against p-vectors):

$$\dot{\gamma}(t) = -d\mathcal{H}(\gamma(t))\lrcorner\pi_{\gamma(a)}, \qquad (13.13)$$

---

[2] In the literature one also finds the opposite convention, without the $-$.

Then (13.13) in symplectic coordinates is precisely (11.37) (no t is appearing explicitly in $\mathscr{H}$).

Remarks 13.9 Interpreting π as operator $T_a^* \mathbb{F} \ni \eta_a \mapsto \pi_a(\ , \eta_a) \in T_a \mathbb{F}$ and ω as operator $T_a \mathbb{F} \ni X_a \mapsto \omega_a(\ , X_a) \in T_a^* \mathbb{F}$, by computation we see that:

$$\pi_a \circ \omega_a = id_{T_a \mathbb{F}}, \qquad \omega_a \circ \pi_a = id_{T_a^* \mathbb{F}}.$$

In other words the two maps are one the inverse of the other.   ∎

13.2.1 Hamiltonian Noether Theorem on Symplectic Manifolds

The flow $\varphi^{(X_f)}$ of X_f on $\mathbb{F}$ (not $F(\mathbb{V}^{n+1}))^3$ still defines (completely) canonical active transformations for any value of the parameter. We will say $\varphi^{(X_f)}$ is the **local one-parameter group di active canonical transformations generated by** f.

Immediately, all properties of the Poisson bracket shown in Proposition 12.26 hold. In particular the closure of the symplectic form, $d\omega = 0$, is equivalent to the *Jacobi identity*. Thus the discussion of Sect. 12.3 on the symmetries and the constants of motion can be rephrased, in an identical but simpler fashion, on a symplectic manifold equipped with a Hamiltonian function.

An important simplification is that now the 1-parameter group of local diffeomorphisms on M describing the solutions of Hamilton's equations—i.e. the (autonomous) *Hamiltonian flow* of Sect. 12.2.2 – becomes a one-parameter group of canonical active transformations with generating function precisely $-\mathscr{H}$. In this context, therefore, the temporal evolution is a local 1-parameter group of active canonical transformations.

Theorem 13.10 *Consider the flow $\varphi^{(W)}$ of the symplectic dynamic vector field W on the symplectic manifold $(\mathbb{F}, \omega)$, described in symplectic local charts by (13.8) in terms of the C^2 (Hamiltonian) function $\mathscr{H} : \mathbb{F} \to \mathbb{R}$. Then*

$$\varphi^{(W)} = \varphi^{(X_{-\mathscr{H}})}.$$

Proof It suffices to note $W = -X_{\mathscr{H}} = X_{-\mathscr{H}}$.   □

A further simplification concerns Proposition (12.28), which now assumes a perhaps more enticing form.

---

[3] For which reason we changed notation.

Proposition 13.11 *Take a C^∞ connected symplectic manifold $(\mathbb{F}, \omega)$ with smooth ω. Consider:*

(i) the Lie algebra $V(\mathbb{F})$ of C^∞ vector fields on $\mathbb{F}$ with Lie bracket given by the commutator,

(ii) the Lie algebra $C^\infty(\mathbb{F})$ of smooth $\mathbb{R}$-valued functions on $\mathbb{F}$ with the Poisson bracket.

The following hold.

(1) The map:

$$L : C^\infty(\mathbb{F}) \ni f \mapsto X_f \in V(\mathbb{F}) \tag{13.14}$$

is a homomorphism of Lie algebras.

(2) If $Z_{C^\infty(\mathbb{F})}$ denotes the centre of the Lie algebra $C^\infty(\mathbb{F})$, then

$$Z_{C^\infty(\mathbb{F})} = Ker(L) = \{g \in C^\infty(\mathbb{F}) \mid g(x) = c \text{ for some } c \in \mathbb{R} \text{ and every } x \in \mathbb{F}\}.$$

Proof The proof is essentially the simplified version of Proposition 12.28. In particular, the functions in $Ker(L)$ cannot depend on time (as t is not defined on $\mathbb{F}$), so they can only be constant.   □

Finally, we can pass to Noether's theorem for autonomous Hamiltonian systems on the symplectic phase space. Theorem 12.41 can be stated and proved in the following simplified version.

Theorem 13.12 (Autonomous Hamiltonian Noether Theorem) *Consider the Hamiltonian physical system described on the symplectic manifold $(\mathbb{F}, \omega)$ by the dynamic vector field $W = X_{-\mathscr{H}}$ with C^2 Hamiltonian $\mathscr{H} : \mathbb{F} \to \mathbb{R}$.*
Let $\varphi^{(X_f)}$ be the local one-parameter group of active canonical transformations associated with some $f \in C^2(\mathbb{F})$. The following are equivalent.

(1) The function f is a first integral, i.e.:

$$W(f) = 0 .$$

(2) We have:

$$[X_f, W] = 0 ,$$

in other words $\varphi^{(X_f)}$ maps solutions to Hamilton's equations (the integral curves of W) to solutions.

(3) We have:

$$\{f, \mathscr{H}\} = 0 .$$

(4) The Hamiltonian is invariant to order one under $\varphi^{(X_f)}$. In other words $\varphi^{(X_f)}$ is a **Hamiltonian symmetry**.

Proof Computing,

$$X_f(\mathcal{H}) = \{f, \mathcal{H}\} = -X_{\mathcal{H}}(f) = W(f)$$

and the left side is just the derivative at zero with respect to the parameter of $\mathcal{H} \circ \varphi^{(X_f)}$. Hence $\mathcal{H}$ is $\varphi^{(X_f)}$-invariant to order one if and only if (1), (2) and (3) hold.
□

13.2.2   The Action of the Galilean Group on Phase Space

The formulation for autonomous Hamiltonian systems on the symplectic phase space allows us to treat in full the action of the Galilean group on phase space $\mathbb{F}$, and discuss the relationship between Hamiltonian symmetries and conserved quantities in particular.

Consider a Hamiltonian system of N point particles without constraints and masses $m_1, \ldots, m_N$, under a conservative force with potential energy $\mathcal{U} = \mathcal{U}(\mathbf{x}_1, \ldots, \mathbf{x}_N)$ of class C^2, where $\mathbf{x}_i$ is the position vector of the i-th point in the inertial frame $\mathcal{R}$, with respect to an origin that we fix once and for all in the rest space of the frame. We will decompose all vectors along a given right-handed orthonormal basis in the frame. Suppose the potential energy only depends on the distances $||\mathbf{x}_i - \mathbf{x}_j||$, $i \neq j$. Using in the obvious way the Cartesian coordinates of every $\mathbf{x}_j$ and their momenta $\mathbf{p}_j$ (the impulses in $\mathcal{R}$) to define the symplectic structure on the C^∞ manifold $\mathbb{F} \equiv \mathbb{R}^{6N}$, the Hamiltonian in $\mathcal{R}$ will be

$$\mathcal{H}(\mathbf{x}_1, \ldots, \mathbf{x}_N, \mathbf{p}_1, \ldots, \mathbf{p}_N) = \sum_{j=1}^{N} \frac{\mathbf{p}_j^2}{2m_j} + \mathcal{U}(\mathbf{x}_1, \ldots, \mathbf{x}_N). \tag{13.15}$$

With these hypotheses the potential energy is invariant under the *active* action of the Galilean group G: *at time* $\tau \in \mathbb{R}$ the group acts on phase space by

$$S_{(c,\vec{c},\vec{v},R)}^{(\tau)} : \mathbb{F} \ni (\mathbf{x}_1, \ldots, \mathbf{x}_N, \mathbf{p}_1, \ldots, \mathbf{p}_N) \mapsto (\mathbf{x}_1', \ldots, \mathbf{x}_N', \mathbf{p}_1', \ldots, \mathbf{p}_N') \in \mathbb{F}$$

where $(\mathbf{x}_i, \mathbf{p}_i)$ is mapped to $(\mathbf{x}_i', \mathbf{p}_i')$ as follows:

$$\begin{cases} x'^{k}_{i} = c^k + (\tau - c)v^k + \sum_{j=1}^{3} R^k{}_j x_i^j (\tau - c), & k = 1, 2, 3 \\ p'^{k}_{i} = m_i v^k + \sum_{j=1}^{3} R^k{}_j p_i^j (\tau - c), & k = 1, 2, 3 \end{cases} \quad i = 1, \ldots, N.$$

$$\tag{13.16}$$

Above, $c \in \mathbb{R}$, $c^k \in \mathbb{R}$ and $v^k \in \mathbb{R}$ are constants, and the $R^k{}_j$ define a 3×3 orthogonal matrix. These coefficients orderly define the element[4] $(c, \vec{c}, \vec{v}, R) \in G$, x_i^k and p_i^k are the three Cartesian components of each $\mathbf{x}_1, \ldots, \mathbf{x}_N$ and $\mathbf{p}_1, \ldots, \mathbf{p}_N$ with respect to the right-handed orthonormal axes chosen in $\mathscr{R}$. Finally:

$$(\mathbf{x}_1(t), \ldots, \mathbf{x}_N(t), \mathbf{p}_1(t), \ldots, \mathbf{p}_N(t)) := \varphi_{t-\tau}^{(X_{-\mathscr{H}})}(\mathbf{x}_1, \ldots, \mathbf{x}_N, \mathbf{p}_1, \ldots, \mathbf{p}_N) \,. \tag{13.17}$$

In other words

$$\mathbb{R} \ni t \mapsto (\mathbf{x}_1(t), \ldots, \mathbf{x}_N(t), \mathbf{p}_1(t), \ldots, \mathbf{p}_N(t)) \in \mathbb{F}$$

is the solution (assumed complete) to Hamilton's equations with Hamiltonian (13.15) and initial conditions *at time* τ:

$$(\mathbf{x}_1(\tau), \ldots, \mathbf{x}_N(\tau), \mathbf{p}_1(\tau), \ldots, \mathbf{p}_N(\tau)) := (\mathbf{x}_1, \ldots, \mathbf{x}_N, \mathbf{p}_1, \ldots, \mathbf{p}_N) \,.$$

Remarks 13.13 Note that $S^{(\tau)}$ then *retains the Hamiltonian dynamics*, and there is a *different* representation of G for every time instant τ. ∎

The map

$$G \ni g \mapsto S_g^{(\tau)} \in Diff(\mathbb{F})$$

for a given τ is a *group representation*, i.e. a homomorphism from G to the diffeomorphism group $Diff(\mathbb{F})$ of $\mathbb{F}$:

$$S_e^{(\tau)} = id \,, \quad S_g^{(\tau)} \circ S_{g'}^{(\tau)} = S_{g \circ g'}^{(\tau)} \,, \quad S_{g^{-1}}^{(\tau)} = (S_g^{(\tau)})^{-1} \,,$$

where $g, g' \in G$ and $e = (0, \vec{0}, \vec{0}, I)$ is the neutral element of G. At last,

$$G \times \mathbb{F} \ni (g, a) \mapsto S_g^{(\tau)}(a) \in \mathbb{F}$$

is jointly C^∞ in its arguments and therefore defines a *smooth action* of the Lie group[5] G on $\mathbb{F}$.

---

[4] See Exercises 3.3 for the details.

[5] G is a (Lie) subgroup of $GL(5, \mathbb{R})$. See (3) in Exercises 3.3.

Let us now introduce the following 10 physical quantities as real functions on $\mathbb{F}$, for $k = 1, 2, 3$:

(1) the k-th component of the system's total impulse in $\mathscr{R}$:

$$P^k := \sum_{j=1}^{N} \mathbf{p}_j \cdot \mathbf{e}_k ,$$

(2) the k-th component of the system's total angular momentum in $\mathscr{R}$ with respect to the origin:

$$L^k := \sum_{j=1}^{N} \mathbf{x}_j \wedge \mathbf{p}_j \cdot \mathbf{e}_k ,$$

(3) the k-th component of the system's boost at time τ in $\mathscr{R}$:

$$K^k(\tau) := \sum_{j=1}^{N} (\tau \mathbf{p}_j - m_j \mathbf{x}_j) \cdot \mathbf{e}_k , \tag{13.18}$$

(4) the system's Hamiltonian $\mathscr{H}$ in $\mathscr{R}$.

By a rather boring computation we could verify the facts listed below; in practice this means finding the tangent vectors to the orbits of the action on $\mathbb{F}$ of the various subgroups, and check they coincides with the corresponding Hamiltonian vector fields. Notice the $-$ sign in front of most generators of the 1-parameter groups of active canonical transformations.

(1) The action on $\mathbb{F}$ of the 1-parameter group of active canonical transformations $\{\varphi_\epsilon^{(-X_{P^k})}\}_{\epsilon \in \mathbb{R}}$ coincides with the action by $S^{(\tau)}$ of the subgroup of G made of **rigid displacements along the axis $\mathbf{e}_k$:**

$$\mathbf{x}_j' := \mathbf{x}_j + \epsilon \mathbf{e}_k , \quad \mathbf{p}_j' := \mathbf{p}_j , \quad \epsilon \in \mathbb{R} ;$$

(2) The action on $\mathbb{F}$ of the 1-parameter group of active canonical transformations $\{\varphi_\epsilon^{(-X_{L^k})}\}_{\epsilon \in \mathbb{R}}$ coincides with the action by $S^{(\tau)}$ of the subgroup of G made of **positive rigid rotations around $\mathbf{e}_k$:**

$$\mathbf{x}_j' := R_{\mathbf{e}_k, \epsilon} \mathbf{x}_j , \quad \mathbf{p}_j' := R_{\mathbf{e}_k, \epsilon} \mathbf{p}_j , \quad \epsilon \in \mathbb{R} ;$$

(3) The action on $\mathbb{F}$ of the 1-parameter group of active canonical transformations $\{\varphi_\epsilon^{(-X_{K^k(\tau)})}\}_{\epsilon \in \mathbb{R}}$ coincides with the action by $S^{(\tau)}$ of the subgroup of G made of **pure Galilean transformations along $\mathbf{e}_k$ at time τ:**

$$\mathbf{x}_j' := \mathbf{x}_j + \tau \epsilon \mathbf{e}_k , \quad \mathbf{p}_j' := \mathbf{p}_j + \epsilon m_j \mathbf{e}_k , \quad \epsilon \in \mathbb{R} ;$$

(4) The action on $\mathbb{F}$ of the one-parameter group of active canonical transformations $\{\varphi_\epsilon^{(X_{\mathscr{H}})}\}_{\epsilon \in \mathbb{R}}$ coincides with the action by $S^{(\tau)}$ of the subgroup of G made of **temporal displacements**:[6]

$$\mathbf{x}'_j := \mathbf{x}_j(\tau - \epsilon), \quad \mathbf{p}'_j := \mathbf{p}_j(\tau - \epsilon), \quad \epsilon \in \mathbb{R}$$

(where $\mathbf{x}_j = \mathbf{x}_j(\tau)$ and $\mathbf{p}_j = \mathbf{p}_j(\tau)$ for the τ that defines $S^{(\tau)}$, in agreement with (13.17)).

Next we shall compare two Lie algebras: the Lie algebra of generating functions with the Poisson bracket, and the Lie algebra of associated Hamiltonian vectors with the commutator.

We recall the **Ricci symbol**, for $a, b, c \in \{1, 2.3\}$:

$$\epsilon_{abc} := \begin{cases} \epsilon_{abc} := 1 \text{ if } (a, b, c) \text{ is a cyclic permutation of } (1, 2, 3), \\ \epsilon_{abc} := -1 \text{ if } (a, b, c) \text{ is a } non - cyclic \text{ permutation of } (1, 2, 3), \\ \epsilon_{abc} := 0 \text{ if } (a, b, c) \text{ is not a permutation of } (1, 2, 3). \end{cases}$$

Finally, we set:[7]

$$\epsilon_{ab}{}^c := \epsilon^{ab}{}_c := \epsilon_{abc}.$$

Regarding vector fields, a straightforward computation produces the following independent relationships (we will omit all $-$ signs in the vector fields appearing in the 1-parameter groups: we will write X_{pk} and not $-X_{pk}$ etc.):

$$[X_{\mathscr{H}}, X_{\mathscr{H}}] = 0, \quad [X_{\mathscr{H}}, X_{pk}] = 0, \quad [X_{\mathscr{H}}, X_{L^k}] = 0,$$

$$[X_{\mathscr{H}}, X_{K^k(t)}] = X_{pk}, \quad [X_{ph}, X_{pk}] = 0,$$

$$[X_{K^h(t)}, X_{pk}] = 0,$$

$$[X_{L^k}, X_{ph}] = \sum_{r=1}^{3} \epsilon_{kh}{}^r X_{pr}, \quad [X_{L^k}, X_{K^h}] = \sum_{r=1}^{3} \epsilon_{kh}{}^r X_{K^r},$$

$$[X_{L^k}, X_{L^h}] = \sum_{r=1}^{3} \epsilon_{kh}{}^r X_{L^r},$$

$$[X_{K^k(\tau)}, X_{K^h(\tau)}] = 0.$$

---

[6] The temporal *displacement* is an *active* transformation, and in some sense the inverse of temporal *evolution*.

[7] From a tensorial point of view these definitions are coherent provided we use orthonormal bases, as presently.

Although we cannot explain more at this juncture, the above commutators characterise the *Lie algebra* of the Galilean group (better: the Lie algebra of the identity component of G, which means taking in G only $SO(3)$ rather than $O(3)$).

Let us now pass to the (Poisson) bracket relationships between the generating functions of 1-parameter groups of canonical transformations. Setting $M := \sum_{j=1}^{N} m_j$ we find

$$\{\mathcal{H}, \mathcal{H}\} = 0, \quad \{\mathcal{H}, P^k\} = 0, \quad \{\mathcal{H}, L^k\} = 0,$$

$$\{\mathcal{H}, K^k(\tau)\} = P^k, \quad \{P^h, P^k\} = 0,$$

$$\{K^h(\tau), P^k\} = -M\delta^{hk},$$

$$\{L^k, P^h\} = \sum_{r=1}^{3} \epsilon^{kh}{}_r P^r, \quad \{L^k, K^h\} = \sum_{r=1}^{3} \epsilon^{kh}{}_r K^r, \quad \{L^k, L^h\} = \sum_{r=1}^{3} \epsilon^{kh}{}_r L^r,$$

$$\{K^k(\tau), K^h(\tau)\} = 0.$$

(13.19)

Remarks 13.14 With the exclusion of the Hamiltonian, the true generating functions would be the functions in the Poisson brackets with a $-$, according to Definition 12.32, and since $-X_f = X_{-f}$. We will ignore that sign because we have dropped it in the corresponding vector fields. In this way we will use the same convention when comparing objects. ∎

The brackets of the generating functions are the same as those between the Hamiltonian vector fields, except for the second row. In the Poisson bracket the zero is replaced by the constant $-M\delta^{hk}$. This is allowed by Proposition 13.11 precisely because, by definition, this constant belongs to the kernel of L in (13.14). The term $-M\delta^{hk}$ (for $h = k$) is called **central charge** since it commutes with all other generators. Analogous constants often appear when one represents other Lie groups on phase space. The interesting aspect is that this term contains the total mass M of the system of point particles.

Remarks 13.15

(1) The commutators of the 10 Hamiltonian vector fields are linear combinations the fields themselves. This means that the real vector space generated by these 10 elements is a Lie subalgebra of $V(\mathbb{F})$ (isomorphic to the Lie algebra of the Galilean Lie group). The Lie algebra (with Poisson bracket) generated by the 10 generating functions, instead, is not isomorphic to the Galilean Lie algebra because of the equation involving the central charge. That said, if we add the constant map 1 to the other ten, it will commute with everything, and the now eleven elements define a Lie subalgebra of $C^\infty(\mathbb{F})$ for the Poisson bracket. The extra generator is clearly in the *centre* of the Lie algebra. This is called a **central extension** of the Galilean Lie algebra, since it is constructed enlarging the Galilean Lie algebra by an element commuting with everything.

(2) The appearance of a central charge proportional to the system's mass at the level of the Lie algebra of generating functions is reminiscent of the same phenomenon in the quantum formulation (see for example [Mor18]) when one studies unitary representations of the Galilean group. The multiple of the mass creates an insurmountable obstruction to the existence of such representations which, by *Bargmann's superselection rule*, must be *projective unitary*.   ∎

Let us analyse the meaning of the relationships involving the Hamiltonian:

$$\{\mathscr{H}, \mathscr{H}\} = 0, \quad \{\mathscr{H}, P^k\} = 0, \quad \{\mathscr{H}, L^k\} = 0, \quad \{\mathscr{H}, K^k(\tau)\} = P^k.$$

$$(13.20)$$

The first 3 correspond to the conservation of the total mechanical energy, of the total impulse and of the total angular momentum along the system's Hamiltonian evolution, by Theorem 13.12. In reality also the last one, involving the boost in the k-th direction, is a conservation law, but we must remember the K^k defined in (13.18) depends on the time parameter τ, and hence we must view it as

$$K^k : \mathbb{R} \times \mathbb{F} \to \mathbb{R}.$$

In other words, identifying τ with t, we are speaking of a quantity defined on phase spacetime, even if we refer to a particular decomposition $F(\mathbb{V}^{n+1}) = \mathbb{R} \times \mathbb{F}$. Now, (13.18) implies (identifying $\tau = t$):

$$\frac{\partial K^k}{\partial t} = P^k$$

so $\{K^k(t), \mathscr{H}\} + P^k = 0$ reads:

$$\{K^k(t), \mathscr{H}\} + \frac{\partial K^k}{\partial t} = 0.$$

This is a conservation law because we can write as:

$$\frac{d}{dt} K^k(t, \mathbf{x}_1(t), \dots, \mathbf{x}_N, \mathbf{p}_1(t), \dots, \mathbf{p}_N(t)) = 0,$$

where $\mathbb{R} \ni t \mapsto (\mathbf{x}_1(t), \dots, \mathbf{x}_N, \mathbf{p}_1(t), \dots, \mathbf{p}_N(t)) \in \mathbb{F}$ is a generic solution to Hamilton's equations for $\mathscr{H}$.

From the point of view of generating functions there is also the constant map given by the central charge M. Clearly this is a first integral. In fact it trivially commutes with $\mathscr{H}$. The constant is also invariant under all other 9 Galilean subgroups, since it commutes with their generators. This fact, albeit algebraically obvious, physically tells us the mass of the physical system is truly an invariant: under rotations, displacements and change of inertial frame.

The meaning of (13.20) in terms of *Hamiltonian symmetries* is similar to what we saw in general on phase spacetime. The first three in (13.20) correspond to $\mathcal{H}$ being invariant (at least to first order) under the corresponding 1-parameter groups of canonical transformations in the sense of Theorem 13.12. The fourth one cannot refer back to that theorem, where the functions *do not* depend on time, and requires a deeper analysis involving the *formal invariance* of $\mathcal{H}$ under the canonical transformations generated by K^k. We have already discussed that interpretation for the Hamiltonian formulation on phase spacetime in (2) Examples 12.43, regarding the generating function $f := -K^k$ (i.e. with $\mathbf{n} := \mathbf{e}_k$).

To finish, note that the Galilean group, represented on $\mathbb{F}$ by (13.16), also acts on physical quantities defined on $\mathbb{F}$ other than the Hamiltonian. If $f : \mathbb{F} \to \mathbb{R}$ is a physical quantity, the Galilean group acts on it by

$$\mathcal{S}_g^{(\tau)}(f) := f \circ \mathcal{S}_{g^{-1}}^{(\tau)}, \quad g \in G. \tag{13.21}$$

In this way (note in particular the inverse g^{-1} on the right) it is easy to see that:

$$\mathcal{S}_e^{(\tau)} = id, \quad \mathcal{S}_g^{(\tau)} \circ \mathcal{S}_{g'}^{(\tau)} = \mathcal{S}_{g \circ g'}^{(\tau)}, \quad \mathcal{S}_{g^{-1}}^{(\tau)} = (\mathcal{S}_g^{(\tau)})^{-1},$$

where e is the neutral element of G. Hence we have a *representation* of G over functions (physical quantities) on phase space. $\mathcal{S}^{(\tau)}$ is an *active* action on the physical quantities. The meaning of $f' := \mathcal{S}_g^{(\tau)}(f)$ is the following: when evaluated on state a, which the group transforms into $a' = \mathcal{S}_g^{(\tau)}(a)$, it 'cancels' the group action:

$$f'(a') = f(a).$$

If, for instance, we "rotate" the system and then we "rotate" the measuring instruments too, the latter's measurements are the same as the measurements of the "unrotated" instruments on the "unrotated" system.

Examples 13.16

(1) Consider the physical system described in the text and the three functions on $\mathbb{F} = \mathbb{R}^{6N}$

$$X^k(\mathbf{x}_1, \ldots, \mathbf{x}_n, \mathbf{p}_1, \ldots, \mathbf{p}_N) := \frac{1}{M} \sum_{j=1}^{N} m_j \mathbf{x}_j, \quad k = 1, 2, 3. \tag{13.22}$$

Clearly these are the three components of the position vector of the centre of mass. Supposing the Hamiltonian flow is globally defined (i.e. a one-parameter group of active canonical transformations), we want to prove

$$X^k \circ \varphi_t^{(W)} = X^k + \frac{t}{M} P^k, \tag{13.23}$$

where $\varphi^{(W)} = \varphi^{(X-\mathscr{H})}$ is the Hamiltonian flow describing the temporal evolution on $\mathbb{F}$. The above formula tells how X^k evolves in time[8] when we evaluate the equation on a state described by a representative point in phase space. We first note that, by definition of K^k,

$$X^k = \frac{1}{M}(t\,P^k - K^k(t))$$

so using the second and third equations in (13.20) we find

$$\{-\mathscr{H}, X^k\} = \frac{1}{M}P^k .$$

From the definition of Poisson bracket, and $W = X_{-\mathscr{H}}$, the above can be understood as

$$\left.\frac{d}{dt}\right|_{t=0} X^k \circ \varphi_t^{(W)} = \frac{1}{M}P^k .$$

In other words, renaming t to h and composing with $\varphi_t^{(W)}$:

$$\left.\frac{d}{dh}\right|_{h=0} X^k \circ \varphi_h^{(W)} \circ \varphi_t^{(W)} = \frac{1}{M}P^k \circ \varphi_t^{(W)} ,$$

which by $\varphi_h^{(W)} \circ \varphi_t^{(W)} = \varphi_{t+h}^{(W)}$ becomes

$$\frac{d}{dt}X^k \circ \varphi_t^{(W)} = \frac{1}{M}P^k \circ \varphi_t^{(W)} = \frac{1}{M}P^k .$$

In the last passage we used $P^k \circ \varphi_t^{(W)} = P^k$ since P^k is a first integral. Given any $a \in \mathbb{F}$, evaluating at a we have:

$$\frac{d}{dt}X^k \circ \varphi_t^{(W)}(a) = \frac{1}{M}P^k(a) ,$$

and integrating in t produces (13.22).

(2) Consider the Galilean group G on $\mathbb{F}$ and let us focus on the one-parameter subgroup of rotations, say around the j-th axis. We know such 1-parameter group of active canonical transformations is associated with L^j, i.e. it is generated by the Hamiltonian vector field $-X_{L^j}$. Consider three C^1 real maps

---

[8] The reader familiar with Quantum Mechanics will notice that the equation corresponds classically to the *Heisenberg evolution* of the position observable [Mor18].

on phase space $\mathbb{F}$:

$$M^k = M^k(\mathbf{x}_1, \ldots, \mathbf{x}_n, \mathbf{p}_1, \ldots, \mathbf{p}_N), \quad k = 1, 2, 3, \tag{13.24}$$

that satisfy commutation relationships with L^j equal to those of L^k, P^k, K^k in (13.19). For example, the aforementioned X^k satisfy:

$$\{L^j, M^k\} = \sum_{r=1}^{3} \epsilon^{jk}{}_r M^r . \tag{13.25}$$

We want to show that the subgroup acts on M^k as follows:

$$M^k \circ \varphi_{-\theta}^{(-X_{Lj})} = \sum_{h=1}^{3} (R_{\theta,\mathbf{e}_j}^{-1})^k{}_h M^h , \tag{13.26}$$

where $R_{\theta,\mathbf{e}_j}$ is the positive rotation by $\theta \in \mathbb{R}$ around $\mathbf{e}_j$. The relation says we can cancel the effect of the *rotation of the state* $(a \to \varphi_\theta^{(-X_{Lj})}(a))$ by *rotating the vector of the three M^k with the inverse rotation*:

$$M^k(a) = \sum_{h=1}^{3} (R_{\theta,\mathbf{e}_j}^{-1})^k{}_h M^h \left(\varphi_\theta^{(-X_{Lj})}(a) \right) .$$

The rotation action on the M^k in the left-hand side of (13.26) is precisely the one in (13.21). Now, to prove the above formula we fix $a \in \mathbb{F}$ and define the C^1 maps:

$$f^k(\theta) := M^k \circ \varphi_{-\theta}^{(-X_{Lj})}(a), \quad g^k(\theta) := \sum_{h=1}^{3} (R_{\theta,\mathbf{e}_j}^{-1})^k{}_h M^h(a) .$$

Arguing as in the previous example and using (13.25) we find

$$\frac{df^k}{d\theta} = \sum_{r=1}^{3} \epsilon^{jk}{}_r f^r(\theta) .$$

Similarly, for the right-hand side:

$$\frac{dg^k}{d\theta} = \sum_{r=1}^{3} \epsilon^{jk}{}_r g^r(\theta) .$$

Finally $f^k(0) = g^k(0) = M^k(a)$. The usual Uniqueness Theorem for systems of differential equations (here linear) guarantees that:

$$f^k(\theta) = g^k(\theta) \quad k = 1, 2, 3, \quad \theta \in \mathbb{R}.$$

As $a \in \mathbb{F}$ is arbitrary, the above is equivalent to (13.26).

(3) The same procedure shows

$$X^j \circ \varphi_{-\epsilon}^{(-X_{K^k(t)})} = X^j - t\delta^{kj}\epsilon, \quad P^j \circ \varphi_{-\epsilon}^{(-X_{K^k(t)})} = P^j - M\delta^{kj}\epsilon.$$

This expresses how pure Galilean transformations at time t act on the quantities, as per Definition (13.26).

(4) Formula (12.39) says how the Hamiltonian changes under a one-parameter group of active canonical transformations $\varphi^{(X_f)}$ generated by a function $f : \mathbb{R} \times \mathbb{F} \to \mathbb{R}$, which may depend parametrically on time::

$$\mathscr{H}_u'(a) = \mathscr{H}\left(\varphi_{-u}^{(X_f)}(a)\right) - \int_0^u \frac{\partial f}{\partial t}\left(t, \varphi_{s-u}^{(X_f)}(a)\right) ds, \quad a \in \mathbb{F}. \tag{13.27}$$

We showed in (2) Examples 12.43, using $f(t, a) = -K^k(t, a)$, that under pure Galilean transformations the Hamiltonian is *formally invariant*. In case f is one of the 10 generators of G other than K^k, and hence does not depend parametrically on time, the new Hamiltonian is

$$\mathscr{H}_u' = \mathcal{S}_{g(u)}^{(t)}(\mathscr{H})$$

where $g = g(u)$ is the one-parameter subgroup of G whose action $\mathcal{S}_{g(u)}^{(t)}$ on $\mathbb{F}$ coincides with $\varphi_u^{(X_f)}$.

13.2.3 The Action of the Poincaré Group on Phase Space for the Free Particle

In the rest of this section we will rely on Chap. 10. The Lagrangian version of the ensuing discussion is contained in Sect. 10.5.4.

As for the Galilean group, the formulation for autonomous Hamiltonian systems on the symplectic phase space allows to understand in a finer way the (proper orthochronous) *Poincaré group's* action on phase space $\mathbb{F}$, and in particular the relationship between Hamiltonian symmetries and conserved quantities.

Henceforth we shall consider the free relativistic particle of mass $m > 0$. The phase space will then be $\mathbb{F} = \mathbb{R}^3 \times \mathbb{R}^3 \ni (\mathbf{x}, \mathbf{p})$. As usual, the factor $\mathbb{R}^3$ refers to the spatial terms in given Minkowski coordinates $(x^0 = ct, x^1, x^2, x^3)$ while the second to the three spatial components of the particle's four-momentum P. If c denotes the speed of light, the Hamiltonian is:

$$\mathcal{H}(\mathbf{x}, \mathbf{p}) = c\sqrt{\mathbf{p}^2 + m^2 c^2}\,. \tag{13.28}$$

Note $(c^{-1}\mathcal{H}, p^1, p^2, p^3)$ are the components of the four-momentum P in the Minkowski coordinates we are using. $\mathcal{H}(\mathbf{x}, \mathbf{p})$ is formally invariant under the *active* transformations of the proper orthochronous Poincaré group $ISO(1, 3)_+$. This group acts—*at time* $\tau \in \mathbb{R}$—on phase space $\mathbb{F}$ by:

$$M^{(\tau)}_{(\Lambda, U)} : \mathbb{F} \ni (\mathbf{x}, \mathbf{p}) \mapsto (\mathbf{x}', \mathbf{p}') \in \mathbb{F} \quad \text{for } (\Lambda, U) \in ISO(1, 3)_+,$$

where:

$$\begin{cases} x'^k = u^k + \Lambda^k{}_0 s(\tau, \mathbf{x}, u^0) + \displaystyle\sum_{j=1}^3 \Lambda^k{}_j x^j(s(\tau, \mathbf{x}, u^0))\,, & k = 1, 2, 3 \\[3mm] p'^k = \frac{1}{c}\Lambda^k{}_0 \mathcal{H}(\mathbf{p}(s(\tau, \mathbf{x}, u^0))) + \displaystyle\sum_{j=1}^3 \Lambda^k{}_j p^j(s(\tau, \mathbf{x}, u^0))\,, & k = 1, 2, 3\,, \end{cases}$$

$$\tag{13.29}$$

and $\Lambda = [\Lambda^k{}_h]_{k,h=0,1,2,3} \in SO(1, 3)_+$ and $U := (u^0, u^1, u^2, u^3) \in \mathbb{R}^4$. Moreover:

$$(\mathbf{x}(t), \mathbf{p}(t)) := \varphi^{(X_{-\mathcal{H}})}_{t-\tau}(\mathbf{x}, \mathbf{p})\,. \tag{13.30}$$

In other words,

$$\mathbb{R} \ni t \mapsto (\mathbf{x}(t), \mathbf{p}(t)) \in \mathbb{F}$$

is the complete solution to Hamilton's equations with Hamiltonian (13.28) and initial conditions *at time* τ:

$$(\mathbf{x}(\tau), \mathbf{p}(\tau)) := (\mathbf{x}, \mathbf{p})\,,$$

i.e.

$$(\mathbf{x}(t), \mathbf{p}(t)) = \left(\mathbf{x} + (t - \tau)\frac{c^2 \mathbf{p}}{\mathcal{H}(\mathbf{p})}, \mathbf{p}\right)\,, \quad t \in \mathbb{R}\,. \tag{13.31}$$

Finally, $s(\tau, \mathbf{x}, u^0) \in \mathbb{R}$ is found by solving:

$$c\tau = u^0 + \Lambda^0{}_0 s(\tau, \mathbf{x}, u^0) + \sum_{k=1}^3 \Lambda^0{}_k x^k(s(\tau, \mathbf{x}, u^0))\,. \tag{13.32}$$

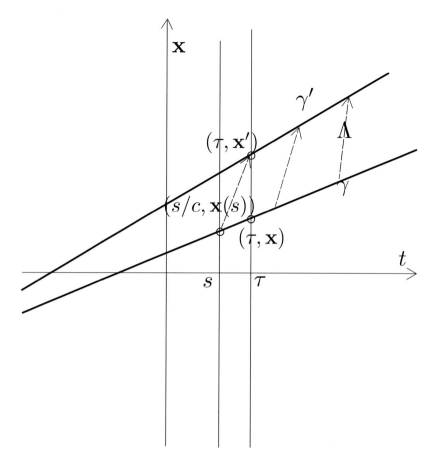

Fig. 13.1 Geometric interpretation of Eq. (13.34) in $\mathbb{R} \times \mathbb{R}^3$

This can always be solved and the solution s is C^∞ in $(\tau, \mathbf{x})$ and linear in u^0, as is immediate from (13.31). If $\Lambda = I$ then $s(\tau, \mathbf{x}, u^0) = c\tau - u^0$.

Remarks 13.17 Expression (13.29) is so scary that is deserves an explanation (Fig. 13.1).

(1) First, the second row reduces to

$$p'^k = \frac{1}{c}\Lambda^k{}_0 \mathcal{H}(\mathbf{p}) + \sum_{j=1}^{3} \Lambda^k{}_j p^j , \quad k = 1, 2, 3 , \tag{13.33}$$

since the solution to (13.31) is trivial in the momentum variable: the momentum stays equal to the its value at the initial instant τ. The Poincaré group's action is just the active action of $SO(1, 3)_+$ on the four-momentum $P \equiv (c^{-1}\mathcal{H}(\mathbf{p}), \mathbf{p})$ read in spatial components, which is what (13.33) is saying, cf. (10.44).

(2) To understand the first row in (13.29) consider the simpler case where $u^0 = u^1 = u^2 = u^3 = 0$, i.e. we work with $SO(1, 3)_+$ instead of $ISO(1, 3)_+$. Then:

$$x'^k = \Lambda^k{}_0 s(\tau, \mathbf{x}) + \sum_{j=1}^{3} \Lambda^k{}_j x^j (s(\tau, \mathbf{x})), \quad k = 1, 2, 3. \tag{13.34}$$

Starting from initial condition $\mathbf{x}$ at time τ with corresponding initial condition $\mathbf{p}$ at that time, let us plot in spacetime the world line γ of the particle evolving from such initial conditions with free Hamiltonian. This will be a geodesic line $\mathbf{x} = \mathbf{x}(t)$ in spatial coordinates. In the figure this is the lower slanted line. The initial condition $\mathbf{p}$ given in this spacetime determines the slope $c^2\mathbf{p}/\mathcal{H}(\mathbf{p})$ of the three components of the line. These components of the four-momentum, rescaled by the positive constant $c^2\mathcal{H}^{-1}$, are just the spatial components of the four-velocity. At an arbitrary instant t let us perform the Lorentz transformation Λ on the corresponding event of γ, thus finding a second world line γ' (upper line) of slope $c^2\mathbf{p}'/\mathcal{H}(\mathbf{p}')$. The Lorentz transformation acts on the events of γ as indicated by the dashed arrows. Clearly Λ also moves the time coordinates of events from γ to γ'. Instead, (13.34) acts *along every temporal fibre* and sends $(\tau, \mathbf{x})$ to $(\tau, \mathbf{x}')$. For that reason we must view $(\tau, \mathbf{x}') \in \gamma'(\mathbb{R})$ as the Λ-transform of an event $(s/c, \mathbf{x}(s)) \in \gamma(\mathbb{R})$ whose time coordinate s/c is different from τ (in the figure, it precedes it). Equation (13.32) for $s = s(\tau, \mathbf{x})$, now simplified to

$$c\tau = \Lambda^0{}_0 s(\tau, \mathbf{x}) + \sum_{k=1}^{3} \Lambda^0{}_k x^k (s(\tau, \mathbf{x})),$$

says exactly that $\Lambda(s, \mathbf{x}(s)) = (c\tau, \mathbf{x}')$.

(3) When we turn on the displacements in $\mathbb{M}^4$ described by $U = (u^0, u^1, u^2, u^3)$, we pass from (13.34) to the second equation in (13.29). Just as for the spatial displacements, there is nothing to say. The temporal displacement associated with u^0 deserves a comment. Above, the event $(s/c, \mathbf{x}(s)) \in \gamma(\mathbb{R})$ must now reach $(\tau, \mathbf{x}')$ under (i) a Lorentz transformation Λ and (ii) a subsequent translation by u^0 along the real line in which $c\tau$ varies. This is precisely what (13.32) requires, hence further explaining why s depends (linearly) on u^0. $\blacksquare$

As in the case of the Galilean group, $M^{(\tau)}$ *retains the Hamiltonian dynamics* and there is a *different* representation of $ISO(1,3)_+$ for each instant $\tau \in \mathbb{R}$. The map:

$$ISO(1,3)_+ \ni g \mapsto M_g^{(\tau)} \in Diff(\mathbb{F})$$

at given τ is a *group representation*, i.e. a homomorphism from $ISO(1,3)_+$ to the diffeomorphism group $Diff(\mathbb{F})$ of $\mathbb{F}$:

$$M_e^{(\tau)} = id \,, \quad M_g^{(\tau)} \circ M_{g'}^{(\tau)} = M_{g \circ g'}^{(\tau)} \,, \quad M_{g^{-1}}^{(\tau)} = (M_g^{(\tau)})^{-1} \,,$$

where $g, g' \in ISO(1,3)_+$ and $e = (I,0)$ is the neutral element of $ISO(1,3)_+$. Finally, the map

$$ISO(1,3)_+ \times \mathbb{F} \ni (g,a) \mapsto M_g^{(\tau)}(a) \in \mathbb{F}$$

is jointly C^∞ in its arguments and therefore defines a *smooth action* of the Lie group[9] $ISO(1,3)_+$ on $\mathbb{F}$.

Next we introduce 10 physical quantities as real functions on $\mathbb{F}$, $k = 1,2,3$. They refer to an inertial frame $\mathscr{R}$ whose Minkowski coordinates are those we used to define phase space and the group representation:

(1) **the k-th component of the system's (relativistic) impulse in $\mathscr{R}$:**

$$P^k := \mathbf{p} \cdot \mathbf{e}_k \,,$$

(2) **the k-th component of the system's (relativistic) angular momentum in $\mathscr{R}$ with respect to the origin:**

$$L^k := \mathbf{x} \wedge \mathbf{p} \cdot \mathbf{e}_k \,,$$

(3) **the Hamiltonian $\mathscr{H} = c\sqrt{\mathbf{p}^2 + m^2 c^2}$ in $\mathscr{R}$,**
(4) **the k-th component of the system's boost at time τ in $\mathscr{R}$:**

$$K^k(\tau) := c\left(\tau \mathbf{p} - \frac{\mathscr{H}\mathbf{x}}{c^2}\right) \cdot \mathbf{e}_k \,. \tag{13.35}$$

---

[9] $ISO(1,3)_+$ is a (Lie) subgroup of $GL(5,\mathbb{R})$ as we saw in Chap. 10.

Remarks 13.18 In order to compare the relativistic boost with the classical (13.18) we must not forget the factor c, which only appears in the former. In fact $c^{-1} K^k \approx K^k_{classical}$ for $||\mathbf{p}|| \ll mc$, since in that case we may approximate $\mathcal{H} = c\sqrt{\mathbf{p}^2 + m^2 c^2}$ by m. ∎

As for the Galilean group, a rather boring computation proves the following facts. Recall that we just need to find the tangent vectors to the orbits in $\mathbb{F}$ of the subgroups and check they coincide with the corresponding Hamiltonian vector fields. Notice the—in front of almost all generators of 1-parameter groups of active canonical transformations.

(1) The action on $\mathbb{F}$ of the 1-parameter group of active canonical transformations $\{\varphi_\epsilon^{(-X_{P^k})}\}_{\epsilon \in \mathbb{R}}$ coincides with the action by $M^{(\tau)}$ of the subgroup of $ISO(1,3)_+$ of **displacements along $\mathbf{e}_k$**:

$$\mathbf{x}'_\epsilon := \mathbf{x} + \epsilon \mathbf{e}_k , \quad \mathbf{p}'_\epsilon := \mathbf{p} , \quad \epsilon \in \mathbb{R} ;$$

(2) The action on $\mathbb{F}$ of the 1-parameter group of active canonical transformations $\{\varphi_\epsilon^{(-X_{L^k})}\}_{\epsilon \in \mathbb{R}}$ coincides with the action by $M^{(\tau)}$ of the subgroup of $ISO(1,3)_+$ of **positive rotations around $\mathbf{e}_k$**:

$$\mathbf{x}'_\epsilon := R_{\mathbf{e}_k, \epsilon} \mathbf{x} , \quad \mathbf{p}'_\epsilon := R_{\mathbf{e}_k, \epsilon} \mathbf{p} , \quad \epsilon \in \mathbb{R} ;$$

(3) The action on $\mathbb{F}$ of the 1-parameter group of active canonical transformations $\{\varphi_\epsilon^{(X_{\mathcal{H}})}\}_{\epsilon \in \mathbb{R}}$ coincides with the action by $M^{(\tau)}$ of the subgroup of $ISO(1,3)_+$ of **temporal displacements**:[10]

$$\mathbf{x}'_\epsilon := \mathbf{x}(\tau - \epsilon) , \quad \mathbf{p}'_\epsilon := \mathbf{p}(\tau - \epsilon) , \quad \epsilon \in \mathbb{R}$$

(where $\mathbf{x} = \mathbf{x}(\tau)$ and $\mathbf{p} = \mathbf{p}(\tau)$ for the τ in $M^{(\tau)}$, as per (13.30));
(4) The action on $\mathbb{F}$ of the 1-parameter group of active canonical transformations $\{\varphi_\epsilon^{(-X_{K^3(\tau)})}\}_{\epsilon \in \mathbb{R}}$ coincides with the action by $M^{(\tau)}$ of the subgroup of

---

[10] The temporal *displacement* is an *active* transformation; in some sense it reverses temporal *evolution*, as in the case of G.

$ISO(1, 3)_+$ of **special Lorentz transformations along** $\mathbf{e}_3$ *at time* τ:

$$
\begin{cases}
x'^3_\epsilon & := \dfrac{(x^3 + c\tau \sinh \epsilon) + \tau \frac{c^2 p^3}{\mathscr{H}(\mathbf{p})}(\cosh \epsilon - 1)}{\cosh \epsilon + \frac{cp^3}{\mathscr{H}(\mathbf{p})} \sinh \epsilon} \\[2ex]
& = \dfrac{x^3 \mathscr{H}(\mathbf{p}) + c^2 \tau (p'^3_\epsilon - p^3)}{\mathscr{H}(\mathbf{p}'_\epsilon)} , \\[2ex]
x'^k_\epsilon & := x^k + \dfrac{p^k}{m_p^2 c^2} \left[\dfrac{\mathscr{H}(\mathbf{p}) x^3}{c} - c\tau p^3 \right] \left[\dfrac{cp^3}{\mathscr{H}(\mathbf{p})} - \dfrac{cp'^3_\epsilon}{\mathscr{H}(\mathbf{p}'_\epsilon)} \right] \\[2ex]
& \quad + c^2 \tau p^k \left[\dfrac{1}{\mathscr{H}(\mathbf{p}'_\epsilon)} - \dfrac{1}{\mathscr{H}(\mathbf{p})} \right] \text{ for } k = 1, 2,
\end{cases}
$$

where $m_p c := \sqrt{m^2 c^2 + (p^1)^2 + (p^2)^2}$, $\mathscr{H}(\mathbf{p}) = c\sqrt{\mathbf{p}^2 + m^2 c^2}$, and

$$
\begin{cases}
p^k \mapsto p'^3_\epsilon := p^3 \cosh \epsilon + c^{-1} \mathscr{H}(\mathbf{p}) \sinh \epsilon , \\
p'^k_\epsilon = p^k \quad \text{for } k = 1, 2.
\end{cases}
$$

We will not write down the analogous special transformations along $\mathbf{e}_1$ and $\mathbf{e}_2$.

As for the Galilean group, let us compare the Lie algebra of generating functions with the Poisson bracket and Lie algebras of Hamiltonian fields with the commutator.

Regarding vector fields, by computation we find the following independent relationships (systematically dropping the—in the vector fields appearing in the 1-parameter groups: for instance we will write X_{pk} and not $-X_{pk}$):

$$
[X_{\mathscr{H}}, X_{\mathscr{H}}] = 0 , \quad [X_{\mathscr{H}}, X_{pk}] = 0 , \quad [X_{\mathscr{H}}, X_{L^k}] = 0 ,
$$

$$
[X_{\mathscr{H}}, X_{K^k(t)}] = c X_{pk} , \quad [X_{ph}, X_{pk}] = 0 ,
$$

$$
[X_{K^h(t)}, X_{pk}] = -\frac{\delta^{hk}}{c} X_{\mathscr{H}} ,
$$

$$
[X_{L^k}, X_{ph}] = \sum_{r=1}^{3} \epsilon_{kh}{}^r X_{pr} , \quad [X_{L^k}, X_{K^h}] = \sum_{r=1}^{3} \epsilon_{kh}{}^r X_{K^r} ,
$$

$$
[X_{L^k}, X_{L^h}] = \sum_{r=1}^{3} \epsilon_{kh}{}^r X_{L^r} ,
$$

$$
[X_{K^k(\tau)}, X_{K^h(\tau)}] = \frac{1}{c^2} \sum_{r=1}^{3} \epsilon_{kh}{}^r X_{L^r} .
$$

Although we cannot explain more in this context, the above characterise the *Lie algebra* of $ISO(1, 3)_+$. They are the same relations found for the Galilean group except for those containing the boost. The last one is actually quite remarkable, for it shows that special Lorentz transformations and rotations are tightly related. This fact has important consequences in Physics, which alas we cannot discuss here.

Let us write the (Poisson) commutators of the generating functions of the 1-parameter groups of canonical transformations:

$$\{\mathcal{H}, \mathcal{H}\} = 0 , \quad \{\mathcal{H}, P^k\} = 0 , \quad \{\mathcal{H}, L^k\} = 0 ,$$

$$\{\mathcal{H}, K^k(\tau)\} = c P^k , \quad \{P^h, P^k\} = 0 ,$$

$$\{K^h(\tau), P^k\} = -\frac{\mathcal{H}}{c} \delta^{hk} ,$$

$$\{L^k, P^h\} = \sum_{r=1}^{3} \epsilon^{kh}{}_r P^r , \quad \{L^k, K^h\} = \sum_{r=1}^{3} \epsilon^{kh}{}_r K^r , \quad \{L^k, L^h\} = \sum_{r=1}^{3} \epsilon^{kh}{}_r L^r ,$$

$$\{K^k(\tau), K^h(\tau)\} = \frac{1}{c^2} \sum_{l=1}^{3} \epsilon^{hk}{}_l L^l .$$

(13.36)

Remarks 13.19 As usual, except for the Hamiltonian, the true generating functions would be those with a—in the Poisson brackets, in agreement with Definition 12.32 and $-X_f = X_{-f}$. We shall ignore the sign, since we have also ignored it in the vector fields, and thus we can compare objects with the same convention. ∎

The brackets of the generating functions are the same as those of the Hamiltonian vector fields. Now the system's mass does not show up, in contrast to what happened for the Galilean group. In particular we no longer have the problem of the central charges, which appeared in the representations in terms of active canonical transformations.

Let us discuss the brackets involving the Hamiltonian:

$$\{\mathcal{H}, \mathcal{H}\} = 0 , \quad \{\mathcal{H}, P^k\} = 0 , \quad \{\mathcal{H}, L^k\} = 0 , \quad \{\mathcal{H}, K^k(\tau)\} = P^k .$$

(13.37)

The first 3, exactly as for the Galilean group, detect the conservation of the total mechanical energy, the total impulse and the total angular momentum along the Hamiltonian evolution, because of Theorem 13.12. As a matter of fact the boost one, too, is a conservation law, but we must remember the K^k in (13.35) depends parametrically on time τ, so we should view it as

$$K^k : \mathbb{R} \times \mathbb{F} \to \mathbb{R} .$$

In other words, identifying τ with t, we are speaking of a quantity defined on phase spacetime, even if we have a specific splitting $F(\mathbb{V}^{n+1}) = \mathbb{R} \times \mathbb{F}$. Now, (13.35) implies (identifying $\tau = t$)

$$\frac{\partial K^k}{\partial t} = c P^k$$

once again, and therefore $\{K^k(t), \mathscr{H}\} + c P^k = 0$ becomes:

$$\{K^k(t), \mathscr{H}\} + \frac{\partial K^k}{\partial t} = 0 .$$

This is a conservation law since we may rephrase it as:

$$\frac{d}{dt} K^k(t, \mathbf{x}(t), \mathbf{p}(t)) = 0 ,$$

where $\mathbb{R} \ni t \mapsto (\mathbf{x}(t), \mathbf{p}(t)) \in \mathbb{F}$ is a generic solution to Hamilton's equations associated with $\mathscr{H}$.

The meaning of (13.37) in terms of *Hamiltonian symmetries* is similar to the general situation on phase spacetime. The first three identities in (13.37) correspond to $\mathscr{H}$ being invariant (at least to order one) under the corresponding 1-parameter groups of canonical transformations by Theorem 13.12. The fourth one does not relate to the theorem, since there the functions *do not* depend on time. This would require talking about the *formal invariance* of $\mathscr{H}$ under the canonical transformations generated by K^k. We have discussed this for the Hamiltonian formulation on phase spacetime in (2) Examples 12.43 regarding the generating function $f := -K^k$ (i.e. $\mathbf{n} := \mathbf{e}_k$) of the Galilean boost. For the Poincaré boost the argument is similar. Actually, if we interpret passively the canonical transformation generated by the boost, we know the Hamiltonian is formally invariant. If we change inertial frame and express the Hamiltonian in terms of the four-momentum's components in the new Minkowski coordinates, the expression will not change. This is because the Hamiltonian, up to a constant factor, is the time component of the four-momentum, which satisfies the coordinate-free formula $g(P, P) = -m^2 c^2$ (see Chap. 10). Consequently: $\mathscr{H} = c P^0 = c \sqrt{\sum_{k=1}^{3} (P^k)^2 + m^2 c^2}$ in any Minkowski coordinates.

13.2.4 The Arnold-Liouville Theorem in a Nutshell

In the context of the Hamiltonian theory on symplectic manifolds $(\mathbb{F}, \omega)$, a **Hamiltonian integrable system** is a collection of n differentiable functions $f_j : \mathbb{F} \to \mathbb{R}$

that is

(a) **functionally independent**, in other words the differentials $df_j|_p$ are linearly independent at every point $p \in \mathbb{F}$,
(b) **involutive**, i.e.:

$$\{f_i, f_j\} = 0 \quad \text{everywhere on } \mathbb{F} \text{ for } i, j = 1, \dots, n, \qquad (13.38)$$

If we choose one f_j to be a Hamiltonian, involutivity forces all the others to be first integrals.

Starting from a Hamiltonian system on a symplectic manifold one can prove that if the Hamiltonian is regular enough then there exists, locally, an involutive set of n independent first integrals (around every state in phase space) [FaMa02].

The theory of integrable systems is an extremely important branch of modern Mathematical Physics, with several consequences in Theoretical Physics [RuSc13]. Here we will merely state the most elementary version of the important *Arnold-Liouville theorem* [Arn92, FaMa02, RuSc13]. This says that under certain conditions integrable systems admit canonical coordinates, the so-called *action-angle variables*, in which Hamilton's equations can be solved completely in a particularly simple way. The action-angle coordinates have played an important role in the transition from Classical to Quantum Mechanics.

Theorem 13.20 (Arnold-Liouville Theorem—Elementary Version) *Consider a Hamiltonian integrable system* $\{f_j\}_{j=1,\dots,n}$ *on a symplectic manifold* $(\mathbb{F}, \omega)$ *and define the level hypersurfaces*

$$\mathbb{F}_{\mathbf{c}} := \{p \in \mathbb{F} \mid f_j(p) = c_j\}, \quad \mathbf{c} = (c_1, \dots, c_n) \in \mathbb{R}^n.$$

If, for some $\mathbf{c} \in \mathbb{R}^n$*, the set* $\mathbb{F}_{\mathbf{c}}$ *is compact and connected, then the following hold.*

(1) $\mathbb{F}_{\mathbf{c}}$ *is an embedded n-dimensional submanifold of* $\mathbb{F}$ *diffeomorphic to the torus*

$$\mathbb{T}^n = \mathbb{S}^1 \times \cdots n \text{ times} \cdots \times \mathbb{S}^1.$$

(2) On an open set $A \supset \mathbb{F}_{\mathbf{c}}$ *there exist canonical coordinates* $\phi_1, \dots, \phi_n, I_1, \dots, I_n$, *called* **action-angle coordinates**, *where* $\phi_k \in (0, 2\pi)$ *and* $I_k \in J_k := (\alpha_k, \beta_k)$ *(whose domain* $U \subset A$ *is* A *up to a zero-measure set, so that each* ϕ_k*, varying in* $[0, 2\pi]$*, describes the k-th circle* $\mathbb{S}^1$ *in* $\mathbb{T}^n$*).*
(3) Taking f_1 *as Hamiltonian function, Hamilton's equations in action-angle coordinates read:*

$$\frac{d\phi_k}{dt} = \omega_k(I_1, \dots, I_n), \quad \frac{dI_k}{dt} = 0 \quad k = 1, \dots, n,$$

for certain differentiable maps $\omega_k : J_1 \times \cdots \times J_n \to \mathbb{R}$ *associated with the* f_k. ∎

Hamilton's equations are evidently completely integrable in the case considered.

13.3 The Symplectic Structure of $F(\mathbb{V}^{n+1})$

We finally return to the general case, with no special assumption on the form of the Hamiltonian nor on the existence of distinguished atlases on $F(\mathbb{V}^{n+1})$. We want to understand the importance of the fibres' symplectic nature in this general setup.

13.3.1 $F(\mathbb{V}^{n+1})$ as Bundle of Symplectic Manifolds

Consider two coordinate systems (U, ψ) and (U', ψ') whose first coordinates t and t' are identified with absolute time up to additive constants. We will suppose that on $U \cap U' \neq \varnothing$ they are related by a canonical transformation. Call $t, x^1, \ldots, x^{2n}$ and $t', x'^1, \ldots, x'^{2n}$ the respective coordinates. Without loss of generality we assume $t = t'$. For a given value $t = t'$, both systems define local coordinates on $\mathbb{F}_t \cap U \cap U'$ if the intersection is non-empty. As $\mathbb{F}_t$ has dimension $2n$, we can define two symplectic forms at $a \in \mathbb{F}_t \cap U \cap U'$, obtained restricting to $T_a \mathbb{F}_t \times T_a \mathbb{F}_t$ the closed 2-forms on $U \cap U'$:

$$\omega := -\frac{1}{2} \sum_{rs} S_{rs} dx^r \wedge dx^s \quad \text{and} \quad \omega' := -\frac{1}{2} \sum_{hk} S_{hk} dx'^h \wedge dx'^k .$$

The natural question is whether the two symplectic forms coincide on $\mathbb{F}_t$. If they do, every phase spacetime naturally turns every phase space at time $\mathbb{F}_t$ into a symplectic manifold (in a differentiable way in t). The symplectic form then is the one coming from the non-temporal coordinates of any natural or canonical system, viewed as coordinates on the fibres $\mathbb{F}_t$. It is fundamental to observe that the two symplectic forms must be compared on pairs of tangent vectors on $\mathbb{F}_t$ and not on $F(\mathbb{V}^{n+1})$. A tangent vector on $\mathbb{F}_t$ at a, in the above coordinates $t, x^1, \ldots, x^{2n}$, looks like:

$$X = X^0 \left.\frac{\partial}{\partial t}\right|_a + \sum_{k=1}^{2n} X^k \left.\frac{\partial}{\partial x^k}\right|_a \quad \text{with } X^0 = 0.$$

Clearly this condition is invariant under $X \in T_a F(\mathbb{V}^{n+1})$ when, in new coordinates $t', x'^1, \ldots, x'^{2n}$, we impose that t and t' are identified with absolute time up to additive constants. In fact, $t = t + c$ implies:

$$X'^0 = \frac{\partial t'}{\partial t} X^0 + \sum_{k=1}^{2n} \frac{\partial t'}{\partial x^k} X^k = 1 X^0 + \sum_{k=1}^{2n} 0 X^k = X^0 .$$

The following theorem shows how $F(\mathbb{V}^{n+1})$ may be understood as a bundle over $\mathbb{R}$ with symplectic fibres, whose symplectic forms are induced by a canonical symplectic form on the standard fibre. Each fibre's symplectic structure also defines the canonical, flow-invariant measure by Liouville's theorem (stated in the theorem's last part).

Theorem 13.21 *Consider a phase spacetime $F(\mathbb{V}^{n+1})$ built on the configuration spacetime $\mathbb{V}^{n+1}$ and viewed as bundle with canonical projection given by absolute time $T : F(\mathbb{V}^{n+1}) \to \mathbb{R}$ and canonical fibre $\mathbb{F}$ (Sect. 11.4). The following hold.*

(1) *Every fibre $\mathbb{F}_\tau$, for generic $\tau \in \mathbb{R}$, has a natural symplectic form Ω_τ, namely the restriction to $T_a\mathbb{F}_\tau \times T_a\mathbb{F}_\tau$, $a \in \mathbb{F}_\tau$, of the closed 2-form*

$$\omega := -\frac{1}{2} \sum_{rs} S_{rs} dx^r \wedge dx^s$$

in natural or canonical local coordinates $t, x^1, \ldots, x^{2n}$ around $a \in \mathbb{F}_\tau$. The restriction does not depend on the coordinates. Hence any natural or canonical coordinate system defines symplectic coordinates on $\mathbb{F}_\tau$.

(2) *The standard fibre $\mathbb{F}$ is a symplectic manifold whose symplectic form Ω satisfies the following properties.*

  (i) *If $g : J' \times \mathbb{F} \to T^{-1}(J')$ is a local trivialisation (g is a diffeomorphism, $J' \subset \mathbb{R}$ an open interval, and $T(g(\tau, a)) = \tau$) then, for any $(\tau, a) \in J' \times \mathbb{F}$:*

$$\Omega_\tau(dg_{(\tau,a)}(0, X), dg_{(\tau,a)}(0, Y)) = \Omega(X, Y), \qquad (13.39)$$

  for any $X, Y \in T_a\mathbb{F}$, where Ω_τ is defined in (1).

  (ii) *If $g^{(\tau)} : \mathbb{F} \ni a \mapsto g(\tau, a) \in \mathbb{F}_\tau$ is the fibre diffeomorphism given by g in (i) for any fibre $\mathbb{F}_\tau$ with $\tau \in J'$, the diffeomorphism preserves the symplectic structure:*

$$\Omega_\tau(dg_a^{(\tau)} X, dg_a^{(\tau)} Y) = \Omega(X, Y), \qquad (13.40)$$

  for any $X, Y \in T_a\mathbb{F}$.

(3) *If $\mathscr{H}$ is a C^1 Hamiltonian defined on a natural or canonical chart (U, ψ) with 2-form ω, a local section $I \ni s \mapsto \gamma(s) \in U$ satisfies Hamilton's equations for $\mathscr{H}$ if and only if:*

$$\langle \dot\gamma, dT \rangle = 1, \qquad (13.41)$$

$$\dot\gamma \lrcorner (\omega - d\mathscr{H} \wedge dT) = 0, \qquad (13.42)$$

where $\dot\gamma$ is the tangent vector of γ.

(4) *Calling $\iota_t : \mathbb{F}_t \to F(\mathbb{V}^{n+1})$ the canonical inclusion, the volume form on $\mathbb{F}_t$*

$$\mu_t := \frac{(-1)^{n(n+1)/2}}{n!} \iota_t^*(\omega \wedge \cdots (n \text{ times}) \cdots \wedge \omega) \qquad (13.43)$$

defines a positive, regular Borel measure by Proposition B.27, still denoted μ_t and called **canonical measure**. *It reduces to the Lebesgue measure*

$$dq^1 \cdots dq^n dp_1 \cdots dp_n$$

by (B.32) in every natural or canonical coordinates around a point of $\mathbb{F}_t$.

(5) $F(\mathbb{V}^{n+1})$ *is an orientable manifold, and the atlases of natural coordinates and of canonical coordinates are coherent. The induced orientation on the fibres $\mathbb{F}_\tau$ by $F(\mathbb{V}^{n+1})$ is the natural one, since $\mathbb{F}_\tau$ is symplectic.*

(6) *Suppose there is a C^2 Hamiltonian dynamic vector field Z as in (12.57) and (12.58). Let $D_{t_1} \subset \mathbb{F}_{t_1}$ be a Borel measurable set of finite measure μ_{t_1} such that $D_t = \Phi^{(Z)}_{t-t_1}(D_{t_1})$ exists for $t \in [t_1, t_2 - t_1]$. Then D_{t_2} has finite measure μ_{t_2} and*

$$\int_{D_{t_2}} d\mu_{t_2} = \int_{D_{t_1}} d\mu_{t_1} . \tag{13.44}$$

The above's infinitesimal version holds as well:

$$\mathcal{L}_Z(\omega - d\mathscr{H} \wedge dT) = 0 . \tag{13.45}$$

Proof First of all, (3) is just Proposition 12.46, in view of (12.59). Let us prove the rest. As for (1), it suffices to show that two coordinate systems, canonical/natural or related by a canonical transformation, induce the same symplectic form on the fibres. Take two such systems where $t' = t + c$ is absolute time τ:

$$\frac{1}{2} \sum_{r,s} S_{rs} dx^r \wedge dx^s = \frac{1}{2} \sum_{h,k,r,s} S_{rs} \frac{\partial x^r}{\partial x'^h} \frac{\partial x^s}{\partial x'^k} dx'^h \wedge dx'^k + \sum_{r,s,k} S_{rs} \frac{\partial x^r}{\partial t'} \frac{\partial x^s}{\partial x'^k} dt' \wedge dx'^k .$$

As the transformation is canonical, the above becomes:

$$\frac{1}{2} \sum_{r,s} S_{rs} dx^r \wedge dx^s = \frac{1}{2} \sum_{h,k} S_{hk} dx'^h \wedge dx'^k + \sum_{r,s,k} S_{rs} \frac{\partial x^r}{\partial t'} \frac{\partial x^s}{\partial x'^k} dt' \wedge dx'^k .$$

In other words:

$$\omega = \omega' - \sum_{r,s,k} S_{rs} \frac{\partial x^r}{\partial t'} \frac{\partial x^s}{\partial x'^k} dt' \wedge dx'^k .$$

When the two sides act on tangent vectors X, Y to $\mathbb{F}_\tau$, the absence of the components $\frac{\partial}{\partial t}$ and $\frac{\partial}{\partial t'}$ gives:

$$\omega_\tau(X, Y) = \omega'_\tau(X, Y) ,$$

proving that the symplectic forms agree on $\mathbb{F}_t$.

Statement (2) can be proved in coordinates using the diffeomorphisms g built from the diffeomorphisms f, as explained in Sect. 11.4. For starters, given $t \in \mathbb{R}$ consider an open interval $J \ni t$ and natural coordinates on it of the type mentioned in Sect. 11.4. Let then $f : J \times \mathbb{F}_t \to T^{-1}(J)$ be the corresponding diffeomorphism, locally trivialising the bundle $F(\mathbb{V}^{n+1})$. If (U, ψ), with $\psi : U \ni a \mapsto (\tau(a), \mathbf{x}(a))$, is one such system, the $\mathbf{x}$ define coordinates on $\mathbb{F}_t$. Using that, f acts as the identity $(\tau, \mathbf{x}) \mapsto (\tau, \mathbf{x})$, where $\mathbf{x}$ in the first pair are the coordinates of a point in $\mathbb{F}_t$, while $(\tau, \mathbf{x})$ in the second pair are the coordinates of a point in $T^{-1}(J)$. In these coordinates (13.39) are trivially true when $\mathbb{F}_t$ has the symplectic form Ω_t induced by the natural coordinates of (U, ψ). Said 2-form does not depend on the coordinates, as shown in (1). Because the aforementioned coordinates cover $T^{-1}(J)$, (13.39) holds everywhere on $T^{-1}(J)$. (13.40) is an easy consequence of the first. Let now $t' \notin J$. We put on $\mathbb{F}_{t'}$ the form $\Omega_{t'}$ and build a diffeomorphism $f' : J' \times \mathbb{F}_t \to T^{-1}(J')$ similar to the previous one, referring to $\mathbb{F}_{t'}$. To finish we have to show the two fibres $\mathbb{F}_t$ and $\mathbb{F}_{t'}$ are identifiable as symplectic manifolds, so to give a unique canonical fibre $\mathbb{F} := \mathbb{F}_t$. For that, as we know from Sect. 11.4, we can define finitely many open intervals $J' = J_1, J_2, \ldots, J_N = J$ so that $J_k \cap J_{k+1} \neq \varnothing$, $t' \in J'$ and on each J_k we perform the above construction. Composing the various fibre diffeomorphisms on each J_k we end up with the diffeomorphism $g_{t'} : \mathbb{F}_t \to \mathbb{F}_{t'}$. As each diffeomorphism that constitutes $g_{t'}$ satisfies (13.40) (replacing $\mathbb{F}$ with $\mathbb{F}_{t_k}$ where $t_k \in J_k$), (13.40) will hold for $g_{t'}$ with $(\mathbb{F}, \Omega)$ replaced by $(\mathbb{F}_t, \Omega_t)$ and $(\mathbb{F}_\tau, \Omega_\tau)$ replaced by $(\mathbb{F}_{t'}, \Omega_{t'})$. Hence the special fibre $\mathbb{F}_t$ chosen at the start can be seen as the canonical fibre $\mathbb{F}$ of the bundle, including its symplectic form $\Omega := \Omega_t$. Let us show this. As we know, the local trivialisations on each $J' \ni t'$ are the diffeomorphisms $g : J' \times \mathbb{F} \ni (\tau, a) \mapsto f'(\tau, g_{t'}(a)) \in T^{-1}(J')$. For these g we have $T(g(\tau, a)) = \tau$ if $\tau \in J'$ and $a \in \mathbb{F}_{t'}$. We also have (13.39), immediately, by definition and because these properties hold for f' with respect to $\mathbb{F}_{t'}$ and $g_{t'}$ satisfies (13.40). Finally (13.40) for g is an easy consequence of (13.39).

Claim (4) follows by computation.

Claim (5) is proved by noting that the Jacobian determinant between canonical or natural coordinates is the determinant of the transformation between the non-temporal coordinates (this is immediate if we expand the determinant along the first row, that of time). The non-temporal coordinates, at given τ, define symplectic coordinates on $\mathbb{F}_\tau$, so the above determinant equals 1 (Proposition 13.6), in particular it is positive. On one hand this shows the canonical coordinate atlas or the natural coordinate atlas induce an orientation, the same on phase spacetime. On the other, it proves that this orientation induces on the fibres the symplectic orientation.

(6) The exterior derivative of the Poincaré-Cartan form:

$$\omega' := \omega - d\mathscr{H} \wedge dT$$

is well defined everywhere on $F(\mathbb{V}^{n+1})$ even if $\mathscr{H}$ depends on the charts. Formula (13.45) is proved as follows, using Cartan's magic formula (B.30):

$$\mathcal{L}_Z(\omega - d\mathscr{H} \wedge dT) = d\left(Z\lrcorner(\omega - d\mathscr{H} \wedge dT)\right) + Z\lrcorner d(\omega - d\mathscr{H} \wedge dT)$$

$$= d0 + Z\lrcorner d0 = 0.$$

Let us prove (13.44). By definition of Lie derivative and (13.45) we have

$$((\Phi_\tau^{(Z)})^*\omega')(a) = \omega'(a),$$

whenever the left side is defined. But in our case it is defined, for $a \in D_1$ if $\tau \in [0, t_2 - t_1]$. Moreover, in (13.43) we can replace ω with ω' since:

$$\iota_t^*\omega = \iota_t^*\omega',$$

where $\iota_t : \mathbb{F}_t \to F(\mathbb{V}^{n+1})$ is the canonical inclusion. That is because ι_t^* kills all summands in the expansion of the product (13.43) that contain dT, as $\iota_t^* dT = 0$. So, the volume form (Definition B.24) is, replacing ω with ω':

$$\mu_t = \frac{(-1)^{n(n+1)/2}}{n!} \iota_t^*(\omega \wedge \cdots (n \text{ times}) \cdots \wedge \omega)$$

$$= \frac{(-1)^{n(n+1)/2}}{n!} \iota_t^*\omega \wedge \cdots (n \text{ times}) \cdots \wedge \iota_t^*\omega$$

$$= \frac{(-1)^{n(n+1)/2}}{n!} \iota_t^*\omega' \wedge \cdots (n \text{ times}) \cdots \wedge \iota_t^*\omega'$$

$$= \frac{(-1)^{n(n+1)/2}}{n!} \iota_t^*(\omega' \wedge \cdots (n \text{ times}) \cdots \wedge \omega').$$

Now (13.44) reads

$$\int_{D_{t_2}} \omega' = \int_{D_{t_1}} \omega'.$$

Proposition B.30 allows us to rephrase the above as:

$$\int_{D_{t_1}} (\Phi_{t_2-t_1}^{(Z)})^*\omega' = \int_{D_{t_1}} \omega',$$

which is true (thus proving (13.45)) because $(\Phi_{t_2-t_1}^{(Z)})^*\omega' = \omega'$ on D_{t_1} as shown earlier. $\qquad\square$

13.3.2   A More General Notion of Phase Spacetime

Theorem 13.21 uncovers a broader perspective, freeing us from the construction of the bundle $F(\mathbb{V}^{n+1})$ over the spacetime of configurations.

We can think of phase spacetime F^{2n+1}, in a completely general way, as a bundle over $\mathbb{R}$, under the canonical projection identified with absolute time $T : F^{2n+1} \to \mathbb{R}$ and with standard fibre being a symplectic manifold $(\mathbb{F}, \Omega)$ (necessarily of dimension $2n$), with a preferred collection of local trivialisations $\{(J_i, g_i)\}_{i \in I}$ (hence, $J_i \subset \mathbb{R}$ is an open interval and $g_i : J_i \times \mathbb{F} \to T^{-1}(J_i)$ is a diffeomorphism with $T(f_i(\tau, a)) = \tau$ for any $\tau \in J_i$ and $a \in \mathbb{F}$). If the fibre at t is $\mathbb{F}_t := T^{-1}(\{t\})$ and $g_i^{(t)} : \mathbb{F} \ni a \mapsto g_i(t, a) \in \mathbb{F}_t$ is the diffeomorphism from the standard fibre and the fibre at t with trivialisation (J_i, g_i), $t \in J_i$, then the above family is such that

(i) $\bigcup_{i \in I} U_i = \mathbb{R}$,
(ii) if $t \in U_j \cap U_k$ and $a \in \mathbb{F}_t$, then:

$$\Omega\left((dg_{ja}^{(t)})^{-1}X, (dg_{ja}^{(t)})^{-1}Y\right) = \Omega\left((dg_{ka}^{(t)})^{-1}X, (dg_{ka}^{(t)})^{-1}Y\right) =: \Omega_t(X, Y),$$

for any $X, Y \in T_a\mathbb{F}_t$.

Thus every fibre has a symplectic form Ω_t induced by Ω, irrespective of the trivialisation.

It is important to stress the above structure is less rigid than that of $F(\mathbb{V}^{n+1})$, since there might not be a special family of local coordinates corresponding to natural coordinates. Nonetheless, all of Hamilton's theory can be formulated in this context. Observe, first, that there exists a privileged atlas. It consists of all coordinates adapted to the fibres of F^{2n+1}, i.e. local charts (U, ψ) with $\psi : U \ni a \mapsto (t(a), x^1(a), \dots, x^{2n}(a))$ of which we further require two things: $T(a) = t(a) + c$ for some constant c; secondly, defining the closed form

$$\omega := -\frac{1}{2} \sum_{i,j=1}^{2n} S_{ij} dx^i \wedge dx^j,$$

we also have

$$\omega(X, Y) = \Omega_\tau(X, Y) \quad \text{if } X, Y \in T_a\mathbb{F}_\tau \subset T_aF^{2n+1} \text{ with } T(a) = \tau.$$

Part (4) in Theorem 12.48 (note the proof only requires what we have assumed thus far) says that the local coordinate transformations of this type are exactly the canonical ones. Yet, we did not make any topological assumption on the domain U, nor have we distinguished between natural and canonical coordinates, since such distinction is not possible in general in our setup. Local coordinate systems defined thus will still be called **canonical coordinate systems** (beware they are not canonical coordinates on a symplectic manifold, but canonical coordinates on a more complicated structure). At present Hamilton's dynamics can be formulated assuming that in each canonical coordinate system (U, ψ), with $\psi : U \ni a \mapsto (t(a), x^1(a), \dots, x^{2n}(a))$, there is a 1-form h (at least C^1) such that

$$dh \wedge dT = 0, \tag{13.46}$$

and also that, when passing to (U', ψ'):

$$\omega - h \wedge dT = \omega' - h' \wedge dT =: v|_{U \cap U'} \tag{13.47}$$

on $U \cap U'$ if non-empty. Therefore we have a global 2-form v that locally reduces to the above expression. Hamilton's equations for a section $\gamma = \gamma(t)$ are then

$$\langle \dot{\gamma}, dT \rangle = 1, \tag{13.48}$$

$$\dot{\gamma} \lrcorner v = 0. \tag{13.49}$$

Let us examine the meaning of (13.46). In any canonical coordinate system (U, ϕ) on U, even if not the one relative to h, with $\phi : U \ni a \mapsto (t(a), x^1(a), \dots, x^{2n}(a))$, where $h = h_0 dt + \sum_{k=1}^{2n} h_k dx^k$, the above condition is:

$$\sum_{i,k=1}^{2n} \left(\frac{\partial h_i}{\partial x^k} - \frac{\partial h_k}{\partial x^i} \right) dx^k \wedge dx^i \wedge dt = 0 \quad \text{for } i, k = 1, \dots, 2n.$$

Applying $\frac{\partial}{\partial t} \lrcorner$, we obtain:

$$\frac{\partial h_i}{\partial x^k} = \frac{\partial h_k}{\partial x^i} \quad \text{for } i, k = 1, \dots, 2n.$$

This implies the previous relation, and hence is equivalent to (13.46). So, in analogy to the core part of the proof of Theorem 12.9, if U is connected and locally simply connected then (13.46) implies that there exists a function $\mathscr{H} = \mathscr{H}(t, x^1, \dots, x^{2n})$, defined up to additive functions of time, for which:

$$\frac{\partial \mathscr{H}}{\partial x^k} = h_k .$$

This holds in any fibre-adapted coordinates on U. Supposing the coordinates are also symplectic, the above equations for γ take the standard form:

$$\langle \dot{\gamma}, dT \rangle = 1, \tag{13.50}$$

$$\dot{\gamma} \lrcorner (\omega - d\mathscr{H} \wedge dT) = 0. \tag{13.51}$$

Equations (13.48) and (13.49) are well defined on F^{2n+1} even if the Hamiltonian is not defined: just having the 1-forms h satisfying (13.47) is enough. The construction also guarantees Liouville's Theorem, in its version (5) in Theorem 13.21, since the proof only needs the Poincaré-Cartan form and the above equations.

An important point is checking whether (13.47) are compatible with (13.46). The former implies immediately

$$(h' - h) \wedge dT = \omega - \omega' \, .$$

Then (13.46) forces

$$0 = (dh' - dh) \wedge dT = d\left[(h' - h) \wedge dT\right] = d(\omega - \omega') = 0$$

on $U \cap U'$. In other words

$$d\omega - d\omega' = 0 \, .$$

But we know this is true, since ω and ω' are closed.

Chapter 14
Complement: Elements of the Theory of Ordinary Differential Equations

In this chapter we introduce the elementary theory of *dynamical systems* in terms of *systems of ordinary differential equations*, also on differentiable manifolds.

From the physical viewpoint, the mathematical tools and results we will discuss are fundamental to make sense of the idea of *determinism* of Classical Mechanics. Indeed, one of the goals will be proving a generalised version of Theorem 3.11, one with weaker assumptions and stronger conclusion. That proof will confirm the success of Newtonian Dynamics in describing and predicting physical phenomena within Classical Mechanics.

We will make use of the following notation.

(a) Throughout the chapter $\mathbb{K}$ denotes one of the fields $\mathbb{R}$, $\mathbb{C}$. In either case $\mathbb{K}^n$ is the vector space over $\mathbb{K}$ of dimension n built in the standard way as direct sum of n copies of $\mathbb{K}$.

(b) If $\Omega \subset \mathbb{K}^m$ is an open set, $C^k(\Omega, \mathbb{K}^n)$ indicates the complex vector space of continuous maps from Ω to $\mathbb{K}^m$ (where the codomain $\mathbb{K}^n$ can be either $\mathbb{R}^n$ or $\mathbb{C}^n$, irrespective of what the domain $\mathbb{K}^m$ is) admitting all partial derivatives up to order k included. In case $\mathbb{K} := \mathbb{C}$ we refer to the underlying real vector space

$$\mathbb{C} = \mathbb{R} + i\mathbb{R} \tag{14.1}$$

and the partial derivatives are computed in real coordinates. For example $f \in C^2(\mathbb{C}^2; \mathbb{C}^n)$ means the following: call

$$g_k(x_1, y_1, x_2, y_2) := Ref_k(z_1, z_2) \quad \text{and} \quad h_k(x_1, y_1, x_2, y_2) := Imf_k(z_1, z_2)$$

the real and imaginary parts of the complex components of $f(z_1, z_2)$, $k = 1, \ldots, n$, with $z_1 = x_1 + iy_1$ and $z_2 = x_2 + iy_2$. Then g_k and h_k are $C^2(\Omega)$ where Ω is an open subset of $\mathbb{R}^4$ under (14.1). ∎

V. Moretti, *Analytical Mechanics*, La Matematica per il 3+2 150,
https://doi.org/10.1007/978-3-031-27612-5_14

14.1 Systems of Differential Equations

The fundamental notion is that of *system of differential equations of order one in normal form*.

Let $\Omega \subset \mathbb{R} \times \mathbb{K}^n$ ($N = 1, 2, \ldots$, given) be a non-empty open set and $\mathbf{f} : \Omega \to \mathbb{K}^n$ a function. A **system of ordinary differential equations (ODE system) of order one in normal form** is an equation:

$$\frac{d\mathbf{x}}{dt} = \mathbf{f}(t, \mathbf{x}(t)) \,, \tag{14.2}$$

of which we seek the **solutions**, i.e. the differentiable maps of the form $\mathbf{x} : I \to \mathbb{K}^n$, with $I \subset \mathbb{R}$ open interval, that solve (14.2) on I.

The system is called **ordinary differential equation (ODE) of order one in normal form** if $n = 1$. It is called **autonomous** if $\mathbf{f}$ does not depend explicitly on the variable t. Finally, if $\mathbf{x} : I \to \mathbb{K}^n$ is a solution of (14.2):

(1) the set $\{(t, \mathbf{x}(t)) \in \mathbb{R} \times \mathbb{K}^n \mid t \in I\} \subset \Omega$ is called **integral curve** of the solution;
(2) the set $\{\mathbf{x}(t) \in \mathbb{K}^n \mid t \in I\}$ is the **orbit** of the solution.

14.1.1 Reduction to Order One

One can similarly define a system of ODEs of order $k > 1$ in normal form as:

$$\frac{d^k\mathbf{x}}{dt^k} = \mathbf{f}\left(t, \mathbf{x}(t), \frac{d\mathbf{x}}{dt}, \ldots, \frac{d^{k-1}\mathbf{x}}{dt^{k-1}}\right) \,, \tag{14.3}$$

where the unknown solution $\mathbf{x} = \mathbf{x}(t)$ is required to be differentiable up to order k on some interval $I \ni t$. The *normality* condition expresses the demand that the highest derivative appears, *alone*, in the left-hand side. When treating Newtonian Dynamics, as we saw in Chap. 3, we arrive at a system of differential equations of order two in normal form, by applying Newton's second law to a system of point particles (keeping in account the cases of Sect. 3.3, including the pure equations of motion). The equation of motion of the point particles forming the system is a solution to the system of differential equations. Systems of differential equations arise in several contexts in Physics for describing the evolution of certain systems; they also crop up in other areas such as Biology, Chemistry and Biomathematics (population dynamics). In general these are ODE systems of order different from two. Order 2 is typical of Physical Mechanics, and more generally of Physics.

Studying the case $k = 1$ is not restrictive because we can always reduce to it starting from any $k \in \mathbb{N}, k > 1$. For this, consider system (14.3) with $k > 1$. Define the new variables

$$\mathbf{y}_r := \frac{d^r \mathbf{x}}{dt^r}, \quad \text{for } r = 1, 2, \ldots, k - 1$$

In these variables system (14.3), of order k, is equivalent to the first-order system

$$\frac{d\mathbf{X}}{dt} = \mathbf{g}\left(t, \mathbf{X}(t)\right) \tag{14.4}$$

in the variable $\mathbf{X} = (\mathbf{x}, \mathbf{y}_1, \ldots, \mathbf{y}_{k-1}) \in \mathbb{R}^{kn}$, where

$$\mathbf{g} : (t, (\mathbf{x}, \mathbf{y}_1, \ldots, \mathbf{y}_{k-1})) \mapsto (\mathbf{y}_1, \mathbf{y}_2, \ldots, \mathbf{f}(t, \mathbf{x}(t), \mathbf{y}_1(t), \ldots, \mathbf{y}_{k-1}(t))) .$$

Thus system (14.4) explicitly reads:

$$\frac{d\mathbf{y}_{k-1}}{dt} = \mathbf{f}(t, \mathbf{x}(t), \mathbf{y}_1(t), \ldots, \mathbf{y}_{k-1}(t)) ,$$

$$\frac{d\mathbf{y}_{k-2}}{dt} = \mathbf{y}_{k-1} ,$$

$$\cdots = \cdots$$

$$\frac{d\mathbf{y}_1}{dt} = \mathbf{y}_2 ,$$

$$\frac{d\mathbf{x}}{dt} = \mathbf{y}_1 .$$

It becomes then evident that any solution $\mathbf{X} = \mathbf{X}(t)$ to (14.4), with $t \in I$, defines a solution $\mathbf{x} = \mathbf{x}(t)$ to (14.3), with $t \in I$, by extracting the first component of the function $\mathbf{X}$; conversely, every solution $\mathbf{x} = \mathbf{x}(t)$ of (14.3), with $t \in I$, defines a solution of (14.4) given by $\mathbf{X} = (\mathbf{x}(t), d\mathbf{x}/dt, \ldots, d^{k-1}\mathbf{x}/dt^{k-1})$ with $t \in I$.

Referring to the aforementioned solutions $\mathbf{x} = \mathbf{x}(t)$ and $\mathbf{X} = \mathbf{X}(t)$, with $t \in I$, observe that $\mathbf{x} = \mathbf{x}(t)$ satisfies the $k - 1$ conditions, for $t_0 \in I$,

$$\mathbf{x}(t_0) = \mathbf{x}_0 \ , \ \frac{d\mathbf{x}}{dt}(t_0) = \mathbf{x}_0^{(1)} \ , \ldots, \ \frac{d\mathbf{x}^{k-1}}{dt^{k-1}}(t_0) = \mathbf{x}_0^{(k-1)} .$$

if and only if $\mathbf{X} = \mathbf{X}(t)$ satisfies

$$\mathbf{X}(t_0) = (\mathbf{x}_0, \mathbf{x}_0^{(1)} \ldots, \mathbf{x}_0^{(k-1)}) .$$

14.1.2 The Cauchy Problem

We pass to the second fundamental definition, that of *Cauchy problem*, or *initial value problem*, which introduces the concept of *initial conditions*.

Given $\mathbf{f} : \Omega \to \mathbb{K}^n$, $\Omega \subset \mathbb{R} \times \mathbb{K}^n$ open and non-empty, the **Cauchy problem** for an ODE system of order one in normal form:

$$\frac{d\mathbf{x}}{dt} = \mathbf{f}\,(t, \mathbf{x}(t))\ , \tag{14.5}$$

consists in finding the solutions to (14.5) that further satisfy a given **initial condition**

$$\mathbf{x}(t_0) = \mathbf{x}_0\ ,$$

where $(t_0, \mathbf{x}_0) \in \Omega$. In other words we seek the differentiable maps $\mathbf{x} : I \to \mathbb{K}^n$, with $I \subset \mathbb{R}$ open interval containing t_0, that solve (14.5) on I and satisfy $\mathbf{x}(t_0) = \mathbf{x}_0$ as well.

Remarks 14.1 We may similarly formulate the Cauchy problem for a system of order $k > 1$,

$$\frac{d^k\mathbf{x}}{dt^k} = \mathbf{f}\left(t, \mathbf{x}(t), \frac{d\mathbf{x}}{dt}, \dots, \frac{d^{k-1}\mathbf{x}}{dt^{k-1}} \right)\ ,$$

by looking for solutions that satisfy $k - 1$ **initial conditions**:

$$\mathbf{x}(t_0) = \mathbf{x}_0\ ,\ \frac{d\mathbf{x}}{dt}(t_0) = \mathbf{x}_0^{(1)}\ ,\dots,\ \frac{d\mathbf{x}^{k-1}}{dt^{k-1}}(t_0) = \mathbf{x}_0^{(k-1)}\ .$$

By what was explained in Sect. 14.1.1, this Cauchy problem is completely equivalent to a Cauchy problem for a differential equation of order one: there is a 1-1 correspondence between the solutions to the Cauchy problem of order k and the solutions to the associated Cauchy problem of order 1. For this reason, in the sequel we shall only study the problems of existence ad uniqueness for initial value problems of order one. ∎

14.1.3 First Integrals

A useful notion, especially in Physics, in the theory of differential equations is that of *first integral*, which we have used throughout the book in a number of situations and that now we shall define formally for completeness.

Definition 14.2 (First Integral) Given $\mathbf{f} : \Omega \to \mathbb{K}^n$, $\Omega \subset \mathbb{R} \times \mathbb{K}^n \times \cdots \times \mathbb{K}^n$ non-empty and open, consider the ODE system (14.3). A function $F : \Omega \to \mathbb{K}$ is called **first integral** of the system if for any solution $\mathbf{x} : I \to \mathbb{K}^n$ to (14.3), $I \subset \mathbb{R}$, we have:

$$F\left(t, \mathbf{x}(t), \frac{d\mathbf{x}}{dt}, \ldots, \frac{d^{k-1}\mathbf{x}}{dt^{k-1}}\right) = c$$

for some constant $c \in \mathbb{K}$, usually dependent of the solution considered. $\diamond$

14.2 Preparatory Notions and Results for the Existence and Uniqueness Theorems

To prove an existence and uniqueness theorem for the Cauchy problem (14.5) with initial condition $\mathbf{x}(t_0) = \mathbf{x}_0$, when the function $\mathbf{f}$ is regular enough (locally Lipschitz in $\mathbf{x}$), we introduce the Banach space of continuous maps on a compact set K with values in $\mathbb{K}^n$, $C^0(K; \mathbb{K}^n)$, and we will prove the fixed-point theorem for contractions on metric spaces. From that result, after transforming the initial value problem into an integral equation involving a certain contraction on a closed subset in $C^0(K; \mathbb{K}^n)$, we shall prove the Existence and uniqueness theorem for the Cauchy problem, in its local and global forms.

14.2.1 The Banach Space $C^0(K; \mathbb{K}^n)$

Normed spaces are special kinds of metric spaces with a richer structure.

We remind that a vector space V over the field $\mathbb{K}$ ($\mathbb{C}$ or $\mathbb{R}$, as we said at the beginning), is called **normed** if there is a function, called **norm**, $|| \cdot || : V \to \mathbb{R}$ satisfying the following properties:

(N1) **positive definiteness**: for any $\mathbf{v} \in V$ we have $||\mathbf{v}|| \geq 0$, with $\mathbf{v} = 0$ if $||\mathbf{v}|| = 0$,:

(N2) **homogeneity**: $||\lambda \mathbf{v}|| = |\lambda| ||\mathbf{v}||$ for any $\lambda \in \mathbb{K}$ and $\mathbf{v} \in V$,

(N3) **triangle inequality**: $||\mathbf{v} + \mathbf{u}|| \leq ||\mathbf{v}|| + ||\mathbf{u}||$ for any $\mathbf{u}, \mathbf{v}$ in V.

Remarks 14.3

(1) (N2) implies $||\mathbf{0}|| = 0$. Hence any norm satisfies: $||\mathbf{v}|| = 0$ if and only if $\mathbf{v} = \mathbf{0}$. If (N1) is weakened to **positive semi-definiteness**: $||\mathbf{v}|| \geq 0$ (without asking for $||\mathbf{v}|| = 0$ to force $\mathbf{v} = \mathbf{0}$), and we keep (N2) and (N3), then $|| \cdot ||$ is called a **semi-norm** on V.

(2) Every normed space V with norm $|| \cdot ||$ is naturally a metric space, hence it has a metric topology with distance:

$$d(\mathbf{u}, \mathbf{v}) := ||\mathbf{u} - \mathbf{v}||, \quad \forall \mathbf{u}, \mathbf{v} \in V.$$

(3) Just like metric spaces, normed spaces are Hausdorff. Hence in normed spaces limits are unique. ∎

Examples 14.4

(1) The simplest normed spaces are built on the vector spaces $\mathbb{C}^n$ or $\mathbb{R}^n$ with norm $|| \cdot ||$ being the standard Euclidean norm:

$$||(c_1, \cdots, c_n)|| := \sqrt{|c_1|^2 + \cdots + |c_n|^2}$$

where $| \cdot |$ is the modulus on $\mathbb{C}$ or $\mathbb{R}$.

(2) If K is a compact topological space (say, a closed and bounded subset of $\mathbb{R}^n$ or $\mathbb{C}^n$), $C^0(K; \mathbb{K})$ denotes the set of *continuous* maps defined on K with values in $\mathbb{K}^n$, which typically is $\mathbb{R}^n$ or $\mathbb{C}^n$. We can turn it into a vector space over $\mathbb{K}$, by defining the linear operations by

$$(\lambda f)(x) := \lambda f(x), \quad (f + g)(x) := f(x) + g(x)$$

$$\text{for any } f, g \in C^0(K; \mathbb{K}^n), \lambda \in \mathbb{K} \text{ and } x \in K.$$

The space $C^0(K; \mathbb{K}^n)$ becomes normed under the norm

$$||f||_\infty := \sup_{x \in K} ||f(x)||.$$

The above is well defined, since for any $f \in C^0(K, \mathbb{K}^n)$, the function $K \ni x \mapsto ||f(x)||$ is continuous (as composite of continuous maps) and hence bounded (as defined on a compact set). Properties (N1), (N2) and (N3) for $|| \cdot ||_\infty$ hold trivially due to the similar properties enjoyed by $|| \cdot ||$ pointwise, and the known properties of the least upper bound.

Recall that if X is a metric space with distance d, a sequence $\{x_n\}_{n \in \mathbb{N}} \subset X$ is called **Cauchy sequence** if the **Cauchy condition** holds: *for any $\epsilon > 0$ there exists $N_\epsilon \in \mathbb{N}$ such that $d(x_n, x_m) < \epsilon$ whenever $n, m > N_\epsilon$.*

It should be well known that any sequence $\{x_n\}_{n \in \mathbb{N}} \subset X$ admitting limit $x \in X$ is Cauchy. In fact, in these hypotheses, for any $\eta > 0$ we have $d(x_n, x_m) \le d(x_n, x) + d(x_m, x) < 2\eta$ if $n > M_\eta$ and $m > M_\eta$ for some $M_\eta \in \mathbb{N}$. But then for any given $\epsilon > 0$, letting $N_\epsilon := M_\eta$ with $\eta = \epsilon/2$, the sequence $\{x_n\}_{n \in \mathbb{N}}$ satisfies the Cauchy condition. We recall that a metric space is called **complete** if the converse holds, i.e. every Cauchy sequence converges in the space.

Definition 14.5 A normed vector space V is called a **Banach space** if it is complete as a metric space. ◇

Recall that $\mathbb{R}$ and $\mathbb{C}$ are complete [Apo91I], so they are elementary instances of Banach spaces. For this reason the space $\mathbb{K}^n$ of Example 14.4.**1** is complete. The proof is easy. Consider a sequence $\{\mathbf{u}_m\}_{m\in\mathbb{N}} \subset \mathbb{K}^n$ where $\mathbf{u}_m = (v_{1m}, \ldots, v_{nm})$. If $\{\mathbf{u}_m\}_{m\in\mathbb{N}} \subset \mathbb{K}^n$ is Cauchy, so are the sequences $\{v_{im}\}_{m\in\mathbb{N}} \subset \mathbb{K}$, for any given $i = 1, 2, \ldots, n$, since $|v_{ip} - v_{iq}| \leq ||\mathbf{u}_p - \mathbf{u}_q||$. If $v_i \in \mathbb{K}$ is the limit of $\{v_{im}\}_{m\in\mathbb{N}}$, it is immediate to see that $\mathbf{u}_n = (v_{1m}, \ldots, v_{nm}) \to (v_1, \ldots, v_n) =: \mathbf{u}$ as $n \to +\infty$, since

$$||\mathbf{u}_m - \mathbf{u}||^2 = |v_{1m} - v_1|^2 + \cdots + |v_{nm} - v_n|^2 \to 0, \quad \text{as } m \to +\infty.$$

The next theorem will be very useful in the theory of differential equations.

Theorem 14.6 *If K is a compact space, the normed space $C^0(K; \mathbb{K}^n)$ with norm $||\cdot||_\infty$ is a Banach space.*

Proof Let $\{f_n\}_{n\in\mathbb{N}} \subset C^0(K; \mathbb{K}^n)$ be a Cauchy sequence. The goal is proving there exists $f \in C^0(K)$ such that $||f_n - f||_\infty \to 0$ as $n \to +\infty$. Since $\{f_n\}_{n\in\mathbb{N}}$ is Cauchy, for any given $x \in K$, also the sequence of vectors in $\mathbb{K}^n$: $\{f_n(x)\}_{n\in\mathbb{N}} \subset \mathbb{K}^n$ is Cauchy. In fact:

$$||f_n(x) - f_m(x)|| \leq \sup_{z\in K} ||f_n(z) - f_m(z)|| = ||f_n - f_m||_\infty < \epsilon, \quad \text{as } n, m < N_\epsilon.$$

In this way, as $\mathbb{K}^n$ is complete, we pointwise define the map:

$$f(x) := \lim_{n\to+\infty} f_n(x).$$

We want to show $||f_n - f||_\infty \to 0$ as $n \to +\infty$ and $f \in C^0(K; \mathbb{K}^n)$. As the sequence $\{f_n\}_{n\in\mathbb{N}}$ is Cauchy, for any $\epsilon > 0$ there exists N_ϵ such that, if $n, m > N_\epsilon$

$$||f_n(x) - f_m(x)|| < \epsilon, \quad \text{for all } x \in K.$$

On the other hand, by definition of f, for a *given* $x \in K$ and for any $\epsilon'_x > 0$, there exists N_{x,ϵ'_x} such that, if $m > N_{x,\epsilon'_x}$ then $||f_m(x) - f(x)|| < \epsilon'_x$. Using both facts, if $n > N_\epsilon$

$$||f_n(x) - f(x)|| \leq ||f_n(x) - f_m(x)|| + ||f_m(x) - f(x)|| < \epsilon + \epsilon'_x$$

where we chose $m > \max(N_\epsilon, N_{x,\epsilon'_x})$. In the end, if $n > N_\epsilon$, then

$$||f_n(x) - f(x)|| < \epsilon + \epsilon'_x, \quad \text{for any } \epsilon'_x > 0.$$

But $\epsilon'_x > 0$ is arbitrary, so the above inequality holds for $\epsilon'_x = 0$ too, possibly as equality. In this way we lose the dependency on x. Overall, for any $\epsilon > 0$ there exists $N_\epsilon \in \mathbb{N}$ such that, if $n > N_\epsilon$

$$||f_n(x) - f(x)|| \le \epsilon, \quad \text{for } all\, x \in K. \tag{14.6}$$

Hence $\{f_n\}$ converges to f *uniformly*. By a well-known basic theorem [Apo91I] (the proof is recalled below for ease of reference) f is continuous, being the uniform limit of continuous maps. Therefore $f \in C^0(K; \mathbb{K}^n)$.

Since (14.6) holds for any $x \in K$, it holds if we take the supremum over x in K: for any $\epsilon > 0$ there exists $N_\epsilon \in \mathbb{N}$ such that $\sup_{x \in K} ||f_n(x) - f(x)|| < \epsilon$, if $n > N_\epsilon$. In other words

$$||f_n - f||_\infty \to 0, \quad \text{as } n \to +\infty.$$

This ends the proof.   □

Lemma 14.7 *Let X be a topological space and $g_n : X \to \mathbb{K}^n$ continuous maps that converge uniformly to $g : X \to \mathbb{K}^n$ as $n \to +\infty$. Then g is continuous on X.*

Proof Fix $x \in X$. For any $\epsilon > 0$, there exists $n > 0$ such that $||g(z) - g_n(z)|| < \epsilon/3$ for any $z \in X$, by uniform convergence. At the same time g_n is continuous at x, so there exists a neighbourhood $G_x \subset X$ of x such that, $||g_n(x) - g_n(y)|| < \epsilon/3$ for any $y \in G_x$. Putting everything together, for any given $\epsilon > 0$ there is a neighbourhood G_x of x such that, if $y \in G_x$:

$$||g(x) - g(y)|| < ||g(x) - g_n(x)|| + ||g_n(x) - g_n(y)||$$
$$+ ||g_n(y) - g(y)|| < \epsilon/3 + \epsilon/3 + \epsilon/3 = \epsilon.$$

As $x \in X$ is arbitrary, we have proved g is continuous on X.   □

At last we prove the following elementary fact.

Proposition 14.8 *If Y is a non-empty close subset in a complete metric space X with distance d, Y is a complete metric space for the restriction of d to $Y \times Y$.*

Proof Clearly Y with d restricted to $Y \times Y$ is a metric space. If $\{y_n\}_{n \in \mathbb{N}} \subset Y$ is a Cauchy sequence, it is Cauchy also in the metric space X. But X is complete, so there exists $y \in X$ with $\lim_{n \to +\infty} y_n = y$. As Y is closed, it contains its limit points, so $y \in Y$. We have shown that any Cauchy sequence in Y converges in Y, and the proof ends.   □

14.2.2 Fixed-Point Theorem in Complete Metric Spaces

In a metric space there are special maps that play a crucial role in the proof of the Existence and uniqueness theorem for the Cauchy problem. We are talking about the so-called *contractions*.

Definition 14.9 Let X be a metric space with norm d. A map $G : X \rightarrow X$ is called **contraction**, if there exists a real number $\lambda \in [0, 1)$ such that:

$$d(G(x), G(y)) \leq \lambda d(x, y) \quad \text{for any pair } (x, y) \in X \times X . \tag{14.7}$$

$\diamond$

Remarks 14.10

(1) We explicitly *exclude* the value $\lambda = 1$.
(2) Formula (14.7) immediately implies that: *every contraction is continuous* in the metric topology of X. In fact, (14.7) implies that if $x \rightarrow y$ then $G(x) \rightarrow G(y)$.
(3) The definition holds in normed spaces, using the distance induced by the norm. The function G in that case *is not* required to be linear. ∎

We conclude with the (Banach-Caccioppoli) *fixed-point theorem*, which has countless consequences in Mathematics and its applications (especially Physics).

Theorem 14.11 (Fixed-Point Theorem) *Let X be a (non-empty) complete metric space and $G : X \rightarrow X$ a contraction. There exists a unique element $z \in X$, called* **fixed point**, *such that:*

$$G(z) = z . \tag{14.8}$$

Proof *Existence of the fixed point.* Consider the sequence, where $x_0 \in X$ is arbitrary, defined recursively by $x_{n+1} = G(x_n)$. The goal is to prove it is Cauchy. As X is complete the sequence will have a limit $x \in X$. We will show that the latter is a fixed point.

Without loss of generality we suppose $m \geq n$. If $m = n$ trivially $d(x_m, x_n) = 0$, and for $m > n$ we use repeatedly the triangle inequality to obtain:

$$d(x_m, x_n) \leq d(x_m, x_{m-1}) + d(x_{m-1}, x_{m-2}) + \cdots + d(x_{n+1}, x_n) . \tag{14.9}$$

Consider the generic term on the right: $d(x_{p+1}, x_p)$. Then:

$$d(x_{p+1}, x_p) = d(G(x_p), G(x_{p-1})) \leq \lambda d(x_p, x_{p-1}) = \lambda d(G(x_{p-1}), G(x_{p-2}))$$

$$\leq \lambda^2 d(x_{p-1}, x_{p-2})$$

$$\leq \cdots \leq \lambda^p d(x_1, x_0) .$$

For any $p = 1, 2, \ldots$, therefore, $d(x_{p+1}, x_p) \leq \lambda^p d(x_1, x_0)$. Inserting that in the right-hand side of (14.9) gives the estimate:

$$d(x_m, x_n) \leq d(x_1, x_0) \sum_{p=n}^{m-1} \lambda^p = d(x_1, x_0)\lambda^n \sum_{p=0}^{m-n-1} \lambda^p \leq \lambda^n d(x_1, x_0) \sum_{p=0}^{+\infty} \lambda^p$$

$$\leq d(x_1, x_0) \frac{\lambda^n}{1 - \lambda}$$

where we used the fact that $\sum_{p=0}^{+\infty} \lambda^p = (1 - \lambda)^{-1}$ if $0 \leq \lambda < 1$. Hence:

$$d(x_m, x_n) \leq d(x_1, x_0) \frac{\lambda^n}{1 - \lambda} . \tag{14.10}$$

But since $|\lambda| < 1$ we have $d(x_1, x_0)\lambda^n/(1 - \lambda) \to 0$ as $n \to +\infty$. Therefore $d(x_m, x_n)$ can be made as small as we like by choosing the lesser between large enough m and n. Immediately, $\{x_n\}_{n\in\mathbb{N}}$ is Cauchy. As X is complete, $\lim_{n\to+\infty} x_n = x \in X$ for some x. Now, G is continuous, being a contraction:

$$G(x) = G\left(\lim_{n\to+\infty} x_n\right) = \lim_{n\to+\infty} G(x_n) = \lim_{n\to+\infty} x_{n+1} = x ,$$

so $G(x) = x$ as required.

Uniqueness of the Fixed Point Assume x and x' satisfy $G(x) = x$ and $G(x') = x'$. Then

$$d(x, x') = d(G(x), G(x')) \leq \lambda d(x, x') .$$

If $d(x, x') \neq 0$, dividing the first and last terms by $d(x, x')$ would give the inequality $1 \leq \lambda$, which is impossible by hypothesis. Hence $d(x, x') = 0$, and then $x = x'$ since the distance is positive definite.

$\square$

14.2.3 (Locally) Lipschitz Functions

The last ingredient to apply the above mathematical machinery to the Existence and uniqueness theorem for the Cauchy problem are *(locally) Lipschitz functions*. The notion's purpose is to specify the regularity of the function in the right-hand side of the Cauchy problem's ODE.

Definition 14.12 Let $p \geq 0$ and $n, m > 0$ be given integers, and $\Omega \subset \mathbb{K}^p \times \mathbb{K}^n$ non-empty and open.

$\mathbf{F} : \Omega \to \mathbb{K}^m$ is called **locally Lipschitz (in the variable $\mathbf{x} \in \mathbb{K}^n$ for $p > 0$)**, if for any $q \in \Omega$ there is a constant $L_q \geq 0$ such that, on an open set $O_q \ni q$:

$$||\mathbf{F}(\mathbf{z}, \mathbf{x}) - \mathbf{F}(\mathbf{z}, \mathbf{x}')|| \leq L_q ||\mathbf{x} - \mathbf{x}'||, \quad \text{for any pair } (\mathbf{z}, \mathbf{x}), (\mathbf{z}, \mathbf{x}') \in O_q. \quad (14.11)$$

If $O_q = \Omega$, $\mathbf{F}$ is called **Lipschitz (in the variable $\mathbf{x} \in \mathbb{K}^n$ for $p > 0$)**. $\diamond$

Remarks 14.13 In (14.11) the pairs $(\mathbf{z}, \mathbf{x})$, $(\mathbf{z}, \mathbf{x}')$ have *equal* first element $\mathbf{z}$. ∎

Examples 14.14

(1) We will use the usual notation $\mathbf{x} = (x^1, \dots, x^n)$, $\mathbf{z} = (z^1, \dots, z^p)$; for $\mathbf{F} : \Omega \to \mathbb{R}^m$ with $\Omega = A \times B$, $A \subset \mathbb{R}^p$ and $B \subset \mathbb{R}^n$ non-empty and open, we write $\mathbf{F}(\mathbf{z}, \mathbf{x}) = (F^1(\mathbf{z}, \mathbf{x}), \dots, F^m(\mathbf{z}, \mathbf{x}))$. The class of locally Lipschitz maps is very large. For example, the next proposition holds.

Proposition 14.15 $\mathbf{F} : \Omega \to \mathbb{R}^m$ *with Ω as above is certainly locally Lipschitz in $\mathbf{x}$ if, for any $\mathbf{z} \in A$, all maps $B \ni \mathbf{x} \mapsto F^k(\mathbf{z}, \mathbf{x})$ are differentiable and the partial derivatives, as $(\mathbf{z}, \mathbf{x})$ varies, are continuous*[1] *on Ω.*

Proof Consider $q = (\mathbf{z}_0, \mathbf{x}_0) \in \Omega$ and let $B_{\mathbf{x}_0} \subset \mathbb{R}^n$ and $B'_{\mathbf{z}_0} \subset \mathbb{R}^p$ be open balls of positive finite radius centred at $\mathbf{x}_0$ and $\mathbf{z}_0$ respectively, and such that $\overline{B'_{\mathbf{z}_0}} \times \overline{B_{\mathbf{x}_0}} \subset \Omega$. If $\mathbf{x}_1, \mathbf{x}_2 \in \overline{B_{\mathbf{x}_0}}$, the segment $\mathbf{x}(t) = \mathbf{x}_1 + t(\mathbf{x}_2 - \mathbf{x}_1)$ is contained in $\overline{B_{\mathbf{x}_0}}$ for $t \in [0, 1]$, and moreover $\mathbf{x}(0) = \mathbf{x}_1$ and $\mathbf{x}(1) = \mathbf{x}_2$. If $\mathbf{z} \in B'_{\mathbf{z}_0}$, and for any given $k = 1, 2, \dots, m$, the Intermediate Value Theorem for $[0, 1] \ni t \mapsto f(t) := F^k(\mathbf{z}, \mathbf{x}(t))$ says that $f(1) - f(0) = 1 f'(\xi)$ for some $\xi \in (0, 1)$ depending on k and $\mathbf{z}$. This, written using F^k, is

$$F^k(\mathbf{z}, \mathbf{x}_2) - F^k(\mathbf{z}, \mathbf{x}_1) = F^k(\mathbf{z}, \mathbf{x}(1)) - F^k(\mathbf{z}, \mathbf{x}(0))$$

$$= \sum_{j=1}^{n} (x_2^j - x_1^j) \left. \frac{\partial F^k}{\partial x^j} \right|_{(\mathbf{z}, \mathbf{x}(\xi))},$$

---

[1] Since a function of several variables is differentiable on an open set if the partial derivatives exist and are continuous on the open set, the request is the same as asking that

$$\Omega \ni (\mathbf{z}, \mathbf{x}) \mapsto \left. \frac{\partial F^k}{\partial x^j} \right|_{(\mathbf{z}, \mathbf{x})}, \quad \text{for } k = 1, 2, \dots, m \text{ and } j = 1, \dots, n,$$

exist and are continuous on Ω.

Hence, by the Schwarz inequality:

$$\left| F^k(\mathbf{z}, \mathbf{x}_2) - F^k(\mathbf{z}, \mathbf{x}_1) \right| \leq \sqrt{\sum_{j=1}^{n} |x_2^j - x_1^j|^2} \sqrt{\sum_{i=1}^{n} \left| \frac{\partial F^k}{\partial x^i} \right|_{(\mathbf{z}, \mathbf{x}(\xi))}^2}$$

$$\leq ||\mathbf{x}_2 - \mathbf{x}_1|| \sqrt{\sum_{i=1}^{n} \left| \frac{\partial F^k}{\partial x^i} \right|_{(\mathbf{z}, \mathbf{x}(\xi))}^2} .$$

As the derivatives in the last term are continuous on the compact set $\overline{B'_{\mathbf{z}_0}} \times \overline{B_{\mathbf{x}_0}}$, there will be $M_k \in [0, +\infty)$ such that, on $\overline{B'_{\mathbf{z}_0}} \times \overline{B_{\mathbf{x}_0}}$,

$$\sqrt{\sum_{i=1}^{n} \left| \frac{\partial F^k}{\partial x^i} \right|_{(\mathbf{z}, \mathbf{x})}^2} \leq M_k .$$

Therefore, on the open neighbourhood of q, $B'_{\mathbf{z}_0} \times B_{\mathbf{x}_0}$, we have the estimate, for any k

$$\left| F^k(\mathbf{z}, \mathbf{x}_2) - F^k(\mathbf{z}, \mathbf{x}_1) \right| \leq M_k ||\mathbf{x}_2 - \mathbf{x}_1|| .$$

If $M := \sqrt{\sum_{k=1}^{m} M_k^2}$, on the open neighbourhood of q, $B'_{\mathbf{z}_0} \times B_{\mathbf{x}_0}$:

$$||\mathbf{F}(\mathbf{z}, \mathbf{x}_2) - \mathbf{F}(\mathbf{z}, \mathbf{x}_1)|| \leq M ||\mathbf{x}_2 - \mathbf{x}_1|| .$$

This inequality proves what we wanted.   □

Note that the thesis holds if we replace $\mathbb{R}^n$ and $\mathbb{R}^m$ respectively with $\mathbb{C}^n$ and $\mathbb{C}^m$ and require, putting $\mathbf{x} = \mathbf{u} + i\mathbf{v} \equiv (\mathbf{u}, \mathbf{v}) \in \mathbb{C}^n \equiv \mathbb{R}^n + i\mathbb{R}^n$ ($\mathbf{u}$ and $\mathbf{v}$ being the real and imaginary parts of $\mathbf{x}$), that the partial derivatives $\frac{\partial}{\partial u^i} Re F^k$, $\frac{\partial}{\partial u^i} Im F^k$, $\frac{\partial}{\partial v^i} Re F^k$, $\frac{\partial}{\partial v^i} Im F^k$ exist and are continuous in all variables $\mathbf{z}, \mathbf{x}$. The proof is left as an exercise.

(2) From the previous example it follows in particular that any $\mathbf{F} \in C^1(\Omega, \mathbb{R}^m)$ is locally Lipschitz.
(3) With the usual conventions $\mathbf{x} = (x^1, \dots, x^n)$, $\mathbf{z} = (z^1, \dots, z^p)$ and considering a map $\mathbf{F} : \Omega \to \mathbb{R}^m$ with $\Omega \subset \mathbb{R}^p \times \mathbb{R}^n$, we can generalise the above remarks to (not just *locally*) Lipschitz maps. From the above discussion, $\mathbf{F} \in C^1(\Omega, \mathbb{R}^m)$ is Lipschitz in $\mathbf{x} \in \mathbb{R}^n$ if restricted to bounded open sets of the form $A \times C$ with $A \subset \mathbb{R}^p$, $C \subset \mathbb{R}^n$, $\overline{A \times C} \subset \Omega$ and C convex. If so, in fact, the first derivatives of the components are certainly bounded on the compact set $\overline{A \times C}$, being continuous; moreover the convexity allows to choose $\mathbf{x}_1$ and $\mathbf{x}_2$ arbitrarily in C, ensuring that the segment between them is entirely contained in C. The

above argument, based on the Intermediate Value Theorem, proves that the map is Lipschitz in $\mathbf{x} \in C$.

If Ω is convex (say, $\Omega = \mathbb{R}$) and all first derivatives of $\mathbf{F} \in C^1(\Omega, \mathbb{R}^n)$ are bounded on Ω, then $\mathbf{F}$ is Lipschitz on Ω.

(4) The continuous map $f(x) = |x|$, from $\mathbb{R}$ to $\mathbb{R}$, is locally Lipschitz in x, even though *it is not differentiable at* $x = 0$. Let us prove it. For $x \neq 0$, there is an open neighbourhood of the point where the map is C^∞, so what we said in the previous example holds. Consider a neighbourhood of $x = 0$, $I := (-\epsilon, \epsilon)$ (with $\epsilon > 0$). If $x_1, x_2 \in I$ suppose, without loss of generality $|x_2| \geq |x_1|$, so $|x_2| = |x_2 - x_1 + x_1| \leq |x_2 - x_1| + |x_1|$ and then $|x_2| - |x_1| \leq |x_2 - x_1|$. Consequently $|\, |x_2| - |x_1| \,| \leq |x_2 - x_1|$, meaning

$$|f(x_2) - f(x_1)| \leq |x_2 - x_1|, \qquad \text{for any pair } x_1, x_2 \in I \,.$$

(5) Instead, the continuous map from $\mathbb{R}$ to $\mathbb{R}$:

$$f(x) := \begin{cases} \sqrt{x} & \text{if } x \geq 0, \\ 0 & \text{if } x < 0, \end{cases} \tag{14.12}$$

is not locally Lipschitz (in x clearly). If we exclude $x = 0$ from the domain, the function is obviously locally Lipschitz by Example 3, since it is C^∞. The problem is then the behaviour around $x = 0$. Let us show that given any open neighbourhood I of 0 there exists no constant $c \in \mathbb{R}$ such that, if $x_1, x_2 \in I$, $|f(x_2) - f(x_1)| \leq c|x_2 - x_1|$. Let us assume $I := (-\epsilon, \epsilon)$, with $\epsilon > 0$, since if c does not exist for these intervals, it cannot exist for any neighbourhood.[2] Let us proceed *by contradiction* and suppose there is a constant $c \in \mathbb{R}$ such that $|f(x_2) - f(x_1)| \leq c|x_2 - x_1|$ for any pair $x_1, x_2 \in (-\epsilon, \epsilon)$. Pick $x_2 := z \in (0, \epsilon)$ and $x_1 = -z$. The inequality $|f(x_2) - f(x_1)| \leq c|x_2 - x_1|$ reads $\sqrt{z} \leq 2cz$. Therefore $2c \geq 1/\sqrt{z}$ for any $z > 0$ arbitrarily small. This is impossible irrespective of how we fix c.

14.3   Existence and Uniqueness Theorems for the Cauchy Problem

At this juncture we are ready to state and prove the theorems on the existence and uniqueness of solutions to a Cauchy problem, both in local and global form.

---

[2] Any open neighbourhood of 0 contains an interval of the type we said!

14.3.1   Theorem of Local Existence and Uniqueness for the Cauchy Problem

We are ready to prove that on a small enough neighbourhood of the initial conditions the Cauchy problem admits a unique solution if the map $\mathbf{f}$ in the ODE is continuous and locally Lipschitz in $\mathbf{x}$.

Theorem 14.16 (Local Existence and Uniqueness)   *Let* $\mathbf{f} : \Omega \to \mathbb{K}^n$ *be continuous and locally Lipschitz in the variable* $\mathbf{x} \in \mathbb{K}^n$ *on the open set* $\Omega \subset \mathbb{R} \times \mathbb{K}^n$. *Consider the Cauchy problem for* $(t_0, \mathbf{x}_0) \in \Omega$:

$$
\begin{cases}
\dfrac{d\mathbf{x}}{dt} = \mathbf{f}(t, \mathbf{x}(t)), \\
\mathbf{x}(t_0) = \mathbf{x}_0.
\end{cases}
\tag{14.13}
$$

There exist an open interval $I \ni t_0$ *and a unique map* $\mathbf{x} : I \to \mathbb{K}^n$ *solving (14.13). Moreover, this map is necessarily in* $C^1(I; \mathbb{K}^n)$.

Proof For starters every solution $\mathbf{x} = \mathbf{x}(t)$ of (14.13) is necessarily C^1. In fact it is continuous as differentiable, but then directly from $\frac{d\mathbf{x}}{dt} = \mathbf{f}(t, \mathbf{x}(t))$ we conclude $\frac{d\mathbf{x}}{dt}$ must be continuous, since the equation's right-hand side is continuous in t, by composition of continuous maps.

Let us pass to the existence and uniqueness. Suppose that $\mathbf{x} : I \to \mathbb{K}^n$ is differentiable and solves (14.13). Keeping into account the Fundamental Theorem of Calculus and integrating (14.13) (the derivative of $\mathbf{x}(t)$ is continuous), $\mathbf{x} : I \to \mathbb{K}^n$ must satisfy

$$
\mathbf{x}(t) = \mathbf{x}_0 + \int_{t_0}^{t} \mathbf{f}(\tau, \mathbf{x}(\tau))\, d\tau, \quad \text{for any } t \in I.
\tag{14.14}
$$

Vice versa, if $\mathbf{x} : I \to \mathbb{K}^n$ is continuous and satisfies (14.14) then by the Fundamental Theorem of Calculus ($\mathbf{f}$ is continuous), $\mathbf{x} = \mathbf{x}(t)$ is differentiable and satisfies (14.13).

We conclude the continuous maps $\mathbf{x} = \mathbf{x}(t)$ defined on an open interval $I \ni t_0$ that solve the integral equation (14.14) are precisely the solutions of (14.13) defined on the same interval I. Instead of solving (14.13) we will solve the equivalent integral problem (14.14). □

Proof of Existence Fix, once and for all, an open set $Q \ni (t_0, \mathbf{x}_0)$ with compact closure and $\overline{Q} \subset \Omega$. Taking $\overline{Q}$ small enough we can use that $\mathbf{f}$ is locally Lipschitz in $\mathbf{x}$. In the sequel:

(i) $0 \le M := \max\{\|\mathbf{f}(t, \mathbf{x})\| \mid (t, \mathbf{x}) \in \overline{Q}\}$;
(ii) $L \ge 0$ is the constant for which $\|\mathbf{f}(t, \mathbf{x}) - \mathbf{f}(t, \mathbf{x}')\| \le L\|\mathbf{x} - \mathbf{x}'\|$ if $(t, \mathbf{x}), (t, \mathbf{x}') \in \overline{Q}$;
(iii) $B_\epsilon(\mathbf{x}_0) := \{\mathbf{x} \in \mathbb{K}^n \mid \|\mathbf{x} - \mathbf{x}_0\| \le \epsilon\}$ for $\epsilon > 0$.

Consider a closed interval $J_\delta = [t_0 - \delta, t_0 + \delta]$ with $\delta > 0$ and consider the Banach space $C^0(J_\delta; \mathbb{K}^n)$ of continuous maps $\mathbf{X} : J_\delta \to \mathbb{K}^n$. Let us define the function G sending a map $\mathbf{X}$ to a new function $G(\mathbf{X})$ defined by

$$G(\mathbf{X})(t) := \mathbf{x}_0 + \int_{t_0}^t \mathbf{f}(\tau, \mathbf{X}(\tau)) \, d\tau, \quad \text{for any } t \in J_\delta.$$

Note $G(\mathbf{X}) \in C^0(J_\delta; \mathbb{K}^n)$ if $\mathbf{X} \in C^0(J_\delta; \mathbb{K}^n)$ by the continuity properties of the integral function (as a function of the upper endpoint, when the integrand is continuous). Let us show G is a contraction on the closed ball[3] of radius $\epsilon > 0$ inside $C^0(J_\delta; \mathbb{K}^n)$:

$$\mathcal{B}_\epsilon^{(\delta)} := \{\mathbf{X} \in C^0(J_\delta; \mathbb{K}^n) \mid \mathbf{X}(t) \in B_\epsilon(\mathbf{x}_0), \forall t \in J_\delta\},$$

if $0 < \delta < \min\{\epsilon/M, 1/L\}$ and $\delta, \epsilon > 0$ are so small that $J_\delta \times B_\epsilon(\mathbf{x}_0) \subset Q$. *In the sequel we will pick $\epsilon > 0$ and $\delta > 0$ so that $J_\delta \times B_\epsilon(\mathbf{x}_0) \subset Q$.* If $\mathbf{X} \in \mathcal{B}_\epsilon^{(\delta)}$, assuming $t \geq t_0$:

$$||G(\mathbf{X})(t) - \mathbf{x}_0|| = \left|\left| \int_{t_0}^t \mathbf{f}(\tau, \mathbf{X}(\tau)) \, d\tau \right|\right| \leq \int_{t_0}^t ||\mathbf{f}(\tau, \mathbf{X}(\tau))|| \, d\tau \leq \int_{t_0}^t M \, d\tau \leq \delta M.$$

Supposing $t \leq t_0$ we obtain the same final estimate, since we have to flip the sign of the integrals in the last two inequalities. Hence $G(\mathcal{B}_\epsilon^{(\delta)}) \subset \mathcal{B}_\epsilon^{(\delta)}$ if $\delta > 0$ is so small that $\delta < \epsilon/M$ (respecting the constraint $J_\delta \times B_\epsilon \subset Q$). If $\mathbf{X}, \mathbf{X}' \in \mathcal{B}_\epsilon^{(\delta)}$ we have, for any $t \in J_\delta$ with $t \geq t_0$:

$$G(\mathbf{X})(t) - G(\mathbf{X}')(t) = \int_{t_0}^t \left[\mathbf{f}(\tau, \mathbf{X}(\tau)) - \mathbf{f}(\tau, \mathbf{X}'(\tau)) \right] d\tau,$$

$$\left|\left| G(\mathbf{X})(t) - G(\mathbf{X}')(t) \right|\right| \leq \left|\left| \int_{t_0}^t \left[\mathbf{f}(\tau, \mathbf{X}(\tau)) - \mathbf{f}(\tau, \mathbf{X}'(\tau)) \right] d\tau \right|\right|$$

$$\leq \int_{t_0}^t \left|\left| \mathbf{f}(\tau, \mathbf{X}(\tau)) - \mathbf{f}(\tau, \mathbf{X}'(\tau)) \right|\right| d\tau.$$

But in our assumptions the Lipschitz condition

$$||\mathbf{f}(t, \mathbf{x}) - \mathbf{f}(t, \mathbf{x}')|| < L||\mathbf{x} - \mathbf{x}'||$$

---

[3] We may write $\mathcal{B}_\epsilon^{(\delta)} = \{\mathbf{X} \in C^0(J_\delta; \mathbb{K}^n) \mid ||\mathbf{X} - \mathbf{X}_0||_\infty \leq \epsilon\}$, where $\mathbf{X}_0$ is here the constant *function* $\mathbf{x}_0$ on J_δ. Hence $\mathcal{B}_\epsilon^{(\delta)}$ is the *closed* ball of radius ϵ centred at $\mathbf{X}_0$ in the Banach space $C^0(J_\delta; \mathbb{K}^n)$.

holds, so for $t \geq t_0$:

$$\left\| G(\mathbf{X})(t) - G(\mathbf{X}')(t) \right\| \leq L \int_{t_0}^{t} \left\| \mathbf{X}(\tau) - \mathbf{X}'(\tau) \right\| d\tau \leq \delta L \| \mathbf{X} - \mathbf{X}' \|_{\infty} .$$

Finally, taking the supremum of the left-hand side:

$$\left\| G(\mathbf{X}) - G(\mathbf{X}') \right\|_{\infty} \leq \delta L \| \mathbf{X} - \mathbf{X}' \|_{\infty} .$$

For $t \leq t_0$ the result is the same. We conclude that if $\delta < 1/L$ (together with the above) then $G : \mathcal{B}_{\epsilon}^{(\delta)} \to \mathcal{B}_{\epsilon}^{(\delta)}$ is a contraction on the closed set $\mathcal{B}_{\epsilon}^{(\delta)}$. The latter is closed in a complete complete metric space, so Proposition 14.8 holds and $\mathcal{B}_{\epsilon}^{(\delta)}$ is in turn a complete metric space. By Theorem 14.11 there will be a fixed point of G, i.e. a continuous function $\mathbf{x} = \mathbf{x}(t) \in \mathbb{K}^n$ with $t \in J_{\delta}$, solving (14.14) by definition of G. The restriction of that map to the open interval $I := (t_0 - \delta, t_0 + \delta)$ is therefore a solution to problem (14.13). □

Proof of Uniqueness Consider, on the above interval $I := (t_0 - \delta, t_0 + \delta)$, another solution $\mathbf{x}' = \mathbf{x}'(t)$ of the integral equation (14.14), a priori different from $\mathbf{x} = \mathbf{x}(t)$. By construction, for any closed interval $J_{\delta'} := [t_0 - \delta', t_0 + \delta']$ with $0 < \delta' < \delta$ the function $G : \mathcal{B}_{\epsilon}^{(J_{\delta'})} \to \mathcal{B}_{\epsilon}^{(J_{\delta'})}$ is a contraction and $\mathbf{x}' = \mathbf{x}'(t)$ is a fixed point. (In particular, since the restriction $\mathbf{x}'$ to $J_{\delta'}$ lives in the complete metric space $\mathcal{B}_{\epsilon}^{(J_{\delta'})}$, from the above inequality we obtain $\|G(\mathbf{x}') - \mathbf{x}_0\|_{\infty} \leq \delta' M < \epsilon$, since $G(\mathbf{x}') = \mathbf{x}'$.) Another fixed point is the restriction $\mathbf{x} = \mathbf{x}(t)$ to the interval $J_{\delta'}$. By uniqueness the two solutions must coincide on $[t_0 - \delta', t_0 + \delta']$. But δ' is completely arbitrary in $(0, \delta)$, so the two solutions coincide on $I = (t_0 - \delta, t_0 + \delta)$. □

The following lemma, proved during the proof of the previous result, will be useful in the sequel.

Lemma 14.17 *Referring to the assumptions of Theorem 14.16, let $Q \ni (t_0, \mathbf{x}_0)$ be an open set with compact closure and $\overline{Q} \subset \Omega$, where $\mathbf{f}$ is Lipschitz in $\mathbf{x}$ with Lipschitz constant $L \geq 0$. Define $M := \max\{\|\mathbf{f}(t, \mathbf{x})\| \mid (t, \mathbf{x}) \in \overline{Q}\}$ and $B_{\epsilon}(\mathbf{x}_0) := \{\mathbf{x} \in \mathbb{K}^n \mid \|\mathbf{x} - \mathbf{x}_0\| \leq \epsilon\}$. If $\delta, \epsilon > 0$ are so small that:*

(i) $0 < \delta < \min\{\epsilon/M, 1/L\}$,
(ii) $[t_0 - \delta, t_0 + \delta] \times B_{\epsilon}(\mathbf{x}_0) \subset Q$,

then Theorem 14.16 holds on $I := (t_0 - \delta, t_0 + \delta)$, and the solution satisfies $(t, \mathbf{x}(t)) \in I \times B_{\epsilon}(\mathbf{x}_0)$ for any $t \in I$.

Remarks 14.18

(1) Asking the map to be locally Lipschitz is not necessary for the *existence* of a solution. A theorem due to Peano and extended by Picard proves the existence of a solution for any Cauchy problem with $\mathbf{f}$ continuous, even if not Lipschitz.

The classical example is the Cauchy problem with $\Omega = \mathbb{R} \times \mathbb{R}$,

$$\begin{cases} \dfrac{dx}{dt} = f(x), \\ x(0) = 0. \end{cases}$$

where f is as in (14.12), so not locally Lipschitz in any neighbourhood of $(t, x) = (0, 0)$. Nonetheless the C^1 map

$$x(t) := \begin{cases} t^2/4 & \text{if } t \geq 0, \\ 0 & \text{if } t < 0, \end{cases} \tag{14.15}$$

solves the Cauchy problem. There are other existence (not uniqueness) theorems based on other forms of the fixed-point theorem, which need weaker assumptions than being locally Lipschitz. Examples include *Schauder's fixed-point theorem*, itself an infinite-dimensional generalisation of the more famous *Brouwer fixed-point theorem*.

(2) The Lipschitz condition is typically necessary to guarantee the solution's *uniqueness*. For example, the previous Cauchy problem has, besides (14.15), also the solution $x(t) = 0$ for any $t \in \mathbb{R}$. ∎

Examples 14.19

(1) The first example discusses a formal way to find the solutions to a class of differential equations called *separable*. The procedure is very useful in practice but normally it is valid only locally. Consider a (normal) differential equation with **separable variables**:

$$\frac{dy}{dx} = f(x)g(y(x)) \tag{14.16}$$

where the functions $f = f(x)$ and $g = g(y)$, respectively defined the open intervals I and J in $\mathbb{R}$, are continuous, and suppose g is locally Lipschitz. If $y = y(x)$ satisfies the equation on I or an open subinterval of I, then it must be C^1 on the interval by construction. Fix the initial condition $y(x_0) = y_0$ with $x_0 \in I$ and $y_0 \in J$. With the assumptions on f and g there is a unique local solution $y = y(x)$ (on an open interval containing x_0) for (14.16) with initial condition $y(x_0) = y_0$. If $g(y_0) \neq 0$, the function $1/g(y)$ is well defined, has constant sign equal to that of $g(y_0)$, and is continuous on an open interval containing y_0. As long as $1/g(y(x))$ is well defined on the solution's domain, we will have:

$$\frac{dy}{dx}\frac{1}{g(y(x))} = f(x). \tag{14.17}$$

Integrating term by term (the integrands are continuous hence integrable):

$$\int_{x_0}^{x} \frac{dy}{dx'} \frac{1}{g(y(x'))} dx' = \int_{x_0}^{x} f(x') dx' . \qquad (14.18)$$

Finally, changing variable in the left-hand side of (14.18) gives:

$$\int_{y_0}^{y(x)} \frac{dy'}{g(y')} = \int_{x_0}^{x} f(x') dx' \qquad (14.19)$$

This identity provides the solution to the Cauchy problem in implicit form. It is fundamental to observe that if $g(y_0) = 0$, the procedure cannot be used. Still, it is clear that the constant solution $y(x) = y_0$ for any $x \in \mathbb{R}$, is the only one solving the Cauchy problem (and is moreover maximal). If we drop the hypothesis that g is locally Lipschitz, the solution's uniqueness (including that of any other solution, for different initial conditions) is no longer guaranteed.

If we do not fix an initial condition, from equation (14.16) we deduce the identity:

$$\int \frac{dy}{g(y)} = \int f(x) dx + C , \qquad (14.20)$$

valid for the equation's solutions, and as long as the two sides exist, where C is an arbitrary constant. Formula (14.20) implicitly determines the class of solutions as $C \in \mathbb{R}$ varies. The parameter C is fixed by the initial conditions, but the solution's (local) uniqueness must be discussed separately using the uniqueness theorem.

(2) Consider the ODE $\frac{dy}{dx} = xy^2$ with initial condition $y(0) = 1$. We can apply the method of separation of variables described above:

$$\int_{1}^{y} \frac{dy'}{y'^2} = \int_{0}^{x} x' dx'$$

so:

$$-\frac{1}{y(x)} + \frac{1}{1} = \frac{x^2}{2} - 0 ,$$

and then:

$$y(x) = \frac{2}{2 - x^2} .$$

This function is defined for $x \in \mathbb{R} \setminus \{\pm\sqrt{2}\}$ and is the unique solution to the Cauchy problem on a neighbourhood of $x = 0$. Actually, by Theorem 14.23

proved below, the restriction to $(-\sqrt{2}, \sqrt{2})$ is the only *maximal* solution: any other solution around $x = 0$ is a restriction of this one. The points $x = \pm\sqrt{2}$ lie outside the solution's domain, where the initial condition was given at $x = 0$. Yet there is no problem whatsoever in assigning the initial condition at $x = \sqrt{2}$, thus studying a *new* Cauchy problem, independent of the former, and obtaining another solution (which exists and is unique, as the Existence and uniqueness theorem holds). If we impose initial condition $y(\sqrt{2}) = 1$ on the same equation $\frac{dy}{dx} = xy^2$, we can find the solution using the previous methods. If we want to use (14.20):

$$\int \frac{dy}{y^2} = \int x\,dx + C \,,$$

i.e.:

$$-\frac{1}{y} = \frac{x^2}{2} + C \,.$$

Then:

$$y(x) = \frac{2}{2C - x^2} \,.$$

Imposing the initial condition $x(\sqrt{2}) = 1$ gives:

$$1 = \frac{2}{2C - 2}$$

from which $C = 2$ and therefore:

$$y(x) = \frac{2}{4 - x^2} \,.$$

This solution is well defined at any $x \in (-2, +2)$. By Theorem 14.23, proved below, this function is the only *maximal* solution: it extends any other solution around $x = \sqrt{2}$.

(3) Consider the differential equation:

$$\frac{dy}{dx} = y^{1/3} \,.$$

The right-hand side is C^1 everywhere on $\mathbb{R} \setminus \{0\}$, so it is locally Lipschitz. But the map is not locally Lipschitz around the origin. Hence, choosing initial conditions $y(x_0) = y_0 \neq 0$ the Existence and uniqueness theorem holds, but we may not have uniqueness if we choose $y(x_0) = 0$. Using (14.20), we find the implicit solutions:

$$\int y^{-1/3} dy = \int dx + C,$$

so:

$$y(x) = \pm \sqrt{\left(\frac{2}{3}(x+C)\right)^3}, \tag{14.21}$$

where the sign in the right-hand side must coincide with that of y_0 by continuity. Immediately, any initial condition $y(0) = y_0 \neq 0$ determines only one value

$$C = C_0 := \frac{3}{2}(y_0^2)^{1/3}.$$

Therefore

$$y(x) = \pm \sqrt{\left(\frac{2}{3}x + (y_0^2)^{1/3}\right)^3},$$

for $x > -C_0$, is a local solution (the only one) of the Cauchy problem. As a matter of fact, by Theorem 14.23 proved below, this is the *maximal* solution: any solution around $x = 0$ is a restriction of it.

The family of solutions (14.21) also contains the solution fulfilling the pathological initial condition $y(0) = 0$, i.e. $y(x) = \sqrt{\left(\frac{2}{3}x\right)^3}$ if $x \geq 0$. We can extend it by setting $y(x) = 0$ if $x < 0$, thus generating a $C^1(\mathbb{R})$ solution. (As expected, since the uniqueness theorem fails) this is *not* the only solution; the constant solution $y(x) = 0$, for any $x \in \mathbb{R}$, solves the Cauchy problem.

14.3.2 A Condition for First Integrals

We will show how the Existence and uniqueness theorem implies a necessary and sufficient condition for a function to be a first integral (Definition 14.2). We state the theorem for first-order systems, and leave to the reader the (obvious) extension to the general case.

Theorem 14.20 *Let* $\mathbf{f} : \Omega \to \mathbb{K}^n$ *be continuous and locally Lipschitz in* $\mathbf{x} \in \mathbb{K}^n$ *on the open set* $\Omega \subset \mathbb{R} \times \mathbb{K}^n$. *An everywhere differentiable function* $F : \Omega \to \mathbb{K}$ *is a first integral for the ODE system*

$$\frac{d\mathbf{x}}{dt} = \mathbf{f}(t, \mathbf{x}(t))$$

if and only if:

$$\mathbf{f}(t, \mathbf{x}) \cdot \nabla_{\mathbf{x}} F|_{(t,\mathbf{x})} + \frac{\partial F}{\partial t}\bigg|_{(t,\mathbf{x})} = 0, \quad \text{for any } (t, \mathbf{x}) \in \Omega. \tag{14.22}$$

Proof Suppose (14.22) holds. If $\mathbf{x} = \mathbf{x}(t)$, with $t \in I$ an open interval in $\mathbb{R}$, solves the system, then

$$\frac{d}{dt} F(t, \mathbf{x}(t)) = \mathbf{f}(t, \mathbf{x}) \cdot \nabla_{\mathbf{x}} F|_{(t,\mathbf{x}(t))} + \frac{\partial F}{\partial t}\bigg|_{(t,\mathbf{x}(t))} = 0,$$

so $I \ni t \mapsto F(t, \mathbf{x}(t))$ is constant on every solution.
Vice versa, if a differentiable F is a first integral, then

$$\frac{d}{dt} F(t, \mathbf{x}(t)) = \mathbf{f}(t, \mathbf{x}) \cdot \nabla_{\mathbf{x}} F|_{(t,\mathbf{x}(t))} + \frac{\partial F}{\partial t}\bigg|_{(t,\mathbf{x}(t))} = 0,$$

on any solution. On the other hand by Theorem 14.16, for any point $(t_0, \mathbf{x}_0) \in \Omega$ there is a solution to the ODE system that satisfies $\mathbf{x}(t_0) = \mathbf{x}_0$ and then:

$$\mathbf{f}(t_0, \mathbf{x}_0) \cdot \nabla_{\mathbf{x}} F|_{(t_0,\mathbf{x}_0)} + \frac{\partial F}{\partial t}\bigg|_{(t_0,\mathbf{x}_0)} = 0.$$

This ends the proof because $(t_0, \mathbf{x}_0) \in \Omega$ is arbitrary.   □

14.3.3   Global Existence and Uniqueness Theorem for the Cauchy Problem

Now we shall prove the existence and uniqueness of "global" solutions. First of all we prove the following proposition.

Proposition 14.21 *Let $\mathbf{f} : \Omega \to \mathbb{K}^n$ be continuous and locally Lipschitz in $\mathbf{x} \in \mathbb{K}^n$ on the open set $\Omega \subset \mathbb{R} \times \mathbb{K}^n$ and consider the Cauchy problem for $(t_0, \mathbf{x}_0) \in \Omega$:*

$$\begin{cases} \dfrac{d\mathbf{x}}{dt} = \mathbf{f}(t, \mathbf{x}(t)), \\ \mathbf{x}(t_0) = \mathbf{x}_0. \end{cases} \tag{14.23}$$

If $\mathbf{x}_1 : I_1 \to \mathbb{K}^n$ and $\mathbf{x}_2 : I_2 \to \mathbb{K}^n$ solve (14.23) and are defined on the open intervals $I_1 \ni t_0$ and $I_2 \ni t_0$ respectively, then

$$\mathbf{x}_1(t) = \mathbf{x}_2(t), \quad \text{if } t \in I_1 \cap I_2.$$

Proof By Theorem 14.16 the two solutions will coincide on the interval I containing t_0 where the local uniqueness theorem holds. If $I \supset I_1 \cap I_2$ there is nothing to prove. So suppose $I \subsetneq I_1 \cap I_2$. To fix ideas we will work on a right neighbourhood of t_0 (the argument for left neighbourhoods is similar). So we know that for a certain $t_1 > t_0$, $\mathbf{x}_1(t) = \mathbf{x}_2(t)$ if $t_0 \le t < t_1$. Let $t_2 > t_1$ be the right end of $I_1 \cap I_2$, possibly $t_2 = +\infty$. We want to show $\mathbf{x}_1(t) = \mathbf{x}_2(t)$ if $t_0 \le t < t_2$.
Let

$$s := \sup\{T \in \mathbb{R} \mid t_0 \le T < t_2 , \quad \mathbf{x}_1(t) = \mathbf{x}_2(t) \text{ if } t_0 \le t < T.\}$$

This supremum exists because the set is non-empty, and upper bounded since $s < +\infty$ if $t_2 < +\infty$. If $s < t_2$ (so s would be finite even if $t_2 = +\infty$) we would have, by continuity, $\mathbf{x}_1(s) = \mathbf{x}_2(s)$, because the solutions are continuous and coincide at all points preceding s. Setting up the initial value problem for $\mathbf{f}$ with initial condition $\mathbf{x}(s) = \mathbf{x}_1(s)$, there would be an open interval $I_s \ni s$ where the Cauchy problem has a unique solution. Consequently on the right of s we would still have $\mathbf{x}_1(t) = \mathbf{x}_2(t)$. But this is impossible by definition of s. We conclude $s = t_2$. By definition of s, therefore, $\mathbf{x}_1(t) = \mathbf{x}_2(t)$ if $t_0 \le t < t_2$. □

Let us pass to the *maximal solutions* of a Cauchy problem. The following definition clarifies what we mean by that. Naively, a solution is maximal when it cannot be extended to another solution of the same Cauchy problem. Note that a solution may be maximal even if not defined on $(-\infty, +\infty)$.

Definition 14.22 Let $\mathbf{f} : \Omega \to \mathbb{K}^n$ be defined on the open set $\Omega \subset \mathbb{R} \times \mathbb{K}^n$ and consider the Cauchy problem for $(t_0, \mathbf{x}_0) \in \Omega$:

$$\begin{cases} \dfrac{d\mathbf{x}}{dt} = \mathbf{f}(t, \mathbf{x}(t)), \\ \mathbf{x}(t_0) = \mathbf{x}_0. \end{cases} \tag{14.24}$$

A solution $\mathbf{x} : I \to \mathbb{K}^n$ to (14.24), with $I \ni t_0$ open interval, is called a **maximal solution** if there exists no other solution $\mathbf{x}' : I' \to \mathbb{K}^n$ to (14.24), with I' open interval, such that:

(i) $I' \supsetneq I$,
(ii) $\mathbf{x}(t) = \mathbf{x}'(t)$ if $t \in I$.

A maximal solution $\mathbf{x} : I \to \mathbb{K}^n$ is called:

(1) **complete**, if $I = (-\infty, +\infty)$;
(2) **future-complete**, if $I = (a, +\infty)$ for $a \in \mathbb{R} \cup \{-\infty\}$;
(3) **past-complete**, if $I = (-\infty, b)$ for $b \in \mathbb{R} \cup \{+\infty\}$.

These definitions extend trivially to Cauchy problems for systems of order $n > 1$. ◇

The existence of maximal solutions is a direct consequence of Zorn's Lemma, only assuming that problem (14.24) admits at least one solution (irrespective of its local or global uniqueness). Consider the collection C of all solutions $\mathbf{x}_J : J \to \mathbb{K}^n$ to (14.24), where $J \ni t_0$ is any open interval where a solution exists. This set is partially ordered by asking $\mathbf{x}_J \leq \mathbf{x}_{J'}$ if and only if $J \subset J'$ and $\mathbf{x}_{J'}|_J = \mathbf{x}_J$. With this partial order relation, totally ordered subsets of C admit least upper bound (the solution defined on the union of all solutions' domains). From Zorn's Lemma (or the Well-ordering Principle, or the Axiom of Choice) C has a maximal element, i.e. problem (14.24) has a maximal solution. Note that it is far from obvious, though, that if there are two maximal solutions then they coincide. That being said, if $\mathbf{f}$ is continuous and locally Lipschitz, the following result ensures existence and uniqueness of maximal solutions, as consequence of Proposition 14.21.

Theorem 14.23 (Global Existence and Uniqueness)   *Let* $\mathbf{f} : \Omega \to \mathbb{K}^n$ *be continuous and locally Lipschitz in* $\mathbf{x} \in \mathbb{K}^n$ *on the open set* $\Omega \subset \mathbb{R} \times \mathbb{K}^n$, *and consider the Cauchy problem for* $(t_0, \mathbf{x}_0) \in \Omega$:

$$\begin{cases} \dfrac{d\mathbf{x}}{dt} = \mathbf{f}(t, \mathbf{x}(t)), \\ \mathbf{x}(t_0) = \mathbf{x}_0. \end{cases} \tag{14.25}$$

Then there exists a unique maximal solution $\mathbf{x} : I \to \mathbb{K}^n$ *of class* C^1.
Furthermore, if $\mathbf{x}_J : J \to \mathbb{K}^n$, *with* $J \ni t_0$ *open interval, is another solution to (14.25), then*

(1) $J \subset I$,
(2) $\mathbf{x}(t) = \mathbf{x}_J(t)$ *if* $t \in J$.

Proof The collection $\mathscr{S}$ of the domains of all solutions to (14.25) is non-empty (Theorem 14.16) and every solution $\mathbf{x}_J : J \to \mathbb{K}^n$, with $J \ni t_0$ open interval, is completely determined by its domain by Proposition 14.21: if two solutions of (14.25) exist on the same open interval J they must coincide. Then let $I :=$ $\cup_{J \in \mathscr{S}} J$. It is open as union of open sets; moreover, being a union of connected sets with a common point (t_0), it is also connected and contains that point. As is well known from the topology of $\mathbb{R}$, open intervals are the open connected sets in $\mathbb{R}$. By construction I is then an open interval containing t_0. Now define the function on I:

$$\mathbf{x}_I(t) := \mathbf{x}_J(t), \quad \text{if } t \in J \text{ and } J \in \mathscr{S}.$$

This $\mathbf{x}_I$ is well defined since Proposition 14.21 implies that if $t \in J \cap J'$ with $J, J' \in \mathscr{S}$, then $\mathbf{x}_J(t) = \mathbf{x}_{J'}(t)$. Hence $\mathbf{x}_I|_J = \mathbf{x}_J$ for any $J \in \mathscr{S}$ and therefore $\mathbf{x}_I$ solves the Cauchy problem if restricted to any $J \in \mathscr{S}$, whence on the whole I. We conclude $\mathbf{x}_I$ is a solution of our Cauchy problem, and by construction it is maximal. Regarding the uniqueness of maximal solutions, if there were two maximal solutions of (14.25), they would coincide on their domains' intersection by Proposition 14.21.

But the domains themselves must coincide, otherwise one solution (or both) would not be maximal (we could extend it beyond its domain using the other maximal solution). Every solution (maximal or not) must be C^1 since $\mathbf{f}$ is continuous. □

Here is an immediate consequence of the theorem.

Proposition 14.24 *Under the hypotheses of Theorem 14.23, let* $\mathbf{x} : I \to \mathbb{K}^n$ *be the maximal solution to* (14.25). *If* $t_1 \in I$ *and* $\mathbf{x}_1 := \mathbf{x}(t_1)$ *then* $\mathbf{x} : I \to \mathbb{K}^n$ *is the maximal solution of the Cauchy problem with the same equation as* (14.25) *and initial condition* $\mathbf{x}(t_1) = \mathbf{x}_1$.

Proof If $\mathbf{x} : I \to \mathbb{K}^n$ were not maximal for the new problem then it would arise as restriction of a solution defined on an interval larger than I. This solution would prolong the solution of (14.25), which cannot be by assumption. □

Theorem 14.23 has an immediate consequence regarding higher-order Cauchy problems. The result is expressed through the following corollary, which in particular also proves Theorem 3.11.

Proposition 14.25 *Let* $I \subset \mathbb{R}$ *be an open interval,* $D \subset \mathbb{K}^m \times \cdots \times \mathbb{K}^m$ *a non-empty open set and set* $\Omega = I \times D$. *Consider* $\mathbf{f} : \Omega \to \mathbb{K}^m$ *and the Cauchy problem*

$$
\begin{cases}
\dfrac{d^n\mathbf{x}}{dt^n} = \mathbf{f}\left(t, \mathbf{x}(t), \dfrac{d\mathbf{x}}{dt}, \ldots, \dfrac{d^{n-1}\mathbf{x}}{dt^{n-1}}\right), \\
\mathbf{x}(t_0) = \mathbf{x}_0, \ \dfrac{d\mathbf{x}}{dt}(t_0) = \mathbf{x}_0^{(1)}, \ \cdots, \ \dfrac{d^{n-1}\mathbf{x}}{dt^{n-1}}(t_0) = \mathbf{x}_0^{(n-1)}
\end{cases}
\tag{14.26}
$$

where $(t_0, \mathbf{x}_0, \mathbf{x}_0^{(1)}, \ldots, \mathbf{x}_0^{(n-1)}) \in \Omega$ *is arbitrary. Put* $\mathbf{X} := (\mathbf{x}, \mathbf{x}^{(1)}, \ldots, \mathbf{x}^{(n-1)}) \in \Omega$ *and suppose:*

(i) $\mathbf{f}$ *is continuous on* Ω,
(ii) $I \times D \ni (t, \mathbf{X}) \mapsto \mathbf{f}(t, \mathbf{X}) \in \mathbb{K}^m$ *is locally Lipschitz in* $\mathbf{X} \in D$ *(in particular, if* $\mathbb{K} = \mathbb{R}$, *this is true when the derivatives of the components of* $\mathbf{f}$ *with respect to all components of* $\mathbf{X} \in D$ *are continuous on* Ω).

Then there exists a unique maximal solution to (14.26) *of class* C^n.

Proof Let us pass to the first-order system equivalent to (14.26)

$$
\begin{cases}
\dfrac{d\mathbf{X}}{dt} = \mathbf{F}(t, \mathbf{X}(t)), \\
\mathbf{X}(t_0) = \mathbf{X}_0.
\end{cases}
$$

as we explained in Sect. 14.1.1. From that discussion, easily, the function $I \times D \ni (t, \mathbf{X}) \mapsto \mathbf{F}(t, \mathbf{X})$ is continuous and locally Lipschitz in $\mathbf{X}$ if so is $I \times D \ni (t, \mathbf{X}) \mapsto \mathbf{f}(t, \mathbf{X})$ (in particular, if $\mathbb{K} = \mathbb{R}$, the Lipschitz condition holds if the derivatives of the components of $\mathbf{f}$ with respect to every component of $\mathbf{X} \in D$ are continuous on Ω). Now we can apply Theorem 14.23 and end the proof. □

14.3.4   Linear Differential Equations

Consider an ODE of order n, on $\mathbb{K} = \mathbb{C}$ or $\mathbb{R}$ indifferently,

$$p_0(t)\frac{d^n x}{dt^n} + p_1(t)\frac{d^{n-1}x}{dt^{n-1}} + \cdots + p_n(t)x = q(t) , \tag{14.27}$$

where the $p_k : \mathbb{R} \to \mathbb{K}$ and $q : \mathbb{R} \to \mathbb{K}$ are continuous, while the unknown function $x = x(t)$ is $\mathbb{K}$-valued and differentiable n times. The above ODE is an **inhomogeneous linear differential equation**. The non-homogeneity is due to the presence of the so-called **source term**: the function q. On the open set $U \subset \mathbb{R}$ where $p_0(t) \neq 0$, we can write the above equation in normal form

$$\frac{d^n x}{dt^n} + P_1(t)\frac{d^{n-1}x}{dt^{n-1}} + \cdots + P_n(t)x = Q(t) , \tag{14.28}$$

having put $P_k(t) := p_k(t)/p_1(t)$ and $Q(t) := q(t)/p(t)$ on U. The points $t \in \mathbb{R}$ where $p_1(t) = 0$ are called **singular points** and must be studied separately.

If $J \subset U$ is an open interval, consider the Cauchy problem of order n on $\Omega = J \times \mathbb{K}$ given by (14.28) together with initial conditions, if $t_0 \in J$,

$$x(t_0) = x_0 , \quad \frac{dx}{dt}(t_0) = x_0^{(1)} , \quad \cdots , \quad \frac{d^{n-1}x}{dt^{n-1}}(t_0) = x_0^{(n-1)} , \tag{14.29}$$

where $x_0, x_0^{(1)}, \ldots x_0^{(n-1)} \in \mathbb{K}$.

With the technique of Sect. 14.1.1, the above problem can always be reduced to a first-order problem on $\Omega := J \times \mathbb{K}^n$ of the form:

$$\begin{cases} \dfrac{d\mathbf{x}}{dt} = -A(t)\mathbf{x}(t) + \mathbf{b}(t), \\ \mathbf{x}(t_0) = \mathbf{x}_0. \end{cases} \tag{14.30}$$

where: $J \ni t_0$, $\mathbf{x}(t) := (x(t), x^{(1)}(t), \cdots , x^{(n-1)}(t))^t$. Moreover, $\mathbf{x}_0 := (x_0, x_0^{(1)}, \cdots , x_0^{(n-1)})^t$, and for any $t \in I$, $A(t) : \mathbb{K}^n \to \mathbb{K}^n$ is an $n \times n$ matrix with coefficients in $\mathbb{K}$ whose last row is the vector $(P_n(t), \ldots , P_1(t))$, while the coefficients on the other rows are 0 or -1 according to the $n-1$ relations:

$$\frac{dx}{dt} = x_1 ,$$

$$\frac{dx_1}{dt} = x_2 ,$$

$$\cdots = \cdots$$

$$\frac{dx_{n-2}}{dt} = x_{n-1} .$$

Finally, the column vector $\mathbf{b}(t)$ has all zero coefficients except for the last one, which equals $Q(t)$.

Problem (14.30) is well posed since the map on the right

$$\mathbf{f}(t, \mathbf{x}) := -A(t)\mathbf{x} + \mathbf{b}(t)$$

is continuous and certainly locally Lipschitz in our hypotheses: we can take as Lipschitz constant $\sup_{t \in J'} \|A(t)\|$, on any set $J' \times B_\epsilon(\mathbf{x}_0)$, where $J' \ni t_0$ is an open interval with closure in J and $B_\epsilon(\mathbf{x}_0) \subset \mathbb{K}^n$ is an open ball centred at $\mathbf{x}_0$ of arbitrary radius $\epsilon > 0$. In conclusion, there always is a unique maximal solution.

Referring to the first-order problem (14.30) on $\Omega = J \times \mathbb{K}^n$ (and the equivalent problem of order n), let us prove the maximal solution is defined for any t in the whole J. *Therefore if $J = \mathbb{R}$, the maximal solution is complete as well.*

Proposition 14.26 *The maximal solution to the first-order Cauchy problem (14.30) on $\Omega = J \times \mathbb{K}^n$ (and the equivalent Cauchy problem of order n) is defined for any t in the whole J.*

Proof Consider initially a homogeneous system, where the Cauchy problem is

$$(14.31) \qquad \begin{cases} \dfrac{d\mathbf{x}}{dt} = \mathbf{f}(t, \mathbf{x}(t)), \\ \mathbf{x}(t_0) = \mathbf{x}_0. \end{cases}$$

and

$$\mathbf{f}(t, \mathbf{x}) := -A(t)\mathbf{x}, \quad \text{with } (t, \mathbf{x}) \in \Omega := J \times \mathbb{K}^n,$$

and $J \ni t \mapsto A(t)$ is a continuous matrix-valued function.

Let $(\alpha, \omega) \subset J$ denote the domain of the maximal solution. Define, in terms of Lemma 14.17, $Q = (t_0 - \Delta, t_0 + \Delta) \times B_E(\mathbf{x}_0)$ with $E, \Delta > 0$ so that

$$J \supset [t_0 - \Delta, t_0 + \Delta] \supset (t_0 - \Delta, t_0 + \Delta).$$

The Lipschitz constant on Q can be chosen to be $L = \max_{t \in [t_0 - \Delta, t_0 + \Delta]} \|A(t)\|$ (if $L = 0$ the discussion becomes trivial, so we will assume $L > 0$ henceforth), while the maximum of $\|\mathbf{f}\|$ clearly is $M = LE$.

If ω is not the right endpoint of J, we can, staying within J, choose Δ so large that $J \supset [t_0 - \Delta, t_0 + \Delta]$ and $t_0 + \Delta > \omega$. Consider then the maximal solution $\mathbf{x} : (\alpha, \omega) \to \mathbb{K}^n$. By Lemma 14.17, if $t_1 \in (\alpha, \omega)$ the maximal solution must be defined on $(t_1 - \delta, t_1 + \delta)$ and take values in $B_\epsilon(\mathbf{x}(t_1))$ as long as $0 < \epsilon < E$ and $\delta < \min(1/L, \epsilon/M)$ i.e. $\delta < (\epsilon/E)(1/L)$. Fix $t_1 < \omega$ so that $|\omega - t_1| < 1/(2L)$. Taking ϵ close enough to E, we can choose δ so small that $\delta < (\epsilon/E)(1/L)$, and also $|t_1 - \omega| < \delta$. This means $t_1 + \delta > \omega$, so the maximal solution, surely defined on $(t_1 - \delta, t_1 + \delta)$, is also defined on a *right neighbourhood* of ω. The latter then cannot be the right end of the maximal interval where we have the solution. The only possibility is that ω coincides with the right endpoint of J. A similar argument shows that α is the left endpoint of J. Hence the maximal solution is defined on the entire J.

Consider at last the inhomogeneous problem:

$$\begin{cases} \dfrac{d\mathbf{x}}{dt} = \mathbf{f}(t, \mathbf{x}(t)), \\ \mathbf{x}(t_0) = \mathbf{x}_0. \end{cases} \qquad (14.32)$$

where

$$\mathbf{f}(t, \mathbf{x}) := -A(t)\mathbf{x} + \mathbf{b}(t), \quad \text{with } (t, \mathbf{x}) \in \Omega := J \times \mathbb{K}^n,$$

with $J \ni t \mapsto A(t)$ matrix-valued and $J \ni t \mapsto \mathbf{b}(t) \in \mathbb{K}^n$ vector-valued. Both are continuous.

The direct computation (left as exercise) shows that for $t \in J$,

$$\mathbf{x}(t) = G(t)^{-1} \left(\mathbf{x}(t_0) + \int_{t_0}^t G(\tau)\mathbf{b}(\tau)\, d\tau \right) \qquad (14.33)$$

solves (on J) problem (14.32). The $n \times n$ matrix $G(t)$ with coefficients in $\mathbb{K}$ is the transpose of the matrix whose k-th column, as t varies, is the maximal solution on J of the associated homogeneous problem

$$\begin{cases} \dfrac{d\,\mathbf{g}_{(k)}}{dt} = -A(t)^t\,\mathbf{g}_{(k)}(t), \\ \mathbf{g}_{(k)}(t_0) = \boldsymbol{\delta}_{(k)}. \end{cases} \qquad (14.34)$$

where $\boldsymbol{\delta}_{(k)}$ is the k-th column of the identity matrix. By construction the solution to the inhomogeneous problem we have found is defined on J, and this ends the proof. □

14.3.5 Structure of the Solution Set of a Linear Equation

Now we would like to examine the structure of the set of all maximal solutions to equation (14.28). Consider then the Cauchy problem of order n on $\mathbb{R} \times \mathbb{K}^n$, with homogeneous linear equation:

$$\frac{d^n x}{dt^n} + P_1(t)\frac{d^{n-1}x}{dt^{n-1}} + \cdots + P_n(t)x = 0, \qquad (14.35)$$

where the P_k are continuous on J, and generic initial conditions, if $t_0 \in J$,

$$x(t_0) = x_0 , \quad \frac{dx}{dt}(t_0) = x_0^{(1)} , \cdots , \frac{d^{n-1}x}{dt^{n-1}}(t_0) = x_0^{(n-1)} , \qquad (14.36)$$

where $x_0, x_0^{(1)}, \ldots x_0^{(n-1)} \in \mathbb{K}$.

Consider two solutions defined on J to (14.35): x and x'. By the equation's linearity clearly if $a, b \in \mathbb{K}$ the map $J \ni t \mapsto ax(t) + bx'(t)$ is still a solution. Immediately, the set of solutions to (14.35) defined on the entire J is a *vector space* over the field $\mathbb{K}$, with linear operations defined "pointwise", as follows.

$$\text{If } a, b \in \mathbb{K} \text{ and } x, x' : J \to \mathbb{K} \text{ solve } (14.35) : (ax + bx')(t)$$

$$:= ax(t) + bx'(t) , \text{ for any } t \in J. \qquad (14.37)$$

We aim to determine the dimension of this vector space. Fix once and for all a point $t_0 \in J$. By the uniqueness theorem, any solution to (14.35) is determined bijectively by its initial data (14.36). Further, the set of possible initial data $(x_0, x_0^{(1)}, \ldots, x_0^{(n-1)})$ fills $\mathbb{K}^n$. If x and x' satisfy the respective initial conditions:

$$x(t_0) = x_0 , \quad \frac{dx}{dt}(t_0) = x_0^{(1)} , \cdots , \frac{d^{n-1}x}{dt^{n-1}}(t_0) = x_0^{(n-1)}$$

and

$$x'(t_0) = x_0' , \quad \frac{dx'}{dt}(t_0) = x_0^{\prime(1)} , \cdots , \frac{d^{n-1}x'}{dt^{n-1}}(t_0) = x_0^{\prime(n-1)} ,$$

the solution $x'' := ax + bx'$ satisfies both sets of conditions (and is the only one doing so):

$$x''(t_0) = ax_0 + bx_0' , \quad \frac{dx''}{dt}(t_0) = ax_0^{(1)} + bx_0^{\prime(1)} , \cdots , \frac{d^{n-1}x''}{dt^{n-1}}(t_0) = ax_0^{(n-1)} + bx_0^{\prime(n-1)} .$$

This immediately proves that the map associating $(x_0, x_0^{(1)}, \ldots, x_0^{(n-1)}) \in \mathbb{K}^n$ with the unique maximal solution to the Cauchy problem (14.35) and (14.36) is a homomorphism of vector spaces. Since initial data and solutions are in 1-1 correspondence, the homomorphism is an isomorphism and therefore preserves the dimensions of the vector spaces. We have in other words proven the following proposition.

Proposition 14.27 *The set of maximal solutions to a homogeneous linear ODE on an interval $J \subset \mathbb{R}$*

$$\frac{d^n x}{dt^n} + P_1(t)\frac{d^{n-1}x}{dt^{n-1}} + \cdots + P_n(t)x = 0 \,,$$

where the P_k are continuous on J, is a vector space over $\mathbb{K}$ under (14.37) and has dimension equal to the order n.

Let us now pass to study the solution set of the *non*-homogeneous linear ODE:

$$\frac{d^n x}{dt^n} + P_1(t)\frac{d^{n-1}x}{dt^{n-1}} + \cdots + P_n(t)x = Q(t) \,, \qquad (14.38)$$

where the P_k and Q are, as usual, continuous on J. The set of all solutions $s : J \to \mathbb{K}$ can be written as

$$\{s = s_0 + x \mid x : J \to \mathbb{K} \text{ solves } (14.35)\} \,,$$

where $s_0 : J \to \mathbb{K}$ is any given solution to (14.38). (Once again we have used the definition of pointwise sum of maps $(s_0 + x)(t) := s_0(t) + x(t)$ for any $t \in J$.) In fact, if s and s_0 solve (14.38), by linearity their difference solves (14.35). Vice versa, if s_0 solves (14.38) and x solves (14.35), again by linearity $s_0 + x$ solves (14.38). So we have proved the following proposition.

Proposition 14.28 *The set of maximal solutions of the inhomogeneous linear ODE on the interval $J \subset \mathbb{R}$*

$$\frac{d^n x}{dt^n} + P_1(t)\frac{d^{n-1}x}{dt^{n-1}} + \cdots + P_n(t)x = Q(t) \,,$$

where the Q and P_k are continuous on J, has the following structure:

$$\{s = s_0 + x \mid x : J \to \mathbb{K} \text{ solves } (14.35)\} \,,$$

where $s_0 : J \to \mathbb{K}$ is an arbitrary solution to the homogeneous equation (14.38).

Remarks 14.29 All of this generalises easily to a Cauchy problem of order n with ODE

$$p_0(t)\frac{d^n \mathbf{x}}{dt^n} + p_1(t)\frac{d^{n-1}\mathbf{x}}{dt^{n-1}} + \cdots + p_n(t)\mathbf{x} = \mathbf{q}(t) \qquad (14.39)$$

in $\mathbf{x} = \mathbf{x}(t) \in \mathbb{K}^m$, for any $t \in U \subset \mathbb{R}$ open. Above, for any $t \in U$, the $p_k(t)$ are $m \times m$ *matrices* in $\mathbb{K}$, $\mathbf{q}(t)$ is a column *vector* in $\mathbb{K}^m$, and as $t \in U$ varies all functions are continuous. If on U we have $\det p_0(t) \neq 0$, the equation can be put in normal form

$$\frac{d^n \mathbf{x}}{dt^n} + P_1(t)\frac{d^{n-1}\mathbf{x}}{dt^{n-1}} + \cdots + P_n(t)\mathbf{x} = \mathbf{Q}(t) \,, \qquad (14.40)$$

where $P_k(t) := (p_0(t))^{-1} p_k(t)$ and $\mathbf{Q}(t) := (p_0(t))^{-1} \mathbf{q}(t)$. The Cauchy problem can be set up on $\Omega = J \times \mathbb{K}^m$ where $J \subset U$ is an open interval. We reduce to a first-order problem on $J \times \mathbb{K}^{n \cdot m}$ with the usual procedure, so the results discussed earlier hold. ∎

Examples 14.30

(1) Consider **homogeneous linear ODEs of order n with constant coefficients**:

$$\frac{d^n x}{dt^n} + a_1 \frac{d^{n-1} x}{dt^{n-1}} + \cdots + a_n x(t) = 0. \tag{14.41}$$

We view it as defined on $(t, x) \in \mathbb{R} \times \mathbb{C}$. The case $(t, x) \in \mathbb{R} \times \mathbb{R}$ arises by restriction. The coefficients $a_1, a_2, \ldots a_{n-1} \in \mathbb{C}$ are fixed numbers. The ODE can be written as:

$$\frac{d^n x}{dt^n} = -a_1 \frac{d^{n-1} x}{dt^{n-1}} - \cdots - a_n x(t). \tag{14.42}$$

Here is an elementary but important observation. Because the maps on the right are differentiable (the left-hand side exists!), we can differentiate the above term by term and find that every solution is of class C^{n+1}. As the procedure can be iterated without end, we conclude that *every solution to (14.41) is actually C^∞*.

The next goal is to write explicitly all maximal solutions. As the solution set is a vector space of dimension n, the generic solution will be a linear combination of n linearly independent particular solutions. The strategy we adopt is as follows.

(i) First of all we seek the maximal solutions, $x : \mathbb{R} \to \mathbb{C}$, of the inhomogeneous first-order equation

$$\left(\frac{d}{dt} - b \right) x(t) = f(t), \tag{14.43}$$

where $f : \mathbb{R} \to \mathbb{C}$ is a given continuous map, $b \in \mathbb{C}$ a constant, and we have used the notation, for any constant a:

$$\left(\frac{d}{dt} + a \right) x(t) := \frac{dx}{dt} + ax(t).$$

(ii) Then we show that (14.41) can be written as:

$$\left(\frac{d}{dt} - b_1 \right) \left(\frac{d}{dt} - b_2 \right) \cdots \left(\frac{d}{dt} - b_n \right) x(t) = 0, \tag{14.44}$$

where $b_1, b_2, \ldots, b_n$ are the solutions in $\mathbb{C}$, possibly with multiplicity, of the algebraic equation in $\chi \in \mathbb{C}$:

$$\chi^n + a_1 \chi^{n-1} + \cdots + a_{n-1} \chi + a_n = 0 . \tag{14.45}$$

(iii) Knowing the general solution to (14.43), i.e. the collection of all solutions for any possible initial conditions, and applying the formula recursively to each element as we shall explain, we eventually find the full solution set to (14.43).

Let us start from (i). From Example 4 we know that the solutions of (14.43) are $s(t) = s_0(t) + x(t)$ with $t \in \mathbb{R}$, where s_0 is an arbitrary particular solution of (14.43) and x roams the solution set of the homogeneous equation:

$$\left(\frac{d}{dt} - b \right) x(t) = 0 . \tag{14.46}$$

Immediately (for instance by separating variables) the latter's solutions are, for any $c \in \mathbb{C}$,

$$x(t) = c e^{bt} , \quad t \in \mathbb{R} .$$

It is manifest that we have a 1-dimensional vector space (over $\mathbb{C}$). By direct computation, a particular solution of (14.43) is

$$s_0(t) = e^{bt} \int_{t_0}^t f(\tau) e^{-b\tau} \, d\tau , \quad t \in \mathbb{R} , \tag{14.47}$$

where $t_0 \in \mathbb{R}$ is an arbitrary point. In the end, given any $t_0 \in \mathbb{R}$, all solutions to (14.43) arise for any $c \in \mathbb{C}$ by:

$$x_c(t) = c e^{bt} + e^{bt} \int_{t_0}^t f(\tau) e^{-b\tau} \, d\tau , \quad t \in \mathbb{R} , \tag{14.48}$$

Let us pass to item (ii). Letting $I = \frac{d^0}{dt^0}$ be the identity operator, consider the set $\mathcal{D}$ of differential operators

$$a_0 \frac{d^n}{dt^n} + a_1 \frac{d^{n-1}}{dt^{n-1}} + \cdots + a_n \frac{d^0}{dt^0} : C^\infty(\mathbb{R}; \mathbb{C}) \to C^\infty(\mathbb{R}; \mathbb{C})$$

as the numbers $a_0, \ldots, a_n \in \mathbb{C}$, and the order $n = 0, 1, 2, \ldots$, vary. Easily, $\mathcal{D}$ is a *commutative ring* with sum, for any $D, D' \in \mathcal{D}$

$$(D + D')(f) := D(f) + D'(f) , \quad \text{for any } f \in C^\infty(\mathbb{R}; \mathbb{C}),$$

product given by composition: for any $D, D' \in \mathcal{D}$

$$(DD')(f) := D(D'(f)),$$

additive neutral element given by the zero operator $O : C^\infty(\mathbb{R}; \mathbb{C}) \to C^\infty(\mathbb{R}; \mathbb{C})$, $(Of)(t) = 0$ for any $t \in \mathbb{R}$, $f \in C^\infty(\mathbb{R}; \mathbb{C})$, and multiplicative unit given by the operator $I = \frac{d^0}{dt^0}$.

If $\mathcal{P}(\mathbb{C})$ is the polynomial ring in the variable χ over $\mathbb{C}$, the map

$$F : a_0\chi^n + a_1\chi^{n-1} + \cdots + a_n \mapsto a_0\frac{d^n}{dt^n} + a_1\frac{d^{n-1}}{dt^{n-1}} + \cdots + a_n\frac{d^0}{dt^0}x(t)$$

is a bijective homomorphism of unital rings. Using that $\frac{d^n}{dt^n}\frac{d^m}{dt^m} = \frac{d^{n+m}}{dt^{n+m}}$, is its immediate to show that F preserves multiplication, addition and $F(1) = I$, so it is a homomorphism of unital rings. The surjectivity is straightforward, while for the injectivity it suffices to show that for $p \in \mathcal{P}(\mathbb{C})$, if $F(p)$ is the zero operator of $\mathcal{D}$, then p is the zero polynomial. Indeed, if $p(\chi) = a_0\chi^n + a_1\chi^{n-1} + \cdots + a_n$ and $F(p) = a_0\frac{d^n}{dt^n} + a_1\frac{d^{n-1}}{dt^{n-1}} + \cdots + a_n\frac{d^0}{dt^0}$ is the zero operator, since $(a_0\frac{d^n}{dt^n} + a_1\frac{d^{n-1}}{dt^{n-1}} + \cdots + a_n\frac{d^0}{dt^0})(f) = 0$ and choosing as f the constant 1, we obtain $a_n = 0$. Choosing as f the constant map equal to t and iterating, we find $a_{n-1} = 0$ and so on, ending in p being the zero polynomial as required. All in all F, as claimed, is an *isomorphism of unital rings*.

By the *Fundamental Theorem of Algebra* every (non-constant) polynomial of $p \in \mathcal{P}(\mathbb{C})$,

$$p(\chi) = \chi^n + a_1\chi^{n-1} + \cdots + a_{n-1}\chi + a_n$$

splits as product of elementary polynomials:

$$p(\chi) = (\chi - b_1)\cdots(\chi - b_n)$$

where $b_1, b_2, \ldots, b_n$ are the n solutions of $p(\chi) = 0$. Under the isomorphism F we arrive at decomposition (14.44).

Now to (iii). The initial ODE (14.41) is equivalent to (14.44):

$$\left(\frac{d}{dt} - b_1\right)\left(\frac{d}{dt} - b_2\right)\cdots\left(\frac{d}{dt} - b_n\right)x(t) = 0,$$

i.e.

$$\left(\frac{d}{dt} - b_1\right)\left[\left(\frac{d}{dt} - b_2\right)\cdots\left(\frac{d}{dt} - b_n\right)x(t)\right] = 0.$$

Keeping in account the general solution (14.48) of (14.46), we conclude that, for $c_1 \in \mathbb{C}$:

$$\left(\frac{d}{dt} - b_2\right) \cdots \left(\frac{d}{dt} - b_n\right) x(t) = c_1 e^{b_1 t} .$$

Iterating, we obtain

$$\left(\frac{d}{dt} - b_3\right) \cdots \left(\frac{d}{dt} - b_n\right) x(t) = c_2 e^{b_2 t} + c_1 e^{b_2 t} \int_{t_0}^{t} e^{b_1 \tau} e^{-b_2 \tau} d\tau .$$

If $b_1 \neq b_2$, gathering the constants in the same exponentials,

$$\left(\frac{d}{dt} - b_3\right) \cdots \left(\frac{d}{dt} - b_n\right) x(t) = k_2 e^{b_2 t} + k_1 e^{b_1 t} .$$

If instead $b_1 = b_2$, the direct calculation gives

$$\left(\frac{d}{dt} - b_3\right) \cdots \left(\frac{d}{dt} - b_n\right) x(t) = k_2 e^{b_1 t} + k_1 t \, e^{b_1 t} .$$

In either case k_1 and k_2 are arbitrary constants in $\mathbb{C}$. Iterating until in the left-hand side there remains $x = x(t)$ only, one can prove as exercise the recipe for the general solution to (14.41)

$$\frac{d^n x}{dt^n} + a_1 \frac{d^{n-1} x}{dt^{n-1}} + \cdots + a_n x(t) = 0 .$$

it goes as follows. We seek the complex roots of the polynomial

$$\chi^n + a_1 \chi^{n-1} + \cdots + a_{n-1} \chi + a_n ,$$

called the associated **characteristic polynomial**. Solutions that coincide are collected:
b_1 with multiplicity m_1, b_2 with multiplicity m_2,..., b_k with multiplicity m_k, so that $1 \leq m_j \leq n$ and $m_1 + \cdots + m_k = n$. The general solution to (14.41) then becomes, depending on the n constants $C_{j,r_j} \in \mathbb{C}$, $j = 1, 2, \ldots, k$ and $r_j = 1, \ldots, m_j$:

$$x(t) = \sum_{j=1}^{k} \sum_{r_j=1}^{m_j} C_{j,r_j} t^{r_j - 1} e^{b_j t} , \quad \text{with } t \in \mathbb{R} . \tag{14.49}$$

In practice: if a root $b \in \mathbb{C}$ of the characteristic polynomial has multiplicity 1 (i.e. there are no other roots equal to it) it contributes to the general solution only the exponential e^{bt}. If it has multiplicity 2, it contributes two functions: e^{bt} and $t \, e^{bt}$. If it has multiplicity m, it contributes m functions e^{bt}, $t e^{bt}$, $t^2 e^{bt}$,..., $t^{m-1} e^{bt}$. Each of these n solutions participates thus to the general solution, and the single contributions are multiplied by n arbitrary complex numbers C_{j,r_j} and then added.

We can justify (14.49) for the general solution to (14.41) from another viewpoint. It is easy to prove that the n particular solutions $\mathbb{R} \ni t \mapsto t^{r_j-1} e^{b_j t}$, obtained varying the solutions of the characteristic equation $p(\chi) = 0$ and remembering the multiplicities, are linearly independent. *These solutions then form a basis of the vector space of solutions to the initial homogeneous linear ODE, which we know has dimension n. Hence the solution's general form (14.49) is nothing but the generic element of the vector space of solutions.*

(2) It is clear that the above method applies to (14.41) when the coefficients $a_0, a_1, \ldots, a_n$ are all *real*, and the initial conditions are *real*: $x(t_0) = x_0$, $\frac{dx}{dt}(t_0) = x_0^{(1)}, \ldots, \frac{d^{n-1}x}{dt^{n-1}}(t_0) = x_0^{(n-1)}$. If so the unique solution to the corresponding Cauchy problem must be real, and this is the case also for *inhomogeneous* linear ODEs

$$\frac{d^n x}{dt^n} + a_1 \frac{d^{n-1}x}{dt^{n-1}} + \cdots + a_n x(t) = f(t),$$

as long as the continuous map f is real.

In fact, if $x = x(t)$ is any such solution, its conjugate $\overline{x} = \overline{x(t)}$ satisfies the ODE with coefficients (and function f) conjugated. But as these are real, $\overline{x}$ will satisfy the same ODE as x. It will satisfy the initial conditions of x too, since these are expressed by real numbers. Therefore $\overline{x}(t) = x(t)$ for any t by the uniqueness theorem. This just says that $x(t)$ is a real number for any t.

The fact that both the coefficients a_k and the initial conditions are real numbers does not guarantee the solutions of $p(\chi) = 0$ will be real. However, if the initial conditions are real, the Cauchy problem's solutions will be real as said before. To write them explicitly, for the homogeneous equation, it is convenient to remember in (14.49) the Euler identity:

$$e^{b_j t} = e^{\alpha_j t} \left(\cos(\beta_j t) + i \sin(\beta_j t) \right)$$

for the *complex solutions* $b_j = \alpha_j + i\beta_j$ of $p(\chi) = 0$, with $\alpha_j, \beta_j \in \mathbb{R}$. If the characteristic polynomial is real, the complex solutions of $p(\chi) = 0$ only show up with the same multiplicity in conjugate pairs. If so, each *pair* of *complex-conjugate* solutions $b_j = \alpha_j + i\beta_j$ and $b_{j'} = \alpha_j - i\beta_j$ (of the same multiplicity $m_j = m_{j'}$) contributes to the general solution of the ODE, respectively:

$$t^{r_j-1} e^{\alpha_j t} C_{j,r_j} \left(\cos(\beta_j t) + i \sin(\beta_j t) \right)$$

$$\text{and} \quad t^{r_j-1} e^{\alpha_j t} C_{j',r_{j'}} \left(\cos(\beta_j t) - i \sin(\beta_j t) \right).$$

Combining these we may write the overall contribution as:

$$t^{r_j-1} e^{\alpha_j t} \left(A_{j,r_j} \cos(\beta_j t) + B_{j,r_j} \sin(\beta_j t) \right), \tag{14.50}$$

where A_{j,r_j} and B_{j,r_j} are coefficients defined by:

$$A_{j,r_j} = C_{j,r_j} + C_{j',r_{j'}} \quad \text{and} \quad B_{j,r_j} = i(C_{j,r_j} - C_{j',r_{j'}}) .$$

If the full solution is real, the coefficients A_{j,r_j} and B_{j,r_j} must be real: the imaginary parts must cancel out in order for (14.50) to be a real solution. We conclude that if (1) *all coefficients of the ODE are real* and (2) *the Cauchy data are real*, then the general solution to (14.41) has the form (14.49), where now the constants C_{j,r_j} are *real* regarding the contribution of the real characteristic roots b_j, while the possible complex-conjugate characteristic roots $b_j, b_{j'}$ provide (14.50) where A_{j,r_j} and B_{j,r_j} are *real*.

(3) For the sake of an example, consider the homogeneous linear ODE with real constant coefficients:

$$\frac{d^4x}{dt^4} - 2\frac{d^3x}{dt^3} + 2\frac{d^2x}{dt^2} - 2\frac{dx}{dt} + x = 0 ,$$

which we wish to solve for any possible *real* initial conditions: $x(0) = x_0$, $\frac{dx}{dt}(0) = x_0^{(1)}$, $\frac{d^2x}{dt^2}(0) = x_0^{(2)}$, $\frac{d^3x}{dt^3}(0) = x_0^{(3)}$. We know that solutions must be real and arise as above. The characteristic polynomial factorises as:

$$p(\chi) = (\chi - 1)^2(\chi + i)(\chi - i) .$$

Hence we have the real root 1 with multiplicity 2 and two conjugate roots $\pm i$ (with zero real part and imaginary part 1). The first roots contributes to the general solution the particular solutions e^t and te^t. The two conjugate roots contribute the particular solutions $\sin t$ and $\cos t$. The *real* general solution is then:

$$x(t) = C_1 e^t + C_2 te^t + A \cos t + B \sin t .$$

The real constants $C_1, C_2, A, B \in \mathbb{R}$, for the time being completely arbitrary, are determined imposing the 4 (real) initial conditions $x(0) = x_0$, $\frac{dx}{dt}(0) = x_0^{(1)}$, $\frac{d^2x}{dt^2}(0) = x_0^{(2)}$, $\frac{d^3x}{dt^3}(0) = x_0^{(3)}$.
If instead we are interested in the *complex* general solution to the ODE, we have:

$$x(t) = C_1 e^t + C_2 te^t + C_3 e^{it} + C_4 e^{-it}$$

where now C_1, C_2, C_3, C_4 live in $\mathbb{C}$.

(4) Consider at last an *inhomogeneous* linear equation with constant coefficients:

$$\frac{d^n x}{dt^n} + a_1 \frac{d^{n-1}x}{dt^{n-1}} + \cdots + a_n x(t) = f(t) . \tag{14.51}$$

where $f : J \to \mathbb{C}$ is continuous on the open interval $J \subset \mathbb{R}$. From Proposition 14.28, and remembering Example 1, we know that the solution set of (14.51) has the form:

$$x(t) = x_0(t) + \sum_{j=1}^{k} \sum_{r_j=1}^{m_j} C_{j,r_j} t^{r_j-1} e^{b_j t}, \quad \text{with } t \in J \tag{14.52}$$

where the C_{j,r_j} are arbitrary constants in $\mathbb{C}$, b_j (with $j = 1, \ldots, k$ and $\sum_{j=1}^{k} m_j = n$) are roots (in $\mathbb{C}$) of the characteristic polynomial of the homogeneous equation (14.41) and m_j is the multiplicity of each solution. Finally, $x_0 : J \to \mathbb{C}$ is a particular solution (arbitrarily fixed) of the inhomogeneous equation (14.51). How do we find such solution? There exist various procedures, to be found in textbooks on differential equations. Here we will just remark that there always exists a standard way to construct such a solution: it is based on formula (14.47), which gives a particular solution to (14.46), since equation (14.51), as we saw in Example 1, can be written as:

$$\left(\frac{d}{dt} - b_1\right)^{m_1} \left(\frac{d}{dt} - b_2\right)^{m_1} \cdots \left(\frac{d}{dt} - b_k\right)^{m_k} x(t) = f(t). \tag{14.53}$$

so:

$$\left(\frac{d}{dt} - b\right)^{m} := \left(\frac{d}{dt} - b\right) \cdots m \text{ times} \cdots \left(\frac{d}{dt} - b\right).$$

Repeatedly applying (14.47) will give a particular solution, as required. For instance, consider an equation of order two:

$$\frac{d^2 x}{dt^2} + a_1 \frac{dx}{dt} + a_2 x(t) = f(t) \tag{14.54}$$

and let $b_1 \neq b_2$ be the roots of the characteristic polynomial of the homogeneous equation. We can write (14.54) as:

$$\left(\frac{d}{dt} - b_1\right) \left(\frac{d}{dt} - b_2\right) x = f.$$

Then formula (14.47) implies that if x satisfies:

$$\left(\frac{d}{dt} - b_2\right) x = e^{b_1 t} \int_{t_0}^{t} f(\tau) e^{-b_1 \tau} d\tau,$$

for a given $t_0 \in J$, then x satisfies (14.51) as well. We may use (14.47) again, where now the f on the right is the map t defined by the right-hand side of the above identity. Thus we find the particular solution of (14.54):

$$x_0(t) = e^{b_2 t} \int_{t_1}^{t} e^{(b_1 - b_2)\tau_1} \left(\int_{t_0}^{\tau_1} e^{-b_1 \tau} f(\tau) d\tau \right) d\tau_1 ,$$

where $t_1 \in J$ is another arbitrary point. By construction, in fact, if we apply $\left(\frac{d}{dt} - b_1 \right) \left(\frac{d}{dt} - b_2 \right)$ to the right-hand side we find f as required. The formula stays valid even if $b_1 = b_2$, since the assumption $b_1 \neq b_2$ has not been used. The procedure extends directly to equations of any order.

Exercises 14.31

(1) Let $\mathbf{f} : \mathbb{R} \times D \to \mathbb{K}^n$ be continuous and locally Lipschitz in $\mathbf{x} \in \mathbb{K}^n$, with $D \subset \mathbb{K}^n$ non-empty and open. Suppose that, for a given $T \in \mathbb{R}$,

$$\mathbf{f}(t - T, \mathbf{x}) = \mathbf{f}(t, \mathbf{x}) , \quad \forall (t, \mathbf{x}) \in \mathbb{R} \times D .$$

Pick any $k \in \mathbb{Z}$. Show that for $(t_0, \mathbf{x}_0) \in \Omega$, the map $\mathbf{x} : (\alpha, \omega) \to \mathbb{K}^n$ is a maximal solution of:

$$\begin{cases} \dfrac{d\mathbf{x}}{dt} = \mathbf{f}(t, \mathbf{x}(t)) \\ \mathbf{x}(t_0) = \mathbf{x}_0. \end{cases} ,$$

if and only if $\mathbf{x}'(t) := \mathbf{x}(t - kT)$ with $t \in (\alpha + kT, \omega + kT)$ is a maximal solution of

$$\begin{cases} \dfrac{d\mathbf{x}'}{dt} = \mathbf{f}\left(t, \mathbf{x}'(t)\right) , \\ \mathbf{x}'(t_0 - kT) = \mathbf{x}_0. \end{cases} ,$$

(2) Let $\mathbf{f} : D \to \mathbb{K}^n$ be locally Lipschitz, with $D \subset \mathbb{K}^n$ non-empty and open. Fix any $T \in \mathbb{R}$. Prove that $\mathbf{x} : (\alpha, \omega) \to \mathbb{K}^n$ is a maximal solution of the autonomous Cauchy problem, for $(t_0, \mathbf{x}_0) \in \Omega$:

$$\begin{cases} \dfrac{d\mathbf{x}}{dt} = \mathbf{f}(\mathbf{x}(t)) \\ \mathbf{x}(t_0) = \mathbf{x}_0. \end{cases} ,$$

if and only if $\mathbf{x}'(t) := \mathbf{x}(t - T)$ with $t \in (\alpha + T, \omega + T)$ is a maximal solution of the autonomous problem

$$\begin{cases} \dfrac{d\mathbf{x}'}{dt} = \mathbf{f}\left(\mathbf{x}'(t)\right) , \\ \mathbf{x}'(t_0 - T) = \mathbf{x}_0. \end{cases} ,$$

(3) Consider a point particle of mass $m > 0$ constrained to a smooth horizontal plane at rest in an inertial frame. Suppose the particle undergoes a force, with $g > 0$:

$$\mathbf{F}(\mathbf{x}) := \kappa r \mathbf{e}_r - mg\mathbf{e}_z \,,$$

where we have introduced cylindrical coordinates r, φ, z with origin O on the plane $z = 0$. Study the motion in the two cases $\kappa > 0$ and $\kappa < 0$. In case $\kappa < 0$ find the equation of motion (maximal solution) with initial conditions $\mathbf{x}(0) = \mathbf{0}$, $\mathbf{v}(0) = v(\mathbf{e}_x - \mathbf{e}_y)$.

(4) Consider a point particle of mass $m > 0$ constrained to a smooth horizontal plane at rest in an inertial frame. Suppose the particle undergoes a force, with $\kappa < 0, \beta > 0, g > 0,$

$$\mathbf{F}(\mathbf{x}, \mathbf{v}) := \kappa r \mathbf{e}_r - \beta \mathbf{e}_z \wedge \mathbf{v} - mg\mathbf{e}_z \,,$$

where we have introduced cylindrical coordinates r, φ, z with origin O on the plane, say of equation $z = 0$. Study the particle's motion.

(5) Consider a point particle of mass $m > 0$ constrained to a smooth horizontal plane at rest in an inertial frame. Suppose the particle undergoes a force, with $\kappa, \gamma > 0, g > 0,$

$$\mathbf{F}(\mathbf{x}, \mathbf{v}) := \kappa r \mathbf{e}_r - \gamma \mathbf{v} - mg\mathbf{e}_z \,,$$

where we have introduced cylindrical coordinates r, φ, z with origin O on the plane $z = 0$. Study the particle's motion for $\kappa > 0$ and for $\kappa < 0$.

14.3.6 Completeness of Maximal Solutions

Now we shall deal with technical matters about the completeness of maximal solutions in cases where the global Existence and uniqueness theorem 14.23 holds.

For starters we will prove a proposition that explains what happens to an incomplete maximal solution. The intuitive idea is that a maximal solution $\mathbf{x} : I \to \mathbb{K}^n$ may be incomplete if, for large values of the parameter t approaching the boundary of I, the integral curve $\{(t, \mathbf{x}(t)) \,|\, t \in I\}$ tends to "exit" the domain Ω, i.e. it converges to some point on the boundary $\mathcal{F}(\Omega)$, where Ω is the domain of the function $\mathbf{f}$ in (14.25). It can also happen that for large t, when getting close to the boundary of I, the integral curve does not converge at all.

Examples 14.32

(1) The first example is trivial. Consider the Cauchy problem on $\Omega = (-1, 1) \times \mathbb{R}$:

$$\begin{cases} \dfrac{dx}{dt} = 1, \\ x(0) = 0. \end{cases}$$

The maximal solution is obviously $x : (-1, 1) \ni t \mapsto t$. It is incomplete, and as $t \to +1$, $(t, x(t)) \to (1, 1) \in \mathcal{F}(\Omega)$. This example's incompleteness is somewhat artificial.

(2) The second example is less contrived and the result somehow unexpected. Consider $\Omega = \mathbb{R}^2$ and on it the Cauchy problem

$$\begin{cases} \dfrac{dx}{dt} = x^2, \\ x(0) = 1. \end{cases}$$

Separating variables we obtain the solution:

$$x(t) = \frac{1}{1 - t}, \qquad t \in (-\infty, 1).$$

This is the only solution, clearly maximal, to the Cauchy problem. Uniqueness is guaranteed by the Global Uniqueness Theorem since the hypotheses holds trivially. The solution is not future-complete, and as $t \to 1^-$ the curve $(t, x(t))$ does not converge to a point. As a matter of fact it diverges.

(3) Consider at last a more complicated example on $\Omega = (0, +\infty) \times \mathbb{R}^2$

$$\begin{cases} \dfrac{dx}{dt} = -\dfrac{y}{t^2}, \\ \dfrac{dy}{dt} = \dfrac{x}{t^2}, \\ x(2/\pi) = 1, \quad y(2/\pi) = 0. \end{cases}$$

On Ω, the system's right-hand side defines a C^1 function for which global Existence and uniqueness theorem holds. A solution is

$$\begin{cases} x(t) = \sin(1/t), \\ y(t) = \cos(1/t), \end{cases} \quad t \in (0, +\infty). \tag{14.55}$$

The solution, clearly maximal, is unique by the aforementioned theorem. In this case though the solution's limit as $t \to 0^+$ does not tend to any value, nor does it diverge. We can nonetheless find a sequence of points t_k in $(0, +\infty)$ that tend to 0^+ as $k \to +\infty$ and such that $(t_k, \mathbf{x}(t_k))$ approaches the boundary $\mathcal{F}(\Omega)$ as $k \to +\infty$. For instance, choose any $t_0 \in (0, +\infty)$ and define $t_k =$

$t_0/(1 + 2\pi k t_0)$ (so that $1/t_k = 2\pi k + 1/t_0$), $k = 1, 2, \ldots$. Hence $\mathbf{x}(t_k) = \mathbf{x}(t_0)$ *always* for any $k = 1, 2, \ldots$, and therefore the limit of $(t_k, \mathbf{x}(t_k))$ exists and equals $(0, \mathbf{x}(t_0)) \in \mathcal{F}(\Omega)$. The existence of such a sequence was precluded in the previous example.

We can prove the following general proposition.

Proposition 14.33 *In the hypotheses of Theorem 14.23 for* $\mathbf{f} : \Omega \to \mathbb{K}^n$, *let* $\mathbf{x} : I \to \mathbb{K}^n$, *with* $I = (\alpha, \omega)$, *be a maximal solution to problem (14.25).*

(1) If $\mathbf{x}$ *is not future-complete,* $\{t_k\}_{k \in \mathbb{N}} \subset I$ *is such that* $t_k \to \omega$ *as* $k \to +\infty$, *and there exists* $\lim\limits_{k \to +\infty} \mathbf{x}(t_k) = \mathbf{x}_\omega \in \mathbb{K}^n$, *then* $(\omega, \mathbf{x}_\omega) \in \mathcal{F}(\Omega)$.

(2) If $\mathbf{x}$ *is not past-complete,* $\{t_k\}_{k \in \mathbb{N}} \subset I$ *is such that* $t_k \to \alpha$ *as* $k \to +\infty$, *and there exists* $\lim\limits_{k \to +\infty} \mathbf{x}(t_k) = \mathbf{x}_\alpha \in \mathbb{K}^n$, *then* $(\alpha, \mathbf{x}_\alpha) \in \mathcal{F}(\Omega)$.

Proof Let us prove (1) only, since (2) is completely similar. We will use Lemma 14.17. By contradiction, assume the solution is not future-complete and that, for the limit of the sequence considered, $(\omega, \mathbf{x}_\omega) \in \Omega$. Consider an open set $Q \subset \Omega$ such that $(\omega, \mathbf{x}_\omega) \in Q$ and $\overline{Q} \subset \Omega$. Choose Q so that on $\overline{Q}$ the map $\mathbf{f}$ is Lipschitz with constant $L \geq 0$, i.e. $\|\mathbf{f}(t, \mathbf{x}) - \mathbf{f}(t, \mathbf{x}')\| \leq L\|\mathbf{x} - \mathbf{x}'\|$ if $(t, \mathbf{x}), (t, \mathbf{x}') \in \overline{Q}$. Then put $M := \max\{\|\mathbf{f}(t, \mathbf{x})\| \mid (t, \mathbf{x}) \in \overline{Q}\}$. To fix ideas take $\overline{Q}$ of the form

$$[\omega - \Delta, \omega + \Delta] \times B_E(\omega),$$

where $\Delta > 0$ and $B_E(\omega)$ is the closed ball in $\mathbb{K}^n$ centred at ω with radius $E > 0$. Since $t_k \to \omega$ and $\mathbf{x}(t_k) \to \mathbf{x}_\omega$ as $k \to +\infty$, choosing k very large we can find small enough $\delta > 0$ and $\epsilon > 0$ so they satisfy the three conditions:

(1) $|t_k - \omega| < \delta$,
(2) $[t_k - \delta, t_k + \delta] \times B_\epsilon(\mathbf{x}(t_k)) \subset [\omega - \Delta, \omega + \Delta] \times B_E(\mathbf{x}_\omega)$,
(3) $\delta < \min\{\epsilon/M, 1/L\}$

(above, $B_\epsilon(\mathbf{x}(t_k))$ is the closed ball of radius ϵ centred at t_k). If $M = 0, \mathbf{f} \equiv 0$ on Q, and starting from the first t_k for which $(t_k, \mathbf{x}(t_k)) \in Q$ the orbit is constant, so $\omega = +\infty$ and the solution is future-complete against the assumption. Therefore $M > 0$. Here, to satisfy (3), it suffices to take $\epsilon = 2\delta M$ and then choose $0 < \delta < 1/(2L)$. We may pick $\delta > 0$ so small that (1) and (2) hold; this can always be done since $(t_k, \mathbf{x}(t_k))$ can be made arbitrarily close to $(\omega, \mathbf{x}_\omega)$ by hypothesis. The first condition warrants $\omega \in (t_k - \delta, t_k + \delta)$. The other two, by Lemma 14.17, ensure the existence of the solution to problem

$$\begin{cases} \dfrac{d\mathbf{x}'}{dt} = \mathbf{f}\left(t, \mathbf{x}'(t)\right), \\ \mathbf{x}'(t_k) = \mathbf{x}(t_k). \end{cases} \tag{14.56}$$

defined on the whole interval $(t_k - \delta, t_k + \delta)$. By construction this solution is a prolongation of the original solution $\mathbf{x} = \mathbf{x}(t)$. In fact it satisfies the same ODE, it agrees with $\mathbf{x} = \mathbf{x}(t)$ at $t = t_k$ and it is defined also on the right of ω, since $\omega \in (t_k - \delta, t_k + \delta)$. This cannot be, because $\mathbf{x} : (\alpha, \omega) \to \mathbb{K}^n$ is maximal. □

Proposition 14.33 has important consequences regarding the completeness of a Cauchy problem's maximal solutions in presence of additional conditions. We list these consequences in one result.

Proposition 14.34 *Suppose* $\mathbf{f} : \Omega \to \mathbb{K}^n$ *satisfies the assumptions of Theorem 14.23, with* $\Omega = \mathbb{R} \times D$ *and* $D \subset \mathbb{K}^n$ *non-empty and open. If* $\mathbf{x} : I \to \mathbb{K}^n$, *with* $I = (\alpha, \omega)$, *is a maximal solution to problem (14.25) the following facts hold.*

(1) If the orbit of $\mathbf{x} : I \to \mathbb{K}^n$ *is contained in a compact set* $K \subset D$ *for* $t \geq T \in I$, *then* $\mathbf{x} : I \to \mathbb{K}^n$ *is future-complete.*

(2) If the orbit of $\mathbf{x} : I \to \mathbb{K}^n$ *is contained in a compact set* $K \subset D$ *for* $t \leq T \in I$, *then* $\mathbf{x} : I \to \mathbb{K}^n$ *is past-complete.*

(3) If $D = \mathbb{K}^n$ *and* $\sup_{(t,\mathbf{x}) \in \Omega} ||\mathbf{f}(t, \mathbf{x})|| < +\infty$ *then* $\mathbf{x} : I \to \mathbb{K}^n$ *is complete.*

(4) If $\mathbf{f}$ *does not depend on* t *(the associated ODE system is autonomous) and the support of* $\mathbf{f}$ *is compact (in* D*), then* $\mathbf{x} : I \to \mathbb{K}^n$ *is complete.*

Proof

(1) Suppose the solution is not future-complete, so $I = (\alpha, \omega)$ with $\omega \in \mathbb{R}$. Consider a sequence $\{t_k\}_{k \in \mathbb{N}} \subset (T, \omega)$ with $t_k \to \omega$ as $k \to +\infty$. Since $K \supset \{\mathbf{x}(t_k)\}_{k \in \mathbb{N}}$ is compact in a metric space, it is sequentially compact as well, so there is a convergent subsequence $\{\mathbf{x}(t_{k_p})\}_{p \in \mathbb{N}}$ to some $\mathbf{x}_\omega \in K$. By Proposition 14.33 $(\omega, \mathbf{x}_\omega) \in \partial\Omega$. But $\Omega = \mathbb{R} \times D$ implies $\mathbf{x}_\omega \in \mathcal{F}(D)$, which is impossible since $\mathbf{x}_\omega \in K \subset D$ and $\mathcal{F}(D) \cap D = \varnothing$ as D is open. Hence $\omega = +\infty$.

(2) The argument is much the same as that of case (1).

(3) Suppose the solution is not future-complete, so $I = (\alpha, \omega)$ with $\omega \in \mathbb{R}$. Consider closed balls $B_k \subset \mathbb{K}^n$ centred at the origin with radii $k = 1, 2, \ldots$. We claim that for any $k \in \mathbb{N}$ there exists $t_k \in (\alpha, \omega)$, with $t_k > t_{k-1}$ and $\mathbf{x}(t_k) \notin B_k$. If not, starting from some $T \in (\alpha, \omega)$ the orbit would be confined to compact sets B_k, and from (1) $\omega = +\infty$. Therefore, as $\{||\mathbf{x}(t_k)||\}_{k \in \mathbb{N}}$ is unbounded and since $t_k - t_0 < \omega - t_0 < +\infty$, we conclude that the difference quotient $(x^1(t_k) - x^1(t_0))/(t_k - t_0)$ of the first component of $\mathbf{x} = \mathbf{x}(t)$ (or its real part if $\mathbb{K} = \mathbb{C}$) must diverge as $k \to +\infty$. The Mean Value Theorem (applied separately to each component of the vector $\mathbf{x}(t_k)$, of which there is a finite number) says there is a subsequence τ_{k_j}, with $t_0 \leq \tau_{k_j} \leq t_{k_j}$, such that the set $\{||\frac{d\mathbf{x}}{dt}(\tau_{k_j})||\}_{j \in \mathbb{N}}$ is unbounded. Directly from the ODE $d\mathbf{x}/dt = \mathbf{f}(t, \mathbf{x}(t))$ we conclude $\mathbf{f}$ cannot be bounded. This contradicts the assumptions, and therefore $\omega = +\infty$. A similar proof, using (2), shows $\alpha = -\infty$.

(4) Call $supp_D\mathbf{f}$ the support of $\mathbf{f}$ in the topology of D, i.e. the closure of $\{\mathbf{x} \in D \mid \mathbf{f}(\mathbf{x}) \neq 0\}$ in the induced topology. By definition $supp_D\mathbf{f} \subset D$ and $supp_D\mathbf{f}$

is compact.[4] If $\mathbf{f}(\mathbf{x}_0) = 0$, then a solution, *hence the only solution*, of the system is $\mathbf{x}(t) = \mathbf{x}_0$ for $t \in \mathbb{R}$. The latter is complete by construction. If $\mathbf{f}(\mathbf{x}_0) \neq 0$, the orbit cannot enter, at some $t_1 \neq t_0$, the set $\{\mathbf{x} \in D \mid \mathbf{f}(\mathbf{x}) = 0\}$ for the same reason. So, the orbit is certainly contained in $\{\mathbf{x} \in D \mid \mathbf{f}(\mathbf{x}) \neq 0\}$ and therefore in the compact set $supp_D \mathbf{f} \subset D$. By (1) and (2) then, the solution is complete.

□

14.4   Comparison of Solutions and Dependency on Initial Conditions and Parameters

In this section we will discuss some technical issues about the qualitative behaviour of the solutions to initial value problems. Essentially we will deal with what happens, mathematically, to solutions when we change "something" in the Cauchy problem. We can change two things: the function $\mathbf{f}$ or the initial condition (or both).

14.4.1   Gronwall Lemma and Consequences

The following very technical result, albeit elementary and apparently useless, is in fact fundamental to the theory of systems of ODEs. Gronwall's Lemma bears important consequences when we study the behaviour *in time* of the solutions to a Cauchy problem (with $\mathbf{f}$ Lipschitz) and we start from different, but "close" initial conditions. The orbits separate exponentially in accordance with the Lipschitz constant. A second corollary is when we consider solutions of *different* initial value problems with the *same* initial condition. In this case, too, as time goes by the orbits diverge from one another exponentially, in a way governed by the Lipschitz constant.

Theorem 14.35 (Gronwall Lemma)   *Let* $h : [t_0, t_1] \rightarrow [0, +\infty)$ *and* $u : [t_0, t_1] \rightarrow [0, +\infty)$ *(with* $t_1 > t_0$*) be continuous maps. If, for some* $a, b \in (0, +\infty)$,

$$h(t) \leq a + b \int_{t_0}^{t} h(s)u(s)ds , \quad \text{for any } t \in [t_0, t_1] \tag{14.57}$$

then

$$h(t) \leq a e^{b \int_{t_0}^{t} u(s)ds} , \quad \text{for any } t \in [t_0, t_1] . \tag{14.58}$$

---

[4] Compactness does not depend on the induced topology, so if $supp_D \mathbf{f}$ is compact in D it is compact in $\mathbb{K}^n$.

Proof Set

$$V(t) := \int_{t_0}^t h(s)u(s)ds \, ,$$

so (14.57) implies

$$\frac{dV}{dt} \leq u(t)(a + bV(t)) \, ,$$

or equivalently

$$\frac{d}{dt}\left(\ln\left|\frac{a+bV(t)}{a}\right| - b\int_{t_0}^t u(s)ds\right) \leq 0 \, .$$

The function in brackets vanishes at $t = t_0$, so it is non-positive if $t > t_0$. Hence

$$\ln\left|\frac{a+bV(t)}{a}\right| \leq b\int_{t_0}^t u(s)ds \, ,$$

and then (as the logarithm's argument is non-negative by construction)

$$a + bV(t) \leq ae^{b\int_{t_0}^t u(s)ds} \, .$$

From (14.57) we then obtain (14.58), ending the proof. □

Gronwall's Lemma produces the following remarkable result.

Theorem 14.36 *Let $\Omega \subset \mathbb{R} \times \mathbb{K}^n$ be non-empty and open, and suppose $\mathbf{f}, \mathbf{g} : \Omega \to \mathbb{K}^n$ are continuous, locally Lipschitz in $\mathbf{x} \in \mathbb{K}^n$. Further assume that:*

(i) $\mathbf{f}$ is Lipschitz in $\mathbf{x}$ everywhere on Ω, with constant K;
(ii) $\|\mathbf{f}(t, \mathbf{x}) - \mathbf{g}(t, \mathbf{x})\| \leq \epsilon$, for some $\epsilon \geq 0$ and any $(t, \mathbf{x}) \in \Omega$.

If $\mathbf{x} : I \to \mathbb{K}^n$ and $\mathbf{y} : J \to \mathbb{K}^n$ respectively solve the Cauchy problems for $\mathbf{f}$ with initial condition $\mathbf{x}(t_0) = \mathbf{x}_0$, and for $\mathbf{g}$ with initial condition $\mathbf{y}(\tau_0) = \mathbf{y}_0$, and we take them on the same finite interval $(\alpha, \beta) \ni t_0, \tau_0$ with $[\alpha, \beta] \subset I \cap J$, then the following estimate holds:

$$\|\mathbf{x}(t) - \mathbf{y}(t)\| \leq (\|\mathbf{x}_0 - \mathbf{y}_0\| + M|t_0 - \tau_0| + \epsilon|\alpha - \beta|) e^{K|t-t_0|} \, ,$$
$$\text{for any } t \in (\alpha, \beta), \tag{14.59}$$

where $M := \max\{\|\mathbf{g}(t, \mathbf{y}(t))\| \mid t \in [\alpha, \beta]\}$.

Proof Let us prove the claim for $t \in [t_0, \beta)$, since on the other semi-interval the argument is similar. We have

$$\mathbf{x}(t) = \mathbf{x}_0 + \int_{t_0}^{t} \mathbf{f}(s, \mathbf{x}(s))\, ds \quad \text{and} \quad \mathbf{y}(t) = \mathbf{y}_0 + \int_{\tau_0}^{t} \mathbf{g}(s, \mathbf{y}(s))\, ds \, .$$

Subtracting term by term:

$$\mathbf{x}(t) - \mathbf{y}(t) = \mathbf{x}_0 - \mathbf{y}_0 + \int_{t_0}^{t} [\mathbf{f}(s, \mathbf{x}(s)) - \mathbf{g}(s, \mathbf{y}(s))]\, ds + \int_{t_0}^{\tau_0} \mathbf{g}(s, \mathbf{y}(s))\, ds \, ,$$

which we can write, adding and subtracting the same integral:

$$\mathbf{x}(t) - \mathbf{y}(t) = \mathbf{x}_0 - \mathbf{y}_0 + \int_{t_0}^{t} [\mathbf{f}(s, \mathbf{x}(s)) - \mathbf{f}(s, \mathbf{y}(s))]\, ds$$

$$+ \int_{t_0}^{t} [\mathbf{f}(s, \mathbf{y}(s)) - \mathbf{g}(s, \mathbf{y}(s))]\, ds + \int_{t_0}^{\tau_0} \mathbf{g}(s, \mathbf{y}(s))\, ds \, .$$

In our assumptions the triangle inequality, together with known integral estimates and the Lipschitz inequality, imply

$$||\mathbf{x}(t) - \mathbf{y}(t)|| \leq ||\mathbf{x}_0 - \mathbf{y}_0|| + K \int_{t_0}^{t} ||\mathbf{x}(s) - \mathbf{y}(s)||\, ds + \epsilon|\beta - \alpha| + M|\tau_0 - t_0| \, .$$

Now, if $\epsilon > 0$:

$$||\mathbf{x}_0 - \mathbf{y}_0|| + M|\tau_0 - t_0| + \epsilon|\beta - \alpha| > 0 \, . \tag{14.60}$$

Hence (14.59) is straightforward from the Gronwall Lemma by using $a := ||\mathbf{x}_0 - \mathbf{y}_0|| + M|\tau_0 - t_0| + \epsilon|\beta - \alpha|$, $t_1 := \beta$, $h(t) := ||\mathbf{x}(t) - \mathbf{y}(t)||$, $u(t) = 1$ identically, and $b := K$. If in our hypotheses $\epsilon = 0$, i.e. $||\mathbf{f}(t, \mathbf{x}) - \mathbf{g}(t, \mathbf{x})|| = 0$ for any $(t, \mathbf{x}) \in \Omega$, then $||\mathbf{f}(t, \mathbf{x}) - \mathbf{g}(t, \mathbf{x})|| \leq \epsilon$ for any $\epsilon > 0$. Then we can use the argument above to obtain that for any given t:

$$||\mathbf{x}(t) - \mathbf{y}(t)|| \leq (||\mathbf{x}_0 - \mathbf{y}_0|| + M|t_0 - \tau_0| + \epsilon|\alpha - \beta|)\, e^{K|t-t_0|}$$

for any $\epsilon > 0$. Since the above holds for any t, that value of t satisfies:

$$||\mathbf{x}(t) - \mathbf{y}(t)|| \leq (||\mathbf{x}_0 - \mathbf{y}_0|| + M|t_0 - \tau_0|)\, e^{K|t-t_0|}$$

for any $\epsilon > 0$, so we have the claim for $\epsilon = 0$ as well. $\square$

There are many situations where the theorem applies. Let us consider two in particular.

(a) Consider a unique system of differential equations ($\mathbf{f} = \mathbf{g}$), but distinct initial conditions $\mathbf{x}_0$, $\mathbf{y}_0$ at the same $t_0 = \tau_0$. The theorem then says that for $t \in (\alpha, \beta)$:

$$||\mathbf{x}(t) - \mathbf{y}(t)|| \leq ||\mathbf{x}_0 - \mathbf{y}_0||e^{K|t-t_0|} .$$

(b) Consider the same initial condition, but two distinct ODE systems, one for $\mathbf{f}$ and one for $\mathbf{g}$, with $||\mathbf{f} - \mathbf{g}||_\infty \leq \epsilon$ on Ω. The theorem says that for $t \in (\alpha, \beta)$:

$$||\mathbf{x}(t) - \mathbf{y}(t)|| \leq \epsilon e^{K|t-t_0|} .$$

14.4.2   Regularity of the Dependency on Cauchy Data and Related Issues

An important problem, especially in physical applications, is the regularity of the maximal solutions of an ODE in terms of the initial conditions and other possible parameters. This issue has been long studied, starting with Poincaré.

Theorem 14.37 *Consider a function* $\mathbf{f} : \Lambda \times \Omega \to \mathbb{K}^n$ *where:*

(i) $\Lambda \subset \mathbb{K}^l$ *is a non-empty open set with elements* λ;
(ii) $\Omega \subset \mathbb{R} \times \mathbb{K}^n$ *is a non-empty open set with elements* $(t, \mathbf{x})$;
(iii) $\mathbf{f}$ *is* $C^0(\Lambda \times \Omega; \mathbb{K}^n)$ *and locally Lipschitz in* $(\lambda, \mathbf{x})$.

If, for $\lambda, t_0, \mathbf{x}_0$ *given and* t *varying, the map*

$$(\lambda, t_0, \mathbf{x}_0, t) \mapsto \mathbf{x}_{\lambda,t_0,\mathbf{x}_0}(t) , \quad \text{with } (\lambda, t_0, \mathbf{x}_0) \in \Lambda \times \Omega, \tag{14.61}$$

is the unique maximal solution to the Cauchy problem

$$\begin{cases} \dfrac{d\mathbf{x}}{dt} = \mathbf{f}(\lambda, t, \mathbf{x}(t)) , \\ \mathbf{x}(t_0) = \mathbf{x}_0 , \end{cases} \tag{14.62}$$

where $t \in I_{\lambda,t_0,\mathbf{x}_0} \subset \mathbb{R}$, *then the following facts hold.*

(1) The set

$$\Gamma := \left\{ (t, \lambda, t_0, \mathbf{x}_0) \mid (\lambda, t_0, \mathbf{x}_0) \in \Lambda \times \Omega , \ t \in I_{\lambda,t_0,\mathbf{x}_0} \right\}$$

is open in $\mathbb{R} \times \Lambda \times \Omega$.
(2) The function (14.61)

$$\Gamma \ni (\lambda, t_0, \mathbf{x}_0, t) \mapsto \mathbf{x}_{\lambda,t_0,\mathbf{x}_0}(t)$$

is of class $C^0(\Gamma; \mathbb{K}^n)$, it admits partial derivative in t and the function

$$\Gamma \ni (\lambda, t_0, \mathbf{x}_0, t) \mapsto \frac{\partial \mathbf{x}_{\lambda, t_0, \mathbf{x}_0}(t)}{\partial t}$$

is also $C^0(\Gamma; \mathbb{K}^n)$.

(3) Strengthening (iii), if $\mathbf{f}$ is $C^k(\Lambda \times \Omega)$ with $k \geq 1$, the maps

$$\Gamma \ni (\lambda, t_0, \mathbf{x}_0, t) \mapsto \mathbf{x}_{\lambda, t_0, \mathbf{x}_0}(t) \quad and \quad \Gamma \ni (\lambda, t_0, \mathbf{x}_0, t) \mapsto \frac{\partial \mathbf{x}_{\lambda, t_0, \mathbf{x}_0}(t)}{\partial t}$$

are $C^k(\Gamma; \mathbb{K}^n)$.

Sketch of Proof The parameter λ may be viewed as a variable exactly as $\mathbf{x}$ by adding to the vector field $\mathbf{f}$ zero components in the space Λ, so that λ does not evolve, and considering the unknown curve to be $t \mapsto (\lambda(t), t, \mathbf{x}(t)) \in \Lambda \times \Omega$ instead of $t \mapsto (t, \mathbf{x}(t)) \in \Omega$. Thus we can absorb λ in the variable $\mathbf{x}$ redefining Ω as $\Lambda \times \Omega$. The regularity properties (2) and (3) of:

$$\Gamma \ni (\lambda, t_0, \mathbf{x}_0, t) \mapsto \frac{\partial \mathbf{x}_{\lambda, t_0, \mathbf{x}_0}(t)}{\partial t}$$

are actually immediate consequences of the similar features of $\mathbf{f}$ and

$$\Gamma \ni (\lambda, t_0, \mathbf{x}_0, t) \mapsto \mathbf{x}_{\lambda, t_0, \mathbf{x}_0}(t) \,,$$

since the ODE reads:

$$\frac{\partial \mathbf{x}_{\lambda, t_0, \mathbf{x}_0}(t)}{\partial t} = \mathbf{f}\left(\lambda, t, \mathbf{x}_{\lambda, t_0, \mathbf{x}_0}(t)\right) \,.$$

From what we have just said (1) follows from [Sid14, Theorem 3.4], (2) follows from [Sid14, Theorem 6.2] (which uses the Gronwall Lemma and Theorem 14.36), and (3) is Corollary 6.3 again in [Sid14]. For (1) and (3) the reader may also consult [Tes12, Section 2.3], Theorem 2.8 in particular. □

14.5　Initial Value Problem on Differentiable Manifolds

The Cauchy problem can be studied also on C^k manifolds $M, k > 1$. It must be recast though in the language of vector fields. In the chapter's sequel we will refer to the Differential Geometry notions introduced in Appendix A.

Definition 14.38 If X is a vector field of class (at least) C^0 on the differentiable manifold M, an **integral curve** $\gamma : I \to M$ of X, where $I \subset \mathbb{R}$ is an open interval, is a C^1 curve (at least) such that

$$\dot{\gamma}(t) = X(\gamma(t)), \quad \text{for any } t \in I. \tag{14.63}$$

$\diamond$

As usual $\dot{\gamma}(t)$ is the tangent vector to γ at the point $\gamma(t)$ where the parameter is t (see Sect. A.6.6). Identity (14.63) in local coordinates (U, ϕ) where

$$X(x^1, \ldots, x^n) = \sum_{i=1}^{n} X^i(x^1, \ldots, x^n) \frac{\partial}{\partial x^i}$$

and where the curve γ is parametrised by $x^i = x^i(t)$ for $i = 1, 2, \ldots, n$, reads

$$\frac{dx^i}{dt} = X^i(x^1(t), \ldots, x^n(t)), \quad \text{for } i = 1, 2, \ldots, n. \tag{14.64}$$

This is just an autonomous ODE on $\Omega = \mathbb{R} \times U$.

The case of X depending on t parametrically can be subsumed in the previous situation in a number of ways. The most direct is to consider the *product manifold* (see Sect. A.6.1) $M' := \mathbb{R} \times M$, and define on it the vector field:

$$X' := \frac{\partial}{\partial u} + X,$$

where $u \in \mathbb{R}$ is the natural coordinate on $\mathbb{R}$. The problem of finding the integral curves of X':

$$\dot{\gamma}'(t) = X'(\gamma'(t)), \quad \text{for any } t \in I, \tag{14.65}$$

reduces in local coordinates $u, x^1, \ldots, x^n$ to the system:

$$\begin{cases} \dfrac{du}{dt} = 1, \\ \dfrac{dx^i}{dt} = X^i(u(t), x^1(t), \ldots, x^n(t)) & \text{for } i = 1, 2, \ldots, n. \end{cases}$$

A less rigid way is to use a bundle (Definition B.4) E, with fibre diffeomorphic to M and the axis $\mathbb{R}$ as base. In natural local coordinates on E adapted to the bundle structure, this returns the above system. The choice is typically suggested by the concrete physical problem. In the rest of this section we will restrict to the *autonomous* case, i.e. when the vector field X does not depend explicitly on the variable t.

14.5.1 Cauchy Problem, Global Existence and Uniqueness

Based on what was said earlier, the Cauchy problem for a vector field X that is at least continuous on the C^k manifold M, $k \geq 0$, can be thought of as the problem of finding the integral curves $\gamma : I \to M$ of X, where $I \subset \mathbb{R}$ is an open interval, that satisfy a certain initial condition $\gamma(t_0) = p_0$:

$$\begin{cases} \dot{\gamma}(t) = X(\gamma(t)), \\ \gamma(t_0) = p_0. \end{cases} \tag{14.66}$$

Definition 14.39 Let X be a continuous (at least) vector field on the differentiable manifold M. Consider the Cauchy problem for $(t_0, p_0) \in \mathbb{R} \times M$:

$$\begin{cases} \dot{\gamma}(t) = X(\gamma(t)), \\ \gamma(t_0) = p_0. \end{cases} \tag{14.67}$$

A solution $\gamma : I \to M$ to (14.67), with $I \ni t_0$ open interval, is called a **maximal solution** to (14.67), or equivalently a **maximal integral curve** of X if there is no other solution $\gamma' : I' \to M$, with I' open interval, to (14.67) satisfying the conditions:

(i) $I' \supsetneq I$,
(ii) $\gamma(t) = \gamma'(t)$ if $t \in I$.

A maximal solution $\gamma : I \to M$ is called:

(1) **complete**, if $I = (-\infty, +\infty)$;
(2) **future-complete**, if $I = (a, +\infty)$ for $a \in \mathbb{R} \cup \{-\infty\}$;
(3) **past-complete**, if $I = (-\infty, b)$ for $b \in \mathbb{R} \cup \{+\infty\}$.

The vector field X is called **complete** when all maximal solutions to (14.67) are complete. ◇

In analogy to $\mathbb{K}^n$, the following global Existence and uniqueness theorem holds, where we ask that X is at least of class C^1.

Theorem 14.40 (Global Existence and Uniqueness) *Let X be C^k a vector field, $k \geq 1$, on a differentiable manifold M. Consider the Cauchy problem for $(t_0, p_0) \in \mathbb{R} \times M$:*

$$\begin{cases} \dot{\gamma}(t) = X(\gamma(t)), \\ \gamma(t_0) = p_0. \end{cases} \tag{14.68}$$

Then there exists a unique maximal solution $\gamma : I \to M$.

Furthermore, if $\gamma_J : J \to M$, with $J \ni t_0$ open interval, is another solution to (14.68), then

(1) $J \subset I$,
(2) $\gamma(t) = \gamma_J(t)$ if $t \in J$.

Proof First, note that in any system of local coordinates, the above Cauchy problem reduces to an autonomous problem on an open set of $\mathbb{R} \times \mathbb{R}^n$. As the field is C^1, the local Lipschitz condition holds and then in any local coordinates the local solution exists and is unique. Every point reached by γ belongs in a system di local coordinates by definition of manifold. For this reason Lemma 14.17 still holds, with same proof, if we replace Ω with $\mathbb{R} \times M$. In view of that, the proof is essentially identical to that of Theorem 14.23.   □

Proposition 14.41 *In the hypotheses of Theorem 14.40 let $\gamma : I_{t_0, p_0} \to M$ be the maximal solution to problem (14.68). The following facts hold.*

(1) If $t_1 \in I_{t_0, p_0}$ and $p_1 := \gamma(t_1)$ then $\gamma : I_{t_0, p_0} \to M$ is also the maximal solution to the problem with the same ODE as (14.68) and initial condition $\gamma(t_1) = p_1$.
(2) If $T \in \mathbb{R}$ then $\gamma_1(t) := \gamma(t - T)$ defined on $I_{t_0, p_0} + T$ (with obvious notation) is the maximal solution to (14.68) with initial condition $\gamma_1(t_0 + T) = p_0$.

Proof

(1) The proof is the same as for Proposition 14.24 with the obvious modifications.
(2) It suffices to observe that differentiating in t is like differentiating in $t - T$, since $\frac{d(t-T)}{dt} = 1$, so γ_1 satisfies the ODE in (14.68) because the variable t is not explicitly present in the right-hand side of (14.68). Moreover γ_1 satisfies the initial condition $\gamma_1(t_0 + T) = p_0$. If γ_1 were not the maximal solution for initial datum $\gamma_1(t_0 + T) = p_0$, i.e. if it were the restriction of a solution with larger domain, then the same would happen to γ, which could then not be maximal for (14.68).

□

14.5.2   Completeness of Maximal Solutions

We can also provide results on the completeness of maximal solutions. First we state a preliminary result.

Proposition 14.42 *If X satisfies the assumptions of Theorem 14.40, let $\gamma : I \to M$, with $I = (\alpha, \omega)$, be a maximal solution to (14.68).*

(1) If γ is not future-complete, there is no sequence $\{t_k\}_{k \in \mathbb{N}} \subset I$ such that $t_k \to \omega$ as $k \to +\infty$ and $\lim_{k \to +\infty} \gamma(t_k) \in M$ exists.
(2) If γ is not past-complete, there is no sequence $\{t_k\}_{k \in \mathbb{N}} \subset I$ such that $t_k \to \alpha$ as $k \to +\infty$ and $\lim_{k \to +\infty} \gamma(t_k) \in M$ exists.

Proof The argument is the same as that of Proposition 14.33, by passing to local coordinates. $\qquad\square$

Proposition 14.42 has important consequences on the completeness of maximal solutions in presence of additional conditions. We gather them in a proposition.

Proposition 14.43 *Assume* X *satisfies the assumptions of Theorem 14.40, and* γ : $I \rightarrow M$, *with* $I = (\alpha, \omega)$, *is a maximal solution to (14.68). Then the following hold.*

(1) If the orbit of $\gamma : I \rightarrow M$ *is contained in a compact set* $K \subset M$ *for* $t \geq T \in I$, *then* $\gamma : I \rightarrow M$ *is future-complete.*
(2) If the orbit of $\mathbf{x} : I \rightarrow M$ *is contained in a compact set* $K \subset M$ *for* $t \leq T \in I$, *then* $\mathbf{x} : I \rightarrow M$ *is past-complete.*
(3) If M *is compact, or more weakly* X *has compact support, then any maximal solution* $\mathbf{x} : I \rightarrow M$ *is complete, i.e.* X *is complete.*

Proof

(1) Suppose the solution is not future-complete, so $I = (\alpha, \omega)$ with $\omega \in \mathbb{R}$. Consider a sequence $\{t_k\}_{k\in\mathbb{N}} \subset (T, \omega)$ with $t_k \rightarrow \omega$ as $k \rightarrow +\infty$. For any $p \in M$, let (U, ϕ) be a local chart with $U \ni p$. For any U take an open set $V \ni p$ such that $\overline{V}$ is compact and completely contained in U. The collection of local charts $(V, \phi|_V)$ defines, by construction, an atlas of M. Since K is compact and covered by the above open sets V, we can extract a finite subcover. There will then be a finite set of local charts $\{(U_i, \phi_i)\}_{i=1,...,N}$ and corresponding finitely many compact sets $\overline{V}_i$, such that $\cup_{i=1}^{N} \overline{V}_i \supset K$ and $\overline{V}_i \subset U_i$. As there are infinitely many points $\gamma(t_k)$, at least one compact set, say $\overline{V}_j$, will contain infinitely many of these points. Hence there is a subsequence $\{\gamma(t_{k_r})\}_{r\in\mathbb{N}} \subset \overline{V}_j$. $\overline{V}_j$ can be taken compact in $\mathbb{R}^n$ because $\overline{V}_j \subset U_j$ and U_j is identified with an open set of $\mathbb{R}^n$ by a homeomorphism ϕ_j. By the sequential compactness of $\mathbb{R}^n$, there exists a subsequence of $\{\gamma(t_{k_r})\}_{r\in\mathbb{N}}$, let us call it $\{\gamma(t_{k_{r_s}})\}_{s\in\mathbb{N}}$, converging in the compact set $\overline{V}_j$. But this means, putting $\tau_s := t_{k_{r_s}}$, that $\tau_s \rightarrow \omega$ as $s \rightarrow +\infty$ and the limit of $\gamma(\tau_s)$ exists in M. This is forbidden by Proposition 14.42. Hence γ must be future-complete. The proof of (2) is essentially identical with the obvious modification.
(3) If X has compact support, any maximal integral curve of X with at least one point outside the support of X must be constant, by the Uniqueness Theorem. Then it is complete by construction. The remaining integral curves have, by definition, orbit in a compact set, namely the support of X. By (1) and (2) these solutions are complete too.

$\qquad\square$

Remarks 14.44 We shall prove, for completeness, a useful compactness result that generalises (3) in Proposition 14.34; it requires some background on Riemannian Geometry, the *Hopf-Rinow Theorem* [doC92] in particular.

Theorem 14.45 *Let* (M, g) *be a* C^{∞} *Riemannian manifold whose geodesics are complete (i.e. defined for all real values of the affine parameter). If* X *is a* C^{∞}

vector field on M such that

$$\sup_{p \in M} g_p \left(X(p), X(p) \right) < +\infty \,,$$

then the maximal solutions to problem (14.68) for X are complete.

Proof Take a maximal solution $\gamma : (\alpha, \omega) \to M$ and, in the theorem's hypotheses, assume by contradiction $\omega < +\infty$ (the other case is analogous). Let t_0 be a point in (α, ω) and consider the map $[t_0, \omega) \ni t \mapsto d(x(t_0), x(t)) =: f(t)$, where d is the distance induced by the metric g on M. It is continuous by construction. If it were bounded we would have an absurd, since the Hopf-Rinow Theorem [doC92] (the manifold is geodesically complete) says that closed metric balls of finte radius are compact, and the solution would be confined to a compact set for $t > t_0$, giving $\omega = +\infty$ by Proposition 14.43. As f is continuous and hence bounded on any interval $[t_0, u]$ with $t_0 < u < \omega$, necessarily there exists a sequence $t_n \to \omega$ for which $d(x(t_0), x(t_n)) \to +\infty$. If $s = s(t)$ is the solution's arclength function for some initial point along the curve, as $t_n \to \omega$

$$\left| \frac{ds}{dt} \right|_{u_n} \right| = \frac{|s(t_n) - s(t_0)|}{t_n - t_0} \geq \frac{d(x(t_0), x(t))}{t_n - t_0} \to +\infty \,,$$

where we have used the Intermediate Value Theorem and the u_n belong in $[t_0, t_n]$. This cannot be, since by definition of arclength,

$$\left(\frac{ds}{dt} \right)^2 = g_{\gamma(t)} \left(X(\gamma(t)), X(\gamma(t)) \right)$$

is bounded by assumption. □

Henceforth we shall no longer need Riemannian manifolds. ∎
Working in local charts, Theorem 14.37 extends easily to differential equations on manifolds, giving the following result (for the C^∞ theory the reader may consult [Lee03]).

Theorem 14.46 *Consider a collection of vector fields $\{X_\lambda\}_{\lambda \in \Lambda}$ on a C^k manifold M, $k \geq 1$, where $\Lambda \subset \mathbb{R}^l$ is non-empty and open and the function $X : (\lambda, p) \mapsto X_\lambda(p)$ is jointly C^k. If*

$$(\lambda, p, t) \mapsto \gamma_{\lambda, p}(t) \,, \quad \text{with } (\lambda, p) \in \Lambda \times M, \tag{14.69}$$

is, for given λ, p and varying t, the only maximal solution to:

$$\begin{cases} \dot{\gamma}(t) = X_\lambda(\gamma(t)) \,, \\ \gamma(0) = p, \end{cases} \tag{14.70}$$

whose (maximal) domain is the open interval $I_{\lambda,p} \subset \mathbb{R}$ containing the origin, then the following hold.

(1) The set $\Gamma_X := \left\{ (t, \lambda, p) \mid (\lambda, p) \in \Lambda \times M, \, t \in I_{\lambda,p} \right\}$ is open in the manifold $\mathbb{R} \times \Lambda \times M$.

(2) The functions:

$$\Gamma_X \ni (\lambda, p, t) \mapsto \gamma_{\lambda,p}(t) \quad \text{and} \quad \Gamma_X \ni (\lambda, p, t) \mapsto \frac{\partial \gamma_{\lambda,p}(t)}{\partial t}$$

are $C^k(\Gamma_X; M)$.

Proof All properties can be proved by working in a local chart of $\mathbb{R} \times \Lambda \times M$ around any $(\lambda, p, t) \in \Gamma_X$, so the theorem is an immediate consequence of Theorem 14.37.
□

Remarks 14.47 Theorem 14.46 generalises to the case where X_λ *also depends on* t parametrically, since it suffices to recast the statement on the product $\mathbb{R} \times M$ by adding a component equal to 1 to the vector field on $\mathbb{R}$, in agreement with the discussion of Sect. 14.5.
∎

14.5.3   One-Parameter Groups of Local and Global Diffeomorphisms

Consider a differentiable manifold M (e.g. $\mathbb{R}^n$) and a C^k vector field X, with $k \geq 1$, on it. Consider the maximal solution $I_p \ni t \mapsto \gamma_p(t)$ to:

$$\begin{cases} \dot{\gamma}(t) = X(\gamma(t)), \\ \gamma(0) = p, \end{cases} \tag{14.71}$$

for any $p \in M$, and build the map:

$$\Gamma_X \ni (t, p) \mapsto \gamma_p(t) =: \phi_t^{(X)}(p), \tag{14.72}$$

where $p \in M$ and $t \in I_p$. By Theorem 14.46 the set $\Gamma_X \subset \mathbb{R} \times M$ is open and looks like

$$\Gamma_X = \bigcup_{p \in M} I_p \times \{p\}, \tag{14.73}$$

where every $I_p \subset \mathbb{R}$ is an open interval such that $I_p \ni 0$.
Finally, by Theorem 14.46 and the Uniqueness Theorem, the following proposition holds.

Proposition 14.48 *Consider a differentiable manifold M, a vector field X of class C^k, with $k \geq 1$, and the map $\phi^{(X)} : \Gamma_X \to M$ defined in (14.72). Then:*

(1) $\Gamma_X \ni (t, p) \mapsto \phi_t^{(X)}(p)$ and $\Gamma_X \ni (t, p) \mapsto \frac{\partial \phi_t^{(X)}(p)}{\partial t}$ are $C^k(\Gamma_X)$.

(2) $\phi_0^{(X)}(p) = p$ for any $p \in M$,

(3) If $(p, t'), (p, t + t') \in \Gamma_X$ or if $(p, t + t') \in \Gamma_X$ and $tt' \geq 0$ then:

$$\phi_t^{(X)}(\phi_{t'}^{(X)}(p)) = \phi_{t+t'}^{(X)}(p) \ .$$

(4) If $(p, t), (p, t'), (p, t + t') \in \Gamma_X$ or if $(p, t + t') \in \Gamma_X$ and $tt' \geq 0$ then:

$$\phi_t^{(X)}(\phi_{t'}^{(X)}(p)) = \phi_{t'}^{(X)}(\phi_t^{(X)}(p)) \ .$$

Proof (1) is part of Proposition 14.48, (2) is immediate from Definition (14.72) of $\phi^{(X)}$. The proof of (3) is based on the property of $\phi^{(X)}$ we may call *flow-invariance of the solutions* to (14.71), in the following sense. □

Lemma 14.49 *Let $\gamma : I_p \to M$ be the unique maximal solution to the initial value problem of the C^k field X, $k \geq 1$, with initial conditions $\gamma(0) = p$. If $t \in I_p$, $t' \in \mathbb{R}$ then:*

$$\phi_t^{(X)}(\gamma(t')) = \gamma(t' + t) \quad \text{assuming } t', t + t' \in I_p \ . \tag{14.74}$$

Proof By definition of $\phi^{(X)}$, if $t' \in I_p$ then $\gamma_1(t) := \phi_t^{(X)}(\gamma(t'))$ is certainly defined for t in an open interval containing 0, and satisfies the ODE in problem (14.71) with initial condition $\gamma_1(0) = \gamma(t')$, irrespective of where defined. Similarly, $\gamma_2(t) := \gamma(t + t')$, defined for $t \in I_p - t'$ (i.e. $t + t' \in I_p$), satisfies the same ODE of (14.71).[5] If finally $t' \in I_p$, then $0 \in I_p - t'$ and γ_2 satisfies the same initial condition $\gamma_2(0) = \gamma(t')$. As the X in (14.71) is at least C^1, the Uniqueness Theorem holds and the two solutions coincide, giving (14.74).

Returning to the proof of (3): if $(p, t'), (p, t + t') \in \Gamma_X$ then $t, t + t' \in I_p$ and the claim coincides with (14.74), which the lemma warrants.

Note that if t and t' have the same sign, asking $(p, t + t') \in \Gamma_X$ implies $(p, t'), (p, t) \in \Gamma_X$, because I_p is an open interval containing 0 and $t + t'$, and t, t' always lie between the former two in the case at hand. (4) follows from (3) since $t + t' = t' + t$. □

The following proposition shows how properties (2), (3) and (4) are *locally* always true, and in particular that ϕ_{-t} is *locally* the inverse of ϕ_t (for t near 0 and p near any given $q \in M$).

---

[5] The equation is autonomous.

Proposition 14.50 *Consider the function* $\phi^{(X)}$ *in (14.72) associated with the C^k vector field X, $k \geq 1$, on the manifold M.*
For any $q \in M$ there exist an open interval $J_q := (-\delta, \delta)$ with small enough $\delta > 0$ and a small enough open neighbourhood of q, $U_q \subset M$, such that the following hold.

(1) If $t \in J_q$ then $\phi_t^{(X)}$ maps U_q to an open set $V_{q,t} := \phi_t^{(X)}(U_q)$ C^k-diffeomorphically.

(2) Restricting to the above pair of open neighbourhoods:

$$\phi_{-t}^{(X)}|_{V_{q,t}} = (\phi_t^{(X)}|_{U_q})^{-1} .$$

(3) If $t, t' \in J_q$ and $p \in U_q$ then the three terms below are well defined, and:

$$\phi_t^{(X)} \circ \phi_{t'}^{(X)}(p) = \phi_{t'}^{(X)} \circ \phi_t^{(X)}(p) = \phi_{t+t'}^{(X)}(p) .$$

Proof (3) For a given $q \in M$, as $\Gamma_X \ni (t, p) \mapsto \phi_t^{(X)}(p)$ is jointly continuous on the open set Γ_X, the pre-image of an open neighbourhood V_q of $q = \phi_0^{(X)}(q)$ is an open neighbourhood F of $(0, q)$. F will contain an open set of the form $(-\epsilon, \epsilon) \times U_q$, where U_q is another open neighbourhood of q, if we choose $\epsilon > 0$ and U_q small. Hence, putting $J_q := (-\delta, \delta)$ where $\delta := \epsilon/2$, if $t, t' \in J_q$ then $t, t', t+t' \in (-\epsilon, \epsilon)$ and therefore $\phi_t^{(X)}(p)$, $\phi_{t'}^{(X)}(p)$ $\phi_{t+t'}^{(X)}(p)$ are well defined if $p \in U_q$. In conclusion (iii) and (iv) in Proposition 14.48 hold.

(1) and (2). With the above J_q and U_q, if $t \in J_q$ in particular $\phi_t^{(X)}(p)$, $\phi_{-t}^{(X)}(p)$ and $\phi_{t-t}^{(X)}(p)$ are well defined, and by (3) in particular:

$$\phi_{-t}^{(X)}\left(\phi_t^{(X)}(p)\right) = \phi_0^{(X)}(p) = p \quad \text{if } p \in U_q.$$

This means $\phi_t^{(X)}|_{U_q} : U_q \to \phi_t^{(X)}(U_q) =: V_{q,t}$ is a 1-1- correspondence and $\phi_{-t}^{(X)}|_{V_{q,t}}$ is the inverse. Given local coordinates on M around p, the above identity shows that the Jacobian matrices of the two sides at p satisfy:

$$J_{\phi_t^{(X)}(p)}^{(\phi_{-t}^{(X)})} J_p^{(\phi_t^{(X)})} = I$$

in the obvious notation, and where I is the identity matrix. That implies $J_p^{(\phi_t^{(X)})}$ is non-singular at p and then $\phi_t^{(X)}$ defines a local diffeomorphism around p. In particular $\phi_t^{(X)}|_{U_q}$ is an open map (Proposition A.25). The consequence is that $V_{q,t}$ is open, as claimed.

$\square$

Let us show that the procedure for constructing the local diffeomorphisms $\phi_t^{(X)}$ from a vector field X can be reversed.

Proposition 14.51 *Consider a C^k manifold M, with $k \geq 1$, together with a map $A \ni (t, x) \mapsto \phi_t(x) \in M$, where $A \subset \mathbb{R} \times M$ is open of the form*

$$A = \bigcup_{p \in M} J_p \times \{p\}, \qquad \textit{and every } J_p \subset \mathbb{R} \textit{ is an open interval such that } 0 \in J_p.$$

(14.75)

Suppose ϕ satisfies the following properties:

(i) the vector field $M \ni p \mapsto \left. \frac{\partial \phi_t(p)}{\partial t} \right|_{t=0}$ is well defined and of class $C^k(M)$,
(ii) $\phi_0(p) = p$ for any $p \in M$,
(iii) $\phi_t(\phi_{t'}(p)) = \phi_{t+t'}(p)$ if $(p, t'), (p, t + t') \in A$.

Then there exists a unique C^k vector field X on M such that:

$$\Gamma_X \supset A \quad \textit{and} \quad \phi_t = \phi_t^{(X)}|_A .$$

Finally, ϕ and $\frac{\partial \phi_t}{\partial t}$ are of class C^k on A.   ◇

Proof Looking at A we deduce $\phi_t(x)$ is well defined for any given $x \in M$ and for t in a neighbourhood of 0. Consider the vector field $X(x) := \frac{\partial}{\partial t}|_{t=0}\phi_t(x)$. It is C^k by (i). From A and properties (ii) and (iii), if $\phi_t(x)$ is defined then $\frac{\partial}{\partial t}\phi_t(x) = X(\phi_t(x))$. By the Global Uniqueness Theorem 14.40, $t \mapsto \phi_t(x)$ is a restriction of the maximal solution of $\frac{d}{dt}\gamma(t) = X(\gamma(t))$ with initial condition $\gamma(0) = x$. Therefore $\phi_t(x) = \phi_t^{(X)}(x)$ whenever the left side is defined. That X is unique follows immediately from the fact that if $\phi_t(p) = \phi_t^{(X)}(p)$ for t in a neighbourhood of 0, then $X(x) := \frac{\partial}{\partial t}|_{t=0}\phi_t(x)$. The last claim in the statement is trivial by the known properties of $\phi^{(X)}$.   □

We may at this point give the following general definition.

Definition 14.52 Consider a C^k function $\phi : A \rightarrow M$, with $k \geq 1$, on a differentiable manifold M, where $A \subset \mathbb{R} \times M$ is open of the form (14.75), and suppose ϕ satisfies (i), (ii) and (iii) in Proposition 14.51.

(1) ϕ is called **one-parameter group of local diffeomorphisms** (of class k) on M, or **local flow of X**, if it has maximal domain, i.e.:

$$\Gamma_X = A \quad \textit{and} \quad \phi = \phi^{(X)}$$

(14.76)

for the vector field X on M determined by ϕ (Proposition 14.51).
(2) Under (14.76), X is called the (infinitesimal) **generator** of ϕ.
(3) Under (14.76), ϕ is said to be a **one-parameter group of (global) diffeomorphisms**, or **global flow of X**, if the variable t is defined on the entire $\mathbb{R}$ irrespective of the point p in $\phi_t^{(X)}(p)$.   ◇

Remarks 14.53 If X is a C^k vector field, $k \geq 1$, on M, the function $\phi := \phi^{(X)}$ from Proposition 14.48 obviously satisfies Definition 14.52. Therefore not only is a one-parameter group of local diffeomorphisms expressible as the flow of a vector field, but every vector field (of class C^k) generates a one-parameter group of local diffeomorphisms.   ∎

To justify Definition (3), observe that any map $\phi_t^{(X)} : M \to M$ in a 1-*parameter group of diffeomorphisms* really defines a (global!) diffeomorphism of M.

Proposition 14.54 *Let X a C^k vector field on M, $k \geq 1$. If $\phi^{(X)}$ is global then any map $\phi_t^{(X)}$ has domain M and is a diffeomorphism of M with*

$$\phi_t^{(X)} \circ \phi_{-t}^{(X)} = \phi_{-t}^{(X)} \circ \phi_t^{(X)} = id_M . \tag{14.77}$$

The maps $\{\phi_t^{(X)}\}_{t \in \mathbb{R}}$ form a commutative (i.e. Abelian) group for the usual composition of maps on M.

Proof Identity (14.77) is immediate from (2), (3) and (4) in Proposition 14.48, so $\phi_t^{(X)} : M \to M$ is a C^k bijection with inverse $\phi_{-t}^{(X)}$, which by assumption is C^k as well. The rest is obvious.   □

Proposition 14.55 *The flow of the vector field X on M is global if and only if X is complete.*

Proof If X is complete, the definition of $\phi^{(X)}$ implies $\phi_t^{(X)}$ is defined for any $t \in \mathbb{R}$. Vice versa if $\phi_t^{(X)}$ is defined for any $t \in \mathbb{R}$ then any maximal solution of the ODE of X has $\mathbb{R}$ as domain by definition of $\phi^{(X)}$; i.e., X is complete.   □

Remarks 14.56 $\phi^{(X)}$ is then a 1-parameter group of (global!) diffeomorphisms on M if, in particular:

(1) M is compact;
(2) M is not compact, but the support of X is compact;
(3) M is not compact, X does not have compact support, but every orbit $\{\phi_t(p) \,|\, t \in I_p\}$ is contained in a compact set K_p. The last situation occurs for example in presence of a *first integral* defined on the entire M, whose *leve sets* are compact (see Sect. 14.5.5);
(4) M is Riemannian and Theorem 14.45 holds.

Needless to say there are many more cases.   ∎

To conclude we have the following result (adapted from [War83]); it complements the local content of Proposition 14.50 with respect to t and p, but refers to the maximal domains of the $\phi_t^{(X)}$. In particular, every $\phi_t^{(X)}$, if not the trivial map on the empty set, is *always* a diffeomorphism from its maximal domain onto its image.

Proposition 14.57 *Let M be a differentiable manifold and X a C^k vector field with $k \geq 1$. If I_p is the domain of the maximal integral solution of X with initial condition $p \in M$ at $t = 0$, define:*

$$D_t := \{p \in M \mid I_p \ni t\} \quad \forall t \in \mathbb{R}.$$

The maps $\phi_t^{(X)}$ and the sets D_t (the maps' maximal domains for given t) enjoy the following properties.

(1) D_t is open (possibly $D_t = \varnothing$).

(2) $\bigcup_{t>0} D_t = M$ and $\bigcup_{t<0} D_t = M$.

(3) $\phi_t^{(X)} : D_t \to D_{-t}$ is a C^k diffeomorphism with inverse $\phi_{-t}^{(X)}$ for any given $t \in \mathbb{R}$.

(4) If $s, t \in \mathbb{R}$ and $p \in M$ are such that $\phi_s^{(X)} \circ \phi_t^{(X)}(p)$ is well defined, then $p \in D_{t+s}$ and

$$\phi_s^{(X)} \circ \phi_t^{(X)}(p) = \phi_{s+t}^{(X)}(p).$$

(5) If $s, t \in \mathbb{R}$ and $st \geq 0$ then

$$\phi_s^{(X)} \circ \phi_t^{(X)}(p) = \phi_t^{(X)} \circ \phi_s^{(X)}(p) = \phi_{s+t}^{(X)}(p) \quad \text{if } p \in D_{s+t}.$$

Proof

(1) If D_t is empty then it is open. Consider D_t containing some point. If $p \in D_t$ then $(t, p) \in \Gamma_X$. Since the domain $\Gamma_X \subset \mathbb{R} \times M$ of $\phi^{(X)}$ is open, there is an open neighbourhood of (t, p) of the form $(t - \delta_p, t + \delta_p) \times N_p \subset \Gamma_X$ for $\delta_p > 0$ small enough, and where $N_p \subset M$ is a small enough open neighbourhood of p. By construction every $q \in N_p$ is contained in D_t since $(t - \delta_p, t + \delta_p) \times \{q\} \subset (t - \delta_p, t + \delta_p) \times N_p \subset \Gamma_X$. Therefore D_t is open because all its points are contained in open neighbourhoods in D_t.

(2) The definition of $\phi^{(X)}$ implies that, if $p \in M$, there is an interval $(0, T_p)$ (respectively $(-T_p, 0)$) with $T_p > 0$ where $\phi_t(p)$ is defined. Therefore $p \in D_t$ for some $t > 0$ (respectively $t < 0$). Consequently: $M \subset \bigcup_{t>0} D_t$ (respectively $M \subset \bigcup_{t<0} D_t$). This proves the claim since obviously $\bigcup_{t>0} D_t \subset M$ (respectively $\bigcup_{t<0} D_t \subset M$).

(3) Suppose $\tau > 0$, since $\tau < 0$ has a similar proof and $\tau = 0$ is trivial. Saying $p \in D_\tau$ is equivalent to saying there exists a curve $\gamma = \gamma(t)$ defined on an open interval containing $[0, \tau]$ such that γ is an integral curve of X through $\gamma(0) = p$. Define $q = \gamma(\tau)$. The curve $\gamma'(t) := \gamma(\tau + t)$ is defined on an open interval containing $[-\tau, 0]$; it is an integral curve of X and it satisfies $\gamma'(-\tau) = q$ and $\gamma'(0) = p$. Therefore $q = \gamma(p) \in D_{-\tau}$ if $p \in D_\tau$. Reading everything in terms of $\phi^{(X)}$, we have shown $\phi^{(X)}(D_\tau) \subset D_{-\tau}$ and $\phi_{-\tau}^{(X)}\left(\phi_\tau^{(X)}(p)\right) = p$ for any $p \in D_\tau$. In particular $\phi_\tau^{(X)} : D_\tau \to D_{-\tau}$ is injective. Swapping τ and $-\tau$

we then obtain $\phi^{(X)}(D_{-\tau}) \subset D_\tau$ and $\phi_\tau^{(X)}\left(\phi_{-\tau}^{(X)}(q)\right) = q$ for any $q \in D_{-\tau}$. Therefore the map $\phi_\tau^{(X)} : D_\tau \to D_{-\tau}$ is surjective, since for any $q \in D_{-\tau}$, if $p_q := \phi_{-\tau}^{(X)}(q)$ we have $\phi_\tau^{(X)}(p_q) = q$. In conclusion $\phi_\tau^{(X)} : D_\tau \to D_{-\tau}$ and $\phi_{-\tau}^{(X)} : D_{-\tau} \to D_\tau$ are inverse bijections. As the latter functions are C^k and defined on open sets, the proof of (3) is finished.

(4) By assumption $\phi_t^{(X)}(p)$ exists, so $(t, p) \in \Gamma_X$. Consider the maximal solution $\gamma = \gamma(\tau) =: \phi_\tau^{(X)}(p)$ to problem (14.71) with initial condition $\gamma(0) = p$ and maximal domain I_p. As $t \in I_p$ by hypothesis, $\gamma = \gamma(\tau)$ with $\tau \in I_p$ is the maximal solution to (14.71) with initial condition *at* $\tau = t$ given exactly by $\gamma(t)$, otherwise the Uniqueness Theorem would be violated. By (2) in Proposition 14.41 the function $\gamma_1(\tau) := \gamma(t+\tau)$, where $\tau \in I_{\gamma(t)} = I_p - t$, will then be the maximal solution of the problem with initial datum $\gamma_1(0) = \gamma(t)$. Said solution can be written, by definition, as $\gamma_1(\tau) = \phi_\tau^{(X)}(\gamma(t))$. Since by assumption $\phi_s^{(X)}(\gamma(t)) = \phi_s^{(X)} \circ \phi_t^{(X)}(p)$ exists, we conclude $s \in I_p - t$ and thence $s + t \in I_p$. We have found that $(s + t, p) \in \Gamma_X$, besides $(t, p) \in \Gamma_X$. The claim then holds by (3) in Proposition 14.48.

(5) If $p \in D_{s+t}$ then $(s + t, p) \in \Gamma_X$. If $st \geq 0$ then the claim holds by (3) and (4) in Proposition 14.48.

$\square$

Remarks 14.58 Evidently, if $0 \leq t < s$ or $s < t \leq 0$ then $D_s \subset D_t$. $\square$

14.5.4 Commuting Vector Fields and Their Local Groups

Consider a differentiable manifold M and two C^k vector fields X and Y on M, with $k \geq 2$. Suppose that for some $p \in M$ and for two open intervals containing the origin I, I', we have:

$$(\phi_t^{(X)} \circ \phi_u^{(Y)})(p) = (\phi_u^{(Y)} \circ \phi_t^{(X)})(p) \quad \text{if } (t, u) \in I \times I'.$$

If, in local coordinates around p, we compute the partial derivative in u at $u = 0$ and then the partial derivative in t at $t = 0$, we find:

$$\sum_{k=1}^n \frac{\partial X^i}{\partial x^k} Y^k(p) = \sum_{k=1}^n \left. \frac{\partial^2 (\phi_u^{(Y)})^i}{\partial u \partial x^k} \right|_{u=0} X^k(p) .$$

The Schwarz Theorem ($\phi_u^{(Y)}$ is C^2 in our set-up) ensures we can swap the derivatives on the right:

$$\sum_{k=1}^{n} \frac{\partial X^i}{\partial x^k} Y^k(p) = \sum_{k=1}^{n} \frac{\partial Y^i}{\partial x^k} X^k(p) \, .$$

In our hypotheses, therefore:

$$[X, Y]_p = 0 \, .$$

We have proved that if two local flows commute at one point, then the Lie bracket of the generators (assumed C^2) is zero at that point. We wish to understand the converse: supposing two C^2 vector fields have zero Lie bracket, can we conclude the local 1-parameter groups commute?

Before stating the result notice that if we have flows $\phi^{(X)}$ and $\phi^{(Y)}$ on M generated by C^k vector fields X and Y, $k \geq 1$, the composite

$$A \ni (u, t, q) \mapsto \phi_u^{(Y)} \circ \phi_t^{(X)}(q) \in M$$

is defined on an open domain $A \subset \mathbb{R} \times \mathbb{R} \times M$, as is easy to show using the continuity of $\phi^{(X)}$ and $\phi^{(Y)}$ and that the functions are defined on open sets. By construction $(0, 0, p) \in A$ for any $p \in M$. Hence $\phi_u^{(Y)} \circ \phi_t^{(X)}(q)$ is defined at $(u, t, q) \in I \times I' \times U$, where I and I' are open intervals containing 0 and U is an open neighbourhood of p.

Theorem 14.59 *Let M be a C^k manifold, $k \geq 2$, and X, Y C^k vector fields such that:*

$$[X, Y] = 0$$

on M. Suppose that, for some $p \in M$ and two open intervals containing the origin I, I', the map $(\phi_u^{(Y)} \circ \phi_t^{(X)})(p)$ if $(t, u) \in I \times I'$ is defined. Then also $(\phi_t^{(X)} \circ \phi_u^{(Y)})(p)$ is defined, and:

$$(\phi_u^{(Y)} \circ \phi_t^{(X)})(p) = (\phi_t^{(X)} \circ \phi_u^{(Y)})(p) \, .$$

Proof Consider, for $t \in I$ given, the TM-valued function

$$I' \ni u \mapsto \psi_{t,p}(u) := \frac{\partial}{\partial t} (\phi_u^{(Y)} \circ \phi_t^{(X)})(p) - X\left((\phi_u^{(Y)} \circ \phi_t^{(X)})(p) \right) \in T_{(\phi_u^{(Y)} \circ \phi_t^{(X)})(p)} M \, .$$

We want to show that it is actually constant. In fact, differentiating in u at $u = u_0$ (which we will not indicate henceforth so the keep the notation readable) and using

Schwarz's Theorem, in local coordinates around $(\phi_{u_0}^{(Y)} \circ \phi_t^{(X)})(p)$ we have

$$\frac{d\psi_{t,p}}{du} = \frac{\partial^2}{\partial t \partial u}(\phi_u^{(Y)} \circ \phi_t^{(X)})(p) - \frac{\partial}{\partial u}X\left((\phi_u^{(Y)} \circ \phi_t^{(X)})(p)\right)$$

$$= \frac{\partial}{\partial t}\sum_i Y^i\left(\phi_t^{(X)}(p)\right)\frac{\partial}{\partial x^i} - \sum_{i,k}\frac{\partial X^i}{\partial x^k}Y^k\left(\phi_t^{(X)}(p)\right)\frac{\partial}{\partial x^i}$$

$$= \sum_{i,k}\frac{\partial Y^i}{\partial x^k}X^k\left(\phi_t^{(X)}(p)\right)\frac{\partial}{\partial x^i} - \sum_{i,k}\frac{\partial X^i}{\partial x^k}Y^k\left(\phi_t^{(X)}(p)\right)\frac{\partial}{\partial x^i}$$

$$= \sum_i [X,Y]^i\left(\phi_t^{(X)}(p)\right)\frac{\partial}{\partial x^i} = 0.$$

We conclude

$$\psi_{t,p}(u) = \psi_{t,p}(0) = \frac{\partial}{\partial t}\phi_t^{(X)}(p) - X\left(\phi_t^{(X)}(p)\right) = \frac{d}{dt}\phi_t^{(X)}(p) - X\left(\phi_t^{(X)}(p)\right) = 0.$$

By the definition of $\psi_{t,p}$ we have then shown that

$$\frac{\partial}{\partial t}(\phi_u^{(Y)} \circ \phi_t^{(X)})(p) = X\left((\phi_u^{(Y)} \circ \phi_t^{(X)})(p)\right).$$

This identity says that, at $u \in I'$ given, the function:

$$I \ni t \mapsto \gamma_{u,p}(t) := (\phi_u^{(Y)} \circ \phi_t^{(X)})(p)$$

satisfies the differential equation:

$$\frac{d}{dt}\gamma_{u,p}(t) = X(\gamma_{u,p}(t)) \quad \text{with initial condition} \quad \gamma_{u,p}(0) = \phi_u^{(Y)}(p).$$

The Uniqueness Theorem for maximal solutions of differential equations on manifolds guarantees that:

$$\gamma_{u,p}(t) = \phi_t^{(X)}\left(\phi_u^{(Y)}(p)\right).$$

Comparing with the definition of $\gamma_{u,p}$ we find:

$$(\phi_u^{(Y)} \circ \phi_t^{(X)})(p) = (\phi_t^{(X)} \circ \phi_u^{(Y)})(p)$$

which is what we wanted.   □

14.5.5   First Integrals and Functionally Independent First Integrals

The notion of *first integral* $F : M \to \mathbb{R}$ for a differentiable vector field X on a differentiable manifold M is a trivial extension of the definition given on $\mathbb{K}^n$:

Definition 14.60 (First Integral on M) Let X be a C^k vector field, $k \geq 1$, on a differentiable manifold M. A function $F : M \to \mathbb{R}$ of class C^k is called a first integral of X when F is constant if restricted to the integral curves of X (the constant depends on the integral curve, in general). $\diamond$

The next result follows immediately.

Proposition 14.61 *Let X be a C^k vector field, $k \geq 1$, on a differentiable manifold M. A function $F : M \to \mathbb{R}$ of class C^k is a first integral of X if and only if*

$$X(F) = 0 , \quad \text{everywhere on } M .$$

If F is a first integral of X, then every set:

$$E_{\alpha,\beta} := \{x \in M \mid \alpha \leq F \leq \beta\}$$

*and in particular every **level set** of F:*

$$E_{\alpha,\alpha}$$

*is **invariant** under X, i.e.:*

$$\phi_t^{(X)}(x) \in E_{\alpha,\beta} \quad \text{if } x \in E_{\alpha,\beta} \text{ for any } t \in \mathbb{R} \text{ such that } \phi_t^{(X)}(x) \text{ is defined.}$$

Proof The proof of the first part, in local coordinates, is the same as for Theorem 14.20. The second part is obvious by definition of first integral. $\square$

The following technical proposition is extremely useful in various constructions.

Proposition 14.62 *Let X a C^k vector field, $k \geq 1$, on a (C^k) manifold M. If $p \in M$ is such that $X(p) \neq 0$, there exists a local chart (U, ϕ), with $p \in U$ and $\phi : U \ni q \mapsto (y^1(q), \ldots, y^n(q))$, on which every integral curve of X in coordinates is of type: $I \ni t \mapsto (y^1 + t, y^2, \ldots, y^n)$ for a corresponding open interval $I \subset \mathbb{R}$. Hence:*

$$X|_U = \frac{\partial}{\partial y^1} .$$

Proof Consider a local chart around p with coordinates $(x^1, \ldots, x^n) \in V \subset \mathbb{R}^n$ so that p corresponds to the origin of $\mathbb{R}^n$. As $X(p) \neq 0$, at least one component of

$$X(p) = \sum_{i=1}^{n} X^i(p) \left. \frac{\partial}{\partial x^i} \right|_p$$

is non-zero. Without loss of generality, and possibly changing the coordinates' names, we assume $X^1(p) \neq 0$. Consider the integral curves of X expressed, in said coordinates on a small enough neighbourhood of p, by: $x^i = x^i(t, x_0^1, \ldots, x_0^n)$ where $(x_0^1, \ldots, x_0^n)$ are the coordinates of the initial point of the curve for $t = 0$, and $t \in (-\delta, \delta)$ with $\delta > 0$ small enough. This function is well defined and of class C^1 by Theorem 14.46. Consider at last the transformation defined around the origin of $\mathbb{R}^n$:

$$(y^1, y^2, \ldots y^n) \mapsto x^i(y^1, 0, y^2, \ldots y^n), \quad \text{for } i = 1, \ldots, n.$$

Keeping in account that, by construction:

$$x^i(t = 0, x_0^1, \ldots, x_0^n) = x_0^i, \quad \text{for } i = 1, \ldots, n,$$

while

$$\left. \frac{\partial x^i(t, x_0^1, \ldots, x_0^n)}{\partial t} \right|_{t=0} = X^i(p), \quad \text{for } i = 1, \ldots, n,$$

we conclude that the transformation's Jacobian matrix at $(t, y^2, \ldots y^n) = (0, 0, \ldots, 0)$ is built as follows:

(i) the first row is $(X^1(p), 0, \ldots, 0)$,

(ii) the first column is $(X^1(p), X^2(p), \ldots, X^n(p))^t$,

(iii) the remaining $(n-1) \times (n-1)$ block, obtained suppressing the first row and column, is the identity matrix I. Expanding along the first row we see that the determinant equals $X^1(p) \neq 0$. But then, by the Implicit Function Theorem [KrPa03], the transformation

$$(y^1, y^2, \ldots y^n) \mapsto x^i(y^1, 0, y^2, \ldots y^n), \quad \text{for } i = 1, \ldots, n.$$

is a local diffeomorphism of class C^k around the origin of $\mathbb{R}^n$. All in all, on a small enough open neighbourhood U of p, the functions $y^i = y^i(x^1, \ldots, x^n)$ define admissible local coordinates on M around p. In this coordinate system, by construction, the integral curves of X look like $I \ni t \mapsto (y^1 + t, y^2, \ldots, y^n)$. $\square$

The above proposition has an immediate corollary on the local existence of $(\dim M) - 1$ functionally independent first integrals for any non-zero vector field in $C^k(M)$.

Theorem 14.63 (Local Existence of $n-1$ Independent First Integrals) *Let X be a C^k vector field, $k \geq 1$, on an n-dimensional (C^k) manifold M. If $p \in M$ is such that $X(p) \neq 0$, there exist a local chart (U, ϕ) with $p \in U$ and $n-1$ functions $F_j : U \to \mathbb{R}$ of class C^k such that:*

(1) the F_j are first integrals of X,
*(2) the F_j are **functionally independent**, i.e. their Jacobian matrix in the local coordinates of U has maximal rank $n-1$.*

Proof Using the coordinate system (U, ϕ) of Proposition 14.62, the $n-1$ functions $F_j : U \ni q \mapsto y^j(q)$ clearly satisfy the two conditions. □

Remarks 14.64

(1) Asking functional independence has nothing to do with the choice of coordinates, and may be expressed saying, for instance, that for any $q \in U$ the $n-1$ vectors of $dF_k(q)$ are linearly independent. We leave the proof to the reader.
(2) The result on the existence of the $n-1$ first integrals is local, and strongly dependent on X not vanishing. The set of points $p \in M$ where $X(p) = 0$ is invariant under the local flow generated by X. There exist isolated points p in that set to which several integral curves of X converge at the same time, or from which they diverge; this appears to violate the Uniqueness Theorem for integral curves of X. For instance, this happens if p is the origin of $M := \mathbb{R}^2$ and $X(x, y) = x\mathbf{e}_x + y\mathbf{e}_y$. That said, none of these curves reaches p, which would happen "when the parameter becomes infinite". ∎

Chapter 15
Complement: The Physical Principles at the Foundations of Special Relativity

In this supplementary chapter we shall construct the spacetime of the theory of Special Relativity in order to justify the mathematical structure that was introduced in axiomatic form in Chap. 10.

15.1 The Classical Perspective'S Crisis

At the end of the nineteenth century, when the equations of Electromagnetism finally reached their final form as the celebrated Maxwell equations, it become clear that the Galilean invariance, as we presented it earlier, could not extend to Electromagnetism.

Maxwell's equations prescribe that in absence of sources, the electromagnetic field in vacuum propagates as *electromagnetic waves* under *D'Alembert's equation*

$$-\frac{1}{c^2}\frac{\partial^2 \Psi}{\partial t^2} + \sum_{i=1}^{3}\frac{\partial^2 \Psi}{\partial (x^i)^2} = 0 \,.$$

In our case Ψ is any component of the electric or magnetic field while t, x^1, x^2, x^3 are the orthonormal Cartesian coordinates of a reference frame. In particular it was understood that light is an electromagnetic wave—and this was one of the triumphs of 1800s Physics, whereby Optics and Electromagnetism could be merged. From the theory of the D'Alembert equation it follows that the parameter c, of the dimensions of a velocity, must be interpreted as the speed of the electromagnetic waves in vacuum irrespective of the direction of motion of the frame under exam: the *speed of light*. Maxwell's theory is single-handedly responsible for the value of c, roughly 300.000 km/s in terms of electrostatic quantities that may be measured independently of the above wave-like propagation.

V. Moretti, *Analytical Mechanics*, La Matematica per il 3+2 150,
https://doi.org/10.1007/978-3-031-27612-5_15

The clash with Galilean invariance is about the following. It is easy to see, by direct substitution and assuming the physical fields are vector (or scalar) fields in absolute space, that the equation's above expression is *not* invariant under a change of frame when the coordinate transformation is an element of the Galilean group. Even Maxwell's equations, which give rise to the wave equation, are not invariant under Galilean transformations.

The simplest proposal that was put forward to interpret that state of affairs is that the wave equation was *only* true in a particular frame, called *aether's* frame. From a physical perspective such a reference frame should be at rest with the medium, that is aether, in which electromagnetic waves propagated. Apart from whether the aether's frame was inertial or not, it still seemed that electromagnetism, in contrast to mechanics, was privileging one special reference frame.

This has a repercussion on the physics of material bodies, which until that moment had been a branch of Classical Mechanics, because electromagnetic waves interact with the bodies that carry charges or electric currents, and apply on them certain forces. Note however that the above does not stand in contrast with the spacetime description of Classical Mechanics, which is independent of the assumption of Galilean invariance. It is for this reason that the spacetime description was kept.

Yet the experimental evidence contradicting the spacetime description's foundations would soon follow. Within such formulation, given that a flash of light should be thought of as an electromagnetic wave with speed c in some particular frame, it immediately follows that this speed should change when measured in another frame. According to this line of thought it seems highly improbable that the Earth is constantly at rest in the aether's frame when circling around the Sun. Therefore we should expect to be able to measure a different speed of light on Earth at each instant. Furthermore, if our planet was not at rest in the aether, the speed of light should not display an isotropic behaviour as prescribed by D'Alembert's equation: it should instead depend on the difference between the light's direction and the direction of the Earth's motion in the aether's frame. By measuring the speed of light on Earth along suitable perpendicular paths, one should obtain distinct values.

Two famous experiments of that sort were indeed conducted in 1887 by A.A. Michelson and E. Morley. On a horizontal plane oriented arbitrarily around a vertical axis, a source S placed at the centre P of the plane produced a beam of light, which was then split in two by a half-silvered mirror. The two beams travelled the same straight distance but in perpendicular directions until they hit two mirrors that reflected them back towards P. At P they were then deflected to S' where an interferometer measured the relative delay of the two beams upon reaching the point were they were recombined. The experiments were slightly modified to account for all possibilities: they were repeated at different times of year and orienting the device along all possible directions, always giving a negative answer. There was never a difference between the separate beams' travel time, and therefore their speed. This *experimental* evidence directly contradicts the classical formulation if we exclude the highly unlikely situation in which the Earth is constantly at rest in the aether.

From 1887 until 1905, when A. Einstein published the paper entitled "On the Electrodynamics of Moving Bodies"[1] that provided the foundations of the theory of Special Relativity, many conjectures were made to explain the negative outcome of Michelson and Morley and to allow Mechanics to coexist with Electromagnetism. Some proposals were based on the aether's partial drag (H. Fizeau) caused by the Earth and still maintained the entire body of Classical Mechanics. Other, more radical theories questioned certain aspects of Classical Mechanics. The latter had been backed by eminent scholars, among whom Lorentz and Poincaré.[2] We shall not deal with all these developments. Instead, we will only consider the theory of Special Relativity, which not only is able to explain the Michelson-Morley experiment (that Einstein apparently was not aware of back in 1905), but has been confirmed so convincingly, so many times, that it is considered most definitely closest to the physical reality than the classical description (which in turn is an approximation of the former in regimes where speeds are "small' compared to c). Suffices to say that modern technology accounts for the effects predicted by Special Relativity. It is using Special Relativity that we design CERN's accelerators, where particles can no longer be treated as classical due to their very high relative speeds (apart from their quantum nature).

15.2   Spacetime and Reference Frames

What allowed (Einstein) to emerge from the classical approach's crisis is the point of view based on the notions of *event* and *spacetime*, which we introduced for Classical Mechanics in Chap. 2.

As we did when formulating Classical Mechanics, we shall assume from now on that spacetime is a smooth manifold[3] of dimension 4 that we will call $\mathbb{M}^4$. With this approach reference frames can be thought of, in general, as physical processes for decomposing spacetime into space plus time, in other words procedures for prescribing the position and time of occurrence of any spacetime event. So we will assume, rather generally, that a *reference frame $\mathscr{R}$ on spacetime $\mathbb{M}^4$ is given by a real three-dimensional Euclidean space $E_{\mathscr{R}} \equiv \mathbb{E}^3$, called rest space of $\mathscr{R}$, and a diffeomorphism*

$$\Phi_{\mathscr{R}} : \mathbb{M}^4 \ni x \to (t_{\mathscr{R}}(x), P_{\mathscr{R}}(x)) \in \mathbb{R} \times E_{\mathscr{R}} ,$$

---

[1] A. Einstein, Ann. d. Physik, 17, (1905).

[2] Poincaré presented a very similar theory to Einstein's relativity, but with subtle philosophical differences based on a *conventionalist* viewpoint. The debate on how different Poincaré's theory is from Einstein's theory is still ongoing.

[3] In the rest of the chapter it would suffice to assume C^2, but eventually we would find a C^∞ structure, actually real analytic.

where $\mathbb{M}^4 \ni x \mapsto t_{\mathscr{R}}(x) \in \mathbb{R}$ *is the time coordinate of* $\mathscr{R}$, *defined up to an additive constant.*

The rest space of a frame $E_{\mathscr{R}}$ then corresponds to the three-dimensional Euclidean physical space at rest with a lab in which the experiments are conducted. Every frame has a temporal axis $\mathbb{R}$ (on which for simplicity we have fixed an arbitrary origin and a unit of measure, even though we could as well view it as a Euclidean space $\mathbb{E}^1$ with no preferred origin).

Physically, we must suppose there is a collection of *ideal rigid rulers* and a collection of *ideal clocks* that are available in every frame. As we saw when discussing Classical Physics in the first two chapters, the rulers' ideality means that any two such have the same length when at rest in the same position in an arbitrary inertial frame, and this is also true if their histories (accelerations included) are different once they return to relative rest in some inertial frame (perhaps different from the initial one). The same criterion holds for ideal clocks: two clocks strike at the same instant when at rest in the same position in an inertial frame, and the same happens after different histories once they return to relative rest in an inertial frame (possibly different from the first). Note that when they meed a second time the clocks may not be synchronised, even if they were initially. Ideal rulers are meant to define the distance employed in every Euclidean space $E_{\mathscr{R}}$, while ideal clocks define the measure of the intervals of time $t_{\mathscr{R}}$ in any frame.

Remarks 15.1 Do ideal clocks exist at all in nature? They do indeed. Atoms themselves are tiny clocks, in the sense that, roughly speaking, they emit radiations at given frequencies, whose period can be used to measure time. These atomic clocks have proved to be the closest approximation of an ideal clock that nature has to offer.   ∎

Classical spacetime and, we will see, relativistic spacetime, admit a metric structure, corresponding to the existence of specific measuring instruments, on top of the topological and differentiable structures. The introduction of the aforementioned reference frames implies that there exist two supposedly distinct metric structures, coming from measuring "time intervals" and measuring "space intervals" between event pairs with respect to an (arbitrary) frame.

In Classical Mechanics these measurements of space and of time between two events are, as we saw in the first chapters, *absolute*: they do not depend on the *reference frame*. Indeed, we know for instance that the time lapse between two events, such as the creation and decay of a particle, does not depend on the frame in which we measure it. Similarly, the spatial distance of two simultaneous events does not depend on the frame.

Let us review a few metric features of the spacetime of Classical Physics, which were addressed in the first chapters.

(a) As for the time coordinate of the various frames, Classical Mechanics makes a requirement based on experimental evidence. Up to additive constants (reflecting the fact we might choose a different time origin in each frame), $t_{\mathscr{R}} = t_{\mathscr{R}'}$ for any pair of frames $\mathscr{R}$ and $\mathscr{R}'$. Equivalently, there exists an *absolute time* function (see Chap. 2) $T : \mathbb{M}^4 \to \mathbb{R}$, defined up to additive constants, that is identified with the time coordinate of every frame.

(b) In Classical Mechanics the rest spaces of the various frames are all identified as affine metric spaces (see Chap. 2) with the spacetime hypersurfaces

$$\Sigma_t = \{p \in \mathbb{M}^4 \mid T(p) = t\}\,.$$

These are three-dimensional Euclidean spaces. Every Σ_t is the *absolute space at time t*. Hence all distances and angles between events p, $p' \in \Sigma_t$ measured in different frames coincide with the absolute measurements with respect to the metric structure of Σ_t.

As we shall see shortly in physical terms (the mathematics was discussed in Chap. 10), the revolution brought on by Relativity has shown that the metric structures of space and time are actually *relative to the reference frame*, but at the same time they partake in an *absolute* spacetime metric structure that has specific symmetry properties governed by the *Poincaré group*. This is true at least until we neglect the relativistic description of the gravitational interaction.

The fallout of the novel relativistic point of view, in which spacetime acquires a geometry, has been incredibly fertile in Physics, and has impacted in a fundamental way the entire development of twentieth century Physics. Relativity Theory has built, jointly with Quantum Mechanics, the *language itself* and the *paradigm* of an entire century's worth of research in Theoretical Physics.

Much has been speculated regarding a possible discontinuous structure of spacetime, or it not being a 4-dimensional manifold, at *Planck's scales*, where some sort of Quantum Theory should manifest itself. There exist several proposals in this respect: *String Theory* in its different declinations, *Loop Quantum Gravity*, and other approaches based on *Noncommutative Geometry*. It is then important to recall that recent observations of so-called γ-bursts made with the Fermi Gamma-ray Space Telescope have pushed the threshold for the existence of quantum gravity phenomena (such as Lorentz symmetry violations) well below the Planck scale.[4]

15.2.1 The Synchronisation Problem

The concept of temporal coordinate $t_{\mathscr{R}}$ of a given frame $\mathscr{R}$ given above, in absence of the classical axiom of absolute time, needs further clarifications. As mentioned, the physical procedure to define a frame's time coordinate is to place an ideal clock at rest at every point of the frame's rest space and then *synchronise at a distance* all the clocks. In other words it means choosing, by some criterion, a time origin for

---

[4] A. A. Abdo ed al., *A limit on the variation of the speed of light arising from quantum gravity effects*, Nature 462, 331–334 (19 November 2009).

each clock by pairing them up. The time struck by a clock when it finds itself at a certain event defines the instant of occurrence of that event in the frame.

The physical demands for synchronisation are:

(a) it must *last* once it has been performed on a pair of clocks;
(b) it must be a *transitive* property (if A is synchronised with B and B with C, then A is synchronised with C), *symmetric* (if A is in sync with B then B is in sync with A), and *reflexive* (any clock A is synchronised with itself).

A priori, we can concoct procedures to synchonise two clocks at rest at distinct points P and Q in $E_{\mathscr{R}}$. Let us examine three possibilities.

(1) take the clock at Q, move it to P, synchonise with the clock at P and then return it to Q;
(2) take a third clock $\mathscr{O}$ in non-accelerated motion in $\mathscr{R}$ along the axis of the segment PQ. When $\mathscr{O}$ passes through P we synchonise it with the clock at P. Then we synchonise the clock at Q with $\mathscr{O}$ when $\mathscr{O}$ reaches Q;
(3) we send towards Q a signal with given speed in $\mathscr{R}$ from the clock at P. Then we synchonise the clock at Q with the one at P keeping into account the distance of P to Q and the signal's travel time to reach Q.

Clearly, in Classical Physics, assuming the existence of absolute time as measured by ideal clocks forces these synchronisation procedures to be all equivalent, because all clocks are measuring (up to additive constants) absolute time. However, we *do not* wish to go back to the classical picture because we know does not reflect the facts. Hence we cannot a priori assume that the three synchronisation procedures (or others) really satisfy (a) and (b). And even if they do, we cannot be sure that they produce the same synchronisation relation. In other words we cannot suppose that clocks synchronised with one procedure are synchronised with another.

Let us expand a bit more on the above processes.

In (1), assuming $\mathscr{R}$ is inertial, there could be the following problem. By moving the clock from Q to P with an initial acceleration and final deceleration, we end up altering the clock's functioning due to the internal forces that are generated.[5] We may suppose, as is the case with "well-built" clocks, that the issues occur during the acceleration (and deceleration) but the clocks continue working in the same way (this makes sense by comparing clocks) after some post-acceleration rebound time. Alas, this fact raises another problem regarding the synchronisation, which involves the period following the acceleration. Without a precise model of how clocks work, this procedure sounds problematic, since it depends on the type of clock. Think of the extreme case of an elementary clock given by an isochronous pendulum in a

---

[5] We are speaking of forces and accelerations in the classical framework, which may seem incoherent with the general discussion. But note that from the experimental viewpoint Classical Physics works exceptionally well when the speeds at stake are small compared to the speed of light. If we displace the clocks using tiny accelerations and velocities, we may retain the classical scheme. Such an approximation is actually a very useful guiding principle that Relativistic Physics must obey: all relativistic laws must reduce to the classical ones when the velocities are small.

gravitational field: the period of small oscillations varies if the support accelerates. At any rate we know that with well-built clock in regimes called, a posteriori, *non-relativistic*, no major issue occurs, so this method might work well for a certain type of clock. It would still be necessary to verify experimentally that (a) and (b) hold.

Case (2), when using a clock in inertial motion, avoids the problems caused by the accelerations, but assumes that the speed differences of clocks synchronised in the same position have no physical relevance, and this too is not obvious. This criterion actually allows to synchronise clocks in relative motion, and it can be proved, under further mathematical hypotheses, that (2) leads directly to the classical formulation of spacetime, provided *(a) and (b) hold.*

Method (3) requires more information in order to be deployed. To know the signal's speed we must measure it, which in turn means we would prior have to synchronise two clocks. Therefore the value of the signal's speed must be known to us by means other than the synchronisation. This speed may further depend on the frame's relative motion against the medium in which it propagates (which may generate an anisotropic velocity).

These three procedures (but we could concoct others) can be employed to synchronise clocks as long as (a) and (b) are experimentally validated.

15.3   The Fundamental Physical Postulates of Special Relativity

The theory of Special Relativity is founded on two or three (sometimes four) physical postulates, depending on the point of view. We shall adopt three postulates and list them below together with commentary.

15.3.1   Constancy of the Speed of Light

Einstein chose (3) as synchronisation criterion to formulate the theory of Special Relativity. The speed in question is that of light in vacuum. Its value can be measured beforehand along a loop by deviating light with mirrors. This procedure *does not necessitate two previously synchronised clocks*, because one is enough as long as we know the distance travelled by light along the circuit. This choice is made by a corresponding principle.

RS1. Constancy of the Speed of Light *There exist reference frames in which, by a corresponding process of synchronisation at a distance, the speed of light in vacuum assumes the same value, independently of the direction of propagation and of the frame.*

Remarks 15.2

(1) Part of the physical content of the postulate is precisely the requirement that synchronisation (3), when using light as a physical means to synchronise, satisfies conditions (a) and (b). Einstein's postulate also implies, though, that the speed of light along *closed paths* at rest with the postulate's frames is constant and *independent of the frame*. This can be tested experimentally, irrespectively of the synchronisation procedure, because it needs a unique clock. The invariance has been verified over and over through experiments not dissimilar to those of Michelson-Morley,[6] even for non-inertial systems (such as the Earth). The value of the speed of light to be used in Einstein's synchronisation is therefore provided irrespective of the procedure itself, as it should, and postulate **RS1** implicitly tells us how to assess it.

(2) As regards the experimental validity of conditions (a) and (b) we can be more precise. The experiments say that the notion of synchronisation in **RS1** does not change in time once it has been imposed. Moreover, the reflexive property is trivially true, *whereas symmetry and transitivity follow from the experimental evidence that the speed of light is constant along closed paths* in inertial frames.

(3) It is important to observe that there may a priori be, and indeed there are, synchronisation procedures other than the one of **RS1** that are physically equally reasonable: they obey (a) and (b), and imply the invariance of the speed of light along loops, which as we said is an experimentally proven fact. These notions lead to an *alternative* description to that of Special Relativity.[7] It could be that one such alternate procedure is completely equivalent to Einstein's one: two clocks at rest and in sync according to Einstein are so for some other method possibly realised in a different way. One example is the procedure (1) above, provided we perform it extremely slowly (and supposing we clarify what "extremely slowly" means without using synchronisation). Or it could be that a different synchronisation is physically admissible (it abides by (a) and (b) and forces constant speed of light along loops), but is not equivalent to Einstein's: two clocks at rest and in sync for Einstein are not in sync for another physically admissible procedure. In any case, once the theory has been set up with Einstein's synchronisation, the other procedures, whether equivalent or not, must be modelled within the theory, i.e. on spacetime, using the geometrical language of the case.

(4) In formulating **RS1** it is understood that a unit of time and a unit of space have been chosen for all frames. The units of measure will still respect **RS1** if they are multiplied by the same factor, possibly different from frame to frame.   ∎

---

[6] See for example R.S. Shankland et al., *Rev. Mod. Phys.* 27 no.2, 167–178 (1955).

[7] S. Liberati, S. Sonego and M. Visser, *Faster-than-c Signals, Special Relativity and Causality*, Ann. Phys. 298, 2002.

15.3.2   Principle of Inertia

The next principle concerns inertial frames, which are defined exactly as in Chap. 3. We will assume that the notion of isolated point particle is *absolute*, i.e. frame-independent: if a particle is isolated in a frame it must be isolated in every other frame.

RS2. Principle of Inertia *The class of reference frames in **RS1** coincides with the class of **inertial frames** defined by requesting that in each such, all isolated point particles move at constant speed.*

15.3.3   Principle of Relativity

The last of Einstein's principles extends Galilean relativity to all of Physics, electromagnetic phenomena included.

RS3. Principle of relativity. *The laws of Physics have the same form in any inertial frame.*

Remarks 15.3 The principle implies that it is never possible to select a preferred reference frame in the class of inertial frames using experiments, since we can replicate the result of an experiment obtained is some frame in any other frame if we prepare the apparatus in the same way.

15.4   From the Postulates of Special Relativity to the Poincaré Group

What we want to do in this section is find the transformation law for the orthonormal Cartesian coordinates of two inertial frames. For that we will use principles **RS1** and **RS2**, plus other mathematical or physical assumptions to be introduced where needed. We shall not use **RS3**, whose role is important in successive developments of the theory, as we have implicitly done in Chap. 10. There, we saw that physical laws are expressed in an intrinsic geometric form, or as physicists would say, in *covariant form*. The latter is one among several mathematical ways to assert the validity of **RS3**.

Remarks 15.4 From now on the universal value of the speed of light in vacuum will be indicated by c.

For starters let us define a family of coordinate systems that will be useful in the sequel. Consider a frame $\mathscr{R}$ in $\mathbb{M}^4$. Fix an orthonormal coordinate system

$$\psi_{\mathscr{R}} : E_{\mathscr{R}} \ni P \mapsto (x^1(P), x^2(P), x^3(P)) \in \mathbb{R}^3$$

on the rest space $E_{\mathscr{R}}$ of $\mathscr{R}$. The global system induced on M^4

$$\Psi_{\mathscr{R}} : \mathbb{M}^4 \ni x \mapsto (ct_{\mathscr{R}}(x), \psi_{\mathscr{R}}(P_{\mathscr{R}}(x))) =: (x^0(x), x^1(x), x^2(x), x^3(x)) \in \mathbb{R}^4$$

is called **system of Minkowski coordinates on** $\mathscr{R}$. It is the same name used in Chap. 10, which is not a problem since eventually these coordinate systems will coincide with the ones defined previously.

Take an inertial frame $\mathscr{R}$. In Minkowski coordinates associated with $\mathscr{R}$, the evolution of a point particle, called its **history** or **world line**, will be described in these coordinates by 4 differentiable maps (C^2 at least)

$$x^\alpha = x^\alpha(u), \quad \alpha = 0, 1, 2, 3, \tag{15.1}$$

where $u \in (a, b)$ is a parameter satisfying $dx^0/du \neq 0$ for any $u \in (a, b)$. Henceforth Greek indices will take any of the values $0, 1, 2, 3$, whereas Roman letters may be $1, 2, 3$. The point particle's velocity in $\mathscr{R}$ will have components

$$v^a = \frac{dx^a}{dt} = c \frac{\frac{dx^a}{du}}{\frac{dx^0}{du}}. \tag{15.2}$$

The definition is independent of the choice of parameter u if the curve is reparametrised by a new u' such that $du'/du > 0$. The velocity's squared modulus in $\mathscr{R}$ equals

$$v^2 = c^2 \sum_{a=1}^{3} \frac{(U^a)^2}{(U^0)^2} \quad \text{where } U^\alpha := \frac{dx^\alpha}{du}. \tag{15.3}$$

The theory should reduce to the classical setup for small velocities, namely when all velocities are admissible. Therefore we will assume the set $\mathcal{C}_{x,\mathscr{R}}$ of tangent vectors at event x to the world lines of point particles passing through x contains the vectors of $T_x \mathbb{M}^4$ associated with motions of sufficiently small velocities in $\mathscr{R}$ and arbitrary direction. Because of (15.3), $\mathcal{C}_{x,\mathscr{R}}$ will contain an *open cone* with axis $U^\alpha = 0$. In other words, for some constant $v_0 > 0$ having the dimensions of a velocity,

$$\left\{ U \in T_x \mathbb{M}^4 \,\middle|\, 0 < \sum_{i=1}^{3} (U^i)^2 < \frac{v_0^2}{c^2} (U^0)^2 \right\} \subset \mathcal{C}_{x,\mathscr{R}} \quad \text{where } U^\alpha := \frac{dx^\alpha}{du}. \tag{15.4}$$

At the same time, all motions happening at the speed of light and in any direction should be admissible. Hence we want all velocities up to the speed of light to be admissible, so that

$$\left\{ U \in T_x \mathbb{M}^4 \,\middle|\, 0 < \sum_{i=1}^{3} (U^i)^2 \le (U^0)^2 \right\} \subset \mathcal{C}_{x,\mathcal{R}} \quad \text{where } U^\alpha := \frac{dx^\alpha}{du}.$$

$$(15.5)$$

15.4.1   RS1 and RS2 Recast in Minkowski Coordinates

Next we wish to present the postulates of Relativity in mathematical form and use the language of Minkowski coordinates, the aim being of finding the most general transformation law between the Minkowski coordinates of two arbitrary inertial frames. We first focus on a point particle evolving with constant velocity. That is because the Principle of inertia forces isolated point particles to move with constant velocity in any inertial frame. Then we will discuss the important case of light particles.

To begin with let us give a general mathematical characterisation of uniform motion in terms of the first and second derivatives of the components of the curve describing the particle's motion in spacetime with respect to an inertial frame. Recall that $dx^0/du \ne 0$ holds for world lines (15.1) written in Minkowski coordinates. Immediately, we will have uniform linear motion in $\mathcal{R}$ if and only if

$$\frac{\frac{dx^\alpha}{du}}{\frac{dx^0}{du}} = \text{constant}$$

for any $\alpha = 1, 2, 3$. This is equivalent to the derivatives in u of the above quotients being zero on (a, b). Let us compute the derivatives:

$$\frac{d}{du}\left(\frac{\frac{dx^\alpha}{du}}{\frac{dx^0}{du}} \right) = \left(\frac{dx^0}{du} \right)^{-2} \left(\frac{d^2x^\alpha}{du^2} \frac{dx^0}{du} - \frac{d^2x^0}{du^2} \frac{dx^\alpha}{du} \right). \quad (15.6)$$

Remember that $dx^0/du \ne 0$ by hypothesis, and moreover if the motion is linear and uniform in $u \in (a, b)$, then $dx^j/du = c^j dx^0/du$ for certain constants c^j. If $c^j \ne 0$ for some $j = 1, 2, 3$ then $dx^0/du = (c^j)^{-1} dx^j/du$, and since the motion is linear and uniform in $u \in (a, b)$, i.e. the left-hand side of (15.6) vanishes, we have:

$$\frac{d^2x^\alpha}{du^2} \frac{dx^j}{du} - \frac{d^2x^j}{du^2} \frac{dx^\alpha}{du} = 0 \quad \text{for all } u \in (a, b).$$

If the motion is uniform and linear and $c^j = 0$ for some $j = 1, 2, 3$, then $dx^j/du = 0$ on (a, b) because $dx^j/du = c^j dx^0/du$. If so, the above relation is once again trivial. Hence for uniform linear motions for $u \in (a, b)$,

$$\frac{d^2x^\alpha}{du^2} \frac{dx^\beta}{du} - \frac{d^2x^\beta}{du^2} \frac{dx^\alpha}{du} = 0 \quad \text{for all } u \in (a, b) \text{ and } \alpha, \beta = 0, 1, 2, 3.$$

$$(15.7)$$

Conversely, if the above relationships hold, putting $\beta = 0$ and dividing by $(dx^0/du)^2 \neq 0$ gives

$$\frac{d}{du} \left(\frac{\frac{dx^\alpha}{du}}{\frac{dx^0}{du}} \right) = 0$$

on (a, b) so the motion is linear and uniform. In summary: *the motion is linear and uniform in the inertial frame $\mathscr{R}$ if and only if (15.7) holds.*

With the idea of writing in coordinates the first two of Einstein's postulates, let us now explain how to express that the velocity of a point particle has modulus equal to c. Keeping (15.3) in account we introduce the matrix

$$\eta := \begin{bmatrix} -1 & 0 & 0 & 0 \\ 0 & 1 & 0 & 0 \\ 0 & 0 & 1 & 0 \\ 0 & 0 & 0 & 1 \end{bmatrix}.$$

$$(15.8)$$

The condition that $v^2 = c^2$ becomes, due to (15.3),

$$\sum_{\alpha,\beta=0}^{3} \eta_{\alpha\beta} \frac{dx^\alpha}{du} \frac{dx^\beta}{du} = 0.$$

$$(15.9)$$

Now take a second inertial frame $\mathscr{R}'$ with Minkowski coordinates y^0, y^1, y^2, y^3. As spacetime is a manifold, the two coordinate systems will be related by a differentiable bijection from $\mathbb{R}^4$ to $\mathbb{R}^4$ with differentiable inverse. We will formally indicate it by

$$y^\alpha = y^\alpha(x^0, x^1, x^2, x^3) \quad \text{where } (x^0, x^1, x^2, x^3) \in \mathbb{R}^4 \text{ and } \alpha = 0, 1, 2, 3.$$

The notion of isolated point particle does not depend on a frame. We shall also assume of having isolated point particles at any place and time (for short, but finite, time intervals) with any possible velocities (within the above bounds). In this scenario the linear uniform motion of a point particle in an inertial frame can always be viewed as being due to the point particle being isolated. We conclude that the linear uniform motion of a point particle in one inertial frame must be described as linear uniform motion in any other inertial frame.

This applies in particular to motions at the speed of light, which can be thought of as motions of light signals.

We shall suppose that the parametrisations of the curve $x^\alpha = x^\alpha(u)$ describing the particle's history in $\mathscr{R}$, which allows to express the velocity via definition (15.2) due to $dx^0/du \neq 0$, are well suited to describe its history also in the coordinates of $\mathscr{R}'$, $y^\alpha = y^\alpha(u)$. Put otherwise, we assume $dy^0/du \neq 0$ and therefore that the velocity in $\mathscr{R}'$ is still expressible by (15.2) using y^α instead of x^α.

We then have the following two demands on the transformation

$$y^\alpha = y^\alpha(x^0, x^1, x^2, x^3) \quad \text{where } (x^0, x^1, x^2, x^3) \in \mathbb{R}^4 \text{ and } \alpha = 0, 1, 2, 3.$$

(a) *By virtue of* **RS2** *the motion of a point particle with world line* $\rho : (a, b) \ni u \mapsto \rho(u) \in \mathbb{M}^4$ *is linear uniform in* $\mathscr{R}'$ *(for a certain time interval) if and only if it is linear uniform in* $\mathscr{R}$ *(for a corresponding interval):*

$$\frac{d^2 y^\alpha}{du^2} \frac{dy^\beta}{du} - \frac{d^2 y^\beta}{du^2} \frac{dy^\alpha}{du} = 0 \,\forall \alpha, \beta$$

$$\text{if and only if} \quad \frac{d^2 x^\alpha}{du^2} \frac{dx^\beta}{du} - \frac{d^2 x^\beta}{du^2} \frac{dx^\alpha}{du} = 0 \,\forall \alpha, \beta \tag{15.10}$$

where ρ *is given in Minkowski coordinates of* $\mathscr{R}$ *by the differentiable maps* $x^\alpha = x^\alpha(u)$, *subject to (15.5) at every event it reaches, and it is given in* $\mathscr{R}'$ *by the differentiable maps* $y^\alpha = y^\alpha(u)$, $\alpha = 0, 1, 2, 3$.

(b) *In particular, by* **RS1**, *a point particle with world line* $\rho : (a, b) \ni u \mapsto \rho(u) \in \mathbb{M}^4$ *travels at the speed of light in* $\mathscr{R}'$ *if and only if it travels at the speed of light in* $\mathscr{R}$. *That is,*

$$\sum_{\alpha,\beta=0}^{3} \eta_{\alpha\beta} \frac{dy^\alpha}{du} \frac{dy^\beta}{du} = 0 \quad \text{if and only if} \quad \sum_{\alpha,\beta=0}^{3} \eta_{\alpha\beta} \frac{dx^\alpha}{du} \frac{dx^\beta}{du} = 0,$$

$$\tag{15.11}$$

using the same conventions as above.

15.4.2   Finding the Poincaré Transformations and the Affine Structure of $\mathbb{M}^4$

The initial problem of determining the general transformation law between the Minkowski coordinates of two inertial frames is therefore mathematically the same as finding the most general C^∞ diffeomorphism of $\mathbb{R}^4$

$$y^\alpha = y^\alpha(x^0, x^1, x^2, x^3) \quad \text{where } (x^0, x^1, x^2, x^3) \in \mathbb{R}^4 \text{ and } \alpha = 0, 1, 2, 3,$$

that satisfies the above (a) and (b) for any world line ρ. We shall next deduce two useful mathematical consequences of (a) and (b), which will eventually give us the answer we want.

Discussion of (a). From

$$y^\alpha(u) = y^\alpha(x^0(u), x^1(u), x^2(u), x^3(u))$$

we obtain

$$\frac{d^2 y^\alpha}{du^2} \frac{dy^\beta}{du} - \frac{d^2 y^\beta}{du^2} \frac{dy^\alpha}{du} = \sum_{\pi,\gamma} \frac{\partial y^\alpha}{\partial x^\pi} \frac{\partial y^\beta}{\partial x^\gamma} \left(\frac{d^2 x^\pi}{du^2} \frac{dx^\gamma}{du} - \frac{d^2 x^\gamma}{du^2} \frac{dx^\pi}{du} \right)$$

$$+ \sum_{\pi,\gamma,\rho} \frac{dx^\pi}{du} \frac{dx^\gamma}{du} \frac{dx^\rho}{du} \left(\frac{\partial^2 y^\alpha}{\partial x^\pi \partial x^\gamma} \frac{\partial y^\beta}{\partial x^\rho} - \frac{\partial^2 y^\beta}{\partial x^\pi \partial x^\gamma} \frac{\partial y^\alpha}{\partial x^\rho} \right) . \qquad (15.12)$$

so by (7), at every event x

$$0 = \sum_{\pi,\gamma,\rho} U^\pi U^\gamma U^\rho \left(\frac{\partial^2 y^\alpha}{\partial x^\pi \partial x^\gamma}\Big|_x \frac{\partial y^\beta}{\partial x^\rho}\Big|_x - \frac{\partial^2 y^\beta}{\partial x^\pi \partial x^\gamma}\Big|_x \frac{\partial y^\alpha}{\partial x^\rho}\Big|_x \right) . \qquad (15.13)$$

where the components $U^\alpha = dx^\alpha/du$ of $U \in T_x \mathbb{M}^4$ refer to the uniform linear motion of a point particle and are evaluated when the corresponding world line crosses x. The right-hand side of (15.13) can be viewed, for x and α, β given, as a polynomial $P_{(x)}^{(\alpha\beta)}[U^0, U^1, U^2, U^3]$ in the U^α. It vanishes on an open set by (15.5) (actually, the weaker condition (15.4) is enough), so it vanishes everywhere, i.e. all its coefficients are zero. In particular, for third-order coefficients:

$$\frac{\partial^3 P_{(x)}^{(\mu\nu)}}{\partial U^\alpha \partial U^\beta \partial U^\gamma} = 0 .$$

Computing the left-hand side gives the identities:

$$\frac{\partial^2 y^\mu}{\partial x^\alpha \partial x^\beta} \frac{\partial y^\nu}{\partial x^\gamma} + \frac{\partial^2 y^\mu}{\partial x^\beta \partial x^\gamma} \frac{\partial y^\nu}{\partial x^\alpha} + \frac{\partial^2 y^\mu}{\partial x^\gamma \partial x^\alpha} \frac{\partial y^\nu}{\partial x^\beta} - \frac{\partial^2 y^\nu}{\partial x^\alpha \partial x^\beta} \frac{\partial y^\mu}{\partial x^\gamma}$$

$$- \frac{\partial^2 y^\nu}{\partial x^\beta \partial x^\gamma} \frac{\partial y^\mu}{\partial x^\alpha} - \frac{\partial^2 y^\nu}{\partial x^\gamma \partial x^\alpha} \frac{\partial y^\mu}{\partial x^\beta} = 0$$

valid at any event $x \in M^4$ and for any $\mu, \nu, \alpha, \beta, \gamma$ in $\{0, 1, 2, 3\}$. Multiplying by $\frac{\partial x^\gamma}{\partial y^\kappa}$ we find:

$$\frac{\partial^2 y^\mu}{\partial x^\alpha \partial x^\beta} \frac{\partial y^\nu}{\partial x^\gamma} \frac{\partial x^\gamma}{\partial y^\kappa} + \frac{\partial^2 y^\mu}{\partial x^\beta \partial x^\gamma} \frac{\partial y^\nu}{\partial x^\alpha} \frac{\partial x^\gamma}{\partial y^\kappa} + \frac{\partial^2 y^\mu}{\partial x^\gamma \partial x^\alpha} \frac{\partial y^\nu}{\partial x^\beta} \frac{\partial x^\gamma}{\partial y^\kappa}$$

$$- \frac{\partial^2 y^\nu}{\partial x^\alpha \partial x^\beta} \frac{\partial y^\mu}{\partial x^\gamma} \frac{\partial x^\gamma}{\partial y^\kappa} - \frac{\partial^2 y^\nu}{\partial x^\beta \partial x^\gamma} \frac{\partial y^\mu}{\partial x^\alpha} \frac{\partial x^\gamma}{\partial y^\kappa} - \frac{\partial^2 y^\nu}{\partial x^\gamma \partial x^\alpha} \frac{\partial y\mu}{\partial x^\beta} \frac{\partial x^\gamma}{\partial y^\kappa} = 0 \, .$$

Now we add over the indices that repeat twice, so $\sum_\gamma \frac{\partial y^\nu}{\partial x^\gamma} \frac{\partial x^\gamma}{\partial y^\kappa} = \frac{\partial y^\nu}{\partial y^\kappa} = \delta^\nu_\kappa$ and also $\sum_\gamma \frac{\partial y^\mu}{\partial x^\gamma} \frac{\partial x^\gamma}{\partial y^\kappa} = \frac{\partial y^\mu}{\partial y^\kappa} = \delta^\mu_\kappa$, obtaining

$$\frac{\partial^2 y^\mu}{\partial x^\alpha \partial x^\beta} \delta^\nu_\kappa + \sum_\gamma \frac{\partial^2 y^\mu}{\partial x^\beta \partial x^\gamma} \frac{\partial y^\nu}{\partial x^\alpha} \frac{\partial x^\gamma}{\partial y^\kappa} + \sum_\gamma \frac{\partial^2 y^\mu}{\partial x^\gamma \partial x^\alpha} \frac{\partial y^\nu}{\partial x^\beta} \frac{\partial x^\gamma}{\partial y^\kappa}$$

$$- \frac{\partial^2 y^\nu}{\partial x^\alpha \partial x^\beta} \delta^\mu_\kappa - \sum_\gamma \frac{\partial^2 y^\nu}{\partial x^\beta \partial x^\gamma} \frac{\partial y^\mu}{\partial x^\alpha} \frac{\partial x^\gamma}{\partial y^\kappa} - \sum_\gamma \frac{\partial^2 y^\nu}{\partial x^\gamma \partial x^\alpha} \frac{\partial y^\mu}{\partial x^\beta} \frac{\partial x^\gamma}{\partial y^\kappa} = 0 \, .$$

Taking $\nu = \kappa$ and adding from 0 to 3:

$$4\frac{\partial^2 y^\mu}{\partial x^\alpha \partial x^\beta} + \sum_\gamma \frac{\partial^2 y^\mu}{\partial x^\beta \partial x^\gamma} \delta^\gamma_\alpha + \sum_\gamma \frac{\partial^2 y^\mu}{\partial x^\gamma \partial x^\alpha} \delta^\gamma_\beta - \frac{\partial^2 y^\mu}{\partial x^\alpha \partial x^\beta}$$

$$- \sum_{\gamma, \nu} \frac{\partial^2 y^\nu}{\partial x^\beta \partial x^\gamma} \frac{\partial y^\mu}{\partial x^\alpha} \frac{\partial x^\gamma}{\partial y^\nu} - \sum_{\gamma, \nu} \frac{\partial^2 y^\nu}{\partial x^\gamma \partial x^\alpha} \frac{\partial y^\mu}{\partial x^\beta} \frac{\partial x^\gamma}{\partial y^\nu} = 0 \, ,$$

and therefore

$$5\frac{\partial^2 y^\mu}{\partial x^\alpha \partial x^\beta} - \sum_{\gamma, \nu} \frac{\partial^2 y^\nu}{\partial x^\beta \partial x^\gamma} \frac{\partial y^\mu}{\partial x^\alpha} \frac{\partial x^\gamma}{\partial y^\nu} - \sum_{\gamma, \nu} \frac{\partial^2 y^\nu}{\partial x^\gamma \partial x^\alpha} \frac{\partial y^\mu}{\partial x^\beta} \frac{\partial x^\gamma}{\partial y^\nu} = 0 \, .$$

Hence, we have deduced from (a) the following useful fact at any event $x \in \mathbb{M}^4$:

$$\frac{\partial^2 y^\mu}{\partial x^\alpha \partial x^\beta}\Big|_x = \Psi_\alpha(x) \frac{\partial y^\mu}{\partial x^\beta}\Big|_x + \Psi_\beta \frac{\partial y^\mu}{\partial x^\alpha}\Big|_x \tag{15.14}$$

where

$$\Psi_\kappa(x) := \frac{1}{5} \sum_{\gamma, \nu} \frac{\partial^2 y^\nu}{\partial x^\gamma \partial x^\kappa}\Big|_x \frac{\partial x^\gamma}{\partial y^\nu}\Big|_x \, . \tag{15.15}$$

Discussion of (b). Observe that

$$\sum_{\mu,\nu} \eta_{\mu\nu} \frac{dy^{\mu}}{du} \frac{dy^{\nu}}{du} = \sum_{\mu,\nu,\pi,\kappa} \left(\eta_{\mu\nu} \frac{\partial y^{\mu}}{\partial x^{\pi}} \frac{\partial y^{\nu}}{\partial x^{\kappa}} \right) \frac{dx^{\pi}}{du} \frac{dx^{\kappa}}{du} . \qquad (15.16)$$

Thus (b) reads

$$\sum_{\mu,\nu} \eta_{\mu\nu} U^{\mu} U^{\nu} = 0 \quad \text{if and only if} \quad \sum_{\mu,\nu,\pi,\kappa} \eta_{\mu\nu} \frac{\partial y^{\mu}}{\partial x^{\pi}}|_x \frac{\partial y^{\nu}}{\partial x^{\kappa}}|_x U^{\pi} U^{\kappa} = 0 .$$

$$(15.17)$$

In particular,

$$\eta'_x(U, U) = 0 \quad \text{if} \quad \eta(U, U) = 0 , \qquad (15.18)$$

where the vector $U \in T_x \mathbb{M}^4$, of components $U^{\kappa} = dx^{\kappa}/du$ in the basis at x corresponding to the Minkowski coordinates of $\mathcal{R}$, is subject to (15.5), and the quadratic form η'_x is

$$\eta'_x(U, U) := \sum_{\mu,\nu,\pi,\kappa=0}^{3} \eta_{\mu\nu} \frac{\partial y^{\mu}}{\partial x^{\pi}}|_x \frac{\partial y^{\nu}}{\partial x^{\kappa}}|_x U^{\pi} U^{\kappa} , \qquad (15.19)$$

In particular, $\eta(U, U) = 0$ and hence $\eta'_x(U, U) = 0$ whenever

$$(U^0)^2 = \sum_{a=1}^{3} (U^a)^2 .$$

A vector of this kind can be found by putting $U^0_V = |V|$ and $U^a_V = V^a$, where $|V| := \sqrt{\sum_{a=1}^{3}(V^a)^2}$. Let us write η'_x using the symmetric matrix of coefficients $c_{00}, c_{0a} = c_{a0}, c_{ab} = c_{ba}$ for $a, b = 1, 2, 3$.
Then $\eta'_x(U_V, U_V) = 0$ becomes:

$$c_{00}|V|^2 + 2 \sum_{a=1}^{3} |V| V^a c_{0a} + \sum_{a,b=1}^{3} V^a V^b c_{ab} = 0$$

for $V^a \in \mathbb{R}$. That is,

$$2 \sum_{a=1}^{3} |V| V^a c_{0a} + \sum_{a,b=1}^{3} V^a V^b (c_{ab} + c_{00}\delta_{ab}) = 0$$

for any $V^a \in \mathbb{R}$, $a = 1, 2, 3$. This is possible only if $c_{ab} = -c_{00}\delta_{ab}$ and $c_{0a} = 0$. Indeed, choosing the column vector of entries $V^a \in \mathbb{R}$ and then the column vector of entries $-V^a \in \mathbb{R}$, we have:

$$2\sum_{a=1}^{3} |V| V^a c_{0a} + \sum_{a,b=1}^{3} V^a V^b (c_{ab} + c_{00}\delta_{ab}) = 0$$

and at the same time

$$-2\sum_{a=1}^{3} |V| V^a c_{0a} + \sum_{a,b=1}^{3} V^a V^b (c_{ab} + c_{00}\delta_{ab}) = 0$$

for any $V^a \in \mathbb{R}$. But then the following hold independently:

$$\sum_{a=1}^{3} |V| V^a c_{0a} = 0 \quad \text{and} \quad \sum_{a,b=1}^{3} V^a V^b (c_{ab} + c_{00}\delta_{ab}) = 0$$

for any $V^a \in \mathbb{R}$ and $a = 1, 2, 3$. The former forces $c_{0a} = 0$ for any $a = 1, 2, 3$. The latter, given the symmetry of the coefficients $c_{ab} + c_{00}\delta_{ab}$, implies[8] $c_{ab} + c_{00}\delta_{ab} = 0$, i.e.: $c_{ab} = -c_{00}\delta_{ab}$. All in all, calling $\lambda(x)$ the coefficient $-c_{00}$, we have:

$$\eta'_x = \lambda(x)\eta , \tag{15.20}$$

where $\mathbb{M}^4 \ni x \mapsto \lambda(x) \in \mathbb{R}$ is an arbitrary function. This map is C^∞ because, thinking the quadratic forms as real 4×4 matrices,

$$2\lambda(x) = \operatorname{tr}(\eta'_x) = \sum_{\mu,\nu,\pi,\kappa=0}^{3} \delta^{\pi\kappa} \eta_{\mu\nu} \frac{\partial y^\mu}{\partial x^\pi}\Big|_x \frac{\partial y^\nu}{\partial x^\kappa}\Big|_x ,$$

where the trace is the usual trace of 4×4 matrices. The right-hand side is C^∞ by assumption, and $\lambda(x)$ is strictly positive by (15.20) and by Sylvester's theorem [Nor86], whereby the signature of a quadratic form is invariant under congruences. In fact η and η_x are related by a congruence because of (15.19), so they have the same signature $(-1, +1, +1, +1)$. This would not be possible by (15.20) if $\lambda(x)$

---

[8] If an $n \times n$ symmetric matrix A satisfies $u^t A u = 0$ for all $u \in \mathbb{R}^n$, putting $u = x + y$ we find $x^t A x + y^t A y + x^t A y + y^t A x = 0$, i.e. $x^t A y + y^t A x = 0$. But $A^t = A$, so $y^t A x = x^t A y$ as well, and therefore $x^t A y = 0$ for any $x, y \in \mathbb{R}^n$, meaning $A = 0$.

was not strictly positive. We conclude that (b) implies

$$\sum_{\mu,\nu=0}^{3} \eta_{\mu\nu} \frac{\partial y^{\mu}}{\partial x^{\pi}}\Big|_x \frac{\partial y^{\nu}}{\partial x^{\kappa}}\Big|_x = \lambda(x)\eta_{\pi\kappa} , \tag{15.21}$$

where $\mathbb{M}^4 \ni x \mapsto \lambda(x) \in \mathbb{R}$ is a strictly positive C^{∞} map.
Finally, under (a) and (b) let us differentiate in x^{ρ} Eq. (15.21) and express the second derivatives of y^{μ} with (15.14). We obtain

$$\sum_{\mu,\nu} \eta_{\mu\nu} \left[\left(\Psi_{\rho} \frac{\partial y^{\mu}}{\partial x^{\rho}} + \Psi_{\rho} \frac{\partial y^{\mu}}{\partial x^{\pi}} \right) \frac{\partial y^{\nu}}{\partial x^{\kappa}} + \left(\Psi_{\kappa} \frac{\partial y^{\nu}}{\partial x^{\rho}} + \Psi_{\rho} \frac{\partial y^{\nu}}{\partial x^{\kappa}} \right) \frac{\partial y^{\mu}}{\partial x^{\pi}} \right] = \frac{\partial \lambda}{\partial x^{\rho}} \eta_{\pi\kappa} .$$

Using (15.21), the above reduces to

$$\Psi_{\pi} \eta_{\rho\kappa} + \Psi_{\kappa} \eta_{\pi\rho} + 2\Psi_{\rho} \eta_{\pi\kappa} = \frac{\partial \ln \lambda}{\partial x^{\rho}} \eta_{\pi\kappa} .$$

To conclude, multiply both sides by $\eta_{\kappa\sigma}$ and sum over κ, remembering that $\sum_{\kappa=0}^{3} \eta_{\alpha\kappa}\eta_{\kappa\sigma} = \delta_{\alpha\sigma}$:

$$\Psi_{\pi} \delta_{\rho\sigma} + \sum_{\kappa=0}^{3} \Psi_{\kappa} \eta_{\pi\rho}\eta_{\kappa\sigma} + 2\Psi_{\rho} \delta_{\pi\sigma} = \frac{\partial \ln \lambda}{\partial x^{\rho}} \delta_{\pi\sigma} . \tag{15.22}$$

We look at this as a system in the unknowns Ψ_{ρ} and $\partial \ln \lambda/\partial x^{\rho}$. Choose $\pi = \rho = 0$ and $\sigma = 1, 2, 3$, so the system gives $\Psi_{\sigma} = 0$ for $\sigma = 1, 2, 3$. Now assuming $\sigma = 0, \pi = \rho = 1$, the system forces $\Psi_0 = 0$. At last, as the maps Ψ_{ρ} are zero, (15.22) returns $\frac{\partial \ln \lambda}{\partial x^{\rho}} = 0$ for $\rho = 0, 1, 2, 3$. So we conclude that the only solution to (15.22), for any $x \in \mathbb{M}^4$ and $\kappa = 0, 1, 2, 3$, is:

$$\Psi_{\kappa}(x) = 0 , \tag{15.23}$$

$$\frac{\partial \ln \lambda}{\partial x^{\rho}} = 0 . \tag{15.24}$$

From (15.15), since the Jacobian matrix of the transformation from x^0, x^1, x^2, x^3 to y^0, y^1, y^2, y^3 is invertible, the above solutions give, for any $\mu = 0, 1, 2, 3$,

$$\frac{\partial y^{\mu}}{\partial x^{\nu}}\Big|_x = \text{constant} , \tag{15.25}$$

$$\lambda(x) = \text{constant} . \tag{15.26}$$

Putting all back together the Minkowski transformation of inertial frames $\mathscr{R}$ and $\mathscr{R}'$ mapping the coordinates x^0, x^1, x^2, x^3 to the y^0, y^1, y^2, y^3 has the form:

$$y^\mu = b^\mu + \sum_{\nu=0}^{3} L^\mu{}_\nu x^\nu \quad \mu = 0, 1, 2, 3, \qquad (15.27)$$

where the 4 coefficients $b^\mu \in \mathbb{R}$ are free and the 16 coefficients $L^\mu{}_\nu \in \mathbb{R}$ obey

$$\sum_{\mu,\nu=0}^{1} \eta_{\mu\nu} L^\mu{}_\pi L^\nu{}_\kappa = \lambda\, \eta_{\pi\kappa} ,$$

where $\lambda > 0$ is a constant. Note that if we redefine the units of space and time of $\mathscr{R}$ by a common dilation factor $\lambda^{1/2}$ (which does not contradict the postulates of Relativity as said in (4), Remarks 15.2), we may reduce to the transformation with coefficients $\Lambda^\mu{}_\nu := \lambda^{-1/2} L^\mu{}_\nu$. In fact, let us define new Minkowski coordinates on $\mathscr{R}$ by $\bar{x}^\nu := \sqrt{\lambda} x^\nu$, so (15.27) implies:

$$y^\mu = b^\mu + \sqrt{\lambda} \sum_\nu \Lambda^\mu{}_\nu x^\nu , \qquad (15.28)$$

that is:

$$y^\mu = b^\mu + \sum_\nu \Lambda^\mu{}_\nu \bar{x}^\nu .$$

Writing once again *without bar* the coordinates on $\mathscr{R}$, the Minkowski systems are related by **Poincaré transformations**:

$$y^\mu = b^\mu + \sum_{\nu=0}^{3} \Lambda^\mu{}_\nu x^\nu \quad \mu = 0, 1, 2, 3, \qquad (15.29)$$

where the $b^\mu \in \mathbb{R}$ are arbitrary by definition, and the matrices Λ of coefficients $\Lambda^\mu{}_\nu$ satisfy, by construction, the **Lorentz relation**:

$$\Lambda^t \eta \Lambda = \eta . \qquad (15.30)$$

Conversely, suppose $\mathscr{R}$ is inertial with Minkowski coordinates x^0, x^1, x^2, x^3 and $\mathscr{R}'$ another frame with Minkowski coordinates y^0, y^1, y^2, y^3 given by (15.29) and (15.30). It is straightforward to see that $\mathscr{R}'$ must be inertial, because by linearity the above linear maps preserve linear uniform motions and the notion of isolated point particle is frame-independent. Furthermore, (15.30) immediately guarantees, by (15.9), that a point particle in motion at the speed of light in $\mathscr{R}$ is seen in $\mathscr{R}'$ as travelling at the speed of light.

In summary, we have proved the following theorem:

Theorem 15.5 *Assume that, on* $\mathbb{M}^4$:

(a) *the Principle of inertia **RS2** holds;*
(b) *the speed of light c is constant (**RS1**), in particular if we adopt Einstein's synchronisation;*
(c) *there is an isolated point particle at every event in spacetime for a finite time interval (in some inertial frame);*
(d) *the velocity of point particles in inertial frames is allowed to have components in any direction and orientation, and modulus contained in $[0, c]$ at least, as per (15.5).*

Then the following hold.

(1) *The transformation law between the Minkowski coordinates of two inertial frames is linear.*
(2) *More precisely (up to rescaling the units of space and time in inertial frames under the above hypotheses), (15.29), called Poincaré transformations, where $b^\alpha \in \mathbb{R}$ and the matrix $\Lambda^\alpha{}_\beta$, $\alpha, \beta = 0, 1, 2, 3$ satisfies (15.30), are exactly the transformations of Minkowski coordinates of the inertial frames.*

Remarks 15.6

(1) In order to have $\lambda = 1$ we have exploited the possibility of dilating or shrinking the units of the single inertial frames (without affecting the value of the speed of light). This raises the next physical issue. Take three inertial frames $\mathscr{R}_l$, $l = 1, 2, 3$. Let λ_{ij} be the coefficients λ coming from (15.28) when $y^0, \ldots, y^3$ refer to frame i and $x^0, \ldots, x^3$ to frame j. By construction:

$$\lambda_{ij} = \lambda_{ik}\lambda_{kj}, \quad \text{for } i, j, k \in \{1, 2, 3\}. \tag{15.31}$$

Choose $i = j = k$ and then $i = j$, so to obtain in particular

$$\lambda_{ii} = 1, \quad \lambda_{ik} = \lambda_{ki}^{-1}. \tag{15.32}$$

Under our hypotheses, we cannot still conclude $\lambda_{ik} = 1$, i.e that the units can be fixed *once and for all in every inertial frame*, in order to always use the coordinate transformation (15.29) for *every* pair of inertial frames. A rather weak hypothesis to grant that is asking that for any pair $\mathscr{R}_i, \mathscr{R}_j$ of distinct inertial frames there exists a third inertial frame $\mathscr{R}_k$ such that $\lambda_{ki} = \lambda_{kj}$. By (15.31) and the second relation in (15.32) we immediately obtain $\lambda_{ij=1}$. It is possible to show that the existence of such an $\mathscr{R}_k$ follows from assuming the isotropy of physical laws formulated in inertial frames. That said, we shall not address the general issue of isotropy and will directly assume instead that one can fix all coefficients $\lambda = 1$ for any pair of inertial frames.

(2) Consider a point particle at rest in an inertial frame $\mathscr{R}$. Let us describe its world line by the time coordinate x^0 of the frame. The spatial coordinates will always stay fixed, say (x_0^1, x_0^2, x_0^3). Now take another inertial frame $\mathcal{R}'$. The world line of the point particle, still parametrised by x^0, will be

$$y^\alpha = y^\alpha(x^0, x_0^1, x_0^2, x_0^3) \quad x^0 \in (a, b).$$

We expect the map $(a, b) \ni x^0 \mapsto y^0 = y^0(x^0, x_0^1, x_0^2, x_0^3) \in \mathbb{R}$ to be *increasing*. That is so not to invert the temporal sequence of physical phenomena when changing inertial frame. This request selects, among the coordinate transformations (15.29), those that further satisfy

$$\Lambda^0{}_0 > 0.$$

The above, as we saw in Chap. 10, corresponds to declaring that only *orthochronous* Poincaré transformations make physical sense. ∎

As the transformations of Minkowski coordinates (15.29) are linear, we can put on $\mathbb{M}^4$ an affine structure for which the Minkowski coordinates are special types of Cartesian coordinates. Furthermore, we can equip the space of displacements of $\mathbb{M}^4$ with an *indefinite inner product* given by the matrix η in any possible Minkowski Cartesian coordinates. *It is precisely the fact that Minkowski coordinates are related by Poincaré transformations, which obey (15.30), that guarantees that the inner product's definition is intrinsic and does not depend on the particular coordinates chosen.*

At this point it becomes an easy task to verify that we can proceed axiomatically as in Chap. 10. The physical explanation that we gave *a posteriori*, in Chap. 10, regarding the metric and kinematic structure associated with inertial frames, corresponds precisely to the physical interpretation we started from here in order to arrive at Theorem 15.5.

Appendix A
Elements of Topology, Analysis, Linear Algebra and Geometry

We will review in this appendix notions of great importance in Classical Mechanics and more generally Mathematical Physics: the basics of Differential Geometry. Let us begin with recalling elementary notions of point-set Topology and Mathematical Analysis in $\mathbb{R}^n$.

A.1 Review of Elementary Topology

We will review here some facts from point-set Topology. A pair $(X, \mathscr{T})$, where X is a set and $\mathscr{T}$ a collection of subsets of X, is called a **topological space** if the following facts hold:

(i) $\varnothing, X \in \mathscr{T}$,
(ii) the union (of any cardinality, including uncountable) of elements of $\mathscr{T}$ is an element of $\mathscr{T}$,
(iii) the finite intersection of elements of $\mathscr{T}$ is an element of $\mathscr{T}$.

The sets of $\mathscr{T}$ are called **open sets**, and $\mathscr{T}$ is the **topology** on X. The **closed sets** of X are by definition the complements in X of open sets.
The **closure** $\overline{S}$ of a set $S \subset X$ is by definition the closed intersection of all closed subsets containing S:

$$\overline{S} := \bigcap \{C \mid X \setminus C \in \mathscr{T}, C \supset S\}.$$

If $A \subset X$, then $\mathcal{F}(A)$ denotes the **boundary** of A, i.e. the set of points $p \in X$ such that every open set $V \ni p$ satisfies $V \cap A \neq \varnothing$ and $V \cap (X \setminus A) = \varnothing$. Similarly, $Int(A)$ denotes the **interior** of A, i.e. the set of points $p \in A$ for which there exists an open set V such that $p \in V \subset A$. Finally, $Ext(A) := Int(X \setminus A)$

indicates the **exterior** of A. If $p \in X$, where $(X, \mathscr{T})$ is a topological space, an **open neighbourhood** of p is a set $A \in \mathscr{T}$ containing p.

On $\mathbb{R}^n$ there is a natural topology called **Euclidean topology** or **standard topology**, whose open sets are, apart from the empty set, arbitrary unions of open balls with any centre and radius. The elementary topological notions on $\mathbb{R}^n$ studied in Analysis courses are subcases of the above general concepts. In particular, the closed subsets of $\mathbb{R}^n$, in the above sense, are exactly those containing their boundary points, meaning closed for the standard topology of $\mathbb{R}^n$ (such a characterisation of closed sets is by the way completely general and holds in any topological space). The closure of $S \subset \mathbb{R}^n$ coincides with the union of S and its accumulation points. We finally recall that a subset in $\mathbb{R}^n$ is called **bounded** if there is a ball of finite radius that contains it.

A **basis** of the topology $\mathscr{T}$ of $(X, \mathscr{T})$ is a collection of open sets $\mathcal{B} \subset \mathscr{T}$ such that every non-empty open set of $\mathscr{T}$ is a union of elements of $\mathcal{B}$. In the Euclidean topology the open balls of any radius and centre forma a basis.

If $A \subset X$ is a set and $(X, \mathscr{T})$ a topological space, we can define on A a topology $\mathscr{T}_A$ using $\mathscr{T}$. Setting $\mathscr{T}_A := \{B \cap A \mid B \in \mathscr{T}\}$, it is easy to check that $(A, \mathscr{T}_A)$ is a topological space. $\mathscr{T}_A$ is the **topology induced (by $\mathscr{T}$) on** A.

If $(X, \mathscr{T})$ and $(X', \mathscr{T}')$ are topological spaces, a map $f : X \to X'$ is called **continuous** if $f^{-1}(A) \subset \mathscr{T}$ whenever $A \in \mathscr{T}'$ (i.e. the pre-image of an open set in the codomain is open in the domain). If $(X, \mathscr{T})$ and $(X', \mathscr{T}')$ are topological spaces and $p \in X$, a map $f : X \to X'$ is called **continuous at** p if for any open neighbourhood B of $f(p)$ there exists an open neighbourhood A of p such that $f(A) \subset B$. it is easy to see that $f : X \to X'$ is continuous if and only if it is continuous at every point $p \in X$. If $(X, \mathscr{T})$ and $(X', \mathscr{T}')$ are topological spaces, a map $f : X \to X'$ is called a **homeomorphism** when it is continuous, bijective and the inverse is continuous.

It can be proved that if the above spaces are $X = \mathbb{R}^n$ and $X' = \mathbb{R}^m$, or subsets thereof, with the respective Euclidean topologies, the classical definitions of continuous map and continuous map at a point are equivalent to the general ones given above.

A topological space $(X, \mathscr{T})$ is said to be **Hausdorff**, or T_2, if for any pair of distinct points $p, q \in X$ there exist open neighbourhoods $U \ni p$ and $V \ni q$ with $U \cap V = \varnothing$.

A topological space $(X, \mathscr{T})$ is **second countable** if it possesses a basis $\mathcal{B} \subset \mathscr{T}$ made by countably many elements. $\mathbb{R}^n$ with the standard topology is both Hausdorff (as every metric space) and second countable (a countable basis is formed by open balls with rational radii and centres of rational coordinates).

A set $S \subset X$ in a topological space $(X, \mathscr{T})$ is called:

(a) **compact** if any collection $\{B_i\}_{i \in I} \subset \mathscr{T}$ such that $\cap_{i \in I} B_i \supset S$ admits a *finite* subcollection, i.e. $F \subset I$ finite, such that $\cap_{i \in F} B_i \supset S$. (In the standard topology of $\mathbb{R}^n$, by the famous **Heine-Borel theorem**, compact sets are precisely the closed and bounded sets);

(a1) **relatively compact** if its closure is compact;

(b) **disconnected** if there exists open sets $A, A' \in \mathcal{T}$ such that $A \cap S \neq \varnothing$, $A' \cap S \neq \varnothing$, $A \cup A' \supset S$, but $(A \cap S) \cap (A' \cap S) = \varnothing$; **connected** if it is not disconnected. (For example, in the standard topology of $\mathbb{R}$, the set $S := (0, 1) \cup [1, 2)$ is connected while $S := (0, 1) \cup (1, 2)$ is disconnected);

(b1) **path-connected** if for any pair $p, q \in S$ there is a continuous map $\gamma : [0, 1] \to S$, called a path, such that $\gamma(0) = p$ and $\gamma(1) = q$; path-connected sets are connected, and in $\mathbb{R}^n$ an open connected set is always path-connected (with continuous and C^∞ paths);

(b2) **simply connected** if it is connected, path-connected and for any pair of points $p, q \in S$ and any pair of paths $\gamma_1, \gamma_2 : [0, 1] \to S$ with $\gamma_i(0) = p$, $\gamma_i(1) = q$ there exists a continuous map $\Gamma : [0, 1] \times [0, 1] \to S$ with $\Gamma(r, 0) = p$, $\Gamma(r, 1) = q$ for any $r \in [0, 1]$ and $\Gamma[0, s] = \gamma_1(s)$ and $\Gamma(1, s) = \gamma_2(s)$. In other words γ_1 can be continuously deformed into γ_2 whilst staying in S. In $\mathbb{R}^n$ a simply connected open set satisfies the simple connectedness property with Γ smooth provided γ_1 and γ_2 are smooth. There is an equivalent definition of simple connectedness based on the notion of *homotopy group*, which we have never used.

A topological space is called **locally compact** if every point has an open neighbourhood with compact closure. $\mathbb{R}^n$ with the Euclidean topology is locally compact.

Compact sets are closed in Hausdorff spaces (such as $\mathbb{R}^n$), and closed subsets in compact sets are in turn always compact.

Continuous maps transform compact sets into compact sets, (path-)connected sets into (path-)connected sets. Homeomorphisms further map simply connected sets to simply connected sets. Functions between topological spaces that map open sets to open sets are called **open maps**.

It is straightforward to see that a subset $S \subset X$ in a topological space is compact, connected, simply connected if and only if it is, respectively, a compact, connected, simply connected topological space for the topology induced by X.

If $(X, \mathcal{T}_X)$ and $(Y, \mathcal{T}_Y)$ are topological spaces, $X \times Y$ has a natural topology, called **product topology**, whose open sets are the unions of all possible Cartesian products $A \times A'$ with $A \in \mathcal{T}_X$ and $A' \in \mathcal{T}_Y$. The standard topology of $\mathbb{R}^{n+m}$ is just the product topology of $\mathbb{R}^n$ and $\mathbb{R}^m$. The canonical projections $\pi_1 : X \times Y \ni (x, y) \mapsto x \in X$ and $\pi_2 : X \times Y \ni (x, y) \mapsto y \in Y$ are continuous and open maps.

A function $f : X \times Y \to Z$, where $(Z, \mathcal{T}_Z)$ is another topological space, is called **jointly continuous** if it is continuous for the domain's product topology.

A.2   Integrals of Limits and Derivatives

Due to the elementary nature of this book we will address the matter referring only to the *Riemann integral* and not the *Lebesgue integral*, for which the ensuing results can be proved with much weaker assumptions using *Lebesgue's dominated convergence theorem*.

Consider a function, at least continuous, $f : I \times K \to \mathbb{R}$, where $K \subset \mathbb{R}^n$ is a compact set (typically the closure of an open bounded set) and $I = (a, b)$ a real open interval. We ask (1) when the integral function

$$F(t) := \int_K f(t, x) d^n x \quad \text{for } t \in I \tag{A.1}$$

is continuous and (2) when it is differentiable in t if f is. The integral can be a Riemann integral if we assume the compact set K is Peano-Jordan measurable[1] as already mentioned. The Riemann integral coincides with the Lebesgue integral in view of the integrand's continuity in x. We can drop the request of K being measurable if the integral is understood in Lebesgue sense, because compact sets are Lebesgue measurable. With that observation, all the proofs that follow remain valid (even though the results could be deduced directly from *Lebesgue's dominated convergence theorem*).

The following elementary facts hold.

Theorem A.1 *Let $f : I \times K \to \mathbb{R}$ be a continuous map, where $K \subset \mathbb{R}^n$ is compact (Peano-Jordan measurable) and $I = (a, b)$ an open interval. If $t_0 \in I$ then:*

$$\lim_{t \to t_0} \int_K f(t, x) d^n x = \int_K \lim_{t \to t_0} f(t, x) d^n x = \int_K f(t_0, x) d^n x .$$

In particular the function $F : I \to \mathbb{R}$ defined in (A.1) is continuous.

Proof Consider a closed interval $J = [t_0 - c, t_0 + c] \subset I$ for some $c > 0$ and restrict f to the compact set (as Cartesian product of compact sets) $J \times K$. Then the restricted f is uniformly continuous as continuous on a compact set: for $\epsilon > 0$ there are $\delta > 0, \delta' > 0$ such that

$$|f(t, x) - f(t', x')| < \epsilon \text{ if } |t - t'| < \delta \text{ and } ||x - x'|| < \delta' \text{ with } t, t' \in J \text{ and } x, x' \in K.$$

In particular

$$|f(t, x) - f(t', x)| < \epsilon \quad \text{if } |t - t'| < \delta \text{ with } t \in J \text{ and } x \in K.$$

Consequently, if $M := \int_K 1 d^n x$ (certainly finite since K is compact),

$$\left| \int_K f(t, x) d^n x - \int_K f(t_0, x) d^n x \right| \leq \int_K |f(t, x) - f(t_0, x)| d^n x$$

$$< M\epsilon \quad \text{for } |t - t_0| < \delta.$$

The claim follows immediately.   □

---

[1] That is, the boundary of K must have zero Lebesgue measure.

Let us pass to the analogous question of whether we can swap the integral in x and the derivative in t.

Theorem A.2 *Let $f : I \times K \to \mathbb{R}$ be a continuous map such that $\frac{\partial f(t,x)}{\partial t}$ exists and is continuous on $I \times K$, where $K \subset \mathbb{R}^n$ is compact (Peano-Jordan measurable) and $I = (a, b)$ an open interval. If $t_0 \in I$ then:*

$$\exists \left.\frac{d}{dt}\right|_{t_0} \int_K f(t, x) d^n x = \int_K \frac{\partial f(t_0, x)}{\partial t} d^n x \, .$$

At last, $F : I \to \mathbb{R}$ defined in (A.1) is $C^1(I)$.

Proof Consider $t_0 \in I$ and an interval $I' = (-c, c)$ for some $c > 0$ so that $t_0 \pm c \in I$. The map $g : I' \times K \mapsto \mathbb{R}$ defined by

$$g(h, x) := \frac{f(t_0 + h, x) - f(t_0, x)}{h} \quad \text{for } h \neq 0 \text{ and } x \in K, \text{ or}$$

$$g(0, x) := \frac{\partial f(t, x)}{\partial t} \quad \text{for } x \in K \tag{A.2}$$

is well defined and continuous as we will show below. We can then apply the previous theorem to g and conclude

$$\exists \left.\frac{d}{dt}\right|_{t_0} \int_K f(t, x) d^n x = \lim_{h \to 0} \int_K \frac{f(t_0 + h, x) - f(t_0, x)}{h} d^n x$$

$$= \int_K \lim_{h \to 0} \frac{f(t_0 + h, x) - f(t_0, x)}{h} d^n x$$

$$= \int_K \frac{\partial f(t_0, x)}{\partial t} d^n x \, .$$

Since the last integrand is continuous, the final claim in the previous theorem ensures that

$$I \ni t \mapsto \frac{d}{dt} \int_K f(t, x) d^n x = \int_K \frac{\partial f(t, x)}{\partial t} d^n x$$

is continuous. Hence F in (A.1) is $C^1(I)$.

To finish the proof we only need to show $g : I' \times K \mapsto \mathbb{R}$, defined in (A.2), is continuous. The only thing to prove is that if $x_0 \in K$ then

$$\lim_{(h,x) \mapsto (0,x_0)} g(h, x) = g(0, x_0) \, ,$$

since the other cases are obvious. The mean value theorem, applied to every given $x \in K$, implies that there exists $h'_x \in [-h, h]$ such that:

$$g(h, x) = \frac{f(t_0 + h, x) - f(t_0, x)}{h} = \frac{\partial f(t_0 + h'_x, x)}{\partial t}.$$

Te map $\frac{\partial f(t,x)}{\partial t}$ is (jointly) continuous at (t_0, x_0) by assumption: for any $\epsilon > 0$ there exist $\delta, \delta' > 0$ such that

$$\left| \frac{\partial f(t, x)}{\partial t} - \frac{\partial f(t_0, x_0)}{\partial t} \right| < \epsilon$$

provided $|t - t_0| < \delta$ and $||x - x_0|| < \delta'$. Immediately

$$|g(h, x) - g(0, x_0)| = \left| g(h, x) - \frac{\partial f(t_0, x_0)}{\partial t} \right| = \left| \frac{\partial f(t_0 + h'_x, x)}{\partial t} - \frac{\partial f(t_0, x_0)}{\partial t} \right| < \epsilon$$

if $|h| < \delta$ and $||x - x_0|| < \delta'$, since $|(t_0 + h'_x) - t_0| = |h'_x| \leq |h| < \delta$. We have shown g is continuous at the points $(0, x_0)$, thus ending the proof.   □

A.3   Series of Vector-Valued Functions

Consider a series of vector-valued functions:

$$\sum_{n=0}^{+\infty} F_n(x), \tag{A.3}$$

where $F_n : A \to \mathbb{R}^m$ and $A \subset \mathbb{R}^n$. The convergence is studied in the topologies induced by the Euclidean norm of $\mathbb{R}^n$ on the common domain A and by the Euclidean norm of $\mathbb{R}^m$ on the common codomain. We will write $|| \cdot ||$ for both norms. Here is a first result.

Proposition A.3 *If the functions F_n in (A.3) are continuous and the series' convergence is uniform, the sum of the series is a continuous function on A.*

Proof Put $S(x) := \sum_{n=0}^{+\infty} F_n(x) = \lim_{N \to +\infty} S_N(x)$, where $S_N(x) := \sum_{n=0}^{N} F_n(x)$ is the N-th partial sum. Simply applying the triangle inequality gives:

$$||S(x) - S(y)|| \leq ||S(x) - S_N(x)|| + ||S_N(x) - S_N(y)|| + ||S_N(y) - S(y)||.$$

Fix $x \in A$. For a given $\epsilon > 0$ we can choose N so that $||S_N(x) - S(x)|| < \epsilon/3$ and $||S_N(y) - S(y)|| < \epsilon/3$ at the same time, due to uniform convergence. For such N, since S_N is continuous in x, we can find $\delta > 0$ for which $||S_N(x) - S_N(y)|| < \epsilon/3$ if

$||x - y|| < \delta$. Putting everything together we have proved that $||x - y|| < \delta$ implies $||S(x) - S(y)|| < \epsilon$, which is indeed the claim given that $x \in A$ is completely arbitrary. □

Another useful result is the following.

Proposition A.4 *Let $A \subset \mathbb{R}^n$ be open. Suppose the terms F_n of series (A.3) are continuously differentiable with respect to the component x^k of $x = (x^1, \ldots, x^n)$. If the series converges on A and the series of partial derivatives with respect to x^k converges uniformly on A, then:*

(1) $S(x)$ admits continuous partial derivative in x^k on A;
(2) the derivative equals:

$$\frac{\partial S}{\partial x^k} = \sum_{n=0}^{+\infty} \frac{\partial F}{\partial x^k} ,$$

everywhere on A.

Proof Fix $x_0 \in A$ and consider an open neighbourhood given by an n-cube Q of edge $2\delta > 0$ centred at x_0, entirely contained in A. Define the continuous (by the previous proposition) map

$$H(x) := \sum_{n=0}^{+\infty} \frac{\partial F_n}{\partial x^k} . \tag{A.4}$$

If $x := (x^1, \ldots, x^k + h, \ldots, x^n) \in Q$ with $h \geq 0$, it follows that:

$$\int_{x_0^k}^{x_0^k+h} H(x) dx^k = \sum_{n=0}^{+\infty} \int_{x_0^k}^{x_0^k+h} \frac{\partial F_n}{\partial x^k} dx^k = \sum_{n=0}^{+\infty} (F_n(x) - F_n(x_0)) = S(x) - S(x_0) .$$

$$\tag{A.5}$$

We will prove the first identity at the end of this argument, while the second and third ones are obvious from the fundamental theorem of calculus and the definition of S, respectively. Assuming the chain of identities does hold, computing the derivative in h at $h = 0$ of the first and last terms and using the fundamental theorem of calculus we find:

$$H(x_0) = \frac{\partial S}{\partial x^k}\bigg|_{x_0} .$$

By definition of H and the arbitrariness of $x_0 \in A$ the claim follows. To finish we only need to justify the first identity in (A.5). Observe that the convergence of the series defining H in (A.4) is uniform by hypothesis, so

$$\lim_{N \to +\infty} \sup_Q \left| H(x) - \sum_{n=0}^{N} \frac{\partial F_n}{\partial x^k} \right| = 0$$

and then

$$\left| \int_{x_0^k}^{x_0^k+h} H(x) dx^k - \sum_{k=0}^{N} \int_{x_0^k}^{x_0^k+h} \frac{\partial F_n}{\partial x^k} dx^k \right| \leq \sup_Q \left| H(x) - \sum_{n=0}^{N} \frac{\partial F_n}{\partial x^k} \right| \int_{x_0^k}^{x_0^k+h} 1 dx^k \to 0$$

as $N \to +\infty$, as we wanted. $\qquad\qquad\qquad\qquad\qquad\qquad\qquad\qquad\qquad\qquad\square$

A.4   Deformation of Curves

Proposition A.5 *Let $\Omega \subset \mathbb{R}^n$ be an open set and $\gamma : I \to \Omega$ a continuous map, where $I = [a, b]$ with $a < b$. If $\eta : I \to \mathbb{R}^n$ is another continuous map there exists $\delta > 0$ such that $\gamma(t) + \alpha \eta(t) \in \Omega$ if $t \in I$ and $|\alpha| < \delta$.*

Proof We assume η is not the zero function, for otherwise the proof would be trivial. For any $t \in I$ there is an open ball B_t centred at $\gamma(t)$ with finite radius such that $B_t \subset \Omega$. The union of such balls covers the compact set $\gamma(I)$ (continuous image of a compact set). By definition of compact set we can extract a finite number of such balls from the covering $\{B_t\}_{t \in I}$ and still have a (now finite) covering of $\gamma(I)$. Let $B \subset \Omega$ denote the union of the finite covering. Clearly B is a bounded set as finite union of bounded sets, so its boundary ∂B is compact (as closed and bounded). By construction, moreover, $\gamma(I) \cap \partial B = \varnothing$. The distance function $\partial B \times I \ni (p, t) \mapsto d(p, \gamma(t))$ is continuous and defined on a compact set (product of compact sets), so it admits minimum d at some $(p_0, t_0) \in B \times I$. Since $\gamma(t_0) \neq p_0$, due to $\gamma(I) \cap \partial B = \varnothing$, we must have $d = d(p_0, \gamma(t_0)) > 0$. If $D := ||\eta(t_1)|| > 0$ is the maximum value of the non-zero continuous map $I \ni t \mapsto ||\eta(t_1)||$ on the compact set I, we define $\delta = d/D$. We conclude that if $|\alpha| < \delta$ then $\gamma(t) + \alpha \eta(t) \in B \subset \Omega$. In fact, if there existed $\alpha_0 \geq 0$ (for $\alpha_0 \leq 0$ the argument is similar) with $\alpha_0 < \delta$, for which $\gamma(t) + \alpha \eta(t) \notin B$ for some $t \in I$, letting $\alpha_1 = \sup\{\beta \in \mathbb{R} \,|\, \gamma(t) + \alpha \eta(t) \in B, \text{ if } 0 \leq \alpha \leq \beta\}$ (this exists, finite, since the set is non-empty and upper bounded by α_0), we would obtain $\gamma(t) + \alpha_1 \eta(t) = p \in \partial B$. But that is impossible because it would mean $d(p, \gamma(t)) = ||\alpha_1 \eta(t)|| < \delta D = d = \min\{d(p, \gamma(u)) \,|\, p \in \partial B, u \in I\}$. $\quad\square$

A.5   Symmetric Operators on Finite-Dimensional Real Vector Spaces

We recall that given a linear operator $A : S \rightarrow S$ on a vector space S over $\mathbb{K} = \mathbb{R}$ or $\mathbb{C}$, a vector $\mathbf{v} \in S \setminus \mathbf{0}$ is an **eigenvector** of A with **eigenvalue** $\lambda \in \mathbb{K}$ whenever $A\mathbf{v} = \lambda\mathbf{v}$. If $\lambda \in \mathbb{K}$ is an eigenvalue, the associated **eigenspace** S_λ is the vector space spanned by the eigenvectors with eigenvalue λ. The number $\dim(S_\lambda)$ (which might be infinite) is the **geometric multiplicity** of λ.

Remarks A.6 In this section we shall only refer to $\mathbb{K} = \mathbb{R}$ and any vector space V will have finite dimension.   ∎

Consider a real vector space V of *finite dimension*, equipped with a positive-definite inner product $\cdot : V \times V \ni (\mathbf{u}, \mathbf{v}) \rightarrow \mathbf{u} \cdot \mathbf{v} \in \mathbb{R}$ (Definition 1.9). As customary, $||\mathbf{v}|| := \sqrt{\mathbf{v} \cdot \mathbf{v}}$, $\mathbf{v} \in V$, is the **standard norm** associated with the inner product.

An **orthonormal set** is a collection of mutually orthogonal, unit vectors. Non-vanishing pairwise othogonal vectors are *linearly independent*. Indeed, if $\sum_{k=1}^n c_k \mathbf{v}_k = 0$, taking the inner product with $\mathbf{v}_{k_0}$ produces $c_{k_0}||\mathbf{v}_{k_0}||^2 = 0$, so $c_{k_0} = 0$. As a consequence an orthonormal set $\{\mathbf{v}_i\}_{i=1,...,n}$ of cardinality $n = \dim(V)$ is a basis of V, called **orthonormal basis**, and furthermore:

$$\mathbf{u} = \sum_{i=1}^{\dim(V)} (\mathbf{u} \cdot \mathbf{v}_i)\mathbf{v}_i , \quad ||\mathbf{u}||^2 = \sum_{i=1}^{\dim(V)} (\mathbf{u} \cdot \mathbf{v}_i)^2 \quad \text{for any } \mathbf{u} \in V$$

as the reader easily proves.

The **orthogonal subspace** to $E \subset V$ is $E^\perp := \{\mathbf{u} \in V \mid \mathbf{u} \cdot \mathbf{v} = 0\}$. Observe that $E^\perp$ is a subspace even if E is not. If $M \subset V$ is a subspace we have a **direct decomposition** $V = M \oplus M^\perp$, which means that any $\mathbf{v} \in V$ decomposes as a sum $\mathbf{v} = \mathbf{u} + \mathbf{u}' =: \mathbf{u} \oplus \mathbf{u}'$ of for unique vectors $\mathbf{u} \in M$, $\mathbf{u}' \in M^\perp$. The proof is left as exercise [Axl16, Nor86].

Definition A.7 Given a linear operator $A : V \rightarrow V$ on the real vector space V, the **adjoint** to A (or **transposed**) is the operator $A^t : V \rightarrow V$ determined by asking

$$\mathbf{v} \cdot A^t\mathbf{u} = A\mathbf{v} \cdot \mathbf{u} \quad \text{for any pair } \mathbf{u}, \mathbf{v} \in V.$$

We call A **symmetric** if $A^t = A$. With other words, $\mathbf{u} \cdot A\mathbf{v} = A\mathbf{u} \cdot \mathbf{v}$ for any vectors $\mathbf{u}, \mathbf{v} \in V$.
A symmetric operator A is **positive semi-definite** if $\mathbf{u} \cdot A\mathbf{u} \geq 0$ for any $\mathbf{u} \in V$, and **positive definite** if $\mathbf{u} \cdot A\mathbf{u} > 0$ for any $\mathbf{u} \in V \setminus \{\mathbf{0}\}$.   ◇

The existence of A^t is proved by picking an orthonormal basis $\{\mathbf{u}_k\}_{k=1,...,\dim V}$ and noting that $\mathbf{v} \cdot A^t\mathbf{u} = A\mathbf{v} \cdot \mathbf{u}$ forces the explicit expression $A^t\mathbf{v} = \sum_k (\mathbf{v} \cdot A\mathbf{u}_k)\mathbf{u}_k$, for any $\mathbf{v} \in V$. Regarding uniqueness, if A^t, $A^{\prime t}$ are both adjoints then $\mathbf{u} \cdot (A^t - A^{\prime t})\mathbf{v} =$

$(A - A)\mathbf{u} \cdot \mathbf{v} = 0$, and choosing $\mathbf{u} := (A^t - A^{\prime t})\mathbf{v}$ leads to $||(A^t - A^{\prime t})\mathbf{v}||^2 = 0$ for any $\mathbf{v} \in V$, so $A^t = A^{\prime t}$.

When A is symmetric, eigenvectors with distinct eigenvalues are orthogonal. Indeed, suppose $A\mathbf{u} = \lambda\mathbf{u}$ and $A\mathbf{v} = \mu\mathbf{v}$. Then:

$$0 = \mathbf{u} \cdot A\mathbf{v} - A\mathbf{u} \cdot \mathbf{v} = \mathbf{u} \cdot \mu\mathbf{v} - \lambda\mathbf{u} \cdot \mathbf{v} = (\mu - \lambda)\mathbf{u} \cdot \mathbf{v}.$$

This immediately proves for symmetric operators the general result whereby eigenvectors of linear operators with distinct eigenvalues are *linearly independent*.

There is an important existence result regarding eigenvector bases for symmetric operators on finite-dimensional real vector spaces.

Proposition A.8 *Let* $A : V \to V$ *be a symmetric linear operator on a finite-dimensional real vector space* V *with positive-definite inner product* $\cdot : V \times V \ni (\mathbf{u}, \mathbf{v}) \to \mathbf{u} \cdot \mathbf{v} \in \mathbb{R}$.

(1) *The eigenvalues* λ *of* A *are the roots of the characteristic polynomial* $\mathbb{C} \ni z \mapsto \det(\mathcal{A}-zI)$ *in* $\mathbb{C}$, *where* $\mathcal{A} \in M(\dim(V), \mathbb{R})$ *is the matrix* $[\mathbf{v}_i \cdot A\mathbf{v}_j]_{i,j=1,\dots,\dim(V)}$ *of* A *in any given orthonormal basis* $\{\mathbf{v}_i\}_{i=1,\dots,\dim(V)}$ *of* V.

(2) *There exists an orthonormal basis* $B \subset V$ *of eigenvectors of* A. *Each eigenvalue of* A *has at least one eigenvector in* B. *The elements of* B *corresponding to the same* λ *form an orthogonal basis of the eigenspace* V_λ. *The matrix of* A *in the basis* B *is diagonal,* $\mathrm{diag}(\lambda_1, \dots, \lambda_{\dim(V)})$, *where the numbers* λ_j *are eigenvalues of* A, *and each eigenvalue* λ *appears precisely with its geometric multiplicity.*

(3) A *is positive (semi-)definite if and only if all eigenvalues are positive (respectively, non-negative).* A *is positive definite if and only if it is positive semi-definite and non-singular (* $A\mathbf{v} = \mathbf{0}$ *implies* $\mathbf{v} = \mathbf{0}$).

Proof Start with (1). Choose an orthonormal basis $\{\mathbf{v}_i\}_{i=1,\dots,\dim(V)}$ and represent A by the matrix $\mathcal{A} = [\mathbf{v}_i \cdot A\mathbf{v}_j]_{i,j=1,\dots,\dim(V)}$. As $A^t = A$, the matrix $\mathcal{A}$ satisfies $\mathcal{A}^t = \mathcal{A}$ where now $\mathcal{A}^t$ is the transpose matrix of $\mathcal{A} \in M(\dim(V), \mathbb{R})$. If $\lambda \in \mathbb{R}$ solves $A\mathbf{v} = \lambda\mathbf{v}$ for some $\mathbf{v} \neq 0$, the $\mathbb{C}^n$ vector made of the (real!) components of $\mathbf{v}$ in the above basis is non-zero, and kills $\mathcal{A} - \lambda I$. Consequently $\det[\mathcal{A} - \lambda I] = 0$. We claim that the roots $\lambda \in \mathbb{C}$ of the equation, which exist by the Fundamental theorem of Algebra, are all real, and for each one there exists a vector $\mathbf{v}_\lambda \in V \setminus \{0\}$ such that $A\mathbf{v}_\lambda = \lambda\mathbf{v}_\lambda$. In this way the eigenvalues (real, by definition) of A are precisely the complex roots of $\det[\mathcal{A} - \lambda I] = 0$, irrespective of the orthonormal basis $\{\mathbf{v}_i\}_{i=1,\dots,\dim(V)}$.

We know $\lambda \in \mathbb{C}$ solves $\det[\mathcal{A} - \lambda I] = 0$ if and only if there is an $\mathbf{x} \in \mathbb{C}^n \setminus \{0\}$ such that $\mathcal{A}\mathbf{x} = \lambda\mathbf{x}$. Hence

$$\lambda\overline{\mathbf{x}}^t\mathbf{x} = \overline{\mathbf{x}}^t\mathcal{A}\mathbf{x} = (\overline{\mathbf{x}}^t\mathcal{A}\mathbf{x})^t = \mathbf{x}^t\mathcal{A}^t\overline{\mathbf{x}} = \mathbf{x}^t\mathcal{A}\overline{\mathbf{x}} = \mathbf{x}^t\overline{\mathcal{A}\mathbf{x}} = \overline{\overline{\mathbf{x}}^t\mathcal{A}\mathbf{x}} = \overline{\lambda\overline{\mathbf{x}}^t\mathbf{x}} = \overline{\lambda}\overline{\mathbf{x}}^t\mathbf{x}$$

(where the overline is complex conjugation on each component). As $\overline{\mathbf{x}}^t\mathbf{x} \neq \mathbf{0}$, we conclude $\overline{\lambda} = \lambda$ and therefore the roots of the characteristic polynomial of $\mathcal{A}$ are real. Finally, conjugating $A\mathbf{x} = \lambda\mathbf{x}$ we obtain $A\overline{\mathbf{x}} = \lambda\overline{\mathbf{x}}$, and so $A(\mathbf{x}+\overline{\mathbf{x}}) = \lambda(\mathbf{x}+\overline{\mathbf{x}})$ and $Ai(\mathbf{x}-\overline{\mathbf{x}}) = \lambda i(\mathbf{x}-\overline{\mathbf{x}})$. Consider the real components of $\mathbf{x}+\overline{\mathbf{x}}$ and $i(\mathbf{x}-\overline{\mathbf{x}})$. At least one set must be non-zero, otherwise $\mathbf{x} = \mathbf{0}$ and this was excluded. Call $\mathbf{x}_\lambda \in \mathbb{R}^{\dim(V)} \setminus \{\mathbf{0}\}$ this set. If the components $\mathbf{v}_\lambda \in V$ with respect to the chosen basis are precisely $\mathbf{x}_\lambda$, then $A\mathbf{v}_\lambda = \lambda\mathbf{v}_\lambda$. Hence each complex root of $\det[A-\lambda I] = 0$ is real and is an eigenvalue of A.

Let us show (2). Take an eigenvalue $\lambda \in \mathbb{R}$ of A and $\mathbf{v}_\lambda$ a corresponding eigenvector. Using $A = A^t$, if $\mathbf{u} \cdot \mathbf{v}_\lambda = 0$ then $A\mathbf{u} \cdot \mathbf{v}_\lambda = \lambda\mathbf{u} \cdot \mathbf{v}_\lambda = 0$, so $A(\{\mathbf{v}_\lambda\}^\perp) \subset \{\mathbf{v}_\lambda\}^\perp$. The restriction $A|_{\{\mathbf{v}_\lambda\}^\perp} : \{\mathbf{v}_\lambda\}^\perp \to \{\mathbf{v}_\lambda\}^\perp$ is a symmetric operator, so it will admit an eigenvector $\mathbf{v}_{\lambda_2}$ with eigenvalue λ_2, which might coincide with λ. Now we repeat the procedure with the symmetric linear map $A|_{\{\mathbf{v}_\lambda,\mathbf{v}_{\lambda_2}\}^\perp} : \{\mathbf{v}_\lambda,\mathbf{v}_{\lambda_2}\}^\perp \to \{\mathbf{v}_\lambda,\mathbf{v}_{\lambda_2}\}^\perp$, thus obtaining a third eigenvector $\mathbf{v}_{\lambda_3}$ of A, orthogonal to the span of $\mathbf{v}_\lambda$, $\mathbf{v}_{\lambda_2}$, with eigenvalue λ_3 (which may coincide with λ or λ_1). Iterating, we end up with a collection of $\dim(V)$ pairwise orthogonal eigenvectors. The procedure stops because the latter are linearly independent (being orthogonal) and their number is the dimension of V. Once normalised, they give an orthonormal basis B of eigenvectors on V. Each eigenvalue λ has one element of B as eigenvector. By contradiction, if an eigenvalue λ_0 did not have an eigenvector in B, then every eigenvector corresponding to λ_0 would be orthogonal to, hence linearly independent of B, which would then not be a basis. At last, take the elements of B associated with one eigenvalue λ. They are pairwise orthogonal, so linearly independent, and must generate the eigenspace V_λ, since the remaining vectors in B are orthogonal to V_λ and therefore linearly independent of V_λ. The last statement in (2) is straightforward from the above properties of B.

Now (3). Write A in the basis B of part (2) as a diagonal matrix $A = \mathrm{diag}(\lambda_1, \ldots, \lambda_{\dim(V)})$, where the diagonal elements are the (possibly not distinct) eigenvalues. Clearly $\mathbf{u} \cdot A\mathbf{u} = \sum_{j=1}^{\dim(V)} \lambda_j(u^j)^2$, where the real numbers u^j are the components of $\mathbf{u}$ in the basis. The claim is now immediate.   □

Corollary A.9 *If $A \in M(n, \mathbb{R})$ is a symmetric matrix, there exists a matrix $R \in O(n)$ such that $RAR^t = \mathrm{diag}(\lambda_1, \ldots, \lambda_n)$, where the λ_j are the eigenvalues of A, each appearing with multiplicity equal to the dimension of its eigenspace.*

Proof It follows directly from the previous argument. The matrix R represents basis change between the canonical basis of $\mathbb{R}^n$ and the orthogonal basis of eigenvectors from (2) in the previous proposition.   □

Definition A.10 An operator $P : V \to V$ satisfying $PP = P$ and $P^t = P$ is called an **orthogonal projector**. We also say that P **projects onto** the image subspace $P(V)$.   ◇

Any subspace $M \subset V$ admits a unique orthogonal projector $P_M : V \to V$ onto M. In fact, from $P_M P_M = P_M$, $P_M^t = P_M$ and $P_M(V) = M$ it is easy to deduce that P_M acts on the orthogonal decomposition $V = M \oplus M^\perp$ by $P_M(\mathbf{u}\oplus\mathbf{v}) = \mathbf{u}\oplus\mathbf{0}$. Therefore P_M is fully determined by M.

Symmetric operators on finite-dimensional real spaces satisfy a spectral decomposition. It is the simplest instance of the *Spectral theorem for self-adjoint operators on complex Hilbert spaces* (see [Mor18] for example).

Theorem A.11 *Let V be a finite-dimensional real vector space with a positive-definite inner product. A symmetric linear operator $A : V \to V$ decomposes uniquely as:*

$$A = \sum_{\lambda \in \sigma(A)} \lambda P_\lambda , \qquad (A.6)$$

where

(1) $\sigma(A) \subset \mathbb{R}$ is a finite set,
(2) the operators $P_\lambda : V \to V$:

> *(a) are non-zero orthogonal projectors,*
> *(b) $P_\lambda P_\mu = 0$ for $\lambda \neq \mu$,*
> *(c) $\sum_{\lambda \in \sigma(A)} P_\lambda = I$.*

Then:

(i) $\sigma(A)$ is the set of eigenvalues of A,
(ii) each P_λ is the orthogonal projector onto the eigenspace V_λ of A associated with $\lambda \in \sigma(A)$.

Formula (A.6) is the **spectral decomposition** *of A, and $\sigma(A)$ is the* **spectrum** *of A.*

Proof In agreement with Proposition A.8 we take an orthonormal basis of V made of eigenvectors of A, whose vectors will be denoted $\mathbf{u}_{\lambda,\alpha_\lambda}$, and $A\mathbf{u}_{\lambda,\alpha_\lambda} = \lambda \mathbf{u}_{\lambda,\alpha_\lambda}$ and $\alpha_\lambda = 1, 2, \ldots, \dim(V_\lambda)$ for any eigenvalue λ of A. Set

$$P_\lambda^{(A)} : V \ni \mathbf{v} \mapsto \sum_{\alpha_\lambda=1}^{\dim V_\lambda} (\mathbf{u}_{\lambda,\alpha_\lambda} \cdot \mathbf{v})\mathbf{u}_{\lambda,\alpha_\lambda} \in V .$$

An easy calculation shows that each $P_\lambda^{(A)}$ is an orthogonal projector onto V_λ (hence it is non-zero) and that (A.6) holds, where $\sigma(A) \subset \mathbb{R}$ is the (finite) eigenvalue set of A. In particular $P_\lambda^{(A)} P_\mu^{(A)} = 0$ when $\lambda \neq \mu$, because eigenvectors of symmetric operators with distinct eigenvalues are orthogonal. Moreover, $\sum_{\lambda \in \sigma(A)} P_\lambda^{(A)} = I$ since $\{\mathbf{u}_{\lambda,\alpha_\lambda}\}_{\lambda \in \sigma(A), \lambda_\alpha=1,\ldots,\dim V_\lambda}$ is an orthonormal basis. We have thus shown that there is a pair $(\sigma(A), \{P_\lambda\}_{\lambda \in \sigma(A)})$ fulfilling (A.6), (1), (2), (i), (ii).
Now let us prove that a pair $(\sigma(A), \{P_\lambda\}_{\lambda \in \sigma(A)})$ satisfying (A.6), (1), (2) must necessarily also satisfy (i) and (ii).
Proof that (A.6), (1), (2) Imply (i) Choose $\lambda \in \sigma(A)$ and set $\mathbf{v} = P_\lambda \mathbf{u}$, where $\mathbf{u} \in V$ is such that $\mathbf{v} \neq \mathbf{0}$. This $\mathbf{u}$ must exist for otherwise $P_\lambda = 0$, which is forbidden by (a). By definition of projector, $P_\lambda \mathbf{v} = P_\lambda P_\lambda \mathbf{u} = P_\lambda \mathbf{u} = \mathbf{v}$. Using (A.6) and (b), we deduce $A\mathbf{v} = \sum_{\mu \in \sigma(A)} \mu P_\mu \mathbf{v} = \sum_{\mu \in \sigma(A)} \mu P_\mu P_\lambda \mathbf{v} = \lambda P_\lambda \mathbf{v} = \lambda \mathbf{v}$. Hence $\sigma(A)$ is

made of eigenvalues of A. On the other hand we claim that each eigenvalue λ of A must belong to $\sigma(A)$. Suppose $\lambda \notin \sigma(A)$ is an eigenvalue with eigenvector $\mathbf{v}$. Then (A.6) implies $\lambda \mathbf{v} = A\mathbf{v} = \sum_{\mu \in \sigma(A)} \mu P_\mu \mathbf{v} = \sum_{\mu \in \sigma(A)} \mu \mathbf{v}_\mu$, where each $\mathbf{v}_\mu := P_\mu \mathbf{v}$ is either null or an eigenvector with eigenvalue $\mu \neq \lambda$, as we saw above. If $\lambda \neq 0$, the identity $-\lambda \mathbf{v} + \sum_{\mu \in \sigma(A)} \mu \mathbf{v}_\mu = \mathbf{0}$ cannot hold, since eigenvectors with distinct eigenvalues are orthogonal and so linearly independent. If, vice versa, $\lambda = 0$, the identity reduces to $\sum_{\mu \in \sigma(A)} \mu P_\mu \mathbf{v} = \mathbf{0}$, where every μ is non-zero ($\mu \neq \lambda = 0$). Therefore (eigenvectors with distinct eigenvalues are linearly independent) $P_\mu \mathbf{v} = 0$ for any $\mu \in \sigma(A)$. But again, this cannot be, because (c) would force the (non-zero) eigenvector $\mathbf{v}$ to satisfy $\mathbf{0} \neq \mathbf{v} = I\mathbf{v} = \sum_{\mu \in \sigma(A)} P_\mu \mathbf{v} = \mathbf{0}$. We conclude that our λ cannot exist, so $\sigma(A)$ contains the eigenvalues of A and hence is their set.

Proof that (A.6), (1), (2) Imply (ii) We know $P_\lambda(V) \subset V_\lambda$ because $A P_\lambda \mathbf{u} = \lambda P_\lambda \mathbf{u} = \lambda P_\lambda P_\lambda \mathbf{u}$. To finish it suffices to show that any eigenvector $\mathbf{v}$ with eigenvalue λ satisfies $P_\lambda \mathbf{v} = \mathbf{v}$. If so, in fact, $\mathbf{v} \in P_\lambda(V)$ and then $P_\lambda(V) \supset V_\lambda$; hence P_λ projects exactly onto V_λ, so it coincides with $P_\lambda^{(A)}$. If $\mathbf{v}$ is the λ-eigenvector of A, by (A.6) $\lambda \mathbf{v} = A\mathbf{v} = \sum_{\mu \in \sigma(A)} \mu P_\mu \mathbf{v}$, and then $\lambda(P_\lambda \mathbf{v} - \mathbf{v}) + \sum_{\mu \neq \lambda} \mu P_\mu \mathbf{v} = \mathbf{0}$. If $\lambda \neq 0$ it follows that $P_\lambda \mathbf{v} - \mathbf{v} = 0$, otherwise $P_\lambda \mathbf{v} - \mathbf{v} \neq \mathbf{0}$ would be a λ-eigenvector and $\lambda(P_\lambda \mathbf{v} - \mathbf{v}) + \sum_{\mu \neq \lambda} \mu P_\mu \mathbf{v} = \mathbf{0}$ would give a non-trivial linear combination of eigenvectors with distinct eigenvalues, so linearly independent. All in all, $\lambda \neq 0$ implies $P_\lambda \mathbf{v} = \mathbf{v}$, ending the proof. In the case $\lambda = 0$ the identity $\lambda \mathbf{v} = \sum_{\mu \in \sigma(A)} \mu P_\mu \mathbf{v}$ still implies $\sum_{\mu \neq 0} \mu P_\mu \mathbf{v} = \mathbf{0}$, and then $P_\mu \mathbf{v} = 0$ for $\mu \neq 0$. Because of (c) we have $\mathbf{v} = I\mathbf{v} = \sum_{\mu \in \sigma(A)} P_\mu \mathbf{v} = P_0 \mathbf{v} + \sum_{\mu \neq 0} P_\mu \mathbf{v} = P_0 \mathbf{v} = P_\lambda \mathbf{v}$, really concluding the argument. $\square$

A.6   Elements of Differential Geometry

The most general and powerful mathematical tool apt to describe the general properties of spacetime, of three-dimensional physical space and of the abstract space in which we describe the physical systems of classical theories, is the notion of *differentiable manifold*. In essence, this is a set of arbitrary objects, called by the generic name of *points*, that can be locally covered by coordinate frames identifying the points with n-tuples of $\mathbb{R}^n$. For completeness we recall below the definition of a differentiable manifold (Definition 1.22) and the observations already made in Chap. 1.

Definition A.12 A **differentiable manifold of dimension** n and **class** C^k, for given $n \in \{1, 2, 3, \cdots\}$ and $k \in \{1, 2, ..., \} \cup \{\infty\}$, is a locally Euclidean, Hausdorff and second countable topological space M, whose elements are called **points**, equipped with a **differentiable structure of class** C^k and **dimension** n. The differentiable structure is a collection of n-dimensional local charts $\mathcal{A} = \{(U_i, \phi_i)\}_{i \in I}$ satisfying the following:

(i) $\cup_{i \in I} U_i = M$;
(ii) the local charts in $\mathcal{A}$ must be pairwise C^k *compatible*;

(iii) $\mathcal{A}$ is **maximal** with respect to (ii): if (U, ϕ) is an n-dimensional local chart on M compatible with every chart of $\mathcal{A}$, then $(U, \phi) \in \mathcal{A}$. ◇

Remarks A.13

(1) Every local chart (U, ϕ) on a differentiable manifold M allows to map, in a 1-1 way, an n-tuple of real numbers $(x_p^1, \cdots, x_p^n) = \phi(p)$ to every point p of U. The elements of the n-tuple are the **coordinates** of p in the chart (U, ϕ). The points in U are therefore in 1-1 correspondence with the n-tuples of $\phi(U) \subset \mathbb{R}^n$. A local chart with domain M is called a **global chart** or **global coordinate system**.

(2) The two demands on the *type* of topology (valid for the standard topology of $\mathbb{R}^n$) are technical. The Hausdorff property warrants the uniqueness of solutions to problems based on differential equations on M (this is necessary in Physics when the equations describe the evolution of physical systems). The existence of a countable basis guarantees that integration over M works properly. That the local charts are, by definition, local homeomorphisms corresponds to the intuitive request that M "is continuous just like" $\mathbb{R}^n$ around every point, though classical counterexamples show how the Hausdorffness of $\mathbb{R}^n$ is not passed to M by a local homeomorphisms, so it must be imposed additionally.

(3) A collection of local charts $\mathcal{A}$ on the differentiable manifold M satisfying (i) and (ii) but perhaps not (iii) is called an **atlas** on M **of dimension** n and **class** C^k. It is easy to prove that for any atlas $\mathcal{A}$ on M there is a unique maximal atlas extending it, i.e. a unique differentiable structure that contains it. This structure is said to be **induced by the atlas**. Note that two atlases on M with mutually compatible charts induce the same differentiable structure on M. Hence to assign a differentiable structure it is enough to give an atlas, one of the many that induce it.

(4) There may exist several inequivalent differentiable structures on the same second countable Hausdorff space.[2] This happens in dimension ≥ 4. Therefore, to define a differentiable manifold N of dimension $n \geq 4$ it is not enough to specify the set N only, even if the appropriate topology was defined. Exceptions are the cases where N is a subset in a higher-dimensional manifold M in which N is embedded, as explained below.

(5) One can prove that if $1 \leq k < \infty$, we can forego some charts in the differentiable structure (an infinite number!), in such a way that the resulting collection is still an atlas with $k = \infty$. One can consider *real analytic* manifolds (in symbols, C^ω), where the maps $\phi \circ \psi^{-1}$ and $\psi \circ \phi^{-1}$ are real analytic. ∎

An important analytical result, with multiple consequences in Differential Geometry, is the following.

---

[2] Technically speaking: *non-diffeomorphic* differentiable structures. For more details refer to Differential Geometry lectures.

Theorem A.14 (Inverse Function Theorem)   *Let $D \subset \mathbb{R}^n$ be open and non-empty and $f : D \to \mathbb{R}^n$ a C^k map, with $k = 1, 2, \ldots, \infty$ given. If the Jacobian matrix of f, evaluated at $p \in D$, has non-zero determinant then there exist open neighbourhoods $U \subset D$ of p and V of $f(p)$ such that: (i) $f|_U : U \to V$ is bijective, (ii) the inverse $f|_U^{-1} : V \to U$ is of class C^k.*

Remarks A.15 If $U \cap V \neq \varnothing$, the k-compatibility of local charts (U, ϕ) and (V, ψ) implies that the Jacobian matrix of the transition function $\phi \circ \psi^{-1}$, being invertible, has everywhere non-zero determinant. Vice versa, if $\phi \circ \psi^{-1} : \psi(U \cap V) \to \phi(U \cap V)$, where the domain and codomain are open in $\mathbb{R}^n$, is bijective, of class C^k, with non-zero Jacobian determinant on $\psi(U \cap V)$, then $\psi \circ \phi^{-1} : \phi(U \cap V) \to \psi(U \cap V)$ is C^k, so the two local charts are k-compatible. The proof is left as Exercise A.21.5.   ∎

A.6.1   Product Manifolds

Given differentiable manifolds M and N of dimension m and n respectively, both of class C^k, $k \geq 1$, we can build a third differentiable manifold of class C^k and dimension $m + n$ on the set $M \times N$ with the product topology (still Hausdorff and second countable). This manifold is called *product manifold* of M and N and is denoted $M \times N$. Its differentiable structure, called *product structure*, is obtained as follows. If (U, ϕ) and (V, ψ) are local charts on M and N respectively, it is immediate that

$$U \times V \ni (p, q) \mapsto (\phi(p), \psi(q)) =: \phi \oplus \psi(p, q) \in \mathbb{R}^{m+n} \qquad (A.7)$$

is a local homeomorphism. Moreover, if (U', ϕ') and (V', ψ') are two other local charts on M and N respectively, each k-compatible with the corresponding previous one, the charts $(U \times V, \phi \oplus \psi)$ and $(U' \times V', \phi' \oplus \psi')$ are clearly k-compatible. Finally, as (U, ϕ) and (V, ψ) vary inside the differentiable structures of M and N, the charts $(U \times V, \phi \oplus \psi)$ define an atlas on $M \times N$. The differentiable structure it generates is by definition the product structure.

Definition A.16 Given two differentiable manifolds M and N, of dimension m and n respectively, both of class C^k, with $k \geq 1$, the **product manifold** $M \times N$ is the manifold $M \times N$ with product topology and differentiable structure induced by the local charts $(U \times V, \phi \oplus \psi)$ defined in (A.7), when (U, ϕ) and (V, ψ) vary in the differentiable structures of M and N.   ◇

A.6.2 Differentiable Maps

Since a differentiable manifold is locally indistinguishable from $\mathbb{R}^n$, the differentiable structure allows to make sense of the notions of a *differentiable map* and a *differentiable curve* defined on a set other than $\mathbb{R}^n$ or a subset, provided it has differentiable structure.

Definition A.17 Let M, N be differentiable manifolds of dimensions m, n and class C^p and C^q $(1 \leq p, q \leq \infty)$.

(1) A continuous map $f : M \rightarrow N$ is of **class C^k** $(0 \leq k \leq p, q$ and possibly $k = \infty)$ if $\psi \circ f \circ \phi^{-1}$ is C^k as map from $\mathbb{R}^m$ to $\mathbb{R}^n$, for any local charts $(U, \phi), (V, \psi)$, respectively in M and N. In this case we write $f \in C^k(M; N)$.

(2) A **curve of class C^k** in N $(0 \leq k \leq q$, possibly $k = \infty)$ is a map $\gamma : I \rightarrow N$, where $I \subset \mathbb{R}$ is an interval different from a point (which may contain either endpoints), that is the restriction of a C^k map $\gamma' : J \rightarrow N$ for some open interval $J \supset I$.

(3) A k-**diffeomorphism** $f : M \rightarrow N$ between manifolds M, N is a C^k bijection with C^k inverse. If M and N are related by a k-diffeomorphism f we say they are k-**diffeomorphic** (under f).

(4) If $f : M \rightarrow N$ is such that $f|_U : U \rightarrow f(U) = V$ is a k-diffeomorphism for open sets $U \subset M$ and $V \subset N$ with the obvious differentiable structures induced by M and N respectively, then f is called a **local k-diffeomorphism** (from U to V). $\diamond$

Remarks A.18

(1) It is customary to speak of *differentiable functions* and *differentiable curves*, with no further specification, when they are of the same class C^k as the manifolds on which they are defined.

(2) We have included differentiable maps of class C^0 which, as should be clear, refers to mere continuous functions, and to homeomorphisms when we say 0-diffeomorphisms. Obviously any k-diffeomorphism is a homeomorphism as well, so for example there cannot exist diffeomorphisms between $\mathbb{S}^2$ and $\mathbb{R}^2$ (or any non-empty open subset), since the source is compact and the target is not. This shows, as mentioned earlier, that there cannot be global charts on $\mathbb{S}^2$.

(3) It is easy to prove that for $f : M \rightarrow N$ to be C^p it suffices that the various $\psi \circ f \circ \phi^{-1}$ are C^k for all local charts $(U, \phi), (V, \psi)$ in two atlases, respectively on M and N, without having to check the condition for *all* possible local charts of the manifolds.

(4) If $f : M \rightarrow N$ is a differentiable map (of class C^k) and we have local charts $(U, \phi), (V, \psi)$ respectively in N and M, the function $\psi \circ f \circ \phi^{-1}$ is called **representation in coordinates** of f.

(5) A function $f : M \times N \rightarrow L$, with M, L, N differentiable manifolds, is said (jointly) **of class C^k** if it is C^k for the product structure of the domain. ∎

A.6.3 Embedded Submanifolds and Non-singular Maps

A fundamental notion is that of *embedded submanifold*. $\mathbb{R}^n$ is an embedded submanifold of $\mathbb{R}^m$ for $m > n$. In canonical coordinates $x^1, \cdots, x^m$ of $\mathbb{R}^m$, $\mathbb{R}^n$ is identified with the subset determined by the conditions $x^{n+1} = \cdots = x^m = 0$ while the first n coordinates of $\mathbb{R}^m$, $x^1, \cdots, x^n$, are identified with the standard coordinates of $\mathbb{R}^n$. The idea is now to generalise, locally, this situation using local coordinates and manifolds N and M instead of $\mathbb{R}^n$ and $\mathbb{R}^m$.

Definition A.19 Let M be a C^k manifold ($k \geq 1$) of dimension $m > n$. An **embedded n-dimensional submanifold of M of class C^k**, N, is a differentiable manifold (of dimension n and class C^k) built as follows.

(1) N is a subset of M with the induced topology of M.
(2) The differentiable structure of N is generated by an atlas $\{(U_i, \phi_i)\}_{i \in I}$ where:

   (i) $U_i = V_i \cap N$ and $\phi_i = \psi_i|_{V_i \cap N}$ for some local chart (V_i, ψ_i) on M;
   (ii) in the coordinates $x^1, \cdots, x^m$ associated with (V_i, ψ_i), the set $V_i \cap N$ is given by $x^{n+1} = \cdots = x^m = 0$ and the remaining coordinates $x^1, \cdots, x^n$ are local coordinates associated with ϕ_i. $\diamondsuit$

Remarks A.20

(1) The topology of N is, by construction, still Hausdorff and second countable.
(2) The structure of embedded submanifold on N, assuming it can be defined, is uniquely determined by the differentiable structure of M as prescribed by Exercise **4** below.
(3) There is a weaker version of submanifold called an *immersed submanifold*. A C^k manifold N ($k \geq 1$) of dimension n is **immersed** in the C^k manifold M ($k \geq 1$) of dimension $m > n$, if $N \subset M$ and for any $p \in N$ there exist a local chart (U, ψ) on N with coordinates $y^1, \ldots, y^n$ and a corresponding local chart (V, ϕ) on M with coordinates $x^1, \ldots, x^m$ such that $V \supset U \ni p$, U is described in V by $x^{n+1} = \ldots = x^m = 0$ and on U the other coordinates $x^1, \ldots, x^m$ coincide with $y^1, \ldots, y^m$ respectively. In practice the difference from an embedded submanifold is that now we do not ask $N \cap V = U$ but only that $V \supset U$, and the topology on N might not be the induced topology of M. An immersed manifold $N \subset M$ is therefore locally embedded in M, where *locally* refers to neighbourhoods of points $p \in N$: there is always an open neighbourhood $U \subset N$ of $p \in N$ such that $U \subset M$ is embedded in M.
(4) Suppose $\gamma : \mathbb{R} \ni t \mapsto \gamma(t) \in M$ is a C^k curve (Definition A.36) that densely fills out, without self-intersecting, a two-dimensional C^k torus M. The image $N = \gamma(\mathbb{R})$ is then an *immersed submanifold* in M, but it is *not* embedded in M, when we put on it the differentiable structure of $\mathbb{R}$ given by the parameter t seen as global coordinate on N. If we restrict to small enough intervals of t, though, we obtain embedded submanifolds. ∎

Exercises A.21

(1) Prove that $\mathbb{S}^2$ in (2), Examples 1.23, is an embedded submanifold in $\mathbb{R}^3$ of dimension 2.

(2) Is the set $N \subset \mathbb{R}^2$ obtained joining the unit circle centred at $(0, 1)$ with the x^1-axis an embedded submanifold in $\mathbb{R}^2$? Does it admit a differentiable structure of dimension 1?

(3) Consider the conical surface C in $\mathbb{R}^3$ of equation $0 \leq x^3 = \sqrt{(x^1)^2 + (x^2)^2}$. Show that C *cannot* be embedded in $\mathbb{R}^3$ as 2-dimensional submanifold, but it has a 2-dimensional differentiable structure. What happens if we consider $C^* := C \setminus \{(0, 0, 0)\}$ instead of C?

(4) Prove that if M is a differentiable manifold, and $N \subset M$, with the induced topology, admits two distinct atlases $\{(U_i, \phi_i)\}_{i \in I}$ and $\{(U'_j, \phi'_j)\}_{j \in J}$ satisfying (2) in Definition A.19, then the two atlases' charts are compatible.

(5) Let (U, ϕ) and (V, ψ) be local charts on M such that $U \cap V \neq \varnothing$ and $\phi \circ \psi^{-1} : \psi(U \cap V) \to \phi(U \cap V)$ is bijective, C^k ($k = 1, 2, \ldots, \infty$) and with non-zero Jacobian determinant on $\psi(U \cap V)$. Show that $\psi \circ \phi^{-1} : \phi(U \cap V) \to \psi(U \cap V)$ is C^k, and hence the two local charts are k-compatible.

One criterion to establish whether a subset in a manifold M is an embedded submanifold is based on the concept of *non-singular function*, which has several other applications.

Definition A.22 Let $f : M \to M'$ be of class C^k, $k \geq 1$, with M and M' differentiable manifolds (in particular $M' = \mathbb{R}$) of dimension m and m' respectively.

(1) f is said to be **non-singular at** $p \in M$ if there exist a local chart (U, ϕ) in M with $U \ni p$, and a local chart (V, ψ) in N with $V \ni f(p)$ such that the Jacobian matrix

$$\left[\frac{\partial(\psi \circ f \circ \phi^{-1})^j}{\partial x^i}\right]_{i=1,\ldots,m,\ j=1,\ldots m'}\Bigg|_{\phi(p)}$$

has maximal rank. In other words the linear map associated with the above matrix is injective if $m \leq m'$ and surjective if $m \geq m'$.

(2) f is called **non-singular** if it is non-singular everywhere on M.

(3) $v \in f(M)$ is called a **regular value** of f if f is non-singular at every $p \in f^{-1}(\{v\})$. $\diamond$

Remarks A.23 It should be clear, remembering how Jacobian matrices transform when changing charts, that the regularity of f at p does not depend on the choice of charts (U, ϕ) and (V, ψ): if the condition in the definition holds for one choice of local charts with domains containing p and $f(p)$ respectively, then it holds for all.   ∎

Here are two technical properties, elementary but important, of non-singular maps.

Proposition A.24 *If $f : M \to N$ is a non-singular map at $p \in M$ of class C^k with $k \geq 1$ and $\dim M = \dim N$, then f defines a local C^k diffeomorphism from an open neighbourhood of p to an open neighbourhood of $f(p)$.*

Proof Obvious from Theorem A.14 when working in local charts, remembering that in this case the Jacobian matrix is square and non-singular. □

Proposition A.25 *If $f : M \to N$ is a non-singular C^k map with $k \geq 1$ and $\dim M = \dim N$ then f is open, i.e. $f(A)$ is open in N if A is open in M.*

Proof For any $p \in M$ there exist open neighbourhoods $U \ni p$ in M and $V \ni f(p)$ in N such that $f|_U : U \to V$ is a C^k diffeomorphism, by Theorem A.14. In particular $f|_U$ is a homeomorphism. Any open subset of U is then mapped to an open subset of V. If $p \in A$ there will exist an open neighbourhood $U_p \subset A$ of p such that $f(U_p) \subset f(A)$ is an open neighbourhood of $f(p)$. Hence $f(A) = \cup_{p \in A} f(U_p)$ is a union of open sets, and so open. □

We mention, without proof (see for example [doC92, Wes78]), an important theorem based on non-singular functions that allows to say when a subset of a differentiable manifold admits an embedded submanifold structure.

Theorem A.26 (Regular-Value Theorem) *Let M and M' be C^k manifolds, $k \geq 1$, of dimension m and m' respectively, with $m > m'$, and $f : M \to M'$ a C^k map. Consider the set:*

$$N_v := \{ p \in M \mid f(p) = v \} .$$

If v is a regular value of f then the following hold.

(1) The set N_v is a C^k embedded submanifold in M of dimension $n := m - m'$.
(2) Consider two local charts:

$$\phi : U \ni q \mapsto (x^1, \ldots, x^m) \in \mathbb{R}^m \quad \text{in } M \quad \text{and}$$

$$\psi : V \ni r \mapsto (y^1, \ldots, y^{m'}) \in \mathbb{R}^{m'} \quad \text{in } M'$$

and the representation in coordinates $y^j = (\psi \circ f \circ \phi^{-1})^j (x^1, \ldots, x^m)$ for $j = 1, \ldots m'$.
If the $m' \times m'$ matrix of elements:

$$\left. \frac{\partial y^j}{\partial x^i} \right|_{\phi(p)} , \quad j = 1, \ldots, m' \text{ and } i = n+1, n+2, \ldots, m$$

*is non-singular and $p \in N_v$, then the first n coordinates $x^1, \ldots, x^n$ in U
define coordinates on $U \cap N_v$ for the differentiable structure of N_v as embedded
submanifold in M, around $p \in N_v$.*

Remarks A.27

(1) The claim in (2) can be made for any m' of the $x^1, \ldots, x^m$ whose corresponding $m' \times m'$ matrix is non-singular. The other n coordinates define,
locally, admissible coordinates on N. The proof is obvious if we keep into
account that if $\phi : U \ni p \mapsto (x^1(p), x^2(p), \ldots, x^m(p)) \in \mathbb{R}^m$ are local
coordinates on M, so are their permutations, for instance $\phi_1 : U \ni p \mapsto$
$(x^2(p), x^1(p), \ldots, x^m(p)) \in \mathbb{R}^m$.

(2) In Analytical Mechanics (Sect. 7.2.1) the regular-value theorem was used to
define the manifold $N_v = \mathbb{V}^{n_c+1}$ as embedded into $\mathbb{V}^{3N+1}$. In that case $M =$
$\mathbb{V}^{3N+1}$, $M' = \mathbb{R}^c$, $n_c := 3N - c$ and the function $f : M \to M'$ was given
via the c constraint functions $f^j : \mathbb{V}^{3N+1} \to \mathbb{R}$. Requiring that the constraints
are functionally independent corresponds to asking that we work on the regular
value $v = (0, \ldots, 0) \in \mathbb{R}^c$. Beware that the dimension of N_v is $n = n_c + 1$,
while in Sect. 7.2.1 n_c was called n.

(3) If $f : M \to \mathbb{R}$ is an everywhere non-singular C^k map on the m-dimensional
C^k manifold M, by the regular-value theorem the sets:

$$\Sigma_t := \{p \in M \mid f(p) = t\},$$

are C^k embedded submanifold of dimension $m - 1$. By construction $\Sigma_t \cap \Sigma_{t'} =$
$\varnothing$ if $t \neq t'$. Finally, $\cup_{t \in f(M)} \Sigma_t = M$. In this situation one says M is a **foliated
manifold** and its **leaves** are the submanifolds Σ_t with $t \in f(M)$. ∎

Examples A.28

(1) Exercise A.21.**1** is solved immediately applying Theorem A.26. The unit sphere
$\mathbb{S}^2$ in $\mathbb{R}^3$ has equation $f(x^1, x^2, x^3) = 0$ where $f(x^1, x^2, x^3) := \sum_{i=1}^3 (x^i)^2$. f
is differentiable, the rectangular matrix of coefficients $\partial f / \partial x^i$ is $(2x^1, 2x^2, 2x^3)$
and never vanishes on $\mathbb{S}^2$ (where it equals $(2, 2, 2)$). Consequently its rank is 1.

(2) The regular-value theorem cannot be applied directly in Exercise A.21.**3** for C
since the function $f(x^1, x^2, x^3) := x^3 - \sqrt{(x^1)^2 + (x^2)^2}$ is not differentiable
at $(0, 0, 0)$, i.e. at the vertex of the cone. If we use the function $g(x^1, x^2, x^3) :=$
$(x^3)^2 - ((x^1)^2 + (x^2)^2)$ the theorem still cannot be invoked, because the rank
of the row matrix of elements $\partial g / \partial x^i$ is zero at the origin.

(3) Call x, y, z the standard coordinates of $\mathbb{R}^3$ and consider the surface S image
of $z = g(x, y)$, with $(x, y) \in D$ open in the plane $z = 0$ and f of
class C^1. This surface is a C^1 embedded submanifold of dimension 2 in
$D \times \mathbb{R}$ with the standard differentiable structure. Moreover the coordinate
system x, y belongs to the submanifold's differentiable structure. Here is the
proof. Put on $D \times \mathbb{R}$ coordinates $y^1 = x^1, y^2 := x^2, y^3 := z - g(x, y)$:
this coordinate transformation is indeed invertible, differentiable and with

differentiable inverse, as one proves easily, so the y^1, y^2, y^3 belong to the differentiable structure of $D \times \mathbb{R}$ (and define a local chart for the differentiable structure of $\mathbb{R}^3$). S is determined by $y^3 = 0$ and is therefore an embedded submanifold by definition.

Alternatively, we can use the regular-value theorem, considering the map $f(x, y, z) := z - g(x, y)$ with $(x, y, z) \in D \times \mathbb{R}$, which describes S by $f(x, y, z) = 0$: in Cartesian coordinates the Jacobian matrix of f is the row vector $(\partial f/\partial x, \partial f/\partial y, 1) \neq 0$.

(4) When studying surfaces in $\mathbb{R}^3$ (or equivalently in $\mathbb{E}^3$), referring to orthonormal coordinates $\mathbf{x} := (x, y, z)$, it is customary to define a **regular surface** as the image S of a map of class (at least) C^1, $\mathbf{x} = \mathbf{x}(u, v)$ with $(u, v) \in D$ open in $\mathbb{R}^2$, so that

$$\frac{\partial \mathbf{x}}{\partial u} \wedge \frac{\partial \mathbf{x}}{\partial v} \neq \mathbf{0}, \quad \text{for any } (u, v) \in D . \tag{A.8}$$

A regular surface defined in this way may not be an embedded submanifold of $\mathbb{R}^3$ of dimension 2 (and class C^1). However, if $S \subset \mathbb{R}^3$ is a two-dimensional embedded submanifold of class (at least) C^1, then S is a union of regular surfaces.

Let us prove these claims. For the first it suffices to consider the upright cylinder S in $\mathbb{R}^3$ with base a *bouquet of 4 circles* $x = \cos(2\theta) \cos\theta$, $y = \cos(2\theta) \sin\theta$. The equations of S are $x = \cos(2u) \cos u$, $y = \cos(2u) \sin u$, $z = v$, for $(u, v) \in (\pi/4, 9\pi/4) \times \mathbb{R}$. A direct calculation shows that condition (A.8) holds on the domain we took. Yet, the curve determining S on the plane $z = 0$, i.e. the bouquet, self-intersects several times at the origin. The topology induced by $\mathbb{R}^3$ on S is such that no point on the z-axis has a neighbourhood in S homeomorphic to $\mathbb{R}^2$. As for the second claim, suppose $S \subset \mathbb{R}^3$ is a C^1 embedded submanifold of dimension 2. If $p \in S$, there is an open neighbourhood $U \ni p$ covered by coordinates y^1, y^2, y^3 of the C^1 structure of $\mathbb{R}^3$, such that $S \cap U$ is described by $y^1 = 0$. $u = y^2, v = y^3$ are local coordinates on S. The fact that y^1, y^2, y^3 are compatible with the standard Cartesian coordinates $\mathbf{x} = (x^1, x^2, x^3)$ of $\mathbb{R}^3$ implies that the Jacobian matrix of elements $\frac{\partial x^i}{\partial y^j}$ has non-zero determinant. This is equivalent to saying that the three vectors $\frac{\partial \mathbf{x}}{\partial y^j}$, $j = 1, 2, 3$ are everywhere linearly independent. But this would be impossible if

$$\frac{\partial \mathbf{x}}{\partial y^2} \wedge \frac{\partial \mathbf{x}}{\partial y^3} = 0 ,$$

so in the end we must have:

$$\frac{\partial \mathbf{x}}{\partial u} \wedge \frac{\partial \mathbf{x}}{\partial v} \neq 0$$

everywhere.

Exercises A.29

(1) Prove the statement in Remarks A.15:
   Assuming $U \cap V \neq \varnothing$, the k-compatibility of the local charts (U, ϕ) and (V, ψ) implies that the Jacobian matrix of $\phi \circ \psi^{-1}$, being invertible, has non-zero determinant everywhere.
(2) Prove that if $f : M \to \mathbb{R}$ is non-singular at $p \in M$ for the local chart (U, ϕ), then it is non-singular for any other local chart around p.
(3) Show that the definition of non-singular function at a point does not depend on the chosen charts.

A.6.4 Tangent and Cotangent Spaces

Take a C^k manifold M of dimension n $(k \geq 1)$. Consider every space $C^k(M)$ as a vector space over the field $\mathbb{R}$, where linear combinations are defined by

$$(af + bg)(p) := af(p) + bg(p) , \quad \text{for any } p \in M$$

for $a, b \in \mathbb{R}$ and $f, g \in C^k(M)$. Given a point $p \in M$, a **derivation** at p is an $\mathbb{R}$-linear map $L_p : C^k(M) \to \mathbb{R}$ satisfying the *Leibniz rule*:

$$L_p(fg) = f(p)L_p(g) + g(p)L_p(f) , \quad \text{for any } f, g \in C^k(M). \tag{A.9}$$

Clearly a linear combination of derivations at p, $aL_p + bL'_p$ $(a, b \in \mathbb{R})$ where

$$(aL_p + bL'_p)(f) := aL_p(f) + bL'_p(f) , \quad \text{for any } f, g \in C^k(M),$$

is still a derivation. Hence derivations at p form a vector space over $\mathbb{R}$, which we denote by $\mathcal{D}^k_p$. Every local chart (U, ϕ) with $U \in p$ automatically defines n derivations at p in the following way. If $x^1, \ldots, x^n$ are the coordinates of ϕ, we define the derivation with respect to the k-th coordinate by:

$$\left. \frac{\partial}{\partial x^k} \right|_p : f \mapsto \left. \frac{\partial f \circ \phi^{-1}}{\partial x^k} \right|_{\phi(p)} , \quad \text{for any } f, g \in C^1(M). \tag{A.10}$$

The n derivations at p, $\left. \frac{\partial}{\partial x^k} \right|_p$, are *linearly independent*: if 0 is the zero derivation and $c^1, c^2, \cdots, c^n \in \mathbb{R}$ are such that:

$$\sum_{k=1}^{n} c^k \left. \frac{\partial}{\partial x^k} \right|_p = 0 ,$$

then choosing a differentiable map that coincides with the coordinate map x^l on an open neighbourhood of p (whose closure is contained in U) and that vanishes outside, then asking

$$\sum_{k=1}^{n} c^k \left.\frac{\partial}{\partial x^k}\right|_p f = 0$$

implies $c^l = 0$. Since we can choose l as we like, every coefficient c^r is zero for $r = 1, 2, \ldots, n$. In the end the n derivations $\left.\frac{\partial}{\partial x^k}\right|_p$ form a basis of a subspace in $\mathcal{D}_p^k$ of dimension n (actually, it can be proved that when $k = \infty$ this subspace coincides with $\mathcal{D}_p^\infty$ itself). Changing local chart and using (V, ψ) with $V \ni p$ and coordinates $y^1, \ldots, y^n$, the new derivations in the new coordinates are related to the old ones by the relationships

$$\left.\frac{\partial}{\partial y^i}\right|_p = \sum_{k=1}^{n} \left.\frac{\partial x^k}{\partial y^i}\right|_{\psi(p)} \left.\frac{\partial}{\partial x^k}\right|_p . \tag{A.11}$$

The proof of this is straightforward from the definitions given. Since the Jacobian matrix of coefficients $\left.\frac{\partial x^k}{\partial y^i}\right|_{\psi(p)}$ is bijective by definition of local charts, we conclude that the subspace of $\mathcal{D}_p^k$ spanned by the $\left.\frac{\partial}{\partial y^i}\right|_p$ coincides with the span of the $\left.\frac{\partial}{\partial x^k}\right|_p$. This subspace is therefore an *intrinsic* object.

Definition A.30 (Tangent Space) Given a differentiable manifold of dimension n and class C^k $(k \geq 1)$, consider a point $p \in M$.

The vector subspace of derivations at $p \in M$ spanned by the n derivations $\left.\frac{\partial}{\partial x^k}\right|_p$, with $k = 1, 2, \ldots, n$, in any local coordinates (U, ϕ) with $U \ni p$, is called the **tangent space to M at p** and is denoted by $T_p M$. Its elements are called **tangent vectors at p to M** or **contravariant vectors at p**. ◇

We recall that if V is a vector space over $\mathbb{R}$, the space V^* of linear maps from V to $\mathbb{R}$ is called **dual space** of V. If the dimension of V is finite, so is that of V^* and they coincide. In particular, if $\{e_i\}_{i=1,\ldots,n}$ is a basis of V, the **dual basis** of V^* is the basis of V^*, $\{e^{*j}\}_{j=1,\ldots,n}$, determined by linearity and by:

$$e^{*j}(e_i) = \delta_i^j , \quad \text{for } i, j = 1, \ldots, n .$$

If $f \in V^*$ and $v \in V$, we use the notation:

$$\langle v, f \rangle := f(v)$$

and the left-hand side denotes the **pairing** of v and f.

Definition A.31 (Cotangent Space) Given a C^k manifold of dimension n $(k \geq 1)$, consider a point $p \in M$.

The dual space of T_pM is called **cotangent space to M at p** and is denoted by T_p^*M. Its elements are called **cotangent vectors at p**, **covariant vectors at p** or **1-forms** at p. For any basis of elements $\left.\frac{\partial}{\partial x^k}\right|_p$ in T_pM, the n elements of the dual basis of T_p^*M are written $dx^i|_p$. By definition:

$$\left\langle \frac{\partial}{\partial x^k}|_p, dx^i|_p \right\rangle = \delta_k^i \ .$$

$\diamond$

A.6.5   Covariant and Contravariant Vector Fields on Manifolds

Now we can define differentiable vector fields on a manifold M.

Definition A.32 If M is a differentiable manifold of class C^k and dimension n, a **vector field of class C^r**, or **contravariant vector field** of class C^r, $r = 0, 1, \ldots, k$, is the assignment of a vector $v(p) \in T_pM$ for any $p \in M$ so that, for any local chart (U, ϕ) with coordinates $x^1, \ldots, x^n$ where

$$v(q) = \sum_{i=1}^{n} v^i(x_q^1, \ldots, x_q^n) \left.\frac{\partial}{\partial x^i}\right|_q \ ,$$

the n functions $v^i = v^i(x^1, \ldots, x^n)$ are of class C^r on $\phi(U)$. A **covector field of class C^r** or **covariant vector field** of class C^r with $r = 0, 1, \ldots, k$ is the assignment of a covector $\omega(p) \in T_pM$ for any $p \in M$, so that, for any local chart (U, ϕ) with coordinates $x^1, \ldots, x^n$ for which

$$\omega(q) = \sum_{i=1}^{n} v_i(x_q^1, \ldots, x_q^n) \left. dx^i \right|_q \ ,$$

the n functions $\omega_i = \omega_i(x^1, \ldots, x^n)$ are of class C^r on $\phi(U)$. $\diamond$

Remarks A.33

(1) We will show later that the notion of tangent vector for affine spaces, and in particular $\mathbb{R}^n$, coincides with the standard notion.

(2) Let $v \in T_p M$ and consider two local charts (U, ϕ) and (V, ψ) with $U \cap V \ni p$ and respective coordinates $x^1, \ldots, x^n$ and $x'^1, \ldots, x'^n$. We must have:

$$v = \sum_{i=1}^n v^i \left. \frac{\partial}{\partial x^i} \right|_p = \sum_{j=1}^n v'^j \left. \frac{\partial}{\partial x'^j} \right|_p ,$$

and therefore

$$\sum_i^n v^i \left. \frac{\partial}{\partial x^i} \right|_p = \sum_{j,i=1}^n v'^j \left. \frac{\partial x^i}{\partial x'^j} \right|_{\psi(p)} \left. \frac{\partial}{\partial x^i} \right|_p ,$$

from which

$$\sum_{i=1}^n \left(v^i - \sum_{j=1}^n \left. \frac{\partial x^i}{\partial x'^j} \right|_{\psi(p)} v'^j \right) \left. \frac{\partial}{\partial x^i} \right|_p = 0 .$$

As the derivations $\left. \frac{\partial}{\partial x^i} \right|_p$ are linearly independent, we conclude that the transformation rule of the components of a vector in $T_p M$, when we change coordinates, is

$$v^i = \sum_{j=1}^n \left. \frac{\partial x^i}{\partial x'^j} \right|_{\psi(p)} v'^j . \tag{A.12}$$

The same procedure gives the similar formula for covariant vectors

$$\omega_i = \sum_{j=1}^n \left. \frac{\partial x'^j}{\partial x^i} \right|_{\psi(p)} \omega'_j , \tag{A.13}$$

when

$$\omega = \sum_{i=1}^n \omega_i \, dx^i \Big|_p = \sum_{j=1}^n \omega'_j \, dx'^j \Big|_p .$$

∎

The following has an obvious proof, but is very important in view of the applications.

Proposition A.34 *If M is a C^k manifold, $k \geq 1$, of dimension n, prescribing a contravariant vector field X of class C^r or a covariant vector field ω of class C^r ($r \geq 0$) is completely equivalent to prescribing n-tuples of C^r functions $\{X^i_{(r)}\}_{i=1,\ldots,n}$*

or, respectively, $\{\omega_{(r)j}\}_{j=1,\dots,n}$ (an n-tuple for any local chart $(U_{(r)}, \phi_{(r)})$ in an (arbitrary) given atlas of M) so that, when we change charts in the atlas:

$$X^i_{(r)}\left(\phi_{(r)}(p)\right) = \sum_{j=1}^{n} \frac{\partial x^i_{(r)}}{\partial x^j_{(r')}}\Bigg|_{\phi_{(r')}(p)} X^j_{(r')}\left(\phi_{(r')}(p)\right), \tag{A.14}$$

or, respectively,

$$\omega_{(r')i}\left(\phi_{(r')}(p)\right) = \sum_{j=1}^{n} \frac{\partial x^j_{(r')}}{\partial x^i_{(r)}}\Bigg|_{\phi_{(r)}(p)} \omega_{(r)j}\left(\phi_{(r)}(p)\right), \tag{A.15}$$

for any point $p \in M$.

A.6.6 Differentials, Curves and Tangent Vectors

Consider a scalar field $f : M \to \mathbb{R}$ of class C^r on the n-dimensional C^k manifold M. Let us explicitly suppose that $k \geq r > 1$. If we give, for any chart in an atlas, the n functions $\{\frac{\partial f}{\partial x^i}\}$ it is immediate to see that the conditions of Proposition A.34 are respected, so we have indeed defined a covariant vector field.

Definition A.35 Consider a scalar field $f : M \to \mathbb{R}$ of class C^r on the n-dimensional C^k manifold M, and assume $k \geq r > 1$. The **differential** of f, df, is the covariant vector field of class C^{r-1} given, in every local chart (U, ψ), by:

$$df|_p = \sum_{i=1}^{n} \frac{\partial f}{\partial x^i}\Bigg|_{\psi(p)} dx^i|_p.$$

$\diamondsuit$.

Consider a C^r *curve* on the C^k manifold M, i.e. a C^r map $(r = 0, 1, \dots, k)$ $\gamma : I \to M$, where $I \subset \mathbb{R}$ is an open interval viewed as differentiable submanifold of $\mathbb{R}$. Assume explicitly $r > 1$. If $p \in \gamma(I)$, we can define the *tangent vector* to γ at p as, if $\gamma(t_p) = p$:

$$\dot{\gamma}(p) := \sum_{i=1}^{n} \frac{dx^i}{dt}\Bigg|_{t_p} \frac{\partial}{\partial x^i}\Bigg|_p,$$

in any local chart defined around p. The definition does *not* depend on the chosen chart. In fact, if we had defined

$$\dot{\gamma}'(p) := \sum_{j=1}^{n} \frac{dx'^{j}}{dt}\bigg|_{t_p} \frac{\partial}{\partial x'^{j}}\bigg|_{p} ,$$

referring to a second system of coordinate around p, under (A.12) we would have found:

$$\dot{\gamma}(p) = \dot{\gamma}'(p) .$$

We may then give the following definition.

Definition A.36 A **curve of class** C^r, $r = 0, 1, \ldots, k$, in the C^k manifold M of dimension n, $k \geq 1$, is a C^r map $\gamma : I \to M$, where $I \subset \mathbb{R}$ is an open interval (seen as embedded submanifold of $\mathbb{R}$). A curve of class C^k is simply called **differentiable**.
If $r \geq 1$, the **tangent vector** to γ at $p = \gamma(t_p)$ for some $t \in I$, is the vector $\dot{\gamma}(p) \in T_p M$ defined by

$$\dot{\gamma}(p) := \sum_{i=1}^{n} \frac{dx^{i}}{dt}\bigg|_{t_p} \frac{\partial}{\partial x^{i}}\bigg|_{p} , \qquad (A.16)$$

in any local chart around p.   ◇.

A.6.7   Affine and Euclidean Spaces as Differentiable Manifolds

Every affine space $\mathbb{A}^n$ admits a natural structure of differentiable manifold (of class C^∞) given by the collection of mutually compatible global *Cartesian coordinates* as per Definition 1.3. Let us recall that such a system is built as follows. We fix a point $O \in \mathbb{A}^n$, called *origin* of the coordinates, and a basis $\mathbf{e}_1, \ldots, \mathbf{e}_n$ of the space of displacements V, called *coordinate axes*. By varying $P \in \mathbb{A}^n$, the components $((P - O)^1, \ldots, (P - O)^n)$ of any vector $P - O$ in the chosen basis defines a bijection $f : \mathbb{A}^n \to \mathbb{R}^n$ allowing to identify the points of $\mathbb{E}^n$ with the points of $\mathbb{R}^n$. This correspondence between points P and n-tuples $((P - O)^1, \ldots, (P - O)^n)$, is bijective. It is injective by (i) in Definition 1.1, and surjective because the domain of $(P, O) \mapsto P - O \in V$ coincides, by definition, with the entire $\mathbb{A}^n \times \mathbb{A}^n$, and now O is kept fixed.

The Euclidean topology of $\mathbb{R}^n$ induces via f a topology on $\mathbb{A}^n$ (defining open sets in $\mathbb{A}^n$ as pre-images of open sets in $\mathbb{R}^n$) that renders it homeomorphic to $\mathbb{R}^n$, hence Hausdorff and second countable. It is easy to see that this topology does not depend on the choice of O nor on the basis of V. Further, the function f defines by itself a

C^∞ atlas on $\mathbb{A}^n$ and equips this space with the differentiable structure it generates. By Exercises 1.7.**1** and **2**, when we pass to different Cartesian coordinates, $(\mathbb{E}^n, g)$, the functions $f \circ g^{-1}$ and $g \circ f^{-1}$ are linear and non-homogeneous hence smooth. In conclusion, all global charts given by Cartesian coordinates are pairwise compatible, and as such they induce the same C^∞ structure. This is generated by the affine structure only, without arbitrary choices, and in this sense it is *natural*.

Considering an affine space $\mathbb{A}^n$ as a differentiable manifold, the obvious question is which relationship there is between the tangent spaces $T_P\mathbb{A}^n$, $P \in \mathbb{A}^n$, and the space of displacements V. For that we have the following theorem.

Theorem A.37 *Let $\mathbb{A}^n$ be an affine space with space of displacements V. Fix $P \in \mathbb{A}^n$ and consider $T_P\mathbb{A}^n$. The following holds.*

(1) If $(\mathbb{A}^n, f)$ is a Cartesian coordinate system of origin O and axes $\mathbf{e}_1, \cdots, \mathbf{e}_n \in V$, let $x^1, \ldots, x^n$ be the coordinate functions of the global chart. Then the map:

$$\chi_P : V \ni \sum_{i=1}^n v^i \mathbf{e}_i \mapsto \sum_{i=1}^n v^i \left.\frac{\partial}{\partial x^i}\right|_P \in T_P\mathbb{A}^n , \tag{A.17}$$

is a vector space isomorphism.

(2) If χ'_P is a similar isomorphism arising for another Cartesian system, then

$$\chi_P = \chi'_P ,$$

and in this sense the isomorphism χ_P is natural.

(3) Given $P \in \mathbb{A}^n$, the unique map $\chi_P^* : T_P^*\mathbb{A}^n \to V^*$ that satisfies:

$$\langle \mathbf{v}, \chi^* \omega_P \rangle = \langle \chi_P \mathbf{v}, \omega_P \rangle \quad \text{for any } \mathbf{v} \in V \text{ and } \omega_p \in T_P^*\mathbb{A}^n,$$

is an isomorphism (also natural) that identifies $T_P^*\mathbb{A}^n$ with V^*.

Proof (1) Since both V and $T_p\mathbb{A}$ have finite dimension n, the linear map χ_P of (A.17), given by identifying the components of a vector in two bases of the two spaces, is an isomorphism. (2) By Exercise 1.7 **1**, if $(\mathbb{A}^n, f)$ is a system of Cartesian coordinates $x^1, \cdots, x^n$ and $(\mathbb{A}^n, g)$ another system with coordinates $x'^1, \cdots, x'^n$, origin O' and axes $\mathbf{e}'_1, \ldots, \mathbf{e}'_n$, such that

$$\mathbf{e}_i = \sum_j B^j{}_i \mathbf{e}'_j , \tag{A.18}$$

then the function $g \circ f^{-1}$ reads, in coordinates:

$$x'^j = \sum_{i=1}^n B^j{}_i (x^i + b^i), \tag{A.19}$$

where $(O - O') = \sum_i b^i e_i$. Then we put:

$$\chi'_P : V \ni \sum_{j=1}^{n} v'^j e'_j \mapsto \sum_{i=j}^{n} v'^j \left.\frac{\partial}{\partial x'^j}\right|_P \in T_P \mathbb{A}^n , \tag{A.20}$$

From (A.19) we have:

$$\left.\frac{\partial}{\partial x^i}\right|_P = \sum_j B^j{}_i \left.\frac{\partial}{\partial x'^j}\right|_P . \tag{A.21}$$

By (A.18) and (A.21) the matrix that changes bases from $\{e'_j\}$ to $\{e_i\}$ in the left-hand sides of (A.17) and (A.20) is the same that we have between the bases $\{\left.\frac{\partial}{\partial x'^j}\right|_P\}$ and $\{\left.\frac{\partial}{\partial x^i}\right|_P\}$ on the right. Immediately, then, $\chi_P = \chi'_P$.

(3) The request that defines χ_P^* can be written as:

$$(\chi_P^* \omega_P)(\mathbf{v}) = \omega_P(\chi_P(\mathbf{v})) \quad \text{for any } \mathbf{v} \in V \text{ and } \omega_p \in T_P^* \mathbb{A}^n,$$

from which it is clear that χ_P^* is well defined and linear. Furthermore, χ_P^* is injective (hence surjective since domain and codomain have the same dimension), because if $\chi_P^* \omega_P = 0$ were the zero map, then $\omega_P(\chi_P(\mathbf{v})) = 0$ for any $\mathbf{v} \in V$ and consequently $\omega_P = 0$ since χ_P is onto. We conclude χ_P^* is a vector space isomorphism. $\square$

Remarks A.38

(1) The map χ_P identifies vectors of V with vectors in the tangent space $T_P \mathbb{A}^n$. The former are not associated with any point of $\mathbb{A}^n$, and in the classical literature they are called *free vectors,*, whereas the latter are associated with the point $P \in \mathbb{A}^n$ and go by the name of **applied vectors**. This distinction only makes sense in affine spaces. On generic differentiable manifolds there only exist applied vectors.

(2) By the theorem just proved, if $(\mathbb{A}^n, f)$ is a Cartesian coordinate system with origin O and axes $e_1, \cdots, e_n \in V$, letting $x^1, \ldots, x^n$ be the coordinate functions of the global chart, we have a canonical identification (due to the existence of the isomorphism χ_P):

$$\mathbf{e}_i \equiv \left.\frac{\partial}{\partial x^i}\right|_P . \tag{A.22}$$

In particular, the tangent vector to a curve $P = P(t)$ parametrized in Cartesian coordinates by $x^i = x^i(t)$, which in Differential Geometry is

$$\mathbf{t}(t) = \sum_i \frac{dx^i}{dt} \left.\frac{\partial}{\partial x^i}\right|_{P(t)} ,$$

can be written as:

$$\mathbf{t}(t) = \sum_i \frac{dx^i}{dt} \mathbf{e}_i \ .$$

Actually when we use the first notation we are thinking of a tangent vector as applied, whilst when using the second we are thinking it as being free. ∎

The above theorem clearly holds in Euclidean spaces $\mathbb{E}^n$, in which case any orthonormal coordinate system alone forms an atlas for the natural differentiable structure of $\mathbb{E}^n$.

The existence of an inner product on the space of displacements has an important consequence. There is an identification—i.e. a *canonical isomorphism*[3] of vector spaces χ, that identifies the space of displacements V with the dual V^*. If, for simplicity, we write $(\ |\ ) : V \times V \to \mathbb{R}$ for the inner product on V, the map

$$\chi : V \ni \mathbf{v} \mapsto (\mathbf{v}|\cdot) \in V^*$$

is well defined (i.e. $(\mathbf{v}|\cdot)$ is truly an element of V^*, due to the inner product's linearity in the right argument). This map, moreover, is evidently linear, because the inner product is linear on the left, and one can easily prove that it is injective. In fact, if $(\mathbf{v}|\cdot) = 0$, where 0 is the zero map, then $||\mathbf{v}||^2 = (\mathbf{v}|\mathbf{v}) = 0$ and hence $\mathbf{v} = 0$. Since V and V^* have the same dimension, the map $V \ni \mathbf{v} \mapsto (\mathbf{v}|\cdot) \in V^*$ is a vector space isomorphism. Note that what we said has nothing to do with the assumption that V is the space of displacements of $\mathbb{E}^n$, as it could be any vector space of dimension n. Returning to Euclidean spaces and considering the differentiable structure of $\mathbb{E}^n$, we easily conclude that for any $P \in \mathbb{E}^n$ the spaces $T_P \mathbb{E}^n$ and $T_P^* \mathbb{E}^n$ are canonically isomorphic by what we have said and by the existence of the isomorphisms χ_P and χ_P^* of Theorem A.37: namely, $\chi_P^{*-1} \circ \chi \circ \chi_P^{-1} : T_P \mathbb{E}^n \to T_P^* \mathbb{E}^n$ is a vector space isomorphism by construction.

Remarks A.39 It is important to note that in reality, irrespective of the differentiable manifold M, from the above discussion $V = T_p M$ and $V^* = T_p^* M$ are canonically identified whenever we have an inner product $(\ |\ )_p$ at p.

---

[3] In naive terms an isomorphism of two algebraic structures is *natural* or *canonical* when it can be defined only using the algebraic structures and without making any choice. For instance, two vector spaces of the same dimension are isomorphic under infinitely many isomorphisms, obtained by choosing and identifying the elements of two bases. These isomorphisms *are not* natural, since the vector space structure is not enough to define them, but one needs to selects bases. Conversely, to define the isomorphism we are talking about between a vector space and its dual, when by vector space we mean the algebraic structure "vector space plus inner product", we do not need any further information apart from the algebraic structure itself and the concept of dual space. The same goes for the isomorphisms χ_P and χ_P^* introduced in Theorem A.37, which only use affine and differentiable structures. The non-intuitive formalisation of these notions is provided by *Category Theory*, which we will not be concerned about.

Appendix B
AC : Advanced Topics in Differential Geometry

We review in the sequel more advanced notions of Differential Geometry: the notions of *pushforward*, *pullback*, *Lie derivative*, *submanifold* and *fibre bundle*, all of which have useful applications to Physics, especially in Lagrangian and Hamiltonian Analytical Mechanics. We will then pass to the basics of integration on manifolds. ∎

B.1 Differentiation on Manifolds and Related Notions

B.1.1 Pushforward and Pullback

Let M and N be differentiable manifolds (of class C^1 at least), of dimensions m and n respectively, and $f : N \to M$ a differentiable map (at least C^1). Given a point $p \in N$ consider local charts (U, ϕ) in N and (V, ψ) in M, respectively around p and $f(p)$. Call $(y^1, \ldots, y^n)$ the coordinates thus defined in U and $(x^1, \ldots, x^m)$ those in V. Let us further define $f^k(y^1, \ldots, y^n) = x^k(f \circ \phi^{-1})$ for $k = 1, \ldots, m$. Then we call:

(1) **pushforward** (or **differential**) the map $df_p : T_pN \to T_{f(p)}M$ given in coordinates by:

$$df_p : T_pN \ni \sum_{i=1}^{n} u^i \left.\frac{\partial}{\partial y^i}\right|_p \mapsto \sum_{j=1}^{m} \left(\sum_{i=1}^{n} \left.\frac{\partial f^j}{\partial y^i}\right|_{\phi(p)} u^i \right) \left.\frac{\partial}{\partial x^j}\right|_p, \qquad (B.1)$$

V. Moretti, *Analytical Mechanics*, La Matematica per il 3+2 150,
https://doi.org/10.1007/978-3-031-27612-5

(2) **pullback** the map $f_p^* : T_{f(p)}^* M \to T_p^* N$ given in coordinates by:

$$f_p^* : T_{f(p)}^* M \ni \sum_{j=1}^{m} \omega_j dx^j|_{f(p)} \mapsto \sum_{i=1}^{n} \left(\sum_{j=1}^{m} \frac{\partial f^j}{\partial y^i} \bigg|_{\phi(p)} \omega_j \right) dy^i|_p. \quad \text{(B.2)}$$

Immediately, the definitions *do not depend on the coordinates around p and f(p)*. The pushforward is also denoted by $f_{*p} : T_p N \to T_{f(p)} M$.

Also easy to see is that df_p and f_p^* may be defined equivalently, though implicitly, by:

$$(df_p X)(g)|_{f(p)} = X(g \circ f)|_p \quad \text{for any } X \in T_p N \text{ and } g : M \to \mathbb{R} \text{ (of class } C^1)$$

and

$$\langle Y, f_p^* \omega \rangle = \langle df_p Y, \omega \rangle \quad \text{for any } Y \in T_p N \text{ and } \omega \in T_{f(p)}^* M .$$

Remarks B.1 Clearly $f : N \to M$ is *non-singular* (Definition A.22) at $p \in N$ if and only if: $df_p : T_p N \to T_{f(p)} M$ is injective when dim $N \leq$ dim M, or surjective when dim $N \geq$ dim M. If moreover dim $N =$ dim M and f is non-singular at p, i.e. $df_p : T_p N \to T_{f(p)} M$ is bijective, then f defines a local diffeomorphism (with the same differentiability class as f) between open neighbourhoods of p and $f(p)$. ∎

B.1.2   Lie Derivative of a Vector Field

Let us now define the notion of *Lie derivative*. Let X be a C^1 vector field on the C^k manifold M of dimension n, with $k \geq 2$. Consider the one-parameter group of local diffeomorphisms $\phi^{(X)}$ (see Sect. 14.5.3) they generate. If we let $\phi^{(X)}$ act on $Y(q)$ as q varies around some given point, for t small enough we will have, locally, another differentiable vector field depending on p:

$$Y_t'(p) := (d\phi_{-t}^{(X)})_q Y(q) \quad \text{where } p = \phi_{-t}^{(X)}(q).$$

The choice of $-t$ instead of t is only conventional and corresponds to the fact that we are letting the group generated by $-X$ act, rather than the one of X, since $\phi_{-t}^{(X)} = \phi_t^{(-X)}$. Now let us focus on the field $M \ni p \mapsto Y_t'(p)$. It depends on the parameter t that varies around the origin. Keeping fixed the point p (not q!) at which the image field is evaluated, we want to understand the rate of variation of $Y_t'(p)$ with respect to t, at $t = 0$. In other words we want to compute:

$$\lim_{t \to 0} \frac{1}{t} \left(Y_t'(p) - Y_0'(p) \right) = \lim_{t \to 0} \frac{1}{t} \left((d\phi_{-t}^{(X)})_q Y(q) - Y(p) \right)$$

$$= \lim_{t \to 0} \frac{1}{t} \left((d\phi_{-t}^{(X)})_{\phi_t^{(X)}(p)} Y(\phi_t^{(X)}(p)) - Y(p) \right)$$

(where we used $q = \phi_t^{(X)}(p)$). The final limit can be written as a derivative:

$$\frac{d}{dt}\Big|_{t=0} (d\phi_{-t}^{(X)})_{\phi_t^{(X)}(p)} Y(\phi_t^{(X)}(p)) \,.$$

The **Lie derivative** of the vector field Y at p along the vector field X is the vector of $T_p M$:

$$\mathcal{L}_X|_p Y := \frac{d}{dt}\Big|_{t=0} (d\phi_{-t}^{(X)})_{\phi_t^{(X)}(p)} Y(\phi_t^{(X)}(p)) \,. \tag{B.3}$$

The meaning of $\mathcal{L}_X|_p Y$ should then be clear: it measures the rate of variation (at $t = 0$) of a vector field Y *at a given point p*, when the vector field is acted upon by the group of local diffeomorphisms generated by the vector field $-X$. Let us emphasise again that the $-$ sign is conventional, and some authors choose the other sign.

In local coordinates around p, the one-parameter group $\phi_t^{(X)}$ transforms $(x^1, \ldots, x^n)$ into $(x_t^1, \ldots, x_t^n)$, so in coordinates:

$$\left(\mathcal{L}_X|_p Y \right)^i := \frac{d}{dt}\Big|_{t=0} \sum_{j=1}^n \frac{\partial x_{-t}^i}{\partial x^j}\Big|_{(x^1(p),\ldots,x^n(p))} Y^j((x_t^1(p), \ldots, x_t^n(p))) \,.$$

The explicit calculation gives:

$$\left(\mathcal{L}_X|_p Y \right)^i := \frac{d}{dt}\Big|_{t=0} \sum_{j=1}^n \frac{\partial x_{-t}^i}{\partial x^j}\Big|_{(x^1(p),\ldots,x^n(p))} Y^j((x^1(p), \ldots, x^n(p))) + \sum_{j=1}^n \delta_j^i \frac{d}{dt}\Big|_{t=0}$$

$$Y^j((x_t^1(p), \ldots, x_t^n(p)))$$

$$= \sum_{j=1}^n \left(-\frac{\partial X^i}{\partial x^j} Y^j + X^k \frac{\partial Y^i}{\partial x^k} \right)\Big|_{(x^1(p),\ldots,x^n(p))} \,,$$

where we used the fact that:

$$\frac{d}{dt}\Big|_{t=0} x_{-t}^i(x^1(p), \ldots, x^n(p)) = -X^i(x^1(p), \ldots, x^n(p)) \,.$$

All in all, we have found:

$$\mathcal{L}_X Y = [X, Y] = -\mathcal{L}_Y X , \tag{B.4}$$

where the **commutator** or **Lie bracket** $[X, Y]$ of the C^2 vector fields X and Y is the C^1 vector field, in local coordinates:

$$[X, Y](p) = \sum_{j=1}^{n} \left(X^j \frac{\partial Y^i}{\partial x^j} - Y^j \frac{\partial X^i}{\partial x^j} \right) \Bigg|_{(x^1(p),\dots,x^n(p))} \frac{\partial}{\partial x^i} \Bigg|_p , \tag{B.5}$$

or, as we did earlier:

$$(\mathcal{L}_X|_p Y)^i = \sum_{j=1}^{n} \left(X^j \frac{\partial Y^i}{\partial x^j} - \frac{\partial X^i}{\partial x^j} Y^j \right) \Bigg|_{(x^1(p),\dots,x^n(p))} . \tag{B.6}$$

Intrinsically, $[X, Y]$ is the unique vector field, seen as differential operator, satisfying:

$$[Y, X](f) = Y(X(f)) - X(Y(f)) \quad \text{for any function } f \in C^1(M).$$

By direct computation one can check the following properties of the commutator, seen as a map associating C^{k-1} vector field $[X, Y]$ with pairs of C^k vector fields X and Y on the C^r manifold, $r \geq k$. If X, Y, Z are C^k vector fields on a C^r manifold, $r \geq k$, then:

(i) **skew-symmetry**: $[X, Y] = -[Y, X]$,
(ii) $\mathbb{R}$-**bilinearity**: $[aX + bZ, Y] = a[X, Y] + b[Z, Y]$ and $[Y, aX + bZ] = a[Y, X] + b[Y, Z]$ for any $a, b \in \mathbb{R}$,
(iii) **Jacobi identity**: $[X, [Y, Z]] + [Y, [Z, X]] + [Z, [X, Y]] = 0$.
In the Jacobi identity the vector fields are assumed C^2, and the 0 on the right is the zero vector field. Finally, if $\cdot$ is the pointwise product:
(iv) $[X, f \cdot Y] = f \cdot [X, Y] + X(f) \cdot Y$ for any function $f \in C^1(M)$.

The commutator turns the space of smooth vector fields into a well-known object.

Definition B.2 A (real) **Lie algebra** is a real vector space V together with a map, called **Lie bracket**,

$$[,] : V \times V \ni (x, y) \mapsto [x, y] \in V$$

satisfying three properties:

(i) skew-symmetry: $[x, y] = -[y, x]$,
(ii) $\mathbb{R}$-bilinearity: $[ax + bz, y] = a[x, y] + b[z, y]$ and $[y, ax + bz] = a[y, x] + b[y, z]$ for any $a, b \in \mathbb{R}$,
(iii) Jacobi identity: $[x, [y, z]] + [y, [z, x]] + [z, [x, y]] = 0$.

The **centre** of V is the subspace

$$Z_V := \{x \in V \mid [x, y] = 0 \quad \forall y \in V\}.$$

If $(V, [,])$ and $(V', [,]')$ are (real) Lie algebras, a linear map $f : V \to V'$ is called a **homomorphism of Lie algebras** if, furthermore:

$$[f(x), f(y)]' = f([x, y]), \quad \text{for } x, y \in V.$$

An **isomorphism of Lie algebras** is a bijective homomorphism of Lie algebras. ◇.

It is clear that the vector space $V(M)$ of C^∞ vector fields on M (with the usual vector space structure, given by $(aX + bY)(p) = aX(p) + bY(p)$ for any $p \in M$, $a, b \in \mathbb{R}$ and $X, Y \in V(M)$) becomes a real Lie algebra when we equip it with the commutator of vector fields. Demanding smoothness is merely due to the fact that the commutator action on vector fields typically lowers by 1 the differentiability class. Hence, for the commutator $[X, Y]$ of C^k vector fields still to be C^k, we need $k = \infty$.

B.2 Immersion of Tangent Spaces for Embedded Submanifolds

If $N \subset M$ is a differentiable submanifold embedded in the differentiable manifold M (both C^k for some $k > 0$) and $n < m$ are the respective dimensions of N and M, the points $p \in N$ become points of M under the map ι that identifies N with a subset of M. The same, then, holds for the vectors of $T_p N$, which can naturally be seen as vectors in $T_p M$ using the pushforward $d\iota_p$, as we set out to explain in detail. Choose a local chart around p, (U, ϕ) in N and a similar local chart (V, ψ) around p in M. Suppose $\phi : U \ni q \mapsto (y^1(q), \ldots, y^n(q))$ while $\psi : V \ni q \mapsto (x^1(q), \ldots, x^m(q))$. The identity map $\iota : N \to M$ defining N as submanifold in M is differentiable and

$$\psi \circ \iota \circ \phi^{-1} : (y^1, \ldots, y^n) \mapsto (x^1(y^1, \ldots y^n), \ldots, x^m(y^1, \ldots y^n)).$$

The pushforward $d\iota_p : T_p N \to T_{f(p)} M$ is, as earlier said, defined in those coordinates by:

$$d\iota_p : \sum_{i=1}^{n} v^i \left.\frac{\partial}{\partial y^i}\right|_p \mapsto \sum_{k=1}^{m} \left(\sum_{i=1}^{n} \left.\frac{\partial x^k}{\partial y^i}\right|_p v^i\right) \left.\frac{\partial}{\partial x^k}\right|_p, \tag{B.7}$$

We know that $d\iota_p$ does not depend on the local charts around p, (U, ϕ) and (V, ψ): if we had started with other local charts we would have obtained the same function $d\iota_p$. By virtue of this we claim $d\iota_p$ is injective.

Choosing coordinates in M adapted to N, i.e a chart (V, ψ) with

$$\psi : V \ni q \mapsto (x^1(q), \ldots, x^m(q)) \in \mathbb{R}^m$$

so that $N \cap U$ corresponds to points of coordinates $x^{n+1} = \cdots = x^m = 0$, the first n coordinates $y^1 = x^1, \ldots y^n = x^n$ define a local chart on N. This is true by definition of embedded submanifold. With this choice of local coordinates on N and M, the explicit expression of $d\iota_p$ is trivial:

$$d\iota_p : \sum_{i=1}^{n} v^i \left.\frac{\partial}{\partial y^i}\right|_p \mapsto \sum_{i=1}^{n} v^i \left.\frac{\partial}{\partial x^i}\right|_p .$$

Evidently in this representation the map $d\iota_p$ is injective. Therefore $d\iota_p$ identifies $T_p N$ with a subspace in $T_p M$.

For any local charts around p, in N and M respectively, the basis vectors $\left.\frac{\partial}{\partial y^i}\right|_p \in T_p N$ identify, under $d\iota_p$, with vectors of $T_p M$. In this sense we may slightly improperly write:

$$\left.\frac{\partial}{\partial y^i}\right|_p = \sum_{k=1}^{m} \left.\frac{\partial x^k}{\partial y^i}\right|_p \left.\frac{\partial}{\partial x^k}\right|_p . \tag{B.8}$$

B.3 Tangent and Cotangent Bundles, Fibre Bundles and Sections

The simples instances of fibre bundles are the *tangent bundle* and the *cotangent bundle* of a differentiable manifold M. Before getting to the general definition let us introduce these two objects.

Consider a C^k manifold M of dimension n, with $k \geq 2$, and the set:

$$TM := \{(p, v) \mid p \in M, v \in T_p M\} .$$

It is possible to put on it a natural structure of differentiable manifold of dimension $2n$ and class $k - 1$; the differentiable manifold thus obtained (still denoted TM) is called **tangent bundle** of M. The differentiable structure is the unique one with atlas the following collection of local charts induced by the differentiable structure of M. For any local chart (U, ϕ) in the differentiable structure $\mathcal{A}_M$ of M, with $\psi : U \ni p \mapsto (x^1(p), \ldots, x^n(p)) \in \mathbb{R}^n$, we put

$$TU := \{(p, v) \in TM \mid p \in U\}$$

and define the 1-1 map

$$T\psi : TU \ni (p, v) \mapsto (x^1(p), \ldots, x^n(p), v^1, \ldots, v^n) \in \mathbb{R}^{2n} ,$$

$$\text{where } v = \sum_k v^k \left.\frac{\partial}{\partial x^k}\right|_p .$$

Each pair $(TU, T\psi)$ on TM, built from a local chart (U, ψ) on M, is called **natural local chart on** TM. Let us put on TM the topology generated by the pre-images of open sets in $\mathbb{R}^{2n}$ under the maps $T\psi$. This topology makes TM Hausdorff, second countable, locally homeomorphic to $\mathbb{R}^{2n}$ (the local homeomorphisms are precisely the $T\psi$). It is easy to see that the set of pairs $T\mathcal{A}_M := \{(TU, T\psi) \mid (U, \psi) \in \mathcal{A}_M\}$ is a C^{k-1} atlas on TM. Let us explain why $k - 1$. If $(TU, T\psi)$ and $(TV, T\phi)$ are local charts of the atlas, $T\psi : (p, v) \mapsto (x^1, \ldots, x^n, \dot{x}^1, \ldots \dot{x}^n)$ and $T\phi : (p, v) \mapsto (y^1, \ldots, y^n, \dot{y}^1, \ldots \dot{y}^n)$, then on $TU \cap TV$ (assuming not empty) we have, in the obvious notation:

$$y^i = y^i(x^1, \ldots, x^n) , \tag{B.9}$$

$$\dot{y}^i = \sum_{j=1}^n \frac{\partial y^i}{\partial x^j} \dot{x}^j . \tag{B.10}$$

The presence of the Jacobian matrix in the second equation lowers by 1 the differentiability class of TM.

The differentiable structure $\mathcal{A}_{TM}$ generated by the atlas $T\mathcal{A}_M$ turns TM into a C^{k-1} manifold of dimension $2n$ called the **tangent bundle** of M. The *surjective* map $\Pi : TM \ni (p, v) \mapsto p \in M$ is C^{k-1} and goes by the name of **canonical projection**. Note that if (U, ψ) is a local chart on M, then $TU = \Pi^{-1}(U)$. The manifold M is the **base** of the tangent bundle. For $p \in M$ the tangent space $T_p M = \Pi^{-1}(p)$, which is an embedded submanifold in TM, is called the **fibre** of TM at (or over) the point $p \in M$.

The **cotangent bundle** T^*M is defined in a completely analogous manner by defining a natural structure of C^{k-1} manifold of dimension $2n$ on the set:

$$T^*M := \{(p, \omega_p) \mid p \in M, \ \omega_p \in T_p^*M\} .$$

For any local chart in the differentiable structure $\mathcal{A}_M$ of M, (U, ϕ), with $\psi : U \ni p \mapsto (x^1(p), \ldots, x^n(p)) \in \mathbb{R}^n$, we let

$$T^*U := \{(p, \omega_p) \in T^*M \mid p \in U\}$$

and define the 1-1 map

$$T^*\psi : T^*U \ni (p, \omega_p) \mapsto (x^1(p), \ldots, x^n(p), (\omega_p)_1, \ldots, (\omega_p)_n) \in \mathbb{R}^{2n} \,,$$

where $\omega_p = \sum_k (\omega_p)_k \, dx^k \Big|_p$.

The atlas on T^*M made of all charts $(T^*U, T^*\psi)$ with $(U, \psi) \in \mathcal{A}_M$, defines the C^{k-1} differentiable structure on the $2n$-manifold T^*M. Each of the above pairs $(T^*U, T^*\psi)$ on T^*M, built from a local chart (U, ψ) on M, is called a **natural local chart on** T^*M. Consider the pair of local charts $(T^*U, T^*\psi)$ and $(T^*V, T^*\phi)$ on T^*M, where $T^*\psi : (p, v) \mapsto (x^1, \ldots, x^n, \tilde{x}_1, \ldots \tilde{x}_n)$ and $T^*\phi : (p, v) \mapsto (y^1, \ldots, y^n, \tilde{y}_1, \ldots \tilde{y}_n)$. On $T^*U \cap T^*V$ (if non-empty) the following relations hold (in the obvious notation):

$$y^i = y^i(x^1, \ldots, x^n) \,, \tag{B.11}$$

$$\tilde{y}_i = \sum_{j=1}^{n} \frac{\partial x^j}{\partial y^i} \tilde{x}_j \,. \tag{B.12}$$

The *surjective* map $\Pi : T^*M \ni (p, v) \mapsto p \in M$ is C^{k-1} and is called **canonical projection**. Notice that if (U, ψ) is a local chart on M, then $T^*U = \Pi^{-1}(U)$. M is the **base** manifold of the cotangent bundle. For $p \in M$, the cotangent space $T_p^*M = \Pi^{-1}(p)$, an embedded submanifold in T^*M, is called **fibre** of T^*M at (or over) the point $p \in M$.

Remarks B.3 We focus on TM even though what we are about to say holds for T^*M as well. First of all, all fibres T_pM are diffeomorphic, since they are vector spaces of the same dimension and the vector space isomorphisms are diffeomorphisms for the differentiable structure of the fibres induced by that of TM. However, there exist infinitely many diffeomorphisms between any pair T_pM, T_qM: none of these is more natural that the others. Hence the fibres are all diffeomorphic, but not canonically; consequently, they are non-canonically diffeomorphic to $\mathbb{R}^n$. We will call F the generic fibre viewed as an "abstract fibre". If we fix $p \in M$, there exist a neighbourhood U of p and a diffeomorphism $f_U : U \times F \to \Pi^{-1}(U)$ such that $\Pi(f_U(q, v)) = q$ for any $q \in U$ and $v \in F$. This U can be chosen as domain of the local chart (U, ψ) of M and the diffeomorphism f_U as map $(T\psi)^{-1}$ seen above. We have shown that TM is *locally diffeomorphic* to $M \times F$.　∎

The definition of *fibre bundle*, or *bundle* for short, arises by generalising the previous constructions and keeping into account the above comment.

Definition B.4 (Fibre Bundle and Sections)　A **fibre bundle** E, or simply a **bundle**, is given by:

(i) a differentiable manifold of dimension n, still called E,
(ii) a second differentiable manifold M of dimension $m < n$ called **base**,

(iii) a surjective differentiable map $\Pi : E \to M$ called **canonical projection**,

(iv) a third differentiable manifold F called **standard fibre**, of dimension $n - m$.

E is required to be **locally diffeomorphic** to $M \times F$ in the following sense: for any $p \in M$ there exist an open neighbourhood U and a diffeomorphism $f : U \times F \to \Pi^{-1}(U)$ with $\Pi(f(x, y)) = x$ for any $x \in U$ and $y \in F$. The diffeomorphism f is called a **local trivialisation** of the bundle.

Furthermore, we say that

(1) a **local section** of E is a differentiable map $s : M_0 \to E$ with $\Pi(s(x)) = x$ for any $x \in M_0$, where $M_0 \subset M$ is open. A local section is a **global section**, or just a **section** of E, if $M_0 = M$.

(2) A system of local coordinates (U, ψ) on E, where:

$$\psi : U \ni p \mapsto (x^1(p), \dots, x^m(p), x^{m+1}(p), \dots, x^n(p)) \in \mathbb{R}^n$$

is called **adapted (to the fibres)**, if there is a local chart $(\Pi(U), \phi)$ on M such that

$$\phi : \Pi(p) \mapsto (x^1(p), \dots, x^m(p)) \in \mathbb{R}^m \quad \forall p \in U .$$

(3) A bundle E with base M, fibre F and canonical projection Π is said to be **trivialisable** if it is diffeomorphic to $M \times F$. More precisely, there is a diffeomorphism $f : M \times F \to E$ such that $\Pi(f(x, y)) = x$ for any pair $(x, y) \in M \times F$. Such diffeomorphism, if it exists, is a **global trivialisation** of the bundle. $\diamondsuit$

N.B. In the above definition we are supposing that the differentiability degrees of functions and manifolds are coherent.

Remarks B.5

(1) Evidently if the local chart

$$\psi : U \ni p \mapsto (x^1(p), \dots, x^m(p), x^{m+1}(p), \dots, x^n(p)) \in \mathbb{R}^n$$

is adapted to the fibres of $\Pi : E \to M$, for any given $p \in \Pi(U)$, the restriction

$$\psi_p : U \cap \Pi^{-1}(\{p\}) \ni q \mapsto (x^{m+1}(q), \dots, x^n(q)) \in \mathbb{R}^{n-m}$$

is a local chart on the fibre $\Pi^{-1}(\{p\})$.

(2) The canonical projection Π satisfies the non-singularity requirement that $d\Pi : T_pE \to T_{\Pi(p)}M$ is surjective, i.e. it has maximal rank. The proof is immediate by differentiating in x the condition $\Pi(f(x, y)) = x$ in coordinates, and observing that in terms of linear maps, where J is the matrix of partial derivatives of f_U with respect to the components of x: $d\Pi J = I$. Hence $d\Pi$ is surjective, as it has a right inverse. The regular-value theorem guarantees that every fibre $F_x := \Pi^{-1}(\{x\})$ is an embedded submanifold of E of dimension $n - m$.

(3) The local trivialisation $f : U \times F \to \Pi^{-1}(U)$ maps a submanifold of the domain to a submanifold of the image diffeomorphically. Therefore for any given $x \in U$, the submanifold F in the domain is transformed diffeomorphically into $f(x, F) = F_x$ (the last fact comes directly from $\Pi(f(x, y)) = x$). In other words $F \ni y \mapsto f(x, F) \in F_x$ is a diffeomorphism, so all fibres F_x in a bundle are diffeomorphic to the standard fibre F.

(4) It can be proved that every fibre bundle with base $\mathbb{R}$ is trivialisable, even if there is no canonical way to see it as a Cartesian product in general. Therefore spacetime, the spacetime of configurations, the spacetime of kinetic states and the phase spacetime, seen as bundles over the temporal axis as base, are trivialisable bundles. Yet there is no unique way to see them as Cartesian products. This, physically, corresponds to the impossibility of choosing a frame once and for all in a canonical way. ■

B.4 Theory of Differential Forms and Integration on Differentiable Manifolds

Let us move to the basics of the theory of integration on differentiable manifolds. The objects to be integrated are known as *differential forms*, and are defined in the next section.

B.4.1 p-Forms and p-Vectors

Let V be a real vector space of finite dimension n, with dual vector space V^* of linear maps $f : V \to \mathbb{R}$. The symbol $\Lambda^p(V^*)$ will indicate the vector space of **totally skew-symmetric multilinear maps** $f : V^p \to \mathbb{R}$ where, from now on, $V^p := V \times \cdots \times V$, p times. In other words f is linear in each argument:

$$f(\mathbf{v}_1, \mathbf{v}_2, \ldots, \mathbf{v}_{k-1}, \alpha\mathbf{v}_k + \beta\mathbf{v}'_k, \mathbf{v}_{k+1}, \ldots, \mathbf{v}_p)$$

$$= \alpha f(\mathbf{v}_1, \mathbf{v}_2, \ldots, \mathbf{v}_{k-1}, \mathbf{v}_k, \mathbf{v}_{k+1}, \ldots, \mathbf{v}_p) + \beta f(\mathbf{v}_1, \mathbf{v}_2, \ldots, \mathbf{v}_{k-1}, \mathbf{v}'_k, \mathbf{v}_{k+1}, \ldots, \mathbf{v}_p),$$

for any $\mathbf{v}_r, \mathbf{v}'_k \in V, r = 1, \ldots, p$ and $\alpha, \beta \in \mathbb{R}$, and it satisfies the property:

$$f(\mathbf{v}_1, \mathbf{v}_2, \ldots, \mathbf{v}_{k-1}, \mathbf{v}_k, \mathbf{v}_{k+1}, \ldots, \mathbf{v}_{h-1}, \mathbf{v}_h, \mathbf{v}_{h+1}, \ldots, \mathbf{v}_p)$$

$$= -f(\mathbf{v}_1, \mathbf{v}_2, \ldots, \mathbf{v}_{k-1}, \mathbf{v}_h, \mathbf{v}_{k+1}, \ldots, \mathbf{v}_{h-1}, \mathbf{v}_k, \mathbf{v}_{h+1}, \ldots, \mathbf{v}_p),$$

for any $h, k = 1, \ldots, p$ and $\mathbf{v}_h, \mathbf{v}_k \in V$. We will call $\Lambda^p(V^*)$ the **space of p-forms** on V, and its elements will be called **p-forms**, p being the **degree** of the p-form. A **form** is a p-form for some p.

The vector space structure of $\Lambda^p(V^*)$ is the natural one induced on functions by the addition and the multiplication by scalars. We will finally write $\Lambda(V^*)$ for the sum of all spaces $\Lambda^p(V^*)$ for $p = 0, 1, \ldots$, assuming $\Lambda^0(V^*) := \{0\}$, $\Lambda^1(V^*) := V^*$, $\Lambda^p(V^*) = \{0\}$ if $p > n$, where 0 is the zero map on the corresponding domain (the null vector of the trivial vector space if $p = 0$). Asking $\Lambda^p(V^*) = \{0\}$ if $p > n$ is not compulsory, as it turns out, since it is easy to show that, apart from the zero map, there are no skew-symmetric multilinear functionals with p arguments when p exceeds the dimension of V (i.e. that of V^*).

Now we want to introduce on $\Lambda(V^*)$ an operation $\wedge$ on p-forms called **exterior product**. Recall that the set of bijections $\sigma : \{1, 2, \ldots, N\} \to \{1, 2, \ldots, N\}$ is a group under composition called **symmetric group on N objects** and written $\mathcal{P}(N)$. The maps in $\mathcal{P}(N)$ are the **permutations of degree N**. If $f \in \Lambda^p(V^*)$ and $g \in \Lambda^q(V^*)$, we define the multilinear map $f \wedge g : V^{p+q} \to \mathbb{R}$ by:

$$(f \wedge g)(\mathbf{v}_1, \mathbf{v}_2, \ldots, \mathbf{v}_{p+q})$$

$$:= \frac{1}{p!q!} \sum_{\sigma \in \mathcal{P}(p+q)} \eta_\sigma f(\mathbf{v}_{\sigma^{-1}(1)}, \ldots, \mathbf{v}_{\sigma^{-1}(p)}) g(\mathbf{v}_{\sigma^{-1}(p+1)}, \ldots, \mathbf{v}_{\sigma^{-1}(p+q)}),$$

$$(\text{B.13})$$

for any $\mathbf{v}_1, \ldots, \mathbf{v}_{p+q} \in V$ (assuming $f \wedge g = 0$ if $f \in \Lambda^0(V^*)$ or $g \in \Lambda^0(V)$). Above, $\eta_\sigma = \pm 1$ is the **sign** of the permutation σ. By definition η_σ equals $+1$ if σ can be written as composite of an even number of swaps of two objects, and it equals -1 if σ can be written as composite of an odd number of swaps of two objects.[1]

Remarks B.6 There is (alas) another definition of exterior product for $f \in \Lambda^p(V^*)$ and $g \in \Lambda^q(V^*)$:

$$(f \hat{\wedge} g)(\mathbf{v}_1, \mathbf{v}_2, \ldots, \mathbf{v}_{p+q})$$

$$:= \frac{1}{(p+q)!} \sum_{\sigma \in \mathcal{P}(p+q)} \eta_\sigma f(\mathbf{v}_{\sigma^{-1}(1)}, \ldots, \mathbf{v}_{\sigma^{-1}(p)}) g(\mathbf{v}_{\sigma^{-1}(p+1)}, \ldots, \mathbf{v}_{\sigma^{-1}(p+q)}),$$

$$(\text{B.14})$$

used in the literature, although in Symplectic Geometry (B.13) is more common. With the new definition all the properties of the following theorem continue to hold. The relationship between the two definitions, which anyway produce *naturally*

---

[1] In other words $\sigma = \sigma_1 \circ \sigma_2 \circ \cdots \circ \sigma_k$ and $\eta = (-1)^k$, where each σ_k swaps two elements in $\{1, 2, \ldots, N\}$ and equals the identity on the complementary subset of cardinality $N - 2$. Decomposing σ into swaps as above is always possible, not in a unique way, but the permutation's sign is uniquely defined.

isomorphic exterior algebras $\Lambda(V^*)$, is

$$f \hat{\wedge} g = \frac{p!q!}{(p+q)!} f \wedge g .$$

One should always make sure to check which definition a book is using. A comparison of the two can be found in [War83]. ■

One can prove the following important theorem.

Theorem B.7 *If V is real vector space of finite dimension n, the function, called* **exterior product** *or* **wedge product***, associating with any pair of p-forms $f \in \Lambda(V^*)$, $g \in \Lambda(V^*)$ the multilinear map $f \wedge g$ of (B.13) is an operation on $\Lambda(V^*)$. More precisely, $f \wedge g \in \Lambda^{p+q}(V^*)$. Furthermore, the wedge product enjoys the following properties:*

associativity: $(f \wedge g) \wedge h = f \wedge (g \wedge h)$ *for any $f, g, h \in \Lambda(V^*)$;*
skew-symmetry: $g \wedge f = (-1)^{rs} f \wedge g$ *if $f \in \Lambda^r(V^*)$, $g \in \Lambda^s(V^*)$,*
linearity in the right argument: $(\alpha f + \beta g) \wedge h = \alpha(f \wedge h) + \beta(g \wedge h)$ *for any $f, g, h \in \Lambda(V^*)$ and $\alpha, \beta \in \mathbb{R}$.*

The exterior product also satisfies the following: if $\{\mathbf{e}^{*k}\}_{k=1,...,n} \subset V^*$ is the dual basis of $\{\mathbf{e}_k\}_{k=1,...,n} \subset V$, then the collection of elements:

$$\mathbf{e}^{*k_1} \wedge \cdots \wedge \mathbf{e}^{*k_p} \quad \text{where } 1 \le k_1 < k_2 < \ldots < k_p \le n,$$

is a basis for $\Lambda^p(V)$. Hence if $f \in \Lambda^p(V^*)$:

$$f = \sum_{1 \le k_1 < k_2 < \ldots < k_p \le n} f_{k_1 \cdots k_p} \mathbf{e}^{*k_1} \wedge \cdots \wedge \mathbf{e}^{*k_p}$$

$$= \sum_{k_1, k_2, \ldots, k_p = 1}^{n} \frac{1}{n!} f_{k_1 \cdots k_p} \mathbf{e}^{*k_1} \wedge \cdots \wedge \mathbf{e}^{*k_p} \tag{B.15}$$

where

$$f_{k_1 \cdots k_p} := f(\mathbf{e}_{k_1}, \ldots, \mathbf{e}_{k_p}) . \tag{B.16}$$

Remarks B.8

(1) Notice that the skew-symmetry together with the linearity in the right argument immediately imply linearity in the left argument.
(2) The dimension of $\Lambda^p(V^*)$, by the last claim in the theorem, equals $\binom{n}{p}$.
(3) The symbol $\wedge$ is the same we have used for the cross product on 3-dimensional vector spaces equipped with inner product. This is not accidental, as the two notions are actually related, as we shall briefly explain. First of all, the presence of an inner product identifies V^* with V—under the natural isomorphism $\chi :$ $V \to V^*$ discussed at the end of Sect. A.6.7—and so we can define the exterior

product of vectors of V and not only of V^*, viewing vectors in V as elements of V^*. It can also be proved, again via the isomorphism χ, that for $p = k$ and $p = n - k$ the spaces $\Lambda^p(V^*)$, having the same dimension, are naturally isomorphic. For $n = 3$ this means in particular that there is a natural isomorphism between the space of 2-forms, generated by the wedge products $\mathbf{u} \wedge \mathbf{v}$, and V itself. This isomorphism gives the cross product. ∎

Let us introduce the **contraction** between vectors and p-forms. If $X \in V$ we define the linear map:

$$X \lrcorner : \Lambda^p(V^*) \to \Lambda^{p-1}(V^*) \quad \text{with } p \geq 1, \tag{B.17}$$

called **contraction**, as the unique linear extension of the following rule, for any 1-forms $\theta_1, \theta_2, \ldots, \theta_p \in \Lambda^1(V^*) = V^*$:

$$X \lrcorner \theta_1 \wedge \cdots \wedge \theta_p := \langle X, \theta_1 \rangle \theta_2 \wedge \cdots \wedge \theta_p + (-1)\langle X, \theta_2 \rangle \theta_1 \wedge \theta_3 \wedge \cdots \wedge \theta_p$$

$$+ \cdots + (-1)^{q+1}\langle X, \theta_q \rangle \theta_1 \wedge \cdots \wedge \theta_{q-1} \wedge \theta_{q+1} \wedge \cdots \wedge \theta_p$$

$$+ \cdots + (-1)^{p+1}\langle X, \theta_q \rangle \theta_1 \wedge \cdots \wedge \theta_{p-1}.$$

Using the linearity of $X \lrcorner$, the definition allows to find $X \lrcorner \omega$ for any $\omega \in \Lambda^p(V^*)$, since we can decompose ω as linear combination of wedge products of 1-forms and then use the above formula.

All definitions and results hold if we replace V with the dual V^*. In that case the elements of $\Lambda^p(V)$ (when seen as skew-symmetric multilinear functionals on $V^* \times \cdots (p \text{ times}) \cdots \times V^*$) are called p-**vectors**.

B.4.2 Differential Forms

Let us pass to a C^k ($k \geq 1$) manifold M of dimension n. A **differential p-form** ω of class $C^{k'}$, with $k' \leq k$, is by definition a map sending points $q \in M$ to p-forms ω_q, so that, in any local coordinates $x^1, \ldots, x^n$ around $q \in M$ where:

$$\omega_q = \sum_{k_1, \ldots, k_p = 1}^{n} (\omega_q)_{k_1 \cdots k_p} dx^{k_1}|_q \wedge \cdots \wedge dx^{k_n}|_q ,$$

the functions $q \mapsto (\omega_q)_{k_1 k_2 \cdots k_p}$ are $C^{k'}$. Given a differential p-form (of class $C^{k'}$ with $k' \geq 1$), its **differential** or **exterior derivative** $d\omega$ is the differential $(p + 1)$-form (of class $C^{k'-1}$) given in any local coordinates by:

$$(d\omega)_q = \sum_{k_0, k_1, \ldots, k_p = 1}^{n+1} \frac{\partial (\omega_q)_{k_1 \cdots k_p}}{\partial x^{k_0}} dx^{k_0}|_q \wedge dx^{k_1}|_q \wedge \cdots \wedge dx^{k_n}|_q . \tag{B.18}$$

One can show that the definition is well posed (it does not depend on the coordinates chosen), and the operator d satisfies the following properties.

Theorem B.9 *The exterior derivative operator d, defined on the differential forms of an n-dimensional C^k manifold M, $k \geq 2$, is characterised by the following properties.*

(1) If ω is a 0-form, i.e. a C^1 function, then $d\omega$ coincides with the differential of the function.

(2) For any C^2 differential form ω of any degree p:

$$d d\omega = 0 .$$ (B.19)

(3) For any pair of C^1 differential forms, ω_1 of degree p and ω_2 of degree q:

$$d(\omega_1 \wedge \omega_2) = (d\omega_1) \wedge \omega_2 + (-1)^p \omega_1 \wedge d\omega_2 .$$ (B.20)

Definition B.10 A differential form ω satisfying $\omega = d\omega'$ for some differential form ω' is called **exact**, and **closed** if $d\omega = 0$.   ◇

By the previous theorem exact differential forms are closed, not conversely. An important result due to Poincaré states that, locally, all closed differential forms are exact as well.

Let us remind that a set U in $\mathbb{R}^n$ is **star-shaped** with respect to $p \in U$ when the straight segment joining p to any other point in U is entirely contained in U. Now we can state the famous *Poincaré Lemma* [Wes78].

Theorem B.11 (Poincaré Lemma) *Consider a manifold M of class C^2 at least and a differential form ω on M of class C^1 at least. If ω is closed:*

$$d\omega = 0 ,$$

then for any local chart (U, ψ) such that $\psi(U) \subset \mathbb{R}^n$ is open, connected and star-shaped with respect to some point, there exists a differential form ω' on U such that:

$$\omega|_U = d\omega' .$$

Remarks B.12

(1) Clearly the above theorem shows that any closed form of class at least C^1 is exact sufficiently close to any given point of the manifold on which it is defined. This is immediate from the fact that the open balls in $\mathbb{R}^n$ are star-shaped with respect to their centres.

(2) For 1-forms, the Poincaré Lemma, recast in terms of vector fields on domains in $\mathbb{R}^n$, shows that a C^2 vector field $\mathbf{V} = \sum_{i=1}^n V^i(\mathbf{x})\mathbf{e}_i$ on an open, star-shaped set $U \subset \mathbb{R}^n$ ($\{\mathbf{e}_1, \ldots, \mathbf{e}_n\}$ is the canonical basis of $\mathbb{R}^n$) is the gradient of a function

$f \in C^2(U)$ if, on U:

$$\frac{\partial V^i}{\partial x^j} = \frac{\partial V^j}{\partial x^i} \quad , i, j = 1, \ldots, n .\tag{B.21}$$

In case $n = 3$, the statement reduces to the known fact that a (C^1) vector field on a star-shaped open set is conservative (the gradient of a scalar field) if its curl is identically zero on the set. Conversely, any C^1 gradient field of some function must fulfil (B.21) on its domain (open but not necessarily star-shaped), because exact 1-forms are closed.

(3) The assumption that U is star-shaped in the Poincaré Lemma may be weakened (keeping the other hypotheses) to asking that the open connected set U (not necessarily a local chart's domain) is **contractible** to some point $x_0 \in U$. In other words, there should be a jointly continuous map $F : [0, 1] \times U \to U$ such that $F(0, x) = x$ for any $x \in U$ and $F(1, x) = x_0$ for any $x \in U$. Intuitively, one should be able to shrink U continuously to the point x_0; for example, a finite open ball $U \subset \mathbb{R}^n$ can be shrunk to its centre x_0.

(4) If the ω in the Poincaré Lemma is a 1-form, we can weaken even more the assumption that U is star-shaped (the other assumptions staying) and just say the open connected set U (not necessarily a local chart's domain) should be simply connected. Remark (2) generalises to this case, and allows to prove Theorem 4.15 in particular.

(5) If U is simply connected and ω a (C^1) closed one-form on U, from elementary Analysis lectures it is known that we may write $\omega = df$ where, for a given point $p_0 \in U$:

$$f(p) = \int_{\gamma_{p_0 p}} \omega := \int_0^1 \sum_{i=1}^n \omega_i(x(t)) \frac{dx^i}{dt} dt .$$

and the line integral is computed along any C^1 path $\gamma_{p_0 p} : [0, 1] \to U$, between p_0 and p and contained in U, expressed in local coordinates on U by $x^i = x^i(t)$, $i = 1, \ldots, n$. ∎

A map $f : M \to N$, of class C^1 at least, defines a **pullback action on differential k-forms** as follows. Let ω be a differential n-form on N. Then $(f^*\omega)_p := f_p^*\omega \in \Lambda^n(T_p M)$, for $p \in M$, defines a differential n-form on M called $f^*\omega$:

$$f_p^*\omega := (\omega \circ f)_p \quad \text{if } \omega \text{ has degree } 0;\tag{B.22}$$

and

$$(f_p^*\omega)(X_1, \ldots, X_n) = \omega(df_p X_1, \ldots, df_p X_n) \quad \text{if } X_1, \ldots, X_n \in T_p M$$

$$\text{and } \omega \text{ has degree } n.\tag{B.23}$$

With this definition the following identities hold, for any $a, b \in \mathbb{R}$ and ω_1, ω_2 differential forms (in the first case of the same degree):

$$f^*(a\omega_1 + b\omega_2) = af^*\omega_1 + bf^*\omega_2 , \tag{B.24}$$

and

$$f^*(\omega_1 \wedge \omega_2) = (f^*\omega_1) \wedge (f^*\omega_2) . \tag{B.25}$$

If ω is a generic p-form, at least C^1, on N, then:

$$df^*\omega = f^*d\omega . \tag{B.26}$$

B.4.3 Lie Derivative of a p-Form

As a last topic we consider the **Lie derivative of differential forms**, extending the definition for vector fields of Sect. B.1.2. The extension of the Lie derivative of a vector field X (at least C^1 in what follows) on the differentiable manifold M is given in terms of the following requests. (In the sequel differential forms will be at least C^1 if not specified otherwise.)

(1) If ω is a differential p-form, then $\mathcal{L}_X\omega$ is a differential p-form (C^0 at least). Moreover if X and ω vanish outside the open set $U \subset M$ then $\mathcal{L}_X\omega$ is zero outside the same set.

(2) If ω_1 and ω_2 are differential forms and $a, b \in \mathbb{R}$ then:

$$\mathcal{L}_X (a\omega_1 + b\omega_2) = a\mathcal{L}_X\omega_1 + b\mathcal{L}_X\omega_2 .$$

(3) The Lie derivative of a 0-form (i.e. a scalar field $f : M \to \mathbb{R}$) coincides with the action of X on f:

$$\mathcal{L}_X f := X(f) .$$

(4) $\mathcal{L}_X$ is a derivation with respect to the pairing of V and V^*. In other words, if θ is a 1-form and Y a vector field then:

$$\mathcal{L}_X \langle Y, \theta \rangle = \langle \mathcal{L}_X Y, \theta \rangle + \langle Y, \mathcal{L}_X\theta \rangle .$$

(5) $\mathcal{L}_X$ is a derivation: if ω_1 and ω_2 are differential forms then

$$\mathcal{L}_X (\omega_1 \wedge \omega_2) = (\mathcal{L}_X\omega_1) \wedge \omega_2 + \omega_1 \wedge \mathcal{L}_X\omega_2 . \tag{B.27}$$

(6) If f is a 0-form (C^2 at least) then

$$\mathcal{L}_X df = dX(f) \, .$$

It is easy to show that the above completely determine the Lie derivative's action on any differential form of any degree. Furthermore, $\mathcal{L}_X$ can be computed as a limit exactly as on vectors:

$$\mathcal{L}_X|_p \omega := \left. \frac{d}{dt} \right|_{t=0} (\phi_t^{(X)*})_{\phi_t^{(X)}(p)} \omega(\phi_t^{(X)}(p)) \, . \tag{B.28}$$

The meaning of $\mathcal{L}_X|_p \omega$, therefore, is similar to what we have seen for vector fields: it measures the rate of variation (at $t = 0$) of a differential form ω at a given point p, when the form is acted upon by the group of local diffeomorphisms generated by the vector field X. The sign of X is now positive in $\phi_t^{(X)*}$, instead of negative as for the vector field case where we had $d\phi_t^{(X)}$ in the similar expression. The reason has to do with the fact that $\phi^{(X)}$ acts on 1-forms under the pullback and not the differential, and the spaces are so to speak exchanged: if $f : M \to N$ is a differentiable map, then $df_r : T_r M \to T_{f(r)} N$ whereas $f_r^* : T_{f(r)}^* N \to T_r^* M$. Notice that if we change the convention chosen for vector fields and insist on the above, for coherence we must change sign in the Lie derivative of forms.

In particular, for a 1-form θ, in local coordinates:

$$(\mathcal{L}_X|_p \theta)_i = \sum_{j=1}^n \left(\frac{\partial X^j}{\partial x^i} \theta_j + X^j \frac{\partial \theta_i}{\partial x^j} \right) \Bigg|_{(x^1(p),...,x^n(p))} \, . \tag{B.29}$$

If ω is a p-form (at least C^1) it is not hard to prove the following relationship between Lie derivatives and contractions, known as **Cartan's magic formula**:

$$\mathcal{L}_X \omega = d (X \lrcorner \omega) + X \lrcorner d\omega \, . \tag{B.30}$$

Using the notation $\iota_X(\omega) := X \lrcorner \omega$, the formula reads:

$$\mathcal{L}_X = d\iota_X + \iota_X d \, ,$$

where $\mathcal{L}_X$ is understood as the Lie derivative of forms.

B.4.4 Integral of Top Forms and Volume Forms on Oriented Manifolds

Let start introducing the subject by working in local charts (U, ψ), with $\psi : U \ni p \mapsto (x^1(p), \ldots, x^n(p)) \in \mathbb{R}^n$, of a differentiable manifold M of dimension n, assumed C^∞. All differential forms will be at least of class C^0. Keeping (B.15) into account, a differential n-form ω on U, i.e. a differential form of top degree, can be written as:

$$\omega|_U = \alpha dx^1 \wedge \cdots \wedge dx^n ,$$

where the function $\alpha : U \to \mathbb{R}$ is uniquely determined by $\omega|_U$ when we have (U, ψ). If we consider another local coordinate system (V, ϕ), with $\phi : V \ni p \mapsto (y^1(p), \ldots, y^n(p)) \in \mathbb{R}^n$ and assume $U \cap V \neq \varnothing$, then:

$$\omega|_V = \beta dy^1 \wedge \cdots \wedge dy^n ,$$

and we ask how the functions α and β are related on $U \cap V$. In this intersection:

$$\omega = \alpha dx^1 \wedge \cdots \wedge dx^n = \sum_{j_1,\ldots,j_n=1}^{n} \frac{\partial x^1}{\partial y^{j_1}} dy^{j_1} \wedge \cdots \wedge \frac{\partial x^n}{\partial y^{j_n}} dy^{j_n}$$

$$= \sum_{j_1,\ldots,j_n=1}^{n} \frac{\partial x^1}{\partial y^{j_1}} \cdots \frac{\partial x^n}{\partial y^{j_n}} dy^{j_1} \wedge \cdots \wedge dy^{j_n}$$

$$= \sum_{\sigma \in \mathcal{P}(n)} \frac{\partial x^1}{\partial y^{\sigma(1)}} \cdots \frac{\partial x^n}{\partial y^{\sigma(n)}} \eta_\sigma dy^1 \wedge \cdots \wedge dy^n ,$$

where we used that $dy^{j_1} \wedge \cdots \wedge dy^{j_n} = 0$ if the indices repeat; if they do not repeat, they take the values of a permutation σ on $1, 2, \ldots, n$. Rearranging $dy^{j_1} \wedge \cdots \wedge dy^{j_n}$ we also have:

$$dy^{\sigma(1)} \wedge \cdots \wedge dy^{\sigma(n)} = \eta_\sigma dy^1 \wedge \cdots \wedge dy^n ,$$

where $\eta_\sigma \in \{-1, 1\}$ is the sign of the permutation σ. Know properties of the determinant say that:

$$\sum_{\sigma \in \mathcal{P}(n)} \eta_\sigma \frac{\partial x^1}{\partial y^{\sigma(1)}} \cdots \frac{\partial x^n}{\partial y^{\sigma(n)}} = \det \left[\frac{\partial x^i}{\partial y^j} \right]_{i,j=1,\ldots n} .$$

The conclusion is:

$$\omega = \alpha dx^1 \wedge \cdots \wedge dx^n = \alpha \det \left[\frac{\partial x^i}{\partial y^j} \right]_{i,j=1,\dots n} dy^1 \wedge \cdots \wedge dy^n .$$

Hence, if (U, ψ), with $\psi : U \ni p \mapsto (x^1(p), \dots, x^n(p)) \in \mathbb{R}^n$ and (V, ϕ), with $\phi : V \ni p \mapsto (y^1(p), \dots, y^n(p)) \in \mathbb{R}^n$ are local charts with $U \cap V \neq \varnothing$, the top-degree differential form ω can be written, on U and V respectively:

$$\omega|_U = \alpha dx^1 \wedge \cdots \wedge dx^n , \quad \omega|_V = \beta dy^1 \wedge \cdots \wedge dy^n$$

where, if $p \in U \cap V$:

$$\beta(p) = (\det J)(p)\alpha(p) , \quad J(p) := \left[\frac{\partial x^i}{\partial y^j} \right]_{i,j=1,\dots n} \Bigg|_{(y^1(p),\dots,y^n(p))} . \tag{B.31}$$

This transformation rule is reminiscent of the transformation of an integral in several variables when we change coordinates; this analogy allows to make sense of the integral of a differential form of top degree.

Remarks B.13

(1) As $\psi \circ \phi^{-1} : \phi(U \cap V) \to \psi(U \cap V)$ is at least C^1 with inverse at least C^1, its Jacobian matrix must have non-zero determinant. Hence $\det J \neq 0$, where J is the Jacobian matrix of $\psi \circ \phi^{-1}$. If $U \cap V$ and so $(\phi(U \cap V))$ is connected, since $\det J$ is continuous and so preserves connectedness, it must map $\phi(U \cap V)$ to a connected subset of $(-\infty, 0) \cup (0, +\infty)$. Therefore if $U \cap V$ is connected the sign of J is always positive or negative on $\phi(U \cap V)$.
(2) The Jacobian determinant's sign for the inverse transformation $\phi \circ \psi^{-1} : \psi(U \cap V) \to \phi(U \cap V)$ is the same as $\det J$, since the two determinants' product equals the determinant of the identity matrix, i.e. 1. ∎

Consider an open set $W \subset \overline{W} \subset U \cap V$ with $\overline{W}$ compact. If W is a Borel set in M so is $\psi(W)$ in $\mathbb{R}^n$, since $\psi^{-1} : \psi(U) \to U$ (the image U and the domain $\psi(U)$ are open) is continuous and hence Borel measurable. Therefore we can use the Lebesgue measure on $\psi(W)$. As ψ is a homeomorphism, $\overline{\psi(W)} = \psi(\overline{W})$ is compact, hence with finite Lebesgue measure. Suppose $(\det J)(p) > 0$ for $p \in W$. If we recall the variable-change formula in integrals, we have the identity:

$$\int_{\phi(W)} \beta \circ \phi^{-1} dy^1 \cdots dy^n = \int_{\phi(W)} \alpha \circ \phi^{-1} (\det J) dy^1 \cdots dy^n$$

$$= \int_{\phi(W)} \alpha \circ \phi^{-1} |\det J| dy^1 \cdots dy^n = \int_{\psi(W)} \alpha \circ \phi^{-1} \circ \phi \circ \psi^{-1} dx^1 \cdots dx^n ,$$

where the integrand functions are continuous under our hypotheses, hence absolutely integrable on compact sets, implying the integrals exist and are finite. In other terms:

$$\int_{\phi(W)} \beta \circ \phi^{-1} dy^1 \cdots dy^n = \int_{\psi(V)} \alpha \circ \psi^{-1} dx^1 \cdots dx^n .$$

All this suggests that we *define*, if the top form ω is at least C^0 and *provided we work in local coordinates for which the Jacobian determinant of the change of variables stays positive*:

$$\int_W \omega := \int_{\psi(W)} \alpha \circ \psi^{-1} dx^1 \cdots dx^n , \qquad (B.32)$$

for any open W with compact closure contained in the domain of a local chart where the coordinates on the right are defined. The right-hand side, in fact, does not depend on the local chart (U, ψ) containing $\overline{W}$.

Now we have two problems. The first is how to properly handle the request that the Jacobian determinant of a coordinate change must be positive; the second is to provide a procedure for "glueing" the integrals on the various domains with local coordinates, in order to define the integral of a top form on a generic set that may not be covered by a unique local chart. Let us begin with the first problem and a corresponding definition.

Definition B.14 Let M be a manifold of class C^k, $k \geq 1$.

(1) An **oriented atlas** on M, if existent, is an atlas (of class C^k) $\mathcal{O} = \{(U_i, \psi_i)\}_{i \in I}$ such that, if $U_i \cap U_j \neq \varnothing$, the Jacobian matrix of $\psi_i \circ \psi_j^{-1}$ on $\psi_j(U_i \cap U_j)$ has positive determinant everywhere.
(2) An **orientation** is a *maximal* oriented atlas $\mathcal{O} = \{(U_i, \psi_i)\}_{i \in I}$, i.e. an oriented atlas containing every local chart (of class C^k) (V, ϕ) of M for which the Jacobian determinant of $\phi \circ \psi_j^{-1}$ on $\psi_j(V \cap U_j)$ is positive for any $j \in I$ such that $V \cap U_j \neq \varnothing$.
(3) M is called **orientable** if it admits an oriented atlas.
(4) An **oriented differentiable manifold** is a pair $(M, \mathcal{O})$, where M is an oriented differentiable manifold and $\mathcal{O}$ an orientation. if so, a local chart (U, ψ) on M is called **oriented** if it belongs to $\mathcal{O}$.
(5) If M and M' are oriented differentiable manifolds of the same dimension with respective orientations $\mathcal{O}$ and $\mathcal{O}'$, a diffeomorphism $f : M \to M'$ is said to **preserve the orientation** when $\psi' \circ f : f^{-1}(U') \to \mathbb{R}^n$ defines a local chart in $\mathcal{O}$ for any local chart $(U', \psi') \in \mathcal{O}'$. $\diamond$

The next result is not hart to prove [KoNo63].

Proposition B.15 *Let M be an orientable C^k manifold, $k \geq 1$. Then*

(1) *if M is connected then there are two possible orientations.*
(2) *Every oriented atlas is a subset of a unique orientation.*

Remarks B.16 There exist differentiable manifolds that are not orientable, the simplest example being the 2-dimensional manifold called **Möbius strip**. It is constructed from the square $\{(x, y) \in \mathbb{R}^2 \mid x \in (-1, 1), y \in (-1, 1)\}$ in $\mathbb{R}^2$ and adding the top edge, i.e. the points $(1, y)$ with $y \in (-1, 1)$. Using local coordinates, $(-1, y)$ is identified with $(1, -y)$ on the bottom edge, for each $y \in (-1, 1)$.   ■

Regarding the integration of top differential forms, from now on we shall refer to oriented differentiable manifolds and will only consider local charts belonging to the orientation. In this way we will not have problems with the sign of the Jacobian determinant when changing coordinate systems.

Let pass to the glueing issue, recalling first some notions related to *paracompactness*. Given a set X, a **cover** of a subset $Y \subset X$ is a collection of subsets $\{U_i\}_{i \in I}$, $U_i \subset X$, such that $\bigcup_{i \in I} \supset Y$. If $\{V_j\}_{j \in J}$ is another cover of Y, we call it **refinement** of $\{U_i\}_{i \in I}$, if for any $j \in J$ there exists $i(j) \in I$ such that $V_j \subset U_{i(j)}$.

Definition B.17 If X is a topological space, a cover $\{U_i\}_{i \in I}$ of $Y \subset X$ is **locally finite**, if for any $y \in Y$ there is an open set $V_y \ni y$ such that $U_i \cap V_y \neq \varnothing$ only for a *finite* number of elements $i \in I$.   ◇

Definition B.18 A topological space X is **paracompact** if every open cover of X admits a locally finite open refinement.   ◇

The following fundamental result holds [KoNo63].

Proposition B.19 *Let M be a topological space that is:*

(i) Hausdorff,
(ii) locally compact (every point has an open neighbourhood with compact closure),
(iii) second countable.

Then M is paracompact.
In particular, every differentiable manifold is paracompact.

Remarks B.20 The theorem implies in particular that any locally finite open cover $\{U_i\}_{i \in I}$ of a differentiable manifold (Hausdorff and second countable) is at most countable. In fact, any point $p \in M$ has an open neighbourhood V_p intersecting a finite number of U_i. Clearly $\{V_p\}_{p \in M}$ is an open cover of M. As the topology of M is second countable, we can extract a countable subcover $\{V_{p_n}\}_{n \in \mathbb{N}}$. Now, V_{p_1} will have non-empty intersection with certain elements $\{U_j\}_{j \in I_1}$ where I_1 is finite, V_{p_2} will have non-empty intersection with elements $\{U_j\}_{j \in I_2}$ where I_2 is finite, and so forth. Each U_i of $\{U_i\}_{i \in I}$ must intersect at least one element of $\{V_{p_n}\}_{n \in \mathbb{N}}$ since the collection covers M. Consequently $I \subset \bigcup_{n \in \mathbb{N}} I_n$ cannot be more than countable.   ■

The last ingredient are *partitions of unity* and a corresponding theorem. Recall that if M is a topological space and $f : M \to \mathbb{R}$, the **support** of f, written supp(f), is the closure of the set $\{p \in M \mid f(p) \neq 0\}$.

Definition B.21 If M is a C^k manifold, $k \geq 1$, and $\mathcal{O} := \{U_i\}_{i \in I}$ is a locally finite open cover of M, a **partition of unity subordinated to** $\mathcal{O}$ is a collection of C^k maps $\{f_i\}_{i \in I}$ with $f_i : M \to \mathbb{R}$ such that:

(i) $\mathrm{supp}(f_i)$ is compact and $\mathrm{supp}(f_i) \subset U_i$ for any $i \in I$,
(ii) $f_i(p) \geq 0$ for any $p \in M$ and $i \in I$,
(iii) for any $p \in M$ then (as $\mathcal{O}$ is locally finite, only finitely many indices contribute to the sum on a neighbourhood of p):

$$\sum_{i \in I} f_i(p) = 1 .$$

◇

The concluding theorem is the following [KoNo63].

Theorem B.22 *Let M be a differentiable manifold and $\mathcal{O} := \{U_i\}_{i \in I}$ a locally finite open cover of M such that $\overline{U}_i$ is compact for any $i \in I$. Then there exists a partition of unity subordinated to $\mathcal{O}$.*

Consider a differentiable manifold M of class C^k with $k \geq 1$, connected, oriented, with orientation $\mathcal{O}$. Any point $p \in M$ has an oriented local chart (U, ψ) with $U \ni p$. We can restrict U so that we remain in $\mathcal{O}$ and $\overline{U}$ is compact and contained in a local chart. Hence there is an atlas of oriented local charts with domains having compact closures. The paracompactness of M ensures we can refine the atlas and make it locally finite. We can put on each open set thus obtained the coordinates inherited from the initial cover. Further, the charts' domains in the final cover have compact closure, since each closure is contained in a compact set. Eventually, on M there is a locally finite cover $\{U_i\}_{i \in I}$ made of open sets with compact closure $\overline{U}_i$ contained in the domain of an oriented local chart. By Remark B.20 I is at most countable. We will call **standard cover** of M one such cover.

By the previous theorem we can find a partition of unity subordinated to any standard cover.

Now we are ready to define the integral of a top form.

Definition B.23 Let M a C^k manifold of dimension n, with $k \geq 1$, connected and oriented. Let $N \subset M$, possibly $N = M$, be a Borel subset.
An n-form of class C^h with $h \geq 0$ is called integrable on N if, for any partition of unity $\{f_i\}_{i \in I}$ subordinated to a corresponding standard cover $\{U_i\}_{i \in I}$, the series

$$\sum_{i \in I} \int_{U_i} f_i \chi_N \omega$$

converges absolutely, where $\chi_N(x) = 1$ if $x \in N$ and $\chi(x) = 0$ otherwise, and the integrals on the right are those of (B.32).

In this case one defines (the summation order is irrelevant, by absolute convergence) **integral of ω on N** the real number:

$$\int_N \omega := \sum_{i \in I} \int_{U_i} f_i \chi_N \omega . \tag{B.33}$$

$\diamond$

Definition B.24 A C^h differential form of top degree μ, with $h \geq 0$, that in every oriented local chart (U, ϕ), with $\phi : U \ni p \mapsto (x^1(p), \dots x^n(p))$, takes the form

$$\mu = \alpha \, dx^1 \wedge \cdots \wedge dx^n \quad \text{where } \alpha(p) \geq 0 \text{ if } p \in U$$

is called a **volume form** on M.

$\diamond$

What we must now prove is that Definition (B.33) does not depend on the partition of unity.

Proposition B.25 *Referring to Definition B.23, if ω is integrable on N then*

$$\sum_{i \in I} \int_{U_i} f_i \chi_N \omega = \sum_{j \in J} \int_{U_i} g_j \chi_N \omega$$

irrespective of the partitions of unity $\{f_i\}_{i \in I}$ and $\{g_j\}_{j \in J}$ subordinated to corresponding standard covers. Therefore Definition (B.33) is well posed.

Proof Call $\{U_i\}_{i \in I}$ and $\{V_j\}_{j \in J}$ the standard covers of $\{f_i\}_{i \in I}$ and $\{g_j\}_{j \in J}$ respectively. It is straightforward to show that $\{U_i \cap V_j\}_{(i,j) \in I \times J}$ is a standard cover and $\{f_i \cdot g_j\}_{(i,j) \in I \times J}$ is a partition of unity subordinated to it. Hence the series

$$\sum_{(i,j) \in I \times J} \int_{U_i \cap V_j} f_i g_j \chi_N \omega$$

converges absolutely and may be computed one sum at a time:

$$\sum_{i \in I} \left(\sum_{j \in J} \int_{U_i \cap V_j} f_i g_j \chi_N \omega \right) = \sum_{(i,j) \in I \times J} \int_{U_i \cap V_j} f_i g_j \chi_N \omega$$

$$= \sum_{j \in J} \left(\sum_{i \in I} \int_{V_j \cap U_i} f_i g_j \chi_N \omega \right) . \tag{B.34}$$

Since g_j is defined everywhere on M and vanishes outside V_j for any given $i \in I$, we may write

$$\sum_{j \in J} \int_{U_i \cap V_j} f_i g_j \chi_N \omega = \sum_{j \in J} \int_{U_i} f_i g_j \chi_{N \cap U_i} \omega .$$

Each integral in the series on the right corresponds to an integral on $\mathbb{R}^n$ (precisely, on the image of U_i in $\mathbb{R}^n$ under the local chart defined on an open neighbourhood of $\overline{U_i}$). The Lebesgue dominated convergence theorem used for the Lebesgue measure on $\mathbb{R}^n$ immediately says:

$$\sum_{j \in J} \int_{U_i} f_i g_j \chi_{N \cap U_i} \omega = \int_{U_i} f_i \left(\sum_{j \in J} g_j \right) \chi_{N \cap U_i} \omega = \int_{U_i} f_i \chi_{N \cap U_i} \omega .$$

Proceeding in the same way for the integrals in

$$\sum_{i \in I} \int_{V_j \cap U_i} f_i g_j \chi_N \omega ,$$

(B.34) becomes:

$$\sum_{i \in I} \int_{U_i} f_i \chi_N \omega = \sum_{j \in J} \int_{U_i} g_j \chi_N \omega$$

which is what we wanted.   □

Proposition B.26 *In Definition B.23 if the Borel set N (or M itself) is compact, or ω (of top degree and class C^h, $h \geq 0$) vanishes outside a compact set K, then ω is integrable on N.*

Proof Suppose N compact. Fix a partition of unity subordinated to a standard cover as in (B.32). If $p \in N$ there is an open neighbourhood O_p of p meeting a finite number of supports of maps f_i in the partition of unity. As N is compact we can extract a finite subcover of N made of neighbourhoods O_p. Therefore only finitely many supports meet N. That is to say, only finitely many terms in series (B.32) will be non-zero. Each one of these is finite, since ω is bounded on each U_i being continuous on $\overline{U_i}$, which is compact hence has finite Lebesgue measure in $\mathbb{R}^n$. In conclusion (B.32) always converges absolutely (it is a finite sum) irrespective of the partition of unity subordinated to an arbitrary standard cover.   □

To finish let us discuss the relationship between *positive Borel measures* and volume forms, referring to [Rud78] for the measure-theoretical notions we will use.

Let us in particular recall that a positive measure $\mu : \mathcal{B}(X) \to [0, +\infty) \cup \{+\infty\}$ is called **Borel** if it acts on the *Borel σ-algebra* $\mathcal{B}(X)$ of a locally compact Hausdorff space X. A differentiable manifold is definitely Hausdorff and locally compact.

A Borel measure is called **regular** if the following two conditions hold:

(i) If $K \in \mathcal{B}(X)$ is compact then $\mu(K) < +\infty$, and if $E \in \mathcal{B}(X)$ then:

$$\mu(E) = \sup\{\mu(K) \,|\, K \subset E \,,\, K \text{ compact}\} \,;$$

(ii) if $E \in \mathcal{B}(X)$ then:

$$\mu(E) = \inf\{\mu(A) \,|\, A \supset E \,,\, A \text{ open}\} \,.$$

Then the following result holds, where $C_c^0(M; \mathbb{C})$ is the complex vector space of complex-valued continuous maps on M with compact support, and $C_c^0(M)$ is the real vector space of real-valued continuous maps on M with compact support.

Proposition B.27 *Let M be an oriented C^h manifold of dimension m, with $h \geq 1$. If ω is a volume form on M the following hold.*

(1) There exists a unique regular, positive Borel measure $\mu_\omega : \mathcal{B}(M) \to [0, +\infty) \cup \{+\infty\}$ such that:

$$\int_M g\omega = \int_M g\,d\mu_\omega \,, \quad \forall g \in C_c^h(M) \,, \quad h \geq 0 \,. \tag{B.35}$$

(2) Take $E \in \mathcal{B}(M)$. If $g \in C^h(M)$, with $h \geq 0$, then $g\omega$ is integrable on E if and only if g is integrable in μ_ω over E, and in such case:

$$\int_E g\omega = \int_E g\,d\mu_\omega \,. \tag{B.36}$$

(3) ω is integrable on $E \in \mathcal{B}(M)$ if, for a partition of unity $\{f_i\}_{i \in I}$ subordinated to a corresponding standard cover $\{U_i\}_{i \in I}$, the series

$$\sum_{i \in I} \int_{U_i} f_i \chi_E \omega$$

converges absolutely. In other words if the condition holds for one partition of unity then is holds for every partition of unity.

Proof

(1) Consider the map:

$$C_c^0(M; \mathbb{C}) \ni g \mapsto I(g) := \int_M Re(g)\omega + i \int_M Im(g)\omega \,.$$

By the definition of the integrals on the right, since ω is a volume form, we have $I(g) \geq 0$ if $g(x) \geq 0$ for any $x \in M$. As I is linear, the *Riesz representation theorem* [Rud78] implies that there is a regular positive Borel measure μ_ω such that

$$I(g) = \int_M g \, d\mu_\omega , \quad \forall g \in C_c^0(M; \mathbb{C}) , \tag{B.37}$$

that reduces to (B.35) if g is real-valued. The same theorem says that a regular Borel measure fulfilling (B.37) is unique. This identity is equivalent to (B.35) by decomposing g into real and imaginary parts.

(2) Suppose the form $g\omega$ is integrable on E and consider a partition of unity $\{f_i\}_{i \in I}$ subordinated to a standard cover $\{U_n\}_{n \in \mathbb{N}}$. Consider only an infinite cover (necessarily countable, by Remarks B.20), since the finite case is just a simplified version of what follows. By definition:

$$\int_E g\omega = \sum_{n \in \mathbb{N}} \int_{U_n} g f_n \chi_E \omega = \sum_{n \in \mathbb{N}} \int_{U_n} g f_n \chi_E d\mu_\omega ,$$

where the series converges absolutely, i.e.

$$\sum_{n \in \mathbb{N}} \int_{U_n} |g f_n \chi_E| d\mu_\omega < +\infty .$$

Given that

$$\sum_{n=0}^{N} |g f_n \chi_E| = \sum_{n=0}^{N} |g \chi_E| f_n \to |g \chi_E| \quad \text{pointwise as } N \to +\infty, \tag{B.38}$$

where the sequence is non-decreasing, the *monotone convergence theorem* [Rud78] implies $g \chi_E$ (Borel measurable as product of a continuous map and a Borel-measurable map) is integrable for the positive measure μ_ω. Remembering that

$$\sum_{n=0}^{N} |g f_n \chi_E| = \sum_{n=0}^{N} |g \chi_E| f_n \leq |g \chi_E| ,$$

the *Dominated convergence theorem* [Rud78] finally gives us:

$$\sum_{n \in \mathbb{N}} \int_{U_n} g f_n \chi_E d\mu_\omega = \int_E g \, d\mu_\omega .$$

This proves (B.36), since the left-hand side is $\int_E g\omega$ by definition. Conversely, assume g is integrable over E with respect to μ_ω. Since

$$\sum_{n=0}^{N} g f_n \chi_E \to g\chi_E \quad \text{pointwise as } N \to +\infty$$

in whichever way we rearrange the series (the convergence is absolute by (B.38)) and because

$$\left| \sum_{n=0}^{N} g f_n \chi_E \right| \le \sum_{n=0}^{N} |g\chi_E f_n| \le |g\chi_E|,$$

the Dominated convergence theorem tells that

$$\sum_{n\in\mathbb{N}} \int_{U_n} g f_n \chi_E \omega = \sum_{n\in\mathbb{N}} \int_{U_n} g f_n \chi_E d\mu_\omega = \int_E g d\mu_\omega$$

for any partition of unity $\{f_n\}_{n\in\mathbb{N}}$ subordinated to a standard cover $\{U_n\}_{n\in\mathbb{N}}$. Since the sum of the series, for a given partition of unity, does not depend on the ordering, the convergence is absolute and then $g\omega$ is integrable on E by definition, because the partition of unity was arbitrary. The definition of $\int_E g\omega$ finally gives (B.36).

(3) If the series

$$\sum_{i\in I} \int_{U_i} f_i \chi_N \omega$$

converges absolutely, then with the same argument used in (2), g is integrable on E in μ_ω, and therefore $g\omega$ is integrable on E, by (2).

□

Remarks B.28 The above proposition holds if ω is a volume form. Item (3) in particular is false, in general, when ω has top degree but is *not* a volume form. ∎

B.4.5 Integral of Forms on Submanifolds

Up to this point we have seen how to define the integral of a differential form of top degree on a differentiable manifold or a subset. We can finally consider the case of a submanifold N of dimension n embedded in a differentiable manifold M of dimension $m > n$. First, if M and N are orientable, the latter inherits an orientation from the former, simply by describing it locally as the locus where the

$m - n$ coordinates of a local chart of M vanish, working in oriented local charts on M. This produces an atlas of N that in turn gives an orientation.

Definition B.29 Consider a C^k ($k \geq 1$) embedded submanifold N of dimension n in an oriented C^k manifold M of dimension $m > n$ (same differentiability class). Suppose N is oriented with orientation induced by that of M, and consider a C^h differential form ω on M of degree n, $h \geq 1$.
Assuming the right-hand side exists, the **integral of** ω **on** $E \subset N$ (a Borel subset) is:

$$\int_E \omega := \int_E \iota^* \omega \,, \tag{B.39}$$

where $\iota : N \to M$ is the immersion $\iota : N \ni p \mapsto p \in M$ and the right-hand side is given by Definition B.23. $\diamond$

We can finally state a useful proposition, whose elementary proof is left to the reader as exercise.

Proposition B.30 *Let M and M' be connected and oriented C^k manifolds of dimension m, with $k \geq 1$, and $N \subset M$ an embedded C^k submanifold of dimension $n < m$ with the induced orientation.*
If $f : M \to M'$ is an orientation-preserving diffeomorphism, η a C^k n-form on M' and $E \subset M$ a Borel set, then η is integrable on $f(E)$ if and only if $f^* \eta$ is integrable on E. If so:

$$\int_{f(E)} \eta = \int_E f^* \eta \,. \tag{B.40}$$

Applying the notion of Lie derivative of a differential n-form we have an immediate corollary.

Proposition B.31 *Under the same hypotheses of Proposition B.30 with $M' = M$, if $\varphi^{(Y)}$ is the one-parameter group of local diffeomorphisms generated by the vector field Y, E is compact and $\varphi_s^{(Y)}(E)$ is defined for small enough $s > 0$, then:*

$$\frac{d}{ds}\bigg|_{s=0} \int_{\varphi_s^{(Y)}(E)} \eta = \int_E \mathcal{L}_Y \eta \,. \tag{B.41}$$

B.4.6 Manifolds with Boundary and the Stokes-Poincaré Theorem

We wish to define *manifolds with boundary*. Consider the closed half-space:

$$\mathbb{H}^n := \{(x^1, \ldots, x^n) \in \mathbb{R}^n \mid x^1 \geq 0\} \,. \tag{B.42}$$

The hyperplane of equation $x^1 = 0$ is the **boundary** of $\mathbb{H}^n$ and denoted $\partial \mathbb{H}^n$. Consider now an open subset $U \subset \mathbb{H}^n$ in the induced topology of $\mathbb{R}^n$. A function f defined on U is said to be **of class** C^k when there exist an open set V with $V \supset U$ and a C^k function F defined on V such that $F|_U = f$. Now we can give the general definition keeping the above into account.

Definition B.32 A C^k **manifold with boundary** M of dimension n is a Hausdorff, second-countable topological space equipped with a collection of **local charts** $\mathcal{A} := \{(U_i, \psi_i)\}_{i \in I}$ that satisfy the following conditions:

   (i) $U_i \subset M$ is open for any $i \in I$ and $\bigcup_{i \in I} U_i = M$;
   (ii) $\psi_i : U_i \to \mathbb{H}^n$, for any $i \in I$, is a homeomorphism when restricted to its image $\psi_i(U_i)$;
   (iii) for any $i, j \in I$, if $U_i \cap U_j \neq \varnothing$, the functions

$$\psi_i \circ \psi_j^{-1} : \psi_j(U_i \cap U_j) \to \psi_j(U_i \cap U_j)$$

   are C^k (in the above sense);
   (iv) the collection $\{(U_i, \psi_i)\}_{i \in I}$ is maximal with respect to (i)–(ii).

A point $p \in M$ is a **boundary point** if there is a local chart $(U, \psi) \in \mathcal{A}$ where $\psi(p) \in \partial \mathbb{H}^n$. The subset $\partial M \subset M$, called **boundary** of M, is the set of boundary points of M. The **interior** of M is the set $M^\circ := M \setminus \partial M$. $\diamond$

If M is a C^k manifold with boundary of dimension n, immediately ∂M is a C^k manifold (without boundary) of dimension $n - 1$, whose differentiable structure is induced by that of M. Similarly, M° is a C^k manifold (without boundary) of dimension $n - 1$, whose differentiable structure is induced by that of M.

For a manifold with boundary one can define in the obvious way tangent spaces, cotangent spaces and orientation. If M is oriented, ∂M and M° acquire an automatic orientation from M.

Examples B.33

(1) A closed ball $B \subset \mathbb{R}^n$ is a manifold with boundary, whose boundary is the topological boundary $\partial B = \mathcal{F}B$.
(2) More generally, let $A \subset M$ be open in a C^k manifold M of dimension n. If, for any $p \in \mathcal{F}(A)$, there exist an open neighbourhood $U \ni p$ and a map $g : U \to \mathbb{R}$ of class C^k such that $dg_p \neq 0$ and $A \cap U = \{p \in U \mid g(p) > 0\}$, then A is a manifold with boundary of class C^k and dimension $n - 1$, and moreover $\partial A = \mathcal{F}(A)$.

We are ready to state the *Stokes-Poincaré Theorem* [KoNo63, Wes78].

Theorem B.34 (Stokes-Poincaré) *Let M be an n-dimensional oriented, compact C^k manifold with boundary, $k \geq 1$. If ω is an $(n-1)$-form on M of class C^k, then:*

$$\int_M d\omega = \int_{\partial M} \omega \,, \tag{B.43}$$

where the orientation of ∂M is induced by M.
If M is compact without boundary, or non-compact without boundary but ω has compact support, the right-hand side is zero.

Remarks B.35

(1) The theorem applies in particular when the manifold with boundary on which we integrate comes from a larger manifold without boundary of class at least C^1 and dimension n, restricting to the closure of an open set A, as we explained in (2) of Examples B.33. Here the theorem holds even if $\mathcal{A}$ is not exactly a C^1 submanifold of dimension $n-1$, but the union of similar manifolds intersecting along submanifolds of dimension $n-2$ or less, belonging to the boundary of A. Think of a cube in $\mathbb{R}^3$: the submanifolds of dimension $3-2 = 1$ are the edges. Everything works so long as ω is sufficiently regular (for instance, if ω is the restriction of a C^1 form defined on the ambient manifold).

(2) The classical Stokes theorem, divergence theorem and Gauss-Green theorem, but also the fundamental theorem of calculus, can be seen as subcases of the above result. ∎

When computing integrals of n-forms on n-submanifolds, especially if we apply the Stokes-Poincaré Theorem, the following is often useful.

Proposition B.36 *Let $N \subset M$ be an embedded submanifold of dimension $n < m := dim(M)$ in the differentiable manifold M, both at least C^1. If ω is an n-form such that $Z \lrcorner \omega = 0$ for any non-zero $Z \in T_p N \subset T_p M$, then $(\iota^* \omega)_p = 0$, where $\iota : N \ni x \mapsto x \in M$ is canonical inclusion.*

Proof Consider local coordinates $x^1, \ldots, x^n, x^{n+1}, \ldots x^m$ on M defined around $p \in N \subset M$, for which $x^1, \ldots, x^n$ give coordinates on N, around p, when $x^{n+1} = x^{n+2} = \cdots = x^m = 0$. Let us choose these coordinates so that $\left. \frac{\partial}{\partial x^1} \right|_p = Z$ (always possible if $Z \neq 0$, by Proposition 14.62). In these coordinates, as $\iota^* \omega$ is of top degree on N, there will exist a real map, say α, for which:

$$\omega = \iota^* \omega = \alpha dx^1 \wedge \cdots \wedge dx^n \,.$$

The condition $Z \lrcorner \omega = 0$ in coordinates $x^1, \ldots, x^m$ reads:

$$Z \lrcorner \alpha(p) dx^1 \wedge \cdots \wedge dx^n|_p = 0;$$

i.e.

$$\frac{\partial}{\partial x^1}\bigg|_p \lrcorner \alpha(p)dx^1 \wedge \cdots \wedge dx^n|_p = 0;$$

and so:

$$\alpha(p)dx^2 \wedge \cdots \wedge dx^n|_p = 0.$$

Since $dx^1 \wedge \cdots \wedge dx^n \neq 0$, we conclude $\alpha(p) = 0$, which implies the claim:

$$(\iota^*\omega)_p = \alpha(p)dx^1 \wedge \cdots \wedge dx^n|_p = 0.$$

$\square$

Appendix C
Solutions and/or Hints to Suggested Exercises

C.1 Exercises for Chap. 1

Exercises 1.2

1.2.1. *Solution.* Let $\mathbf{u} \in V$. By (1) in Definition 1.1 there must exist $R \in \mathbb{A}^n$ such that $R - P = \mathbf{u}$. Using (2), $\mathbf{u} + (P - P) = (R - P) + (P - P) = R - P = \mathbf{u}$. As the additive neutral element $\mathbf{0}$ in V is unique, $P - P = \mathbf{0}$.

1.2.2. *Solution.* Using (2) in Definition 1.1 we have: $((Q + \mathbf{u}) + \mathbf{v}) - Q = [((Q+\mathbf{u}) + \mathbf{v}) - (Q+\mathbf{u})] + [(Q+\mathbf{u}) - Q]$. By definition the first two summands on the right are respectively: $\mathbf{v}$ and $\mathbf{u}$. Hence $((Q+\mathbf{u}) + \mathbf{v}) - Q = \mathbf{u} + \mathbf{v}$. By definition, then: $(Q + \mathbf{u}) + \mathbf{v} = Q + (\mathbf{u} + \mathbf{v})$.

1.2.3. *Solution.* Using (2) in Definition 1.1 we have: $P - Q + Q - P = P - P = \mathbf{0}$ and the claim follows.

1.2.4. *Solution.* Using (2) in Definition 1.1 we have:

$$(P + \mathbf{u}) - (Q + \mathbf{u}) = [(P + \mathbf{u}) - P] + [P - (Q + \mathbf{u})]$$
$$= [(P + \mathbf{u}) - P] + [P - Q] + [Q - (Q + \mathbf{u})] \,.$$

Applying the definition of $P + \mathbf{u}$ and $Q + \mathbf{u}$ on the last side, and remembering the previous exercise:

$$(P + \mathbf{u}) - (Q + \mathbf{u}) = \mathbf{u} + [P - Q] - \mathbf{u} = P - Q \,.$$

Exercises 1.7

1.7.1. *Solution.* By construction

$$\mathbf{e}_i = \sum_j B^j{}_i \mathbf{e}'_j \,,$$

where the matrix B of coefficients $B^j{}_i$ is non-singular. If $\mathbf{v} \in V$ decomposes along the above bases as $\mathbf{v} = \sum_i v^i \mathbf{e}_i = \sum_j v'^j \mathbf{e}'_j$, expressing $\mathbf{e}_i$ in terms of the other $\mathbf{e}'_j$ we find:

$$\sum_{i,j} v^i B^j{}_i \mathbf{e}'_j = \sum_j v'^j \mathbf{e}'_j .$$

Then

$$\sum_j \left(v'^j - \sum_{i=1}^n B^j{}_i v^i \right) \mathbf{e}'_j = 0 .$$

As the $\mathbf{e}'_j$ are linearly independent, we obtain the *transformation rules* of the components of $\mathbf{v}$:

$$v'^j = \sum_{i=1}^n B^j{}_i v^i . \tag{C.1}$$

From these it is easy to recover the transformation law between the Cartesian systems f and g. Consider a point $P \in \mathbb{A}^n$. By the properties of affine spaces we may write:

$$P - O' = (P - O) + (O - O'). \tag{C.2}$$

The components of $P - O, x^1, \cdots, x^n$, in the basis $\{\mathbf{e}_i\}_{i=1,\ldots,n}$, are the coordinates of P in the global chart $(\mathbb{A}^n, f)$, while those of $P - O', x'^1, \cdots, x'^n$, in $\{\mathbf{e}'_j\}_{j=1,\ldots,n}$ are the coordinates of P in the other global chart $(\mathbb{A}^n, g)$. By (C.1) we have, immediately from (C.2):

$$(P - O')'^j = \sum_{i=1}^n B^j{}_i (P - O')^i = \sum_{i=1}^n B^j{}_i (P - O)^i + \sum_{i=1}^n B^j{}_i (O - O')^i ,$$

and then

$$x'^j = \sum_{i=1}^n B^j{}_i (x^i + b^i),$$

where $(O - O') = \sum_i b^i \mathbf{e}_i$.

1.7.2. Solution. $f \circ g^{-1}$ is just the relation expressing the coordinates $x^1, \cdots, x^n$ in terms of the coordinates $x'^1, \cdots, x'^n$. It arises by inverting (1.4). By definition of inverse matrix, $\sum_j (B^{-1})^k{}_j B^j{}_i = \delta^k{}_i$ where $\delta^k{}_i$ is the usual *Kronecker delta*, defined as $\delta^k{}_i = 0$ if $k \neq i$ and $\delta^k{}_k = 1$. Multiplying (1.4) by $(B^{-1})^k{}_j$ and adding over j gives:

$$\sum_j (B^{-1})^k{}_j x'^j = \sum_{i,j} (B^{-1})^k{}_j B^j{}_i (x^i + b^i) = \sum_{i,j} \delta^k{}_i (x^i + b^i) = x^k + b^k,$$

and (1.5) follows.

1.7.3. Let $O_1 \in \mathbb{A}^n$ and $O_2 \in \mathbb{A}^m$ be the origins of two Cartesian coordinate systems and $\{e_1, \cdots, e_n\} \subset V_1$ and $\{f_1, \cdots, f_m\} \subset V_2$ the bases in the spaces of translations. Call $x_1^1, \cdots, x_1^n$ and $x_2^1, \cdots, x_2^m$ the associates Cartesian coordinates on $\mathbb{A}^n$ and $\mathbb{A}^m$. If $x_1^1, \cdots, x_1^n$ are the coordinates of $P \in \mathbb{A}^n$, then $P - O_1 = \sum_j x_1^j e_j$ and so:

$$\psi(P) = \psi\left(O_1 + \sum_j x_1^j e_j\right) = \left[\psi\left(O_1 + \sum_j x_1^j e_j\right) - \psi(O_1)\right] + \psi(O_1).$$

The map in brackets $d\psi(\sum_j x_1^j e_j)$ is linear by hypothesis, so:

$$\psi(P) = \sum_j x_1^j d\psi(e_j) + \psi(O_1).$$

Therefore

$$\psi(P) - O_2 = \sum_j x_1^j d\psi(e_j) + (\psi(O_1) - O_2) = \sum_j x_1^j L(e_j) + c,$$

where $c := \psi(O) - O_2 \in V_2$. We can decompose all vectors in the basis $f_1, \cdots, f_m \subset V_2$:

$$\sum_i (\psi(P) - O_2)^i f_i = \sum_j x_1^j \sum_i L^i(e_j) f_i + \sum_i c^i f_i.$$

By definition $x_2^i := (\psi(P) - O_2)^i$, and putting $L^i{}_j := d\psi^i(e_j)$, the above identity becomes:

$$\sum_i \left[x_2^i - \left(\sum_j L^i{}_j x_1^j + c^i\right)\right] f_i = 0.$$

Since the f_i are linearly independent, (1.6) is proven. That a transformation $\psi : \mathbb{A}^n \to \mathbb{A}^m$ is affine if of the form (1.6) in Cartesian coordinates is immediate.

1.7.**4**. We have:

$$\psi(P(t)) = \psi(P(t)) - \psi(P) + \psi(P) = \psi(P + t\mathbf{u}) - \psi(P) + \psi(P)$$
$$= d\psi(t\mathbf{u}) + \psi(P) = \psi(P) + t d\psi(\mathbf{u}) .$$

Exercises 1.14

1.14.**1**. *Solution*. Observe that $0 \le s(\mathbf{u} + a\mathbf{v}, \mathbf{u} + a\mathbf{v})$, i.e.:

$$0 \le s(\mathbf{u}, \mathbf{u}) + as(\mathbf{u}, \mathbf{v}) + as(\mathbf{v}, \mathbf{u}) + a^2 s(\mathbf{v}, \mathbf{v}) = s(\mathbf{u}, \mathbf{u}) + 2as(\mathbf{u}, \mathbf{v}) + a^2 s(\mathbf{v}, \mathbf{v})$$

for any $a \in \mathbb{R}$ and $\mathbf{u}, \mathbf{v} \in V$. This is possible for any $a \in \mathbb{R}$ only if the polynomial $P(a) = a^2 s(\mathbf{v}, \mathbf{v}) + 2as(\mathbf{u}, \mathbf{v}) + s(\mathbf{u}, \mathbf{u})$ has no real roots, so: $4s(\mathbf{u}, \mathbf{v})^2 - 4s(\mathbf{u}, \mathbf{u})s(\mathbf{v}, \mathbf{v}) \le 0$. This is the claim.

1.14.**2**. *Solution*. Taking the inner product with $\mathbf{e}_j$ of both sides of $\mathbf{v} = \sum_{k=1} v^k \mathbf{e}_k$, and recalling $\mathbf{e}_k \cdot \mathbf{e}_j = \delta_{kj}$, we find $v^j = \mathbf{v} \cdot \mathbf{e}_j = \mathbf{e}_j \cdot \mathbf{v}$. Finally, $\mathbf{u} \cdot \mathbf{v} = \left(\sum_j u^j \mathbf{e}_j\right) \cdot \left(\sum_k v^k \mathbf{e}_k\right) = \sum_{k,j} u^j v^k \mathbf{e}_k \cdot \mathbf{e}_j = \sum_{k,j} u^j v^k \delta_{k,j} = \sum_j u^j v^j$

1.14.**3**. *Solution*. Observe first that $R R^t = I$ and $R^t R = I$ separately imply R and R^t are invertible. (Just compute the determinant, use that the determinant of a product equals the product of the determinants, and $\det I = 1$.) We now have two cases. (a) If $R R^t = I$ then multiplying by R on the right and by R^{-1} on the left gives $R^t R = I$. But $(R^t)^t = R$, so $R \in O(n)$ implies $R^t \in O(n)$. (b) If, vice versa, $R^t \in O(n)$, i.e. $R^t (R^t)^t = I$, we can recast this as $R^t R = I$. Applying R^{-1} on the right and R on the left produces $R R^t = I$. Hence $R^t \in O(n)$ implies $R \in O(n)$. These facts also show that $R R^t = I$ and $R^t R = I$ are equivalent conditions.

1.14.**4**. *Solution*. Putting $||\mathbf{u}|| := \sqrt{\mathbf{u} \cdot \mathbf{u}}$ we have: $||\mathbf{u} + \mathbf{v}||^2 = ||\mathbf{u}||^2 + ||\mathbf{v}||^2 + 2\mathbf{u} \cdot \mathbf{v}$. Therefore $||\mathbf{u} + \mathbf{v}||^2 - ||\mathbf{u}||^2 - ||\mathbf{v}||^2 = 2\mathbf{u} \cdot \mathbf{v}$. The Cauchy-Schwarz inequality implies

$$\left| ||\mathbf{u} + \mathbf{v}||^2 - ||\mathbf{u}||^2 - ||\mathbf{v}||^2 \right| \le 2||\mathbf{u}|| \, ||\mathbf{v}|| .$$

In particular:

$$||\mathbf{u} + \mathbf{v}||^2 - ||\mathbf{u}||^2 - ||\mathbf{v}||^2 \le 2||\mathbf{u}|| \, ||\mathbf{v}|| ,$$

so:

$$||\mathbf{u} + \mathbf{v}||^2 \le 2||\mathbf{u}|| \, ||\mathbf{v}|| + ||\mathbf{u}||^2 + ||\mathbf{v}||^2 = (||\mathbf{u}|| + ||\mathbf{v}||)^2$$

and this forces:

$$||\mathbf{u} + \mathbf{v}|| \le ||\mathbf{u}|| + ||\mathbf{v}|| .$$

The triangle inequality for the Euclidean distance follows immediately from the definition and the previous relation.

Exercises 1.16

1.16.1. *Hint.* Let $||\mathbf{u}|| := \sqrt{\mathbf{u} \cdot \mathbf{u}}$ be the norm associated with the inner product. The latter's symmetry and linearity imply:

$$(\mathbf{u} \pm \mathbf{v}) \cdot (\mathbf{u} \pm \mathbf{v}) = \mathbf{u} \cdot \mathbf{u} + \mathbf{v} \cdot \mathbf{v} \pm 2(\mathbf{u} \cdot \mathbf{v}),$$

so

$$\mathbf{u} \cdot \mathbf{v} = \frac{1}{4}||\mathbf{u} + \mathbf{v}||^2 + \frac{1}{4}||\mathbf{u} - \mathbf{v}||^2.$$

But $d(O, O + \mathbf{w}) := ||\mathbf{w}||$, and the claim follows immediately.

1.16.2. *Solution.* Clearly the conservation of the inner product implies the conservation of distances, so we need only show the converse implication. From the (real symmetric) inner product's properties

$$(\mathbf{u}|\mathbf{v})_1 = \frac{1}{2}[(\mathbf{u} + \mathbf{v}|\mathbf{u} + \mathbf{v})_1 - (\mathbf{u} - \mathbf{v}|\mathbf{u} - \mathbf{v})_1],$$

so the inner product of $P - Q$ and $P' - Q$, with $R \in \mathbb{E}_1^3$ such that $R - Q = (P - Q) + (P' - Q)$, equals

$$(P - Q|P' - Q)_1 = \frac{1}{2}[(R - Q|R - Q)_1 - (P - P'|P - P')_1]$$

$$= \frac{1}{2}[d_1(R, Q)^2 - d_1(P, P')^2].$$

But ϕ is affine:

$$\phi(R) - \phi(Q) = d\psi(R - Q) = d\psi(P - Q) + d\psi(P' - Q)$$

$$= (\phi(P) - \phi(Q)) + (\phi(P') - \phi(Q)).$$

As ϕ preserves distances,

$$(P - Q|P' - Q)_1 = \frac{1}{2}[d_1(R, Q)^2 - d_1(P, P')^2]$$

$$= \frac{1}{2}[d_2(\phi(R), \phi(Q))^2 - d_2(\phi(P), \phi(P'))^2].$$

The last side is precisely $(\phi(P) - \phi(Q)|\phi(P') - \phi(Q))_2$.

1.16.3. *Solution.* If ϕ is given in coordinates by (1.14), in the orthonormal bases associated to those distances the matrix of the linear map $d\phi$ is the matrix R of coefficients $R^i{}_j$. This is orthogonal, so the inner products are preserved. From the previous exercise ϕ preserves distances. Conversely, suppose ϕ is affine and preserves distances, hence inner products by the previous exercise. In orthonormal coordinates the affine map ϕ (Exercise 1.7.3) is given by:

$$x_2^i = \sum_{j=1}^{n} R^i{}_j x_1^j + b^i ,$$

so in the same bases used to define the Cartesian coordinates, the linear map $d\phi$: $V_1 \to V_2$ reads:

$$w_2^i = \sum_{j=1}^{n} R^i{}_j w_1^j$$

where $\mathbf{w}_2 = d\phi(\mathbf{w}_1)$, $\mathbf{w}_1 \in V_1$ and $\mathbf{w}_2 \in V_2$. As the bases are orthonormal, the conservation of the inner products is

$$\sum_{i=1}^{n} v_2^i u_2^i = \sum_{j=1}^{n} v_1^j u_1^j .$$

Using the above form of $d\phi$:

$$\sum_{i=1}^{n} v_1^k R^i{}_k R^i{}_j u_1^j = \sum_{j=1}^{n} v_1^j u_1^j$$

i.e.

$$\sum_{i,j,k=1}^{n} v_1^k (R^i{}_k R^i{}_j - \delta_{kj}) u_1^j = 0 .$$

Choosing all coefficients v_1^k and u_1^j zero except for one pair, and varying pairs in all possible ways, we find:

$$\sum_{i=1}^{n} R^i{}_k R^i{}_j = \delta_{kj}$$

for any k and j. In matrix form

$$R^t R = I .$$

By Exercise 1.14.3 this is equivalent to $R \in O(n)$.

   1.16.4. *Solution.* Let us prove (i), (ii) and (iii). If $\phi : \mathbb{E}_1^n \to \mathbb{E}_2^n$ preserves distances it must be affine by Theorem 1.12, and has form (1.14) in orthonormal coordinates. As non-homogeneous linear maps are C^∞, ϕ is C^∞. Orthogonal matrices are invertible, so the inverse of (1.14) is:

$$x_1^j = c^j + \sum_{i=1}^n (R^{-1})^j{}_i x_2^i, \quad \text{with } c^j = \sum_{i=1}^n (R^{-1})^j{}_i b^i.$$

But the inverse of an orthogonal matrix is orthogonal by the previous exercise, so the inverse of ϕ is affine and preserves distances, in particular it is C^∞. Hence ϕ is a diffeomorphism. The composite of affine isometries, in orthonormal coordinates $x_2^i = \sum_{j=1}^n R^i{}_j x_1^j + b^i$ and $x_3^i = \sum_{j=1}^n R'^i{}_j x_2^j + b'^i$ trivially gives the transformation:

$$x_3^i = \sum_{j=1}^n (R'R)^i{}_j x_1^j + \left(b'^i + \sum_{j=1}^n R'^i{}_j b^j \right)$$

Orthogonal matrices for a group, so $R'R$ is orthogonal and the above is an isometry by Exercise 1.16.**3**.

The proof of (iv) is straightforward because, by Exercise 1.16.**3**, the matrix of $d\phi$ is orthogonal when written with respect to the bases giving the orthonormal coordinates in which we write ϕ.

1.16.**5**. *Solution*. By the previous exercise the set of affine isometries on $\mathbb{E}^n$ is closed under composition and inversion, and the identity map is clearly an isometry. Hence, for the (associative) composition of maps, the affine isometries of a given Euclidean space form a group. As affine maps are isomorphisms, the group is a subgroup of the group of isomorphisms of $\mathbb{E}^n$.

1.16.**6**. *Solution*. If f is an affine isometry it is obviously an isometry. For the converse it suffices to show that f is an affine map. We need a preliminary result.

Lemma *Let V_1 and V_2 be vector spaces of finite dimension n with (real symmetric) inner products $(\cdot|\cdot)_1$ and $(\cdot|\cdot)_2$ and norms $\|\cdot\|_1$ and $\|\cdot\|_2$ respectively. A map $\phi : V_1 \to V_2$ is linear and preserves the inner product if and only if:*

(1) $\phi(0_1) = 0_2$, where 0_i is the zero vector of V_i, $i = 1, 2$;
(2) $\|\phi(\mathbf{v}) - \phi(\mathbf{u})\|_2 = \|\mathbf{v} - \mathbf{u}\|_1$ for any $\mathbf{u}, \mathbf{v} \in V_1$.

Proof Clearly, if ϕ is linear and preserves the inner product, then (1) and (2) hold. Let us prove that under (1) and (2) ϕ preserves the inner product. For any $\mathbf{w} \in V_1$

$$\|\phi(\mathbf{w})\|_2 = \|\phi(\mathbf{w}) - 0_2\|_2 = \|\phi(\mathbf{w}) - \phi(0_1)\|_2 = \|\mathbf{w} - \mathbf{0}\|_1 = \|\mathbf{w}\|_1.$$

But $\|\phi(\mathbf{u}) - \phi(\mathbf{v})\|_2^2 = \|u - v\|_1^2$ can be written:

$$\|\phi(\mathbf{u})\|_2^2 + \|\phi(\mathbf{v})\|_2^2 + 2(\phi(\mathbf{u})|\phi(\mathbf{v}))_2 = \|\mathbf{u}\|_1^2 + \|\mathbf{v}\|_1^2 + 2(\mathbf{u}|\mathbf{v})_2$$

so using $||\phi(\mathbf{w})||_2 = ||\mathbf{w}||_1$ we find $(\phi(\mathbf{u})|\phi(\mathbf{v}))_2 = (\mathbf{u}|\mathbf{v})_2$, which is what we wanted. To finish, let us prove ϕ is linear. Let $\{\mathbf{e}_i\}_{i=1,\dots,n}$ be an orthonormal basis on V_1. By the conservation of the inner product, $\{\phi(\mathbf{e})_i\}_{i=1,\dots,n}$ is an orthonormal basis on V_2. Hence for any $\mathbf{v} \in V_1$, splitting $\phi(\mathbf{v})$ along $\{\phi(\mathbf{e})_i\}_{i=1,\dots,n}$ using the conservation of the inner product:

$$\phi(\mathbf{v}) = \sum_{i=1}^{n}(\phi(\mathbf{e}_i)|\phi(\mathbf{v}))\phi(\mathbf{e}_i) = \sum_{i=1}^{n}(\mathbf{e}_i|\mathbf{v})\phi(\mathbf{e}_i) .$$

The linearity in $\mathbf{v}$ of $(\mathbf{e}_i|\mathbf{v})$ implies immediately ϕ is linear.   □

Now we return to the proof that an isometry is affine. Fix $O \in \mathbb{E}_1^n$. Any $\mathbf{v} \in V_1$ can be written, uniquely, as $\mathbf{v} = P_{\mathbf{v}} - O$ for some $P_{\mathbf{v}} \in \mathbb{E}_1^n$. In particular $\mathbf{0}_1 = O - O$. Define $\phi : V_1 \to V_2$ so that $\phi(\mathbf{v}) := f(P_{\mathbf{v}}) - f(O)$. Thus $\phi(\mathbf{0}_1) = f(O) - f(O) = \mathbf{0}_2$, and

$$||\phi(\mathbf{u}) - \phi(\mathbf{v})||_2^2 = ||(f(P_{\mathbf{u}}) - f(O)) - (f(P_{\mathbf{v}}) - f(O))||_2^2 = ||f(P_{\mathbf{u}}) - f(P_{\mathbf{v}})||_2^2,$$

but the last side equals:

$$d_2(f(P_{\mathbf{u}}), f(P_{\mathbf{v}}))^2 = d_1(P_{\mathbf{u}} - P_{\mathbf{v}})^2 = ||\mathbf{u} - \mathbf{v}||_1^2 .$$

Therefore (1) and (2) hold. Applying the lemma, ϕ is linear (and preserves the inner products). By definition of ϕ and its linearity:

$$f(P_{\mathbf{v}}) - f(P_{\mathbf{u}}) = f(P_{\mathbf{v}}) - f(O) - (f(P_{\mathbf{u}}) - f(O))$$
$$= \phi(\mathbf{u}) - \phi(\mathbf{v}) = \phi(\mathbf{u} - \mathbf{v}) = \phi(P_{\mathbf{u}} - P_{\mathbf{v}}) .$$

But $P_{\mathbf{w}}$ ranges over $\mathbb{E}_1^n$ when $\mathbf{w} \in V_1$, so the above reads:

$$f(P) - f(Q) = \phi(P - Q) , \quad \text{for any } P, Q \in \mathbb{E}^n,$$

where ϕ is a linear map. This implies f is translation-invariant and affine, with $df = \phi$.

C.2   Exercises for Chap. 2

Exercises 2.11

2.11.**1**. *Solution.* To prove

$$\sigma_{\mathscr{R}} : \mathbb{V}^4 \ni p \mapsto (T, x^1, x^2, x^3) \in \mathbb{R}^4$$

defines a chart compatible with the differentiable structure of $\mathbb{V}^4$, by Exercise A.21.**5** it is enough to exhibit around any point of $\mathbb{V}^4$ a local chart of the differentiable structure of $\mathbb{V}^4$, with local coordinates y^1, y^2, y^3, y^4 such that the Jacobian matrix of $x^i = x^i(y^1, y^2, y^3, y^4)$ has everywhere non-zero determinant. Consider $p \in \mathbb{V}^4$ and $\Sigma_t \ni p$. As Σ_t is an embedded submanifold in $\mathbb{V}^4$, there is a local chart (U, ϕ) with $\mathbb{V}^4 \supset U \ni p$ and $\phi : q \mapsto (z^0(q), z^1(q), z^2(q), z^3(q))$ such that $U \cap \Sigma_t$ is given by $z^0 = 0$. At last, $\phi|_{U \cap \Sigma_t} : q \mapsto (z^1(q), z^2(q), z^3(q))$ defines a local chart on $U \cap \Sigma_t$. In coordinates z^i the absolute time function T must satisfy $\partial T / \partial z^0 \neq 0$ (since $\partial T / \partial z^i = 0$ if $i = 1, 2, 3$ by construction and T is non-singular). The Jacobian determinant, at p, of the transformation $T = T(z^0, z^1, z^2, z^3)$, $z^i = z^i$ ($i = 1, 2, 3$), is exactly $\partial T / \partial z^0 \neq 0$. By Theorem 1.25, restricting the domain U around p to a smaller U', the map $(z^0, z^1, z^2, z^3) \mapsto (T(z^0, z^1, z^2, z^3), z^1, z^2, z^3)$ is differentiable, bijective with differentiable inverse. All in all, for any $p \in \mathbb{V}^4$, there is a local chart (U', ψ) in the differentiable structure of $\mathbb{V}^4$ with $\mathbb{V}^4 \supset U' \ni p$ and such that $\psi : q \mapsto (y^0(q), y^1(q), y^2(q), y^3(q))$ where $(y^0(q), y^1(q), y^2(q), y^3(q)) = (T(q), z^1(q), z^2(q), z^3(q))$. Clearly $\Sigma_t \cap U'$ is determined by $y^0 = t$ and y^1, y^2, y^3 are local coordinates on Σ_t. The x^1, x^2, x^3 are global coordinates on Σ_t that are compatible with the differentiable structure since they are defined by $\Pi_{\mathscr{R}} \lceil_{\Sigma_t} : \Sigma_t \to E^3_{\mathscr{R}}$ (a diffeomorphism as Euclidean isometry). The Jacobian matrix of $x^i = x^i(y^1, y^2, y^3)$, $i = 1, 2, 3$, then has determinant J different from zero. Computing, the Jacobian determinant of $T = y^0, x^i = x^i(y^1, y^2, y^3)$, $i = 1, 2, 3$, still has determinant $J \neq 0$. This was the claim.

2.11.**2**. *Solution.* Fix an event $p \in \mathbb{V}^4$ and consider the coordinates for $\sigma_{\mathscr{R}}$ and $\sigma_{\mathscr{R}'}$. The first coordinate $t(p)$ in $(t(p), x^1(p), x^2(p), x^3(p))$ corresponds to the value of $T(p)$. Changing frame, $t'(p)$ must coincide with $T(p)$ up to an additive constant c in the definition of absolute time T. This proves (2.2). Consider now an event $p \in \Sigma_t$. The coordinates $(x'^1(p), x'^2(p), x'^3(p))$ are related to $(x^1(p), x^2(p), x^3(p))$ by $\Pi_{\mathscr{R}'} \lceil_{\Sigma_t} \circ (\Pi_{\mathscr{R}} \lceil_{\Sigma_t})^{-1}$, which is an isometry as composite of isometries (see Exercise 1.16.**4**) by Definition 2.5. Applying Exercise 1.16.**3** we obtain (2.2). The differentiability of $R^i{}_j(t)$ and $c^j(t)$ is a straightforward consequence of condition (2) in Definition 2.5 if we choose properly the point P: for example, if $P \in E^3_{\mathscr{R}}$ is the origin of the unprimed coordinates, from (2.2) the $c^j(t)$ are differentiable when we write the world line of P in primed coordinates.

2.11.**3**. *Solution.* (i) Let us represent spacetime as $\mathbb{R}^4$ with coordinates t, x^1, x^2, x^3. Consider the collection of curves labelled by triples $(x'^1, x'^2, x'^3) \in \mathbb{R}^3$ and parametrised by $t' \in \mathbb{R}$ as follows:

$$x^j(t') = \sum_{i=1}^{3} (R(t' - c)^{-1})^j{}_i (x'^i - c^i(t' - c)), \quad j = 1, 2, 3. \tag{C.3}$$

Immediately, these world lines satisfy the Definition 2.5, and by construction t', x'^1, x'^2, x'^3 are orthonormal coordinates moving with this frame $\mathscr{R}'$. By construction, Eq. (2.2) transform the Cartesian coordinates of $\mathscr{R}$ into those on the above $\mathscr{R}'$.

(ii) Evidently, if (2.2) transform coordinates inside $\mathscr{R}$, since $\Pi_{\mathscr{R}}$ is the same in source and target, there cannot be temporal dependence in (2.2). If, vice versa, the maps in (2.2) do not depend on time, the world lines of points at rest in (C.3) are the same world lines of points at rest in $\mathscr{R}$ labelled by $(x'^1, x'^2, x'^3) \in \mathbb{R}^3$, possibly for another time origin. Hence $\mathscr{R} = \mathscr{R}'$.

Exercises 2.18

2.18.1. *Solution.* First of all let us write down the unit basis of polar coordinates in terms of the Cartesian ones:

$$\mathbf{e}_r = \cos\phi\, \mathbf{e}_x + \sin\phi\, \mathbf{e}_y \,, \tag{C.4}$$

$$\mathbf{e}_\varphi = -\sin\phi\, \mathbf{e}_x + \cos\phi\, \mathbf{e}_y \,, \tag{C.5}$$

with inverse relations

$$\mathbf{e}_x = \cos\phi\, \mathbf{e}_r - \sin\phi\, \mathbf{e}_\varphi \,, \tag{C.6}$$

$$\mathbf{e}_y = \sin\phi\, \mathbf{e}_r + \cos\phi\, \mathbf{e}_\varphi \,. \tag{C.7}$$

After picking an origin O, by definition of $\mathbf{e}_r$

$$\mathbf{x}(t) = P(t) - O = x(t)\, \mathbf{e}_x + y(t)\, \mathbf{e}_y = r(t)\, \mathbf{e}_r(t) \,, \tag{C.8}$$

where we made explicit that $\mathbf{e}_r$ depends on time, since r and ϕ depend on time when the point moves. The time derivative of $P(t) - O$ should be computed remembering this. In the sequel, we follow the convention (going back to Newton) whereby the time derivative is denoted by a dot over the variable to be differentiated. Then

$$\mathbf{v}(t) = \dot{\mathbf{x}}(t) = \dot{r}(t)\, \mathbf{e}_r(t) + r(t)\, \dot{\mathbf{e}}_r(t) \,. \tag{C.9}$$

Assuming $\mathbf{e}_r$ and $\mathbf{e}_\varphi$ in (C.4) and (C.5) depend on time and differentiating in t, then using (C.6) and (C.7), we find

$$\dot{\mathbf{e}}_r = \dot{\phi}\, \mathbf{e}_\varphi \,, \tag{C.10}$$

$$\dot{\mathbf{e}}_\varphi = -\dot{\phi}\, \mathbf{e}_r \,, \tag{C.11}$$

so for example

$$\ddot{\mathbf{e}}_r = \ddot{\phi}\, \mathbf{e}_\varphi - \dot{\phi}^2\, \mathbf{e}_r \,,$$

$$\ddot{\mathbf{e}}_\varphi = -\dot{\phi}^2\, \mathbf{e}_\varphi - \ddot{\phi}\, \mathbf{e}_r \,.$$

Inserting (C.10) and (C.11) in (C.9) gives

$$\mathbf{v} = \dot{r}\, \mathbf{e}_r + r\dot{\phi}\, \mathbf{e}_\varphi \, .$$

Differentiating in time and using (C.10) and (C.11) we find the acceleration

$$\mathbf{a} = (\ddot{r} - r\dot{\phi}^2)\, \mathbf{e}_r + (r\ddot{\phi} + 2\dot{r}\dot{\phi})\, \mathbf{e}_\varphi \, .$$

2.18.2. *Solution.* Going about as in the previous exercise, we have

$$\mathbf{x}(t) = P(t) - O = x(t)\, \mathbf{e}_x + y(t)\, \mathbf{e}_y + z(t)\, \mathbf{e}_z = r(t)\, \mathbf{e}_r(t) \, . \qquad (C.12)$$

Moreover

$$\mathbf{e}_r = \sin\theta \cos\phi\, \mathbf{e}_x + \sin\theta \sin\phi\, \mathbf{e}_y + \cos\theta\, \mathbf{e}_z \, , \qquad (C.13)$$

$$\mathbf{e}_\theta = \cos\theta \cos\phi\, \mathbf{e}_x + \cos\theta \sin\phi\, \mathbf{e}_y - \sin\theta\, \mathbf{e}_z \, , \qquad (C.14)$$

$$\mathbf{e}_\varphi = -\sin\phi\, \mathbf{e}_x + \cos\phi\, \mathbf{e}_y \, . \qquad (C.15)$$

Observe that the triple is orthonormal, so if we have two vectors the third is their cross product (special attention should be paid to the sign). In order to invert the above transformation, drawing a picture suggests that

$$\mathbf{e}_z = \cos\theta\, \mathbf{e}_r - \sin\theta\, \mathbf{e}_\theta \, .$$

It is convenient to use the radial unit vector on $z = 0$:

$$n := \sin\theta\, \mathbf{e}_r + \cos\theta\, \mathbf{e}_\theta \, .$$

To find $\mathbf{e}_x$ and $\mathbf{e}_y$ we use the same formulas we employed for the polar coordinates, replacing $\mathbf{e}_r$ by n:

$$\mathbf{e}_x = \cos\phi\, n - \sin\phi\, \mathbf{e}_\varphi \, ,$$

$$\mathbf{e}_y = \sin\phi\, n + \cos\phi\, \mathbf{e}_\varphi \, .$$

Putting everything together:

$$\mathbf{e}_x = \sin\theta \cos\phi\, \mathbf{e}_r - \cos\theta \cos\phi\, \mathbf{e}_\theta - \sin\phi\, \mathbf{e}_\varphi \, , \qquad (C.16)$$

$$\mathbf{e}_y = \sin\theta \sin\phi\, \mathbf{e}_r + \cos\theta \sin\phi\, \mathbf{e}_\theta + \cos\phi\, \mathbf{e}_\varphi \, , \qquad (C.17)$$

$$\mathbf{e}_z = \cos\theta\, \mathbf{e}_r - \sin\theta\, \mathbf{e}_\theta \, . \qquad (C.18)$$

This can be obtained by noting that both triples are orthonormal, so related by an orthogonal map. Therefore the matrix representing (C.16)–(C.18) is the transpose of the matrix of (C.13)–(C.15).

By differentiating in t formulas (C.13)–(C.15) then using (C.16)–(C.18), an involved calculation produces:

$$\dot{\mathbf{e}}_r = \dot{\theta}\, \mathbf{e}_\theta + \dot{\phi} \sin\theta\, \mathbf{e}_\varphi \,, \tag{C.19}$$

$$\dot{\mathbf{e}}_\theta = -\dot{\theta}\, \mathbf{e}_r + \dot{\phi} \cos\theta\, \mathbf{e}_\varphi \,, \tag{C.20}$$

$$\dot{\mathbf{e}}_\varphi = -\dot{\phi} \sin\theta\, \mathbf{e}_r - \dot{\phi} \cos\theta\, \mathbf{e}_\theta \,. \tag{C.21}$$

Differentiating in t (C.12) and using (C.19) gives

$$\mathbf{v} = \dot{r}\, \mathbf{e}_r + r\dot{\theta}\, \mathbf{e}_\theta + r\dot{\phi} \sin\theta\, \mathbf{e}_\varphi \,.$$

As for the acceleration, differentiating the velocity and using (C.19)–(C.21) gives:

$$\mathbf{a} = (\ddot{r} - r\dot{\theta}^2 - r\dot{\phi}^2 \sin^2\theta)\, \mathbf{e}_r + (r\ddot{\theta} + 2\dot{r}\dot{\theta} - r\dot{\phi}^2 \sin\theta \cos\theta)\, \mathbf{e}_\theta$$
$$+ (r\ddot{\phi} \sin\theta + 2\dot{r}\dot{\phi} \sin\theta + 2r\dot{\phi}\dot{\theta} \cos\theta)\, \mathbf{e}_\varphi \,.$$

2.18.3. *Hint.* The exercise has actually been solved in the form of Exercise 2.18.1, since the z-axis does not add any complication because $\mathbf{e}_z$ is independent of t and:

$$\mathbf{x}(t) = P(t) - O = x(t)\, \mathbf{e}_x + y(t)\, \mathbf{e}_y + z(t)\, \mathbf{e}_z = r(t)\, \mathbf{e}_r(t) + z(t)\, \mathbf{e}_z \,.$$

2.18.4. The explanation is based on the fact that in orthonormal coordinates the Levi-Civita connection has zero coefficients, so (2.11) automatically gives the acceleration. Hence the particle's acceleration is just the second covariant derivative along the motion:

$$\mathbf{a} = \nabla_{\dot{P}} \dot{P} \,.$$

In generic coordinates, the above is exactly (2.11). Clearly, then, the result holds not only in cylindrical or spherical coordinates, but in any curvilinear coordinates.

Exercises 2.18

2.21.**1.** *Solution.* Take orthonormal coordinates in which Π is the plane $z = 0$. Then the curve's equations, by hypothesis, are $x = x(s)$, $y = y(s)$, $z = 0$. Hence $\mathbf{t}(s) = dx/ds\,\mathbf{e}_x + dy/ds\,\mathbf{e}_y$. By assumption $\mathbf{n}$ is defined (Γ cannot be reparametrised as a line segment in any subinterval of (a, b)), and by construction $\mathbf{n}$ belongs to Π because it is a linear combination of $\mathbf{e}_x$ and $\mathbf{e}_y$. Therefore $\mathbf{b}(s) = \mathbf{t}(t) \wedge \mathbf{n}(s) = \mathbf{e}_z$, perpendicular to Π and constant.

2.21.**2.** *Solution.* Let $P = P(s)$ denote the curve with $s \in (a, b) \ni c$. Consider the map $f(s) := \mathbf{b} \cdot (P(s) - P(c))$ of s. As $\mathbf{b}$ is constant in s, the derivative equals $f'(s) = \mathbf{b} \cdot \mathbf{t}(s) = 0$, so the map is constant. Actually it is zero since $f(c) = 0$. We conclude $\mathbf{b} \cdot (P(s) - P(c)) = 0$ for any $s \in (a, b)$, but the set $\{P \in \mathbb{E}^3 \mid \mathbf{b} \cdot (P - P(c))\} = 0$ is the plane through $P(c)$ normal to $\mathbf{b} \neq 0$. Therefore Γ belongs to the plane.

2.21.**3**. *Solution.* Note, first of all, that the last identities in each equation follow from the definition of Darboux vector and taking the cross products, since $\mathbf{t}, \mathbf{n}, \mathbf{b}$ is a right-handed orthonormal basis. So let us concentrate on the first identities. (2.27) is the definition of $\mathbf{n}$ and need not be proved, so let us prove (2.25). As $\mathbf{t}, \mathbf{n}, \mathbf{b}$ is an orthonormal basis,

$$d\mathbf{n}/ds = (d\mathbf{n}/ds \cdot \mathbf{t})\mathbf{t} + (d\mathbf{n}/ds \cdot \mathbf{n})\mathbf{n} + (d\mathbf{n}/ds \cdot \mathbf{b})\mathbf{b} \,.$$

Computing the components separately,

$$d\mathbf{n}/ds \cdot \mathbf{t} = d/ds(\mathbf{n} \cdot \mathbf{t}) - \mathbf{n} \cdot d\mathbf{t}/ds = 0 - \rho^{-1}(\mathbf{n} \cdot \mathbf{n}) = -\rho^{-1} \,.$$

For the **n**-component we have:

$$d\mathbf{n}/ds \cdot \mathbf{n} = (1/2)d/ds\mathbf{n} \cdot \mathbf{n} = (1/2)d/ds1 = 0 \,.$$

For the **b**-component:

$$d\mathbf{n}/ds \cdot \mathbf{b} = d\mathbf{n}/ds \cdot (\mathbf{t} \wedge \mathbf{b}) =: -\tau^{-1} \quad \text{by definition.}$$

Gathering all up, we find (2.25). Let us pass to (2.26). Since $\mathbf{t}, \mathbf{n}, \mathbf{b}$ are orthonormal,

$$d\mathbf{b}/ds = (d\mathbf{b}/ds \cdot \mathbf{t})\mathbf{t} + (d\mathbf{b}/ds \cdot \mathbf{n})\mathbf{n} + (d\mathbf{b}/ds \cdot \mathbf{b})\mathbf{b} \,.$$

Let us compute the components.

$$d\mathbf{b}/ds \cdot \mathbf{t} = d/ds(\mathbf{b} \cdot \mathbf{t}) - \mathbf{b} \cdot d\mathbf{t}/ds = 0 - \rho^{-1}(\mathbf{b} \cdot \mathbf{n}) = 0 \,.$$

For **b**:

$$d\mathbf{b}/ds \cdot \mathbf{b} = (1/2)d/ds(\mathbf{b} \cdot \mathbf{b}) = (1/2)d/ds1 = 0 \,.$$

For **n**:

$$d\mathbf{b}/ds \cdot \mathbf{n} = 0 - \mathbf{b} \cdot d\mathbf{n}/ds = -\mathbf{b} \cdot (-\mathbf{t}/\rho - \mathbf{b}/\tau) = (\mathbf{b} \cdot \mathbf{b})/\tau = 1/\tau \,.$$

All in all, we recover (2.26).

Exercises 2.30

2.30.**1**.*Solution.* Let us write the unit vectors of the basis of $\mathscr{R}$ in terms of the basis of $\hat{\mathscr{R}}$. By construction $\mathbf{e}_3 \cdot \hat{\mathbf{e}}_3 = \cos\theta$, $\mathbf{e}_3 \cdot \hat{\mathbf{e}}_1 = \sin\theta \ \sin\phi$ and $\mathbf{e}_3 \cdot \hat{\mathbf{e}}_2 = -\sin\theta \ \cos\psi$, so

$$\mathbf{e}_3 = \sin\theta \ \sin\psi \ \hat{\mathbf{e}}_1 - \sin\theta \ \cos\psi \ \hat{\mathbf{e}}_2 + \cos\theta \ \hat{\mathbf{e}}_3 \ .$$

Now to $\mathbf{e}_1$. First, we split it along $\mathbf{N}$, $\mathbf{e}_3$, $\mathbf{N}'$ where $\mathbf{N}' := \mathbf{e}_3 \wedge \mathbf{N}$. By construction $\mathbf{e}_1 = \cos\phi \ \mathbf{N} + \sin\phi \ \mathbf{N}'$, and from $\mathbf{N} = \cos\psi \ \hat{\mathbf{e}}_1 + \sin\psi \ \hat{\mathbf{e}}_2$ we have

$$\mathbf{e}_1 = \cos\phi \ (\cos\psi \ \hat{\mathbf{e}}_1 + \sin\psi \ \hat{\mathbf{e}}_2) + \sin\phi \ \mathbf{e}_3 \wedge (\cos\psi \ \hat{\mathbf{e}}_1 + \sin\psi \ \hat{\mathbf{e}}_2)$$

In other words:

$$\mathbf{e}_1 = \cos\phi \ (\cos\psi \ \hat{\mathbf{e}}_1 + \sin\psi \ \hat{\mathbf{e}}_2) + \sin\phi \ \left(\sin\theta \ \sin\psi \ \hat{\mathbf{e}}_1 - \sin\theta \ \cos\psi \ \hat{\mathbf{e}}_2\right.$$
$$\left. + \cos\theta \ \hat{\mathbf{e}}_3\right) \wedge (\cos\psi \ \hat{\mathbf{e}}_1 + \sin\psi \ \hat{\mathbf{e}}_2) \ ,$$

so in the end:

$$\mathbf{e}_1 = (\cos\phi \cos\psi - \sin\phi \sin\psi \cos\theta) \ \hat{\mathbf{e}}_1 + (\cos\phi \sin\psi + \sin\phi \cos\theta \cos\psi) \ \hat{\mathbf{e}}_2$$
$$+ \sin\phi \sin\theta \ \hat{\mathbf{e}}_3 \ .$$

The expression for $\mathbf{e}_2$ comes from taking the cross product $\mathbf{e}_3 \wedge \mathbf{e}_1$

$$\mathbf{e}_2 = -(\sin\phi \cos\psi + \cos\phi \sin\psi \cos\theta) \ \hat{\mathbf{e}}_1 + (-\sin\phi \sin\psi + \cos\phi \cos\theta \cos\psi) \ \hat{\mathbf{e}}_2$$
$$+ \cos\phi \sin\theta \ \hat{\mathbf{e}}_3$$

2.30.**2**. *Hint.* As the matrices are orthogonal, it suffices to transpose the matrix of the previous exercise.

2.30.**3**. *Hint.* Using the decomposition rule for ω, $\omega_{\hat{\mathscr{R}}|\mathscr{R}}$ is the sum of three ω vectors. Let $\mathscr{R}_2$ be the frame with moving trihedron $\hat{\mathbf{e}}_3$, $\mathbf{N}$, $\hat{\mathbf{e}}_3 \wedge \mathbf{N}$, and $\mathscr{R}_3$ the one with $\mathbf{e}_3$, $\mathbf{N}$, $\mathbf{e}_3 \wedge \mathbf{N}$. Then:

$$\omega_{\mathscr{R}|\hat{\mathscr{R}}} = \omega_{\mathscr{R}|\mathscr{R}_3} + \omega_{\mathscr{R}_3|\mathscr{R}_2} + \omega_{\mathscr{R}_2|\hat{\mathscr{R}}} \ ,$$

i.e., by definition of Euler angles:

$$\omega_{\mathscr{R}|\hat{\mathscr{R}}} = \dot{\phi} \ \mathbf{e}_3 + \dot{\theta} \mathbf{N} + \dot{\psi} \ \hat{\mathbf{e}}_3 \ ,$$

where $\mathbf{N} = \cos\psi \ \hat{\mathbf{e}}_1 + \sin\psi \ \hat{\mathbf{e}}_2$. Writing the unit vectors $\hat{\mathbf{e}}_i$ in terms of the $\mathbf{e}_k$ with the formulas of the previous exercise, we find the final expression for $\omega_{\mathscr{R}|\hat{\mathscr{R}}}$.

Exercises 2.32

2.32.**1**. *Solution*. Fix a frame $\mathscr{R}_D$ at rest with D and a frame $\mathscr{R}_{D'}$ at rest with D'. By construction: $\omega_{\mathscr{R}_D}|_{\mathscr{R}} = \frac{d\Theta}{dt}\mathbf{e}_z$ and $\omega_{\mathscr{R}_{D'}}|_{\mathscr{R}} = \frac{d\theta}{dt}\mathbf{e}_z$. If $P \in D$ and $P' \in D'$ are points moving with the two discs then

$$\mathbf{v}_P|_{\mathscr{R}} = \mathbf{v}_O|_{\mathscr{R}} + \mathbf{v}_P|_{\mathscr{R}_D} + \omega_{\mathscr{R}_D}|_{\mathscr{R}} \wedge (P - O) = 0 + 0 + \frac{d\Theta}{dt}\mathbf{e}_z \wedge (P - O)\,.$$

Similarly:

$$\mathbf{v}_{P'}|_{\mathscr{R}} = \mathbf{v}_{O'}|_{\mathscr{R}} + \mathbf{v}_{P'}|_{\mathscr{R}_{D'}} + \omega_{\mathscr{R}_{D'}}|_{\mathscr{R}} \wedge (P' - O') = 0 + 0 + \frac{d\theta}{dt}\mathbf{e}_z \wedge (P' - O')\,.$$

The rolling condition $\mathbf{v}_Q|_{\mathscr{R}} = \mathbf{v}_Q|_{\mathscr{R}}$, where Q is the contact point of the two discs, reads

$$\frac{d\Theta}{dt}\mathbf{e}_z \wedge R\mathbf{e}_x = \frac{d\theta}{dt}\mathbf{e}_z \wedge (-r)\mathbf{e}_x\,.$$

That is to say:

$$R\frac{d\Theta}{dt}\mathbf{e}_y = -r\frac{d\theta}{dt}\mathbf{e}_y\,,$$

so

$$R\frac{d\Theta}{dt} = -r\frac{d\theta}{dt}\,.$$

This ODE can be integrated to

$$R\Theta(t) = r\theta(t) + c\,,$$

where the constant c can be found evaluating the angles at the initial time (or any other instant).

2.32.**2**. *Solution*. Fix a frame $\mathscr{R}_D$ at rest with D and a frame $\mathscr{R}_{D'}$ at rest with D'. By construction: $\omega_{\mathscr{R}_D}|_{\mathscr{R}} = \frac{d\Theta}{dt}\mathbf{e}_z$ and $\omega_{\mathscr{R}_{D'}}|_{\mathscr{R}} = \frac{d\theta}{dt}\mathbf{e}_z$. If $P \in D$ and $P' \in D'$ are points moving with the discs,

$$\mathbf{v}_P|_{\mathscr{R}} = \mathbf{v}_O|_{\mathscr{R}} + \mathbf{v}_P|_{\mathscr{R}_D} + \omega_{\mathscr{R}_D}|_{\mathscr{R}} \wedge (P - O) = 0 + 0 + \frac{d\Theta}{dt}\mathbf{e}_z \wedge (P - O)\,.$$

Analogously:

$$\mathbf{v}_{P'}|_{\mathscr{R}} = \mathbf{v}_{O'}|_{\mathscr{R}} + \mathbf{v}_{P'}|_{\mathscr{R}_{D'}} + \omega_{\mathscr{R}_{D'}}|_{\mathscr{R}} \wedge (P' - O')$$

$$= \frac{d\phi}{dt}\mathbf{e}_z \wedge (O' - O) + 0 + \frac{d\theta}{dt}\mathbf{e}_z \wedge (P' - O')\,;$$

i.e.

$$\mathbf{v}_{P'}|_{\mathscr{R}} = \frac{d\phi}{dt}\mathbf{e}_z \wedge (O' - O) + \frac{d\theta}{dt}\mathbf{e}_z \wedge (P' - O').$$

We have used that O' rotates about O on the plane $z = 0$ and is determined by the polar angle ϕ. The rolling condition $\mathbf{v}_Q|_{\mathscr{R}} = \mathbf{v}_Q|_{\mathscr{R}}$, where Q is the contact point of the discs, is

$$\frac{d\Theta}{dt}\mathbf{e}_z \wedge R\mathbf{e} = \frac{d\phi}{dt}\mathbf{e}_z \wedge (R + r)\mathbf{e} - \frac{d\theta}{dt}\mathbf{e}_z \wedge r\mathbf{e}$$

where $\mathbf{e}$ is the unit vector of $Q - O$. In other words:

$$\left(R\frac{d\Theta}{dt} - (R + r)\frac{d\phi}{dt} + r\frac{d\theta}{dt}\right)\mathbf{e}_z \wedge \mathbf{e} = \mathbf{0},$$

and since $\mathbf{e}_z \wedge \mathbf{e} \neq \mathbf{0}$, the above is equivalent to:

$$R\frac{d\Theta}{dt} - (R + r)\frac{d\phi}{dt} + r\frac{d\theta}{dt} = 0.$$

This ODE integrates to

$$R\Theta(t) - (R + r)\phi(t) + r\theta(t) = c,$$

where the constant c can be fixed by evaluating the three angles at the initial time (or any other time).

C.3   Exercises for Chap. 3

Exercises 3.3

    3.3.**1.** *Hint.* Compose the Galilean transformations associated with $(c_2, \vec{c}_2, \vec{v}_2, R_2)$ and $(c_1, \vec{c}_1, \vec{v}_1, R_1)$ in this order, and find the coefficients of the resulting Galilean transformation.

    3.3.**2.** *Hint.* We must check that the composition law is associative, that $(0, \vec{0}, \vec{0}, I)$ is the neutral element and

$$(c, \vec{c}, \vec{v}, R)^{-1} = \left(-c, -R^{-1}(\vec{c} + c\vec{v}), -R^{-1}\vec{v}, R^{-1}\right)$$

the inverse.

3.3.3. *Hint.* Identify vectors $(t, \vec{x}) \in \mathbb{R}^4$ with vectors in $\mathbb{R}^5$ of the form $(1, t, \vec{x})$. Associate with each element $G := (c, \vec{c}, \vec{v}, R)$ in the Galilean group the matrix

$$\Omega_G := \begin{bmatrix} 1 & 0 & \vec{0} \\ c & 1 & 0 \\ \vec{c} & \vec{v} & R \end{bmatrix} .$$

Check that the action of Ω_G on the (column) vector $(1, t, \vec{x})$, Ω_G produces the known action of G on t and $\vec{x}$, whilst leaving untouched the 1. Check that the map $G \mapsto \Omega_G$ is 1-1. Verify Ω_G is non-singular and the composition law for the coefficients G, G' in the Galilean group corresponds to the matrix product of $\Omega_G, \Omega_{G'}$, and the neutral element of the Galilean group is associated with the identity matrix.

Exercises 3.5

3.5.1. *Solution.* Let $\mathbf{e}_x$, $\mathbf{e}_y$, $\mathbf{e}_z$ be axes at the point O and at rest with the ground. Suppose $\hat{O}$ moves with the seat and the axes $\hat{\mathbf{e}}_x$, $\hat{\mathbf{e}}_y$, $\hat{\mathbf{e}}_z$ moving with the seat are applied at $\hat{O}$. The axes $\mathbf{e}_z$, $\hat{\mathbf{e}}_z$ are both vertical, and the wheel lies on the plane $y = \hat{y} = 0$. Observe that by construction $\hat{\mathbf{e}}_x$, $\hat{\mathbf{e}}_z$ stay parallel to $\mathbf{e}_x$, $\mathbf{e}_z$ when the seat moves. Hence $\omega_{\hat{\mathscr{R}}|\mathscr{R}} = 0$. However, $\hat{\mathscr{R}}$ is not inertial. Suppose $\alpha \neq 0$ is the wheel's angular speed (constant). Putting O at the centre of the larger wheel, the coordinate transformation is (assume $\hat{t} = t$): $x = R\cos(\alpha t) + \hat{x}$, $y = \hat{y}$, $z = R\sin(\alpha t) + \hat{z}$. This transformation does not belong to the Galilean group, so $\hat{\mathscr{R}}$ cannot be inertial if $\mathscr{R}$ is.

Exercises 3.17

3.17.1. *Hint.* Newton's second law of motion for the system P, Q is

$$m\mathbf{a}_P = -mg\mathbf{e}_z - \kappa(P - Q) + \boldsymbol{\phi}_P$$

$$m\mathbf{a}_Q = -mg\mathbf{e}_z - \kappa(Q - P) + \boldsymbol{\phi}_Q$$

Parametrising Γ by arclength we determine P by s_P and Q by s_Q; using the formulas in **(1)**, Examples 2.20, the above read:

$$m\mathbf{a}_P = -mg\mathbf{e}_z - \kappa \left(\cos\left(\frac{s_P}{\sqrt{2}}\right) - \cos\left(\frac{s_Q}{\sqrt{2}}\right) \right) \mathbf{e}_x - \kappa \left(\sin\left(\frac{s_P}{\sqrt{2}}\right) - \sin\left(\frac{s_Q}{\sqrt{2}}\right) \right) \mathbf{e}_y$$

$$- \kappa \frac{s_P - s_Q}{\sqrt{2}} \mathbf{e}_z + \boldsymbol{\phi}_P$$

and

$$m\mathbf{a}_Q = -mg\mathbf{e}_z - \kappa \left(\cos\left(\frac{s_Q}{\sqrt{2}}\right) - \cos\left(\frac{s_P}{\sqrt{2}}\right) \right) \mathbf{e}_x - \kappa \left(\sin\left(\frac{s_Q}{\sqrt{2}}\right) - \sin\left(\frac{s_P}{\sqrt{2}}\right) \right) \mathbf{e}_y$$

$$- \kappa \frac{s_Q - s_P}{\sqrt{2}} \mathbf{e}_z + \boldsymbol{\phi}_Q \, .$$

Now, we make use of the formulas of **1**, Examples 2.20, project both equations along the tangent vectors $\mathbf{t}_P$ and $\mathbf{t}_P$ to Γ at P and Q and recall (2.18): we find the pure equations of motion ($\boldsymbol{\phi}_P$ and $\boldsymbol{\phi}_Q$ are normal to Γ by hypothesis, so they do not appear):

$$m\frac{d^2 s_P}{dt^2} = -\frac{mg}{\sqrt{2}} + \frac{\kappa}{\sqrt{2}} \sin\left(\frac{s_P}{\sqrt{2}}\right) \left[\cos\left(\frac{s_P}{\sqrt{2}}\right) - \cos\left(\frac{s_Q}{\sqrt{2}}\right) \right]$$

$$- \frac{\kappa}{\sqrt{2}} \cos\left(\frac{s_P}{\sqrt{2}}\right) \left[\sin\left(\frac{s_P}{\sqrt{2}}\right) - \sin\left(\frac{s_Q}{\sqrt{2}}\right) \right] - \frac{\kappa}{\sqrt{2}}(s_P - s_Q)$$

and

$$m\frac{d^2 s_Q}{dt^2} = -\frac{mg}{\sqrt{2}} + \frac{\kappa}{\sqrt{2}} \sin\left(\frac{s_Q}{\sqrt{2}}\right) \left[\cos\left(\frac{s_Q}{\sqrt{2}}\right) - \cos\left(\frac{s_P}{\sqrt{2}}\right) \right]$$

$$- \frac{\kappa}{\sqrt{2}} \cos\left(\frac{s_Q}{\sqrt{2}}\right) \left[\sin\left(\frac{s_Q}{\sqrt{2}}\right) - \sin\left(\frac{s_P}{\sqrt{2}}\right) \right] - \frac{\kappa}{\sqrt{2}}(s_Q - s_P)$$

3.17.2. *Hint.* Compute the time derivative of the right-hand side of E and use (3.52). E is the mechanical energy of the point particle in the non-inertial frame $\mathscr{R}'$.

C.4   Exercises for Chap. 4

Exercises 4.3

4.3.1. *Solution.* From the definition of $\Gamma_O|_{\mathscr{R}}$ and using relations (2.67) (for O' at rest in $\mathscr{R}'$) we have:

$$\Gamma_O|_{\mathscr{R}}(t) = \Gamma_O|_{\mathscr{R}'}(t) + \sum_{k=1}^{n} m_k(P_k(t) - O(t)) \wedge \left(\boldsymbol{\omega}_{\mathscr{R}'}|_{\mathscr{R}}(t) \wedge (P_k(t) - O'(t)) \right)$$

$$+ \sum_{k=1}^{n} m_k(P_k(t) - O(t)) \wedge \mathbf{v}_{O'}|_{\mathscr{R}}(t) \, .$$

As O', at rest with $\mathscr{R}'$, can be chosen arbitrarily, we choose it so that at time t it coincides with $O(t)$ (the choice is different instant by instant, but this is not forbidden). Hence

$$\boldsymbol{\Gamma}_O|_{\mathscr{R}}(t) = \boldsymbol{\Gamma}_O|_{\mathscr{R}'}(t) + \sum_{k=1}^n m_k (P_k(t) - O(t)) \wedge (\boldsymbol{\omega}_{\mathscr{R}'}|_{\mathscr{R}}(t) \wedge (P_k(t) - O(t))) .$$

4.3.2. *Solution.* By definition of $\boldsymbol{\Gamma}_O|_{\mathscr{R}}$:

$$\boldsymbol{\Gamma}_O|_{\mathscr{R}} = \sum_{k=1}^n m_k (P_k - O') \wedge \mathbf{v}_{P_k}|_{\mathscr{R}} + \sum_{k=1}^n m_k (O' - O) \wedge \mathbf{v}_{P_k}|_{\mathscr{R}} = \boldsymbol{\Gamma}_{O'}|_{\mathscr{R}} + (O - O') \wedge \mathbf{P}|_{\mathscr{R}},$$

where in the last passage we used (4.1).

Exercises 4.16

4.16.1. *Solution.* Let us work in spherical coordinates centred at O. Consider a closed C^1 curve in Ω, say $\Gamma : r = r(s), \theta = \theta(s), \varphi = \varphi(s)$ with $s \in [0, 1]$. Then $\mathbf{x}(s) = r(s) \, \mathbf{e}_r(s)$, so:

$$\frac{d\mathbf{x}}{ds} = \frac{dr}{ds} \, \mathbf{e}_r(s) + r(s) \frac{d\theta}{ds} \mathbf{e}_\theta(s) + r(s) \sin(\theta(s)) \frac{d\phi}{ds} \, \mathbf{e}_\varphi(s) .$$

As $\mathbf{e}_r(s)$ is always perpendicular to $\mathbf{e}_\theta(s)$ and $\mathbf{e}_\varphi(s)$ by construction, immediately, if $\mathbf{F} = F(r) \, \mathbf{e}_r$:

$$\oint_\Gamma \mathbf{F}(r) \cdot d\mathbf{x} = \int_0^1 F(r(s)) \, \mathbf{e}_r \cdot \frac{d\mathbf{x}}{ds} ds = \int_0^1 F(r(s)) \frac{dr}{ds} ds = \int_{r(0)}^{r(1)} \mathbf{F}(r) dr = 0 ,$$

since by hypothesis $r(0) = r(1)$. But Γ is arbitrary, so (1) in Theorem 4.15 implies $\mathbf{F}$ is conservative.

4.16.2. *Solution.* First, the angular momentum $\boldsymbol{\Gamma}_O|_{\mathscr{R}} = (P - O) \wedge \mathbf{v}_P|_{\mathscr{R}}$ satisfies, along the motion, $d\boldsymbol{\Gamma}_O|_{\mathscr{R}}/dt = (P - O) \wedge \mathbf{F} = \mathbf{0}$, because $\mathbf{F}$ is parallel to $P - O$ by hypothesis. Therefore the angular momentum is conserved in time. If $\boldsymbol{\Gamma}_O|_{\mathscr{R}} = \mathbf{0}$, in Cartesian coordinates $\mathbf{x} = (x, y, z)$ moving with $\mathscr{R}$ and centred at O,

$$\frac{d\mathbf{x}}{dt} = \lambda(t)\mathbf{x}(t) ,$$

where $\lambda(t)$ is a continuous map, hence integrable, except perhaps at the times where $\mathbf{x}(t_0) = \mathbf{0}$. In components, this ODE integrates immediately to:

$$\mathbf{x}(t) = e^{\int_0^t \lambda(\tau)d\tau} \mathbf{x}(0) .$$

The motion therefore occurs along an a priori direction, hence on a plane, which is definitely orthogonal to the angular momentum as the latter is the zero vector. For

$\Gamma_O|_{\mathscr{R}} \neq 0$, the condition:

$$\Gamma_O|_{\mathscr{R}} = (P(t) - O) \wedge \mathbf{v}_P|_{\mathscr{R}}(t)$$

says in particular that $P(t) - O$ is always perpendicular to the non-zero constant vector $\Gamma_O|_{\mathscr{R}}$, so $P(t)$ lies on the plane through O and perpendicular to that vector.
Exercises 4.20

4.20.1. *Solution.* (i) In the inertial frame $\mathscr{R}$, the second law of motion is:

$$m\mathbf{a}_P|_{\mathscr{R}} = -mg\,\mathbf{e}_z - k(P - H) + \boldsymbol{\phi}\,.$$

Writing the acceleration $\mathbf{a}_P|_{\mathscr{R}}$ in terms of $\mathbf{a}_P|_{\mathscr{R}'}$ with (2.69), and using the fact that the angular velocity is constant, Newton's law of motion in $\mathscr{R}'$ reads:

$$m\mathbf{a}_P|_{\mathscr{R}'} + 2m\omega\,\mathbf{e}_z \wedge \mathbf{v}_P|_{\mathscr{R}'} + m\omega^2\,\mathbf{e}_z \wedge (\mathbf{e}_z \wedge (P - O)) = -mg\,\mathbf{e}_z - k(P - H) + \boldsymbol{\phi}\,,$$

i.e.:

$$m\mathbf{a}_P|_{\mathscr{R}'} = -2m\omega\,\mathbf{e}_z \wedge \mathbf{v}_P|_{\mathscr{R}'} - m\omega^2\,\mathbf{e}_z \wedge (\mathbf{e}_z \wedge (P - O)) - mg\,\mathbf{e}_z - k(P - H) + \boldsymbol{\phi}\,.$$

Set $\mathbf{x} := P - O = x\mathbf{e}_x$, so that $\mathbf{v}_P|_{\mathscr{R}'} = \dot{x}\,\mathbf{e}_x$ where the dot means time derivative, and $P - H = x\,\mathbf{e}_x - h\,\mathbf{e}_z$. Since the constraint is smooth $\boldsymbol{\phi} = \phi^y\,\mathbf{e}_y + \phi^z\,\mathbf{e}_z$, and the above equation becomes:

$$m\ddot{x}\,\mathbf{e}_x = (m\omega^2 - \kappa)x\,\mathbf{e}_x + (\phi^y - 2m\omega\dot{x})\,\mathbf{e}_y + (\kappa h + \phi^z - mg)\,\mathbf{e}_z\,, \qquad \text{(C.22)}$$

from which we find the pure equation of motion

$$\frac{d^2x}{dt^2} = \left(\omega^2 - \frac{\kappa}{m}\right)x\,, \qquad \text{(C.23)}$$

and two equations for the constraint reaction once we know the motion

$$\phi^y(t) = 2m\omega\frac{dx}{dt}\,, \qquad \text{(C.24)}$$

$$\phi^z = mg - \kappa h\,. \qquad \text{(C.25)}$$

The general solution to the homogeneous linear equation with constant coefficients (C.23) depends on the ratio $m\omega^2/\kappa$, and we have three cases.

(1) $m\omega^2/\kappa < 1$. The roots of the characteristic polynomial are purely imaginary so:

$$x(t) = A \sin\left(\sqrt{\frac{\kappa}{m} - \omega^2}\, t\right) + B \cos\left(\sqrt{\frac{\kappa}{m} - \omega^2}\, t\right) ,$$

with $A, B \in \mathbb{R}$ arbitrary constants.

(2) $m\omega^2/\kappa = 1$. The roots of the characteristic polynomial coincide and are both 0. Therefore:

$$x(t) = Ae^{0t} + Bte^{0t} = A + Bt ,$$

with $A, B \in \mathbb{R}$ arbitrary constants.

(3) $m\omega^2/\kappa > 1$. The roots of the characteristic polynomial are real, so:

$$x(t) = A \exp\left(\sqrt{\omega^2 - \frac{\kappa}{m}}\, t\right) + B \exp\left(\sqrt{\omega^2 - \frac{\kappa}{m}}\, t\right) ,$$

with $A, B \in \mathbb{R}$ arbitrary constants.

(ii) Consider case (1). The initial condition $\mathbf{v}_P|_{\mathscr{R}'}(0) = \mathbf{0}$ immediately gives $A = 0$. Imposing $P(0) - O = x_0\mathbf{e}_x$ with $x_0 > 0$ we find $B = x_0$. Hence the motion with these initial conditions is

$$x(t) = x_0 \cos\left(\sqrt{\frac{\kappa}{m} - \omega^2}\, t\right) , \quad \text{for any } t \in \mathbb{R}.$$

Using (C.24) and (C.25) the constraint reaction, in time, is

$$\boldsymbol{\phi} = -2m\omega x_0 \sqrt{\frac{\kappa}{m} - \omega^2} \sin\left(\sqrt{\frac{\kappa}{m} - \omega^2}\, t\right) \mathbf{e}_y + (mg - \kappa h)\mathbf{e}_z .$$

(iii) and (iv) The second governing equation in our case gives

$$\frac{d}{dt}\bigg|_{\mathscr{R}} [(P(t)-O) \wedge m\mathbf{v}_P|_{\mathscr{R}}(t)] = (P(t) - O) \wedge (-mg\,\mathbf{e}_z - \kappa(P(t) - H) + \boldsymbol{\phi}(t)) ,$$

i.e.

$$\frac{d}{dt}\bigg|_{\mathscr{R}} [(P(t) - O) \wedge m\mathbf{v}_P|_{\mathscr{R}}(t)]$$
$$= x(t)\,\mathbf{e}_x(t) \wedge \left(-mg\,\mathbf{e}_z - \kappa x(t)\,\mathbf{e}_x(t) + h\kappa\,\mathbf{e}_z + \phi^y(t)\,\mathbf{e}_y(t) + \phi^z\,\mathbf{e}_z\right) .$$

The right-hand-side's component along the z-axis is, using (C.24):

$$x(t)\phi^y(t) = 2m\omega x(t)\frac{dx}{dt} = m\omega\frac{d}{dt}x(t)^2 \, .$$

Hence the angular momentum along z is not conserved except for the case when the motion does not satisfy $x(t)^2 = c$ constant, for any $t \in \mathbb{R}$, i.e. $x(t) = \pm c$ for any $t \in \mathbb{R}$. As motions are continuous maps, this means that the only admissible motions are $x(t) = x_0$ for any $t \in \mathbb{R}$.

In case (1): $m\omega^2/\kappa < 1$, only possible if $A = B = 0$. Hence $x(0) = 0$ and $dx/dt(0) = 0$.

In case (2): $m\omega^2/\kappa = 1$, only possible if $A = 0$. Hence $x(0) = x_0$ arbitrary and $dx/dt(0) = 0$.

In case (3): $m\omega^2/\kappa > 1$, only possible if $A = B = 0$. Hence $x(0) = 0$ and $dx/dt(0) = 0$.

  4.20.**2**. *Solution*. (i) Consider the kinetic energy theorem in $\mathscr{R}'$:

$$\frac{d\mathscr{T}|_{\mathscr{R}'}}{dt} = \mathbf{v}_P|_{\mathscr{R}'} \cdot \mathbf{F} \, ,$$

where $\mathbf{F}$ represents all inertial real forces acting on P as seen in $\mathscr{R}'$. $\mathbf{F}$ then equals the right-hand side of (C.22). Explicitly:

$$\frac{d\mathscr{T}|_{\mathscr{R}'}}{dt} = \frac{dx}{dt}\,\mathbf{e}_x \cdot ((m\omega^2 - \kappa)x\,\mathbf{e}_x + (\phi^y - 2m\omega\dot{x})\,\mathbf{e}_y + (\kappa h + \phi^z - mg)\,\mathbf{e}_z) \, .$$

In particular, the Coriolis force ($2m\omega\dot{x}\,\mathbf{e}_y$) and the constraint reaction do not dissipate power, since they contribute nothing to the above balance equation's right side. All in all:

$$\frac{d\mathscr{T}|_{\mathscr{R}'}}{dt} = (m\omega^2 - \kappa)\frac{dx(t)}{dt}x(t) \, .$$

which we may recast as:

$$\frac{d\mathscr{T}|_{\mathscr{R}'}}{dt} = -\frac{d}{dt}\left(\kappa\frac{x^2}{2} - m\omega^2\frac{x^2}{2}\right) \, .$$

Therefore, along the motion the following quantity is conserved:

$$\mathscr{E}|_{\mathscr{R}'} := \mathscr{T}|_{\mathscr{R}'} + \kappa\frac{x^2}{2} - m\omega^2\frac{x^2}{2} \, .$$

The second summand on the right is the usual potential energy of the spring, while the last term is a potential energy due to the centrifugal force, since:

$$-\nabla\left(-m\omega^2\frac{x^2}{2}\right) = (m\omega^2)x\,\mathbf{e}_x = -m\boldsymbol{\omega}_{\mathcal{R}'}|_{\mathcal{R}} \wedge (\boldsymbol{\omega}_{\mathcal{R}'}|_{\mathcal{R}} \wedge (P(t)-O)))$$

is precisely the centrifugal force appearing in $\mathcal{R}'$.

(ii) Let us examine the energy levels in $\mathcal{R}$. In $\mathcal{R}$ we have the spring's force, the constraint reaction and gravity. The conservative forces are the spring's and gravity. The gravitational potential energy mgz is constant because of the constraint on the point particle. All in all the balance theorem of the mechanical energy, defined as $\mathcal{E}|_{\mathcal{R}} = (1/2)m\mathbf{v}_P|_{\mathcal{R}} + (1/2)\kappa(P-H)^2 + mgz$, tells us that:

$$\frac{d\mathcal{E}|_{\mathcal{R}}}{dt} = \mathbf{v}_P|_{\mathcal{R}} \cdot (\phi^y\,\mathbf{e}_y(t) + \phi^z\,\mathbf{e}_z)\,,$$

where $\phi^y(t) = 2m\omega\dot{x}(t)$ by (C.24), while $\mathbf{v}_P|_{\mathcal{R}} \cdot \phi^z\,\mathbf{e}_z = 0$ since the motion occurs on the plane $z = 0$. We further know that:

$$\mathbf{v}_P|_{\mathcal{R}} = \mathbf{v}_P|_{\mathcal{R}'} + \boldsymbol{\omega}_{\mathcal{R}'}|_{\mathcal{R}} \wedge (P(t)-O)$$
$$= \dot{x}(t)\,\mathbf{e}_x(t) + \omega\,\mathbf{e}_z \wedge x(t)\,\mathbf{e}_x(t) = \dot{x}(t)\,\mathbf{e}_x(t) + \omega x(t)\,\mathbf{e}_y(t)\,.$$

Altogether:

$$\frac{d\mathcal{E}|_{\mathcal{R}}}{dt} = (\dot{x}(t)\,\mathbf{e}_x(t) + \omega x(t)\,\mathbf{e}_y(t)) \cdot 2m\omega\dot{x}(t)\,\mathbf{e}_y(t) = m\omega^2\frac{dx(t)^2}{dt}\,.$$

The same result is obtained in the previous exercise: $\mathcal{E}|_{\mathcal{R}}$ is conserved only along the motions where $x(t) = $ constant for any $t \in \mathbb{R}$. The initial conditions warranting such motions were examined in the previous exercise.

4.20.3. *Solution.* Let us use cylindrical coordinates r, φ, z adapted to the surface in the obvious way. Recall

$$P - O = z(P)\,\mathbf{e}_z + R\,\mathbf{e}_r(P)\,, \quad \mathbf{v}_P = \dot{z}(P)\,\mathbf{e}_z + R\dot{\varphi}\,\mathbf{e}_\varphi(P)\,,$$

$$\mathbf{a}_P = -R\dot{\varphi}^2(P)\mathbf{e}_r(P) + R\ddot{\varphi}(P)\mathbf{e}_\varphi(P) + \ddot{z}\,\mathbf{e}_z\,.$$

Newton's second law, for P, is:

$$-MR\dot{\varphi}^2(P)\mathbf{e}_r(P) + MR\ddot{\varphi}(P)\mathbf{e}_\varphi(P) + M\ddot{z}\,\mathbf{e}_z$$
$$= -Mg\,\mathbf{e}_z - \kappa(P-Q) - \gamma(P-O) + \phi^r(P)\,\mathbf{e}_r(P)\,,$$

which we express as

$$- MR\dot{\varphi}^2(P)\mathbf{e}_r(P) + MR\ddot{\varphi}(P)\mathbf{e}_\varphi(P) + M\ddot{z}(P)\,\mathbf{e}_z$$
$$= -Mg\,\mathbf{e}_z - (\kappa + \gamma)(P - O) + \kappa(Q - O) + \phi^r(P)\,\mathbf{e}_r(P)\,. \quad\quad (C.26)$$

For Q we have, similarly:

$$- mR\dot{\varphi}^2(Q)\mathbf{e}_r(Q) + mR\ddot{\varphi}(Q)\mathbf{e}_\varphi(Q) + m\ddot{z}(Q)\,\mathbf{e}_z$$
$$= -mg\,\mathbf{e}_z - (\kappa + \gamma)(Q - O) + \kappa(P - O) + \phi^r(Q)\,\mathbf{e}_r(Q)\,. \quad\quad (C.27)$$

To write everything in the cylindrical basis, remember that:

$$\mathbf{e}_x = \cos\varphi\,\mathbf{e}_r - \sin\varphi\,\mathbf{e}_\varphi \quad\text{and}\quad \mathbf{e}_y = \sin\varphi\,\mathbf{e}_r + \cos\varphi\,\mathbf{e}_\varphi\,.$$

Trivially then $P - O = z(P)\,\mathbf{e}_z + R\,\mathbf{e}_r(P)$, while

$$Q - O = z(Q)\,\mathbf{e}_z + R\,\mathbf{e}_r(Q) = z(Q)\,\mathbf{e}_z + R\cos\varphi(Q)\,\mathbf{e}_x + R\sin\varphi(Q)\,\mathbf{e}_y =$$
$$z(Q)\,\mathbf{e}_z + R\cos\varphi(Q)(\cos\varphi(P)\,\mathbf{e}_r(P) - \sin\varphi(P)\,\mathbf{e}_\varphi(P))$$
$$+ R\sin\varphi(Q)(\sin\varphi(P)\,\mathbf{e}_r(P) + \cos\varphi(P)\,\mathbf{e}_\varphi(P))\,.$$

Hence:

$$Q - O = z(Q)\,\mathbf{e}_z + R(\cos\varphi(Q)\cos\varphi(P) + \sin\varphi(Q)\sin\varphi(P))\,\mathbf{e}_r(P)$$
$$+ R(\sin\varphi(Q)\cos\varphi(P) - \cos\varphi(Q)\sin\varphi(P))\,\mathbf{e}_\varphi(P)\,,$$

i.e.:

$$Q - O = z(Q)\,\mathbf{e}_z + R\cos(\varphi(Q) - \varphi(P))\,\mathbf{e}_r(P) + R\sin(\varphi(Q) - \varphi(P))\,\mathbf{e}_\varphi(P)$$

Inserting in (C.26) and gathering terms along the unit vectors $\mathbf{e}_r(P)$, $\mathbf{e}_\varphi(P)$, $\mathbf{e}_z$ gives the following 3 equations, with slightly altered notation:

$$MR\ddot{\varphi}_P = \kappa R\sin(\varphi_Q - \varphi_P)\,, \quad\quad (C.28)$$

$$M\ddot{z}_P = -Mg - (\kappa + \gamma)z_P + \kappa z_Q\,, \qu\quad (C.29)$$

$$MR\dot{\varphi}_P^2 = (\gamma + \kappa)R - \kappa R\cos(\varphi_Q - \varphi_P) - \phi_P^r\,. \quad\quad (C.30)$$

Similarly, we obtain from (C.27) the equations for Q:

$$m R \ddot{\varphi}_Q = -\kappa R \sin \left(\varphi_Q - \varphi_P \right) , \tag{C.31}$$

$$m \ddot{z}_Q = -M g - (\kappa + \gamma) z_Q + \kappa z_P , \tag{C.32}$$

$$m R \dot{\varphi}_Q^2 = (\gamma + \kappa) R - \kappa R \cos \left(\varphi_Q - \varphi_P \right) - \phi_Q^r . \tag{C.33}$$

We then have the pure equations of motion

$$\ddot{\varphi}_P = \frac{\kappa}{M} \sin \left(\varphi_Q - \varphi_P \right) , \tag{C.34}$$

$$\ddot{\varphi}_Q = -\frac{\kappa}{m} \sin \left(\varphi_Q - \varphi_P \right) , \tag{C.35}$$

$$\ddot{z}_P = -g - \frac{(\kappa + \gamma)}{M} z_P + \frac{\kappa}{M} z_Q , \tag{C.36}$$

$$\ddot{z}_Q = -g - \frac{(\kappa + \gamma)}{m} z_Q + \frac{\kappa}{m} z_P , \tag{C.37}$$

and the constraint equations along the motion:

$$\phi_P^r = (\gamma + \kappa) R - \kappa R \cos \left(\varphi_Q - \varphi_P \right) - M R \dot{\varphi}_P^2 . \tag{C.38}$$

$$\phi_Q^r = (\gamma + \kappa) R - \kappa R \cos \left(\varphi_Q - \varphi_P \right) - m R \dot{\varphi}_Q^2 . \tag{C.39}$$

If, at $t = 0$, $\varphi_P = 0$, $\varphi_Q = \pi/2$ and $\dot{\varphi}_P = \dot{\varphi}_Q = 0$, irrespective of the values of z_P and z_Q, the reactions are $\boldsymbol{\phi}_P = (\gamma + \kappa) R \, \mathbf{e}_x$ and $\boldsymbol{\phi}_Q = (\gamma + \kappa) R \, \mathbf{e}_y$.
To conclude we need to show that $M \dot{\varphi}_P + m \dot{\varphi}_Q$ is conserved along the motion. This can be done directly from (C.34) and (C.35), multiplying the first by M, the second by m and adding. Thus $M \dot{\varphi}_P + m \dot{\varphi}_Q$ has zero time derivative on the solutions of the equations of motion, and therefore is a constant of motion. Another way is to observe $M R \dot{\varphi}_P + m R \dot{\varphi}_Q$ is exactly the z-component of the total angular momentum with respect to O, and at the same time the total momentum of the external forces does not have z-component. Therefore the second governing equation ensures the conservation of $R(M \dot{\varphi}_P + m \dot{\varphi}_Q)$.

C.5   Exercises for Chap. 5

Exercises 5.22
   5.22.1. *Hint.* Pick a triple of axes moving with the system (square + rod + point P) with origin O, one axis extending $C - O$ and one, of unit vector $\mathbf{n}$, perpendicular to the plane on which the square lies (the third axis makes the triple orthonormal and right-handed). The vector $\omega_{S|\mathscr{R}}$ of the rigid system is $\dot{\theta} \mathbf{n}$ for this axis. The equations of motion come from the second governing equation with respect to the pole O. The

above is a principal triple of inertia of the rigid system composed of the square and the rod. It is easy to find the contribution to the inertia tensor or the total angular momentum caused by P. The equation of motion must have the form (we leave the reader to interpret I, A and B)

$$I\frac{d^2\theta}{dt^2} + m||P - O||^2\frac{d^2\theta}{dt^2} = A\sin\theta + B\sin(\theta + \beta)$$

where β is the angle between $P - O$ and $C - O$. The system's total mechanical energy is obviously a first integral.

5.22.**2**. The system has two freedom degrees, say s and φ. The first coordinate is the signed distance travelled by the geometric point of contact C between the ramp and the disc, starting from the initial time configuration; the positive orientation is conventionally the downwards direction. The angle φ is formed by a given ray and the normal line through C to the ramp. The orientation of φ is set so that it increases when s increases. The value of φ is zero when $s = 0$ in the initial configuration. The rolling constraint, as usual, forces $s = R\varphi$. During the motion the freedom degrees reduce to one. However, we do not a priori know when the rolling occurs, so we must use two freedom degrees. The first governing equation has the form (all with respect to the inertial frame $\mathcal{R}$ in which the ramp is at rest and $\mathbf{e}_z$ is vertical)

$$m g \mathbf{e}_z + \boldsymbol{\phi}_C = m\mathbf{a}_G \quad \text{i.e.} \quad mg\sin\alpha + \phi^t = m\frac{d^2s}{dt^2}, \quad mg\cos\alpha + \phi^n = 0.$$
$$\text{(C.40)}$$

where we split $\boldsymbol{\phi}_C$ into a normal part ϕ^n to the ramp and a tangent part ϕ^t, and G is the disc's centre of mass, i.e. it geometric centre. The second governing equation for G is:

$$\mathbf{M}_G = I_G\left(\frac{\boldsymbol{\omega}}{dt}\right) + \boldsymbol{\omega} \wedge I_G(\boldsymbol{\omega}) + m(G - C) \wedge \mathbf{a}_G \quad \text{i.e.} \quad mgR\sin\alpha = \frac{1}{2}R^2\frac{d^2\varphi}{dt^2}.$$
$$\text{(C.41)}$$

At this point we shall consider two physical situations. (1) In case of pure rolling (no slipping) $s = Rd\varphi/dt$. Hence (C.41) gives the pure equation of motion:

$$\frac{3}{2}\frac{d^2s}{dt^2} = g\sin\alpha \quad \text{i.e.} \quad \frac{d^2s}{dt^2} = \frac{2}{3}g\sin\alpha, \quad \text{(C.42)}$$

while (C.40) provide the components ϕ^t and ϕ^n of the constraint reaction at C:

$$\varphi^t = \frac{1}{3}mg\sin\alpha, \quad \varphi^n = -mg\cos\alpha.$$

Therefore, given the static friction coefficient μ, we have rolling without slipping if and only if $|\varphi'| \leq \mu|\varphi''|$, meaning:

$$\frac{1}{3}mg \sin \alpha \leq \mu mg \cos \alpha \quad \text{i.e.} \quad \tan \alpha \leq 3\mu . \tag{C.43}$$

If so, the solution to (C.42) gives the system's motion.

(2) If (C.43) is violated, the contact point C slides on the ramp (i.e. it has non-zero velocity with respect to it). Then the constitutive characterisation ($\mathbf{t}$ is the unit tangent vector to the ramp increasing with s):

$$\phi^t \mathbf{t} = -f|\phi^n|\frac{\mathbf{v}_P}{||\mathbf{v}_P||} , \tag{C.44}$$

together with the last equation in (C.40), allows to express the components of the constraint reaction as:

$$\varphi'' = -mg \cos \alpha , \quad \varphi' = -f| - mg \cos \alpha| = -fmg \cos \alpha , \tag{C.45}$$

where, in any case, $ds/dt > 0$. The penultimate equation in (C.40) plus (C.41) provide the pure equations of motion:

$$g(\sin \alpha - f \cos \alpha) = \frac{d^2s}{dt^2} , \quad 2fg \cos \alpha = R\frac{d^2\varphi}{dt^2} . \tag{C.46}$$

Integrating gives the solution to the problem of motion if $\tan \alpha > 3\mu$.

C.6   Exercises for Chap. 6

Exercises 6.31

6.31.**1.** *Solution.* The differential equations for the motion are

$$\frac{d^2x}{dt^2} = -\frac{2\kappa}{m}x^2 ,$$

$$\frac{d^2y}{dt^2} = -\frac{2\kappa}{m}y^2 ,$$

$$\frac{d^2z}{dt^2} = \frac{4\kappa}{m}z^3 . \tag{C.47}$$

Consider the maximal solution with initial condition $\mathbf{x}(0) = (0, 0, z_0)$ and $\mathbf{v}(0) = (0, 0, 0)$ where $z_0 > 0$. We know this solution exists and is unique since the Existence and uniqueness theorem applies. The first two equations have unique solution $x(t) = 0$ and $y(t) = 0$ for any $t \in \mathbb{R}$, respectively, whereas (C.47) is not integrable elementarily. Nevertheless we can extract sufficient information to prove the claim, using a suitable first integral. Multiplying by dz/dt both sides of (C.47), easy manipulations give:

$$\frac{d}{dz}\left[\left(\frac{dz(t)}{dt}\right)^2 - 2\frac{\kappa}{m}z(t)^4\right] = 0.$$

Therefore

$$2\frac{\kappa}{m}z_0^4 = \left(\frac{dz(t)}{dt}\right)^2 - 2\frac{\kappa}{m}z(t)^4$$

and so:

$$\frac{dz(t)}{dt} = \pm\sqrt{2\frac{\kappa}{m}\left[z(t)^4 - z_0^4\right]}.$$

This can be integrated[1] to:

$$t(z) = \pm\int_{z_0}^{z}\frac{d\zeta}{\sqrt{2\frac{\kappa}{m}\left[\zeta^4 - z_0^2\right]}}.$$ (C.48)

As the integrand is strictly positive, by fixing the sign we can invert the map to obtain two monotone maps: $z = z_\pm(t)$. z_+ is defined for $t \geq 0$, z_- for $t < 0$, and z in either case lives in $[z_0, +\infty)$. The maximal solution, which we know exists and is unique, must coincide piecewise with $z_\pm$. So it is given by: $z := z_-(t)$ if $t < 0$ and $z = z_+(t)$ if $t \geq 0$. It is incomplete since the integral in (C.48) converges to some $\omega \in \mathbb{R}$ as $z \to +\infty$, and to some $\alpha \in \mathbb{R}$ when $z \to -\infty$. This is because as $z \to \pm\infty$:

$$\frac{1}{\sqrt{z^4 - z_0^4}} = O(z^{-2}).$$

Consider an open ball B_ϵ in $\mathbb{R}^3 \times \mathbb{R}^3$ at the origin, $\mathbf{x} = \mathbf{0}$, $\mathbf{v} = \mathbf{0}$, of radius $\epsilon > 0$. We cannot find any open neighbourhood U of the origin in $\mathbb{R}^3 \times \mathbb{R}^3$ so that all initial conditions in V give maximal solutions with orbits contained in B_ϵ. We can in fact pick as initial condition $\mathbf{x}(0) = (0, 0, z_0)$ and $\mathbf{v}(0) = (0, 0, 0)$ where $z_0 > 0$ is so

---

[1] The integral converges as $z \to z_0$ since:

$$\frac{1}{\sqrt{z^4 - z_0^4}} = \frac{1}{\sqrt{z - z_0}\sqrt{(z + z_0)(z^2 + z_0^2)}}$$

and the singularity is caused by the first factor on the right, which is integrable.

small that it belongs to V. The maximal solution exits B_ϵ in finite time since the solution's component $z(t)$ is increasing and tends to $+\infty$ as $t \to \omega$. The same thing happens for $t \to \alpha$: $z(t) \to -\infty$. To sum up, the equilibrium configuration given by the origin of $\mathbb{R}^3$ is not past- nor future-stable.

C.7 Exercises for Chap. 7

Exercises 7.33

7.33.**1**. *Solution.* For starters, let us notice that $\mathscr{R}$ is inertial so we just need to consider real forces; besides, the constraints are holonomic and ideal: we can apply the Lagrangian formalism.

(a) Using the free coordinates ϕ_P and ϕ_Q, the positions of P and Q are determined, in $\mathscr{R}$, by the position vectors:

$$\mathbf{x}_P = R \cos \phi_P \, \mathbf{e}_x + R \sin \phi_P \, \mathbf{e}_y + R\phi_P \, \mathbf{e}_z , \tag{C.49}$$

$$\mathbf{x}_Q = R \cos \phi_Q \, \mathbf{e}_x + R \sin \phi_Q \, \mathbf{e}_y + R\phi_P \, \mathbf{e}_z . \tag{C.50}$$

From these we find the velocities (in $\mathscr{R}$):

$$\mathbf{v}_P = \dot\phi_P \left(-R \sin \phi_P \, \mathbf{e}_x + R \cos \phi_P \, \mathbf{e}_y + R \, \mathbf{e}_z \right) , \tag{C.51}$$

$$\mathbf{v}_Q = \dot\phi_Q \left(-R \sin \phi_Q \, \mathbf{e}_x + R \cos \phi_Q \, \mathbf{e}_y + R \, \mathbf{e}_z \right) . \tag{C.52}$$

The latter give the kinetic energy $\mathscr{T} = \frac{m}{2}\mathbf{v}_P^2 + \frac{m}{2}\mathbf{v}_Q^2$ in $\mathscr{R}$, in terms of the free coordinates ϕ_P, ϕ_Q and the associated dotted coordinates $\dot\phi_P, \dot\phi_Q$:

$$\mathscr{T} = \frac{mR^2}{2}\dot\phi_P^2 \left(\sin^2 \phi_P + \cos^2 \phi_P + 1 \right) + \frac{mR^2}{2}\dot\phi_Q^2 \left(\sin^2 \phi_Q + \cos^2 \phi_Q + 1 \right) .$$

Simplifying, we obtain:

$$\mathscr{T} = mR^2 \left(\dot\phi_P^2 + \dot\phi_Q^2 \right) .$$

Consider first $\gamma = 0$. In this case all active forces are conservative and therefore the system admits a Lagrangian $\mathscr{L}|_{\mathscr{R}} = \mathscr{L} = \mathscr{T} - \mathscr{U}$, where $\mathscr{U} = \mathscr{U}|_{\mathscr{R}}$ is the sum of the potential energies of every conservative force acting on the system. In practice we have a contribution to $\mathscr{U}$ due to the gravitational potential energy:

$$mgz_P + mgz_Q = mgR(\phi_P + \phi_Q)$$

and one due to the potential energy of the spring connecting P and Q:

$$\frac{\kappa}{2}(P - Q)^2 = \frac{\kappa R^2}{2}\left[(\cos\phi_P - \cos\phi_Q)\,\mathbf{e}_x + (\sin\phi_P - \sin\phi_Q)\,\mathbf{e}_y + (\phi_P - \phi_Q)\,\mathbf{e}_z\right]^2 .$$

The latter, after squaring, simplifies to:

$$\frac{\kappa R^2}{2}\left[2(1 - \cos\phi_P\cos\phi_Q - \sin\phi_P\sin\phi_Q) + (\phi_P - \phi_Q)^2\right]$$

$$= \frac{\kappa R^2}{2}\left[2(1 - \cos(\phi_P - \phi_Q)) + (\phi_P - \phi_Q)^2\right] .$$

All in all:

$$\mathscr{U} = mgR(\phi_P + \phi_Q) + \kappa R^2\left[1 - \cos(\phi_P - \phi_Q) + \frac{(\phi_P - \phi_Q)^2}{2}\right] .$$

The system's Lagrangian when $\gamma = 0$ is then:

$$\mathscr{L} = mR^2\left(\dot\phi_P^2 + \dot\phi_Q^2\right) - mgR(\phi_P + \phi_Q)$$

$$- \kappa R^2\left[1 - \cos(\phi_P - \phi_Q) + \frac{(\phi_P - \phi_Q)^2}{2}\right] . \tag{C.53}$$

In case $\gamma = 0$ the non-trivial part of the Euler-Lagrange equations is:

$$\frac{d}{dt}\frac{\partial\mathscr{L}}{\partial\dot q^k} - \frac{\partial\mathscr{L}}{\partial q^k} = 0 , \quad \text{where } q^1 = \phi_P \text{ and } q^2 = \phi_Q.$$

The trivial part, obviously, is the one saying that along the motion the dotted variables are the time derivatives of the undotted ones. So, we end up with the (second-order) system:

$$2mR^2\frac{d^2\phi_P}{dt^2} = -mgR - \kappa R^2\left[\sin(\phi_P - \phi_Q) + (\phi_P - \phi_Q)\right] , \tag{C.54}$$

$$2mR^2\frac{d^2\phi_Q}{dt^2} = -mgR + \kappa R^2\left[\sin(\phi_P - \phi_Q) + (\phi_P - \phi_Q)\right] . \tag{C.55}$$

Let us consider now a viscous force acting on P only. It does not have a potential as it does not depend on position, hence it is neither conservative. Therefore we must use the formulation of the Euler-Lagrange equations with the explicit Lagrangian components of the active forces:

$$\frac{d}{dt}\frac{\partial\mathscr{T}}{\partial\dot q^k} - \frac{\partial\mathscr{T}}{\partial q^k} = \mathscr{Q}_k , \quad \text{where } q^1 = \phi_P \text{ and } q^2 = \phi_Q.$$

Among all forces acting on the points, only the viscous force has no potential, so we may simplify the above equations as follows, using the same Lagrangian:

$$\frac{d}{dt}\frac{\partial \mathscr{L}}{\partial \dot{q}^k} - \frac{\partial \mathscr{L}}{\partial q^k} = \mathscr{Q}_k^{(visc)}, \quad \text{where } q^1 = \phi_P \text{ and } q^2 = \phi_Q, \tag{C.56}$$

where $\mathscr{Q}_k^{(visc)}$ are the Lagrangian components of the viscous force only. The conservative forces have been absorbed by the term $-\mathscr{U}$ in the Lagrangian $\mathscr{L}$. Therefore

$$\mathscr{Q}_k^{(visc)} = \sum_{i=1}^{2} \frac{\partial \mathbf{x}_i}{\partial q^k} \cdot \mathbf{F}_i^{(visc)} = \frac{\partial \mathbf{x}_R}{\partial q^k} \cdot (-\gamma \mathbf{v}_P).$$

All in all, using (C.49) and (C.49) to compute the viscosity's Lagrangian components, only the component with $k = 1$ (i.e. the one relative to ϕ_P) is non-zero, and it equals:

$$\mathscr{Q}_1^{(visc)} = -\gamma \left[\frac{\partial}{\partial \phi_P} (R \cos \phi_P \, \mathbf{e}_x + R \sin \phi_P \, \mathbf{e}_y + R\phi_P \, \mathbf{e}_z) \right]$$
$$\cdot \dot{\phi}_P \left(-R \sin \phi_P \, \mathbf{e}_x + R \cos \phi_P \, \mathbf{e}_y + R \, \mathbf{e}_z \right).$$

A few calculations eventually provide:

$$\mathscr{Q}_1^{(visc)} = -2\gamma R^2 \dot{\phi}_P.$$

In this case the Euler-Lagrange equations become, from (C.56):

$$2m R^2 \frac{d^2\phi_P}{dt^2} = -mgR - \kappa R^2 \left[\sin(\phi_P - \phi_Q) + (\phi_P - \phi_Q) \right] - 2\gamma R^2 \frac{d\phi_P}{dt},$$
$$\tag{C.57}$$

$$2m R^2 \frac{d^2\phi_Q}{dt^2} = -mgR + \kappa R^2 \left[\sin(\phi_P - \phi_Q) + (\phi_P - \phi_Q) \right]. \tag{C.58}$$

(b) The change of variables $\theta = \phi_P - \phi_Q$, $\tau = \phi_P + \phi_Q$ is differentiable, bijective with differentiable inverse. The inverse formulas are: $\phi_P = (\theta + \tau)/2$ and $\phi_Q = (\tau - \theta)/2$, from which it is clear that varying τ and fixing θ gives a rigid rotation of the system by $\tau/2$ composed with a z-translation by $R\tau/2$.
From the above transformation rules we have: $\dot{\phi}_P = (\dot{\theta} + \dot{\tau})/2$ and $\dot{\phi}_Q = (\dot{\tau} - \dot{\theta})/2$. Substituting in (C.53) we obtain the Lagrangian in the new variables:

$$\mathscr{L}(\theta, \tau, \dot{\theta}, \dot{\tau}) = \frac{mR^2}{2} \left((\dot{\theta} + \dot{\tau})^2 + (\dot{\tau} - \dot{\theta})^2 \right) + mgR\tau - \kappa R^2 \left[1 - \cos\theta + \frac{\theta^2}{2} \right],$$

i.e.

$$\mathcal{L}(\theta, \tau, \dot{\theta}, \dot{\tau}) = \frac{mR^2}{2}\left(\dot{\theta}^2 + \dot{\tau}^2\right) + mgR\tau - \kappa R^2\left[1 - \cos\theta + \frac{\theta^2}{2}\right].$$

When $g = 0$ the Lagrangian reduces to:

$$\mathcal{L}(\theta, \tau, \dot{\theta}, \dot{\tau}) = \frac{mR^2}{2}\left(\dot{\theta}^2 + \dot{\tau}^2\right) - \kappa R^2\left[1 - \cos\theta + \frac{\theta^2}{2}\right].$$

In this case the E-L equation for τ is, given that $\mathcal{L}$ does not explicitly depend on τ:

$$\frac{d}{dt}\left(\frac{\partial\mathcal{L}}{\partial\dot{\tau}}\right) = 0.$$

In other words, the quantity:

$$I := \frac{\partial\mathcal{L}}{\partial\dot{\tau}} = mR^2\frac{d\tau}{dt},$$

is conserved in time along the motion (i.e. on the solutions to the Euler-Lagrange equations).

Going back to the free coordinates ϕ_P and ϕ_Q, the quantity L_z takes the form:

$$I := \frac{\partial\mathcal{L}}{\partial\dot{\tau}} = mR^2\left(\frac{d\phi_P}{dt} + \frac{d\phi_Q}{dt}\right).$$

Consider now the angular momentum in $\mathcal{R}$ with respect to O:

$$\boldsymbol{\Gamma}_O|_{\mathcal{R}} = (P - O) \wedge m\mathbf{v}_P + (Q - O) \wedge m\mathbf{v}_Q.$$

The calculation involving (C.49), (C.50), (C.51) and (C.52) immediately gives

$$\mathbf{e}_z \cdot \boldsymbol{\Gamma}_O|_{\mathcal{R}} = mR^2\left(\frac{d\phi_P}{dt} + \frac{d\phi_Q}{dt}\right).$$

Consequently, the conserved quantity I is nothing but the z-component of the system's angular momentum in $\mathcal{R}$ with respect to O.

7.33.2. *Solution.* Observe that $\mathcal{R}$ is an inertial frame so we just need to consider real forces; moreover, the constraints are holonomic and ideal, and we may use the Lagrangian formalism.

(a) Let us find the system's Lagrangian, which will be called $\mathcal{L}$ without mention of the frame $\mathcal{R}$. As all active forces are conservative the Lagrangian will be $\mathcal{L} = \mathcal{T} - \mathcal{U}$, where $\mathcal{T}$ is the kinetic energy in $\mathcal{R}$, and $\mathcal{U}$ is the total potential energy in the same frame. Let us determine the kinetic energy. We write positions and velocities in terms of the free coordinates r_1, φ_1 and r_2, φ_1 and the associated dotted coordinates. Given the cone's equation, the height z_1 of P_1 coincides with the radius r_1 of the same point on $z = 0$ in polar coordinates, trivially

$$P_1 - O = r_1(\cos\varphi_1\, \mathbf{e}_x + \sin\varphi_1\, \mathbf{e}_y + \mathbf{e}_z).$$

and also:

$$P_2 - O = r_1(\cos\varphi_2\, \mathbf{e}_x + \sin\varphi_2\, \mathbf{e}_y + \mathbf{e}_z).$$

Let us focus on P_1. Assuming r_1, φ_1 depend on time, differentiating with respect to time and replacing the dotted variables with the time derivatives of the free coordinates:

$$\mathbf{v}_1 = \dot{r}_1(\cos\varphi_1\, \mathbf{e}_x + \sin\varphi_1\, \mathbf{e}_y + \mathbf{e}_z) + r_1\dot{\varphi}_1(-\sin\varphi_1\, \mathbf{e}_x + \cos\varphi_1\, \mathbf{e}_y).$$

Squaring and using $\cos^2\varphi_1 + \sin^2\varphi_1 = 1$ gives:

$$\mathbf{v}_1^2 = 2\dot{r}_1^2 + r_1^2\dot{\varphi}_1^2.$$

Similarly:

$$\mathbf{v}_2^2 = 2\dot{r}_2^2 + r_2^2\dot{\varphi}_2^2.$$

Finally, the kinetic energy is:

$$\mathscr{T} = \frac{m}{2}\left(2(\dot{r}_1^2 + \dot{r}_2^2) + r_1^2\dot{\varphi}_1^2 + r_2^2\dot{\varphi}_2^2\right). \tag{C.59}$$

The potential energy is the sum of the gravitational potential energy $mgz_1 + mgz_2$ and the spring's potential energy $\kappa(P_1 - R_2)^2/2$. In particular, from the above expression for $P_1 - O$ and $P_2 - O$ we have:

$$(P_1 - P_2)^2 = (r_1\cos v\phi_1 - r_2\cos\varphi_2)^2 + (r_1\sin\varphi_1 - r_2\sin\varphi_2)^2 + (r_1 - r_2)^2,$$

so squaring and using $\cos(a - b) = \cos a \cos b + \sin a \sin b$ we end up with:

$$(P_1 - P_2)^2 = r_1^2 + r_2^2 - 2r_1r_2\cos(\varphi_1 - \varphi_2).$$

All in all:

$$\mathscr{U}(P_1, P_2) = mg(r_1 + r_2) + \frac{\kappa}{2}\left(r_1^2 + r_2^2 - 2r_1r_2\cos(\varphi_1 - \varphi_2)\right), \tag{C.60}$$

and then, by (C.59):

$$\mathcal{L} = \frac{m}{2}\left(2(\dot{r}_1{}^2 + \dot{r}_2{}^2) + r_1^2\dot{\varphi}_1^2 + r_2^2\dot{\varphi}_2^2\right) - mg(r_1 + r_2)$$

$$- \frac{\kappa}{2}\left(r_1^2 + r_2^2 - 2r_1r_2\cos(\varphi_1 - \varphi_2)\right). \qquad (C.61)$$

With this Lagrangian the Euler-Lagrange equations are

$$2m\frac{d^2r_1}{dt^2} = mr_1\left(\frac{d\varphi_1}{dt}\right)^2 - mg - k(r_1 - r_2\cos(\varphi_1 - \varphi_2)), \quad (C.62)$$

$$2m\frac{d^2r_2}{dt^2} = mr_2\left(\frac{d\varphi_2}{dt}\right)^2 - mg - k(r_2 - r_1\cos(\varphi_1 - \varphi_2)), \quad (C.63)$$

$$m\frac{d}{dt}\left(r_1^2\frac{d\varphi_1}{dt}\right) = -\kappa r_1r_2\sin(\varphi_1 - \varphi_2), \qquad (C.64)$$

$$m\frac{d}{dt}\left(r_2^2\frac{d\varphi_2}{dt}\right) = -\kappa r_1r_2\sin(\varphi_2 - \varphi_1). \qquad (C.65)$$

(b) We have $\Phi := (\varphi_1 - \varphi_2)/2$ and $\Theta = (\varphi_1 + \varphi_2)/2$ so $\varphi_1 = \Phi + \Theta$, $\varphi_2 = \Theta - \Phi$, $\dot{\varphi}_1 = \dot{\Phi} + \dot{\Theta}$, $\dot{\varphi}_2 = \dot{\Theta} - \dot{\Phi}$. The Lagrangian (C.61) in the variables then reads:

$$\mathcal{L} = \frac{m}{2}\left[2(\dot{r}_1{}^2 + \dot{r}_2{}^2) + (r_1^2 + r_2^2)\left(\dot{\Phi}^2 + \dot{\Theta}^2\right) + 2(r_1^2 - r_2^2)\dot{\Phi}\dot{\Theta}\right]$$

$$- mg(r_1 + r_2) - \frac{\kappa}{2}\left(r_1^2 + r_2^2 - 2r_1r_2\cos 2\Phi\right).$$

Hence the Lagrangian does not explicitly depend on Θ, meaning that $\partial\mathcal{L}/\partial\dot{\Theta}$ is conserved along the motion due to the Euler-Lagrange equations. Back in initial coordinates:

$$\frac{\partial\mathcal{L}}{\partial\dot{\Theta}} = \frac{m}{2}2\left[(\dot{r}_1{}^2 + \dot{r}_2{}^2)\dot{\Theta} + (\dot{r}_1{}^2 - \dot{r}_2{}^2)\dot{\Phi}\right]$$

$$= \frac{m}{2}\left[(\dot{r}_1{}^2 + \dot{r}_2{}^2)(\dot{\varphi}_1 + \dot{\varphi}_2) + (\dot{r}_1{}^2 - \dot{r}_2{}^2)(\dot{\varphi}_1 - \dot{\varphi}_2)\right].$$

Expanding along the motion,

$$\frac{\partial\mathcal{L}}{\partial\dot{\Theta}} = mr_1^2\frac{d\varphi_1}{dt} + mr_2^2\frac{d\varphi_2}{dt}.$$

Writing $P_i - O$ and the velocities $\mathbf{v}_i$ in terms of the free coordinates as determined at the beginning, easy manipulations show that:

$$\frac{\partial \mathscr{L}}{\partial \dot{\Theta}} = \mathbf{e}_z \cdot \boldsymbol{\Gamma}_0|_{\mathscr{R}} ,$$

where:

$$\boldsymbol{\Gamma}_0|_{\mathscr{R}} = (P_1 - 0) \wedge m\mathbf{v}_1 + (P_2 - 0) \wedge m\mathbf{v}_2$$

is the total angular momentum in $\mathscr{R}$ with respect to O.

7.33.3. *Solution.* (a) We use as free coordinate the x-component of $P - O$ in the non-inertial frame $\mathscr{R}$. Since in the inertial frame there are no non-conservative forces (except for the constraint reaction $\boldsymbol{\phi}$, obviously), it is convenient to use the Lagrangian in the inertial frame $\mathscr{L}|_{\hat{\mathscr{R}}} = \mathscr{T}|_{\hat{\mathscr{R}}} - \mathscr{U}|_{\hat{\mathscr{R}}}$. Let us write the kinetic energy $\mathscr{T}|_{\hat{\mathscr{R}}} = \frac{1}{2}m\mathbf{v}_P^2|_{\hat{\mathscr{R}}}$ in terms of x and $\dot{x}$. The velocity in the non-inertial frame $\mathscr{R}$ is clearly the time derivative of:

$$P - O = x(t)\mathbf{e}_x + \sinh x(t)\mathbf{e}_z ,$$

since the unit vectors $\mathbf{e}_x$ and $\mathbf{e}_z$ are fixed in said frame. Then:

$$\mathbf{v}_P|_{\mathscr{R}} = \dot{x}\mathbf{e}_x + \dot{x}\cosh x\,\mathbf{e}_z .$$

The velocity in $\hat{\mathscr{R}}$ therefore is

$$\mathbf{v}_P|_{\hat{\mathscr{R}}} = \mathbf{v}_P|_{\mathscr{R}} + \boldsymbol{\omega}_{\mathscr{R}}|_{\hat{\mathscr{R}}} \wedge (P - O) = \dot{x}\mathbf{e}_x + \dot{x}\cosh x\,\mathbf{e}_z + \Omega\mathbf{e}_z \wedge (x\mathbf{e}_x + \sinh x\,\mathbf{e}_z) .$$

Computing,

$$\mathbf{v}_P|_{\hat{\mathscr{R}}} = \dot{x}\mathbf{e}_x + \Omega x\mathbf{e}_y + \dot{x}\cosh x\,\mathbf{e}_z . \tag{C.66}$$

Consequently:

$$\mathscr{T}|_{\hat{\mathscr{R}}} = \frac{m}{2}\dot{x}^2 \left(1 + \cosh^2 x\right) + \frac{m}{2}\Omega^2 x^2 . \tag{C.67}$$

There are two contributions to the potential energy $\mathscr{U}|_{\hat{\mathscr{R}}}$. One is due to gravity mgz, the other to the spring: $(\kappa/2)x^2$. All in all:

$$\mathscr{L}|_{\hat{\mathscr{R}}} = \frac{m}{2}\dot{x}^2 \left(1 + \cosh^2 x\right) - mg\sinh x - \frac{1}{2}(\kappa - m\Omega^2)x^2 . \tag{C.68}$$

With this Lagrangian

$$\frac{\partial \mathcal{L}|_{\hat{\mathcal{R}}}}{\partial \dot{x}} = m\dot{x}(1+\cosh^2 x), \quad \frac{\partial \mathcal{L}|_{\hat{\mathcal{R}}}}{\partial x} = m\dot{x}^2 \sinh x \cosh x + (m\Omega^2 - \kappa)x - mg\cosh x.$$

The Euler-Lagrange equations therefore are:

$$\frac{d}{dt}m\left(\frac{dx}{dt}(1+\cosh^2 x)\right) = m\dot{x}^2 \sinh x \cosh x + (m\Omega^2 - \kappa)x - mg\cosh x,$$

giving:

$$m\frac{d^2x}{dt^2}(1+\cosh^2 x) + 2m\left(\frac{dx}{dt}\right)^2 \sinh x \cosh x$$

$$= m\left(\frac{dx}{dt}\right)^2 \sinh x \cosh x + (m\Omega^2 - \kappa)x - mg\cosh x.$$

The final form for the Euler-Lagrange equations is:

$$m\frac{d^2x}{dt^2} = \frac{(m\Omega^2 - \kappa)x - mg\cosh x - m\left(\frac{dx}{dt}\right)^2 \sinh x \cosh x}{1 + \cosh^2 x} \tag{C.69}$$

(b) Directly by definition of $\mathcal{H}$ we have:

$$\frac{d}{dt}\mathcal{H} = \frac{d\dot{x}}{dt}\frac{\partial \mathcal{L}|_{\hat{\mathcal{R}}}}{\partial \dot{x}} + \dot{x}\frac{d}{dt}\frac{\partial \mathcal{L}|_{\hat{\mathcal{R}}}}{\partial \dot{x}} - \frac{d}{dt}\mathcal{L}|_{\hat{\mathcal{R}}}.$$

Using the Euler-Lagrange equations we may recast the above as:

$$\frac{d}{dt}\mathcal{H} = \frac{d\dot{x}}{dt}\frac{\partial \mathcal{L}|_{\hat{\mathcal{R}}}}{\partial \dot{x}} + \dot{x}\frac{\partial \mathcal{L}|_{\hat{\mathcal{R}}}}{\partial x} - \frac{d}{dt}\mathcal{L}|_{\hat{\mathcal{R}}}.$$

Adding and subtracting $\partial \mathcal{L}|_{\hat{\mathcal{R}}}/\partial t$ gives:

$$\frac{d}{dt}\mathcal{H} = \frac{d\dot{x}}{dt}\frac{\partial \mathcal{L}|_{\hat{\mathcal{R}}}}{\partial \dot{x}} + \dot{x}\frac{\partial \mathcal{L}|_{\hat{\mathcal{R}}}}{\partial x} + \frac{\partial \mathcal{L}|_{\hat{\mathcal{R}}}}{\partial t} - \frac{d}{dt}\mathcal{L}|_{\hat{\mathcal{R}}} - \frac{\partial \mathcal{L}|_{\hat{\mathcal{R}}}}{\partial t}.$$

On the other hand, as $\mathcal{L}|_{\hat{\mathcal{R}}} = \mathcal{L}|_{\hat{\mathcal{R}}}(t, x, \dot{x})$ we must have:

$$\frac{d}{dt}\mathcal{L}|_{\hat{\mathcal{R}}} = \frac{d\dot{x}}{dt}\frac{\partial \mathcal{L}|_{\hat{\mathcal{R}}}}{\partial \dot{x}} + \dot{x}\frac{\partial \mathcal{L}|_{\hat{\mathcal{R}}}}{\partial x} + \frac{\partial \mathcal{L}|_{\hat{\mathcal{R}}}}{\partial t}.$$

Inserting this in the previous expression says that on any solution to the Euler-Lagrange equations the following remarkable identity holds:

$$\frac{d}{dt}\mathscr{H}(t, x(t), \dot{x}(t)) = -\frac{\partial}{\partial t}\mathscr{L}|_{\hat{\mathscr{R}}}(t, x(t), \dot{x}(t)) .$$

In our case, (C.68) obviously implies that $\mathscr{L}|_{\hat{\mathscr{R}}}$ does not explicitly depend on time, so

$$\frac{d}{dt}\mathscr{H}(t, x(t), \dot{x}(t)) = -\frac{\partial}{\partial t}\mathscr{L}|_{\hat{\mathscr{R}}}(t, x(t), \dot{x}(t)) = 0 ,$$

and therefore on any solution to the Euler-Lagrange equations $\mathscr{H}$ is constant. Let us make physical sense of $\mathscr{H}$, which from (C.68) equals:

$$\mathscr{H}(x, \dot{x}) = \frac{m}{2}\dot{x}^2 \left(1 + \cosh^2 x\right) + mg \sinh x + \frac{1}{2}(\kappa - m\Omega^2)x^2 . \qquad \text{(C.70)}$$

We will show that $\mathscr{H}$ is the total mechanical energy in the non-inertial frame $\mathscr{R}$. In the inertial frame $\hat{\mathscr{R}}$ the mechanical energy is not conserved: to keep the system rotating one must provide external work. This work is dissipated by the constraint reaction $\boldsymbol{\phi}$. In fact, the constraint reaction $\boldsymbol{\phi}$ in general does work in $\hat{\mathscr{R}}$ since it is normal to the curve, but the curve's orientation is not that of the particle's velocity, because the curve itself moves in $\hat{\mathscr{R}}$. Let us look at the picture in the non-inertial frame $\mathscr{R}$, taking into account the inertial Coriolis force $-m\boldsymbol{\omega}_{\mathscr{R}}|_{\hat{\mathscr{R}}} \wedge \mathbf{v}_P|_{\mathscr{R}}$ and the inertial centrifugal force $-m\boldsymbol{\omega}_{\mathscr{R}}|_{\hat{\mathscr{R}}} \wedge (\boldsymbol{\omega}_{\mathscr{R}}|_{\hat{\mathscr{R}}} \wedge (P - O))$ (other inertial forces are absent since $\boldsymbol{\omega}_{\mathscr{R}}|_{\hat{\mathscr{R}}}$ is constant in time). In the non-inertial frame, $\boldsymbol{\phi}$ does not work because, by construction, it is always orthogonal to the velocity of P in that frame. The other forces are: gravity, which is conservative; the elastic force, conservative; the Coriolis force, which does no work (it is by definition perpendicular to the velocity); finally, the centrifugal force, which we will see is conservative. The total mechanical energy in $\mathscr{R}$, given the kinetic energy plus the three potential energies mentioned, is therefore a constant of motion. We claim it coincides with $\mathscr{H}$. Let us write the equation of motion in Newtonian form in $\mathscr{R}$. If $\mathbf{x} = P - O = x\mathbf{e}_x + \sinh x\mathbf{e}_z$ and hence $\mathbf{v}_P|_{\mathscr{R}} = \dot{x}\mathbf{e}_x + \dot{x}\cosh x\mathbf{e}_z$, we have:

$$m\frac{d^2}{dt^2}\bigg|_{\mathscr{R}}\mathbf{x} = -mg\mathbf{e}_z - \kappa x\mathbf{e}_x - m\Omega\,\mathbf{e}_z \wedge \left(\frac{dx}{dt}\mathbf{e}_x + \frac{dx}{dt}\cosh x\mathbf{e}_z\right)$$

$$- m\Omega^2\,\mathbf{e}_z \wedge (\mathbf{e}_z \wedge (x\mathbf{e}_x + \sinh x\mathbf{e}_z)) + \boldsymbol{\phi} .$$

Expanding,

$$m\frac{d^2}{dt^2}(x\mathbf{e}_x + \sinh x\mathbf{e}_z) = -mg\mathbf{e}_z - \kappa x\mathbf{e}_x + m\Omega^2 x\mathbf{e}_x - m\Omega\frac{dx}{dt}\mathbf{e}_y + \boldsymbol{\phi} . \qquad \text{(C.71)}$$

The last two forces produce no work, being orthogonal to the velocity (which always lies along $\mathbf{e}_x$). The centrifugal force arises by differentiating in x and flipping the sign of the centrifugal potential energy $-\frac{1}{2}m\Omega^2 x^2$, which corresponds to a spring with negative elastic constant. By the theorem on the conservation of the mechanical energy, on any motion the sum of kinetic energy + total potential energy is conserved. The kinetic energy in $\mathscr{R}$ equals

$$\frac{1}{2}m v_P^2|_{\mathscr{R}} = \frac{m}{2}\left(\frac{dx}{dt}\right)^2 \left(1 + \cosh^2 x\right) ,$$

so the previous first integral is:

$$\frac{m}{2}\left(\frac{dx}{dt}\right)^2 \left(1 + \cosh^2 x\right) + mg \sinh x + \frac{1}{2}(\kappa - m\Omega^2)x^2 .$$

This is, on any motion, exactly expression (C.70) for $\mathscr{H}$.

(c) Directly from (C.71):

$$\phi^x = m\frac{d^2 x}{dt^2} + (\kappa - m\Omega^2)x .$$

Using (C.69) to write $md^2 x/dt^2$ in terms of x and its time derivative, on any motion:

$$\phi^x(x,\dot{x}) = (\kappa - m\Omega^2)x + \frac{(m\Omega^2 - \kappa)x - mg\cosh x - m\left(\frac{dx}{dt}\right)^2 \sinh x \cosh x}{1 + \cosh^2 x}$$

i.e., in terms of x and $\dot{x}$:

$$\phi^x(x,\dot{x}) = \frac{(\kappa - m\Omega^2)x \cosh^2 x - mg\cosh x - m\dot{x}^2 \sinh x \cosh x}{1 + \cosh^2 x} . \qquad \text{(C.72)}$$

(d) As $\mathscr{H}$ is conserved in time, for the motion with initial conditions $x(0) = 0$ and $\dot{x}(0) = v > 0$ we have:

$$\mathscr{H}(0, v) = \mathscr{H}(x(t), \dot{x}(t)) , \qquad \text{for any } t.$$

Therefore, from (C.70) we deduce

$$mv^2 = \frac{m}{2}\dot{x}(t)^2 \left(1 + \cosh^2 x(t)\right) + mg \sinh x(t) + \frac{1}{2}(\kappa - m\Omega^2)x(t)^2 ,$$

and then:

$$m\dot{x}(t)^2 = \frac{2mv^2 - 2mg \sinh x(t) + (m\Omega^2 - \kappa)x(t)^2}{1 + \cosh^2 x(t)} .$$

Substituting in (C.72) provides the answer to the problem:

$$\phi^*(x(t)) = \frac{(\kappa - m\Omega^2)x(t)\cosh^2 x(t) - mg\cosh x(t)}{1 + \cosh^2 x(t)}$$

$$+ \frac{2mg\sinh x(t) - 2mv^2 + (\kappa - m\Omega^2)x(t)^2}{(1 + \cosh^2 x(t))^2} \sinh x(t)\cosh x(t) .$$

7.33.4. Solution. (a) The Lagrangian in the inertial frame $\hat{\mathscr{R}}$, as all active forces are conservative, has the general form:

$$\mathscr{L} = \frac{m}{2}(v_P^2 + v_Q^2) - \mathscr{U} \tag{C.73}$$

with obvious notation. The velocities v_P and v_Q refer to $\hat{\mathscr{R}}$. We should write all quantities in the Lagrangian in terms of $\phi, \dot{\phi}, \psi, \dot{\psi}$. First note that $Q - O = Q - P + P - O$ and

$$P - O = L(\cos\phi\, e_x + \sin\phi\, e_z) , \quad Q - P = L(\cos\psi\, e_x + \sin\psi\, e_z) . \tag{C.74}$$

Therefore

$$Q - O = L[(\cos\phi + \cos\psi)e_x + (\sin\phi + \sin\psi)e_z] . \tag{C.75}$$

Hence we can make

$$\mathscr{U} = mg(z_P + z_Q) + \frac{k}{2}(Q - O)^2$$

explicit in terms of the angles:

$$\mathscr{U}(\phi, \psi) = mgL(2\sin\phi + \sin\psi) + \frac{kL^2}{2}\left((\cos\phi + \cos\psi)^2 + (\sin\phi + \sin\psi)^2\right)^2 ,$$

and therefore:

$$\mathscr{U}(\phi, \psi) = mgL(2\sin\phi + \sin\psi) + kL^2(1 + \cos(\phi - \psi)) . \tag{C.76}$$

The above expressions of $P - O$ and $Q - O$ allow to compute the velocities of P and Q in $\mathscr{R}$ in terms of $\phi, \dot{\phi}, \psi, \dot{\psi}$. The velocities in $\hat{\mathscr{R}}$ come from the usual kinematical transformations, keeping in account the frames' relative motion. Assuming the angles vary on the system's motion, differentiating $P - O$ and $Q - O$ in (C.74) and (C.75) in time, and recalling the unit vectors are constant in time in

$\mathscr{R}$, we obtain:

$$\mathbf{v}_P|_{\mathscr{R}} = L(-\dot{\phi}\sin\phi\, \mathbf{e}_x + \dot{\phi}\cos\phi\, \mathbf{e}_z),\tag{C.77}$$

$$\mathbf{v}_Q|_{\mathscr{R}} = -L(\dot{\phi}\sin\phi + \dot{\psi}\sin\psi)\, \mathbf{e}_x + L(\dot{\phi}\cos\phi + \dot{\psi}\cos\psi)\mathbf{e}_z.\tag{C.78}$$

The velocities of P and Q in $\hat{\mathscr{R}}$ arise from:

$$\mathbf{v}_P = \mathbf{v}_P|_{\mathscr{R}} + \boldsymbol{\omega}_{\mathscr{R}}|_{\hat{\mathscr{R}}}\wedge(P-O) = \mathbf{v}_P|_{\mathscr{R}} + \Omega\mathbf{e}_z\wedge(P-O), \quad \mathbf{v}_Q = \mathbf{v}_Q|_{\mathscr{R}} + \Omega\mathbf{e}_z\wedge(Q-O).$$

With the help of (C.74), (C.75), (C.77) and (C.78), we find:

$$\mathbf{v}_P = -\dot{\phi}L\sin\phi\, \mathbf{e}_x + \Omega L\cos\phi\mathbf{e}_y + L\dot{\phi}\cos\phi\, \mathbf{e}_z,\tag{C.79}$$

$$\mathbf{v}_Q = -L(\dot{\phi}\sin\phi + \dot{\psi}\sin\psi)\, \mathbf{e}_x + \Omega L(\cos\phi + \cos\psi)\, \mathbf{e}_y + L(\dot{\phi}\cos\phi + \dot{\psi}\cos\psi)\mathbf{e}_z.$$
$$\tag{C.80}$$

Substituting these in the right-hand side of (C.73) and using (C.76), we find the Lagrangian $\mathscr{L}$ in terms of the chosen coordinates:

$$\mathscr{L} = \frac{mL^2}{2}\left(2\dot{\phi}^2 + \dot{\psi}^2 + 2\dot{\phi}\dot{\psi}\cos(\phi - \psi)\right) + \frac{mL^2\Omega^2}{2}\left[(\cos\phi + \cos\psi)^2 + \cos^2\phi\right]$$

$$- mgL(2\sin\phi + \sin\psi) - kL^2(1 + \cos(\phi - \psi)).\tag{C.81}$$

The Euler-Lagrange equations then are:

$$2mL^2\frac{d^2\phi}{dt^2} + mL^2\cos(\phi - \psi)\frac{d^2\psi}{dt^2} = -mL^2\left(\frac{d\psi}{dt}\right)^2\sin(\phi - \psi)$$

$$- mL^2\Omega^2(2\cos\phi + \cos\psi)\sin\phi$$

$$+ kL^2\sin(\phi - \psi) - 2mgL\cos\phi,\tag{C.82}$$

plus

$$mL^2\frac{d^2\psi}{dt^2} + mL^2\cos(\psi - \phi)\frac{d^2\phi}{dt^2} = -mL^2\left(\frac{d\phi}{dt}\right)^2\sin(\psi - \phi)$$

$$- mL^2\Omega^2(\cos\psi + \cos\phi)\sin\psi$$

$$+ kL^2\sin(\psi - \phi) - mgL\cos\psi.\tag{C.83}$$

(b) The above Euler-Lagrange equations *are not* in normal form: on the left there is a linear combination of the highest derivatives, rather than just one. The system can be recast as:

$$
mL^2
\begin{bmatrix}
2 & \cos(\phi - \psi) \\
\cos(\phi - \psi) & 1
\end{bmatrix}
\begin{bmatrix}
\frac{d^2\phi}{dt^2} \\
\frac{d^2\psi}{dt^2}
\end{bmatrix}
$$

$$
=
\begin{bmatrix}
-mL^2 \left(\frac{d\psi}{dt}\right)^2 \sin(\phi - \psi) - ML^2\Omega^2(2\cos\phi + \cos\psi)\sin\phi \\
+kL^2 \sin(\phi - \psi) - 2mgL\cos\phi \\
\\
-mL^2 \left(\frac{d\phi}{dt}\right)^2 \sin(\psi - \phi) - ML^2\Omega^2(\cos\psi + \cos\phi)\sin\psi \\
+kL^2 \sin(\psi - \phi) - mgL\cos\psi
\end{bmatrix}
$$

If the square matrix on the left is invertible, the system can be put in normal form by multiplying by the inverse matrix. The determinant (apart from mL^2) is $2 - \cos^2(\phi - \psi)$, a strictly positive quantity. Hence the matrix is invertible. Cramer's rule gives:

$$
\begin{bmatrix}
2 & \cos(\phi - \psi) \\
\cos(\phi - \psi) & 1
\end{bmatrix}^{-1}
$$

$$
= (2 - \cos^2(\phi - \psi))^{-1}
\begin{bmatrix}
1 & -\cos(\phi - \psi) \\
-\cos(\phi - \psi) & 2
\end{bmatrix}
$$

The Euler-Lagrange system in normal form is:

$$
\begin{bmatrix}
\frac{d^2\phi}{dt^2} \\
\frac{d^2\psi}{dt^2}
\end{bmatrix}
=
\frac{1}{mL^2(2 - \cos^2(\phi - \psi))}
\begin{bmatrix}
1 & -\cos(\phi - \psi) \\
-\cos(\phi - \psi) & 2
\end{bmatrix}
\times
$$

$$
\times
\begin{bmatrix}
-mL^2 \left(\frac{d\psi}{dt}\right)^2 \sin(\phi - \psi) - mL^2\Omega^2(2\cos\phi + \cos\psi)\sin\phi \\
+kL^2 \sin(\phi - \psi) - 2mgL\cos\phi \\
\\
-mL^2 \left(\frac{d\phi}{dt}\right)^2 \sin(\psi - \phi) - mL^2\Omega^2(\cos\psi + \cos\phi)\sin\psi \\
+kL^2 \sin(\psi - \phi) - mgL\cos\psi
\end{bmatrix}
$$

(c) As the Lagrangian (C.81) does not explicitly depend on time, by Theorem 8.13 the Hamiltonian:

$$
\mathcal{H} = \dot\phi \frac{\partial \mathcal{L}}{\partial \dot\phi} + \dot\psi \frac{\partial \mathcal{L}}{\partial \dot\psi} - \mathcal{L}
$$

is conserved. Explicitly:

$$\mathcal{H} = \frac{mL^2}{2}\left(2\dot{\phi}^2 + \dot{\psi}^2 + 2\dot{\phi}\dot{\psi}\cos(\phi - \psi)\right) - \frac{mL^2\Omega^2}{2}\left[(\cos\phi + \cos\psi)^2 + \cos^2\phi\right]$$

$$+ mgL(2\sin\phi + \sin\psi) + kL^2(1 + \cos(\phi - \psi)). \tag{C.84}$$

Physically, $\mathcal{H}$ is the system's mechanical energy in the non-inertial frame $\mathcal{R}$. In fact, (C.77) and (C.78) immediately imply the kinetic energy in $\mathcal{R}$ is:

$$\mathcal{T}|_{\mathcal{R}} = \frac{mL^2}{2}\left(2\dot{\phi}^2 + \dot{\psi}^2 + 2\dot{\phi}\dot{\psi}\cos(\phi - \psi)\right).$$

In $\mathcal{R}$ we have: the constraint reactions, which do no overall work since the constraints are ideal; the conservative forces of gravity and the spring's, with total potential energy:

$$+ mgL(2\sin\phi + \sin\psi) + kL^2(1 + \cos(\phi - \psi));$$

the Coriolis force, which does no work as perpendicular to the velocity vectors; the centrifugal forces, perpendicular to and stemming from the z-axis, with modulus $m\Omega^2 d_P$ and $m\Omega^2 d_Q$ respectively, where d_P and d_Q are the points' distances to the z-axis. These forces have the same structure of repulsive elastic forces of constant $-m\Omega^2$ attached to the z-axis at points P' and Q' at the same height as the spring's corresponding other end P and Q. The centrifugal forces are therefore conservative, with total potential energy:

$$-\frac{m\Omega^2}{2}\left(d_P^2 + d_Q^2\right) = -\frac{mL^2\Omega^2}{2}\left[\cos^2\phi + (\cos\phi + \cos\psi)^2\right].$$

Adding up all contributions, the total mechanical energy in $\mathcal{R}$ equals:

$$E|_{\mathcal{R}} = \frac{mL^2}{2}\left(2\dot{\phi}^2 + \dot{\psi}^2 + 2\dot{\phi}\dot{\psi}\cos(\phi - \psi)\right) - \frac{mL^2\Omega^2}{2}\left[(\cos\phi + \cos\psi)^2 + \cos^2\phi\right]$$

$$+ mgL(2\sin\phi + \sin\psi) + kL^2(1 + \cos(\phi - \psi)).$$

Therefore it coincides with $\mathcal{H}$ along the motion.

(d) Directly from system (C.82)–(C.83) the 4 constant curves $(\phi(t), \psi(t)) = (\pm\pi/2, \pm\pi/2)$ satisfy equations of motion and initial conditions. Since they can be put in normal form with C^∞ source terms, the solutions are unique, so what we found are the only solutions for the initial conditions given in the hypothesis.

C.8   Exercises for Chap. 10

Exercises 10.61

10.61.**1.** (1) If $o \notin J^+(o') \cup J^-(o')$ there are no C^1 causal curves connecting the points. Hence the line segment from o to o', which exists since $\mathbb{M}^4$ is affine, must be spacelike. Suppose conversely there exists a spacelike segment between o and o'. We may adapt Minkowski coordinates so that o and o' lie on $\mathbf{e}_1$. Any C^1 curve connecting the points will be like $[a, b] \ni s \mapsto (x^0(s), x^1(s), x^2(s), x^3(s))$. As $x^0(a) = x^0(b) = 0$ there must exist a point $c \in [a, b]$ where $\frac{dx^0}{ds}|_c = 0$. If the curve is causal, then $|\dot{x}^0|^2 \geq \sum_{a=1}^{3} |\dot{x}^a|^2$. Hence for $s = s_0$ we have $\dot{x}^\alpha(s_0) = 0$, $\alpha = 0, 1, 2, 3$, contradicting the fact that causal curves have non-zero tangent vector. Hence $o \notin J^+(o') \cup J^-(o')$.

(2) Because $J^+(o') \cup J^-(o')$ is (in Minkowski coordinates on $\mathbb{R}^4$) a closed convex (double) cone with vertex o', if $o \in J^+(o') \cup J^-(o')$ the segment from o' to o must lie on the cone and hence is causal. Vice versa, if there is a causal segment between o and o', by definition $o \in J^+(o') \cup J^-(o')$.

(3) Because $I^+(o') \cup I^-(o')$ is (in Minkowski coordinates on $\mathbb{R}^4$) an open convex cone with vertex o' made of disjoint half-cones, if $o \in I^+(o') \cup I^-(o')$ the segment connecting o' and o lies on the cone and therefore is timelike. Conversely, if there is timelike segment from o to o', by definition $o \in I^+(o') \cup I^-(o')$.

10.61.**2.** *Hint.* Think of the situation where the source emits impulses, relative to proper time, of frequency ν_0. Apply the formula of time dilation to find $\nu = \nu_0 \sqrt{1 - \frac{\Omega^2 r^2}{c^2}}$.

10.61.**3.** *Hint.* Recall that simultaneity is relative. For the driver, the doors do not open and close simultaneously.

10.61.**4.** *Solution.* Let γ be the spacelike segment from o to o', and consider Minkowski coordinates whose origin O is the midpoint of γ and where $\mathbf{e}_1$ is parallel to γ itself. If $2L$ is the length of γ, the piecewise C^1 curve γ' given by the lightlike segments from o to $O + L\mathbf{e}_0$ and from $O + L\mathbf{e}_0$ to o' respectively, has zero total length. This is because the segments are lightlike. We can approximate γ' by a family of spacelike C^1 curves from o to o', on the plane spanned by $\mathbf{e}_0$ and $\mathbf{e}_1$ through O, whose lengths can be made as small as we please.

10.61.**5.** *Hint.* The sum $\mathbf{P}$ of future timelike vectors is future timelike because all are contained in the same half of the light cone. Choosing $\mathcal{R}$ parallel to $\mathbf{P}$, the spatial components of the sum of the momenta will vanish. Letting M be as said, $M^2 c^2 = -g(\mathbf{P}, \mathbf{P})$ does not depend on the Minkowski coordinates. Therefore we may define the four-velocity of the centre of mass by $\mathbf{V} := M^{-1}\mathbf{P}$. Picking any Minkowski coordinates we have:

$$MV^a = \sum_{i=1}^{N} P_i^a = \sum_{i=1}^{N} \frac{m_i v_i^a(t)}{\sqrt{1 - \left(\frac{v_i}{c}\right)^2}}$$

from which (10.99) follows once we recall definition (10.100).

10.61.**6** and **7**. *Hint.* It si only a matter of applying the principle of conservation of the four-momentum, taking care that outside of the interaction events the four-momenta on each world line are constant. Also note that every surface Σ meets each world line only once.

10.61.**8**. *Solution.* Asking $g(\mathbf{V}, \mathbf{F}) = 0$, since $V^0 > 0$, implies $F^0 = (V^0)^{-1} \sum_{a=1}^3 V^a F^a$. On the other hand from $V^0 = \sqrt{c^2 + \sum_a (V^a)^2}$ we obtain $m \frac{dV^0}{d\tau} = \frac{1}{\sqrt{c^2 + \sum_b (V^b)^2}} \sum_{a=1}^3 V^a m \frac{dV^a}{d\tau} = \frac{1}{V^0} \sum_{a=1}^3 V^a m \frac{dV^a}{d\tau} = \frac{1}{V^0} \sum_{a=1}^3 V^a F^a = F^0$.

10.61.**8**. *Hint.* Write the four-velocity's components in the two coordinate systems, using the spatial velocities in each frame.

C.9   Exercises for Chap. 11

Exercises 11.10

11.10.**1**. *Solution.* (a) Directly from the equation of $\mathbb{T}^2$, viewing P as the point particle and $\phi = \phi(t), \theta = \theta(t)$, we have:

$$\mathbf{x} = P - O = (R + r\cos\theta)\cos\phi\,\mathbf{e}_x + (R + r\cos\theta)\sin\phi\,\mathbf{e}_y + r\sin\phi\,\mathbf{e}_z$$

and then:

$$\mathbf{v}_P = (R + r\cos\theta)\dot\phi(-\sin\phi\,\mathbf{e}_x + \cos\phi\,\mathbf{e}_y) + r\dot\theta\cos\theta\,\mathbf{e}_z - r\dot\theta\sin\theta\,(\cos\phi\,\mathbf{e}_x + \sin\phi\,\mathbf{e}_y).$$

In other words:

$$\mathbf{v}_P = -\left[(R + r\cos\theta)\dot\phi\sin\phi + r\dot\theta\sin\theta\cos\phi\right]\mathbf{e}_x$$
$$+ \left[(R + r\cos\theta)\dot\phi\cos\phi - r\dot\theta\sin\theta\sin\phi\right]\mathbf{e}_y + r\dot\theta\cos\theta\,\mathbf{e}_z .$$

Squaring and multiplying by $m/2$ gives the kinetic energy in $\mathscr{R}$, that is:

$$\mathscr{T} = \frac{m(R + r\cos\theta)^2}{2}\dot\phi^2 + \frac{mr^2}{2}\dot\theta^2 .$$

The spring's potential energy is:

$$\frac{k}{2}(P - O)^2 = \frac{k}{2}r^2\sin^2\theta + \frac{k}{2}(R + r\cos\theta)^2 = \frac{k}{2}\left(R^2 + r^2 + 2rR\cos\theta\right) .$$

We may redefine the potential energy by dropping the additive constant without losing information:

$$\mathscr{U}(\theta, \phi) := krR\cos\theta .$$

In case $\gamma > 0$ the viscous force is describes by Lagrangian components:

$$\mathcal{Q}_\phi = \frac{\partial \mathbf{x}}{\partial \phi} \cdot (-\gamma \mathbf{v}_P) = -\gamma (R + r \cos \theta)^2 \dot{\phi} \,,$$

and

$$\mathcal{Q}_\theta = \frac{\partial \mathbf{x}}{\partial \theta} \cdot (-\gamma \mathbf{v}_P) = -\gamma r^2 \dot{\theta} \,.$$

We introduce the partial Lagrangian (it does not account for the viscosity!):

$$\mathcal{L} := \mathcal{T} - \mathcal{U} = \frac{m(R + r \cos \theta)^2}{2} \dot{\phi}^2 + \frac{mr^2}{2} \dot{\theta}^2 - kr R \cos \theta \qquad (C.85)$$

so the non-trivial Euler-Lagrange equations are:

$$\frac{d}{dt} \frac{\partial \mathcal{L}}{\partial \dot{\phi}} - \frac{\partial \mathcal{L}}{\partial \phi} = \mathcal{Q}_\phi \,,$$

$$\frac{d}{dt} \frac{\partial \mathcal{L}}{\partial \dot{\theta}} - \frac{\partial \mathcal{L}}{\partial \theta} = \mathcal{Q}_\theta \,,$$

or explicitly:

$$m \frac{d}{dt} \left((R + r \cos \theta)^2 \frac{d\phi}{dt} \right) = -\gamma (R + r \cos \theta)^2 \frac{d\phi}{dt} \,, \qquad (C.86)$$

$$mr^2 \frac{d^2 \theta}{dt^2} = -mr(R + r \cos \theta) \left(\frac{d\phi}{dt} \right)^2 \sin \theta - \gamma r^2 \frac{d\theta}{dt} + kr R \sin \theta \,. \qquad (C.87)$$

(b) In case $\gamma = 0$, the complete Lagrangian is (C.85). It admits ϕ as cyclic coordinate, so the associated conjugate momentum is a first integral. Hence we have the first integral:

$$p_\phi := \frac{\partial \mathcal{L}}{\partial \dot{\phi}} = m(R + r \cos \theta)^2 \frac{d\phi}{dt} \,.$$

We claim this is the z-component of the angular momentum with respect to O in $\mathcal{R}$. From

$$\mathbf{x} = P - O = (R + r \cos \theta) \cos \phi \, \mathbf{e}_x + (R + r \cos \theta) \sin \phi \, \mathbf{e}_y + r \sin \phi \, \mathbf{e}_z$$

and

$$\mathbf{v}_P = - \left[(R + r \cos\theta)\dot{\phi} \sin\phi + r\dot{\theta} \sin\theta \cos\phi \right] \mathbf{e}_x$$
$$+ \left[(R + r\cos\theta)\dot{\phi}\cos\phi - r\dot{\theta}\sin\theta\sin\phi \right] \mathbf{e}_y + r\dot{\theta}\cos\theta\mathbf{e}_z ,$$

we deduce:

$$\Gamma_O|_{\mathscr{R}} \cdot \mathbf{e}_z = m\mathbf{x} \wedge \mathbf{v}_P \cdot \mathbf{e}_z = m(x v_{Py} - y v_{Px}) = m(R + r\cos\theta)^2\dot{\phi} ,$$

i.e. on any motion:

$$p_\phi = \Gamma_O|_{\mathscr{R}} \cdot \mathbf{e}_z = m(R + r\cos\theta)^2\dot{\phi} . \tag{C.88}$$

Regarding the total mechanical energy, we are under the assumptions of Jacobi's theorem, since in the above coordinates (adapted to $\mathscr{R}$) the Lagrangian does not depend on time explicitly. Therefore it is conserved and coincides with the system's Hamiltonian

$$\mathscr{H}(\phi, \theta, \dot{\phi}, \dot{\theta}) = \mathscr{T} + \mathscr{U} = \frac{m(R + r\cos\theta)^2}{2}\dot{\phi}^2 + \frac{mr^2}{2}\dot{\theta}^2 + kr R \cos\theta . \tag{C.89}$$

(c) We seek the law that relates $\phi, \theta, p_\phi, p_\theta$ to $\phi, \dot{\phi}, \theta, \dot{\theta}$, invert it and make the Hamiltonian in (C.89) explicit in terms of $\phi, \theta, p_\phi, p_\theta$. We know p_ϕ is given by (C.88) while:

$$p_\theta = \frac{\partial\mathscr{L}}{\partial\dot{\theta}} = mr^2\dot{\theta} .$$

Therefore:

$$\dot{\theta} = \frac{1}{mr^2}p_\theta , \tag{C.90}$$

$$\dot{\phi} = \frac{1}{m(R + r\cos\theta)^2}p_\phi . \tag{C.91}$$

In the variables $\phi, \theta, p_\phi, p_\theta$ the Hamiltonian (C.89) reads:

$$\mathscr{H}(\phi, \theta, p_\phi, p_\theta) = \frac{p_\phi^2}{2m(R + r\cos\theta)^2} + \frac{p_\theta^2}{2mr^2} + kr R \cos\theta .$$

Hamilton's equations then are:

$$\frac{dp_\phi}{dt} = \left(-\frac{\partial \mathcal{H}}{\partial \phi} =\right) 0 , \tag{C.92}$$

$$\frac{d\phi}{dt} = \left(\frac{\partial \mathcal{H}}{\partial p_\phi} =\right) \frac{p_\phi}{m(R + r\cos\theta)^2} , \tag{C.93}$$

$$\frac{dp_\theta}{dt} = \left(-\frac{\partial \mathcal{H}}{\partial \theta} =\right) - \frac{p_\phi^2 r \sin\theta}{m(R + r\cos\theta)^3} + krR\sin\theta , \tag{C.94}$$

$$\frac{d\theta}{dt} = \left(\frac{\partial \mathcal{H}}{\partial p_\theta} =\right) \frac{p_\theta}{mr^2} . \tag{C.95}$$

(d) Consider the Euler-Lagrange equations (C.86) and (C.87). If we insert in the second one the initial conditions the right-hand side vanishes. Then the map $\theta(t) = 0$, for any $t \in \mathbb{R}$, satisfies the θ-initial conditions and the second equation, irrespective of the first one. As the system is normal with regular source, what we found is the unique solution. With such $\theta = \theta(t)$ the first equation is:

$$m\frac{d}{dt}\left((R+r)^2\frac{d\phi}{dt}\right) = -\gamma(R+r)^2\frac{d\phi}{dt} ,$$

i.e.:

$$\frac{d^2\phi}{dt^2} + \frac{\gamma}{m}\frac{d\phi}{dt} = 0 .$$

The characteristic equation

$$\chi^2 + \frac{\gamma}{m}\chi = 0$$

has solutions $\chi = 0, -\gamma/m$. The general solution is then:

$$\phi(t) = A + Be^{-\gamma t/m} .$$

The initial conditions say that: $\phi(0) = A + B = 0$ so $A = -B$, and $v = d/dt|_{t=0}\phi(t) = -\gamma Be^{-\gamma t/m}/m|_{t=0}$ implying $B = -mv/\gamma$. The (maximal) solution to the problem, i.e. the motion given by the aforementioned initial conditions, is:

$$\theta(t) = 0 , \quad \phi(t) = \frac{mv}{\gamma}(1 - e^{-\gamma t/m}) , \quad \text{for any } t \in \mathbb{R}.$$

11.10.2. *Solution.* (a) We will use the Lagrangian $\mathcal{L}|_{\hat{\mathcal{R}}}$ relative to the inertial frame $\hat{\mathcal{R}}$. First we should express the velocity of the points in $\hat{\mathcal{R}}$ using the free coordinates x, θ, s. In polar coordinates (r, θ) on the plane orthogonal to $\mathbf{e}_x$ through $O + x\mathbf{e}_x$:

$$\mathbf{e}_r = \cos\theta\mathbf{e}_y + \sin\theta\mathbf{e}_z \quad \mathbf{e}_\theta = -\sin\theta\mathbf{e}_y + \cos\theta\mathbf{e}_z \,.$$

Recall $P - O = x\mathbf{e}_x + \cosh x\mathbf{e}_z$ while $Q - O = x\mathbf{e}_x + r\mathbf{e}_r$. Then:

$$\mathbf{v}_P|_{\mathscr{R}} = \dot{x}\mathbf{e}_x + \dot{x}\sinh x\mathbf{e}_z \,, \quad \mathbf{v}_Q|_{\mathscr{R}} = \dot{x}\mathbf{e}_x + \dot{r}\mathbf{e}_r + r\dot{\theta}\mathbf{e}_\theta \tag{C.96}$$

and therefore:

$$\mathbf{v}_P|_{\hat{\mathscr{R}}} = \mathbf{v}_P|_{\mathscr{R}} + \Omega\mathbf{e}_z \wedge (x\mathbf{e}_x + \cosh x\mathbf{e}_z) = \dot{x}\mathbf{e}_x + \Omega x\mathbf{e}_y + \dot{x}\sinh x\mathbf{e}_z \,,$$

while:

$$\begin{aligned}
\mathbf{v}_Q|_{\hat{\mathscr{R}}} &= \mathbf{v}_Q|_{\mathscr{R}} + \Omega\mathbf{e}_z \wedge (x\mathbf{e}_x + r\mathbf{e}_r) \\
&= \dot{x}\mathbf{e}_x + \dot{r}\mathbf{e}_r + r\dot{\theta}\mathbf{e}_\theta + \Omega x\mathbf{e}_y + \Omega\mathbf{e}_z \wedge (r\cos\theta\mathbf{e}_y + r\sin\theta\mathbf{e}_z) \\
&= \dot{x}\mathbf{e}_x + \dot{r}\mathbf{e}_r + r\dot{\theta}\mathbf{e}_\theta + \Omega x\mathbf{e}_y - \Omega r\cos\theta\mathbf{e}_x \\
&= \dot{x}\mathbf{e}_x + \dot{r}\mathbf{e}_r + r\dot{\theta}\mathbf{e}_\theta + \Omega x(\cos\theta\mathbf{e}_r - \sin\theta\mathbf{e}_\theta) - \Omega r\cos\theta\mathbf{e}_x \,.
\end{aligned}$$

We have found:

$$\mathbf{v}_P|_{\hat{\mathscr{R}}} = \dot{x}\mathbf{e}_x + \Omega x\mathbf{e}_y + \dot{x}\sinh x\mathbf{e}_z \,, \tag{C.97}$$

$$\mathbf{v}_Q|_{\hat{\mathscr{R}}} = (\dot{x} - \Omega r\cos\theta)\mathbf{e}_x + (\dot{r} + \Omega x\cos\theta)\mathbf{e}_r + (r\dot{\theta} - \Omega x\sin\theta)\mathbf{e}_\theta \,. \tag{C.98}$$

Both velocities are written in (different) orthonormal triples of unit vectors. The kinetic energy in $\mathscr{T}_{\hat{\mathscr{R}}}$ then is, recalling $\cosh^2 x - \sinh^2 x = 1$,

$$\begin{aligned}
\mathscr{T}|_{\hat{\mathscr{R}}} &= \frac{m}{2}\left(\mathbf{v}_P|_{\hat{\mathscr{R}}}^2 + \mathbf{v}_Q|_{\hat{\mathscr{R}}}^2\right) \\
&= \frac{m}{2}\left(\dot{x}^2 + \Omega^2 x^2 + \dot{x}^2\sinh^2 x + (\dot{x} - \Omega r\cos\theta)^2 \right. \\
&\qquad \left. + (\dot{r} + \Omega x\cos\theta)^2 + (r\dot{\theta} - \Omega x\sin\theta)^2\right) \,.
\end{aligned}$$

Hence:

$$\begin{aligned}
\mathscr{T}|_{\hat{\mathscr{R}}} &= \frac{m}{2}\left(\dot{x}^2(1 + \cosh^2 x) + \dot{r}^2 + r^2\dot{\theta}^2\right) + \frac{m}{2}\left(2\Omega^2 x^2 + \Omega^2 r^2\cos^2\theta\right) \\
&\quad - m\Omega\dot{\theta}rx\sin\theta + m(\dot{r}x - \dot{x}r)\Omega\cos\theta \,.
\end{aligned}$$

The potential energy is the sum of the gravitational potential energy plus the spring's. Using $P - Q = P - O - (Q - O) = x\mathbf{e}_x + \cosh x \mathbf{e}_z - (x\mathbf{e}_x + r\mathbf{e}_r) = \cosh x \mathbf{e}_z - r\mathbf{e}_r$ and $\mathbf{e}_r \cdot \mathbf{e}_z = \sin\theta$, we obtain:

$$\mathscr{U} = mg\left(z_P + z_Q\right) + \frac{k}{2}(P - Q)^2$$

$$= mg\left(\cosh x + r\sin\theta\right) + \frac{k}{2}\left(r^2 + \cosh^2 x - 2r\cosh x \sin\theta\right).$$

The Lagrangian in $\hat{\mathscr{R}}$ then has form:

$$\mathscr{L}\big|_{\hat{\mathscr{R}}} = \frac{m}{2}\left(\dot{x}^2(1 + \cosh^2 x) + \dot{r}^2 + r^2\dot{\theta}^2\right) + \frac{m}{2}\left(2\Omega^2 x^2 + \Omega^2 r^2 \cos^2\theta\right)$$

$$- m\Omega\dot{\theta}rx\sin\theta + m(\dot{r}x - \dot{x}r)\Omega\cos\theta - mg\left(\cosh x + r\sin\theta\right)$$

$$- \frac{k}{2}\left(r^2 + \cosh^2 x - 2r\cosh x \sin\theta\right).$$

The three Euler-Lagrange equations for the coordinates x, r, θ are:

$$m\frac{d}{dt}\left((1 + \cosh^2 x)\frac{dx}{dt} - r\Omega\cos\theta\right)$$

$$= m\left(\frac{dx}{dt}\right)^2 \sinh x \cosh x + 2m\Omega^2 x - m\Omega\frac{d\theta}{dt}r\sin\theta + m\frac{dr}{dt}\Omega\cos\theta$$

$$- mg\sinh x - k\cosh x \sinh x + kr\sinh x \sin\theta,$$

$$m\frac{d}{dt}\left(\frac{dr}{dt} + x\Omega\cos\theta\right)$$

$$= mr\left(\frac{d\theta}{dt}\right)^2 + m\Omega^2 r\cos^2\theta - m\Omega\frac{d\theta}{dt}x\sin\theta - m\frac{dx}{dt}\Omega\cos\theta - mg\sin\theta$$

$$- kr - k\cosh x \sin\theta,$$

$$m\frac{d}{dt}\left(r^2\frac{d\theta}{dt} - \Omega rx\sin\theta\right) = -m\Omega^2 r^2 \cos\theta \sin\theta - m\frac{d\theta}{dt}\Omega rx\cos\theta$$

$$- m\left(\frac{dr}{dt}x - \frac{dx}{dt}r\right)\Omega\sin\theta$$

$$- mgr\cos\theta + kr\cosh x \cos\theta.$$

(b) The explicit Hamiltonian associated with the Lagrangian $\mathscr{L}|_{\hat{\mathscr{R}}}$, in the coordinates x, r, θ, is:

$$\mathscr{H} = \dot{x}\frac{\partial \mathscr{L}|_{\hat{\mathscr{R}}}}{\partial \dot{x}} + \dot{r}\frac{\partial \mathscr{L}|_{\hat{\mathscr{R}}}}{\partial \dot{r}} + \dot{\theta}\frac{\partial \mathscr{L}|_{\hat{\mathscr{R}}}}{\partial \dot{\theta}} - \mathscr{L},$$

that is:

$$\mathscr{H} = \frac{m}{2}\left(\dot{x}^2(1 + \cosh^2 x) + \dot{r}^2 + r^2\dot{\theta}^2\right) - \frac{m\Omega^2}{2}\left(x^2 + (x^2 + r^2\cos^2\theta)\right)$$

$$+ mg\,(\cosh x + r\sin\theta)$$

$$+ \frac{k}{2}\left(r^2 + \cosh^2 x - 2r\cosh x \sin\theta\right). \tag{C.99}$$

As $\mathscr{L}|_{\hat{\mathscr{R}}}$ does not depend explicitly on time, the Jacobi theorem guarantees that $\mathscr{H}$ is conserved along the Euler-Lagrange solutions. The term

$$\frac{m}{2}\left(\dot{x}^2(1 + \cosh^2 x) + \dot{r}^2 + r^2\dot{\theta}^2\right)$$

is the kinetic energy in $\mathscr{R}$, as is easy to recover from (C.96) directly. The third and fourth summands on the right in (C.99) represent the potential energies of gravity and the spring, as seen before. The second term

$$-\frac{m\Omega^2}{2}\left(x^2 + (x^2 + r^2\cos^2\theta)\right)$$

reads $-\frac{m\Omega^2}{2}d_P^2 - \frac{m\Omega^2}{2}d_Q^2$. $d_P = |x_P|$ is the distance of P to the z-axis and d_Q the distance of Q to the z-axis. d_Q is the length of the projection on $z = 0$ of $Q - O = x\mathbf{e}_x + r\mathbf{e}_r = x\mathbf{e}_x + r\cos\theta\mathbf{e}_y + r\sin\theta\mathbf{e}_z$; hence the square of d_Q equals $x^2 + r^2\cos^2\theta$. We know that

$$-\frac{m\Omega^2}{2}d_P^2 - \frac{m\Omega^2}{2}d_Q^2$$

is the potential energy of the centrifugal force. The Coriolis force does no work in $\mathscr{R}$ as perpendicular to the velocity. The constraint reactions do not add work since they are ideal and independent of time in $\mathscr{R}$. All in all, the right-hand side of $\mathscr{H}$ is precisely the total mechanical energy in the frame $\mathscr{R}$.

(c) Explicitly:

$$p_x = \frac{\partial \mathcal{L}|_{\hat{\mathcal{R}}}}{\partial \dot{x}} = m\dot{x}(1 + \cosh^2 x) - mr\Omega \cos\theta , \tag{C.100}$$

$$p_r = \frac{\partial \mathcal{L}|_{\hat{\mathcal{R}}}}{\partial \dot{r}} = m\dot{r} + mx\Omega \cos\theta , \tag{C.101}$$

$$p_\theta = \frac{\partial \mathcal{L}|_{\hat{\mathcal{R}}}}{\partial \dot{\theta}} = mr^2\dot{\theta} - m\Omega xr \sin\theta . \tag{C.102}$$

The inverse formulas are:

$$\dot{x} = \frac{p_x + mr\Omega \cos\theta}{m(1 + \cosh^2 x)} , \tag{C.103}$$

$$\dot{r} = \frac{p_r - mx\Omega \cos\theta}{m} , \tag{C.104}$$

$$\dot{\theta} = \frac{p_\theta + m\Omega xr \sin\theta}{mr^2} . \tag{C.105}$$

Substituting in (C.99) we find:

$$\mathcal{H} = \frac{(p_x + mr\Omega \cos\theta)^2}{2m(1 + \cosh^2 x)} + \frac{(p_r - mx\Omega \cos\theta)^2}{2m} + \frac{(p_\theta + m\Omega xr \sin\theta)^2}{2mr^2}$$

$$- \frac{m\Omega^2}{2}\left(2x^2 + r^2 \cos^2\theta\right)$$

$$+ mg(\cosh x + r\sin\theta) + \frac{k}{2}\left(r^2 + \cosh^2 x + 2r\cosh x \sin\theta\right) . \tag{C.106}$$

Hamilton's equations $\frac{dq^k}{dt} = \frac{\partial \mathcal{H}}{\partial p_k}$ immediately give:

$$\frac{dx}{dt} = \frac{p_x + mr\Omega \cos\theta}{m(1 + \cosh^2 x)} ,$$

$$\frac{dr}{dt} = \frac{p_r - mx\Omega \cos\theta}{m} ,$$

$$\frac{d\theta}{dt} = \frac{p_\theta + m\Omega xr \sin\theta}{mr^2} .$$

The remaining Hamiltonian equations $\frac{dp_k}{dt} = -\frac{\partial \mathcal{H}}{\partial q^k}$ are:

$$\frac{dp_x}{dt} = \frac{(p_x + mr\Omega \cos \theta)^2}{m(1 + \cosh^2 x)^2} \sinh x \cosh x + \Omega \cos \theta (p_r - mx\Omega \cos \theta)$$
$$- \Omega r \sin \theta \frac{p_\theta + m\Omega xr \sin \theta}{r^2}$$
$$+ 2m\Omega^2 x - mg \sinh x - k \cosh x \sinh x + kr \sinh x \sin \theta ,$$

$$\frac{dp_r}{dt} = \frac{(p_\theta + m\Omega xr \sin \theta)^2}{mr^3} - \Omega \cos \theta \frac{p_x + mr\Omega \cos \theta}{(1 + \cosh^2 x)} - \Omega x \sin \theta \frac{p_\theta + m\Omega xr \sin \theta}{r^2}$$
$$+ m\Omega^2 r \cos^2 \theta - mg \sin \theta - kr - k \cosh x \sin \theta ,$$

$$\frac{dp_\theta}{dt} = - \left(x(p_r - mx\Omega \cos \theta) - r \frac{p_x + mr\Omega \cos \theta}{1 + \cosh^2 x} \right) \Omega \sin \theta$$
$$- \Omega x \cos \theta \frac{p_\theta + m\Omega xr \sin \theta}{r}$$
$$- m\Omega^2 r^2 \cos \theta \sin \theta - mgr \cos \theta + kr \cosh x \cos \theta .$$

Exercises 11.40

11.40.**1**. *Solution.* The proof is identical to that for exponentials of real numbers (defined via power series) and can be found in any Analysis textbook.

11.40.**2**. *Solution.* Consider the map:

$$f(t) := \det e^{tF} , \quad t \in \mathbb{R}$$

and compute

$$f'(t) = \lim_{h \to 0} \frac{1}{h} \left(\det e^{(t+h)F} - \det e^{tF} \right) = \lim_{h \to 0} \frac{1}{h} \left(\det \left(e^{tF} e^{hF} \right) - \det e^{tF} \right)$$
$$= \det e^{tF} \lim_{h \to 0} \frac{1}{h} \left(\det e^{hF} - 1 \right) .$$

Expanding e^{hF} in series up to order one and using Lemma 11.21 we obtain:

$$\lim_{h \to 0} \frac{1}{h} \left(\det e^{hF} - 1 \right) = \operatorname{tr} F .$$

Substituting above we find the normal ODE with C^∞ source term:

$$f'(t) = f(t) \operatorname{tr} F .$$

The function $g(t) := e^{t \, \mathrm{tr} \, F}$ satisfies the same ODE and the same initial condition $f(0) = g(0) = 1$. The Existence and uniqueness theorem then forces $f(t) = g(t)$, as claimed.

11.40.**3** and **4**. *Hint*. Argue in analogy to what was done for the group $Sp(n, \mathbb{R})$. Keep in mind the previous exercise, which implies that $\det e^{tA} = e^{t \, \mathrm{tr} \, A} = e^0 = 1$ if $A = -A^t$.

C.10   Exercises for Chap. 12

Exercises 12.64

12.64.**1**. *Solution*. It suffices to specialise Proposition 12.63 to this case, viewing $M = \Omega$ and $\mathscr{L} : TM \to \mathbb{R}$. If the matrix $a_{ij}(q)$ is everywhere non-singular then the Euler-Lagrange equations can be seen as the normal equation on TM of Proposition 12.63, with right-hand-side term of class C^2.

12.64.**1**. *Solution*. The problem's solutions have the form

$$q(\tau) = A \cosh t + B \sinh t .$$

Imposing $q(T) = Q$ and $q(t) = q$ gives the general form

$$q(\tau) = \frac{q \sinh(\tau - T) + Q \sinh(t - \tau)}{\sinh(t - T)} , \qquad \tau \in \mathbb{R} .$$

Hence:

$$\dot{q}(\tau) = \frac{q \cosh(\tau - T) - Q \cosh(t - \tau)}{\sinh(t - T)} , \qquad \tau \in \mathbb{R} .$$

From the second one we obtain:

$$\frac{\partial q}{\partial \dot{Q}} = \sinh(t - T) ,$$

where $\dot{Q}$ is the velocity at the initial time T of the solution starting at Q and reaching q at time $t > T$. This agrees with the general result (12.98), which shows how the local construction of Hamilton's principal function works. Finally, from the above:

$$\mathscr{L}(q(\tau), \dot{q}(\tau)) = \frac{q^2 \cosh(2(\tau - T)) + Q^2 \cosh(2(t - \tau)) - 2q \, Q \cosh(\tau - (t + T))}{2 \sinh^2(t - T)} .$$

Now we may integrate from T to t and find Hamilton's principal function:

$$S_{T,Q}(t,q) = \frac{(q^2 + Q^2)\cosh(t - T) - 2qQ}{2\sinh(t - T)}.$$

In particular, (12.92) holds around any (t, q) if we choose $T = T_0$ suitably, because:

$$\frac{\partial^2 S_{T,Q}}{\partial q \partial Q} = -\frac{1}{\sinh(t - T)}$$

becomes singular only at $t = T$. Hence, around any $(t, q) \in \mathbb{R}^2$ there always is a complete solution to the Hamilton-Jacobi equation provided with choose the instant $T = T_0$ appropriately.

C.11   Exercises for Complement 14

Exercises 14.31

14.31.**1**. *Hint.* It is enough to prove the claim for $k = \pm 1$ and then iterate k times. Apply the global Existence and uniqueness theorem, together with ($k = 1$)

$$\left.\frac{d\mathbf{x}'}{dt}\right|_{t=s-T} = \left.\frac{d\mathbf{x}}{dt}\right|_{t=s}$$

and that $\mathbf{f}$ is periodic in t of period T.

14.31.**2**. *Sketch of Solution.* Use the previous exercise for $k = 1$, observing that $\mathbf{f}$ is constant in t so periodic of any period T in t.

14.31.**3**. *Solution.* Let us ignore from the start the motion along z, due to the constraint. The plane's constraint reaction is $mg\mathbf{e}_z$. The problem reduces to two dimensions:

$$\frac{d^2x}{dt^2} = m^{-1}\kappa x,$$

$$\frac{d^2y}{dt^2} = m^{-1}\kappa y.$$

This is of order two and normal. Reducing to first order:

$$\frac{d^2v_x}{dt^2} = m^{-1}\kappa x,$$

$$\frac{d^2v_y}{dt^2} = m^{-1}\kappa y,$$

$$\frac{d^2x}{dt^2} = v_x ,$$

$$\frac{d^2y}{dt^2} = v_y .$$

Clearly the right-hand side is C^1, so the Existence and uniqueness theorems apply. Back to order two, the equations decouple, so we can integrate them separately. Let us first view $x = x(t)$ and $y = y(t)$ as complex-valued, and later we will make them real. The characteristic polynomial of both is $\chi^2 - \frac{\kappa}{m} = 0$. The solutions are:

$$b_\pm = \pm\sqrt{\frac{\kappa}{m}} , \quad \text{if } \kappa > 0 ,$$

or

$$b_\pm = \pm i\sqrt{\frac{-\kappa}{m}} , \quad \text{if } \kappa < 0.$$

In case $\kappa > 0$, the law of motion is:

$$x(t) = c_1 e^{\sqrt{\frac{\kappa}{m}}t} + c_2 e^{-\sqrt{\frac{\kappa}{m}}t} , \tag{C.107}$$

$$y(t) = d_1 e^{\sqrt{\frac{\kappa}{m}}t} + d_2 e^{-\sqrt{\frac{\kappa}{m}}t} , \tag{C.108}$$

where c_1, c_2, d_1, d_2 are arbitrary constants in $\mathbb{R}$. They can be fixed by the initial conditions.

In case $\kappa < 0$ the discussion is slightly more complicated. In $\mathbb{C}$, the solutions would be:

$$x(t) = c_1 e^{i\sqrt{\frac{\kappa}{m}}t} + c_2 e^{-i\sqrt{\frac{\kappa}{m}}t} ,$$

$$y(t) = d_1 e^{i\sqrt{\frac{\kappa}{m}}t} + d_2 e^{-i\sqrt{\frac{\kappa}{m}}t} .$$

In general, these are not acceptable because x and y are complex. Let us recall the Euler formula: $e^{i\alpha} = \cos\alpha + i\sin\alpha$. Then the above solutions become:

$$x(t) = (c_1 + c_2)\cos\left(\sqrt{\frac{\kappa}{m}}t\right) + i(c_1 - c_2)\sin\left(\sqrt{\frac{\kappa}{m}}t\right) , \tag{C.109}$$

$$y(t) = (d_1 + d_2)\cos\left(\sqrt{\frac{\kappa}{m}}t\right) + i(d_1 - d_2)\sin\left(\sqrt{\frac{\kappa}{m}}t\right) . \tag{C.110}$$

All real solutions arise by varying the constants $(c_1 + c_2)$, $(d_1 + d_2)$ in $\mathbb{R}$ and $(c_1 - c_2)$, $(d_1 - d_2)$ in $i\mathbb{R}$ (purely imaginary numbers). In summary, the general solution is:

$$x(t) = A_1 \cos\left(\sqrt{\frac{\kappa}{m}}t\right) + A_2 \sin\left(\sqrt{\frac{\kappa}{m}}t\right), \tag{C.111}$$

$$y(t) = B_1 \cos\left(\sqrt{\frac{\kappa}{m}}t\right) + B_2 \sin\left(\sqrt{\frac{\kappa}{m}}t\right). \tag{C.112}$$

with $A_1, A_2, B_1, B_2 \in \mathbb{R}$.

To finish we find the law of motion, for $\kappa < 0$, with initial conditions $\mathbf{x}(0) = \mathbf{0}$ and $\mathbf{v}(0) = v(\mathbf{e}_x - \mathbf{e}_y)$. The first condition, inserted in (C.111) and (C.112), gives: $0 = A_1$ and $0 = B_1$. Differentiating (C.111) and (C.112), with what we found above, produces:

$$v_x(0) = \sqrt{\frac{\kappa}{m}}A_2, \tag{C.113}$$

$$v_y(0) = \sqrt{\frac{\kappa}{m}}B_2. \tag{C.114}$$

But $\mathbf{v}(0) = v(\mathbf{e}_x - \mathbf{e}_y)$, so $v_x(0) = v$ and $v_y(0) = -v$, and then $A_2 = v\sqrt{\frac{m}{\kappa}}$ and $B_2 = -v\sqrt{\frac{m}{\kappa}}$. The required law of motion then is:

$$\mathbf{x}(t) = v\sqrt{\frac{m}{\kappa}}\sin\left(\sqrt{\frac{\kappa}{m}}t\right)(\mathbf{e}_x - \mathbf{e}_y), \quad \text{for } t \in \mathbb{R}.$$

14.31.**4**. *Solution.* Arguing as in the previous exercise, we find:

$$\frac{d^2x}{dt^2} = m^{-1}\kappa x - m^{-1}\beta\frac{dy}{dt},$$

$$\frac{d^2y}{dt^2} = m^{-1}\kappa y + m^{-1}\beta\frac{dx}{dt}.$$

The Existence and uniqueness theorem clearly holds. In contrast to the previous exercise, now the equations are coupled, so it is best to proceed as follows. Define the complex variable $z = x + iy$ and write the equations as:

$$\frac{d^2x}{dt^2} = m^{-1}\kappa x + m^{-1}\beta i\frac{diy}{dt},$$

$$\frac{d^2iy}{dt^2} = m^{-1}\kappa iy + m^{-1}\beta i\frac{dx}{dt}.$$

Adding term by term gives a unique complex equation subsuming the two real equations:

$$\frac{d^2 z}{dt^2} - i\frac{\beta}{m}\frac{dz}{dt} - \frac{\kappa}{m}z = 0 \, . \tag{C.115}$$

Remember that eventually, the solutions in x and y will be the real and imaginary parts of the solution z.
The characteristic equation:

$$\chi^2 - i\frac{\beta}{m}\chi - \frac{\kappa}{m} = 0$$

has solutions:

$$b_\pm = i\frac{\beta}{2m} \pm \frac{1}{2}\sqrt{\frac{4\kappa}{m} - \frac{\beta^2}{m^2}} \, .$$

Let us examine first the case of double root, i.e. $\frac{4\kappa}{m} = \frac{\beta^2}{m^2}$. The general solution, for $a, b \in \mathbb{C}$ arbitrary constants, is:

$$z(t) = (a + bt)e^{i\frac{\beta}{2m}t} \, .$$

Using the Euler formula:

$$z(t) = (a + bt)\left(\cos\left(\frac{\beta}{2m}t\right) + i\sin\left(\frac{\beta}{2m}t\right)\right) \, .$$

Writing $a = A + iB$ and $b = C + iD$ for $A, B, C, D \in \mathbb{R}$, and taking the real and imaginary parts we find the general real solution:

$$x(t) = (A + Ct)\cos\left(\frac{\beta}{2m}t\right) - (B + Dt)\sin\left(\frac{\beta}{2m}t\right)$$

and

$$y(t) = (B + Dt)\cos\left(\frac{\beta}{2m}t\right) + (A + Ct)\sin\left(\frac{\beta}{2m}t\right) \, .$$

When $\frac{4\kappa}{m} \neq \frac{\beta^2}{m^2}$ the solution in $\mathbb{C}$ to (C.115) is:

$$z(t) = e^{i\frac{\beta}{2m}t}\left(C_+ e^{\frac{1}{2}\sqrt{\frac{4\kappa}{m} - \frac{\beta^2}{m^2}}t} + C_- e^{-\frac{1}{2}\sqrt{\frac{4\kappa}{m} - \frac{\beta^2}{m^2}}t}\right) \, ,$$

with $C_\pm \in \mathbb{C}$ arbitrary. We may rewrite it as

$$z(t) = \cos\left(\frac{\beta}{2m}t\right)\left(C_+ e^{\frac{1}{2}\sqrt{\frac{4\kappa}{m} - \frac{\beta^2}{m^2}}t} + C_- e^{-\frac{1}{2}\sqrt{\frac{4\kappa}{m} - \frac{\beta^2}{m^2}}t}\right)$$
$$+ i\sin\left(\frac{\beta}{2m}t\right)\left(C_+ e^{\frac{1}{2}\sqrt{\frac{4\kappa}{m} - \frac{\beta^2}{m^2}}t} + C_- e^{-\frac{1}{2}\sqrt{\frac{4\kappa}{m} - \frac{\beta^2}{m^2}}t}\right).$$

Now we have two cases to consider.
(i) *Case* $\frac{4\kappa}{m} > \frac{\beta^2}{m^2}$. The exponents are real. Writing $C_+ = a + ib$ and $C_- = c + id$, so to have only real constants, we immediately find:

$$x(t) = \cos\left(\frac{\beta}{2m}t\right)\left(ae^{\frac{1}{2}\sqrt{\frac{4\kappa}{m} - \frac{\beta^2}{m^2}}t} + ce^{-\frac{1}{2}\sqrt{\frac{4\kappa}{m} - \frac{\beta^2}{m^2}}t}\right)$$
$$- \sin\left(\frac{\beta}{2m}t\right)\left(be^{\frac{1}{2}\sqrt{\frac{4\kappa}{m} - \frac{\beta^2}{m^2}}t} + de^{-\frac{1}{2}\sqrt{\frac{4\kappa}{m} - \frac{\beta^2}{m^2}}t}\right),$$

$$y(t) = \cos\left(\frac{\beta}{2m}t\right)\left(be^{\frac{1}{2}\sqrt{\frac{4\kappa}{m} - \frac{\beta^2}{m^2}}t} + de^{-\frac{1}{2}\sqrt{\frac{4\kappa}{m} - \frac{\beta^2}{m^2}}t}\right)$$
$$+ \sin\left(\frac{\beta}{2m}t\right)\left(ae^{\frac{1}{2}\sqrt{\frac{4\kappa}{m} - \frac{\beta^2}{m^2}}t} + ce^{-\frac{1}{2}\sqrt{\frac{4\kappa}{m} - \frac{\beta^2}{m^2}}t}\right),$$

where a, b, c, d are arbitrary in $\mathbb{R}$.
(i) *Case* $\frac{4\kappa}{m} < \frac{\beta^2}{m^2}$. Now the exponents are purely imaginary:

$$z(t) = \cos\left(\frac{\beta}{2m}t\right)\left(C_+ e^{i\frac{1}{2}\sqrt{\frac{\beta^2}{m^2} - \frac{4\kappa}{m}}t} + C_- e^{-i\frac{1}{2}\sqrt{\frac{\beta^2}{m^2} - \frac{4\kappa}{m}}t}\right)$$
$$+ i\sin\left(\frac{\beta}{2m}t\right)\left(C_+ e^{i\frac{1}{2}\sqrt{\frac{\beta^2}{m^2} - \frac{4\kappa}{m}}t} + C_- e^{-i\frac{1}{2}\sqrt{\frac{\beta^2}{m^2} - \frac{4\kappa}{m}}t}\right).$$

Use the Euler formula and $C_+ = a + ib$ and $C_- = c + id$, so to have real constants only. Taking the real part of z, the explicit x-component of the solution is:

$$x(t) = (a + c)\cos\left(\frac{\beta}{2m}t\right)\cos\left(\frac{1}{2}\sqrt{\frac{\beta^2}{m^2} - \frac{4\kappa}{m}}t\right)$$
$$+ (c - a)\sin\left(\frac{\beta}{2m}t\right)\sin\left(\frac{1}{2}\sqrt{\frac{\beta^2}{m^2} - \frac{4\kappa}{m}}t\right)$$

$$+ (-b + d) \cos\left(\frac{\beta}{2m}t\right) \sin\left(\frac{1}{2}\sqrt{\frac{\beta^2}{m^2} - \frac{4\kappa}{m}}\,t\right)$$

$$- (b + d) \sin\left(\frac{\beta}{2m}t\right) \cos\left(\frac{1}{2}\sqrt{\frac{\beta^2}{m^2} - \frac{4\kappa}{m}}\,t\right).$$

Taking the imaginary part of z:

$$y(t) = (d + b) \cos\left(\frac{\beta}{2m}t\right) \cos\left(\frac{1}{2}\sqrt{\frac{\beta^2}{m^2} - \frac{4\kappa}{m}}\,t\right)$$

$$+ (d - b) \sin\left(\frac{\beta}{2m}t\right) \sin\left(\frac{1}{2}\sqrt{\frac{\beta^2}{m^2} - \frac{4\kappa}{m}}\,t\right)$$

$$+ (a - c) \cos\left(\frac{\beta}{2m}t\right) \sin\left(\frac{1}{2}\sqrt{\frac{\beta^2}{m^2} - \frac{4\kappa}{m}}\,t\right)$$

$$+ (a + c) \sin\left(\frac{\beta}{2m}t\right) \cos\left(\frac{1}{2}\sqrt{\frac{\beta^2}{m^2} - \frac{4\kappa}{m}}\,t\right).$$

Changing names for the real constants we find, in terms of real constants A, B, C, D:

$$x(t) = A \cos\left(\frac{\beta}{2m}t\right) \cos\left(\frac{1}{2}\sqrt{\frac{\beta^2}{m^2} - \frac{4\kappa}{m}}\,t\right) + B \sin\left(\frac{\beta}{2m}t\right) \sin\left(\frac{1}{2}\sqrt{\frac{\beta^2}{m^2} - \frac{4\kappa}{m}}\,t\right)$$

$$+ C \cos\left(\frac{\beta}{2m}t\right) \sin\left(\frac{1}{2}\sqrt{\frac{\beta^2}{m^2} - \frac{4\kappa}{m}}\,t\right) + D \sin\left(\frac{\beta}{2m}t\right) \cos\left(\frac{1}{2}\sqrt{\frac{\beta^2}{m^2} - \frac{4\kappa}{m}}\,t\right),$$

and

$$y(t) = -D \cos\left(\frac{\beta}{2m}t\right) \cos\left(\frac{1}{2}\sqrt{\frac{\beta^2}{m^2} - \frac{4\kappa}{m}}\,t\right)$$

$$+ C \sin\left(\frac{\beta}{2m}t\right) \sin\left(\frac{1}{2}\sqrt{\frac{\beta^2}{m^2} - \frac{4\kappa}{m}}\,t\right)$$

$$- B \cos\left(\frac{\beta}{2m}t\right) \sin\left(\frac{1}{2}\sqrt{\frac{\beta^2}{m^2} - \frac{4\kappa}{m}}\,t\right) + A \sin\left(\frac{\beta}{2m}t\right) \cos\left(\sqrt{\frac{1}{2}\frac{\beta^2}{m^2} - \frac{4\kappa}{m}}\,t\right).$$

14.31.5. *Hint.* Proceed as in Exercise 3, since the differential equations of the components x and y are decoupled.

C.12 Exercises for Appendix A

Exercises A.21

A.21.1. *Hint.* To build the atlas $\{(U_i, \phi_i)\}_{i \in I}$ on $\mathbb{S}^2$, consider spherical coordinates with varying polar axis z, and using the coordinate $r' := r - 1$ instead of the radial coordinate r.

A.21.2. *Hint.* The answer to both questions is no: N is not locally homeomorphic to $\mathbb{R}^1$ (what happens near the point $(0, 0)$?).

A.21.3. *Hint.* The vertex of the cone $(0, 0, 0)$ does not have, around it, coordinates of the type in (ii), Definition A.19. Nonetheless, one can define a differentiable structure on C using the global chart that projects the points of C onto the plane $x^3 = 0$, with coordinates (x^1, x^2). Excluding the vertex, C^* is an embedded 2-dimensional submanifold in $\mathbb{R}^3$, with global chart obtained as above, projecting onto the cone's base.

A.21.4. *Hint.* Recall that the atlases' local charts are restrictions of compatible local charts on M.

A.21.5. *Solution.* By Theorem 1.25, for any p in the open set $\phi(U \cap V)$ there exists an open neighbourhood $O \subset \phi(U \cap V)$ of p mapped to an open set $O' \subset \psi(U \cap V)$ under $\psi \circ \phi^{-1}$. Moreover, on O', the inverse g to $(\psi \circ \phi^{-1}) \upharpoonright_U$ is well defined and C^k. As the only inverse of $\psi \circ \phi^{-1} : \phi(U \cap V) \to \psi(U \cap V)$ by assumption is $\phi \circ \psi^{-1} : \psi(U \cap V) \to \phi(U \cap V)$, g must coincides with the restriction of the latter to O'. As p varies, the neighbourhoods O' cover $\psi(U \cap V)$, and the restriction of $\phi \circ \psi^{-1} : \psi(U \cap V) \to \phi(U \cap V)$ to a neighbourhood (O') of any point in the domain $\psi(U \cap V)$ is C^k. But then $\phi \circ \psi^{-1} : \psi(U \cap V) \to \phi(U \cap V)$ is C^k.

Exercises A.29

A.29.1. Consider local coordinates $x^1, \ldots, x^n$ on (U, ϕ) and $y^1, \ldots, y^n$ on (V, ψ) with $U \cap V \neq \emptyset$. Since in coordinates $\psi \circ \phi^{-1}$ is $y^i = y^i(x^1, \cdots, x^n)$ and $\phi \circ \psi^{-1}$ is $x^i = x^i(y^1, \cdots, y^n)$, formula

$$id_{\phi(U \cap U)} = (\phi \circ \psi^{-1}) \circ (\psi \circ \phi^{-1})$$

reads:

$$x^i = x^i\left(y^1(x^1, \ldots, x^2), \ldots y^n(x^1, \ldots, x^2)\right), \quad i = 1, \ldots, n. \tag{C.116}$$

Assuming both $\psi \circ \phi^{-1}$ and $\phi \circ \psi^{-1}$ are differentiable (C^1 at least) as prescribed by the charts' compatibility, using the chain rule in (C.116) when differentiating in

x^k produces:

$$\delta^i_k = \sum_{j=1}^{n} \frac{\partial x^i}{\partial y^j} \frac{\partial y^j}{\partial x^k} .$$

This says that the product of the square Jacobian matrices on the right is the identity. Hence either is invertible, and in particular both have non-zero determinant.

A.29.2. *Solution.* Passing from local coordinates $x^1, \ldots, x^n$ on (U, ϕ) around p to local coordinates $y^1, \ldots, y^n$ on (V, ψ) around p, and assuming the passage is non-singular:

$$\frac{\partial f \circ \phi^{-1}}{\partial x^i}\Big|_{\phi^{-1}(p)} = \sum_{j=1}^{n} \frac{\partial y^j}{\partial x^i}\Big|_{\phi^{-1}(p)} \frac{\partial f \circ \psi^{-1}}{\partial y^j}\Big|_{\psi^{-1}(p)} .$$

The matrix of coefficients $\frac{\partial y^j}{\partial x^i}\big|_{\phi^{-1}(p)}$ is non-singular by the previous exercise, so if the row vector of components $\frac{\partial f \circ \psi^{-1}}{\partial y^j}\big|_{\psi^{-1}(p)}$ is non-zero, so is the row vector of components $\frac{\partial f \circ \phi^{-1}}{\partial x^i}\big|_{\phi^{-1}(p)}$.

A.29.3. *Hint.* Generalise the argument of the previous exercise.

References

[AbMa78] Abram, R., Marsden, J.E.: Foundations of Mechanics. Benjamin-Cummings, San Francisco (1978)

[Apo91I] Apostol, T.M.: Calculus, vol. 1, 2nd edn. Wiley, New York (1991)

[Apo19II] Apostol, T.M.: Calculus, vol. 2, 2nd edn. Wiley, New York (1919)

[Arn92] Arnold, V.I.: Mathematical Methods of Classical Mechanics, 2nd edn. Springer, Berlin (1989)

[Axl16] Axler, S.: Linear Algebra Done Right, 3rd edn. Springer, Berlin (2016)

[Bis16] Biscari, P.: Introduzione alla Meccanica Razionale: elementi di teoria con esercizi. Springer, Berlin (2016)

[BRSG16] Biscari, P., Ruggeri, T., Saccomandi, G., Vianello, M.: Meccanica Razionale, 3rd edn. Springer, Berlin (2016)

[Cardin15] F.Cardin, Elementary Symplectic Topology and Mechanics. Springer, Berlin (2015)

[doC92] do Carmo, M.P.: Riemannian Geometry. Birkhäuser, Boston (1992)

[FaMa02] Fasano, A., Marmi, S.: Analytical Mechanics: An Introduction. Oxford Graduate Texts. Oxford University Press, Oxford (2013)

[Gol50] Goldstein, H.: Classical Mechanics. Addison-Wesley, Cambridge (1950)

[Jac98] Jackson, J.D.: Classical Electrodynamics, 3rd edn. Wiley, New York (1998)

[KoNo63] Kobayashi, S., Nomizu, K.: Foundations of Differential Geometry, vol. I. Interscience, New York (1963)

[KrPa03] Krantz, S.G., Parks, H.R.: The Implicit Function Theorem, 2003 edn. Birkhäuser, Basel (2003)

[Lee03] Lee, J.M.: Introduction to Smooth Manifolds. Springer, New York (2003)

[Mal52] Malkin, I.G.: Theory of stability of motion, in *ACE-tr-3352 Physics and Mathematics*. US Atomic Energy Commission, Washington (1952)

[Min19] Minguzzi, E.: Lorentzian causality theory. Living Rev. Relat. **22**(1), 1–202 (2019)

[Mor18] Moretti, V.: Spectral Theory and Quantum Mechanics, 2nd edn. Springer, Berlin (2018)

[Mor19] Moretti, V.: Fundamental Mathematical Structures of Quantum Theory Spectral Theory, Foundational Issues, Symmetries, Algebraic Formulation. Springer, Berlin (2019)

[Nor86] Norman, C.W.: Undergraduate Algebra. Oxford University Press, Oxford (1986)

[ONe83] O'Neill, B.: Semi-Riemannian Geometry with Applications to Relativity. Academic Press, Cambridge (1983)

[Rin06] Rindler, W.: Relativity: Special, General, and Cosmology, 2nd edn. Oxford University Press, Oxford (2006)

[Rud78] Rudin, W.: Real and Complex Analysis, 3rd edn. McGraw-Hill Education, New York (2001)

[RuSc13] Rudolph, G., Schmidt, M.: Differential Geometry and Mathematical Physics. Part I. Manifolds, Lie Groups and Hamiltonian Systems. Springer, Berlin (2013)

[Sid14] Sideris, T.C.: Ordinary Differential Equations and Dynamical Systems. Atlantis Press, Paris (2014)

[Tes12] Teschl, G.: Ordinary Differential Equations and Dynamical Systems. Graduate Studies in Mathematics AMS. American Mathematical Society, Providence (2012)

[Wes78] von Westenholtz, C.: Differential Forms in Mathematical Physics. North-Holland, Amsterdam (1978)

[Wal84] Wald, R.M.: General Relativity. Chicago University Press, Chicago (1984)

[War83] Warner, F.W.: Foundations of Differentiable Manifolds and Lie Groups. Springer, Berlin (1983). Interscience, New York, 1963

Index

Printed in the United States
by Baker & Taylor Publisher Services